ENCYCLOPEDIA OF PHYSICS

EDITED BY

S. FLÜGGE

VOLUME XXI

ELECTRON-EMISSION
GAS DISCHARGES I

WITH 378 FIGURES

SPRINGER-VERLAG
BERLIN · GÖTTINGEN · HEIDELBERG
1956

HANDBUCH DER PHYSIK

HERAUSGEGEBEN VON

S. FLÜGGE

BAND XXI

ELEKTRONEN-EMISSION
GASENTLADUNGEN I

MIT 378 FIGUREN

SPRINGER-VERLAG
BERLIN · GÖTTINGEN · HEIDELBERG
1956

ISBN-13: 978-3-642-45846-0 e-ISBN-13: 978-3-642-45844-6
DOI: 10.1007/978-3-642-45844-6

Softcover reprint of the hardcover 1st edition 1956

Inhaltsverzeichnis.

Seite

Thermionic Emission.

By

Wayne B. Nottingham[1].

With 62 Figures.

Glossary of symbols.

a Thermionic constant, Eq. (9.1).

a Ratio of radii as defined by Eq. (60.6).

a_A, a_B Empirical constants of Eqs. (77.1) and (77.2).

A Universal thermionic constant, Eqs. (5.2) and (18.6).

A Area of conduction specimen, Eq. (65.28).

A Area of electron emitter as used in Eq. (82.5).

A_F Fowler's thermionic constant defined by Eq. (64.24).

A_R Richardson constant, Eq. (50.2).

A_ν Constant in general equation as in Eq. (50.8).

B Symbol introduced into Eq. (26.1) to represent barrier properties.

C_g Grid to emitter capacitance (Sect. 82).

C_p Correction factor to pore-conduction equation to include porosity of specimens, Eq. (65.28).

C_s Correction factor in Eq. (65.9).

C_x Capacitance calculated as in Sect. 82.

D Distance between emitter and collector, Eq. (27.4).

$D(p, B)$ Transmission over emission barrier, Eq. (26.1).

$\overline{D(p, B)}$ Average transmission of electrons from the interior of a solid across a barrier to the outside, Eq. (64.22).

$\overline{D}$ An abbreviation $\overline{D(p, B)}$ (Sect. 64).

e Electron charge, Eq. (5.1).

E Used as abbreviation of E_D in Eq. (64.6).

E Electric intensity, Eq. (27.3).

E_c Electric intensity at critical distance x_c, Eq. (27.7).

E_x Component of electric intensity, Eq. (23.3).

E_a Activation energy defined by Eq. (65.13).

E_D Energy level of donors relative to bottom of conduction band (Sect. 64).

f Fraction of a Maxwellian distribution, Eqs. (33.2) and (34.1).

f_M Maximum fractional distance from an emitter of very high capability to space-charge minimum, Eq. (61.3).

FL Fermi level (Fig. 2).

$F_c(p, V, T)$ Probability of electron absorption by collector depends on momentum, potential and temperature of emitter, Eq. (26.1).

$F(S, a)$ Function given in Eq. (60.4) and computed for Table 10, plotted in Fig. 19.

$F_c(S, kT/\omega)$ Tabulated function applied to theory of electron emission from a cylinder in a retarding field, Eq. (26.9).

$F(V_s, V_T, n)$ Defined by Eq. (62.6)

$F(x)$ Force function, Eq. (27.6).

$f(\eta_c)$ Function of collector region potential difference given by Eq. (46.14) and Table 6.

$F(\eta_c)$ A function applied to the collector space defined by Eqs. (38.4) and (39.5).

g Conduction per unit area, Eq. (65.28).

g_m Transconductance (or mutual-conductance) defined by Eq. (82.5).

G Abbreviated form of $G(x)$ as in Eq. (27.11).

$G(S)$ Electron flow in a retarding field between cylinders, Eq. (26.11), and Tables 2 and 10.

[1] This work was supported in part by the Signal Corps; the Office of Scientific Research, Air Research and Development Command; the Office of Naval Research; the Research Laboratory of Electronics and the Department of Physics of the Massachusetts Institute of Technology.

$G(x)$	Geometrical function relating E to V_c, Eq. (27.3).
G'	Empirical value of constant of Eq. (79.1).
GE-218	Special form of General Electric tungsten wire (Sect. 70).
h	PLANCK's constant, Eq. (15.1).
h^3	Extention in phase space per quantum state (Sect. 15).
i	Emission current as in Eq. (82.8).
I	Electron current density, Eqs. (5.1) and (5.2).
I_c	"Random" electron current density in a cavity, Eqs. (18.2) and (18.4).
I_s	Current density at the surface of a cavity. Same as I_{00}, Eq. (18.1).
I_0	Constant of the SCHOTTKY Equation, Eq. (27.11).
I_0	Current density at zero field as in Eqs. (27.14) and (40.1).
I_{00}	Current density defined by Eqs. (18.5) and (26.10).
I_L	Minimum current density with zero field at emitter computed by Eq. (46.12).
I_m	Maximum possible current density flowing across a diode with zero field at the collector, Eq. (43.2).
I_M	Maximum current density given by Eqs. (46.11) or (58.2).
I'_r	Current density computed by Eq. (52.4).
I_R	Current density under critical condition of zero field at the collector, Eq. (43.1).
I_Θ	Current density with zero field at the emitter and the critical temperature Θ, Eq. (47.1).
k	BOLTZMANN's constant, Eq. (5.1).
K^2	A number defined by Eq. (64.14).
K_L	The LANGMUIR-CHILD constant, Eq. (38.3).
l_p	Average free flight distance in a pore, Eq. (65.17).
l_s	Free flight distance in a semiconductor, Eq. (65.6).
l_{s0}	Free flight distance at some low temperature T_0, Eq. (65.6).
L	Thickness of a test specimen for pore conduction in a high field, Eq. (66.3).
L_M	Designation of straight lines in Figs. 11 and 12.
m	Electron mass, Eq. (5.1).
$\overline{m}$	Average dipole moment per atom, Eq. (73.2).
$\overline{m}_0$	Dipole moment per atom at very low surface concentration.
m_g	Proportionality constant defined by Eq. (82.3).
M_p	Electron mobility in a pore, Eq. (65.4).
$M_{p\,\mathrm{II}}$	Mobility of electrons in a pore over temperature range II, Eq. (65.18).
M_s	Electron mobility in a semiconducting solid, Eq. (65.3).
n	Concentration of free electrons, Eqs. (5.1) and (22.1).
n	Term index in summation of Eqs. (62.3) and (62.5).
n_a	Number of atoms per unit area, Eq. (73.2).
n_c	Concentration of electrons in cavity space, Eq. (18.3).
n_D	Donor concentration used in Sect. 64.
n_1	Adsorbed atoms per unit area in a monolayer, Eq. (75.1).
n_0	Electron concentration at cavity center of Fig. 3.
n_{0m}	Maximum possible electron density at the center of a cavity, Eq. (24.4).
n_s	Concentration of electrons at emitting surface of cavity, Eq. (24.8).
N	A numerical constant defined by Eqs. (38.8) or (46.4).
N_{xc}	Number of electrons that can cross a solid boundary out of a cavity per second per unit area per unit range in energy, Eq. (17.2).
N_{xs}	Number of electrons that can cross a solid boundary into a cavity per second per unit area per unit range in energy, Eq. (17.1).
$N(\varepsilon_x)$	Number of electrons crossing unit area per second per unit range in ε_x, Eq. (16.1).
p	Momentum of an electron, Eq. (26.2).
p_x, p_y, p_z	Momentum component along x, y, z direction.
$\overline{p}_x$	Average momentum component, Eq. (33.3).
p'_x	Momentum component in the x direction of an electron at the limiting part of the barrier, Eq. (26.4).
$\overline{\Delta p_x}$	Average gain in momentum as in Eq. (65.1).
p_r	Radial component of momentum, Eq. (60.1).
p_{rs}	Same as p_r at the surface of the emitter, Eq. (60.1).
p_{rl}	Limiting value of initial radial momentum at the emitter for arrival at a potential point V_s negative, Eq. (60.3).
$p_{\Theta l}$	Limiting value of tangential momentum at the emitter for arrival at a potential point V_s negative, Eq. (60.3).
p_Θ	Tangential component of momentum, Eq. (60.1).
$p_{\Theta s}$	Same as p_Θ but at the surface of the emitter, Eq. (60.1).

P	Abbreviation for contact difference in potential, Eq. (38.10).
P_0	Contact difference in potential at 0° K by extrapolation according to Eq. (38.10) (usually not the true value).
P_T	Contact difference in potential (a function of temperature), Eq. (27.1).
$P(\psi^{\frac{1}{2}})$	Probability integral, Eq. (34.6).
r	Radius of emitter, Eqs. (26.5) and (27.5).
r	Radius between that of the emitter and the collector, Eq. (60.2).
r_s	Radius of emitter in Eq. (60.20).
R	Radius of collector, Eqs. (26.5) and (27.5), or radius at potential minimum as in Eq. (60.3).
R_c	Collector radius as used in Sect. 67.
s	A dimensionless measure of applied voltage defined by Eq. (56.1).
s_R	Dimensionless measure of applied voltage for zero field at collector, Eq. (57.1).
S	Retarding potential in dimensionless units, Eqs. (26.8) and (56.5).
S_a	Slope of line in Fig. 11.
S_b	Slope of line in Fig. 12.
S'	Dimensionless change in potential with zero field at collector as reference, Eq. (57.1).
S'_0	Maximum value of collector potential change in dimensionless units (Fig. 14).
S'_u	Value of S' when current ratio is expressed as u^2, Eq. (58.13).
$S_{z'}$	Slope given by Eq. (52.8) and illustrated in Fig. 13.
T	Temperature on Kelvin scale (see also V_T).
T_a	Activation temperature (Sect. 75).
T_0	Matching temperature for relating thermionic constants, Eq. (50.5).
T_0	A low temperature at which a free flight distance l_{s0} is known, Eq. (65.6).
ΔT	Increase in temperature above the critical value Θ, Eq. (40.2).
u^2	Current ratio as in Eq. (57.3).
u_0^2	Maximum current ratio between space-charge limits, Eq. (57.4).
U^2	Current ratio defined by Eq. (58.3).
v	Average speed of electrons of low concentration, Eq. (65.6).
v	Applied potential, Eq. (27.1).
v''	Applied voltage for current density I_M (Sect. 47).
v'	Applied voltage for current density I_L (Sect. 47).
v^{**}	Voltage computed from observables as in Eq. (47.5).
v_1	Applied voltage for a calculated current, Eq. (52.1).
v_1^*	Defined by Eq. (52.2).
v_j	Defined by Eq. (52.3).
v'_r	Observable applied voltage, Eq. (52.5).
v_R	Applied voltage for zero gradient at the collector (Sect. 57).
v_Θ	Applied voltage at critical condition of zero field at the emitter, Eq. (49.7).
Δv	Observable potential difference over temperature difference of ΔV_T with zero net current (Sect. 86).
$\Delta v'$	Internal potential due to surface charges as in Sect. 86.
V	Magnitude of true retarding potential in the space between the emitter and collector in Eqs. (26.5) and (26.6).
V	Potential with respect to the space-charge minimum of Fig. 3, see Eq. (22.1).
V_l	A constant depending on tube geometry and temperature. See Sect. 9 and Eq. (27.11).
V_c	Potential difference between emitter and collector as in Eq. (27.11) or potential across the collector space as in Eq. (32.1).
V_m	Lowest temperature of range expressed in electron-volts, Eq. (49.1) and (11.3).
V_n	Highest temperature of range expressed in electron-volts, Eq. (49.1) and (11.3).
V_p	Voltage drop over an average pore distance of l_p, Eq. (66.2).
V_r	Retarding potential difference between surface of emitter and surface of collector, Eq. (42.1).
V_s	Potential difference in emitter space as in Eq. (32.1) in Fig. 7.
V_s	Difference in potential between center of cavity and its boundary just outside the conducting surface as in Sect. 24.
V_T	The electron volt equivalent of the energy kT, Eq. (46.9).
V'_T	Electron-volt equivalent of minimum temperature for space-charge minimum of V_T as in Eq. (82.7).
V_{T0}	Electron volt equivalent of T_0 in Eq. (65.6).
V_w	Characteristic temperature expressed in electron volts defined by Eq. (47.12).
V_{z1}, V_{z2}	Critical temperatures expressed in electron volt equivalent, Eq. (53.1).
V_Θ	Electron-volt equivalent of the critical temperature Θ and given by $\Theta/11600$ (Sect. 47).
w	Distance between plane conducting surfaces (Sect. 21, Figs. 1, 2, 3, and 7).

w	Separation of surface charges, Eq. (73.1) (Fig. 34).
W_a	Potential energy difference shown in Fig. 2—Electron affinity of a solid, Eq. (17.4).
W_v	Electron affinity modified by an externally applied potential, Eq. (27.10).
x	Integration variable, Eqs. (26.11) and (60.4).
x	Distance variable as in Sect. 27.
x_1	Characteristic unit distance of space-charge theory, Eqs. (35.1), (35.2) and (38.9).
x_c	Escape distance as in Fig. 5 and Eq. (27.6).
x_c	Distance from potential minimum to collector Fig. 7 and Eq. (38.2).
x_g	Distance from emitter to control grid of idealized triode.
x_0	Location of mirror-image surface with respect to an arbitrary reference as in Eq. (27.8).
X_s	Maximum value of distance from emitter to space-charge minimum, Eq. (46.10).
y	Integration variable, Eqs. (26.12) and (34.3).
z	Ratio of distances as in Eqs. (24.2) and (43.4).
z^2	Ratio of electron concentrations as in Eq. (24.11) or currents as in Eqs. (43.4) and (51.2).
z'^2	Arbitrary ratio of currents as in Eq. (52.4).
z'_1, z'_2	Arbitrary choices of z, Eq. (53.1).
z_R^2	Current ratio defined by Eq. (58.7).
α	$(d\varphi/dV_T)$ or the temperature coefficient of work-function with temperature expressed in its electron-volt equivalent, Eq. (65.21).
γ	Proportionality constant of Eq. (80.2).
δ	Distance from potential minimum to surface at which ψ_s is infinite, Eqs. (24.1) and (43.1).
ε	Kinetic energy when potential energy is zero.
ε	Energy, Eq. (15.1).
ε_0	Permitivity of free space, Eq. (22.1).
ε_x	Kinetic energy associated with motion in the positive x direction, Eq. (16.2).
ε_x	$p_x^2/2m$ or kinetic energy over an image barrier as in Eq. (63.2).
ε'_x	Kinetic energy associated with the x direction of motion in the cavity of Fig. 2.
ε_{x2}	Kinetic energy associated with electron motion in the semiconductor illustrated in Fig. 27.
$\bar{\varepsilon}_x$	Average kinetic energy of an electron, Eqs. (63.1) and (63.2).
$\Delta\varepsilon_x$	Difference in energy between bottom of conduction band in the semiconductor and that of the metal in Fig. 27.
Θ	Temperature critical for zero field at either the collector or emitter with space charge (Sect. 31).
ϑ	Fraction of surface covered by adsorbed atoms.
ϑ_p	Coefficient of thermal emf (thermoelectric power) defined by Eq. (86.5) for a pore in a semiconducting structure.
ϑ_s	Coefficient of thermal emf (thermoelectric power) defined by Eq. (86.3) for semiconducting solid.
λ	Approximately half of the characteristic pore length l_p, Eq. (66.1).
μ	FERMI level measured relative to bottom of conduction band as in Eqs. (16.1) and (64.1).
μ'	A measure of donor concentration and temperature as in Eq. (64.9).
μ_1	FERMI level in the metal illustrated in Fig. 27.
μ_2	FERMI level in a semiconductor illustrated in Fig. 27.
μ_c	FERMI level in cavity (see Fig. 2).
μ_s	FERMI level in solid (see Fig. 2).
$\Delta\mu$	Change in the FERMI level with temperature relative to its value at $T = 0°$ K as used in Eq. (81.4).
$\Delta\mu$	Change in FERMI level with a temperature change of ΔV_T as in Eq. (86.2).
ν	Number of electrons per unit volume.
ν_0	Electron concentration at space-charge minimum, Eq. (35.3).
ν	Temperature exponent in Eq. (50.8).
σ	Integration variable representing energy in dimensionless units as in Figs. 23 and 24 and Eqs. (62.3) and (62.7).
σ	Conductivity as in Eq. (65.2).
σ	Surface charge per unit area, Eq. (73.1).
σ_A	Conductivity over range I of Fig. 37.
σ_B	Conductivity over range II of Fig. 37.
$\sigma_{p\,\mathrm{II}}$	Pore conductivity over temperature range II, Eq. (65.19).
Σ	Change in applied potential in dimensionless units, Eq. (58.5).

$\bar{\tau}$	Average time between inelastic collisions as in Eq. (65.1).
τ_E	Average time of flight across a pore in a strong field, Eq. (66.2).
$\bar{\tau}_p$	Average time for electrons in pores, Eq. (65.4).
$\bar{\tau}_{p\,\mathrm{II}}$	Average time of flight across a pore in temperature range II, Eq. (65.17).
$\bar{\tau}_s$	Average time for electrons in solids, Eq. (65.3).
φ	True work-function and specifically defined by Eq. (19.1).
$\Delta\varphi$	Reduction of true work-function by accelerating field (Sect. 27).
φ_c	True work-function of collector, Eq. (27.2).
φ_F	FOWLER work-function or true work-function at 0° K for an N type semiconductor.
φ_R	RICHARDSON work-function, Eq. (50.2).
φ_s	True work-function of emitter, Eq. (27.2).
φ_ν	Constant in general equation as in Eq. (50.8).
Φ	Work-factor, Eq. (9.1), related to work-function in Sect. 50.
Φ_σ	Activation energy for pore conductivity, Eq. (65.27).
Φ_A	Activation energy of Eq. (77.1).
Φ_B	Activation energy of Eq. (77.1).
χ	Dimensionless measure of distance from the space-charge potential minimum, Eqs. (22.4), (35.2) and (35.4).
χ_a, χ_b	Values associated with the arbitrary choices of ψ_a and ψ_b (Sect. 47).
χ_c	Collector region distance in dimensionless units, Eq. (37.8).
χ_m	Maximum value of χ_s which is 1.806 as in Eq. (37.6).
χ_s	Emitter region distance in dimensionless units, Eqs. (37.5) and (37.6).
χ_{cu}, χ_{su}	Values corresponding to the potentials ψ_{cu} and ψ_{su}, Eq. (58.11).
χ_{sR}	Value corresponding to ψ_{sR} of Eq. (58.9).
ψ	Dimensionless representation of potential differences in units of kT.
ψ_a	Arbitrary choice of ψ_c (Sect. 47).
ψ_b	Arbitrary choice of ψ_c (Sect. 47).
ψ_c	Collector region potential difference in dimensionless units, Eq. (37.7), Fig. 14.
ψ_s	Dimensionless potential difference in the emitter space, Eq. (37.1), Fig. 14.
$\Delta\psi_s$	Change in ψ_s as in Eq. (57.2) and Fig. 14.
ψ_{c0}	Maximum value of collector region potential in dimensionless units, Fig. 14 and Eq. (57.9).
ψ_{cu}	Potential at the collector with reference to the potential minimum when current ratio is u^2, Eq. (58.11).
ψ_{sR}	Maximum value of emitter region potential in dimensionless units (Fig. 14).
ψ_{su}	Potential at the emitter with reference to the potential minimum when current ratio is u^2, Eq. (58.10).
ω	Empirical constant, Eq. (46.4).

A. Scope and objectives.

1. Definitions. The emission of electrons across the boundary surface that separates a heated electronic conductor from an otherwise nonconducting space has become synonomous with the term "thermionic emission". The broadest application of the word thermionic might include the emission of charged atomic or molecular particles that may carry with them either a net positive or a net negative charge. Since these phenomena are so very different basically, this article will be devoted exclusively to the more fundamental aspects of the experimental and the theoretical investigations of the phenomenon of *electron* emission from heated conductors.

Thousands of experiments have been reported in the literature and serve as the basic work material from which an explanation of the phenomenon in terms of the fundamental principles of physics emerges. The studies relate to four surface classifications which are:

1. clean homogeneous surfaces;
2. clean heterogeneous surfaces;
3. simple composite surfaces;
4. complex surfaces.

To clarify these classifications an illustration of each will be given. Single crystal wires of circular cross section have been used as sources of thermionic electrons and provide the nearest approach to the realization of experimental conditions appropriate to theoretical interpretation[1,2,3].

Emission can be observed and investigated in detail from the more important crystallographically homogeneous surfaces. These surfaces must be maintained under such perfect vacuum conditions that an absolutely negligible fraction of a monomolecular layer of impurity is present. Since fundamental studies show that the thermionic emission is dependent not only on the atomic composition of the emitting conductor but also on the crystallographic structure of the exposed surfaces, it is evident that practically all investigations that describe the electron emission from polycrystalline substances yield data characteristic only of the particular specimen. In general such observations measure the electron emission summed over an assembly of *heterogeneous surfaces*, which can never be accurately described. Most of the work reported in the literature of the subject applies to these surfaces.

At a given temperature, the efficiency of electron emission from a given conductor may be altered through many orders of magnitude by the adsorption of polarizable atoms or molecules on an otherwise clean homogeneous or heterogeneous underlying conductor. The presence of a surface-layer coverage having a density even smaller than $^1/_{100}$ part of a monolayer can be observed to alter the emission properties[4-7]. Such surfaces are classified as *simple composite surfaces* if the extent of the coverage is of the order of a monolayer or less.

The last of the above four classifications, the *complex surface*, is by far the most important in terms of its usefulness as a source of electrons and includes as its most important member the *oxide cathode*. The structure represents an emitter which depends on the conduction of a heated metallic support such as nickel or platinum upon which has been placed, after due processing, a layer of alkaline earth oxide crystals. The surfaces of these crystals serve as the emitting areas. These crystals are generally solid solutions of barium oxide and strontium oxide of comparable proportions and sometimes have traces of other substances added. It is not unexpected that such a complex emitter will have properties very difficult to explain.

It is considered within the scope of this chapter to review critically some of the important investigations that yield information relevant to all four of these classifications of thermionic emitters. It is not the purpose of this chapter to give an exhaustive bibliography; only such references as apply specifically to the information given here will be cited. The reader may refer to a recently published bibliography[8] on physical electronics which contains several hundred references to contributions on thermionic emission.

2. Theory and experiment. It is an illusion to believe that the main features of thermionic emission have been worked out theoretically and are in agreement with experiment. In spite of the generality often associated with the thermo-

---

1 M. H. NICHOLS: Phys. Rev. **57**, 297 (1950).
2 G. F. SMITH: Phys. Rev. **94**, 295 (1954).
3 A. R. HUTSON: Phys. Rev. **98**, 889 (1955).
4 J. B. TAYLOR and I. LANGMUIR: Phys. Rev. **44**, 423 (1933).
5 J. A. BECKER: Phys. Rev. **28**, 341 (1926); **33**, 1082 (1929).
6 J. A. BECKER and G. E. MOORE: Phil. Mag. **29**, 129 (1940).
7 W. B. NOTTINGHAM: Phys. Rev. **49**, 78 (1936).
8 W. B. NOTTINGHAM: Bibliography in Physical Electronics. Cambridge, Massachusetts: Addison-Wesley Press 1954.

dynamic interpretation of thermionic emission, emphasis must be given to the fact that this branch of theory cannot be relied upon to give accurate information concerning the current flow across a boundary under experimental conditions that violate the basic assumptions of the theory. The most important assumption made is that the system under consideration can be bounded in a manner that will still permit the actual measurement of an electron emission across a boundary. Thermodynamic considerations can be applied to the electron gas in the interior of a single crystal of conducting substance so cut out as to bound the interior cavity by surfaces of perfect homogeneity as regards their crystal structure. With the combined help of the equations of thermodynamics and electrostatics the time average of the density of electrons can be specified as a function of the coordinates within the cavity. The pressure that these electrons would exert on the surface of the cavity can be computed with confidence and from these quantities one might presume to calculate a current of electrons which would flow from the interior of the conducting specimen across a surface boundary to the region just outside the conductor. To assume that such a calculation of the current would be directly related to an observable emission current is wrong. No theory as general as this can predict the fraction of the electrons which are reflected at the boundary.

Even though usually considered to be less general, the statistical theory of an assembly of free electrons is more capable of giving valid information concerning the thermionic emission process. It will therfore be appropriate to base practically all of the theoretical analysis brought forward in this chapter on the application of the statistics of free electrons to thermionic emission phenomena.

B. Historical highlights.

I. General background.

3. Introduction. It will be the purpose of this section to review briefly some of the main events that have marked both the development of the theory and the understanding of thermionic emission as well as the technological advances which have been made during the past 70 years. Following this qualitative review of events a detailed analysis will be given to review the present state of our understanding with respect both to theory and experiment.

4. Discovery and identification of thermionic emission. That negative electricity escapes from hot filaments was probably first established by THOMAS A. EDISON and later identified by WILLIAM H. PREECE[1] as the "EDISON Effect". The account by PREECE of his own experiments on the EDISON effect does not separate the phenomenon now recognized as thermionic emission from the "blue effect" which was evidently the ionization of the residual gas produced by the electrons as they were accelerated from the negative hot filament to the plate of the diode. The identification of the charge carrier emitted from the hot carbon filament as a particle with a very small mass compared with that of a hydrogen ion and with a charge equal in magnitude but opposite in sign to that of the hydrogen ion was made by J. J. THOMSON[2]. In spite of the important experiments of J. J. THOMSON, general agreement that electron flow through conductors and electron emission from hot surfaces are truly different phenomena

---

[1] W. H. PREECE: Proc. Roy. Soc. Lond. **38**, 219 (1885).

[2] J. J. THOMSON: Phil. Mag. **48**, 547 (1899).

from ionic flow through substances and ionic conduction through gases did not come until about 1914[1].

5. RICHARDSON equations. Thermionic emission is so intimately associated with the phenomenon of electronic conduction in solids that advances in these two fields of scientific investigation are associated. The DRUDE[2] development of the theory of free electrons in metallic conductors paved the way for the better understanding of thermionic emission which marked the contributions of O. W. RICHARDSON. The basic idea of work-function as being a measure of the energy per electron required to transfer charges from the interior of the conductor to the field-free space outside of it is largely due to RICHARDSON[3]. Founded on a very literal classical interpretation of the free electron theory of electronic conduction in metals, RICHARDSON[4] derived his first thermionic emission equation which took the following form:

$$I = n e \left(\frac{k}{2\pi m}\right)^{\frac{1}{2}} T^{\frac{1}{2}} e^{-\frac{e\varphi}{kT}}. \tag{5.1}$$

The recognition of difficulties encountered by the classical free electron theory of conduction and its relation to the specific heat of metals led RICHARDSON[5] and VON LAUE[6] to the "T^2" equation given as follows:

$$I = A T^2 e^{-\frac{e\varphi}{kT}}. \tag{5.2}$$

In its original form this equation depended upon thermodynamic arguments and at the present time it is accepted as the correct expression for electron current flow at a boundary which is in a system maintained under conditions of thermodynamic equilibrium. The mistake generally made in the application of this equation is that the current I is identified as the thermionic emission current one should expect to observe in a laboratory experiment. The current density of Eq. (5.2) is independent of reflection and of other phenomena that may alter the distribution in energy of the electrons which do cross the boundary in an actual emission experiment. A second common error in the application of Eq. (5.2) to thermionic emission is the assumption that the true work-function φ is a constant.

DUSHMAN[7] was one of the first to apply quantum theory to the analysis upon which Eq. (5.2) is based and he deduced that the constant A is a universal constant. His numerical value for this constant was in error by a factor of 2 because it was not known at the time that the "statistical weight" for a free electron should be 2 in order to take into account electron spin.

6. SCHOTTKY mirror-image effect. Of the many important contributions made by SCHOTTKY to the better understanding of thermionic emission, one of the

[1] I. LANGMUIR: Proc. Inst. Radio. Engrs. **3**, 261 (1915). — Gen. Electr. Rev. **18**, 327 (1915). — Trans. Amer. Electrochem. Soc. **29**, 125 (1916), (see p. 134). — The FREDERICK GUTHRIE [Phil. Mag. **46**, 257 (1873)] report "On a Relation Between Heat and Static Electricity" has been interpreted by some readers to indicate thermionic emission. The low temperatures involved in his experiments and the general phenomena reported can more easily be interpreted in terms of ionic production, both positive and negative, than as a clear indication of electron emission.

[2] P. DRUDE: Ann. Phys., Lpz. **1**, 566 (1900); **3**, 369 (1900). — H. A. LORENTZ: Theory of Electrons. New York: G. E. Stechert & Co. 1909.

[3] O. W. RICHARDSON: Phil. Mag. **23**, 263 (1912).

[4] O. W. RICHARDSON: Proc. Cambridge Phil. Soc. **11**, 286 (1902).

[5] O. W. RICHARDSON: Phil. Mag. **23**, 594 (1912).

[6] M. v. LAUE: Jb. Radioakt. u. Elektr. **15**, 205, 257, 301 (1918).

[7] S. DUSHMAN: Phys. Rev. **21**, 623 (1923).

most important was his recognition that the mirror-image force acting on an electron as it escapes from the surface of a good conductor accounts for the major part of the total work which must be supplied for an electron to escape[1]. Since this mirror-image force extends over such a great distance compared with inter-atomic distances, it is possible to counteract it by the application of an externally applied accelerating electric field. This neutralization of forces results in a reduction in the work required for an electron to escape and as a consequence there is an increase in the thermionic emission observed from a constant temperature emitter as the accelerating field is increased. The correctness of this theory and the supporting experimental data leave the "SCHOTTKY Effect" unchallenged. Details of this theory are given in Sect. 27.

7. CHILD-LANGMUIR space charge. A second effect produced by accelerating fields—an important one in the understanding of thermionic emission—was recognized by LANGMUIR[2] and CHILD[3] and others as being accounted for by the phenomenon of space charge.

As the temperature of an emitting conductor increases, the observed current does not increase indefinitely, even though a fixed strongly accelerating positive potential is maintained on the electron collector. If the number of electrons in transit between the emitter and the collector exactly equals the total surface charge maintained on the collector by the external circuit, then the electric field at the emitter becomes zero. Further increases in temperature are followed by very little increase in observed current because of the development of a retarding field at the surface of the emitter produced by space-charge. Even though space charge effects are strongly dependent on electrode geometry and act in the space well outside of the thermionic emitter itself, it is necessary to have a full understanding of their influence. Thermionic emission is an electron flow observed as a current between suitably placed electrodes and the phenomenon of space charge seldom should be neglected in the interpretation of the observations.

8. SOMMERFELD'S free electron theory. SOMMERFELD[4] applied the statistical theories of FERMI and DIRAC to reconstruct the DRUDE free-electron theory of conduction in metals to include the concept of quantization of phase space and the introduction of PAULI's exclusion principle. This SOMMERFELD theory of conduction by free electrons was applied by FOWLER[5] and NORDHEIM[6] to form the basis of the quantum theory derivation of Eq. (5.2). It is from this beginning that we have the means for the better understanding of the mechanism of thermionic emission even though some phases of the theory are incomplete at the present time.

II. Experiments with clean surfaces.

9. Empirical equations and work-factor. As knowledge of the thermionic emission process has developed, it has become more and more evident that little or no theoretical significance can be attributed to the application of the RICHARDSON form of thermionic equation as given by Eq. (5.2) for heterogeneous

[1] W. SCHOTTKY: Phys. Z. **15**, 872 (1914). — Z. Physik **14**, 63 (1923).
[2] I. LANGMUIR: Phys. Rev. **2**, 450 (1913); **21**, 419 (1923).
[3] C. D. CHILD: Phys. Rev. **32**, 498 (1911).
[4] A. SOMMERFELD: Z. Physik **47**, 1 (1928).
[5] R. H. FOWLER: Proc. Roy. Soc. Lond., Ser. A **117**, 549 (1928); **118**, 229 (1928).
[6] L. NORDHEIM: Z. Physik **30**, 177 (1929).

surfaces whether clean or not. The experiments of NICHOLS[1], SMITH[2] and HUTSON[3] have given data on homogeneous clean surfaces which will be shown later in this chapter to yield the only data appropriate for analysis by the RICHARDSON form of equation. This statement is not a denial of the engineering usefulness of some equation by which experimental data on the thermionic emission properties of clean metals can be expressed. The nearly universal application of the RICHARDSON equation for the correlation of the observed data is in some respects unfortunate, since the equation is in fact used purely as an empirical one containing two empirically determined constants. An equation of the following form is a far more useful empirical equation:

$$I = a\,\mathrm{e}^{-\frac{e\Phi}{kT}}. \tag{9.1}$$

In this equation a is a thermionic constant determined by experiment to fit the observed data. This constant depends on the substance from which the electrons are emitted and also on the details of its crystallographic configuration. In general the experimenter describes his specimen as one of some specified composition and as "polycrystalline". The use by others of empirical constants so determined to predict the emission properties of other samples of the same material implies that, averaged over the entire emitting surface, the heterogeneous distribution of crystal surfaces is reproducible. The constant Φ of Eq. (9.1) is referred to as the "work-factor" and again it is dependent on the detailed surface configuration of the specimen.

It is self-evident that Eq. (9.1) is easier to use as an empirical equation than is Eq. (5.2). The computation required to determine the emission current for a particular temperature is easy to make with either equation but the reverse process of determining the temperature at which specified current density may be expected is difficult with Eq. (5.2) but is straightforward with Eq. (9.1). A number[4,5] of tabulations have been made of the empirical constants suitable for use in the RICHARDSON form of empirical equation but no complete tabulation has been made of the constants suitable for use in Eq. (9.1). For this reason, tabulations of both sets of empirical constants are given in Appendix 1 to this article. The equations for conversion from one representation to the other are given in Sect. 50.

The "SCHOTTKY effect"[6], which is the reduction in work-function at constant temperature under the influence of an applied electron-accelerating field at the surface of the emitter, accounts for an increase in emission current density proportional to the factor $\exp(V_c^{\frac{1}{2}}/V_1^{\frac{1}{2}})$. For this expression to hold, the geometric relations between the emitter and the electron collector must be such that the surface field at the emitter is everywhere constant and proportional to the potential difference (V_c) between the emitter and the collector. In this relation the constant (V_1) depends upon the geometry and the temperature of the emitter and can be computed with accuracy [see Eq. (27.11)]. Experimental results seldom show an increase in current with applied potential as predicted by this factor, for two reasons. The "SCHOTTKY effect" refers exclusively to the reduction in that part of the work-function that depends on mirror-image forces. Electrons which escape from heterogeneous polycrystalline surfaces are acted

[1] M. H. NICHOLS: Phys. Rev. **57**, 297 (1940).
[2] G. F. SMITH: Phys. Rev. **94**, 295 (1954).
[3] A. R. HUTSON: Phys. Rev. **98**, 889 (1955).
[4] C. HERRING and M. H. NICHOLS: Rev. Mod. Phys. **21**, 185 (1949).
[5] H. B. MICHAELSON: J. Appl. Phys. **21**, 536 (1950).
[6] W. SCHOTTKY: Phys. Z. **15**, 872 (1914). — Z. Physik **14**, 63 (1923).

upon by additional electrical forces arising from the work-function differences that occur between different crystallographic faces of an otherwise pure surface of the material being investigated. It follows, therefore, that the SCHOTTKY effect experiment may be used to yield information concerning these other forces.

10. Periodic deviations from the "SCHOTTKY effect". A second and more basic cause of the difference between experimental results and the "SCHOTTKY effect" theory relates to the fact that the accelerating field not only lowers the work-function precisely as computed by SCHOTTKY but it alters the *transmission* of the barrier for the impinging electron stream[1-5]. This escape probability and its variation with the electric field is a matter not well understood in all of its detail. Experiments[6,7,8] show that this field-dependent effect results in a very small "periodic" alteration in the integrated transmission probability that is indicated by a periodic deviation with respect to the SCHOTTKY law of emission current. The existence of this periodic deviation effect is the latest of the series of discoveries relevant to the thermionic emission properties of clean surfaces. More details concerning this effect will be presented in Sect. 70.

III. Experiments with composite surfaces (mainly the discoveries).

11. Oxide cathodes. WEHNELT[9] discovered in 1903 that barium oxide and other alkaline earth oxides could be applied to poorly emitting surfaces such as platinum with the result that such a thermionic emitter would yield an emission current many million-fold greater than that characteristic of the base metal in the absence of the oxide. Resulting from extensive research, the oxide cathode has become economically the most important thermionic emitter ever discovered and is the most indispensable member of the group of inventions and discoveries upon which our "Electronic Age" depends. In spite of a half century of research, many of the most fundamental problems related to the phenomenon of thermionic emission from oxide cathodes are understood only in a qualitative manner. Much disagreement exists among workers in this field.

12. Thoriated filaments. LANGMUIR and ROGERS[10] discovered in 1913 that tungsten wire into which a small amount of thoria had been inserted could, by suitable heat treatment, be made to emit a current density of electrons at a given temperature 10^5 times greater than that observed from pure tungsten. Later analysis by LANGMUIR and others showed that this enhanced emission, produced as a result of the creation of a simple composite surface could be related quantitatively to the average dipole moment per unit area created at the surface by the polarizable atoms adsorbed on it. This discovery and the related experiments, including the adsorption of the alkalies and specifically of cesium[11], led to a far better understanding of the properties of thermionic emitters than would have been available to us otherwise.

---

[1] R. L. E. SEIFERT and T. E. PHIPPS: Part I, Phys. Rev. **56**, 652 (1939).
[2] E. GUTH and C. J. MULLIN: Phys. Rev. **59**, 575 (1941); **59**, 867 (1941); **51**, 339 (1942).
[3] C. HERRING and M. NICHOLS: Rev. Mod. Phys. **21**, 185 (1948).
[4] D. W. JUENKER, G. S. COLLADAY and E. A. COOMES: Phys. Rev. **90**, 772 (1953).
[5] S. C. MILLER jr. and R. H. GOOD jr.: Phys. Rev. **92**, 1367 (1953).
[6] W. B. NOTTINGHAM: Phys. Rev. **57**, 935 (1940) (L).
[7] E. G. BROCK, A. L. HOUDE and E. A. COOMES: Phys. Rev. **89**, 851 (1953).
[8] D. W. JUENKER, G. S. COLLADAY and E. A. COOMES: Phys. Rev. **90**, 772 (1953).
[9] A. WEHNELT: Verh. dtsch. phys. Ges. **5**, 255, 423 (1903). — Ann. Phys. **14**, 425 (1904).
[10] I. LANGMUIR and W. ROGERS: Phys. Rev. **4**, 544 (1914).
[11] J. B. TAYLOR and I. LANGMUIR: Phys. Rev. **44**, 423 (1933).

13. Dispenser cathodes. The application of efficient thermionic emitters to new contemporary cathode structures has stimulated the development of an emitter which in many respects is a combination of the oxide cathode and the simple composite surface type[1]. The various embodiments of this type fall under a general classification of dispenser cathodes. A base metal of tungsten or molybdenum, usually sintered, can be prepared with a high degree of porosity so that active material, such as barium, can be diffused through it at a rate suitable for maintaining on the porous structure of the exposed surface an activating layer of polarizable molecules or atoms. Another form of the dispenser cathode has the activating material formed right into the sintered structure. Although these cathodes are of commercial importance and go far to satisfy certain specialized needs for high-current density as electron emitters, this development is of such recent origin that important details concerning the basic emission mechanism still remain to be established by experiment.

The first of these dispenser cathodes to become popular was that proposed by LEMMENS. It is referred to in present-day literature as the "L" cathode. This simple designation has been well received but is no longer quite appropriate because there are new forms of dispenser cathodes that have superseded the "L" cathode. It is therefore considered desirable to use the expression "D" cathode for all of the more recent forms of dispenser cathodes.

C. Theory.

I. Statistical mechanics as a basis for emission equations.

14. Free electrons. The basic concepts needed for the derivation of thermionic emission equations are very elementary and yet they are sufficient for the purpose. One pictures the interior of a conducting crystal as an organized arrangement of atoms characterized by specific interatomic distances which are specifically dependent on the atomic composition and the phase taken on by the crystal, depending upon the temperature and the previous temperature history of the specimen. Each crystal as a whole should be thought of as being electrically neutral within any extended region in the interior. Any excess of charge either positive or negative will be found at the surface only. Quantum theory indicates that most of the electrons that neutralize the positive charge on the atomic nuclei are localized near them and in general contribute nothing to the electrical conductivity of the specimen. The valence electrons associated with these atoms, however, occupy quantum states that extend throughout the entire interior of each isolated crystal and it is to these electrons that the statistical theory of the free electron gas may be applied. The free electron theory as applied to these valence electrons describes their behavior in practically classical terms and finally depends upon experiment to justify the applicability of the simplifying assumptions. It is the purpose of this article to indicate as clearly as possible that the most recent experiments serve to support strongly the concepts of the mechanism of thermionic emission which can be derived from the theory even though they are based on a semiclassical analysis of behavior of valence electrons in a conductor.

15. Three basic assumptions. The first assumption made for the development of this theory is that the inter-electronic forces can be neglected and therefore

[1] A. W. HULL: Phys. Rev. **56**, 86 (1939). — H. J. LEMMENS, M. J. JANSEN and R. LOOSJES: Philips techn. Rev. **11**, 341 (1950). — R. C. HUGHES and P. P. COPPOLA: Phys. Rev. **85**, 388 (1952). — J. Appl. Phys. **23**, 1261 (1952).

the electrons behave as though they were particles of three degrees of freedom. The phase space suitable for representing the behavior of an assembly of electrons can therefore be taken to be a six-dimensional phase space in which a representative point exists for each electron in the assembly. The six bits of information needed to localize this representative point are three coordinates and three components of momentum. The second assumption is that for each quantum state an extension in phase space of size h^3 is needed and that a representative point cannot be localized (nor need it be) more specifically than to indicate that one representative point lies within the quantum-state region. Actually this is not quite the whole story because quantum principles permit two electrons to occupy a single quantum state if their spin vectors are always antiparallel. A factor 2 that appears repeatedly in the equations derived on these assumptions is therefore this weight factor which is thus incorporated into the theory. Already the third postulate has been mentioned, namely, the PAULI Exclusion Principle, which limits the number of electrons in a given quantum state to two with antiparallel spin vectors.

It is the purpose of a statistical theory to find an expression for the distribution of representative points in phase space which is consistent with basic principles of thermodynamics and has associated with it the greatest likelihood of occurrence. The function thus obtained, without the need for introducing any additional assumptions, is the following:

$$f(\varepsilon)\, dx\, dy\, dz\, dp_x\, dp_y\, dp_z = 2\, \frac{dx\, dy\, dz\, dp_x\, dp_y\, dp_z}{h^3} \left[\frac{1}{e^{\frac{\varepsilon - \mu}{kT}} + 1}\right]. \tag{15.1}$$

Some explanation of this equation may make its use and meaning easier to grasp. The energy ε is generally separable into two terms, one of which expresses the kinetic energy of a particle whose representative point lies in a specified region in phase space, and the other term is the potential energy expressible in terms of the coordinates of a particle whose representative point is in that region in phase space. The quantity μ is a constant for a given problem which contains implicitly the concentration of electrons and is a function of the temperature. The fundamental concept that determines the value of this parameter is that the integration of Eq. (15.1) over the entire phase space shall exactly equal the number of electrons in the assembly, that is, the number of free electrons within a crystal, for example. Although this statement defines the manner in which the constant μ is determined, there is a second meaning to the constant which is interesting to note, if it applies to a concentration of electrons of the order of 10^{20} per cm.^{3}. The energy value μ is that to be associated with that quantum state for which the probability of occupancy is exactly one-half.

For electron concentrations less than approximately 10^{19} per cm.^{3} the appropriate value of μ is generally a *negative* number. This statement demands a word of explanation. The simplest application of Eq. (15.1) is made to regions in coordinate space over which there is no change in potential energy. It is therefore sufficient for the present purpose to apply Eq. (15.1) to problems in which the potential energy may be taken to be zero. In that case the energy ε will be the kinetic energy of the electron whose representative point lies in a particular region of phase space. In problems of this kind which occur in connection with the theory of the oxide cathode, the algebraic sign of the quantity μ can be defined as negative, and therefore, it lies below the conduction band in the energy, band system. All of the available quantum states associated with the particular problem for low-density distributions of the electrons are less than half filled if μ is negative.

Note that the extension in phase space $(dx\,dy\,dz\,dp_x\,dp_y\,dp_z)/h^3$ represents the number of quantum states in this extension, since the extension per quantum state is h^3, as mentioned previously. The factor 2 is the double occupancy of a quantum state by the two electrons with antiparallel spin vectors.

Finally, the factor in the square brackets of Eq. (15.1) can be identified by its name the "FERMI factor" which gives a direct means of computing the probability that a given quantum state identified by its energy ε will be occupied. The energy is given explicitly in terms of the momenta and the coordinate values associated with the representative point in phase space. The name given to the quantity μ is the "FERMI level". It is evident at once that if the numerical value of μ is positive, then there can be an energy level ε exactly equal to μ and, as mentioned above, the FERMI factor takes on the value $^1/_2$.

16. The electron flow equation. Although Eq. (15.1) is the basic starting point for all equations relevant to thermionic emission, the following equation which is derived directly from Eq. (15.1) without the introduction of any approximations is the most important equation applicable to thermionic emission.

$$N(\varepsilon_x)\,d\varepsilon_x = 2\,\frac{(2\pi\,m\,kT)}{h^3}\ln\left(1+e^{-\frac{\varepsilon_x-\mu}{kT}}\right)d\varepsilon_x. \tag{16.1}$$

The independent variable in this equation ε_x is defined by

$$\varepsilon_x = \frac{p_x^2}{2m}. \tag{16.2}$$

By the use of Eq. (16.1), the number of electrons $N(\varepsilon_x)\,d\varepsilon_x$ that cross a unit area in unit time with kinetic energy associated with the positive x direction of motion can be computed for the energy range, $d\varepsilon_x$. This equation holds for all values of μ, either positive or negative, and therefore applies to all densities of electrons provided μ is expressed relative to the energy level for which the kinetic energy is zero or, in other words, with respect to the potential energy at the region in space for which the number of electrons crossing a boundary perpendicular to the x direction is being computed. The first application of this formula will be to compute the "random" currents which impinge on various boundaries of a pillbox-like cavity within the interior of a homogeneous crystal.

17. Electrons in a cavity. The pillbox problem is of interest because it is the only example of the application of theory to an experimentally realizable structure for which all of the essential details are easy to describe. The structure visualized is shown in Fig. 1. The cross-hatched solid structure S represents a section through the interior of a single crystal and the cavity within this crystal is represented by C.

The perpendicular distance across that cavity, ab, should be visualized as being not less than 10^{-4} cm. and can very well be any amount larger than this. The requirement that the cavity be esentially pillbox form is necessary because of the need to have the entire interior of the cavity of a single surface structure type. The pillbox has the further advantage that the problem can be handled exactly, even though sufficient electrons exist in the cavity to give an appreciable space-charge field there. The first steps of the discussion can be carried through without the introduction of space charge as a factor of any importance. The space-charge solution of the problem will be introduced later (Sect. 21).

In the energy diagram of Fig. 2 the potential[1] energy of an electron is shown as a function of distance as one progresses in the x direction from A to B. The

[1] See Sect. 27.

region A to a, is the potential in the interior of the solid taken here to be uniform. It will be shown later that the periodicity of the true potential is of no consequence in the thermionic emission theory. The potential of the electron in the space between a and b is shown to be higher than that in the interior of the metal by an amount W_a. This energy difference is the integration of all of the actual forces that act on an electron as it escapes from the metal into the cavity. In the absence of space charge the cavity potential will be constant at distances greater than approximately 10^{-5} cm. from either surface, since the dominant long-range force acting on an electron is the mirror-image force which at this distance has fallen to a negligible amount. Eq. (16.1) may be used to calculate the number of electrons which approach the boundary from the left at a, with energy between ε_x and $\varepsilon_x + d\varepsilon_x$ associated with the x component of the momentum. In the space between a and b the corresponding energy state lies at ε'_x.

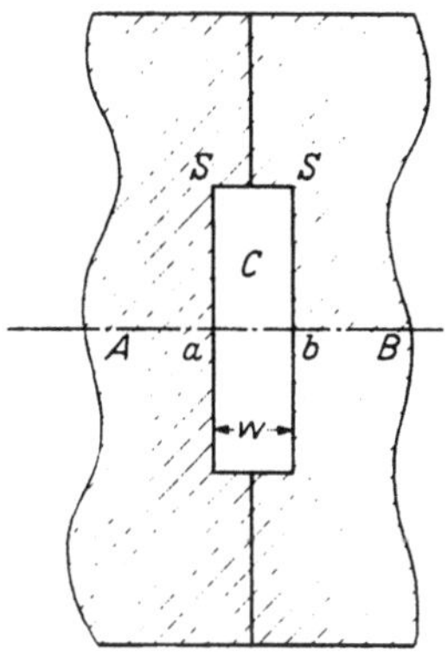

Fig. 1. Cross section of pillbox cavity.

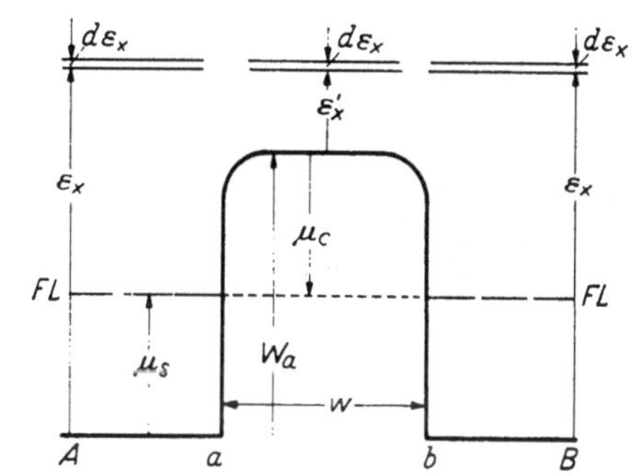

Fig. 2. Potential energy diagram for an electron in the pillbox cavity problem.

For the net current to be zero it is necessary that the current in this band from the left be equal and opposite to the current in the band which approaches surface a from the right. This statement would in general not be true if it were applied to a geometrical configuration in which currents were being observed as electron emission currents in the usual way. An essential part of this analysis is that the entire region surrounding the cavity be at a constant temperature and of course this includes the cavity itself.

In the interior of the crystal the FERMI level, FL, is located at an energy μ_s positive with respect to the potential energy line Aa relative to which the kinetic energy ε_x is referred. The application of Eq. (16.1) shows that there is a simple and yet a necessary condition which must be satisfied if the cavity currents are in statistical equilibrium with the currents flowing in the solid. This condition is that the FERMI level be continuous right through the cavity space. Relative to the potential energy of an electron in the cavity, the FERMI level is *negative*, the amount shown as μ_c. The formal writing of the two equations for the two electron streams serves to illustrate this point and will be used for further development. These equations are the following:

$$N_{xs}\, d\varepsilon_x = \frac{2(2\pi m k T)}{h^3} \ln\left(1 + e^{-\frac{\varepsilon_x - \mu_s}{kT}}\right) d\varepsilon_x, \tag{17.1}$$

$$N_{xc}\, d\varepsilon_x = \frac{2(2\pi m k T)}{h^3} \ln\left(1 + e^{-\frac{\varepsilon'_x - \mu_c}{kT}}\right) d\varepsilon_x. \tag{17.2}$$

It is clear from an inspection of these two equations that the necessary condition for the equality of these two flows of electrons is that the exponents be equal and the following equation may therefore be written:

$$\varepsilon_x - \mu_s = \varepsilon_x' - \mu_c. \tag{17.3}$$

Eq. (17.4) gives additional relations as a result of the reorganization of Eq. (17.3) which are self-evident:

$$\varepsilon_x - \varepsilon_x' = \mu_s - \mu_c = W_a. \tag{17.4}$$

The final rearrangement of this equation is written as follows:

$$-\mu_c = W_a - \mu_s = (\text{true work-function}) \times e = \varphi e. \tag{17.5}$$

This equation stated in words demonstrates the fact that the true work-function expressed in energy units is a direct measure of the location of the FERMI level appropriate to the cavity space outside of a thermionic emitting conductor *when equilibrium exists between the conductor and the space.*

18. The RICHARDSON equation. The integration of Eq. (17.1) between the limits of $\varepsilon_x = W_a$ and ∞ gives a means of calculating the total electron current that impinges on the boundary at a from the interior with the energy range limited, in such a way that any of the electrons included could have gone into the space if reflection effects at the boundary a did not exist. This result is given as Eq. (18.1). The integration of Eq. (17.2) gives the total current which would flow into the conductor from the exterior under conditions of perfect equilibrium. These two currents must be equal. These equations integrated give the following results:

$$I_s = \frac{2e(2\pi m kT)kT}{h^3} e^{-\frac{W_a - \mu_s}{kT}}, \tag{18.1}$$

$$I_c = \frac{2e(2\pi m kT)kT}{h^3} e^{\frac{\mu_c}{kT}}. \tag{18.2}$$

The validity of Eq. (18.2) depends on the assumption that the numerical value of μ_c is not less than $5\,kT$ for an accuracy of better than 1%. For smaller values of μ_c other terms in the power series expansion must be used.

The fact that μ_c is clearly a negative number implies that the electron density in the cavity space is smaller than approximately 10^{19} per cm.³. Under these conditions the statistical theory gives a suitable expression for μ_c which is the following:

$$\mu_c = -kT \ln\left[\frac{2(2\pi m kT)^{\frac{3}{2}}}{n_c h^3}\right]. \tag{18.3}$$

The substitution of this value for μ_c into Eq. (18.2) yields the following:

$$I_c = n_c e \left(\frac{kT}{2\pi m}\right)^{\frac{1}{2}} \tag{18.4}$$

in which n_c is the concentration of electrons in the cavity space near enough to the surface so that space-charge fields can be neglected and yet far enough from the surface so that the mirror-image fields are negligible. This equation is the familiar one from classical mechanics and may be explained in the following terms: $(n_c/2)$ represents the concentration of electrons moving with a component of velocity in any specified direction; $2(kT/2\pi m)^{\frac{1}{2}}$ represents the average of the velocity component of these electrons in a classical distribution; and e is the charge on an electron.

The rewriting of Eq. (18.1) yields at once the RICHARDSON form of the equation so often misused when it is identified with observable thermionic emission.

$$I_s = \frac{4\pi m k^2 e}{h^3} T^2 e^{-\frac{W_a - \mu_s}{kT}}. \tag{18.5}$$

The first factor of this equation may be recognized as the familiar universal thermionic constant A.

$$A = \frac{4\pi m k^2 e}{h^3} = 120 \text{ amp/cm.}^2\, T^2. \tag{18.6}$$

19. The true work-function and its temperature coefficient. An explanation of the reason why Eq. (18.5) cannot be taken as a theoretical prediction of the true functional relation between the observed thermionic current emission density and temperature is in order. Three features of thermionic measurements stand in the way of the direct use of this equation. First, the true work-function is dependent on the temperature; second, the observed energy distribution of emitted electrons is deficient in the low-energy range and, third, the evaluation of current density is seldom correct because of the surface inhomogeneity. More particulars concerning these points are given in the following section.

Eq. (17.5) may be rewritten to express the true work-function in the unit of "electron volts"

$$\text{true work-function} = \varphi = \frac{W_a - \mu_s}{e}. \tag{19.1}$$

In this equation, e is the charge on an electron of 1.6×10^{-19} coulomb and both W_a and μ_s are expressed in the energy unit of joules. Over the temperature range generally used for the determination of a RICHARDSON constant, the true work-function (φ) is not independent of the temperature. For good conductors such as the metals, the FERMI level (μ_s) is a positive quantity and always decreases with an increase in temperature. It is not possible to make a completely generalized statement concerning μ_s if the specimen is a semiconductor except when it applies to the most important of the semiconductor emitters, namely, the oxide cathode. In that case μ_s is generally negative and *increases* in *absolute* value as the temperature is raised (see Sect. 64). At the surface of good conductors the potential energy difference (W_a), (as measured between the "bottom" of the conduction band and an electron at rest at infinity) is of the order of 10 electron volts, whereas for semiconductors the value is close to one electron volt. This quantity is given the name "electron affinity" and depends not only upon the crystal structure of the surface but also on the average dipole moment of any adsorbed atoms. The temperature coefficient of the electron affinity for clean metallic surfaces is in all probability negative. That is, W_a decreases with a temperature increase. Notice, however, that φ may nevertheless increase or decrease with the temperature depending on the relative rates of change of the two terms of Eq. (19.1).

If electropositive adsorbed layers exist on the surface of a good conductor, it is generally true that their dipole moment per unit area will decrease more rapidly than μ_s with increasing temperature. If follows, therefore, since the dipole moment operates to reduce the electron affinity of the composite surface, that with these electropositive layers, W_a will actually *increase* with the temperature. The opposite effect is to be expected for the adsorption of electronegative impurity atoms or molecules.

20. The RICHARDSON equation and its relation to experiment. Eq. (18.5) can be accepted with a great deal of confidence since its validity is derived both from a generalized thermodynamical argument and from the theory of quantum statistics. It is not this equation that is tested in any of the direct experiments usually associated with the measurement of the thermionic emission of electrons from practical specimens.

Since Eq. (18.1) and its equivalent Eq. (18.5) were derived directly from Eq. (17.1) it is equally evident that the validity of Eq. (17.1) in its possible application to the actual thermionic emission problem should be established independently. The uncertainty with regard to the temperature coefficient of the true work-function, which has just been discussed, does not in any way interfere with the experimental determination of the energy distribution of the electrons that are emitted. The prediction made by Eq. (17.1) on the distribution that must be observed to make Eq. (18.5) valid, has been found by direct experiment to be in error. Later in this chapter the detailed experiments by which Eq. (17.1) has been evaluated as a true representation of the energy distribution will be discussed. It is sufficient for the present purposes to state without qualification that the energy distribution predicted by this equation is not found experimentally. The observed distributions found in all experiments capable of yielding direct information in answer to this question show that there is a marked deficiency of low energy electrons. It cannot be stated at present, that the true explanation for this deficiency is a "reflection effect" at the boundary of the surface. The internal consistency in the experimental interpretation that comes very easily from this hypothesis tends to support the view that the deficiency is caused by reflection. An alternative explanation which qualitatively seems less able to correlate the experimental results depends on the assumption that the extraction of the electrons observed as electron emission current disturbs the population of the quantum states near the immediate surface of the conductor. When those electrons capable of escape are taken away, the quantum states will not be refilled continuously, as they would in the pillbox, and a deficiency in slow electrons may be the result.

Finally, a third effect, which is seldom taken into account in the application of the Richardson form of the equation to experimental data, is the lack of uniformity of emission over the surface of the specimen, since in most calculations of current density the total area of the surface is used. If the emission current is observed in the presence of a moderately strong electric field designed specifically to sweep the emitted electrons over to the collector, then the emission is largely dominated by the low work-function areas.

This discussion should serve to justify the statement made above that the current density predicted by Eq. (18.5) should not be identified with thermionic emission as measured by the usual laboratory methods.

II. The density of an electron atmosphere in an enclosed space[1].

21. Introduction. In spite of the uncertainty with regard to the true energy distribution of the electrons emitted from a surface, a solution can be given for the distribution in electron density within a cavity bounded by parallel planes. Even though this problem in itself is somewhat artificial and relates only indirectly to practical needs, the solution can be worked out exactly and is of interest. In its qualitative aspects it bears most directly on the emission properties of hollow cathodes, but, in addition, it may be of importance in connection with the flow of electrons across the cavities within the body of oxide cathodes.

The solution to the "pillbox" problem depends on the assumptions that a cavity is created within the interior of a uniform substance and that two plane

[1] R. H. Fowler: Statistical Mechanics, 2nd edit., p. 366. Cambridge Univ. Press 1936.

conducting surfaces are separated as shown in Fig. 1 by a distance w. It is assumed that this separation between the surfaces is small compared with the linear surface dimensions in order to eliminate the effects of the boundaries that form the periphery of the pillbox. Because of the relatively high ionization potential of atoms making up the emitting surface bounding the cavity, the presence of positive ions can be neglected and at once, it becomes evident that the distribution of electron density within the cavity will be a function only of the distance variable, x, and that the electron density at the midpoint across the cavity will be a minimum. To make the statement that the density of electrons will depend upon the potential according to the BOLTZMANN relation is the equivalent of stating that the FERMI level is continuous from the interior of one conducting surface right through the evacuated space to the interior of the other conductor. Fig. 2 illustrates in a qualitative manner the variation in the motive with distance from the metallic interior at A through the surface boundary at a and across the cavity to the surface at b and then to the interior at B if space charge is neglected. The true potential distribution in the cavity is to be found by the analysis which follows.

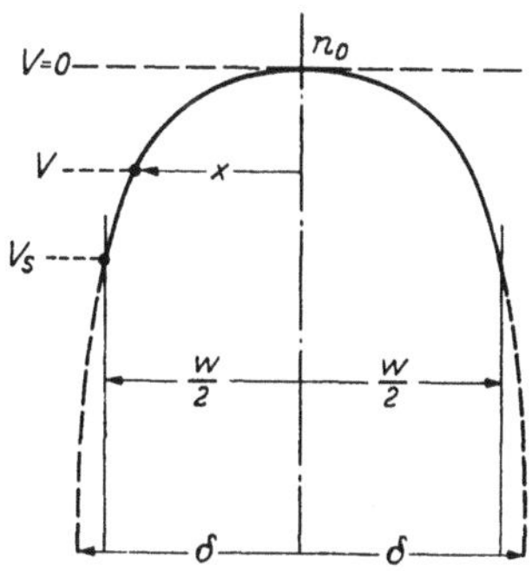

Fig. 3. Electrostatic potential in a cavity with space-charge present.

22. Mathematical formulation of the electron density problem. It is mathematically convenient to measure distances from the motive maximum (see Sect. 27) which lies half-way between the two emitting surfaces. Since the motive is the measure of the potential energy of an electron in this example a motive maximum is a potential minimum in the electrostatic sense. The potential will be expressed as a function of x with reference to its value at this minimum point. This potential, V, at any point x will be positive, since the diagrams of Figs. 2 and 3 represent the variation in the potential energy of an electron which in itself carries a negative charge. Fig. 3 shows the corresponding distribution in electrostatic potential with space charge present. The two basic equations are the following:

$$\frac{d^2 V}{dx^2} = -\frac{n e}{\varepsilon_0}, \tag{22.1}$$

$$n = n_0 e^{\frac{eV}{kT}}. \tag{22.2}$$

In POISSON's relation given as Eq. (22.1), n is the concentration of electrons, e is the electron charge and ε_0 is the permitivity of free space. Eq. (22.2) is the BOLTZMANN relation which expresses the concentration in terms of the potential, V, with respect to the midpoint and the electron density, n_0, at the midpoint. Qualitatively the variation in potential with distance is illustrated in Fig. 3.

It is the purpose of this section not only to outline as briefly as possible the results that one obtains from the simultaneous solution of these two equations, but also to show how these results can be fitted to specific boundary conditions for their application.

These two equations can be brought together and simplified by the introduction of dimensionless variables defined as follows:

$$\psi = \frac{eV}{kT}, \tag{22.3}$$

$$\chi = \frac{x}{x_1} \quad \text{with} \quad x_1^2 = \left[\frac{kT\varepsilon_0}{e^2 n_0}\right]. \tag{22.4}$$

By direct substitution of these equations into Eqs. (22.1) and (22.2) the following relation may be written:

$$\frac{d^2\psi}{d\chi^2} = -e^{\psi}. \tag{22.5}$$

The first step in the solution of this equation yields a relation from which the electric field in the cavity can be calculated and is given as follows:

$$\frac{d\psi}{d\chi} = \sqrt{2}\,(e^{\psi} - 1)^{\frac{1}{2}}. \tag{22.6}$$

The second step in the solution results in the final answers expressed in either of the forms that follow:

$$\chi = \sqrt{2}\,\operatorname{arc\,tan}\sqrt{e^{\psi} - 1}, \tag{22.7}$$

$$\psi = \ln\left(\sec^2\frac{\chi}{\sqrt{2}}\right). \tag{22.8}$$

An inspection of either of these equations shows that the total range for the variable χ, is $0 \lesseqgtr \chi \lesseqgtr (\pi/\sqrt{2})$.

23. Solution to the electron density problem. The solutions given as Eqs. (22.7) and (22.8) may be rewritten in terms of the original variables to give the distance at which any specified potential occurs or to give the potential found at any specified distance. These equations are as follows:

$$x = \left[\frac{2kT\varepsilon_0}{e^2 n_0}\right]^{\frac{1}{2}} \operatorname{arc\,tan}\left(e^{\frac{eV}{kT}} - 1\right)^{\frac{1}{2}}, \tag{23.1}$$

$$V = \frac{kT}{e}\ln\sec^2\left[x\left(\frac{e^2 n_0}{2kT\varepsilon_0}\right)^{\frac{1}{2}}\right]. \tag{23.2}$$

The equation for the electric intensity E_x or the field at any point is written as follows:

$$E_x = -\frac{dV}{dx} = \left[\frac{2kTn_0}{\varepsilon_0}\right]^{\frac{1}{2}} \tan\left[x\left(\frac{e^2 n_0}{2kT\varepsilon_0}\right)^{\frac{1}{2}}\right], \tag{23.3}$$

$$E_x = -\frac{dV}{dx} = \left[\frac{2kTn_0}{\varepsilon_0}\right]^{\frac{1}{2}}\left(e^{\frac{eV}{kT}} - 1\right)^{\frac{1}{2}}. \tag{23.4}$$

24. The fit to boundary conditions. The difficulty with these equations as they now stand is that the concentration at the center of the cavity, that is, n_0, must be known before quantitative calculations can be made. This concentration depends not only on the properties of the emitting surfaces, a and b, but also on the dimensions of the cavity (see Fig. 3). The following steps will serve as a guide to one of the more satisfactory methods of applying these results.

An inspection of the diagram of Fig. 3 indicates that there is a distance, δ, always greater than $w/2$ at which the potential approaches infinity for any given value of the electron concentration, n_0. Eq. (24.1) serves to define this value of δ and the relation may be written as follows:

$$\delta^2 = \frac{kT\varepsilon_0}{e^2 n_0}\cdot\frac{\pi^2}{2}. \tag{24.1}$$

It is convenient for the purpose of calculation to define the ratio of the actual half width $w/2$ of the cavity to this quantity δ as follows:

$$z = \frac{w}{2\delta}. \tag{24.2}$$

It is to be noted that if the concentration, n_0, is very small then δ would be very large and it is therefore clear that the range in the quantity z is from zero to one.

A rewriting of Eq. (24.1) gives an expression for the electron concentration at the center of a cavity.

$$n_0 = z^2 \left[\left(\frac{2}{w}\right)^2 \frac{\pi^2 k \varepsilon_0}{2e^2} T\right]. \tag{24.3}$$

Notice that the maximum possible value of the concentration occurs if z is unity. Define this quantity as n_{0m} and it may be computed in terms of the temperature and the cavity width by the following equation:

$$n_{0m} = 9.4 \times 10^4 \frac{T}{w^2} \text{ electrons per m}^3. \tag{24.4}$$

Let the concentration of electrons at the emitting surface of the cavity be n_s and let the difference in potential between the center of the cavity and the region just outside of the surface be V_s. With these definitions the following relations may be written from Eqs. (22.2) and (22.8):

$$n_0 = n_s \, \mathrm{e}^{-\frac{eV_s}{kT}} = n_s \cos^2\left(\frac{\pi}{2} z\right). \tag{24.5}$$

The final equation which permits the determination of the appropriate value of z in terms of the known parameters of the problem is given as:

$$\frac{n_s}{n_{0m}} = \frac{z^2}{\cos^2\left(\frac{\pi}{2} z\right)}. \tag{24.6}$$

The equation which will serve for the computation of n_s comes directly from the fact that I_s of Eq. (18.1) is $n_s e (kT/2\pi m)^{\frac{1}{2}}$ and may be written as follows:

$$n_s = \frac{2(2\pi m k T)^{\frac{3}{2}}}{h^3} \mathrm{e}^{-\frac{e\varphi}{kT}}. \tag{24.7}$$

For this equation to be exact, the true work-function φ is needed and it is to be identified as $(W_a - \mu_s)/e$. To simplify the calculation, the assembly of fundamental constants may be evaluated and the equation written as follows:

$$n_s = 4.83 \times 10^{21} T^{\frac{3}{2}} \mathrm{e}^{-\frac{e\varphi}{kT}} \text{ (electrons per m}^3). \tag{24.8}$$

For any given problem, a numerical value can be established for the ratio (n_s/n_{0m}) from Eqs. (24.4) and (24.8) and it is only necessary, therefore, to determine the correct value of z to satisfy Eq. (24.6). To make this procedure easy, values of this function have been computed for the range of z from 0.05 to 0.99 and are given in Table 1. Eq. (24.9) and (24.10) serve as the most convenient means of calculating z in the extreme low range and in the extreme high range:

$$z = \left(\frac{n_s}{n_{0m}}\right)^{\frac{1}{2}} \quad \text{for} \quad 0 < z < 0.1, \tag{24.9}$$

$$z = 1 - \frac{2}{\pi}\left(\frac{n_{0m}}{n_s}\right)^{\frac{1}{2}} \quad \text{for} \quad 0.98 < z < 1. \tag{24.10}$$

In terms of the density ratio, Eq. (24.9) applies when the surface density is less than one percent of the maximum possible density at the center of the cavity n_{0m} and Eq. (24.10) applies when the surface density is more than a thousand times greater than the maximum density at the center.

Since the maximum density will have been calculated already from Eq. (24.4), the true density at the center of the cavity is given at once by the following:

$$n_0 = z^2 n_{0m}. \tag{24.11}$$

As an aid to computation, the solutions to this problem may be written in terms of the parameter z and the controllable variables w and T.

The distance x from the midpoint of the cavity at which a specified V is found may be computed as

$$x = \frac{w}{\pi z} \arc\tan \left(e^{\frac{eV}{kT}} - 1\right)^{\frac{1}{2}}. \tag{24.12}$$

The potential at any distance is given by

$$V = \frac{kT}{e} \ln \left(\sec^2 \pi z \frac{x}{w}\right). \tag{24.13}$$

The field may be computed at any distance point or potential by the following equations:

$$E_x = -\frac{dV}{dx} = \left(\frac{2\pi k}{e}\right) \frac{zT}{w} \tan \left(\pi z \frac{x}{w}\right), \tag{24.14}$$

$$E_x = -\frac{dV}{dx} = \left(\frac{2\pi k}{e}\right) \frac{zT}{w} \left(e^{\frac{eV}{kT}} - 1\right)^{\frac{1}{2}}. \tag{24.15}$$

The results of this analysis may be summarized by the following statements:

1. The maximum electron density that can be obtained at the center of a cavity is proportional to the temperature and inversely proportional to the square of the smallest dimension.

2. To attain approximately 90% of this maximum density at the center, the concentration of electrons at the surface of the cavity must be greater by a factor of approximately 200.

3. Further increase in the density of electrons at the surface of the cavity, even though it be many orders of magnitude, cannot increase the density at the center by more than 10%.

4. By the proper control of the porosity of the surface of a cathode, the effective emission current density of the cathode can be made higher than the actual average net current density that crosses the surface boundary which separates the solid from the evacuated space within the pore. This property may have an important bearing on the performance characteristics of suitably constructed oxide cathodes.

III. Field effects with current flow.

25. Three field effects. The three most important field effects that relate to thermionic emission are: 1. the influence of retarding potentials, externally applied; 2. the influence of accelerating fields, externally applied; and 3. the influence of the combined effect of a space charge and external fields. The theoretical aspects of these three field effects will be discussed in this order.

26. The influence of retarding fields. The earliest experiments on thermionic emission established that the observed currents that flow across the space in a diode structure are very dependent upon the applied voltage difference between the electron emitter and the collector. If the emission current density is sufficiently low, so that space-charge effects can be neglected, the two principal regions associated with the applied voltage are: 1. the accelerating field range, and 2. the retarding field range. These ranges are separated quite sharply because of the fact that the current change in the first region is almost negligible, while in the second region the current change goes through many orders of magnitude for each additional volt of applied retarding potential. Superficially, this phenomenon is accounted for as a measure of the energy distribution of the emitted electrons. It is so difficult to satisfy the necessary experimental conditions which will permit the direct interpretation of the results in terms of the electron energy distribution that very few authentic data are available.

The starting point of any theory covering this phase of the subject is Eq. (15.1) From this equation, it is possible to write a general expression for the current received as a function of the applied voltage between the emitter and the collector. This equation is the following:

$$I = e \int_{p_z=-\infty}^{p_z=\infty} \int_{p_y=-\infty}^{p_y=\infty} \int_{p_x=0}^{p_x=\infty} D(p, B)\, F_c(p, V, T)\, n(p)\, dp_x\, dp_y\, dp_z \tag{26.1}$$

$D(p, B)$ = transmission over emission barrier; $F(p, V, T)$ = transmission to and over collector barrier.

$$n(p) = \frac{2 p_x}{h^3 m} \frac{1}{e^{\frac{(p_x^2+p_y^2+p_z^2)/2m-\mu_s}{kT}}+1}. \tag{26.2}$$

Even though Eq. (26.1) gives the current observed at the collector per unit area of the emitter, the two transmission functions $D(p, B)$ and $F(p, V, T)$ must be known to make the calculation. Therefore to apply Eq. (26.1) to experimental results, it is necessary to design the experiment in such a manner that simplifying assumptions with regard to the transmission coefficients are legitimate. Since the crucial part of this theory is dependent on these assumptions, the following paragraphs will deal with the more important aspects of the restrictions needed.

The symbol $D(p, B)$ used for the transmission of electrons over the emission barrier has been chosen to call attention to the dependence of this factor upon the components of momentum (p) with which the electron approaches the barrier and the symbol B includes the properties of the emission surface barrier itself. The simplest known surface barrier is that for a clean, atomically smooth single face of a single crystal of a good conductor. The only experiment reported in the literature which is designed to evaluate this transmission function specifically is that of HUTSON[1].

Although the experimental result obtained by HUTSON does not lend itself to a representation by a simple analytic expression for the transmission coefficient, the results are in such excellent quantitative agreement with an analytic function found empirically by NOTTINGHAM[2] that this expression will best serve the present purpose. These results may be summarized by the following relations:

Clean single crystal surface

I. Range in $(p_x^2/2m) < W_a$

$$D(p, B) = 0. \tag{26.3}$$

II. Range $(p_x^2/2m) \gtrless W_a$

$$D(p, B) = \left(1 - e^{-\frac{p_x'^2/2m}{\omega}}\right) \tag{26.4}$$

where $(p_x'^2/2m) = p_x^2/2m - W_a$.

The interpretation placed on these conditions is that no electron can escape unless it approaches a barrier perpendicular to the x direction with a kinetic

[1] A. R. HUTSON: Phys. Rev. **98**, 889 (1955).
[2] W. B. NOTTINGHAM: Phys. Rev. **49**, 78 (1936).

energy associated with that component of the momentum which is greater than the electron affinity W_a. The usual assumption is that, if this kinetic energy exceeds W_a then its probability of escape is unity. The expression given in Eq. (26.4) indicates that the probability of escape approaches unity only after the *excess* kinetic energy expressed as $(p_x'^2/2m)$ exceeds the empirical constant ω by a considerable factor. The empirical value of ω is 3.05×10^{-20} joules or 0.191 electron volt.

Even though this transmission coefficient given by Eq. (26.4) is in excellent agreement with the experiments of both HUTSON[1] and NOTTINGHAM[2], it yet remains to be proven conclusively that the phenomenon is correctly interpreted as a deficiency of slow electrons because of the "reflection effect" at the boundary. Reflection can be established unambiguously only by the experimental observation of it for very slow electrons which impinge on the external surface of a clean, single crystal face. This direct evaluation is being undertaken by LANGE[3].

If the surface is not uniform, the transmission over it may be far more complex and may even be field-sensitive when the retarding field is small. Polycrystalline surfaces, even though they are clean, generally exhibit different crystal facets for the individual crystals that make up the surface. Since these facets are likely to have different work-functions, local contact-difference in potential fields will make the specification of a suitable choice for the electron affinity W_a difficult, if not impossible. The expression "patch effect" is used to describe this difficulty qualitatively. It will become clear after a discussion of the collection coefficient $F_c(p, V, T)$ that all cathode inhomogeneities become more and more unimportant as the retarding potential increases.

The transmission coefficient through and over the collector barrier is also very difficult to evaluate. Factors that are important may be enumerated: 1. geometrical arrangement of the collector with respect to the emitter and all other elements in the experimental diode such as glass walls, insulators, etc. 2. the surface structure of the collector including non-uniformities in work-function that are derived either from a polycrystalline structure or a non-uniform distribution of polarizable adsorbed atoms, 3. the energy distribution of the electrons that impinge on the collector barrier since there may be a reflection effect at this barrier which is energy-dependent, 4. changes in the surface properties of the collector (specifically, its work-function) which often accompany changes in the current density being received as well as the previous history of the electron bombardment of that surface[4].

If the purpose of the experiment is to determine the energy distribution of the electrons emitted from the cathode, then every means must be adopted to eliminate as many of the complexities enumerated above as possible. Hence, the reader is therefore warned in his future studies that results obtained from experiments in which due consideration has not been given to these factors should be viewed with scepticism.

A superficial examination of the problem suggests immediately that the ideal geometrical arrangement for the determination of an electron energy distribution

[1] A. R. HUTSON: Phys. Rev. **98**, 889 (1955).

[2] W. B. NOTTINGHAM: Phys. Rev. **49**, 78 (1936).

[3] W. J. LANGE: Research in progress, Department of Physics, Massachusetts Institute of Technology, Cambridge, Massachusetts.

[4] W. B. NOTTINGHAM: Phys. Rev. **39**, 183 (1932); **44**, 311 (1933); P. L. COPELAND: Phys. Rev. **57**, 625 (1940).

would be that of a closely-spaced diode formed with parallel planes. It is safe to say that such a diode is practically impossible to achieve since for it to be successful the collector would have to be a single crystal capable of being heated sufficiently to drive off all adsorbed films. Collectors such as these have not been made[1]. The next best solution involves the construction of a diode with coaxial cylinders. This structure, also, is subject to the difficulty of reflection from the collector, unless the collector radius is large compared with the emitter radius. The coaxial diode has the distinct advantage that electron reflections from the collector are minimized because any electron which leaves the filamentary cathode with sufficient energy to have been collected will, in general, find itself repeatedly coming to the collector even though it may have been reflected one time or more. Since the collecting cylinder can be constructed from a refractory metal such as tantalum, the purity of the surface can be insured under proper vacuum conditions. A structural form like this eliminates practically all the objectionable features that might interfere with the correct interpretation of the emission properties of the emitter in the presence of a retarding field. For a cylindrical structure, Eq. (26.1) has been put into usable form by NOTTINGHAM[2].

The emitter transmission function $D(p, B)$ is taken to be that given as Eqs. (26.3) and (26.4). The factor $F_c(p, V, T)$ which represents the probability of electron collection at the collector, has two ranges that can be explained most concisely by the following equations if the z direction is along the axis, the x direction along the radius and the y direction tangent to the emitting filament at the surface point of electron emission.

I. Range:

$$\left.\begin{aligned} p_x'^2 + p_y^2\left(1 - \frac{r^2}{R^2}\right) &< 2meV, \\ F_c(p, V, T) &= 0. \end{aligned}\right\} \tag{26.5}$$

II. Range:

$$\left.\begin{aligned} p_x'^2 + p_y^2\left(1 - \frac{r^2}{R^2}\right) &\geqq 2meV, \\ F_c(p, V, T) &= 1. \end{aligned}\right\} \tag{26.6}$$

In these expressions the potential (V) is defined as the *retarding* potential in the space and is the work required to carry an electron from a point just outside the emitter surface to a point just outside the collector surface. Since it is a necessary condition that the radius of the emitter (r) be small compared with the radius of the collector (R) in order to minimize the effect of reflection at the collector, the condition equation given above (which was derived from an analysis of the electron trajectory between coaxial electrodes) can be approximated by one's neglecting (r^2/R^2) in comparison with unity. It is this condition on the probability of transmission to the collector that establishes the limits of integration not specified explicitly in Eq. (26.7). This equation gives the electron emission current density expected as a function of the temperature and the retarding potential.

$$I = \frac{2e}{h^3}\, e^{-\frac{W_a - \mu_s}{kT}} \int_{p_z=-\infty}^{+\infty} \int^{+\infty} \int^{+\infty} \left(1 - e^{-\frac{p_x'^2}{2m\omega}}\right) \frac{p_x'}{m}\, e^{-\frac{p_x'^2 + p_y^2 + p_z^2}{2mkT}}\, dp_x'\, dp_y\, dp_z. \tag{26.7}$$

---

[1] H. SHELTON: Research Laboratory of Electronics, Massachusetts Institute of Technology, Cambridge, Massachusetts, is experimenting with a single crystal tantalum collector to receive an electron beam collimated by a magnetic field.

[2] W. B. NOTTINGHAM: Phys. Rev. **49**, 78 (1936).

In spite of the apparent complexity of this equation, it can be integrated and put into the usable form indicated by the following equations:

$$S=\frac{eV}{kT}=\frac{11600}{T}V, \tag{26.8}$$

$$I=I_{00}\left[G(S)-F(S,kT/\omega)\right], \tag{26.9}$$

$$I_{00}=\frac{4\pi m k^2 e}{h^3}T^2 e^{-\frac{W_a-\mu_s}{kT}}, \tag{26.10}$$

$$G(S)=\frac{2}{\sqrt{\pi}}S^{\frac{1}{2}}e^{-S}+1-\frac{2}{\sqrt{\pi}}\int\limits_0^{S^{\frac{1}{2}}}e^{-x^2}dx, \tag{26.11}$$

$$F(S,kT/\omega)=\frac{1}{1+\frac{kT}{\omega}}\left[\left(\frac{\omega}{kT}\right)^{\frac{1}{2}}\frac{2}{\sqrt{\pi}}e^{-S\left(1+\frac{kT}{\omega}\right)}\int\limits_0^{\left(\frac{SkT}{\omega}\right)^{\frac{1}{2}}}e^{y^2}dy+1-\frac{2}{\sqrt{\pi}}\int\limits_0^{S^{\frac{1}{2}}}e^{-x^2}dx\right]. \tag{26.12}$$

Since the controllable variables are the retarding potential V and the temperature T, the function expressed in Eq. (26.12) must be worked out in tabular form with sufficient detail so that smooth curves can be drawn for the purpose of interpolation. An inspection of Eqs. (26.11) and (26.12) shows that Eq. (26.12) reduces exactly to Eq. (26.11) as (kT/ω) approaches zero. It is, therefore, not necessary to tabulate the function $G(S)$[1], of since the values of this function for the range S from 0 to 10 will be found in the first row of the Table 2. A graphical representation of Eq. (26.9) is shown in Fig. 4. Three of the curves show the computed result for the three temperatures, 813, 1160, and 1852° K. The value of ω used is the one found experimentally: 3.05×10^{-20} joules (0.191 ev). The fourth curve in the diagram is the universal curve expected for a transmission coefficient $D(p,B)$ of unity for all positive values of p'_x.

Even though retarding potential measurements often have questionable validity for the determination of the true energy distribution of the electrons emitted from a thermionic source, an analysis of these measurements may yield other useful information. The temperature of the emitter may be determined sufficiently accurately to be useful information if the rate of change of emission is determined as a function of the applied voltage for the largest retarding potentials. Observations in this range demand the measurement of the smallest currents possible consistent with the insulation properties of the diode structure. At high retarding fields, cathode inhomogeneity in structure and reflection effects become less and less important. The accuracy of the agreement found by Fan[2] and Hung[3] also supports the view that cathode temperatures may be determined in this manner under specialized geometrical conditions.

An analysis of this problem by Ikeharra[4] has supplied a method of determining the temperature of oxide cathodes by evaluation of higher order harmonics observed in the output of a diode (or its equivalent) while a small range in the retarding potential characteristic is swept by a sinusoidal voltage variation. (For additional discussion see Sects. 60 and 62.)

[1] W. Schottky: Ann. Phys., Lpz. **44**, 1011 (1914). Tabulated by L. H. Germer: Phys. Rev. **25**, 795 (1925).

[2] H. Y. Fan: J. Appl. Phys. **14**, 552 (1943).

[3] C. S. Hung: J. Appl. Phys. **21**, 37 (1955).

[4] S. Ikeharra: J. Appl. Phys. **25**, 725 (1954).

A second use for thermionic measurements in retarding fields is in the quantitative determination of changes in the work-function of the collector. The work of I. LANGMUIR and KINGDON[1] and also of D. B. LANGMUIR[2] may be cited as examples of the use of a retarding potential method for the determination of the variation in the work-function of the collector.

Although the current received by a particular collector operated at a high retarding potential is practically independent of the emission properties of the cathode for a given *applied* potential, the current observed is a very sensitive function of the anode itself. The reason for this sensitivity is that the applied potential measures the difference in energy of the FERMI level of the collector with respect to the FERMI level of the emitter. The motive[3-6] maximum over which the electrons must flow in order to register as a current at the collector is higher than the FERMI level of the collector by its own work-function. Therefore, even though the applied potential be maintained constant, the change in the work-function of the collector can alter the current that flows to it in a very significant manner; while a corresponding change in the emitter is generally of no measurable significance under these conditions. To use this method as a quantitative measure of any change in the work-function of the collector, it is necessary that other fields that might possibly alter the electron trajectories remain unchanged or else some method must be used for normalizing them. This method, which is adaptable for the study of surface conditions and alterations in these surface conditions, should find a wider application than it seems to have at present.

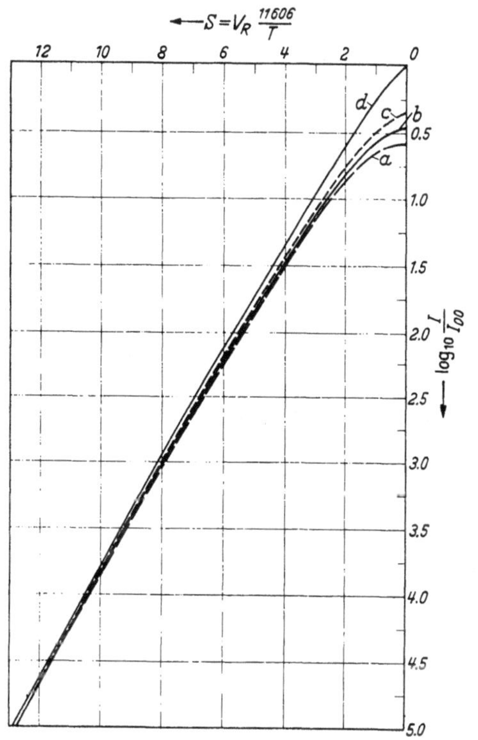

Fig. 4. Computed energy distributions. *a* 813° K with reflection. *b* 1160° K with reflection. *c* 1852° K with reflection. *d* No reflection.

Two additional precautions need to be mentioned in association with the analysis of retarding potential measurements. Gross inhomogeneities on the emitter surface, such as are found on partially coated oxide surfaces, must be considered. Care must be exercised to confine the range of emission current density to that for which space-charge effects are negligible (see Sect. 83). Development of a space-charge limitation first occurs at the collector as the retarding potential is reduced, and then the maximum in the motive for an escaping electron progresses across the diode as the applied potential to the collector is made more positive. This motive maximum is produced by the

[1] I. LANGMUIR and K. H. KINGDON: Phys. Rev. **34**, 129 (1929).

[2] D. B. LANGMUIR: Phys. Rev. **49**, 428 (1936).

[3] See Sect. 27.

[4] I. LANGMUIR and K. H. KINGDON: Proc. Roy. Soc. Lond., Ser. A **107**, 61 (1925).

[5] P. W. BRIDGEMAN: The Thermodynamics of Electrical Phenomena in Metals. New York: Macmillan Co. 1934.

[6] C. HERRING and M. H. NICHOLS: Rev. Mod. Phys. **21**, 185 (1949).

space charge and limits the emission. In simple geometrical cases, a theoretical analysis can be applied to determine the space-charge distribution and, therefore, the emission properties in the presence of space charge. In most practical tubes, it is not profitable to try to work out a detailed theory to be applied to a particular structure.

27. The influence of accelerating fields. In order to discuss the influence of electron accelerating fields on thermionic emission phenomena, it is desirable to treat the problem under conditions such that the total charge of electrons in transit from the emitter to the collector will be negligible in comparison with the net surface charge on the collector. This limitation is the equivalent of stating that the influence of space charge is negligible. It is also necessary to define explicitly the potential field in which the electron is moving. Since mirror-image forces must be considered, the potential is defined in a special way and is given the name "motive" as suggested by I. LANGMUIR[1]. The definition of "motive" may be given in LANGMUIR'S words:

"The motive is thus defined as a scalar quantity whose gradient in any direction and at any point represents the force component per unit charge which must be applied to an electron or an ion to hold it in equilibrium at the given point." The gradient of the motive is the "motive intensity" and is therefore a vector which measures the force per unit charge *on an electron* at the point in the space at which the gradient is taken. It is to be noted that this definition differs from that of "electrostatic potential" and of "electric intensity" in that definitions for these are given as the limit of the time average of the work or force, respectively, per unit charge as the charge approaches zero. It is appropriate to state at this point that even the BOLTZMANN equation for the electron density dependence on motive is always applicable. The potential and the fields that enter into the solution of POISSON'S equation deal only with the "electrostatic" contribution to the motive. It was for this reason that the formulation and discussion of the pillbox problem in Sect. 22 defined the potentials used there as directly applicable at distances greater than 10^{-5} cm. from the surface. This is the distance at which mirror-image forces are usually negligible.

As an electron leaves a conducting surface, it is acted upon by short-range and by long-range forces. The short-range forces dominate for a distance of the order of 1 or 2 interatomic distances (10^{-7} cm. or less) and depend on the surface composition, that is, the orientation of the actual surface with respect to atomic or crystallographic arrangement of the interior and with respect to the adsorbed atoms in the immediate neighborhood of the emission point. Even though there can be no doubt about the existence of these forces, no experiments have been formulated which yield quantitive information other than the knowledge that the integral of all of the forces with respect to distance determines the true work-function. It is this variation in the short-range forces from substance to substance and from surface to surface that accounts for the observed differences in their true work-functions. The most important of the long-range forces was identified by SCHOTTKY[2] as the "mirror-image" force.

The second kind of long-range force arises from cathode inhomogeneity. No practical electron emitter has ever been constructed which is truly free from inhomogeneity. For example, a single crystal of tungsten formed as part of a wire and used in a coaxial diode always exhibits in its surface a variation in the crystal structure which depends on the orientation of the crystal axes with respect

[1] I. LANGMUIR and K. H. KINGDON: Proc. Roy. Soc. Lond., Ser. A **107**, 61 (1925).
[2] W. SCHOTTKY: Phys. Z. **15**, 872 (1914).

to the axis of the wire. For the present purposes it is sufficient to say that the greater the linear dimensions of the areas of inhomogeneity, the greater the distance into space near the cathode that these local field effects will extend. Since extreme differences in work-function are unlikely to exceed one or two electron volts, these very extended fields will be correspondingly weaker the larger the area of each patch of high or low work-function. Neighboring high work-function areas cause local fields to exist over the low work-function areas which inhibit the emission of electrons from the latter. As the externally applied accelerating field is increased, it can be made larger than the local field at a relatively low value of applied potential. After the influence of the local field has been largely wiped out, the emission obtained from the various patch areas is that which would have been available, expressed as emission current density independent of the presence of inhomogeneity. The total emission is the sum of the emissions from the various areas and is therefore dominated by low work-function regions. It will be shown in Sect. 71 that since there is such a wide variation in the current density over the surface of a single-crystal tungsten wire of circular cross section, approximately one-half of the wire emits all of the observed electrons.

Basically, the reason that electron emission increases as an externally applied accelerating field is increased is that more electrons become available for emission as the maximum in the barrier over which the electrons must escape is reduced relative to the FERMI level in the interior of the emitter. Spurious effects can occur which conceivably cause a decrease in emission as the field increases. These spurious effects always reflect lack of stability of the surface conditions, which can be brought about either by the ionic migration in the interior of the solid near its emission surface or by the adsorption on the surface of electronegative gases (cathode poisons) that often arise as a result of the bombardment of the anode or an associated insulator by the electrons. The yield of such poisonous products is often negligible for an electron energy of five volts or less but it continues to increase after onset with the increase in electron energy[1].

There are two ways of evaluating the reduction in true work-function with accelerating field, if one carries through the experiment in such a manner that the emitter surface remains unchanged. These ways are: 1. direct measurement of the actual energy distribution of the electrons, 2. the measurement of the variation in the integrated thermionic emission as a function of the field. The thermionic emission method depends on the validity of certain assumptions which can be explained most easily by a quick review of the detailed theory applicable to this experiment.

The motive function for an electron as it leaves a conducting surface is represented qualitatively by the solid line of Fig. 5a. Corresponding to this motive function, there is a force function per unit charge which is represented by the line in Fig. 5b. This motive intensity is obtained, at least in principle, from the gradient of the motive as defined above. In the presence of an accelerating field, the modified motive function is everywhere lower than the zero field curve and the computation which must be made in order to determine the extent of the lowering is most easily seen in the force-function diagram of Fig. 5b. Over the range in distance from x_0 to x_c the force function in the absence of space charge is diminished by an amount directly proportional to the accelerating potential V_c multiplied by a geometric factor $G(x)$. The accelerating potential

[1] J. D. HOBBS: Energy Dependence of Electron Produced Poisoning of Oxide Cathodes. Master's Thesis, M. I. T., 1954.

is defined as the applied potential v corrected for by the contact difference in potential P_T as shown in Eq. (27.1).

$$V_c = v - P_T. \tag{27.1}$$

The contact difference in potential is defined as the difference between the true work-function of the collector and that of the emitter when these work-functions are expressed in electron-volts. This relation is given as Eq. (27.2).

$$P_T = \varphi_c - \varphi_s. \tag{27.2}$$

The electric intensity (E), shown in Fig. 5b, may be a function of the distance from the emitter surface depending on the geometrical arrangement of the emitter and the collector. The general expression for the intensity is the following:

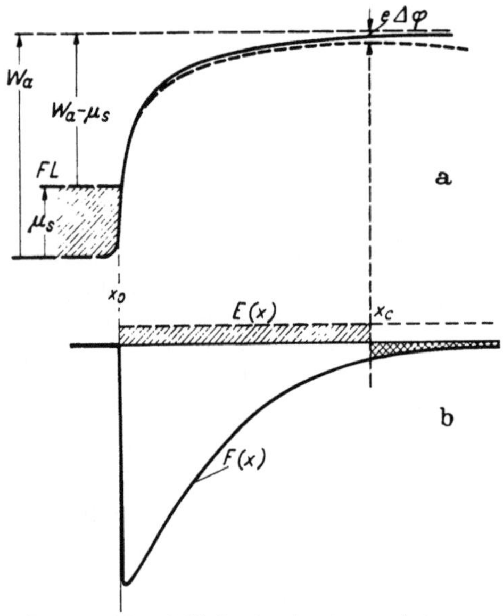

Fig. 5a and b. a) Motive function for an electron as it leaves a conducting surface. b) Motive intensity for an electron.

$$E = V_c\, G(x). \tag{27.3}$$

The specific expression for parallel planes separated by distance D gives the field as independent of x as follows:

$$E = V_c/D. \tag{27.4}$$

For coaxial cylinders of radii r and R, the field depends on x and is

$$E = V_c\left[\frac{1}{\left(1+\frac{x}{r}\right) r \ln\frac{R}{r}}\right]. \tag{27.5}$$

Notice that the field in the neighborhood of an emitting cylinder of radius r is constant for distances from the surfaces which are small compared with the emitter radius.

An inspection of Fig. 5b permits one to write the expression for the reduction in true work-function as a function of the accelerating potential. This result is written as follows:

$$\Delta\varphi = V_c \int_{x=x_0}^{x=x_c} G(x)\,dx + \int_{x=x_c}^{x=\infty} F(x)\,dx. \tag{27.6}$$

It is evident from this equation that if the force function $F(x)$ is known, the reduction in work-function can be computed, since at the critical distance (x_c) the force function and the accelerating field are also related by Eq. (27.7):

$$V_c\, G(x_c) + F(x_c) = E_c + F(x_c) = 0. \tag{27.7}$$

The converse of the above statement is also true: that is, if there is an experimental way of making a direct determination of the change in work-function $\Delta\varphi$ with field, then the force function $F(x)$ can be determined. Both of these procedures have been used and will be illustrated below.

The method followed by SCHOTTKY was to assume the force function per unit charge (motive intensity at zero external field) to be dominated by the mirror image. This function is the following:

$$F(x) = -\frac{e}{16\pi\,(x-x_0)^2\,\varepsilon_0} \text{ newtons per coulomb (or volts per meter)}. \tag{27.8}$$

This function serves as a means of difining the location (x_0) which can be taken as the "perfect" image boundary of the conductor. The original distances used in the above equations were presumably measured from some arbitrary point $x=0$ considered to be at the surface of the emitter. If the long-range forces at some distance from the surface are given in terms of Eq. (27.8), it is as though the actual force were matched to this equation in such a manner that, if the mirror-image function continued in exactly this form down to x_0 it would become infinite. The finiteness of the actual integral (i.e., the electron affinity) establishes that the true forces against which the electron works as it escapes are weaker than the mirror-image force at short range. It has been pointed out by DE BOER[1], LANGMUIR, and others that, in a qualitative manner, the difference between the true motive-intensity function and the mirror-image function should be related to the interatomic spacing of the surface atoms or to the roughness of the surface through which the electron is escaping. For a smooth and closely packed surface, such as the 110 surface of tungsten, the short-range forces are the strongest and the integral of these forces gives a very high electron affinity. It follows directly from Eqs. (27.7) and (27.8) that the escape distance (x_c) is given by Eq. (27.9):

$$x_c = x_0 + \left(\frac{e}{16\pi\,\varepsilon_0\,G(x_c)\,V_c}\right)^{\frac{1}{2}}. \tag{27.9}$$

If the structure of the diode is such that the geometrical function $G(x)$ is dependent on the distance, full knowledge of that function is needed in order to carry through the evaluation indicated by Eq. (27.6). If over the range of distance from the surface to x_c, the value of $G(x)$ is a constant (as is the case for parallel planes and cylinders if the distance is small compared with the radius of curvature of the emitter), it is possible to carry through the integration. The result is the well-known SCHOTTKY equation which shows that the decrease in work-function is directly proportional to the square root of the surface electric intensity. This equation checks with experiment only if the dominating long-range force, that acts upon an electron as it escapes, is the mirror-image force.

The only direct observation of this "SCHOTTKY effect" was made by HUTSON[2] even though *indirect* observations have been made both in association with thermionic emission and with photoelectric emission.

One needs only to refer to Eq. (26.1) to realize that the introduction of this accelerating field effect can best be made by incorporating it into the definition of $D(p, B)$ which expresses the probability that an electron can escape from the emitter. If it is possible to substitute for the electron affinity, W_a, its modified value as given in Eq. (27.10), then the analysis can be carried through very simply:

$$W_a' = W_a - (\Delta\varphi)\,e. \tag{27.10}$$

A necessary condition for the application of SCHOTTKY'S results to an experiment in thermionic emission is that the energy distribution of the electrons as they pass over the limiting barrier at the distance x_c from the surface remains independent of the externally imposed electric field.

It could be assumed from the NOTTINGHAM[3] experiment that the energy distribution of the electrons observed at zero field was one deficient in low-energy electrons, but there was no evidence prior to the HUTSON experiments

[1] J. H. DE BOER: Electron Emission and Adsorption Phenomena, p. 18. Cambridge Univ. Press. 1935.

[2] A. R. HUTSON: Phys. Rev. **98**, 889 (1955).

[3] W. B. NOTTINGHAM: Phys. Rev. **49**, 78 (1936).

that the same energy distribution was applicable to the electrons emitted in the presence of a strong accelerating field. He showed that the distribution was invariant with the applied field. It is this result which could not possibly have been predicted that warrants the use of Eqs. (26.3) and (26.4) for the transmission probability. There is one minor difference, however, and that is that W_v as defined by Eq. (27.10) must be used to establish the ranges for the use of Eqs. (26.3) and (26.4) and that the momentum component (p'_x) used in Eq. (26.4) is defined as the momentum of the particular electron as it passes over the limiting barrier at the distance x_c. With these limitations in mind it is possible to write the SCHOTTKY equation immediately, both in the direct form and in its logarithmic form. These equations are the following:

$$I = I_0 \exp\left\{\frac{e}{k}\left[\frac{e}{4\pi\varepsilon_0}\right]^{\frac{1}{2}}\frac{(G V_c)^{\frac{1}{2}}}{T}\right\}, \tag{27.11}$$

$$\ln I = \ln I_0 + 0.4402\,\frac{(G V_c)^{\frac{1}{2}}}{T}, \tag{27.12}$$

$$\log_{10} I = \log_{10} I_0 + 0.1912\,\frac{(G V_c)^{\frac{1}{2}}}{T}. \tag{27.13}$$

In these equations the current I_0 is a constant of the equation which superficially represents the "observed" emission current at zero field, that is, when the accelerating potential V_c is zero. Experiment shows, as discussed in detail in Sect. 70 that the observed current at zero field is always less than the value predicted by these equations when it is compared with the empirical value of the constant I_0 that best fits the data observed at high field strength.

Undoubtedly, one of the most important influences in this disagreement is the inhomogeneity of all laboratory specimens. The high field data apply more directly to the emission from the low work-function areas, uninhibited by the local fields brought about in these areas by the neighboring high work-function regions.

It was pointed out above that if the change in work-function were known as a function of the accelerating potential, the force-function could be deduced by the proper application of Eq. (27.6). To obtain this change in work-function from thermionic emission directly, it is necessary to make the assumption that the actual energy distribution of the electrons is independent of the applied voltage, but it is not necessary to assume that the emission is a true MAXWELLian distribution. A reflection deficiency of the type described in connection with the derivation of the SCHOTTKY equation and illustrated originally by Eq. (26.4) will not alter the result. The current is related to the decrease of the work-function by an amount $\Delta\varphi$ by Eq. (27.14)

$$I = I_0\, e^{\frac{e\Delta\varphi}{kT}}. \tag{27.14}$$

In this equation the current I_0 is that found at a given temperature before any modification has occurred in the work-function. The actual choice of I_0 is not so very important in the analysis since it is the change in emission that is used.

Eq. (27.14), in its logarithmic form, can be differentiated with respect to V_c to obtain the following result:

$$\frac{d(\Delta\varphi)}{dV_c} = \frac{kT}{e}\,\frac{d(\ln I)}{dV_c}. \tag{27.15}$$

This equation shows that if $\ln I$ is known as a function of the accelerating voltage, then the slope of that curve, multiplied by (kT/e) will give the rate with which the change of work-function shifts with voltage. One needs only to differentiate Eq. (27.6) with respect to the accelerating voltage to obtain the result given in Eq. (27.16)

$$\frac{d(\Delta\varphi)}{dV_c} = \int_{x_0}^{x_c} G\,dx. \tag{27.16}$$

In order to obtain this equation, it was necessary to make use of the equality of the force-function and field as given in Eq. (27.7). Since in most applications the geometrical function $G(x)$ is constant over the important range from x_0 to x_c, the integration can be carried forward and one obtains, by combination with Eq. (27.15), the following value for the critical distance:

$$x_c - x_0 = \frac{kT}{eG}\,\frac{d(\ln I)}{dV_c}. \tag{27.17}$$

Having determined the critical distance x_c from the observed data, one then makes use of Eq. (27.7) to determine the value of the force-function which is effective at that distance.

Although it is possible to deduce the fact that the force-function is of the mirror-image type at applied potentials great enough to overcome the patch fields, it is clearly more practical to use the SCHOTTKY equation first and plot, as is indicated in Eq. (27.13), the logarithm of the current as a function of $(V_c)^{\frac{1}{2}}/T$. If the plot gives a straight line of slope $0.1912G^{\frac{1}{2}}$, the experiment shows that SCHOTTKY's mirror-image force dominates. In general, experiments carried out under suitable conditions yield results that agree well with this equation over a considerable range in surface field. A small systematic variation has been observed which will be discussed in the next section.

28. Periodic deviations from the SCHOTTKY theory prediction. A very detailed study of the increase in thermionic emission with accelerating field by SEIFERT and PHIPPS[1] showed that the so-called "SCHOTTKY Plot" is not a straight line but deviates in a regular manner as a function of the surface electron accelerating field. Since the average slope of the plotted data is that predicted by the SCHOTTKY mirror-image theory, there is a temptation to describe the observed deviation as a "periodic" deviation with reference to this line. Although the effect is extremely small, it has been verified by a number of researchers[2].

The nonlinearity of the SCHOTTKY plot acquires significance mainly because it can be explained only on the basis of a complete wave-mechanics understanding of the behavior of electrons as they escape over a mirror-image barrier in the presence of a strong field. The theories of GUTH and MULLIN[3] have been presented and also discussed in considerable detail by HERRING and NICHOLS[4].

The theoretical explanation of the periodic deviations brings the conclusion that a small reflection effect exists, and that the integrated influence of this reflection is dependent on the location and shape of the mirror-image barrier modified by the external field. Since the extent of reflection effects at the surface of an electron emitter is clearly an open question, the details of the theoretical analysis will be omitted from this presentation. A brief review of the experimental evidence will be given later in Sect. 70.

---

[1] R. L. B. SEIFERT and T. E. PHIPPS: Phys. Rev. **56**, 652 (1939).

[2] W. B. NOTTINGHAM: Phys. Rev. **57**, 935 (1940) (L). — E. G. BROCK, A. L. HOUDE and E. A. COOMES: Phys. Rev. **89**, 851 (1953). — D. W. JUENKER: G. S. COLLADAY and E. A. COOMES: Phys. Rev. **90**, 772 (1953).

[3] E. GUTH and C. J. MULLIN: Phys. Rev. **59**, 575 (1941); **59**, 867 (1941); **51**, 339 (1942).

[4] C. HERRING and M. H. NICHOLS: Rev. Mod. Phys. **21**, 185 (1949).

IV. LANGMUIR'S space-charge theory.

29. Statement of the problem. Early observations showed that for a given diode structure and a particular applied accelerating potential on the collector, the thermionic emission current increased exponentially with the temperature up to a fairly sharp limited current and showed very little further increase as the temperature exceeded this critical value. The limiting current was found to increase approximately as the $\frac{3}{2}$ power of the applied voltage and, of course, the critical temperature at which current limitation took place was voltage-dependent. The first step toward the theoretical understanding of this problem came when CHILD[1] and also LANGMUIR[2] applied the POISSON equation and the equation for current continuity as a pair of simultaneous relations.

Further study indicated that the current limitation was really due to the motive[3] maximum (for electrons) between the emitter and the electron collector because under the conditions of the experiment there were more electrons carrying negative charges in transit between the emitter and the collector than the total of positive charge on the surface of the collector. This motive maximum acts directly as a retarding potential and, therefore, limits the emission. It will be shown in this analysis that the magnitude of this retarding potential is directly proportional to the temperature in excess of the critical temperature at which the surface field at the emitter shifts from an electron accelerating field to zero field. The theory to be presented also may be applied in the retarding potential region near zero field. The correct analysis to this problem therefore involves the solution of three basic equations which are: 1. POISSON'S equation. 2. The BOLTZMANN density relation by which the energy dependence on temperature enters. 3. The equation of continuity of current. LANGMUIR[4] gave one of the first analyses of this problem which will be summarized here[5,6,7].

In the discussion of an electron atmosphere in a cavity (given in Sect. 21) mathematical simplification resulted from the choice of a "pillbox" which permitted the exact computation on the basis of an electron density variation with respect to distance which was a function of a single coordinate. For the same reason this analysis will be applied to a "plain parallel diode" structure. One surface of this diode is assumed to be capable of electron emission and its motive at all points over the surface at a distance of about 10^{-5} cm. differs in energy from the FERMI level in the interior of the emitter by the true work-function of the surface. This work-function must be assumed to be uniform all over the surface.

In the analysis which follows, the potentials and the potential gradients will apply only in the space between the two elements of the diode and will not include the distance over which mirror-image forces are appreciable. The electron collector must be assumed to be a perfect receiver of electrons. Any reflections or secondary emission that might occur at the surface of the electron collector must be assumed to be negligible. With these restrictions, it is obvious that no practical diode can satisfy the detailed requirements necessary for an exact application of the theory to experiment. In spite of this limitation, the theory certainly aids in the interpretation of experimental results.

---

[1] C. D. CHILD: Phys. Rev. **32**, 492 (1911).
[2] I. LANGMUIR: Phys. Rev. **2**, 450 (1913).
[3] See Sect. 27.
[4] I. LANGMUIR: Phys. Rev. **21**, 419 (1923).
[5] P. H. J. A. KLEYNEN: Philips Res. Rep. **1**, 81 (1946).
[6] A. VAN DER ZIEL: Philips Res. Rep. **1**, 97 (1946).
[7] H. F. IVEY: Advances in Electronics, Vol. VI, p. 137. New York: Academic Press 1954.

Even though experiment has established that the distribution in energy of the electrons emitted from smooth surfaces constituting a measurable emission current is deficient in the low-energy group, it would complicate the analysis to attempt to bring this fact into the theory. The strict application of the theory as it will be developed here can be made only after qualitative consideration has been given to all of these special limitations in terms of emitter and collector properties.

30. Uses of space-charge theory. The qualitative aspects of space-charge theory are well known and often used, but application in a truly quantitative manner is not widespread. The coverage of this subject in most text books is incomplete because the incorporation into the theory of the energy distribution of the electrons is usually treated superficially. As the properties of oxide cathodes have become better understood within recent years, more and more research effort has been directed toward the solution of the cathode evaluation problem. It has not been easy to devise generally acceptable methods for cathode evaluation because the usual tests carried out at the normal operating temperature of the emitter border on being destructive of the cathode under test. Even though the tube structure under investigation can withstand the high voltage required to evaluate cathode emission properties by sweeping out the space-charge for a very short period of time, the validity of the results is definitely open to question. The practical use of the emitter is generally under a condition of strong space-charge limitation. An alternate method of cathode evaluation involves an analysis of the emission as a function of the temperature and with relatively small applied voltages over the temperature range below and up to the critical temperature at which space-charge limitation sets in. This method also has its drawbacks, since it does not evaluate the emitter at or even near the normal operating temperature. Indirect methods of cathode evaluation based largely on empirical procedures have been developed and are widely used (see Sects. 80 and 82).

It will be the purpose of the following sections to develop the theory of space charge in a plane parallel diode with the ultimate objective of applying it in a practical manner to the quantitative evaluation of emitter properties in the useful range in temperature and in a completely nondestructive manner.

The development of methods for the application of the theory depends on the combination of three factors not hitherto applied to this problem. The basic formulation of the problem is that of LANGMUIR. His final results were made available in tabular form. Although these tables are a necessary step toward the final application of the theory, the derivation of empirical equations adapted to represent the tabular data with the required accuracy has made it possible to carry LANGMUIR'S results nearer to experimental applicability. Another factor that simplifies the analysis is the introduction of the emission equation given as Eq. (9.1).

The tangibly observable quantities are emission currents, applied voltages, emitter temperature, and the geometrical configuration. The emitter and collector properties are the intangible factors that actually determine the relations between the four observables. One must, therefore, use the space-charge theory to discover the emitter and collector properties that are most consistent with the laboratory observations.

31. Potential curves and the critical temperature. Figs. 6a and 6b apply, respectively, to the application of accelerating fields in the space between the electrodes and that of retarding fields. The straight line (I) in each example represents the distribution in electrostatic potential in the absence of space charge from a point just outside the emitter to the corresponding point in space just outside of the collector. The second line (II) shows qualitatively the potential distribution after the temperature of the emitter has been raised to exactly that critical value designated as Θ at which the gradient of the potential at either of the electrodes is exactly zero. In Fig. 6a the surface charge on the collector is measured directly by the gradient of the potential at that surface and the total of this surface charge is exactly equal to the total volume charge of all of the electrons in transit between the emitter and the collector. Under these conditions no lines of electric intensity extend all the way from the one electrode to the other. In Fig. 6b Curve (I) shows the surface charge on the emitter positive and the number of electrons in the space is negligible. The surface charge on the collector is negative. As the temperature of the emitter increases, its surface charge becomes still more positive and finally is exactly equal to the total charge in the space. At this temperature, Θ, the potential distribution of Curve (II), has a zero gradient at the collector. In both examples (accelerating and retarding), temperatures in excess of Θ produce a potential minimum (motive maximum for electrons) which is located between the two electrodes. Under these circumstances the mathematical theory which is applicable is the same for both figures, since positive charge will be found on both electrode surfaces. The sum of all of these positive charges will exactly equal the total charge of the electrons in the interelectrode space.

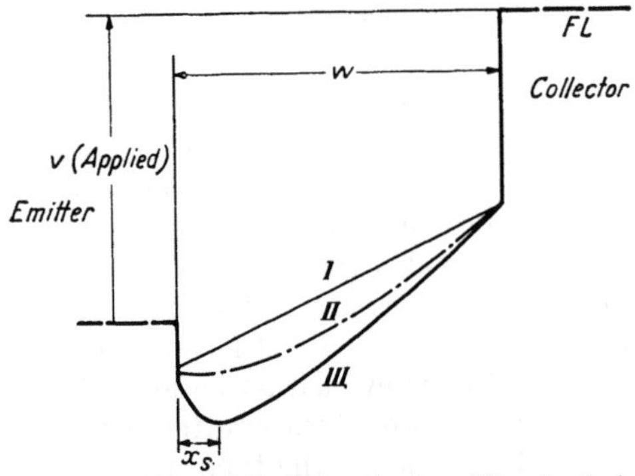

Fig. 6a. Potential distributions for three different cathode temperatures. Constant applied potential; positive temperature coefficient of work-function. I) No space charge. II) Zero field at emitter. III) With space-charge minimum at x_s.

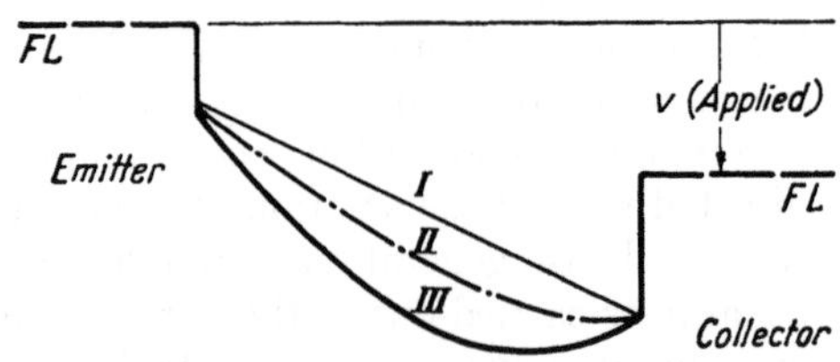

Fig. 6b. Potential distributions for three temperatures. I) No space charge. II) Zero field at the collector. III) With space-charge minimum.

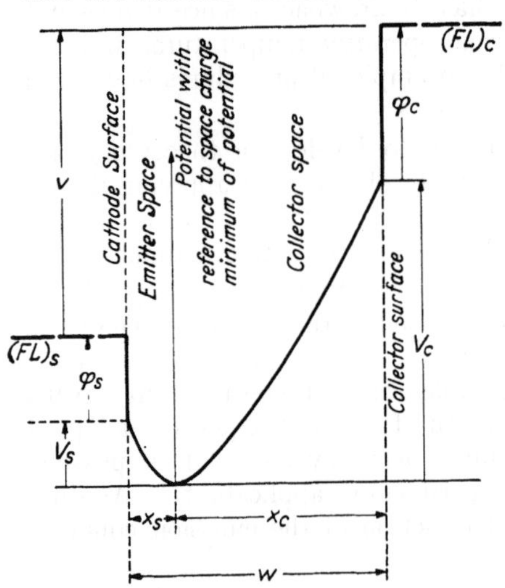

Fig. 7. Electrostatic potential with a space-charge minimum between the emitter and collector with applied accelerating potential.

32. Boundary conditions with positive applied potential. In general, there is more interest in the analysis as it applies to Curve (III) of Fig. 6a. The curve in Fig. 7 represents the electrostatic potential as a function of the distance either way from the potential minimum, which will always be found between the electrodes if the temperature is greater than Θ and the current is limited by space charge. The problem to be solved is that of determining the actual value of the potential as a function of distance which will satisfy the experimentally describable boundary conditions. The requirements are: 1. emitter to collector separation is w; 2. the true work-function of the emitter is φ_s; 3. the true work-function of the collector is φ_c; 4. the applied difference in potential v establishes the FERMI level of the collector with respect to the FERMI level of the emitter; 5. the difference in potential between points just outside the collector and the emitter is related to the applied difference in potential by the following equation:

$$V_c - V_s = v - (\varphi_c - \varphi_s); \tag{32.1}$$

6. the temperature of the emitter is T, which is greater than the critical temperature Θ.

33. Formulation of the general space-charge problem. With the above parameters specified, the density of charge in the immediate neighborhood of the cathode will be given by Eq. (24.7). Notice, however, that this equation applied only to the *ideal cathode* and that no real cathode ever satisfies the necessary ideal conditions.

Subject to the conditions specified above, a potential minimum will be found at a distance x_s away from the cathode and with reference to the potential at this minimum; the point in space just outside of the emitter will be positive V_s. The corresponding point just outside of the collector will be V_c positive with respect to the potential minimum. The plane surface at the distance x_s from the cathode surface divides the interelectrode region into two parts which may be designated as the "emitter space" and the "collector space". The potential function is not symmetrical about this dividing line, because in the emitter space electrons are streaming in both directions from the emitter to the potential minimum and back to the emitter. A small *net* current of electrons, which left the emitter with high energy, stream over the potential minimum and result in a current which flows in one direction only in the collector space.

It is the purpose of this theory to formulate the solution to the problem of satisfying POISSON's equation, the BOLTZMANN relation, and the continuity of current requirement. These three relations expressed in equation form are written as follows:

POISSON's equation:

$$\frac{d^2 V}{dx^2} = \frac{\nu e}{\varepsilon_0}. \tag{33.1}$$

BOLTZMANN relation:

$$\nu = \left(n_0 e^{\frac{eV}{kT}}\right) f. \tag{33.2}$$

Continuity of current:

$$I = \nu \frac{\overline{p}_x}{m} e. \tag{33.3}$$

The new symbols used in these equations may be defined as follows:

ν = number of electrons per unit volume near any plane located at a distance x from the potential minimum.

n_0 = number of electrons per unit volume that would be found in a complete MAXWELLian distribution at the potential minimum.

f = the fraction of a complete MAXWELLian distribution found at any plane between the two electrodes. For further explanation note that for a high-emission cathode, this fraction may approach unity very close to the cathode. The fraction will be exactly one-half at the potential minimum and may approach zero in the immediate neighborhood of the collector, when it is operated at a high positive potential.

I = electron current density in the direction from emitter to collector which is a constant at any plane between the emitter and the collector.

$\bar{p}_x/m$ = average velocity of the electrons that cross any plane in the direction emitter to collector. Notice that this average velocity depends on the temperature of the cathode and the potential, since it involves the actual energy distribution of the electrons. The potential value at the point in question determines the gain in kinetic energy which the electrons have received from the electric field.

Clearly, any theory dependent on the above conditions applies only if any change in external potential occurs very slowly in comparison with the average transit time of an electron as it crosses from the emitter to the collector.

34. Equation from statistical mechanics. In order that there be no misunderstanding with regard to the method by which f is calculated, the basic equation which comes directly from the statistical theory of free particles is the following:

$$f = \frac{\int\limits_{p_z=-\infty}^{p_z=+\infty} \int\limits_{p_y=-\infty}^{p_y=+\infty} \int\limits_{p_x=\pm\sqrt{2meV}}^{p_x=+\infty} e^{-\frac{p_x^2+p_y^2+p_z^2}{2mkT}} dp_x\,dp_y\,dp_z}{(2\pi m k T)^{\frac{3}{2}}}. \tag{34.1}$$

The writing of this equation comes directly from an inspection of the distribution of representative points in momentum space. The proper incorporation of the potential at the space point at which the fraction f is determined, comes in the setting of the limits of integration with respect to p_x. If the region in space for which f is being calculated lies in the emitter space as shown in Fig. 7, then the minus sign is used, whereas if the space location is in the collector space, the plus sign is used. Eq. (34.1) can be simplified and made appropriate for calculation by defining two new quantities as given in Eqs. (34.2) and (34.3) and the substitution of them into Eq. (34.1):

$$\psi = \frac{eV}{kT}, \tag{34.2}$$

$$y^2 = \frac{p_x^2}{2mkT}, \tag{34.3}$$

$$f = \frac{1}{2}\left[1 \pm \frac{2}{\sqrt{\pi}} \int\limits_{y=0}^{y=\psi^{\frac{1}{2}}} e^{-y^2}\,dy\right]. \tag{34.4}$$

$$f = \tfrac{1}{2}\left[1 \pm P(\psi^{\frac{1}{2}})\right]. \tag{34.5}$$

After the integration of Eq. (34.1), the plus sign corresponds to the emitter region. The integral in Eq. (34.4) has been tabulated under the well-known designation "Probability Integral" defined as:

$$P(\psi^{\frac{1}{2}}) = \frac{2}{\sqrt{\pi}} \int\limits_0^{\psi^{\frac{1}{2}}} e^{-y^2}\,dy. \tag{34.6}$$

35. Definition of distance unit x_1. In order to solve simultaneously Eqs. (33.1), (33.2), and (33.3) it is desirable to introduce a unit of distance which is exactly the same as that used in Eq. (22.4) except for an additional numerical factor of $\sqrt{2}$. The similarity of these unit distances can be seen more clearly by the following relation:

$$x_1^2 = 2\left[\frac{kT\varepsilon_0}{e^2 n_0}\right] \frac{e n_0 (kT/2\pi m)^{\frac{1}{2}}}{I}. \tag{35.1}$$

or

$$x_1^2 = \frac{2\varepsilon_0 (kT)^{\frac{3}{2}}}{e\,(2\pi m)^{\frac{1}{2}} I} = 2.370\times 10^{-12}\frac{T^{\frac{3}{2}}}{I} = 2.963\times 10^{-6}\frac{V_T^{\frac{3}{2}}}{I}. \tag{35.2}$$

Note that in Eq. (35.1) the constant current density I is expressible directly in terms of the density of electrons n_0 that can give exactly this current density for those electrons of a complete MAXWELLian distribution which cross a boundary. The actual concentration of electrons at the potential minimum is given by Eq. (33.2) and allows one to write the following:

$$\nu_0 = \tfrac{1}{2} n_0 . \tag{35.3}$$

The direct substitution of these relations in Eq. (33.1) results in the single differential equation that follows:

$$\frac{d^2\psi}{d\chi^2} = -\left[1 \pm P(\psi^{\frac{1}{2}})\right] e^{\psi} . \tag{35.4}$$

In this equation the plus sign is used for the emitter region and the minus sign for the collector region illustrated in Fig. 7. With the unit of distance defined by Eq. (35.2), the distance parameter χ is defined as (x/x_1) and x is measured from the potential minimum.

36. Solution to give electric intensity. Even though an explicit algebraic solution to Eq. (35.4) cannot be formulated, a step toward that solution can be made and the result is the following:

$$\frac{d\psi}{d\chi} = \sqrt{2}\left\{\left[1 \pm P(\psi^{\frac{1}{2}})\right] e^{\psi} - \left(1 \pm \frac{2}{\sqrt{\pi}}\psi^{\frac{1}{2}}\right)\right\}^{\frac{1}{2}} . \tag{36.1}$$

The next step in the complete solution of the problem would normally require the solving of Eq. (36.1). To find this solution explicitly seems to be impossible, but, of course, it can be accomplished by numerical integration. The solution by this method is given in Table 3.

By using a combination of theoretical forms as well as the results recorded in the table, the author has developed empirical equations which represent the results with considerable accuracy. These equations will be given in Sect. 37.

Eq. (36.1) permits the computation of the electric field at any point at which the potential is known. Note that the field is given by the following equation:

$$E = -\frac{dV}{dx} = -\left(\frac{kT}{e\,x_1}\right)\frac{d\psi}{d\chi} . \tag{36.2}$$

In this calculation the temperature is characteristic of the problem and the distance parameter x_1 is computed by Eq. (35.2). The emission current density I carried across the diode must be known.

37. Empirical equations that represent the solution to the LANGMUIR equation. In the emitter region the solution to Eq. (36.1) can be represented with good accuracy by the two equations given here as Eqs. (37.1) and (37.3). Notice that each applies for a specific range.

Range $0 \lessapprox \chi_s \lessapprox 1.4$ (Emitter Region)

$$\psi_s = \frac{\chi_s^2}{2} + 0.6\left(e^{0.32\,\chi_s^{3.17}} - 1\right), \tag{37.1}$$

$$\psi_s = \frac{\chi_s^2}{2} + 0.6\left(10^{0.139\,\chi_s^{3.17}} - 1\right). \tag{37.2}$$

Range $1.4 \lessapprox \chi_s \lessapprox 1.806$ (Emitter Region)

$$\psi_s = 2\ln\left[\cos(\chi_s - 0.235)\right]^{-1}, \tag{37.3}$$

$$\psi_s = 4.605\log_{10}\left[\cos(\chi_s - 0.235)\right]^{-1}. \tag{37.4}$$

Attention may be directed to the fact that as χ_s approaches 1.806, the value of ψ_s approaches infinity. The symbol $\chi_m = 1.806$ will often be used for this maximum value.

The inverse relation to the above equations is also of value. Again, a single equation cannot be used for the entire range but the two given below as Eqs. (37.5) and (37.6) give quite accurate results.

Range $0 \lessgtr \psi_s \lessgtr 2.0$ (Emitter Region)

$$\chi_s^2 = 2\psi_s + 0.72\psi_s^{1.46}. \tag{37.5}$$

Range $2 \lessgtr \psi_s$

$$\chi_s^2 = \left[\arc\tan\left(e^{\psi_s} - 1\right)^{\frac{1}{2}} + 0.235\right]^2. \tag{37.6}$$

The above equations are presented as empirical results that serve as a solution to Eq. (36.1) when the plus signs are used and are therefore applicable to the emitter region. For Eq. (36.1) to apply to the collector region, the negative signs must be used and under these circumstances the following equations give quite accurate results:

(Collector Region)

$$\psi_c = 1.17 \frac{\chi_c^{\frac{4}{3}}}{1 + \frac{1.68}{\chi_c^{0.681}}}, \tag{37.7}$$

$$\chi_c = 0.8888\,\psi_c^{\frac{3}{4}}\left(1 + \frac{1.413}{\psi_c^{0.475}}\right)^{\frac{3}{4}}. \tag{37.8}$$

The numerical constants of the two equations are, respectively, $\left(\frac{9}{4}\sqrt{\pi}\right)^{\frac{2}{3}}$ and $\left(4\sqrt{\pi}/9\right)^{\frac{1}{2}}$.

38. The CHILD-LANGMUIR space-charge equation. As a step toward the writing of a relation that is the equivalent of the CHILD-LANGMUIR equation, Eq. (37.8) squared may be rearranged as follows:

$$1 = \frac{4\sqrt{\pi}}{9}\frac{\psi_c^{\frac{3}{2}}}{\chi_c^2}\left(1 + \frac{1.413}{\psi_c^{0.475}}\right)^{\frac{3}{2}}. \tag{38.1}$$

This equation rewritten in terms of the symbols introduced as Eqs. (34.2) and (35.2) takes on the familiar form as follows:

$$I = \frac{4\varepsilon_0}{9}\left(\frac{2e}{m}\right)^{\frac{1}{2}}\frac{V_c^{\frac{3}{2}}}{x_c^2}\left(1 + \frac{1.413}{\left(\frac{eV_c}{kT}\right)^{0.475}}\right)^{\frac{3}{2}}. \tag{38.2}$$

It is clear from this equation that under many experimental conditions the applied potential may be very nearly equal to that shown in Fig. 7 as V_c and that the distance x_c may be very nearly equal to w. For large values of applied voltage, the last factor of the equation is practically unity and the equation then reverts to that of the CHILD-LANGMUIR equation. The constant of that equation is given as follows:

$$K_L = \frac{4\varepsilon_0}{9}\left(\frac{2e}{m}\right)^{\frac{1}{2}} = 2.334\times 10^{-6}. \tag{38.3}$$

It will be useful to define the ψ_c function of Eq. (38.1) by the symbol $F(\psi_c)$ given as follows:

$$F(\psi_c) = \psi_c\left(1 + \frac{1.413}{\psi_c^{0.475}}\right). \tag{38.4}$$

The true numerical values, that this function was developed to give, can be derived as indicated by Eq. (38.1) from the tabular values of χ_c, since the relation is given as follows:

$$F^{\frac{3}{2}}(\psi_c) = \frac{9}{4\sqrt{\pi}}\chi_c^2 = 1.269\,\chi_c^2. \tag{38.5}$$

Two equations very useful in computation that are the equivalent of Eq. (38.2) are given as Eqs. (38.6) and (38.7):

$$I = K_L \frac{F^{\frac{3}{2}}(\psi_c)}{x_c^2} V_T^{\frac{3}{2}}. \tag{38.6}$$

$$I = N \frac{\chi_c^2}{x_c^2} V_T^{\frac{3}{2}}. \tag{38.7}$$

The numerical value of a constant N may be computed from basic physical quantities as follows:

$$N = \varepsilon_0 \left(\frac{2e}{\pi m}\right)^{\frac{1}{2}} = 2.963 \times 10^{-6}. \tag{38.8}$$

A final and closely related expression which serves to link directly the properties in the emitter region to those in the collector region is given in equation form as:

$$x_1^2 = \frac{x_s^2}{\chi_s^2} = \frac{x_c^2}{\chi_c^2} = N\frac{V_T^{\frac{3}{2}}}{I}. \tag{38.9}$$

Even though it may be said that solutions to the space-charge problem are given by these various equations it is not easy to apply them to laboratory results. There are basically five independent variables which may be listed as follows:

1. The *diode spacing* w. The spacing to the potential minimum is related to w in that the sum of x_s and x_c is always equal to w.

2. The emitter temperature expressed either as T or its equivalent V_T [see Eq. (46.9)].

3. The *zero field* (at the emitter) *electron emission* current density of the emitter. This may be related to the temperature by an empirical equation of the form of Eq. (9.1) in which the two constants a and Φ constitute a description of the emitter.

4. The *applied difference in potential* which is directly controllable by the observer.

5. The *contact difference in potential* including its temperature coefficient, which depends both upon the properties of the emitter and the collector. Even though the temperature variation of the contact potential must be expected to be nonlinear, observations generally apply to such a limited range that it will be assumed at the start of the analysis that the contact difference in potential may be expressed as:

$$P_T = P_0 + V_T \frac{dP}{dV_T}. \tag{38.10}$$

The remaining sections of this theoretical analysis relate in one way or another to the bridging of the gap between the basic space-charge equations and their application to the interpretation of experimentally obtained results.

39. Application of the theory to boundary conditions. The equation which represents the thermionic emission current as a function of the temperature for stable cathodes was given as Eq. (9.1) and is reproduced as follows:

$$I = a\,e^{-\frac{e\Phi}{kT}} = a\,e^{-\frac{\Phi}{V_T}}. \tag{39.1}$$

This equation applies over that range in temperature for which the potential distribution between the emitter and the collector lies between curve (I) and curve (II) of Fig. 6a. At a very critical temperature $T=\Theta$ the potential gradient at the emitter is zero and this critical current I_0 is given by Eq. (39.2).

$$I_0 = a\,\mathrm{e}^{-\frac{e\Phi}{k\Theta}}. \tag{39.2}$$

In order for this critical temperature to be determined with accuracy, the emitter must be very uniform, in regard to both its surface composition and its temperature.

A specific use to which Eq. (38.2) may be applied is that of determining for a given electrode spacing and given applied voltage v between the electrodes, the current I_0 that will flow when the critical temperature Θ is reached. Under these conditions V_c is related to the applied potential according to Eq. (32.1). Note that V_s is zero. The distance x_c is identically equal to the spacing w. With the temperature $T=\Theta$ yet to be determined precisely, a first approximation for the space-charge limited current can be obtained by neglecting the correction term involving the temperature in Eq. (38.2). If the cathode properties are known and specified by Eq. (39.2), it is easy to determine the temperature at which the approximate value of current density will be obtained. This temperature, now identified as Θ' may be used in the correction term of Eq. (38.2) and a corrected value of the current density computed. Having thus determined the most suitable I_0, one may make a very small correction to determine the exact temperature Θ from Eq. (39.2) that wiil yield a self-consistent set of data.

40. Determination of value of potential minimum. As the temperature increases above the critical value Θ, the potential minimum develops and the equation for the current is the following:

$$I = a\,\mathrm{e}^{-\frac{e(\Phi+V_s)}{kT}}. \tag{40.1}$$

If ΔT is the increase in temperature above the critical value Θ then Eqs. (39.2) and (40.1) may be combined to give the appropriate value for the difference in potential between a point just outside of the emitter and the potential minimum illustrated in Fig. 7.

$$V_s = \Phi\frac{\Delta T}{\Theta} - \frac{kT}{e}\ln\left(\frac{I}{I_0}\right). \tag{40.2}$$

An inspection of this equation shows at once that if the current density I did not change at all with an increased temperature above its critical value I_0, then the potential minimum would increase in value in direct proportion to the temperature excess above the critical value Θ. The constant of proportionality is the work-factor Φ. Under well-controlled experimental conditions the current I will exceed I_0 by only a small amount ΔI for a considerable increase in temperature. Under these conditions an approximate form of Eq. (40.2) may be written by which the correction term may be evaluated most readily:

$$V_s \approx \Phi\frac{\Delta T}{\Theta} - \frac{T}{11600}\cdot\frac{\Delta I}{I_0} = \Phi\frac{\Delta T}{\Theta} - V_T\frac{\Delta I}{I_0} \tag{40.3}$$

where V_T is defined as $(T/11600)$.

This equation is useful in its application to experimentally realizable emitters under most circumstances, since the distance to the space-charge minimum x_s is generally smaller than the radius of curvature of the emitter. The steps by which this critical distance is determined follow.

41. Determination of distance to potential minimum. The first step is to compute the value of ψ_s:

$$\psi_s = \frac{e V_s}{k T} = \frac{\Phi e}{k \Theta} \cdot \frac{\Delta T}{T} - \ln \frac{I}{I_0}. \tag{41.1}$$

This value can be computed from observable quantities which are specifically: 1. the work-factor Φ; 2. the critical temperature Θ; 3. the absolute temperature T; 4. the excess temperature ΔT; 5. the current ratio (I/I_0). The corresponding value χ_s may be computed directly from Eq. (37.5) or Eq. (37.6), depending on the numerical value of ψ_s computed from Eq. (41.1). Since the current density and the temperature are known, the correct value for the distance from the cathode surface to the potential minimum may be computed directly from Eq. (35.2) or Eq. (38.9) and the definition of χ_s.

42. Application of theory to the retarding potential problem. In the retarding potential problem illustrated by Fig. 6b and Fig. 8 the magnitude of the retarding potential applied is constant and given by

$$V_r = V_s - V_c. \tag{42.1}$$

The expression that corresponds to Eq. (40.2) of the previous example is given as follows:

$$V_s = \frac{\Delta T}{\Theta} (\Phi + V_r) + V_r - V_T \ln \frac{I}{I_0}. \tag{42.2}$$

The procedure outlined above for the calculation of the distance to the potential minimum can be carried through in an analogous manner.

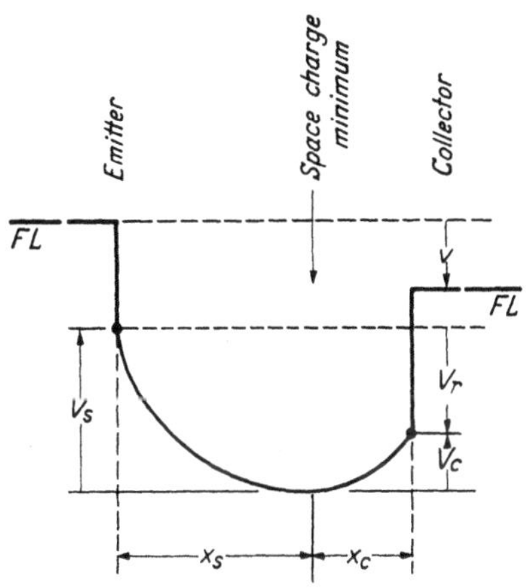

Fig. 8. Electrostatic potential with a space-charge minimum between the emitter and the collector for an applied retarding potential.

An inspection of Fig. 8 and qualitative considerations permit one to conclude that if the retarding potential, V_r, were increased while at the same time the temperature and properties of the cathode were held constant the potential minimum would shift toward the collector until at a particular value of retarding potential the electric field is exactly zero at this surface. If it is assumed that there are no reflection effects at the collector, space charge will have no influence on the emission current for still greater values of retarding potential. It is of some importance to devise a means for determining this critical value of retarding potential at which the onset of space-charge limitation takes place. A review of the "Pillbox Theory" given in Sect. 24 shows that the method of computation should follow much the same procedure.

43. Limiting retarding potential for onset of space charge. Eqs. (37.1) and (37.3) are directly applicable to this problem. The fact that χ_s cannot exceed 1.806 permits one to write two important relations which form the basis for the method of calculation to follow. The numerical factor comes from Eq. (35.2). These relations are given as:

$$(1.806)^2 \times 2.370 \times 10^{-12} = \frac{I_R}{\Theta^{\frac{3}{2}}} \delta^2 = 7.729 \times 10^{-12} \tag{43.1}$$

and

$$7.729 \times 10^{-12} = \frac{I_m}{\Theta^{\frac{3}{2}}} w^2. \tag{43.2}$$

$$9.664 \times 10^{-6} = \frac{I_m}{V_\Theta^{\frac{3}{2}}} w^2 = N \chi_m^2. \tag{43.3}$$

Eq. (38.8) defines N.

In Eq. (43.1) the current (I_R) is the actual current measured at the collector. The distance δ is the measure of the diode spacing for a diode of unlimited emission current capacity which would yield the current, I_R, actually observed. In Eq. (43.2) the current I_m is the maximum current which can flow across a diode of the actual spacing w and still achieve zero potential gradient at the collector. These two equations lead to the next which is

$$z^2 = \frac{w^2}{\delta^2} = \frac{I_R}{I_m}. \tag{43.4}$$

If the emission current capability of the emitter at zero field over the surface of the emitter is defined as I_0, then the current I_R will be given by the following relation:

$$I_R = I_0 \, \mathrm{e}^{-\frac{e V_R}{k\Theta}} = I_0 \, \mathrm{e}^{-\psi_{sR}}. \tag{43.5}$$

The combination of these two equations is the following:

$$\frac{I_0}{I_m} = z^2 \, \mathrm{e}^{\psi_{sR}}. \tag{43.6}$$

Eqs. (37.1) and (37.3) give the relation between ψ_{sR} and χ_{sR}. More accurate data for this relation are given in Table 3E. Note also that the relation between z and χ_{sR} is given as follows:

$$\chi_{sR} = 1.806\, z. \tag{43.7}$$

An abridged table has been prepared which relates the function on the right-hand side of Eq. (43.6) to z. See Table 4.

44. Application of universal function to problem solution. The method by which the universal function can be used will be outlined as follows. Experimental data on cathode properties must be depended upon to yield the numerical value of I_0 as a function of the temperature in terms of the two empirical constants (a) and (Φ) as given in Eq. (39.2). At any specified temperature and for a diode of specified dimensions Eq. (43.2) establishes the current density, I_m. The properties of the cathode and the spacing of the diode permit one to write the current ratio, (I_0/I_m), needed in Eq. (43.6). The plot given in Fig. 9 or the tabular data in Table 4 allow for the determination of the correct value of z to satisfy Eq. (43.6). As soon as z is known, then Eq. (43.4) permits the correct determination of the actual current I_R. A knowledge of this current and the specified temperature will permit the determination of the precise retarding potential at which space-charge limitation will begin. Ideally, for all values of retarding potential greater than the one that satisfies these equations the current received at the collector will be governed by the exponential relation similar to Eq. (43.5), with the current I less than the limiting one associated with the solution to this problem. This method of solution properly adapted also permits the user to determine the temperature at which the onset of space-charge limitation occurs for a given retarding potential V_r.

Table 4 has been computed over the middle range for which an explicit expression for z cannot be written. The following equation will serve as a means of establishing the correct value for the very small values of z.

Range $(I_m/I_0) > 100$

$$z = \frac{\left(\frac{I_0}{I_m}\right)^{\frac{1}{2}}}{1 + 0.8\left(\frac{I_0}{I_m}\right)}. \tag{44.1}$$

For the highest range of z from 0.98 to 1.0 the following equation should be used:

Range $(I_0/I_m) \gtrless 800$

$$z = 1 - \frac{2}{(\pi + 0.47)} \left(\frac{I_m}{I_0}\right)^{\frac{1}{2}} = 1 - 0.554 \left(\frac{I_m}{I_0}\right)^{\frac{1}{2}}. \tag{44.2}$$

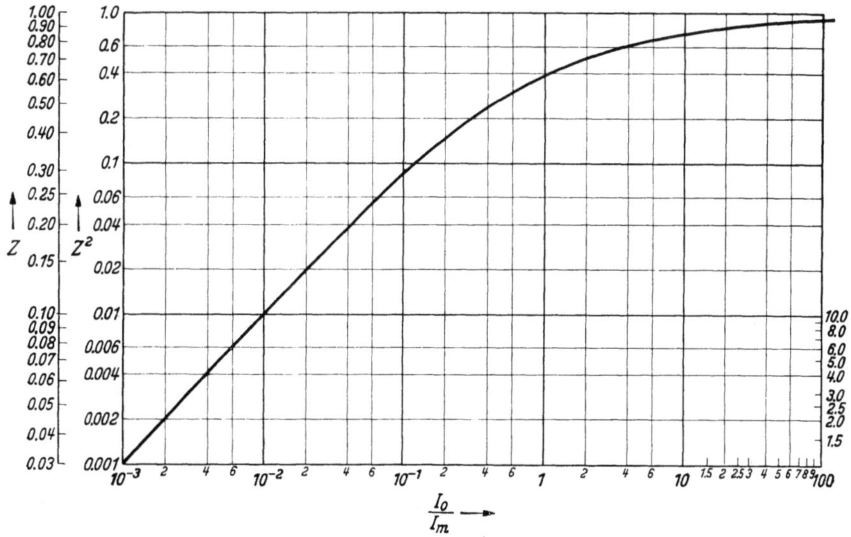

Fig. 9. Plot of data in Table 4 for solution of Eq. (43.6).

V. Idealized use of space-charge method for cathode property determination.

45. Statement of the problem. Even though the oxide cathode is the most important thermionic emitter of electrons, no generally accepted method has been developed for the evaluation of cathode properties under conditions that approach those used in most practical applications. This section will develop the theoretical background for the application of the space-charge equations to cathode evaluation. A brief outline of the fundamental principles involved may facilitate the direct application of the theory.

The following statements will serve to define the symbols used and the experimental conditions directly controllable.

1. The initial development of the theory will apply to a plane-parallel diode with the emitter to collector distance w small compared with other dimensions.

2. The emission properties of the emitter may be expressed by the empirical relation of Eq. (39.1) for which the emission constant a and the work-factor Φ are to be determined.

3. The absolute temperature of the emitter is T and this must be capable of determination either by thermocouple measurements, by pyrometry or by the investigation of the electron energy distribution with very high retarding fields.

4. The temperature and the emission properties of the emitter must be uniform.

5. The collector must have a uniform work-function and be a perfect receiver for the electrons that come to it.

6. The actual current density of the electrons, that cross the diode and therefore constitute the collector current per unit area, will be represented by I and will be directly measurable under any specified conditions.

7. The applied voltage v will be directly measurable and will represent the difference in energy between the FERMI levels of the collector and the emitter. Notice that if there is a sufficient resistance between the emission surface and the point of contact used for the applied potential measurement so that there is an appreciable difference in the FERMI level from the point of contact to the emission surface, then the measured applied potential must be corrected for the "IR" drop. In the formulation of the theory it is assumed that the applied potential v is identical with the true potential difference between the FERMI level of the collector and that of the emitter at its immediate surface.

Although the above statements serve to define the important directly controllable variables, other symbols will be introduced in the development of the theory. Basically, there are four unknowns in the list given above, which are: the emission constant a, the work-factor Φ, the contact difference in potential, and the temperature coefficient of the contact potential. The contact difference in potential P_T is the difference between the true work-functions of the collector and the emitter. In principle, four suitably chosen experiments should permit the evaluation of these four quantities. Although a minimum of four measurements is required, it will be shown that practical results can be obtained more easily from a larger number of measurements. It is, therefore, the purpose of this theoretical discussion to present the principles on which the suitable choice of measurements can be based.

All of the information desired can be obtained by retarding potential methods, if measurements outlined below are made at four or more temperatures. A second avenue of approach depends on a similar set of measurements made with accelerating fields applied. The final method to be discussed depends on a combination of the first two; it involves making observations at only three temperatures with suitably chosen applied potentials that give both retarding potential and accelerating potential data.

a) Emitter evaluation by accelerating potential methods.

46. Basic space-charge equations. The potential distribution between the emitter and the collector is shown by the curve of Fig. 7 for a given applied potential difference v. The relation between the applied potential and the other potentials pertinent to Fig. 7 and analysis of this problem is given in the following equation:

$$v = V_c - V_s + P_T. \tag{46.1}$$

Eq. (38.2) may be rewritten in a more useful form as:

$$I = K_L \frac{F^{\frac{3}{2}}(\psi_c)}{w^2\left(1 - \frac{x_s}{w}\right)^2} (V_T)^{\frac{3}{2}}. \tag{46.2}$$

The following equations serve as definitions of the symbols used here:

$$K_L = \frac{4\varepsilon_0}{9}\left(\frac{2e}{m}\right)^{\frac{1}{2}} = 2.334\times 10^{-6}, \tag{46.3}$$

$$N = \varepsilon_0\left(\frac{2e}{\pi m}\right)^{\frac{1}{2}} = 2.963\times 10^{-6}, \tag{46.4}$$

$$\chi_m = 1.806 \text{ [from (37.4)]}, \tag{46.5}$$

$$\psi_c = \frac{e V_c}{k T}, \tag{46.6}$$

$$F^{\frac{4}{3}}(\psi_c) = \frac{9}{4\sqrt{\pi}}\,\chi_c^2 = 1.2694\,\chi_c^2, \tag{46.7}$$

$$F(\psi_c) = \psi_c\left(1 + \frac{1.413}{\psi_c^{0.475}}\right), \tag{46.8}$$

$$V_T = \frac{kT}{e} = \frac{T}{11600}. \tag{46.9}$$

In order to facilitate the use of these functions, tables of values have been prepared for various useful powers of $F(\psi_c)$. See Table 5. Notice that this function results from a direct application of Eq. (38.2) to the problem.

Inspection of Eq. (46.2) permits the following conclusions to be drawn. At any given temperature V_T and a specified value of ψ_c, the current which can flow across a diode (of emitter to collector separation w) is a function only of the emitter to potential minimum distance x_s and *this is the only variable dependent on the cathode properties.* The minimum current that can flow and still have Eq. (46.2) applicable occurs with zero field at the emitter, that is, with x_s equal to zero. This situation establishes a limiting condition for an emitter of minimum capability at a particular temperature. For a given emitter let this temperature be represented by the symbol Θ.

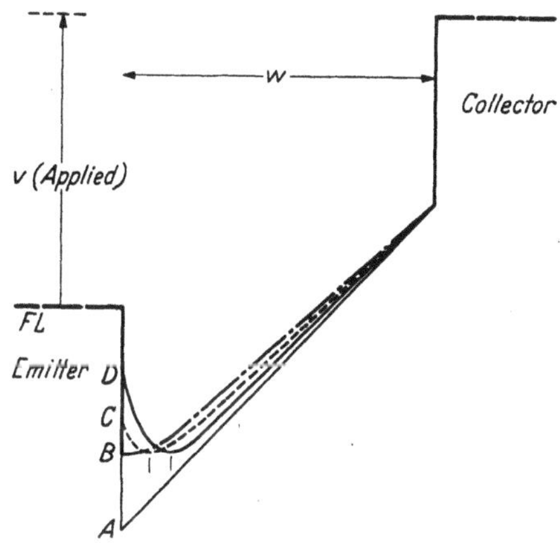

Fig. 10. Potential distributions for four different cathode conditions, constant temperature and applied potential. *A* Very low emission. *B* Critical condition of zero field at emitter. *C* Partial activation. *D* Highest activation.

Fig. 10 has been prepared to show qualitatively the change in the distribution in potential between an emitter held at a constant temperature and a collector maintained at a fixed *applied potential* v. An emitter of very low emission capability is identified by condition A of this figure and the current is not limited by space charge. An activation process may lower the work-function to condition B and the critical situation of zero field at the emitter is shown by this curve. Further lowering of the work-function to conditions C and D results in the development of the space-charge minimum and its progression across the interelectrode space toward the collector.

The above description applied specifically to a change in emitter activation and to no change in the *applied* potential. The analysis of experimental data that will lead to an evaluation of emitter properties can be simplified by taking the potential difference V_c in the collector space as constant instead of the applied potential. Under this condition and with a constant emitter temperature it follows that, as the emission capability increases, the potential difference V_s across the emitter region increases at a slightly lower rate than P_T (the contact difference in potential) in order to permit an increase in current without a change in V_c. This result follows because the emitter region distance x_s increases, and it is evident from Eq. (46.1) that the applied potential v must be increased.

As the activation process continues, Eq. (37.6) becomes a controlling factor and x_s approaches a maximum value of X_s which may be computed by Eq. (46.10), as it will be shown. The following equation may be derived by the elimination of the parameter x_1 between Eqs. (37.6) and (37.8) and it serves as a direct means of making the calculation of

$$X_s = \frac{w}{1 + \frac{2\pi^{\frac{1}{4}}}{3\,(1.806)}\, F^{\frac{3}{4}}(\psi_c)} = \frac{w}{1 + 0.4914\, F^{\frac{3}{4}}(\psi_c)} = \frac{w}{1 + \frac{\chi_c}{\chi_m}}. \tag{46.10}$$

The numerical constant is written out to show the direct connection with Eqs. (37.6) and (37.8).

The maximum current is given by the following equation:

$$I_M = 7.729 \times 10^{-12} \frac{T^{\frac{3}{2}}}{X_s^2} = 9.664 \times 10^{-6} \frac{V_T^{\frac{3}{2}}}{X_s^2}. \tag{46.11}$$

Let the current I_L represent some of the factors in Eq. (46.2) and be expressed as

$$I_L = K_L \frac{F^{\frac{3}{2}}(\psi_c)}{w^2} (V_T)^{\frac{3}{2}} = N \frac{\chi_c^2}{w^2} (V_T)^{\frac{3}{2}}. \tag{46.12}$$

The ratio of this current to the maximum current may be expressed as follows:

$$\frac{I_L}{I_M} = \left(1 - \frac{X_s}{w}\right)^2. \tag{46.13}$$

For any specific value of emitter temperature and collector potential relative to the potential minimum, namely V_c, the current I_L can be computed for a given diode structure by Eq. (46.12). The introduction of the value for the maximum distance X_s as given by Eq. (46.10) in Eq. (46.13) gives an alternative method for the determination of the maximum current as the product of the current I_L and a universal function which has also been computed and recorded n Table 6. This relation is given as

$$I_M = I_L \left(1 + \frac{2.035}{F^{\frac{3}{4}}(\psi_c)}\right)^2 = I_L\, f(\psi_c) = I_L \left(1 + \frac{\chi_m}{\chi_c}\right)^2. \tag{46.14}$$

47. Application of basic equations to cathode evaluation. As the basic equations are applied to determine emitter properties the only two directly controllable and observable variables are the emitter temperature and the applied difference in potential. The unknowns which it is the purpose of the study to evaluate quantitatively may be listed as follows:

1. The work factor Φ.
2. The RICHARDSON work-function φ_R.
3. The thermionic constant a.
4. The critical temperature Θ or its equivalent V_Θ which is the minimum temperature at which a given emitter will operate with the current limited by space charge for a specified value of V_c.
5. The contact difference potential P_T at any temperature.
6. The temperature coefficient of the contact difference in potential which is also the negative of the temperature coefficient of the true work-function of the emitter.
7. The contact potential P_0 which is the value at absolute zero of temperature determined by a linear extrapolation from the observational temperature region to absolute zero. For stable emitters with emission current given by either Eq. (39.1) or the RICHARDSON type of equation, the temperature variation of the work-function is dominated by the first power term. It will be shown in

Sect. 81 that for oxide cathodes higher power terms are needed. The present analysis depends on the use of a suitably chosen average temperature coefficient and a corresponding P_0.

There are still other quantities which can be considered as indirect variables. The analysis depends on the arbitrary choice of two or more values for ψ_c. If it is assumed that this analysis applies to the evaluation of an oxide cathode, or its equivalent, then the choice of values of ψ_c between 15 and 50 will correspond to the application of accelerating potentials such that the energy with which the electrons bombard the collector will lie between 1.5 and 5 electron volts. For a particular choice of ψ_c, Eq. (46.12) permits the preparation of a table of values of I_L as a function of the temperature. In particular these values should be computed for the normal temperature of operation represented by V_n and for a range of lower temperatures. The value of I_M may also be computed directly with the aid of the tabulated function expressed in Eq. (46.14) as $f(\psi_c)$. The analysis which follows will show the method by which the unknown quantities can be related to the directly and indirectly controllable variables.

Eq. (40.2) may be rewritten in terms of the variables presently useful in the analysis, as follows:

$$V_s = \Phi \frac{V_T - V_\Theta}{V_\Theta} - V_T \ln \frac{I}{I_\Theta}. \tag{47.1}$$

In this equation I_Θ is the current density which flows to the collector at the critical temperature Θ when the value of V_s is zero for a particular value of ψ_c. The current I is the current density at any higher temperature for the same value of ψ_c. Eq. (46.1) may be rewritten in terms of the quantities now defined, as follows:

$$v = V_T \left(\psi_c + \frac{dP}{dV_T} - \frac{\Phi}{V_\Theta} + \ln \frac{I}{I_\Theta}\right) + P_0 + \Phi. \tag{47.2}$$

For this equation to hold exactly, the correct value of I should be known. Since its value cannot be determined in advance of a determination of v, an experimental procedure must be used.

With good cathodes, the current I will be very nearly equal to I_M for an appreciable range in temperature above and below the normal operating temperature expressed as V_n. The experimental procedure is to compute I_M as well as I_L for each temperature and the specified value of ψ_c. One establishes by experiment the applied potentials v'' and v' which are observed to be required to give diode current densities of I_M and I_L, respectively. The equation for the I_M characteristic curve is the following:

$$v'' = V_T \left[\psi_c + \frac{dP}{dV_T} - \frac{\Phi}{V_\Theta} + \ln [f(\psi_c)] + \frac{3}{2} \ln \left(\frac{V_T}{V_\Theta}\right)\right] + P_0 + \Phi \tag{47.3}$$

since

$$\ln \frac{I_M}{I_\Theta} = \ln \frac{I_M}{I_L} \left(\frac{V_T}{V_\Theta}\right)^{\frac{3}{2}} = \ln [f(\psi_c)] + \frac{3}{2} \ln \left(\frac{V_T}{V_\Theta}\right). \tag{47.4}$$

Over the high-temperature range the characteristic curve for the I_L data follows an equation identical to Eq. (47.3) except that a lower value of ψ_c is applicable. It is possible to compute this lower value of ψ_c but it does not serve a useful purpose in the present analysis. The more useful result to keep in mind is that a linear relation between an observable quantity and the temperature can be formulated from Eq. (47.3). This relation depends on an observable voltage quantity defined by the following:

$$v^{**} = v'' - V_T \left[\frac{3}{2} \ln \frac{V_T}{V_\Theta} + \ln f(\psi_c)\right]. \tag{47.5}$$

Note that the computation of the correction term depends on an estimate of the critical temperature V_Θ. The means for making this estimate will be described in connection with the discussion of the data presented in Fig. 11.

With this definition of the voltage v^{**}, the following equation may be written:

$$v^{**} = V_T\left[\psi_c + \frac{dP}{dV_T} - \frac{\Phi}{V_\Theta}\right] + P_0 + \Phi. \tag{47.6}$$

At the critical temperature V_Θ, Eq. (47.2) relates to the observations made along the I_L characteristic curve and satisfies the requirement as expressed in the following equation:

$$v'_\Theta = V_\Theta\left(\psi_c + \frac{dP}{dV_T} - \frac{\Phi}{V_\Theta}\right) + P_0 + \Phi. \tag{47.7}$$

A comparison between Eqs. (47.6) and (47.7) shows that the purpose of this analysis is to establish an accurate value for the critical temperature V_Θ. The straight line described by the plotted values of v^{**} as a function of the temperature V_T over the high-temperature range intersects the voltage scale at the zero value of V_T to give a measure of $(P_0 + \Phi)$. At the critical temperature, v^{**} is identical to v'_Θ which establishes the fact that the intersection of the straight line described with the observed I_L characteristic curve, as shown in Fig. 11, determines the exact value of V_Θ to be used.

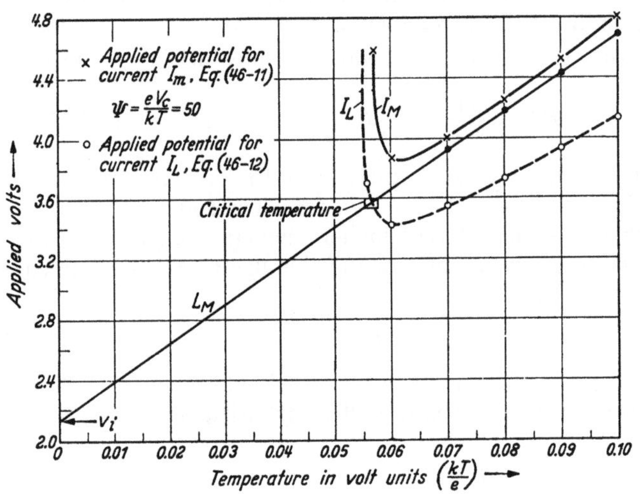

Fig. 11. Applied potentials for emitter evaluation at $\psi_c = 50$.

It is to be anticipated that in carrying out the experiment, observations would not be made at this exact temperature, and yet, if the observations were made at temperatures slightly above and below this temperature, it is easy to interpolate between the observation points and determine the suitable value of V_Θ. The insertion of this value of temperature in Eq. (46.12) gives an accurate measure of the zero field emission capability of the emitter. This result is expressed as follows:

$$I_0 = I_{L\Theta} = N\,\frac{\chi_c^2}{w^2}\,(V_\Theta)^{\frac{3}{2}}. \tag{47.8}$$

The application of these principles can be illustrated best by the numerical data presented in Figs. 11 and 12. For these curves, typical values of all of the constants are assumed in advance so that the curves can be drawn to scale.

The solid and the dashed curves of Figs. 11 and 12 join a set of points that represent, as a function of temperature, the directly observable applied potentials needed so that the computed current densities I_M and I_L are the ones actually observed. Note that Fig. 11 applies to a value of ψ_c of 50 and Fig. 12 is for ψ_c of 16. Before the corrections can be applied that correspond to the application of Eq. (47.5), an estimated value of the critical temperature V_Θ must be determined. A sufficiently accurate choice may be made by drawing a tangent to the smooth curve designated as I_M at the highest temperature and then determining the intersection of this tangent with the dashed curve designated as I_L. This approximate value of the critical temperature, V_Θ, can be used in Eq. (47.5) to locate the points identified in these figures by solid dots. An inspection of Eq. (47.6) shows that a straight line drawn through these points has an intercept v_i on the voltage axis at the zero of temperature. The intersection of the straight line designated L_M, with the dashed experimental curve is indicated by the square □ on each of the figures. The exact value of the critical temperature, V_Θ, is determined by this intersection. It is anticipated that this exact value of the critical temperature will not differ enough from the previously determined approximate value to make it necessary to re-evaluate the potentials computed by Eq. (47.5).

This analysis of experimental data as illustrated by Figs. 11 and 12, should yield a single value of the intersection potential v_i for both choices of ψ_c of 50 and 16. If the initial analysis of the data for the two cases gives slightly different values for v_i, an average value should be taken and new straight lines should be drawn to correspond to the best representation of the entire set of data. A lack of self-consistency will generally indicate a failure to satisfy the uniformity requirements either with respect to emitter temperature or the surface conditions of the emitter and collector.

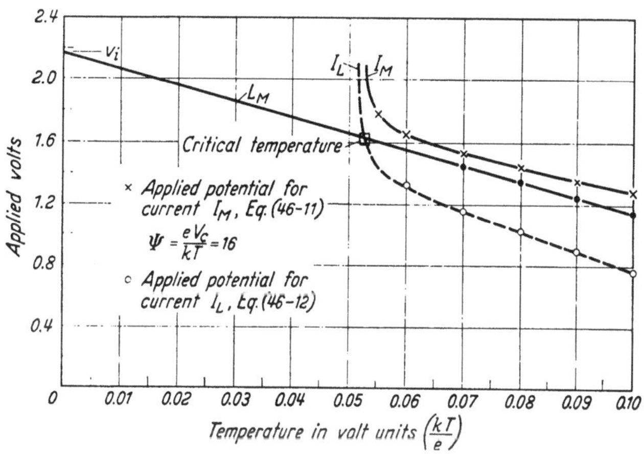

Fig. 12. Applied potentials for emitter evaluation at $\psi_c = 16$.

If the contact potential variation with temperature cannot be represented well enough by Eq. (38.10), then experimental points will not fall exactly on a straight line but may deviate systematically. This may indicate that the best straight line will give a good mean value of the temperature coefficient and the corresponding P_0 that will be useful only over the observational range in temperature.

The slope of the line L_M is given by the terms contained within the square brackets of Eq. (47.6). There are essentially two unknowns in this expression for the slope; they are the temperature coefficient of the contact potential and the work-factor. The acquisition of data at the two chosen values of ψ_c identified here as ψ_a and ψ_b, instead of 50 and 16, permits the determination of the work-factor by the following relation:

$$\Phi = \frac{(S_a - S_b) - (\psi_a - \psi_b)}{\frac{1}{V_{\Theta b}} - \frac{1}{V_{\Theta a}}}. \tag{47.9}$$

Use may be made of Eq. (47.8) and an alternative method of computing the work-factor Φ may be employed as in the following equation:

$$\Phi = \frac{2\ln\frac{\chi_a}{\chi_b} + \frac{3}{2}\ln\frac{V_{\Theta a}}{V_{\Theta b}}}{\frac{1}{V_{\Theta b}} - \frac{1}{V_{\Theta a}}}. \tag{47.10}$$

In this equation the values of χ_a and χ_b are those found in Table 3C corresponding to the selected values of ψ_a and ψ_b. A direct calculation of the thermionic constant expressed as $\ln a$ may be made from either set of data taken at ψ_a or ψ_b. The following equation will be written in terms of the choice of ψ_a and is:

$$\ln a = \ln\left[N \frac{\chi_a^2}{w^2} V_{\Theta a}^{\frac{3}{2}}\right] + \frac{\Phi}{V_{\Theta a}}. \tag{47.11}$$

To facilitate calculation, it will be convenient to define a characteristic "temperature" that in effect describes the diode spacing. This symbol is defined by the following equation:

$$V_w^{\frac{3}{2}} = \frac{w^2}{N \chi_m^2}. \tag{47.12}$$

The substitution of the symbol V_w in Eq. (47.11) gives the form by which the thermionic constant can be computed most easily.

$$\ln a = 2 \ln \frac{\chi_a}{\chi_m} - \frac{3}{2} \ln\left(\frac{V_w}{V_{\Theta a}}\right) + \frac{\Phi}{V_{\Theta a}}. \tag{47.13}$$

48. Determination of contact difference in potential and its temperature coefficient. Since the intercept of the straight lines on the ordinate axis can be determined and identified as v_i, the contact difference in potential may be computed as follows:

$$P_0 = v_i - \Phi. \tag{48.1}$$

Since the slopes S_a or S_b are known, either one can be used in the following equation to determine the temperature coefficient of the contact potential applicable over the observation range

$$-\frac{dP}{dV_T} = \frac{d\varphi}{dV_T} = \psi_a - \left(S_a + \frac{\Phi}{V_{\Theta a}}\right). \tag{48.2}$$

49. Determination of the RICHARDSON work-function and the true work-function. If the work-factor Φ is known, it is easy to calculate the most appropriate value for the RICHARDSON work-function φ_R that will best represent the data over any specified temperature range, as shown in the next section. Assume for this calculation that V_n represents the highest temperature of interest and that V_m corresponds to the lowest temperature of interest. The RICHARDSON work-function is then given by the following equation:

$$\varphi_R = \Phi - (V_n + V_m). \tag{49.1}$$

With the temperature coefficient of the work-function known from Eq. (48.2), the best value for the true work-function is given as follows:

$$\varphi = \varphi_R + V_T\left(\frac{d\varphi}{dV_T}\right). \tag{49.2}$$

50. Determination of the RICHARDSON constant A_R and its relation to the thermionic constant a. When they are considered as empirical equations to represent the temperature-dependence of thermionic emission, the simplified equation and the RICHARDSON form are equally valid. These two equations are written as follows:

$$I = a e^{-\frac{\Phi}{V_T}}, \tag{50.1}$$

$$I = A_R T^2 e^{-\frac{\varphi_R}{V_T}}. \tag{50.2}$$

Since the range of temperature of practical interest is generally less than two to one, it is possible to express relations between the empirical constants such that either equation will represent the experimental data within the accuracy of the temperature scales used. By differentiation it may be shown that the slope of the simplified equation plot is related to the RICHARDSON slope by the following equation:

$$\Phi = \varphi_R + 2 V_T. \tag{50.3}$$

Since the average temperature over the range of interest is $(V_n + V_m)/2$ it is appropriate to take this value for V_T of Eq. (50.3) and the following relation is obtained:

$$\Phi = \varphi_R + V_n + V_m. \tag{50.4}$$

With this relation between the work-factor Φ and the RICHARDSON work-function φ_R, the thermionic constants a and A_R should be chosen so that the two equations give exactly the same values of current density at some intermediate temperature T_0. This temperature should preferably be approximately one-quarter of the range each way from the midtemperature so that the deviations that must exist will be more or less equally divided and there will be two points at which the equations give precisely the same value for the current density. As a result of this matching procedure, one may write the following relation:

$$\ln 10 = \frac{T_n + T_m}{T_0} = 2.3 \tag{50.5}$$

If the matching temperature T_0 is chosen to satisfy this relation, then a convenient means of calculating the thermionic constant a in terms of the RICHARDSON constant A_R follows:

$$a = 10 A_R T_0^2. \tag{50.6}$$

The inverse relation is useful as it applies to the results obtained from the analysis of Sect. 47:

$$A_R = \frac{a}{10 T_0^2}. \tag{50.7}$$

Although the name "work-factor" has been found to be a convenient term for Φ as Eq. (50.1) is applied specifically to thermionic emission data, an equation of this form has been used as a means of expressing the conductivity of semiconductors as a function of the temperature. In the conduction process, there is also the problem of the temperature coefficient of the FERMI level. In the discussion in Sect. 65 it is shown that the temperature exponent of the theoretical equation that best represents results is $-(\frac{3}{4})$ instead of 2 as in the RICHARDSON empirical formula given here as Eq. (50.2) and therefore a more general conversion formula is needed. In the application of Eq. (50.1) to these conduction problems the quantity Φ has acquired the name "activation energy"[1, 2]. Actually, it is no more a direct measure of the true activation energy in the conduction case than it is a measure of the true work-function in the emission case. In both analyses it is simply an empirical constant.

Since data found in the literature are often expressed this way, however, it will be convenient to write general forms for the conversion of the empirical constants representing data according to Eq. (50.1) to suitable choices for the empirical constants in the generalized equation given as follows:

$$I = A_\nu T^\nu e^{-\frac{\varphi_\nu}{V_T}}. \tag{50.8}$$

---

[1] N. B. HANNAY, D. MACNAIR and A. H. WHITE: J. Appl. Phys. **20**, 669 (1949).
[2] R. LOOSJES and H. J. VINK: Philips Res. Rep. **4**, 449 (1949).

This equation gives the same variation in the quantity $\ln I$ with respect to V_T^{-1} as that given by Eq. (50.1) at the temperature V_T if the following condition is satisfied:

$$\varphi_\nu = \Phi - \nu V_T. \tag{50.9}$$

The two equations will give the same value for the quantity I at a chosen temperature if the relation between a and A_ν is given by

$$A_\nu = \frac{a}{e^\nu T_0^\nu}. \tag{50.10}$$

One generally wishes to choose the constants so that they will give comparable results over a specific range in temperature. If the extent of that range can be bracketed by V_n and V_m (or T_n and T_m) which carries the interpretation now that the first of these temperatures is the maximum temperature of the range and the second the minimum temperature of the range, then the best choice for the relations between the empirical constants may be written according to the following two equations:

$$\varphi_\nu = \Phi - \frac{\nu}{2}(V_n + V_m), \tag{50.11}$$

$$T_0 = \tfrac{1}{2}(T_n + T_m). \tag{50.12}$$

b) Emitter evaluation by retarding potential methods.

51. Basic space-charge equations. A general statement covering the objectives of this theory of cathode evaluation is given in Sect. 45 and should be referred to in order that such limitations as necessarily exist will be clearly understood. Oxide cathodes generally operate at a temperature for which the emission current is strongly limited by space charge and, therefore, it is appropriate to present a theory of cathode evaluation which depends on the direct application of space-charge theory to this practical problem. The basic equation has already been worked out by which one can calculate the maximum current which can flow across a diode of given dimensions if the cathode is operated at a specified temperature and the *potential gradient at the collector is exactly zero.* This equation was first presented as Eq. (43.3) and may be written in terms of notation presently useful as follows:

$$I_m = 7.729 \times 10^{-12} \frac{T^{\frac{3}{2}}}{w^2} = 9.664 \times 10^{-6} \frac{V_T^{\frac{3}{2}}}{w^2}. \tag{51.1}$$

For a good oxide cathode this equation gives the observed current when zero field exists at the collector for the temperature range higher than 800° K. For lower temperatures the actual current obtained under the zero field condition is lower than I_m and the ratio of these currents is defined as in Eq. (43.4) as follows:

$$z^2 = \frac{I_R}{I_m}. \tag{51.2}$$

The analysis which follows shows the means by which the correct value of z can be determined. With z known, it is possible to give a quantitative value to the true retarding potential that exists across the space between a point just outside of the emitter and the corresponding point outside of the collector. When this retarding potential is known, the emission capability of the emitter will also be known at the specific temperature at which z has been evaluated.

The current received at the collector can be expressed in terms of cathode properties as follows:

$$I_R = a\, e^{-\frac{\Phi + V_R}{V_T}}. \tag{51.3}$$

In this equation the constant, a, is a measure of an emitter property and is to be determined by the series of experiments to be described. The work-factor is

Φ, and the true retarding potential is V_R. It is to be noted that the symbol V_R is appropriate, since it measures the true difference in potential across the emitter space as described in Sect. 31. In the present example there is no collector space, since the zero gradient of the potential distribution occurs at the collector. The relation between the applied difference in potential v and the true retarding potential V_R is the following:

$$v = P_T - V_R. \tag{51.4}$$

The true contact difference of potential including the temperature variation of the emitter work-function is P_T. See Sects. 45 and 47. These relations permit the following equation to be written:

$$v - V_T \ln I_m = V_T \left[\frac{dP}{dV_T} - \ln a - \ln \frac{1}{z^2}\right] + \Phi + P_0. \tag{51.5}$$

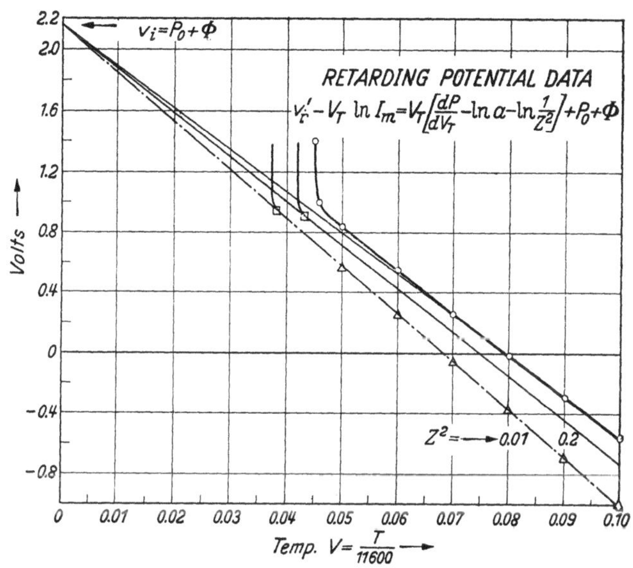

Fig. 13. Emitter evaluation by retarding potential methods.

52. Application of basic equations to cathode evaluation. Over the high-temperature range of approximately 800° to 1200° K for oxide cathodes it is to be expected that Eq. (51.5) can be simplified to read as follows:

$$v_1 - V_T \ln I_m = V_T \left[\frac{dP}{dV_T} - \ln a\right] + \Phi + P_0. \tag{52.1}$$

Over this range z^2 is so near unity that its logarithm can be neglected. The experimental observations grouped on the left-hand side of the equation involve first, the computation of I_m by means of Eq. (51.1) and the observation of the applied voltage v_1 required at each temperature to give the corresponding current density I_m in the test diode. The voltage v_1^* can be defined as in the following equation and a plot made of this voltage as a function of the temperature

$$v_1^* = v_1 - V_T \ln I_m. \tag{52.2}$$

Computed points typical of observational data are shown in Fig. 13. A tangent may be drawn through the points in the high-temperature range and the intercept of this line on the voltage axis may be identified as follows:

$$v_i = \Phi + P_0. \tag{52.3}$$

As the temperature is lowered, the observed points will deviate very gradually as shown in Fig. 13. At each of the known emitter temperatures, observed applied voltages may be determined to give a set of points corresponding to some arbitrarily chosen values of z^2. For example, 0.2 and 0.01 may be selected. The current that is observed in the experiment can be related to the computed current I_m by the following equation:

$$I'_r = z'^2 I_m. \tag{52.4}$$

The reason for the change in notation from I_R to I'_r is that, in the first case, the current flow I_R indicated that particular value for which zero gradient existed at the collector for a known or specified value of z^2. In that case, the true retarding potential between the emitter and the collector would be identical to the emitter potential relative to the space-charge potential minimum which has been identified as V_s. For a given emitter there is only one temperature at which this condition is satisfied if the value of z'^2 differs appreciably from unity. It has already been mentioned that for good oxide cathodes there is a very useful range in temperature for which z^2 is so near unity that its variation is not detectable. For any choice of z'^2 and of temperature such that a potential minimum does *not* exist between the emitter and the collector, the collector current is controlled exclusively by the true retarding potential and is independent of space charge. Let this true retarding potential be represented by the symbol V_r and the corresponding applied potential be identified as v'_r, These two potentials are related by the following equation:

$$v'_r = P_T - V_r. \tag{52.5}$$

The new equation that corresponds to Eq. (51.3) and relates the observed current to the true retarding potential for the whole range for which V_r is greater than V_s, may be written as:

$$I'_r = a\,e^{-\frac{\Phi + V_r}{V_T}}. \tag{52.6}$$

The three equations above may be combined to give the observed relation between the temperature expressed as V_T and the observed applied potential v'_r which yields the observable current I'_r after a definite choice has been made for the value of z'^2

$$v'_r - V_T \ln I_m = V_T \left[\frac{dP_T}{dV_T} - \ln a - \ln \frac{1}{z'^2}\right] + \Phi + P_0. \tag{52.7}$$

It is to be seen that this equation has exactly the same form as Eq. (51.5) and it indicates that observed points will fall on a straight line of slope $S_{z'}$ as defined by the following equation:

$$S_{z'} = \frac{dP}{dV_T} - \ln a - \ln \frac{1}{z'^2}. \tag{52.8}$$

Note that all lines should be expected to have a negative slope increasing in value as z'^2 decreases and all lines intersect at the same voltage point v_i, as given by Eq. (52.3).

Under ideal conditions of emitter uniformity the observed points will fall on the straight lines described for all temperatures in excess of the critical temperature for which the condition of zero gradient at the collector is satisfied. As the temperature is lowered, below the critical value, the observed points break away sharply from the straight lines, as shown in Fig. 13. With these critical temperatures known for two or more values of z' appreciably less than unity, the properties of the emitter may be specified by the methods given in the following section.

53. Interpretation of results. The location of corresponding critical temperatures, collector currents, and z' values leads first to a determination of true retarding potentials applicable in Eq. (51.3). Since this equation involves the two unknowns a and Φ, their numberical values can be determined from data obtained at two temperatures. These temperatures and the corresponding currents may be determined as described in Sect. 52. The next step, to determine the relation between V_R and z, still remains to be completed. Eq. (43.7) gives the relation between χ_s and z and Eq. (37.1) relates ψ_s to χ_s for the smaller values of χ_s. Thus, for values of z less than 0.77 or z^2 less than 0.6, Eq. (37.1) will be found satisfactory for most calculations. The data in Table 3E combined with an accurately prepared plot for interpolation may also be used.

Let the two arbitrary values of z' be identified as z_1' and z_2' with z_1' being the larger. The corresponding values of ψ_s will be ψ_1 and ψ_2 and the corresponding critical temperatures will be V_{z1} and V_{z2}. With these quantities defined, it is possible to solve for the work-factor and the result is the following:

$$\Phi = \frac{\left[\ln\left(\frac{z_1'^2}{z_2'^2}\right) + \psi_1 - \psi_2\right] + \frac{3}{2}\ln\frac{V_{z1}}{V_{z2}}}{\frac{1}{V_{z2}} - \frac{1}{V_{z1}}}. \tag{53.1}$$

Note that the quantity in the square bracket is essentially an arbitrary constant of the problem that depends only on the chosen z values.

The numerical values of this constant for z^2 values of 0.2 and 0.01 is 3.409.

With the work-factor Φ known and the emission current given at a particular temperature such as V_{z1}, the value of the thermionic constant may be expressed as:

$$\ln a = \frac{\Phi}{V_{z1}} + \psi_1 + \ln\left(z_1'^2 I_{m1}\right). \tag{53.2}$$

The slope of the line described by Eq. (52.7) may be expressed by Eq. (52.8) and is a directly measurable quantity. This equation contains only the temperature coefficient of the contact potential as an unknown and therefore it may be evaluated. The temperature coefficient of the work-function is the negative of the quantity (dP/dV_T). The intercept on the voltage axis is a directly measurable quantity and, combined with the work-factor as in Eq. (52.3), serves to evaluate the effective contact difference in potential at the zero of temperature.

54. Criticism of the retarding potential method of cathode evaluation. In Sects. 51 through 53 the theory of cathode evaluation by retarding potential methods has been presented in order to prepare for the use of certain of its elements in the theory which follows. The next sections show the most practical means of evaluating an electron emitter under conditions nearly identical to those under which an oxide cathode is most likely to be used in practical applications.

The principal objection to the retarding potential method of cathode evaluation involves the critical nature of the emission current and temperature measurements at the low end of the temperature scale. The most basic quantity, namely, the work-factor, depends so strongly on these temperature measurements and on the identification of the critical temperature that cathode inhomogeneities can be expected to interfere seriously with the accuracy of the results. Of the three methods of cathode evaluation presented here the retarding potential method is the least satisfactory. The theory presented above, however, is basic to an understanding of the combination method, undoubtedly the best of the three, which will be described in the next section.

c) Emitter evaluation by a combination of retarding and accelerating potential methods.

55. Statement of the problem. Practical electron emitters such as oxide cathodes cannot be expected to be perfect surfaces of uniform emission efficiency. Any method of cathode evaluation which does not assess the cathode in the presence of space charge under conditions similar to those of its practical operation may very well yield misleading results. Evaluation methods that depend only on the drawing of saturation current by means of an accelerating field at the cathode evaluate patches of low work-function area, since these are likely to yield a disproportionate fraction of the emission current. It follows that a method which will permit evaluation of the emitter and which will include information obtained at normal operating temperatures and with very low fields can be expected to yield the most reliable information. It is the purpose of the next few sections to present new methods for emitter evaluation by taking full advantage of the space-charge theory[1].

The quantities to be determined are (1) the work-factor Φ, (2) the thermionic constant a, (3) the average temperature coefficient of the work-function, and (4) the contact difference in potential. These four quantities may be determined by measurements made at a minimum of three different temperatures.

The method to be described depends on measurements made in both the retarding potential and the accelerating ranges. Some previous knowledge of the general properties of the emitter is presupposed in order that the temperatures most suitable for study may be selected. The highest temperature used can be that associated with the normal operation of the emitter. For oxide cathodes this would be approximately 1160° K. Temperatures of approximately 660° and 580° K are generally suitable for the other two temperatures. The analysis requires that these temperatures be known quite accurately and, therefore, thermocouple measurements combined with electron energy distribution measurements should be used in their determination.

The methods of cathode evaluation, described in Sects. 45 to 54, required observations made at five or six different values of temperature and a knowledge of the diode spacing, with considerable accuracy, in order to obtain reliable results. The method to be presented is more practical, since it depends on the accumulation of data, for subsequent analysis, by making a large number of current measurements as a function of applied voltage while the temperature is held constant. It is for this reason that importance is attached to this phase of the theoretical analysis.

To obtain the most accurate results, it is necessary to measure the emission as a function of the applied voltage from currents of approximately 10^{-9} amp/cm.^{2} to values of about 4×10^{-3} amp/cm.^{2} for a diode of approximately 1 mm. spacing. The normal procedure is to cover the range in the accelerating field region to

[1] Although the author read the article by W. R. Ferris [RCA Review **10**, 134 (1949)], some years ago, the present analysis has been developed independently. Since the author's objective is cathode evaluation, it has been necessary to go into far more detail in the theory. It will be shown that Ferris' concept of a "Universal diode characteristic" is practically the equivalent of the "Master Curve" described here. The theory is extended to make it applicable to the evaluation of emitters with any ratio of (I_0/I_R) and a means of adapting it to the cylindrical diode is provided.

an applied voltage of approximately 6 volts. It is necessary for the cathodes to be very stable and the measurements must be reproducible with high accuracy to indicate such stability.

56. Methods of comparison between theory and experiment. A graphical method of analysis is the most appropriate means of comparing the observations with the theory. In this manner the current flow under the various critical conditions may be determined by experiment. The first of these conditions to be discussed will be that associated with zero gradient at the collector for each of the temperatures. This condition may be defined as that for $\psi_c = 0$. Two other critical conditions involve the determination of the collector potential at which zero gradient exists at the emitter for two or more of the lower temperatures chosen for investigation.

The first step in the procedure is to prepare a plot of the logarithm of the emission current as a function of the observed applied potential expressed in the dimensionless unit of (v/V_T). Let this ratio define the quantity, s, as follows:

$$s = \frac{v}{V_T} \,. \tag{56.1}$$

Depending on the polarity of v, the quantity s may be positive or negative.

The emission current density characteristic of the emitter itself, which it is the purpose of this analysis to determine, is given by the following equation:

$$I_0 = a\,e^{-\frac{\Phi}{V_T}} \,. \tag{56.2}$$

With the true retarding potential, defined as V_r, the emission current density which arrives at the collector may be expressed as follows:

$$\ln I = \ln I_0 - \frac{V_r}{V_T} \,. \tag{56.3}$$

Since it is generally more convenient for the graphic analysis to use semilogarithmic paper, Eq. (56.3) may be written in terms of common logarithms as

$$\log_{10} I = \log_{10} I_0 - 0.4343\, S \,. \tag{56.4}$$

In this equation, the dimensionless form of the absolute value of the retarding potential expressed as S is related to the observable quantity defined by Eq. (56.1) by the following relation

$$|S| = -\frac{P_T}{V_T} + \frac{v}{V_T} = -\frac{P_T}{V_T} + s \,. \tag{56.5}$$

Even though the current I_0 and the true contact difference in potential P_T at the temperature T are unknown, it is nevertheless possible to compare the theory with experiment by noting that the observed currents plotted as a function of s on semi logarithmic paper yield a straight line of slope 0.4343. This straight-line relation applies over the entire range for which there is no potential minimum between the emitter and the collector. At a critical value of s, a departure from the straight line can be detected. This departure serves as an approximate indication of the maximum current which can flow across the diode under the condition of zero gradient at the collector. The use of a nonuniform collector surface can cause some uncertainty in the correct identification of the applied potential required to give zero gradient at the collector. The more serious difficulty standing in the way of a simple interpretation of experimental results is the reflection effect at the collector. This effect makes the recorded current at the critical condition of zero gradient less than the true emission current, since additional space charge is created by the reflected electrons as they approach and leave the collector without being recorded. Evidence for this enhanced space charge will be presented in Sect. 83.

To establish a more suitable identification of the applied voltage which would have been correct for $\psi_c = 0$ in the absence of reflection, it is necessary to observe the rise in current with applied voltage through a sufficient range of positive values of ψ_c so that the observed curve can be matched by the theoretical curve. The matching of curves in this region by the technique described in the following sections makes it possible to evaluate accurately the ideal critical current I_R for the zero value of ψ_c. The method also permits an evaluation of I_0 over the lower range in temperature. See Sect. 67 for the detailed application of this theory to experimental data.

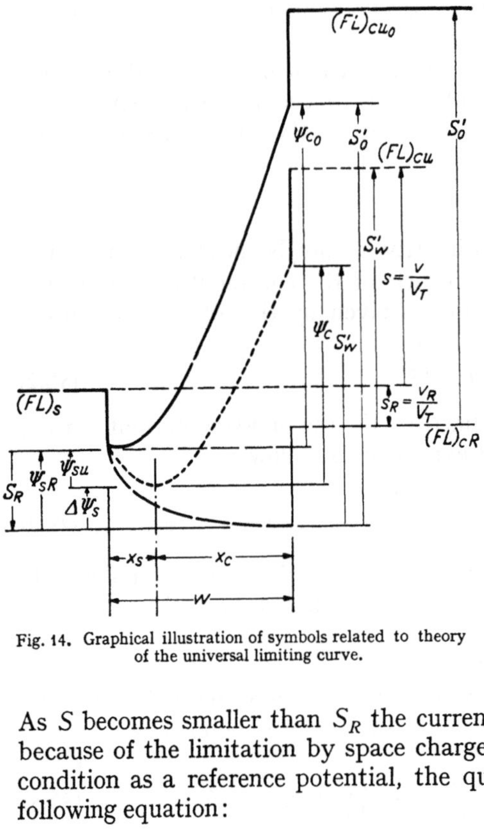

Fig. 14. Graphical illustration of symbols related to theory of the universal limiting curve.

57. The onset of space-charge limitation and the universal limiting curve. For any given temperature and emission property of the emitter, let the critical value of applied voltage for which zero gradient occurs at an ideal collector be identified as v_R and the corresponding value of s as s_R. These quantities are related as in Eq. (56.1). In Sect. 56 the absolute value of the surface potential of the collector relative to the emitter was defined as S and the true retarding potential was equal to S as long as no space-charge minimum existed between the emitter and the collector. At a certain critical valu of s zero field develops at the collector and this value of S is identified in Fig. 14 as S_R. As S becomes smaller than S_R the current increase no longer follows Eq. (56.4) because of the limitation by space charge which then sets in. With this critical condition as a reference potential, the quantity $S' \gtrless 0$ may be defined by the following equation:

$$S' = s - s_R = \frac{v - v_R}{V_T}. \tag{57.1}$$

Although the actual value of s_R will not be known until the analysis is carried further, it is to be expected that a "universal" relationship may be found between the current expressed as a ratio and the value of S' (Sect. 58). Thus the logarithm of the current ratio as a function of S' will create a curve which will serve as a means of determining the location of the exact value of s_R. The relation between the observed current and s determined experimentally may be plotted on semi-logarithmic paper and for comparison with theory may be superpositioned on the plotted set of "Master curves" described in Sect. 58. It will be the purpose of the next few paragraphs to derive this relation from space-charge theory.

The relation between S' and other quantities of interest is best illustrated by Fig. 14. Three curves are shown. The dashed line represents the distribution in potential for $S'=0$, that is, the whole difference in potential in the space is exactly ψ_{sR}. The applied difference in potential expressed in dimensionless units is s_R. As the applied potential is changed to a more positive value by an amount S' as expressed in Eq. (57.1), the potential at the surface of the collector changes exactly the same amount and is related to the change in the emitter region potential $\Delta\psi_s$ and to the change in the collector region potential ψ_c, by the relation given as follows:

$$S' = \Delta\psi_s + \psi_c. \tag{57.2}$$

The intermediate potential distribution to which this equation applies is shown by the dotted curve of Fig. 14. This change in the emitter region potential results in an increase in current from the critical value I_R to the new value I given as

$$\ln (I/I_R) = \Delta\psi_s = \ln u^2. \tag{57.3}$$

Eq. (57.3) serves to define the quantity u^2 which is (I/I_R).

A very important limiting condition of potential distribution is illustrated by the solid line of Fig. 14. A cathode of finite emission capability I_0 has zero field at its surface when the current carried across the diode is exactly equal to I_0. This condition may be related to the previously defined critical situation of zero gradient at the collector by the following equation:

$$\frac{I_0}{I_R} = u_0^2 = \mathrm{e}^{\psi_{sR}}. \tag{57.4}$$

Since the current I_R is directly related to the emission capability I_0, this property of the emitter establishes the zero reference for the quantity S' defined by Eq. (57.2) and illustrated in Fig. 14. With space-charge theory dominating, the maximum permissible value for S' occurs when the current flow across the diode is I_0 and is associated with zero gradient at the emitter. Let this maximum value of S' be represented by S_0'. A very important universal curve relates u_0^2 to S_0', since both of these quantities relate specifically to the emission capability I_0. This relation is to be derived.

From the basic considerations expressed as Eq. (38.9), the following three equations may be written directly:

$$I_0 = N \frac{\chi_{c0}^2}{w^2} (V_T)^{\frac{3}{2}}, \tag{57.5}$$

$$I_R = N \frac{\chi_{sR}^2}{w^2} (V_T)^{\frac{3}{2}}, \tag{57.6}$$

$$\frac{I_0}{I_R} = \frac{\chi_{c0}^2}{\chi_{sR}^2} = u_0^2. \tag{57.7}$$

It is with the help of Eq. (57.7) that the relation may be written that ties together the two critical values of χ as given in the following equation:

$$\chi_{c0} = u_0 \chi_{sR} = \chi_{sR}\, \mathrm{e}^{\psi_{sR}/2}. \tag{57.8}$$

This equation illustrates the plan of calculation by which the desired relation can best be computed. The steps in the calculation and the results obtained are given in Table 7. Note that χ_s and ψ_s combined as in Eq. (57.8) yield a table of values for χ_{c0}. From accurately made plots of the LANGMUIR data of Table 3C or from Eq. (37.7) rewritten below as Eq. (57.9) the corresponding values of ψ_{c0} may be obtained:

$$\psi_{c0} = 1.172 \frac{\chi_{c0}^{\frac{4}{3}}}{1 + \frac{1.68}{\chi_{c0}^{0.681}}}. \tag{57.9}$$

With the knowledge of the appropriate values of ψ_{sR} and ψ_{c0}, the correct value of S_0' may be written as follows:

$$S_0' = \psi_{sR} + \psi_{c0}. \tag{57.10}$$

The desired universal curve is obtained as a plot of the quantity u_0^2 of Eq. (57.7) on semi-logarithmic paper as a function of S_0' of Eq. (57.10). This curve has been plotted on Figs. 16 and 17 and is designated the "Universal Limiting Curve".

The basic purpose for this universal curve is to serve as the means by which the correct value of I_0 may be assigned as it applies to an experimentally measured current-voltage characteristic. It is therefore equally important that a set of master curves he developed to be used in association with this Universal Limiting Curve so that the best value for the current I_R can be established for a particular case. At the same time a most appropriate value of I_0 will be obtained at the intersection of the observed current-voltage characteristic with the Universal Limiting Curve. The next section shows the means for deriving the required master curves.

58. Equations and tables for the master curves. The first step in the derivation leading to the desired set of master curves is to obtain the limiting curve applicable to cathodes so copious in their emission that they become the "ideal emitters" of essentially unlimited emission capability. This master curve will form an upper boundary, illustrated in Figs. 16 and 17, to the set of master curves applicable to emitters of limited emission capability. The first part of this section will develop the theory by which the current voltage characteristic applicable to the "ideal emitter" may be computed.

Even though the emitter has practically unlimited electron emission available in the presence of a retarding field, it was shown in Sect. 43 that there is a maximum current I_m that can flow across a diode of spacing w with an emitter at a temperature V_T and have a zero gradient at the collector. The value of this current is given by the following equation which is essentially the same as Eq. (43.3):

$$I_m = \chi_m^2 N \frac{V_T^{\frac{3}{2}}}{w}. \tag{58.1}$$

For any positive value of ψ_c there is also a maximum current that can flow across a diode from an ideal emitter. This current may be computed directly from the following equation which is derived from the combination of Eqs. (46.10) and (46.11):

$$I_M = \chi_m^2 N \frac{V_T^{\frac{3}{2}}}{w} \left(1 + \frac{\chi_c}{\chi_m}\right)^2. \tag{58.2}$$

These two equations may be combined to define the quantity U^2 as in the following equation:

$$\frac{I_M}{I_m} = U^2 = \left(1 + \frac{\chi_c}{\chi_m}\right)^2. \tag{58.3}$$

The present objective is to express the relationship between U^2 and a variation in the applied voltage with reference to the zero value established when zero gradient occurs at the collector. When ψ_c is zero, U^2 is unity. Voltages expressed in dimensionless units are best illustrated by Fig. 15 which, it will be noted, is very similar in its construction to Fig. 14 except that, in this case, the condition of zero gradient at the emitter is never reached. This figure will also define symbols used in the following equations.

The increase in current expressed in terms of the current ratio U^2 is related to the *change* in the potential across the emitter region by the following equation:

$$U^2 = e^{\Delta\psi_s}. \tag{58.4}$$

With the change in applied potential represented by the symbol Σ, the following equations may be written:

$$\Sigma = \psi_c + \Delta\psi_s = \psi_c + \ln U^2, \tag{58.5}$$

$$\Sigma = \psi_c + \ln\left(1 + \frac{\chi_c}{\chi_m}\right)^2. \tag{58.6}$$

The "master curve" that forms the limiting set is obtained from a plot of the corresponding values of U^2 and Σ; its value is computed by Eq. (58.6). These corresponding values are given in Table 8.

All real emitters have finite emission capability and yet practical emitters, such as oxide cathodes operated at their normal temperature, can be expected to have such high emission capability that their operation is characterized by a z_R^2 value close to 0.998, as defined in the following equation:

$$\frac{I_R}{I_m} = z_R^2. \tag{58.7}$$

Eq. (44.2) and an estimated value for the emission capability of an oxide cathode operating at 1160° K of 10 amp/cm.[2] was used to determine this value of z_R^2 that approaches so near unity. The value of z_R^2 for a good cathode operated at one-half its normal operating temperature will be close to 0.66. These figures are quoted to show that over a considerable range in temperature, practical oxide cathodes of good quality can be expected to approach in their performance very close to that of the ideal emitter of unlimited capability.

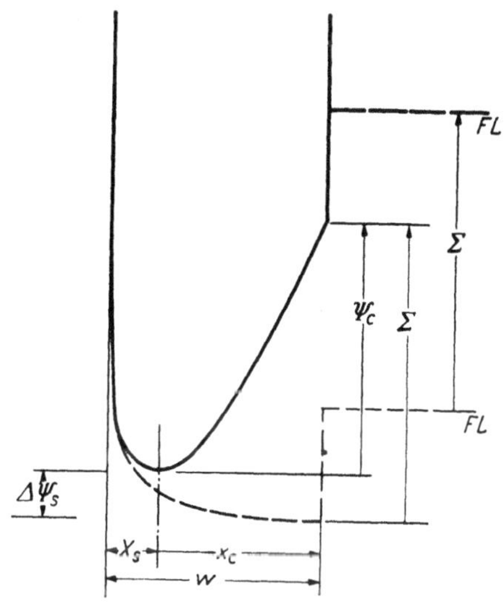

Fig. 15. Potential distribution for an emitter of unlimited emission.

For every value of I_0 expressed in terms of u_0^2, as in Eq. (57.4), there is a corresponding value of z_R^2 and a characteristic current-voltage relation that covers the range of current from I_R to I_0. The parameter u^2 expressed as follows serves as a measure of the current I.

$$\frac{I}{I_R} = u^2. \tag{58.8}$$

Just as it was the purpose of the previous paragraphs to find the relation between U^2 and Σ, the following analysis is designed to show the method of calculating the relation between u^2 and S'. The steps for obtaining this relation are rather involved and yet, because the result is so important, the details are given.

Again Fig. 14 serves as the reference by which the variation of the potentials and the current may be linked in a simple and yet completely correct manner.

From Eq. (57.4) the following relation may be written:

$$\psi_{sR} = \ln u_0^2. \tag{58.9}$$

When the potential difference between the emitter and the collector is exactly ψ_{sR}, the value of S' is zero and the value of ψ_c is also zero. As the potential of the collector is made more positive with respect to this limiting value, a difference in potential of an amount which is symbolized by ψ_{cu} develops between the potential minimum and the collector. Under the same conditions of current flow, identified by a particular value of u, there is also a difference in potential of ψ_{su} between the emitter and the potential minimum. The relation between the measured current I and the emission capability I_0 is given by the following equation:

$$\ln \frac{I_0}{I} = \ln \frac{u_0^2}{u^2} = \psi_{su}. \tag{58.10}$$

Eqs. (38.9) and (57.6) may be combined and solved to form the following basic relation:

$$\chi_{cu} + \chi_{su} = \left(\frac{I}{I_R}\right)^{\frac{1}{2}} \chi_{sR}. \tag{58.11}$$

With a substitution from Eq. (58.8) in Eq. (58.11), it is possible to express the direct correspondence between u, χ_{su} and χ_{cu}. In principle this relation serves to locate the potential minimum between the two electrodes when the particular current ratio u^2 obeys the equation

$$u\,\chi_{sR} - \chi_{su} = \chi_{cu}. \tag{58.12}$$

The corresponding value for ψ_{cu} can be obtained either from Table 3C or from Eq. (57.9).

Since Eq. (57.3) relates u^2 to the change in the emitter space potential as S' increases from zero, Eq. (57.2) may be rewritten as follows:

$$S'_u = \Delta\psi_{su} + \psi_{cu}. \tag{58.13}$$

These equations, from Eq. (58.8) through Eq. (58.13) form the basis for the procedure by which one is able to calculate characteristic curves of current as a function of applied collector potential for emitters of all capability, as measure by the parameter u_0^2 expressed by Eq. (57.4).

The method of calculation involves nine steps and these are outlined very briefly as follows:

1. An arbitrary choice is made of the u_0^2 for which the calculation is to be made. Calculations have been made for a set of selected values of u_0^2 between 2 and 200 as shown in Table 9.

2. Each selection of u_0^2 gives by Eq. (58.9) a specific value of ψ_{sR} and also a corresponding value of χ_{sR}.

3. Corresponding values of ψ_{su} and χ_{su} may be copied from Table 3E for the range $0 < \psi_s \lessapprox 1.0$ to make a good working table for the best choice of u^2 values.

4. It follows from Eq. (58.10) that a suitable set of u^2 values may be obtained from the following relation

$$u^2 = u_0^2\, e^{-\psi_{su}}. \tag{58.14}$$

5. A reference work, such as BARLOW's Tables[1], serves as an excellent aid in converting these u^2 values to u.

6. The operation described by Eq. (58.12) is performed to obtain the corresponding set of χ_{cu} values.

[1] BARLOW's Tables, Edited by L. J. COMRIE, 4th ed. New York: Chemical Publishing Co., 1944.

7. The χ_{cu} values may be converted by means of Table 3C or by Eq. (57.9) to obtain the corresponding ψ_{cu}.

8. Logarithm tables may be used to determine $\ln u^2$ values corresponding to those recorded as step 4.

9. Since $\ln u^2$ is related to $\Delta\psi_{su}$ by Eq. (57.3), all of the data are at hand to perform the addition indicated in Eq. (58.13) and thus to obtain the correct value of S'_u.

The desired curve that relates u^2 to S'_u for a particular value of u_0^2 is thus available. The manner in which the master curves are used as they apply to experiment will be described in the next sections. Typical master curves are plotted in Figs. 16 and 17.

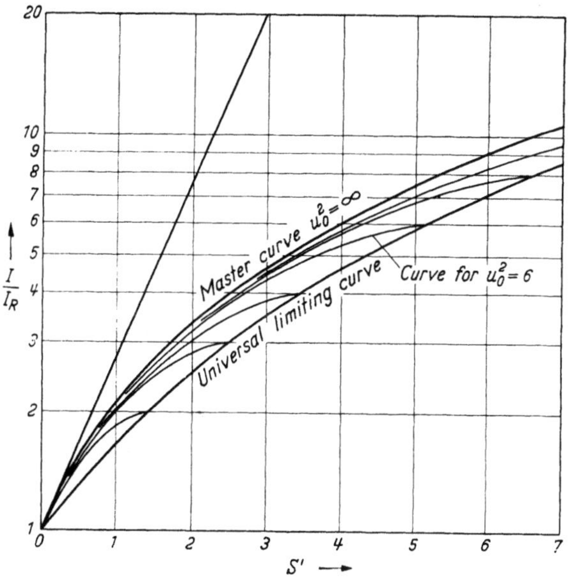

Fig. 16. Master curves for selected values of u_0 from 2 to 8 and the limiting curves.

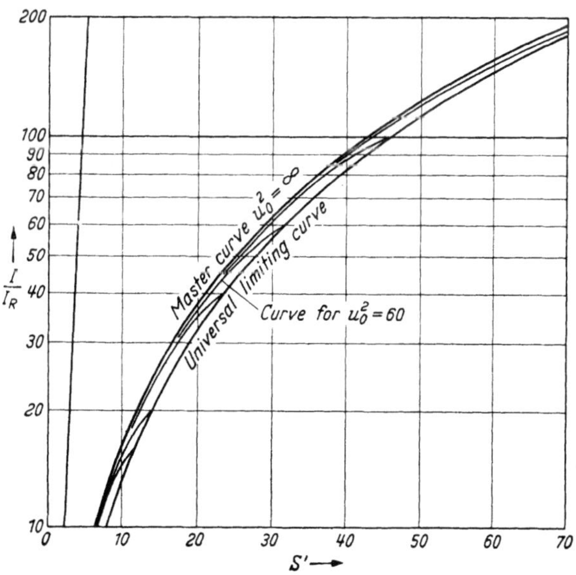

Fig. 17. Master curves for selected values of u_0 from 16 to 200 and the limiting curves.

59. General discussion of the application of theory to experiment. It has been the plan of this treatise to begin at Sect. 14 and to present as much as possible of the theoretical structure of the science of thermionic emission with a minimum of references to specific experimental results. The theories have been developed with the ultimate objective of applying them directly to the experiments to be described in later sections. It now becomes necessary to discuss, in a general way certain specific features of experimental research to justify the development of more specialized features of the theory in advance of the application of the entire theory to the interpretation of experimental research.

Basically, nearly all researches on oxide cathodes are performed with diode tube structures. There are only two practical forms of the diode structure that lend themselves to a complete theoretical analysis: (1) a diode formed between parallel planes and (2) a diode formed by concentric cylinders. The fundamental reason for this limitation in structure is that only these tube structures can yield a uniform field at the emitting surface and at the collecting surface for a fixed applied difference in potential between them. Although the plane parallel electrode system is easier to deal with theoretically, it is not as easy to construct as the concentric cylinder diode. Although both structures are of practical importance, the cylindrical emitter is used more widely and techniques for its construction are more highly developed. It is therefore considered significant to discuss the application of theories already developed to the cylindrical structure.

It is not only important to have the electric fields within the test diode uniform but also to have the temperature of the emitter uniform and known. With the oxide cathode the best way to determine the temperature of the emitter over the wide range for which it is desirable to have its value determined accurately is by means of a thermocouple. A fine tungsten wire may be welded to the interior surface of a cylindrical emitter sleeve and the thermal electromotive force developed relative to a cold junction can be calibrated with satisfactory accuracy. Since the entire length of such a sleeve cannot be maintained at the desired temperature, some means must be chosen to assure that the measured electron emission originates over the uniform section of the emitter. The two methods most commonly used are the "guard-ring" method and the method of limiting the activated area of the emitter. The guard-ring method involves structural requirements far more difficult to satisfy than the other and yet, as a research instrument, diodes created in this manner yield reliable results over a wider range of research conditions than apply if the observer depends on the limiting of the emission area by the application of the oxide coating to just the central portion. The objection to the latter method becomes evident especially in the retarding potential range. At a sufficiently high retarding potential the highly activated and the unactivated regions of the cathode contribute to the observed current in direct proportion to their areas. In the intermediate retarding potential range saturation currents will be drawn from the unactivated area and under many circumstances of test such currents are small enough to be neglected. This lack of certainty about the origin of the observed current must not be overlooked in the interpretation of experimental observations.

In order to carry a research investigation over the widest range in temperature and to the greatest retarding potential one must design the test structure so that leakage currents are reduced to far smaller values than would normally be acceptable in practical tubes.

In the immediate neighborhood of a cylindrical emitting surface, the electron motion and the properties of electron gas are not appreciably different from those at the surface of a plane emitter. At distances away from the emitter comparable with its radius, the fact that a cylinder is used becomes more and more important in the interpretation of experimental data. Therefore, for practical cylindrical structures in which the radius of the collector is greater than that of the emitter by a factor of 1.5 or more, it is desirable to work out the theory by which the observed electron emission collected can be interpreted with respect to the influence of the radial and tangential momenta of the electrons as they leave the emitter. Sect. 60 will present an outline of the theory and indicate some of the methods by which tabulated functions presented there may be used.

Even though a given cylindrical diode structure is created with a well-defined collector to emitter ratio of radii, in the application of space-charge theory to the electron flow, the potential minimum crosses the diode as the collector space potential expressed as ψ_c increases from zero to very moderate values of approximately 10 units of V_T. In order to understand the increase in current expected at the collector region as the potential is made positive, it is necessary to combine the theory for the electron flow in a cylindrical structure with an evaluation of the location of the potential minimum between the emitter and the collector. These relations will be presented in Sect. 61.

It was appropriate to develop the basic theories in the previous sections as they would apply to emitters and collectors which exhibited no reflection effects. The application of these theories to experimental data always seems to lead to unexpected disagreement between theory and experiment if reflection effects are neglected. At this time a theoretical basis for the existence of the reflection effect has not been developed and most of the arguments that one may apply in a classical manner to the transmission of electrons across the boundary between a solid and a vacuum lead one to conclude that reflection cannot be of much importance. Since all of the experimental data known to this writer, including those of FAN[1] and HUNG[2], are completely consistent with the reflection-effect hypothesis—even though in some cases the writers do not interpret their own results in that manner—the theory of the influence of reflection needs to be presented. In the examples for which quantitative data are available, Eq. (26.4) offers the most satisfactory form of expression for the transmission of electrons out of an emitter or into a collector. This form may be assumed and equations may be derived to interpret reflection effects either at the collector or the emitter or at both. These equations will be presented in Sects. 62 and 63 along with a discussion of the influence that a deficiency of slow electrons in the energy distribution will have on the application of space-charge theory. This theory, developed on the basis of LANGMUIR'S equations, depended on the electron energy distribution being MAXWELLian.

Sect. 64 will be devoted to the evaluation of the temperature variation of the FERMI level as it applies to the "N" type semiconductor. This theory is essential to the interpretation of the temperature coefficient of both the work-function and the contact difference in potential.

Sect. 65 will relate to the conductivity properties of the solid structure of the semiconductor typical of oxide-coated cathodes and also to the conductivity through the occluded pores in the coating. After the theoretical background has been developed in Sects. 60 through 66, it will be possible to return to the discussion of the theory of cathode evaluation in cylindrical structures. Sect. 67 will complete the theory of thermionic emission requisite of the explanation and discussion of the experimental results presented in the last division of this treatise.

VI. General theory.

60. Electron flow between concentric cylinders. The principles of conservation of energy and of momentum serve to establish the basic equation for the flow of emission current from an internal electron emitter to a concentric cylindrical collector. To establish this limiting condition the potential difference between a point just outside the surface of the emitter and the potential minimum which occurs at the surface of the collector in the absence of space-charge or at the space-charge minimum, when one exists between the electrodes, is expressed as V_s (see Figs. 8, 10, and 14). The surface of the emitter is thus V_s positive with respect to the cylindrical surface at which the limiting condition for electron arrival is being expressed. This surface either coincides with the collector or lies between the emitter and the collector. The tangential (p_θ) and the radial (p_r)

[1] H. Y. FAN: J. Appl. Phys. **14**, 552 (1943).
[2] C. S. HUNG: J. Appl. Phys. **21**, 37 (1950).

momenta are related to the potential difference and to the initial values of these momenta at the emitter by the following equation:

$$\frac{p_\vartheta^2}{2m} + \frac{p_r^2}{2m} + e V_s = \frac{p_{\vartheta s}^2 + p_{rs}^2}{2m} = \text{const}. \tag{60.1}$$

This equation expresses the conservation of energy and the following expresses the conservation of momentum:

$$r p_\vartheta = r_s p_{\vartheta s}. \tag{60.2}$$

Since the limiting condition is that associated with p_r being zero, the following relation gives the limiting values of the momenta p_{r1} and $p_{\vartheta 1}$ at the emitter for electrons to arrive at the potential point V_s negative with respect to the emitter at a radius of R:

$$2 m e V_s = p_{r1}^2 + \left(1 - \frac{r_s^2}{R^2}\right) p_{\vartheta 1}^2. \tag{60.3}$$

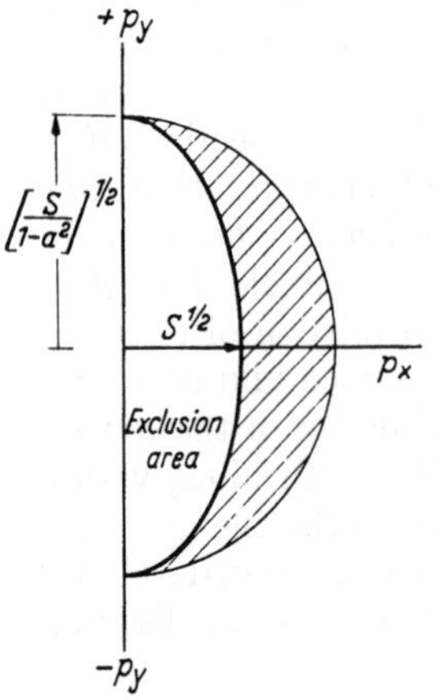

Fig. 18. Momentum space illustration of exclusion area for cylinders of $(1/a)$ ratio of 1.2 as used in Eq. (60.4).

The basic equation for the calculation of current is Eq. (26.7) and for the present purpose it will be best to compute the integral on the basis that the transmission coefficient expressed by Eq. (26.4) is one, or that ω of that equation is zero. The method of integration which must be employed can be illustrated best by the momentum space diagram of Fig. 18.

At the surface of the emitter the radial component of the momentum is interchangeable with the x component used in Eq. (26.7) and the tangential component is the equivalent of the y component of that equation. The exclusion area in the momentum space of Fig. 18 is bounded by an ellipse the equation of which is Eq. (60.3). The integration method found most convenient involved first, the evaluation of the integral in the entire area of positive p_x values outside of the circle drawn with a radius of $(2 m e V_s)^{\frac{1}{2}}$; then, to this value is added the contribution represented by the cross-hatched area of Fig. 18 which lies between the exclusion boundary and the circle. The results of this integration are expressed as follows:

$$I = I_{00} \left\{ \left[1 + \frac{2}{\sqrt{\pi}} \sqrt{\frac{S}{1-a^2}}\, e^{-S/(1-a^2)} - \frac{2}{\sqrt{\pi}} \int_0^{\sqrt{\frac{S}{1-a^2}}} e^{-x^2} dx + \right.\right.$$
$$\left.\left. + \frac{2}{\sqrt{\pi}} \int_{\sqrt{S}}^{\sqrt{\frac{S}{1-a^2}}} \frac{2x^2 \left(1 - \frac{S}{x^2}\right)^{\frac{1}{2}}}{a} e^{-x^2} dx \right] \right\}. \tag{60.4}$$

The two parameters of this equation, S and a, are defined by the following:

$$S = \frac{V_s}{V_T}, \tag{60.5}$$

$$a = \frac{r_s}{R}. \tag{60.6}$$

It is to be noted that when a^2 is small compared with unity, Eq. (60.4) reduces to the essential part of Eq. (26.11) which was derived by SCHOTTKY for the limiting case of a small filament coaxial with respect to a collector of large radius. As a approaches one, Eq. (60.4) becomes identical with the simple BOLTZMANN relation expressed as follows:

$$I = I_{00}\, e^{-S}. \tag{60.7}$$

For arbitrary choices of the ratio of radii, represented by the parameter a, the function shown as Eq. (60.4) has been evaluated by numerical integration for the entire useful range of S from zero to 20. These results are summarized in Table 10.

Smooth curves drawn through plotted points that represent the data given in Table 10 are shown in Fig. 19 for selected values of $(1/a)$ of 1, 1.5, 2.5 and

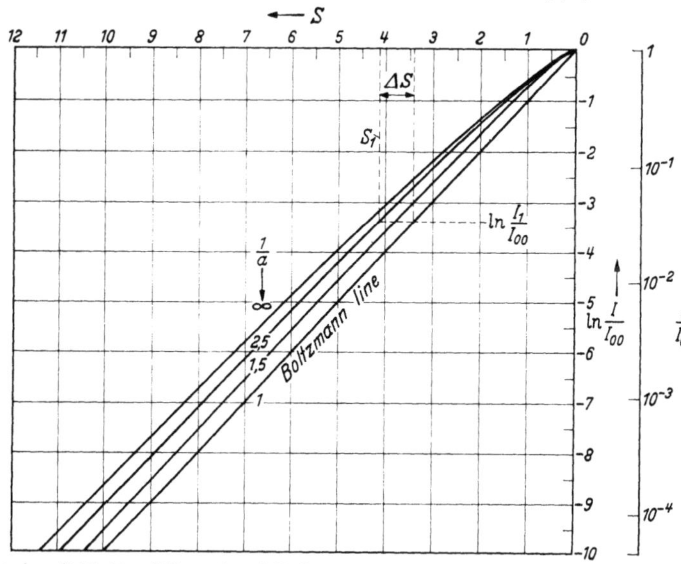

Fig. 19. Plots from Table 10 and illustration of displacement method of conversion to equivalent of parallel planes.

infinity. An inspection of these curves leads to the following conclusions: (1) for small values of $(1/a)$ the lines rapidly become parallel to the BOLTZMANN line, and (2) the larger the values of $(1/a)$ the greater is the displacement of the curve from the BOLTZMANN line.

Mention has been made of the fact that the temperature of the emitter can be determined by an analysis of the electron energy distribution at high retarding

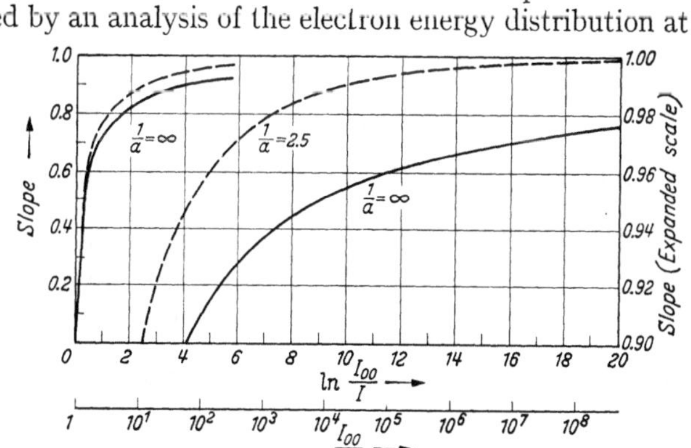

Fig. 20. Slope correction factor from Table 10.

fields applied to a plane parallel structure. These results show that the electron temperature may also be determined in a similar manner for a cylindrical structure if the observed average slope is corrected by the slope factor characteristic of the radius ratio $(1/a)$ of the tube in which the observations were made. In Table 10 the slopes are given for the various values of $(1/a)$ as a function of the quantity $\ln(I/I_{00})$. Smooth curves drawn through the computed points for the two cases of $(1/a)$ of 2.5 and infinity are shown in Fig. 20. These curves are offered by way of illustration of the usefulness of the data in Table 10. For very large values of $(1/a)$ they show graphically that the apparent electron temperature will be 7% higher than the true temperature if the average slope of experimental data is determined for the range of current ratio (I/I_{00}) of 10^{-2} to 10^{-4}. With a ratio of 2.5 the correction will be only 2% for the same observational current range. Since the evaluation of thermionic constants is very sensitive to the correctness of the temperature scale, it is very important that use be made of this tabular information in order to arrive at a correct determination of the temperature from the electron energy distribution observed in a particular structure. In the application of these results gross inhomogeneity of the emitter must be guarded against to avoid misinterpretation of the data. See Sect. 83 for an example.

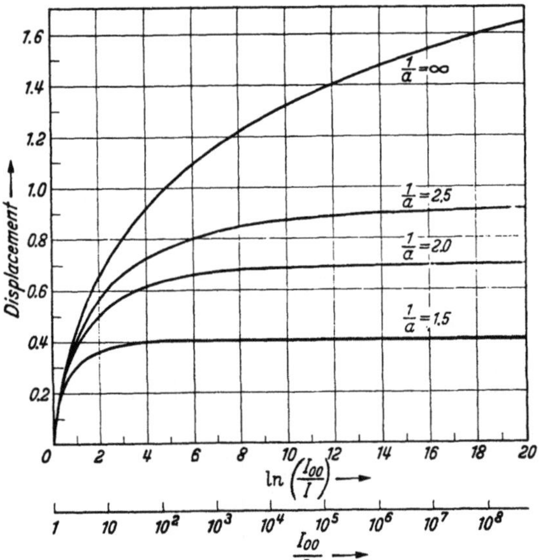

Fig. 21. Displacement for correction to equivalent of parallel planes from Table 10.

For all values of the radii ratio $(1/a)$, the lines illustrated in Fig. 19 are displaced to the left relative to the BOLTZMANN line. Note that at any particular value of S the excess of the current over the corresponding value on the BOLTZMANN line is a direct measure of the contribution to the current that arrived at the collector or at the space-charge minimum because tangential initial momentum contributed to the total energy of the electron. The symmetry of the function represented by Eq. (60.4) shows that an equally valid method for converting observational data taken with a concentric cylindrical structure to its equivalent for a plane parallel structure involves the identification of the S-value at which a particular current is observed as illustrated at point S_1 in Fig. 19. Note that the horizontal displacement shown there as ΔS permits the determination of the applied potential which would yield the same emission current (I_1) from an emitter which had identical properties to that used in the observation but was an equivalent parallel plane diode structure.

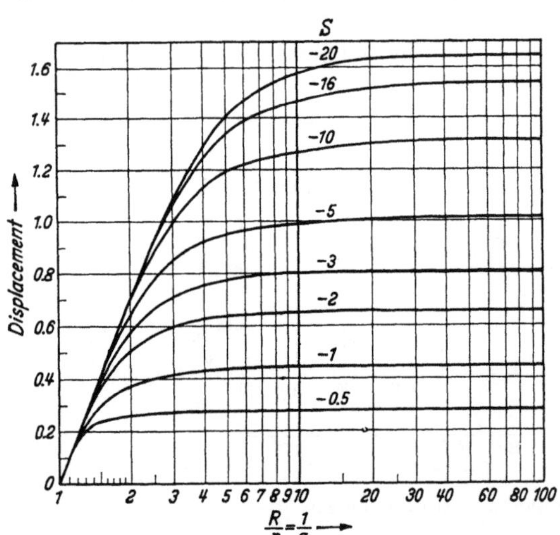

Fig. 22. Displacement as related to shift in space-charge minimum.

For a particular ratio of radii such as 2.5 for example, the displacement ΔS is a function of the ratio (I/I_{00}) given by the separation between the "2.5" line and the BOLTZMANN line. In general, there will be a value of I at which space charge will set in and following that the effective radii ratio will decrease rapidly from 2.5 to a value near to unity. As $(1/a)$ approaches unity, tangential components of momentum will no longer contribute to the initial kinetic energy needed to pass the space-charge minimum. The data derivable from Table 10 for the displacement term can be combined with the knowledge of the location of the space-charge minimum and its movement across the space between the collector and the emitter to correct, with fair accuracy, currents observed in a cylindrical structure to the equivalent plane parallel structure (see Sect. 61). These methods make the space-charge theories developed in the previous sections directly applicable to the cylindrical diode. Fig. 21 has been prepared to illustrate the usefulness of these data recorded in Table 10.

In Fig. 21 the parameter used is the radii ratio $(1/a)$. A second very useful graphical presentation of the data is shown in Fig. 22. For this figure the parameter used is the S of Eq. (60.7) which is the equivalent of $\ln(I/I_{00})$.

61. Space location of space-charge minimum. The need for a simple method of calculating the location of the space-charge minimum as it moves away from the collector surface was demonstrated in the previous section. The distance from the emitter to the space-charge minimum has been defined as x_{su}. With a diode spacing w, the following equation expresses the fractional distance from the emitting surface to the space-charge minimum.

$$f = \frac{x_{su}}{w} = \frac{\chi_{su}}{\chi_{cu} + \chi_{su}}. \tag{61.1}$$

This equation may be combined with Eq. (58.11) to give the following results:

$$f = \frac{\chi_{su}}{u\,\chi_{sR}}. \tag{61.2}$$

The details of the method by which this equation is used will be outlined below. It is of interest to note that for emitters of very high emission capability which have $u_0^2 = (I_0/I_R) > 10^3$, the values of χ_{su} and χ_{sR} are so close to χ_m that Eq. (61.2) is well approximated by the following:

$$f_M = \frac{1}{u}. \tag{61.3}$$

When the value of u_0^2 is much less than 10^3 the fractional distance from the emitter to the space-charge minimum will always be less than $(1/u)$, and it may be computed with accuracy at any particular value of u by the method to be outlined.

For the lower values of I_0, a satisfactory approximation for u_0^2 may be obtained without the complete analysis necessary to determine this quantity accurately by the measurement of the "saturation current". In most practical diode structures used for the evaluation of oxide cathodes such saturation currents will be observed at less than 6 volts applied potential at the collector relative to the emitter. With ψ_{sR} given by Eq. (58.9) the corresponding value of χ_{sR} is obtained from Table 3E. At intermediate values of u between unity and u_0 the appropriate value of ψ_{su} is given by Eq. (58.10) and the corresponding value of χ_{su} may be obtained from Table 3E. With these quantities determined, the fractional distance from the emitter to the space-charge minimum in a plane parallel diode structure is given accurately by Eq. (61.2). Experiments with cylindrical structures indicate that the same formula gives an excellent approximation for the location of the space-charge minimum as it progresses from the collector surface toward the emitter surface as the collector is made more and more positive with respect to the critical potential at which zero gradient is established at the collector. This analysis, combined with the tabular information developed in the previous section goes far to bridge the gab between the exact theories applicable to the plane parallel structure and their use in association with cylindrical diodes.

62. Influence of multiple electron reflection on diode currents. In spite of the fact that electron reflection effects at the boundary between solids and vacuum are not generally accepted as fact, there are no published experimental data that demonstrate conclusively the absence of reflection effects. GERMER[1] interpreted his experiment to indicate that reflection effects are absent, but NOTTINGHAM'S studies[2] show that GERMER'S experimental data are in excellent agreement with the reflection hypothesis. HUTSON'S experiments[3] to be described in more detail in Sect. 72 are most easily explained by the reflection hypothesis and in fact yield a transmission coefficient in excellent agreement with NOTTINGHAM'S. HUNG[4] made a careful study of the electron energy distribution from an oxide cathode source and interpreted the deviation between his observations and those to be expected in the absence of reflection as the result of patchiness of the emitter surface. It is true that patchiness can cause similar discrepancy.

It is the purpose of this section to make available the theoretical background by which reflection can be included in the analysis of experimental data if the observer finds it necessary. Sect. 81 will show that electron emission properties of oxide cathodes, both in the presence and absence of space charge, can best be understood if the reflection hypothesis is integrated with the theoretical interpretation of the experimental data.

The most convenient expressions that best fit the quantitative data available for the transmission and reflection of electrons at a solid surface are repeated from Sect. 26 as follows:

For transmission:

$$D(p'_x) = \left(1 - e^{-\frac{(p'^2_x/2m)}{\omega}}\right). \tag{62.1}$$

For reflection:

$$R(p'_x) = e^{-\frac{(p'^2_x/2m)}{\omega}}. \tag{62.2}$$

In these expressions for transmission and reflection, the quantity $(p'^2_x/2m)$ is the kinetic energy associated with an electron which has a component of momentum p'_x perpendicular to the barrier at the surface of the solid as the electron passes so close to the barrier that the principal forces acting are the mirror-image forces. Although it might be anticipated that the empirical constant ω entering into these equations might very well be expected to be dependent on the externally applied surface field, the researches of HUTSON show that the constant does not change appreciably for moderate field strength.

An emitter in the presence of a strong retarding potential can be expected to deliver across its barrier all of the electrons that would be predicted in the absence of any reflection effect. Over the energy range of 0.2 electron volts and less, strong reflection effects would normally be expected. In the presence of a retarding field, an electron atmosphere is built up outside the emitter with

[1] L. H. GERMER: Phys. Rev. **25**, 795 (1925).
[2] W. B. NOTTINGHAM: Phys. Rev. **49**, 78 (1936).
[3] A. R. HUTSON: Phys. Rev. **98**, 889 (1955).
[4] C. S. HUNG: J. Appl. Phys. **21**, 37 (1950).

exactly that concentration that would be found without reflection effects, because in a region of this kind transmission and reflection exactly compensate so that the normal density is maintained. The situation is different at the collector, since electrons arrive there with a distribution of energy such that many of them cross the collector barrier with a very small energy. These electrons, if reflected, return to the emitter with such high energy that they should be expected to be completely reabsorbed by the emitter surface without multiple reflection. Unless there are unexpected localized energy regions characteristic of the crystal structure of the emitter which would result in a transmission characteristic far more complex than that described by Eq. (62.1), an observer must expect to find a variation in current received with retarding potential changes that will follow the BOLTZMANN relation and therefore seem to indicate an absence of reflection at the collector. This failure to show reflection even though it may exist results from the fact that the energy distribution of the electrons arriving at the collector is independent of the retarding potential. The number of electrons arriving depends on the retarding potential. The fraction reflected will be independent of the retarding potential.

The situation reverses when observations are made in the presence of a small accelerating field. If the minimum energy of the electrons arriving at the collector is in excess of one electron volt and if there are no reflection maxima not described by Eq. (62.1), then practically all of the electron current that arrives at the collector will be absorbed by it. This statement holds true for that range in applied potential for which secondary electron emission effects are negligible.

Even in the absence of space-charge, multiple reflections can occur when the retarding or accelerating fields are very weak. The equations which are to be developed in this section are capable of providing a description of the current-voltage relation to be anticipated under geometrical conditions that involve an equality of transmission and reflection effects at the emitter and the collector boundaries. In the presence of space charge, the problem becomes somewhat more involved, but the same principle is used. Over the retarding potential range and before a space-charge minimum develops, a constant fraction of the electrons that arrive at the collector will be reflected back and reabsorbed by the emitter. As the space-charge minimum develops, the electrons which pass over this minimum will have a higher transmission probability of entering the collector, since the very slowest electrons will be received at the collector with and energy of V_c at the surface. Again, the space-charge which develops in the immediate neighborhood of the emitter will be fully equal to that which would be expected in the absence of a reflection effect, because of the self-compensating action that takes place at the emitter barrier for the electrons passing into space with insufficient energy to go over the space-charge potential minimum. As the value of this potential between the surface of the emitter and the potential minimum as expressed by V_s decreases, the space-charge density will decrease and finally become characteristic of the electron energy distribution which actually comes from the emitter. This distribution will contain an abnormally high proportion of high-energy electrons in comparison with the distribution that would exist in the absence of a reflection effect.

It should be evident from this preliminary discussion of the problem that, if reflection effects are present, they would show in diode studies first because a deviation from the BOLTZMANN distribution should be observed at a smaller current that would be predicted by the computation called for in Sect. 43. Experiments to be described in Sect. 83 will show that such a deviation is found. Many other evidences in these studies of space-charge in diodes are entirely consistent with the reflection hypothesis and the features made available by this hypothesis are needed to explain the observations. The computation which follows applies directly to the circumstance and temperature range for which space-charge effects are negligible.

In the formulation of this problem, there are two ranges of interest, designated as the "retarding range" and the "accelerating range". It is to be anticipated that the equations which are derived and are applicable to each of these ranges individually will yield exactly the same answer at zero field. The voltages that enter into the equation will be designated V_s or V_c. The first will give the potential difference of the emitter surface relative to the collector surface and will therefore represent the actual magnitude of the retarding potential (see Fig. 23). Over the accelerating potential range the symbol will be V_c for the voltage difference between the collector surface and the emitting surface (see Fig. 24). Under this circumstance, the potential will be an accelerating potential for the electrons emitted at the surface. The equation for the retarding range will be developed first.

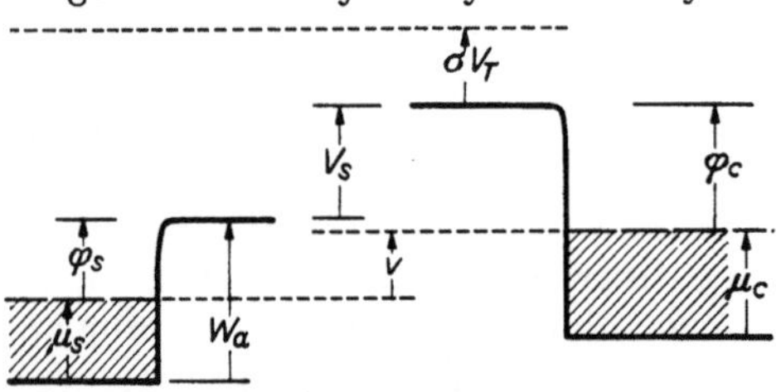

Fig. 23. Symbols used for multiple reflection theory with retarding potentials.

Electrons which are emitted from the surface with sufficient energy to pass across the diode and approach the collector will have a certain probability of acceptance by the collector which is expressed by Eq. (62.1). Those reflected, with a probability expressed by Eq. (62.2), will return to the emitter where, again, there will be a certain probability of acceptance by the emitter given by Eq. (62.1) and a certain probability of reflection to the collector. The final equation for the current delivered from the emitter to the collector must represent the sum of an infinite series which, under most circumstances, converges so rapidly that three or four terms in the series are sufficient to give high accuracy. This series is expressed as follows:

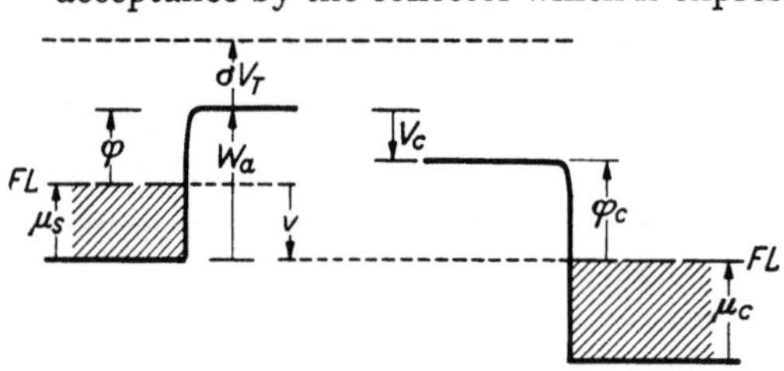

Fig. 24. Symbols used for multiple reflection theory with accelerating potentials.

$$I = I_{00} \sum_{n=0}^{\infty} \int_{\sigma=0}^{\infty} e^{-\left(\frac{V_s}{V_T}+\sigma\right)} \left(1 - e^{-\frac{V_s+\sigma V_T}{\omega}}\right) \left(1 - e^{-\frac{\sigma V_T}{\omega}}\right) e^{-n\left(\frac{V_s+2\sigma V_T}{\omega}\right)} d\sigma. \tag{62.3}$$

In this equation the quantity σ is best defined by Fig. 23 in which it is shown to be the energy of an electron in excess of that needed for entrance across the motive maximum V_s and is expressed in dimensionless units. The quantity I_{00} is defined by the following relation:

$$I_{00} = \frac{4\pi m k^2 e}{h^3} T^2 e^{-\frac{\varphi}{V_T}} = A T^2 e^{-\frac{\varphi}{V_T}}. \tag{62.4}$$

In this equation φ is the *true* work-function (not the Richardson work-function) and the current density expressed is that which impinges on the interior surface for the group of electrons energetic enough to escape. The constant A of this equation is the universal constant from the theory and it should apply to emission from all substances.

Inspection of Eq. (62.3) shows that, for each value of the parameter n, there will be four terms in the final equation. These may be combined algebraically to give the following results:

$$I = I_{00}\, e^{-\frac{V_s}{V_T}} \sum_{n=0}^{\infty} F(V_s, V_T, n) \tag{62.5}$$

with

$$F(V_s, V_T, n) = \frac{e^{-\frac{n V_s}{\omega}}}{\frac{\omega}{V_T} + 4n + 1 + 2n(2n+1)\frac{V_T}{\omega}} - \frac{e^{-(n+1)\frac{V_s}{\omega}}}{\frac{\omega}{V_T} + 4n + 3 + (2n+1)(2n+2)\frac{V_T}{\omega}}. \tag{62.6}$$

The diagram which shows the symbols appropriate to the analysis of the multiple reflection problem with accelerating fields is shown as Fig. 24. The formulation of the problem is exactly analogous to that required for Eq. (62.3) except for the definition of (σV_T) which for the application of an accelerating field measures the kinetic energy associated with the electron motion as it passes over the emission barrier instead of the collection barrier as was illustrated in Fig. 23. With this change of symbolization the new equation is of exactly the same general form and is given as follows:

$$I = I_{00} \sum_{n=0}^{\infty} \int_{\sigma=0}^{\infty} e^{-\sigma}\left(1 - e^{-\frac{\sigma V_T}{\omega}}\right)\left(1 - e^{-\frac{V_c + \sigma V_T}{\omega}}\right) e^{-\frac{n}{\omega}(2\sigma V_T + V_c)}\, d\sigma. \tag{62.7}$$

After integration, the solution is very similar to that given as Eq. (62.5) and is the following:

$$I = I_{00} \sum_{n=0}^{n=\infty} F(V_c, V_T, n). \tag{62.8}$$

Since the function $F(V_c, V_T, n)$ is exactly the same as that given above as Eq. (62.6) except for the substitution of V_c for V_s this function will not be written out in detail. Although in Eqs. (62.1) and (62.2) ω is expressed in energy units of joules, it is far more convenient for computational purposes to express ω in electron volts. The value of this constant which seems to fit the experiments best is 0.191 electron volts or its equivalent 3.06×10^{-18} joules.

Although these relations have been worked out specifically for the plane parallel electrode structure, they are applicable to the cylindrical structure if the radii ratio is close enough to unity so that a large fraction of the electrons reflected from the collector return to the emitter. If the emitter is of very small radius compared with the collector then multiple reflection effects are unimportant even though consideration must be given to the reflection effect at the emission surface. The reason multiple effects are negligible for this geometry is that, after an electron leaves the emitter, there is very small probability that it will return to the emitter without having made many excursions to the collector with a finite probability that it will be accepted there and record as observed current each time it returns. The usual structures for emitter testing with oxide cathodes make it necessary to include in the analysis the concept of multiple reflection. The application of these theories to the analysis of electron flow in the presence of space charge introduces still more complications.

For high values of V_s or V_c, Eq. (62.6) has a value only for the $n=0$ term. The value of the function is given as follows:

$$f(V_c, V_T, 0) = \frac{1}{\frac{\omega}{V_T} + 1}. \tag{62.9}$$

It follows that, for high retarding potentials, the current is given by Eq. (62.10) and for high accelerating potentials it is given by Eq. (62.11) if one neglects the reduction in work-function that results from the SCHOTTKY effect discussed in detail in Sect. 27:

$$I = I_{00}\,\mathrm{e}^{-\frac{V_c}{V_T}}\,\frac{1}{\frac{\omega}{V_T}+1} \tag{62.10}$$

(Range of retarding potentials $V_s > 3\omega$).

$$I = I_{00}\,\frac{1}{\frac{\omega}{V_T}+1} \tag{62.11}$$

(Range of accelerating potentials $V_c > 5\omega$).

Since the observed currents in both the retarding-potential range and the accelerating-potential range are lowered by the same factor which is given by Eq. (62.9), the intersection of the extrapolated retarding-potential line and the saturation-current line in the absence of space-charge will nevertheless give a rather accurate determination of the "zero-field" condition. HUNG[1] observed this intersection point as a function of the temperature. Although he did not point out that he had measured the temperature coefficient of the work-function in this manner, the discussion of this problem in Sect. 81 will indicate that his value of the temperature coefficient of the work-function for oxide cathodes is typical of that expected for a well-activated cathode over the temperature range of his observations.

63. Average energy of electrons transmitted across the mirror-image field reflection barrier. For high retarding potentials, it has already been pointed out that the reflection effect leaves the distribution in electron energy unaltered. However, an analysis of the statistics shows that the average energy associated with the component of momentum perpendicular to a surface across which electrons travel with a MAXWELLian distribution of velocity will be given in the following equation:

$$\bar{\varepsilon}_x = \frac{\overline{p_x'^2}}{2m} = kT = eV_T\,. \tag{63.1}$$

To calculate the average energy associated with the component of momentum p_x' in the presence of a reflection effect boundary one may write the following basic equation:

$$\bar{\varepsilon}_x = \frac{\int\limits_0^\infty \frac{p_x'^2}{2m}\,\frac{p_x'}{m}\,\mathrm{e}^{-\frac{\varepsilon_x}{kT}}\left(1-\mathrm{e}^{-\frac{\varepsilon_x}{\omega}}\right)dp_x'}{\int\limits_0^\infty \frac{p_x'}{m}\,\mathrm{e}^{-\frac{\varepsilon_x}{kT}}\left(1-\mathrm{e}^{-\frac{\varepsilon_x}{\omega}}\right)dp_x'}\,. \tag{63.2}$$

This equation, evaluated, gives the following result:

$$\bar{\varepsilon}_x = kT\,\frac{1+2\frac{\omega}{kT}}{1+\frac{\omega}{kT}}\,. \tag{63.3}$$

[1] C. S. HUNG: J. Appl. Phys. 21, 143 (1950).

If the average energy is expressed in electron volts, the equation takes on the following form:

$$\bar{\varepsilon}_x = V_T \frac{1 + 2\frac{\omega}{V_T}}{1 + \frac{\omega}{V_T}}. \tag{63.4}$$

This equation shows that there is a significant change in the average energy of electrons emitted through a reflection barrier compared with the nonreflection barrier. With the expected value of ω taken as 0.191 electron volts combined with the normal value of V_T of 0.1 electron volts it shows that the average energy under these circumstances is 66% higher than the usual kT value. The difference is still greater if one applies this equation to a lower value of V_T such as 0.05. Under this circumstance the average energy is 80% above the usually accepted value.

The indications are that over the average temperature range important for oxide cathodes, it is to be expected that space-charge effects and other phenomena that depend on the energy distribution of the electrons should be examined in the light of the fact that the average energy of escaping electrons will be significantly higher than the value usually assigned of kT or, in electron units, V_T. As space charge develops with an increase in temperature for a given emitter, the reflection effect alteration of the average energy of the electrons diminishes gradually and finally becomes small enough to be negligible for most practical purposes.

64. Temperature coefficient of the FERMI level in N type semiconductors and the FOWLER thermionic equation. As a starting point for this discussion, Sects. 14, 15 and 16 should be reviewed, since the derivation of the basic equation, Eq. (16.1). depended on that material. This equation is repeated as the following:

$$N(\varepsilon_x)\, d\varepsilon_x = 2\frac{(2\pi m k T)}{h^3} \ln\left(1 + e^{-\frac{\varepsilon_x - \mu}{kT}}\right) d\varepsilon_x. \tag{64.1}$$

In this equation, the variable ε_x is the kinetic energy[1] associated with an electron whose x component of momentum is perpendicular to the surface across which $N(\varepsilon_x)$ electrons cross in the positive direction per unit area per unit time per unit energy range in ε_x. This equation gives the random current which crosses the boundary at the stated energy value with respect to the zero of kinetic energy in the conduction band. If the conductor is reduced to absoiute zero, then the highest occupied electronic state in the conductor is a direct measure of the FERMI level at absolute zero of temperature. It is most convenient to assign the energy at this level its value with respect to the zero of kinetic energy in the conduction band. For all of the common metallic conductors, the FERMI level designated in this equation by μ is positive and decreases in absolute value as the temperature increases. The rate of decrease is dependent on both the electron concentration and the expansion properties of the crystal.

In all of the semiconductors of particular interest as they relate directly to thermionic emission, the concentration of electrons in the conduction band at absolute zero is generally zero and the FERMI level is negative. Although one might consider the approach to absolute zero in the semiconductor problem to yield an indeterminante value for the FERMI level, it is possible to show that

[1] If the temperature-energy kT is expressed in joules as in Eq. (64.1) then ε_x and μ, and later μ' and E, are expressed in joules but if the temperature-energy is expressed in its electron volt equivalent then ε_x, μ, μ' and E are expressed in electron volts.

at the zero of temperature, the FERMI level is exactly halfway between the zero state in the conduction band and the highest occupied impurity level. The energy level diagram of Fig. 25 will serve to relate the quantities mentioned here.

As the temperature increases from absolute zero, the position of the FERMI level changes in a manner that depends on the donor concentration n_D and on the energy level or levels of these donors expressed in the diagram as E_D. Although equations[1] may be worked out by which the temperature variation of the FERMI level can be computed for a multiplicity of donor levels (two such levels are shown in Fig. 25), it will simplify the calculations to treat first the problem for a single, well-defined set of donors. Fig. 26 may be used to describe the situation at an intermediate temperature between absolute zero and the maximum at which structural changes can occur in the semiconductor. This range

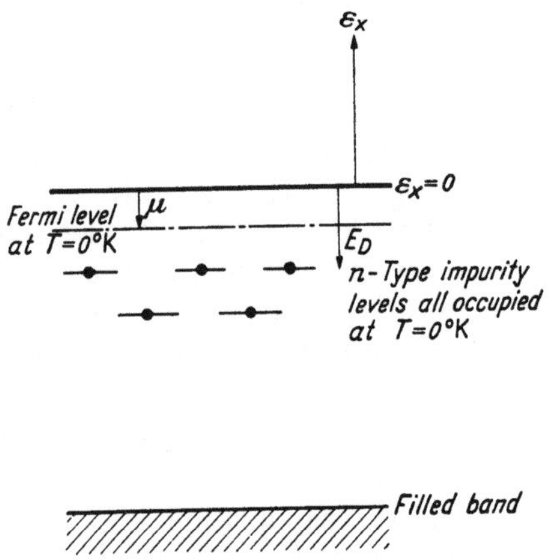

Fig. 25. Energy states in a semiconductor.

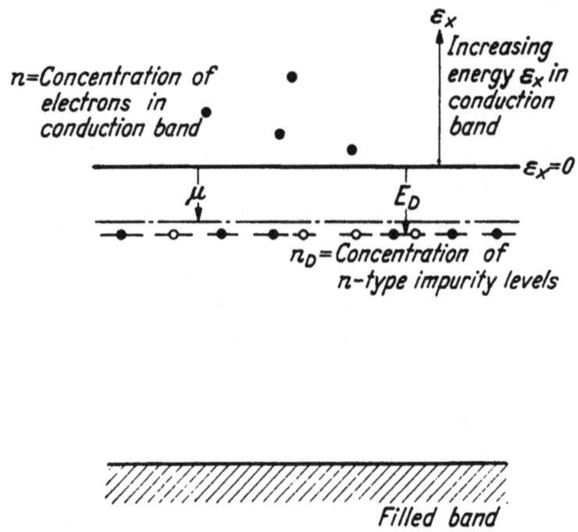

Fig. 26. An N type semiconductor at a high temperature.

in temperature for the oxide cathode lies below approximately 1200° K. As the temperature is increased from absulute zero, electrons transfer from the donor states into the conduction band to maintain a steady state situation at each temperature for which the rate of transfer from the donor states to the conduction band is exactly equalized by the return of electrons from the conduction band to the donor states.

If the experimental evaluation of the theory depended on exact knowledge of the true distribution of the electrons in the conduction band, then a quantum theory analysis of the wave functions suitable for the particular crystals and crystallographic direction of the solid would be demanded before any further progress could be made in the discussion of the temperature variation of the FERMI level. Such details are not necessary for the present purpose. since in the final analysis, the center of interest relates to the random current rather than to the true distribution of the electrons which gives rise to that current.

Since crystals having very different distributions in available quantum states can be formed in intimate contact with each other and under thermal equilibrium conditions there is no net flow of electricity across the boundary, the random current flow in the interiors of the separate crystals must each be exactly that called for by Eq. (64.1) or else the random flow must be exactly zero in one or the other of the crystals. In the application of this principle, the FERMI level is continuous across the boundary and therefore it is to be expected

[1] R. A. HUTNER, E. S. RITTNER and F. K. DUPRÉ: Philips Res. Rep. 5, 188 (1950).

that for dissimilar substances, the location of the energy state $\varepsilon_x = 0$ will be different in the two crystals. The FERMI levels in those crystals expressed relative to the $\varepsilon_x = 0$ of the individual crystal will differ in magnitude. These points are illustrated in Fig. 27.

In Fig. 27 an energy diagram is shown that is applicable to a contact between a metallic conductor and an N type semiconductor. This contact is considered to be in thermodynamic equilibrium with no net current flowing across the boundary layer lying between the planes $A-A$ and $B-B$. These planes are taken sufficiently far inside of each of the conductors so that the presence of the boundary does not influence the random currents within the interiors of the two conductors. In the metal, the FERMI level μ_1 is positive, whereas in the semiconductor its value is negative. The algebraic difference between these FERMI levels is equal to the sum of their absolute values and measures the displacement in energy of the reference levels at the bottoms of each of the conduction bands. Since no conduction takes place in the semiconductor below the level designated as $\varepsilon_{x2} = 0$, all electrons that impinge on the boundary $A-A$ as they move from left to right will be totally reflected without changing energies at the boundary potential and will return into the metal. Thus the net current over this energy band remains zero, in spite of the fact that there is no corresponding current in the semiconductor. The electrons which move from left to right across $A-A$ with energy well above the motive maximum shown in the diagram will pass directly across the boundary layer and enter the semiconductor at its boundary $B-B$. In order to maintain zero net current across these boundaries an exactly equal number of electrons must return from the semiconductor into the metal no matter how complex the energy band structure of either the semiconductor or the metal may be. For that electron group which has energy within the metal greater than $\Delta\varepsilon$ (shown in the figure) and less than that corresponding to the maximum in the motive shown between A and B, there will be a partial transmission from the metal into the semiconductor. Exactly the same reflection and transmission properties must be assigned to this barrier for the electrons which approach from right to left within the semiconductor. It follows that the random current in both the metal and the semiconductor, even in this energy band of incomplete transmission, must be exactly the same and may be computed by the introduction of the correct values of the FERMI level relative to the zero energy state in the individual conductor by Eq. (64.1).

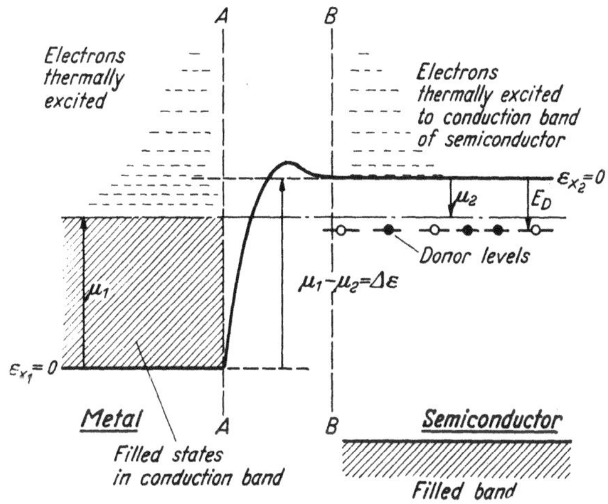

Fig. 27. A contact between a semiconductor and a metal.

This discussion concerning the universality of the random current found in all conductors gives the basic condition upon which the simplifications needed, to avoid the impossible problem of working it out exactly as it would apply to a practical oxide-coated, rests. The exact method would demand a detailed examination of the distribution in quantum states that would come from a wave-mechanical analysis.

Subject only to the restriction that there are no forbidden bands in the energy-level system above $\varepsilon_x = 0$, the random current density in any conductor is given by the following formula:

$$I = 2 \frac{(2\pi m k T)\, e}{h^3} k T \int\limits_{\varepsilon_x=0}^{\infty} \frac{\ln\left(1 - e^{-\frac{\varepsilon_x - \mu}{kT}}\right)}{k T} d\varepsilon_x. \tag{64.2}$$

In all cases of importance to thermionic emission from semiconductors, μ is negative and its absolute value is large compared with kT. Under these circumstances, the exponential part of Eq. (64.2) is small in comparison with unity over the entire range of ε_x from zero to infinity. This fact permits the integration of Eq. (64.2) to give the following result:

$$I = 2 \frac{(2\pi m k T)\, e\, k T}{h^3} e^{\frac{\mu}{kT}}. \tag{64.3}$$

Considered superficially, this equation does not appear to contain the concentration of electrons in the conduction band, since the concentration appears only implicitly in the value of μ.

An ideal free-electron gas at low concentration has a random current expressed by the following relation:

$$I = n\, e \left(\frac{kT}{2\pi m}\right)^{\frac{1}{2}}. \tag{64.4}$$

Since these two equations must yield exactly the same current, they may be equated and in this manner may express the functional relation between the free-electron concentration n and the value of μ required for the electron gas within the semiconductor to be capable of existing in thermal equilibrium with a free-electron gas. The value obtained for μ by this procedure is the following:

$$\mu = -kT \ln \frac{2(2\pi m k T)^{\frac{3}{2}}}{n h^3}. \tag{64.5}$$

In an "N" type semiconductor, the number of electrons which make transitions from the filled valence band into the conduction band at the temperatures encountered with oxide cathodes is so small in comparison with those which transfer from the donor levels to the conduction band that only the latter need to be considered. Since the FERMI factor [Eq. (15.1)] expresses the probability that an electronic state will be occupied, the following equation will give the concentration of electrons as a function of the temperature, the donor concentration, and the energy level of the donors expressed relative to the conduction band reference at $\varepsilon_x = 0$

$$n = n_D \left(1 - \frac{1}{e^{\frac{E-\mu}{kT}} + 1}\right) = \frac{n_D}{1 + e^{-\frac{E-\mu}{kT}}}. \tag{64.6}$$

This equation may also be solved for μ to obtain the following results:

$$e^{-\frac{\mu}{kT}} = \frac{e^{-\frac{E}{kT}}}{\left(\frac{n_D}{n} - 1\right)}. \tag{64.7}$$

Eq. (64.5) and (64.7) may be combined to obtain an explicit expression for the electron concentration as a function only of the temperature for a given donor density (n_D) located in a level at E with respect to the bottom of the conduction band. The equation obtained without the introduction of any approximation is the following:

$$n = \frac{2n_D}{1 + \left[1 + \frac{2n_D h^3 e^{-\frac{E}{kT}}}{(2\pi m k T)^{\frac{3}{2}}}\right]^{\frac{1}{2}}}. \tag{64.8}$$

For the purposes for which these theories are being developed the approximate forms of this equation[1] are not suitable.

Eq. (64.8) may be put into a more useful form for the purpose of computation by expressing the energy E in electrons volts. At the same time one may use to advantage the electron-volt equivalent of the temperature expressed here as V_T defined by Eq. (46.9). It is also convenient to define a quantity μ' by means of the relation shown as:

$$e^{\frac{\mu'}{V_T}} = \frac{2(2\pi m e V_T)^{\frac{3}{2}}}{(n_D/2)\, h^3} = \frac{4(2\pi m e)^{\frac{3}{2}}}{h^3} \cdot \frac{V_T^{\frac{3}{2}}}{n_D}. \tag{64.9}$$

With the above substitutions, Eq. (64.8) may be simplified in its writing as follows:

$$n = \frac{2n_D}{1 + \left[1 + 8\, e^{-\frac{\mu'+E}{V_T}}\right]^{\frac{1}{2}}}. \tag{64.10}$$

In order to obtain a direct expression for the FERMI level as a function of the temperature, Eq. (64.10) can be substituted in Eq. (64.7) and solved for μ. The result is given as follows:

$$\mu = E - V_T \ln 2 + V_T \ln\left\{\left[1 + 8 e^{-\frac{\mu'+E}{V_T}}\right]^{\frac{1}{2}} - 1\right\}. \tag{64.11}$$

The first step in the quantitative use of this equation involves the computation of μ'. Constants in Eq. (64.9) have been combined to permit the easy calculation of μ' by the following relation:

$$\mu' = V_T\left[\ln \frac{1.208 \times 10^{28}}{n_D} - \frac{3}{2} \ln V_T^{-1}\right]. \tag{64.12}$$

In order to assist in the use of these equations, values of the various functions presented here are given in Table 11 for a wide range of values of n_D and E. A very important equation is that for the temperature coefficient of the FERMI level which is given by the differentiation of Eq. (64.11) with respect to V_T. The result of this differentiation is the following:

$$\frac{d\mu}{dV_T} = \frac{\mu - E}{V_T} - \frac{\left(\frac{3}{2} - \frac{E}{V_T}\right) 8 e^{-\frac{\mu'+E}{V_T}}}{2\left[\left(1 + 8 e^{-\frac{\mu'+E}{V_T}}\right)^{\frac{1}{2}} - 1\right]\left[1 + 8 e^{-\frac{\mu'+E}{V_T}}\right]^{\frac{1}{2}}}. \tag{64.13}$$

In reviewing these equations, practically the only assumption which has been made is that the numerical value of $|\mu/V_T|$ be approximately 5 or more.

To satisfy this condition without at the same time having a very small concentration of donor states, the value of (E/V_T) must be of the order of 10. If the value of μ' as computed by Eq. (64.12) can be related to the temperature and

[1] F. SEITZ: The Modern Theory of Solids, p. 188. New York: McGraw-Hill 1948.

be found to give (μ'/V_T) a value greater than 5, then E may have practically any value and the requirement that $|\mu/V_T|$ be greater than 5 will be satisfied.

The combination of symbols described in the following equation occurs so often in the use of all of the equations in this section that it will be used to define a quantity K as follows:

$$K^{-2} = 8\,\mathrm{e}^{-\frac{\mu'+E}{V_T}}. \tag{64.14}$$

In this work, the principal application of Eq. (64.13) comes in the close relation to the temperature coefficient of the work-function. If the electron affinity W_a is practically independent of the temperature, then the temperature coefficient of the FERMI level is a direct measure of the temperature coefficient of the work-function.

This important phase of the basic theory of thermionic emission from oxide cathodes has received very little attention and yet it will be shown in later sections that by making a suitable choice of the donor density n_D and the associated energy E, excellent correlation is found between the temperature variation of the FERMI level and the temperature variation of the contact difference in potential. See Sect. 81 and 83. Eq. (64.13) can be adapted to permit a direct calculation of the temperature coefficient without the previous determination of the numerical value of the FERMI level. The equation written in this form is:

$$\frac{d\mu}{dV_T} = \frac{\ln 2}{2} - \frac{\mu'+E}{2V_T} + \ln\left[(K^2+1)^{\frac{1}{2}} - K\right] - \frac{\left(\frac{3}{2} - \frac{E}{V_T}\right)}{2\left[K^2+1-K(K^2+1)^{\frac{1}{2}}\right]}. \tag{64.15}$$

With this equation it is possible to write expressions for the extreme variation in the temperature coefficient of the FERMI level, since for very low temperatures K is very small, whereas for very high temperatures K is large. In the high-temperature range, for which the approximate expression gives very accurate results, the value of K should be 5 or more. The following equations result from this analysis:

High temperature range:

$$\frac{d\mu}{dV_T} \approx -\left(\frac{\mu'}{V_T} + \frac{3}{2} - \ln 2\right) = -\left(\frac{\mu'}{V_T} + 0.8069\right). \tag{64.16}$$

Low temperature range:

$$\frac{d\mu}{dV_T} \approx -\frac{1}{2}\left(\frac{\mu'}{V_T} + \frac{3}{2} - \ln 2\right) = -\frac{1}{2}\left(\frac{\mu'}{V_T} + 0.8069\right). \tag{64.17}$$

The significance of these two equations stands out strongly after the substitution is made from Eq. (64.12) for the definition of μ'. Again for the two ranges, equations may be written as follows:

High temperature range:

$$\frac{d\mu}{dV_T} \approx -\left[\ln\frac{1.208\times 10^{28}}{N_D} - \frac{3}{2}\ln(V_T^{-1}) + 0.8069\right]. \tag{64.18}$$

Low temperature range:

$$\frac{d\mu}{dV_T} \approx -\frac{1}{2}\left[\ln\frac{1.208\times 10^{28}}{N_D} - \frac{3}{2}\ln(V_T^{-1}) + 0.8069\right]. \tag{64.19}$$

The two main points of this result are: (1) the total range in the variation of the temperature coefficient of the FERMI level can be very little more than a factor of two and (2) the temperature coefficient depends *only* on the concentration of donors and the temperature and is completely independent of the energy level from which the donors arrive. This latter result is of major significance as it relates to the emission properties of oxide cathodes and related semiconductor phenomena.

Various experiments on oxide cathodes which will be discussed in Sect. 81 and 83 indicate that the probable range in *average* temperature coefficient of the work-function as expressed in the units used in these equations lies between 5 and 9 for well-activated, normal oxide cathodes. Since the probable value of E is approximately 0.7 electron volts negative with respect to the bottom of the conduction band, the low temperature range requirement is satisfied with V_T approximately 0.05. With the donor concentration of 5×10^{22} donors per m.^{3}, the value of the temperature coefficient is 4.3. Since the average coefficient can be expected to be approximately 1.5 times greater, this calculation yields an average value of 6.5. An increase in donor concentration by factor of 10 would reduce the average coefficient to 4.2 and a decrease in concentration by a factor of 10 would increase the average coefficient to 8.8. The indication is that for normal oxide cathodes, the expected range of donor concentrations will be 5×10^{21} to 5×10^{23}. It remains to be shown, in the next section, that conductivity data can be used to advantage to estimate the location of the energy level to be associated with these donors.

FOWLER[1] derived an emission equation applicable to the "N-type" semiconductor under specifically limited conditions. Note, first, that the introduction of the BOLTZMANN relation in Eq. (64.3) yields the true RICHARDSON equation, which is Eq. (18.5) and is written here as follows:

$$I = 2\,\frac{(2\pi\, m\, k T)\, e\, k T}{h^3}\, \mathrm{e}^{-\frac{W_a - \mu}{k T}}\,. \tag{64.20}$$

In this equation, W_a is the electron affinity of the semiconductor and, under the conditions of special interest as they relate to oxide cathodes, μ is actually a negative number. The true work-function, which is $(W_a - \mu)$ is, therefore, the numerical sum of these two quantities. The equations given in this section show clearly that μ is a function of the temperature and, therefore, the true work-function will be very temperature dependent. It could be said that the purpose of the FOWLER equation is to take account of the temperature variation of the work-function in a more satisfactory manner than to assume that its average value will be a constant. Since his equation has often been used for the analysis of thermionic emission data, observed with the oxide cathode, it will be worthwhile to derive this formula to show very clearly the restrictions which the user should impose but sometimes does not[2].

An examination of Eq. (64.8) shows, that, under the conditions of a high donor concentration (n_D) and low temperatures, the concentration of electrons in the conduction band can be small in comparison with the number of donors. This condition holds when the principal term in the denominator is very large compared with unity. Under these conditions, the concentration of electrons in the conduction band may be expressed by the following equation:

$$n = n_D^{\frac{1}{2}}\, \frac{\sqrt{2}\,(2\pi\, m\, k T)^{\frac{4}{3}}}{h^{\frac{3}{2}}}\, \mathrm{e}^{\frac{E}{2 k T}}\,. \tag{64.21}$$

[1] R. H. FOWLER: Statistical Mechanics, 2nd edit., p. 402. Cambridge Univ. Press 1936.
[2] J. NAKAI, Y. INUISHI and Y. TSUNG-CHE: J. Phys. Soc. Japan **10**, 437 (1955).

This equation, when combined with the current Eq. (64.4), gives a good approximation for the random current that crosses any boundary within the interior of the semiconducting crystal. This random current, reduced by the BOLTZMANN factor $\exp(-W_a/kT)$, gives the current density in the immediate neighborhood of the crystal surface, when thermodynamic equilibrium applies. Again, there is the problem as to whether or not all electrons which impinge on the boundary from the interior enter the space outside the crystal. In order to include the possibility that some are reflected, the generalized transmission function first introduced in Eq. (26.1) and illustrated further by Eq. (26.4) will be introduced here in the final formulation of the FOWLER emission equation.

$$I = \overline{D(p, B)}\, n_D^{\frac{1}{2}} \left[\frac{e\sqrt{2}\,(2\pi m)^{\frac{1}{4}}\, k^{\frac{5}{4}}}{h^{\frac{3}{2}}}\right] T^{\frac{5}{4}}\, e^{-\frac{W_a-(E/2)}{kT}}. \tag{64.22}$$

The symbol $\overline{D(p, B)}$ is designed to call attention to the fact that the transmission will depend on the electron energy and the barrier itself. An example will shown of the use of this equation in Sect. 81 and 83. In order to assist in computation, the universal constant included in the brackets is evaluated as follows:

$$\frac{e\sqrt{2}\,(2\pi m)^{\frac{1}{4}}\, k^{\frac{5}{4}}}{h^{\frac{3}{2}}} = 1.726 \times 10^{-5}. \tag{64.23}$$

The units are such that, if the donor concentration is expressed in the number per m.³ and the temperature given on the KELVIN scale, then the current density will be expressed in amp/m².

Although Eq. (64.22) can be used as an empirical equation in exactly the same manner as the RICHARDSON form of equation, its most useful purpose is that of determining an estimated value of the donor concentration. Assume that it is possible to measure experimentally the zero-field thermionic emission current density from a semiconducting surface such as an oxide-coated cathode and construct a plot of $\log_{10}(I/T^{\frac{5}{4}})$ as a function of $(1/T)$. The slope of such an experimentally obtained curve will determine $(W_a - 0.5\,E)$. It should be clear from the method by which this equation was derived that this quantity really represents the true work-function at absolute zero. The true work-function is of course temperature-dependent and the negative of the temperature coefficient is given by Eq. (64.19). It is assumed that the temperature coefficient of the electron affinity W_a is negligible. After the FOWLER emission plot has been constructed as described above and the slope determined, any point on the plot will serve to determine the value of the numerical constant defined as follows:

$$A_F = 1.726 \times 10^{-5}\, \overline{D(p, B)}\, n_D^{\frac{1}{2}}. \tag{64.24}$$

An experimentally determined value (Sect. 81) for the empirical constant A_F is 1.5×10^5 or, if the current is expressed in amp/cm.², the number is 15 amp per cm² · deg$^{\frac{5}{4}}$. With a value of $\overline{D}$ of 0.175 (which is not unlikely for the very low temperature range) the donor density n_D is 2.4×10^{21} per m³. The very accurate data of HUNG are used for this analysis in Sect. 81 and there the reflection effect is introduced as indicated in Eq. (62.11) *before* the plot is made according to Eq. (64.22). The equation form then becomes:

$$I_{00} = 1.726 \times 10^{-5}\, n_D^{\frac{1}{2}}\, T^{\frac{5}{4}}\, e^{-\frac{(W_a-(E/2))}{kT}}. \tag{64.25}$$

The FOWLER work-function is given by:

$$\varphi_F = \frac{(W_a - (E/2))}{e}. \tag{64.26}$$

65. Conduction processes in porous semiconductors. LOOSJES and VINK[1] were among the first to call specific attention to the need for a detailed analysis of conduction processes in porous semiconductors to serve as step toward the better understanding of thermionic emission phenomena from oxide cathodes. This section will deal with basic concepts needed for the interpretation of conductivity data applicable to polycrystalline semiconductors.

Practically all experiments show that at very low field strength the conduction current is directly proportional to the average electric field within the test specimen. The basic condition which must be satisfied is that *the energy gained on the average by the electrons between inelastic collisions must be small compared with their thermal energy expressed as kT*. The force acting on an electron in the presence of a field can be expressed as the product $(e E_x)$ and the matter of the sign of the field in relation to the sign of the charge carrier can be omitted for the time being. The average change in momentum is then the force multiplied by the average time between inelastic collisions. This relation is expressed as follows:

$$\overline{\Delta p_x} = e E_x \bar{\tau}. \tag{65.1}$$

If the electron concentration is n and the effective mass that of a free electron, the conductivity may be expressed as follows:

$$\sigma = \frac{\overline{\Delta p_x}}{m E_x} e n = \frac{e \bar{\tau}}{m} e n. \tag{65.2}$$

The factor $(e \bar{\tau}/m)$ is the electron mobility which will be identified as M_s for the electron mobility within the solid structure and as M_p for the electron mobility in the porous structure. These definitions expressed in equation form are the following:

For crystalline structure:

$$M_s = \frac{e \bar{\tau}_s}{m}. \tag{65.3}$$

For pore conduction:

$$M_p = \frac{e \bar{\tau}_p}{m}. \tag{65.4}$$

Studies of the conductivity of oxide-coating material as a function of the temperature and the previous history (which would include activation processes) serve as the best means of gaining information concerning the conduction mechanism. LOOSJES and VINK have identified three ranges in temperature, two of which are associated with the conductivity of the free electrons in the pores of the porous structure. The third is dominated by the conduction through the solid. These two conduction mechanisms may be thought of as parallel conductors. Eq. (65.2) is suitable for the computation of each and depends on a knowledge of the electron concentration n and the average time between inelastic collisions $\bar{\tau}$. These two quantities will differ many orders of magnitude depending on whether one is dealing with the conduction in the solid or in the pore. The theoretical background for these two cases will be presented very briefly in this section. Conductivity in the crystalline structure will be discussed first.

Since Eq. (64.8) or (64.10) may be relied upon for the calculation of the electron density in an N-type semiconductor as a function of temperature, it remains only to discuss $\bar{\tau}_s$ to establish the conductivity characteristic. The use

[1] R. LOOSJES and H. J. VINK: Philips Res. Rep. **4**, 449 (1949).

of either classical or FERMI statistics, as it applies to a free electron concentration of low density, permits one to write an expression for the average speed of electrons in an electron distribution characterized by a specific temperature. The expression for the average speed is the following:

$$\bar{v} = 4\left(\frac{kT}{2\pi m}\right)^{\frac{1}{2}}. \tag{65.5}$$

One of the factors that must enter into the calculation of the time between inelastic collisions is the average speed of the electrons and the time should be inversely proportional to the speed. A second factor which must enter is the distance between collision centers and its dependence on temperature. At the relatively high temperature associated with thermionic emission there can be little doubt that the thermal vibration of the molecules of the structure sets up a disorder of the potential function through which the electrons travel in their random motions. Therefore the average distance of "free flight" must decrease as the temperature increases. Even if quantum mechanics could be applied to solve this problem of the lattice scattering of the electron waves for a pure barium oxide crystal, it is doubtful that the solution would apply more than approximately to the barium-strontium-oxide crystals employed in all thermionic emitters of this classification. One may assume, therefore, for the purposes of the present discussion, that the average "free flight" distance in the solid structure is inversely proportional to the temperature and that an experimental determination of the electron mobility will be depended upon to determine the order of magnitude of the proportionality constant and, in turn, the approximate distance between scattering points. With these assumptions the following equation may be written:

$$\tau_s = \frac{l_s}{\bar{v}} = \frac{l_{s0} T_0}{T} \cdot \frac{1}{4} \cdot \left(\frac{2\pi m}{kT}\right)^{\frac{1}{2}}. \tag{65.6}$$

In this equation the proportionality constant is written as $(l_{s0} T_0)$ with l_{s0} the average free flight distance at some low temperature T_0 above which it is assumed that the free flight distance is inversely proportional to the temperature. Eq. (65.6) may be combined with Eq. (65.3) to obtain the following expression for the mobility:

$$M_s = \left(\frac{\pi e}{8m}\right)^{\frac{1}{2}} l_{s0} \frac{V_{T0}}{V_T^{\frac{3}{2}}} = 2.63 \times 10^5 \frac{V_{T0}}{V_T^{\frac{3}{2}}} \cdot l_{s0}. \tag{65.7}$$

The equation has been put in the above form in order to make it ready for easy numerical calculation. Rather than assume a value of l_{s0} it is of more direct interest to take the value for the mobility given by PELL[1] for electrons in barium-oxide single-crystals which was 5×10^{-4} m.^{2} per v-sec. If it is assumed that this value is appropriate at a temperature V_{T0} of 0.04 (464° K) and the value of l_{s0} is computed it is 3.8×10^{-10} m. Calculations based on the molecular weight and density figures for barium and strontium oxide lead to the conclusion that this distance is exactly equal to the average molecular distance in the solid. If one considers the high temperature at which this mobility measurement was made and the fact that it could very well have been twice as great since there was considerable uncertainty in the exact value to be chosen, the fact that the free flight distance in an ionic crystal is equal to one or two intermolecular distances seems altogether reasonable and justifies the use of Eq. (65.7). With the introduction of these numbers, one obtains a mobility equation which follows:

$$M_s = \frac{4 \times 10^{-6}}{V_T^{\frac{3}{2}}} \text{ m.}^2/\text{v-sec.} \tag{65.8}$$

[1] E. M. PELL: Phys. Rev. **87**, 457 (1952).

This expression for the mobility of electrons in a barium-strontium oxide crystal may be combined as in Eq. (65.2) with the expression for electron density as a function of the temperature given in Eq. (64.10) to express the conductivity of typical crystals used in oxide cathodes so that the prediction of the equation can be compared with experiment and in that manner an appropriate value for the donor energy level E can be derived. The equation so derived is the following:

$$\sigma = \frac{1.82\times 10^{-24}\, C_s\, n_D\, V_T^{-\frac{3}{2}}}{1+\left[1+8\,\mathrm{e}^{-\frac{(\mu'+E)}{V_T}}\right]}\,. \tag{65.9}$$

The constant C_s is introduced to correct for the fact that a porous structure will have a smaller conduction cross section than the same volume of solid crystalline material.

The usual experimental procedure is to plot the logarithm of the conductivity σ as a function of $(1/V_T)$ and use the average slope of this plotted curve as a measure of the "activation energy". The value of E is then often taken as twice the activation energy. The analysis to follow shows that, although this procedure is approximately correct, it is necessary to analyze the results more critically if accuracy is desired.

A quantity K has been defined by Eq. (64.14) and the evaluation of this number indicated that K is generally small for the range of temperature, electron concentration, and energy level E of greatest interest in thermionic emission. With this restriction, Eq. (65.9) may be written as follows:

$$\sigma = \frac{1.82\times 10^{-24}\, C_s\, n_D\, V_T^{-\frac{3}{4}}\, K}{1+K}\,. \tag{65.10}$$

This equation may be modified still further by the introduction of the value of K as expressed in Eq. (64.14) into the numerator of the above equation and the conductivity relation becomes the following:

$$\sigma = \frac{7.07\times 10^{-11}\, C_s\, n_D^{\frac{1}{2}}\, V_T^{-\frac{3}{4}}\, \mathrm{e}^{\frac{E}{2V_T}}}{1+K}\,. \tag{65.11}$$

The main purpose of measuring the conductivity of an oxide cathode as a function of the temperature is to evaluate the effective energy difference E between the bottom of the conduction band and the energy levels occupied by the donors. It should be clear from the above equation that this can be accomplished best by plotting $\ln(\sigma V_T^{\frac{3}{4}})$ as a function of $(1/V_T)$. The form of the equation is the following:

$$\ln(\sigma V_T^{\frac{3}{4}}) = \ln(7.07\times 10^{-10}\, C_s\, n_D^{\frac{1}{2}}) + \frac{E}{2}\, V_T^{-1} - K\,. \tag{65.12}$$

This equation makes use of the fact that the value of K is small. The slope of the straight line obtained from this plot should give the best obtainable value for $(E/2)$, since the temperature variation of K is very small in comparison with $(E/2)$.

In most examples given in the literature, it is the activation energy that has been evaluated and this has been defined by the following relation:

$$\frac{d(\ln\sigma)}{d(V_T^{-1})} = E_a\,. \tag{65.13}$$

By the direct differentiation of Eq. (65.9) one obtains the complete expression for the activation energy which is clearly not independent of the temperature, as the following analysis will show:

$$E_a = \frac{3}{2} V_T - \frac{\frac{3}{2} V_T - E}{2(K^2+1)^{\frac{1}{2}}\left[K + (K^2+1)^{\frac{1}{2}}\right]}. \tag{65.14}$$

Since the principal range of interest is that for which K^2 is very small in comparison with unity, this equation may be simplified as follows:

$$E_a = \frac{E}{2} + \frac{3}{4} V_T - \left(\frac{E}{2} - \frac{3}{4} V_T\right) K. \tag{65.15}$$

In general, the last term of this equation is very small and, therefore, within the accuracy of most available measurements, an improved value for the true energy of the donor state would be given in terms of the activation energy E_a by the following equation:

$$E = 2E_a - \tfrac{3}{2} V_T. \tag{65.16}$$

The range in activation energy reported in the literature and discussed in Sect. 77 lies between -0.25 and -0.4 ev., with some dependence on the state of activation of the test specimen. The only activation energies that are appropriate for comparison with this theory are those taken over such a low temperature range that the electron conductivity through the pores contributes a negligible proportion of the total conductivity of the specimen. An average value of V_T for this low temperature range is 0.05. Eq. (65.16) shows that if the activation energy E_a is -0.31 the effective location of the donor state will be at 0.7 ev below the bottom of the conduction band.

With a very densely packed crystalline structure, the constant C_s of Eq. (65.9) could presumably approach the value of 1.0. Since the equations as they have been developed apply to the conductivity of single crystals, they do not take into account the fact that in a practical oxide cathode less than 50% of the superficial cross section is occupied by crystals. Furthermore, these crystals make contact at very small surface areas compared with the crystal cross section. It is practically impossible to generalize, therefore, about the best choice of C_s to fit a specific test specimen which will permit a direct comparison between theory and experiment. Under the circumstances, the best procedure is to evaluate the energy E which measures the location of the donor levels by means of the temperature variation in the conductivity over the low temperature range, as has been described in this section. The best means of determining the most suitable concentration of donors has been demonstrated in a previous section to depend on a measurement of the temperature coefficient of the contact difference in potential, which in turn, is the negative of the temperature coefficient of the true work-function. With the donor density and the energy level known, observations of the conductivity analyzed according to Eq. (65.9) yield a direct value for C_s. Again, the published material for comparison with theory is very meager, but the indications are that a value of C_s of the order of 0.05 to 0.1 is needed to be consistent with experimental values unless the mobility of the electrons in the single crystals of barium-strontium oxide is smaller than the PELL[1] value of 5×10^{-4} m.^{2} per v-sec. Since a choice of C_s within the range described here seeems reasonable, confidence should be placed in these equations and in this analysis until better data become available to justify the development of an alternative theoretical analysis.

[1] E. M. PELL: Phys. Rev. 87, 457 (1952).

The conductivity of the open structure between the particles of an oxide cathode will become more and more important as the temperature increases. The theory to be developed in the following paragraphs shows that the conduction through the pores will be directly proportional to the electric field within the porous structure provided that the energy gained from the field in the time of one free flight of the electron across the pore is small compared with the thermal energy V_T (or kT). The analysis will show that at very high field strength comparable to that associated with pulse emission studies, the conduction will increase as the square root of the field as LOOSJES and VINK[1] have shown experimentally. The conductivity in the pore will be given by Eq. (65.2) in which the density of the electrons within the pore will be relatively uniform over the lower temperature range but will become non-uniform in a manner that can best be evaluated by the theory presented here in Sect. 21 to 24, inclusive. This development of space charge within the pore influences both the determination of $\bar{\tau}_p$ and the value of electron density n. These factors control the conductivity-temperature function in the high-temperature range, and it will be different from the function followed over the low-temperature range of pore conductivity. These phenomena account very satisfactorily for the two high-temperature ranges in conductivity first reported by HANNAY, MACNAIR, and WHITE[2] and identified by LOOSJES and VINK as ranges II and III.

In the application of Eq. (65.2) to the conductivity of the pores, the computation of the average time of flight of an electron will depend mainly on the electron speed and the effective path length for the low-field condition. As the intensity of the electric field increases, the energy gained by an electron in one free flight across the pore can eventually exceed its temperature energy under which condition the time of flight is greatly shortened and is field-dependent. It is only necessary to formulate the problem for these two extremes, if one realizes that between these extremes the average time of flight can take on any intermediate value.

If the temperature range is that in which pore conduction predominates and measurement shows that the current is directly proportional to the field, sufficient indication is given that the energy gained per free flight is small enough in comparison to the thermal energy and that the time of flight computed by the following equation is the proper choice.

$$\bar{\tau}_{p\,\mathrm{II}} = \left(\frac{\pi m}{8e}\right)^{\frac{1}{2}} \frac{l_p}{V_T^{\frac{1}{2}}}. \tag{65.17}$$

In this equation, the number II is applied as a subscript on $\bar{\tau}_p$ to indicate that this equation applies over the temperature below approximately 0.09 ev equivalent (1044° K). In this equation l_p corresponds to the free-flight distance averaged over all of the electron trajectories within the pore. This free-flight distance is, in general, smaller than the average dimension of the pore but is the same order of magnitude. For example, if the pore were spherical, then l_p would be approximately 64% of the diameter. The relation given in Eq. (65.17) can be used to calculate the mobility for this low-field condition and the result is given as follows:

$$M_{p\,\mathrm{II}} = \left(\frac{\pi e}{8m}\right)^{\frac{1}{2}} \frac{l_p}{V_T^{\frac{1}{2}}} = 2.63 \times 10^5 \frac{l_p}{V_T^{\frac{1}{2}}}. \tag{65.18}$$

[1] R. LOOSJES and H. J. VINK: Philips Res. Rep. **4**, 449 (1949).
[2] N. B. HAANNAY, D. MACNAIR and A. H. WHITE: J. Appl. Phys. **20**, 669 (1949).

This equation combined with the evaluation of the electron density gives the conductivity of the pore

$$\sigma_{p\,\mathrm{II}} = n_0\, e \left(\frac{\pi e}{8m}\right)^{\frac{1}{2}} \frac{l_p}{V_T^{\frac{1}{2}}} = 4.21\times 10^{-14}\, n_0 \frac{l_p}{V_T^{\frac{1}{2}}}. \tag{65.19}$$

Eq. (24.7) is the most suitable one for the computation of the electron density in the immediate neighborhood of the surface boundary of the pore. This equation, with the universal constants evaluated, and the temperature expressed in its electron-volt equivalent, is the following:

$$n_s = 6.04\times 10^{27}\, V_T^{\frac{3}{2}}\, \mathrm{e}^{-\frac{\varphi}{V_T}}. \tag{65.20}$$

In this equation, the symbol φ represents the *true work-function*. The true work-function depends to some extent on the state of activation, principally through the alteration of its average temperature coefficient which decreases as the concentration of donors increases. A formula which applies as a satisfactory approximation for a limited range in temperature for a well-activated oxide cathode is the following:

$$\varphi = \varphi_0 + \alpha\, V_T. \tag{65.21}$$

For these oxide cathodes the range in φ_0 is 0.9 to 1.2 and for α it is 5 to 9.

Over the range of conductivity characteristic below 1000° K the pore concentration of electrons n_0 is practically equal to the electron concentration at the surface of the pore. The temperature at which there is an appreciable difference between these concentrations depends on the pore size and can be determined quite well by the theory presented in Sect. 24. There the space charge relations were computed for a pillbox like pore with the small dimension w between parallel planes. Qualitative considerations indicate that the use of l_p, which is defined as the *effective* average distance of free flight within the pore, is the equivalent to the use of w in Eq. (24.4). With this condition considered, the following equation gives the maximum density of electrons which one can expect in the central region of a pore of this size:

$$n_{0m} = 1.09\times 10^{9}\, \frac{V_T}{l_p^2}. \tag{65.22}$$

The actual density n_0 in the central region of a pore the boundaries of which are characterized by a definite value of the true work-function is related to this maximum possible value for the density through the parameter z expressed by:

$$n_0 = n_{0m}\, z^2. \tag{65.23}$$

Since both n_s and n_{0m} can be computed from a knowledge of the temperature, the pore size, and the true work-function, the expression given as Eq. (24.6) permits the direct evaluation of z^2 from the ratio of these two concentrations with the help of Table 4 or of the curve shown as Fig. 9. Since the main purpose of this calculation is to estimate the upper limit for the temperature range identified by II, Eq. (24.9) serves as a satisfactory means of computing z, since it is only for z values less than 0.1 that n_0 and n_s are essentially equal. Eq. (24.5) shows that the reduction in concentration at the center of the pore under this condition is 2.5%. Since n_s increases very rapidly with the temperature after this particular temperature has been reached, the transition from conduction region II to III is quite sharp. With this value of z as the arbitrary criterion for the determination of the transition temperature, the following relation may be written:

$$1.8\times 10^{-21} = l_p^2\, V_T^{\frac{1}{2}}\, \mathrm{e}^{-\frac{\varphi}{V_T}}. \tag{65.24}$$

If the characteristic pore dimension l_p is 2.5×10^{-6} m. (2.5 microns), this equation is satisfied at the value of V_T of 0.0833 (970° K). This calculation is mentioned at this point to indicate that the above equation is in excellent agreement with the temperature limit that marks the separation between regions II and III found by LOOSJES and VINK. The variation of the critical temperature with activation will be discussed in Sect. 77.

The introduction of Eq. (65.20) into Eq. (65.19) completes the calculation of the expression for the pore conductivity as a function of the pore dimension, the temperature, and the true work-function for that range of temperature which lies within range II:

$$\sigma_{p\,\mathrm{II}} = 2.54\times10^{14}\, V_T\, \mathrm{e}^{-\frac{\varphi}{V_T}}\, l_p. \tag{65.25}$$

If the true work-function is known only through its approximate representation given as Eq. (65.21), then:

$$\sigma_{p\,\mathrm{II}} = 2.54\times10^{14}\, \mathrm{e}^{-\alpha}\, V_T\, \mathrm{e}^{-\frac{\varphi_0}{V_T}}\, l_p. \tag{65.26}$$

A plot of $\ln \sigma_{p\,\mathrm{II}}$ as a function of V_T^{-1} yields an activation energy Φ_σ which is related to φ_0 by Eq. (50.11) which yields:

$$\varphi_0 = \Phi_\sigma - \frac{V_n + V_m}{2} \tag{65.27}$$

in which V_n and V_m are the voltage equivalents of the maximum and minimum temperatures of range II.

Although Eq. (65.25) represents the conductivity of a pore of characteristic dimension l_p, the quantity usually measured is the conduction per unit area, as exemplified by the researches of LOOSJES and VINK. The following equation serves to describe this function in which g is the conduction of a specimen of thickness L over which a potential difference of V is applied:

$$\frac{g}{A} = 2.54\times10^{14}\, \frac{V}{L}\, C_p\, V_T\, \mathrm{e}^{-\frac{\varphi}{V_T}}\, l_p. \tag{65.28}$$

Two additional symbols are introduced in this equation, which are A, for the area, and C_p, for a constant introduced in this equation to account for some of the structural factors similar to those considered, just as the corresponding constant, C_s, was introduced in Eq. (65.9). Before this theoretical equation is compared with experimental results, an estimate must be made for the value of C_p. The higher the degree of porosity, the greater the value of C_p. Even though the known porosity might be 50%, the choice of a value of 0.5 would be too large for two reasons. First, the conduction through the pores will follow a devious route, even though it might be direct from pore to pore. Second, some conduction will be through solid structure in series with the pores. This resistance will reduce the conduction. It is estimated, therefore, that the value of C_p to be used for comparison with experiment should be 0.1 to 0.4 for the usual coating structure.

A question may be raised as to why Eq. (65.20) is used for the electron density instead of Eq. (64.21), since it was the latter that led to the FOWLER equation for thermionic emission. The latter formula is not appropriate, even though it was used by NAKAI, INUISHI and TSUNG-CHE[1]. The purpose of Sect. 64 was to show that the temperature variation of the FERMI level could be incorporated n the theory of the FOWLER emission equation under the restriction that the

[1] J. NAKAI, Y. INUISHI and Y. TSUNG-CHE: J. Phys. Soc. Japan **10**, 437 (1955).

range of temperature for which it is applied be sufficiently low. The simplified equation used as a substitute for the exact equation, namely, Eq. (64.10), is thus limited in its range of application. It is therefore more appropriate to use Eq. (65.20), even though it involves the true work-function and this, in turn, may have to be represented by its approximate form given as Eq. (65.21).

66. Conductivity with high internal field strength. Experiments have been carried out by LOOSJES and VINK[1] and by others[2] in which sufficiently high fields were applied to an oxide coating to pass beyond the linear part of the conduction characteristic. The first deviation in the linearity should appear when the energy gained from the field is comparable with the temperature energy V_T or (kT). The experiments of LOOSJES and VINK show that, in the lower temperature range which is still high enough so that pore conductivity may be separated from crystal conductivity, the conduction current is proportional to the square root of the total voltage applied across the specimen. The following equation indicates the conditions under which this result may be anticipated.

$$\left(\frac{e\,E_x\,l_p}{2m\,\lambda\,l_p}\right)\tau^2 + \left(\frac{8e\,V_T}{\pi\,m\,l_p^2}\right)^{\frac{1}{2}}\tau = 1\,. \tag{66.1}$$

An explanation of this equation will be given. The average time of flight τ enters the quadratic equation in two terms. The first term relates to the energy acquired by an electron in traveling a distance λ in the presence of a field of electric intensity E_x. A suitable value for λ will be close to one half of the characteristic pore dimension defined in the previous section and represented by l_p.

The second term comes from the average thermal speed, as in Eq. (65.17). Two extremes may be considered, depending on the value of $(E_x\,l_p)$. In the low-field range for which the first term of this equation is very small, the free-flight time may be computed by the second term to obtain the result already given as Eq. (65.17). As E_x increases, the solution to the quadratic equation above gives an ever decreasing value of the flight time τ, and more and more importance must be attached to the first term of this equation. Finally, when it becomes large compared to the second term, the average flight time τ_E becomes:

$$\tau_E = \left(\frac{2m\,\lambda}{e\,E_x}\right)^{\frac{1}{2}} = \left(\frac{2m\,\lambda\,l_p}{e\,V_p}\right)^{\frac{1}{2}}. \tag{66.2}$$

Here, V_p is the drop in potential over a single "effective" pore distance. Note that the flight time is inversely proportional to $E_x^{\frac{1}{2}}$ (or $V_p^{\frac{1}{2}}$) and proportional to a characteristic distance $(l_p\,\lambda)^{\frac{1}{2}}$. At moderately low temperatures, this distance should be approximately 0.4 of the average pore dimension. As the temperature increases and/or the field increases, additional complications enter into the computation of the flight time and the conduction equation.

Assume first that the field is not quite strong enough to wipe out all of the effects of space charge, but nearly sufficient to do so. Under this condition, Eq. (65.2) may be combined with Eq. (65.20) to yield an equation for the pore conduction per unit area with a moderately strong field and not too high a temperature. The result is the following:

$$\frac{g}{A} = 5.75\times 10^{14}\,C_p\,\lambda^{\frac{1}{2}}\left(\frac{V}{L}\right)^{\frac{1}{2}} V_T^{\frac{3}{4}}\,\mathrm{e}^{-\frac{\varphi}{V_T}}. \tag{66.3}$$

This equation is similar in many respects to that given as Eq. (65.28) and is the one most suitable for comparison with experiment, as will be shown in Sect. 77.

---

[1] R. LOOSJES and H. J. VINK: Philips Res. Rep. 4, 449 (1949).

[2] See footnote 1, p. 91.

The conduction per unit area is g/A for an applied voltage difference of V over a specimen of thickness L. Thus, the average field within the specimen is V/L. The quantity C_p must be estimated, since it depends on the actual physical structure of the specimen investigated. In general, the porosity of practical cathodes is close to 0.5, although somewhat smaller and larger values than this have often been studied. As in the previous section, it is indicated that the conduction through pores may follow devious routes and in some circumstances may even have to be considered to be in series with a very small thickness of the solid structure. The crystalline solid has a higher resistance and therefore interferes slightly with the conduction attributed to the pores alone. These considerations lead to the establishment of a probable range for C_p of 0.1 to 0.4. Applications will be made of this equation in the discussion of experiments with oxide cathodes given in Sect. 84.

An inspection of Eq. (66.1) shows that the maximum value of the time of free flight is associated with the lowest value of the internal field. As the field increases, the first term of the equation becomes more and more important until it finally dominates, under which condition the time of free flight is inversely proportional to the square root of the field. At some particular field, the two terms will contribute equally, that is, the value of each will be exactly 0.5. The relation between the field and the temperature for this special circumstance is given in the following equation:

$$\frac{E_x l_p}{V_T} = \frac{V_p}{V_T} = \frac{8}{\pi} \frac{\lambda}{l_p} \approx 1.3. \tag{66.4}$$

In this equation, the product $(E_x l_p)$ is the potential drop over a characteristic pore length and is symbolized by V_p. Since electrons enter the pore at various positions along its x direction, the distance λ should be approximately one half of the characteristic distance l_p. Subject to this assumption, a numerical value of the ratio is approximately 1.3. It follows, therefore, that if the potential V_p is just 1.3 V_T, then the conduction will be exactly half of that calculated by Eq. (65.28). Again, inspection of Eq. (66.1) shows that under this condition of (V_p/V_T) of 1.3, the conduction will be 70% of that calculated by Eq. (66.3). The fact that this analysis of the variations in conduction with field is in good agreement with the experiments of Loosjes and Vink will be shown in Sect. 77.

Implicit in the derivation of Eq. (66.3) was the assumption that the concentration of electrons in the immediate neighborhood of the pore surface at which they enter will be independent of the electric intensity superimposed on the space within the pore. This condition is satisfied if the emission current into the pore is limited by the space-charge within it. As the field is increased, space-charge limitations will gradually be removed and the supply of electrons needed to maintain the anticipated current density will be too small. If the only influence of the field were that of shortening the transit time as indicated in Eq. (66.2), then the conductivity would be inversely proportional to the square root of the field and the conduction directly proportional to $E_x^{\frac{1}{2}}$. After the supply of available electrons decreases, a "saturation" effect will set in gradually and the conduction will rise even less rapidly than is indicated by Eq. (66.3). Nakai, Inuishi and Tsung-che[1] attributed the lack of complete saturation of the conduction current to a lowering of the work-function on the interior of the pore surface by the Schottky effect (see Sect. 27). The application of this theory depends on the assumption that the emission increase results *only* from the cancellation of mirror-image forces by the external field. The total lack of uni-

[1] J. Nakai, Y. Inuishi and Y. Tsung-che: J. Phys. Soc. Japan **10**, 437 (1955).

formity of field over the surface of the interior of the pore makes this assumption inappropriate. More details concerning these experiments will be given in Sect. 79.

The application of these equations to the interpretation of the experiments of LOOSJES and VINK yields a value of the characteristic pore length of 2.8×10^{-6} m. and C_p of 0.18 for the lower temperature range. The presence of space-charge over the high-temperature range interferes with the direct application of this theory. As space-charge effects set in, the effective pore length decreases, since this length depends on the distance of free flight and the presence of space-charge limits it. As the temperature is increased, and the externally applied electric intensity E_x is maintained constant, the product $(E_x\, l_p)$ which defines the quantity V_p decreases, and therefore the range in applied field for which the conduction increases linearly with the voltage is extended enough to be clearly measurable, and is shown by LOOSJES and VINK.

Even though the theory is capable to indicating quantitatively the dependence of pore conduction on field and temperature in this high field, high-temperature range, its application would depend upon the choice of pore structures not realizable in actual experiments. It is therefore sufficient to show by this qualitative discussion, the relation between theory and observation for these experimental conditions.

67. Emitter evaluation in a diode having concentric cylinders. In the previous sections, space-charge theory as it applies to thermionic emission observations in diodes of parallel-plane construction has been formulated and the results of the theory adapted for direct application to experiment either in equation form or tabular form. The theory has also been developed for the flow of electrons from a cylindrical emitter across the space to a cylindrical collector for all ratios of collector to emitter radii. In order to determine the property of an emitter, it is necessary to combine these two divisions of thermionic emission theory in the manner that is to be outlined here.

A concentric cylinder diode can be constructed with practically any arbitrarily chosen ratio of radii. A practical choice for test purposes would be (R_c/r_s) of 2.5. The collector radius is defined as R_c and the emitter radius r_s. For the idealized application of the theory, all parts of the emitter that deliver electrons to the collector must have a single value of the true work-function at the temperature of test and the emitter must be uniform in its temperature. This condition can be realized in the laboratory by the use of the "guard ring" principle of tube construction. In a properly constructed diode having practical dimensions there will be a sufficient range of applied voltage over which the potential at the surface of the collector will be so negative with respect to that at the emitter that the presence of space-charge will not interfere with the flow of electrons from the emitter to the collector. An estimate of the maximum current density which can flow without space-charge interference may be computed by Eq. (51.1). The use of this equation may be illustrated best by a numerical example in which the effective diode spacing is taken as 1.0 mm. and the extremes in temperature are 580 and 1160° K. The corresponding current densities are respectively: 10.8 and 30.5 microamp/cm.². With these values serving as an indication of the maximum current flow without space-charge interference, there will be a range of lower current values extending through many orders of magnitude for which curves of the type shown in Fig. 19 will be directly applicable.

A well-activated oxide cathode operating at 1160° K should be capable of an electron emission density of 10 amp/cm.² which is a factor of 3×10^5 greater than the approximate current density required to develop the zero field condition at the collector. Reference to Fig. 21 shows that the displacement of the observed points over the entire range of retarding potentials negative with respect to the critical value will be close to 0.9.

Fig. 28 has been prepared to illustrate the use of the displacement theory in combination with "master curve $u_0^2 = \infty$" shown in Figs. 16 and 17. In Fig. 28 the current density values that would be obtained in an experiment are plotted on semi-logarithmic paper as a function of the applied voltage in

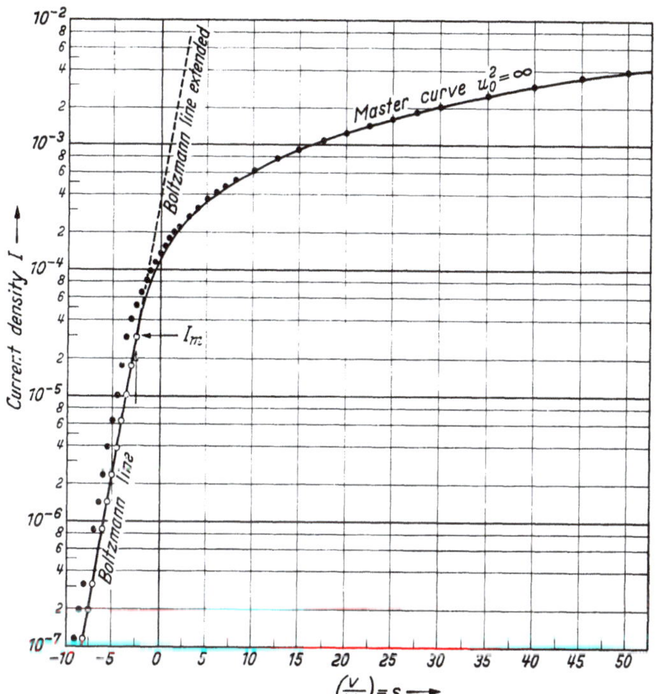

Fig. 28. The application of space-charge theory and the displacement method to evaluate an emitter in a concentric cylindrical diode.

dimensionless units defined by (v/V_T). The current density value designated I_m marks the maximum value of the current density expected at the onset of space-charge limitation. The displacement theory calls for a displacement to the right of 0.9 for each of the observed points for which the current density is less than I_m. These displaced points are shown by open circles. The purpose of this displacement is to establish the proper location of the straight line designated the "BOLTZMANN Line" which is the correct extrapolation of the "master curve" for values of u^2 defined as (I/I_R) which are less than unity. This is the range of current ratio for which there is no space-charge minimum between the emitter and the collector. The analysis shows that this displacement process converts experimental data obtained in a cylindrical diode to a set of points appropriate to represent the emission properties of the same cathode had they been determined in a parallel-plane diode. For the data shown in Fig. 28 the BOLTZMANN Line crosses the critical current value at -2.5 on the "s" scale. This corresponds to an applied potential after correction of -0.25 volts. The actual applied potential for this current would be -0.34 volts. As the actual applied voltage is made more positive than this -0.34 volts, a potential minimum develops first very close to the collector and then moves with each increase in applied voltage across the space until it approaches very close to the emitter surface. As this transition takes place, the tangential components of momentum with which the electrons left the emitter become less and less effective as contributors to the total kinetic energy required for an electron to pass the potential minimum and ultimately register as collector current.

Fig. 22 may be referred to as the quantitative means by which the correct displacement may be computed as the space-charge minimum moves across the diode from the collector to the emitter. Notice that at 2.5 for $(1/a)$ the displacement indicated for the higher values of S is 0.9 as mentioned above. In this example, the effective value of $(1/a)$ is 1.7 when the potential minimum lies halfway between the emitter and the collector. The corresponding displacement is 0.5. When the fractional distance from the emitter to the space-charge minimum has been reduced to 0.1, the effective value of $(1/a)$ is 1.15 and the displacement is reduced to 0.1. Since Eq. (61.3) relates the maximum of this fractional distance to the quantity u, it follows that for current densities greater than approximately 100 I_m no displacement correction is required and the observed points should follow the master curve.

This fitting process illustrated by Fig. 28 shows that the master curve combined with the BOLTZMANN line establishes a unique value for the current and voltage conditions that apply to the equivalent plane parallel diode that has precisely the properties and characteristics of the cylindrical diode investigated.

This introduction to the method of analysis applicable to data taken with a cylindrical structure applies specifically to that high temperature range over which the properties of the emitter are practically those of one which has unlimited emission capability. As the temperature is lowered and the potential minimum as it forms at the collector surface becomes equal to or less than V_T (that is the equivalent of kT) the emitter capability begins to be important and this capability as measured in terms of z_R^2 in Eq. (58.7) may be established with the required accuracy for most purposes. Further discussion of this phase of the theory will be postponed to Sect. 83 in which its application will be made to actual laboratory observations made on cylindrical diodes having well-activated and stabilized thermionic properties.

68. Thermionic emission theory. Concluding remarks. The theories presented in Sects. 14 to 66 are the necessary background for the interpretation of experimental researches in thermionic emission. Much of the theoretical work presented here may be found scattered in the literature and therefore it forms a body of uncoordinated information. It is one of the present objectives of this writing, not only to coordinate, but to present the subject in a manner that will permit its use as effective reference material without the demand on the reader that he master all the phases of the subject at one time. Its greatest usefulness depends on one's being able to open the study at the particular section of special interest and derive from it the information desired.

The phases of the theory that have been purposely omitted and yet may be relevant to the detailed understanding of thermionic emission phenomena include the analysis of electronic band structure in solids; the corresponding calculations at the surface boundaries and the present-day theories of the transmission of electrons over and through barriers as these relate to the waves associated with the interaction of electrons with atoms that form solid structures. The more important experiments on thermionic emission reported in the present-day literature and during the past twenty-five years require for their explanation only those phases of the theory that are presented here. Additional experiments that relate specifically to the emission of electrons from single crystals and their absorption by single crystals are likely to demand additional theoretical develop ment. With these exceptions, it is hoped that the theories presented here will be sufficiently comprehensive to satisfy all of the important needs of those who come to this source for information related specifically to the theoretical problems of thermionic emission.

D. Applications of theory to experiments on thermionic emission.

I. General discussion.

69. Emitter classification. Most thermionic emitters may be grouped under either of two general classifications. The first of these groups to be discussed in detail is that of the metallic electron emitter which includes specifically thermionic emission from pure metals or metals coated with polarizable atoms adsorbed to the surface of the pure metal to an extent equal to or less than one monatomic layer. The emission properties of the second classification depend upon the adherence of a complex non-metallic structure, generally thousands of atom layers thick, and best represented in its most useful form by the oxide-coated cathode. Under special circumstances an oxide cathode or its equivalent may be built upon a pure metal base by the condensation of molecules from an external source or, by migration over the surfaces of a sintered, porous structure of metallic crystallites from an interior source. Specifically, the dispenser cathode is an example of this emitter type. The properties of such emitters fall between the extremes of the metallic emitter and the oxide-coated cathode.

The feature that distinguishes these emitter classifications is the *concentration* of free electrons in the interior of the conductor at the emitting surface. The fact that metallic conductors improve in their electrical conductivity as the temperature is decreased even to extremely low values indicates that the FERMI level is positive with respect to the bottom of the conduction band and that in most cases the concentration of free electrons in this conduction band is practically equal to the concentration of atoms. In tungsten, for example, there are approximately 6.3×10^{28} atoms per cubic meter, and with the number of free electrons in the conduction band being approximately equal to this value, the FERMI level is located positive with respect to the bottom of the conduction band 5.7 ev. Since the true work-function of tungsten, as will be shown in Sect. 72, is between 4.4 and 5.3 ev depending on the crystallographic configuration of the exposed surface, the electron affinity (W_a) is close to 10.5 ev again depending on the exposed crystallographic surface. Over the entire operating range of temperature, the concentration of free electrons in the interior of tungsten crystals will remain high and not change significantly. The equation most suitable for the calculation of the FERMI level as a function of the temperature is given in Appendix 3.

Eq. (18.5) shows that the emission properties of a specific surface do not depend on the absolute values of the electron affinity (W_a) and the FERMI level (μ_s) but depend on the difference between these quantities which is, by definition, the true work-function. It is expected that for all metallic surfaces free from adsorbed layers, that the electron affinity will decrease slightly as the temperature increases. Furthermore, the value of the FERMI level also decreases and therefore depending only on the relative rates of decrease, the true work-function may increase or decrease with the temperature. In all known cases of metallic emitters free from adsorbed layers these changes in the true work-function with temperature are relatively small.

The demand for increased electron emission densities for the same input of power, that is, the same temperature, stimulates the search for a practical means of lowering the true work-function. The adsorption onto the surface of a tungsten crystal of a fraction of a monolayer (approximately 0.7) of electropositive polarizable atoms such as thorium or cesium results in an effective lowering of the electron affinity in direct proportion to the *average dipole moment per unit area.* This method of increasing electron emission efficiency is made use of in all practical devices that depend on thoriated tungsten ribbons or filaments as their electron emission sources. For metals there is no evidence to indicate that practical electron emitters can be improved in their efficiency by the introduction into the interior of alloying impurities which are capable of increasing the concentration of free electrons and in that manner appreciably alter the value of μ_s.

Electron emission from poor conductors of which the oxide-coated cathode is the most important example is predominantly characterized by factors that in most respects are exactly opposite to those important for metallic emitters. First, the electron affinity (W_a) is low, that is, generally less than 1 ev. Secondly, the free electron concentration in the interior of the crystal near its surface is generally lower by a factor of 10^6 and at very low temperatures the concentration of free electrons approaches zero. Under these low temperature conditions, the FERMI level will lie almost exactly halfway between the bottom of the conduction band and the highest occupied electronic level within the crystal. Such levels are shown schematically in Figs. 25, 26 and 27. As the temperature is increased, the magnitude of μ increases very appreciably and this change in the value of μ therefore *increases* the value of the true work-function and decreases the emission compared with the emission that would have been realized had there been no change in the FERMI level.

With these poor conductors, the number of free electrons in the conduction band is therefore dependent not only on the concentration of donor centers (which are probably oxygen vacancies in the oxide cathode), but also it is very dependent on the temperature. The equation which relates the value of the FERMI level to the concentration of donors and to the temperature is Eq. (64.11). An oxide cathode in a thoroughly deactivated state would have no excess of barium over oxygen and the FERMI level would fall at an energy value approximately 2 ev below the bottom of the conduction band. The introduction of a barium excess of as little as 1 part in 10^7 would result in a shifting in the FERMI level to a value halfway between the energy E_D as in Fig. 25 and the bottom of the conduction band if the temperature is sufficiently low. The introduction of such a very small impurity content into an otherwise perfect ionic crystal does not change the electron affinity but does make a very significant reduction in the true work-function because of the important reduction in the magnitude of the FERMI level value μ.

An increase in the concentration of donors by a factor of 100 or more will not change the electron energy level E_D associated with a particular donor type such as, for example, an oxygen vacancy. The really significant factor that is influenced by the increase in the concentration of the donors is the temperature coefficient of the FERMI level. The exact value of this coefficient may be computed by Eq. (64.13) as was done in the preparation of the data given in Table 11 for selected values of donor concentration and donor energy level.

This discussion may be summarized by the statement that the high emission efficiency of oxide cathodes is achieved by the use of a non-metallic emitting surface having a very low electron affinity and the introduction into the crystalline structure of this emitter of a sufficiently high concentration of donor centers to reduce the temperature coefficient of the FERMI level. With this combination the work-function has an acceptably small value at the normal operating temperature. Any changes in the crystalline structure such as those associated with the addition of a high concentration of strontium to form a solid solution of barium-strontium oxide crystals can presumably make some change in the donor level relative to the bottom of the conduction band or it may make a change in the maximum donor concentration that can be incorporated into the solid structure of the crystal.

The experimental evidence which forms the basis of these introductory remarks will be presented in the later sections of this treatise. These researches will be discussed in the order of increasing complexity which is first, emission from pure tungsten; second, the emission from tungsten in the presence of cesium vapor; third, the emission from thoriated tungsten; and fourth, the emission and conduction properties of oxide cathodes.

70. General studies of thermionic emission from tungsten. Of all of the pure metals which have been studied to determine their thermionic emission properties, tungsten has received the most attention. Because of the economic importance of tungsten in the lamp industry, the major metalurgical problems associated with the production of high-quality filaments were solved many years ago. It will be shown in Sect. 71 that there is no unique value of the true work-function that can be associated with any crystalline material without the specification of the particular crystallographic surface across which the electrons are escaping into the evacuated space. Reproducibility of average emission properties from tungsten is attained only after sufficient heat treatment in a completely inert atmosphere (high vacuum or inert gas) to recrystallize the microcrystals into much larger ones and at the same time drive out and evaporate off occluded impurities which had been purposely introduced in order to facilitate the production of the wire.

The evaluation of the thermionic emission properties of polycrystalline wire is basically an engineering problem rather than a problem in physics. The conventional method of expressing the results in a useful form is to quote the experimentally determined empirical constants suitable for use in the RICHARDSON type of equation [Appendix I and Eq. (50.2)]. A practical choice of these constants is $\varphi_R = 4.5$ ev and $A_R = 60$ amp per cm.^{2} degree^2. NICHOLS[1] has measured the average RICHARDSON constants for various samples of wire and compared his results with those previously obtained on similar samples less well defined in terms of their impurity content. The expected average values of these constants have been obtained by computations based on the known variation of true work-function with crystallographic direction and assumptions with regard to the probable distribution of the particular elemental areas associated with specific crystallographic orientations over the entire surface of the specimen.

The empirically determined work-functions expressed as φ_R are suitable only for the calculation of the average electron emission density as a function of the temperature at an applied potential which is nominally zero. If the average electric intensity at the surface of the emitter can be computed from geometrical considerations then the SCHOTTKY relation given as Eq. (27.13) may be used for the calculation of the average current density obtainable in the presence of a moderately strong accelerating field.

[1] M. H. NICHOLS: Phys. Rev. **78**, 158 (1950).

Theoretical relations applicable in the field of surface chemistry often demand a knowledge of the work-function of the surface at which the reaction takes place. The use of the empirical quantity φ_R for this work-function should be avoided if accuracy is of importance. In all probability no part of the real surface will have a true work-function of this value since some areas conceivably of the order of 50% of the superficial surface may have a slightly lower work-function while the remaining area can very well have a work-function higher than the thermionic average by as much as 1 electron volt. Since work-functions are generally measured at high temperatures, the surface will often be "self cleaning" and give a result which is characteristic of a clean surface even though it may not be an homogeneous one. As the temperature is lowered, polarizable atoms or molecules condense on the surface very quickly unless the partial pressure of condensable gases is maintained at extremely low values (of the order 10^{-13} mm. or less). These adsorbed gases may form a monolayer which will be quite stable under well-defined conditions and give reproducible results. It must be emphasized that reproducibility is not a reliable indication of the absence of surface contamination.

A detailed analysis of thermionic emission data by SEIFERT and PHIPPS[1] aimed at a more accurate determination of the RICHARDSON work-function of tungsten led them to the discovery of the *periodic deviation effect*. If the best of experimental conditions are maintained, measurements show that the rise in average emission current from a polycrystalline test specimen does not follow precisely along the straight line predicted by the SCHOTTKY theory [Eq. (27.13)] but shows small periodic deviations from this line. This effect has been studied most extensively by COOMES and his collaborators[2] and the theory has been developed by GUTH and MULLIN[3] and by HERRING and NICHOLS[4]. Even though this periodic deviation effect is so small that it is insignificant as it relates to practical problems of thermionic emission, it is an effect so well established by experiment that any detailed theory for the transmission of electrons from the interior of the metal across an uncontaminated barrier into a vacuum must include a satisfactory explanation of this phenomenon and at the same time not introduce assumptions that conflict with other known properties of the emitter.

In order to illustrate the major points of this section by carefully taken experimental data, Fig. 29 has been prepared. Two sets of data are shown on the figure. The solid circles apply to data taken at 1373° K while the open circles show results at 1790° K. The logarithm of the observed current is plotted as a function of the square root of the electric field at the surface of the emitter. At a field of approximately 6.4 kv per cm., the currents differ by a factor of exactly 10^4. With the corresponding displacement vertically, the two curves cross each other at this field and they can, therefore, be displayed on the same figure. This method of display permits the graphic illustration of a number of points important for the understanding of thermionic emission from clean, polycrystalline tungsten in the presence of a strong applied field. After a brief description of the method of measurement, the points illustrated by these data will be explained.

A long filament of GE 218 wire was polished to remove all die marks and scratches and its diameter was measured by an interferometer method to be

[1] R. L. E. SEIFERT and T. E. PHIPPS: Phys. Rev. **56**, 652 (1939).
[2] R. J. MUNICK, W. B. LABERGE and E. A. COOMES: Phys. Rev. **80**, 887 (1950). — D. W. JUENKER, G. S. COLLADAY jr., and E. A. COOMES: Phys. Rev. **90**, 772 (1953).
[3] E. GUTH and C. J. MULLIN: Phys. Rev. **61**, 339 (1942) and earlier papers.
[4] C. HERRING and M. H. NICHOLS: Rev. Mod. Phys. **21**, 185 (1949).

4.24×10^{-3} cm. The filament was mounted accurately on the axis of a set of five cylinders, each having an inside diameter of 2 cm. The center collector was 3 cm. in length and the currents measured correspond to the arrival of electrons at this collector. The tube was so constructed that applied voltages up to 10000 volts could be used without appreciable leakage or spurious current. At this voltage, the surface field was 0.76×10^6 volts per cm. which was sufficient to give a very measurable field emission with the filament at room temperature. This contribution due to field emission is clearly evident in curve *a* of Fig. 29 at a field strength of 0.2×10^6 volts per cm. The deviation of the three highest points from the extrapolated straight line, shown on this figure, corresponded quite precisely with the field emission observed at room temperature. For the high temperature data, shown as curve *b*, the field emission over the range shown was not important since the measured thermionic currents at the higher temperature were 10000 times greater than those observed for curve *a*.

Eqs. (27.5) and (27.13) were used to calculate the field at the surface of the emitter and determine the slope of the two straight lines, shown in Fig. 29, associated with curves *a* and *b*. These lines are both drawn to fit the observed data points in the very high field range. This choice is quite arbitrary and depends on the assumption that even with these strong fields, the penetration of electrons through the modified mirror-image barrier is of minor significance. An inspection of the location of the data points in comparison with these lines shows that over the range of field of 4×10^4 volts per cm. and higher, the deviation of the plotted points with respect to the straight line is practically independent of the temperature of the emitter. At lower fields the curves show quite graphically that the deviation of the observed points is very measurably greater the lower the temperature. This fact alone indicates that space charge is playing no important part in the control of the electron flow from the emitter to the collector.

If the periodic deviations are neglected and an approximate average slope is determined, the departure from the theoretical SCHOTTKY lines is close to 8% It is evident from these data that average slopes that might be used to represent experimental data according to the so-called "SCHOTTKY plot" will deviate more and more from the theoretical slope the lower the field and the lower the temperature.

These deviations are consistent with the explanation that long range surface fields are produced because of the non-uniformity of the work-function exhibited by the specimen of tungsten used for this study. Undoubtedly, the crystals were large and exhibited a variation in work-function depending on the crystallographic direction. The minimum value was approximately 4.3 and the maximum was about 5.3 volts. Although patch theories have been worked out[1] for special geometrical arrangements of patches, it is hardly worthwhile to do more than indicate that a choice of patches may be made which is capable of accounting for characteristics similar to those shown in Fig. 29. Although the patch effect is important, such an explanation leaves much to be desired and serves relatively little purpose since the details concerning the specimen used for observations are generally lacking.

These data may be taken as typical and represent thermionic emission from polycrystalline tungsten wire at two temperatures. It is conventional, as has already been explained, to use the intercepts of straight lines of the type shown in Fig. 29 as appropriate data for the RICHARDSON equation. The limitations as well as the usefulness of the empirical constants A_R and φ_R have been discussed.

[1] J. A. BECKER: Rev. Mod. Phys. **7**, 95 (1935). — W. B. NOTTINGHAM: Phys. Rev. **49**, 78 (1936). — C. HERRING and M. H. NICHOLS: Rev. Mod. Phys. **21**, 185 (1949).

Computation shows that the empirical constants as given in Appendix I may be used to compute the intercepts expected for these two temperatures. These observed data show a deviation in current density observed of only 20% from that expected for curve *a* and 40% for curve *b*. Lack of agreement with the average predicted current density for tungsten of this magnitude is typical.

A careful examination of the data shown in Fig. 29 serves to establish the fact that the data points deviate systematically from a straight line. Clearly, these deviations are small and stand out strongly only when accurately taken data permit the direct plotting of the deviation from the theoretical straight line as a function of the field. The data shown in this figure, in addition to other data taken at the same time at other temperatures provide the information on which maxima and minima in the deviation can be indicated. In this figure the minima are indicated by arrows pointing downward and the maxima by arrows pointing upwards. One striking fact that came from these observations is that the fractional magnitude of these deviations is practically independent of the temperature and the location of the maxima and minima also seems to be practically independent of the temperature. These results were reported very briefly by NOTTINGHAM[1].

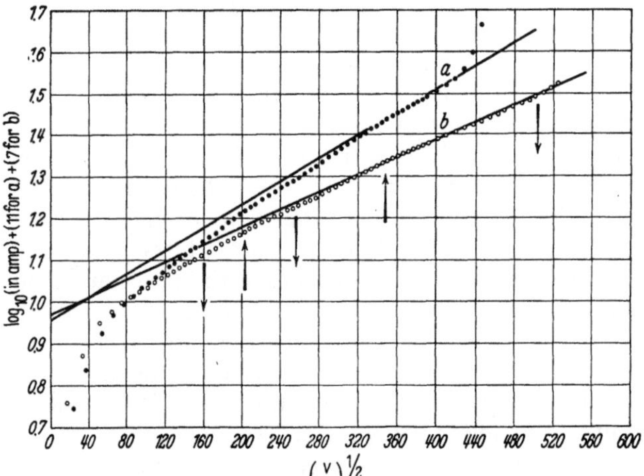

Fig. 29. Electron emission from tungsten as a function of field at two temperatures. Curve *a* for 1373° K; *b* for 1790° K. Straight lines represent SCHOTTKY mirror-image theory. Maxima of periodic deviation ↑; Minima ↓.

The work of DUSHMAN and others[2] led to the erroneous conclusion that the experimentally determined value of the thermionic constant A_R should be 60.2 amps per cm.²-degree² and therefore agreed exactly with the value of the theoretical constant which he computed. This value is now recognized to have been in error by a factor of 2. The factors that enter into the correct calculation of the thermionic constant A are given in Eq. (18.6). Contributing to this difference between the observational value of 60 and the computed value of 120 are three factors not previously taken into account. These are:

1. For tungsten, practically all the emission comes from approximately 50% of the superficial area of the emitting filament;

2. The effective temperature coefficient of the true work-function of the important emitting areas is not only uncertain as regards its actual value but even the sign, remains in doubt;

3. The observed deficiency in the number of slow electrons emitted from the surface must also be given consideration.

The combination of these last two factors would seem to indicate that the average temperature coefficient of the true work-function of the strongly emitting areas is small and negative so that the combination of all factors gives an empirical value of the constant A_R of 60 amps per cm.²-degree².

[1] W. B. NOTTINGHAM: Phys. Rev. **57**, 935 (1940).
[2] S. DUSHMAN: Rev. Mod. Phys. **2**, 381 (1930).

II. Emission from single crystals.

71. Thermionic emission properties of single crystals of tungsten. The pioneering work of MÜLLER[1] showed by the use of his electron projection microscope that the field emission properties of a tungsten single crystal were very dependent on crystallographic direction. Shortly before MÜLLER's work, thermionic electrons were projected in a similar manner from a filamentary source to a surrounding cylindrical fluorescent screen and thus indicated at least qualitatively the variation in the emission properties of a single crystal of tungsten as a function of crystallographic direction. Following a series of researches by JOHNSON and SHOCKLEY[2], NELSON[3], and MARTIN[4], NICHOLS[5] made the first quantitative measurements of the thermionic emission from a single crystal of tungsten as a function of its crystallographic direction.

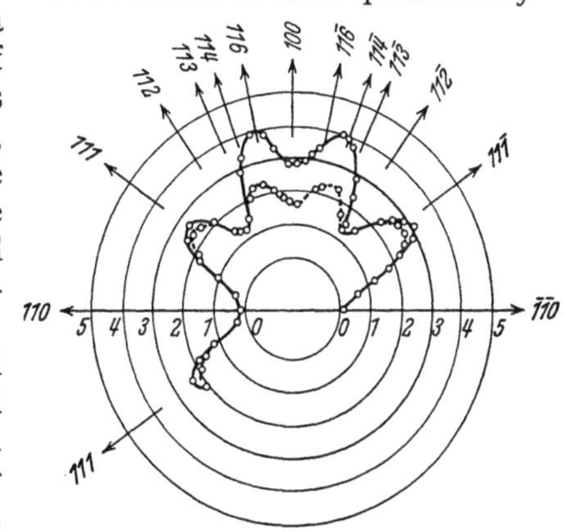

Fig. 30. Polar diagram of thermionic emission current as a function of angle as observed by NICHOLS at 1880° K, applied surface field of 60 kv per cm. Solid line shows emission from a cleaner surface after the crystal was heated to 2500° K.

In the NICHOLS experiment, a polished tungsten filament in which a single crystal 9 cm. long had been grown was mounted on the axis of a cylindrical electron collector system. The filament was supported on a frame with bearings accurately aligned so that the filament could be rotated about its own axis through more than 180°. A slot in the collector system defined an opening through which a beam of electrons could be delivered to a shielded collector for measurement. Cold working of polycrystalline tungsten wire previous to its first heating tends to line up the 110 direction of the crystallites along the axis of the wire. Such wires invariably grow crystals with this orientation. The polar diagram shown as Fig. 30 is a reproduction of the one presented by NICHOLS. The diagram shows weak emission areas in the 110 and the 112 direction. Maxima occur in the 111 and 116 directions. The minimum in the 100 direction is sometimes not as pronounced as that found by NICHOLS.

Table 71.1.

Direction	NICHOLS		SMITH	
	φ_R	A_R	φ_R	A_R
(111)	4.39	35	4.38	52
(112)	4.69	125	4.65	120
(116)	4.39	53	4.29	40
(100)	4.56	117	4.52	105
(110)[7]	4.68	15	4.58	8

The NICHOLS tube was reconstructed by SMITH[6] and very extensive additional measurements were made on a new crystal. These measurements included a study of the emission as a function of surface field. Although the SMITH measurements were in substantial agreement with NICHOLS, analysis showed that the electron current observed with the crystal oriented with the 110 face in direct line with the measurement slot was almost entirely spurious. Work-function difference evaluations by

[1] E. W. MÜLLER: Z. Physik **106**, 541 (1937); **108**, 668 (1938).
[2] R. P. JOHNSON and W. SHOCKLEY: Phys. Rev. **49**, 439 (1936).
[3] R. B. NELSON: M. I. T. PhD Thesis 1938.
[4] S. T. MARTIN: Phys. **56**, 947 (1939).
[5] M. N. NICHOLS: Phys. Rev. **57**, 297 (1940).
[6] G. F. SMITH: Phys. Rev. **94**, 295 (1954).
[7] Emission in this direction is spurious and the corresponding A_R and φ_R do not represent the RICHARDSON constants truly characteristic of this surface direction.

HUTSON (Sect. 72) show that the true work-function associated with this direction is so high that the currents measured by NICHOLS and by SMITH were correctly identified by SMITH to be the result of secondary effects and not a valid measurement of the primary emission in that direction. The results of both of these studies are best summarized by the Table 71.1.

72. Velocity analysis of thermionic emission from single crystals of tungsten. A study[1] that required a homo-energetic beam of electrons indicated that an investigation of the electron energy distribution from a hot filament by means of a semicircular velocity analyzer could yield an accurate determination of the contact difference in potential between the emitter surface and the interior surface of the analyzer. Provided that the local work-function differences over the surface of the interior of the analyzer were sufficiently small and the dimensions of such patch areas small compared to the distance from the surface to the electron beam, the average work-function of the analyzer interior serves as a suitable reference. HUTSON[2] designed a tube similar to the NICHOLS tube in that the single crystal tungsten filament could be rotated about its own axis so that the electron emission derived from well-defined crystallographic surfaces could be made to enter the aperture of the analyzer. The filament was heated by means of a pulse current source so that there would be no drop in potential over the filament during the measuring half cycle. Electron accelerating voltages in the range 100 to 2000 volts were applied between the filament and the co-axial cylinder. The slot in the cylinder subtended a very small angle at the filament and therefore permitted electrons to enter it from a well-defined area. A carefully designed electron-optical system provided a means for slowing down the electrons so that they entered the 180° velocity analyzer with an energy of about 3 ev. Depending on the strength of the uniform magnetic field, a precisely known energy group passed around the analyzer and out the exit slit to the shielded electron collector. The data shown in Fig. 31 are typical of the results obtained. In this figure the current observed is plotted on a logarithmic scale as a function of the kinetic energy, expressed as V_p, of the electrons at the emission surface. The current and energy values are not shown in this diagram since they were normalized by vertical and horizontal displacements in order to match the three sets of observed data which were taken with the extremes in applied voltage of 150 to 2000 volts.

It was possible to establish that no change had occurred in the average work-function of the interior of the analyzer during the time required for a set of measurements, and therefore the horizontal displacements required to bring the various curves similar to those shown in Fig. 31 into alignment are a direct measure of the changes in the true work-function with electric field. The observed changes for three of the important crystallographic directions are plotted in Fig. 32. The SCHOTTKY theory of the change in true work-function with field was presented in detail in Sect. 27 and Eq. (27.11) shows that the change in work-function should be proportional to the square root of the electric intensity at the surface of the emitter.

In Fig. 32 the work-function change is plotted as a function of the square root of the surface field and the straight lines are drawn with the theoretical slope required by the SCHOTTKY theory. This study therefore stands as the first direct observation of the SCHOTTKY effect reduction in work-function by the influence of an external field. All previous verifications of the SCHOTTKY theory

[1] W. B. NOTTINGHAM: Phys. Rev. **55**, 203 (1939).
[2] A. R. HUTSON: Phys. Rev. **98**, 889 (1955).

have depended on the assumption not previously established that the energy distribution of the electrons transmitted across the maximum of the potential barrier remained independent of the applied field. HUTSON's experiments demonstrate this independence separately from the direct observation of the shift in the work-function which is established by these data to be exactly that predicted by the SCHOTTKY theory within the experimental error. Fig. 33 illustrates the fact that the energy distribution at a constant temperature of 2000° K and a constant accelerating anode voltage is independent of the crystallographic

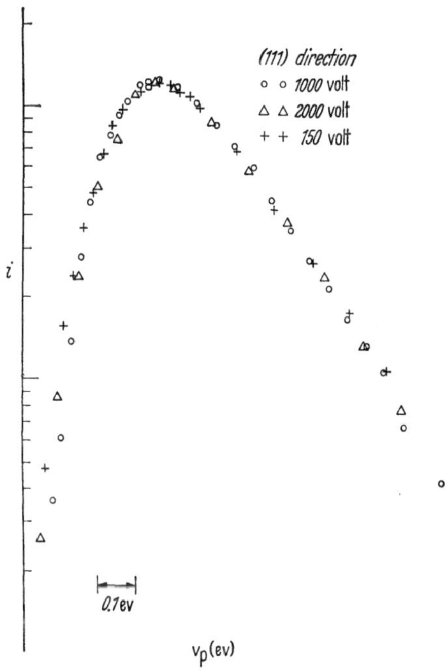

Fig. 31. Superposition of energy distributions obtained by HUTSON at 2000° K for three different anode potentials and electron emission in the 111 direction. Initial energy of electron is V_p in ev.

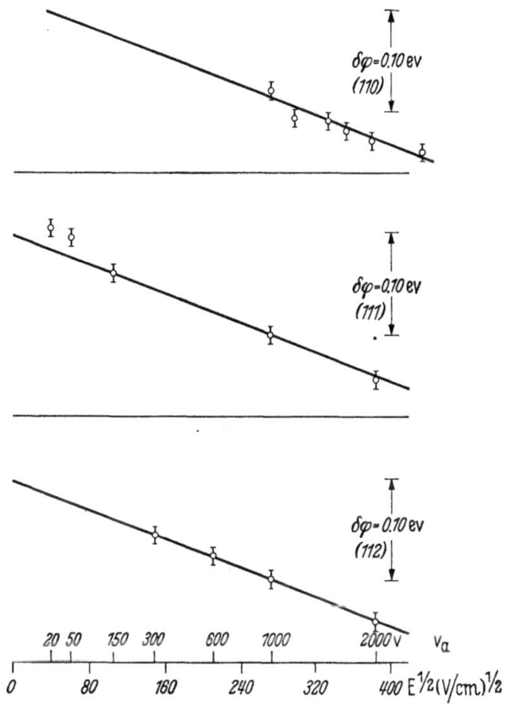

Fig. 32. Change of work-function with applied field as observed by HUTSON from the relative voltage displacements required to superpose distributions illustrated by Fig. 31. Solid lines are drawn with a theoretical slope required by SCHOTTKY's mirror-image theory.

direction and furthermore that there is a very marked deficiency of slow electrons. This result is indicated by the lack of agreement between the observed points illustrated there and the solid line which represents the results that would have been obtained had the electrons been emitted from the hot filament with a perfect MAXWELLian distribution.

Electron energy distributions determined by the retarding potential method by NOTTINGHAM[1] indicated a similar departure from a MAXWELLian distribution for electrons emitted from tungsten filaments and from thoriated tungsten filaments. Patch effects on the tungsten filaments interferred with the accurate reproducibility of the retarding potential curve but did not interfere with the results obtained with thoriated filaments. An empirical relation discussed in Sect. 26 and given by Eq. (26.4) served as an accurate representation of the retarding potential data. This relation would predict the distribution function

[1] W. B. NOTTINGHAM: Phys. Rev. **49**, 78 (1936).

represented by the dotted line of Fig. 33. Since the method by which the electron energy distribution was determined by NOTTINGHAM was inherently less accurate than that of HUTSON. There is no significant disagreement between these studies.

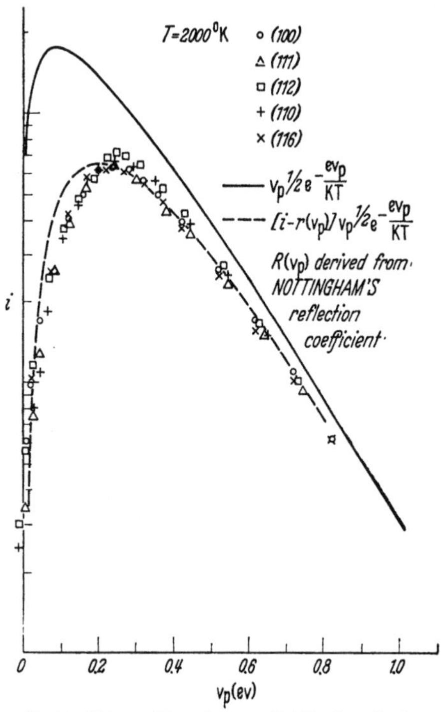

Fig. 33. Superposition of energy distributions obtained by HUTSON for the selected directions shown at maxima and minima of the polar diagram. Anode potential 1000 volts. Solid line computed for a true MAXWELLian distribution. Dotted line computed according to the empirical equation of NOTTINGHAM which describes the observed deficiency of slow electrons. See Sects. 26 and 75.

The retarding potential measurements made by GERMER[1] are often cited as proof that electrons are emitted from tungsten with a MAXWELLian distribution. Since GERMER'S experimental error was large, he was not justified in drawing any other conclusion, and yet NOTTINGHAM showed that his data and the GERMER data are in excellent agreement. There exist in the literature, therefore, reports of only three experiments on electron energy distribution that have been taken with suitable geometrical conditions and freedom from surface contamination required for these studies. The results of these three researches are in agreement.

Far more difficult is the direct measurement of the temperature coefficient of the contact difference in potential. HUTSON'S experiments yielded some information on this subject and indicated both positive and negative temperature coefficients depending on the crystallographic direction investigated. These results combined with the direct determination of the contact difference in potential between the various crystal faces permitted HUTSON to prepare the data contained in Table 72.1.

Table 72.1.
HUTSON'S *estimates of the true work-functions for a tungsten crystal in the temperature range 1700 to 2000° K.*

Direction	φ in ev	φ in ev
111	$4.3 + (3\times 10^{-5})\,T$	$4.3 + 0.36\,V_T$
112 (Note 1)	$4.57 - (5\times 10^{-5})\,T$	$4.57 - 0.6\,V_T$
112 (Note 2)	$4.66 - (8\times 10^{-5})\,T$	$4.66 - 0.9\,V_T$
116 (Note 3)	$4.20 + (3\times 10^{-5})\,T$	$4.2 + 0.36\,V_T$
116 (Note 3)	$4.31 + (3\times 10^{-5})\,T$	$4.31 + 0.36\,V_T$
100	$4.44 - (2\times 10^{-5})\,T$	$4.44 - 0.24\,V_T$
110 (Note 4)	$5.09 \pm$ (uncertain)	

Note 1: Assume no effect of spurious current similar to that found for 110 orientation

Note 2: Based on observed temperature coefficient.

Note 3: Range thought to be result of heat treatment.

Note 4: Best estimate for crystal at 2000° K.

Since HUTSON was not prepared to make a direct determination of the absolute value of any of the RICHARDSON work-functions with an accuracy comparable with that of NICHOLS and of SMITH, those given in Table 72.1 depend on the acceptance of the NICHOLS and SMITH work-function determined for the 111 direction.

Because of the fact that no theory has been devised to account for the observed deficiency of slow electrons emitted thermionically from hot surfaces, experimentally unsupported criticism has followed the publication of the experimental

[1] L. H. GERMER: Phys. Rev. 25, 795 (1925).

results that show the deficiency. The latest is that offered by SMITH[1] in criticism of the work of HUTSON. SMITH makes the hypothesis that, for reasons unknown, the resolution of HUTSON'S velocity analyzing equipment was a factor of 10 poorer than the value claimed by HUTSON. All attempts to identify a cause for this lack of resolution assumed by SMITH have failed and independent evidence by LANGE[2] indicates that his analyzer, which was similar in construction to that of HUTSON, has practically the resolution expected from it according to calculations similar to those applied by HUTSON. LANGE'S experiment was not designed specifically to evaluate the energy distribution of the electrons emitted from a hot filament, but when used for that purpose indicated a distribution very similar to that reported by HUTSON.

LANGE'S experiment was designed to measure the reflection of slow electrons as they enter, from the outside, into a clean, single crystal of tantalum. It is premature to report LANGE'S results in any detail, but the present indications seem to be that very slow electrons are *not reflected* as they approach a single crystal which is at room temperature. This result suggests that possibly the important difference between the reception and the emission of electrons is that LANGE'S receiver operated at a low temperature while all emitters studied have operated at high temperatures. The experiment is to be modified to permit the evaluation of the electron absorption properties of the target when it is at a high temperature sufficient to excite the vibrational states of the crystal in a manner comparable with those in the high temperature electron emitters studied.

III. Modification of electron affinity by polar layers.

73. Influence of polarizable adsorbed layers on thermionic emission. Since a tungsten filament can be raised to a very high temperature (3000° K) for a short period of time without excessive evaporation, the surface can be freed of adsorbed impurities. Under moderately good vacuum conditions the thermionic emission can be measured accurately at a test temperature of the order of 1400° K. If the partial pressure of gases capable of adsorbing on the filament at this temperature is of the order of 10^{-10} mm. or higher, the thermionic emission from the test filament usually decreases with the time. This phenomenon often described as "emitter poisoning" is characteristic of the adsorption onto the filament of electro-negative gases such as oxygen. Since the electro-negative adsorbed layers are difficult to control and reproduce, they have not been investigated in terms of their influence on thermionic emission to the extent that adsorbed layers of electro-positive material have been studied.

The adsorption of films on the surface of a single crystal of tungsten and its influence on the thermionic emission is exhibited dramatically by the researches of MARTIN[3]. A large single crystal of tungsten was cut and polished to form a hemispherical cap which could be heated by radiation and electron bombardment from a filament mounted in a cavity in this cap. This structure was located at the center of a spherical bulb coated with a phosphor so that the pattern of light observed at the surface of the phosphor served as a qualitative measure of the distribution of electron emission over the surface of the tungsten emitter. It was not possible to control nor measure the quantity of adsorbed material on the tungsten, but it was possible to establish that the adsorption and emission

[1] G. F. SMITH: Phys. Rev. **100**, 1115 (1955).
[2] W. J. LANGE: Unpublished researches at M. I. T.
[3] S. T. MARTIN: Phys. Rev. **56**, 947 (1939).

properties of the surface varied in an extremely complex manner and yet maintained symmetries that were determined by the known crystallographic directions of the single crystal specimen. These studies of MARTIN's demonstrated the futility of attempting to interpret, in much more than a qualitative manner, the influence of adsorbed films on thermionic emission. For this reason, Sect. 74 will give a very brief review of the researches of TAYLOR and LANGMUIR[1] which was the latest and most comprehensive study of the properties of cesium films on tungsten. Sect. 75 will deal with some of the properties of tungsten when activated by a partial layer of thorium as produced on a thoriated tungsten filament.

The forces that act on an electron as it leaves the interior of a metallic conductor and escapes across the conducting boundary into free space are complex and understood in some detail only after the very short-range forces become reduces in importance in comparison with the long-range mirror-image force. The integral of all of the forces that act on an electron as it escapes is measured by the quantity W_a known as the "electron affinity". The true work-function of a surface as discussed in Sect. 19 is not the electron affinity itself, but the difference between two quantities both measured from the same reference and given by Eq. (19.1). Additional discussion will be found in Sect. 27.

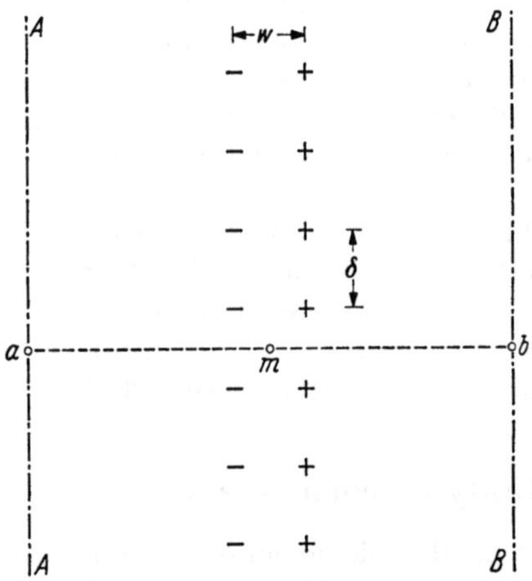

Fig. 34. Infinite plane of dipoles for computation of difference in potential from a to b.

The adsorption of polarizable atoms in the form of a film less than one mono-molecular layer in thickness can have no influence on the location of the FERMI level μ_s, but can act directly to reduce (or even increase) the average value of the electron affinity W_a. The effect of a dipole layer can be evaluated quantitatively by first considering that the problem is closely related to that of computing the difference in potential between the sheets of charge that reside on the surfaces of a parallel plane capacitor with a separation between the surface charges of w and a charge per unit area of σ. The following equation expresses that difference in potential:

$$\Delta V = \frac{w\sigma}{\varepsilon_0}. \tag{73.1}$$

If the system of charges is considered to be the equivalent of the grouping shown in Fig. 34, then an electron at point a in the plane AA will experience practically no net force acting on it due to the presence of the dipoles if the distance $a-m$ is large compared with the average distance between the dipoles shown in the figure as δ. The net force on the electron at point b also vanishes under these same conditions. However, as the electron approaches close to the sheet of dipoles, it is acted upon by forces which, because of the negative charge accelerate the electron across the space. In other words, it would have fallen through a difference in potential expressible by an equation similar to Eq. (73.1) as follows:

$$\Delta V = \frac{\overline{m}\, n_a}{\varepsilon_0}. \tag{73.2}$$

[1] J. B. TAYLOR and I. LANGMUIR: Phys. Rev. 44, 423 (1933).

In this equation $\bar{m}$ is the average dipole moment per atom and is defined as the average displaced charge within the polarized atomic or molecular unit multiplied by the average distance of displacement. The quantity n_a is the number of adsorbed molecules per unit area. On the basis of this reasoning, it is an accurate statement to say that the reduction in the true work-function that results from the adsorption of an electro-positive substance to the extent of one monolayer or less is directly proportional to the average dipole moment per unit area that results from the adsorption of this layer.

The arrangement of charges shown in Fig. 34 suggests at once that the adsorbed dipoles should generally repel each other and therefore act in their motion over the surface of a crystal as a two-dimensional gas. When the average distance between dipoles is large compared with atomic spacings, then the dipole moment per atom or molecule should be independent of their concentration. As the number on the surface increases, however, the dipoles react upon each other and reduce the dipole moment per atom. As it will be shown in Sects. 74 and 75, this fact accounts for the linear reduction in work-function with increased concentration when a small fraction of the surface is covered.

With larger concentrations, the dipole moment per atom decreases to such an extent that a maximum dipole moment per unit area occurs with approximately 70% of the surface covered and not when a complete monolayer is formed.

For a given concentration of adsorbed atoms, it is to be anticipated that the dipole moment per atom will decrease as the temperature increases. Because direct measurements of the temperature coefficient of the contact potential are difficult, although not impossible, there exist no quantitative data on the net increase in work-function with temperature. Indirect measurements by NOTTINGHAM[1] would indicate a maximum of 28.5×10^{-5} volts per degree for thorium on tungsten. This factor, when converted to the units used in Eq. (38.10), gives the value of (dP/dV_T) of 3.3. The thermionic constant A_R for surfaces activated by electropositive fractional monolayers is low, and this fact is a direct indication that this positive temperature coefficient of the work-function is influencing the emission-*versus*-temperature relation.

The opposite effect is observed when the surface is contaminated by electronegative atoms or molecules. Again, the first direct indication comes from the fact that the empirical A_R values for those contaminated surfaces are very high compared with the theoretical A given by Eq. (18.6).

74. The properties of cesium films on tungsten. The properties of fractional monolayers of cesium adsorbed to tungsten have been studied experimentally and analytically in the greatest detail by TAYLOR and LANGMUIR[2]. The principal tube used for their study had two "well-aged" tungsten filaments, each of 30 cm. total length mounted close to the axis of a three-electrode collector system. The central 10 cm. sections of these filaments were straight and 10 cm. of filament at each end were wound in a tight spiral so that the center section would heat very uniformly. The collectors were formed by evaporation onto the glass wall of the tube so that the entire structure except for the filaments themselves could be maintained accurately at a pre-assigned temperature in the range from about 270 to 300° K. The concentration of cesium in the vapor phase was controlled by the regulation of this temperature.

---

[1] W. B. NOTTINGHAM: Phys. Rev. **49**, 78 (1936).

[2] J. B. TAYLOR and I. LANGMUIR: Phys. Rev. **44**, 423 (1933). — I. LANGMUIR: J. Amer. Chem. Soc. **54**, 2798 (1932).

LANGMUIR and KINGDON[1] showed that all cesium atoms which strike a high-temperature tungsten filament leave as positive ions and the measurement of this ion current is a direct indication of the arrival rate of cesium atoms at the filament. As the temperature of the test filament is reduced, atoms tend to adsorb on the filament until a sufficient layer is built up so that the rate of evaporation from the surface is exactly equal to the rate of arrival of atoms at the surface. For a wide range of surface concentrations that approached closely to a monolayer, TAYLOR and LANGMUIR determined rates of evaporation of atoms, positive ions and electrons as a function of the temperature. LANGMUIR based his theoretical analysis on the assumption that 99.5% of the entire surface of his "aged" filaments was 110 surface. This hypothesis was well supported by the self-consistency of the data, but was never verified by actual observation of the emission pattern by the electron projection tube method later used so successfully by JOHNSON and SHOCKLEY and by MARTIN[2] to show that the filament actually had developed an etched structure such that only 110 surfaces were exposed. ROBINSON[3] made a vigorous attempt to follow the techniques of TAYLOR and LANGMUIR to produce a test specimen of uniform surface structure, but found it impossible.

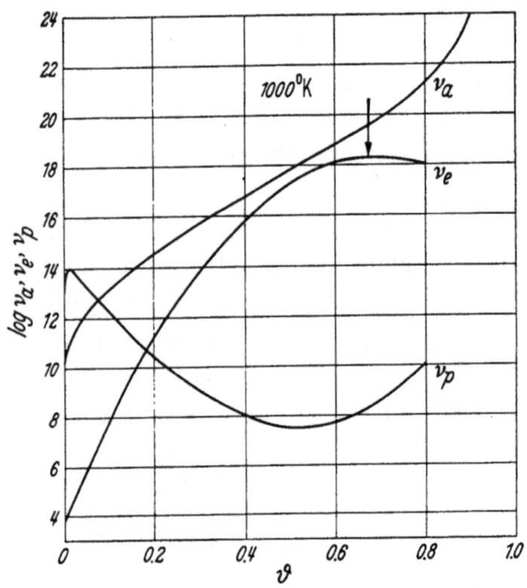

Fig. 35. Atom (ν_a), ion (ν_p), and electron (ν_e) evaporation rates at 1000° K with zero-field as observed by TAYLOR and LANGMUIR as a function of the fraction of the surface covered ϑ.

In spite of the uncertainty in the true nature of the underlying crystallographic structure of the tungsten specimen studied by TAYLOR and LANGMUIR, there can be no doubt concerning the broader aspects of their results. These will be summarized very briefly by the following statements.

1. The adsorption of cesium on a tungsten surface reduces the work-function in direct proportion to the number of cesium atoms adsorbed for coverages less than approximately 15% of the number of cesium sites available to a monolayer of adsorbed atoms. LANGMUIR's determination gives 4.8×10^{14} atoms per cm.² for complete coverage.

2. A maximum reduction in average work-function of approximately 3 volts takes place if the surface covered is 0.67 of a monolayer. This maximum occurs because the dipole moment per atom decreases at a greater and greater rate so that in spite of an increase in cesium concentration above this critical value, the average dipole moment per unit area actually decreases.

3. The relative rates of atom evaporation, electron evaporation and positive ion evaporation at 1000° K are best illustrated by Fig. 35. In this figure, the fraction of the surface covered is represented by the symbol ϑ and the evaporation rates are expressed in terms of units per cm.² per sec. This figure shows that, for very small concentrations the ion evaporation is very large compared with

[1] I. LANGMUIR and K. H. KINGDON: Science, Lancaster, Pa. **57**, 52 (1923). — Proc. Roy. Soc. Lond., Ser. A **107**, 61 (1925).

[2] R. P. JOHNSON and W. SHOCKLEY: Phys. Rev. **49**, 436 (1936). — S. T. MARTIN: Phys. Rev. **56**, 947 (1939).

[3] C. S. ROBINSON: Unpublished research. Dept. of Physics, M. I. T.

either atom evaporation or electron evaporation. As the surface coverage increases, ion evaporation drops to practically negligible proportions, whereas the atom evaporation continues to increase and as a monolayer is approached, the atom evaporation increases so rapidly that at this high filament temperature it is practically impossible to form even a small part of a second layer of cesium on the surface. Electron evaporation is shown to have a maximum at a ϑ value of 0.67.

These studies of TAYLOR and LANGMUIR have contributed more than any other single investigation to an understanding of the properties of adsorbed films on an underlying metallic structure. The results are best accounted for by the statement that the adsorption of an electropositive layer modifies the electron affinity in direct proportion to the dipole moment per unit area. The adsorbed layer reduces the work-function and permits the thermionic emission of enormously more electrons than would otherwise obtain. This is one of the clearest examples of catalytic action since the cesium itself assists in the "reaction" without in any way being used up in the process.

75. Some properties of thoriated tungsten emitters. In the early days of lamp manufacture, improved ductility of tungsten was obtained by the introduction of thorium oxide to the extent of about one percent along with the tungsten powder before sintering and wire production. Before heating to a high temperature, wires containing thorium oxide are practically indistinguishable by inspection from pure tungsten wires. Upon the first high-temperature heating, however, there is a very great difference in the average crystal size of the heated wire. Pure tungsten invariably develops large crystals that generally extend as single crystals across the entire diameter of the wire, while thoriated tungsten develops many small crystals per wire diameter that seem never to grow to develop large crystals in later heating. Hundreds of pockets of thorium oxide are formed in the interstices between the tungsten crystals in a centimeter of wire length. LANGMUIR and ROGERS[1] discovered that, after a suitable sequence of heating, thoriated tungsten wires yielded tens of thousands greater electron emission density at a favorable operating temperature in comparison with pure tungsten at that same temperature. The main features of the phenomena of thorium oxide reduction and diffusion to the surface were studied by LANGMUIR[2] and BRATTAIN and BECKER[3].

From these and other studies it has become clear that thorium oxide is reduced to some extent by the evoluation of oxygen in excess of the thorium when the filament is flashed to a high temperature, preferably between 2800 and 3000° K. After a flash of the order of one minute at this high temperature, the concentration of thorium averaged over the entire surface of the tungsten is practically negligible and the emission properties of the specimen are very close to those of pure tungsten. In order to activate this filament, it must be maintained in an evacuated tube from which all active gases have been removed, either by pumping or by the "getter action" of some material such as barium. For quantitative researches, it is necessary to reduce the partial pressure of these active gases to less than 10^{-14} mm.

If the filament is maintained, preferably with no applied electron accelerating field, at a temperature of 2100° K, thermionic emission tests made at some fixed test temperature such als 1400° K or less show that the emission capability

[1] I. LANGMUIR and W. ROGERS: Phys. Rev. **4**, 544 (1914).
[2] I. LANGMUIR: Phys. Rev. **22**, 357 (1923).
[3] W. H. BRATTAIN and J. A. BECKER: Phys. Rev. **43**, 428 (1933).

of the filament will increase through many orders of magnitude and finally stabilize after approximately 30 min. of activation at a value which is independent of further activation heating

For a quantitative investigation of the activation procedure, it is more instructive to choose a lower activation temperature such as 1900° K and a test temperature of 1200° K or less. After the filament has been aged by repeating the activation procedure a few times, it may be flashed to a high temperature for 30 sec. and a detailed study of the rise in emission with time of activation may be made. Since a decrease in true work-function as a result of activation is a direct measure of the average dipole moment per unit area that results from the presence of the thorium on the tungsten surface, the following equation gives the relation between the observed emission current and the time:

$$\ln I = \ln I_0 + \frac{e}{\varepsilon_0 k T_s} \int_{t=0}^{t=t} \overline{m}_0 \left[1 - f(n_a/n_1)\right] \left(\frac{d n_a}{d t}\right)_{T_a} dt. \qquad (75.1)$$

This equation is written in this form in order to bring out certain features of the phenomena involved. Consider first the time rate of increase in the density of adsorbed atoms represented by $(dn_a/dt)_{T_a}$. For a temperature of activation T_a as low as 1900° K, the evaporation of thorium atoms is practically negligible until the surface concentration builds up to 0.8 of a monolayer. Under that condition this factor is controlled by the diffusion of thorium from a large number of relatively active spots between crystals out of which the thorium atoms stream and spread (because of their mutual repulsive forces) over the entire surface of the thoriated tungsten filament. For the most accurate results, the thorium atoms should be given an opportunity to disperse uniformly over the entire emitter by holding it at a temperature of approximately 1400° K for a considerable period of time after each activation period before the emission observations are made at 1200° K. Other factors in the equation include $\overline{m}_0$ which is defined as the average dipole moment per thorium atom when the concentration is less than 5% of a monolayer and the second factor of the equation is written in this form to indicate that the average dipole moment per atom is known to decrease as the concentration increases. In other words, the function $f(n_a/n_1)$ would start out for small concentrations to be practically zero, but as the concentration increases, this function will gradually increase until it approaches the concentration of a monolayer specified as n_1. It follows from this discussion that if the arrival rate of thorium atoms is initially constant, and if the function $f(n_a/n_1)$ is small compared with 1, then $\ln I$ should be a linear function of the time. Initially it can be assumed that the evaporation rate can be neglected in comparison with the rate of arrival of thorium atoms from the interior, and that this rate of arrival is governed by an exponential temperature relation. Studies of the initial rise in emission current for small surface concentrations should permit a determination of the dipole moment $\overline{m}_0$. The exact rate of arrival at a given activation temperature T_a could differ from sample to sample, but it would be expected that the dipole moment per atom would be reproducible. Tests of this kind have not been made with all factors of the problem under control.

Having once determined a suitable value for $\overline{m}_0$ and the rate of arrival for a given low temperature of activation T_a, then it should be possible to determine more specifically the interaction function $f(n_a/n_1)$.

As regards this process of activation, it is sufficient to say that a surface concentration of approximately 0.7 of a monolayer yields the highest emission current at a given test temperature and that the addition of extra thorium atoms actually results in a lower emission as measured at zero field since, because of the interaction of one dipole on its neighbor, the average dipole moment per unit area decreases with an increase in thorium concentration above the 0.7 value.

Fundamental to the understanding of the physics of thermionic emission is knowledge concerning the energy distribution of the electrons emitted. NOTTINGHAM[1] developed a circuit for the pulse heating of a thoriated filament in order to study its emission properties as a function of applied voltage without the results being distorted by the presence of a drop in potential over the length of the filament that delivered current to the collector. Extensive and detailed research yielded information that indicated an unmistakable departure from the anticipated MAXWELLian distribution. By means of analytic methods described in detail in Sect. 26, it was possible to show that, over a very wide range in activation and in temperature, the empirical equation given as Eq. (26.8) and discussed in Sect. 26, was an excellent representation of the data. This result is illustrated by Fig. 36, which is a composite of electron energy distributions for 9 states of activation which covered a very wide range of emission capabilities. The test temperature was 1160° K.

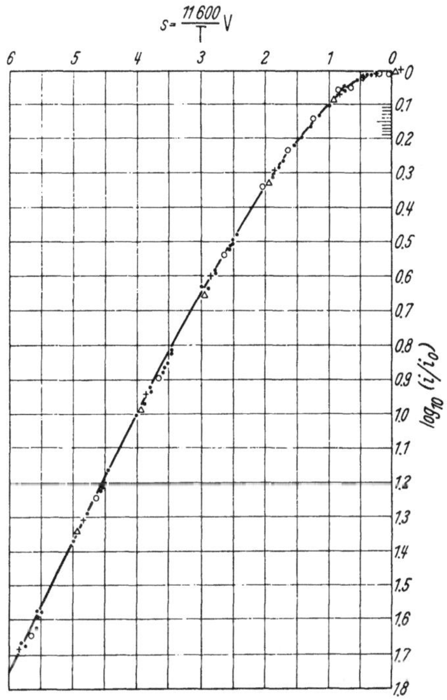

Fig. 36. Composite of the electron energy distributions observed by NOTTINGHAM for nine states of activation. Circle, $\vartheta = 0.032$; Cross, $\vartheta = 0.18$; Triangle, $\vartheta = 0.9$. Other states of activation not identified. Solid line computed with $\omega = 3.05 \times 10^{-20}$ joules or 0.191 ev. i_0 = current at zero-field. $T = 1160°$ K.

The extent of this deficiency in slow electrons is best illustrated by the curves of Fig. 4, which apply to three selected temperatures. The statement has been made by HERRING and NICHOLS[2] that the observed deficiency in slow electrons can be accounted for on the basis of a hypothetical distribution of nonuniformities. In reply to this suggestion, it may be stated that it seems highly improbable that a single sample of thoriated filament wire in so many different states of activation would have effectively the same distribution of nonuniformity and furthermore, that all samples investigated with the same degree of care as the one under present discussion show the same energy distribution. All that can be said is that all of the known data at present capable of giving an answer to this question of the electron energy distribution yield the same answer and further more, this answer is in excellent agreement with HUTSON's data, discussed in Sect. 72.

The influence of patches and the details of their development can be studied by observing the thermionic emission with an accelerating field. Typical data covering this phase of the subject are given in NOTTINGHAM's paper. The most significant conclusion that can be drawn from these studies is that although the long-range forces between the parallel dipoles formed by the adsorption of thorium atoms are repulsive forces, the experimental evidence indicates that there are strong, short-range attractive forces which permit patch areas of large

[1] W. B. NOTTINGHAM: Phys. Rev. **49**, 78 (1936).
[2] C. HERRING and M. H. NICHOLS: Rev. Mod. Phys. **21**, 185 (1949).

groups of thorium atoms to accumulate together and form local regions of high thorium density separated by regions of very low density. In other words, the "surface phase postulate" of LANGMUIR that "all the properties of an adsorbed film on an underlying surface of given composition are uniquely determined by ϑ and T" is not fulfilled with thorium and, in all probability, is not as perfectly satisfied for cesium on tungsten as TAYLOR and LANGMUIR[1] thought.

IV. The oxide-coated cathode.

76. General considerations. The oxide-coated cathode is by far the most important thermionic emitter of electrons. The fact that the application of an alkaline earth oxide to a supporting metallic conductor converted it into an amazingly efficient electron source was discovered in 1903 by WEHNELT[2]. In spite of the fact that over fifty years of research have been applied to the investigation of phenomena associated with the oxide cathode, a comprehensive and generally accepted theory of its operation has not been developed.

Within the past ten years, the results of many researches carried on under superior conditions have yielded so much new material that a fairly clear picture seems to be emerging, and it will be the purpose of the following sections to integrate as many of these results as possible into a unified theory. See Sect. 69 for a discussion of the basic difference between the oxide emitter and metallic emitters.

A very complete analysis of the chemistry of the oxide cathode has been prepared by RITTNER[3]. His study has been made with the background assumption that the presence of an excess of barium within the crystal of the oxide cathode is necessary for its activation. In spite of the fact that the results of WOOTEN, RUEHLE and MOORE[4] show no correlation between their measurements of "Thermionic Performance" and the free barium found either in the coating itself or in the tube under test, the theory to be presented here also depends on the excess barium hypothesis. The application of the equations developed in Sect. 65 shows that the concentration of excess barium need not be more than a few parts per million and that a variation in this concentration has a more direct bearing on the temperature coefficient of the true work-function than it does on the work-function derived as an empirical constant from the RICHARDSON formula (see Sect. 81).

RITTNER's analysis of the chemistry of the oxide cathode shows that pure nickel reacts with barium and strontium oxide to such an extremely small extent that it is the impurities intentionally introduced into the nickel base that are are the most efficient reducing agents in practical cathode nickel to develop and maintain the needed excess barium. It would carry this discussion too far afield to go into all the details of the thermo-chemistry involved. It must be summarized simply by stating that reducing agents such as aluminum, silicon, magnesium and titanium are favored in practical applications. Good production procedure demands a combination of elements so that the initial activation can proceed rapidly and possibly exhaust a strongly acting reducing agent such as

---

[1] J. B. TAYLOR and I. LANGMUIR: Phys. Rev. **44**, 423 (1933). — I. LANGMUIR: J. Amer. Chem. Soc. **54**, 2798 (1932).

[2] A. WEHNELT: Verh. dtsch. phys. Ges. **5**, 255, 423 (1903). — Ann. Phys. **14**, 425 (1904).

[3] E. S. RITTNER: Philips Res. Rep. **8**, 184 (1953).

[4] L. A. WOOTEN, A. E. RUEHLE and G. E. MOORE: J. Appl. Phys. **26**, 44 (1955). — L. A. WOOTEN, G. E. MOORE and W. G. GULDNER: J. Appl. Phys. **26**, 937 (1955). — G. E. MOORE, L. A. WOOTEN and J. MORRISON: J. Appl. Phys. **26**, 943 (1955).

magnesium within the first few hours of cathode operation. The excess barium thus produced would soon leave the crystals in the course of life unless some slower process of BaO reduction continued the maintenance of the required excess barium concentration. Silicon and titanium alloyed into the base nickel seem to serve this function. An excessive use of silicon results in the development of an interface layer identified by x-ray analysis[1] as barium ortho-silicate. Although the formation of the compound results in the liberation of barium, the interface thus developed may have an extremely high resistance which is in general very objectionable in the practical use of such cathodes.

Practical experience teaches that a solid solution of barium-strontium carbonate in proportions close to 50—50 by weight forms one of the best sources for oxide cathode material. These crystals are prepared in a finely divided form mixed with a binder and sprayed on to the base metal. Under moderately good vacuum conditions the carbonate breaks down at a temperature of approximately 1200 to 1300° K with the evoluation of carbon dioxide and carbon monoxide. The most detailed descriptions concerning manufacturing procedure available in the literature are given by HERRMANN and WAGENER[2]. Even though the discussion and details given in that book are very helpful, it must be recognized that manufacturers of oxide cathodes have developed the art as it applies particularly to their own environment and generally maintain their procedures as "company confidential".

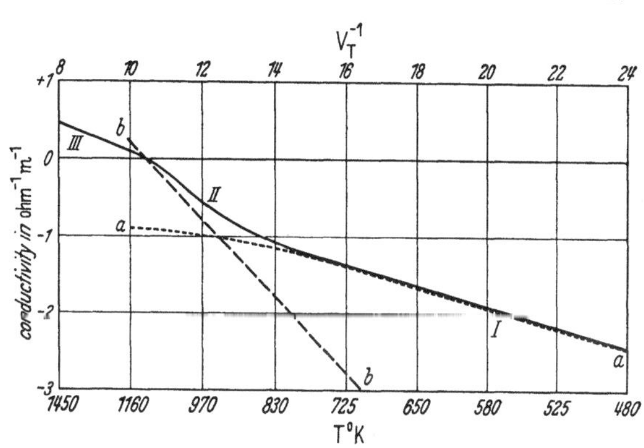

Fig. 37. Conductivity of a typical barium-strontium coating as computed by theories of Sect. 65. Range I, semiconductor conductivity within crystals. Range II, pore conductivity becomes more important than crystal conductivity. Range III pore conductivity inhibited by space charge. Curve *aa*, semiconductor conduction alone, Curve *bb*, pore conduction alone without space charge.

77. General discussion of the electrical properties of the coating. LOOSJES and VINK[3] were probably the first to present the most acceptable generalized description of the electrical properties of an oxide coating. Their description and data apply more specifically to the uniform and regular features of the coating than they do to the distorted and complex situations associated with cathodes in which a non-uniform distribution of the donor centers exists. The maintenance of high electron emission currents always develops some nonuniformity in the coating because of the high internal electric field. With well-activated coatings the internal field within the semiconductor coating is often so small as the electron current is drawn that the ionic flow of the impurity center in the presence of this field can be neglected. Although it will be necessary to discuss the evidence which is directly related to the properties of cathodes with a nonuniform distribution of donors, it will be best to postpone that complication until after the basic properties have been described.

The conductivity characteristic of an oxide cathode may be divided into three ranges, which are identified in Fig. 37 by the numbers I, II and III. The

[1] A. S. EISENSTEIN: Adv. Electronics **1**, 24 (1948).
[2] G. HERRMANN and S. WAGENER: The Oxide-Coated Cathode, vol. 1. London: Chipman & Hall 1951.
[3] R. LOOSJES and H. J. VINK: Philips Res. Rep. **4**, 449 (1949).

very low temperature range (I) is dominated by the mechanism of electronic conductivity through the solid structure of the crystalline semiconductor of which the coating is made. In parallel with these conducting paths through the crystals and their contacts with each other is a second means of conduction. This additional mechanism is the conduction by electrons which can escape from the crystalline surfaces into the pores between crystals and thus carry current by the unimpeded flow of the electrons across the pores. Approximate expressions for the electron mobility associated with these two mechanisms were presented as Eqs. (65.8) and (65.18). These formulae show that, if the average free-flight distance of an electron in a pore is three microns, then at 1000° K the mobility of an electron in the pore is well over 16000 times greater than the mobility for electrons in the crystalline solid. It will be shown that, at about 1000° K the expected ratio of electron density in the pore compared to that of the free electrons within the crystal is also about 16000 to one but in the reverse manner such that the product of the mobility and the electron density for these two mechanisms results in an equality. When all of the properties of a porous conductor are considered, these two means of electrical conductivity are equally important at about 900° K. As the temperature is lowered, the density of electrons within the pores falls so rapidly in comparison with the decrease in density of the free electrons within the crystals that the semiconducting mechanism of the crystals predominates.

This recognition of the fact that the conduction over the ranges I and II arises from these two parallel mechanisms allows the observed curve to be resolved into its two components identified in Fig. 37 as *aa* and *bb*. The empirical results expressed by the approximate linearity of these two components as one plots the logarithm of the conduction as a function of $(1/T)$, can be represented by the equation forms given as follows:

$$\sigma_A = a_A \mathrm{e}^{-\frac{e\Phi_A}{kT}} = a_A \mathrm{e}^{-\frac{\Phi_A}{V_T}}, \tag{77.1}$$

$$\sigma_B = a_B \mathrm{e}^{-\frac{e\Phi_B}{kT}} = a_B \mathrm{e}^{-\frac{\Phi_B}{V_T}}. \tag{77.2}$$

These equations are of exactly the same form as a simplified thermionic equation first presented as Eq. (9.1). They also serve the same purpose which was to yield two numbers by which a set of observation data points can be correlated. When the basic electronic mechanism responsible for the conduction behavior can be identified, the generalized conversion equations of Sect. 50 are most useful.

The empirical constants Φ_A and Φ_B are expressed in energy units of "electron volts". By analogy with certain chemical reactions, these quantities have often been referred to as "activation energies". In view of the detailed theory already presented in Sect. 65, it is to be expected that the activation energy Φ_B could very well be identified as the work-factor associated with the surfaces of the pores. More details on this point are given in Sect. 86. In order to relate the activation energy Φ_A to such details of a specific electronic conductor as the donor concentration and energy level, it is necessary to identify the conducting mechanism. If it is that of conduction through the N-type semiconductor, the equation form given as Eq. (65.11) is the best one to use. In the low temperature range, experiment can give a good value of Φ_A for conduction through the crystalline structure, and since Eq. (65.11) involves the temperature in a manner such that $\nu = -\frac{3}{4}$, then Eq. (50.11) may be applied to evaluate the location of the donor level. The result is given as follows:

$$E = -2\left[\Phi_A + \tfrac{3}{8}(V_n + V_m)\right]. \tag{77.3}$$

In this equation the highest temperature of the range over which Φ_A has been evaluated is V_n in its electron volt equivalent and the lowest temperature of the range is V_m in the same units.

Conduction curves which have all of the characteristics of the calculated curve shown in Fig. 37 by the solid line have been observed[1] with oxide cathodes. Fig. 38 presents the experimental data of two of these observers and the theoretical curve of Fig. 37 is superimposed. A reader familiar with the difficulty associated with the experimental determination of the conductivity of the oxide coating as a function of temperature will not be surprised at the relatively large differences in the experimentally determined characteristics, and he will recognize that the theoretical curve shown is in satisfactory agreement with the experimental determinations. It is therefore in order to explain in considerable detail the model on which the theoretical curve depended so that the necessary precautions may be taken to avoid the drawing of unwarranted conclusions from experimental data.

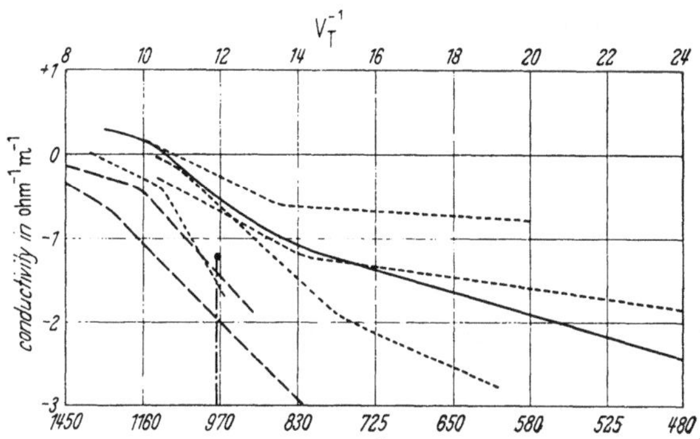

Fig. 38. The comparison between theory and observation. Observations by LOOSJES and VINK shown by dotted lines. Observations by HANNAY, MACNAIR and WHITE shown by dashed lines and the dot-dash line shows range covered also by Fig. 40. Circle shows highest state of activation. Theory curve of Fig. 37 shown as solid line.

The theoretical calculations for the curves in Fig. 37 assume properties of the oxide coating structure tabulated as follows:

1. The oxide coating is a porous structure of small crystallites in contact with each other to give an average density of approximately 50% of that which would be obtained for the solid crystalline structure of the composition used. Thus the total volume of the pores is approximately one-half of that of the coating as a whole.

2. The conductivity and the emission properties are to be computed on the basis that the electrons are in thermal equilibrium with a concentration of donors of 5×10^{22} per m.^{3} located at an energy level 0.7 ev negative with respect to the bottom of the conduction band.

3. At very low temperatures, the conductivity of the coating is attributed to the electronic conductivity of the crystals only, since the concentration of electrons in the pores between the crystals is completely negligible in that temperature range. This conductivity for the actual coating will be taken to be a factor of 10 lower than that computed by the theory for this system of donors in order to correct for the fact that the true average cross-sectional area of the conduction paths will be reduced because of the presence of the voids and because of the fact that the crystals will make contact with each other over surface areas small compared with the crystal cross-sections. In addition, the conduction paths will be longer because the electrons will have to follow devious routes. It is not possible to evaluate these corrections separately and therefore they are all lumped together and approximated by this factor of 10. This factor was introduced into the theoretical equations as C_s of Eq. (65.9).

[1] R. LOOSJES and H. J. VINK: Philips Res. Rep. **4**, 449 (1949). — N. B. HANNAY, D. MAC NAIR and A. H. WHITE: J. Appl. Phys. **20**, 669 (1949). — R. FORMAN: Phys. Rev. **96**, 1479 (1954). — J. R. YOUNG: J. Appl. Phys. **23**, 1129 (1952). — J. NAKAI, Y. INUISHI and Y. TSUNG-CHE: J. Phys. Soc. Japan **10**, 437 (1955).

4. A measurement of the temperature coefficient of the contact difference in potential, to be described in more detail in Sects. 81 and 83, served to establish the magnitude of the donor density to be associated with a well-activated oxide cathode. It must be assumed that a slightly higher density is to be found close to the interior surfaces of the pores. Since the exact density actually located there in an experimentally observable structure of the oxide cathode is unknown, it is assumed that the temperature coefficient of the true work-function at the surfaces of the pores and averaged over the important temperature range is (6.3). This coefficient (α) is defined in the units described by Eq. (65.21). The value of φ_0, associated with that same equation is taken as 1.05 ev and may very legitimately be compared with tabulated values of the RICHARDSON work-function as observed and reported in the literature for well-activated oxide cathodes.

5. The above choices of donor density, of donor energy level and true work-function make it necessary to ascribe the constant value of 0.875 ev to the electron affinity. This quantity, designated as W_a, is defined as the energy difference between an electron found in a level at the bottom of the conduction band and the energy of an electron at rest just outside of the emitting surface.

6. The effective pore length l_p of Eq. (65.17) is taken as 2.5×10^{-6} m. It is to be expected that the average linear dimension of the pore will be slightly greater than this figure and will probably be of the order of 3 to 3.3 microns.

In the above tabulation, all of the necessary quantities have been specified in order to use Eq. (65.9) for a direct computation of the contribution to the coating conductivity ascribed to electronic conduction through the crystals. The results of this calculation are shown in Fig. 37 by curve *aa*. A comparison between this calculated curve and the composite curve shows that over the range of temperature below 725° K conductivity is entirely dominated by this component. Note that curve *aa* is not a straight line and when it is combined with the conductivity component attributed to the pores, the apparent straight line range extends to higher temperatures. Note also that the "activation energy" Φ_A for the temperature range identified as the LOOSJES-VINK range I is 0.315 ev. The conversion formula given either as Eq. (77.3) or Eq. (65.16) permits the evaluation of the energy level of the donors to be 0.7 ev.

As the temperature increases, range II is clearly evident. LOOSJES and VINK in their analysis extrapolate linearly from range I and deduce what they consider a suitable representation of pore conductivity by analysis of the difference between the actually observed curve and the extrapolated straight line. It is evident from Fig. 37 that theory indicates a curve for the range *aa* instead of a straight line and therefore over region II the conductivity of the electrons through the pores plays a still more important part than was attributed to it by LOOSJES and VINK. Eq. (65.26), derived from theoretical considerations and with the constants called for in the tabulation above, serves as the means for the calculation of the pore conductivity as shown in Fig. 37 by the line *bb*. Note that a line drawn tangent to the composite curve in the range II would have yielded an apparent activation energy of 0.84. This value, it will be noted, has no obvious significance at all. If the correct procedure is followed, however, of establishing the location of line *bb* by subtracting the contribution really due to conduction through the crystalline structure, then the "activation energy" associated with this line is 1.14 ev. Eq. (65.27) furnishes the means by which this activation energy can be converted to obtain the correct value of φ_0.

It is more difficult to obtain precision in a theoretical calculation that applies to range III. It is easy to make use of Eqs. (65.20), (65.22) and (65.23) to establish certain limits for the conductivity expected in the pores after space charge within them becomes an important limiting factor. The curve shown in Fig. 37 has been drawn to fall more or less midway between those computable limits. Again, it is possible to calculate a so-called "activation energy" for this range and the value obtained is 0.43 ev. This result is in good agreement with the values reported by HANNAY, MACNAIR and WHITE. This reduced rate of increase in conductivity with temperature is the result of the building up of space charge within the pores and the shortening of the time of free flight. The term "activation energy" has no physical significance under these conditions except to express that rate as an empirical constant.

An analysis has been made of the electronic structure of barium oxide by HENSLEY[1]. His energy level diagram is shown as Fig. 39. It is to be noted that he does not include a donor state nearer to the conduction band than 1.4 ev. Practically all experiments on the conductivity of well-activated oxide cathodes show "activation energies" for the low-temperature range between 0.25 and 0.4 ev. The analysis presented here in Sect. 65 indicates that there must therefore be a donor state close to 0.7 ev below the conduction band. The presence of this state may be difficult to establish by methods that are completely independent of thermionic studies since the important concentration range for this donor state lies below a few parts per million. It is probable that the presence of other states such as those shown by HENSLEY is indicated by the higher values of activation energy associated with poorly activated cathodes.

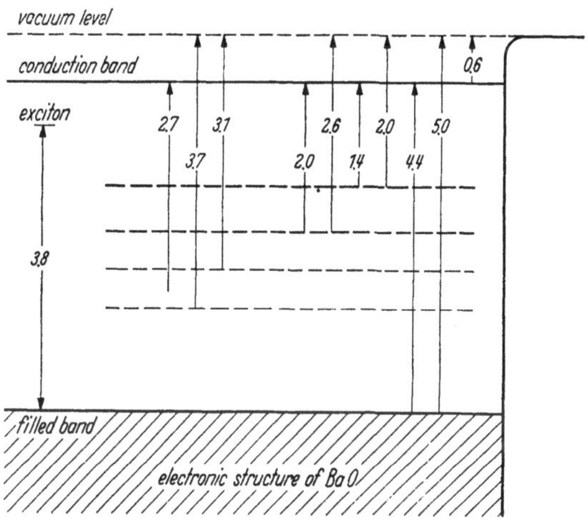

Fig. 39. Energy level diagram for barium oxide as prepared by HENSLEY. Energies expressed in ev.

It has not been possible to establish unambiguously the nature of the important donor center. The choice seems to be either it is an oxygen vacancy or an interstitial barium atom. Experiments on electron emission which are interpreted in terms of the local depletion of the donor concentration have been interpreted by FROST[2] to indicate that the activation energy for the diffusion process of the donor is about 0.4 ev. Since this energy is so low and the concentration of donors needed for activation so small, it seems quite likely that an electron in an oxygen vacancy may be the important donor level. The resolution of this problem of donor identification will probably have to depend on a theoretical analysis of the properties of excess barium as it is incorporated into the crystal in this way.

Although most of the conduction properties of oxide cathodes are accounted for quantitatively by the application of the LOOSJES and VINK model of the porous conductor, the experimental observation of electron mobility as a function

[1] E. B. HENSLEY: Report on the M. I. T. 15th Annual Conference on Physical Electronics, Cambridge, Mass., p. 18, 1955.

[2] H. B. FROST: Transient Changes in Oxide Cathodes. ScD Thesis, M. I. T. 1954; also H. B. FORST: Report on the M. I. T. Annual Physical Electronics Conference, Cambridge, Mass., p. 1, 1955.

of temperature as reported by FORMAN[1] contributes additional facts in favor of this model. The direct observation of mobilities through the interpretation of conductivity combined with the measurement of the HALL-effect emf is subject to many difficulties. The observed value of mobility over the high-temperature range was about 30 m.^{2} per v-sec. With FORMAN's estimated mean-free path within the pores of his test sample as 10 microns, Eq. (65.18) gives a mobility anticipated for the pore conduction model of about 10 m.^{2} per v-sec. at 1000° K. This direct observation of electron mobility in the high-temperature range and its corresponding fall to much lower values as the temperature is lowered serves as strong evidence in favor of the LOOSJES and VINK pore-conduction hypothesis. A very high magneto-resistive effect has also been observed by FORMAN in oxide cathode test samples.

78. Experiments on coating conductivity. The importance of measurements of coating conductivity as a basis for the development of a theory of the oxide cathode was recognized by REIMANN and MURGOCI and also by BECKER and SEARS[2]. Their method depended upon the imbedding into the coating of probes by which the conductivity could be deduced from potential and current measurements. This method has been used by many other investigators and has been found to be extremely difficult in that the processed coating very seldom activates in a normal manner to result in a really active cathode. The method used by LOOSJES and VINK[3] depended on the direct measurement of conductivity across two similar coatings that were processed in a normal manner and pressed together for conduction measurements. It was on the basis of these results that they formulated their conduction theory. They made observations on coatings in different stages of activation and results are summarized in Fig. 38 by the three curves shown there by dotted lines. The theoretical curve, shown there as a solid line, represents these data quantitatively better than it does the data of HANNEY, MACNAIR and WHITE[4].

These experimenters developed a tube of unusual design in which the oxide coating was deposited upon a magnesium oxide base which, in turn, was heated by an internally located tungsten heater. Four electrical connections were made to the coating by means of platinum wires mounted in grooves cut in the ceramic. These wires served as connections and potential leads so that the conductivity could be measured by the observation of the drop in potential over a well-defined length of the coating of about one centimeter. Since the coating could not be activated in the usual manner, a heat treatment in an atmosphere of methane served as a reducing agent to bring about activation. Many observations were made at a standard temperature of approximately 970° K and the conductivity was found to vary over three orders of magnitude as it depended on the activation and aging process. These points are represented by the vertical dot-dash line of Fig. 38. The most favorable state of activation reported gave a conductivity of 6×10^{-2} ohm^{-1} m.$^{-1}$. This state of activation is represented by the circle shown in the figure. In view of the difficulty of accomplishing full activation by this process, it is not surprising that the observed maximum conductivity is approximately a factor of four below that obtained by LOOSJES and VINK at the same temperature.

---

[1] R. FORMAN: Report on the M. I. T., Annual Physical Electronics Conference, Cambridge, Mass., p. 10, 1955; and Phys. Rev. **96**, 1479 (1954).

[2] A. L. REIMANN and R. MURGOCI: Phil. Mag. **9**, 440 (1930). — J. A. BECKER and R. W. SEARS: Phys. Rev. **38**, 2193 (1931).

[3] See ref. 1, p. 117.

[4] See ref. 1, p. 117.

In addition to the extensive measurements made at 970° K, others were made over the high temperature range extending up to about 1400° K. These results are summarized by the dashed lines of Fig. 38. The indications are that the coating material was not fully activated at the 0.7 ev donor level and therefore because of the high temperature coefficient of the work-function, the conductivity in the pores of the structure was low, and the electronic conductivity in the crystals also sufficiently low so that it did not show its contribution over the range of temperature studied . In spite of this lower state of activation, clear evidence exists in these studies for the space-charge limitation on conductivity as the temperature increases above 1160° K.

In addition, HANNAY, MACNAIR and WHITE [1] measured electron emission at the same temperature of 970° K at which conductivity measurements were made as mentioned above. Under all conditions that corresponded to a uniform distribution of activating centers throughout the specimen, exact proportionality existed between conduction and emission and the results are reproduced here as Fig. 40. Although HANNAY, MACNAIR and WHITE interpreted their results to prove that in their experiments the conductivity through the crystalline structure was more important than the conductivity through the pores, it will be shown in Sect. 86 that their constant proportionality factor between emission current and conductivity is in satisfactory agreement with the one computed from the theories presented here. It will then be shown that these experiments do not conflict with the pore conduction model of the oxide cathode.

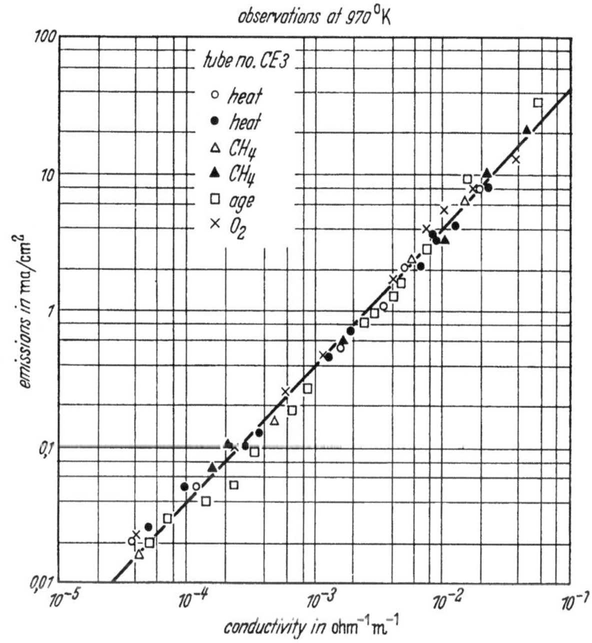

Fig. 40. Linear relation between electron emission and coating conductivity as observed by HANNAY, MACNAIR and WHITE at 970° K.

79. An analysis of emission current measurement methods. Not only is the theory of the oxide cathode a very controversial subject, but an equally unresolved problem is that of the evaluation of cathode-electron emission. It will be the purpose of this section to review some of the more important methods of cathode emission evaluation and discuss the basic principles which must be considered in the interpretation of data obtained by the various laboratory procedures used. Although references will be made to specific examples of the uses to which these various methods have been put, the references to these examples will, of necessity, be very limited.

Five methods of emission evaluation will be described. These methods are tabulated as follows and will be discussed in that order

1. SCHOTTKY plot (*dc* or pulse method) and RICHARDSON plot evaluation;
2. Deviation form the $\frac{3}{2}$ power space charge line;

[1] See ref. 1, p. 117.

3. Analysis of retarding potential data and the use of FOWLER's equation;

4. Indirect methods including harmonic analysis and changes in transconductance;

5. Voltage current characteristics with a detailed evaluation by space-charge theory.

The theory applicable ot the SCHOTTKY concept that the work-function of an electron emitter is reduced through the application of an external field was discussed in detail in Sect. 27. This theory depended on the fact that for good conductors such as tungsten a most important long range force that acts on an electron as it leaves is the "mirror-image" force. The fact that the electron affinity of an oxide seems to be only 0.9 ev indicates that the simple mirror-image concept probably does not apply. The direct application of the mirror-image theory led to the development of Eq. (27.13) which depended on the assumption that space charge could be neglected. This equation reformulated to express the temperature in its electron volt equivalent (V_T) is written as follows:

$$\mathrm{Log}_{10} I = \mathrm{Log}_{10} I_0 + 1.65 \times 10^{-5} \frac{G'^{\frac{1}{2}}}{V_T} (V_c)^{\frac{1}{2}}. \tag{79.1}$$

In spite of the fact that the form of this equation depends on the mirror-image concept it is often applied to the interpretation of experimental data observed with oxide cathodes for which I is the current density, I_0 is an empirical constant obtained from the fitting of the equation to experimental data. The numerical constant comes directly from the theory by way of Eq. (27.13). In that equation, the constant G represented the geometrical factor by which the potential difference between the emitter surface and the collector surface (V_c) needed to be multiplied by in order to obtain the electric field intensity near the surface of the emitter. Eq. (27.4) and (27.5) served as examples for the calculation of G. In Eq. (79.1) the constant G' must be looked upon as an empirical constant rather than one derivable directly from the geometry.

Numerous examples[1] of the application of this equation to emission currents from oxide cathodes lead to the result that G' is always larger than the true geometrical constant G by a factor of 3 to 10. At least four factors contribute to this lack of agreement between G' and G. First, the surfaces of most examples to which this equation has been applied have many sharp points and edges that correspond to the sintered crystallite structure of which the cathode surface is made. Secondly, the very low work-function areas which at a given temperature contribute a large fraction of the observed emission current do not experience a uniform field over the whole emission area and therefore no one simple geometrical factor can be applied. Furthermore if these surfaces are within pores their emission may be space-charge limited within such lccalized areas. Thirdly, in the neighborhood of low work-function areas the surrounding high work-function areas inhibit the emission because of the local contact difference in potential field and this effect is reduced as the field is increased. Fourth, the mirror-image concept may not apply.

In view of the above restrictions, the method by which Eq. (79.1) is usually applied involves the plotting of $\mathrm{Log}_{10} I$ as a function of $V_c^{\frac{1}{2}}$ or better yet, as a junction of $(V_c^{\frac{1}{2}}/V_T)$. The advantage in the second choice is that if the value of G' remains constant as the temperature is varied then the slopes of the plotted

[1] C. S. HUNG: J. Appl. Phys. **21**, 37 (1950). — T. ARIZUMI and S. NARITA: J. Phys. Soc. Japan, **6**, 118 (1951). — J. NAKAI, Y. INUISHI and Y. TSUNG-CHE: J. Phys. Soc. Japan, **10**, 437 (1955). — L. S. NERGAARD: RCA-Rev. **13**, 464 (1952).

curves will all have the same value or vary in a systematic way as for example, in the experiments of HUNG[1]. With either method of plotting, the observed currents invariably fall below the line represented by Eq. (79.1) as the applied voltage is lowered. Typical data are illustrated in Fig. 41. In some examples departure from the straight line results from the influence of space charge, whereas in other examples where the observations are made at such a low temperature that space charge cannot possibly influence the result, the departure from the straight line is attributed to "patch effect", that is, a lack of homogeneity of the emitting surface.

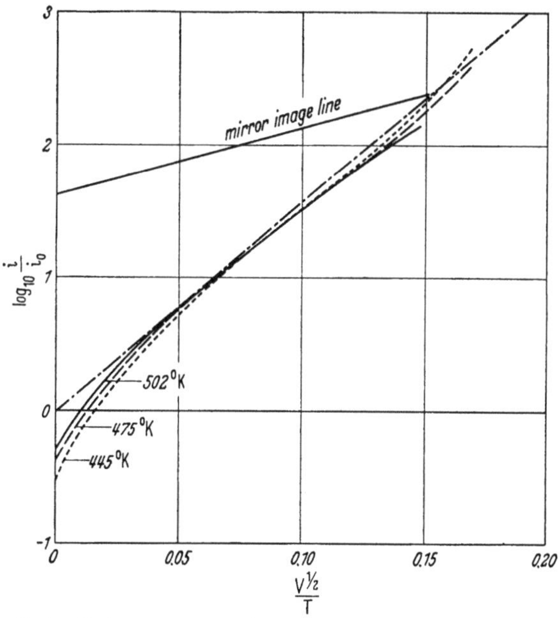

Fig. 41. Emission from an oxide cathode as a function of field as observed by HUNG at three temperatures. Note disagreement with mirror-image slope and other evidence for patch effect. Slope of dot-dash line is 3.1 times slope of SCHOTTKY line.

The straight line extrapolation from the high voltage range in spite of the obvious departure serves to establish the empirical value of I_0 which in turn is often taken to be the "zero-field" current density appropriate for further analysis according to the RICHARDSON or the FOWLER equation. An example of such a plot is shown in Fig. 42. This plot serves to determine the two other empirical constants A_R and φ_R or A_F and φ_F.

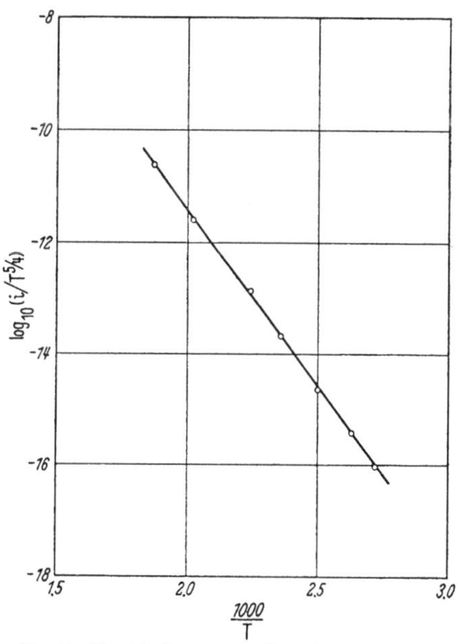

Fig. 42. Empirical representation of HUNG's observations as fitted to the FOWLER equation. $\varphi_F = 1.25$ ev.

Basically this method of analysis yields four empirical constants by which the emission properties of the particular cathode under investigation can be described in the high field region by the RICHARDSON type of equation given as Eq. (50.2) or the FOWLER Equation (64.22) and the SCHOTTKY form given as Eq. (79.1).

Although this method of data analysis has been used more extensively than others it has not been successful as a means for distinguishing between cathodes as they are used and therefore does not serve as a reliable method of emission evaluation. The objections to this method of cathode study deserve some elaboration. Most applications of oxide cathodes involve their use under the condition that the emission is limited by space charge. The theory presented in Sect. 58

[1] See ref. 1, p. 122.

shows that if the electron emission under all conditions of normal tube operation is adequate to maintain a potential difference across the "emitter space" between the emitter surface and the potential minimum created by space charge equal to or greater than V_T or (kT) then in terms of tube performance emitters of widely different saturation emission capability are indistinguishable from each other. It follows, therefore, that cathode evaluation by the "SCHOTTKY plot—RICHARDSON plot" method yields information which is neither well adapted to the quantitative understanding of the emission process of the cathode as a whole nor does it lead to the identification of processes most likely to give the desirable characteristics required of these emitters in practical applications.

An experimental difficulty arises in the use of these methods because the principal voltage range over which investigations must be carried out extends from approximately 20 to 1000 volts or more depending on the tube geometry and the maximum temperature at which the test is to be made. Since well-activated cathodes at normal operating temperatures can yield as much as 10 amps per cm.^{2}, an electron collector at 1000 volts may be required to dissipate kilowatts of energy. Since very few structures are capable of this accomplishment under continuous operation, a pulse method of test is used so that advantage can be taken of the heat capacity of the collector in order to handle this very high power. Microsecond pulses repeated thirty times per second increase the collector's capacity for handling instantaneous power by over thirty thousand fold. This method of cathode evaluation if it is to be applied for the range of normal operating temperature requires powerful, accurately calibrated and controlled pulse equipment. To cover the entire range of study both *dc* equipment and pulse equipment are needed. The drawing of the needed large currents through the cathode coating creates a sufficiently high electric intensity within the coating to result in ionic migration or even cathode destruction in the form of sparking which in turn changes the properties of the emitter as a result of the test procedure itself. Still another difficulty is associated with the creation of decomposition products at the surface of the collector. The electro-negative elements of this decomposition have a poisoning effect as they condense upon the surface of the emitter and neutralize the effectiveness of the donor centers near the surface sufficiently to cause a very objectionable change in cathode properties and thus reduce the emission.

Most of the reported values for the RICHARDSON work-function Φ_R, obtained by this procedure, fall between 0.94 and 1.1 ev. The RICHARDSON thermionic constant A_R usually obtained falls within the range 0.01 and 0.5 amp per cm.^{2}-degree^2.

80. Deviations from the three halves power line. The second most popular method for cathode evaluation, aimed principally at an understanding of the fundamental properties of the emitter, depends on observations taken over a range of applied potential for which the current is limited by space charge. The cathode is then evaluated in terms of a departure from the "three-halves power space-charge line". The two methods of analysis of data usually used will be described in this section and a critical discussion of the results that may be obtained this way will be presented. A third very closely related method not known to be in present use is described in Sect. 82.

Coordinate paper is available which has a linear scale across the abscissa and a two-thirds power scale along the ordinate. This paper is used, as illustrated in Figs. 43 and 44, by one's plotting the applied voltage along the linear scale and the corresponding observed current along the two-thirds power scale. Plots of carefully taken observed data according to this method yield fairly straight lines as long as the current is limited by space charge, but never yield truly straight lines because the three halves power "law" is at best a rather poor approximation as will be shown later in this section.

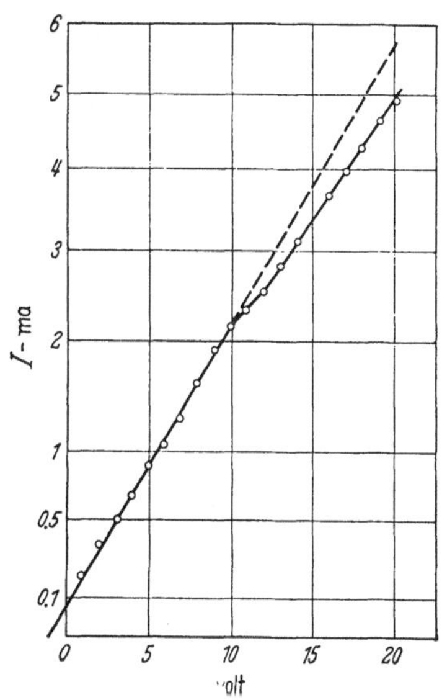

Fig. 43. Observations of NERGAARD[1] plotted on $\frac{2}{3}$ power paper illustrating distortion caused by "high-speed 10-volt effect".

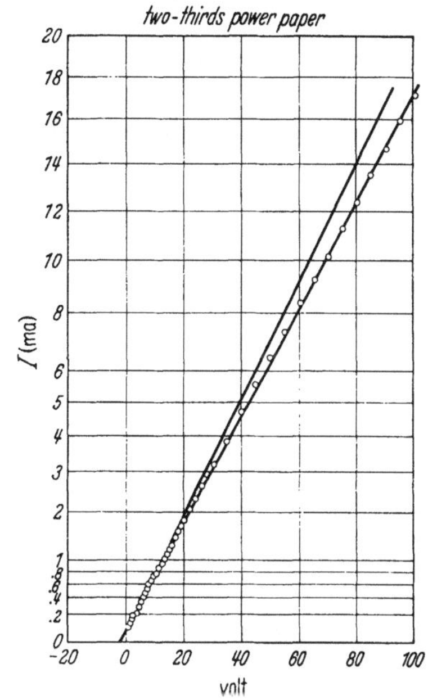

Fig. 44. NERGAARD's use of $\frac{2}{3}$ power paper with an extension into the higher voltage range.

The second method of analysis has been developed at the Bell Telephone Laboratories and is described very briefly by HANNAY, MACNAIR and WHITE[2]. Fig. 45 reproduced from that paper, serves as the best means of describing their method, which involves the following steps.

The diode under test receives a single, one-half cycle voltage pulse applied to the electron collector with respect to the emitter and derived from the 60-cycle power line. In four milliseconds, the applied potential rises to a maximum and then falls to zero in the next four milliseconds. The current, conducted through the tube, produces a small IR drop through a calibrated resistance and after amplification is applied to an oscilloscope in a manner to make the vertical deflection of the spot an accurate measure of the current through the test diode. A special sweep circuit has been designed so that at each instant of time the horizontal deflection of the oscilloscope beam is directly proportional to the three-halves power of the applied voltage. On the

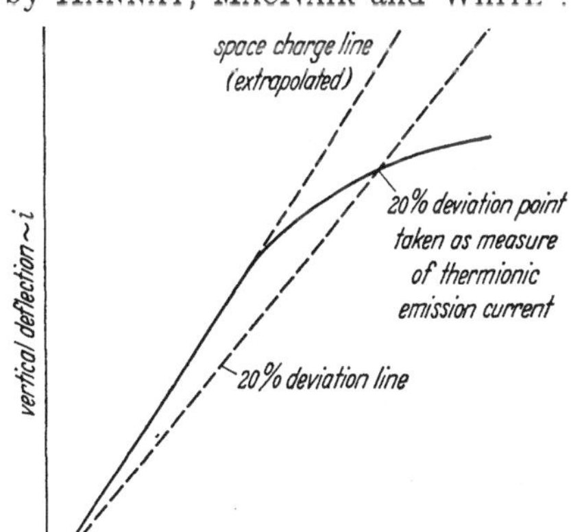

Fig. 45. Emission evaluation by HANNAY, MACNAIR and WHITE by an oscilloscope display which utilized a $\frac{3}{2}$ power voltage axis.

[1] L. S. NERGAARD: RCA-Rev. **13**, 464 (1952). Curves are reproductions of NERGAARD's, Figs. 14 and 15.

[2] See ref. 1, p. 117.

face of the oscilloscope is a ruled coordinate system that is illustrated by Fig. 45. Drawn across the coordinates are straight lines which differ in slope by 20%. The observer uses this equipment by his adjusting the horizontal deflection amplitude until the lower part of the curve produced by the electron beam exactly coincides with the socalled "space-charge line". This adjustment is made simultaneously with the selection of a suitable resistance in the cathode circuit so that within the range of the displayed figure on the oscilloscope, the observed characteristic departs from the upper straight line and intersects the lower one. The current value at this intersection is identified by these observers as a measure of the emission property of the test diode.

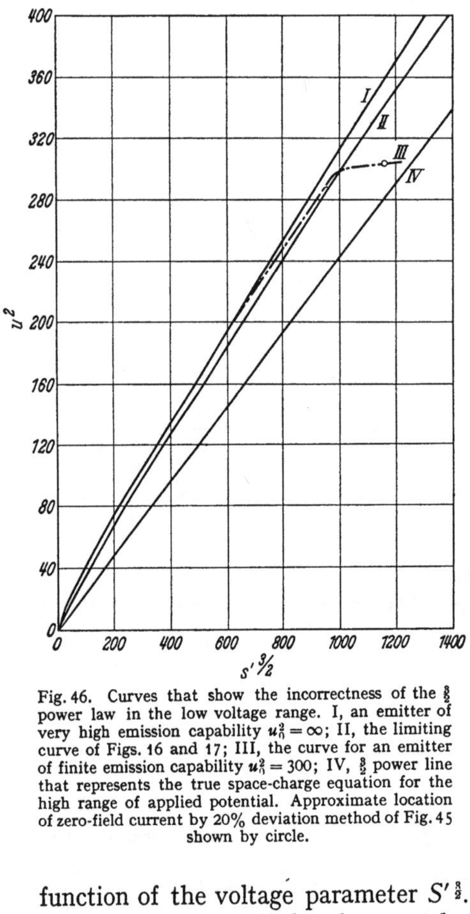

Fig. 46. Curves that show the incorrectness of the $\frac{3}{2}$ power law in the low voltage range. I, an emitter of very high emission capability $u_0^2 = \infty$; II, the limiting curve of Figs. 16 and 17; III, the curve for an emitter of finite emission capability $u_0^2 = 300$; IV, $\frac{3}{2}$ power line that represents the true space-charge equation for the high range of applied potential. Approximate location of zero-field current by 20% deviation method of Fig. 45 shown by circle.

These two methods of study are basically the same and therefore are treated together in this discussion. In Sect. 58 the theory for the computation of "master curves" was developed in detail. The characteristic parameter u_0^2 was defined as the ratio of the true zero-field emission capability of the emitter to the emission current observed under that critical condition for the onset of space-charge limitation which is zero-field at the surface of the collector. It was shown there that over the entire range of the master curves, corresponding to various values of u_0^2 from approximately 5 to infinity, they cannot be distinguished from each other until a sufficiently high accelerating field is applied to the collector to reduce the value of the space-charge minimum to V_T or less. Fig. 46 has been prepared to illustrate the application of the correct theory to the interpretation of data plotted according to the three-halves power method.

From computed data found in Tables 7, 8 and 9, three curves are plotted as a function of the voltage parameter $S'^{\frac{3}{2}}$. The precise relation between this parameter and the applied voltage and temperature is given by Eq. (57.1). The algebraic expression which relates the value of U^2 to the quantity χ_c was derived as Eq. (58.3) as it applied to the emitter of unlimited capability and therefore forms the basis for curve I of the figure. This equation is repeated as follows:

$$U^2 = \left(1 + \frac{\chi_c}{\chi_m}\right)^2. \tag{80.1}$$

The assumption that the three-halves power law will give the same emission current at very high values of applied potential is the equivalent of writing the following relation:

$$U^2 = \gamma\,(\psi_c)^{\frac{3}{2}}. \tag{80.2}$$

In this equation, the proportionality factor γ determines the slope of the line drawn between the origin of the diagram in Fig. 46 and any specified point on line I. If it is assumed that the value of ψ_c is practically equal to S' over the important range to which this analysis is usually applied, this approximation is satisfactory. These equations may be combined with Eq. (37.8) to yield the following relation:

$$\gamma = \frac{4\sqrt{\pi}}{9} \cdot \frac{1}{(1.806)^2} \cdot \left(1 + \frac{1.413}{\psi_c^{0.475}}\right)^{\frac{3}{2}} = 0.2422 \left(1 + \frac{1.413}{\psi_c^{0.475}}\right)^{\frac{3}{2}}. \tag{80.3}$$

This equation shows that the slope of the line which would represent the true properties of a diode in the very high range over which the three-halves power law holds with accuracy is 0.2422. Line IV is drawn in Fig. 46 with this slope.

Over an important portion of the range in S' to which this method of analysis is often used, the average slope of the *true curve* is nearly 30% greater than the line that represents the three-halves power law. This discrepancy does not usually cause much concern because many experimenters are satisfied to obtain a linear plot and to not inquire into the detailed relation between the observed slope and that which would be computed by theory from a knowledge of the dimensions of the diode and the temperature of the emitter.

Another phenomenon which interferes with the usefulness of this method of analysis has been given the name "high-speed, 10-volt effect". According to MATHESON and NERGAARD[1], this effect was first discovered in 1935 by H. NELSON and is easily understood qualitatively as being the result of secondary emission of electrons from the composite surface that exists on the electron collector of practically all test diodes used to study the properties of the oxide cathode. This effect is well illustrated by Figs. 43 and 44. Over the low range of applied voltage the variation in emission current should follow a very nearly straight line relation even though it does not have the correct slope as long as the emission is really limited by a space-charge minimum greater than V_T. In the range of applied voltage between 8 and 12 volts, slow secondary electrons are emitted from the collector and increase the space-charge in the neighborhood of the collector enough so that the observer needs to apply additional voltage to the extent of one or two volts or more in order to re-establish the field in the rest of the tube that would have existed there had it not been for this additional group of slow-moving electrons in the immediate neighborhood of the electron collector.

This effect is a property of the collector and can be expected to be different in detail from tube to tube independent of the emitter properties under investigation. It makes the application of space-charge theory to oxide emitter properties difficult to interpret with an applied voltage greater than 10 volts. The fact that very useful data may be obtained in the low-voltage range will be demonstrated in Sect. 83.

There is still another effect which must be guarded against. The bombardment of the electron collector by high-energy electrons produces objectionable decomposition products which have not been identified with certainty but do react in an unfavorable manner at the surface of an oxide cathode emitter. This effect was first reported by HEADRICK and LEDERER[2] and investigated by others[3,4]. Even though JACOBS' identification of the reaction products is not generally accepted, practically all experiments indicate that the onset occurs at approximately 6 volts electron energy and that the lower the temperature of the emitter,

[1] R. R. MATHESON and L. S. NERGAARD: RCA-Rev. **12**, 258 (1951).

[2] L. B. HEADRICK and E. A. LEDERER: Phys. Rev. **50**, 1094 (1936).

[3] H. JACOBS: J. Appl. Phys. **17**, 596 (1946).

[4] J. D. HOBBS: M. I. T. Thesis "Energy Dependence of Electron Produced Poisoning of Oxide Cathodes". 1954.

the easier it is to detect the influence of this poisoning. The higher the electron energy, the greater is the supply of poison[1] and therefore it is possible to explain the results reported by NERGAARD[2] which indicated that the higher the temperature of the oxide cathode, the higher the energy required to produce sufficient poison to give a measurable reaction at the surface of the cathode.

The current ratio represented by the symbol u^2 was defined in Eq. (57.3). The ordinate scale in Fig. 46 is therefore essentially a current scale with the unit of current that which flows across the test diode at the specified temperature and creates zero field at the surface of the collector. In most examples, this unit of current represented by the symbol I_R is indistinguishable from the maximum current that can flow across such a diode which was represented by the symbol I_m and given quantitatively for the plane parallel diode by Eq. (51.1). The theoretical evaluation of the relation between the radial dimensions of the elements of a diode made with concentric cylinders and the "effective spacing" represented by w of Eq. (51.1) has not been worked out but experiment shows that for a ratio of radii of 2.5, the effective value of w is 0.83 mm. when the actual difference in the radii for the true spacing is 1.0 mm. (see Sect. 83). As the ratio of radii approaches one, the effective spacing should approach the true spacing. This result came from an experiment applied to the low-voltage range and seemed to be independent of the temperature. The LANGMUIR[3] formula, which would be applicable at best to the very high voltage range would give 0.68 mm. for this structure. Really accurate fittings to the theoretical curves in the low-voltage range have not been made except for studies with cylindrical structures and reported in Sect. 83. The ideal structure for the application of theory is the plane parallel one and an excellent design has been fabricated by the Bell Telephone Laboratories and is described in detail in Sect. 84. Test diodes are often made with oval electron collectors instead of cylindrical ones. Although superficially, the dependence of observed current on applied potential may follow so close to the theoretical curves over the low-voltage range that an "effective-spacing" can be established by experiment, the prediction of theory is that such a diode would be unsuitable for the general purpose of emitter evaluation except under very specialized conditions.

The equation which yields the data for the plotting of the "limiting curve" II of Fig. 46 is the combination of Eqs. (57.7) and (57.10). These equations serve as the basis for Table 7. If the experimental data are not distorted by spurious effects, then observed points should follow curve I very closely until approximately half of the maximum zero-field emission from the cathode is obtained. A very gradual deviation from curve I takes place and continues, as shown by curve III, until this curve intersects the limiting curve II. This intersection establishes the true value of the zero-field current for this particular emitter. Since the emission increases very slowly after zero-field has been established, the choice of the 20% point as is Bell Laboratory practice and illustrated in Fig. 45 and 46 is not very much in error.

81. Analysis by retarding potential data and the application of FOWLER'S and RICHARDSON'S equations. The study of thermionic emission from an oxide cathode in the transition range from retarding potentials through to accelerating potentials by HUNG[4] will be used exclusively as the basis for this discussion, since it is one

[1] See ref. 4, p. 127.

[2] L. S. NERGAARD: RCA-Rev. **13**, 464 (1952).

[3] I. LANGMUIR and K. T. COMPTON: Rev. Mod. Phys. **3**, 191 (1931) p. 245.

[4] C. S. HUNG: J. Appl. Phys. **21**, 37 (1950). For greater detail see M. I. T. Thesis, Department of Physics 1948.

of the most carefully executed studies reported in the literature. His data are shown in Fig. 47. The diode used for this study contained a cylindrical coated cathode 0.2 cm. in diameter and 4.5 cm. long concentric with a cylindrical collector system which had a central section one cm. long insulated from the rest so that the current to this collector came from the most uniform part of the emitter. The temperature of the emitter was determined by a calibrated thermocouple. The collector was movable so that evaporation products normally found on such collectors were eliminated. The ratio of the collector to the emitter diameter was four. The spacing between collector and emitter was three mm.

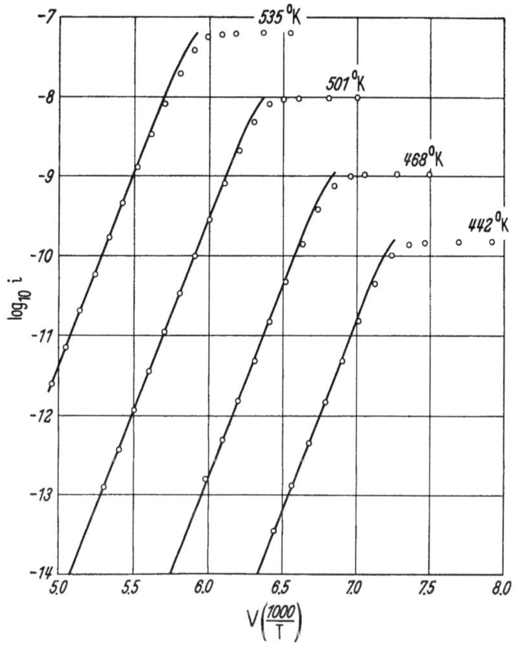

Fig. 47. Current measurements by HUNG as a function of applied potential at four temperatures. Termination of theoretical lines locates zero-field point.

In Fig. 47, experimental results are compared with the solid lines which are the theoretical curves that show the rise in current as a function of applied voltage as computed for the tube geometry used. An idealized FERMI distribution of emitted electrons is assumed (see Sect. 60 and Table 10). The application of the space-charge equations indicates that over the temperature range covered the effect of space charge is negligible even very close to zero-field. HUNG's evidence is very clear that there was a deficiency of electron current at zero field and he interpreted his results on the hypothesis that a lack of uniformity of work-function over the emitting surface was responsible for this deficiency. The detailed theory of multiple reflection as it applies to a diode has been presented in Sect. 62. One of the results of that analysis was that, over a wide range in the retarding potential part of the characteristic, the shape of the curve is unaltered by the reflection effect since every observed point is reduced in current value by exactly the same fraction. Furthermore, this same fractional reduction applies also on the accelerating side [see Eqs. (62.10) and (62.11)]. Although it cannot be taken as a proof of the reflection effect, detailed calculations show that HUNG's experimental points are in excellent agreement with the curve computed by the multiple reflection equations and it is not necessary to make an ad-hoc assumption concerning patchiness to explain his results. Patch structure has a far more important influence on the emission observed with strong accelerating fields. Zero-field observations made at 563° and 610° K show still more deviation from theory which is contrary to prediction from patch theory but this result would be expected because of the space charge created by the extra multiply reflected electrons. Further evidence for this effect will be given in Sect. 83.

If the reflection hypothesis is assumed, then the procedure used by HUNG for establishing the applied potential required to create zero field in the space between the emitter and the collector is precisely the one that should be used. His table of values of the contact potential as a function of the temperature serves as a direct observation of the temperature change of the work-function. These results are shown in Fig. 48.

Since the change in measured current as a function of applied voltage is so rapid as the field changes from retarding to accelerating, there can be little doubt about the choice of the emission value as a function of the temperature that characterized his specimen. HUNG plotted his data in accordance with the FOWLER emission equation derived in Sect. 64 and given there as Eq. (64.22). It was shown in that section that FOWLER attributes the entire change in the work-function with temperature as being due to variation of the FERMI level and assumes implicitly that the electron affinity W_a changes so little with temperature that its contribution may be neglected. Until precise measurements are made with special attention directed toward the experimental determination of the temperature variation of W_a, it may be assumed that the electron affinity decreases very slightly as the temperature increases, but that the amount of this decrease is offset by the very considerable increase in the work-function because of the temperature variation of the FERMI level.

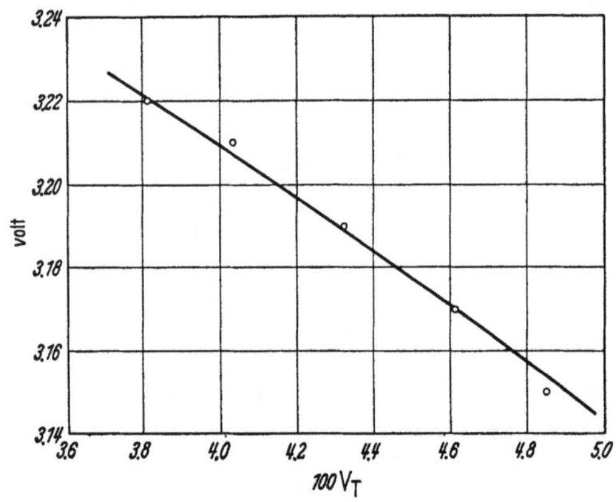

Fig. 48. Contact difference in potential as observed by HUNG as a function of the temperature. Solid line is the theoretical curve for donor density n_D of 3×10^{21} per m.^{3} at a level 0.6 ev below the conduction band.

In the analysis that follows, the basic experimental data depend entirely on the results of HUNG's study reported in much greater detail in his thesis than in the published paper. None of this analysis was made by HUNG because he did not accept the fact that his data are in good agreement with the reflection hypothesis. The reader will find that the analysis presented here demands for its excellent agreement with theory the recognition of Eq. (62.11) as a suitable means of correcting the observed current to compute the value I_{00} which is the current density which would have been observed at zero field in the absence of any deficiency of slow electrons which Sects. 72 and 75 have shown.

The curves of Fig. 47 show that the measured emission current is very accurately represented by the theoretical curves applicable to this geometry and characterized by the temperature for each curve as measured by a calibrated thermocouple. Consistent with the theories presented in Sect. 62, the applied potential for zero field and the zero-field emission can be established with high accuracy in spite of the deviation from the theoretical non-reflection curve. The curve computed by the theory of multiple reflection does pass through the experimental points, but since the remaining part of this analysis depends only on the knowledge of the zero-field emission and the measured contact difference in potential as well as its variation with temperature, the needed data are all illustrated by Fig. 47. HUNG has reported five contact potential measurements instead of just the four shown in this figure, and zero-field emission current measurements at seven temperatures. It will be these values that are used in the rest of this analysis.

The five observed contact potential measurements of Fig. 48 are joined together by the theoretical curve which depends for its location in this figure on the emitter work-function value of 1.225 ev. This value is obtained from an analysis of the emission data according to the method described in detail later in this section. The line also depends on the choice of a collector work-function at a temperature somewhat above room temperature of 4.629 ev. The choice of donor state concentration and its average energy level with respect to the bottom of the conduction band will be discussed.

The observations of zero-field current made by HUNG at seven different temperatures were tabulated in terms of the corresponding current densities and then, with the help of Eq. (62.11), the corresponding value of current I_{00} was calculated for each temperature. These values represent the current density which would have been observed in the absence of any reflection effect and it is this current density which is appropriate for analysis according to the FOWLER equation given as Eq. (64.25). The plot of these points is shown as Fig. 49. All of the points lie on such an excellent straight line that its slope can be determined with high accuracy. The theory which led to Eq. (64.25) established the fact that this slope yields an accurate measure of the *true* work-function at absolute zero. That is the advantage of the FOWLER equation compared with the RICHARDSON T^2 equation which, it will be shown, does not yield the *true* work-function at absolute zero, but gives one determined by a linear extrapolation which would be valid only in case the true work-function were a linear function of the temperature over the entire range. The values obtained from this experimental study are recorded in equation form as follows:

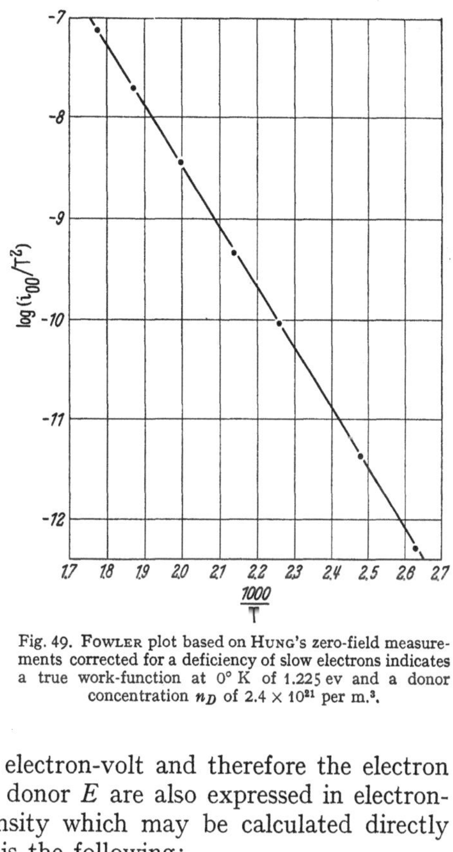

Fig. 49. FOWLER plot based on HUNG's zero-field measurements corrected for a deficiency of slow electrons indicates a true work-function at 0° K of 1.225 ev and a donor concentration n_D of 2.4 × 10²¹ per m.³.

$$\varphi_F = W_a - \frac{E}{2} = 1.225 \text{ ev}, \qquad (81.1)$$

$$1.726 \times 10^{-5}\, n_D^{\frac{1}{2}} = (8.5 \pm 0.2) \times 10^5. \qquad (81.2)$$

Note that the unit in Eq. (81.1) is the electron-volt and therefore the electron affinity W_a and the energy level of the donor E are also expressed in electron-volts instead of joules. The donor density which may be calculated directly from the result expressed in Eq. (81.2) is the following:

$$n_D = (2.4 \pm 0.4) \times 10^{21} \text{ per m.}^3. \qquad (81.3)$$

This analysis yields a direct measure of the donor density which may be used in association with the theories presented in Sect. 64 to compute the location of the FERMI level at each temperature. The change in FERMI level with temperature with reference to its value at absolute zero may be computed by Eq. (64.11) written as follows:

$$\varDelta\mu = \mu - \frac{E}{2} = \frac{E}{2} - V_T \ln 2 + V_T \ln\left\{\left[1 + 8\,\mathrm{e}^{-\frac{u'+E}{V_T}}\right]^{\frac{1}{2}} - 1\right\}. \qquad (81.4)$$

Although this equation gives the correct change in the FERMI level over the entire range of temperature even beyond that for which the FOWLER equation gives a perfect linear log-plot, the approximation used in deriving his equation may be applied and the function suitable for accurate computation over the low temperature range is the following:

$$\Delta\mu = \mu - \frac{E}{2} = \frac{V_T}{2}\left[\ln 2 - \ln\frac{1.208\times 10^{28}}{n_D} + \frac{3}{2}\ln(V_T^{-1})\right]. \tag{81.5}$$

Since these equations were used for the computation of Table 11, the data found there for a donor density of 3×10^{21} per m.³ and a choice of E of -0.6 ev yielded directly the theoretical curve represented by the solid line of Fig. 50. This line represents the true work-function as a function of the temperature starting at its measured value of 1.225 ev at absolute zero and carrying through to the range of observed contact difference in potential. The points reproduced in this figure come from Fig. 48. Since the data used for this calculation came directly from Table 11 and were not recomputed for the slightly smaller donor density of 2.4×10^{21}, the fit to the data is not quite as good as it would have been if a new curve had been computed with the donor energy level taken at -0.7 ev. These data therefore place the donor energy level at (0.7 ± 0.1)ev. This value, combined with the experimental value of the work-function at absolute zero as given in Eq. (81.1) indicates an electron affinity W_a of 0.875 ev.

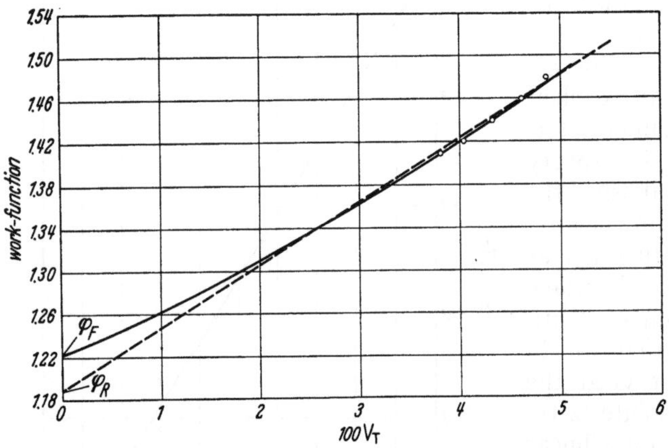

Fig. 50. True work-function as a function of the temperature with φ_F determined by plot of Fig. 49 and experimental points determined by HUNG's observation of the temperature variation of the contact potential. Solid line, the theoretical curve for $n_D = 3\times 10^{21}$ and $E = -0.6$ ev. RICHARDSON plot, Fig. 51. Average temperature coefficient of the work-function over observational range determined by dashed straight line.

Even though this analysis according to the FOWLER equation gives all of the information as a result of the very skillful and direct observation of the electron emission properties of the specimen of an oxide cathode studied by HUNG, it is important to show additional relations by carrying through the analysis according to the RICHARDSON T^2 equation. The same basic data for the emission current density I_{00}, which would have been observed in the absence of reflection, were used to form the plot shown as Fig. 51. The experimental points fall on an excellent straight line since it was demonstrated in Fig. 50 that over the range of observation the temperature variation of the work-function is dominated by the linear term and the average slope is 5.8.

The direct analysis of Fig. 51 gives the following results:

$$\varphi_R = 1.19\,\text{ev}, \tag{81.6}$$

$$A_R = A_H\,e^{\left(\frac{d\varphi}{dV_T}\right)} = 3390. \tag{81.7}$$

Note that the straight line shown in Fig. 50 starting at the RICHARDSON work-function value of 1.19 and drawn with a slope of 5.8 is in excellent agreement with HUNG's observed data on the contact potential changes of his specimen. With this relation, $(d\varphi/dV_T) = 5.8$, the value of the thermionic constant expressed here as A_H to indicate that it applied specifically to HUNG's data is thus determined experimentally as follows:

$$A_H = 112 \times 10^4 \text{ amp/m.}^2\text{-degree}^2. \tag{81.8}$$

This value is in such excellent agreement with the theoretical value of A which is 120×10^4 amp/m.²-degree² that there is very little room for improvement in the application of theory to experiment as it applies to these data.

This demonstration of the application of theory to experimental data should remove the doubt often expressed as to whether or not the RICHARDSON equation or the FOWLER equation should be used for the analysis of emission data as it applies to semiconductors. The answer is that both equations apply perfectly well and give different answers only in that the fundamental theories ask different questions. Thus the difference in the work-functions expressed as Eqs. (81.1) and (81.6) does not in any sense represent an error in their determination since the φ_R represents the *effective* work-function at absolute zero on the basis that a linear extrapolation from that point will give a satisfactory representation of the true work-function in the operating range. Clearly, the most convenient formula for the expression of the true work-function in the operating range as applied to this specimen is based on the RICHARDSON value and given as follows:

$$\varphi = 1.19 + 5.8\,V_T. \tag{81.9}$$

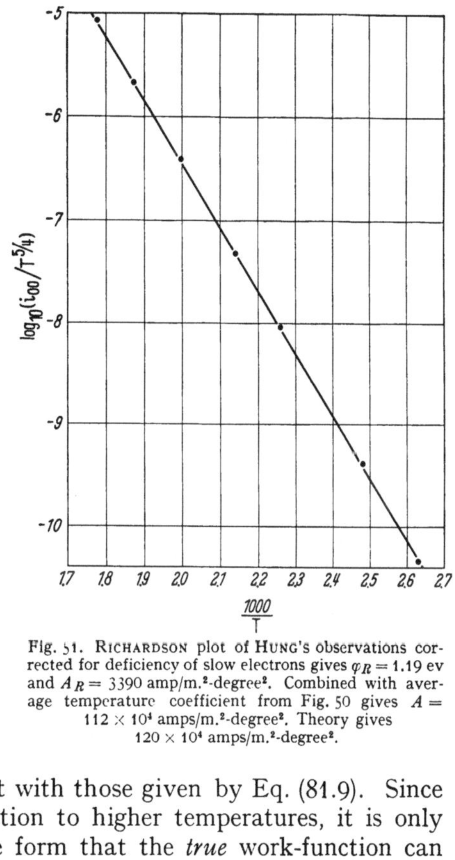

Fig. 51. RICHARDSON plot of HUNG's observations corrected for deficiency of slow electrons gives $\varphi_R = 1.19$ ev and $A_R = 3390$ amp/m.²-degree². Combined with average temperature coefficient from Fig. 50 gives $A = 112 \times 10^4$ amps/m.²-degree². Theory gives 120×10^4 amps/m.²-degree².

The values of the true work-function, calculated from its basic definition ($\varphi = W_a - \mu$) by means of Eq. (64.11) and the experimentally determined value of W_a of 0.875, are in excellent agreement with those given by Eq. (81.9). Since Eq. (81.9) cannot be used for extrapolation to higher temperatures, it is only by means of Eq. (64.11) in its complete form that the *true* work-function can be computed for the whole range in temperature. With the *true* work-function known the emission current in the high range can be computed from the low range data.

Too often in the literature one finds the statement: "Thermionic emission measurements are not capable of deciding the *correct* power of T to be used in emission formulae". Such a statement confuses the issue since it is here demonstrated that it is not a matter of the correctness of the choice of the power of T for the representation but the choice must be made entirely in terms of the information desired. If the emitter is an N-type semiconductor and the range of investigation is that for which the FOWLER equation applies, then its use is appropriate and the basic information obtained rewarding. Both forms are needed for the correct analysis and use of data and *neither* should be used for the empirical representation of data since Eq. (9.1) is far more useful for the purpose.

82. Indirect methods of cathode evaluation including transconductance changes and its theory. Industrial laboratories have need for methods of cathode evaluation that are reliable and simple even though often qualitative, since their tests must apply to production tubes that serve a variety of functions. The tube types include not only simple diodes but complex multigrid structures. With very few exceptions, these practical tubes depend on the flow of electrons across the space-charge minimum for their performance to be within specification limits. Sect. 55 through 59 and especially Sect. 58 develop the application of space-charge theory to the observable emission properties of a cathode to show that no important observable differences will occur as the result of major differences in the available supply of electrons as long as the conditions of test do not reduce the retarding effect of the potential minimum created by space charge near the cathode to a value less than V_T as expressed in electron volts or kT as expressed in joules. The fundamental rule therefore the for evaluation of emission capability is that the test must involve, in one manner or another, an observation which depends upon the tube operation with the space-charge potential minimum *less* than V_T. Two specific test procedures are described here, although there must be many others that will satisfy this condition. The first will relate to the testing of diodes and the second to the testing of grid-controlled tubes.

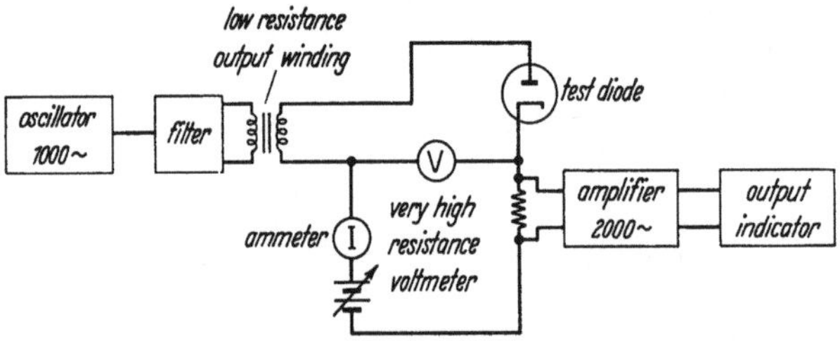

Fig. 52. Block diagram for circuit used for cathode evaluation based on the generation of the second harmonic of 1000 cycles in the test diode.

The diode test depends upon the fact that a circuit element, that has both a nonlinear and nonsymmetrical current voltage characteristic with respect to a chosen operating condition, is a generator of harmonics including even multiples of a pure sine-wave input frequency. A block diagram of the test method is indicated in Fig. 52. The diode under test can be biased to deliver an arbitrarily chosen emission current to the collector that is within the normal operating range for the diode with the current limited by space charge. The circuit provides for a modulation of this bias by a sine wave of controllable amplitude and *high purity*. Since the current-voltage relation in the diode is nonlinear and may be represented by a power series expansion, there will be a second harmonic component of current which can be amplified by means of a narrow band amplifier and the output observed on an oscilloscope or else rectified to produce a direct current proportional to the amplitude of the *second* harmonic. This signal may be observed with a meter or used to actuate a recorder. For a pre-set bias voltage, observations may be made as the temperature of the emitter is lowered. The second harmonic will decrease very slowly until the emission capability of the cathode approaches that critical value at which the space-charge minimum is equal to or less than V_T. Under this condition, the voltage-current relation becomes much more linear and the harmonic output decreases to a minimum and at the same time its phase changes. This dip in the harmonic output serves as a quick index of the fact that the space-charge minimum has decreased to the order of V_T. The heating current required to maintain this condition is a practical measure that can be related to predetermined standards. The time required for the cathode to drop in temperature after the heating power has been removed may also serve as a measure of the critical temperature. The lower the temperature at which this minimum in second harmonic content occurs, the higher the emission capability of the cathode.

An alternate way of conducting this test requires the setting of the cathode temperature at some arbitrarily chosen value. The applied voltage v of Fig. 52 is increased gradually until the second harmonic output indicator shows a minimum. The observer may record both the critical applied voltage and the corresponding space current as well as the temperature of the emitter. This test serves as an excellent method of determining a current which is very close to the zero-field current capability of the emitter at the specified temperature. Such a test can be made as nearly automatic as desired to reduce the time required for it to a very small fraction of a second after temperature equilibrium has been achieved. With a standardized tube structure and the application of the theories presented here, tests made at two or more temperatures can be analyzed to yield quantitative data concerning many detailed properties of the emitter. Tests made at a single temperature can be used very effectively for production control.

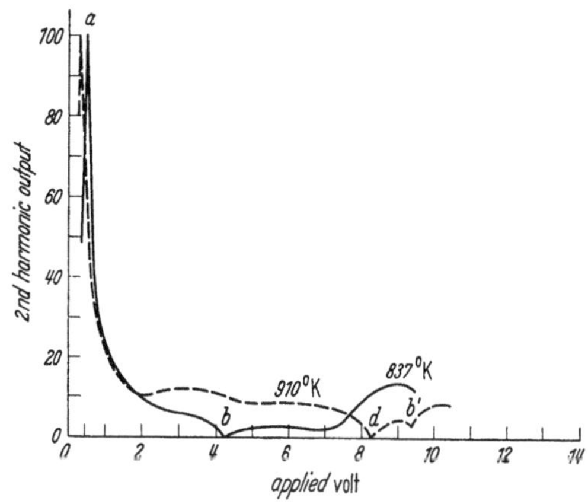

Fig. 53. Harmonic output from test diode as a function of applied volts for two temperatures. Solid line, 837° K; dashed line 910° K. Onset of space charge at collector shown at A; zero-field at emitter shown at B. Minimum at D results "high-speed 10-volt effect". Zero-field at emitter at B'.

Results that may be obtained by the use of this method of second harmonic generation can be illustrated best by the presentation of specific laboratory data. The emitter was an indirectly heated cylinder concentric with respect to a collector with a spacing of 1 mm. and a ratio of radii of 2.5. The central 12 mm. section of the emitter was coated with a standard barium-strontium oxide and on each end beyond this coating the nickel sleeve extended an additional 10 mm. The diode had been processed according to standard methods and had been operated at a temperature of 1160° K for one thousand hours with a space-charge limited current of 55 ma. Additional point-by-point data which show the voltage current properties of this diode will be presented in Sect. 83. Fig. 53 shows the second harmonic output as a function of the applied voltage for two different temperatures. These temperatures were determined by thermocouple measurements of the emitter.

With the emitter maintained at 837° K, the second harmonic was measured as a function of the applied voltage and the gain of the analyzer was adjusted to give an output of 100 units at the maximum located at 0.45 volts (see a of Fig. 53). This maximum was an accurate measurement of the applied potential at which zero gradient occurred at the collector and thus represented the onset of space-charge limitation as discussed in detail in Sect. 57. As the applied potential was made more positive, the harmonic output decreased approximately as the inverse square root of this change in potential. Over this range, the emission current increased with the applied voltage in accordance with the theoretical curves of Figs. 16 and 17. The sharp minimum (b) shown at 4.2 volts was a direct indication of the fact that the space-charge minimum had moved across the diode and was practically coincident with the emitter. The emission current observed with this applied potential was an excellent measure of the "zerofield" thermionic emission from this cathode at this temperature. The change in potential of 3.75 volts thus observed yields additional information concerning the properties of the emitter.

As the temperature was lowered, the minimum shifted to lower values of applied potential and conversely as the temperature was raised, it moved to higher values. At each temperature, the current flow at the corresponding critical applied potential served as an excellent basis for the determination of the emission properties.

Observed data are shown for the emitter operating at 910° K and two minima are evident, with one at 8.28 volts and the other at 9.3. Detailed study showed that the 9.3 minimum is the one that characterized the emission property of the cathode and the 8.28 minimum was a clear indication of the electron energy required to develop the space-charge distortion effect known as the "high-speed 10-volt effect" described in Sect. 80. Detailed studies not yet completed indicate that at higher electron energies of about 20 to 30 volts, additional disturbances in the space-charge situation caused by secondaries from the collector appear. Fig. 54 is presented to show both the sensitivity of this method of analysis and phenomena which observers of thermionic emission must understand in order not to report misleading data. Note first the two prominent minima (d and f) at 7.6 and 9.7 volts applied potential. These are the direct result of the "high-speed 10-volt effect" and it will be noted that the location of the first with respect to the condition of zero-field at the collector occurs at exactly the same potential difference of 8 volts as that observed at the lower temperature and shown in Fig. 53. The feature of additional interest and importance is the small sharp minimum (c) at zero. Detailed study showed that this minimum was a direct result of the thermionic emission obtained at this temperature (1160° K) from the bare nickel parts of the emitter outside of the central section which was coated. At this temperature and this applied voltage and for all applied voltages more negative than this, the "bare" nickel part of the emitter, because of its larger area, contributed more current to the collector than the central portion. These data are presented in this brief form to stress the need for test diodes designed with guard rings or diodes that use some other method for insuring that the observed emission originates at that part of the emitter which is being investigated.

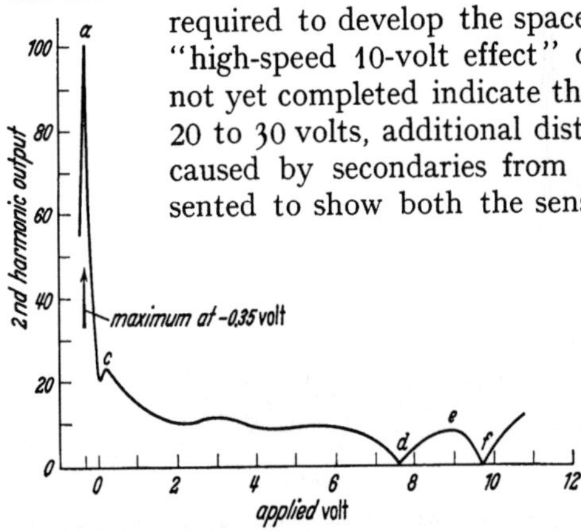

Fig. 54. Harmonic output of test diode at 1160° K. Onset of space charge with zero-field at collector at A. Influence of emission from uncoated nickel at C. Structure due to "high-speed 10 volt effect" at D, E, and F.

Most grid-controlled vacuum tubes have an electron collector usually called its "plate" which receives a current from the emitter depending upon the potential applied to the control grid. Since the control grid itself receives a negligible part of the emission current, practically all the current which passes the space-charge minimum is received on one or more of the electron collectors. The following relation is basic to the theory concerning the space-charge control of emission discussed in detail in Sect. 57:

$$I = I_0 \, \mathrm{e}^{-\varphi_{s\,u}} = I_0 \, \mathrm{e}^{-\frac{V_s}{V_T}} . \tag{82.1}$$

In this equation, I_0 is the zero-field emission current density capability of the emitter at the temperature V_T. The potential in the immediate neighbourhood of the emitter surface is V_s positive with respect to the potential minimum located a short distance from the emitter. The dimensionless expression for this quantity was used almost exclusively in Sects. 57 and 58 and is given as ψ_{su}. It may be assumed that the change in plate current of interest is either equal to dI or directly proportional to it for all small variations of V_s. The variation in plate current may be expressed by the following equation:

$$\frac{dI}{dV_s} = -\frac{I_0\,\mathrm{e}^{-\frac{V_s}{V_T}}}{V_T}. \tag{82.2}$$

If the potentials of other electrodes in the tube are held constant, then *small* variations in the grid potential will make a directly proportional change in the potential of the space-charge minimum. Since the grid potential is usually expressed with respect to the emitter, the making of the grid more positive results in a decrease in the value of V_s and thus accounts for the negative sign used in the following equation:

$$d\,V_s = -\,m_g\,d\,V_g. \tag{82.3}$$

Here, V_g is the applied potential of the grid with respect to the emitter and m_g is the proportionality constant always less than unity and greater than zero. In Eq. (40.3) it was shown that the value of V_s could be represented with considerable accuracy by the following very simple formula:

$$V_s = \Phi\,\frac{V_T - V_\Theta}{V_\Theta}. \tag{82.4}$$

In this equation, Φ is the work-factor; the temperature expressed in its electron volt equivalent is V_T; and the temperature for which the space-charge minimum is exactly zero for a fixed set of applied conditions is given by V_Θ. The substitution of these equations into Eq. (82.2) gives the following result when it is remembered that definition of I_0 is given by Eq. (39.2):

$$g_m = \frac{dI}{dV_g}\,A = m_g\,\frac{a\,\mathrm{e}^{-\frac{\Phi}{V_\Theta}}}{V_T}\,A. \tag{82.5}$$

Since the definition of a transconductance (or mutual-conductance) g_m is the variation in plate current per unit variation in grid potential expressed as $(dI/dV_g)\,A$, this equation shows that the transconductance for an emitter of area A and an ideal structure would show a very small increase as the temperature is lowered from its normal operating value to one closer to the critical value V_Θ. This result is not observed in a practical triode because there is no single value of V_Θ applicable to the entire effective area of the emitter.

Since an equation of the form of Eq. (82.5) expresses some of the most basic features of transconductance principles and since no theory including them is found in the textbooks on the theory of vacuum tubes, an illustration of its quantitative usefulness will be given in this paragraph. A plane surface emitter of 0.5 cm.² may be designed into a troide having an emitter to grid spacing of 5×10^{-3} cm. The grid structure is such that its effective capacity at the emitter plane is 6% of the value that it would have if it were a parallel plane instead of a grid structure. With the grid operated sufficiently negative, so that it receives no appreciable part of the emission current the plate of the troide may be sufficiently positive so that the current (i) delivered from emitter to the plate is 25×10^{-3} amps. The minimum temperature V_Θ of a well-activated oxide cathode that can meet the current demand may be computed by the following relation:

$$i = A\,a\,\mathrm{e}^{-\frac{\Phi}{V_\Theta}}. \tag{82.6}$$

A typical emitter will have $a = 3 \times 10^{10}$ amps per m.^{2} (or 3×10^6 amps/cm.^{2}) and $\Phi = 1.26$ volt. Computation shows that the critical value V_Θ is 7.03×10^{-2} or $\Theta = 816°$ K. The following equation may be used to compute the minimum temperature for satisfactory operation for a given emitter as expressed in terms of the critical temperature (V_Θ) and the work-factor:

$$V_T' = V_\Theta \left(1 + \frac{V_\Theta}{\Phi - V_\Theta}\right). \tag{82.7}$$

The substitution of numbers yields the value 7.45×10^{-2} ev or 865° K. No important change in the transconductance will be observed in a triode operated as described if the temperature of an emitter having these specifications is raised above this value V_T'. The next step for the calculation of m_g of Eq. (82.5) demands a knowledge of the distance from the emitter to the space-charge minimum. For all temperatures above V_T' Eq. (46.11) may be used for calculating the distance to the space-charge minimum. This equation, arranged for direct application is the following:

$$x_s = 3.109 \times 10^{-3} \frac{V_T^{\frac{3}{4}}}{I^{\frac{1}{2}}} = 3.109 \times 10^{-3} \frac{V_T^{\frac{3}{4}}}{(i/A)^{\frac{1}{2}}}. \tag{82.8}$$

Let the distance from the control grid to the emitter be x_g and let the capacitance which one can calculate from the known area and this distance for parallel planes be defined as C_x. By a suitably chosen set of measurements the true capacitance between the grid and the emitter may be determined experimentally. Let its value be C_g. The ratio of these two capacitances (C_g/C_x) is a measure of the effectiveness of the grid in controlling a charge distribution in the immediate neighborhood of the emitter. The fact that the space-charge minimum is at a fractional distance (x_s/x_g) from the emitter makes the modulation of the space-charge minimum by the grid possible. These considerations combine to give an expression for m_g of Eq. (82.5), which is:

$$m_g = \frac{x_s}{x_g} \cdot \frac{C_g}{C_x}. \tag{82.9}$$

These equations may be used to yield the following formula for transconductance:

$$g_m = 3.109 \times 10^{-3} \left[\frac{C_g}{x_g C_x}\right] \cdot \frac{(i/A)^{\frac{1}{2}}}{V_T^{\frac{1}{4}}}. \tag{82.10}$$

In this equation the factor in the square bracket depends on the geometrical configuration of the triode. The current i is the actual emission current which is grid-controlled. Since Eq. (82.10) applies specifically to a structure of parallel planes, the capacitance C_x may be expressed by the following equation:

$$C_x = \frac{\varepsilon_0 A}{x_g}. \tag{82.11}$$

The proportionality factor ε_0 is the permitivity of free space. This equation substituted into Eq. (82.10) yields the following:

$$g_m = 3.109 \times 10^{-3} \frac{C_g}{\varepsilon_0} \frac{(i/A)^{\frac{1}{2}}}{V_T^{\frac{1}{4}}} = 3.511 \times 10^8 C_g \frac{(i/A)^{\frac{1}{2}}}{V_T^{\frac{1}{4}}}. \tag{82.12}$$

In this equation C_g is the true capacitance of that part of the grid which modulates the space-charge minimum and therefore is much less than the input capacitance of an actual triode since the latter includes the coupling of not only the grid wires with respect to all other parts of the structure, but the supports and the lead-in wires as well. A close-spaced triode of high transconductance has been constructed with the following specifications:

$i = 25 \times 10^{-3}$ amps; $A = 0.5 \times 10^{-4}$ m.^{2}; $C_g = 0.5 \times 10^{-12}$ farad (Estimated)
$x_g = 5 \times 10^{-5}$ m.; $V_T = 0.1$ ev.

With these numbers inserted into Eq. (82.12) the computed transconductance g_m is 7000 microamps per volt which agrees well with laboratory measurements. The quantity least accurately known is C_g, which was estimated from the known input capacitance rating for this triode of 2×10^{-12} farad.

This numerical example has been included here to illustrate the usefulness of the theories of thermionic emission in the presence of space charge which have been presented. It also illustrates the fact that, as the temperature is reduced from its normal value to one that is *between* V_Θ and V_T' as determined by Eq. (82.7), the distance x_s which could be calculated accurately by Eq. (82.8) for the higher temperatures, shrinks very rapidly to zero. Note that this shrinking of x_s reduces m_g to zero as indicated in Eq. (82.9). Since the basic equation for transconductance is Eq. (82.5), it is evident that in this critical range in temperature, transconductance falls rapidly to zero because of the m_g factor.

Although an equivalent of the above theory of transconductance is not widely known judging by the books on tube design, the engineers in the tube industry use a test for emitter quality control which is related to it.

The better the emitter, as measured in terms of the lowering of the work-factor Φ or the increase in the constant a, the lower the critical temperature V_Θ for a given set of operating conditions. These conditions are determined by the potentials applied to all elements of the tube except the heater, which supplies power to maintain the temperature of the emitter. It is not customary to make a direct determination of V_Θ but an indirect measure of it comes by the making of a quantitative determination of the time which elapses between the cutting off of the normal power and the time that the transconductance begins to fall very rapidly. A second method is to permit the heating power to be decreased slowly and determine the minimum value at which the tube under test will maintain a specified fraction of its normal transconductance. A third and less quantitative test involves the measurement of the transconductance at the normal operating temperature and a remeasurement of g_m with a 10% reduction in heater voltage. This method may be used to eliminate very poor emitters but will not distinguish between exceptionally good emitters and only fair ones if the less active surfaces are still able to maintain the potential minimum of the amount V_T over the major part of the emitter surfaces even at the reduced heater power.

83. Voltage-current characteristics with a detailed evaluation of emission according to space-charge theory. Precision measurements that yield quantitative data on electron emission from oxide cathodes have been discussed in detail in Sect. 81. In order to avoid space-charge effects, it was necessary to hold the temperature to less than 600° K. This temperature is so low compared with the normal operating temperature of 1160° K that experiments were undertaken to test the applicability of the space-charge equations as well as the emission equations covering a wide range of temperature for diodes of very simple construction.

The "ideal" diode for scientific study can be designed and built with a few main features which will be mentioned in the following tabulation.

1. The emitter and the collector should be parallel planes.
2. The temperature of each should be measurable by means of thermocouples.

3. A convenient design for the emitter could be a small cylindrical cap, a few millimeters in diameter similar in general construction to that used by the Bell Telephone Laboratories and described in Sect. 84. The part of the collector to which the measured electrons flow should be a pure nickel disc of small diameter guarded by a surrounding sheet of pure nickel to form the parallel plane structure and yet limit the measured emission current to the central part of the emitter.

4. The collector should be capable of being heated to any temperature at least equal to that of the emitter if desired.

5. For fundamental studies, especially of the cathode nickel, all of the parts of the test diode except the emitter should be of the purest nickel available.

6. Leakage current to the collector should be eliminated so that electron currents as small as 10^{-14} amps may be measured with accuracy.

Since the Bell Telephone Laboratories design is the nearest to this ideal, details concerning it will be given in Sect. 84 even though no measurements made with this diode are available for analysis.

The diode on which the test measurements reported in this section were carried out was one produced by the Raytheon Manufacturing Company[1]. It was built with pure nickel parts except for the emitter. The essential dimensions are given in the following tabulation:

Table 83.1.

	Length		Coated band		Outside diameter		Inside diameter	
Emitter. . . .	27 mm.	1.06 in.	12 mm.	0.47 in.	1.15	0.045 in.		
Coated					1.35	0.053 in.		
After exhaust .					1.30	0.051 in.		
Collector . . .	19 mm.	0.75 in.					3.35 mm.	0.132 in.

Amer. Soc. Testing Materials 2, 883 (1952) Designation B 270-52T.

Although a number of specimens built according to this design were investigated in sufficient detail to show that the theoretical analysis developed in the previous sections applied, the data to be presented in this section were taken with great care on one of these diodes which was known from previous measurements to be typical. Before the test was made, it had been operated at the normal operating temperature of 1160° K for 1000 hours with sufficient positive collector potential to give an emission current of about 50 ma which corresponded to a current density of approximately 0.1 amp per cm.^{2}. Throughout the entire set of measurements the emitter temperatures were set and monitored by thermocouple measurements since the emitter had welded to its interior surface a fine tungsten wire. The properties of the tungsten-to-nickel junction for the particular alloy used as the emitter had been studied to obtain the calibration. During a test, the heater power input was maintained constant to better than 0.1%. Voltages were set with potentiometer accuracy and the currents measured with calibrated meters which gave reproducible results to 1%. The emission properties of the cathode were perfectly stabilized and the total range of applied voltage never exceeded 6 volts. All currents were accurately reproducible independent of the temperature sequence used for the investigation. In order to maintain

[1] Raytheon Manufacturing Company, Newton, Massachusetts, U.S.A.

emitter stability during a particular run, a 4 volt applied potential to the collector was maintained at all times, except for the short interval of time required for each individual reading. There were no instabilities that indicated that this precaution was really necessary. After a change in temperature, the electron emission took up its final value as quickly as the temperature could be adjusted. Thereafter, it remained perfectly constant. Because of this reproducibility, it is thought that the physical structure, including the concentration and distribution of donors, remained sufficiently constant over the entire range of time and temperature required for the experiment, so that the analysis, carried out on the basis that all measurements apply to the emitter in a particular state of activation, is justified.

Detailed measurements were made at 9 different temperatures ranging from 689 to 1160° K. Currents were measured as a function of applied voltage both for retarding and accelerating potentials. Over the low temperature range the 6-volt applied potential was sufficient to move the space-charge minimum to the emitter surface and with the help of the curves shown as Figs. 16 and 17, it was possible to determine the zero field emission current with accuracy. On the basis of these data the electron concentration in the immediate neighborhood of the emitter could be computed for the higher temperatures with sufficient accuracy for use in the quantitative determination of the contact difference in potential over the entire temperature range investigated. This evaluation will be illustrated by Fig. 56 and will be discussed in more detail later in this section. The important result is the quantitative determination of the *donor density* in the neighborhood of the emitting surfaces and the determination of the electronic *energy level* associated with the donor. This final step in the analysis comes by the application of the FOWLER and also the RICHARDSON thermionic equations to determine the electron affinity of the emitter and compare the overall results with thermionic emission theory. This paragraph serves to outline the analytic procedure to be followed.

For the diode structure described above, space-charge influences the measured current as it develops to give zero gradient at the collector at about 6×10^{-6} amps for the lowest temperature and about 20×10^{-6} amps for the highest temperature. Since the leakage current was less than 10^{-9} amps, the current flow could be studied as a function of the applied potential over the retarding field range to cover a current ratio of nearly 10^4 without the interference of space charge. Three of the 9 curves have been chosen for Fig. 55 since they serve not only to illustrate the method of analysis but also the complications that arise which must be understood in order not to misinterpret laboratory data.

Observations made with the emitter at 760° K serve as the basis for the plotted points of curve I. The solid line is the theoretical curve which should be expected for this geometry and this temperature. It will be noted that the plotted points follow this line very accurately for currents below 1 microampere. The two points designated *a* and *b* fall below the theoretical line as the result of an influence of space charge at a smaller observed current value than would have been expected if all electrons that impinge on the collector were absorbed by it. This result is a direct indication that electron reflection is taking place at the collector and must be given consideration in the detailed analysis of the observations.

As the applied potential is made more positive than 0.6 volt, the current rise follows that anticipated from space-charge theory for this geometry and temperature as the space-charge minimum moves across the diode and is located at the emitter surface when the measured current is 220×10^{-6} amps. This method of analysis is illustrated by Fig. 28.

In order to show the relation between the analysis of the point-by-point observations of the current as a function of the voltage compared to the second harmonic method, an additional curve has been drawn in on this figure. This curve shown by the dotted line is a measure of the second harmonic generated in this diode as a function of the applied potential and corresponds to the detailed exploration over the top of the sharp peak shown in Figs. 53 and 54 at *a*. Although

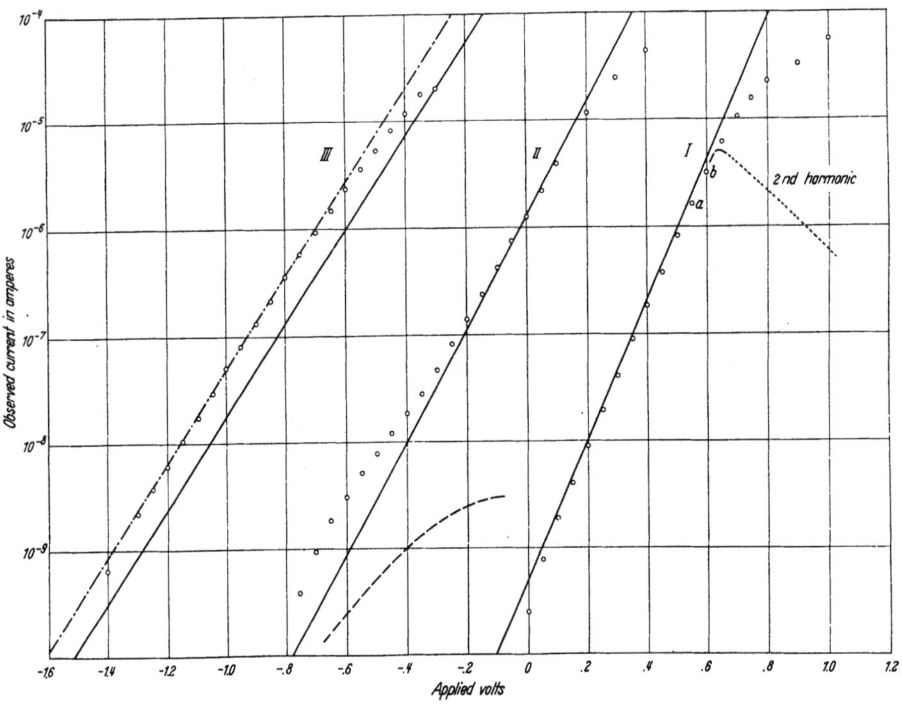

Fig. 55. Retarding potential curves for three temperatures. I. 760° K, dotted line shows second harmonic analysis. II. 956° K, dashed line, emission from bare nickel. III. 1160° K, solid line, emission from coating. Dot-dash line, total emission including bare nickel.

in Fig. 55 the deviation from the solid line was interpreted as the onset of space-charge, it is very clear from the second harmonic curve that the second derivative of the current curve is changing very rapidly at this point.

The data points associated with curve II were observed at a temperature of 956° K. A superficial analysis of the plotted points would indicate that the electron temperature was much higher than the thermocouple temperature. The theoretical curve that represents the true variation in the emission from the oxide coating is again represented by the solid line which is drawn at the theoretical slope for this temperature and geometry. The fact that so many points lie above the predicted line is a direct indication that a contribution to the observed current of major importance is coming from the uncoated areas of the nickel sleeve. This uncoated area seems to have a work-function almost exactly the same as the collector. These work-functions are probably 2.5 and 2.6 respectively. The zero field emission from the uncoated nickel is approximately 3×10^{-9} amps at this temperature. The influence of space-charge is clearly evident for the reading at 0.2 volt under which condition the current flowing is 12×10^{-6} amps.

Again the rise in current follows the curve given by space-charge theory as the space-charge minimum moves across from the collector to the emitter. At this temperature the measured current increases over 1000 fold and the data presented in Table 7 indicate that zero field emission would be observed at 12 to 13 volts applied potential. This prediction was later verified by experiment with the second harmonic method of analysis described in Sect. 82. The complication introduced into the analysis by the existence of the "high-speed 10-volt effect" interfered with the exact determination of the applied voltage required to force the potential minimum to coincide with the emitter.

Curve III which represents observations at 1160° K yields a result which would seem to indicate that the emission changed with applied potential exactly as one would expect for a MAXWELLian distribution characterized by this temperature. The line that represents this distribution is shown as dot-dash. The result which is not so obvious is that the observed distribution follows this line because there is adequate emission from the entire emitter which includes both the coated part and the uncoated part and therefore at each point the currents are larger by about a factor of three than the currents which legitimately belong to the emission properties of the coated part of the emitter. The solid line drawn parallel to the dot-dash line represents the emission from the oxide cathode after the total current has been corrected to subtract out the spurious current. Direct point-by-point observations and also the harmonic analysis illustrated by Fig. 54 shows that the emission from the uncoated area saturates at approximately zero potential. As more and more positive potential is applied, the current is predominantly that emitted from the oxide cathode and it follows very accurately the master curve described by the data of Table 8 after proper consideration is given over the low voltage range to the fact that the observations were made in a cylindrical diode instead of a plane parallel diode. An applied potential of approximately 180 volts would have brought the potential minimum over to the emitter to yield a direct observation of the zero-field emission of this emitter. Under these conditions, the collector would have been required to dissipate about 70 watts. Pulse equipment, not available at the time these measurements were made, would be required to verify this prediction.

On the basis of these analyses, the results presented quantitatively in Table 83.2 summarize the necessary information by means of which an evaluation of the emission is possible.

Table 83.2.

Heater volts	T° K	100 V_T	V_T^{-1}	s_R	$i_R \times 10^6$	i_0 [1]	u_0^2	z^2	Obs. [2] I_m	Cal. [2] I_m	ψ_{SR}	$s_R+\psi_{SR}$	P_T	I_{00}
2.1	689	5.938	16.84	14.9	6.8	30×10^{-6}	4.41	0.52	0.261	0.229	1.8	16.7	0.993	2.35
2.3	729	6.281	15.92	13.6	8.0	110×10^{-6}	13.75	0.718	0.223	0.249	2.7	16.0	1.003	8.0
2.5	760	6.549	15.27	11.4	9.5	220×10^{-6}	23.1	0.780	0.243	0.265	3.36	14.76	0.967	20
3.0	837	7.212	13.87	8.2	12.5	1800×10^{-6}	144	0.910	0.275	0.306	5.04	13.24	0.956	133
3.5	897	7.729	12.94	6.0	15.0	$*6.3\times10^{-3}$	427	0.947	0.317	0.340	6.0	12.00	0.927	
4.0	956	8.237	12.14	3.9	17.5	$*20\times10^{-3}$	1143	0.968	0.362	0.374	6.91	10.81	0.894	
4.5	1003	8.642	11.57	2.7	19.1	$*43\times10^{-3}$	2250	0.977	0.391	0.402	7.64	10.34	0.894	
5.0	1051	9.056	11.04	1.3	21.2	$*91\times10^{-3}$	4670	0.984	0.431	0.429	8.21	9.51	0.861	
6.3	1160	10.00	10	−2.0	25.0	$*380\times10^{-3}$	15200	0.990	0.505	0.505	9.42	7.42	0.742	

[1] High temperature values * obtained by extrapolation according to Eq. (9.1) based on temperature range 689 to 837° K.

[2] Calculated on basis that equivalent plane-parallel diode had spacing $W = 0.78\times10^{-3}$ m.; actual spacing was 1.0×10^{-3} m. between cylinders; range in equivalent spacing observed was $(0.8\pm0.02)\ 10^{-3}$ m.

Based on the observed current at zero field for the four lowest temperatures and a detailed analysis of the effect of multiple reflection developed in Sect. 62, the column headed "I_{00}" represents the average current density which these theories indicate would have been observed in the complete absence of reflection at both the emitter and the collector. These data were used in the FOWLER emission equation given as (64.25) to determine the true work-function at absolute zero and the donor concentration that would yield these results. The graphical analysis is represented by Fig. 56. The FOWLER work-function computed from the slope of this curve is expressed as follows:

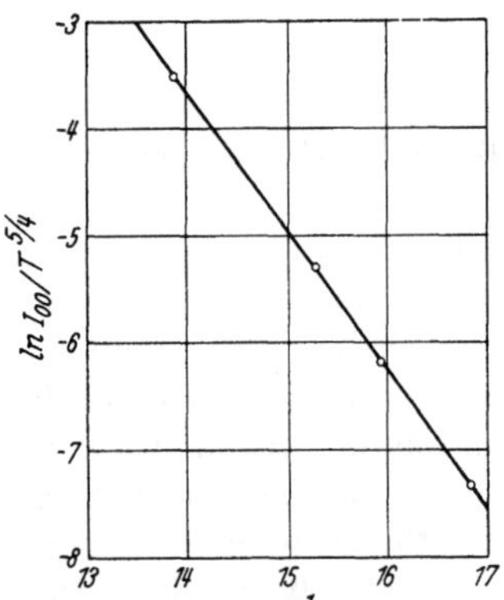

Fig. 56. Plot of emission data according to the FOWLER equation.

$$\varphi_F = W_a - \frac{E}{2} = 1.29\ \text{ev}. \tag{83.1}$$

These data also yield a FOWLER thermionic constant which determines the donor density as follows:

$$n_D = \left[\frac{1.85\times 10^{6}}{1.726\times 10^{-5}}\right]^2 = 1.15\times 10^{22}\ \text{per m.}^3. \tag{83.2}$$

If the electron affinity W_a is taken to be 0.88 ev as indicated by HUNG's experiment discussed in detail in Sect. 81, the energy level of the donor is -0.82 ev with respect to the bottom of the conduction band.

With this information available it is possible to compare the computed temperature variation of the work-function with the contact difference in potential observed. With the true work-function at absolute zero known to be 1.29 ev and the donor concentration and donor level known, it is possible to use the data of Table 11 to obtain the true work-function over the entire range in temperature. This result is shown by the solid line in Fig. 57. The choice of an average collector work-function of 2.58 ev places the observed contact difference in potential found at 956° K exactly on the theoretical true work-function line. If it is assumed that the true work-function of the collector is independent of the temperature of the emitter, then all the other nine points of observed contact difference in potential fall as shown in Fig. 57. Even though the agreement between theory and experiment is not exact, it is considered to be practically within experimental error, because the analysis was complicated by the fact that the diode under test was a cylindrical diode without guard rings, and that the correct introduction of the multiple

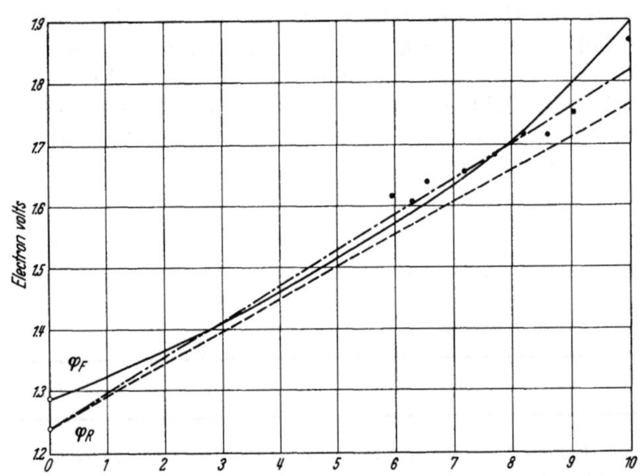

Fig. 57. True work-function with temperature dependence. Open circles, φ_F from FOWLER plot of Fig. 56; φ_R from RICHARDSON plot of Fig. 58. Solid line, theoretical curve computed for n_D of 10^{22} and $E = -0.8$; solid circles, observed contact difference in potential with average collector work-function assumed 2.58 ev; dot-dash line, best straight line through φ_R and experimental points; dotted line, based on choice of temperature coefficient according to RICHARDSON equation.

reflection theory especially as it applies to very low temperature range was not easy to carry through with exactness.

The next step in the analysis involved the application of the RICHARDSON formula to the same data shown in Fig. 56. The RICHARDSON plot is given in Fig. 58. The theories discussed in detail in Sect. 81 show that the slope of this line should give the RICHARDSON work-function which is the one found at absolute zero by linear extrapolation from the observational range. The value obtained for φ_R from this plot is 1.24 ev. This point is plotted in Fig. 57 and identified by φ_R. The straight line drawn from this φ_R point through the data points is shown by a dot-dash line in Fig. 57. The slope of this line is seen to be 5.8 and represents the apparent average temperature coefficient of the work-function in dimensionless units:

Computation based on the data of Fig. 58 gives the RICHARDSON constant A_R a value of 0.63×10^4 amp per m.^{2}-degree^2. If it is assumed that the average temperature coefficient of the work-function can be derived from this RICHARDSON constant and its relation to the theoretical constant A of Eq. (18.6), then the temperature coefficient would be 5.25. The dotted line of Fig. 57 is drawn with this slope. It is evident that if the true work-function of the collector were taken to be 2.54 ev instead of the value chosen of 2.58 ev, the observed points would be in very good agreement with the theoretical line. An inspection of Table 11 as it applies to a donor density of 3×10^{22} shows that a choice of a slightly higher donor density than the one required by Eq. (83.2) would yield slightly better overall self-consistency.

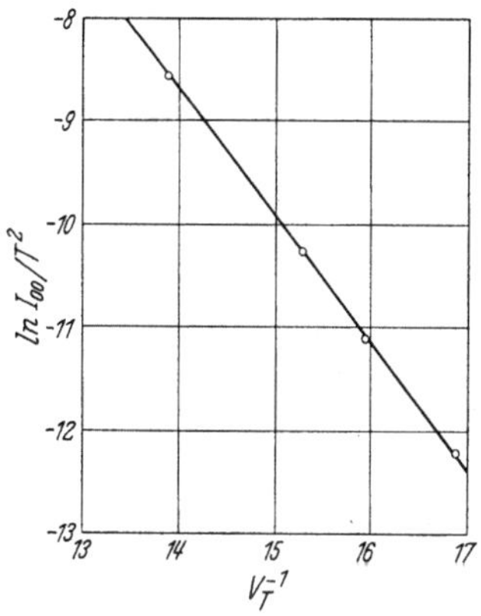

Fig. 58. Plot of emission data according to the RICHARDSON equation.

Since this method of emission analysis has been presented as it was carried through for the first time, the results may be summarized by the statement that the emission properties of this oxide cathode are in excellent agreement with those anticipated by the theories presented here. All factors including reflection effects, correct space-charge theory and analysis of electron flow between concentric cylinders were taken into account. The neglect of any one of these factors would result in a disagreement with theory that would have been outside of the experimental error.

84. Description of Bell Telephone Laboratories M-1949 planar diode. The data presented in Sect. 83 serve to forecast the results which may be obtained by a very careful analysis of voltage, current and temperature relations in diodes having oxide cathodes. The use of a cylindrical structure in contrast to a planar structure is distinctly disadvantageous in that far more effort is required for the analysis of the data. It is therefore important for the advancement of the better understanding of oxide cathodes to design a test structure which is feasible to make and at the same time more adaptable in terms of the theoretical analysis.

The Bell Telephone Laboratories planar diode M-1949 so nearly represents the ideal structure that a description of it is presented here in lieu of the fact that these details are not available in the scientific literature. The following paragraphs are quoted from information supplied by Mr. H. E. KERN of the Bell Telephone Laboratories.

A planar type vacuum tube diode has been designed and built at the Bell Telephone Laboratories for studying the fundamental properties of oxide-coated cathodes. A pictorial view of the diode structure is shown in Fig. 59.

The important structural and material features of this diode are listed below:

1. The cathode is a hat-shaped detail which can be machined from solid rod or formed from sheet material of thicknesses ranging from 0.060 inch to 0.001 or 0.002 inch.

2. The cathode hat is located directly over a hole (not shown) in the top surface of a hollow rectangular cathode heater box and is welded in place along the brim of the cathode hat. The hole in the cathode heater box facilitates removal of gases from beneath the cathode hat.

3. The emission coating is deposited on the top of the cathode hat as indicated by the dotted area. The coated area is 0.05 cm.².

4. The anode is a hollow rectangular box of exactly the same dimensions as the cathode heater box with the exception of the small hole required for the cathode outgassing.

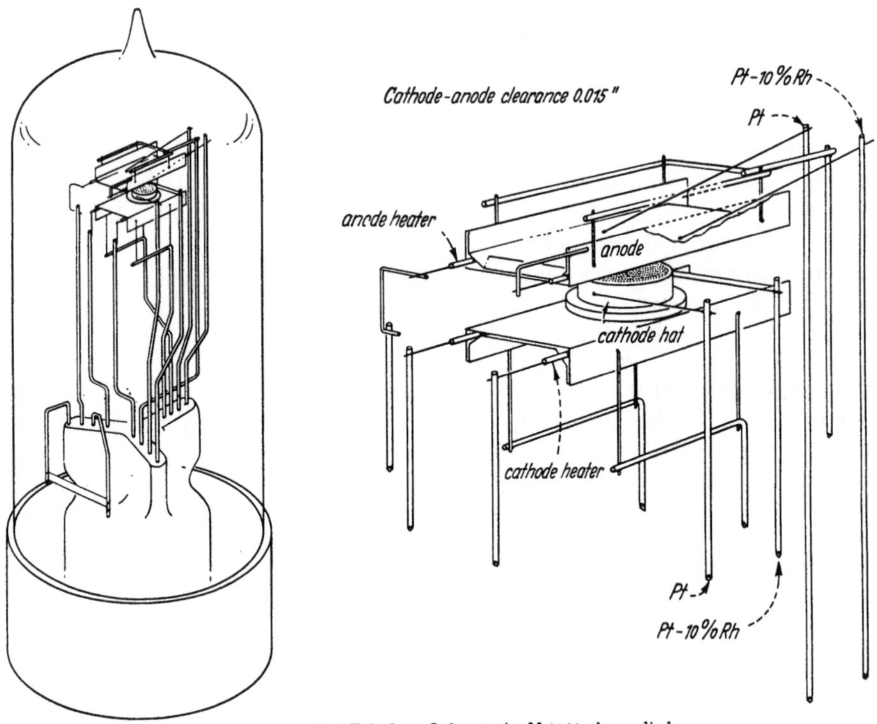

Fig. 59. Bell Telephone Laboratories M-1949 planar diode.

5. A heater is contained in the anode box so that the anode may be heated independently through the same temperature range as the cathode.

6. The anode and cathode heaters are identical and are aluminum-oxide coated-tungsten heaters (RCA-FH-484H).

7. Platinum and platinum 10% rhodium (0.002 inch diameter) thermocouple wires are separately welded to both the anode box and the cathode hat as shown. An average cathode or anode temperature reading is obtained as a result of the physical separation of the 0.002 inch diameter, Pt and Pt-10% Rh thermocouple wires. This average reading is perhaps of greater interest than a temperature measurement at a single point.

8. The 0.002 inch diameter thermocouple wires are welded to 0.020 inch diameter similar Pt and Pt-10% Rh stem into the soft glass press. In this way reference junctions within the glass envelope are avoided and more accurate temperature measurements can be made.

9. The anode-cathode assembly is mounted on a 12-lead soft glass tripod stem. The dimensions of the completed mount are such that the anode-cathode assembly is approximately 2 inches away from the ultimate seal-in area. This positioning minimizes the possibility of oxidation of parts during seal-in.

10. No ceramics or micas (possible sources of contaminating gases) are contained in this diode.

11. All parts with the exception of the cathode hat and the tungsten heaters are made from high purity vacuum melted nickel.

Some of the advantages of the structure described above are:

1. Fundamental properties of the oxide cathode can be determined over a wide range of temperatures with an exact knowledge of the true temperature of the cathode.

2. The anode can be thoroughly outgassed by means of the anode heater before the carbonates on the cathode are decomposed. During this anode heating, the cathode temperature remains less than the maximum temperature (400 to 450° C) it attained during prior glass bake-out. For example, with the anode at 1000° C true temperature, the temperature of the cathode does not exceed 250° C.

3. A complete temperature *vs.* time record for both anode and cathode can be obtained during all process steps for possible correlation with the emission and life history of the tube.

4. Because of the relatively small coating area and close anode to cathode spacing, life tests can be run at *dc* cathode current densities as high as 500 ma/cm.^{2} without causing excessive anode heating.

5. Cathode current density can be varied over a wide range under conditions of constant anode temperature by making use of the anode heater.

6. A more thorough study can be undertaken of anode deactivation phenomena as related to anode materials, anode processing, and anode operating temperature.

85. Relation between emission properties and conduction properties of uniform oxide-coated emitters. The experiments of HANNAY, MACNAIR and WHITE[1] and also YOUNG[2] serve as examples of researches in which emitters of uniform structural properties were investigated to determine the relation between conduction and emission at an arbitrarily chosen temperature for various states of emitter activation. The HANNAY, MACNAIR and WHITE data, observed at 970° K, are best represented by Fig. 40. An examination of this figure yields the ratio of emission current density to conductivity expressed as follows:

$$\frac{I}{\sigma_{p\,\mathrm{II}}} = 4\times 10^3 \text{ amp-ohm/meter}. \tag{85.1}$$

Results tabulated by YOUNG when expressed in the same units range from 3×10^3 to 25×10^3 amp-ohm per meter for values applicable to studies at 1000° K. The higher value applied to YOUNG's best specimen.

The temperature chosen for both of these studies was sufficiently high to make the conduction mechanism through the coating be dominated by pore conduction in contrast to the direct conduction through the crystals. Eq. (65.28) may be adapted to give an expression for the conductivity through the pores as follows:

$$\sigma_{p\,\mathrm{II}} = 2.54\times 10^{14}\, V_T\, e^{-\frac{\varphi}{V_T}}\, C_p\, l_p. \tag{85.2}$$

In this expression, the true work-function is expressed as φ; the temperature in its electron-volt equivalent is V_T; and the constants C_p and l_p are determined by the physical structure of the emitter. In both experiments the emission current density figures used were intended to be as accurate a measure of the zero-field emission as it was possible to obtain without detailed analysis. In neither case were the geometrical configurations suitable for an exact evaluation of this current density. The equation that will best represent the emission current density is of the basic type used for the analysis of zero-field data given in Sects. 81 and 83. This equation comes directly from Eq. (18.5) with the modification, which experiment seems to call for, to account for the deficiency in slow electrons. The need for this factor was demonstrated in Sects. 81 and 83. The equation follows:

$$I = \frac{120\times 10^4\times 1.35\times 10^8\, V_T^2\, e^{-\frac{\varphi}{V_T}}}{1+\frac{0.191}{V_T}}. \tag{85.3}$$

[1] N. B. HANNAY, D. MACNAIR and A. H. WHITE: J. Appl. Phys. **20**, 669 (1949).
[2] J. R. YOUNG: J. Appl. Phys. **23**, 1129 (1952).

In this equation the true work-function is expressed as φ as in Eq. (85.2). These two equations combine to yield the following:

$$\frac{I}{\sigma_{p\,\mathrm{II}}} = \frac{63.8\times 10^{-2}\cdot V_T}{\left(1+\frac{0.191}{V_T}\right) C_p\, l_p}. \tag{85.4}$$

This analysis shows that a comparison between theory and experiment can be valid only under the condition that the true work-function applicable in Eq. (85.2) and Eq. (85.3) be the same. In turn this means that averaged over the entire coating, the electron affinity at the surfaces of the pores and the donor concentration found there must be essentially equal to the corresponding quantities at the emission surface. This demands that operation at a constant temperature without the drawing of emission current should precede any measurements. It is also clear that the introduction of poisoning impurities such as oxygen without a period of coating conditioning can increase the true work-function of the emitting surface without at the same time influencing to the same extent the true work-function found at the surfaces of the pores. The experiments of HANNAY, MACNAIR and WHITE are consistent with this view.

The most important factors in Eq. (85.4) that influence the emission current to conductivity ratio are the geometrical constants C_p and l_p. The first of these, as indicated in Sects. 65 and 77, depends on the detailed nature of the porosity of the specimen. Without this information it is not possible to do more than estimate the value of this constant for a given experimental specimen and for the present purposes the estimated value of C_p will be 0.3. Similarly, techniques for the determination of the most suitable value of l_p have not been developed beyond the estimation stage. It is thought that the range in l_p is likely to be from 2 microns to 8 microns. With the assumption that the product of these two constants is 1.5×10^{-6}, the computation of the ratio expressed by Eq. (85.4) gives a value of 11×10^3 amp-ohm per m. for comparison with experiment. This figure is well within the range of observation made by YOUNG and is less than a factor of 3 larger than that observed by HANNAY, MACNAIR and WHITE as illustrated by Fig. 40.

86. Thermoelectric effects in an oxide coating. The fact that the temperature coefficient of the FERMI level in an oxide cathode is large indicates that the maintenance of a temperature gradient across a coating will result in the production of a measurable difference in potential between suitably placed electrodes. This difference in potential has been observed by YOUNG[1] and found to be directly proportional to the temperature gradient. Although the potential developed depends on the average temperature at which the observation is made, the high temperature electrode is always more positive than the low temperature contact. A typical curve as observed by YOUNG is shown as Fig. 60. The experiment involved the application of a barium-strontium oxide coating to two nickel caps. The coated surfaces were held in contact with each other mechanically, face to face, and each cap could be heated by a suitable internal heater. Thermocouples were used to measure temperatures and temperature differences. Not only could the thermal emf be measured as a function of average temperature, but also the electrical conductivity of the coatings. Fig. 61 presents results obtained by YOUNG for the same specimen as was used for the data of Fig. 60. A comparison between Fig. 61 and Fig. 37 indicates that the LOOSJES and VINK ranges I and II are covered by this study. The very low temperature range therefore is dominated by conductivity in the crystalline substance and the very

[1] J. R. YOUNG: J. Appl. Phys. **23**, 1129 (1952).

high temperature range by conductivity of electrons that build an electron atmosphere in the pores between the crystals. The conductivity data show no indication of important space-charge effects within the pores since range III (see Sects. 77 and 78) is not present. The equations developed in Sects. 64 and 65 may be adapted to the discussion of these thermoelectric effects.

If a temperature gradient is maintained across a single crystal of a substance which has a temperature coefficient of the FERMI level when measured with respect to the bottom of the conduction band, then in the absence of any current flow potentiometer methods permit the measurement of the difference in energy of the FERMI levels at the two points of contact between which the temperature difference exists. In YOUNG's experiment observations made in the very low temperature range permitted a direct determination of the emf per unit temperature difference averaged over the specimen. Since it is presumed that the tempera-

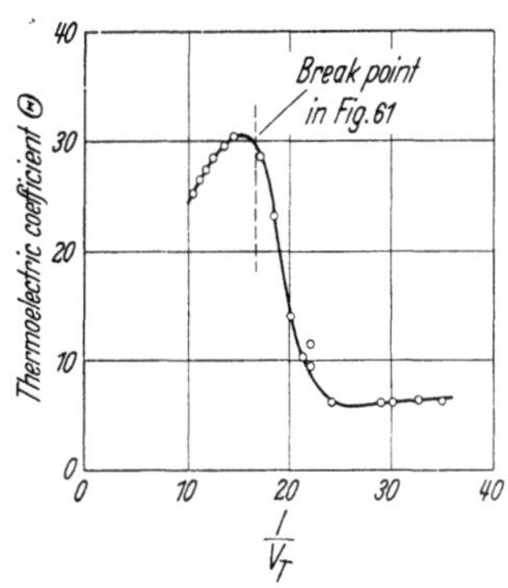

Fig. 60. Thermoelectric coefficient as observed by YOUNG.

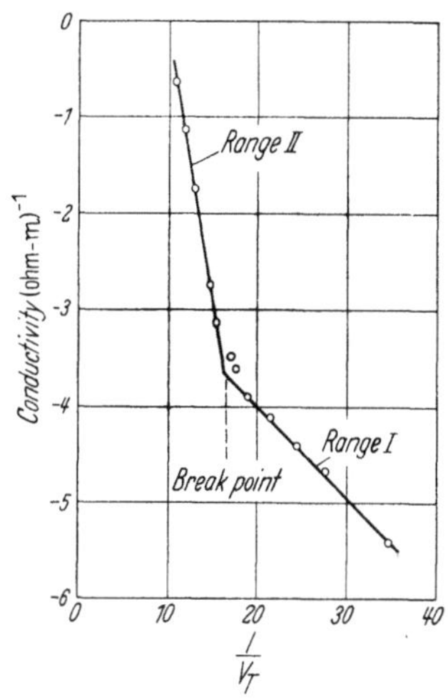

Fig. 61. Conductivity data as observed by YOUNG on same specimen as Fig. 60.

ture difference existed over the same thickness of specimen as that which gave rise to the emf, it was not necessary to know exactly the thickness in order to acquire significant data for comparison with theory.

Eq. (64.3) may be used as the starting point for the derivation of an equation which should relate an observable thermoelectric emf to the temperature gradient maintained across a single crystal of activated barium-strontium oxide. The use of "activated" indicates that the theory is applicable only to a crystal in which there are a sufficient number of donor centers close enough to the bottom of the conduction band so that the location of the FERMI level is dominated by the free electron concentration in the conduction band obtained by transitions from the donors and that contributions to the electron density from the valence band are negligible. In the formulation of the theory it is assumed that over a very small distance within the crystal there is a difference in temperature of (ΔV_T) and that over this same distance there is an observable potential difference (Δv). This observable potential difference would have to be measured by potentiometer methods with the help of a potential indicator of the electrometer type since in the very low temperature range the resistance of a practical specimen as it applies to the oxide cathode may be very high. Any appreciable current flow may interfere with the accurate determination of the thermal emf.

The analysis that follows shows that the observable emf may be thought of as being derived from two phenomena. Since the observable emf represents a displacement of the FERMI level in the high temperature region, relative to the location of the FERMI level in the lower temperature region, the presence of a temperature variation of the FERMI level will contribute a term to the thermal emf. An analysis of Eq. (64.8) as well as purely qualitative considerations show that the concentration of free electrons in the high temperature region will be slightly greater than the concentration of free electrons in the low temperature region. Furthermore, these electrons in the high temperature region will be moving with a greater average velocity as they cross the imaginary boundary between the low temperature and the high temperature region. In order that the time average of the current flowing across this boundary be exactly zero, a true internal field will develop which basically is derived from a small excess of a negative surface charge on the low temperature end of the crystal and a corresponding surface charge of positive electricity at the high temperature end of the crystal. There will also be a distribution of surface charges along the sides of the crystal so that the interior field will be uniform and, in the absence of magnetic fields, the field will be in the same direction as the temperature gradient. This latter contribution to the observable emf will be designated by $(\Delta v_s')$. With these definitions and the use of Eq. (64.3) a current density of the electrons that flow across the boundary from the low to the high temperature regions is given as follows:

$$I = 2\frac{2\pi\, m\, e^3}{h^3}\, V_T^2\, e^{\frac{\mu}{V_T}}. \tag{86.1}$$

Eq. (86.2) expresses the current density in the opposite direction under the steady state condition of *zero net current* across the boundary that separates the high and low temperature regions.

$$I = 2\frac{2\pi\, m\, e^3}{h^3}\,(V_T + \Delta V_T)^2\, e^{\frac{\mu+\Delta\mu}{(V_T+\Delta V_T)}}\, e^{\frac{\Delta v'}{(V_T+\Delta V_T)}}. \tag{86.2}$$

Clearly these equations are valid only in the extreme case of ΔV_T approaching zero. The equations are nevertheless useful since the coefficient of the thermal emf may be defined by the following relation:

$$\vartheta_s = \left(\frac{\Delta v}{\Delta V_T}\right)_{\Delta V_T \to 0} = \left(\frac{\Delta\mu + \Delta v'}{\Delta V_T}\right)_{\Delta V_T \to 0}. \tag{86.3}$$

To solve for the expression of Eq. (86.3), the logarithms of Eq. (86.1) and Eq. (86.2) are set equal to each other and the algebraic solution yields the following result:

$$\vartheta_s = \frac{\Delta v}{\Delta V_T} = -2 + \frac{\mu}{V_T}. \tag{86.4}$$

In the use of this equation, it must be remembered that it applies only for values of μ which are negative, and that the negative sign indicated for the expression as a whole results from the fact that the current carrier is an electron and that the high temperature region is the more positive in the electrostatic sense. In the plot of the theoretical curve as it is compared with YOUNG's experiment, the absolute value of the thermoelectric coefficient will be plotted as a function of $(1/V_T)$.

Note that this expression is identical in basic form to that derived by HENSLEY[1] and does not represent a precisely straight line plot since the location of the FERMI level (μ) is an implicit function of the temperature. Over any short range in temperature the true value of μ may be represented as a linear function of the

[1] E. B. HENSLEY: J. Appl. Phys. 33, 1122 (1952).

temperature but to do so hardly serves any immediate purpose since Table 11 makes available accurately computed values of μ for a wide range of donor densities and donor energy levels.

It was demonstrated in Sects. 17 and 18 that in the absence of space-charge the potential energy of an electron in a pore differs from that of an electron in the bottom of a conduction band by the electron affinity W_a. Exactly the same principles that are illustrated above serve to give the following equation for the thermoelectric coefficient appropriate to a pore across which there is a small temperature gradient. This equation is the following:

$$\vartheta_p = \frac{\Delta v}{\Delta V_T} = -2 - \frac{W_a - \mu}{V_T}. \tag{86.5}$$

The evidence presented in Sects. 81 and 83 indicate that a well-activated oxide cathode might very well have a donor density of 3×10^{22} donors per m.³ at an energy level of approximately -0.6 ev and an electron affinity of 0.9 ev.

The two solid lines of Fig. 62 represent the values of the thermo-electric coefficient as a function of $(1/V_T)$ as computed by Eqs. (86.4) and (86.5)

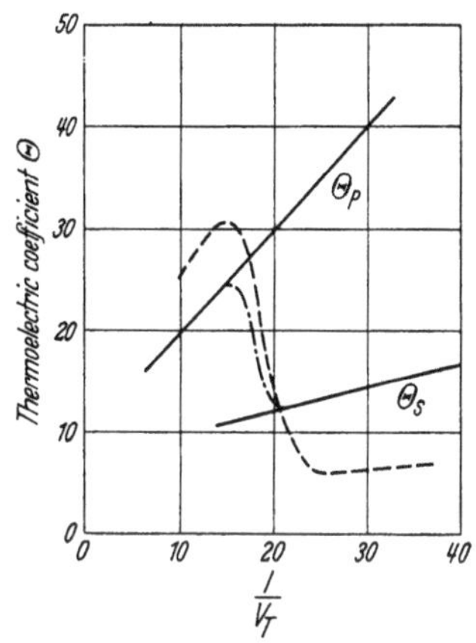

Fig. 62. Comparison between theory and experiment. ϑ_P thermoelectric coefficient for pores; ϑ_s same for crystalline material; dot-dash line, approximate intermediate curve; dotted line, YOUNG's experimental data.

It is an open question as to whether or not either of these computed curves can be expected to represent the results obtained in an experiment similar to that of YOUNG. In the first place over the low temperature region the conduction of heat from the nickel contact at high temperature to the nickel contact at low temperature will be almost entirely that conducted through the multitude of crystals which are sintered together to give contact areas of considerable variability. Of necessity, this results in a nonuniform distribution of temperature gradient, and yet it is not completely unreasonable that for small total temperature differences this nonuniformity will average out satisfactorily. The second difficulty is related to the almost certain lack of similarity between the crystals in direct contact with the nickel surface and those in the interior of the test specimen. Although no obvious interface structure may be present, the thermoelectric behavior of this region might be sufficiently different from the rest of the specimen that it would disturb the results. Finally, in the intermediate temperature range there will be conflict between the two sources of thermal emf since the pores will tend to generate a larger emf than the surrounding solid structure. If the net current averaged over the entire specimen is exactly zero, then in the neighborhood of each pore there will be a circulation of current. This circulation will be in the direction that electrons will flow through the crystalline substance from the low temperature region to the high temperature region, and they will flow in the opposite direction in the pores. The transition region indicated by the dot-dash curve of Fig. 62 will be very dependent on the pore structure and the conduction properties of the specimen under study. No attempt will be made here to derive equations applicable to this intermediate range. The observed results of YOUNG presented in Fig. 60 are reproduced by the dotted line of Fig. 62.

Although qualitative agreement between theory and experiment is evident, the overall picture needs to be clarified. The experiment is not an easy one and therefore it should be duplicated with attention directed toward improvement in the activation of the specimen. The indications are that YOUNG's specimen was not as well activated as would be desirable. Although YOUNG's experiment used potentiometer methods for the measurement of the thermal emf, a more sensitive indicator of higher internal resistance than a conventional galvanometer should be used. In spite of this criticism, the experiments of YOUNG contributed valuable information to support the LOOSJES and VINK model of coating conductivity at a time when this theory was less firmly established than at present.

87. Coating activation and related problems. The crystals that constitute an oxide cathode coating are generally formed following the decomposition of a solid solution of barium-strontium carbonate. Some manufacturers include calcium as well as barium and strontium with a proportion of calcium of approximately 5%. Conversion to the oxide takes place rapidly at temperatures above 1150° K and it is not unusual to use temperatures even above 1450° K for very quick conversion. Since the results of detailed studies are not available, production procedures that yield specific advantageous properties are not well understood, nor is there general agreement concerning some of the most basic factors. Enough of the theory has been presented in the previous sections to indicate that a porous structure can be created to have a higher electrical conductivity than can be expected from a crystalline structure having the same total weight of material and the same density of donors. The desired lowering of the coating resistance depends upon the development of pores of optimum dimensions. If the pore size is too large, the presence of space-charge within the pores limits their effectiveness as conductors and if the pores are too small crystalline material interferes with the free flow of the electrons and the resistance of a cathode may increase.

Porosity also plays a part in the liberation of the decomposition products and with the physical strength of the coating. Coating strength in turn relates to coating adherence which is a problem of great practical importance much in need of scientific investigation.

Although the emission properties of an oxide cathode immediately after its formation are generally good enough to yield current densities far in excess of those that would have been available from the base metal, an activation process is necessary to create a really efficient electron emitter. Much evidence points to the fact that activation results from a creation of an oxygen deficiency of approximately 10 to 20 parts per million averaged over the crystalline structure of the coating. As far as the experiments go, the term "oxygen deficiency" is completely analogous to "barium excess". The difference in concept relates specifically to the location of these donors. Excess barium within the crystal in all probability will find itself in normal barium sites and the oxygen atoms that are missing will leave defects within the body of the crystal which serve as trapping centers for the electrons normally associated with the excess barium atom. This trapping of electrons near negative ion sites is well known as it applies to the alkali halides. In potassium chloride, for example, electrons from the excess potassium atoms are trapped near chlorine vacancies and give rise to an absorption band for visible light in the green. A potassium chloride crystal which would otherwise be perfectly clear and transparent appears colored when viewed in white light because of this absorption band. These trapped electrons are often known as "color centers".

Since the crystal of a barium-strontium oxide cathode has an ionic structure, it is very likely that the donors in an activated cathode are the oxygen vacancies which are capable of delivering one electron per donor to the conduction band by the absorption of energy of approximately 0.7 ev. Direct evidence for this particular transition in an activated oxide cathode is not available although the conduction and emission properties of well-activated cathodes serve as indirect indications of the presence of a donor level at this energy.

The difficulties encountered in the production of the necessary concentration of donors when pure nickel is used as a base metal and great care is exercised to eliminate reducing reactions by the presence of impurities made available from neighboring structures, indicates strongly that the most important chemical reactions responsible for the formation of oxygen vacancies depends on there being present in the base metal reducing agents such as: magnesium, silicon, titanium and manganese[1]. The oxide reduction takes place at the boundary layer between the base metal and the oxide, and if the reaction takes place too rapidly, the excess barium which must be absorbed into proper lattice sites within the crystals in order to be useful may migrate over the surfaces of the crystals and after evaporation from the cathode, deposit on insulators and other structural parts of practical tubes in an objectionable manner. In many cases the compound formed by the oxygen and the reducing agent has a very high resistivity unless this compound itself can hold within it enough donors so that it may be characterized as an "*n*" type semiconductor.

The drawing of an electron emission current during the activation process serves the purpose of creating within the coating itself an electric field which tends to maintain a high concentration gradient of donors within the interior of the coating and inhibit their loss at the exterior surface. With a very small concentration of a very active reducing agent such as magnesium, a high concentration of donors can be created in the first few minutes of the activation schedule and by suitable heat treatment and aging activate the entire coating in a few hours. At least in theory, this quick, initial activation is important from a practical point of view. The ultimate life of an emitter, however, depends on the maintenance of the required donor density throughout the cathode for a very long period of time. The presence of excess donor density in the immediate neighborhood of the vacuum surface of the emitter results in excess barium evaporation. It is therefore necessary to attain a suitable compromise such that the rate of production of excess barium is exactly balanced by its loss either as a result of evaporation or as a result of the neutralization of the oxygen vacancies by the absorption of oxygen or other electronegative elements brought to the surface from the external environment. Constructional and vacuum techniques determine the availability of these poisoning agents. It follows therefore that the achievement of truly long-life reliable cathodes depends on the maintenance of a suitable rate of production of oxygen vacancies over a long period of time and the minimizing of their loss.

Closely associated with this problem of activation is the stability of the emitter under high current drain conditions. Only a few of the many phenomena that depend on current drain can be mentioned here. It has been established by WAYMOUTH[2] that in many examples of moderately high silicon content nickel

[1] E. S. RITTNER: Philips Res. Rep. **8**, 184 (1953).
[2] J. F. WAYMOUTH: J. Appl. Phys. **22**, 80 (1951).

the interface of barium ortho-silicate ($2\,BaO \cdot SiO_2$) will have a low resistance through many hours of emitter life if an electron current is maintained continuously to keep an electric field applied at all times of cathode operation. Such a field inhibits the temperature diffusion of donors out of the interface compound. Without this flow of current, a high internal resistance may develop which could very well interfere with the practical usefulness of such an emitter. The base metal used for emitters in applications for which intermittent service is demanded should contain a minimum of silicon and have substituted in its place reducing agents less prone to develop high resistance in the absence of current flow.

Attention may be directed to the main structure of the oxide cathode, and since one realizes that at the normal operating temperature there is considerable mobility of the atoms that make up such crystals, there is a strong tendency for the donors (probably oxygen vacancies) to become more or less uniformly distributed throughout the interior of each of the crystals if no electron current is flowing. The maintenance of emission current results in an internal electric field within the crystals which interferes with the randomness of temperature diffusion and results in a redistribution of donors with a concentration gradient which in effect balances the electric field. Thus, the higher the emission current, the higher the field and the higher the concentration gradient. In the extreme case of high pulse-emission, the concentration gradient can be so great that the surface crystals increase in work-function so much that a drop in emission may take place. If the emission is too low to maintain a space-charge potential minimum at the surface of the emitter of kT or approximately 0.1 volt for a practical cathode, such an emitter in a vacuum tube may not be sufficiently space-charge limited to maintain in the desired modulation properties.

This increase in work-function is basically the result of the increase in the temperature coefficient of the true work-function as the donor density decreases. The rapidity with which the work-function and its temperature coefficient changes with the reduction in donor concentration is well illustrated by the data in Table 11.

With well-activated cathodes, emission current densities of approximately 5 amps per $cm.^2$ may be obtained for reasonable periods of time without prohibitive reduction in donor density at the emission surface. Unless the cathode is well activated and the vacuum environment free from electronegative elements, excessive field develops near the superficial surface which in turn aggravates the situation in that the stronger the field, the greater the difficulty in maintaining a sufficient concentration near the surface.

Although this concept of donor migration in the presence of field dates back at least to the early work of BECKER[1], the most complete recent treatment is that of NERGAARD[2].

These problems of donor concentration, donor distribution and donor mobility in the presence of fields are likely to be the subject of many researches in the near future since these phenomena combined with the control of porosity determine the important emission properties of oxide cathodes.

88. Dispenser cathodes. Dispenser cathodes have been developed to meet the demand for high emission current density under environmental conditions

---

[1] J. A. BECKER: Phys. Rev. **34**, 1323 (1929).
[2] L. S. NERGAARD: RCA Rev. **13**, 464 (1952).

that would often be considered unfavorable to high emission efficiency. One of the earliest dispenser cathode structures was that proposed by HULL[1]. In this example the emitting surface was molybdenum which received the activating material by evaporation from a closely woven "stocking" of molybdenum wire which contained as a filler a fuzed barya-alumina eutectic (70% BaO, 30% Al_2O_3 by weight). Barium oxide is evaporated from the dispenser and condensed on the molybdenum surfaces presumably with a sufficient concentration of excess barium to maintain a well-activated emission surface for thousands of hours. The thickness of the coatings used was never established by direct experiment, but the properties were so similar to those of a conventional oxide cathode that it may be presumed that a sufficiently thick layer of barium oxide existed to create a multi-layer emitter surface with semiconducting properties.

The development of the "*L*" cathode by LEMMENS, JANSEN and LOOSJES[2] took advantage of a technique by which a porous tungsten plug could be inserted into a molybdenum cup into which had been placed barium-strontium carbonate which was thus sealed into position behind the plug. High temperature operation permitted the conversion of the carbonate to the oxide and the removal of the gaseous products of this reaction. The barium oxide and barium diffuses through the porous tungsten and activates the surface.

HUGHES and COPPOLA[3] have described still another version of the dispenser cathode. In the preparation of this emitter the principle active constituent is barium carbonate which is fuzed with a small amount of calcium carbonate and aluminum oxide (Al_2O_3). After fusion, this material is ground to a fine powder and sintered under high pressure with a powdered refractory material. Typical of this matrix material is the tungsten 25%, molybdenum 75% alloy. The favorable properties resulting from the introduction of calcium were first described by LEVI[4].

Not unrelated to these emitters are those described by MACNAIR, LYNCH and HANNAY[5]. In this example, a plug is formed on a pressed nickel base and contains particles of nickel pressed in special dies with barium-strontium carbonate particles. These molded cathodes, after suitable activation, have an emission efficiency very comparable with well-activated sprayed oxide cathodes.

Still another dispenser cathode has been described by BALAS, DEMPSEY, and REXER[6]. In this case, nickel powder is pressed and sintered to form a cathode matrix. This matrix is filled by water soluble compounds which finally precipitate the barium-strontium carbonate needed for the eventual conversion to barium-strontium oxide. The material contained within the matrix migrates to the surface and activates it. Observations indicate that these emitters have high efficiency and are generally far more able to deliver continuously high current density. They also withstand the temporary admission of oxygen to their surfaces.

Most of the observations with dispenser cathodes are interpreted on the assumption that the surface should be classified as "simple composite". It

---

[1] A. W. HULL: Phys. Rev. **56**, 86 (1939).
[2] H. J. LEMMENS, M. J. JANSEN and R. LOOSJES: Philips Techn. Rev. **11**, 341 (1950).
[3] R. C. HUGHES and P. P. COPPOLA: Phys. Rev. **85**, 388 (1952). — J. Appl. Phys. **23**, 1261 (1952). — Proc. Inst. Radio Engrs. **44**, 351 (1956).
[4] R. LEVI: J. Appl. Phys. **26**, 639 (1955).
[5] D. MACNAIR, R. T. LYNCH and N. B. HANNAY: J. Appl. Phys. **24**, 1335 (1953).
[6] W. BALAS, J. DEMPSEY and E. F. REXER: J. Appl. Phys. **26**, 1163 (1955).

is assumed that a mono-molecular layer of barium oxide reduces the true work-function of an otherwise clean base metal with the resulting high emission efficiency. It should be considered an open question as to whether or not this interpretation is entirely correct since in some cases oxidation of the base metal is a necessary step in the production of an efficient emitter. If this oxide is some 10 to 20 atom layers thick and can be converted into a suitable "*n*" type semiconductor by the presence of the barium oxide, then the concepts associated with the "complex composite" surface may be more appropriate for the explanation of the emission properties.

In order to understand the physics of the emission process for these dispenser cathodes, research data not available in the literature are needed. The general discussion of Sect. 69 indicates broadly the nature of the data wanted. Specifically one may ask the question: can a polarizable monolayer of molecules reduce the work-function as much as is observed? The alternative is that the dispenser cathode should be described as a complex composite surface, and with this statement, it is implied that the surface film is thick enough to have taken on many of the properties of an "*n*" type semiconductor. If this second picture represents the facts, then high emission efficiency results from a very great reduction in the electron affinity and by having present within the complex layer a high concentration of donors. The temperature coefficient of the FERMI level can then be maintained small enough so that a satisfactory true work-function at operating temperature is obtained.

It is not easy to outline the researches which would give an unambiguous answer as to the nature of the surface film on a dispenser cathode. Practically all of the techniques that have been described in Sects. 73 through 83 may be used to yield electrical data that serve to describe the emission properties of these cathodes. Direct measurements of the true work-function over a wide range of temperature should be most helpful. A determination of the temperature coefficient of the work-function as a function of the temperature could very well be useful for the identification of the emission mechanism. Still additional properties may be discovered by the use of a dispenser cathode as a receiver of electrons. In that manner additional direct determinations may be made of the variation of the true work-function with temperature and with activation over a very wide range of these variables. After the electrical properties of various specific types of dispenser cathodes have been determined, they could be compared to the properties of ribbons of the same base metal upon which barium-strontium oxide had been deposited from exterior sources much after the manner of the work of MOORE and ALLISON[1]. They found that a condensation of approximately 10 to 20 molecular layers of barium oxide was required to convert the properties of a tungsten ribbon to those very analogous to a conventional oxide-coated cathode. It seems unlikely that such a large number of layers can be formed by diffusion and molecular migration from the interior of a dispenser cathode. Yet even though the surface may not be covered with that number of layers, if the extent of the coverage is appreciably in excess of a monolayer, the emission properties may be dominated by phenomena associated with the complex layer structure in contrast to the simple composite type. If oxidation of the base metal beyond a monolayer is a fundamental requirement for high efficiency then theories applicable to complex-composite surfaces should apply.

---

[1] G. E. MOORE and H. W. ALLISON: Phys. Rev. **77**, 246 (1950).

89. Concluding remarks and acknowledgements. This treatise has been prepared with the objective that it might serve as a suitable source of basic physics as it relates to thermionic emission. The introduction of many cross-references and the presentation of a glossary of symbols are devices which it is hoped will assist the reader in the acquisition of information in the particular phases of this subject of thermionic emission in which he is most interested. In spite of its length, it touches only lightly on many branches of the subject while at the same time it attempts to discuss with thoroughness, and it is hoped with rigor some of the more important phases of the subject.

Criticism may be made because so much of the writing has been allotted to the analysis and discussion of *space-charge problems*. The justification for this emphasis in an article on thermionic emission is that a thorough mastering of this subject and its intelligent use is needed for the evaluation of thermionic emission and its application in engineering. It is hoped that this better understanding of space-charge theory in its relation to thermionic emission will ultimately assist in the improvement of practical vacuum tubes not only in their performance characteristics but also in their reliability both on a short time and a long time basis.

For similar reasons the *oxide cathode* well deserves the very best effort that any author writing on the subject of thermionic emission can give to it. Millions upon millions of tubes that depend for their usefulness on oxide cathodes have been produced and for some considerable time in the future they will still be needed. It is hoped, therefore, that this treatise will serve as a guide in the formulation of specific means by which the basic problems can be attacked and ultimately solved. The acquisition of new and more reliable information may very well show that the theoretical analysis presented here is not adequate since it does depend on simplifying assumptions. The reader will note that the main effort has been to try to bring together in a simplified manner both experimental and theoretical studies basic to the analysis of thermionic emission from oxide cathodes as it is understood at this time.

Apologies are in order for the very superficial handling of such an important subject as the dispenser cathode. If more time were available for the preparation of this treatise there is no doubt that existing experimental data could be organized and presented in a far better manner so as to establish more clearly the basic phenomena involved.

A number of my colleagues, acquaintances and assistants have aided immeasurably in the production of this treatise. Specifically I wish to acknowledge the help given me by Mr. JAMES CARDELL and many of his associates at the Raytheon Manufacturing Company whose studies on the properties of cathode nickel have been most helpful. In addition, this group has made available innumerable test diodes on which studies of cathode material could be undertaken.

Among my students and former students who have helped by their suggestions and criticisms, I wish to mention in particular ANDREW R. HUTSON, WILLIAM J. LANGE and HAYWOOD SHELTON. Without the help of the computing division of the Research Laboratory of Electronics under the directorship of Miss ELIZABETH J. CAMPBELL it would have been impossible for me to have prepared the tables, many of which are being made available in this treatise for the first time. It is hoped that this choice of tabular information and its method of presentation will prove useful. Dr. HELEN L. THOMAS, also of the Research Laboratory of Electronics, assisted in many phases of the editing and the production. Finally I wish to acknowledge the great assistance given me in the preparation of this manuscript by my assistant, LAWRENCE E. SPRAGUE, who prepared practically all of the drawings. Thanks and credit are due EVELINE VAN BERKUM who helped me with much of the original draft and editorial work. My appreciation has been earned by Miss ANNE F. CRIMMINGS who prepared all of the final manuscript and helped most effectively in many aspects of the final production.

Table 1. *Density of an electron atmosphere in an enclosed space. (See Sect. 24 for details.)* n_s = electron concentration just outside of surface. n_{0m} = maximum possible density at center of cavity [Eq. (24.4)]. n_0 = true density at center of cavity. $z^2 = n_0/n_{0m}$ [see Eqs. (24.2) and (24.11)]. $V_T = T/11600$ = electron volt equivalent of temperature. V_s = true potential difference in volts between surface and center of cavity.

z	z^2	n_s/n_{0m}	n_s/n_0	V_s/V_T
0.05	25×10^{-4}	2.515×10^{-3}	1.00619	0.00617
0.10	1×10^{-4}	10.25×10^{-3}	1.02508	0.02477
0.15	225×10^{-4}	23.79×10^{-3}	1.05764	0.05604
0.20	400×10^{-4}	44.22×10^{-3}	1.10561	0.10040
0.25	625×10^{-4}	73.22×10^{-3}	1.17158	0.15835
0.30	900×10^{-4}	113.4×10^{-3}	1.25957	0.23077
0.35	1225×10^{-4}	0.1685	1.37555	0.31885
0.40	16×10^{-2}	0.24445	1.52781	0.42383
0.45	20.25×10^{-2}	0.3502	1.72957	0.54787
0.50	25×10^{-2}	0.5000	2.0	0.69315
0.55	30.25×10^{-2}	0.7171	2.3707	0.86318
0.60	36×10^{-2}	1.042	2.8946	1.06281
0.65	42.25×10^{-2}	1.5475	3.6627	1.29821
0.70	49×10^{-2}	2.378	4.8527	1.57954
0.75	56.25×10^{-2}	3.841	6.8287	1.92114
0.80	64.0×10^{-2}	6.701	10.4699	2.3485
0.82	6.724×10^{-1}	8.636	12.8431	2.5528
0.84	7.056×10^{-1}	11.41	16.1729	2.7833
0.86	7.396×10^{-1}	15.54	21.017	3.0453
0.88	7.744×10^{-1}	22.06	28.481	3.3492
0.90	8.100×10^{-1}	33.10	40.861	3.7102
0.91	8.281×10^{-1}	41.71	50.363	3.9193
0.92	8.464×10^{-1}	53.88	63.654	4.1535
0.93	8.649×10^{-1}	71.83	83.050	4.4194
0.94	8.836×10^{-1}	99.77	112.917	4.7267
0.95	9.025×10^{-1}	146.6	162.462	5.0904
0.96	9.216×10^{-1}	233.7	253.60	5.5358
0.97	9.409×10^{-1}	424.0	450.59	6.1104
0.98	9.604×10^{-1}	973.4	1013.53	6.9212
0.99	9.801×10^{-1}	3968.2	4048.7	8.3062

Table 2. *Values of $F(S, kT/\omega)$ for various values of S and kT/ω [see Eq. (26.12)].*

kT/ω \ S	$S\to0$	0.5	1	1.5	2	3	4	5	6	8	10
0	1.0	8.01^{-1}	5.72^{-1}	3.91^{-1}	26.1^{-2}	11.2^{-2}	4.6^{-2}	18.6^{-3}	7.38^{-3}	11.3^{-4}	17^{-5}
0.06	0.943	7.47	5.25	3.51	23.1	9.5	3.78	14.3	5.62	8.4	11.8
0.12	0.893	6.98	4.83	3.18	20.5	8.14	3.12	11.7	4.35	6.05	7.9
0.20	0.833	6.41	4.31	2.81	17.7	6.75	2.45	8.85	3.20	3.99	5.00
0.40	0.714	5.30	3.42	2.09	12.6	4.39	1.47	5.00	1.75	1.96	2.33
0.60	0.625	4.47	2.75	1.62	9.12	2.69	0.974	3.21	1.00	1.12	1.25
0.80	0.556	3.86	2.28	1.29	7.23	2.25	0.70	2.29	0.66	0.751	0.873
1.0	0.500	3.38	1.90	1.05	5.13	1.73	0.538	1.62	0.502	0.578	0.672
1.2	0.455	2.95	1.63	0.872	5.00	1.41	0.401	1.28	0.450	0.461	0.552
1.4	0.417	2.62	1.40	0.739	4.00	1.19	0.331	1.06	0.40	0.386	0.470
1.6	0.385	2.34	1.23		3.37						
2.0	0.337	1.92	0.959		2.61	0.718					
3.0	0.250	1.29	0.611		1.54	0.453	0.143				
5.0	0.167	0.735	0.325		0.879						

Note 1: The row for $kT/\omega = 0$ is $G(S)$ since the first term of Eq. (26.12) reduces to $2S^{\frac{1}{2}}e^{-S}/\sqrt{\pi}$ in the limit. A more complete table of values is given in Table 10.

Note 2: The superscript at the top of each column indicates the power of ten by which the numbers in that column should be multiplied.

Table 3E (Emitter Space). *Numerical solution*[1] *to the* LANGMUIR *equation for epace charge in the Emitter Space* [*Eq. (36.1)*].

ψ_s	χ_s	χ_s/χ_m	$(\chi_s/\chi_m)^2$	φ_s	χ_s	χ_s/χ_m	$(\chi_s/\chi_m)^2$
0.005	0.09864	0.05462	0.002983	1.3	1.242	0.6877	0.4731
0.006	0.1080	0.05980	0.003575	1.35	1.258	0.6966	0.4853
0.007	0.1165	0.06451	0.004160				
0.008	0.1244	0.06888	0.004746	1.4	1.273	0.7049	0.4970
0.009	0.1317	0.07292	0.005316	1.45	1.288	0.7132	0.5086
				1.5	1.302	0.7209	0.5197
0.010	0.1387	0.07680	0.005899	1.6	1.330	0.7364	0.5424
0.011	0.1454	0.08051	0.006481	1.7	1.356	0.7508	0.5638
0.012	0.1517	0.08400	0.007055				
0.013	0.1578	0.08738	0.007634	1.8	1.380	0.7641	0.5837
0.014	0.1636	0.09059	0.008204	1.9	1.402	0.7763	0.6028
0.015	0.1692	0.09369	0.008778	2.0	1.423	0.7879	0.6208
				2.1	1.444	0.7996	0.6392
0.02	0.1947	0.10781	0.01162	2.2	1.462	0.8095	0.6552
0.025	0.2169	0.1201	0.01442				
0.03	0.2369	0.1312	0.01721	2.3	1.480	0.8195	0.6714
0.04	0.2721	0.1507	0.02270	2.4	1.497	0.8289	0.6871
0.045	0.2879	0.1594	0.02541	2.5	1.513	0.8378	0.7018
				2.6	1.527	0.8455	0.7150
0.05	0.3027	0.1676	0.02809	2.7	1.541	0.8533	0.7282
0.06	0.3302	0.1828	0.03342	2.8	1.555	0.8610	0.7413
0.07	0.3553	0.1967	0.03869	2.9	1.568	0.8682	0.7539
0.08	0.3783	0.2095	0.04387				
0.09	0.3999	0.2214	0.04902	3.0	1.580	0.8749	0.7652
				3.1	1.590	0.8804	0.7751
0.1	0.4201	0.2326	0.05411	3.2	1.602	0.8870	0.7867
0.15	0.5070	0.2807	0.07879	3.3	1.611	0.8920	0.7956
0.2	0.5777	0.3199	0.10231	3.4	1.621	0.8976	0.8057
0.25	0.6385	0.3535	0.1250	3.5	1.631	0.9031	0.8155
0.3	0.6923	0.3833	0.1469	3.6	1.639	0.9075	0.8235
0.35	0.7396	0.4095	0.1677	3.8	1.655	0.9164	0.8398
0.4	0.7835	0.4338	0.1882	4.0	1.670	0.9247	0.8551
0.45	0.8238	0.4561	0.2081	4.2	1.683	0.9319	0.8683
0.5	0.8605	0.4765	0.2270	4.4	1.694	0.9380	0.8799
0.55	0.8952	0.4957	0.2457	4.6	1.705	0.9441	0.8913
0.6	0.9277	0.5137	0.2639	4.8	1.715	0.9496	0.9017
				5.0	1.724	0.9546	0.9112
0.65	0.9581	0.5305	0.2814				
0.70	0.9871	0.5466	0.2987	5.5	1.742	0.9646	0.9305
0.75	1.014	0.5615	0.3152	6.0	1.756	0.9723	0.9455
0.80	1.039	0.5753	0.3311	6.5	1.767	0.9784	0.9572
0.85	1.064	0.5891	0.3471	7.0	1.776	0.9834	0.9670
0.9	1.088	0.6024	0.3630	7.5	1.783	0.9873	0.9746
0.95	1.109	0.6141	0.3771				
1.0	1.131	0.6262	0.3921	8.0	1.788	0.9900	0.9802
				9.0	1.795	0.9939	0.9878
1.1	1.171	0.6484	0.4203	10.0	1.799	0.9961	0.9921
1.15	1.189	0.6584	0.4335	12.0	1.803	0.9983	0.9967
1.2	1.208	0.6689	0.4473	14.0	1.805	0.9994	0.9989
1.25	1.225	0.6783	0.4602	16.0	1.805	0.9994	0.9899

[1] Based on tables computed by P. H. J. A. KLEYNEN: Philips Res. Rep. **1**, 81 (1946).

Note 1: See Eqs. (37.1), (37.3), (37.5) and (37.6) for empirical equations for these data and the means for extrapolation.

Note 2: The limiting value of χ_s is $\chi_m = 1.806$.

Note 3: Definitions: $\psi_s = V_s/V_T = e V_s/kT$ and $\chi_s^2 = (x_s/x_1)^2$ with $(x_1)^2$ given by Eq. (35.2).

Table 3C (Collector Region). *Numerical solution*[1] *to the* LANGMUIR *equation for space charge in the Collector Region [Eq. (36.1)].*

ψ_c	χ_c	χ_c/χ_m	$(\chi_c/\chi_m)^2$
0.01	0.1440	0.07973	0.006359
0.02	0.2053	0.11368	0.01292
0.03	0.2529	0.1400	0.01961
0.04	0.2933	0.1624	0.02637
0.05	0.3293	0.1823	0.03323
0.06	0.3621	0.2005	0.04019
0.07	0.3924	0.2173	0.04721
0.08	0.4209	0.2331	0.05433
0.09	0.4477	0.2479	0.06144
0.10	0.4733	0.2621	0.06868
0.15	0.5866	0.3248	0.1055
0.20	0.6841	0.3788	0.1435
0.25	0.7714	0.4271	0.1825
0.30	0.8515	0.4715	0.2223
0.35	0.9262	0.5128	0.2630
0.40	0.9965	0.5518	0.3044
0.45	1.063	0.5886	0.3464
0.50	1.127	0.6240	0.3894
0.60	1.247	0.6905	0.4767
0.70	1.359	0.7525	0.5663
0.80	1.465	0.8112	0.6579
0.90	1.567	0.8677	0.7527
1.00	1.663	0.9208	0.8480
1.10	1.756	0.9723	0.9455
1.20	1.846	1.0221	1.0449
1.40	2.018	1.1174	1.248
1.60	2.181	1.208	1.458
1.80	2.336	1.293	1.673
2.00	2.486	1.377	1.895
2.20	2.630	1.456	2.121
2.40	2.769	1.553	2.351
2.60	2.904	1.608	2.585
2.80	3.036	1.681	2.826
3.00	3.164	1.752	3.069
3.20	3.290	1.822	3.319
3.40	3.413	1.890	3.571
3.60	3.533	1.956	3.827
3.80	3.651	2.022	4.087
4.00	3.767	2.086	4.351
4.50	4.049	2.242	5.026

ψ_c	χ_c	χ_c/χ_m	$(\chi_c/\chi_m)^2$
5.00	4.320	2.392	5.722
5.50	4.583	2.538	6.440
6.00	4.838	2.679	7.176
6.50	5.086	2.816	7.931
7.00	5.328	2.950	8.703
7.50	5.564	3.081	9.491
8.00	5.796	3.209	10.300
9.00	6.245	3.458	11.957
10.0	6.680	3.699	13.68
11.0	7.101	3.932	15.46
12.0	7.510	4.158	17.29
13.0	7.909	4.379	19.18
14.0	8.298	4.595	21.11
15.0	8.680	4.806	23.10
16.0	9.053	5.013	25.13
18.0	9.780	5.415	29.32
20.0	10.48	5.803	33.67
25.0	12.16	6.733	45.34
30.0	13.74	7.608	57.88
35.0	15.24	8.439	71.21
40.0	16.68	9.236	85.30
45.0	18.08	10.011	100.22
50.0	19.43	10.76	115.74
60.0	22.02	12.19	148.7
70.0	24.50	13.57	184.0
80.0	26.88	14.88	221.5
90.0	29.17	16.15	260.9
100.0	31.40	17.39	302.3
150.0	41.78	23.13	535.2
200.0	51.25	28.38	805.3
300.0	68.50	37.93	1439
400.0	84.28	46.67	2178
500.0	99.04	54.84	3007
600.0	113.06	62.60	3919
700.0	126.5	70.04	4906
800.0	139.4	77.19	5958
900.0	151.9	84.11	7074
1000.0	164.1	90.86	8256

[1] Based on tables computed by W. R. FERRIS: RCA-Rev. **10**, 134 (1949).

Note 1: See Eqs. (37.7) and (37.8) for empirical equations for these data and the means for extrapolation.

Note 2: The limiting value of χ_s is $\chi_m = 1.806$.

Note 3: Definitions: $\psi_c = V_c/V_T = eV_c/kT$ and $\chi_c^2 = (x_c/x_1)^2$ with $(x_1)^2$ given by Eq. (35.2).

Table 4. *Emitter region potential and its relation to emitter properties and current flow. (See Sect. 43 and 44 and Fig. 9.)*

ψ_s	z	z^2	u_0	u_0^2	$(I_0/I_m)^{\frac{1}{2}}$	I_0/I_m
0.02	0.1078	0.01162	1.0100	1.0202	0.1089	0.01185
0.025	0.1201	0.01442	1.0126	1.0253	0.1216	0.01478
0.03	0.1312	0.01721	1.0151	1.0305	0.1332	0.01773
0.04	0.1507	0.02271	1.0202	1.0408	0.1538	0.02364
0.05	0.1676	0.02809	1.0253	1.0513	0.1718	0.02953
0.06	0.1828	0.03342	1.0304	1.0618	0.1884	0.03549
0.07	0.1967	0.03869	1.0356	1.0725	0.2037	0.04150
0.08	0.2095	0.04389	1.0408	1.0833	0.2181	0.04755
0.09	0.2214	0.04902	1.0460	1.0942	0.2316	0.05364
0.10	0.2326	0.05410	1.0513	1.1052	0.2445	0.05979
0.15	0.2807	0.07879	1.0779	1.1618	0.3026	0.09154
0.2	0.3199	0.1023	1.1052	1.2214	0.3534	0.1249
0.25	0.3535	0.1250	1.1331	1.2840	0.4006	0.1605
0.3	0.3833	0.1469	1.1619	1.3499	0.4453	0.1983
0.35	0.4095	0.1677	1.1913	1.4191	0.4879	0.2380
0.4	0.4338	0.1882	1.2214	1.4918	0.5299	0.2808
0.45	0.4561	0.2080	1.2523	1.5683	0.5711	0.3262
0.5	0.4765	0.2270	1.2840	1.6487	0.6118	0.3743
0.55	0.4957	0.2457	1.3165	1.7333	0.6526	0.4259
0.6	0.5137	0.2639	1.3499	1.8221	0.6935	0.4809
0.7	0.5466	0.2988	1.419	2.014	0.7758	0.6018
0.8	0.5753	0.3310	1.492	2.226	0.8584	0.7368
0.9	0.6024	0.3629	1.568	2.460	0.9448	0.8927
1.0	0.6262	0.3921	1.619	2.718	1.032	1.066
1.1	0.6484	0.4204	1.733	3.004	1.124	1.263
1.2	0.6689	0.4474	1.822	3.320	1.219	1.485
1.3	0.6877	0.4729	1.915	3.669	1.317	1.735
1.4	0.7049	0.4969	2.014	4.055	1.420	2.015
1.5	0.7209	0.5197	2.117	4.482	1.526	2.329
1.6	0.7364	0.5423	2.226	4.953	1.639	2.686
1.7	0.7508	0.5637	2.340	5.474	1.757	3.086
1.8	0.7641	0.5838	2.460	6.050	1.879	3.532
1.9	0.7763	0.6026	2.586	6.686	2.007	4.029
2.0	0.7879	0.6208	2.718	7.389	2.142	4.587
2.1	0.7996	0.6394	2.858	8.166	2.285	5.221
2.2	0.8095	0.6553	3.004	9.025	2.432	5.914
2.3	0.8195	0.6716	3.158	9.974	2.588	6.699
2.4	0.8289	0.6871	3.320	11.023	2.752	7.574
2.5	0.8378	0.7019	3.490	12.18	2.924	8.549
2.6	0.8455	0.7149	3.669	13.46	3.102	9.623
2.7	0.8533	0.7281	3.857	14.88	3.292	10.834
2.8	0.8610	0.7413	4.055	16.44	3.491	12.19
2.9	0.8682	0.7538	4.263	18.17	3.701	13.70
3.0	0.8749	0.7654	4.482	20.09	3.922	15.38
3.1	0.8804	0.7751	4.712	22.20	4.148	17.21
3.2	0.8870	0.7868	4.953	24.53	4.393	19.30
3.3	0.8920	0.7957	5.207	27.11	4.644	21.57
3.4	0.8976	0.8057	5.474	29.96	4.913	24.14
3.5	0.9031	0.8156	5.755	33.12	5.197	27.01
3.6	0.9075	0.8236	6.050	36.60	5.490	30.14
3.8	0.9164	0.8398	6.686	44.70	6.127	37.54
4.0	0.9247	0.8551	7.389	54.60	6.833	46.69
4.2	0.9319	0.8684	8.166	66.69	7.610	57.91
4.4	0.9380	0.8798	9.025	81.45	8.465	71.66
4.6	0.9441	0.8913	9.974	99.48	9.416	88.67
4.8	0.9496	0.9017	11.023	121.51	10.47	109.6

Table 4. (Continued.)

ψ_s	z	z^2	u_0	u_0^2	$(I_0/I_m)^{\frac{1}{2}}$	I_0/I_m
5.0	0.9546	0.9113	12.18	148.4	11.63	135.2
5.5	0.9646	0.9304	15.64	244.7	15.09	227.7
6.0	0.9723	0.9454	20.09	403.4	19.53	381.4
6.5	0.9784	0.9573	25.79	665.1	25.23	636.7
7.0	0.9834	0.9671	33.12	1096.6	32.57	1060.5
7.5	0.9873	0.9748	42.52	1808.0	41.98	1762.4
8	0.9900	0.9801	54.60	2981.0	54.05	2921.7
9	0.9939	0.9878	90.02	8103.0	89.47	8004.0
10	0.9961	0.9922	148.4	22026.0	147.8	21854.0
12	0.9983	0.9966	403.4	0.16275×10^6	402.7	0.1622×10^6
14	0.9994	0.9988	1096.6	1.2026×10^6	1096.0	1.201×10^6
16	0.9994	0.9988	2981.0	8.8861×10^6	2979.0	8.875×10^6

Note 1: ψ_s and χ_s from Table 3E. $z = (\chi_s/\chi_m)$ from Eq. (43.7). $u_0^2 = I_0/I_R = e^{\psi_s}$ from Eq. (43.5). $(I_0/I_m) = z^2 e^{\psi_s}$ from Eq. (43.6).

Table 5. *Collector region potential and its relation to emitter properties and current Flow. [Use with Eq. (46.2) and related equations.]*

ψ_c	χ_c^2	$[F(\psi_c)]$	$[F(\psi_c)]^{\frac{3}{4}}$	$[F(\psi_c)]^{\frac{3}{2}}$
0.01	0.02074	0.08849	0.16227	0.02633
0.02	0.04215	0.1420	0.23132	0.05351
0.03	0.06396	0.1875	0.28494	0.08119
0.04	0.08602	0.22846	0.33045	0.1092
0.05	0.1084	0.26654	0.37094	0.1376
0.06	0.1311	0.3025	0.40792	0.1664
0.07	0.1540	0.3369	0.44215	0.1955
0.08	0.1772	0.3698	0.47424	0.2249
0.09	0.2004	0.4015	0.50438	0.2544
0.10	0.2240	0.4324	0.53320	0.2843
0.15	0.3441	0.5757	0.66091	0.4368
0.20	0.4680	0.7067	0.77078	0.5941
0.25	0.5951	0.8295	0.86914	0.7554
0.30	0.7251	0.9462	0.95937	0.9204
0.35	0.8578	1.0584	1.0436	1.089
0.40	0.9930	1.1672	1.1229	1.261
0.45	1.130	1.2716	1.1975	1.434
0.50	1.270	1.3748	1.2696	1.612
0.60	1.555	1.5736	1.4050	1.974
0.70	1.847	1.7649	1.5312	2.345
0.80	2.146	1.9542	1.6505	2.724
0.90	2.455	2.1334	1.7652	3.116
1.00	2.766	2.3101	1.8738	3.511
1.10	3.084	2.4840	1.9786	3.915
1.20	3.408	2.6550	2.0799	4.326
1.40	4.072	2.9896	2.2735	5.169
1.60	4.757	3.3163	2.4574	6.039
1.80	5.457	3.6338	2.6319	6.927

Table 5. (Continued.)

ψ_c	χ_c^2	$[F(\psi_c)]$	$[F(\psi_c)]^{\frac{3}{4}}$	$[F(\psi_c)]^{\frac{3}{2}}$
2.00	6.180	3.9482	2.8009	7.845
2.20	6.917	4.2560	2.9631	8.780
2.40	7.667	4.5583	3.1196	9.732
2.60	8.433	4.8559	3.2711	10.70
2.80	9.217	5.1538	3.4205	11.70
3.00	10.011	5.4463	3.5651	12.71
3.20	10.824	5.7367	3.7068	13.74
3.40	11.649	6.0243	3.8453	14.79
3.60	12.482	6.3072	3.9799	15.84
3.80	13.330	6.5908	4.1134	16.92
4.00	14.190	6.8708	4.2438	18.01
4.50	16.394	7.5657	4.5618	20.81
5.00	18.662	8.2486	4.8672	23.69
5.50	21.004	8.9243	5.1633	26.66
6.00	23.406	9.5926	5.4507	29.71
6.50	25.867	10.255	5.7306	32.84
7.00	28.388	10.911	6.0033	36.04
7.50	30.958	11.559	6.2690	39.30
8.00	33.594	12.205	6.5299	42.64
9.00	39.000	13.483	7.0363	49.51
10.0	44.622	14.749	7.5260	56.64
11.0	50.424	16.002	8.0006	64.01
12.0	56.400	17.241	8.4611	71.59
13.0	62.552	18.473	8.9107	79.40
14.0	68.857	19.696	9.3493	87.41
15.0	75.342	20.914	9.7796	95.64
16.0	81.957	22.115	10.198	104.0
18.0	95.648	24.516	11.018	121.4
20.0	109.83	26.885	11.807	139.4
25.0	147.87	32.782	13.700	187.7
30.0	188.79	38.587	15.482	239.7
35.0	232.26	44.294	17.170	294.8
40.0	278.22	49.966	18.794	353.2
45.0	326.89	55.637	20.372	415.0
50.0	377.52	61.236	21.891	479.2
60.0	484.88	72.357	24.809	615.5
70.0	600.25	83.426	27.604	762.0
80.0	722.53	94.400	30.285	917.2
90.0	850.89	105.26	32.863	1080.1
100.0	985.96	116.16	35.378	1252.0
150.0	1745.6	169.97	47.074	2216.0
200.0	2626.6	223.44	57.741	3334.0
300.0	4692.3	328.57	77.175	5956.0
400.0	7103.1	433.22	94.958	9017.0
500.0	9808.9	537.20	111.58	12451.0
600.0	12783.0	640.95	127.39	16227.0
700.0	16002.0	744.48	142.52	20313.0
800.0	19432.0	847.43	157.06	24667.0
900.0	23074.0	950.20	171.14	29290.0
1000.0	26929.0	1053.3	184.89	34184.0

Note 1: Tabular form of Eq. (46.2) comes from relation $[F(\psi_c)]^{\frac{3}{2}} = 1.2694\ \chi_c^2 = (\frac{9}{4}\sqrt{\pi})\ \chi_c^2$.
Note 2: Use Eq. (37.8) or (46.5) for $\psi_c > 1000$.

Table 6. *Relation between maximum and minimum current flow for a given collector region potential. [See Eq. (46.14).]*

ψ_c	$f(\psi_c)$	$\ln f(\psi_c)$	$f^{\frac{2}{3}}(\psi_c)$	$(\frac{2}{3}) \ln f(\psi_c)$
1.0	4.3515	1.47052	2.6654	0.98035
2.0	2.9810	1.09226	2.0713	0.72818
3.0	2.4674	0.90316	1.8260	0.60211
4.0	2.1890	0.78344	1.6859	0.52230
5.0	2.0110	0.69863	1.5932	0.46576
6.0	1.8861	0.63451	1.5266	0.42301
7.0	1.7929	0.58383	1.4758	0.38922
8.0	1.7204	0.54256	1.4358	0.36171
9.0	1.6621	0.50808	1.4032	0.33872
10.0	1.6139	0.47865	1.3759	0.31910
12.0	1.5389	0.43107	1.3329	0.28738
14.0	1.4827	0.39386	1.3003	0.26257
16.0	1.4389	0.36388	1.27455	0.24259
18.0	1.4035	0.33897	1.2536	0.22598
20.0	1.3744	0.31802	1.2362	0.21201
25.0	1.3191	0.27695	1.2028	0.18463
30.0	1.2802	0.24702	1.1790	0.16468
35.0	1.2511	0.22402	1.1611	0.14935
40.0	1.2283	0.20563	1.1469	0.13709
45.0	1.2098	0.19046	1.1354	0.12697
50.0	1.1946	0.17781	1.1259	0.11854
60.0	1.1708	0.15769	1.1109	0.10513
70.0	1.1529	0.14228	1.0995	0.09485
80.0	1.1389	0.13006	1.0906	0.08671
90.0	1.1277	0.12018	1.0834	0.08012
100.0	1.1183	0.11181	1.0774	0.07454
150.0	1.0883	0.08462	1.0580	0.05641
200.0	1.0717	0.06925	1.0473	0.04617
300.0	1.0534	0.05202	1.0353	0.03468
400.0	1.0433	0.04239	1.0287	0.02826
500.0	1.0368	0.03614	1.0244	0.02409
600.0	1.0322	0.03169	1.0214	0.02113
700.0	1.0288	0.02839	1.0191	0.01893
800.0	1.0261	0.02577	1.0173	0.01718
900.0	1.0239	0.02362	1.0159	0.01575
1000.0	1.0221	0.02186	1.0147	0.01457

Note 1: $f(\psi_c) = \left(1 + \frac{\chi_m}{\chi_c}\right)^2$ (Eq. (46.14).

Table 7. *The universal limiting curve of Figs. 16 and 17 is a plot of u_0^2 as a function of S'. (See Sect. 57.)*

			u_0	$u_0 \chi_{sR}$		u_0^2	S'
ψ_{sR}	χ_{sR}	$\psi_{sR}/2$	$e^{(\psi_{sR}/2)}$	χ_{co}	ψ_{co}	$e^{\psi_{sR}}$	$\psi_{sR} + \psi_{co}$
0.02	0.1947	0.01	1.0101	0.1967	0.0180	1.0202	0.038
0.04	0.2721	0.02	1.0202	0.2776	0.0362	1.0408	0.076
0.06	0.3302	0.03	1.0305	0.3403	0.0533	1.0618	0.113
0.08	0.3783	0.04	1.0408	0.3937	0.0703	1.0833	0.150
0.10	0.4201	0.05	1.0513	0.4417	0.0871	1.1052	0.187
0.15	0.5070	0.075	1.0779	0.5465	0.1322	1.1618	0.282
0.20	0.5777	0.10	1.1052	0.6385	0.1750	1.2214	0.375
0.25	0.6385	0.125	1.1332	0.7235	0.223	1.2840	0.473
0.30	0.6923	015	1.1618	0.8043	0.270	1.3499	0.570
0.40	0.7835	0.2	1.2214	0.9570	0.372	1.4918	0.772
0.50	0.8605	0.25	1.2840	1.1049	0.483	1.6487	0.983
0.60	0.9277	0.3	1.3499	1.252	0.605	1.8221	1.205
0.693	0.985	0.347	1.414	1.393	0.732	2	1.425
0.80	1.039	0.4	1.4918	1.550	0.883	2.226	1.683

Table 7. (Continued.)

			u_0	$u_0\chi_s R$		u_J^2	S'
$\psi_s R$	$\chi_s R$	$\psi_s R/2$	$e^{(\psi_s R/2)}$	χ_{c0}	ψ_{c0}	$e^{\psi_s R}$	$\psi_s R+\psi_{c0}$
1.00	1.131	0.5	1.6487	1.865	1.22	2.718	2.220
1.099	1.169	0.549	1.732	2.025	1.382	3	2.481
1.2	1.208	0.6	1.8221	2.201	1.627	3.320	2.827
1.386	1.267	0.693	2.000	2.534	2.068	4	3.454
1.4	1.273	0.7	2.014	2.564	2.110	4.055	3.51
1.6	1.330	0.8	2.226	2.961	2.690	4.953	4.29
1.792	1.375	0.896	2.449	3.369	3.33	6	5.12
1.8	1.380	0.9	2.460	3.395	3.37	6.050	5.17
2.0	1.423	1.0	2.718	3.868	4.19	7.389	6.19
2.079	1.439	1.040	2.828	4.071	4.55	8	6.63
2.303	1.480	1.151	3.162	4.680	5.69	10	7.99
2.4	1.497	1.2	3.320	4.970	6.28	11.023	8.68
2.773	1.550	1.386	4.000	6.200	8.92	16	11.69
2.8	1.555	1.4	4.055	6.306	9.15	16.44	11.95
2.996	1.580	1.498	4.472	7.065	10.91	20	13.91
3.2	1.602	1.6	4.953	7.935	13.09	24.53	16.29
3.6	1.639	1.8	6.050	9.916	18.40	36.60	22.00
3.689	1.645	1.844	6.325	10.40	19.77	40	23.46
4.0	1.670	2.0	7.389	12.34	25.59	54.60	29.59
4.094	1.675	2.047	7.746	12.98	27.56	60	31.6
4.4	1.694	2.2	9.025	15.288	35.0	81.45	39.4
4.605	1.704	2.303	10.000	17.04	41.0	100	45.6
5.0	1.724	2.5	12.18	20.998	55.7	148.4	60.7
5.298	1.730	2.649	14.142	24.47	70.0	200	75.3
5.5	1.742	2.75	15.64	27.245	81.5	244.7	87.0
6.0	1.756	3.0	20.09	35.278	118	403.4	124.0
7.0	1.770	3.5	33.12	58.821	240	1096.6	247.0
8	1.788	4.0	54.60	97.625	488	2981	496
9	1.795	4.5	90.02	161.59	975	8103	984
10	1.799	5	148.4	266.97	1944	22026	1954
12	1.803	6	403.4	727.33	7490	0.16275×10^6	7500
14	1.805	7	1096.6	1979.36	28800	1.2026×10^6	28800
16	1.805	8	2981	5380.71	110000	8.8861×10^6	110000

Table 8. *Master curve for emitter of unlimited capability $u_0^2=\infty$. (See Sect. 58 and Figs. 16 and 17.)*

ψ_c	U^2	Σ	ψ_c	U^2	Σ
0.02	1.2403	0.23535	35.0	89.09	39.490
0.04	1.3512	0.34099	40.0	104.8	44.652
0.06	1.4412	0.42548	45.0	121.2	49.797
0.10	1.5928	0.56549	50.0	138.3	54.929
0.20	1.9011	0.8424	60.0	174.1	65.160
0.40	2.4080	1.2788	70.0	212.2	75.358
0.60	2.8577	1.6500	80.0	252.3	85.531
1.00	3.6895	2.3055	90.0	294.2	95.684
1.40	4.483	2.9003	100.0	338.0	105.823
2.0	5.648	3.7313	150.0	582.4	156.37
3.0	7.573	5.0246	200.0	863.1	206.76
5.0	11.506	7.4429	300.0	1515.0	307.32
7.0	15.60	9.7473	400.0	2272.0	407.73
10.0	22.08	13.095	500.0	3118.0	508.04
12.0	26.61	15.281	600.0	4045.0	608.31
14.0	31.30	17.444	700.0	5047.0	708.53
16.0	36.15	19.588	800.0	6113.0	808.72
20.0	46.28	23.835	900.0	7244.0	908.89
25.0	59.80	29.091	1000.0	8439.0	1009.04
30.0	74.10	34.305			

Note 1: $U^2=\left(1+\dfrac{\chi_c}{\chi_m}\right)^2$ [Eq. (58.3)]. Note 2: $\Sigma=\psi_c+\ln U^2$ [Eq. (58.5)].

Table 9. *Master curve for emitter of limited capability*

$u_0^2 \to$ $\psi^s R$ $\chi^s R$	2 0.693 0.985		3 1.099 1.169		4 1.386 1.267		6 1.792 1.375		8 2.079 1.439	
ψ_s	u^2	S'	u^2	S'	u^2	S'	u^2	S'	u^2	S'
0	2	1.425	3	2.509	4	3.456	6	5.11	8	6.629
0.04	1.922	1.126	2.882	2.113	3.843	2.999	5.765	4.5418	7.686	5.9495
0.10	1.810	0.929	2.714	1.836	3.619	2.651	5.429	4.1157	7.238	5.4393
0.15	1.721	0.802	2.582	1.661	3.443	2.436	5.164	3.824	6.886	5.0995
0.2	1.637	0.6918	2.456	1.506	3.275	2.244	4.912	3.568	6.550	4.7895
0.3	1.482	0.508	2.222	1.238	2.963	1.908	4.445	3.127	5.926	4.2424
0.4	1.341	0.3513	2.011	1.015	2.681	1.624	4.022	2.739	5.362	3.7684
0.5	1.213	0.2163	1.820	0.817	2.426	1.376	3.639	2.401	4.852	3.344
0.6	1.097	0.098	1.646	0.6413	2.195	1.134	3.293	2.094	4.390	2.960
0.8	0.899		1.348	0.3459	1.797	0.7723	2.696	1.569	3.594	2.309
1.0	0.736		1.104	0.1035	1.472	0.4615	2.207	1.133	2.943	1.764

Table 10. *Electron current in a retarding field for cylinders of various radii ratios from 1 to ∞ (Sect. 60).*

$$\frac{1}{a} = \infty$$

S	$F_1(S)$	$\ln F_1(S)$	$\log_{10} F_1(S)$	Slope	Disp.
0	1.0000	0	0	0	0
0.5	0.8012	−0.2216	−0.0962	0.608	0.2784
1.0	0.5724	−0.5579	−0.2423	0.728	0.4421
1.5	0.3916	−0.9374	−0.4071	0.787	0.5626
2.0	0.2615	−1.3414	−0.5826	0.826	0.6586
3.0	0.1116	−2.1927	−0.9523	0.872	0.8073
4.0	4.601×10^{-2}	−3.0788	−1.3371	0.898	0.9212
5.0	1.857×10^{-2}	−3.9864	−1.7313	0.9157	1.0136
6.0	0.7383×10^{-2}	−4.9086	−2.1318	0.9281	1.0914
7.0	0.2905×10^{-2}	−5.8413	−2.5368	0.9370	1.1587
8.0	0.1134×10^{-2}	−6.7820	−2.9454	0.9444	1.2180
9.0	4.398×10^{-4}	−7.7291	−3.3567	0.9501	1.2709
10.0	1.697×10^{-4}	−8.6812	−3.7702	0.954	1.3188
12.0	0.2498×10^{-4}	−10.5974	−4.6024	0.962	1.4026
14.0	3.632×10^{-6}	−12.526	−5.4400	0.967	1.474
16.0	0.5234×10^{-6}	−14.463	−6.2812	0.970	1.537
18.0	7.488×10^{-8}	−16.407	−7.1255	0.974	1.593
20.0	1.066×10^{-8}	−18.357	−7.9723	0.977	1.643

$$\frac{1}{a} = 5.0$$

S	$F(S, a)$	$\ln F(S, a)$	$\log_{10} F(S, a)$	Slope	Disp.
0	1.0000	0	0	0	0
0.5	0.7979	−0.2257	−0.0980	0.619	0.2743
1.0	0.5668	−0.5678	−0.2466	0.740	0.4322
1.5	0.3854	−0.9535	−0.4141	0.800	0.5465
2.0	0.2557	−1.3639	−0.5923	0.839	0.6361
3.0	0.1077	−2.2280	−0.9676	0.886	0.7720
4.0	4.384×10^{-2}	−3.1272	−1.3581	0.911	0.8728
5.0	1.747×10^{-2}	−4.0475	−1.7578	0.9286	0.9525
6.0	0.6858×10^{-2}	−4.9824	−2.1638	0.9406	1.0176
7.0	0.2664×10^{-2}	−5.9278	−2.5744	0.9499	1.0722
8.0	0.1027×10^{-2}	−6.8811	−2.9884	0.9566	1.1189
9.0	3.935×10^{-4}	−7.8405	−3.4051	0.9622	1.1595
10.0	1.500×10^{-4}	−8.8050	−2.8240	0.9667	1.1950

u_0^2 = selected values. (See Sect. 58 and Figs. 16 and 17.)

10 2.303 1.480		20 2.996 1.580		40 3.689 1.645		60 4.094 1.675		100 4.605 1.704		200 5.298 1.730	
u^2	S'	u^2	S^2	u^2	S'	u^2	S'	u^2	S'	u^2	S'
10	8.003	20	13.906	40	23.489	60	31.644	100	45.55	200	75.298
9.608	7.2527	19.22	12.9059	38.43	22.0588	57.65	29.9544	96.08	43.5651	192.2	72.2585
9.048	6.6926	18.00	12.0559	36.19	20.7688	54.29	28.2943	90.48	40.707	181.0	68.193
8.607	6.2826	17.21	11.3255	34.43	19.8089	51.64	27.0443	86.47	39.5598	172.1	66.2481
8.187	5.9126	16.37	10.8454	32.75	18.9089	49.12	25.8942	81.87	37.4051	163.7	63.398
7.408	5.2726	14.82	9.8459	29.63	17.3188	44.45	23.7743	74.08	35.0	148.2	58.4985
6.703	4.7226	13.41	8.946	26.81	15.8587	40.22	21.8743	67.03	32.3851	134.1	54.0986
6.065	4.2276	12.31	8.2104	24.26	14.538	36.39	20.1443	60.65	29.9051	121.3	49.9
5.488	3.7776	10.97	7.3651	21.95	13.3287	32.93	18.4944	54.88	27.6051	109.7	46.3
4.493	3.001	8.99	6.0761	17.97	11.2187	26.96	15.2443	44.93	23.5151	89.9	39.5
3.679	2.353	7.36	4.981	14.72	9.3992	22.07	13.2342	36.79	20.0252	73.6	34.2

Table 10. (Continued.)

S	$F(S,a)$	$\ln F(S,a)$	$\log_{10} F(S,a)$	Slope	Disp.
		$\frac{1}{a} = 5.0$ (Continued)			
12.0	0.2154×10^{-4}	−10.7454	−4.6667	0.9733	1.2546
14.0	3.059×10^{-6}	−12.6973	−5.5144	0.9784	1.3027
16.0	0.4307×10^{-6}	−14.6578	−6.3658	0.9821	1.3422
18.0	6.026×10^{-8}	−16.6247	−7.2200	0.9848	1.3753
20.0	0.8386×10^{-8}	−18.5967	−8.0764	0.9872	1.4033
		$\frac{1}{a} = 4.0$			
0	1.0000	0	0	0	0
0.5	0.7960	−0.2281	−0.0991	0.625	0.2719
1.0	0.5635	−0.5736	−0.2491	0.747	0.4264
1.5	0.3818	−0.9628	−0.4182	0.807	0.5372
2.0	0.2524	−1.3768	−0.5979	0.847	0.6232
3.0	0.1056	−2.2482	−0.9764	0.893	0.7518
4.0	4.266×10^{-2}	−3.1544	−1.3699	0.918	0.8456
5.0	1.688×10^{-2}	−4.0817	−1.7727	0.9358	0.9183
6.0	0.6582×10^{-2}	−5.0234	−2.1816	0.9472	0.9766
7.0	0.2541×10^{-2}	−5.9753	−2.5950	0.9566	1.0247
8.0	9.731×10^{-4}	−6.9351	−3.0119	0.9631	1.0649
9.0	3.705×10^{-4}	−7.9007	−3.4312	0.9685	1.0993
10.0	1.404×10^{-4}	−8.8712	−3.8527	0.9725	1.1288
12.0	0.1994×10^{-4}	−10.8230	−4.7004	0.9789	1.1770
14.0	2.801×10^{-6}	−12.7857	−5.5528	0.9836	1.2143
16.0	0.3904×10^{-6}	−14.7562	−6.4085	0.9869	1.2438
18.0	5.410×10^{-8}	−16.7324	−7.2668	0.9893	1.2676
20.0	0.7464×10^{-8}	−18.7132	−8.1270	0.9915	1.2868
		$\frac{1}{a} = 3.0$			
0	1.0000	0	0	0	0
0.5	0.7916	−0.2336	−0.1015	0.640	0.2664
1.0	0.5562	−0.5866	−0.2548	0.762	0.4134
1.5	0.3739	−0.9837	−0.4272	0.823	0.5163
2.0	0.2452	−1.4056	−0.6105	0.863	0.5944
3.0	0.1010	−2.2926	−0.9956	0.907	0.7074
4.0	4.022×10^{-2}	−3.2135	−1.3956	0.933	0.7865

Table 10. (Continued.)

S	$F(S, a)$	$\ln F(S, a)$	$\log_{10} F(S, a)$	Slope	Disp.
		$\frac{1}{a} = 3.0$ (Continued)			
5.0	1.568×10^{-2}	-4.1552	-1.8046	0.9492	0.8448
6.0	0.6034×10^{-2}	-5.1104	-2.2194	0.9609	0.8896
7.0	0.2299×10^{-2}	-6.0752	-2.6384	0.9687	0.9248
8.0	8.701×10^{-4}	-7.0469	-3.0604	0.9747	0.9531
9.0	3.276×10^{-4}	-8.0239	-3.4847	0.9795	0.9761
10.0	1.228×10^{-4}	-9.0050	-3.9108	0.9829	0.9950
12.0	0.1710×10^{-4}	-10.9764	-4.7670	0.9883	1.0236
14.0	2.361×10^{-6}	-12.9564	-5.6269	0.9916	1.0436
16.0	0.3242×10^{-6}	-14.9420	-6.4892	0.9940	1.0580
18.0	4.433×10^{-8}	-16.9316	-7.3533	0.9957	1.0684
20.0	0.6045×10^{-8}	-18.9240	-8.2186	0.9968	1.0760
		$\frac{1}{a} = 2.5$			
0	1.0000	0	0	0	0
0.5	0.7867	-0.2399	-0.1042	0.658	0.2601
1.0	0.5486	-0.6005	-0.2608	0.780	0.3995
1.5	0.3658	-1.0058	-0.4368	0.839	0.4942
2.0	0.2379	-1.4357	-0.6235	0.878	0.5643
3.0	9.651×10^{-2}	-2.3381	-1.0154	0.922	0.6619
4.0	3.788×10^{-2}	-3.2735	-1.4217	0.947	0.7265
5.0	1.458×10^{-2}	-4.2280	-1.8362	0.9610	0.7720
6.0	0.5544×10^{-2}	-5.1949	-2.2561	0.9724	0.8051
7.0	0.2090×10^{-2}	-6.1704	2.6708	0.9787	0.8296
8.0	7.836×10^{-4}	-7.1516	-3.1059	0.9840	0.8484
9.0	2.924×10^{-4}	-8.1373	-3.5340	0.9878	0.8627
10.0	1.088×10^{-4}	-9.1263	-3.9635	0.9902	0.8737
12.0	0.1495×10^{-4}	-11.1107	-4.8253	0.9941	0.8893
14.0	2.043×10^{-6}	-13.1011	-5.6897	0.9962	0.8989
16.0	0.2782×10^{-6}	-15.0950	-6.5557	0.9977	0.9050
18.0	3.779×10^{-8}	-17.0912	-7.4226	0.9984	0.9088
20.0	0.5127×10^{-8}	-19.0887	-8.2901	0.9990	0.9113
		$\frac{1}{a} = 2.0$			
0	1.0000	0	0	0	0
0.5	0.7773	-0.2520	-0.1094	0.688	0.2480
1.0	0.5335	-0.6284	-0.2729	0.812	0.3716
1.5	0.3501	-1.0494	-0.4557	0.870	0.4506
2.0	0.2244	-1.4943	-0.6490	0.908	0.5057
3.0	8.859×10^{-2}	-2.4238	-1.0526	0.947	0.5762
4.0	3.397×10^{-2}	-3.3824	-1.4690	0.968	0.6176
5.0	1.282×10^{-2}	-4.3567	-1.8921	0.9794	0.6433
6.0	0.4795×10^{-2}	-5.3402	-2.3192	0.9876	0.6598
7.0	0.1783×10^{-2}	-6.3296	-2.7489	0.9911	0.6704
8.0	6.607×10^{-4}	-7.3222	-3.1800	0.9941	0.6778
9.0	2.442×10^{-4}	-8.3173	-3.6122	0.9963	0.6827
10.0	0.9014×10^{-4}	-9.3141	-4.0451	0.9974	0.6859
12.0	0.1224×10^{-4}	-11.3104	-4.9120	0.9988	0.6896
14.0	1.660×10^{-6}	-13.3086	-5.7799	0.9993	0.6914
16.0	2.249×10^{-6}	-15.3077	-6.6480	0.9998	0.6923
18.0	3.044×10^{-8}	-17.3074	-7.5165	0.9999	0.6926
20.0	0.4121×10^{-8}	-19.3073	-8.3851	1.0	0.6927

Table 10. (Continued.)

S	$F(S,a)$	$\ln F(S,a)$	$\log_{10} F(S,a)$	Slope	Disp.
			$\frac{1}{a} = 1.5$		
0	1.0000	0	0	0	0
0.5	0.7678	−0.2642	−0.1147	0.8543	0.2358
1.0	0.4960	−0.7012	−0.3045	0.8942	0.2988
1.5	0.3142	−1.1577	−0.5028	0.9320	0.3423
2.0	0.1953	−1.6330	−0.7092	0.9640	0.3670
3.0	7.357×10^{-2}	−2.6095	−1.1333	0.9866	0.3905
4.0	2.731×10^{-2}	−3.6006	−1.5637	0.9950	0.3994
5.0	1.008×10^{-2}	−4.5971	−1.9965	0.9977	0.4029
6.0	0.3714×10^{-2}	−5.5956	−2.4296	0.9990	0.4044
7.0	0.1367×10^{-2}	−6.5952	−2.8643	0.9996	0.4048
8.0	5.035×10^{-4}	−7.5949	−3.2984	0.9998	0.4051
9.0	1.851×10^{-4}	−8.5946	−3.7326	0.9999	0.4054
10.0	0.6810×10^{-4}	−9.5946	−4.1669	1.0000	0.4054
12.0	9.216×10^{-6}	−11.595	−5.0354	1.0000	0.405
14.0	1.247×10^{-6}	−13.595	−5.9042	1.0000	0.405
16.0	0.1688×10^{-6}	−15.595	−6.7728	1.0000	0.405
18.0	2.284×10^{-8}	−17.595	−7.6414	1.0000	0.405
20.0	0.3091×10^{-8}	−19.595	−8.5100	1.0000	0.405
			$\frac{1}{a} = 1.0$		
0	1.0	0	0	1.0	0
0.5	0.6065	−0.5	−0.2171		
1.0	0.3679	−1.0	−0.4343		
1.5	0.2231	−1.5	−0.6514		
2.0	0.1353	−2.0	−0.8686		
3.0	4.979×10^{-2}	−3.0	−1.3029		
4.0	1.832×10^{-2}	−4.0	−1.7372		
5.0	0.6738×10^{-2}	−5.0	−2.1715		
6.0	0.2479×10^{-2}	−6.0	−2.6058		
7.0	9.119×10^{-4}	−7.0	−3.0401		
8.0	3.355×10^{-4}	−8.0	−3.4744		
9.0	1.234×10^{-4}	−9.0	−3.9087		
10.0	0.4540×10^{-4}	−10.0	−4.3429		
12.0	6.144×10^{-6}	−12.0	−5.2115		
14.0	0.8315×10^{-6}	−14.0	−6.0801		
16.0	0.1125×10^{-6}	−16.0	−6.9487		
18.0	1.523×10^{-8}	−18.0	−7.8173		
20.0	0.2061×10^{-8}	−20.0	−8.6859		

Table 11. *The* FERMI *level and its temperature coefficient for selected donor concentration and energy level.* (Sect. 64.)

$n_D = 10^{18}$

V_T^{-1}	μ'	$E=-0.6$	$E=-0.8$	$E=-0.9$	$E=-1.0$	$E=-1.1$	$E=-1.2$	$E=-1.4$	$E=-1.6$
		μ							
8	2.512	−2.425	−2.425	−2.425	−2.425	−2.425	−2.425	−2.425	−2.425
10	1.976	−1.907	−1.907	−1.907	−1.907	−1.907	−1.907	−1.907	−1.911
12	1.624	−1.566	−1.566	−1.566	−1.566	−1.566	−1.567	−1.576	−1.616
14	1.375	−1.326	−1.326	−1.326	−1.327	−1.329	−1.336	−1.384	−1.468
16	1.191	−1.148	−1.148	−1.149	−1.153	−1.166	−1.194	−1.278	−1.375
18	1.049	−1.010	−1.012	−1.017	−1.034	−1.067	−1.105	−1.205	−1.305
20	0.936	−0.902	−0.907	−0.925	−0.960	−0.998	−1.051	−1.151	−1.251
22	0.844	−0.813	−0.831	−0.865	−0.907	−0.956	−1.006	−1.106	−1.206
24	0.769	−0.741	−0.780	−0.820	−0.870	−0.920	−0.970	−1.070	−1.170
26	0.705	−0.682	−0.743	−0.789	−0.839	−0.889	−0.939	−1.039	−1.139

Table 11. (Continued.)

$n_D = 10^{18}$ (Continued)

V_T^{-1}	$\frac{d\mu'}{dV_T}$	$E=-0.6$	$E=-0.8$	$E=-0.9$	$E=-1.0$	$E=-1.1$	$E=-1.2$	$E=-1.4$	$E=-1.6$
		$\frac{d\mu}{dV_T}$							
8	21.596	−20.902	−20.902	−20.902	−20.902	−20.902	−20.902	−20.894	−20.885
10	21.261	−20.598	−20.568	−20.568	−20.567	−20.564	−20.557	−20.478	−19.896
12	20.987	−20.294	−20.293	−20.290	−20.280	−20.244	−20.117	−18.621	−14.456
14	20.756	−20.063	−20.056	−20.030	−19.916	−19.468	−18.099	−13.335	−10.980
16	20.556	−19.861	−19.813	−19.595	−18.654	−16.252	−13.489	−10.790	−10.126
18	20.379	−19.680	−19.370	−18.062	−15.058	−12.384	−10.891	−10.045	−9.883
20	20.221	−19.498	−17.972	−14.661	−11.921	−10.508	−10.085	−9.818	−9.772
22	20.078	−19.262	−15.012	−11.891	−10.444	−9.966	−9.791	−9.706	−9.694
24	19.948	−18.785	−12.309	−10.472	−9.908	−9.720	−9.658	−9.631	−9.628
26	19.828	−17.623	−10.809	−9.916	−9.672	−9.598	−9.576	−9.568	−9.565

$n_D = 3 \times 10^{18}$

V_T^{-1}	μ'	μ							
8	2.375	−2.288	−2.288	−2.288	−2.288	−2.288	−2.288	−2.288	−2.288
10	1.866	−1.797	−1.797	−1.797	−1.797	−1.797	−1.797	−1.799	−1.809
12	1.532	−1.475	−1.475	−1.475	−1.475	−1.476	−1.478	−1.497	−1.557
14	1.297	−1.247	−1.248	−1.248	−1.250	−1.255	−1.270	−1.336	−1.427
16	1.122	−1.079	−1.080	−1.082	−1.092	−1.115	−1.151	−1.242	−1.340
18	0.988	−0.949	−0.953	−0.964	−0.992	−1.032	−1.0776	−1.175	−1.275
20	0.881	−0.847	−0.860	−0.888	−0.929	−0.975	−1.024	−1.123	−1.223
22	0.795	−0.764	−0.796	−0.836	−0.883	−0.932	−0.982	−1.082	−1.182
24	0.723	−0.698	−0.753	−0.799	−0.842	−0.897	−0.947	−1.047	−1.147
26	0.663	−0.646	−0.720	−0.769	−0.818	−0.868	−0.918	−1.018	−1.118

V_T^{-1}	$\frac{d\mu'}{dV_T}$	$\frac{d\mu}{dV_T}$							
8	20.497	−19.804	−19.804	−19.803	−19.803	−19.803	−19.802	−19.794	−19.750
10	20.162	−19.469	−19.469	−19.468	−19.466	−19.459	−19.438	−19.210	−17.596
12	19.889	−19.196	−19.193	−19.202	−19.154	−19.048	−18.700	−15.756	−12.206
14	19.658	−18.963	−18.942	−18.866	−18.551	−17.504	−15.365	−11.440	−10.031
16	19.457	−18.759	−18.616	−18.041	−16.216	−13.511	−11.502	−9.878	−9.495
18	19.281	−18.567	−17.748	−15.450	−12.550	−10.783	−9.950	−9.418	−9.317
20	19.123	−18.341	−15.393	−12.243	−10.473	−9.720	−9.415	−9.246	−9.220
22	18.980	−17.938	−12.487	−10.427	−9.613	−9.313	−9.204	−9.151	−9.144
24	18.849	−16.971	−10.654	−9.607	−9.137	−9.135	−9.097	−9.080	−9.078
26	18.729	−15.083	−9.735	−9.235	−9.083	−9.037	−9.024	−9.018	−9.018

$n_D = 10^{19}$

V_T^{-1}	μ'	μ							
8	2.224	−2.137	−2.137	−2.137	−2.137	−2.137	−2.138	−2.138	−2.152
10	1.746	−1.676	−1.676	−1.677	−1.677	−1.677	−1.677	−1.682	−1.706
12	1.432	−1.374	−1.374	−1.375	−1.375	−1.377	−1.383	−1.422	−1.498
14	1.211	−1.161	−1.162	−1.163	−1.168	−1.181	−1.207	−1.287	−1.382
16	1.047	−1.004	−1.006	−1.013	−1.033	−1.066	−1.108	−1.203	−1.302
18	0.921	−0.883	−0.892	−0.914	−0.951	−0.995	−1.043	−1.141	−1.241
20	0.821	−0.787	−0.814	−0.851	−0.896	−0.944	−0.994	−1.093	−1.193
22	0.740	−0.712	−0.762	−0.807	−0.855	−0.904	−0.954	−1.054	−1.154
24	0.673	−0.654	−0.725	−0.773	−0.822	−0.872	−0.922	−1.022	−1.122
26	0.616	−0.612	−0.696	−0.745	−0.795	−0.845	−0.895	−0.995	−1.095

Table 11. (Continued.)

$n_D = 10^{19}$ (Continued)

V_T^{-1}	$\frac{d\mu'}{dV_T}$	$E=-0.6$	$E=-0.8$	$E=-0.9$	$E=-1.0$	$E=-1.1$	$E=-1.2$	$E=-1.4$	$E=-1.6$
		$\frac{d\mu}{dV_T}$							
8	19.293	−18.600	−18.600	−18.599	−18.599	−18.597	−18.594	−18.568	−18.426
10	18.958	−18.265	−18.264	−18.261	−18.253	−18.231	−18.162	−17.500	−14.985
12	18.685	−17.991	−17.987	−17.954	−17.856	−17.535	−16.652	−13.016	−10.450
14	18.454	−17.757	−17.688	−17.448	−16.611	−14.739	−12.517	−9.963	−9.181
16	18.253	−17.544	−17.104	−15.753	−13.283	−11.164	−9.955	−9.041	−8.842
18	18.077	−17.315	−15.364	−12.580	−10.531	−9.512	−9.052	−8.760	−8.706
20	17.919	−16.945	−12.532	−10.316	−9.305	−8.889	−8.723	−8.630	−8.615
22	17.776	−16.095	−10.436	−9.247	−8.799	−8.634	−8.575	−8.545	−8.541
24	17.645	−14.379	−9.344	−8.766	−8.571	−8.507	−8.486	−8.477	−8.476
26	17.525	−12.266	−8.810	−8.535	−8.451	−8.427	−8.419	−8.416	−8.415

$n_D = 3 \times 10^{19}$

V_T^{-1}	μ'	μ							
8	2.087	−2.000	−2.000	−2.000	−2.000	−2.000	−2.000	−2.001	−2.005
10	1.636	−1.567	−1.567	−1.567	−1.567	−1.568	−1.569	−1.582	−1.624
12	1.340	−1.283	−1.283	−1.284	−1.285	−1.291	−1.304	−1.362	−1.448
14	1.132	−1.083	−1.084	−1.088	−1.099	−1.122	−1.157	−1.245	−1.342
16	0.978	−0.935	−0.941	−0.956	−0.986	−1.026	−1.071	−1.168	−1.268
18	0.860	−0.822	−0.842	−0.874	−0.916	−0.963	−1.012	−1.111	−1.211
20	0.766	−0.735	−0.778	−0.820	−0.867	−0.916	−0.966	−1.066	−1.166
22	0.690	−0.668	−0.734	−0.781	−0.830	−0.879	−0.929	−1.029	−1.129
24	0.627	−0.619	−0.701	−0.750	−0.799	−0.849	−0.899	−0.999	−1.099
26	0.574	−0.583	−0.674	−0.724	−0.774	−0.824	−0.874	−0.974	−1.074

V_T^{-1}	$\frac{d\mu'}{dV_T}$	$\frac{d\mu}{dV_T}$							
8	18.194	−17.501	−17.501	−17.500	−17.498	−17.494	−17.485	−17.408	−17.020
10	17.860	−17.166	−17.163	−17.155	−17.131	−17.062	−16.871	−15.416	−12.405
12	17.586	−16.891	−16.863	−16.782	−16.513	−15.751	−14.250	−10.889	−9.289
14	17.355	−16.652	−16.452	−15.844	−14.300	−12.190	−10.519	−8.958	−8.504
16	17.155	−16.415	−15.327	−13.241	−11.019	−9.629	−8.911	−8.397	−8.266
18	16.978	−16.086	−12.837	−10.504	−9.214	−8.617	−8.350	−8.182	−8.150
20	16.820	−15.387	−10.448	−9.055	−8.463	−8.223	−8.127	−8.073	−8.065
22	16.677	−13.900	−9.097	−8.400	−8.141	−8.046	−8.011	−7.995	−7.992
24	16.546	−11.887	−8.429	−8.094	−7.982	−7.945	−7.933	−7.927	−7.927
26	16.426	−10.209	−8.095	−7.935	−7.887	−7.873	−7.868	−7.867	−7.867

$n_D = 10^{20}$

V_T^{-1}	μ'	μ							
8	1.936	−1.850	−1.850	−1.850	−1.850	−1.850	−1.850	−1.853	−1.864
10	1.516	−1.446	−1.446	−1.447	−1.447	−1.449	−1.454	−1.482	−1.546
12	1.240	−1.182	−1.183	−1.185	−1.190	−1.204	−1.228	−1.302	−1.395
14	1.046	−0.997	−1.001	−1.011	−1.032	−1.066	−1.107	−1.201	−1.299
16	0.903	−0.861	−0.876	−0.902	−0.940	−0.985	−1.032	−1.130	−1.230
18	0.793	−0.758	−0.795	−0.835	−0.880	−0.928	−0.978	−1.077	−1.177
20	0.706	−0.680	−0.742	−0.788	−0.836	−0.886	−0.936	−1.036	−1.136
22	0.635	−0.624	−0.704	−0.753	−0.802	−0.852	−0.902	−1.002	−1.102
24	0.577	−0.585	−0.675	−0.724	−0.774	−0.824	−0.874	−0.974	−1.074
26	0.528	−0.556	−0.651	−0.701	−0.751	−0.801	−0.851	−0.951	−1.051

Table 11. (Continued.)

$n_D = 10^{20}$ (Continued)

V_T^{-1}	$\frac{d\mu'}{dV_T}$	$E=-0.6$	$E=-0.8$	$E=-0.9$	$E=-1.0$	$E=-1.1$	$E=-1.2$	$E=-1.4$	$E=-1.6$
		$\frac{d\mu}{dV_T}$							
8	16.990	−16.297	−16.295	−16.293	−16.288	−16.274	−16.242	−16.243	−15.018
10	16.656	−15.961	−15.949	−15.923	−15.846	−15.634	−15.118	−12.708	−10.187
12	16.382	−15.682	−15.590	−15.344	−14.645	−13.253	−11.543	− 9.208	− 8.308
14	16.151	−15.424	−14.837	−13.552	−11.608	− 9.966	− 8.946	− 8.073	− 7.824
16	15.951	−15.107	−12.857	−10.694	− 9.198	− 8.399	− 8.002	− 7.715	− 7.647
18	15.774	−14.493	−10.372	− 8.859	− 8.130	− 7.800	− 7.663	− 7.562	− 7.544
20	15.616	−13.172	− 8.794	− 8.006	− 7.681	− 7.549	− 7.496	− 7.467	− 7.462
22	15.473	−11.282	− 7.997	− 7.614	− 7.471	− 7.420	− 7.401	− 7.391	− 7.390
24	15.342	− 9.644	− 7.600	− 7.416	− 7.355	− 7.335	− 7.328	− 7.325	− 7.325
26	15.222	− 8.574	− 7.390	− 7.302	− 7.276	− 7.268	− 7.266	− 7.265	− 7.265

$n_D = 3 \times 10^{20}$

V_T^{-1}	μ'	μ							
8	1.799	−1.712	−1.712	−1.712	−1.713	−1.713	−1.714	−1.722	−1.746
10	1.406	−1.336	−1.337	−1.338	−1.340	−1.344	−1.356	−1.404	−1.482
12	1.149	−1.091	−1.093	−1.098	−1.111	−1.134	−1.167	−1.252	−1.347
14	0.968	−0.919	−0.929	−0.948	−0.979	−1.019	−1.064	−1.160	−1.260
16	0.835	−0.794	−0.824	−0.858	−0.901	−0.948	−0.997	−1.096	−1.196
18	0.732	−0.702	−0.757	−0.801	−0.848	−0.897	−0.947	−1.047	−1.147
20	0.651	−0.636	−0.712	−0.760	−0.809	−0.858	−0.908	−1.008	−1.108
22	0.585	−0.590	−0.678	−0.727	−0.777	−0.827	−0.877	−0.977	−1.077
24	0.531	−0.557	−0.652	−0.701	−0.751	−0.801	−0.851	−0.951	−1.051
26	0.486	−0.532	−0.630	−0.680	−0.729	−0.779	−0.829	−0.929	−1.029

V_T^{-1}	$\frac{d\mu'}{dV_T}$	$\frac{d\mu}{dV_T}$							
8	15.892	−15.198	−15.194	−15.187	−15.170	−15.131	−15.039	−14.431	−12.729
10	15.557	−14.860	−14.825	−14.748	−14.536	−14.025	−13.052	−10.438	− 8.726
12	15.284	−14.569	−14.309	−13.718	−12.485	−10.897	− 9.551	− 8.087	− 7.563
14	15.052	−14.260	−12.946	−11.209	− 9.567	− 8.494	− 7.886	− 7.378	− 7.235
16	14.852	−13.745	−10.550	− 8.916	− 7.993	− 7.525	− 7.295	− 7.129	− 7.184
18	14.675	−12.638	− 8.678	− 7.756	− 7.332	− 7.141	− 7.057	− 7.004	− 6.993
20	14.517	−10.892	− 7.686	− 7.227	− 7.039	− 6.963	− 6.932	− 6.915	− 6.913
22	14.374	− 9.252	− 7.192	− 6.970	− 6.888	− 6.858	− 6.847	− 6.842	− 6.841
24	14.244	− 8.143	− 6.934	− 6.828	− 6.793	− 6.781	− 6.777	− 6.776	− 6.775
26	14.124	− 7.475	− 6.787	− 6.737	− 6.722	− 6.717	− 6.716	− 6.715	− 6.715

$n_D = 10^{21}$

V_T^{-1}	μ'	μ							
8	1.648	−1.562	−1.562	−1.562	−1.563	−1.565	−1.586	−1.587	−1.634
10	1.285	−1.216	−1.218	−1.220	−1.226	−1.238	−1.260	−1.328	−1.415
12	1.048	−0.991	−0.998	−1.010	−1.033	−1.067	−1.107	−1.199	−1.296
14	0.882	−0.835	−0.858	−0.888	−0.927	−0.972	−1.019	−1.117	−1.216
16	0.759	−0.724	−0.774	−0.815	−0.861	−0.909	−0.959	−1.058	−1.158
18	0.665	−0.646	−0.719	−0.766	−0.814	−0.864	−0.913	−1.013	−1.113
20	0.591	−0.594	−0.680	−0.729	−0.778	−0.828	−0.878	−0.978	−1.078
22	0.530	−0.557	−0.650	−0.700	−0.750	−0.800	−0.850	−0.949	−1.049
24	0.481	−0.530	−0.626	−0.676	−0.726	−0.776	−0.826	−0.926	−1.026
26	0.439	−0.508	−0.606	−0.656	−0.706	−0.756	−0.806	−0.906	−1.006

Table 11. (Continued.)

$n_D = 10^{21}$ (Continued)

V_T^{-1}	$\frac{d\mu'}{dV_T}$	$E=-0.6$	$E=-0.8$	$E=-0.9$	$E=-1.0$	$E=-1.1$	$E=-1.2$	$E=-1.4$	$E=-1.6$
		$\frac{d\mu}{dV_T}$							
8	14.688	−13.992	−13.979	−13.956	−13.903	−13.777	−13.518	−12.232	−10.220
10	14.353	−13.646	−13.534	−13.304	−12.766	−11.797	−10.563	−8.542	−7.543
12	14.080	−13.318	−12.605	−11.485	−10.033	−8.794	−7.955	−7.128	−6.840
14	13.848	−12.849	−10.546	−9.011	−7.918	−7.302	−6.965	−6.686	−6.608
16	13.648	−11.887	−8.498	−7.498	−6.979	−6.722	−6.596	−6.505	−6.484
18	13.471	−10.285	−7.324	−6.809	−6.576	−6.471	−6.425	−6.396	−6.390
20	13.313	−8.696	−6.735	−6.482	−6.379	−6.338	−6.321	−6.312	−6.311
22	13.171	−7.596	−6.431	−6.309	−6.264	−6.248	−6.242	−6.239	−6.239
24	13.040	−6.928	−6.260	−6.202	−6.183	−6.176	−6.174	−6.174	−6.173
26	12.920	−6.531	−6.153	−6.125	−6.117	−6.114	−6.114	−6.113	−6.113

$n_D = 3\times10^{21}$

V_T^{-1}	μ'	μ							
8	1.511	−1.425	−1.425	−1.426	−1.428	−1.433	−1.441	−1.478	−1.543
10	1.175	−1.107	−1.110	−1.117	−1.130	−1.152	−1.184	−1.265	−1.357
12	0.957	−0.901	−0.917	−0.939	−0.972	−1.012	−1.056	−1.152	−1.250
14	0.804	−0.761	−0.802	−0.840	−0.883	−0.930	−0.979	−1.077	−1.177
16	0.691	−0.666	−0.733	−0.778	−0.826	−0.874	−0.924	−1.024	−1.124
18	0.604	−0.603	−0.686	−0.734	−0.783	−0.833	−0.883	−0.983	−1.083
20	0.536	−0.560	−0.652	−0.701	−0.751	−0.801	−0.851	−0.951	−1.051
22	0.480	−0.529	−0.625	−0.675	−0.725	−0.775	−0.825	−0.925	−1.025
24	0.435	−0.505	−0.603	−0.653	−0.703	−0.753	−0.803	−0.903	−1.003
26	0.397	−0.486	−0.585	−0.635	−0.685	−0.735	−0.785	−0.885	−0.985

V_T^{-1}	$\frac{d\mu'}{dV_T}$	$\frac{d\mu}{dV_T}$							
8	13.589	−12.889	−12.850	−12.785	−12.638	−12.333	−11.784	−10.041	−8.413
10	13.254	−12.521	−12.215	−11.693	−10.776	−9.640	−8.603	−7.281	−6.693
12	12.981	−12.092	−10.682	−9.360	−8.190	−7.381	−6.877	−6.395	−6.228
14	12.750	−11.298	−8.546	−7.455	−6.807	−6.446	−6.251	−6.090	−6.029
16	12.550	−9.877	−7.117	−6.520	−6.218	−6.069	−5.996	−5.943	−5.932
18	12.373	−8.316	−6.389	−6.082	−5.947	−5.887	−5.860	−5.844	−5.840
20	12.215	−7.180	−6.006	−5.860	−5.801	−5.762	−5.768	−5.762	−5.761
22	12.072	−6.479	−5.800	−5.730	−5.704	−5.695	−5.691	−5.690	−5.689
24	11.941	−6.061	−5.674	−5.641	−5.630	−5.626	−5.625	−5.624	−5.624
26	11.821	−5.806	−5.587	−5.571	−5.566	−5.565	−5.564	−5.564	−5.564

$n_D = 10^{22}$

V_T^{-1}	μ'	μ							
8	1.361	−1.275	−1.277	−1.280	−1.286	−1.297	−1.316	−1.374	−1.454
10	1.055	−0.988	−0.999	−1.014	−1.038	−1.071	−1.110	−1.199	−1.295
12	0.856	−0.805	−0.839	−0.872	−0.912	−0.956	−1.003	−1.100	−1.200
14	0.718	−0.687	−0.748	−0.791	−0.838	−0.886	−0.935	−1.034	−1.134
16	0.615	−0.610	−0.691	−0.738	−0.787	−0.836	−0.886	−0.986	−1.086
18	0.537	−0.560	−0.651	−0.700	−0.750	−0.799	−0.849	−0.949	−1.049
20	0.476	−0.526	−0.621	−0.671	−0.720	−0.770	−0.820	−0.920	−1.020
22	0.426	−0.500	−0.597	−0.647	−0.697	−0.747	−0.797	−0.897	−0.997
24	0.385	−0.479	−0.578	−0.628	−0.678	−0.728	−0.778	−0.878	−0.978
26	0.351	−0.462	−0.562	−0.612	−0.662	−0.712	−0.762	−0.862	−0.962

Table 11. (Continued.)

$n_D = 10^{22}$ (Continued)

V_T^{-1}	$\frac{d\mu'}{dV_T}$	$E=-0.6$	$E=-0.8$	$E=-0.9$	$E=-1.0$	$E=-1.1$	$E=-1.2$	$E=-1.4$	$E=-1.6$
		$\frac{d\mu}{dV_T}$							
8	12.388	−11.668	−11.547	−11.359	−10.988	−10.376	−9.572	−7.980	−6.910
10	12.050	−11.229	−10.463	−9.583	−8.556	−7.656	−6.989	−6.229	−5.905
12	11.772	−10.534	−8.486	−7.426	−6.688	−6.225	−5.944	−5.680	−5.588
14	11.552	−9.267	−6.860	−6.216	−5.854	−5.656	−5.549	−5.461	−5.436
16	11.340	−7.776	−5.982	−5.651	−5.485	−5.403	−5.364	−5.335	−5.328
18	11.166	−6.648	−5.535	−5.371	−5.297	−5.264	−5.249	−5.240	−5.238
20	11.020	−5.944	−5.293	−5.213	−5.181	−5.167	−5.162	−5.159	−5.159
22	10.872	−5.521	−5.148	−5.130	−5.096	−5.090	−5.088	−5.087	−5.087
24	10.740	−5.261	−5.050	−5.031	−5.025	−5.023	−5.022	−5.022	−5.022
26	10.626	−5.094	−4.975	−4.966	−4.963	−4.962	−4.962	−4.962	−4.962

$n_D = 3 \times 10^{22}$

V_T^{-1}	μ'	μ							
8	1.223	−1.138	−1.144	−1.152	−1.166	−1.187	−1.216	−1.290	−1.378
10	0.945	−0.882	−0.906	−0.931	−0.965	−1.004	−1.048	−1.142	−1.239
12	0.765	−0.724	−0.777	−0.817	−0.861	−0.908	−0.956	−1.054	−1.154
14	0.639	−0.627	−0.703	−0.749	−0.797	−0.846	−0.895	−0.995	−1.095
16	0.547	−0.566	−0.655	−0.703	−0.752	−0.802	−0.852	−0.952	−1.052
18	0.476	−0.525	−0.620	−0.669	−0.719	−0.769	−0.819	−0.919	−1.019
20	0.421	−0.496	−0.593	−0.643	−0.693	−0.743	−0.793	−0.893	−0.993
22	0.376	−0.474	−0.572	−0.622	−0.672	−0.722	−0.772	−0.872	−0.972
24	0.339	−0.456	−0.555	−0.605	−0.655	−0.705	−0.755	−0.855	−0.955
26	0.308	−0.441	−0.541	−0.591	−0.641	−0.691	−0.741	−0.841	−0.941

V_T^{-1}	$\frac{d\mu'}{dV_T}$	$\frac{d\mu}{dV_T}$							
8	11.284	−10.524	−10.210	−9.806	−9.184	−8.428	−7.684	−6.561	−5.934
10	10.950	−9.916	−8.610	−7.686	−6.888	−6.298	−5.892	−5.448	−5.260
12	10.680	−8.836	−6.803	−6.104	−5.659	−5.388	−5.225	−5.072	−5.019
14	10.446	−7.413	−5.714	−5.334	−5.124	−5.010	−4.948	−4.897	−4.882
16	10.252	−6.254	−5.156	−4.964	−4.869	−4.822	−4.798	−4.782	−4.778
18	10.068	−5.513	−4.860	−4.765	−4.723	−4.704	−4.695	−4.690	−4.689
20	9.920	−5.064	−4.687	−4.641	−4.622	−4.615	−4.612	−4.610	−4.610
22	9.772	−4.789	−4.573	−4.551	−4.543	−4.541	−4.539	−4.538	−4.538
24	9.636	−4.611	−4.489	−4.478	−4.475	−4.473	−4.473	−4.473	−4.473
26	9.508	−4.489	−4.420	−4.415	−4.413	−4.413	−4.413	−4.413	−4.413

Appendix 1.

Thermionic constants.

The following empirical equations may be used with the tabulated constants to calculate approximate current densities in amp/cm.² from the sources listed.

$$I = A_R T^2 e^{-\frac{e\varphi_R}{kT}}, \qquad I = A_R T^2 e^{-\frac{b}{T}}, \qquad I = A_R T^2 10^{-\frac{b'}{T}},$$

$$I = a e^{-\frac{e\Phi}{kT}}, \qquad I = a e^{-\frac{\beta}{T}}, \qquad I = a\, 10^{-\frac{\beta'}{T}}.$$

Note 1: T_m and T_n are minimum and maximum temperatures.
Note 2: T_0 computed from T_m and T_n by Eq. (50.5)
Note 3: Φ computed from φ_R by Eq. (50.3).
Note 4: a computed from A_R by Eq. (50.6).

Source	T_m	T_n	T_0	A_R	Φ_R	b	b'	a	Φ	β	β'	Ref.
C	1300	2200	1520	30	4.34	50340	21870	70×10^7	4.64	53820	23390	a
	1300	2200	1520	15	4.38	50810	22080	35×10^7	4.68	54290	23590	b
βFe	1040	1180	1010	26	4.48	51970	22580	27×10^7	4.68	54290	23590	c
γFe	1180	1680	1240	1.5	4.21	48840	21220	2.3×10^7	4.46	51740	22490	c
Mo	1300	2100	1480	115	4.37	50690	22020	250×10^7	4.69	54400	23640	d
	1350	2000	1450	55	4.15	48140	20920	155×10^7	4.44	51500	22380	e
Ni	1300	1700	1300	30	4.61	53480	23230	50×10^7	4.86	56380	24490	c
	1150	1700	1240	50	5.24	60780	26410	80×10^7	5.49	63680	27670	f
Pt	1700	2100	1650	32	5.32	61710	26810	90×10^7	5.65	65540	28480	g
Ta	1200	2000	1390	52	4.19	48600	21120	100×10^7	4.47	51850	22530	h
W	1400	2400	1650	72	4.52	52430	22780	200×10^7	4.85	56260	24440	i
	1200	2000	1390	60	4.51	52320	22730	120×10^7	4.79	55560	24140	j
W+Th	1200	2000	1390	3	2.63	30510	13255	6×10^7	2.91	33760	14670	k
"L" (W+Ba)	1300	1700	1300	1	1.8	20880	9070	2×10^7	2.06	23900	10380	l
	1300	1700	1300	15	2.0	23200	10080	20×10^7	2.26	26220	11390	l
La+LaB_6	1100	1900	1300	29	2.66	30860	13410	50×10^7	2.92	33870	14720	m
BaSrO	600	1200	780	0.5	1.0	11600	5040	3×10^6	1.16	13460	5846	n

a) A. L. Reimann: Proc. Phys. Soc. Lond. **50**, 496 (1938).
b) A. Braun and G. Busch: Helv. phys. Acta **20**, 33 (1947).
c) H. B. Wahlin: Phys. Rev. **61**, 509 (1942).
d) R. W. Wright: Phys. Rev. **60**, 465 (1941).
e) L. A. DuBridge and W. W. Roehr: Phys. Rev. **42**, 56 (1932).
f) A. B. Cardwell: Phys. Rev. **76**, 125 (1949).
g) L. V. Whitney: Phys. Rev. **50**, 1154 (1936).
h) M. D. Fiske: Phys. Rev. **61**, 513 (1942).
i) A. L. Reimann: Phil. Mag. **25**, 834 (1938).
j) W. B. Nottingham: Phys. Rev. **47**, 806 (1935).
k) S. Dushman and J. Ewald: Phys. Rev. **29**, 857 (1927).
l) H. J. Lemmens, M. J. Janson and R. Loosjes: Philips Rech. Rev. **11**, 341 (1950).
m) J. M. Lafferty: J. Appl. Phys. **22**, 299 (1951).
n) Reliable thermionic constants for the oxides are particularly difficult to obtain. The ones proposed here depend on emission measurements made on selected diodes constructed by the Raytheon Manufacturing Company.

Appendix 2.

Some useful equations from statistical theory of free electrons.

Particle current $I(\varepsilon_x)$ in x-direction per unit energy range of ε_x (Fermi Statistics)

$$I(\varepsilon_x)\,d\varepsilon_x = \frac{4\pi m k T}{h^3}\ln\left(1+e^{-\frac{(\varepsilon_x-\mu)}{kT}}\right)d\varepsilon_x .$$

Energy of Fermi *level*:

For high concentrations of electrons:

$$\mu = \frac{h^2}{2m}\left(\frac{3n}{8\pi}\right)^{\frac{2}{3}}\left\{1-\frac{\pi^2}{12}\left[\frac{kT}{\frac{h^2}{2m}\left(\frac{3n}{8\pi}\right)^{\frac{2}{3}}}\right]^2+\cdots\right\}.$$

For

$$T=0,\qquad \mu=\frac{h^2}{2m}\left(\frac{3n}{8\pi}\right)^{\frac{2}{3}}.$$

For low concentrations of electrons:

$$\mu = -kT\ln\left[\frac{2}{n h^3}\,(2\pi m k T)^{\frac{3}{2}}\right].$$

Field Emission.

By

R. H. Good jr. and Erwin W. Müller.

With 39 Figures.

I. Introduction*.

1. Scope and objectives. The emission of electrons from the surface of a conductor into a vacuum under the influence of a high electric field is an elementary effect comparable with thermionic emission, photoelectric emission, or secondary emission. From the standpoint of technical significance, field emission ranges definitely as the last one. This is due to technical difficulties still arising in the application of extremely high fields. However, the theoretical understanding is at least as complete as that of the other mechanisms of electron emission. If one considers as the purpose of a cathode the emission of as many electrons as possible from the almost inexhaustible reservoir inside a metal, a field emitter surpasses all other cathodes because current densities of millions of amps per cm^2 are easily obtained. Field emission has therefore deservedly become an attractive subject of research as can be seen in the historical development. The quantum mechanical theory of field emission from metals is well established and has been confirmed experimentally. From such experiments the field emission microscope emerged and opened the way for investigating various surface phenomena by field emission such as adsorption on different crystal planes, surface migration and desorption. As an entirely new effect, field ion emission was discovered and used to study the behavior of surfaces under fields up to 500 million volts per cm. We will confine ourselves to the discussion of field emission of electrons and ions into a vacuum or a gas of very low pressure. More complex effects which may possibly include field emission such as high vacuum breakdown, and certain kinds of arc discharge will not be considered here, nor will the so-called internal field emission which initiates the electrical breakdown of insulating crystals.

2. Historical development. Field emission was certainly occurring in many early experiments where high voltage was applied to sharply pointed cathodes or narrow gaps in evacuated tubes. The first one to describe this phenomenon in detail was R. W. Wood[1], in 1897, who reported enthusiastically what one would call today a kind of fire-works in his discharge tube. For experimental physicists who would rather like a more reproducible and reliable effect, field emission turned out to be quite discouraging for almost half a century afterwards. Lilienfeld[2] in 1922 ran an x-ray tube with a pointed cold cathode drawing quite erratically several milliamperes by a strong electric field, and he also noted the activation of the cathode by caesium. In 1923, Schottky[3] made the first attempt

* In this article the following abbreviations will be used: F-N theory = Fowler-Nordheim theory; FEM = field emission microscope.

[1] R. W. Wood: Phys. Rev. **5**, 1 (1897).

[2] J. E. Lilienfeld: Phys. Z. **23**, 506 (1922).

[3] W. Schottky: Z. Physik **14**, 63 (1923).

to explain field emission from cold metals, proposing a connection with thermionic emission with his potential trough model. An external field reduces the height of the potential barrier at the metal surface. The crest is located at a distance x_0 from the surface where the image force equals the field force on the electron $\frac{e^2}{4x_0^2}=eF$ ($-e$ is the electronic charge, F the field strength). Image force and external field together lower the potential barrier by

$$\Delta\varphi=\frac{e^2}{4x_0}+eF\,x_0$$

and by eliminating x_0 SCHOTTKY obtains

$$\Delta\varphi=e\sqrt{eF}$$

for the decrease of the work function. At relatively low fields, this decrease in thermionic work function has been experimentally established to be in full agreement with SCHOTTKY'S prediction. Field emission from cold cathodes was now assumed to be due to the complete reduction of the surface potential barrier, which would take place at $1.4\cdot10^8$ volts/cm for tungsten with a work function of $\varphi=4.5$ e-volts.

The preceding experiments, however, required only a field of some 10^6 volts/cm. SCHOTTKY explained this discrepancy by assuming that field emission originated from submicroscopic protuberances on the cathode which increased locally the field by a field enhancement factor estimated to lie between ten and one hundred.

SCHOTTKY'S theory of 1923 stimulated quite extensive research for the following seven years. ROTHER[1] obtained the high field by extremely small clearances between cathode and anode. With tantalum and iridium spheres as cathodes, field strengths up to $8\cdot10^6$ volts/cm were obtained. Most of the following investigators worked with cathodes having a small radius of curvature. GOSSLING'S[2] observations were made with fine wire loops, a brush of wires, or a single electro-chemically etched point. The measured current-voltage characteristics were plotted as log i against the square root of the applied voltage, but the straight lines which were to be expected according to SCHOTTKY'S theory could never be obtained. A further consequence of this theory is a considerable dependence upon temperature below the critical field of zero work function. However, GOSSLING showed the field currents to be independent of temperature in the wide range from 300° K to the onset of thermionic emission at 1500° K. GOSSLING came to the general conclusion that the SCHOTTKY theory could not describe the experimental results and that a better explanation might be found by the application of quantum theory rather than classical theory.

At the same time a research group around MILLIKAN came to similar results. MILLIKAN and EYRING[3] worked with a fine tungsten wire in the axis of a cylindrical anode and under very good vacuum conditions. The fields went up to $5\cdot10^6$ volts per cm after the cathode had been conditioned by drawing large field currents. From the appearance of some luminous regions on the anode, it was concluded that the field emission originated only from a few minute spots at the cathode surface, and the conditioning was due to the smoothing of these submicroscopic protrusions by ion bombardment from the residual gas. Temperature independence was again observed up to 1000° K. A most important discovery was made by MILLIKAN and LAURITSEN[4]. They found that the measured current-voltage

[1] F. ROTHER: Ann. Phys., Lpz. **81**, 317 (1926).
[2] B. S. GOSSLING: Phil. Mag. **1**, 609 (1926).
[3] R. A. MILLIKAN and C. F. EYRING: Phys. Rev. **27**, 51 (1926).
[4] R. A. MILLIKAN and C. C. LAURITSEN: Phys. Rev. **33**, 598 (1929).

characteristics could be described by $i = A \cdot e^{-B/F}$, so that a plot of $\log i$ versus $1/F$ gave a straight line over the whole accessible range. They also tried to incorporate field emission and thermionic emission in one formula by extending the Richardson equation to

$$i = A\,(T + c\,F)^2 \exp\left(\frac{-b}{T + c\,F}\right)$$

where A, b, and c are constants. This was subsequently criticized by de Bruyne[1] who produced experimental evidence that in the range below the onset of thermionic emission, field emission was completely independent of temperature. The effect found by Millikan and Lauritsen in the intermediate range of 1000 to 1500° K was apparently due to some activation of the tungsten cathode by thorium, giving rise to thermionic emission enhanced by the Schottky effect. The complete absence of a temperature influence was also shown by Piersol[2], who immersed a carefully evacuated and sealed off field emission tube into liquid air without noting a change in emission.

With the event of Sommerfeld's electron theory of metals, time was ripe to explain field emission in the light of the quantum theory. The first clear picture of the mechanism was obtained by Fowler and Nordheim[3] in 1928. Assuming the Fermi-Dirac statistics for the energy distribution of the metal electrons, Fowler and Nordheim calculated the number of electrons arriving from the inside at the surface potential barrier in each range of energy. They solved the Schrödinger equation to find the fraction of electrons that penetrate the potential hump at the surface. When the product of the number of electrons arriving at the potential barrier and the probability of penetration is integrated over all electron energies, a formula for the current density results:

$$j = \frac{4\sqrt{\mu\varphi}}{\mu + \varphi}\,\frac{e^3 F^2}{8\pi h\varphi}\,e^{-\frac{8\pi\sqrt{2m}\,\varphi^{\frac{3}{2}}}{3heF}}$$

where $-e$ is the electron charge, h Planck's constant, μ the energy of the Fermi level inside the metal relative to the bottom of the conduction band, and φ its thermionic work function. The surface barrier assumed for the derivation of this fundamental Fowler-Nordheim equation is simply a potential step at zero external field, and a triangular hump with a high field applied. Nordheim[4] refined the calculation further to include the deformation of the potential barrier by Schottky's image force, which resulted in a slightly decreased field strength necessary for the same current density.

The physical picture of the emission phenomena, according to Fowler and Nordheim, is as follows: in thermionic emission, electrons which have taken up enough thermal energy pass over the barrier; in field emission, electrons with energies below the Fermi level tunnel through the narrow potential barrier. The Fowler-Nordheim formula corresponds to Millikan and Lauritsen's plot of $\log i$ versus $1/F$ except for the experimentally hard to observe term of F^2 before the exponential function. It does not agree with Millikan and Lauritsen's generalized formula for both thermionic and field emission, which as an empirical equation had been verified only for small fields at high temperatures and for high fields at low temperatures, but not in the intermediate region. The Fowler-Nordheim formula predicts appreciable field currents at field strengths

[1] N. A. De Bruyne: Phys. Rev. **35**, 172 (1930).
[2] R. J. Piersol: Phys. Rev. **31**, 441 (1928).
[3] R. H. Fowler and L. Nordheim: Proc. Roy. Soc. Lond., Ser. A **119**, 173 (1928).
[4] L. Nordheim: Proc. Roy. Soc. Lond., Ser. A **121**, 626 (1928).

about five times smaller than SCHOTTKY's theory, so that the assumption of smaller field enhancement factors is sufficient to achieve agreement with the experimental field strengths.

After the establishment of the wave mechanical theory, STERN, GOSSLING, and FOWLER[1] used the results to obtain from some measured characteristics the field enhancement factor for a given cathode with a known work function. For an electrolytically etched tungsten point, the actual field was found to be 2.68 times larger than the measured field. They derived also a method of explaining the change of the ordinate intersection in a FOWLER-NORDHEIM plot ($\log i/F^2$ versus $1/F$) which occurs if the cathode is activated by an electro-positive contamination such as sodium. Though they obtained quantitatively the proper thickness for a sodium monolayer from field emission measurements, their assumption about the potential distribution in such a film seems not to be permissible.

Various attempts to confirm experimentally the FOWLER-NORDHEIM theory (in the following referred to as F-N theory) were made during the thirties, but no real progress could be achieved because of the impossibility of producing a defined cathode surface. Irregular protrusions were found by CHAMBERS[2] to be responsible for field emission from wires. The hope of several investigators like BEAMS[3] and later QUARLES[4] and MOORE[5] of obtaining a really smooth surface by using liquid mercury were not fulfilled. The field to start an arc discharge, requiring an electron emission of about 1 amp/cm², went up to only $1.8 \cdot 10^6$ volts per cm after repeated distillation. Microsecond pulses were used to prevent distortion of the liquid surface by the field. In spite of all precautions the existence of a field enhancement factor of at least 15 remained to be explained. Experiments to find the dependence upon the work function as predicted by the F-N theory gave quite contradictory results. AHEARN[6] used a thoriated tungsten cathode the work function of which could be controlled by varying the activation and measuring the thermionic emission. However, field emission from the same cathode turned out to be independent of the thermionic work function. The assumption had to be made that field emission originated again from small protrusions only, which did not change their activation as did the bulk surface for thermionic emission.

MÜLLER[7], on the other hand, activating a tungsten point cathode with evaporated electropositive substances, preferably barium, found a dependence upon the third power of the work function rather than the predicted $\frac{3}{2}$ power. Here again protrusions were responsible, produced either by sharp edged crystallites of the condensed material or by arcing in disruptive breakdown discharges. A year before WEHNELT and SCHILLING[8] produced pictures of an emitting edge by means of an electron microscope, demonstrating that the emission originated from quite randomly appearing and disappearing single spots along the edge.

In full agreement with the assumption of the F-N theory that the field electrons come directly from the top of the FERMI distribution, MÜLLER could not detect any cooling effect due to field emission. The origin of the field electrons at the FERMI level had been demonstrated before by HENDERSON and coworkers[9].

[1] T. E. STERN, B. S. GOSSLING and R. H. FOWLER: Proc. Roy. Soc. Lond., Ser. A **124**, 699 (1929).
[2] C. C. CHAMBERS: J. Franklin Inst. **218**, 463 (1934).
[3] J. W. BEAMS: Phys. Rev. **44**, 803 (1933).
[4] L. R. QUARLES: Phys. Rev. **48**, 260 (1935).
[5] D. H. MOORE: Phys. Rev. **50**, 344 (1936).
[6] A. J. AHEARN: Phys. Rev. **50**, 238 (1936).
[7] E. W. MÜLLER: Z. Physik **102**, 734 (1936).
[8] A. WEHNELT and W. SCHILLING: Z. Physik **98**, 286 (1936).
[9] J. E. HENDERSON and R. E. BADGLEY: Phys. Rev. **38**, 590 (1931).

They applied the method of retarding potentials to measure the energy of the field electrons but neither their original cylindrical arrangement nor an electrode system with an emitting point and concentric spheres which was used later by Dahlstrom, Mackenzie and Henderson[1], as well as by Müller[2], was suitable for measuring the true narrow velocity distribution in the range below the Fermi level.

A period of new productivity started with the introduction of the field emission microscope by Müller[3] in 1937. He observed that at high temperatures surface migration on a clean metal point tends to produce a very perfectly rounded and almost hemispherical tip which is smooth down to the lattice steps of atomic dimensions. An especially simple device was obtained by arranging a fluorescent screen at anode potential opposite to the tip. Field electrons leaving the tip almost radially project a greatly enlarged picture of the cathode surface on the screen. The patterns obtained with W, Mo, Ni, and Cu indicated the differences in current density at the various planes of the single crystal emitter and demonstrated convincingly that there were no irregular protrusions. Adsorption and desorption of oxygen, barium, and thorium, their dependence upon the crystallographic structure of the substrate, and the surface migration at elevated temperatures could be shown. The granular structure of the same films at small coverages was attributed to the visualization of individual atoms. Field induced deformation of the tungsten hemisphere at high temperatures was found. The low emission of the 011, 112, and 001 planes of the body centered cubic lattice was ascribed to an intrinsically high work function of these planes. The measurement of the tungsten tip radius with an oil immersion microscope showed for the first time field emission occurring at a field as high as $35 \cdot 10^6$ volts/cm, in agreement with the F-N theory and without requiring any field enhancement factor. A current density of 10^8 amps/cm^2 through a partially thoriated tip was experimentally established.

When the commercial electron microscope became available, Haefer[4] used it to measure the emitter geometry with improved resolution and made more accurate field calculations. The validity of the F-N theory was confirmed within an experimental error then believed to be only 15% in the field, and the dependence upon the $\frac{3}{2}$ power of work function was proven as well by measuring the characteristics of W-tips coated with Ba, K, and Cs in a field emission microscope.

Benjamin and Jenkins[5] adopted Müller's field emission microscope (FEM) to study again field emission from clean W, Mo, and Ni, as well as adsorption layers of Ba, Th, and Na, using sealed off tubes. The change in tip geometry due to field induced surface migration at elevated temperatures was emphasized. Müller[6] measured then the actual velocity distribution of field electrons and determined the resolution of the FEM. Recently an increasing number of investigators have used the FEM to study surface phenomena and to extend the knowledge of the emission process itself. One major contribution is the application of pulse techniques by Dyke[7] and collaborators. This permits the current densities to be substantially increased, reaching the range of space charge limitation and final breakdown by arcing, and promises technical applications with reasonable lifetimes [3].

[1] R. K. Dahlstrom, K. V. Mackenzie and J. E. Henderson: Phys. Rev. **48**, 484 (1935).
[2] E. W. Müller: Z. Physik **102**, 734 (1936).
[3] E. W. Müller: Z. Physik **106**, 541 (1937); **108**, 668 (1938).
[4] R. Haefer: Z. Physik **116**, 604 (1940).
[5] M. Benjamin and R. O. Jenkins: Proc. Roy. Soc. Lond., Ser. A **176**, 262 (1940); **180**, 225 (1942).
[6] E. W. Müller: Z. Physik **120**, 261, 270 (1943).
[7] W. P. Dyke and J. K. Trolan: Phys. Rev. **89**, 799 (1953).

New aspects came into field emission microscopy when it was found that adsorbed positive ions could be torn off by reversed fields in the 10^8 volts/cm range[1]. This led to the general phenomenon of field ion emission and the design of a field ion microscope[2]. A resolution of 2.7 Å is now obtainable and electric fields up to $5.7 \cdot 10^8$ volts/cm, where tungsten evaporates even at liquid hydrogen temperature, can now be handled experimentally[3].

II. Theory of field emission.

3. Electron gas statistics. In their theory of field emission FOWLER and NORDHEIM[4, 5], following SOMMERFELD, assume that the conduction electrons in a metal form a gas of free particles which have FERMI-DIRAC statistics. An outline of the properties of the gas which are needed to understand the field emission process is given below. (See also Reference [9] and this Encyclopedia, Vol. XIX.)

A small volume v inside the metal will be considered. Within v the electrons move freely. Since they are FERMI-DIRAC particles only wave functions antisymmetric with respect to interchange of particles arise. The possible states of a single particle within v may be divided into groups at energies ε_i containing altogether g_i states and populated by n_i electrons. Here i is the serial number of the group. The zero point for measuring energy will be specified later. Since n_i FERMI-DIRAC particles can be distributed among g_i states in $\frac{g_i!}{n_i!\,(g_i - n_i)!}$ ways, the relative probability of a specific arrangement of the electrons into groups is

$$w = \prod_i \frac{g_i!}{n_i!\,(g_i - n_i)!}\,.$$

The equilibrium distribution is found by maximizing w with respect to changes in the populations. This is equivalent to maximizing $\log w$ which, as long as the populations are so large that the STIRLING approximation

$$\log n! = n \log n - n$$

is valid, can be written in the form

$$\log w = \sum_i \left[g_i \log g_i - n_i \log n_i - (g_i - n_i) \log (g_i - n_i)\right].$$

Therefore at equilibrium the change in this quantity with respect to an arbitrary variation in the populations is zero:

$$-\,\delta \log w = \sum_i \left[\log n_i - \log (g_i - n_i)\right] \delta n_i = 0\,.$$

If the volume v is isolated as far as the flow of energy or particles is considered, the populations cannot all be varied independently and the auxiliary conditions

$$\delta \sum_i n_i = \sum_i \delta n_i = 0\,,$$

$$\delta \sum_i \varepsilon_i n_i = \sum_i \varepsilon_i\, \delta n_i = 0$$

---

[1] E. W. MÜLLER: Naturwiss. **29**, 533 (1941).

[2] E. W. MÜLLER: Z. Physik **131**, 136 (1951).

[3] E. W. MÜLLER: Z. Elektrochem. **59**, 372 (1955). — Z. Naturforsch. **11**a, 88 (1956). — J. Appl. Phys. **27**, 474 (1956).

[4] R. H. FOWLER and L. W. NORDHEIM: Proc. Roy. Soc. Lond., Ser. A **119**, 173 (1928).

[5] L. W. NORDHEIM: Proc. Roy. Soc. Lond., Ser. A **121**, 626 (1928). — Phys. Z. **30**, 177 (1929).

must be imposed. These three equations can be solved for the equilibrium distributions by using the method of LAGRANGE's undetermined multipliers. Multiplying the last two by α and β and adding all three, one finds

$$\sum_i \left[\log \frac{n_i}{g_i - n_i} + \alpha + \beta\,\varepsilon_i\right] \delta n_i = 0.$$

In this sum α and β may be chosen to make two of the coefficients of the δn_i zero—then the other δn_i can be varied independently so their coefficients are zero also. In general one may conclude that

$$\log \frac{n_i}{g_i - n_i} + \alpha + \beta\,\varepsilon_i = 0, \quad n_i = \frac{g_i}{e^{\alpha + \beta\varepsilon_i} + 1},$$

which is the FERMI-DIRAC distribution formula.

The physical significance of the constants α, β can be illustrated by considering a neighboring volume v' interacting with v. For v' let the states be divided into groups at energies ε_i' containing g_i' states and n_i' electrons. The relative probability of an arrangement of the compound system is

$$w = \prod_i \frac{g_i!}{n_i!\,(g_i - n_i)!} \prod_j \frac{g_j'!}{n_j'!\,(g_j' - n_j')!},$$

If energy and particles can flow between v and v' but the compound system is isolated, then the equilibrium distribution is to be found by maximizing $\log w$ with the auxiliary conditions that the total energy and number of particles is conserved:

$$-\delta \log w = \sum_i \{[\log n_i - \log (g_i - n_i)]\,\delta n_i + [\log n_i' - \log (g_i' - n_i')]\,\delta n_i'\} = 0,$$

$$\delta \sum_i (n_i + n_i') = \sum_i (\delta n_i + \delta n_i') = 0,$$

$$\delta \sum_i (\varepsilon_i n_i + \varepsilon_i' n_i') = \sum_i (\varepsilon_i\,\delta n_i + \varepsilon_i'\,\delta n_i') = 0.$$

If undetermined multipliers α and β are used as before, the result is evidently

$$n_i = \frac{g_i}{e^{\alpha+\beta\varepsilon_i} + 1}, \qquad n_i' = \frac{g_i'}{e^{\alpha+\beta\varepsilon_i'} + 1},$$

so each system comes to a FERMI-DIRAC equilibrium distribution and the same parameters α, β apply to each. It is seen that β will be the same for any two systems that are in equilibrium even though they are in energy contact, so it is associated with the temperature. Also α, since it is the same for two systems in equilibrium even though they may exchange particles, must be associated with the chemical potential. One may take as the statistical definitions of these quantities

$$\alpha = -\frac{\zeta}{kT},$$

$$\beta = \frac{1}{kT},$$

where ζ is the chemical potential, k is BOLTZMANN's constant, and T is the absolute temperature. The energy ζ is called the FERMI energy. The distribution formula then takes the form

$$n_i = \frac{g_i}{e^{\frac{\varepsilon_i - \zeta}{kT}} + 1}$$

for any volume v inside the metal, with the understanding that ζ and T have the same value throughout the metal at equilibrium.

The number of states g_i can be conveniently written in terms of a momentum range by using the well-known theorem that in each volume h^3 of phase space there are two states for a free particle of spin $\frac{1}{2}$ (here h is PLANCK's constant). This means that the number of states with momenta in the range $dp_x\,dp_y\,dp_z$ is $\frac{2v}{h^3}\,dp_x\,dp_y\,dp_z$ and that

$$d\,n = \frac{2v}{h^3}\,\frac{dp_x\,dp_y\,dp_z}{e^{\frac{\varepsilon-\zeta}{kT}}+1} \tag{3.1}$$

is the number of electrons within volume v and with momenta within the indicated range. The energy and momentum are of course related as follows:

$$p_x^2 + p_y^2 + p_z^2 = 2m(\varepsilon - V) \tag{3.2}$$

where m is the mass of an electron and V is the potential energy within v. One can choose v small so that V is constant within it and the free-particle statistics used above apply. This equation for the number of electrons is the fundamental formula for the properties of the electron gas.

4. Potential barrier. The field emission process can be treated by considering a one-dimensional potential which is assumed to have the same effect on a conduction electron as the actual metal. This potential was originally proposed by SCHOTTKY[1] and was used by NORDHEIM[2] in his modification of the original FOWLER-NORDHEIM theory.

There are three contributions to this effective potential:

1. Within the metal the potential energy has some constant value $-W_a$ relative to zero when the electron and metal are separated. This energy is lower than the FERMI energy ζ in metals. Its actual value is immaterial in the ordinary discussions of the theory, as will be shown below.

2. An electric field F is applied to draw the electrons out of the metal. The origin will be chosen on the metal surface and the positive x-axis perpendicular to the surface and out of the metal. This field will be constant as far as the surface barrier effects are concerned and so gives a contribution of $-eFx$ to the potential energy. It is presumed that the free charges cause this field to be neutralized within the metal and so it applies only outside. The zero point is chosen so this contribution to the potential is zero when the applied field is zero.

3. An electron outside a metal is attracted to the metal as a result of the charge it induces on the surface. The force on an electron of charge $-e$ is $-\frac{e^2}{4x^2}$ and the energy of position is $-\frac{e^2}{4x}$. The potential energy is chosen to be zero when the electron is far from the metal. This energy also applies only to the electrons outside the metal.

These three contributions give

$$\left.\begin{aligned} V(x) &= -W_a && \text{where} \quad x<0, \\ &= -eFx - \frac{e^2}{4x} && \text{where} \quad x>0 \end{aligned}\right\} \tag{4.1}$$

for the effective potential energy.

[1] W. SCHOTTKY: Z. Physik **14**, 63 (1923).
[2] L. W. NORDHEIM: Proc. Roy. Soc. Lond., Ser. A **121**, 626 (1928).

This potential is not expected to have significance at the point $x=0$. The main features of field emission can be understood from the shape of the potential outside the metal only. In Fig. 1, a plot of the effective potential is given for a representative case. The maximum value of the potential is found by differentiation to be at the point

$$x_0 = \frac{1}{2}\sqrt{\frac{e}{F}}, \tag{4.2}$$

and the corresponding maximum value of $V(x)$ is

$$V_{\max} = -\sqrt{e^3 F}. \tag{4.3}$$

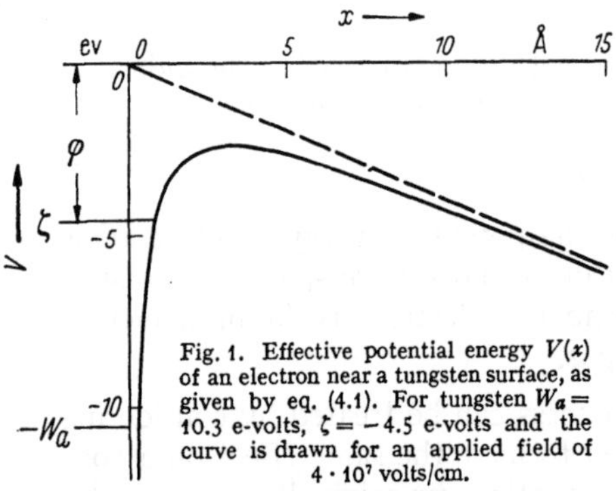

Fig. 1. Effective potential energy $V(x)$ of an electron near a tungsten surface, as given by eq. (4.1). For tungsten $W_a = 10.3$ e-volts, $\zeta = -4.5$ e-volts and the curve is drawn for an applied field of $4 \cdot 10^7$ volts/cm.

Under practical conditions this maximum remains above the Fermi energy.

5. Wave mechanical emission formula. In their theory of field emission Fowler and Nordheim take the electron gas model of the conduction electrons and the effective potential model of the metal and make the additional physical assumption that the electrons inside the metal remain essentially at equilibrium in spite of those that are escaping from the metal. Then the current of electrons is found by integrating, over all electron energies, the equilibrium flux of electrons incident on the surface times the probability that an electron penetrate the barrier. This integral can conveniently be written in terms of the x-part of the energy W, defined by

$$\left.\begin{aligned} W &= \varepsilon - \frac{p_y^2}{2m} - \frac{p_z^2}{2m} \\ &= \frac{p_x^2}{2m} + V(x). \end{aligned}\right\} \tag{5.1}$$

[All energies will be measured from the same reference as the effective potential $V(x)$.] Then, if $N(W)\,dW$ is the number of electrons with x-part of their energy within dW incident on the surface per second per unit area and $D(W)$ is the probability of penetration of the barrier, the product $P(W)\,dW$ gives the number within dW that emerge from the metal per second per unit area

$$P(W)\,dW = D(W)\,N(W)\,dW \tag{5.2}$$

and j, the electric current per unit area, is

$$j = e\int_{-W_a}^{\infty} P(W)\,dW. \tag{5.3}$$

The function $N(W)$ is called the supply function and $D(W)$ is called the transmission coefficient.

The supply function is easily found from the equilibrium distribution formula for the electrons, eq. (3.1). The number of electrons per second per unit area moving in the x-direction and with x-momentum within dp_x is found by multiplying the number per unit volume with momenta within $dp_x\,dp_y\,dp_z$ by the x-velocity and integrating over all p_y and p_z. The result is

$$\int_{p_y=-\infty}^{\infty}\int_{p_z=-\infty}^{\infty} \frac{p_x}{m}\,\frac{2}{h^3}\,\frac{dp_x\,dp_y\,dp_z}{e^{\frac{\varepsilon-\zeta}{kT}}+1}.$$

However, according to eq. (5.1), $p_x\,dp_x$ is equivalent to $m\,dW$. If this replacement is made, the quantity becomes the flux of electrons with x-energy within dW incident on the surface or, by definition, $N(W)\,dW$:

$$N(W)\,dW = \frac{2}{h^3}\,dW \int\limits_{-\infty}^{\infty}\int\limits_{-\infty}^{\infty} \frac{dp_y\,dp_z}{e^{\frac{W-\zeta}{kT}+\frac{p_y^2+p_z^2}{2mkT}}+1},$$

where eq. (5.1) has been used to write ε in terms of W in the denominator. The double integral can be easily evaluated if polar coordinates are introduced

$$p_y = \varrho\cos\vartheta,$$

$$p_z = \varrho\sin\vartheta$$

for then it is found that

$$\begin{aligned} N(W) &= \frac{2}{h^3}\int\limits_{\varrho=0}^{\infty}\int\limits_{\vartheta=0}^{2\pi} \frac{\varrho\,d\varrho\,d\vartheta}{e^{\frac{W-\zeta}{kT}+\frac{\varrho^2}{2mkT}}+1} \\ &= -\frac{4\pi m k T}{h^3}\log\left(e^{\frac{W-\zeta}{kT}} + e^{-\frac{\varrho^2}{2mkT}}\right)\Big|_0^{\infty} \\ &= \frac{4\pi m k T}{h^3}\log\left(1+e^{-\frac{W-\zeta}{kT}}\right). \end{aligned} \tag{5.4}$$

The transmission coefficient is to be found from a quantum mechanical discussion of the electrons incident on the barrier. Corresponding to the energy-momentum connection of eq. (5.1) the time-independent SCHRÖDINGER equation for the motion of one of the electrons in the x direction is

$$\frac{d^2u}{dx^2} + \frac{2m}{\hbar^2}\left[W - V(x)\right]u = 0,$$

where $u(x)$ is the wave function and $\hbar$ is $h/2\pi$. It would be awkward to carry through an exact solution of this equation but the properties of the solutions can be studied easily in a WKB approximation. In this approximation the transmission coefficient is

$$D(W) = e^{-\int\limits_{x_1}^{x_2}\sqrt{\frac{8m}{\hbar^2}[V(x)-W]}\,dx}$$

where x_1 and x_2 are the zeros of the radicand, chosen so that $x_1 < x_2$. This formula is valid only when W is appreciably less than $V_{\max}$; this is the range required for field emission. It is a specialization of a more general formula, valid for all W, which has been discussed by, for example, KEMBLE[1] and by MILLER and GOOD[2]. Substituting for $V(x)$ from eq. (4.1) and evaluating the roots one finds that

$$-\log D = \int\limits_{x_1}^{x_2}\sqrt{\frac{8m}{\hbar^2}\left(-eFx + |W| - \frac{e^2}{4x}\right)}\,dx,$$

where

$$\left.\begin{matrix}x_1\\x_2\end{matrix}\right\} = \frac{|W|}{2eF}\left(1 \mp \sqrt{1-\frac{e^3F}{W^2}}\right)$$

[1] EDWIN C. KEMBLE: The Fundamental Principles of Quantum Mechanics, first edit., Sec. 21j. New York: McGraw-Hill Book Co. 1937.

[2] S. C. MILLER jr. and R. H. GOOD jr.: Phys. Rev. 91, 174 (1953).

(W is always negative). This integral was calculated by NORDHEIM[1]; his result was improved by BURGESS, KROEMER, and HOUSTON[2]. The expression for the roots suggests introducing the parameter

$$y = \frac{\sqrt{e^3 F}}{|W|} \tag{5.5}$$

and a new integration variable

$$\xi = \frac{2eF}{|W|}\, x,$$

so that

$$-\log D = \frac{\sqrt{m\,|W|^3}}{\hbar e F} \int\limits_{1-\sqrt{1-y^2}}^{1+\sqrt{1-y^2}} \sqrt{-\xi^2 + 2\xi - y^2}\, \frac{d\xi}{\sqrt{\xi}}\,.$$

The substitution $\eta = \sqrt{\xi}$ brings this to a standard form for an elliptic integral:

$$-\log D = \frac{2\sqrt{m|W|^3}}{\hbar e F} \int\limits_b^a \sqrt{(a^2-\eta^2)(\eta^2-b^2)}\, d\eta\,,$$

where

$$b = \sqrt{1 - \sqrt{1-y^2}}\,,$$
$$a = \sqrt{1 + \sqrt{1-y^2}}\,.$$

An application of BYRD and FRIEDMAN'S tables[3] yields

$$-\log D = \frac{4\sqrt{m|W|^3}\,a}{3\hbar e F}\left[\frac{(a^2+b^2)}{2}\, E(k) - b^2 K(k)\right],$$

where

$$k^2 = \frac{a^2 - b^2}{a^2}$$

and K and E are the complete elliptic integrals of the first and second kinds:

$$K(k) = \int\limits_0^{\pi/2} \frac{d\varphi}{\sqrt{1 - k^2 \sin^2\varphi}}\,,$$

$$E(k) = \int\limits_0^{\pi/2} \sqrt{1 - k^2 \sin^2\varphi}\, d\varphi\,.$$

Rewriting in terms of the parameter y, one obtains finally for the transmission coefficient

$$D(W) = e^{-\frac{4\sqrt{2m|W|^3}}{3\hbar e F} v(y)}\,, \tag{5.6}$$

where

$$v(y) = 2^{-\frac{1}{2}} \sqrt{1+\sqrt{1-y^2}}\left[E(k) - \left(1 - \sqrt{1-y^2}\right) K(k)\right], \tag{5.7}$$

$$k^2 = \frac{2\sqrt{1-y^2}}{1+\sqrt{1-y^2}}\,, \tag{5.8}$$

---

[1] L. W. NORDHEIM: Proc. Roy. Soc. Lond., Ser. A **121**, 626 (1928).

[2] R. E. BURGESS, H. KROEMER and J. M. HOUSTON: Phys. Rev. **90**, 515 (1953).

[3] PAUL F. BYRD and MORRIS D. FRIEDMAN: Handbook of Elliptic Integrals for Engineers and Physicists. Formulas 217.09 and 361.27. Berlin: Springer 1954.

and y is given by eq. (5.5). The function $v(y)$ has been evaluated for representative values of y by BURGESS, KROEMER, and HOUSTON[1]. Their results are given in Table 1.

The number of electrons within dW that emerge from the metal per second per unit area is found by combining eqs. (5.4) and (5.6) according to eq. (5.2):

$$P(W)\,dW = \frac{4\pi m k T}{h^3}\, e^{-\frac{4\sqrt{2m|W|^3}}{3\hbar e F} v\left(\frac{\sqrt{e^3 F}}{|W|}\right)} \log\left(1 + e^{-\frac{W-\zeta}{kT}}\right) dW. \tag{5.9}$$

The field emission electrons have energies in the neighborhood of $W=\zeta$. It is permissible then to approximate the exponent in the transmission coefficient by the first two terms in a power series expansion at $W=\zeta$. One finds directly that

$$-\frac{4\sqrt{2m|W|^3}}{3\hbar e F} v\left(\frac{\sqrt{e^3 F}}{|W|}\right) \approx -c + \frac{W-\zeta}{d}, \tag{5.10}$$

where

$$c = \frac{4\sqrt{2m\varphi^3}}{3\hbar e F} v\left(\frac{\sqrt{e^3 F}}{\varphi}\right), \tag{5.11}$$

$$d = \frac{\hbar e F}{2\sqrt{2m\varphi}\, t\left(\frac{\sqrt{e^3 F}}{\varphi}\right)}, \tag{5.12}$$

$$t(y) = v(y) - \frac{2}{3} y \frac{dv(y)}{dy}, \tag{5.13}$$

and $\varphi = -\zeta$ is the work function. Numerical values of the function $t(y)$ are also given in Table 1—it is seen that it is a slowly varying function, ranging only between 1.00 and 1.11. If numerical values of m, e, $\hbar$, are inserted here and if φ is expressed in e-volts and F in volts/cm, then c and d are given by

$$c = \frac{6.83\times 10^7\, \varphi^{\frac{3}{2}}}{F} v\left(3.79\times 10^{-4} \frac{F^{\frac{1}{2}}}{\varphi}\right), \tag{5.14}$$

$$d = \frac{9.76\times 10^{-9}\, F}{\varphi^{\frac{1}{2}}\, t\left(3.79\times 10^{-4} \frac{F^{\frac{1}{2}}}{\varphi}\right)} \text{ e-volts}. \tag{5.15}$$

Table 1. *Values of the functions $v(y)$, $s(y)$, $t(y)$, defined by eqs. (5.7), (5.8), (8.2), (5.13).*

y	$v(y)$	$s(y)$	$t(y)$
0	1.0000	1.0000	1.0000
0.05	0.9948	0.9995	1.0011
0.1	0.9817	0.9981	1.0036
0.15	0.9622	0.9958	1.0070
0.2	0.9370	0.9926	1.0111
0.25	0.9068	0.9885	1.0157
0.3	0.8718	0.9835	1.0207
0.35	0.8323	0.9777	1.0262
0.4	0.7888	0.9711	1.0319
0.45	0.7413	0.9637	1.0378
0.5	0.6900	0.9554	1.0439
0.55	0.6351	0.9464	1.0502
0.6	0.5768	0.9366	1.0565
0.65	0.5152	0.9261	1.0631
0.7	0.4504	0.9149	1.0697
0.75	0.3825	0.9030	1.0765
0.8	0.3117	0.8903	1.0832
0.85	0.2379	0.8770	1.0900
0.9	0.1613	0.8630	1.0969
0.95	0.0820	0.8483	1.1037
1	0	0.8330	1.1107

In calculating the numerical values here and below, DUMOND and COHEN's[2] recommendations for the values of the atomic constants have been followed. Also the discussion will be restricted in this section to the low temperature limit of eq. (5.9). The effect of higher temperatures is treated in Sect. 6. For low enough temperatures it is seen that

$$\left.\begin{aligned} kT\log\left(1+e^{-\frac{W-\zeta}{kT}}\right) &= 0 \quad &&\text{when}\quad W>\zeta,\\ &= \zeta - W \quad &&\text{when}\quad W<\zeta. \end{aligned}\right\} \tag{5.16}$$

If eqs. (5.10) and (5.16) are substituted into eq. (5.9), the result is

$$\left.\begin{aligned} P(W) &= 0 \quad &&\text{when}\quad W>\zeta,\\ &= \frac{4\pi m}{h^3}\, e^{-c+\frac{W-\zeta}{d}} (\zeta - W) \quad &&\text{when}\quad W<\zeta \end{aligned}\right\} \tag{5.17}$$

[1] R. E. BURGESS, H. KROEMER and J. M. HOUSTON: Phys. Rev. **90**, 515 (1953).
[2] J. W. M. DUMOND and E. R. COHEN: Rev. Mod. Phys. **25**, 691 (1953).

and this is the low temperature limit of the distribution of the emitted electrons. By differentiation one finds easily that the peak of $P(W)$ occurs at the energy $W = \zeta - d$ and has the value

$$P_{\max} = \frac{4\pi m d}{h^3} e^{-(c+1)}.$$

From a numerical discussion of the function one finds that the width of the peak at half this maximum value is $2.44d$. These considerations were first made by Richter[1].

The total electric current flowing is found by integrating the distribution given by eq. (5.17) over all energies according to eq. (5.3). Ordinarily the lowest energy $-W_a$ is far below the Fermi energy ζ so the lower limit may be taken at $-\infty$. The results are as follows:

$$\left.\begin{aligned} j &= e \int_{-\infty}^{\zeta} \frac{4\pi m}{h^3} e^{-c + \frac{W-\zeta}{d}} (\zeta - W)\, dW = \frac{4\pi m e d^2}{h^3} e^{-c} \\ &= \frac{e^3 F^2}{8\pi h \varphi t^2\left(\frac{\sqrt{e^3 F}}{\varphi}\right)} e^{-\frac{4\sqrt{2m}\,\varphi^{\frac{3}{2}}}{3\hbar e F} v\left(\frac{\sqrt{e^3 F}}{\varphi}\right)}. \end{aligned}\right\} \tag{5.18}$$

If φ in e-volts and F in volts/cm are inserted the result is

$$j = \frac{1.54\times 10^{-6} F^2}{\varphi t^2\left(3.79\times 10^{-4}\frac{F^{\frac{1}{2}}}{\varphi}\right)} e^{-6.83\times 10^{7}\frac{\varphi^{\frac{3}{2}}}{F} v\left(3.79\times 10^{-4}\frac{F^{\frac{1}{2}}}{\varphi}\right)} \text{ amp/cm}^2. \tag{5.19}$$

Some numerical values of the common logarithm of the current density for various fields and work functions are given in Table 2. A more extensive set of values has been calculated by Dolan[2] using eq. (5.19) but with t set equal to unity. This simplification is often made because t^{-2} varies only between 1.00 and 0.81, contributing between 0 and -0.09 to $^{10}\text{Log}\, j$.

Table 2. *Common logarithm of the current density j in amps/cm² for various values of the work function φ in e-volts and the electric field F in volts/cm according to eq. (5.19).*

$\varphi=2.0$		$\varphi=4.5$		$\varphi=6.3$	
F	$^{10}\text{Log}\, j$	F	$^{10}\text{Log}\, j$	F	$^{10}\text{Log}\, j$
1.0×10^7	2.98	2.0×10^7	−3.33	2.0×10^7	−12.90
1.2	4.45	3.0	1.57	4.0	−0.88
1.4	5.49	4.0	4.06	6.0	3.25
1.6	6.27	5.0	5.59	8.0	5.34
1.8	6.89	6.0	6.62	10.0	6.66
2.0	7.40	7.0	7.36	12.0	7.52
2.2	7.82	8.0	7.94	14.0	8.16
2.4	8.16	9.0	8.39	16.0	8.65
2.6	8.45	10.0	8.76	18.0	9.04
		12.0	9.32	20.0	9.36

6. Field emission at higher temperatures. The approximation used for the logarithmic term in eq. (5.16) restricts the derivation of the previous section to zero temperature. At finite temperatures there are electrons above the Fermi

[1] G. Richter: Z. Physik **119**, 406 (1942).
[2] W. W. Dolan: Phys. Rev. **91**, 510 (1953).

energy incident on the barrier and these are more likely to penetrate because of the increase in transmission coefficient with increase in energy. For smal temperatures the emission is only slightly changed from the emission at zero temperature, but at higher temperatures the character of the emission changes completely and the primary contribution to the current comes from electrons above rather than below the FERMI energy. The high temperature limit is the thermionic emission in which the current consists mainly of electrons which have energies above the peak of the barrier V_{max}. In the following paragraphs the first correction to the cold emission formula for increasing temperature is given.

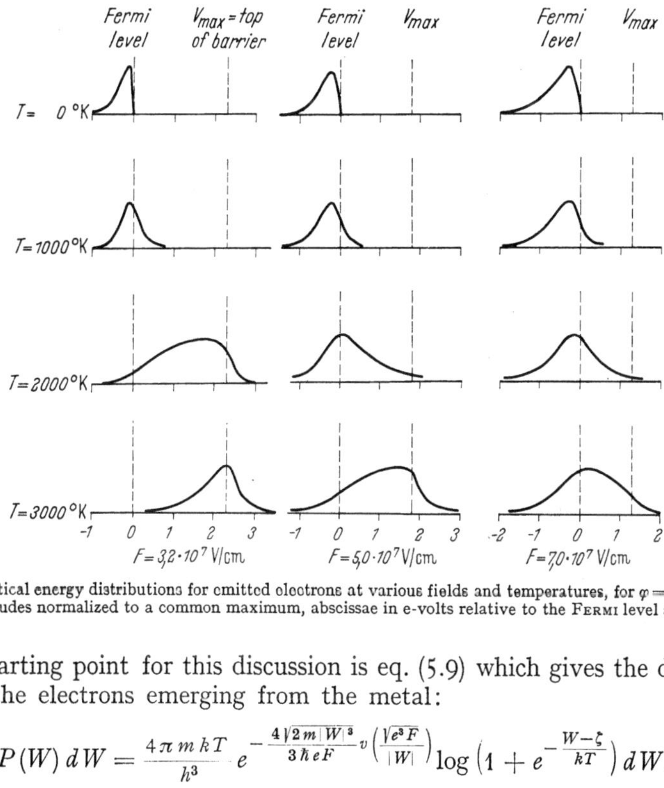

Fig. 2. Theoretical energy distributions for emitted electrons at various fields and temperatures, for $\varphi = 4.5$ e-volts, with amplitudes normalized to a common maximum, abscissae in e-volts relative to the FERMI level at zero.

The starting point for this discussion is eq. (5.9) which gives the distribution in W of the electrons emerging from the metal:

$$P(W)\,dW = \frac{4\pi m k T}{h^3}\, e^{-\frac{4\sqrt{2m}\,|W|^{\frac{3}{2}}}{3\hbar e F}\, v\left(\frac{\sqrt{e^3 F}}{|W|}\right)} \log\left(1 + e^{-\frac{W-\zeta}{kT}}\right) dW.$$

If numerical values of the constants are introduced the equation reads

$$\left.\begin{aligned} P(W)\,dW = 8.70\times 10^{24}\, T\, e^{-6.83\times 10^{7}\frac{|W|^{\frac{3}{2}}}{F}\, v\left(3.79\times 10^{-4}\frac{F^{\frac{1}{2}}}{|W|}\right)} \times \\ \times \log\left(1 + e^{-1.16\times 10^{4}\frac{W+\varphi}{T}}\right) dW\,\mathrm{sec}^{-1}\,\mathrm{cm}^{-2}, \end{aligned}\right\} \tag{6.1}$$

where W and φ are in e-volts and F is in volts/cm. This formula is applicable only when $W < V_{max}$ because of the limitations of the transmission coefficient approximation. DOLAN and DYKE[1] have made plots of the distribution in W of the emitted electrons for a few values of temperature and field, and for work function $\varphi = 4.5$ e-volts. Some of their results are shown in Fig. 2. They use the supply function of eq. (5.4), the transmission coefficient of eq. (5.6) for energies W below V_{max}, and unit transmission for energies W above V_{max}. They

[1] W. W. DOLAN and W. P. DYKE: Phys. Rev. 95, 327 (1954).

apply therefore eq. (6.1) when $W < V_{max}$ and the same relation except with the transmission coefficient exponential replaced by unity when $W > V_{max}$. This is not entirely satisfactory because the transmission coefficient is about $\frac{1}{2}$ when $W = V_{max}$ and makes a gradual transition to unity as W increases above that value but the distribution curves are sensitive to this difference only near $W = V_{max}$. Dolan and Dyke distinguish between three types of emission which are illustrated by their curves: temperature emission at high temperatures and low fields as described by the Richardson equation, field emission at low temperatures and high fields as described by the Nordheim eq. (5.18), and temperature-and-field or T-F emission at temperatures and fields intermediate between the two extremes.

When the expansion at the Fermi energy of eq. (5.10) is introduced into eq. (5.9) the formula for the distribution of emitted electrons becomes

$$P(W) = \frac{4\pi m k T}{h^3} e^{-c + \frac{W-\zeta}{d}} \log\left(1 + e^{-\frac{W-\zeta}{kT}}\right). \tag{6.2}$$

When $W > \zeta$ a first approximation to the logarithm is

$$\log\left(1 + e^{-\frac{W-\zeta}{kT}}\right) \approx e^{-\frac{W-\zeta}{kT}},$$

and when both this and the expansion at the Fermi level apply the distribution is

$$P(W) = \frac{4\pi m k T}{h^3} e^{-c + (W-\zeta)\left(\frac{1}{d} - \frac{1}{kT}\right)}. \tag{6.3}$$

This result is of interest in itself because the exponential tail of the distribution above the Fermi energy can be observed directly. Also from eq. (6.3) one can make an estimate of the range of temperature and field in which the emitted current is mainly from below the Fermi energy. The requirement is that the exponential in eq. (6.3) should be strong enough to make $P(W)$ appreciably smaller as the energy increases toward V_{max}. In symbols one has

$$(V_{max} - \zeta)\left(\frac{1}{d} - \frac{1}{kT}\right) < -1$$

or, replacing $V_{max} - \zeta$ by $\varphi - \sqrt{e^3 F}$,

$$\frac{1}{kT} > \frac{1}{d} + \frac{1}{\varphi - \sqrt{e^3 F}}.$$

For ordinary operating conditions the second term on the right can be disregarded compared to the first so that the condition is simply $d > kT$ or, as a consequence of eq. (5.12),

$$F > \frac{2\sqrt{2m\varphi}\, kT\, t\left(\frac{\sqrt{e^3 F}}{\varphi}\right)}{\hbar e}. \tag{6.4}$$

If one substitutes numerical values and puts $t \approx 1$ this condition becomes

$$F > 8.83 \times 10^3 \varphi^{\frac{1}{2}} T. \tag{6.5}$$

It marks a boundary of the field emission phenomenon; beyond it, the current is carried mainly by electrons with energies above the Fermi level and a different theory applies.

However, as long as eq. (6.4) is fulfilled the total emitted current can be found by integrating over the distribution given in eq. (6.2):

$$j(T) = e\int_{-\infty}^{\infty} P(W)\,dW = \frac{4\pi m e k^2 T^2}{h^3}\, e^{-c}\int_0^{\infty} \xi^{\frac{kT}{d}-1}\log\left(1+\frac{1}{\xi}\right)d\xi,$$

where the integration variable has been changed to $\xi = e^{\frac{W-\zeta}{kT}}$. Then an integration by parts brings the integral to a standard form[1]:

$$\left.\begin{aligned} j(T) &= \frac{4\pi m e k T d}{h^3}\, e^{-c}\int_0^{\infty}\frac{\xi^{\frac{kT}{d}-1}}{1+\xi}\,d\xi \\ &= \frac{4\pi m e k T d}{h^3}\, e^{-c}\,\frac{\pi}{\sin\frac{\pi k T}{d}} \\ &= j(0)\,\frac{\frac{\pi k T}{d}}{\sin\frac{\pi k T}{d}}, \end{aligned}\right\}\quad (6.6)$$

where $j(0)$ is the zero-temperature current density of eq. (5.18). Numerically one finds that

$$\left.\begin{aligned} \frac{\pi k T}{d} &= 2.77\times 10^4\,\frac{T\varphi^{\frac{1}{2}}}{F}\times \\ &\quad\times t\left(3.79\times 10^{-4}\,\frac{F^{\frac{1}{2}}}{\varphi}\right). \end{aligned}\right\}\quad (6.7)$$

If an expansion for small temperature is made, the expression for the current density becomes

$$j(T) = j(0)\left[1+\frac{1}{6}\left(\frac{\pi k T}{d}\right)^2+\cdots\right]. \quad (6.8)$$

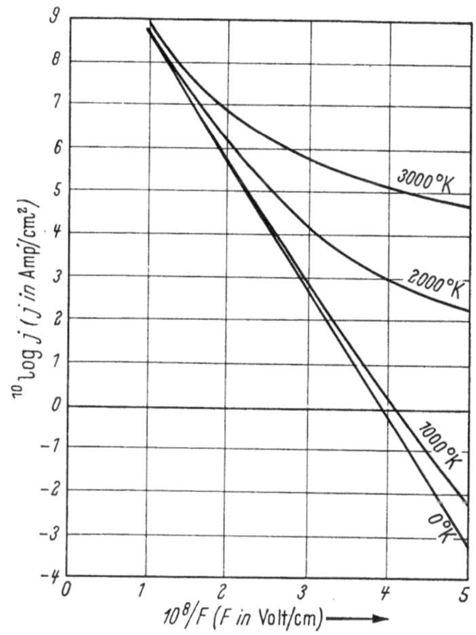

Fig. 3. Computed current density j versus reciprocal field $1/F$ at various temperatures, for $\varphi = 4.5$ e-volts.

For example, if the field is 4×10^7 volt/cm on tungsten ($\varphi = 4.5$ e-volts) at room temperature, eq. (6.7) gives $\frac{\pi k T}{d} = 0.4$ and eq. (6.8) yields $j(T) = 1.03\, j(0)$. On the other hand if the temperature is increased to 1000° K one finds $\frac{\pi k T}{d} = 1.5$ and, from eq. (6.6), $j(T) = 1.5\, j(0)$.

Electron emission in the range specified by eq. (6.4) has been studied by GUTH and MULLIN[2] using a series expansion method to evaluate the current integral. Also DOLAN and DYKE[3] have integrated their distribution curves numerically, obtaining results outside the range of eq. (6.4). Some of their results are reproduced in Fig. 3. MURPHY and GOOD[4] have studied the electron emission problem from a general point of view which gives thermionic emission and field emission as limiting cases. They give a more detailed discussion of the values of temperature and field for which eq. (6.6) is valid. CUTLER and GOOD[5] have considered higher order terms in the expansions made to obtain eq. (6.6) and have found analytic expressions for the corresponding higher order corrections to the current formula.

[1] See, for example, H. B. DWIGHT: Tables of Integrals, revised edit., formula 856.2, New York: The Macmillan Co. 1947.
[2] EUGENE GUTH and CHARLES J. MULLIN: Phys. Rev. **61**, 339 (1942).
[3] W. W. DOLAN and W. P. DYKE: Phys. Rev. **95**, 327 (1954).
[4] E. L. MURPHY and R. H. GOOD jr.: Phys. Rev. **102**, 1464 (1956).
[5] P. H. CUTLER and R. H. GOOD jr.: Bull. Amer. Phys. Soc. **1**, 279 (1956).

III. Experimental results.

7. Dependence on field strength. The first step of the experimental verification of the F-N theory is the determination of the relation between field strength and current density from the directly measurable voltage-current graph. At the present time the field can only be calculated from the geometry of the electrodes using potential theory, which restricts the possible arrangements to simple forms. The most convenient device of concentric cylinders, the inner one—a fine wire—being the field emitter, turns out to be impractical, since the wire is never smooth enough for a homogeneous emission. One has therefore to rely on the etched and heat polished point cathode as developed by MÜLLER[1]. The exact geometrical form of it may be determined by electron microscopy, and the density distribution of field emission over the tip can be checked in the FEM. MÜLLER[2] and HAEFER[3] used a hyperboloid of revolution to approximate the tip geometry, and BECKER[4]

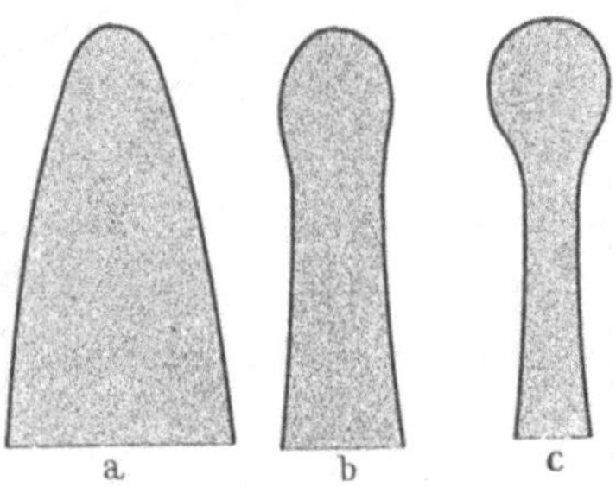

Fig. 4a.

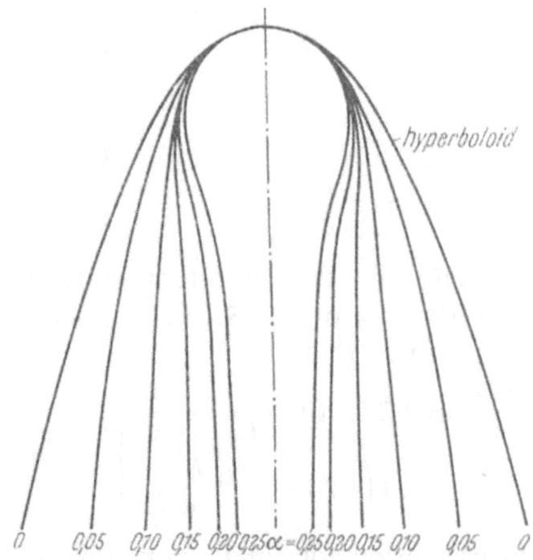

Fig. 4b.

Fig. 4a and b. Typical shapes of cathode tips: a, as observed in an electron microscope and b, as described by equipotential surfaces obtained by superposing a hyperboloidal and a spherical field (the fractional contribution α of the spherical field is indicated).

used a paraboloid; either of these can roughly describe some of the forms as found by electron microscopy. A much wider variety of the actually occurring forms, however, can be covered by fitting the profile of the emitter with a suitable equipotential surface produced by two core bodies. DYKE and co-workers[5] use a sphere mounted on an orthogonal cone, but their method requires some experimentation with the proper combination of three parameters to fit the given emitter profile. DRECHSLER and HENKEL[6] superimpose the potential of a sphere onto the potential of a hyperboloid, the center of the sphere coinciding with the center of the sphere of curvature at the apex of the hyperboloid. Fig. 4 gives some of the equipotential surfaces they obtain, described by only one form factor α, where α is simply the contribution of the spherical field as a fraction of one and $1-\alpha$ the contribution of the hyperboloidal field. The maximum field strength at the apex is then

$$F_0 = \alpha F_{\text{sphere}} + (1-\alpha) F_{\text{hyp}} = V\left(\frac{\alpha}{r} + \frac{2(1-\alpha)}{r \log \frac{4l}{r}}\right)$$

(V = anode potential, l = distance tip-anode, r = radius of curvature in the apex, $l \gg r$).

[1] E. W. MÜLLER: Z. Physik **106**, 132 (1937).
[2] E. W. MÜLLER: Z. Physik **108**, 668 (1938).
[3] R. HAEFER: Z. Physik **116**, 604 (1940).
[4] J. A. BECKER: Bell Syst. Techn. J. **30**, 907 (1951).
[5] W. P. DYKE, J. K. TROLAN, W. W. DOLAN and G. BARNES: J. Appl. Phys. **24**, 570 (1953).
[6] M. DRECHSLER and E. HENKEL: Z. angew. Phys. **6**, 341 (1954).

In Fig. 5 the emitter shape has been calculated for different form factors and with always the same field strength F_0 at the apex. One notices that the radius of curvature varies by more than a factor of two. The field drops with increasing polar angle ϑ, measured from the apex, as shown in Fig. 6, causing an even steeper drop of current density. Fig. 7 has been calculated by DRECHSLER under the assumption of the validity of the F-N equation for $F_0 = 4 \cdot 10^7$ volts/cm, and a homogeneous work function of $\varphi =$ 4.50 e-volts. DYKE and co-workers use a similar curve to determine an average current density $j = i/A$, where i is the ex-

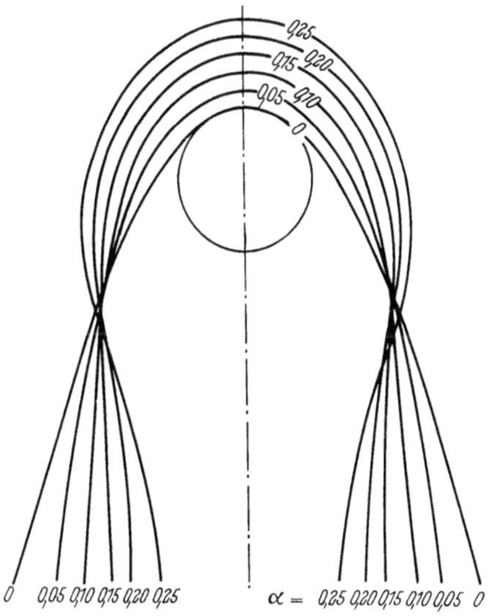

Fig. 5. The equipotential surfaces of Fig. 4b, adjusted to the same apex field strength F_0.

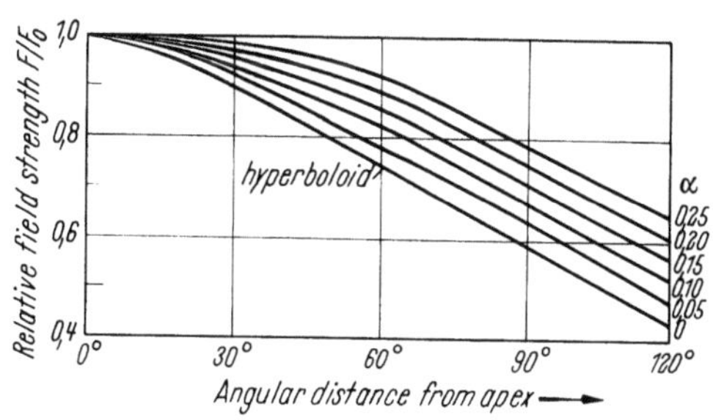

Fig. 6. Decrease of relative field strength with angular distance from apex.

perimental current. The effective emitting area A increases slightly with the field strength. The most important error occurs when the tip shape is determined by the electron microscope. There is an uncertainty in the magnification and the micrographs give only profiles, which depend on the direction of viewing, of the

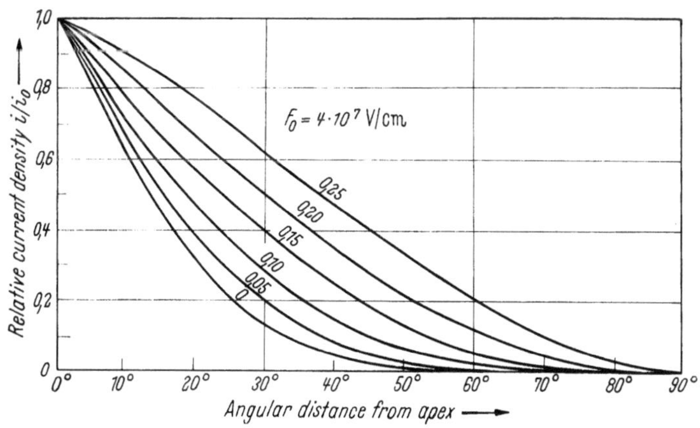

Fig. 7. Decrease of relative current density with angular distance from apex.

true emitter geometry. Within an error of 15% in the ratio of field to voltage the experimental data for W tips are found to agree with the current densities predicted by the F-N theory in a range up to almost 10^7 amp/cm². From there on to higher fields, space charge begins to be effective, as will be shown below. Already HAEFER's[1] experiments give clear evidence that the image force contribution as included in the F-N theory is essential, predicting a 20 to 40% lower field for the

[1] R. HAEFER: Z. Physik **116**, 604 (1940).

same current. DYKE and TROLAN[1] point out that the accuracy of field determination in HAEFER'S work is overestimated and that only by accidental compensation of several errors was the close fit to the F-N theory obtained. In their experiments they monitored the emitting surface for cleanness with the FEM and determined the field more carefully. Also the complicated current density distributions of the different crystallographic planes was taken into account by summing over the single areas. Their data give full evidence for the correctness of the F-N theory within 15% of the field [3].

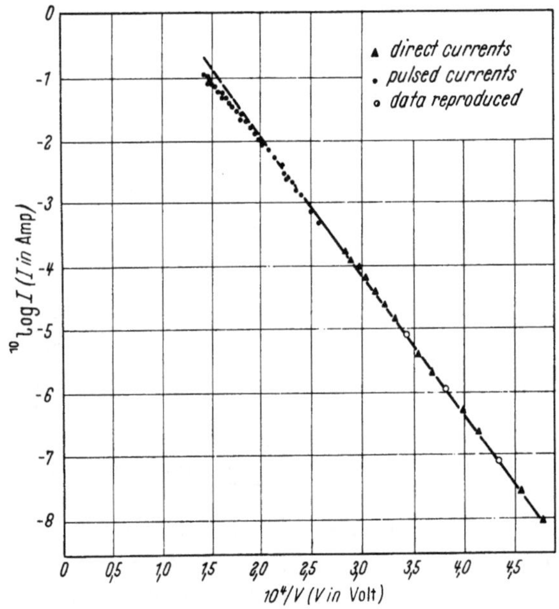

Fig. 8. Logarithm of current versus reciprocal voltage for combined pulse and direct current operation of a tungsten emitter with space charge deviation at large currents.

Beginning at current densities of about $6 \cdot 10^6$ amps/cm², the current no longer rises as fast with the applied voltage as would be expected from the F-N theory. As has been shown by STERN, GOSSLING, and FOWLER[2], space charge should finally limit the current to become approximately proportional to the $\frac{3}{2}$ power of applied voltage. One finds that at a current density of 10^7 amps/cm² the electrons contained in a sheet of $2 \cdot 10^{-6}$ cm thickness would reduce the effective field at the surface by 1%, enough to explain the deviation from the F-N curve as observed by DYKE and co-workers[1] (Fig. 8). According to their observation the space charge limitation starts at the apex of the tip emitter, where field and current density are maximal. In this area the further increase of current density with voltage is small, and the increase in total current comes mainly from the previously less emitting tip areas with higher work functions or lower local field at greater apex angles. However, a part of the deviations may be ascribed to the inaccuracy of the classical image force model when the entire barrier lies within one or two lattice distances from the surface (LEWIS[3]).

The pre-exponential term F^2 in the F-N equation (5.18) cannot be easily deduced from the voltage-current graphs. If one restricts measurements to a small field range, perhaps between 25 and $50 \cdot 10^6$ volts/cm, a slightly better straight line will be obtained with a plot of $\mathrm{Log}\, j/F^2$ versus $1/F$ rather than simply $\mathrm{Log}\, j$ versus $1/F$, but both plots deviate because of the image force correction if one covers a wider range.

8. Dependence on work function. According to the F-N equation the work function is the only material constant affecting field emission. In a plot $\mathrm{Log}\, j/F^2$ versus $1/F$, the slope of the approximately straight line is obtained by taking the derivative of the F-N equation (5.19)

$$\text{slope} = \frac{d\,{}^{10}\mathrm{Log}\,(j/F^2)}{d\,(1/F)} = -2.96 \times 10^7 \varphi^{\frac{3}{2}} s \left(3.79 \times 10^{-4} \frac{F^{\frac{1}{2}}}{\varphi}\right), \tag{8.1}$$

[1] W. P. DYKE and J. K. TROLAN: Phys. Rev. **89**, 799 (1953). — J. P. BARBOUR, W. W. DOLAN, J. K. TROLAN, E. E. MARTIN and W. P. DYKE: Phys. Rev. **92**, 45 (1953).

[2] T. E. STERN, B. S. GOSSLING and R. H. FOWLER: Proc. Roy. Soc. Lond., Ser. A **124**, 699 (1929).

[3] T. J. LEWIS: Proc. Phys. Soc. Lond. B **68**, 938 (1955).

where

$$s(y) = v(y) - \frac{y}{2} \frac{dv(y)}{dy} . \tag{8.2}$$

Some numerical values of $s(y)$ are given in Table 1, page 187.

In the field range that is accessible for d.c. field emission $s(y)$ varies only a few per cent. If one takes as an example a tungsten field emitter with a hemispherical tip of $2 \cdot 10^{-5}$ cm radius and 4.5 e-volts work function, a field of $26 \cdot 10^6$ volts/cm will produce a current of $2.5 \cdot 10^{-9}$ amps or 1 amp/cm², and $s(y)$ is 0.967. The same cathode will emit 25 ma or 10^7 amps/cm² at a field of $64 \cdot 10^6$ volts/cm, and $s(y)$ is then 0.921. The deviation from the straight line will amount to only 4.6% in slope over the whole range. Thus the slope is mainly determined by the work function.

For an experimental check, it is practical to eliminate the uncertainty of the voltage-field strength calibration by using only one emitter with a given geometry and changing its work function by the application of adsorption layers. Since it is impossible to measure thermionic work functions on the small area of a field emitter, it is preferable to prepare layers which are known to give reproducible data on macroscopic thermionic cathodes. The F-N plots obtained by HAEFER[1] Fig. 9 for either heavy coatings of Ba, K, and Cs with the work functions of the bulk adsorbed material or for "monolayers" with minimum work functions have slopes whose $\frac{2}{3}$ power is proportional to the work functions known to occur at such layers. Additional adsorption of contaminations, especially oxygen, can greatly influence the actual work functions, and the uneven distribution on the different crystallographic planes of the substrate, formation of crystallites and clusters and a possible field dependence of the dipole moment of such layers makes the exact confirmation of the $\varphi^{\frac{3}{2}}$ dependence difficult.

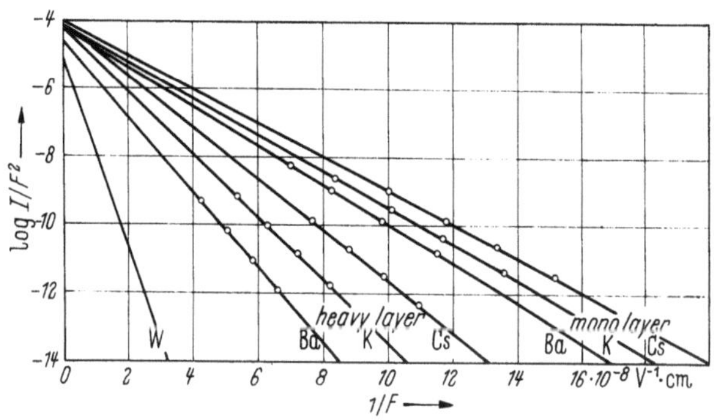

Fig. 9. Current-field characteristics for tungsten, covered with a monolayer or a heavy coating of Ba, K, Cs.

The crystallographic patterns as obtained from clean metals in the field emission microscope were explained as resulting from differences in work functions of the individual crystal planes[2]. Since thermionic data for tungsten single crystals were measured by NICHOLS[3] and SMITH[4], one can correlate the field current density distribution with the thermionic work functions. Except for the 011-plane of tungsten, the differences of the various planes amount to a few tenths of a volt only, so that the experiments themselves, as well as the analysis of the data, require special precautions. The simplest approach is taken by WILKINSON[5], who measures the light output of the different areas in the image of an FEM with a photomultiplier. Since no precautions are taken to avoid light

[1] R. HAEFER: Z. Physik **116**, 604 (1940).
[2] E. W. MÜLLER: Z. Physik **106**, 541 (1937).
[3] M. H. NICHOLS: Phys. Rev. **57**, 297 (1940).
[4] G. F. SMITH: Phys. Rev. **94**, 295 (1954).
[5] M. K. WILKINSON: J. Appl. Phys. **24**, 1203 (1953).

scattering, the maximal contrast, defined as the ratio of apparent current densities in the 116 and 011 planes, is only 35. DYKE and co-workers[1] apply a photographic method and use an aluminum backed screen, which raises the contrast immediately to 1600. However, as MÜLLER points out, there is still a considerable illumination of the dark screen areas by x-rays and secondary electrons inside the tube. MÜLLER[2] therefore uses a collector behind a small hole in the screen in order to measure the current density directly. This raises the contrast between the most strongly emitting 116-plane and the low emitting 011-plane to 10^4 or even 10^5 at lower fields. In a recent paper MÜLLER[3] obtains a full agreement with SMITH's latest thermionic data for all crystallographic planes except 011. For the latter plane the $\varphi^{\frac{3}{2}}$ relation of the F-N theory gives a work function as high as 5.70 to 5.99 e-volts, depending on the degree of frozen-in thermal disorder of that plane. In SMITH's thermionic measurements the weak emission of the 011-plane was almost completely covered by scattered secondaries. The unexpectedly high field emission work function is in accordance with an observation of thermal ionization of aluminum (ionization energy 5.96 e-volts) on a hot tungsten filament. Since all other metals investigated so far exhibit a similar contrast in their field emission microscope patterns, such high work functions may occur generally on the close packed crystal planes. The conventional methods for measuring work functions are not adequate for low emitting surface areas.

9. Velocity distribution. Though the retarding potential method as applied first by HENDERSON and co-workers[4] showed clearly that the field emission originates from energy levels at or below the FERMI level, the true, very narrow energy distribution cannot be observed with their cylindrical arrangement. As RICHTER[5] points out, the field distortions at a possible fine structure of the cathode surface, as well as in the apertures of the perforated anode, produce a tangential velocity component which is then missing in the normal component as measured at the retarding potential collector. If the irregularity on the cathode is represented by a small hemisphere with the radius r sitting on a plane, an emitted electron will travel a distance of the order of r in the original direction normal to the hemisphere before leaving the locally distorted field. By graphical integration RICHTER obtains, for an electron emitted at an angle of 45° to the plane, the tangential energy $W_t = 0.48\, e r F$, where F is the undistorted field above the plane. With $F = 40 \cdot 10^6$ volts/cm, and $r = 10^{-6}$ cm, the tangential energy is 20 e-volts, which would be missing in a plane retarding potential arrangement. In a cylindrical arrangement this loss occurs only for electrons having a velocity component parallel to the axis, while the tangential component in the plane perpendicular to the axis is less by the ratio of the squares of cathode and anode radii. This is true for all tangential components in the spherical arrangement of MÜLLER[6], and the remaining unexpected width of the velocity distribution, still more than 2.5 volts, is due to the distortion in the anode perforation. By using the smooth field emitter of the field emission microscope, a small hole in the bottom of the concentric anode cup and a field free space between this hole and

---

[1] W. P. DYKE, J. K. TROLAN, W. W. DOLAN and F. J. GRUNDHAUSER: J. Appl. Phys. **25** 106 (1954).

[2] E. W. MÜLLER: Z. Physik **120**, 261 (1943). — M. DRECHSLER and E. W. MÜLLER: Z. Physik **134**, 208 (1953).

[3] E. W. MÜLLER: J. Appl. Phys. **26**, 732 (1955).

[4] J. E. HENDERSON and R. E. BADGLEY: Phys. Rev. **38**, 590 (1931). — J. E. HENDERSON and R. K. DAHLSTROM: Phys. Rev. **55**, 473 (1939).

[5] G. RICHTER: Z. Physik **119**, 406 (1942).

[6] E. W. MÜLLER: Z. Physik **102**, 734 (1936).

a second aperture of the anode, MÜLLER[1] obtains finally the velocity distribution in exact agreement with the theoretical width. Fig. 10 gives an improved experimental arrangement[2], Fig. 11 shows schematically the potential distribution in the tube, and Fig. 12 presents the results of the measurements as obtained by differentiating the recorded graph of collector current versus retarding potential for two different cathode temperatures. The temperature tail of the FERMI distribution shows up clearly, and the half width of 0.40 e-volts for the emission at room temperature corresponds closely to the theoretical width for $F = 35 \cdot 10^6$ volts/cm. A small deviation at the high energy side is due to the beginning of secondary emission (reflection) when the incoming electrons strike the collector with velocity of more than a few tenths of a volt. With the cathode at room temperature, where the temperature tail of the FERMI distribution is negligible, the onset of collector current is almost sharp and occurs at a voltage with respect to cathode equal to the work function of the collector material.

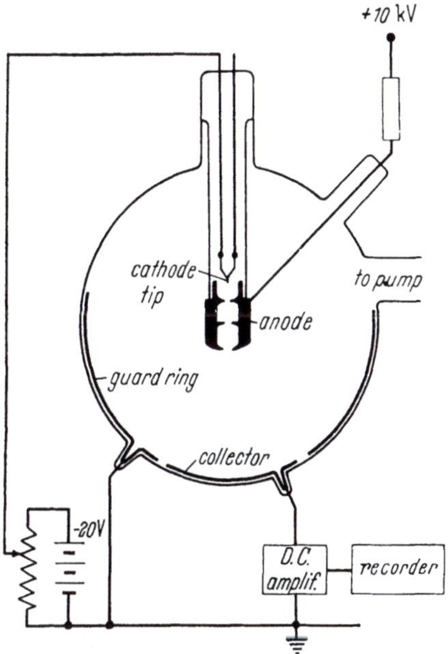

Fig. 10. Retarding potential tube for measuring the energy distribution of field electrons.

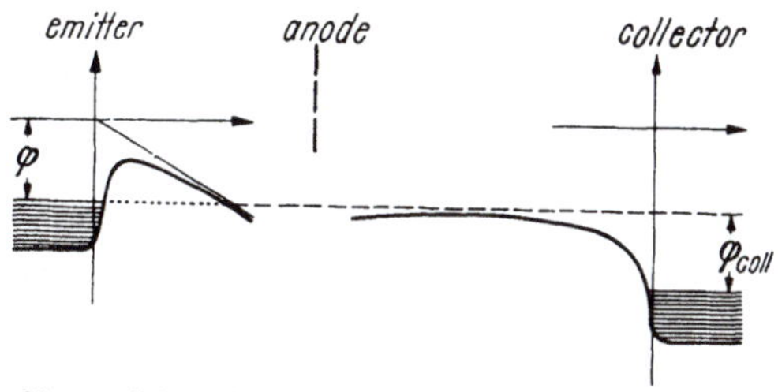

Fig. 11. Schematic diagram of potential distribution for the retarding potential method.

With the accuracy now obtainable in measuring velocity distributions, the temperature of the emitter surface can be derived from an analysis of the current tail above the FERMI level. This would be a worthy project because of the present uncertainty about heat effects connected with field emission. Also, details in the energy band structure may show up in the distribution curve.

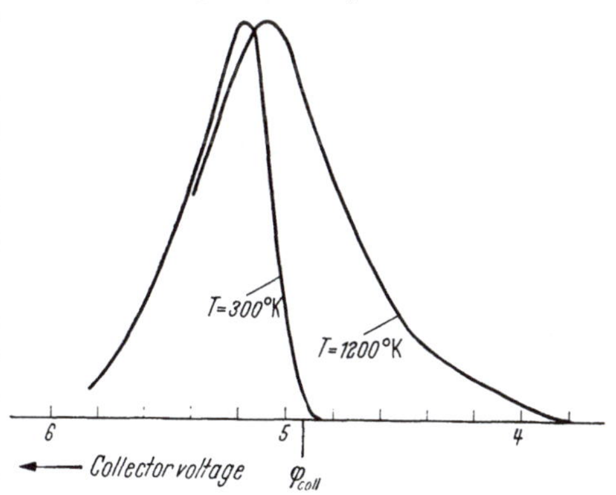

Fig. 12. Measured velocity distribution of field emission from tungsten, $F = 38 \times 10^6$ volts/cm, $T = 300°$ K and 1200° K, ordinates normalized.

10. Temperature dependence. For low temperatures the temperature dependence of field emission is small, since the electrons come primarily from below the FERMI level. This is demonstrated by GOMER and HULM[3] who insert a field emission tube with a tantalum cathode into a bath of liquid helium. No change in the

[1] E. W. MÜLLER: Z. Physik **120**, 261 (1943).
[2] E. W. MÜLLER and K. BAHADUR: Phys. Rev. **102**, 624 (1956).
[3] R. GOMER and J. K. HULM: J. Chem. Phys. **20**, 1500 (1952).

current can be detected, within an accuracy of 0.2%, when the tip is cooled to 2.2° K or warmed up to 300° K. It may be noted that the transition to the superconductive state at 4.2° K does not show up in the field emission either. The magnetic field from the emission current itself is estimated to be too low for quenching superconduction, since the applied current densities are below 80 amps/cm² and range down to 0.1 amp/cm².

At temperatures above 1000° K, the temperature tail of the supply function $N(W)$ becomes increasingly important, and since the transmission function $D(W)$ increases also, electron emission becomes strongly dependent on both the variables, temperature and field. This region cannot be properly investigated by *dc* methods because of the progressive deformation of the hot cathode under the influence of field induced surface migration. It becomes experimentally accessible with microsecond pulse techniques as applied by Dolan and Dyke[1]. They call emission which originates mainly at energy levels between the Fermi level and the peak of the barrier "temperature and field emission", abbreviated T-F emission.

Measuring T-F emission is important for the evaluation of heat effects connected with field emission. Three mechanisms have to be considered: resistive heating of the cathode, heating by the Nottingham process, and cooling by emission of thermally excited electrons. The destruction of very fine cathode tips was attributed by Müller[2] to excessive resistive heat. Haefer[3] calculated the amount of Joule heat developed in a part of the tip assumed to be cylindrical and found that the melting point of tungsten could be reached easily. An improved computation of the resistively heated cone emitter is given by Dolan, Dyke, and Trolan[4]. They consider the resistive heat generation and heat conduction and neglect radiation and temperature dependence of resistivity and specific heat. The differential equation can be solved by expansion in series. Computation of the steady state is relatively simple and leads to a formula $T_{max} = 9.5 \cdot 10^{-4} j^2 r^2$ for the temperature (in °C) reached at the tungsten tip with a hemispherical end of radius r on a cone with a half angle of 11°. The cone angle makes relatively little difference to the results in terms of current density for a given temperature, not more than a factor of three between half angles of 5° and 20°. The rise time in which the steady state temperature is closely approached is not more than 10^{-5} sec. Within one microsecond about one fourth of the final temperature will be reached, allowing twice as large a current density as can be sustained for the steady state with *dc* operation. The result of these computations agrees very well with the experimental determination of the maximal current density that can be obtained in microsecond pulses without destruction of the emitter.

Applying a flat top microsecond high voltage pulse, Dyke and co-workers observe a corresponding flat top current pulse as long as the current density is not too high. At a certain current density or field strength, however, the current no longer remains constant during the pulse duration, but increases steadily. A further increase of field by only 1% results in the initiation of a low impedance arc and a subsequent destruction of the emitter. An arc current of 46 amps is reported which melted the cathode tip down to an irregular ball of more than 15 micron radius. The observed anomalous current rise by about a factor of six just before the initiation of the arc is attributed to T-F emission caused by resistive heating to more than 3000°. At this temperature or at the melting point,

---

[1] W. W. Dolan and W. P. Dyke: Phys. Rev. **95**, 327 (1954).

[2] E. W. Müller: Z. Physik **106**, 132 (1937).

[3] R. Haefer: Z. Physik **116**, 604 (1940).

[4] W. W. Dolan, W. P. Dyke and J. K. Trolan: Phys. Rev. **91**, 1054 (1953).

evaporation of tungsten is thought to provide enough ions in the high current density space in front of the emitter to neutralize the space charge and initiate the explosive arc breakdown. This mechanism has not yet been discussed in detail. Since the concentration of tungsten vapor at the melting point is only 10^{14} cm^{-3}, one can assume that the evaporated tungsten might not suffice to produce enough ions for space charge neutralization. Larger quantities of tungsten can be brought into the electron beam by field evaporation (page 223). Also the field forces will tear the surface of the tip instantaneously after melting.

The large contribution of T-F emission at extreme current densities just before arc initiation requires the consideration of heat effects connected with the emission. The energy distribution curves for field and T-F emission as presented in Fig. 2 in Sect. 6 give an indication of the heat effects which are to be expected. The cathode will be cooled by emitting electrons from an energy level above the FERMI level. Inversely, NOTTINGHAM[1] has proposed also that emission from below the FERMI level should heat the cathode as vacancies are filled up by electrons dropping from higher levels. While the cooling process is evident and well-known from thermionic emission, it is not quite clear whether the NOTTINGHAM process is actually effective or at what distance from the emitting surface the heat will be generated. FLEMING and HENDERSON[2], using a sharply pointed W-Ta thermocouple as emitter, fail to find any heat effect amounting to more than 0.002 e-volt per emitted electron. The energy distribution according to Fig. 2 in Sect. 6 suggests a heat developed by the NOTTINGHAM process of several tenths of a volt for low temperature field emission. At higher temperatures and intermediate fields, about $50 \cdot 10^6$ volts/cm, emission cooling prevails, but at the extreme fields and current densities just before arc initiation, it is not effective enough to balance the resistive heating, as the experiments of DYKE and co-workers clearly indicate.

From the microsecond pulse experiments and the explanation of arc initiation by resistive heating, it would appear that the highest experimentally obtainable current densities are limited to about 10^8 amps/cm^2. This is not necessarily true since MÜLLER[3] reports about the same density obtained by *dc*. He limits the emission to only a small fraction of the total tip surface. A monolayer of thorium for instance can be arranged in such a manner that it covers only the small area comprising the two 111 planes, while the rest of the surface has a high work function and contributes negligibly to the total current. This arrangement can be obtained in a field emission microscope by proper heat treatment of a thoriated tungsten wire or better with much larger differences in work functions by tearing off a monolayer film of thorium with such a field strength, that only the 111 planes keep the thorium adsorbed. (See Sect. 17, on field desorption.) In one case the two 111 planes had a diameter of 120 Å and the *dc* current passing through each of the two patches could be raised for a short time to 0.5 ma, or a density of more than $2 \cdot 10^8$ amps/cm^2. This is still only a small fraction of the total electron supply to the internal surface. If larger pulse currents could be obtained from such a cathode with an extremely small effective area and hence not too much space charge limitation, a new breakdown mechanism might occur when the metal lattice of positive ions is deprived of the neutralizing conduction electrons.

11. Various effects related to field emission. The failure to apply field emission successfully for technical purposes is mainly due to limitations in lifetime. Field

[1] W. B. NOTTINGHAM: Phys. Rev. **59**, 906 (1941).
[2] G. M. FLEMING and J. E. HENDERSON: Phys. Rev. **59**, 907 (1941).
[3] E. W. MÜLLER: Z. Physik **106**, 541 (1937).

emission microscopy reveals that the erratic and unstable performance in a clean, dust-free tube is caused by cathode sputtering. MÜLLER[1] measures the sputtering rate of a monolayer of thorium on tungsten in argon, nitrogen, and hydrogen. For the first two gases it takes only $t \approx \frac{10^{-8}}{ip}$ seconds to remove the thorium film (i = emission current in amps, p = gas pressure in mm mercury) while the effect in hydrogen is 500 times slower. The bare tungsten surface sputters at about the same rate. Although only tips of radii between 2000 and 3000 Å were used, the observed rates apply approximately to all field emitters if they are adjusted to the same current density. The lifetime of an emitter depends of course on the amount of change one considers as admissible. Practically, irregularities which are dangerous in a low impedance circuit build up after removal of 10 to 100 monolayers only. Since the curvature of the field lines close to strongly emitting protrusions cannot be followed by the impinging ions, the sputtering proceeds mainly at the circumferences of the protrusions, making them sharper. The current at constant applied voltage increases gradually until the heat development at some spots accelerates the building up by field supported surface migration and results in a sudden vacuum arc breakdown [*6*]. Thus one can assume the lifetime to be of the order of $t \approx \frac{10^{-6}}{ip}$ seconds. If one maintains a vacuum of 10^{-8} mm in a technical tube, the lifetime at only 1 ma current would not exceed one day. Though it is not difficult to obtain initially a much higher vacuum in a sealed off tube[2], the diffusion of helium from the atmosphere through the glass envelope is serious. The helium pressure builds up to about 10^{-6} mm in a sealed off FEM made of Pyrex glass within a month. On the other hand, commercial FEMs with soft glass envelopes have been working satisfactorily after years of storage. For a really safe performance it is best either to pump the tube continuously by an ionization pump or to smooth the tip by surface migration at elevated temperatures, which has also the advantage of evaporating adsorbed contaminations. While occasional flashing of the cathode is standard procedure in field emission microscopy, continous heating and pulse techniques have been used recently by DYKE[3] to draw T-F currents of several hundred milliamperes over several hundred hours. This will make field emission again attractive for the technical solution of cathode problems where large current densities and small dimensions or transit times are essential.

Other technical problems connected with field emission are the high vacuum breakdown in high voltage tubes[4] on the one hand, and the breakdown in very small gaps which occurs below the minimum sparking potential in the presence of a gas[5] on the other hand and a variety of conditions in between as occurring for instance in gas filled rectifier tubes. Most of these breakdown phenomena are not pure field emission effects, since the total voltage is often more determining than the cathode field. Also the study of formative time delays reveals the occurrence of complex processes. As to the cathode phenomena, it appears that three different mechanisms are to be considered. First, a vacuum arc might be initiated by heating up a small protrusion which carries a large field emission current as investigated by DYKE and co-workers. Second, field emission might be enhanced greatly by the proximity of positive ions, either fixed on insulating

---

[1] E. W. MÜLLER: Z. Physik **106**, 132 (1937).
[2] D. ALPERT: J. Appl. Phys. **24**, 860 (1953).
[3] W. P. DYKE: Proc. Inst. Radio Engrs. **43**, 162 (1955).
[4] J. G. TRUMP and R. J. VAN DE GRAAFF: J. Appl. Phys. **18**, 327 (1947).
[5] L. H. GERMER and F. E. HAWORTH: Phys. Rev. **73**, 1121 (1948).

particles sitting on the cathode (MALTER effect[1,2]) or during their transit time when approaching the cathode[3]. Third, one finds from field emission microscopical observations that extremely small insulating particles on the cathode may decrease the potential barrier at the surface to such an extent that the current density becomes exceedingly large[4].

Field emission from semiconductors has not yet been investigated to a large extent. The high current density required to produce measurable currents from microscopic tips will cause a large voltage drop inside the emitter if the conductivity is low. APKER and TAFT[5] find the emission from a cadmium sulfide tip extremely sensitive to radiation that excites photoconductivity in the crystal. Besides cadmium sulfide other semiconductors like germanium and silicon would be very interesting. However, it is difficult to produce clean cathode tips from these Low-melting materials.

IV. Field emission microscopy.

12. Principle of field emission microscope. In the past fifteen years almost every experimental work on field emission makes use of the field emission microscope (FEM). It was originally invented by MÜLLER[6] in order to obtain the directional distribution of field emission from a sharp point. A hemispherical fluorescent screen at anode potential is arranged around the emitter. The most important feature of the design is the perfection of the hemispherically pointed tip which has to be smooth down to the lattice steps of atomic dimensions. Field electrons leaving the tip surface with low initial energy start to follow paths along the lines of force in essentially radial directions. With tip radii usually between 10^{-5} and 10^{-4} cm, magnifications, essentially the ratio of screen distance and tip radius, in the order of 10^5 to 10^6 are obtained. Fig. 13 shows one of the many designs for such a microscope tube. The shape of the anode, a ring in front of the tip or a conductive coating on the wall, is not of great importance for the image formation. The tube may be permanently connected with the pumping system in order to evaporate degassing substances onto the tip for adsorption

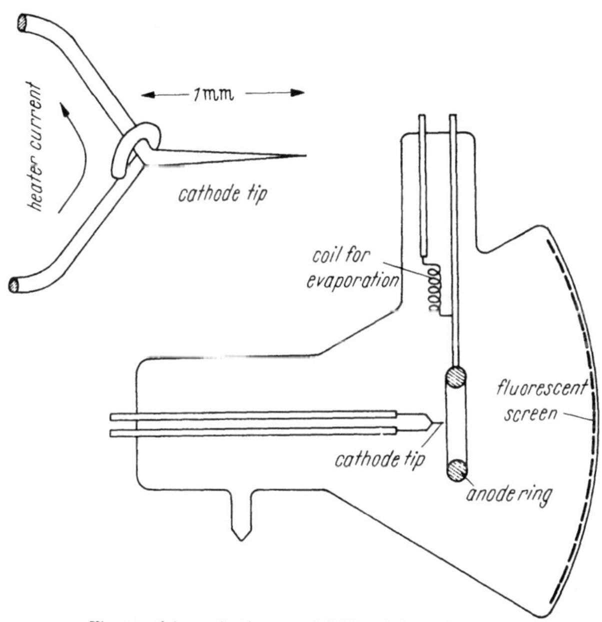

Fig. 13. Schematic diagram of field emission microscope.

[1] L. MALTER: Phys. Rev. 50, 48 (1936).
[2] J. MÜHLENPFORDT: Z. Physik 108, 698 (1938).
[3] W. S. BOYLE, P. KISLIUK and L. H. GERMER: J. Appl. Phys. 26, 720 (1955).
[4] E. W. MÜLLER: Z. Elektrochem. 59, 372 (1955).
[5] L. APKER and E. TAFT: Phys. Rev. 88, 1037 (1952).
[6] E. W. MÜLLER: Phys. Z. 37, 838 (1936). — Z. Physik 106, 541 (1937).

studies or to let in various gases at low pressures. For other purposes it may be advantageous to seal the tube off and to produce an ultra high vacuum by means of a getter or a BAYARD-ALPERT[1] ionization gauge. For quantitative work and for larger currents the fluorescent screen should have a definite potential, provided either by a transparent conductive coating on the glass wall underneath the screen or by an aluminum backing. The emission currents used for adsorption studies are in the microamp range, but under good vacuum conditions the currents can be raised to several hundred microamps, allowing the image on the screen to be projected to some meters diameter in a large auditorium, and permitting motion pictures of surface migration phenomena and single shot microsecond pictures to be taken.

So far field emitters for the FEM have been made only from metals with high melting points. The heater loop should be of the same material as the emitter itself to avoid alloying. The tip wire is etched chemically or electrolytically to the desired cone[2,3]. Tungsten and molybdenum can be etched in a $NaNO_2$ or NaOH melt, tantalum only in the latter or in aqueous HF. Applying a few volts a c or d c between the dipped emitter assembly and the etch bath, which might be a melt as well as an aqueous solution, accelerates the process considerably. Platinum tips can be produced by electrolytic etching as an anode in molten NaCl. Tungsten points can also be obtained easily by the oxidation of the end of the wire in a gas flame, the volatile oxide leaving a sharp point behind. In order to produce extremely fine points, the oxidation of a pre-etched tip can be carried out in the final FEM tube itself under a reduced oxygen pressure[4]. Another method, convenient for resharpening a point which has become too blunt by prolonged heating during use in the FEM, is to apply cathode sputtering[5], preferably by the heavier noble gases or mercury. The positive ions produced in the neighborhood of the tip when field electron current is drawn in the presence of some microns of the gas will drop mainly on the lateral part of the tip and remove cathode metal from there. With skillfull handling the latter two methods yield the finest points with radii of 100 Å or less.

The smoothing of these roughly sharpened points is achieved by surface migration[4,6] during a heat treatment in high vacuum. Becoming evident at about $\frac{1}{4}$ of the absolute melting temperature, surface migration allows the atoms on the rough tip to rearrange themselves, reducing the free surface energy. Perfect smoothness is achieved after a few minutes at about $\frac{2}{3}$ of the absolute melting temperature, provided, however, that no oxides or other contaminations still stable at this temperature are present. While no difficulties are encountered for the highest melting metals, it is extremely hard to achieve satisfactory surfaces with chemically more active metals like zirconium, titanium, or iron.

The determination of the tip radius, as required for measuring the magnification, current density, or surface migration rates, can be done by means of a conventional electron microscope after removing the emitter from the FEM. The accuracy is about 10% for the larger radii and much less for the extremely fine points. For controlling small changes in radius which occur during surface migration experiments and for determining the approximate size of the tip, it is more convenient to derive the radius from the anode voltage that is required

[1] R. T. BAYARD and D. ALPERT: Rev. Sci. Instrum. **21**, 571 (1950). — Cf. vol. 13 of this Encyclopedia, article of D. ALPERT.

[2] E. W. MÜLLER: Z. Physik **106**, 541 (1937).

[3] M. BENJAMIN and R. O. JENKINS: Proc. Roy. Soc. Lond., Ser. A **176**, 262 (1940).

[4] E. W. MÜLLER: Z. Physik **108**, 668 (1938).

[5] E. W. MÜLLER: Z. Physik **106**, 132 (1937).

[6] E. W. MÜLLER: Z. Physik **126**, 642 (1949).

for a given current. Assuming the validity of the F-N theory, the work function of the emitter, and the geometrical shape of a hyperboloid, MÜLLER[1] derives $r = \frac{0.85}{\varphi^2} V^{\frac{3}{4}}$ for the tip radius in Å, if φ is expressed in e-volts and V, the anode voltage for 10^{-5} amps emission, in volts. Though there do occur tips whose shape can be described closely as a hyperboloid, a better approximation can be made by a combination of hyperboloidal and spherical fields as shown by DRECHSLER and HENKEL[2] (see also Sect. 7). The result of their computation is shown in Fig. 14. One has to know the form factor α (contribution of the spherical field as a fraction of one), which can be estimated or measured from the drop in current density with increasing angular distance from the apex. A pure hyperboloidal field ($\alpha = 0$) gives a sharp drop while a bulb shaped tip ($\alpha = 0.25$) has a wide angle picture, showing emission even at 90° from the apex. Sharper tips tend to have smaller form factors; an overall formula $r = \frac{0.15}{\varphi^2} V^{\frac{3}{2}}$ is recommended for evaluating the radius when α is unknown.

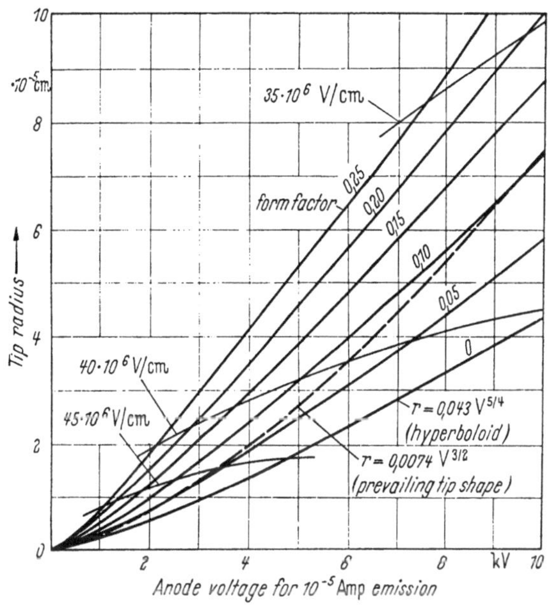

Fig. 14. Chart for determining the radius of a tungsten tip from the anode voltage required for 10^{-5} amps emission.

The FEM is particularly suited for observations in a wide range of temperatures. Elevated temperatures are obtained by heating the cathode loop electrically. Heat conduction in the short cone is sufficient to prevent any large temperature difference between the base of the cone where the temperature can be measured and the fine emitting tip. If ΔT is the temperature difference between the point of measurement of temperature T and tip 0.5 mm distant, a calculation gives $\Delta T = 2°$ K at $T = 1200°$ K and $\Delta T = 86°$ K at $T = 2800°$ K for a tungsten tip with a 5° half cone angle. Temperatures above red heat are measured pyrometrically, above 260° C by infrared photography and comparison with the radiation of a long tungsten wire of known temperature-resistivity relationship. Particularly for low temperature work, with the FEM immersed in a cooling bath, it is convenient to fasten fine potential leads to a central section of the cathode heater loop, and to determine the temperature again from the resistance [*4*].

13. Resolution of the field emission microscope. The most important feature of a microscope is the resolution. For the field emission microscope it is defined as the minimum distance between two objects on the tip which produce separate images on the screen. It can be calculated by finding the image size Δy_s on the screen of an object of width Δy_t on the tip and minimizing the image size Δy_s

[1] E. W. MÜLLER: Z. Physik **126**, 642 (1949).
[2] M. DRECHSLER and E. HENKEL: Z. angew. Phys. **6**, 341 (1954).

with regard to Δy_t; then the resolution δ is the distance between two objects which produce images separated by the distance $\Delta y_{s\,\min}$.

There are three factors which contribute to the broadening of a beam through a spot on the tip: the divergence of the electron orbits due to the tube magnification, the velocities parallel to the surface which the electrons retain when they emerge from the metal, and the diffraction of an electron wave passing through the emitting aperture[1-4].

The first effect is described by the magnification factor M, where $\Delta y_{sm} = M\Delta y_t$ is the spot width which would be observed on the screen if there were no transverse velocity or diffraction effects. If the tip and screen are approximated by parts of concentric spheres of radii r and R, then the magnification is written as $M = \frac{R}{\beta r}$ where the image compression factor β depends on the entire geometry, especially the support of the tip. It can be determined by measuring the angle $45°/\beta$ between the centers of the 011 and the 001 spots on the screen as seen from the tip.

The effect of the electron velocities parallel to the surface can be estimated by using the root-mean-square transverse velocity $\bar{v}_p$. Referring to the distribution of electrons with momentum given by eq. (3.1) one finds that

$$(\bar{v}_p)^2 = \frac{1}{m^2}\int\limits_{-\infty}^{\infty}\int\limits_{-\infty}^{\infty}\frac{(p_y^2+p_z^2)\,dp_y\,dp_z}{e^{\frac{\varepsilon-\zeta}{kT}}+1}\Bigg/\int\limits_{-\infty}^{\infty}\int\limits_{-\infty}^{\infty}\frac{dp_y\,dp_z}{e^{\frac{\varepsilon-\zeta}{kT}}+1}$$

$$= \frac{1}{m^2}\int\limits_{0}^{\infty}\frac{\varrho^3\,d\varrho}{e^{\frac{W-\zeta}{kT}+\frac{\varrho^2}{2mkT}}+1}\Bigg/\!\!\Bigg/\int\limits_{0}^{\infty}\frac{\varrho\,d\varrho}{e^{\frac{W-\zeta}{kT}+\frac{\varrho^2}{2mkT}}+1},$$

where polar coordinates ϱ, ϑ have been introduced in the $p_y\,p_z$ plane and ε has been eliminated according to eq. (5.1). The average here depends parametrically on the x-energy W; it is appropriate to evaluate it at $W=\zeta-d$ which, as discussed in Sect. 5, is the energy at which the maximum number of electrons emerge. The average velocity is then given by

$$\left.\begin{aligned}(\bar{v}_p)^2 &= \frac{1}{m^2}\int\limits_{0}^{\infty}\frac{\varrho^3\,d\varrho}{e^{-\frac{d}{kT}+\frac{\varrho^2}{2mkT}}+1}\Bigg/\!\!\Bigg/\int\limits_{0}^{\infty}\frac{\varrho\,d\varrho}{e^{-\frac{d}{kT}+\frac{\varrho^2}{2mkT}}+1}\\ &= \frac{1}{m^2}\int\limits_{0}^{\sqrt{2md}}\varrho^3\,d\varrho\Bigg/\int\limits_{0}^{\sqrt{2md}}\varrho\,d\varrho\\ &= \frac{d}{m}\end{aligned}\right\}\qquad(13.1)$$

where, as a further approximation, the denominators have been replaced by unity when $\varrho < \sqrt{2md}$ and by infinity when $\varrho > \sqrt{2md}$. The additional width of the spot on the screen due to the transverse velocity only is given simply by $\Delta y_{sv} = 2\bar{v}_p\tau$ where τ is the transit time to the screen. The potential drop from

[1] M. Benjamin and R. O. Jenkins: Proc. Roy. Soc. Lond., Ser. A **176**, 262 (1940).
[2] E. W. Müller: Z. Physik **120**, 270 (1943).
[3] F. Ashworth: Adv. Electronics **3**, 1 (1951).
[4] R. Gomer: J. Chem. Phys. **20**, 1772 (1952).

tip to screen takes place mostly near the tip so the electrons travel mainly with the final velocity and

$$\tau = R\sqrt{\frac{m}{2eV}}, \tag{13.2}$$

where V is the potential difference between tip and screen.

The diffraction broadening can be estimated from the wave mechanical uncertainty principle. As an electron passes through the spot Δy_t the uncertainty in its y-momentum is greater than $\hbar/2\Delta y_t$, so the y-velocity will be about $\hbar/2m\,\Delta y_t$ and

$$\Delta y_{sd} = 2\,\frac{\hbar}{2m\,\Delta y_t}\,\tau$$

is the diffraction contribution to the image width.

Postponing for the moment the detailed substitution for these quantities, the total width of the spot on the screen is given by

$$(\Delta y_s)^2 = (\Delta y_{sm})^2 + (\Delta y_{sv})^2 + (\Delta y_{sd})^2 = M^2(\Delta y_t)^2 + 4(\bar{v}_p)^2\tau^2 + \frac{\hbar^2}{m^2(\Delta y_t)^2}\tau^2.$$

The minimum value of this expression as a function of Δy_t is

$$(\Delta y_s)^2_{\min} = \frac{2\hbar\,\tau M}{m} + 4(\bar{v}_p)^2\tau^2,$$

so the resolution δ is

$$\delta = \frac{(\Delta y_s)_{\min}}{M} = \sqrt{\frac{2\hbar\tau}{mM} + \frac{4(\bar{v}_p)^2\tau^2}{M^2}}. \tag{13.3}$$

Here one can put in the results found above for $M, \bar{v}_p, d$ (setting $t=1$) so that

$$\delta^2 = \sqrt{\frac{2}{meV}}\,\hbar\beta r + \frac{\hbar F\beta^2 r^2}{\sqrt{2m\varphi}\,V}. \tag{13.4}$$

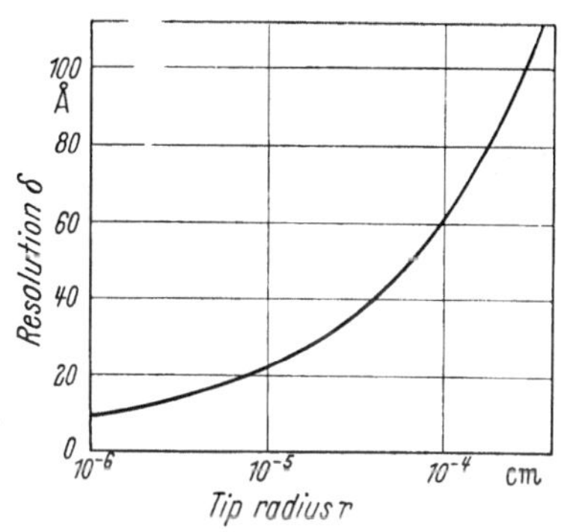

Fig. 15. Resolution of a typical field emission microscope, computed for $\varphi = 4.5$ e-volts and an emission of 10^{-5} amps.

For practical purposes the field at the tip and the applied voltage are related as follows:

$$\frac{F}{V} = \frac{k}{r},$$

where k is roughly constant. This permits the resolution to be written finally as

$$\delta = \sqrt{\frac{\hbar k\beta^2 r}{\sqrt{2m\varphi}}\left(1 + \frac{2}{k\beta}\sqrt{\frac{\varphi}{eV}}\right)}. \tag{13.5}$$

A typical set of operating conditions is: $k=\frac{1}{5}$, $\beta=1.5$, $r=10^{-5}$ cm, $\varphi=4.5$ e-volt, $V=2500$ volt. These conditions give for the resolution $\delta = 23$ Å. These arguments may be verified experimentally by observing the minimum size of spot which appears on the screen; this will be the resolution times the magnification. For a distance to the screen of 5 cm and otherwise operating conditions as above, this predicts a minimum spot size of 0.8 mm, which agrees with the experiments[1]. Fig. 15 shows how the resolution varies with tip radius for a fixed total current of 10^{-5} amps.

[1] E. W. MÜLLER: Z. Physik **120**, 270 (1943).

ROSE[1] extended similar considerations to hemispherical projections on an otherwise smooth tip of relatively large radius. With a protrusion radius of 10^{-7} cm a resolution of 3 Å can be expected for an object on the top of the protrusion.

14. Emission from clean surfaces. Field emission microscopy reveals immediately whether the observed surface is actually clean; the criterion is that the emission pattern should be independent of the annealing temperature, except perhaps for small changes in the relative sizes of the dark areas. Only recently high vacuum experimentalists became aware of the fact that almost all experiments dealing with allegedly clean surfaces and their reactions were actually made on surfaces covered with adsorption layers (see references [*2*], [*4*], and [*6*]). Even at a vacuum of 10^{-8} mm Hg, it takes only minutes to build up a monolayer of contaminations on a freshly cleaned metal surface[2]. However, where modern ultra high vacuum techniques[3] are applicable, it is no problem to reduce the pressure of residual gases far below 10^{-10} mm. The field emitter can be cleaned by heating, provided the oxides or other impurities possess a higher vapor- or decomposition-pressure than the metal, so that they disappear before the tip is blunted too much by surface migration. Tungsten, molybdenum, rhenium tantalum, niobium, platinum, and nickel are easy to manipulate because of the volatility of their oxides and the high degree of purity in which they are available. The presence of impurities can often be observed by their tendency to form an enriched surface state after annealing for some time. This is particularly typical for carbon and silicon on tungsten, molybdenum and nickel and can be used to determine the activation heat of bulk diffusion by measuring rates of appearance at the surface after flashing the tip. Annealing in oxygen or hydrogen [*6*] can remove such impurities.

Only a few planes of low index number appear on a clean tip (Fig. 16 and 17). Electron microscope studies confirm, as far as the limited resolution allows, the conclusion drawn from experiments with evaporation of Ba or W from one side, that the tip consists of small flat regions at the lowest index directions, merging without sharp edges into spheroidal regions which contain all other possible indices. MÜLLER[2] assumed the shape of a tungsten tip after long annealing to be a dissolution form, since a continuous increase in radius was observed. However, this is true only for the more hyperboloidal form, while the bulk form obtained on very slim cones remains comparatively steady and cannot be far from the equilibrium form, which might be spheroidal as well. Experimentally one finds an increase of the size of the flats at lower annealing temperatures[4] in agreement with HERRING's[5] theoretical consideration. The tip can certainly not assume the equilibrium form of an isolated crystal, since its shape results from a competition between the migration of the atoms towards the shank due to the gradient of free surface energy and the opposite motion of matter due to the temperature gradient. MÜLLER uses the increase of tip radius with prolonged heating to measure the heat of activation of tungsten on its own lattice (106500 cal per mol)[2]. At lower temperatures the changes in tip radius are due to surface migration, while at very high temperatures, bulk diffusion could be significant. HERRING finds that the blunting rate results in a $1/r^3$ dependence when volume

---

[1] D. J. ROSE: J. Appl. Phys. **27**, 215 (1956).
[2] E. W. MÜLLER: Z. Physik **126**, 642 (1949).
[3] D. ALPERT: J. Appl. Phys. **24**, 860 (1953).
[4] J. A. BECKER: Bell Syst. Techn. J. **30**, 907 (1951).
[5] C. HERRING: Structure and Properties of Solid Surfaces, edit. by R. GOMER and C. S. SMITH. University of Chicago Press 1953; also: J. Appl. Phys. **21**, 301 (1950).

diffusion is effective and a $1/r^4$ dependence when surface migration predominates. Experiments show that the blunting rate becomes extremely low when a radius of 6 to $8 \cdot 10^{-5}$ cm is reached. A second method of measuring activation energies for surface migration on tungsten is to produce a surface deformation and determine the rate of re-establishment of the rounded form. The deformation can be obtained either by evaporating the same material onto the tip[1] or by exposing the heated tip to a high positive or negative field and freezing in the built up form, as described in detail by BENJAMIN and JENKINS[2] and by BECKER[3]. For the migration of tungsten evaporated on tungsten activation energies between 80000 and 106500 cal/mole are observed on various crystal planes[1]. They agree sufficiently with data calculated with a method developed by STRANSKI[4]. A remarkably low activation energy of only 30000 cal/mole for the surface migration on the closely packed 011 plane suggests the visualization of individual W atoms[1].

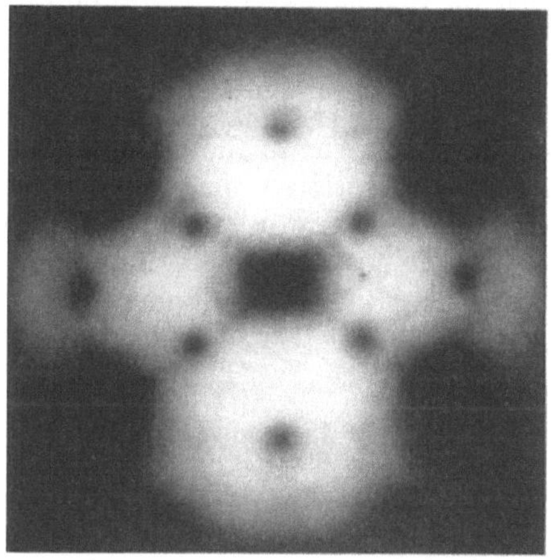

Fig. 16. Field emission pattern of clean tungsten ($r = 2.5 \times 10^{-5}$ cm, $V = 3500$ volts, $i = 10^{-5}$ amps).

The increase of radius occurs on a smooth tip by gradual dissolution of the edges of the predominant lattice planes. This can be shown indirectly by observing adsorbed material like SrO or Al_2O_3 on tungsten. When the tip is heated to about 1400° K, collapsing rings, representing edges of 011 planes and tagged by the adsorbate, can be seen in the emission pattern [6]. Without interference of an adsorbate, the emission of the single lattice steps is too weak to appear in the great contrast between the dark 011 plane and the bright surroundings, unless one takes special precautions[5]. The best method is to observe with

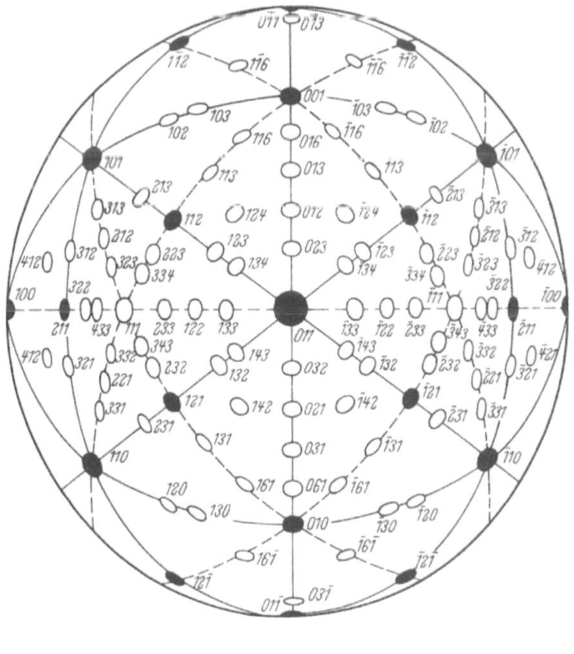

Fig. 17. Crystallographic map for field emission pattern of a cubic crystal.

[1] E. W. MÜLLER: Z. Physik **126**, 642 (1949).
[2] M. BENJAMIN and R. O. JENKINS: Proc. Roy. Soc. Lond., Ser. A **176**, 262 (1940).
[3] J. A. BECKER: Bell Syst. Techn. J. **30**, 907 (1951).
[4] I. N. STRANSKI and R. SUHRMANN: Ann. Physik **1**, 153 (1947).
[5] E. W. MÜLLER: J. Appl. Phys. **26**, 732 (1955).

micro-second pulses [1]. Above $5 \cdot 10^6$ amp/cm² the contrast due to different work functions is much diminished and the collapsing rings can be observed at tip temperatures above 1700° K. A low duty cycle minimizes field induced migration, so that the thermal surface migration rate can be measured quite accurately. 73000 cal/mole is obtained for the heat of activation on tungsten and 60000 cal/mole on tantalum.

The spherical shape of the annealed clean tip suggests that the appearance of the dark areas in the crystallographic pattern is almost entirely due to the high work function of the low index planes (Sect. 8). The relative degree to which these planes show up is not a function of the crystal structure alone. While W and Mo give exactly the same patterns with the predominant 011 plane and pronounced 112 and 001 planes, the latter ones show up only very weakly on

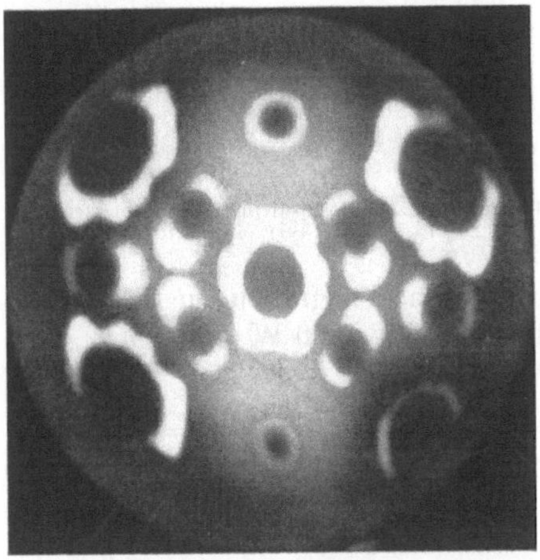

Fig. 18. Tungsten condensed at 1290° K on a tungsten tip. Sharp edges are built up around the predominant planes.

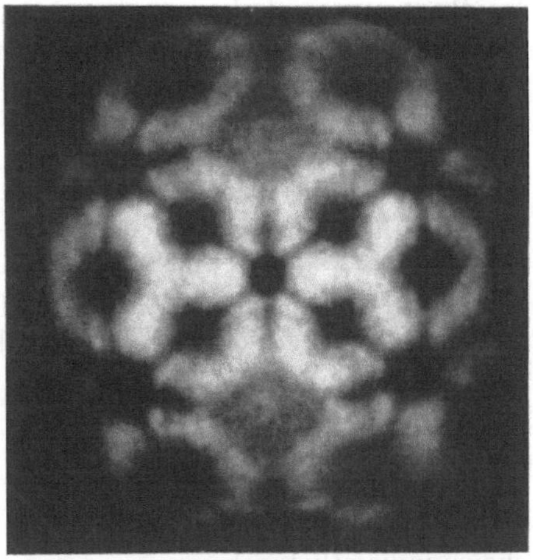

Fig. 19. Barium at low degree of coverage on a tungsten tip.

Ta and Nb tips with the same body centered cubic lattice. Face centered cubic crystals like Pt and Ni [*6*] [2] exhibit 111 and 001 most predominantly, while the hexagonal Re has the lowest emission in the close packed (01$\bar{1}$0) plane [3].

15. Adsorption phenomena. Electron emission methods can be effective tools for studying adsorption phenomena. This is particularly true for field emission microscopy which takes advantage of the change in work function connected with adsorption and of the wide accessible range from liquid helium temperature to above 2000° K. Beautiful pictorial illustrations of the anisotropy of adsorption on single crystals, sometimes under well-defined conditions of cleanness, are easily obtainable in a large magnification (see Fig. 19, 20 and References [*6*], [*7*], and [*8*]). However, only in a few cases has it been possible to get quantitative information on the parameters of the adsorption isotherm. What one observes directly is the specific pattern on the screen, the brightness of which can be, with some precautions, converted into current density at the emitting surface or into effective work function. The determination of surface density or degree of covering for the various crystal planes and the geometrical arrangement of the adsorbate on an atomic scale remains the goal of an interpretation which is still quite vague at the present time.

[1] J. K. Trolan, J. P. Barbour, E. E. Martin and W. P. Dyke: Phys. Rev. **100**, 1646 (1955).

[2] R. Gomer: J. Chem. Phys. **21**, 293 (1953).

[3] G. Barnes: Phys. Rev. **97**, 1579 (1955).

Changes in work function occurring with the adsorption of a substance different from the substrate metal can be described as arising from a dipole layer. Electropositive atoms like Ba repel electrons at the surface, while electronegative atoms like oxygen attract electrons, the excess charge being often only a fraction of one electron per adsorbed atom. The resulting change in potential across this dipole sheet as appearing in macroscopic contact potential measurements or in thermionic emission is $\Delta\varphi/e = 2\pi\Theta M$, where M is the dipole moment per unit area at full coverage and Θ the actual fraction of coverage. In the case of field emission, however, only the potential distribution in the barrier close to the surface is significant, so that the discreteness of the adsorbed atoms results in a slight decrease in $\Delta\varphi$, particularly noticeable at low degrees of coverage and high fields[1].

The determining exponent in the F-N equation contains the factor $\varphi^{\frac{3}{2}}/F$, so that a change in current density may

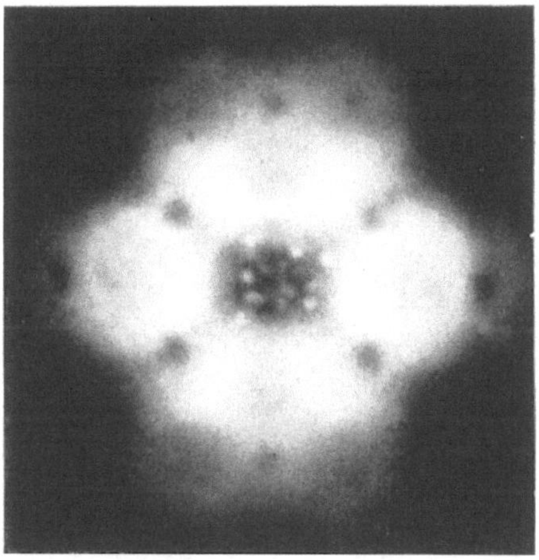

a

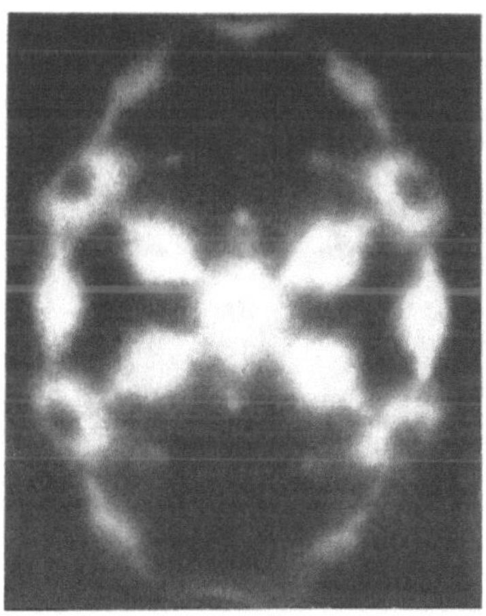

b

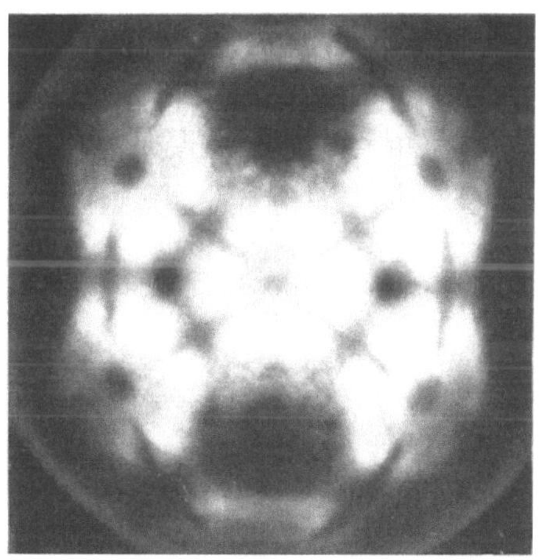

c

Fig. 20a—c. Caesium adsorbed on tungsten tip. The field emission microscope is immersed in liquid air to prevent surface migration. (a) Very low coverage, probably individual Cs atoms on 011 plane in center. (b) About half a monolayer of Cs adsorbed. (c) Monolayer of Cs with minimum work function.

be due either to a local field enhancement or to a decrease in work function. In some cases the adsorption of a substance occurs either in patches, clusters, or small crystallites or it may cause the substrate to rearrange itself into a sharply edged form rather than the smooth form of the uncovered surface, as soon as the temperature is high enough to allow surface migration. Overgrowths of new phases or compounds oriented to the substrate structure may occur as well. Such protuberances give not only an enhanced current density but also an additional local magnification due to the divergence of the lines of force

[1] R. Gomer: J. Chem. Phys. **21**, 1869 (1953).

above the adsorbate. While in some cases the local field enhancement can easily be recognized by the large contrast of the emitting particles or ridges, it is not always possible to distinguish this effect from a locally reduced work function.

Field emission microscopy of adsorption layers started with simple observations of the arrangement of barium and thorium on tungsten points[1], establish-

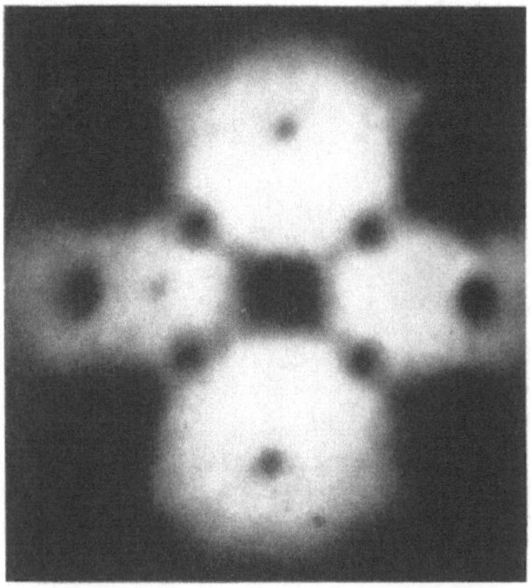

Fig. 21 a.

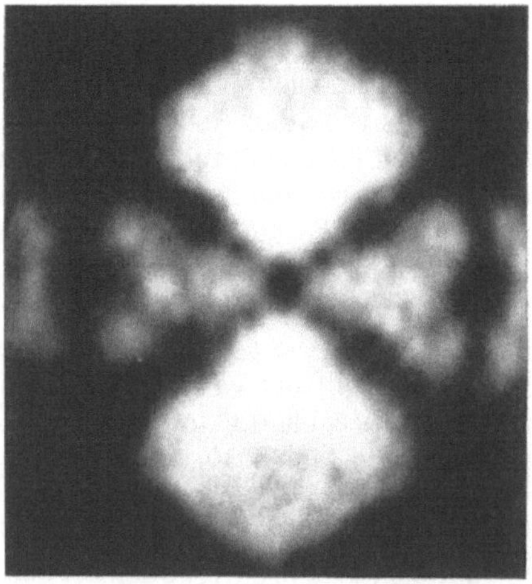

Fig. 21 b.

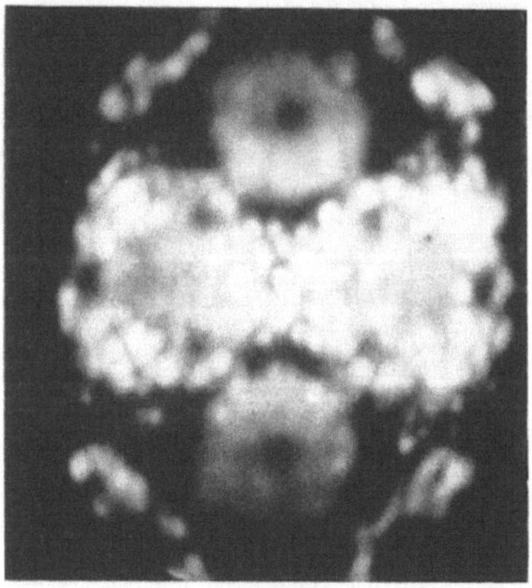

Fig. 21 c.

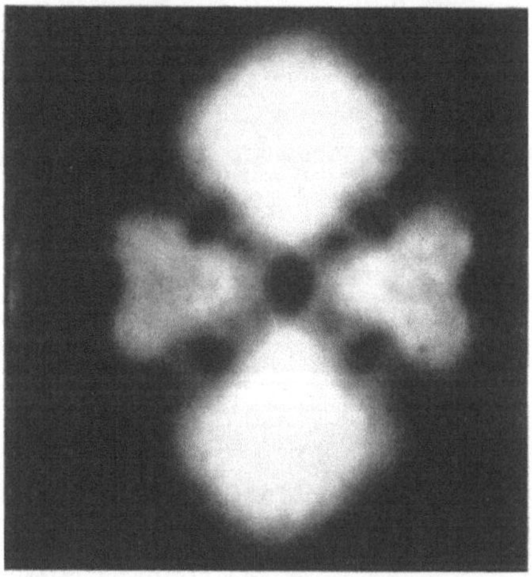

Fig. 21 d.

ing clearly the significance of surface migration, which was not yet unanimously recognized by then. More data were subsequently gathered by BENJAMIN and JENKINS[2]. STRANSKI and SUHRMANN[3] as well as ASHWORTH [1], explain the observed properties of Ba, Cs, Th, Na, and Cu films on the basis of their detailed consideration of the geometrical structure of the various tungsten planes and the fit of the adsorbed atoms. However, that geometrical structure is not the only determining feature is shown by comparative adsorption of carbon on

---

[1] E. W. MÜLLER: Z. Physik **106**, 541 (1937); **108**, 668 (1938).

[2] M. BENJAMIN and R. O. JENKINS: Proc. Roy. Soc. Lond., Ser. A **180**, 225 (1942).

[3] I. N. STRANSKI and R. SUHRMANN: Ann. Physik **1**, 169 (1947).

W and Mo tips (References [6] and [7]). The lattice constants differ only by 0.6%, but the adsorption patterns are quite different. For such experiments a twin tube is most suitable, that is a field emission microscope with two independent cathode tips, one of W and the other of Mo, both working under the same vacuum conditions and evaporated simultaneously from one source equidistant

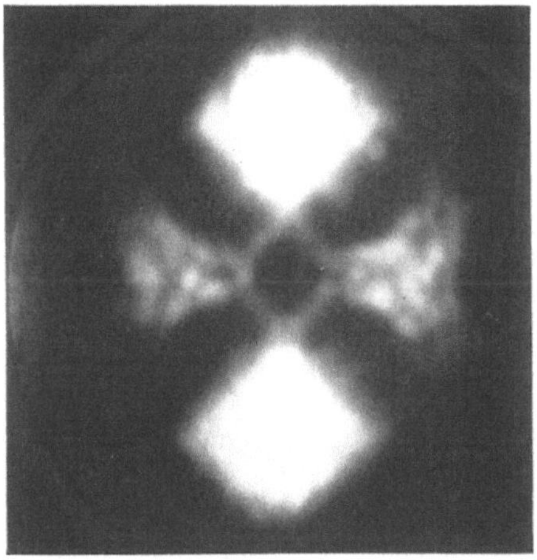

Fig. 21e.

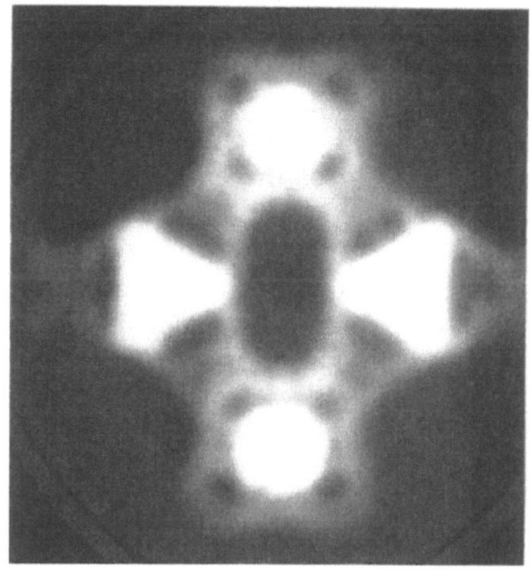

Fig. 21f.

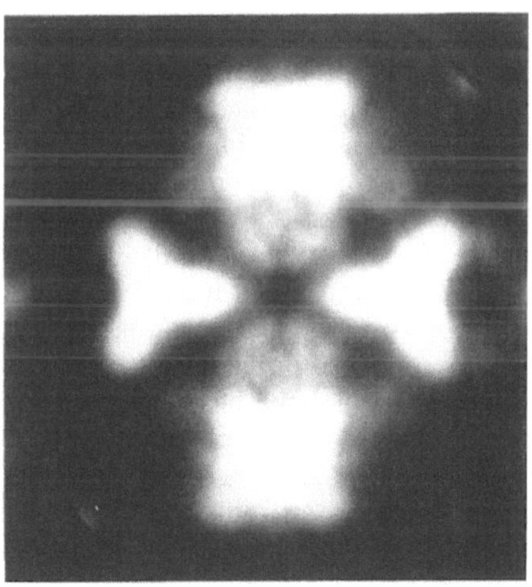

Fig. 21g.

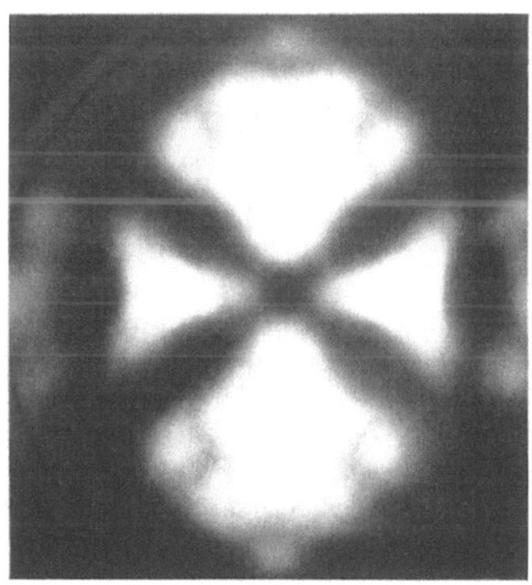

Fig. 21h.

Fig. 21 a—h. Oxygen adsorbed on tungsten tip, $r \approx 10^{-5}$ cm. The anode voltages are given for 10^{-6} amp emission. (a) Clean tungsten surface, 2410 volts. (b) tip exposed for about 30 seconds to 5×10^{-8} mm of oxygen, 2800 volts. (c) Adsorption film almost saturated after exposure to 10^{-5} mm of oxygen, 3860 volts. (d) Oxygen film partly desorbed by heating the tip for one minute at 1000° K in high vacuum, 3680 volts. (e) Further desorption for one minute at 1400° K, picture taken at room temperature and 3330 volts. (f) Desorption continued at 1700° K, 3200 volts. (g) Desorption continued at 1900° K, 2850 volts. (h) Desorption continued at 2100° K, 2470 volts. One minute annealing at 2200° K is required to re-establish the clean tungsten surface.

to each tip, or exposed to the same gas for building up the adsorption layer. A detailed study of surface migration of barium on the W tip is made by DRECHSLER[1].

[1] M. DRECHSLER: Z. Elektrochem. **58**, 327, 340 (1954).

The adsorption system that attracted the interest of several experienced workers from the beginning of field emission microscopy to the present time and that still leaves many questions open is oxygen on tungsten[1], [2], [4], [6]. If one admits oxygen at a constant pressure of $5 \cdot 10^{-8}$ mm, the gradual formation of the adsorption layer occurs slowly enough to be followed in detail (Fig. 21). The emission decreases first in the regions of the 111 zones, then around 112 and 111. When more oxygen deposits the region around 001 becomes much darker. At a constant anode voltage the total emission drops by about five orders of magnitude while the adsorption film is built up. Knowing the pressure, the number of impinging O_2 molecules can be calculated. With APKER's flash filament method[2] one determines further the amount of oxygen that is actually adsorbed on the surface of the entire cathode loop. The ratio of both numbers per unit area gives the sticking probability of oxygen molecules impinging on tungsten. It is of the order of $\frac{1}{10}$ and remains constant until about $6 \cdot 10^{14}$ oxygen atoms per cm² occupy the surface. This may be considered as a monolayer. Further admission of oxygen results in continued adsorption with a rapidly decreasing sticking equivalent to two or three monolayers. Unfortunately no distinction can be made for evaluating the degree of coverage on individual crystal planes. The contribution to $\Delta\varphi$ of a single adsorbed atom may be different on the various planes. The surface migration can be studied by immersing the FEM into liquid helium[3]. Oxygen produced in the tube by thermal decomposition of CuO covers only the one side of the tip showing towards the gas source, since no reflection of the O_2 molecules from the cold walls is possible. If less than about a monolayer of oxygen is deposited, the adsorption film starts to spread over the whole tip only after it has been heated to 450° K. GOMER [4] determines the activation energy to be about 30000 cal/mole, while MÜLLER[4], using field desorption for the removal of oxygen from the very tip of the emitter, finds only half as much probably because his experiments were made on a surface which still contained some strongly bound oxygen in deeper sites[5]. The thin layer formed at helium temperature does not seem to differ from that formed at room temperature, suggesting that no appreciable activation energy is required for the bond. In a second adsorption layer surface mobility is already detectable at 40° K. More information is obtained by observing the oxygen covered tip at higher temperatures in a very high vacuum. Heating to 700 or 1000° K produces apparently the agglomeration of small oxide crystallites superimposed on the main pattern. Whether the many details that can be observed at prolonged heating at higher temperatures are due to formation and epitaxy of an oxide or simply a progressive desorption of the second physically bound molecular layer and subsequently of the first chemically bound atom layer is still debated [2], [4]. Field ion microscopy (Sect. 20) reveals that the bright areas on 111 planes in Fig. 21f—h are caused by built up edges rather then by a low work function. The final traces of adsorbed oxygen disappear only after some heating above 2000° K. No other adsorption system has been studied in such a wide temperature range. The desorption of the oxygen film can also be carried out by chemical reduction with hydrogen, which attacks the film at about 1200° K beginning at the 116 planes. At 1500° K the reduction is complete within two minutes. In these experiments the FEM is immersed in liquid air to bind water as the

[1] E. W. MÜLLER: Z. Physik **106**, 541 (1937). — Z. Elektrochem. **59**, 372 (1955).
[2] L. APKER: Industr. Engng. Chem. **40**, 846 (1948).
[3] R. GOMER and J. K. HULM: J. Amer. Chem. Soc. **75**, 4114 (1953).
[4] E. W. MÜLLER: Z. Elektrochem. **59**, 372 (1955).
[5] R. GOMER: Private communication.

reaction product. Numerous other adsorbates have been studied on tungsten surfaces, and a few adsorption patterns have been published. Co, U, and Zr (Fig. 22) exhibit peculiar surface migration effects [7]. UO_2, CuO_2, Al_2O_3 (Fig. 23) and BeO give beautiful epitaxy patterns with preferences for the 112,

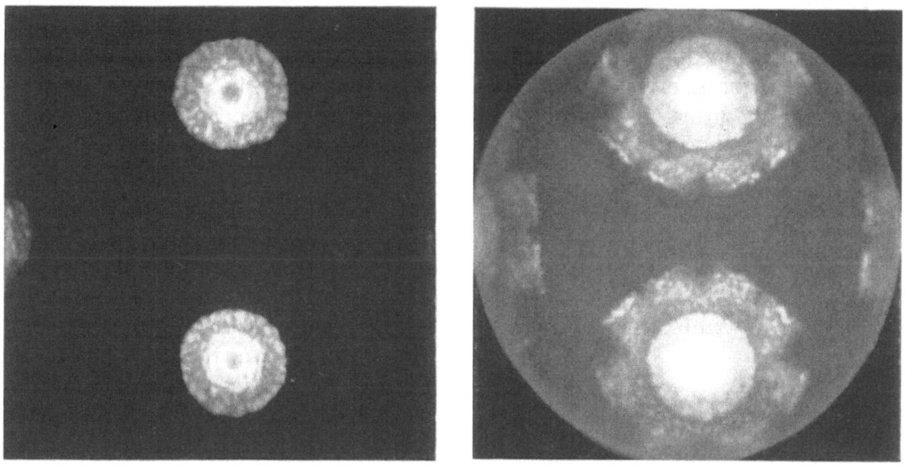

a b

Fig. 22 a and b. Zirconium adsorbed on tungsten. (a) Arrangement around the 001 planes after brief annealing at 1600° K. (b) Spreading by surface migration after one minute at 1300° K.

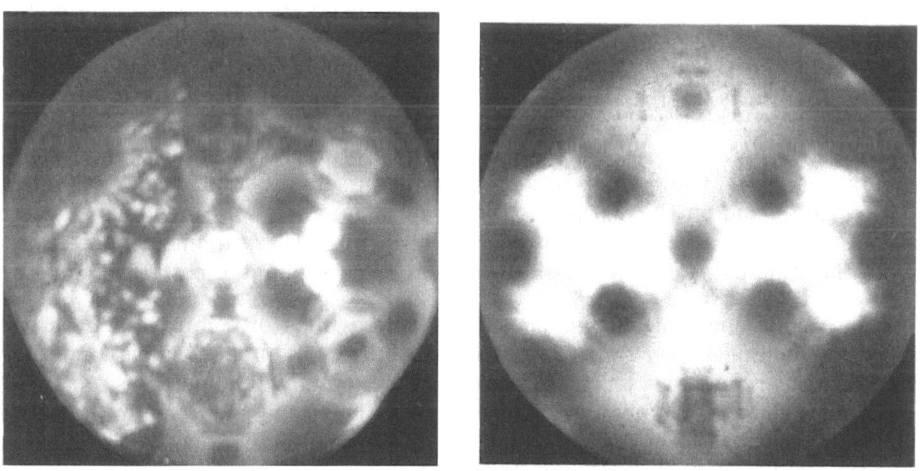

a b

Fig. 23 a and b. Aluminum oxide evaporated on to a tungsten tip. (a) Aluminum oxide was condensed on the left side of the tip and then spread at 1600° K within 10 seconds. Some crystallites remain on the left half while a thin film covers the right half of the pattern. (b) Very small amount of aluminum oxide with characteristic epitaxis on and around the 001 planes.

013, and 001 planes respectively. More or less clean surfaces of Mo, Pt, Ni, Fe, and Ti, some with identified contaminations have been presented. A particularly interesting substance is carbon [6], Fig. 24, which can be deposited on tungsten by decomposing organic substances or by evaporation from a carbon filament. Above 1000° K surface migration spreads the carbon crystallites over the surface and rearranges the tungsten substrate to form 3 to 8 parallel ridges surrounding the 011 planes and sometimes the 112 planes, while the 334 planes appear as dark regions. The latter ones are so characteristic for carbon that they may be used to indicate the presence of minute amounts of organic substances such as pump oils with vapor pressures of 10^{-8} mm Hg. Presumably the 334 spots are due to epitaxy of a flat graphitic arrangement of carbon atoms with a misfit of 5%. A part of the carbon diffuses into the interior and after the tip has been flashed at 2800° K to produce a clean surface, it can be brought out again by an extended annealing process at lower temperatures, with a rate corresponding to an activation energy of 110000 cal/mole. This is not necessarily the energy step for bulk diffusion of carbon in tungsten but rather the bulk diffusion of tungsten itself which is required to build up the characteristic ridges. Adsorbed carbon can be removed in the 1100 to 1200° K temperature range by hydrogen or oxygen. Since graphite in bulk may be oxydized at lower temperatures, Klein[1] suggests that the breaking of W—O bonds has to occur before the reaction can start. Large amounts of carbon deposited on W or Mo emitters form a carbide tip with a typical hexagonal pattern [6]. Schleicher[2] suspects that also his iron tip might have been nonmetallic but rather Fe_3O_4 with sufficient conductivity to sustain the required high current density for producing a picture.

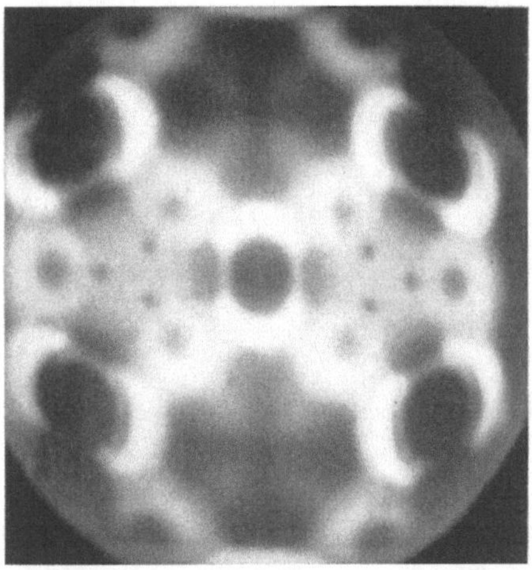

Fig. 24. Carbon adsorbed on tungsten, after annealing at 1200° K, with characteristic enhanced emission around the 011 and 112 planes and with dark 334 planes.

In principle field emission microscopy could be applied to the study of some more surface reactions besides the oxidation of carbon or the reduction of an oxygen film. The decomposition of NH_3 with adsorbed nitrogen remaining chemically bound on the W tip was tudied by Korb [6]. Only preliminary tests have been made with CO and NH_3 on Ni and Pt tips. These are more interesting from the standpoint of catalysis but yield extremely complex patterns.

16. Atomic and molecular objects. Adsorption films of comparatively large atoms or molecules may show a granular structure. Müller[3] assumes that the finest spots on the screen represent blurred images of individual particles protruding from the smooth substrate. Such granulation is observed not only with elements which decrease the work function due to their electropositive character, like the alkali and alkali earth metals, and also zirconium, thorium, and uranium, but as well with substances like sulfur, selenium, barium oxide, and many others which occur in molecular form. Evaporated upon the cold tip, these substances condense in irregular small clusters and crystallites perhaps 10 to 50 Å in size. After some heating, surface migration spreads the material more evenly so that all crystal planes of the same structure assume the same granulation density. Müller[4] gives the following arguments for his assumption that individual atoms

---

[1] R. Klein: J. Chem. Phys. **21**, 1177 (1953); **22**, 1406 (1954).
[2] H. W. Schleicher: Z. Naturforsch. **7**a, 471 (1952).
[3] E. W. Müller: Z. Physik **108**, 668 (1938).
[4] E. W. Müller: Z. Naturforsch. **5**a, 473 (1950).

are visualized: 1. A barium film on tungsten shows the granulation only if the overall degree of coverage is low enough to insure an average distance between the particles in the order of the resolving power. This occurs when the average work function indicates a degree of coverage Θ of about 10% of a monolayer, corresponding to an average distance of 20 Å. BECKER[1] finds that for all values of Θ from 1.0 to 0.1 the emission comes largely from clusters. This discrepancy still needs to be explained. It appears to be difficult to produce really pure barium deposits. Small traces of contaminations suppress surface migration spread and promote clustering. 2. A known average degree of covering can be obtained in a twin tube[2] as shown schematically in Fig. 25. One tip is placed five times as far from a point source of barium as the other. If one evaporates enough barium to deposit just a monolayer on the nearby tip, the more distant tip receives $\frac{1}{25}$ of a monolayer. In a second experiment, one produces an equally thin layer on the near tip as indicated by the similarity of appearance, and obtains consequently on the distant tip $\frac{1}{625}$ of a monolayer. Assuming $2.5 \cdot 10^{14}$ barium atoms/cm² for the monolayer (the exact value is not known and depends on the crystalline structure of the substrate) the average distance between two atoms on the distant tip is in the order of 160 Å, which makes it easy to count the spots which appear on the screen. Actually one finds about one half of the expected number. The missing atoms are probably adsorbed on sites where they do not protrude sufficiently. 3. One can estimate the relative intensity of the electron beam emitted through a single adsorbed barium atom, by determining through numerical integration the transmission coefficient of the potential barrier in front of an adsorbed atom. The shape of the barrier is obtained from a model and includes the field distortion due to the atom and the image force. Multiplying the transmission coefficient by the supply function $N(W)$ and distributing the total current over the area of the image disk on the screen, gives the current density in the image, which can be compared with the current density in the bare crystal plane. The ratio of current densities is found to be 2.5 for a single tungsten atom sitting on a closely packed 011 tungsten plane, and 31 for a barium atom at the same place; this is about what one observes.

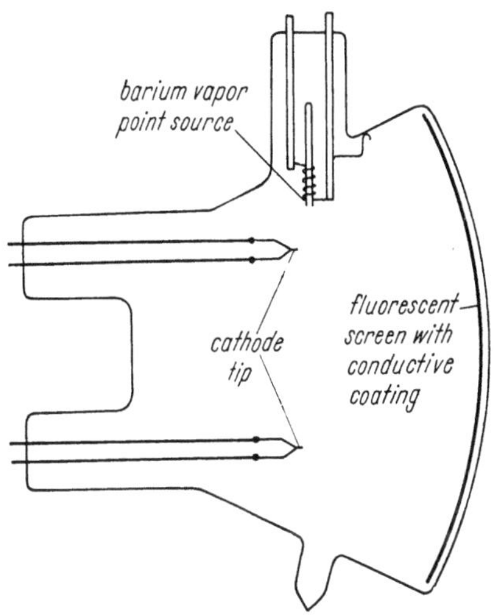

Fig. 25. Twin tube producing definite degrees of coverage.

Actually barium at low degrees of coverage does not show up on the 011 plane of tungsten, probably because it is mobile at room temperature and migrates immediately to adsorption sites along the lattice steps around that plane, where it is bound more strongly and protrudes less far, so that the observed contrast is slightly less than calculated. Tungsten atoms on the 011 plane are just barely visible and begin to migrate above 800° K with an activation heat of 30000 cal/mole as is to be expected for single atoms[3]. If one assumes that a barium spot consists of a small cluster[1] filling an area of about 100 Å diameter rather than containing only one atom, this adsorption island would be equivalent to

[1] J. A. BECKER: Bell Syst. Techn. J. **30**, 907 (1951).
[2] M. DRECHSLER and E. W. MÜLLER: Z. Physik **132**, 195 (1952).
[3] E. W. MÜLLER: Z. Physik **126**, 642 (1949).

a piece of a monolayer and would have correspondingly a work function of less than 2 e-volts. The local field is almost of the same size as in the bare substrate and the local current density should be at least 10^4 times enhanced. One can produce such low work function patches by using a high positive field to strip a monolayer from all parts of the tip except the 116 planes with barium or the 111 planes with thorium. These spots appear indeed with the expected large contrast which is entirely different from what one observes in the granulated pictures described above. It seems therefore clear that the field emission microscope visualizes individual atoms.

The condensation of medium sized molecules on the tip yields again a granular structure. However, a very peculiar situation occurred when MÜLLER[1] evaporated small quantities of phthalocyanine onto the tip, Fig. 26. Besides single spots intensely bright doublets and quadruplet patterns appeared. They acted definitely as a unit, suddenly changing into another shape, rotating, librating, or disappearing under the impact of an ion, or evaporating finally from the heated tip. This behavior makes it highly plausible that the patterns are caused by individual molecules. Since the structure of the phthalocyanine molecule as known by x-ray analysis is very similar to the observed quadruplets, it was assumed that one can see directly a structural image of the molecule, although it is only 10 by 10 Å in size. The enhanced resolution was explained as made possible by the local field distortion in front of the molecule adsorbed on a protrusion, causing a beam spread and an additional magnification perhaps by a factor of ten. While GOMER and SPEER[2] confirm these observations and explain the similarity of the image by assuming field ejection of the π-electrons of the molecule, further experiments by MÜLLER [6] and WOLF[3] and HAEFER[4] establish clearly that the same patterns can be obtained from many substances having a quite different structure. Trifold symmetric molecules like triphenylene produce quadruplets, as do acetone or carbon dioxide when applied at a higher pressure. Finally BECKER[5] [2] finds the same multiple units after producing a heavy adsorption layer of oxygen on a tungsten tip and baking the tube. He interprets the doublets again as structural images of diatomic oxygen molecules, and the quadruplets correspondingly as O_4 complexes which rest on or protrude above the first two oxygen adsorption layers and which share electrons with the atoms in these layers.

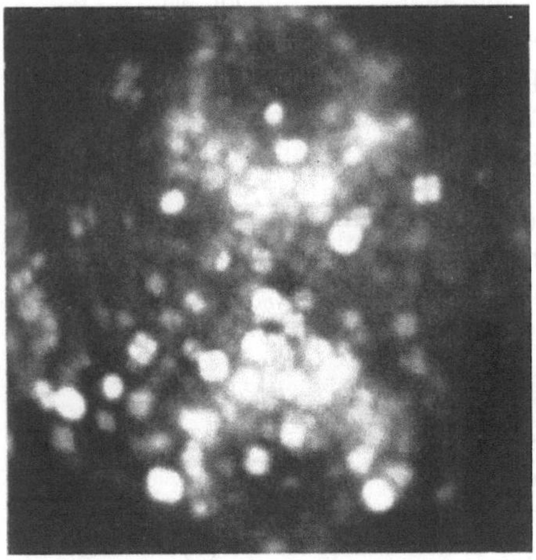

Fig. 26. Pattern produced by condensing a small quantity of phthalocyanine on a tungsten tip.

With this confusing situation the interpretation of these "molecular images" is quite uncertain. It is necessary to explain the appearance of only a few typical patterns from such a wide variety of substances on a more common basis than

[1] E. W. MÜLLER: Naturwiss. **37**, 333 (1950). — Z. Naturforsch. **5**a, 473 (1950).
[2] R. GOMER and D. A. SPEER: J. Chem. Phys. **21**, 73 (1953).
[3] P. WOLF: Z. angew. Phys. **6**, 529 (1954).
[4] R. HAEFER: Acta phys. Austriaca **8**, 105 (1953).
[5] J. A. BECKER and R. G. BRANDES: J. Chem. Phys. **23**, 1323 (1955). — J. Appl. Phys. **27**, 221 (1956).

the assumption of a resembling image. The fact that both parts of a doublet and usually also the four parts of a quadruplet have the same intensity suggests a scattering mechanism which would split the electron beam coming from the interior of the cathode while it is passing through the molecular field of the adsorbed particle. It is felt that the local field enhancement due to a protrusion would not suffice to explain the large intensity of the patterns, since the dimensions particularly of oxygen atoms are too small compared with the extension of the potential barrier. The experimental procedure required for producing the oxygen patterns does not exclude entirely the possibility of forming extremely small tungsten oxide crystallites (in the 10 to 20 Å range). A potential well model[1] according to Fig. 27a shows a medium sized molecule or a small crystal adsorbed at a metal surface. The important feature is a forbidden energy band in the particle at the FERMI level, above which are unoccupied levels. The external field tilts the potential well as in Fig. 27b, so that the empty levels can be filled by metal electrons tunneling in. This reservoir provides then the electrons for the external field emission. The two potential barriers which are to be penetrated here are together considerably smaller than the one in front of the plain metal surface at the same external field, so the current density is considerably enhanced. If the crystal is sufficiently thick, the outer potential peak falls below the FERMI level. With certain arrangements of the energy bands the current density may then become very large, leading to the evaporation of the particle and to arc initiation. This mechanism might be responsible for some kinds of high vacuum breakdown. This tilted band mechanism, explaining the large intensity of the patterns, holds for small crystallites as well as for molecules. The actual nature of the particles is still doubtful. HAEFER assumes them to be crystallites of quite considerable size, and MÜLLER is also inclined to assume very small crystallites or a $W_n - O_m$ complex in the case of the oxygen patterns. However, the following experimental observations by MÜLLER and WOLF might be considered in favor of an interpretation as scattering images of individual molecules: 1. No steady or almost continuous growth or dissolution is observed, while small crystallites of barium with somewhat different fourfold patterns do show these effects, when surface migration, condensation, and re-evaporation experiments are made. 2. Crystals of the organic substances are known to evaporate above 200° C. However, the field emission patterns are stable at tip temperatures beyond 600° C, due to strong molecular adsorption. 3. Patterns appear also when the substance is condensed upon a tip at liquid air temperature, where no surface migration should allow recrystallization. 4. The cross-section for being hit by an impinging ion can be estimated to be 100 Å² only from the lifetime of a quadruplet under ion bombardment. 5. No shadow casting can be observed when barium or copper is evaporated onto the tip from the side. 6. Impinging barium or copper atoms produce rotations and motions of the particles.

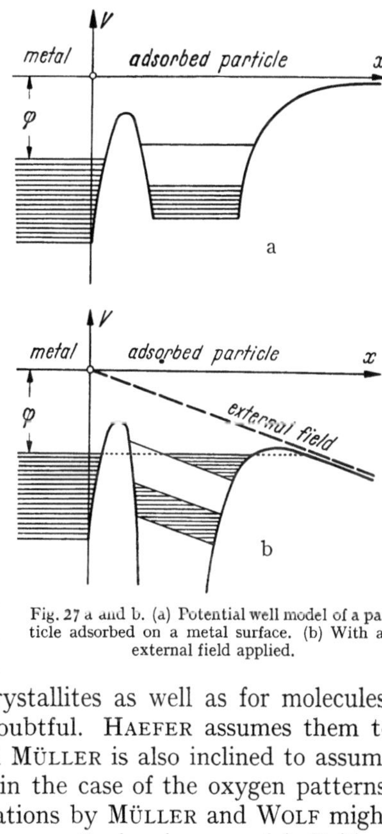

Fig. 27 a and b. (a) Potential well model of a particle adsorbed on a metal surface. (b) With an external field applied.

[1] E. W. MÜLLER: J. Appl. Phys. 24, 1414 A (1953). — Z. Elektrochem. 59, 372 (1955).

The atoms do not have enough inertia to move crystallites. 7. The two or four parts of a multiplet have equal intensity, suggesting the splitting of an electron beam inside a molecular unit. 8. Operation of the microscope with hydrogen ions shows that in the center of each doublet or quadruplet electron pattern of phthalocyanine lies a small particle which is not larger than 15 Å, the estimated resolution of the hydrogen field ion microscope under such conditions.

V. Field emission of positive ions.

17. Field desorption. Emission of positive ions may occur from a metal surface if a very strong electric field is applied in such a direction as to make the metal

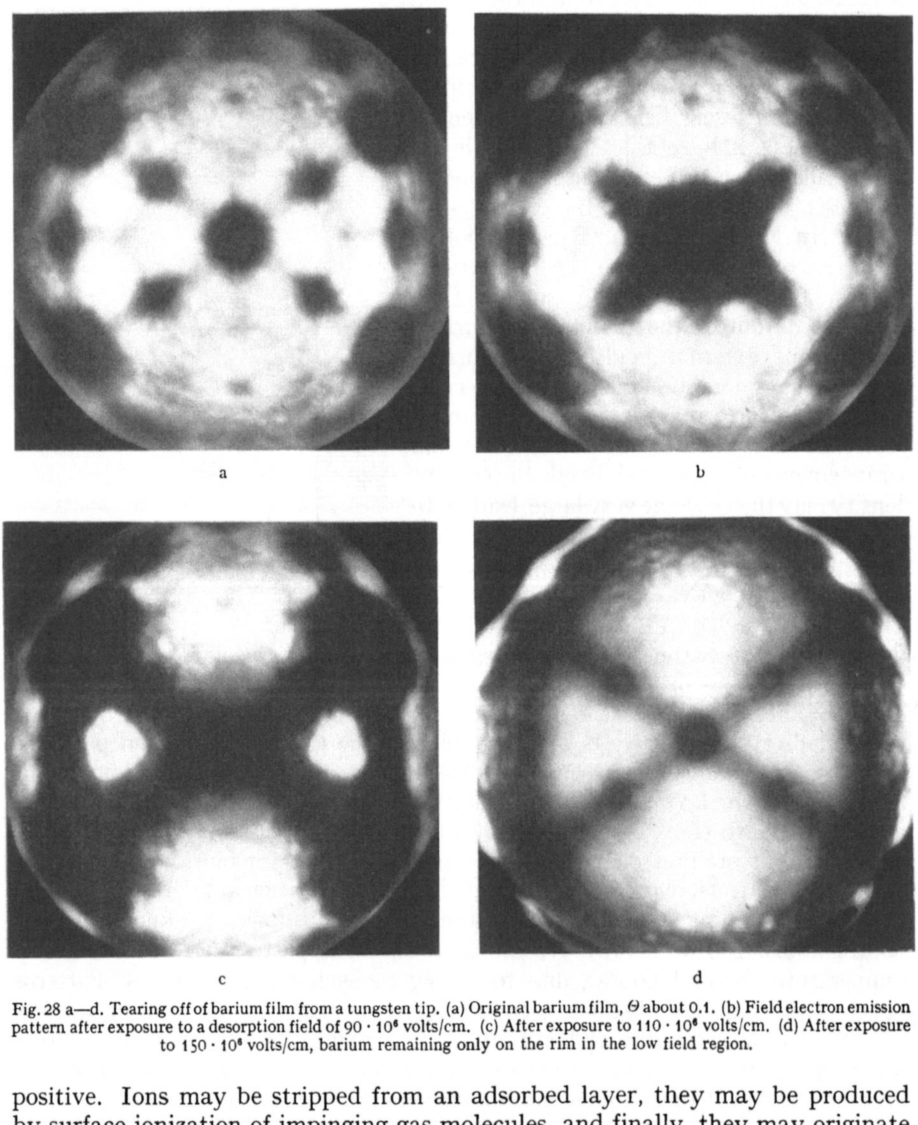

Fig. 28 a—d. Tearing off of barium film from a tungsten tip. (a) Original barium film, Θ about 0.1. (b) Field electron emission pattern after exposure to a desorption field of $90 \cdot 10^6$ volts/cm. (c) After exposure to $110 \cdot 10^6$ volts/cm. (d) After exposure to $150 \cdot 10^6$ volts/cm, barium remaining only on the rim in the low field region.

positive. Ions may be stripped from an adsorbed layer, they may be produced by surface ionization of impinging gas molecules, and finally, they may originate in the metal surface itself by a kind of field supported evaporation. The supply is naturally more limited than in the case of electron emission. Because of the extremely high field strengths, a point electrode with a well-defined surface as used in the field emission microscope is almost mandatory. The performance of experiments in a field emission microscope with reversed polarity has further the advantage of immediately showing the effect of the positive field in the electron picture of the surface after the voltage has been reversed again. This technique was introduced by MÜLLER[1] who investigated the field desorption (tearing off) of barium from a tungsten surface in some detail, Fig. 28. The barium film is characterized by its average degree of coverage, as derived from the work function with the help of the relation between Θ and φ found thermionically by BECKER[2]. If for instance a surface with $\Theta = 0.35$ is exposed at room temperature to a positive field of $100 \cdot 10^6$ volts/cm, one does not notice any change of the field electron emission pattern. However, after an exposure to $102 \cdot 10^6$ volts/cm for three seconds, barium is stripped off the 011 vicinity. This stripped area increases when a stronger positive field is applied. At $110 \cdot 10^6$ volts/cm the barium film remains only in the 111 and 001 regions, at $125 \cdot 10^6$ volts/cm the desorption takes place also at 111, and finally at about $150 \cdot 10^6$ volts/cm the film is completely desorbed except from the shank of the tip where the field is weaker. These tearing off field strengths were obtained by calibrating the clean *W* tip with field electron emission, using the radius formula on page 203, with the same limitations of absolute accuracy. However, the relative data for the stripping from the different crystal planes of the substrate are as accurate as the voltage measurement, about 1%. Fig. 29 gives the desorption field which is required in the vicinity of 011 as a function of degree of coverage and with the temperature as a parameter. If the temperature is high enough so that barium migrates on all planes, it will be desorbed completely at the critical field for desorption on the 011 edges, where it comes off most easily.

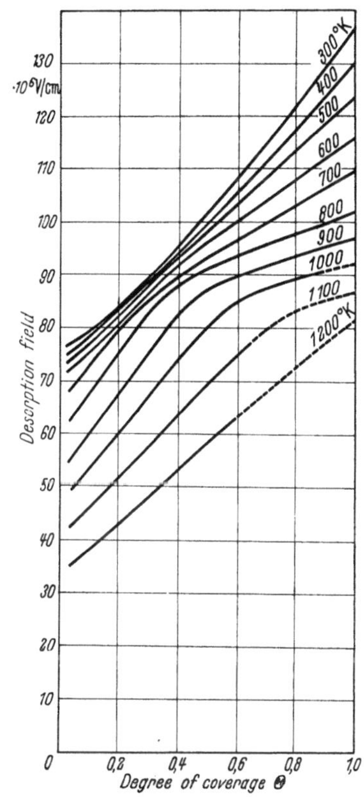

Fig. 29. Threshold field strength for desorption of barium from the vicinity of the 011 plane of tungsten.

The tearing off of thorium from a tungsten surface requires much stronger fields. It starts again in the vicinity of 011 with about $240 \cdot 10^6$ volts/cm at low degrees of coverage, but does not increase with Θ as rapidly as in the case of barium. However, on other planes, particularly on 111 and 116, the binding forces seem to be much larger, (Fig. 30). If for instance $360 \cdot 10^6$ volts/cm are applied to the tip originally covered with $\Theta = 0.85$, the thorium remains in the form of single islands on the 111 and 116 planes only. Because of the low work function of these islands compared with the surrounding tungsten, large electron currents can be drawn from such a tip, with d.c. current densities well above 10^8 amps/cm^2. Fig. 31 gives desorption field strengths for stripping thorium from the 011 vicinity as a function of temperature and degree of coverage[3]. When

[1] E. W. MÜLLER: Naturwiss. **29**, 533 (1941).
[2] J. A. BECKER: Trans. Faraday Soc. **28**, 151 (1932).
[3] E. W. MÜLLER: Phys. Rev. **102**, 618 (1956).

surface migration is possible at above 800° K, the whole adsorption layer comes off at the critical field of the 011 vicinity, and the desorption field is independent of Θ from 0 to 0.7.

Other more qualitative experiments on stripping adsorbed layers by strong positive fields have been made with lithium, sodium, and caesium. For all the

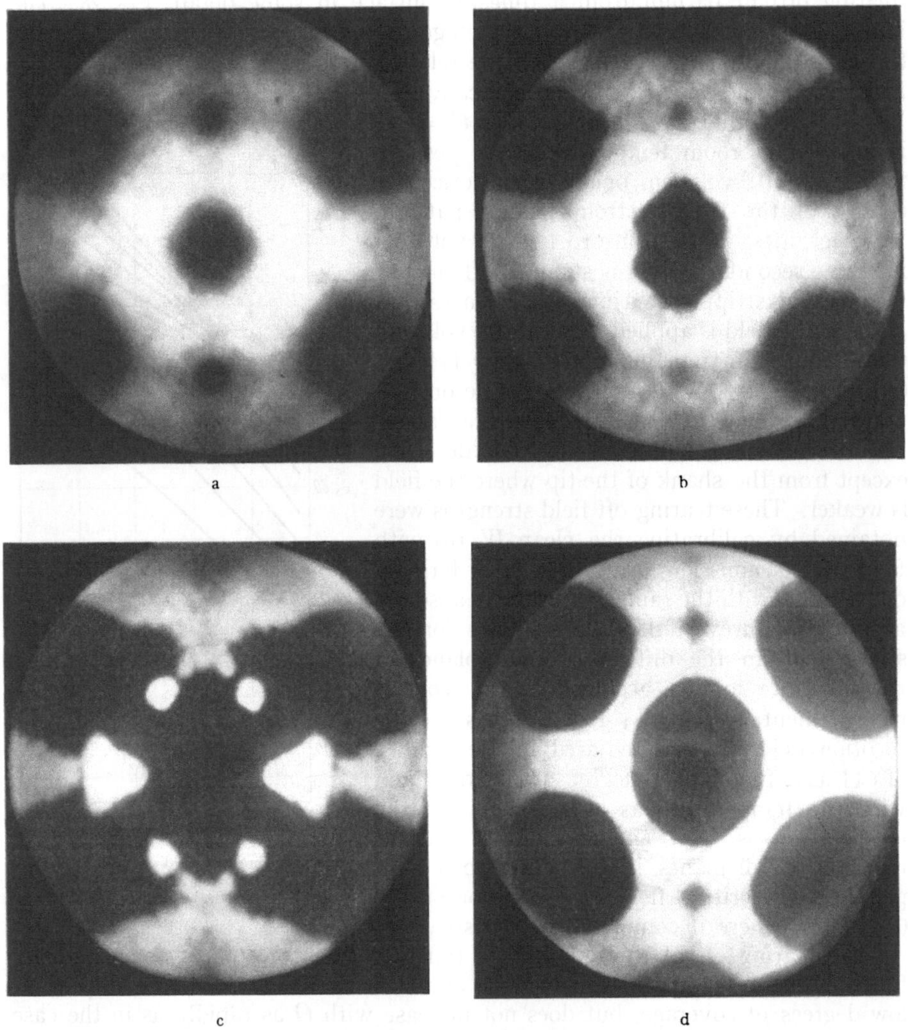

Fig. 30 a—d. Tearing off of a thorium film from a tungsten tip. (a) Original thorium film, Θ about 0.6. (b) Field electron emission pattern after exposure to a desorption field of $270 \cdot 10^6$ volts/cm. (c) Desorption field increased to $340 \cdot 10^6$ volts/cm, thorium remaining on the 111 and 116 planes only. (d) Original thorium film after exposure to desorption field of $220 \cdot 10^6$ volts/cm at 700° K.

alkali and alkali earth metals, which come off at relatively low field strengths, the effect can be easily observed by applying an alternating voltage to the field emission microscope. The peak of the positive half wave produces the tearing off field, while the negative half wave gives the electron picture of the surface. With a proper current limiting resistance in the circuit, one observes the tearing off progressing over the different crystal planes, as the a.c. voltage is slowly increased.

Field desorption can be understood as an evaporation of positive ions of charge ne over a potential barrier. The height Q of the barrier is the binding energy of the ion Q_0 decreased by the SCHOTTKY depression $\sqrt{n^3 e^3 F}$ of the potential peak:

$$Q = Q_0 - \sqrt{n^3 e^3 F}\,. \tag{17.1}$$

The binding energy can be computed from the thermionic cycle

$$Q_0 = \Lambda + V_I - n\varphi, \tag{17.2}$$

where Λ is the heat of evaporation of the neutral atom and V_I is the ionization energy. The quantities on the right are known from other experiments. The time of evaporation over the barrier is

$$\tau = \tau_0 e^{\frac{Q}{kT}}, \tag{17.3}$$

where τ_0 is the vibration time of the adsorbed particle, about 10^{-13} sec. The desorption field which strips the surface within time τ is then

$$F_{\text{des}} = \frac{1}{n^3 e^3}\left(Q_0 - kT\log\frac{\tau}{\tau_0}\right)^2. \tag{17.4}$$

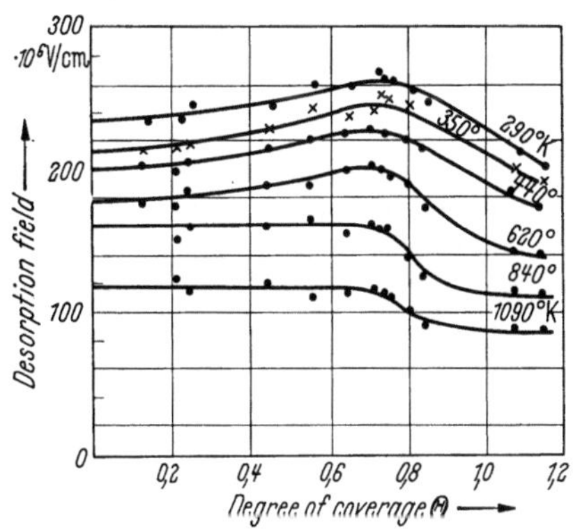

Fig. 31. Threshold field strength for desorption of thorium from the vicinity of the 011 plane of tungsten.

For barium on tungsten DE BOER's energy considerations[1] suggest doubly charged ions. (They may at least occur in the instant of tearing off.) The heat of evaporation of the neutral atom is not well known for $\Theta \approx 0$. MOORE and ALLISON[2] give $\Lambda = 3.68$ e-volts for $\Theta = 0.3$ and note a poor fit already between $\Theta = 0.3$ and $\Theta = 0.2$. From DOBRETZOW and MOROZOW's determination[3] one obtains $\Lambda = 7.1$ e-volts at $\Theta = 0$, which seems to be too large. If one assumes $\Lambda = 5.1$ e-volts and $n = 2$, eq. (17.4) gives with $V_I = 15.2$ e-volts, $\varphi = 5.0$ e-volts (vicinity of 011 plane) and $Q_0 = 10.3$ e-volts the following results for an evaporation time of $\tau = 3$ sec:

$$F_{\text{des}} = 78 \cdot 10^6 \text{ volts/cm at } 300^\circ \text{ K},$$

$$F_{\text{des}} = 45 \cdot 10^6 \text{ volts/cm at } 1200^\circ \text{ K}.$$

This is in excellent agreement with the observed data. F. KIRCHNER and H. KIRCHNER[4] make a similar consideration, assuming singly charged ions. However, this would require Λ to be 1.6 e-volts only.

In the case of thorium on tungsten LANGMUIR[5] measured $\Lambda = 7.7$ e-volts, but the ionization potential is unknown. With the assumption $V_I = 5$ e-volts one obtains under the same conditions as above

$$F_{\text{des}} = 333 \cdot 10^6 \text{ volts/cm at } 300^\circ \text{ K},$$

$$F_{\text{des}} = 167 \cdot 10^6 \text{ volts/cm at } 1090^\circ \text{ K}.$$

---

[1] J. H. DE BOER: Elektronenemission und Adsorptionserscheinungen, p. 72. Leipzig 1937.
[2] G. E. MOORE and H. W. ALLISON: J. Chem. Phys. **23**, 1609 (1955).
[3] L. N. DOBRETZOW and G. A. MOROZOW: Phys. Z. Sowjet. **9**, 352 (1936).
[4] F. KIRCHNER and H. KIRCHNER: Z. Naturforsch. **10**a, 394 (1955).
[5] I. LANGMUIR: J. Franklin Inst. **217**, 543 (1934).

Both values are 40% higher than the experimental data. This is a satisfactory agreement in view of the limitations of the image force picture, particularly if one considers a local field enhancement factor at the lattice steps.

The tearing off field increases with increasing degree of coverage because the applied external field must also balance the opposite field which is produced at the considered ion by the dipoles of the neighbors. This dipole field F_{dipole} can be found by adding the contributions of a square array of dipoles with moments μ and having a lattice constant $a \cdot \Theta^{-\frac{1}{2}}$

$$F_{\text{dipole}} = \frac{9.05\,\mu\,\Theta^{\frac{3}{2}}}{a^3}. \tag{17.5}$$

The dipole moment μ as a function of Θ can be found from the experimental determination of the decrease of thermionic work function. For barium, measurements of BECKER[1] and others[2] can be approximated by the formula $\Delta\varphi = B\Theta^{\frac{1}{4}}$. B for barium equals 2.9 e-volts, for thorium 1.9 e-volts. Since a sheet of Θ/a^2 dipoles per unit area with moment μ changes the work function by $\Delta\varphi = \frac{2\pi\mu\Theta e}{a^2}$, (only half of the dipole effective in producing F_{dipole} contributes to $\Delta\varphi$), we obtain the desorption field

$$F_{\text{des}}(\Theta) = F_{\text{des}}(0) + \frac{9.05\,B\,\Theta^{\frac{3}{4}}}{2\pi a e}. \tag{17.6}$$

Since no accurate data are known, we assume for barium as well as for thorium, a number of $2.5 \cdot 10^{14}$ adsorbed particles/cm² when $\Theta = 1$. The resulting desorption fields are, after adjusting only $F_{\text{Th des}}(0)$ to the experimental value,

$$F_{\text{Ba des}} = 78 \cdot 10^6 + 66 \cdot 10^6\,\Theta^{\frac{3}{4}}\ \text{volts/cm},$$

$$F_{\text{Th des}} = 240 \cdot 10^6 + 43 \cdot 10^6\,\Theta^{\frac{3}{4}}\ \text{volts/cm}.$$

These figures lie within the experimental error of the measurements. A more detailed interpretation would have to include the polarizability of the adsorbed particles. This does not seem worth while at the present time because of the uncertainty in the polarizability and the spacing.

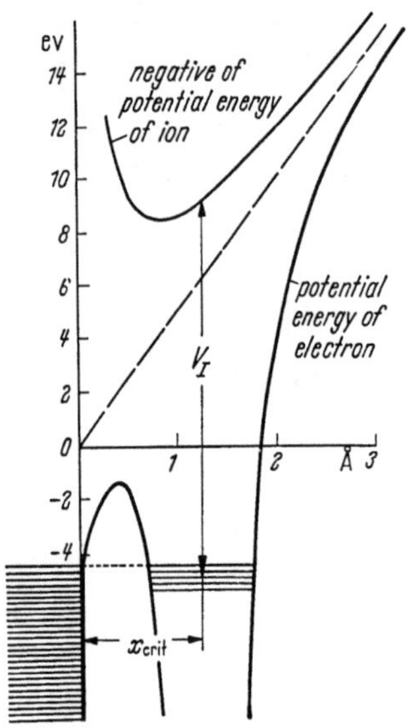

Fig. 32. Potential scheme for an oxygen atom adsorbed on a tungsten surface at $500 \cdot 10^6$ volts/cm.

Field desorption may also yield valuable information on adsorption forces in other cases. However, the interpretation is more difficult than for electropositive adsorbates. A saturated oxygen film[3] shows the first effect of a slightly decreased work function after an application of $310 \cdot 10^6$ volts/cm, while the grainy structure of the adsorbed molecules disappears only above $420 \cdot 10^6$ volts/cm. The chemisorbed layer in close contact with the tungsten substrate comes off the 011 vicinity at about $480 \cdot 10^6$ volts/cm, and only at $500 \cdot 10^6$ volts/cm is the desorption completed. According to Fig. 32, the ionization of an oxygen atom takes place if the field is high enough to lift the ground level above the FERMI level in the substrate, taking into account the ion image force. After the oxygen film has been torn off the cap, one can by gradual heating observe the surface migration of oxygen from the still covered shank of the tip, which

[1] J. A. BECKER: Trans. Faraday Soc. **28**, 151 (1932).

[2] J. W. RYDE and N. L. HARRIS: In A. L. REIMANN: Thermionic Emission, p. 164. New York: John Wiley & Sons 1934.

[3] E. W. MÜLLER: Z. Elektrochem. **59**, 372 (1955).

occurs with an activation energy of 14000 cal/mole. The fact that even oxygen can be removed by field desorption suggests the use of this method for cleaning surface contaminations from a field emitter without applying heat.

At fields of about $500 \cdot 10^6$ volts/cm the surface of tungsten evaporates at room temperature[1]. The rate of field evaporation can be measured by determining the increase of tip radius after exposure to the field. A more accurate technique is to observe the dissolution of the surface in a helium operated field ion microscope. The concentric edges of 011 net planes collapse at a rate of 5 layers per sec. at $570 \cdot 10^6$ volts/cm when the tip is cooled by liquid hydrogen, while for the same evaporation rate the fields at room temperature and at 900° K are $510 \cdot 10^6$ volts/cm and $400 \cdot 10^6$ volts/cm, respectively. Lacking a more convincing theory one can apply again eq. (17.4) with $\Lambda = 8.78$ e-volts[2], $V_I =$ 7.98 e-volts[3], and $\varphi_{011} = 6.0$ e-volts. For a desorption time of 3 sec at room temperature one finds $F = 693 \cdot 10^6$ volts/cm. As in the case of thorium desorption this is 40% greater than the experimental value. These considerations give also the observed time and temperature dependences.

18. Field ion emission with gas supply. While field desorption of adsorbed matter gives only a small pulse of ions (in the range of 10^{-15} coulombs for the usual tip size), a continuous field ion current can be obtained by the ionization of gas molecules impinging on the emitter[4]. The current is the product of the supply function (the number of molecules arriving at the emitter per unit of time) and the efficiency of surface ionization. The number of molecules striking the tip is largely determined by the attraction of the induced dipoles in the inhomogeneous field. A sphere of capture with radius r_c can be defined extending as far as the attraction is equal to the centrifugal force of a tangentially approaching molecule:

$$-\alpha F \frac{dF}{dr} = \frac{m v^2}{r_c} \tag{18.1}$$

(α = polarizability of molecule). Setting $\frac{m v^2}{2} = kT$ and using for the description of the field the semi-empirical formula

$$F = \frac{7.75 \cdot V r_0^{\frac{2}{3}}}{r^{\frac{4}{3}}} \tag{18.2}$$

(V = applied voltage, r_0 = tip radius in cm, F in volts/cm) one obtains

$$r_c = \frac{4 \alpha^{\frac{3}{8}} V^{\frac{3}{4}} r_0^{\frac{1}{2}}}{k^{\frac{3}{8}} T^{\frac{3}{8}}} .$$

Outside the sphere of capture the molecules are assumed to move as in an ordinary gas of a pressure p. The number of molecules striking the sphere of capture and hence the tip per second is then

$$\left.\begin{aligned} z &= \frac{p}{\sqrt{2\pi m k T}}\, 4\pi r_c^2 , \\ z &= \frac{80\, p\, \alpha^{\frac{3}{4}} V^{\frac{3}{2}} r_0}{m^{\frac{1}{2}} k^{\frac{5}{4}} T^{\frac{5}{4}}} \end{aligned}\right\} \tag{18.3}$$

and the field ion current is obtained by multiplying z by the ion charge. This result is in fairly good agreement with the measured current-voltage characteristics

[1] E. W. Müller: Phys. Rev. **102**, 618 (1956).
[2] Landolt-Börnstein: 3. Erg.-Bd., Teil 3, p. 2711. Berlin 1936.
[3] W. Finkelnburg und W. Humbach: Naturwiss. **42**, 35 (1955).
[4] E. W. Müller: Z. Physik **131**, 136 (1951).

in the high field range where all striking molecules will be ionized[1] (Fig. 33). The observed currents are about 5 times larger than the calculated ones. This is due to secondary electrons which are not separated, and also due to the approximation made in the field formula eq. (18.2).

The ionization of the approaching molecules occurs at a certain distance above the surface. If the field is extremely high, the potential wall surrounding the molecule will become narrow enough to allow autoionization by tunneling[2]. With a slightly smaller field, autoionization in space will be negligible, but when the molecule approaches the metal surface of the emitter, the image potential will provide an additional decrease of the potential wall[3,4]. However, there is a sharp limit placed for the ionization when the ground level of the approaching molecule sinks below the Fermi level in the emitter, because a tunneling electron

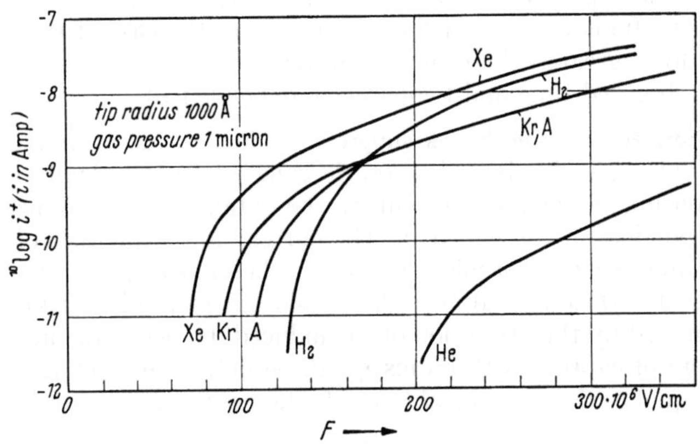

Fig. 33. Current-field characteristics for field ion emission in various gases.

would not find an unoccupied level in the metal. The result of a crude numerical computation[5,6] for the probability P that an electron tunnels while the atom approaches from d Å to $(d-1)$ Å with a speed determined by the field and the polarizability is given in Fig. 34. It was assumed that the electron strikes the inside wall of the potential funnel with a frequency of $\nu = 1.5 \cdot 10^{16}$ sec.$^{-1}$ for argon and $2.4 \cdot 10^{16}$ sec.$^{-1}$ for helium. The potential of the ion was approximated by $\frac{-e}{r}$. The total effective potential on the electron is then $V(x) = -\frac{e}{d-x} + Fx - \frac{e}{4x} + \frac{e}{d+x}$. Here d is the distance of the ion from the surface, x the distance of the electron from the surface. The third term is due to the attraction of the electron by its image in the metal, the fourth term is due to the repulsion from the image of the ion in the metal. The penetration probability $D(d)$ was obtained by graphical integration;

$$D(d) = e^{-\sqrt{\frac{8em}{\hbar^2}}\int\limits_{x_1}^{x_2}\left(V(x) - Fd + V_I - \frac{e}{4d}\right)^{\frac{1}{2}} dx}$$

where V_I is the ionization potential. The probability P of being ionized is finally obtained by multiplying $D(d)$ by the frequency of the bound electron and the

[1] E. W. Müller: Pittsburgh Field Emission Symposium, 1954.
[2] E. W. Müller: Z. Physik **131**, 136 (1951).
[3] F. Kirchner: Naturwiss. **41**, 136 (1954).
[4] M. G. Inghram and R. Gomer: J. Chem. Phys. **22**, 1279 (1954).
[5] E. W. Müller, Z. Naturforsch. **11** a, 88 (1956).
[6] E. W. Müller and K. Bahadur: Phys. Rev. **102**, 624 (1956).

transit time $\frac{10^{-8}}{F\sqrt{\alpha/m}}$ required to travel 1 Å. The ionization probability is mainly determined by the ionization potential, but is also sensitive to the special assumptions about the shape of the potential close to the ion. The experimental current versus field strength curves, Fig. 33, give the onset of the ion current at fields roughly proportional to $V_I^{\frac{3}{2}}$. This result is in agreement with the calculations. Since the field drops only slowly even at 100 Å distance from the emitter, an ionization probability of 10^{-2} per Å interval means already auto-ionization before the molecule comes really close to the surface. This occurs above $200 \cdot 10^6$ volts/cm for argon and above $450 \cdot 10^6$ volts/cm for helium

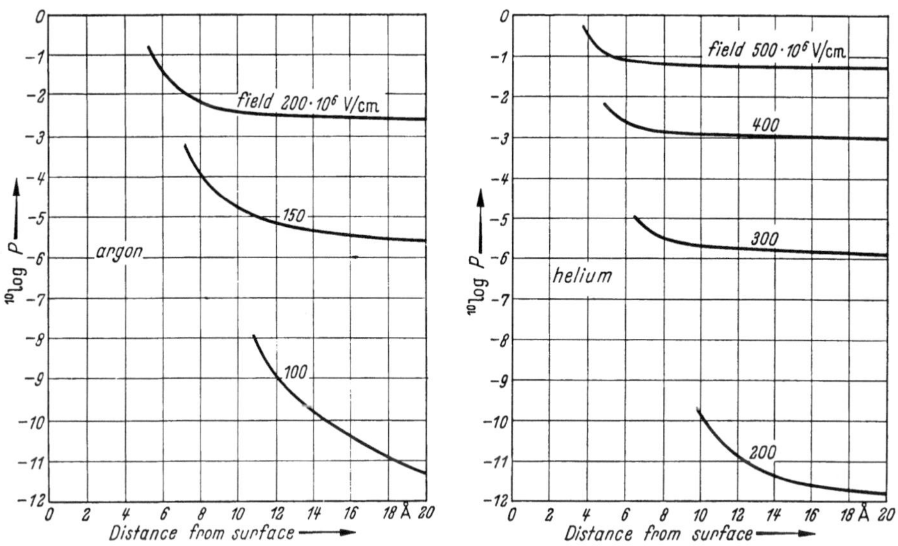

Fig. 34 a and b. (a) Probability of field ionization for argon atoms approaching a tungsten surface. (b) Same for helium.

according to Fig. 34. Observation of the blurring of the field ion microscope image gives evidence that this actually happens at about these fields.

The ion current for a given field can be obtained by integrating the ionization probability P over the distance and multiplying the integral by the supply function. However, if at low fields $\int_0^\infty P\,dx$ is smaller than one, the molecules that have not been ionized will strike the emitter surface and will bounce back and pass through the ionization zone again. The particles are reflected at random angles, and so the rebounding molecules have, on the average, larger tangential components than the incident molecules. Therefore, they stay longer in the zone of high ionization probability. Whether they can leave the sphere of capture at all or whether they have to return to pass the ionization zone again and again until they are finally ionized depends on the unknown accommodation coefficient and on the ratio of $\frac{3}{2}kT$ to $\frac{\alpha}{2}F^2$. So one has to expect a more pronounced negative temperature effect in the low field range, which is actually observed. Not only does the ion current increase when the whole tube is immersed in liquid air, but also one observes a fading out of the image when the tip of a field ion microscope is heated to some hundred degrees above room temperature.

This would be unexplainable if all ions were produced at some distance from the tip before the molecules touch the surface.

The ionization zone can be localized by measuring the velocity distribution of the ions. INGHRAM and GOMER[1] investigate the ions in a mass spectrometer which has a peak width of 20 volts. (Some of the field strengths quoted by these authors are too high by a factor of almost two.) The width broadens when the field is increased, indicating clearly that at least some of the ions are produced at a certain distance from the tip. In measuring the velocity distribution a higher resolution can be obtained by a retarding potential method[2]. The tube shown in Fig. 10 is first used to measure field electron emission in order to show that it gives the theoretical distribution within a few hundredths of a volt. This measurement gives also the work function of the collector φ_{coll}; it is the voltage at which electrons start to enter the collector according to the potential diagram Fig. 11. Field ions produced after about 1 micron of gas has been admitted and after the voltages have been reversed can reach the collector only if it is negative by $V_{\text{coll}} = V_I - \varphi_{\text{coll}}$ (Fig. 35). Fig. 36 gives sample recordings of collector currents, where $\varphi_{\text{coll}} = 5.7$ e-volts. Actually the current measured at the collector has the direction of entering electrons. These electrons are produced when ions that do not reach the collector fall back and strike the outside surface of the accelerating electrode and when collisions occur in the gas. This electron current begins to drop at V_{coll} when ions start to contact the collector. For argon $V_{\text{coll}} = 10.0$ volts, the value which would be expected according to the potential diagram Fig. 35. In the same way the onset of ion current for xenon starts at $V_{\text{coll}} = 6.5$ volts, as expected. The distribution curves that can be obtained by differentiating the collector current recordings show clearly that there is a minimum distance from the emitter beyond which the field ions are produced. This distance is determined by the field and the ionization potential so that the ground level is just above the FERMI level. The curves show further that at increasing fields the ionization zone extends to a larger distance from the emitter surface.

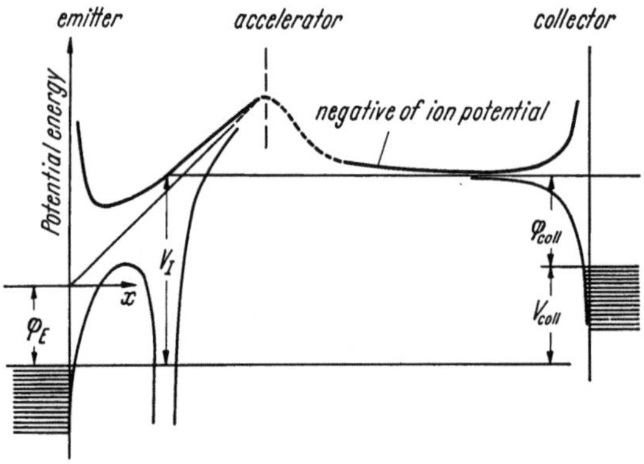

Fig. 35. Potential scheme for measuring energy distribution of field ion emission.

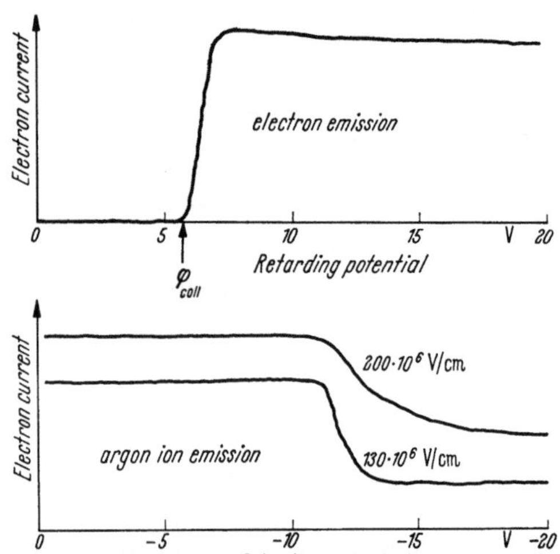

Fig. 36. Recordings of collector currents, as a function of voltage, in retarding potential tube.

[1] M. G. INGHRAM and R. GOMER: J. Chem. Phys. **22**, 1279 (1954).
[2] E. W. MÜLLER and K. BAHADUR: Phys. Rev. **102**, 624 (1956).

19. Field ion mass spectroscopy. A gas supplied field ion emitter provides an ion source for mass spectroscopy with quite remarkable properties. While the conventional ion source uses electron collision and contains a hot cathode, a field ion emitter can be run with fields high enough to autoionize the molecules before they even strike the surface. If the gas pressure is kept well below one micron to prevent collisions between ions and gas molecules, there is not much chance for fragmentation. On the other hand at lower fields one can obtain some fragmentation when the molecules break up after having touched the surface and go back through the ionization zone. INGHRAM and GOMER[1] have shown that hydrogen yields molecule ions as well as protons, and also that methyl alcohol is fragmented. They localize the origin of the different fragments by studying the velocity distribution in the ion beam within the obtainable resolution of about 20 volts in a conventional mass spectrometer. A simpler apparatus[2,3] utilizing the point shape of the ion source can be made by focussing the divergent ion beam with an electrostatic lens system onto a distant screen and separating the masses by a transverse magnetic field. A mass resolution of about 80 can be easily obtained, enough to show that O_2, N_2, and CH_4 do not split into atoms or fragments. A full utilization of mass spectroscopic analysis of fragmentation as a function of field and tip temperature, combined with accurate retarding potential measurements to locate the origin of the ions, will probably yield valuable information on fragmentation processes occurring on metal surfaces. The application to gas analysis is also promising because of the small fragmentation, although selective effects due to different polarizabilities and possible permanent dipole moments of the different constituents of the sample will have to be considered [*4*].

20. Field ion microscopy. The field ion microscope[4] forms an image of the tip surface in the field emission microscope with ions instead of electrons. Not only is supplementary information to be expected because of the different medium for image production, but also an increased resolution can be foreseen. The resolving power is determined by the transverse velocity component v_p of the emitted particles and by the diffraction effect exactly as considered in Sect. 13 for the case of electrons. Replacing in eq. (13.3) the magnification M by $\frac{R}{\beta r}$ and the transit time τ by $R\sqrt{\frac{m}{2eV}}$ one obtains for the resolution

$$\delta = \sqrt{\frac{2\hbar\beta r}{\sqrt{2emV}} + \frac{2\beta^2 r^2 m \overline{v}_p^2}{eV}}\,. \tag{20.1}$$

Compared with electron emission the square of the diffraction contribution, the first term under the root, is smaller by a factor of at least 100 because of the larger mass of the ion and the higher voltage required for field ion emission. The second term will also be smaller if the transverse energy $\frac{m\overline{v}_p^2}{2}$ is of about the same size as for electrons. As considered in Sect. 18, the ions produced from incoming gas molecules will not have more transverse energy than kT, which is at room temperature about 4 times smaller than $\frac{m\overline{v}_p^2}{2}$ for field electrons. This would give a very much better resolution for the ion microscope. However, if

[1] M. G. INGHRAM and R. GOMER: J. Chem. Phys. **22**, 1279 (1954). — Z. Naturforsch. **10**a, 863 (1955).
[2] E. W. MÜLLER and K. BAHADUR: Phys. Rev. **102**, 624, (1956).
[3] E. W. MÜLLER: Pittsburgh Field Emission Symposium, 1954.
[4] E. W. MÜLLER: Z. Physik **131**, 136 (1951).

relatively low fields are applied, as is necessary to provide ionization close to the surface only instead of autoionization in space, the majority of ions is produced after the gas molecules bounce back from the surface. Then they will have a randomly directed velocity according to their kinetic energy $\frac{m v^2}{2} = \frac{3}{2} kT + \frac{\alpha}{2} F^2$. This gives for the resolution

$$\delta = \sqrt{\frac{2\hbar\beta r}{\sqrt{2emV}} + \frac{\pi^2\beta^2 r^2 (3kT + \alpha F^2)}{8eV}}. \tag{20.2}$$

Neglecting the diffraction effect and using the semi-empirical eq. (18.2) for the field-voltage relation one finds finally the following formula for the resolution of the field ion microscope:

$$\delta = \pi\beta r^{\frac{2}{3}} \sqrt{\frac{3kT + \alpha F^2}{eF}}. \tag{20.3}$$

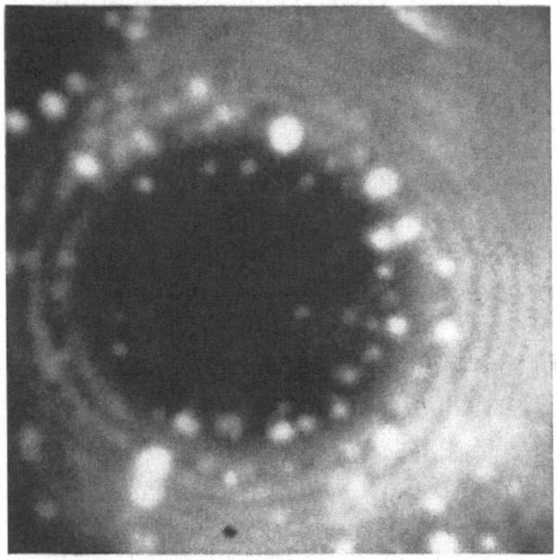

Fig. 37. Helium on image of a tungsten tip, radius 1000 Å. Vicinity of 011 plane with individual lattice steps. Resolution about 5 Å.

For a given tip radius the polarizability of the molecule is now the determining factor. This makes helium probably the most suitable gas with $\alpha = 0.2 \cdot 10^{-24}\,\text{cm}^3$ only. If we choose $r = 1000$ Å and $F = 325 \cdot 10^6$ volts/cm, V will be 20000 volts and the resolution is expected to be $\delta = 5.6$ Å. With $r = 250$ Å, V is only 6500 volts, and $\delta = 2.5$ Å, both at room temperature.

Actually eq. (20.3) describes the resolution obtainable in a layer parallel to and some 5 to 10 Å above the surface, where the ions originate. How much of the atomic structure of the surface shows up in the local field distribution so far above the surface is hard to figure out quantitatively. One can expect to see the edges of single lattice steps, while it might be well at the limit of contrast to resolve the fine field ripple that can be expected by the individual atoms constituting the edge.

The expected resolution is fully confirmed by experiments, which were begun with hydrogen. Images of the tungsten tip can also be obtained with a variety of other gases. Helium has the smallest polarizability and the highest ionization potential and so is most advantageous in respect to resolution[1]. The small currents obtainable with helium can be balanced by the application of a higher pressure. Up to 25 microns were used without too much loss of definition due to scattering and without a gaseous discharge breakdown. Fig. 37 gives a section of the tip surface with the lattice steps surrounding the 011 plane, and a resolution of about 5 Å. A pearl chain structure of some lattice steps is just barely indicated, which may be due to the resolution of the individual atoms along the edge. If the field is increased to above $450 \cdot 10^6$ volts/cm, the lattice steps decrease continuously in diameter and collapse at the center of 011 and 112, indicating removal of tungsten atoms by field evaporation. The data at the end of Sect. 17 were derived from such observations. DRECHSLER *et al*[2,3]

[1] E. W. MÜLLER and K. BAHADUR: Phys. Rev. **102**, 624 (1956).
[2] M. DRECHSLER and G. PANKOW: Proc. Conf. Electr. Microscopy, London 1954.
[3] M. DRECHSLER, G. PANKOW and R. VANSELOW: Z. phys. Chem. **4**, 249 (1955).

have also obtained ion pictures with various gases and with cesium and mercury vapor. They interpreted some structures that were obtained by exposing the heated emitter to fields near $200 \cdot 10^6$ volts/cm as spirals due to screw dislocations. Multiple lattice steps were observed with the limited resolution of the hydrogen ion microscope.

Eq. (20.3) does not predict an essential improvement of resolution when the temperature is decreased, since at room temperature $3kT$ is already much smaller than αF^2. However, if the temperature is lowered enough to accomodate

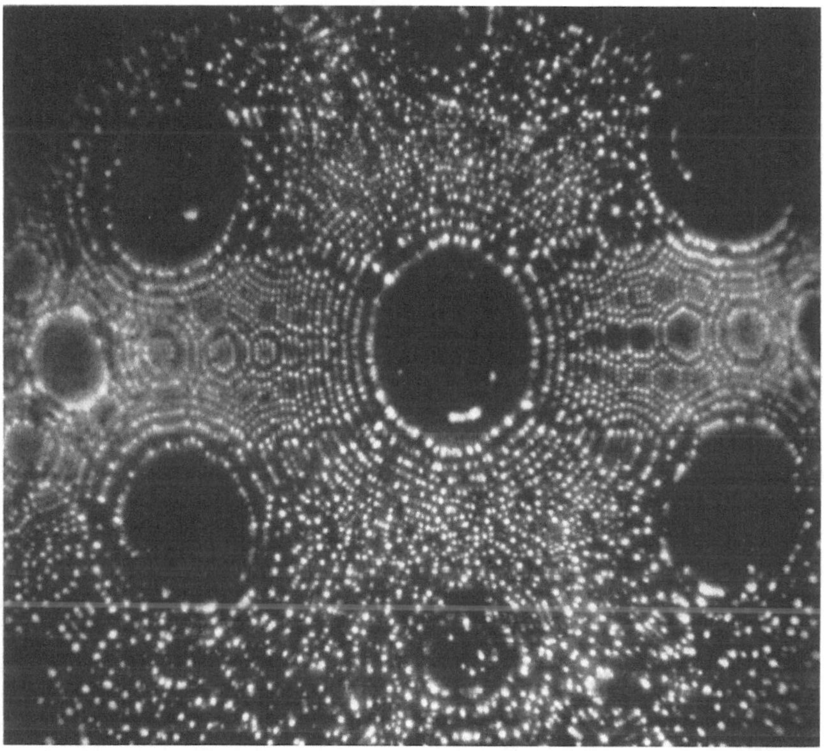

Fig. 38. Helium ion image of a tungsten tip of radius 950 Å, with 011 plane in center and the four dark 112 planes around. Helium gas pressure 0.2 micron, 19000 volts, after annealing at 2000 °K and field desorption at 22000 volts, at 21 °K.

the impinging gas molecules completely to the tip temperature, the rebounding molecules and subsequently the ions will have a kinetic energy of kT_{tip}/only, so that eq. (20.3) is reduced to

$$\delta = \pi \beta r^{\frac{2}{3}} \sqrt{\frac{3kT_{\text{tip}}}{eF}} . \tag{20.4}$$

The theoretical resolution for $r = 1000$ Å and $T = 60$ °K will then be 1.5 Å. Experiments with a specially built tube which allowed the tip to be cooled to liquid hydrogen temperature confirmed the expectation[1]. Apparently sufficient accomodation for helium on tungsten can be obtained below 60 °K, solid nitrogen temperature. Fig. 38 shows the central area on a tungsten tip. All lattice steps of the 011 net planes are visible, and in the regions where the local field is at its optimum value (the very small tip has a slightly irregular shape) the individual tungsten

[1] E. W. MÜLLER: Z. Naturforsch. **11**a, 88 (1956). — J. Appl. Phys. **27**, 474 (1956).

atoms in a protruding position along the steps are resolved. Two adjacent lattice sites are very rarely occupied in such a manner as to have both atoms protrude equally. However, on several occasions such neighboring tungsten atoms with 2.74 Å separation have been photographed with perfect resolution. Scattering discs as small as 1.6 Å were obtained, so that it seems possible to observe the atomic arrangement in surface compounds of sufficient strength like tungsten

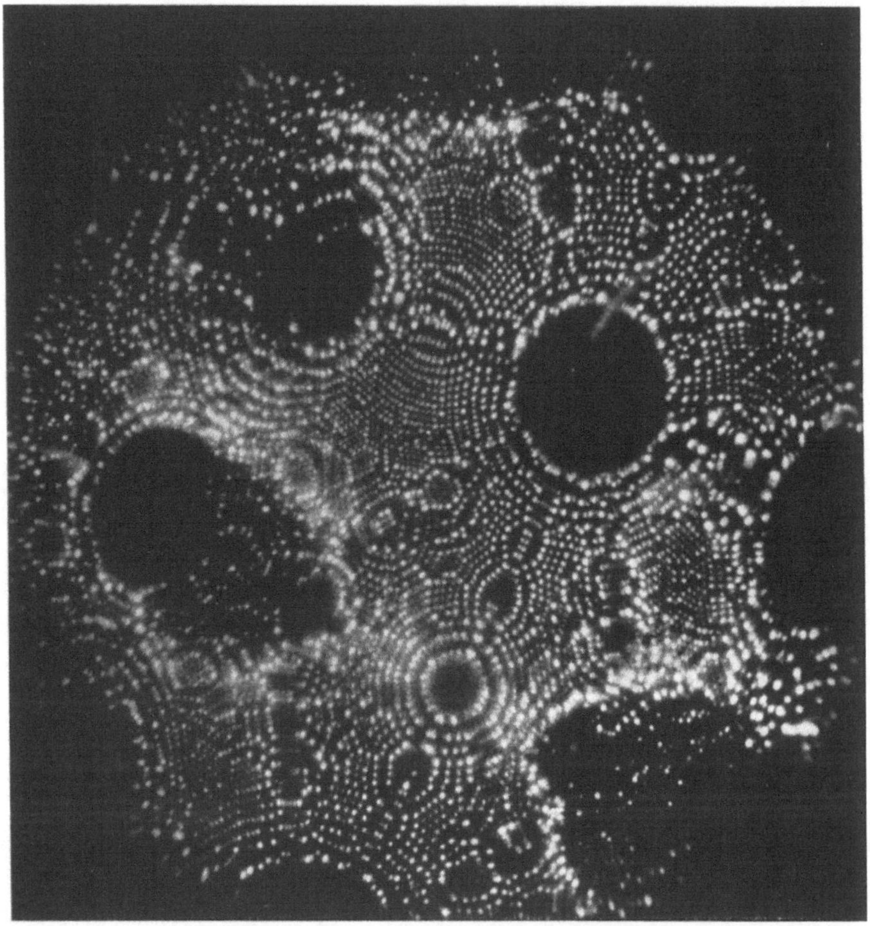

Fig. 39. Helium ion image of a rhenium tip, with close packed hexagonal lattice, radius about 800 Å, after annealing at 2500 °K and some field evaporation at liquid hydrogen temperature. The dark area closest to the center is a 1010 net plane.

carbide or nitride. Various kinds of dislocations which are not removed by the large field stresses (of the order of 1000 kg/mm^2) can be observed. While an annealed surface turns out to have a considerable frozen in thermal disorder, an ideally smooth and clean surface with perfect high index planes can be produced by field evaporation. On such surfaces the initial stage of cathode sputtering or the damage done by impinging individual ions or fast atoms may be studied in detail. At the present time cooling with liquid hydrogen seems to provide optimum conditions, not only because of increased intensity, but also because the rebounding atoms cannot hop higher than perhaps 5 Å, that is just to the inner border of the ionization zone, before they drop back to the surface due to dipole attraction. Thus the majority of the ions are produced in a thin layer quite close to the surface, giving a high resolution.

A further improvement can be obtained by returning to the original concept of field desorption, where the ions are actually produced at the surface and start there with transverse energies of kT only. To do this one operates the field ion microscope with high voltage pulses instead of d.c. The voltage peaks of a suitable repetition rate tear off the adsorbed ions and produce the image, while in the time between two pulses the sites at the surface are being filled up again by adsorption from the gas. Some difficulties arising from the extremely low intensity of the image have still to be overcome.

References.

[1] ASHWORTH, F.: Field Emission Microscopy. Adv. Electronics, Vol. 3, 1. New York: Academic Press 1951. This is the first review article on the subject, with several original contributions. It discusses the use of the field emission microscope in the study of adsorption on specific crystal planes.

[2] BECKER, J. A.: Adsorption on Metal Surfaces and its Bearing on Catalysis. Adv. Catalysis **7** (1955). This comprehensive paper contains a careful discussion of adsorption phenomena as observed by the field emission microscope, particularly a detailed description and interpretation of oxygen on tungsten.

[3] DYKE, W. P., and W. W. DOLAN: Field Emission. Adv. Electronics, Vol. 8. New York: Academic Press 1956. This succinct presentation is particularly valuable for its detailed account of the many contributions of the Linfield College group to the more quantitative knowledge of field emission phenomena made possible by pulse techniques.

[4] GOMER, R.: Field Emission Microscopy and Some Applications to Catalysis and Chemisorption. Adv. Catalysis **7** (1955). Here is an excellent survey of the standard methods, including detailed techniques. Adsorption experiments with oxygen on tungsten are extended down to liquid helium temperatures. The interpretation of the oxygen on tungsten system differs from that in reference [2].

[5] JENKINS, R. O.: Field Emission of Electrons. Rep. Progr. Phys. **9**, 177 (1942/43). This is a comprehensive account of the development of knowledge in field emission from WOOD's first observation to the beginning of field emission microscopy.

[6] MÜLLER, E. W.: Feldemission. Ergebn. exakt. Naturw. **27**, 290 (1953). This extensive review article contains many original results in field emission microscopy and field ion emission. The reference list is complete up to 1952.

[7] MÜLLER, ERWIN W.: Field Emission Microscopy. Mod. Res. Techn. Phys. Metallurgy. Amer. Soc. for Metals, Cleveland, Ohio **1953**, 33. This is a brief summary with emphasis on field emission microscopy for metallurgists.

[8] MÜLLER, ERWIN W.: Field Emission Microscopy. Article in W. G. BERL, Physical Methods in Chemical Analysis, Vol. 3. New York: Academic Press 1956. This paper contains an account of the latest results in the field emission microscopy of a wide variety of adsorption systems, and includes a description of techniques and a reference list.

[9] SOMMERFELD, A., and H. BETHE: Article in GEIGER-SCHEEL's Handbuch der Physik, Vol. 24/2, p. 333. Berlin: Springer 1933. This is the fundamental article in the study of the electron theory of metals. It includes an account of the FOWLER-NORDHEIM theory of field emission.

Sekundärelektronen-Emission fester Körper bei Bestrahlung mit Elektronen.

Von

R. Kollath.

Mit 50 Figuren.

A. Einführung.

1. Der Grundvorgang der Sekundäremission. Treffen Elektronen auf feste Körper auf, so treten aus den festen Körpern wiederum Elektronen aus (Fig. 1). Diese Erscheinung der Sekundärelektronen-Emission (SE) wurde schon um 1900 entdeckt und ist seither Gegenstand vieler Untersuchungen gewesen, insbesondere seit sich die Technik für diesen Gegenstand zu interessieren begann. Man kann sich auf den Standpunkt stellen, daß es sich grundsätzlich einer Feststellung entziehe, wie weit solche „sekundären" Elektronen ursprünglich dem Metallverband angehörten und wie weit es sich um irgendwie gestreute Elektronen des auftreffenden Strahls, um Primärelektronen (P), handelt. Die genauere Untersuchung der Erscheinung hat aber gute Gründe dafür geliefert, mit dem Namen „Sekundärelektronen" (S) nur einen gewissen Teil der von dem Festkörper ausgehenden Elektronen mit ganz bestimmten Eigenschaften zu bezeichnen. Daß zum mindesten ein Teil der „sekundären" Elektronen nicht primäre Elektronen sein können, geht schon daraus hervor, daß in vielen Fällen mehr „sekundäre" Elektronen den Festkörper verlassen, als primäre aufgetroffen sind.

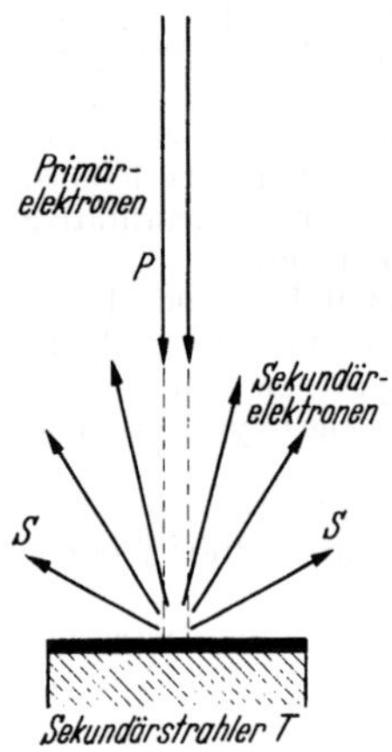

Fig. 1. Grunderscheinung der SE (schematisch).

2. Zur Gliederung des Stoffes. So einfach dieser Emissionsvorgang rein äußerlich zu beschreiben ist, so kompliziert und zum Teil noch ungeklärt sind die Grundvorgänge dabei. Bei dieser Sachlage erscheint es angebracht, zunächst die experimentell gefundenen Erscheinungen und Gesetzmäßigkeiten der SE darzustellen (Teil B—D). Anschließend werden im Teil E einige anderweitige Erfahrungen über die Bewegung von Elektronen in festen Körpern bis zu Energien von einigen keV wiedergegeben. Teil F gibt eine Zusammenstellung der Ansätze zur theoretischen Deutung. Ungeachtet der Schwierigkeiten bei der Aufklärung der Grundvorgänge der SE hat diese Emissionsart heute bereits eine erhebliche Bedeutung für die Technik erlangt.

3. Historisches. Ein kurzer geschichtlicher Überblick dürfte für das Verständnis und die Beurteilung des derzeitigen Standes unserer Kenntnis nützlich sein. Nach der Entdeckung der SE durch Austin und Starke [*9*], [*10*] wurden bis zum Beginn des ersten Weltkrieges schon einige wesentliche Eigenschaften der S aus Metallen festgestellt, besonders durch die Messungen von v. Baeyer [*17*], [*18*], [*19*] und Gehrts [*20*]. In und nach dem ersten Weltkrieg begann sich auch ein technisches Interesse an der SE zu regen, denn sie machte sich in Elektronenröhren störend bemerkbar. Ferner tauchte schon 1919 in einer Patentschrift von SLEPIAN [29] der Gedanke auf, die SE zur Stromverstärkung auszunutzen; die praktische Durchführung scheiterte damals jedoch daran, daß noch keine Stoffe mit genügend großer SE bekannt waren.

Wesentliche Fortschritte wurden zwischen 1925 und 1930 gemacht: BECKER [46], [47] führte die ersten sorgfältigen Messungen der Energieverteilung der S durch; ferner wurde von SUHRMANN [78] und bei der RCA gleichzeitig die große SE der Alkaliphotoschichten entdeckt, und schließlich setzte sich in diesen Jahren, wesentlich auf Grund der Arbeiten von FARNSWORTH [56], [57] die Erkenntnis durch, daß nur unter besten Vakuumbedingungen zuverlässige, d. h. reproduzierbare SE-Messungen sich durchführen lassen. Ausgedehnte und sehr zuverlässige Meßreihen stammen in der Folgezeit von WARNECKE [124], [158],]221] und von BRUINING (vgl. [193]). Die Kenntnis von der SE wurde durch zahlreiche experimentelle Arbeiten immer weiter ausgedehnt und es erschienen auch theoretische Betrachtungen zur Erklärung der Eigenschaften der S und zur Erfassung der Vorgänge bei ihrer Entstehung. Gerade hierbei stellte sich aber heraus, wieviel experimentelles Material noch fehlt, besonders über das Verhalten von Elektronen in Metallen im Energiegebiet bis zu einigen keV. Es wird noch einer erheblichen Arbeit und besonders enger Zusammenarbeit zwischen Experiment und Theorie bedürfen, um über die Erscheinung der SE endgültige Aufklärung zu gewinnen. Immerhin scheinen sich schon verschiedene grundsätzliche Gesichtspunkte abzuzeichnen. Seit der ersten Übersicht von LANGMUIR und JONES erschienen zusammenfassende Darstellungen von: R. KOLLATH [4], R. WARNECKE [8], MCKAY [3] und H. BRUINING [1], [2], die den Fortschritt auf experimentellem Gebiet und den Übergang von rein phänomenologischer Betrachtungsweise zu exakten theoretischen Untersuchungen zeigen, aber bei dem heutigen Stand noch immer keine abschließende Darstellung der SE fester Körper zu geben vermochten; das gleiche gilt daher auch von dem vorliegenden Bericht, wenn auch seit der letzten Übersicht von BRUINING immerhin wieder einige bemerkenswerte Tatsachen hinzugekommen sind.

Die Schwierigkeiten, die einem endgültigen Verständnis der SE entgegenstehen, lassen sich vielleicht am besten durch Gegenüberstellung der SE und anderer Elektronenemissionsprozesse (Glühemission, lichtelektrische Emission, Feldemission) verständlich machen[1]. Bei der Glühemission haben wir es nur mit den quasifreien Elektronen des Elektronengases im Metall zu tun; die FERMI-Verteilung dieses Elektronengases und ihre Temperaturabhängigkeit hat — jedenfalls bei reinen Metallen — ein quantitatives Verständnis der Erscheinungen ermöglicht. Bei der Feldemission hat die wellenmechanische Behandlung (Tunneleffekt) ebenfalls zu einer befriedigenden quantitativen Deutung der Erscheinungen geführt, da auch hier die Leitungselektronen wesentlich am Emissionsvorgang beteiligt sind. Beim Photoeffekt liegen die Dinge bereits komplizierter (was theoretisch schon in der Unterscheidung zwischen Volumen- und Oberflächeneffekten zum Ausdruck kommt); hier haben wir es mit einem primären Auslösevorgang zu tun (Energiezuführung von außen durch die Lichtquanten) und einem davon getrennten Vorgang des Elektronenaustritts. Aber die auf den Festkörper auftreffenden Lichtquanten besitzen nur einen vernachlässigbar kleinen Impuls

[1] Die S-Auslösung durch positive Ionen wird in diesem Zusammenhang *nicht* behandelt; sie stößt auf die gleichen, vielleicht sogar auf noch größere experimentelle und theoretische Schwierigkeiten wie die Auslösung durch Elektronen. Über das vorliegende Material vgl. den Artikel von P. F. LITTLE in diesem Bande.

und ihre Energie ist von der Größenordnung der Austrittsarbeit; an der langwelligen Grenze kann es sich demnach bei Metallen nur um Leitungselektronen handeln, wobei außerdem an dem Emissionsvorgang nur eine dünne Oberflächenschicht (wenige Atomlagen) beteiligt ist. Beim Photoeffekt läßt sich also der Vorgang noch gut übersehen. Bei der SE liegt die Sache wesentlich schwieriger. Die Energie der P reicht im allgemeinen dafür aus, auch wesentlich fester gebundene Elektronen im Metall zu befreien. Es ist also zunächst die Grundfrage zu klären, an welche Elektronen im Metall anteilig die Energie der P abgegeben wird, wobei die Größe und die Wahrscheinlichkeit der Energieverluste von Elektronen von einigen hundert eV Energie beim Durchgang durch Festkörper eine wesentliche Rolle spielen. Nicht weniger schwierig ist dann die Frage nach der Wechselwirkung (Streuung, Energieverlust) der energetisch angehobenen Elektronen mit dem Gitter und den Elektronen des Metalls auf ihrem Wege zur Oberfläche, der je nach Material und Energie der P bis zu 100 Atomlagen betragen kann. Über Streuung und Energieverluste von Elektronen im Energiebereich von etwa 10 bis zu einigen hundert eV gibt es aber bisher kaum einigermaßen sichere Daten. Für Isolatoren würde sich diese Frage wahrscheinlich leichter beantworten lassen als für Metalle (zunächst nur gebundene Elektronen, relativ große freie Weglängen der ins Leitungsband gehobenen S), unglücklicherweise stoßen aber Versuche an Isolatoren auf erhebliche experimentelle Schwierigkeiten, wenn man wirklich eindeutige Aussagen gewinnen will.

Diese kurzen Andeutungen mögen genügen; sie sollen vor allem begreiflich machen, warum trotz intensiver, jahrzehntelanger Arbeit vieler Wissenschaftler auf diesem Gebiet immer noch keine abschließende, in allen Teilen befriedigende Darstellung möglich ist.

B. Die Energieverteilung der Sekundärelektronen.

4. Allgemeine Form der Energieverteilung. Wir betrachten die allgemeine Form der Energieverteilung von Elektronen, die ein fester Körper bei Beschuß mit P emittiert, am Beispiel der Fig. 2. Abszisse ist die Energie der emittierten Elektronen in eV, Ordinate die relative Anzahl $N(E)$ von Elektronen, deren Energie im Energiebereich zwischen E und $E+dE$ liegt; die Fläche zwischen der Kurve und der Abszissenachse ist dann ein Maß für die Gesamtzahl der austretenden Elektronen.

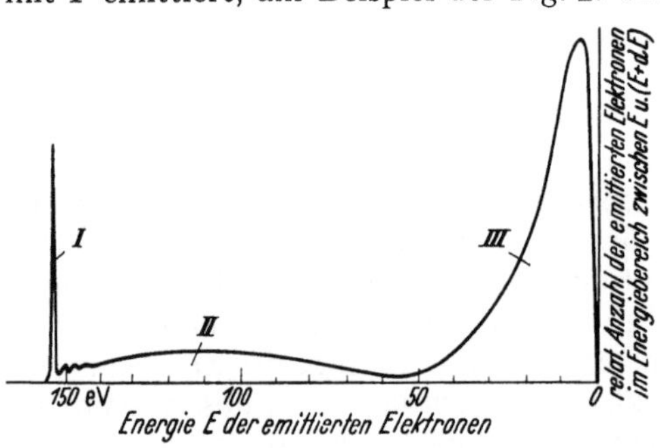

Fig. 2. Allgemeine Form der Energieverteilung (schematisch).

Der Kurvenverlauf von links nach rechts betrachtet, zeigt folgendes:

a) Ein Teil der ausgelösten Elektronen (Gruppe I) besitzt noch die volle Primärenergie, wie das scharfe Maximum bei 155 eV erkennen läßt; sie werden allgemein als „elastisch gestreute P" bezeichnet;

b) eine zweite Gruppe (II) tritt mit Energien zwischen 150 und etwa 50 eV aus. Man kann sie mit solchen P identifizieren, die auf Grund von Streuprozessen „rückdiffundierten"[1] und dabei mehr oder weniger große Energieverluste erlitten;

[1] Diese Bezeichnung geht auf Lenard zurück.

c) unterhalb von etwa 50 eV steigt mit kleiner werdender Energie die Zahl der austretenden Elektronen stark an, wobei Elektronen mit Energien von wenigen eV besonders häufig sind. Die große Mehrzahl dieser zur Gruppe III gehörenden Elektronen kann man ihren Eigenschaften nach (wie später noch deutlicher werden wird), offenbar *nicht* als P deuten: sie bilden die eigentlichen „S", von denen im weiteren hauptsächlich gesprochen werden soll.

Natürlich ist eine wirkliche, individuelle Unterscheidung der S von den P nicht möglich; es läßt sich ferner oft keine scharfe Grenze zwischen den „rückdiffundierenden" und den „eigentlichen" S ziehen. Diese Frage hat aber in dem hier vornehmlich interessierenden Bereich der Primärenergien (bis etwa 1500 eV) keine große Bedeutung, weil die eigentlichen S zahlenmäßig den größten Teil der insgesamt ausgelösten Elektronen bilden, und weil sich die Energieverteilungskurve auf der Grenze zwischen II und III der Abszisse stark nähert (vgl. hierzu z. B. Fig. 10 und 13). Über die „rückdiffundierenden" Elektronen wird in Teil C und E noch einiges gesagt werden.

5. Methoden zur Messung der Energieverteilung. Bei der Messung einer Energieverteilung von langsamen Elektronen sind einige grundsätzliche Vorbedingungen zu beachten, wenn man einwandfreie Daten erhalten will.

Der Raum vor dem Sekundärstrahler T (vgl. Fig. 1) muß feldfrei sein, damit die von *T* ausgehenden Elektronen in ihrer Laufrichtung unbeeinflußt bleiben. Falls für die Energieanalyse Felder gebraucht werden, müssen sie nach Größe und Richtung genau bekannt sein; Kontaktpotentiale, magnetische Streufelder o. ä. müssen also kompensiert werden. Da auch Raumladungen vor der Platte *T* die Elektronenbahnen stören, sind bei den Messungen nur kleine Stromdichten zulässig. Ferner sollte der Primärstrahl durch den Meßvorgang möglichst wenig beeinflußt werden und zwar bezüglich seiner geometrischen Definition, seiner Auftreffstelle und seiner Energie; dies läßt sich im allgemeinen befriedigend durchführen, weil die Energie der P wesentlich größer ist als die der S.

Die eben aufgezählten Grundbedingungen sind offenbar bei verschiedenen Arbeiten über die Energieverteilung der S oft nicht genügend beachtet worden; denn nur so ist es zu verstehen, daß die Resultate verschiedener Arbeiten über die Energieverteilung auch an ein und demselben Material erheblich differieren.

Zunächst sollen die verwendeten Meßmethoden zusammengestellt und soweit erläutert werden, wie es für die Beurteilung der Resultate notwendig erscheint.

Im Gebiet langsamer Elektronen werden zur Zeit für die Energieanalyse verwendet:

a) das longitudinale elektrische Feld, dessen Gebrauch schon auf LENARD [*6*], [*7*] zurückgeht (Gegenfeldmethode);

b) das transversale elektrische Feld, aus elektronenoptischen Gründen am besten mittels eines Zylinderkondensators in der von HUGHES und ROJANSKY [*119*] angegebenen Form;

c) das longitudinale magnetische Feld, dessen Eigenschaften erstmalig von BUSCH [*35*] untersucht wurden;

d) das transversale Magnetfeld, das zuerst von RAMSAUER [*25*] für die Energieanalyse langsamer Elektronen verwendet wurde.

Wie sich im folgenden zeigen wird, liefert keine der genannten Methoden direkt die Energieverteilung, vielmehr bedürfen die Meßwerte in allen Fällen einer Umrechnung [*147*]. Für die Beurteilung der Genauigkeit sowie der Vor- und Nachteile der einzelnen Methoden bei Untersuchungen an S ist die Dispersion der verschiedenen Anordnungen von Bedeutung, d. h. die Größe der Energieaufspaltung als Funktion der Energie (vgl. den Schluß dieser Ziffer).

Benutzt wurde in den weitaus meisten Fällen die Methode a) wegen ihrer Einfachheit, ferner in vielen Fällen die Methode d); nach der Methode c) wurde bisher nur eine Arbeit durchgeführt; die Methode b) scheint im Gebiet der SE bisher noch nicht verwendet worden zu sein; der Grund ist wohl darin zu suchen, daß bei der SE ausheizbare Apparaturen benötigt werden und der Aufbau nach b) etwas sperrig ist; grundsätzlich läßt sich aber Methode b) nach unveröffentlichten Erfahrungen des Verfassers mit gutem Erfolg auch auf diesem Gebiet verwenden.

Zu a). Die Gegenfeldmethode, auch als Methode des elektrischen Verzögerungsfeldes bezeichnet, wird gewöhnlich in der Ausführungsform wie in *Fig. 3a* benutzt. Die P vom Glühdraht *Gl* treffen auf die Sekundärstrahlerplatte *T*, die entstehenden S laufen radial zum Auffänger *K*. Legt man *K* gegenüber *T*

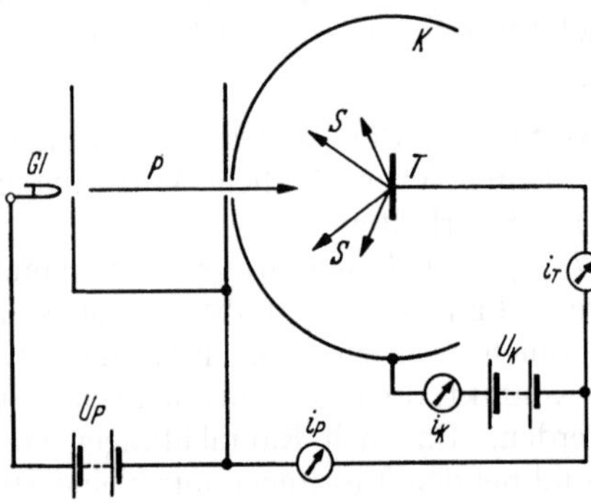

Fig. 3a. Gegenfeldmethode mit „Zentralfeld" (schematisch).

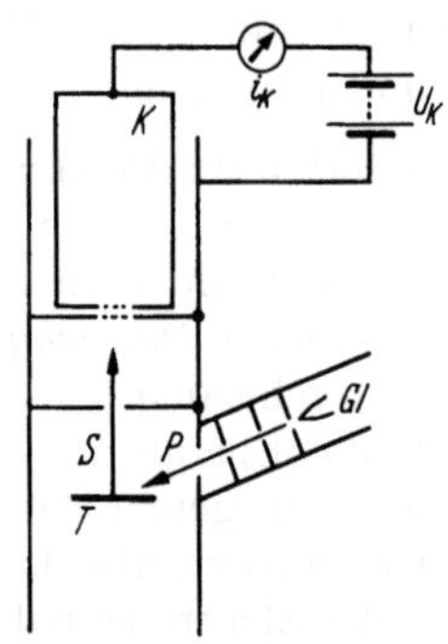

Fig. 3b. Gegenfeldmethode zur Untersuchung einer bestimmten Austrittsrichtung (nach Langenwalter [*133*]).

nacheinander auf verschieden großes negatives Potential (Spannung U_K), so werden nur diejenigen S den Auffänger *K* erreichen, deren Energie zur Überwindung der Potentialdifferenz zwischen *T* und *K* ausreicht. Trägt man den von *K* abfließenden S-Strom i_K als Funktion von U_K auf, so erhält man die „Gegenfeldkurve", auch „Gegenspannungskurve" oder „Anlaufkurve" genannt *(Fig. 4a)*. Die gesuchte Energieverteilung wird aus dieser Meßkurve durch Differentiation erhalten *(Fig. 4b)*. Will man speziell die Energieverteilung derjenigen S untersuchen, die unter einem bestimmten Austrittswinkel φ (gerechnet gegen die Normale auf der Oberfläche) gestartet sind, so kann man eine Anordnung nach Art der *Fig. 3b* benutzen; man kann aber in diesem Fall auch eine der anderen Methoden verwenden, bei denen sowieso jeweils nur eine vorgegebene Austrittsrichtung untersucht werden kann.

Beim Übergang von einer Integralkurve (Gegenfeldkurve) zu einer Differentialkurve wird man keine so große Genauigkeit bzw. Empfindlichkeit erwarten können wie bei direkter Messung einer Differentialkurve (Methode b, c, d). Man kann die Differentiation allerdings gleich in die Messung selbst verlegen, indem man der Gegenspannung U_K eine kleine konstante Wechselspannung ΔU_K überlagert und die Größe des entstehenden Wechselstroms Δi_K als Ordinate über U_K aufträgt. Der wesentliche Einwand gegen die Methode a) in der Form der Fig. 3a liegt in der Schwierigkeit, einwandfreie Feldverhältnisse zu schaffen. (Vgl. hierzu z. B. Johnson [*338*], der aus diesem Grunde seinen Energieverteilungsmessungen nach dieser Methode nur qualitativen, orientierenden Wert beimißt.) Bei kritischer Handhabung ist diese Methode aber zum mindesten für qualitative Orientierung brauchbar (vgl. A. Beckers kritische Diskussion der Gegenfeldmethode [*46*]); sie hat dann den Vorteil, *alle* S zu erfassen und gestattet infolgedessen die Verwendung kleiner Stromdichten (keine Raumladungen!).

Zu b). Da diese Methode bisher zur Messung der Energieverteilung von S nicht verwendet wurde, sei hier nur der Vollständigkeit halber in *Fig. 5* eine entsprechende Anordnung wiedergegeben. Richtungsfokussierung tritt nach $\pi/\sqrt{2}$ entsprechend einem Winkel von 127,5° ein [*119*]. Die Energieverteilung erhält man aus dem gemessenen Auffängerstrom i_K durch Auftragung von i_K/E_s über E_s. Diese Methode bietet als einzige der genannten 4 die Möglichkeit, den Einfallswinkel α der P und den Austrittswinkel φ der S unabhängig voneinander zu variieren. (E_s ist durch das Feld im Kondensator bestimmt.)

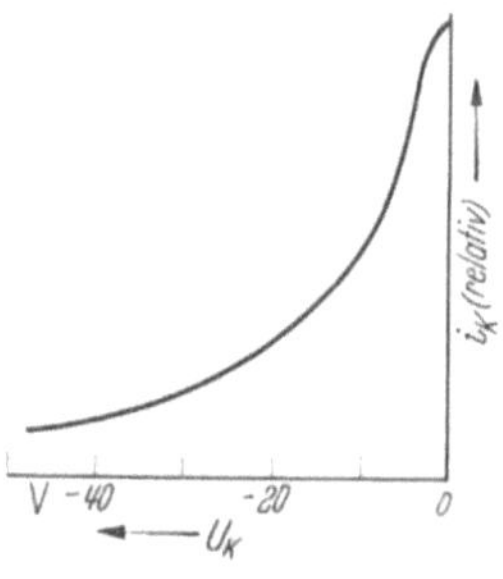

Fig. 4a. Gegenspannungskurve.

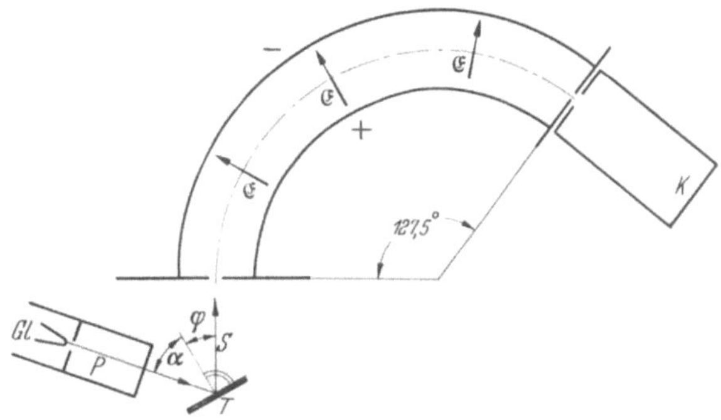

Fig. 5. Methode des transversalen elektrischen Feldes ($\mathfrak{E}$ = elektrische Feldstärke, φ = Austrittswinkel der S aus T, α = Einfallswinkel der P auf T).

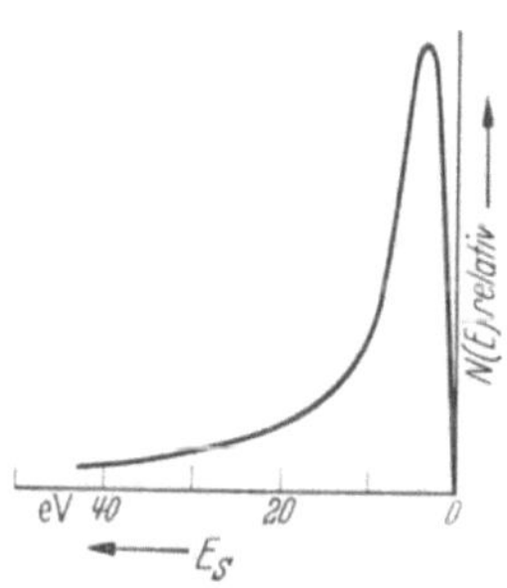

Fig. 4b. Energieverteilung.

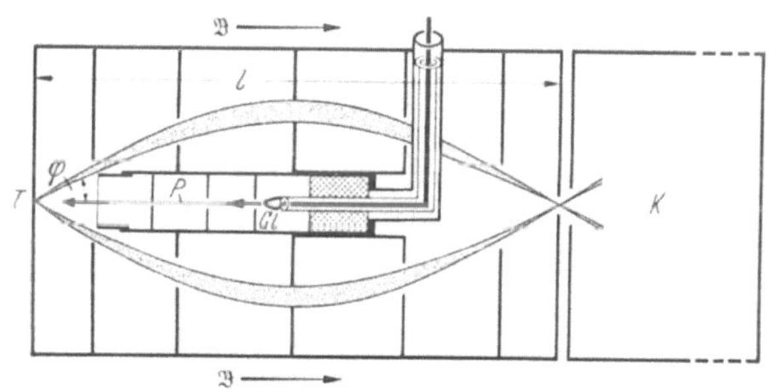

Fig. 6. Prinzipskizze der Methode des longitudinalen Magnetfeldes (nach KOLLATH [*258*]).

Zu c). Die Methode des longitudinalen Magnetfeldes wurde in ihrer Eignung für die Messung der Energieverteilung langsamer Elektronen von KOLLATH [*293*] geprüft und dann auf die Untersuchung einer ganzen Reihe von Metallen und Metallegierungen angewendet [*344*]. Treten Elektronen in ein homogenes Magnetfeld etwas geneigt zur Richtung der magnetischen Feldlinien ein, so winden sie sich in Schraubenlinien um die Feldlinien herum (gleichzeitige lineare Bewegung *in* und Kreisbewegung *senkrecht* zur Richtung des Magnetfeldes). Bei zylindersymmetrischer Anordnung beschreiben dann die Abstände der Elektronen von der Symmetrieachse der Anordnung eine sin-Linie (*Fig. 6*). Alle Elektronen, die mit gleicher Energie von einem bestimmten Punkt T der Achse ausgehen, werden in einer festen Entfernung l (= Länge des Analysators) wieder in einem Achsenpunkt gesammelt, wo sich die Eintrittsöffnung des Auffangkäfigs K befindet. Durch Änderung der Flußdichte B des Magnetfeldes erhält man nacheinander Elektronen verschiedener Energie. Die Energieverteilungskurve ergibt sich aus dem gemessenen Käfigsstrom i_K durch Auftragung von i_K/E_s über $E_s \sim B^2$. Da die Mitte des Analysators nicht anderweitig benötigt wird, kann dieser Raum zur Unterbringung der P-Quelle *Gl* ausgenutzt werden, so daß die P längs der Achse von rechts nach links laufend auf die Platte *T* auftreffen; da die P in Richtung der magnetischen Feldlinien laufen, bleibt die Auftreffstelle der P bei einer Änderung der Flußdichte während der Messung unbeeinflußt. Durch die Zylindersymmetrie der Anordnung werden außerdem alle S, die unter einem bestimmten, durch die Apparaturdimensionen festgelegten Austrittswinkel φ die Platte *T* verlassen, für die Messung ausgenutzt (Intensität!); im Gegensatz dazu ist bei der Methode des transversalen Magnetfeldes nur ein wesentlich kleinerer Raumwinkel verwendbar, deshalb muß dort mit wesentlich kleineren Elektronenströmen gearbeitet werden, um Raumladungen vor *T* zu

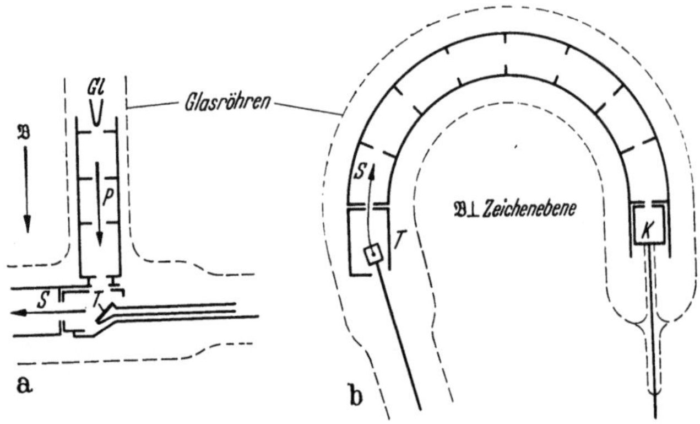

Fig. 7 a u. b. Methode des transversalen Magnetfeldes in der von HARWORTH [*128*] benutzten Form.

vermeiden. Der Hauptnachteil der Methode des longitudinalen Magnetfeldes in der oben beschriebenen Form liegt darin, daß der Einfallswinkel α der P gar nicht und der Austrittswinkel φ der S nur durch einen weitgehenden Umbau der Anordnung geändert werden können.

Zu d). Das Verfahren des transversalen Magnetfeldes, das von RAMSAUER in seiner Wirkungsweise genau untersucht und dann zur Messung der Energieverteilung von Photoelektronen verwendet wurde [*25*], ist so allgemein bekannt, daß eine kurze Beschreibung in diesem Zusammenhang genügen dürfte *(Fig. 7)*. Die Elektronen von der Platte *T*, ausgelöst durch P, die in Fig. 7b senkrecht zur Zeichenebene auf *T* auftreffen, beschreiben in einem homogenen Magnetfeld einen Halbkreis (Richtungsfokussierung!), dessen Radius durch mehrere Blenden vorgeschrieben ist. Die Methode kann in zwei verschiedenen Abarten verwendet werden: entweder wird bei konstantem Magnetfeld der Auffangkäfig längs des Kreisdurchmessers verschoben, oder es wird mit feststehendem Käfig gearbeitet und die Flußdichte des Magnetfeldes variiert (übliche Ausführungsform). Im ersten Fall erhält man aus dem gemessenen Käfigstrom i_K als Funktion des Kreisdurchmessers D die Energieverteilungskurve durch Auftragung von i_K/D über D^2 (das proportional zu E ist), im zweiten Fall durch Auftragung von i_K/E über B^2 (das proportional zu E ist). Wegen des geringen zur Messung gelangenden Teilstroms ist erhebliche Vorsicht bezüglich eventuell entstehender Raumladungen im Raum vor *T* geboten; ferner ist darauf zu achten, daß die P, wie in Fig. 7a angegeben, in Richtung des magnetischen Flusses laufen, damit sich der Auftreffpunkt der P nicht während der Messung verschiebt. Der experimentelle Aufwand hält sich in erträglichen Grenzen; die Methode hat ferner den Vorteil, daß sich der Auftreffwinkel der P und der Austrittswinkel der S durch Drehung von T verändern lassen, allerdings nicht unabhängig voneinander: eine Drehung der Quelle um T ist nämlich nicht ohne weiteres möglich, weil dann die P schief zum Feld laufen würden; der Auftreffwinkel der P und der Austrittswinkel der S bilden also zusammen eine Summe von 90°. Diese Einschränkung fällt übrigens bei der Methode des elektrischen Transversalfeldes fort; es ergibt sich bei näherer Betrachtung also, daß gerade die beiden Methoden, die sich zur Energieanalyse der S besonders gut eignen, bisher wenig oder gar nicht herangezogen worden sind.

Bei den beschriebenen Methoden hängt das Energie-Auflösungsvermögen in verschiedener Weise von der Energie ab. Während für das Gegenfeld die von der Messung erfaßte Energiebreite für alle Energien konstant ist, hängt sie bei den drei anderen Verfahren von der Energie ab [*147*]. Die Überlegenheit der Differentialmethoden gegenüber der Gegenfeldmethode beruht gerade darauf, daß bei der *gemessenen* Kurve der Bereich kleiner Energien auseinandergezogen wird, der bei der SE besonders interessiert, und daß gleichzeitig im Bereich größerer Energien (vgl. oben) die Empfindlichkeit gegenüber Abweichungen vom glatten Kurvenverlauf stark erhöht wird[1]. So wird es verständlich, daß bei den Methoden b) bis d) noch Maxima bei größeren Energien einwandfrei gemessen werden, die bei Umrechnung auf die Energieverteilung völlig „eingeebnet“ und deshalb von der Gegenfeldmethode gar nicht erfaßt werden (vgl. hierzu Ziff. 7).

6. Übersicht über die bisher durchgeführten Messungen der Energieverteilung. Im Vordergrund des Interesses steht natürlich die Energieverteilung der eigentlichen S mit Energien bis etwa 50 eV und ihre Abhängigkeit von verschiedenen Parametern (Energie der P, Material des Sekundärstrahlers, seine Struktur und Oberflächenbeschaffenheit usw.). Soweit hierzu experimentelles Material vorliegt, wird es anschließend zusammengestellt.

Darüber hinaus macht aber die Energieverteilung bei genauerer Untersuchung der Elektronen, die mehr oder weniger große Geschwindigkeitsverluste erlitten haben (also im Energiegebiet zwischen 50 eV und der Primärelektronen-Energie E_P), Aussagen über die Größe und Häufigkeit der Geschwindigkeitsverluste der P im Plattenmaterial und über die Zahl der „rückdiffundierenden“ Elektronen; Angaben hierüber werden gebraucht, um das Zustandekommen der SE zu erklären. Sie werden im Zusammenhang mit der Diskussion über die Frage der Entstehung der S behandelt (vgl. Teil C, E, F).

Vermutlich ist die Energieverteilung der S eine viel schärfere Sonde für die Richtigkeit theoretischer Vorstellungen über die Entstehung der SE als die relative *Zahl* der aus der Platte austretenden Elektronen, die sog. Ausbeute. Tatsächlich liegen bisher aber nur wenige zuverlässige Untersuchungen der Energieverteilung von S vor, während Ausbeute-Messungen unter den verschiedensten Bedingungen in großer Zahl durchgeführt sind, wie sich im Teil C zeigen wird. Das hängt vermutlich damit zusammen, daß eine Messung der Energieverteilung experimentell wesentlich schwieriger ist als eine Ausbeutemessung. In Tabelle 1 sind Untersuchungen der Energieverteilung seit 1925 zusammengestellt.

Das gesamte Material der Tabelle 1 ist wesentlich lückenhafter, als es bei flüchtiger Betrachtung den Anschein hat; das wird bei der nachfolgenden Diskussion der Parameter klar werden, die möglicherweise auf die Form der Energieverteilung Einfluß nehmen: abgesehen von verschiedenen ausführlichen Messungen der Energieverteilung als Funktion der Primärenergie E_P und einer

[1] Das hängt zusammen mit dem Verlauf der Energieverteilungskurve der S, die sich bei größeren Energien der Abszisse wieder stark annähert.

unter weitgehend vergleichbaren Bedingungen durchgeführten Untersuchung einiger Metalle, Legierungen und Halbleiterschichten bei verschiedener Wärmebehandlung wurden quantitativ nur noch wenige spezielle Fälle behandelt. Dazu kommen noch einige Messungen der Energieverteilungskurve bei Isolatoren, die aber höchstens qualitative Aussagen gestatten. Selbst eine so grundsätzliche Frage wie die Temperaturabhängigkeit ist durchaus noch nicht abschließend geklärt.

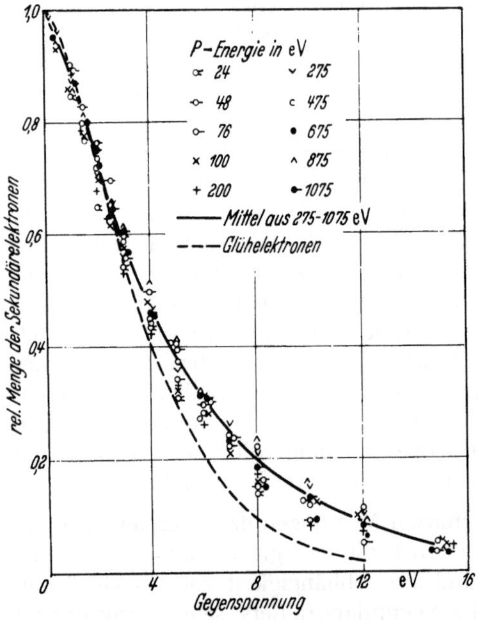

Fig. 8a. S-Gegenspannungskurve bei verschiedener Energie der P an Ruß nach BECKER [46] (die gestrichelte Kurve ist bei der Ordinate 0,7 willkürlich auf die ausgezogene Kurve gelegt).

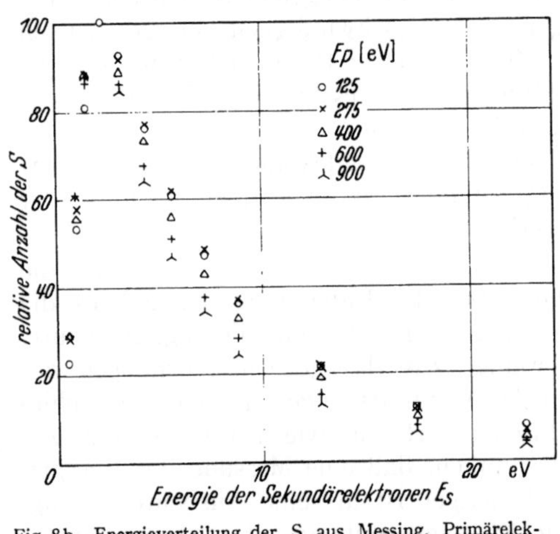

Fig. 8b. Energieverteilung der S aus Messing, Primärelektronen-Energie E_P als Parameter (nach KOLLATH [344]).

7. Energieverteilung der Sekundärelektronen aus Metallen als Funktion verschiedener Parameter. Von der *Energie der P* ist die Energieverteilung im Bereich zwischen 100 und 1000 eV im wesentlichen unabhängig, wie als erster BECKER [*46*] einwandfrei für C (Ruß) und Pt und KOLLATH [*344*] an Messing und Be-Aufdampfschichten nachwies. Aus den Fig. 8a, b läßt sich auf eine eindeutige Zuordnung der Primärenergie und der Lage der Meßpunkte (bei fester S-Energie) nicht ohne weiteres schließen, immerhin scheint ein geringer systematischer Gang angedeutet zu sein. Ob man diesen geringen Gang aber der Energieverteilung der *eigentlichen* S zuschreiben darf oder ob er durch andere Ursachen bedingt ist, bleibt ungewiß (vgl. die ausführliche Diskussion bei KOLLATH [*344*]).

Über die Energieverteilung als Funktion des *Einfallwinkels der P und des Austrittswinkels der S* ist bisher kaum etwas bekannt. Zwar sind die Untersuchungen nach Tabelle 1 bei vielen verschiedenen Einfalls- und Austrittswinkeln durchgeführt worden, ein direkter Vergleich zwischen den Arbeiten verschiedener Autoren ist aber sogar am gleichen Material wegen der verschiedenen Materialbeschaffenheit und wegen Abweichungen auf Grund von unkontrollierbaren Fehlerquellen bei den Messungen nicht möglich. Die bisher einzige systematische Untersuchung von KUSHNIR und FRUMIN [*294*] an Mo und Ag ergibt, daß sich die Lage des Maximums der Energieverteilungskurve (wahrscheinlichste Energie) mit steigendem Austrittswinkel der S nach größeren Energien hin verschiebt. Eine Nachprüfung und Ausdehnung dieser Messungen wäre bei der Wichtigkeit einer solchen Aussage für die theoretische Behandlung sehr zu begrüßen.

Die nach Tabelle 1 *stark unterschiedlichen Angaben über die wahrscheinlichste Energie der S* ($E_{s_{max}}$) veranlaßten KOLLATH zu einer genauen *Vergleichs*untersuchung verschiedener Materialien unter möglichst gleichen Bedingungen. Dafür wurde die Anordnung nach Fig. 6 in geeigneter Weise modifiziert (Fig. 9). In Fig. 10 ist schraffiert der Bereich angegeben, innerhalb dessen sämtliche Meßpunkte für die Metalle Ta, W, Mo, Ag, Zn (in kompakter Schicht), ferner für Aufdampfschichten von Be, Mg, Cs und schließlich für die Legierungen Messing, CuBe, NiBe zu finden waren; die Versuchsröhre einschließlich der Schichten wurde vorher wie üblich mehrere Stunden bei etwa 350 bis 400° C ausgeheizt[1]. Nach dieser zusammenhängenden Meßreihe variiert die wahrscheinlichste Energie der S nur in einem kleinen Bereich zwischen etwa 1,4 und 2,3 eV. Wenn die Messungen zunächst auch nur für einen Einfallswinkel der P von 0° und einen Austrittswinkel der S von etwa 30° gelten, so bestehen doch sehr starke Bedenken gegen die Richtigkeit von Angaben über die Lage des Maximums, die über etwa 4 eV hinausgehen[2]. Irgendwelche Fehlerquellen wie Kontaktpotentiale oder Raumladungen wirken sich in der Regel dahingehend aus, daß die *langsamsten* Elektronen „verschwinden", daß also eine Verschiebung der gesamten Energieverteilung nach größeren Energien vorgetäuscht wird[3].

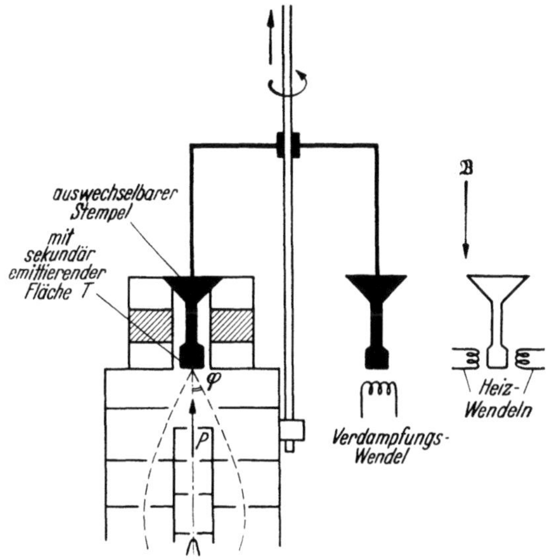

Fig. 9. Anordnung zum unmittelbaren Vergleich der S-Energieverteilung verschiedener Materialien (nach KOLLATH [258]).

Den Messungen nach der Gegenfeldmethode mit „Zentralfeld" kann nur qualitativer Wert beigemessen werden. Damit verringert sich das quantitativ brauchbare Material der Tabelle 1 um die mit einem Sternchen bezeichneten Beiträge.

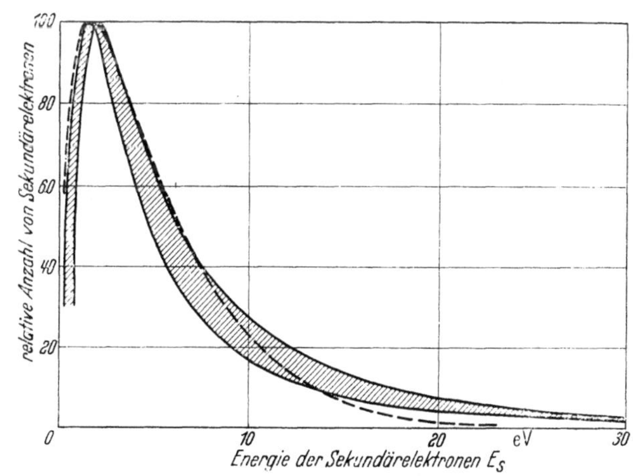

Fig. 10. S-Energieverteilung für verschiedene Materialien, vgl. Text (— — —: MAXWELL-Verteilung mit Maximum bei 1,75 eV zum Vergleich).

[1] Die leicht verdampfbaren Substanzen Zn und Messing wurden nur bei etwa 250—300° C ausgeheizt, die Cs-Schicht nach dem Aufdampfen gemessen.

[2] Die Messungen von HAWORTH werden im nächsten Absatz behandelt; sie sind nicht ohne weiteres mit diesem Befund vergleichbar.

[3] Dies gilt für die differentiellen Methoden; bei der Gegenfeldmethode ist es schwer, solche Einflüsse auf die Energieverteilung abzuschätzen und was schlimmer ist, sich experimentell durch entsprechende Kontrollmessungen Gewißheit zu verschaffen, daß sie nicht vorliegen.

Tabelle 1. *Zusammenstellung von Messungen der Energieverteilung von Sekundärelektronen (seit 1925)*

Material	Autor	Jahr	Methode[1]	(Einfalls-) Winkel in Grad	(Austritts-) Winkel in Grad	Primärenergie E_p in eV	$E_{s\,max}$ in eV	Zitat	Bemerkungen
Ag (kompakt)	FARNSWORTH	1928	Gegenf.	0		0 bis 50		[68]	* „Zentralfeld“
	CHILINSKY	1932	trans. M.				8	[99]	* Raumladungen
	LANGENWALTER	1935	Gegenf.	70	0	10 bis 100	2 bis 3	[133]	
	RUDBERG	1936	trans. M.	45	45	155	5,4	[152]	nur 1 Übersichtskurve
	KUSHNIR und FRUMIN	1941			variabel			[294]	
	KOLLATH	1947	long. M	0	30	400	1,5 bis 2,1	[344]	je nach Vorbehandlung
Al (kompakt)	BRINSMADE	1927	trans. M.	45	45	36 bis 176	5 bis 6	[59]	* Primärstromstärke nicht angegeben (Raumladung ?)
Au	RUDBERG	1936	trans. M.	45	45	155	5,4	[152]	nur 1 Übersichtskurve
	KOLLATH	1952	trans. E.	$\alpha+\varphi=90$		150	2 bis 3	unveröffentlicht	Au-Draht
Be (aufgedampft)	KOLLATH	1947	long. M.	0	30	85 bis 900	1,9 bis 2,2	[344]	je nach Behandlung weitere Maxima vgl. S. 244
C (Ruß)	BECKER	1925	Gegenf.	30	0	24 bis 2075	} 2	[46], [47]	
	BECKER	1925	Gegenf.	30		24 bis 2075	}		
(Graphit)	SHARMAN	1927	trans. M.				5	[66]	* Raumladung
Cu (kompakt)	SHARMAN	1927	trans. M.				15	[66]	* Raumladung
	FARNSWORTH	1928	Gegenf.	0		0 bis 50		[68]	* „Zentralfeld“
	RUDBERG	1936	trans. M.	45	45	155	3,3	[152]	nur 1 Übersichtskurve
Cs (aufgedampft auf Ag)	KOLLATH	1947	long. M.	0	30	400	2	[344]	
Fe	FARNSWORTH	1928	Gegenf.	0		0 bis 50		[68]	* „Zentralfeld“

Mg (aufgedampft)	KOLLATH	1947	long. M.	0	30	400	1,5 bis 1,9	[344]	je nach Behandlung
Mo (kompakt)	SOLLER HAWORTH KUSHNIR und FRUMIN KOLLATH	1930 1935 1941 1947	trans. M. trans. M. long. M.	45 45 0	45 45 variabel 30	7,5 bis 100 150 400	4 bis 20 3 1,8 bis 1,9	[93] [128] [294] [344]	* Raumladung weitere Maxima vgl. S. 244
Nb	HAWORTH	1936	trans. M.	45	45	147	4,5	[145]	
Ni	FARNSWORTH	1928	Gegenf.	0		0 bis 50		[68]	* „Zentralfeld"
Pd	LANGENWALTER	1935	Gegenf.	70	0	10 bis 100	2	[133]	
Pt	BECKER SHARMAN LANGENWALTER	1925 1927 1935	Gegenf. trans. M. Gegenf.	30 70	0 0	 10 bis 100	 14 2 bis 3	[46] [66] [133]	* Raumladung
Ta	KOLLATH	1947	long. M.	0	30	400	2,0 bis 1,9	[344]	je nach Behandlung
Wo	KOLLATH	1947	long. M.	0	30	400	2,1 bis 2,2	[344]	je nach Behandlung
Zn (aufgedampft)	KOLLATH	1947	long. M.	0	30	400	1,8	[344]	
Messing CuBe-Legierung NiBe-Legierung	KOLLATH KOLLATH KOLLATH	1947 1947 1947	long. M. long. M. long. M.	0 0 0	30 30 30	125 bis 900 400 400	302 1,4 bis 1,9 1,5 bis 1,7	 [344] [344]	je nach Behandlung weitere Maxima vgl. S. 244
Cs-Oxyd Cs—Sb	PIATNITZKY KOLLATH	1938 1947	Gegenf. long. M.	0 0	 30	50 bis 300 400	2 2,0	[211] [344]	* „Zentralfeld", Photoschicht Photoschicht
BaO—SrO	JOHNSON	1948	Gegenf.	0		1250	1	[351]	* „Zentralfeld", Oxydkathode
MgF_2 NaCl NaCl	GEYER VUDINSKY GEYER	1942 1939 1942	Gegenf. Gegenf. Gegenf.	0 0 0		300 bis 1500	1 1 1	[311] [346] [311]	* „Zentralfeld" (Raumladung?) aufgedampft mit verschiedener Dicke

[1] Gegenf. = Gegenfeldmethode; trans. M. = transversales Magnetfeld; long. M. = longitutinales Magnetfeld; trans. E. = transversales elektrisches Feld.

Bemerkenswert ist an der Verteilungskurve nach Fig. 10: das Kurvenmaximum liegt bei etwa 2 eV, die wahrscheinlichste Energie der S ist also um etwa eine Zehnerpotenz größer als die der Glühelektronen. Dabei läßt sich mit Sicher-

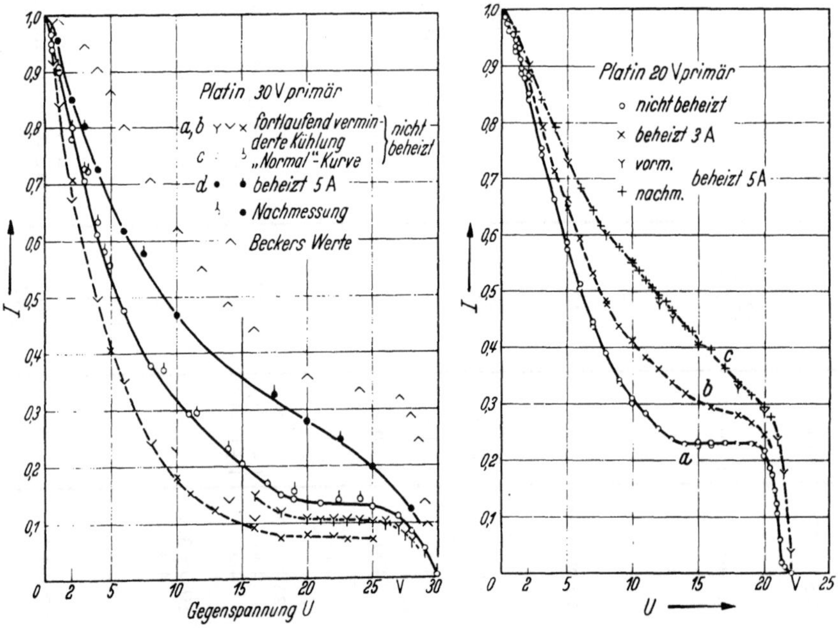

Fig. 11. Einfluß der Gasbeladung auf die Gegenspannungskurve der S nach LANGENWALTER [*133*].

heit sagen, daß der Kurvenverlauf bis zu einigen Zehntel eV herab die angegebene Form hat, daß insbesondere die Lage des Maxiumms nicht durch Fehler der Meßmethoden vorgetäuscht sein kann, denn die Methode des transversalen Magnetfeldes liefert auch noch im Bereich einiger Zehntel eV einwandfreie Resultate [*83*] und die Zuverlässigkeit der Methode des longitudinalen Magnetfeldes wurde bis zu etwa 0,5 eV herab durch Vergleichsmessungen an Photoelektronen direkt geprüft ([*293*], S. 79).

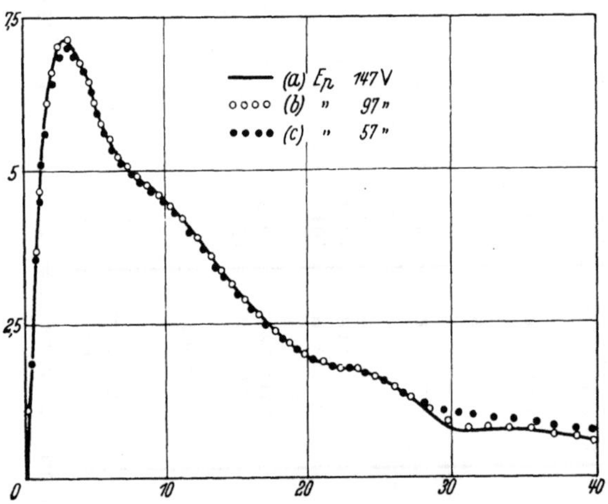

Fig. 12a. S-Energieverteilung an stark entgastem Molybdän nach HAWORTH [*128*].

Der Entgasungszustand des SE-Strahlers ist von erheblichem Einfluß auf die Energieverteilung, wie LANGENWALTER [*133*] durch Messungen (Gegenfeldmethode, Fig. 3b) an ausgeheizten und absichtlich gasbeladenen Flächen zeigen konnte (Fig. 11). In diesem Zusammenhang wesentlich ist vor allem die Feststellung, daß vergleichbare Messungen nur an ausgeheizten Schichten, d. h. mit ausheizbaren Apparaturen zu erhalten sind.

Charakteristisch für den oberflächlich entgasten Zustand ist es anscheinend nach Fig. 10, daß sich die Energieverteilungskurven verschiedener Metalle nur durch kleine Abweichungen in der Lage des Maximums unterscheiden; in Fig. 10

ist zum Vergleich eine MAXWELLsche Verteilungskurve ($\sim\sqrt{E}\,e^{-E/kT}$) mit einem Maximum bei 1,8 eV aufgetragen.

Erhebliche individuelle Verschiedenheiten der einzelnen Metalle treten in ihren Energieverteilungskurven erst dann zutage, wenn man sie einer wesentlich intensiveren Wärme-, d. h. *Glühbehandlung* unterzieht: Neben dem bisher einzigen

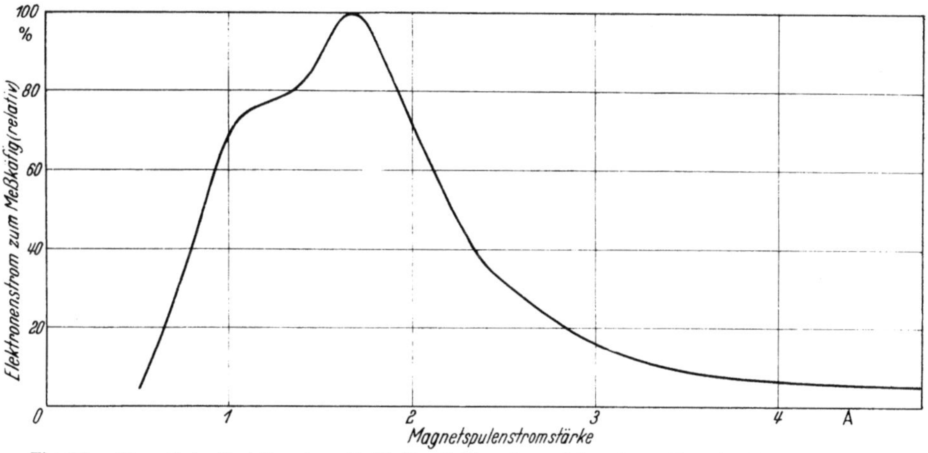

Fig. 12b. „Magnetische Verteilungskurve" (direkte Meßkurve) zur S-Energieverteilung in Fig. 12c nach KOLLATH [*344*]; vgl. Text.

Maximum bei etwa 2 eV treten dann ein oder mehrere, verschieden stark ausgeprägte „Nebenmaxima" bei größeren Energien auf, eine Tatsache, die 1935 zuerst von HAWORTH [*128*] an Molybdän entdeckt und dann auch an Niob [*145*]

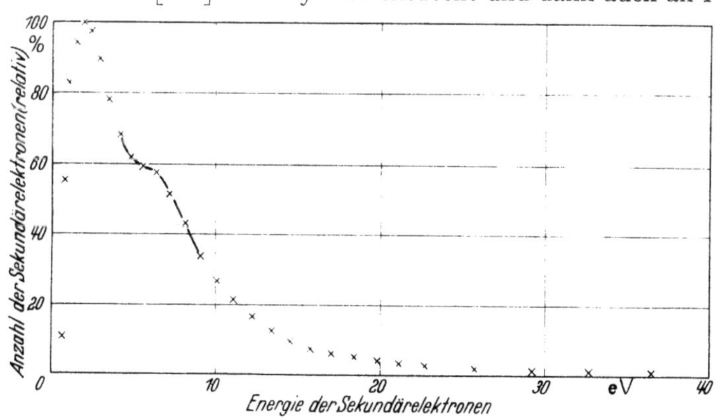

Fig. 12c. S-Energieverteilung an einer ausgeglühten Aufdampfschicht von Be auf Mo nach KOLLATH [*344*].

nachgewiesen wurde. Später erbrachte KOLLATH [*344*] den Nachweis, daß es sich um eine allgemeine Eigenschaft metallischer Schichten handelt, die besonders ausgeprägt auch bei Beryllium und seinen Legierungen auftritt, wenn man diese Materialien einer Glühbehandlung unterwirft. *Fig. 12a* zeigt die Meßergebnisse von HAWORTH an Mo: Dem allgemeinen Kurvenverlauf sind mehrere Maxima aufgeprägt, deren Lage von HAWORTH zu etwa 11, 24, 36 eV angegeben wird. KOLLATH konnte das Vorhandensein zum mindesten des ersten, am schärfsten ausgeprägten Nebenmaximums bestätigen[1].

In der Energieverteilungskurve treten die Nebenmaxima vielleicht nicht überzeugend genug hervor, deshalb ist in *Fig. 12b* und c eine Messung an einer

[1] Die Maxima für W und Mo wurden neuerdings von G. A. HARROWER bestätigt, vgl. S. 282.

Be-Aufdampfschicht direkt (12b) als „magnetische Verteilungskurve" und nach Umrechnung in die zugehörige Energieverteilung (12c) wiedergegeben. In der magnetischen Verteilungskurve tritt das entsprechende Nebenmaximum bei 1,7 A deutlich in Erscheinung (das Hauptmaximum liegt bei etwa 1,1 A), und ein zweites Nebenmaximum ist bei 2,6 A angedeutet. Die zugehörige Energieverteilungskurve ist an der Stelle des ersten Nebenmaximums (6,5 eV) nur noch etwas ausgebuchtet, das zweite Nebenmaximum ist völlig verwischt, es wäre bei etwa 18—20 eV zu suchen.

Das mehr oder weniger deutliche Hervortreten der Nebenmaxima hängt erheblich von der Stärke der Glühbehandlung bezüglich Temperatur und Zeitdauer ab. Die Sorgfalt, mit der HAWORTH die Glühbehandlung durchführte,

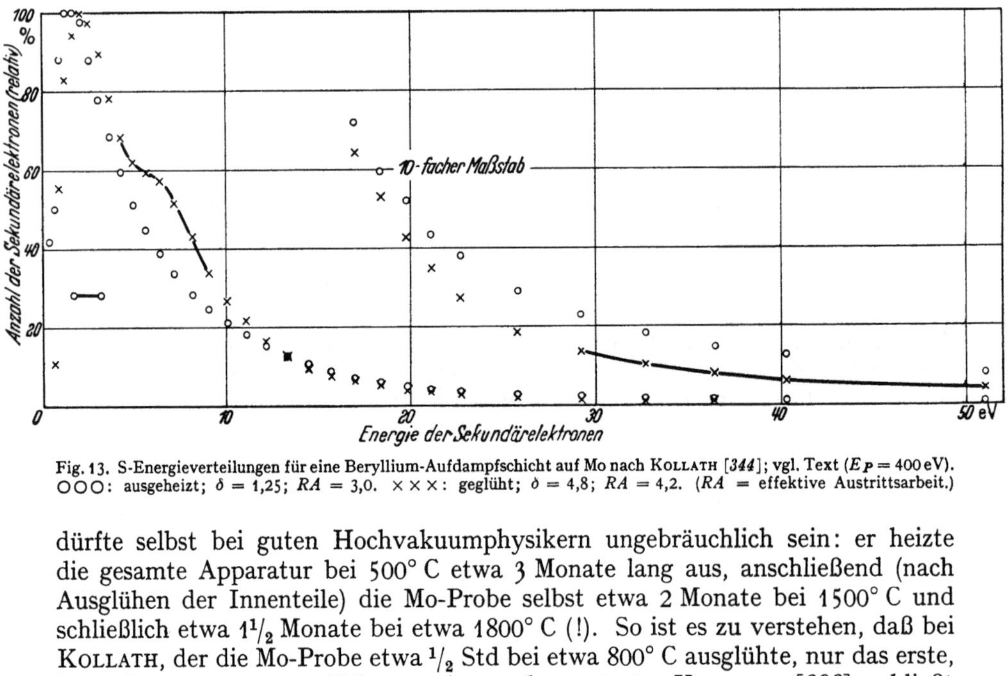

Fig. 13. S-Energieverteilungen für eine Beryllium-Aufdampfschicht auf Mo nach KOLLATH [*344*]; vgl. Text ($E_P = 400$ eV). ○○○: ausgeheizt; $\delta = 1{,}25$; $RA = 3{,}0$. ×××: geglüht; $\delta = 4{,}8$; $RA = 4{,}2$. (RA = effektive Austrittsarbeit.)

dürfte selbst bei guten Hochvakuumphysikern ungebräuchlich sein: er heizte die gesamte Apparatur bei 500° C etwa 3 Monate lang aus, anschließend (nach Ausglühen der Innenteile) die Mo-Probe selbst etwa 2 Monate bei 1500° C und schließlich etwa $1^1/_2$ Monate bei etwa 1800° C (!). So ist es zu verstehen, daß bei KOLLATH, der die Mo-Probe etwa $^1/_2$ Std bei etwa 800° C ausglühte, nur das erste, besonders ausgeprägte Nebenmaximum hervortrat. KOLLATH [*293*] schließt daraus, daß die Glühbehandlung nicht einfach zur Reinigung der *Oberfläche* bzw. zur Vertreibung einer oberflächlichen Gashaut dient, sondern zur Austreibung *im* Metall okkludierter Gase; diese sollen die Nebenmaxima soweit „einebnen", daß nach der üblichen Ausheizung davon noch nichts zum Vorschein kommt. Diese Auffassung wird auch dadurch gestützt, daß beim Stehenlassen der Apparatur an der Pumpe die Nebenmaxima erst im Verlauf von Tagen bis Wochen wieder verschwinden[1]. In der theoretischen Behandlung der SE wird man einen solchen Einfluß okkludierter Gase nur schwer berücksichtigen können; umso wichtiger ist es, *die* Energieverteilung zu *messen*, die dem eigentlichen Metallverband zugehört. Vermutlich hängt die Lage der Nebenmaxima mit der Lage der Energiebänder im Metall zusammen, worauf schon FRÖHLICH [*101*] hingewiesen hat; aus der relativen Höhe der Maxima läßt sich vielleicht darauf

[1] Man kann sich leicht ausrechnen, daß selbst bei einem Vakuum von 10^{-6} bis 10^{-7} mm Hg (und selbst, wenn der Akkomodationskoeffizient wesentlich kleiner als 1 ist) bei Zimmertemperatur nur Zeiten von der Größenordnung Sekunden bis Minuten notwendig sind, um eine Gleichgewichtsgasbeladung von etwa einer Atomlage auf der Metalloberfläche herzustellen.

schließen, in welchem Maße die verschiedenen Bänder an der Entstehung der S beteiligt sind. Ähnliche Erscheinungen sind eigentlich in noch ausgeprägterer Weise bei Halbleitern und Isolatoren zu erwarten; doch sind die bisher einzigen Energieverteilungsmessungen, bei denen von derartigen Erscheinungen berichtet wird (GEYER [*311*]) nicht recht überzeugend. Was aber auch der Grund für das

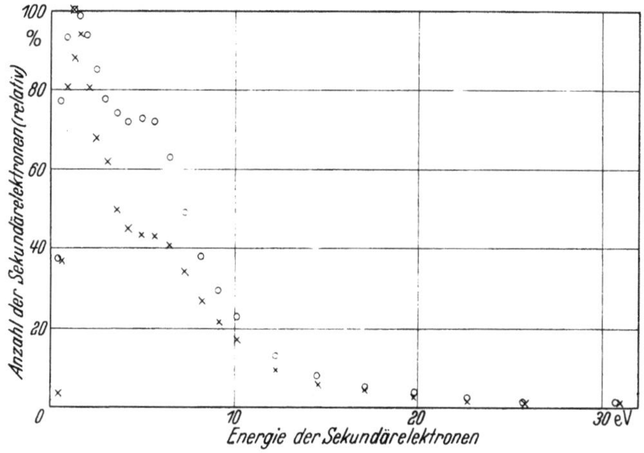

Fig. 14a. S-Energieverteilungen für eine Kupfer-Berylliumlegierung vor und nach „Formierung“ nach KOLLATH [*344*]; vgl. Text ($E_P = 400$ eV). ○○○: ausgeheizt; $\delta = 5{,}48$; $RA = 3{,}8$. ×××: geglüht; $\delta = 4{,}73$; $RA = 4{,}7$. (RA = effektive Austrittsarbeit).

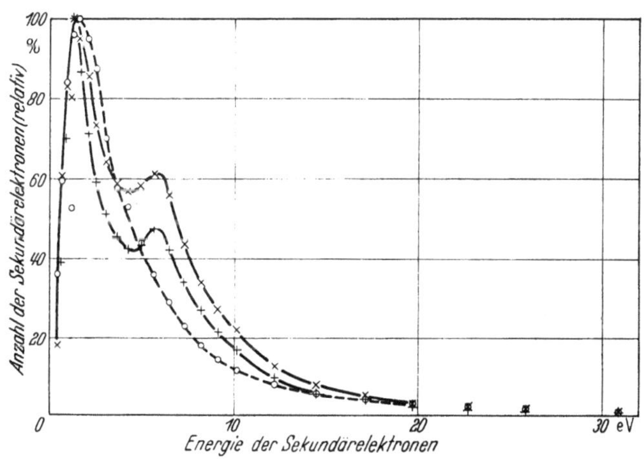

Fig. 14b. S-Energieverteilungen für eine Nickel-Berylliumlegierung vor und nach „Formierung“ nach KOLLATH [*344*]; vgl. Text ($E_P = 400$ eV). ○○○: ausgeheizt = 2,14; $RA = 3{,}7$. ×××: geglüht (Probe 1); $\delta = 4{,}08$, $RA = 4{,}5$; Probe 2: +++ $\delta = 4{,}13$; $RA = 4{,}5$. (RA = effektive Austrittsarbeit.)

Auftreten der Nebenmaxima bei Glühbehandlung und für die zeitliche Rückbildung dieser Maxima sein mag, die Aufklärung dieser Erscheinung wird zum endgültigen Verständnis der SE wesentlich beitragen.

Wie schon weiter oben erwähnt, treten solche Nebenmaxima in wesentlich ausgeprägterer Form beim Be und seinen Legierungen auf. Fig. 13 zeigt die Energieverteilung für eine auf Mo aufgedampfte Be-Schicht: Nach dem Aus*heizen* im Ofen erhält man den üblichen glatten Verlauf mit nur einem Maximum bei 1,8 eV, nach dem Aus*glühen* tritt ein deutliches Nebenmaximum bei etwa 6,5 eV auf. Am deutlichsten zeigte sich diese Erscheinung bisher an „formierten“ CuBe- und NiBe-Schichten ([*293*], auch [*287*], [*296*]) vgl. (Fig. 14a, b). Hingewiesen sei schließlich noch darauf, daß auch für die Energieverteilung mit mehreren

Nebenmaxima die weiter oben besprochene Unabhängigkeit der Kurvenform von der Energie der P (für $100 \leq E_p \leq 900$ eV) gilt.

8. Energieverteilung der Sekundärelektronen aus Halbleitern und Isolatoren. Zur Zeit kann man sehr wenig quantitative Aussagen über die Energieverteilung bei Halbleitern und Isolatoren machen, obgleich schon eine ganze Anzahl von Messungen vorliegt.

Das hat seinen Grund in den experimentellen Schwierigkeiten, mit denen diese Messungen verbunden sind. Die Hauptschwierigkeit besteht darin, Aufladungserscheinungen an der Oberfläche und im Innern von Isolatoren bzw. Halbleitern zu vermeiden oder unter Kontrolle zu halten. Mit abnehmender Leitfähigkeit wird den üblichen Methoden insofern eine Grenze gesetzt, als sich die Messungen nicht mit beliebig kleinen Stromstärken ausführen lassen; die Stromdichten müßten aber mit der Leitfähigkeit immer weiter abnehmen, um Aufladungseffekte vernachlässigen zu können. Erwärmt man aber solche Schichten, um Aufladungen zu vermeiden (vgl. z. B. VUDINSKY [*246*]), dann hat man eben nicht mehr den Isolator, an dem man messen wollte.

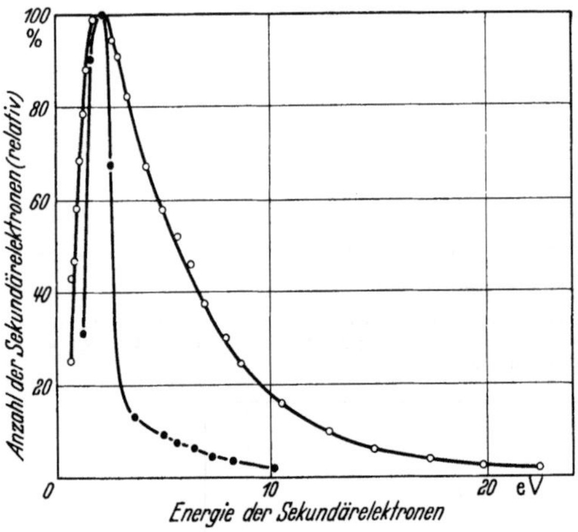

Fig. 15. S-Energieverteilung für eine Caesium-Aufdampfschicht auf Silber und für eine Caesium-Antimonphotoschicht auf Silber nach KOLLATH [*344*]. ○○○: Cs auf Ag; $\delta = 4{,}7$; $RA = 2$. ●●●: Cs-Sb-Photoschicht; $\delta = 2{,}18$; $RA = 1{,}5$.

Eine Energieverteilungskurve der S für eine CsSb-Photoschicht wurde von KOLLATH [*293*] im longitudinalen Magnetfeld gemessen (Fig. 15). Der Vergleich mit einer mittleren Energieverteilungskurve für ausgeheizte Metalle ergibt: bei etwa gleicher Lage des Maximums (etwa 2 eV)[1] fällt die CsSb-Kurve nach größeren Energien hin wesentlich steiler ab. Einen ähnlich steilen Kurvenverlauf beobachteten PJATNITZKY an der CsO-Photoschicht auf Ag [*211*] und JOHNSON an BaSr-Oxydkathoden [*353*]. JOHNSON hat übrigens die Verschiebung des Maximums mit steigender Temperatur nach kleineren Energien hin *nicht* bestätigen können, die von MORGULIS und NAGORSKY [*207*] sowie von POMERANTZ [*341*] bei Oxydkathoden gefunden wurde. Die Abweichung der Verteilungskurven von Photoschichten und Oxydkathoden gegen die Kurvenform für Metalle (auch Cs!) ist so stark, daß sie auch bei Messungen nach der Gegenfeldmethode (PJATNITZKY und JOHNSON benutzten ein „Zentralfeld") noch deutlich zum Ausdruck kommt[2]. Vorsichtig muß man bei der Beurteilung von Meßergebnissen an gut isolierenden Schichten sein, wenn die Messungen auf das Vorhandensein von Aufladungen an der Oberfläche oder im Innern des Isolators hindeuten. Die Beobachtung einer größeren Anzahl von Nebenmaximis an Aufdampfschichten von MgF_2 und NaCl (GEYER [*311*]) ist sehr interessant; umso mehr sollte die Realität dieser Maxima nachgeprüft werden. Übereinstimmend geben mehrere Autoren die Lage des Maximums für NaCl bei einer Energie ≤ 1 eV an [*246*], [*311*], eins der wenigen Resultate, die als gesichert gelten können. Es scheint so, als ob für die Energieverteilungsmessungen an Isolatoren neue Wege beschritten werden müssen.

---

[1] Der steile Abfall der Kurve unterhalb etwa 2 eV, kann *nicht* vorgetäuscht sein; denn eine unmittelbar folgende Messung der Photoelektronen-Energieverteilung an der gleichen Schicht zeigte noch ein deutliches Maximum bei merklich kleinerer Energie.

[2] Vgl. hierzu die kritischen Ausführungen von JOHNSON ([*338*], S. 1070).

C. Die Anzahl der Sekundärelektronen (Ausbeute).

9. Grundsätzliche Betrachtungen. Von allen Erscheinungen der SE wurde die *Anzahl* von S, die aus einer Oberfläche bei Elektronenbeschuß herauskommen, am häufigsten und unter erheblicher Variation aller Versuchsbedingungen untersucht. Dabei hat sich gezeigt, daß der Strom der von der Platte emittierten Elektronen i_K bei sonst gleichen Bedingungen stets proportional zum Strom i_P der auftreffenden P ist. Wir können also i_K auf i_P beziehen und gewinnen damit den Begriff der „Ausbeute“: $\delta = i_K/i_P$. Diese Ausbeute hängt ebenso wie die bereits behandelte Energieverteilung von vielen Parametern ab, worüber in den anschließenden Abschnitten berichtet wird.

Hierbei ist natürlich δ als *Mittelwert* beim Auftreffen sehr vieler P zu verstehen: $\delta = i_K/i_P$ gibt an, wieviel Elektronen im *Mittel* von einem P aus einer Fläche herausgeschlagen werden. Bei der technischen Anwendung der SE, z. B. im Vervielfacher, interessiert dabei durchaus auch die Frage, wieviel Elektronen im Einzelprozeß von einem P erzeugt werden, bzw. wie sich unter sonst gleichen Bedingungen die Anzahl der von *einem* P ausgelösten Elektronen statistisch verteilt (Verteilung der differentiellen δ-Werte). Diese Frage wurde z. B. von HAYNER [*129*], KURRELMEYER und HAYNER [*173*], ZIEGLER [*161*], [*162*] sowie SHOCKLEY und PIERCE [*216*] am Schroteffekt der SE untersucht mit dem Resultat, daß die statistischen Schwankungen beim SE-Prozeß im Mittel klein sind gegenüber denjenigen, die der P-Strom hervorruft. Daher sind die Stromschwankungen, die zusätzlich z. B. in einem SE-Vervielfacher (Multiplier) hervorgerufen werden, stets klein gegen die Schwankungen des Primärstroms. Wenn im folgenden von der „Ausbeute“ gesprochen wird, so ist stets der Mittelwert im Sinne der obigen Definition gemeint.

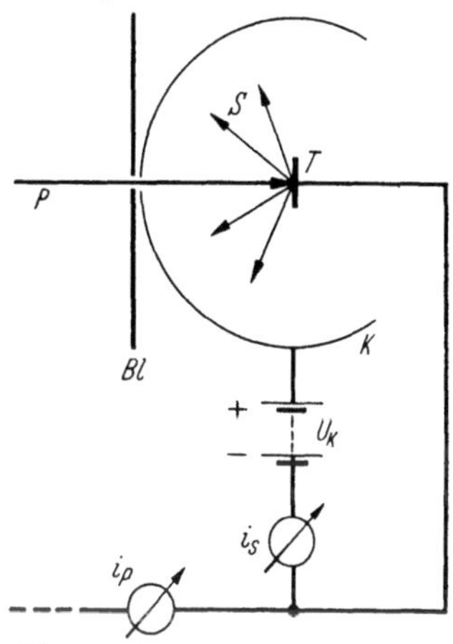

Fig. 16: Anordnung zur Messung der Ausbeute δ (schematisch).

Die *Messung* von δ scheint auf den ersten Blick eine einfache Aufgabe zu sein: In Fig. 16 kommen von links die P und lösen an dem Sekundärstrahler T Elektronen aus. Legt man den Auffänger K, der T umgibt, auf ein positives Potential gegenüber T (Spannung U_K), so werden die an T ausgelösten Elektronen nach K gezogen und fließen als Sekundärstrom i_K ab[1]. Ein zweiter Strommesser mißt die Gesamtmenge der P, die durch die Blende Bl in die aus T und K bestehende Meßanordnung einfallen; i_K/i_P ist dann die Ausbeute δ.

Ein kurzer Rückblick auf den allgemeinen Verlauf der Energieverteilungskurven der von T ausgehenden Elektronen (Fig. 2) zeigt aber, daß die Dinge in Wirklichkeit sehr viel komplizierter liegen, wenn man nur die „eigentlichen S“ (Fläche III in Fig. 2) bei der Messung erfassen will. Die dafür notwendigen Maßnahmen bzw. Korrekturen hängen unter anderem auch von der Geometrie der Anordnung und von dem Material und der Oberflächenbeschaffenheit der Elektroden T, K wesentlich ab. Die Größe der Korrekturglieder läßt sich aber nicht mit Sicherheit angeben. Man darf daher die Genauigkeit beim *Vergleich* der Ausbeutewerte verschiedener Autoren nicht überschätzen, während bei ein und derselben Anordnung sich eine erhebliche Genauigkeit der *Einzel*messungen erreichen läßt. Die Meßmethoden und die bei der Messung auftretenden Schwierigkeiten sollen deshalb hier *vor* der Besprechung der Versuchsergebnisse diskutiert werden.

---
[1] In Fig. 16 mit i_s bezeichnet.

10. Apparaturen zur Messung der Ausbeute, Auswertungsfragen. Als Beispiel für die *Anordnungen*, die *zur Messung der Ausbeute* verwendet werden und die im Prinzip durchweg der Fig. 16 entsprechen, wird in Fig. 17 die Meßanordnung von BRUINING und DE BOER [*189*] wiedergegeben; mit dieser Anordnung wurde eine große Anzahl von Materialien untersucht. Ein abschmelzbarer Glaskörper, der mit einem Getter *Ge* — oder mit einem Absorptionskohlegefäß versehen ist (gute Vakuumbedingungen!), trägt unten auf einem Quetschfuß die eigentliche Meßeinrichtung. Sie besteht aus der Elektronenquelle (Glühdraht *Gl*), einem Blendensystem (WEHNELT-Elektrode *W* und Anode *A*); in Richtung des Primärstrahls folgt dann der Auffangkäfig *K* für die S und schließlich der Sekundärstrahler *T*. Natürlich darf der P-Strahl nicht direkt auf den Käfig *K* auftreffen; darüber hinaus sollen auch keine S aus dem Auffänger *K* durch die Primärstrahlöffnung entkommen bzw. keine an der Anode durch P ausgelösten S durch diese Öffnung hindurchtreten oder nach *K* gelangen. Das kann in einfacher Weise durch elektronenoptische Mittel erreicht werden, z. B. durch eine weitere Blende zwischen *A* und *K* auf stark negativem Potential gegen die Anode. Eine solche Blende mit geeigneter Öffnung bildet bei genügend negativem Potential eine Sperre für die (langsamen) S in beiden Richtungen, während sie — abgesehen von ihrer elektronenoptischen Wirkung als Einzellinse — die P ungehindert hindurchläßt [*202*], [*235*]. Die Platte *T* kann nach oben angehoben und aus den Glühwendeln F_1 und F_2 mit verschiedenen Materialien bedampft werden. Der Auffänger *K* ist kugelförmig ausgebildet, um aus Gegenspannungskurven im „Zentralfeld" Schlüsse auf die Energie der S ziehen zu können (vgl. oben Ziff. 5).

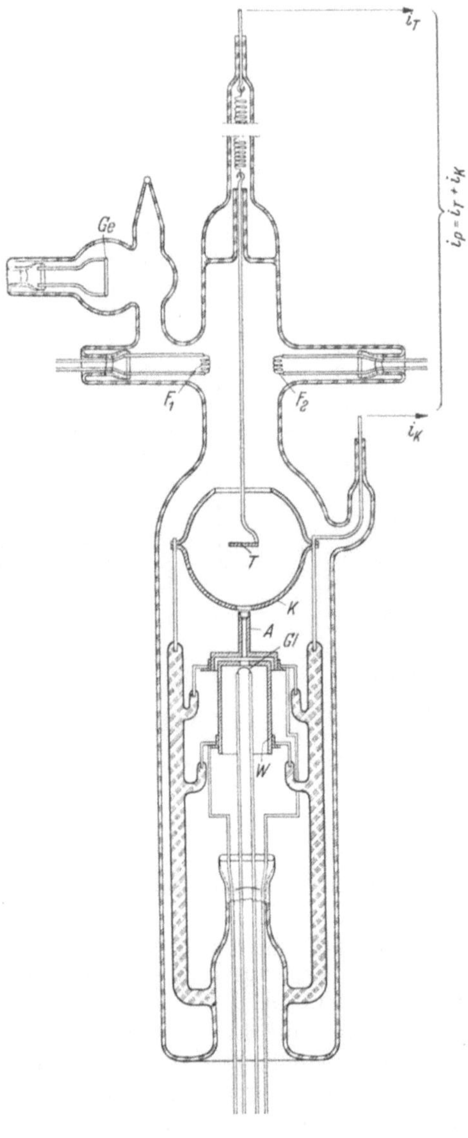

Fig. 17. Anordnung zur Messung der Ausbeute δ in abgeschmolzener Röhre (nach BRUINING und DE BOER [*189*]; vgl. Text).

Gemessen werden mit dieser Anordnung bei positivem bzw. negativem Potential von *K* gegen *T* zwei der folgenden Größen: Elektronenstrom von der Platte i_T, Elektronenstrom vom Auffänger i_K, Gesamtstrom $i_P = i_T + i_K$ (die vorgegebene Anodenspannung U_A zwischen *Gl* und *T* bestimmt dabei die Auftreffenergie[1] der P). Die Größen i_K, i_T und i_P sind in Fig. 18 als Funktion von U_K

[1] Die Austrittsenergie der Glühelektronen ist im Mittel vernachlässigbar klein.

schematisch dargestellt. i_P ist der Gesamtstrom, der in das System K, T eintritt und sollte daher konstant, speziell *un*abhängig von der Spannung U_K sein (jedenfalls für nicht zu große U_K). Der Elektronenstrom i_T ist bei negativem U_K nahezu gleich i_P, fällt in der Nähe von $U_K = 0$ steil ab und nähert sich bei großen positiven U_K-Werten einem konstanten Wert, dessen Höhenlage von der Größe der Ausbeute abhängt (in Fig. 18 ist ein Ausbeutewert $\delta \approx 1{,}6$ zugrunde gelegt).

Die Ausbeute δ ist bei genügend positivem U_K gleich dem Quotienten i_K/i_P, bzw., was offenbar dasselbe ist, gleich dem Quotienten $(i_P - i_T)/i_P$; das ist die meist benutzte Form der Auswertung. Man kann aber auch einen angenäherten Wert für δ durch Vergleich der i_T-Werte bei genügend negativem und genügend positivem U_K gewinnen:

$$\delta \sim \frac{i_{T-} - i_{T+}}{i_{T-}}$$

(vgl. Fig. 18). Beide Auswertungsverfahren ergeben nur dann denselben Wert, wenn $i_{T-} = i_P$ ist. Abweichungen zwischen i_{T-} und i_P werden im allgemeinen deswegen vorhanden sein, weil ja beim Auftreffen der P nicht nur die langsamen, „eigentlichen“ S an der Platte ausgelöst werden, sondern weil auch noch elastisch reflektierte und rückdiffundierende Elektronen von T ausgehen, wie die Energieverteilung der austretenden Elektronen zeigt (Fig. 2). Wie kompliziert die Angelegenheit wird, wenn man das Vorhandensein der reflektierten und diffundierenden Elektronen in die Auswertung einbeziehen will[1], um einen „wahren“ δ-Wert der eigentlichen S zu gewinnen, soll die folgende Betrachtung zeigen.

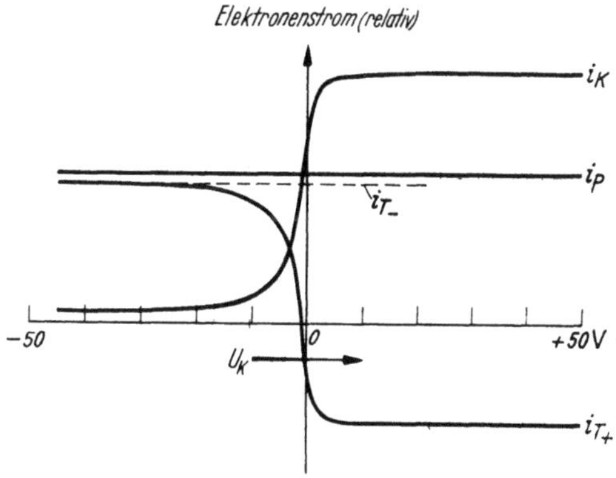

Fig. 18. Primärelektronenstrom i_P, Auffängerstrom i_K und Plattenstrom i_T als Funktion der Spannung U_K zwischen K und T (schematisch).

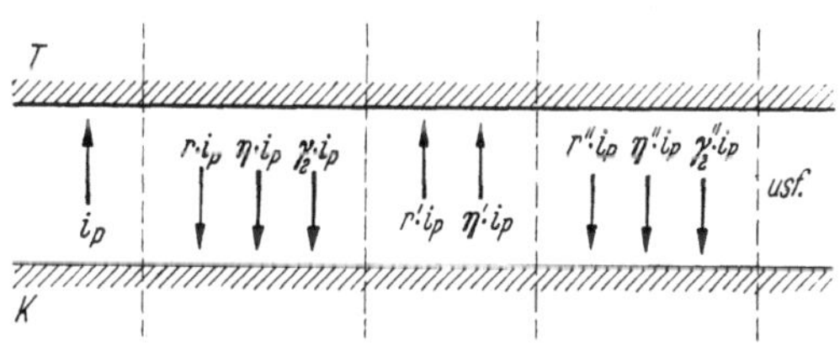

Fig. 19. Elektronenübergang zwischen zwei Elektroden T und K (K etwa 50 V positiv gegen T; schematisch). P-Strom: i_P; elastische reflektierte Elektronen: Faktor $r, r', r'', \ldots$; rückdiffundierende Elektronen: Faktor $\eta, \eta', \eta'', \cdots$; S: Faktor $\gamma_2, \gamma_2', \gamma_2'', \ldots$

„*Wahre*“ *Ausbeute.* Von dem gemessenen Primärelektronenstrom $i_P = (i_T + i_K)$, der in Fig. 19 auf T auftrifft, wird ein Teil $(r \cdot i_P)$ elastisch reflektiert, ein anderer Teil $(\eta \cdot i_P)$ diffundiert zurück und ferner werden $(\gamma_2 \cdot i_P)$ S ausgelöst. Bei ihrem Auftreffen auf den positiv geladenen Auffänger K werden die $(\gamma_2 \cdot i_P)$ dort zurückgehalten, während von den $(r \cdot i_P)$ und den $(\eta \cdot i_P)$ wieder ein Bruchteil $(r' \cdot i_P)$ reflektiert wird, ein Bruchteil $(\eta' \cdot i_P)$ rückdiffundiert usw.: die nach K fließende Ladung ist also, streng genommen, die Summe einer alternierenden Reihe. Die Glieder dieser Reihe nehmen aber schnell ab, da die Koeffizienten r und η für alle Materialien wesentlich kleiner als 1 sind (vgl. E). Man wird daher den wahren S-Strom i_s nach K in genügender Näherung erhalten, wenn man nur die ersten Glieder der Reihe berücksichtigt[2].

Aus dem gemessenen Auffängerstrom $i_K = i_P . (r + \eta + \gamma_2)$ ergibt sich dann der Strom der „eigentlichen“ S zu: $i_s^* = i_K - i_P\,(r + \eta)$. Es bestehen nun

[1] Und das ist für die Theorie der SE notwendig!

[2] Man vergleiche hierzu auch die Überlegungen von Bruining [2] auf S. 6 und 7.

verschiedene Möglichkeiten, i_i^* eine Bezugsgröße zuzuordnen, und man erhält dementsprechend *verschiedene Definitionen für die „wahre“ Ausbeute* der S:

Wir beziehen i_s^* auf i_P und erhalten mit $\delta = i_K/i_P$:

$$\frac{i_s^*}{i_P} = \delta - (r + \eta). \tag{10.1}$$

Bei Vernachlässigung von r und η ergibt sich hieraus (in nullter Näherung) der üblicherweise gewonnene Wert $\gamma_0 = \delta$, in erster Näherung (Vernachlässigung von r)

$$\gamma_1 = \delta - \eta. \tag{10.1a}$$

Diese Definition benutzt STERNGLASS [*404*]. Berücksichtigen wir aber r *und* η, dann ist von dem gemessenen δ noch der Summand $(r + \eta)$ abzuziehen und wir erhalten in zweiter Näherung γ_2.

HACHENBERG und BRAUER [*400*] definieren einen wirksamen Primärstrom, indem sie von i_P den Strom der elastisch reflektierten Elektronen $r \cdot i_P$ abziehen, Sie erhalten damit als Ausbeute:

$$\gamma_3 = \frac{\delta - (r + \eta)}{1 - r}.$$

Bei unseren bisherigen Überlegungen zur Ausbeute haben wir stillschweigend den Satz von der Erhaltung der Ladung benutzt: $i_P = i_K + i_T$. Nun soll uns die Betrachtung der *Energieerhaltung* neue Gesichtspunkte eröffnen. Für den hier behandelten Fall besagt der Energiesatz, daß sich die eingestrahlte (Primär-) Energie vollständig wiederfinden muß in der emittierten Energie und in der Energie, die im Sekundärstrahler verbleibt:

$$E_{\text{ein}} = E_{\text{aus}} + E_{\text{abs}}.$$

Die ausgestrahlte Energie wird im wesentlichen von den emittierten Elektronen mitgeführt (E_{em}), der Rest z.B. von elektromagnetischen Wellen (ΔE_{aus}):

$$E_{\text{aus}} = E_{\text{em}} + \Delta E_{\text{aus}}. \tag{10.2}$$

In unserem Zusammenhang interessiert in erster Linie die Größe von ΔE_{aus} relativ zu E_{ein}. Bei den hier benutzten E_p-Werten ist die Entstehung von Wellenstrahlung nachgewiesen (vgl. Ziff. 24), $\Delta E_{\text{aus}}/E_{\text{ein}}$ dürfte kleiner sein als 10^{-2}.

Die je Zeiteinheit eingestrahlte Energie ist $E_{\text{ein}} = i_P \overline{E}_p$, wenn $\overline{E}_p$ die mittlere Energie ist, die pro P eingestrahlt wird. (Dimension: erg/Ladung.) Entsprechend ist

$$E_{\text{em}} = i_s^* \overline{E}_s + i_d \overline{E}_d + i_r \overline{E}_r, \tag{10.3}$$

wobei $\overline{E}_s$ die je S, $\overline{E}_d$ die je rückdiffundierendes Elektron, $\overline{E}_r$ die je reflektiertes Elektron vom Sekundärstrahler emittierte mittlere Energie ist. $\overline{E}_s$, $\overline{E}_d$, $\overline{E}_r$ lassen sich offenbar aus der Energieverteilungskurve berechnen. Es ist $E_{\text{em}} = \int_0^{E_p} E\,dN(E)$; E_{em} ist also bei vorgegebener Energieverteilung bekannt. In dem Ausdruck (10.3) steckt aber wegen der ungenauen Definition von i_s^*, i_d und i_r (vgl. Ziff. 4) eine gewisse Willkür.

Damit erhält man:

$$i_p \overline{E}_p = i_s^* \overline{E}_s + i_d \overline{E}_d + i_r \overline{E}_r + \Delta E_{\text{aus}} + E_{\text{abs}} \tag{10.4}$$

oder mit

$$i_s^* = \gamma_2 i_p; \quad i_d = \eta\, i_p; \quad i_r = r i_p;$$
$$i_p \overline{E}_p = \gamma_2 i_p \overline{E}_s + \eta\, i_p \overline{E}_d + r \cdot i_p \overline{E}_r + \Delta E_{\text{aus}} + E_{\text{abs}}.$$

Mangels näherer Kenntnisse über Betrag und Bedeutung von E_{abs} fassen wir diese Größe mit $i_s^* \overline{E}_s$ zusammen[1]: $i_s^* \hat{E}_s = i_s^* \overline{E}_s + E_{\text{abs}}$. $i_s^* \hat{E}_s$ ist die Energie, die im Primärstrahl zugeführt werden müßte, damit ein Sekundärstrom i_s^* resultiert, wenn keine rückdiffundierenden und keine reflektierten Elektronen auftreten und auch anderweitig keine Energie emittiert wird. $i_s^* \hat{E}_s$ entspricht also einem bestimmten Primärstrom, bei dem jedes Elektron nach unserer Übereinkunft die Energie $\overline{E}_p$ mit sich führen sollte. Dieser Strom ist offenbar

$$i_p^* = i_p\left(1 - \eta \frac{\overline{E}_d}{\overline{E}_p} - r \frac{\overline{E}_p}{\overline{E}_r}\right) - \frac{\Delta E_{\text{aus}}}{\overline{E}_p}. \tag{10.5}$$

Auf diesen Strom sollte man sich bei der Definition der Ausbeute beziehen:

$$\gamma_4 = \frac{i_s^*}{i_p^*}. \tag{10.6}$$

Man erhält dann:

$$\left.\begin{aligned} \gamma_4 &= \frac{i_s^*}{i_p} \cdot \frac{i_p}{i_p^*} = \gamma_2 \frac{1}{1 - \overline{\chi}\,\eta - \overline{\varrho}\, r - \varepsilon}, \\ \text{mit} \qquad \overline{\chi} &= \frac{\overline{E}_d}{\overline{E}_p}; \quad \overline{\varrho} = \frac{\overline{E}_r}{\overline{E}_p}; \quad \varepsilon = \frac{\Delta E_{\text{aus}}}{i_p \overline{E}_p}. \end{aligned}\right\} \tag{10.7}$$

Vernachlässigt man in (10.7) ε gegen die übrigen Größen (s. oben) und setzt man $\overline{\varrho} = 1$, dann ergibt sich:

$$\gamma_4 = \frac{\gamma_2}{1 - (r + \overline{\chi}\,\eta)} \tag{10.8}$$

oder[2] mit (10.1):

$$\gamma_4 = \frac{\delta - (r + \eta)}{1 - (r + \overline{\chi}\,\eta)}. \tag{10.9}$$

$\overline{\chi}$ ist die von STERNGLASS als $\overline{K}$ eingeführte Größe, für die er die Formel $\overline{K} = 0{,}15 + 2 \cdot 10^{-3} Z$ angibt (Z = Ordnungszahl). $\overline{\chi}$ ist also ungefähr 0,5.

Im Abschnitt C ist stets nur der *gemessene Wert der Ausbeute* $\delta = i_K/i_p$ gemeint, wenn nicht ausdrücklich auf Korrektionen dieses Meßwertes hingewiesen wird.

Es gibt noch *einige andere Meßanordnungen* zur Bestimmung der Ausbeute. Man kann Trioden in Dynatronschaltung benutzen (Gitter positiv, Anode negativ gegen Kathode), um δ für das Anodenmaterial zu bestimmen. Verschiedene Autoren haben viel Mühe darauf verwendet, Meßmethoden zu ersinnen, um aus Triodenmessungen *quan*titative Ausbeutewerte zu gewinnen [*39*], [*44*], [*71*], [*72*], [*109*]. Eine spezielle Methode wurde von TRELOAR [*156*] ausgearbeitet, um δ für drahtförmige Materialien zu messen. Ausführliche Beschreibungen dieser Methoden findet man bei BRUINING [*2*].

Für die *Messung der Ausbeute an Isolatoren*[3] wurde vielfach ebenfalls die oben genauer beschriebene „statische" Meßmethode nach Fig. 17 benutzt, unter anderem von GEYER [*312*]. Versuche, die hierbei zusätzlich auftretenden Schwierigkeiten (vgl. Ziff. 8) zu umgehen, (Verwendung sehr dünner Schichten, Erhöhung der Leitfähigkeit des Materials durch Erwärmung oder Ultrarotbestrahlung) sind wenig befriedigend. Man hat deshalb andere, „dynamische" Meßmethoden entwickelt, deren Prinzip übereinstimmend darauf hinausläuft, nur

---

[1] $\hat{E}_s$ ist die Energie, die man im Mittel dem Material zuführen muß, um ein S der Energie $\overline{E}_s$ zu erzeugen. $\hat{E}_s$ ist identisch mit $\overline{E}_0$ bei STERNGLASS [404], vgl. hierzu auch Teil F.

[2] Betrachtungen über die absorbierte Energie, wie sie zur Ableitung von γ_4 führten, sind nicht mehr zu umgehen, wenn man die Ausbeute an dünneren Folien mißt.

[3] In sinngemäßer Übertragung gilt das folgende auch für Halbleiter.

mit kurzzeitigen P-*Impulsen* zu arbeiten und in der Zeit zwischen zwei Impulsen die Oberfläche des bestrahlten Materials wieder auf einen festen Ausgangszustand zurückzuführen. SALOW [*265*] erreicht das dadurch, daß er neben der Impulsbestrahlung die Oberfläche dauernd mit Elektronen aus einer zweiten Elektronenquelle auf einen festen Potentialwert auflädt HEIMANN und GEYER [*254*] dadurch, daß sie die Energie der P in kurzen Impulsen auf den gewünschten Wert bringen, und dazwischen die Oberfläche mit einer bestimmten Ausgangsenergie der P beschießen. Bei HEIMANN und GEYER sind also sozusagen die beiden Elektronenquellen von SALOW in einer vereinigt. Impulsmethoden zur Messung der Ausbeute wurden ferner von SCHERER [*267*], JOHNSON [*353*] und POMERANTZ [*341*] verwendet. Bei den „dynamischen" Methoden werden zwar die Oberflächen-Aufladungen unter Kontrolle genommen, das Problem der Aufladungen im Inneren eines Isolators bleibt jedoch ungelöst.

11. Allgemeines Aussehen und charakteristische Daten der „Ausbeutekurve". Schon frühzeitig [*17*] bis [*20*] wurde erkannt, daß die Ausbeute δ in charakteristischer Weise von der Energie E_P der P abhängt (Fig. 20): Die sog. „*Ausbeutekurve*" steigt von kleinen E_P her bis zu einem Maximum an, das bei E_P-Werten von einigen 100 eV liegt, und fällt dann nach größeren E_P wieder langsam ab. Diesen allgemeinen Verlauf der Ausbeutekurve kann man sich in folgender Weise verständlich machen: mit wachsender Energie der P werden immer mehr S im Material erzeugt. Mit wachsender Energie der P dringen diese aber auch immer tiefer ins Material ein und damit wächst auch die Tiefe, in welcher die S erzeugt werden, so daß ihr Weg zur Oberfläche immer länger wird. In Analogie zur Absorption langsamer Elektronen in Gasen wird man (mangels ausreichender Kenntnis) auch hier ein *e*-Gesetz für die Absorption annehmen. Die Ausbeute muß daher schließlich wieder abfallen, also bei irgendeinem Wert der P-Energie ein Maximum überschreiten.

Die Maximalwerte δ_m liegen für reine Metalle zwischen 0,5 und 2, bei Nichtleitern wurden zum Teil Maximalwerte bis zu $\delta_m = 6$ gefunden und schließlich gibt es einige komplizierter aufgebaute Schichten, bei denen δ_m-Werte bis zu 12 erreicht werden. In der Literatur hin und wieder angegebene wesentlich größere δ-Werte beruhen auf Erscheinungen, die mit der Feldemission sehr dünner Schichten zusammenhängen („MALTER-Effekt" vgl. Ziff. 19). Ausbeutekurven wurden für viele Materialien gemessen und es wurde stets qualitativ der in Fig. 20a wiedergegebene charakteristische Verlauf gefunden; Unterschiede zwischen den Kurven bestehen nur im Quantitativen, in vielen Fällen reicht es daher aus, einige besonders markante Werte der Ausbeutekurve zu kennen. Zu diesen gehören vor allem der Maximalwert δ_m und die zugehörige P-Energie $E_{P\max}$; ferner — falls der maximale Ausbeutewert über 1 liegt — die Abszissenlage der beiden Punkte, in denen die Ausbeutekurve den Wert 1 annimmt (E_P^{I}, E_P^{II}).

Abgesehen von der Festlegung des ungefähren Verlaufs der Ausbeutekurve spielen die Werte E_P^{I} und besonders E_P^{II} eine gewisse Rolle in technischen Röhren. Wenn ein isolierter Metallkörper von P getroffen wird und eine Elektrode auf positiverem Potential (wir beziehen alle Potentiale auf das Kathodenpotential als Nullpunkt) als E_P^{II} in der Röhre vorhanden ist, gibt es nur zwei Gleichgewichtszustände für das Potential des Metallkörpers: sein Potential stellt sich auf Kathodenpotential ein oder es nimmt denjenigen Wert an, der E_P^{II} entspricht; welches von den beiden stabilen Potentialen sich einstellt, hängt dabei von dem Potential ab, das der isolierte Metallkörper zufällig hatte, als die ersten P ihn erreichten. Das kann man sich durch Betrachtung der Ausbeutekurve in Fig. 20a leicht klar machen: bei einer Aufladung von z. B. +1000 V gegen Kathode verlassen mehr S den Metallkörper als P auftreffen ($\delta > 1$); sein Potential wird sich also nach größeren Werten verschieben. Bei einer Aufladung von z. B. $+2000$ V gehen dagegen weniger S von dem Metallkörper weg, als P auftreffen ($\delta < 1$); sein Potential wird sich also nach kleineren Werten verschieben. In beiden Fällen strebt die Aufladung des Metallkörpers dem zu E_P^{II} gehörenden Potentialwert zu: die Potentiallage des Metallkörpers im Punkt E_P^{II} ist *stabil*. Eine ganz entsprechende Überlegung für den Punkt E_P^I führt zu dem Schluß, daß die Potentiallage eines isolierten Metallkörpers im Punkt E_P^I *labil* ist: bei zufälligen Anfangswerten $> E_P^I$ strebt das Potential dem Punkt E_P^{II} zu, bei zufälligen Anfangswerten $< E_P^I$ dem Kathodenpotential, das also wiederum eine *stabile* Gleichgewichtslage darstellt.

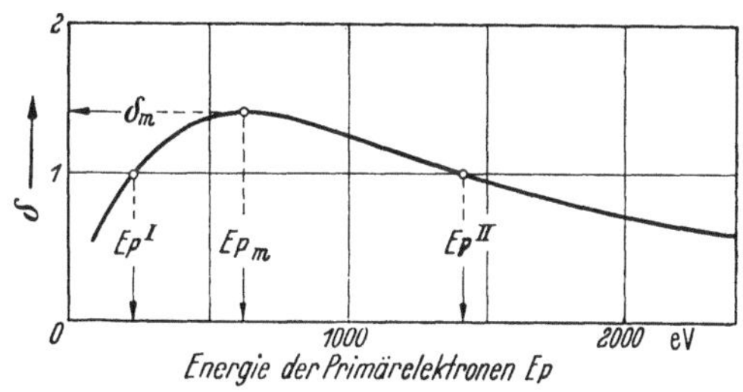

Fig. 20a. „Ausbeutekurve“ $\delta = f(E_P)$: allgemeines Aussehen und charakteristische Größen.

Liegt das höchste Potential der Anordnung („Absaug-Elektrode“) zwischen E_P^I und E_P^{II}, so lädt sich ein isolierter Metallkörper auf Kathodenpotential oder auf ein Potential in der Nähe desjenigen der „Absaugelektrode“ auf (er kann sich nicht positiv gegen die „Absaugelektrode“ aufladen, weil er dann die S von den anderen Elektroden einzufangen beginnt und damit sogleich wieder negativer wird). Liegt schließlich das höchste Potential in der Röhre zwischen Kathodenpotential und E_P^I, so kann sich der isolierte Metallkörper nur auf Kathodenpotential aufladen.

Diese Gesetzmäßigkeiten werden übrigens auch bei den „dynamischen“ Methoden zur Messung der Ausbeute an Isolatoren (vgl. Ziff. 10) ausgenutzt: Man hält durch geeigneten Elektronenbeschuß die Oberfläche des Isolators entweder auf Kathodenpotential fest oder auf einem geeigneten Potential entsprechend dem Punkt E_P^{II} der Ausbeutekurve, auf jeden Fall also auf einem definierten Ausgangspotential. Die Auftreffenergie der P ist dann definiert durch die Differenz zwischen diesem stabilen Wert des Oberflächenpotentials und demjenigen Potentialwert, auf den die elektronenemittierende Kathode für die kurze Zeit des Impulses gebracht wird.

Wie weitgehend bei Kenntnis der oben angegebenen charakteristischen Größen der Verlauf der Ausbeutekurve festgelegt ist, läßt sich an Hand von Fig. 20b beurteilen. In dieser Figur sind eine Anzahl von Ausbeutewerten für verschiedene Stoffe eingetragen, wobei die Ordinaten auf den maximalen Ausbeutewert, die Abszissen auf die zu $\delta_{\max}$ gehörende P-Energie $E_{P\max}$ bezogen sind. Die Punkte geben die direkt gemessenen δ-Werte nach einer Zusammenstellung von Bruining [*2*] wieder, die Kreise sind einer Zusammenstellung von Sternglass [*404*] entnommen, der die gemessenen Ausbeuten bezüglich der rückdiffundierenden P korrigierte. Da das Maximum der Ausbeutekurve bei vielen Stoffen sehr flach ist, dürfte die Auftragung in dieser Form mit einer gewissen Willkür behaftet sein, sie zeigt aber, daß im großen und ganzen die Ausbeutekurven sämtlich den gleichen charakteristischen Verlauf besitzen.

Die Steilheit des Kurvenabfalls bei $E_P > E_{P\max}$ sowie die Steilheit des Kurvenanstiegs bei kleinen Primärenergien scheinen ebenfalls charakteristische Größen der Ausbeutekurve zu sein (vgl. Ziff. 16). Falls man sich nach Fig. 20b auf den Standpunkt stellt, daß die Ausbeutekurven für verschiedene Materialien, auf das Ausbeutemaximum bezogen, sich tatsächlich durch ein- und dieselbe Kurve darstellen lassen, so bestehen bestimmte Beziehungen zwischen Lage und Höhe des

Ausbeutemaximums einerseits und der Steilheit des Kurvenanstiegs bzw. Abfalls links und rechts vom Maximum andererseits; die Frage scheint aber noch nicht genügend geklärt, daher wurden diese Daten unter den charakteristischen Daten in Fig. 20a nicht mit aufgeführt.

Ausbeutekurven wurden in Abhängigkeit von einer großen Anzahl von Parametern gemessen. Die Ergebnisse dieser Untersuchungen werden in den folgenden Abschnitten von C zusammengestellt, wobei in den Ziff. 12 bis 17 die Messungen an Metallen, in Ziff. 18 bis 20 die Ausbeutemessungen an Halbleitern und Isolatoren behandelt werden.

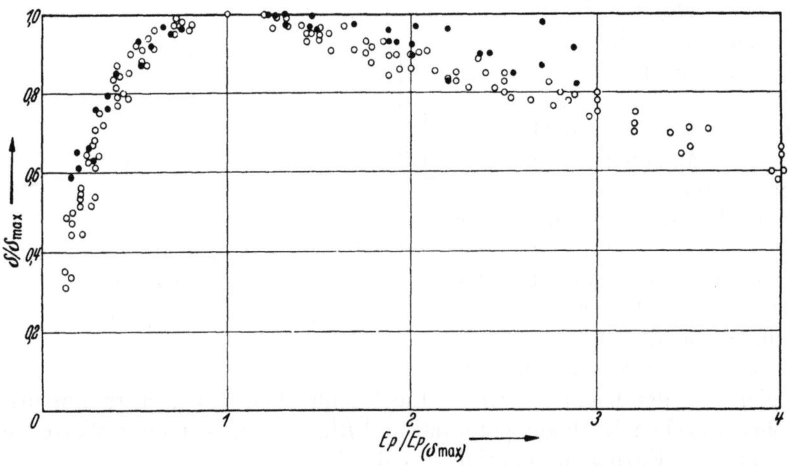

Fig. 20b. „Normierung" der Ausbeutekurven auf $\delta_{max} = 1$ und $E_{P\delta max} = 1$. Gemessene δ-Werte für verschiedene Materialien nach BRUINING [2]: ● ● ●. Korrigierte Werte (δ — η) für Li, Mg, Al, Si, Ti, Fe, Cu, Ni, Ga, Ge, Rb, Nb, Mo, Pt nach STERNGLASS [404]: ○○○.

12. Abhängigkeit der Ausbeute vom Reinheitszustand des SE-Strahlers. *Verunreinigungen sowie Gasbeladung* der untersuchten Metallschichten haben einen erheblichen Einfluß auf die Größe der Ausbeute, was schon frühzeitig erkannt wurde [*24*], [*26*]. Verläßliche und wirklich dem betreffenden Stoff zugehörende Ausbeutewerte können daher nur nach längerem Ausheizen der gesamten Apparatur und speziell durch längeres Glühen der Materialproben bei möglichst hohen Temperaturen erhalten werden. Zu den *Verunreinigungen*, die die Ausbeute für Metalle stark erhöhen, gehören vor allem Fettdämpfe [*96*]. Durch lange Erwärmung der untersuchten Metalle auf hohe Temperatur lassen sich, wenigstens bei den hochschmelzenden Metallen, Verunreinigungen so weit entfernen, daß schließlich ein „Endzustand" der Ausbeutekurven erreicht wird, der sich auch bei weiterer Wärmebehandlung nicht mehr ändert und der daher dem reinen Metall zugeschrieben werden kann (vgl. z. B. AHEARN [*95*], COPELAND [*96*], FARNSWORTH [*48*], KHLEBNIKOW [*199*], WARNECKE [*124*]. Im allgemeinen liegt dabei die Ausbeutekurve im „Endzustand" tiefer als bei nicht entgastem Material (vgl. Fig. 21), doch hat WARNECKE *während* des Ausheizvorganges Bewegungen der Ausbeutekurve oft zu größeren *und* zu kleineren Werten am gleichen Material beobachtet. Eine Abnahme der Ausbeute bei Einwirkung von Sauerstoff wurde z. B. von TRELOAR [*157*] und MCKAY [*316*] beobachtet, während SUHRMANN und KUNDT [*243*] keine meßbare Änderung der Ausbeute für Kupfer-, Silber- und Goldschichten feststellen konnten, wenn diese einer Wasserstoffatmosphäre ausgesetzt wurden.

Die eben beschriebenen Erscheinungen hängen offenbar eng mit der Frage zusammen, wie weit die SE-Ausbeute durch *Adsorption* von neutralen Atomen und

Molekülen oder von Ionen[1] an der Oberfläche des SE-Strahlers beeinflußt wird, und damit auch mit der Frage nach dem *Einfluß der Austrittsarbeit.* Wegen der Größe der Energie der S (vgl. Fig. 10) wird man a priori nicht erwarten können, daß die Austrittsarbeit für die SE eine so entscheidende Rolle spielt wie z. B. für die Glühemission; tatsächlich lassen sich die Ausbeutekurven *verschiedener* Metalle nicht eindeutig den zugehörigen Werten der Austrittsarbeiten zuordnen. Man hat andererseits die Möglichkeit, den Einfluß der Austrittsarbeit auf die Ausbeute ohne wesentliche Änderung anderer Parameter *an ein und demselben Metall* dadurch zu untersuchen, daß man die Austrittsarbeit dieses Metalls durch Adsorption von Fremdatomen verändert, z. B. die des Wolframs verkleinert durch Niederschlagen von Caesium[2]. Auf diese Weise untersuchten SIXTUS (Th auf W [*84*]), TRELOAR (Ba und Th auf W [*187*]), DE BOER und BRUINING (Ba auf W [*229*], COOMES (Th auf W [*227*]), MCKAY (Na auf W [*316*]), den Einfluß einer Änderung der Austrittsarbeit. Bei den Versuchen wurden die Austrittsarbeiten nach einer der bekannten Methoden für sich bestimmt. Alle Autoren[3] fanden eine systematische Zunahme der Ausbeute bei Erniedrigung der Austrittsarbeit *(Fig. 22a)*, bzw. bei stetiger Vergrößerung der Schichtdicke ein Maximum der Ausbeute für den kleinsten Wert der Austrittsarbeit (Fig. 22b)[4]. Für Bedeckungen von weniger als einer Atomlage erhielten sowohl SIXTUS als auch TRELOAR empirische Beziehungen von der Form: $\log\delta = A - B\cdot e\Phi$ (A und B Konstante, $e\Phi$ = Austrittsarbeit). Bei Adsorption von Sauerstoff auf W hat TRELOAR [*157*] eine Erniedrigung des δ-Wertes von 1,31 (reines W) auf 1,06 (W mit Sauerstoffbedeckung) festgestellt, was der gleichzeitig gemessenen Erhöhung der Austrittsarbeit um 1,78 eV entspricht.

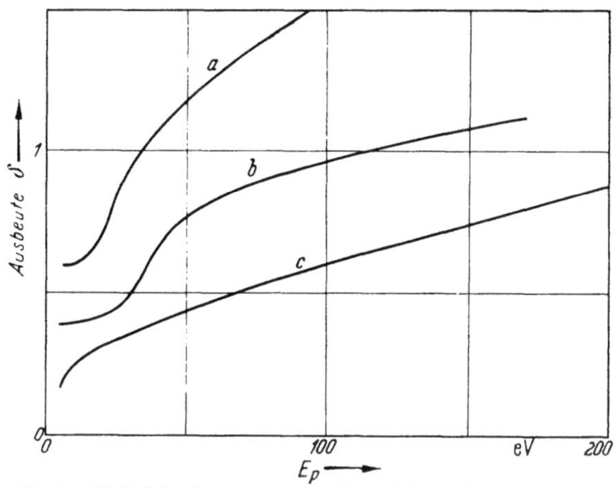

Fig. 21. Einfluß des Entgasungszustandes auf die Ausbeute von W (nach AHEARN [*95*]). *a* Nach Ausheizung bei weniger als 900° K; *b* nach einer Stunde Ausheizen bei 1300° K; *c* nach einer Stunde Ausheizen bei 1700° K („Endzustand").

[1] Vgl. hierzu besonders DE BOER: Electron Emission and Adsorption Phenomena. Cambridge (England): University Press 1935.

[2] Hierbei wäre zunächst zu prüfen, ob an einer etwa monoatomaren Schicht von Fremdatomen eine merkliche Anzahl von S ausgelöst werden kann. BRUINING und DE BOER [*229*] kommen auf Grund eingehender Überlegungen und Versuche zu dem Schluß, daß Schichten bis zu monoatomarer Dicke von sich aus nur einen vernachlässigbar kleinen Beitrag zur Ausbeute liefern können, jedenfalls für P-Energien, die größer als 50 eV sind. Es muß jedoch auf den Befund von FARNSWORTH [*144*] hingewiesen werden, daß in einer zweiatomaren Schicht von Silber 70% der auftreffenden Elektronen Zusammenstöße erleiden bei Energien bis zu 300 eV, ferner darauf, daß auch die rückdiffundierenden (langsameren) Elektronen sowie eventuell ein Teil der schnelleren S die Schicht vom Metall her kommend durchdringen und dabei eventuell Elektronen aus der Oberflächenschicht zusätzlich befreien können.

[3] Mit Ausnahme von COOMES; man vergleiche dazu die ausführliche kritische Diskussion von MCKAY auf S. 73 seines zusammenfassenden Berichts [*3*].

[4] In Versuchen von AFANASJEWA und TIMOFEEV [*165*] (Kalium auf Au, Ag, Pt) wurde mit wachsender Schichtdecke ebenfalls ein Ausbeutemaximum gefunden, das aber nicht an der Stelle kleinster Austrittsarbeit lag.

Die relativ geringe Ausbeuteerhöhung bei starker Erniedrigung der Austrittsarbeit (z. B. nach McKAY nur 60% größere δ-Werte bei einer Austrittsarbeitserniedrigung, die ein Anwachsen des Glühemissionsstroms um den Faktor 10^6 zur Folge gehabt hätte) läßt sich zwanglos mit der relativen Größe der Energie der S im Vergleich zur Energie der Glühelektronen verstehen. Man kann aus diesen Resultaten wohl schließen, daß auch eine sehr starke Erniedrigung der Austrittsarbeit die Ausbeute nicht um erhebliche Beträge vergrößern kann (im Grenzfall $e\Phi \rightarrow 0$ geht $\delta_m \rightarrow D \leq 3$ für W nach TRELOAR [*187*]).

Wenn man die eben gewonnenen Erkenntnisse nun auf die *Gasbeladung* von Oberflächen und die dadurch bewirkte Änderung von δ anwendet, so läßt

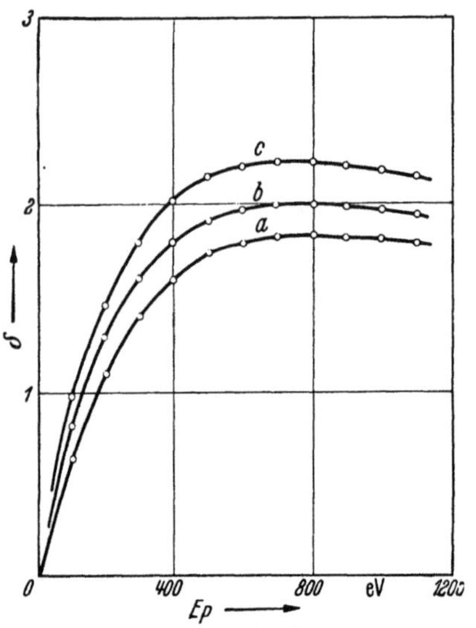

Fig. 22a. Ausbeutekurve an W bei verschiedenem Bedeckungsgrad mit Thorium (nach SIXTUS [*84*]). Effektive Austrittsarbeiten für Kurve *a*: 4,52 eV, *b*: 3,30 eV; *c*: 2,63 eV.

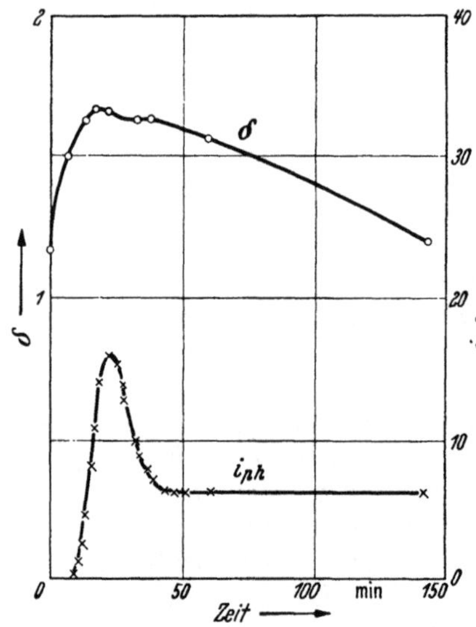

Fig. 22b. Ausbeute δ und Photostrom i_{ph} von Molybdän, das zeitlich zunehmend mit Ba bedeckt wird ($E_P = 200$ eV) nach DE BOER und BRUINING [*229*].

sich sicher ein wesentlicher Teil dieser Änderung deuten. Es ist aber nicht zu übersehen, daß so große Änderungen wie z.B. in Fig. 21 auf diesem Weg nicht restlos zu erklären sind. Nach 1 Std Ausglühen bei 1300° K sollte jedenfalls keine oberflächliche Gasbeladung mehr zurückbleiben können; wenn nach einer weiteren Stunde Ausglühen bei 1700° K trotzdem noch eine erhebliche Änderung der Ausbeute festgestellt wurde, so muß man andere Gründe dafür suchen, vielleicht in einer Richtung, wie sie bereits in Ziff. 7 dieses Berichts bei Besprechung der Energieverteilung der S angedeutet wurde.

Versuche mit *dickeren Schichten* eines Materials auf einem anderen bei gleichzeitiger Messung der Ausbeute wurden von COPELAND [*86*], [*106*], [*127*] und HASTINGS [*253*] angestellt. In Fig. 23 ist die Änderung der Ausbeutekurve bei wachsender Bedeckung von Aluminium[1] mit Platin wiedergegeben. Man sieht, daß bei kleinen P-Energien die Ausbeutekurve sich derjenigen einer dicken Pt-Schicht wesentlich schneller annähert als bei größeren P-Energien. Aus solchen Versuchen lassen sich gewisse Schlüsse auf die Eindringtiefe der P ziehen, wenn die Dicke der aufgedampften Schicht gleichzeitig gemessen wird (vgl. Ziff. 23).

Während sich bei den hochschmelzenden Metallen und den Edelmetallen kompakte Schichten mit reiner Oberfläche herstellen lassen, ist dies bei vielen

[1] Kompakte Al-Platte, also mit Oxydschicht; vgl. weiter unten.

leicht schmelzbaren und bei den elektropositiven Metallen nicht möglich: kompakte Schichten von z. B. Erdalkalien bedecken sich sofort bei ihrer Herstellung an Luft mit einer so beständigen Oxydschicht, daß durch Ausheizen keine Entfernung dieser Schicht mehr möglich ist. Die an solchen kompakten Metallschichten gemessenen Ausbeutewerte können also nicht dem reinen Metall zugeschrieben werden, sie werden in diesem Bericht in Ziff. 19 behandelt. Um für diese Stoffe trotzdem die Ausbeuten zu messen, die dem *reinen* Metall zugehören, kann man zwei Wege einschlagen: man kann Aufdampfschichten von ihnen im Hochvakuum herstellen, oder man muß im Hochvakuum die Oxydschicht dieser Metalle durch Kathodenzerstäubung „abtrommeln", wobei die Frage zu stellen ist, ob die SE-Ausbeute von Aufdampf- bzw. „abgetrommelten" Schichten gleich derjenigen des kompakten Metalls gesetzt werden darf. BRUINING und DE BOER [*189*] fanden *keinen* Unterschied zwischen der Ausbeute aufgedampfter Schichten von Kupfer und Silber und der Ausbeute derselben Metalle in kompakter Form; BRUINING [*284*] stellte ferner für kompaktes Aluminium, dessen oberflächliche Oxydschicht durch Kathodenzerstäubung im Argon „abgetrommelt" war, die gleiche Ausbeute fest wie für die Aluminium-Aufdampfschicht. Vermutlich dürfen wir diese Befunde auch auf andere Metalle verallgemeinern und können also die Ausbeuten, die an Aufdampfschichten gewonnen sind, den betreffenden reinen Metallen zuordnen.

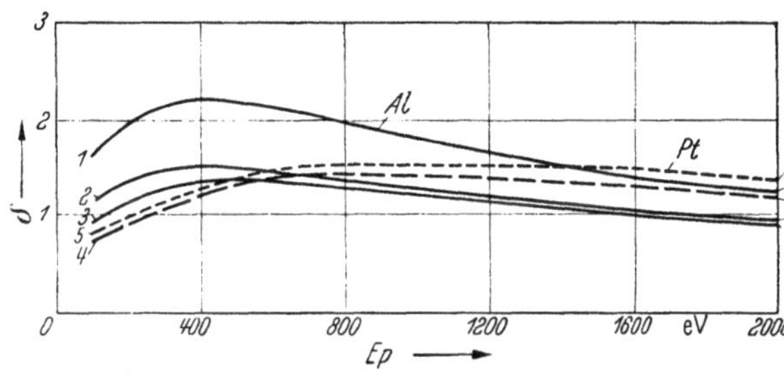

Fig. 23. Änderung der Ausbeutekurve beim Aufdampfen von Pt auf eine kompakte Al-Schicht nach COPELAND [*117*]. Kurve 1: Al ohne Pt. Kurve 2 bis 4: Pt-Aufdampfschicht wachsender Dicke. Kurve 5: Sehr dicke Pt-Schicht („Endzustand").

13. Abhängigkeit der Ausbeute vom Einfallswinkel der Primärelektronen, vom Bearbeitungs-, vom Ordnungs- und vom Kristallisationszustand. Die bisherigen Betrachtungen bezogen sich auf senkrechten Einfall der P in das SE-Material. Zahlreiche Autoren haben aber auch die *Ausbeute bei schiefem Einfall der P* untersucht und stets eine größere Ausbeute gemessen als für senkrechten Einfall [*34*], [*48*], [*54*], [*138*], [*181*], [*196*], [*204*]. Dieses Resultat ist, jedenfalls qualitativ, bei Betrachtungen von Fig. 24 auch verständlich: Bei senkrechtem Einfall der P ist der Absorptionsweg für die S bis zur Oberfläche gleich der mittleren Eindringtiefe $\bar{x}$ der P, bei schiefem Einfall der P unter dem Winkel Θ gegen das Einfallslot haben die S bis zur Oberfläche dagegen nur den Weg $(\bar{x} \cdot \cos\Theta)$ zurückzulegen bei gleicher mittlerer Eindringtiefe $\bar{x}$ der P[1]. Setzt man nach BRUINING [*138*] für die Absorption der S ein Exponentialgesetz an:

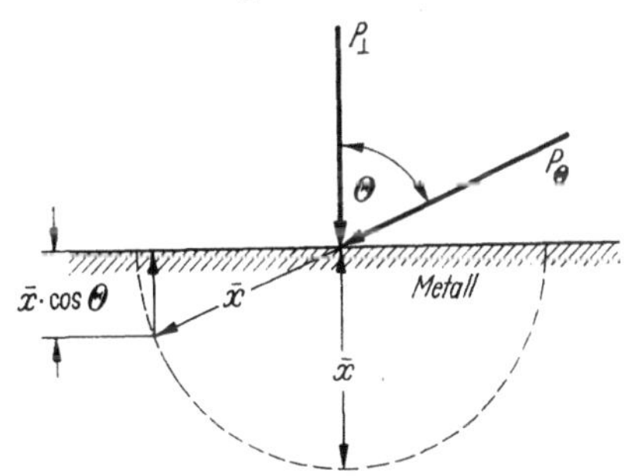

Fig. 24. Absorptionsweg $\bar{x}$ der S bei senkrechtem und ($\bar{x}$. cos Θ) bei schiefem Einfall der P (Einfallswinkel Θ).

$$\delta_0 = \Delta\, e^{-\alpha x}, \qquad (13.1)$$

[1] Der Gedankengang beruht auf der Annahme der Isotropie der Richtungsverteilung der S sofort nach dem Erzeugungsakt.

wobei $\delta_0 = \delta$ für $\Theta = 0$, Δ = Zahl der von einem P erzeugten S im Metall, α = Absorptionskoeffizient, so gilt für den Einfallswinkel Θ entsprechend:

$$\delta_\Theta = \Delta\, e^{-\alpha \bar{x} \cos\Theta}. \tag{13.1a}$$

Durch Division erhält man aus (13.1) und (13.1a):

$$\ln\left(\frac{\delta_\Theta}{\delta_0}\right) = \alpha \bar{x}(1 - \cos\Theta) = \mathrm{const}\,(1 - \cos\Theta). \tag{13.2}$$

Mit dieser Formel läßt sich die gefundene Abhängigkeit der Ausbeute vom Einfallswinkel gut darstellen. Fig. 25 zeigt die Ausbeutekurve an einer polierten

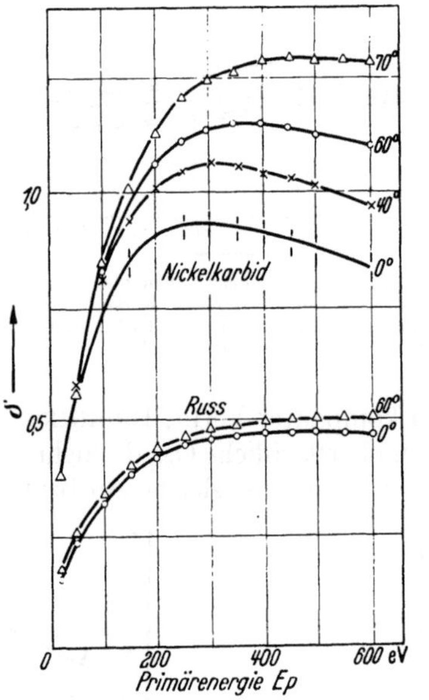

Fig. 25. Ausbeutekurve als Funktion der Primärenergie bei Nickelkarbid und Ruß (nach BRUINING [*138*]) (Einfallswinkel als Parameter).

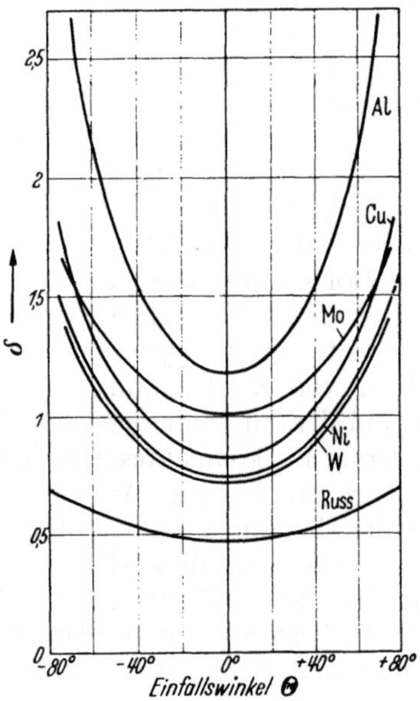

Fig. 26. Ausbeute bei E_P = 2,5 keV als Funktion des Einfallswinkels für Al, Cu, Ni, Mo, W und Ruß (nach MÜLLER [*181*]).

Nickelkarbidschicht und an einer Rußschicht mit dem Einfallswinkel Θ als Parameter nach BRUINING [*138*]. Fig. 26 zeigt die Ausbeute bei fester Primärenergie (E_P = 2500 eV) als Funktion des Einfallswinkels Θ für eine Reihe von Metallen und ebenfalls für eine Rußschicht. Besonders bemerkenswert ist in Fig. 25 bei der Nickelkarbidfläche die schwache Abhängigkeit vom Einfallswinkel bei kleinen Primärenergien, die Verlagerung von δ_m nach größeren E_P mit wachsendem Einfallswinkel und der immer schwächer werdende Abfall der δ-Kurve nach größeren Energien hin (für $\Theta = 90°$ sollte die Ausbeutekurve kein *Maximum* mehr überschreiten; vgl. hierzu die Diskussion von BRUINING [*2*], S. 103). Das ist auch zu erwarten, denn bei sehr geringer Eindringtiefe von der Größenordnung 1 bis 2 Atomlagen ist die Absorption der S klein und damit spielt der Einfallswinkel keine wesentliche Rolle mehr; mit größer werdender Primärenergie wirkt sich dagegen die Absorption der S immer stärker aus.

Hingewiesen sei noch auf den relativ geringen Einfluß des Einfallswinkels der P bei einer Rußfläche und den geringen Ausbeutewert an dieser Fläche im ganzen. Zur Erklärung möge die schematische Fig. 27 dienen, die einer Arbeit von BRUINING entnommen ist [*192*]: an einer einigermaßen glatten Oberfläche ist der Einfallswinkel genügend definiert und kann sich daher so auswirken, wie weiter oben an Hand von Fig. 24 besprochen; eine Rußfläche ist dagegen aus kleinen Flocken porös aufgebaut, so daß einerseits viele S keinen Ausweg aus dem „Labyrinth" finden (kleine Ausbeute), andererseits aber auch kein definierter Einfallswinkel der P mehr existiert und damit auch keine wesentliche Abhängigkeit der Ausbeute von der Einfallsrichtung der P. Ganz entsprechende Unterschiede in der Ausbeute bei glatten und porösen Flächen ergeben sich auch an

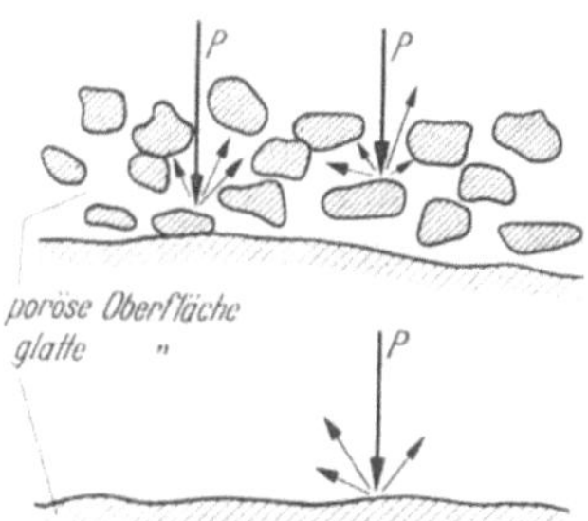

Fig. 27. Zum Verständnis der geringen Ausbeute an rauhen (porösen) Oberflächen gegenüber glatten (nach BRUINING [*192*]).

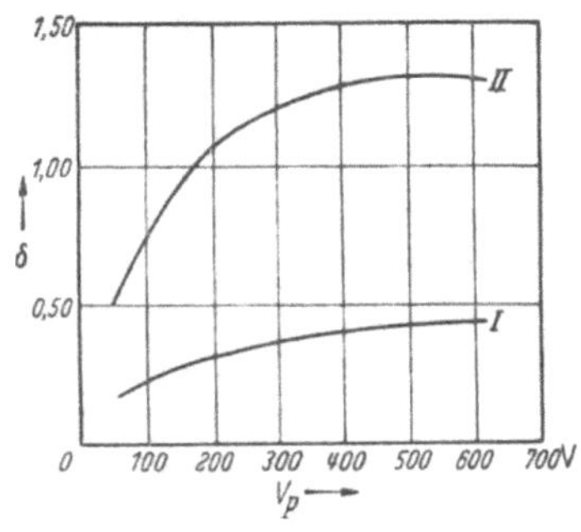

Fig. 28. Ausbeutekurve an einer blanken und einer „schwarzen" Ni-Schicht (nach BRUINING [*192*]; vgl. Text).

Metallschichten, die bei der Herstellung absichtlich „optisch schwarz" gemacht werden, was z. B. durch Aufdampfen in einer Gasatmosphäre[1] erreicht wird. *Fig. 28* zeigt die Ausbeutekurve einer solchen „schwarzen" Nickelschicht (I) und zum Vergleich die Ausbeutekurve einer blanken Nickelplatte (II) nach BRUINING [*192*]. In Fortsetzung dieser Gedankengänge hat JONKER vorgeschlagen, Oberflächen von Elektroden in technischen Röhren mechanisch aufzurauhen, um ihre SE zu verkleinern[2].

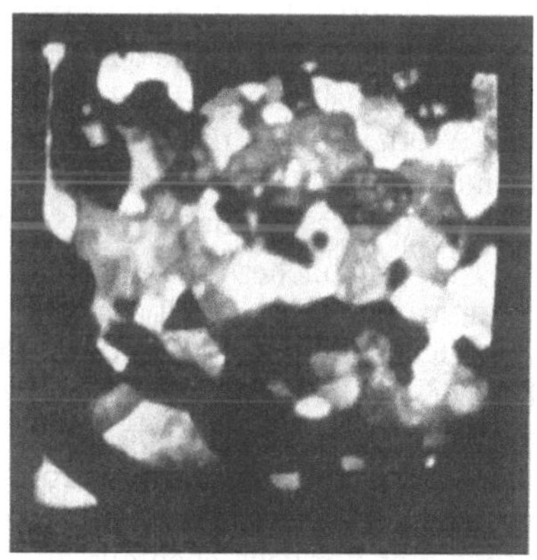

Fig. 29. Strukturbild von siliziertem Eisenblech, sichtbar gemacht durch Abtastung der SE-Ausbeute der Kristallite (nach KNOLL und THEILE [*232*]).

Es sollen in diesem Zusammenhang eine Reihe von Arbeiten erwähnt werden, die sich mit dem Einfluß der *Kristallstruktur* von Metallschichten auf die Ausbeute befassen [*48*], [*68*], [*82*], [*89*], [*105*], [*113*], [*282*]. Da die Austrittsarbeiten verschiedener Einkristallflächen etwas verschieden sind, wie ja schon von der Glühemission her bekannt ist, wird man erwarten, daß auch die SE-Ausbeuten an verschiedenen Kristallflächen desselben Metallstücks sich etwas unterscheiden. Dies wurde auch von KNOLL und THEILE [*232*] durch Abtastung eines Stücks silizierten Eisenblechs direkt sichtbar gemacht (Fig. 29). Keine Änderung der Ausbeute konnte am CURIE-Punkt bei Nickel und Eisen festgestellt werden (WOOLDRIDGE [*277*], [*278*], TRELOAR [*217*]). Der Einfluß des „Ordnungszustandes einer Schicht" auf die Ausbeute wird in Arbeiten von KOLLATH [*201*], [*203*], SUHRMANN und KUNDT [*243*], [*244*], [*323*], WOOLDRIDGE und HARTMANN [*279*] sowie MOROZOW [*301*] untersucht. KOLLATH glaubt, daß die starke Zunahme der Ausbeute von Be-Aufdampfschichten bei Glühbehandlung nicht

[1] Am besten Edelgase.
[2] J. L. H. JONKER: Philips techn. Rev. **3**, 211.

ausschließlich auf die Oxydation des Be zurückgeführt werden kann, sondern daß auch „Ordnungsvorgänge" in der Schicht beteiligt sein müssen. SUHRMANN und KUNDT erhalten an Schichten von Cu, Ag, Au, die bei 83° K aufgedampft wurden („ungeordnete Schicht") eine Erhöhung der Ausbeute um 30%, wenn diese Schichten auf Zimmertemperatur erwärmt wurden („geordnete Schicht"). WOOLDRIDGE und HARTMANN finden keine Ausbeuteänderung an einer Cu_3Au-Legierung bei Erwärmung bis 400° C. MOROZOW beobachtet eine Änderung der Ausbeute bis zu 10% für Pb, Sb, Bi beim Übergang vom festen zum flüssigen Zustand, erklärt aber seinen Befund mit einer Änderung der Leitfähigkeit und nicht durch eine Veränderung des „Ordnungszustandes".

14. Einfluß der Temperatur auf die Ausbeute bei Metallen. In Ziff. 12 wurde genauer auseinandergesetzt, daß Verunreinigungen der Oberflächen erheblichen Einfluß auf den Ausbeutewert haben. Temperaturänderungen des Sekundärstrahlers können z. B. seinen Gasbeladungszustand ändern und eine Änderung der Ausbeute mit der Temperatur hervorrufen, wenn die Röhre selbst und ihre Metallteile nicht gründlich entgast werden. Von diesen Temperatureinflüssen soll hier nicht die Rede sein, sondern nur von einer Temperaturabhängigkeit der Ausbeute eines reinen (entgasten) Metalls. Wegen der grundsätzlichen Wichtigkeit gerade einer eventuellen Temperaturabhängigkeit der SE für die theoretische Behandlung wurden schon 1937 in einem zusammenfassenden Bericht des Verfassers [*4*] die bis dahin vorhandenen Messungen der Ausbeute als Funktion der Temperatur in einer Tabelle zusammengestellt, aus der eindeutig hervorging, daß die SE-Ausbeute für *Metall*schichten von der Temperatur *nicht* abhängt. Inzwischen sind noch eine ganze Anzahl weiterer Arbeiten erschienen (TRELOAR und LANDON [*217*], COOMES [*227*], KUSHNIR [*234*], MCKAY [*316*], MOROZOW [*300*], WOOLDRIDGE [*277*], [*278*]), die das frühere Ergebnis über die Temperatur-*Un*abhängigkeit der Ausbeute an Metallschichten bestätigen. Speziell zeigten MOROZOW und WOOLDRIDGE, daß der Temperaturkoeffizient der SE für Co, Fe und Mo kleiner sein muß als der Temperaturkoeffizient der linearen Ausdehnung dieser Metalle (also z. B. für Mo: $< 5 \cdot 10^{-6}$ pro Grad).

Eine erhebliche Abnahme der Ausbeute mit der Temperatur glaubt neuerdings STERNGLASS [*395*] an Pt, Ta und Mo gefunden zu haben mit Temperaturkoeffizienten von 5 bis $7 \cdot 10^{-4}$ pro Grad. Da STERNGLASS aber eine ganz aus Messing bestehende Röhre mit mehreren gedichteten Ansätzen[1] verwendet, haben vermutlich die eingangs erwähnten Verunreinigungen der Oberfläche einen Gang der Ausbeute mit der Temperatur vorgetäuscht[2]. Die Abhängigkeit der Energieverteilung (Gegenfeldkurven) der S von der Temperatur in der Arbeit von STERNGLASS gibt übrigens ebenfalls zu starken Bedenken Anlaß; über einen ähnlichen Temperatureinfluß berichtete seinerzeit REICHELT [*264*] für W-Flächen, während KOLLATH [*291*] an der Energieverteilung (Gegenfeldkurve) der S aus einer geglühten Be-Aufdampfschicht keinerlei meßbare Veränderung bis zu einer Temperatur von 800° C nachweisen konnte. Immerhin wurde schon am Schluß von Ziff. 6 darauf hingewiesen, daß speziell zur Frage der Temperaturabhängigkeit der Energieverteilung der S *direkte* Messungen (nach Methode b bis d, vgl. Ziff. B 5), natürlich an gründlich entgasten Oberflächen, noch fehlen.

15. Elastisch reflektierte und rückdiffundierende Elektronen. Bei der Besprechung der Meßmethoden in Ziff. 10 wurde die Frage behandelt, inwieweit es

[1] *Nicht* mit Fett gedichteten.

[2] Die Messungen der *rückdiffundierenden* Elektronen (das *eigentliche* Ziel der Arbeit von STERNGLASS) werden durch die obige Kritik nicht notwendig berührt (vgl. Ziff. 15), weil die rückdiffundierenden Elektronen eine wesentlich größere Energie haben.

möglich ist, aus den gemessenen Ausbeutewerten δ auf eine „wahre Ausbeute" der S zurückzuschließen: es ergab sich in erster Näherung der Ausdruck (10.9), in dem außer dem gemessenen Ausbeutewert δ noch die relative Zahl $r i_p$ der elastisch reflektierten Elektronen, die Zahl ηi_p der rückdiffundierenden Elektronen und schließlich noch eine Größe $\bar{\chi}$ auftritt, die mit der Energieverteilung der rückdiffundierenden Elektronen zusammenhängt. Im folgenden ist zusammengestellt, was man über diese drei Größen weiß.

Für Primärenergien $E_P < 10$ eV treten nur *elastisch reflektierte Elektronen* (Bereich I in Fig. 2) auf [*19*], [*37*], [*191*], [*320*]; die Energiegrenze hängt dabei noch vom Material und vom Gasbeladungszustand der Oberfläche ab. Der Reflexionskoeffizient r liegt für verschiedene Materialien zwischen 0,05 und 0,4, bei Gasbeladung eventuell noch etwas höher. Wesentlich ist, daß der Beitrag, den die elastisch reflektierten Elektronen zur Ausbeute liefern, mit wachsender E_P schnell klein wird und daher, jedenfalls im Bereich des Maximums der Ausbeute und darüber, im allgemeinen vernachlässigt werden kann.

Über *die rückdiffundierenden P* (Bereich II in Fig. 2), liegt etwas experimentelles Material vor, wenn auch gerade im E_P-Bereich um das Ausbeutemaximum herum die Frage nach der Menge der rückdiffundierenden Elektronen und nach ihrer Energieverteilung noch nicht als befriedigend geklärt bezeichnet werden kann.

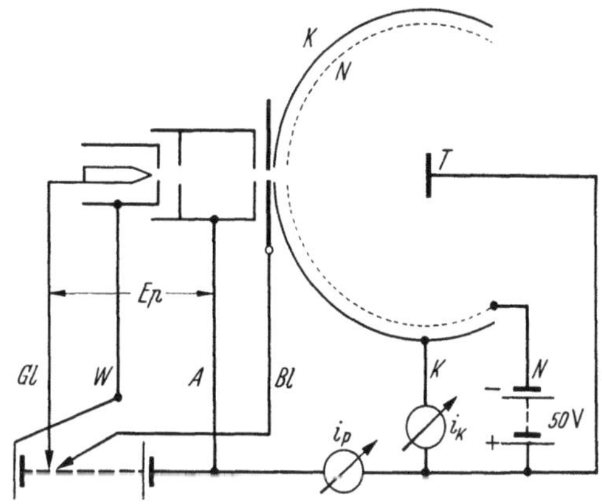

Fig. 30. Prinzip einer Anordnung und Schaltung zur Messung der Rückdiffusion η.

Wir betrachten zunächst an Hand von Fig. 30 eine Versuchsanordnung, mit welcher der Anteil der rückdiffundierenden Elektronen an der Gesamtzahl der austretenden gemessen werden kann. Eine Anordnung dieser Art haben STEHBERGER [*77*] und STERNGLASS [*395*] benutzt. Sie unterscheidet sich von der üblichen Anordnung zur Messung der Ausbeute (vgl. Fig. 16 und 17) durch ein kugelförmiges Netz N im Inneren von K aus möglichst feinen Drähten (möglichst kleiner Bedeckungsgrad!). Gibt man N ein negatives Potential gegenüber T und K z. B. -50 eV im E_P-Bereich von einigen 100 eV aufwärts, so hält das Netz N alle Elektronen zurück, deren Energie kleiner ist als 50 eV, läßt also nur Elektronen nach K gelangen, deren Energie größer ist als 50 eV, d. h. die sog. rückdiffundierenden P[1]. Dann ist $\eta = i_K/i_P$. Bei diesen Messungen ist folgendes zu beachten: Zwar werden sowohl die S vom Sekundärstrahler T als auch am Käfig K ausgelöste S durch das negative Potential von N zurückgehalten; von den auf K auftreffenden Elektronen diffundiert aber ein Bruchteil η' zurück und am Netz N werden zusätzlich S ausgelöst, von denen ein nicht leicht zu bestimmender Bruchteil ebenfalls nach K gezogen wird. Man wird für N daher ein Material mit möglichst geringem δ wählen[2] und Gasbeladungen vermeiden, da sonst der niedrige δ-Wert erheblich heraufgesetzt werden kann (vgl. Ziff. 12); durch den zum Teil streifenden Einfall der Elektronen auf die Drähte des Netzes kommt ohnehin noch eine gewisse Vergrößerung von δ hinzu.

Charakteristisch ist, zunächst im großen und ganzen gesehen, daß die so gefundenen Werte von η bei $E_P < 1000$ eV jedenfalls unter 0,3 liegen, während

[1] Die Betrachtung gilt wegen des Durchgriffes nur näherungsweise.
[2] Zum Beispiel Russ, vgl. Tabelle 2.

sie mit weiter wachsender E_P bis etwa 0,5 anwachsen, wie die Zusammenstellung der bisher gemessenen η-Werte in Fig. 31 zeigt. Da die Werte der Ausbeute δ in diesem Energiebereich mit steigenden E_P kleiner werden und unter 1 sinken, kann man aus dem allgemeinen Verlauf der η-Kurven bereits schließen, daß der „wahre δ-Wert", d. h. die Ausbeute an „eigentlichen" S, nach großen E_P hin auf sehr kleine Beträge abfällt. Zu der Darstellung in Fig. 31 sei im einzelnen noch folgendes bemerkt. Nach PALLUELL [348] nähern sich die η-Kurven mit wachsender P-Energie konstanten Endwerten, die recht gut mit den Werten von BOTHE bei 680 keV [357 a] übereinstimmen, und auch die Werte von TRUMP und VAN DE GRAAFF [349] lassen sich qualitativ diesem Bild einordnen (wenn man den Kurvenabfall unterhalb $2 \cdot 10^5$ eV unberücksichtigt läßt). Die η-Werte scheinen also für $E_P \geq 10^4$ eV konstante Werte anzunehmen, die sich entsprechend ihrer Kern-

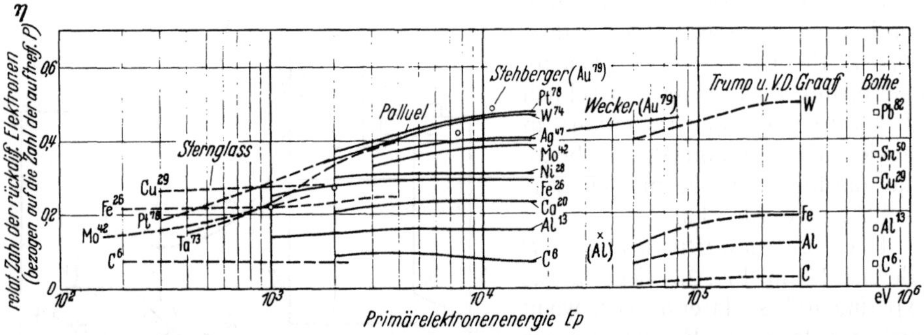

Fig. 31. Relative Anzahl η der rückdiffundierenden Elektronen bezogen auf die Anzahl der primär auftreffenden als Funktion der Primärenergie E_P nach Messungen verschiedener Autoren (vgl. Text).

ladungszahl zwanglos ordnen lassen (nicht so gut ihrer Dichte nach!). Komplizierter wird das Bild, wenn man zu kleineren E_P-Werten übergeht. Nach STERNGLASS [395] fallen die η-Kurven für $Z > 30$ (Z = Ordnungszahl) nach kleinen Energien hin wesentlich steiler ab als die η-Kurven für Elemente mit $Z < 30$; Messungen von STEHBERGER an Au [77] stützen dieses Resultat. Es ergeben sich für P-Energien von einigen 100 eV bei niedrigen Ordnungszahlen zum Teil wesentlich größere η-Werte als bei höheren Ordnungszahlen. STERNGLASS gibt dafür eine Erklärung, indem er die Zahl der in äußeren Schalen der betreffenden Elemente gebundenen Elektronen mit einer Bindungsenergie kleiner als $E_P/2{,}7$ zur Charakterisierung heranzieht. Bezüglich eingehenderen Studiums gerade dieser Frage sei auf die Originalarbeit von STERNGLASS verwiesen.

Um über die Größe $\bar{\chi}$, die in diesem Zusammenhang interessiert, Aussagen zu erhalten, brauchen wir noch die Energieverteilung der rückdiffundierenden Elektronen bei verschiedenen E_P. Da zur Bestimmung von $\bar{\chi}$ eine Mittelwertbildung bezüglich der Energieverteilung der rückdiffundierenden Elektronen vorzunehmen ist[1], spielt der genaue quantitative Verlauf der Energieverteilung keine sehr entscheidende Rolle. Fig. 32 soll daher einer *Übersicht* dienen, ohne Anspruch auf Vollständigkeit zu erheben. Die Energieverteilungskurven zeigen folgendes: die allgemein auftretenden Maxima verschieben sich bei konstanten E_P um so mehr nach kleineren Energiewerten, je kleiner die Ordnungszahl des betreffenden Materials ist. Der Abstand der Maxima nimmt mit kleiner werdenden E_P zu (Fig. 32a—c). Bei den Energieverteilungen für $E_P \leq 2$ keV (Fig. 32c nach STERNGLASS [395]) fallen die relativ großen Ordinaten auf, mit der die Energieverteilungskurven links einmünden; es wäre wichtig, diesen Befund mit exakten

[1] Vgl. Ziff. 10, Gl. (10.7).

Methoden nachzuprüfen, also nicht im „zentralen" Gegenfeld wie bei STERNGLASS (vgl. hierzu den Verlauf der Energieverteilungskurven nach Fig. 10 dieses Berichts im Zusammenhang mit Fig. 31). Wenn auch bei der Unsicherheit des Verlaufs der Energieverteilungskurven in Fig. 32c eine quantitative Auswertung dieser Kurven für die Größe $\bar{\chi}$ nicht gerechtfertigt erscheint[1], so läßt sich doch mit ziemlicher Sicherheit sagen, daß $\bar{\chi}$ nicht weit von dem Wert 0,5 entfernt sein kann und daß auch der Materialeinfluß keine wesentliche Änderung des Wertes $\bar{\chi} = 0{,}5$ bedingt.

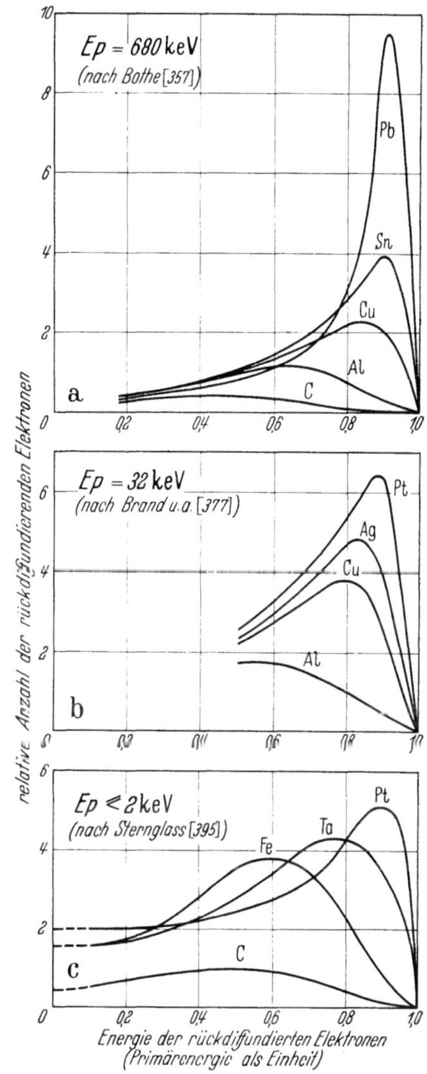

Fig. 32a—c. Beispiele für Energieverteilungskurven der rückdiffundierenden Elektronen bei verschiedenen Primärenergien; Material als Parameter (vgl. Text).

16. Zusammenhänge zwischen der Ausbeute und dem periodischen System. Schon seit den ersten systematischen Untersuchungen der SE-Erscheinung hat es nicht an Versuchen gefehlt, irgendwelche charakteristische Daten der Ausbeutekurve mit der Stellung des zugehörigen Materials im periodischen System in Zusammenhang zu bringen. Bei Verwendung der Höhe oder der Abszissenlage des Ausbeutemaximums kam man dabei zu keinen eindeutigen Ergebnissen. Dagegen konnte COPELAND [*100*], [*117*] und später WARNECKE [*158*] einen gewissen Gang der Steilheit des Abfalls der Ausbeutekurve nach größeren Energien hin mit der Ordnungszahl feststellen in dem Sinne, daß mit steigender Ordnungszahl dieser Abfall der Ausbeutekurve immer schwächer wird; allerdings fiel dabei schon auf, daß im einzelnen starke Schwankungen zwischen benachbarten Elementen auftraten. GEYER [*311*], [*312*] berichtet über Zusammenhänge die bei seinen eigenen Messungen und denen anderer Autoren (BRUINING und DE BOER, COPELAND) zwischen der Anfangssteilheit der Ausbeutekurve und der Ordnungszahl auftreten; MCKAY konnte bei Prüfung dieser Frage an einem erweiterten Material die Beziehungen jedoch nicht bestätigen (man vergleiche hierzu seine kritische Diskussion in [*3*], S. 70ff.).

In neuerer Zeit hat STERNGLASS [*404*] über sehr ins einzelne gehende Zusammenhänge zwischen korrigierten Ausbeutekurven und dem periodischen System berichtet. STERNGLASS korrigiert zunächst die gemessenen δ-Werte bezüglich der rückdiffundierenden Elektronen nach Gl. (10.1a); er trägt dann $\ln[(\delta - \eta)/E_P]$ über verschiedenen Potenzen von E_P auf und findet einen *linearen*

[1] STERNGLASS gibt die Formel an: $\bar{\chi} = 0{,}45 + 2 \cdot 10^{-3} \cdot Z$, vgl. Ziff. 10.

Abfall speziell bei der Auftragung über $\sqrt{E_P}$ im Bereich $200 < E_P < 1500$ eV. Er trägt dann den reziproken Wert der Neigung β über der Ordnungszahl auf und erhält damit die Darstellung der Fig. 33: in der Anordnung der $1/\beta$-Werte über der Ordnungszahl tritt der Schalenaufbau des Atoms deutlich zutage. Näheres über die theoretischen Überlegungen von STERNGLASS, die dieser Auswertungsform zugrunde liegen, findet man im Teil F dieses Berichtes. Zu der von STERNGLASS durchgeführten Korrektur vergleiche man die grundsätzlichen Überlegungen der Ziff. 10. Abgesehen davon mahnt die Streuung der in der Literatur angegebenen δ- und η-Werte zur Vorsicht bei der quantitativen Bewertung: bezüglich der δ-Werte sei auf die Ansicht von MCKAY [3], S. 69, bezüglich der η-Werte auf Fig. 31 und die zugehörige Diskussion hingewiesen.

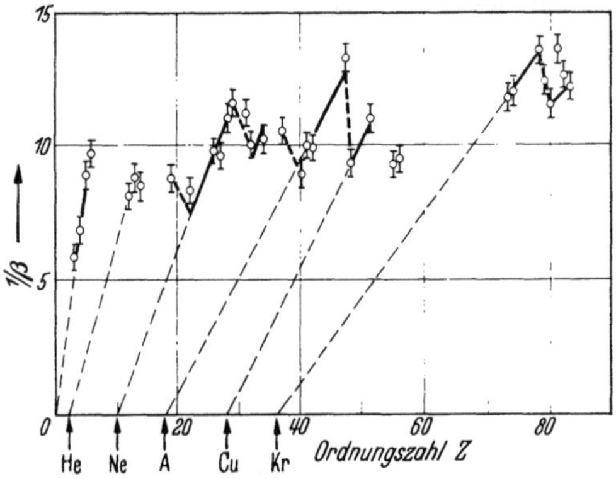

Fig. 33. Zusammenhang zwischen SE-Ausbeute und Ordnungszahl des SE-Strahlers (nach STERNGLASS [404], vgl. Text).

17. Materialzusammenstellung über Ausbeutemessungen an reinen Metallen. Zum Abschluß der Diskussion der Ausbeuteuntersuchungen an reinen Metallen ist in Tabelle 2 das gesamte Material nach steigender Ordnungszahl Z zusammengestellt. Man findet in den Kolonnen die charakteristischen Daten der Ausbeutekurven (vgl. Fig. 20a, S. 255): den gemessenen Maximalwert $\delta_{\max}$ und die zugehörige Primärenergie $E_{P\max}$; ferner — falls die Ausbeutekurve den Wert $\delta = 1$ überschreitet — die Abszissenlage der Punkte, in denen $\delta = 1$ wird (E_P^{I} und E_P^{II}); die Zahlen in der letzten Kolonne beziehen sich auf das Literaturverzeichnis. Es sei darauf hingewiesen, daß es sich bei den Tabellenwerten um die Ausbeutewerte bei *senkrechtem Einfall der P* handelt. Die in Klammern gesetzten Zahlen sind unsicher. Das „größer als"-Zeichen bedeutet, daß die Kurve nicht bis zu genügend großen E_P verfolgt wurde, um genauere Zahlenangaben zu machen. In Fig. 34 sind schließlich noch einige Ausbeutekurven für in der Praxis öfters gebrauchte Metalle in ihrem Gesamtverlauf wiedergegeben, gleichzeitig als Beispiel für den Verlauf von Ausbeutekurven im einzelnen.

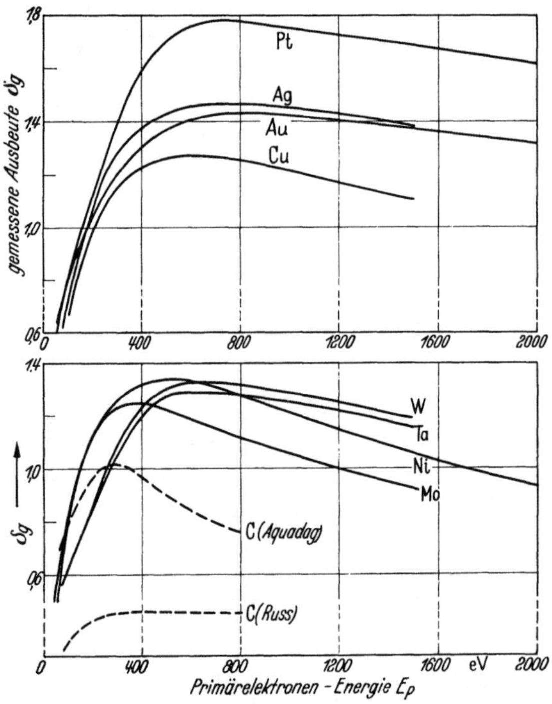

Fig. 34. Ausbeutekurven von einigen Metallen und Kohlenstoff nach Messungen verschiedener Autoren (vgl. Tabelle 2).

Tabelle 2. *Ausbeutewerte für reine Metalle.*

Element *Z*	Element Chemisches Zeichen	Maximale Ausbeute δ_{max}	$E_{P\,max}$ in eV	E_P^{I} in eV	E_P^{II} in eV	Autoren
3	Li	0,5	85	—	—	[*189*],
4	Be	0,5	200	—	—	[*189*], [*202*], [*215*]
5	B	1,2	150	100	450	[*339*]
6	C {Graphit	1,0	300	—	—	[*193*], [*368*]
	C {Ruß	0,45	500	—	—	
12	Mg	0,95	300	—	—	[*48*], [*189*], [*199*]
13	Al	0,95	300	—	—	[*189*]
14	Si	1,1	250	90	650	[*339*]
19	K	0,7	200	—	—	[*165*], [*235*], [*289*]
22	Ti	0,9	280	—	—	[*192*]
26	Fe	1,3	(400)	120	1400	[*51*], [*158*]
27	Co	1,2	(500)	(200)		[*217*], [*249*]
28	Ni (blank)	$1{,}3_5$	550	150	1750	[*51*], [*89*], [*158*], [*192*] [*265*], [*370*]
28	Ni (schwarz)[1]	(0,5)	>600	—	—	[*192*]
29	Cu	1,3	600	200	1500	[*58*], [*158*], [*323*]
31	Ga	1,55	500	100	>1500	[*372*]
32	Ge	1,2	400	200	900	[*339*], [*401*]
34	Se	$1{,}3_5$	400	110	>1200	[*390*]
37	Rb	0,9	350	—	—	[*165*]
40	Zr	1,1	350	175	(600)	[*192*]
41	Nb (Cb)	1,2	375	175	1100	[*158*]
42	Mo	1,25	375	150	1300	[*51*], [*107*], [*158*]
46	Pd	$>1{,}3$	>250	120		[*48*]
47	Ag	$1{,}4_7$	800	150	>2000	[*158*], [*189*]
48	Cd	$1{,}1_4$	450	300	700	[*324*]
50	Sn	$1{,}3_5$	500	150	>1500	[*301*]
51	Sb	1,3	500	200	>1200	[*390*]
55	Cs	$0{,}7_2$	400	—	—	[*189*], [*198*], [*235*]
56	Ba	$0{,}8_3$	400	—	—	[*168*], [*189*]
73	Ta	1,3	600	250	>2000	[*158*]
74	W	$1{,}3_5$	650	250	>1500	[*58*], [*75*], [*95*], [*158*]
78	Pt	1,8	700	150	>2000	[*96*]
79	Au	$1{,}4_5$	800	150	>2000	[*58*], [*96*], [*117*], [*127*], [*158*]
80	Hg	1,3	600	350	>1200	[*372*]
81	Tl	1,7	650	70	>1500	[*390*]
82	Pb	1,1	500	250	1000	[*301*], [*372*]
83	Bi	$1{,}1_5$	550	200	1600	[*301*], [*372*]
90	Th	1,1	800			[*168*]

18. Ausbeute von zusammengesetzten Photoschichten und Oxydkathoden. Als man sich gegen Ende der zwanziger Jahre für die technischen Aussichten der SE-Erscheinung zu interessieren begann, stießen nahezu gleichzeitig SUHRMANN [*78*] sowie ZWORYKIN und Mitarbeiter [*164*] auf die große Ausbeute der Alkaliphotoschichten, speziell der Cs-Photoschichten auf oxydiertem Ag als Unterlagemetall[2]. Trotz vieler Untersuchungen ist der Mechanismus der großen SE-Ausbeute dieser Photoschichten noch nicht befriedigend geklärt[3]. Eine ausführlichere Diskussion der Ergebnisse dieser Untersuchungen[4] gibt McKAY [*3*], hier

[1] Ni (blank): Ni-Schicht aufgedampft im Vakuum; Ni (schwarz): Ni-Schicht aufgedampft in Edelgasatmosphäre.

[2] Die Herstellungsvorschriften für diese Schichten sind heutzutage geläufig; sie können in der einschlägigen Literatur nachgelesen werden.

[3] Das ist durchaus verständlich, wenn man bedenkt, daß selbst für die SE an reinen Metallen noch keine umfassende Deutung besteht (vgl. Teil F). Literatur: [*118*, [*130*], [*141*], [*142*], [*143*], [*148*], [*170*], [*186*, [*198*], [*218*], [*224*].

[4] Die zum Teil nicht widerspruchsfrei sind.

seien nur einige Einzelheiten erwähnt. Die Photoausbeute und die SE-Ausbeute müssen bei diesen Schichten nicht parallel gehen [*134*], [*159*]. Die notwendige Menge von Cs-Atomen an der Schichtoberfläche ist für größte SE-Ausbeute geringer als für die maximale Photoausbeute [*155*], [*164*]. Die Art des Unterlagemetalls ist von gewissem Einfluß auf die erreichbaren Höchstausbeuten [*164*], ferner in gewissen Grenzen auch der Oxydationszustand vor dem Aufbringen des Cs [*160*]. In seiner Diskussion des vorliegenden Versuchsmaterials (s. oben) schreibt McKAY [*3*], S. 115/116: „Die Photoempfindlichkeit gegenüber sichtbarem Licht hat mit den SE-Eigenschaften der Schichten nichts zu tun. Die S stammen aus dem Innern der Oxydschicht. BRUINING glaubt, daß die freien Metallpartikel nur dazu dienen, um die Leitfähigkeit der Schicht zu verbessern,

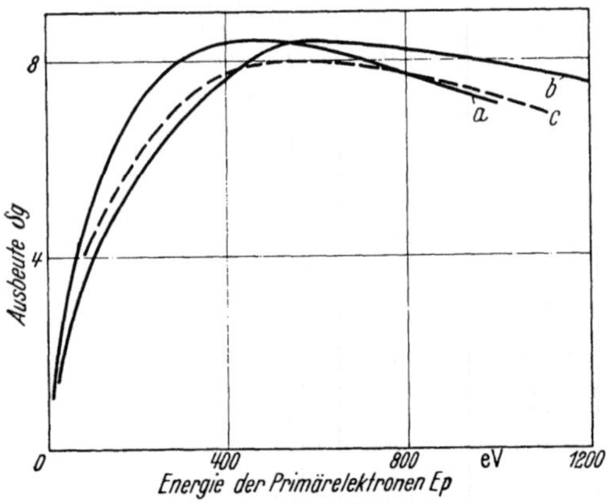

Fig. 35. Ausbeutekurven an Cs-Photoschichten. *a* nach ZWORYKIN u. a. [*164*]; *b* nach WEISS [*159*] (beide Cs_2O auf Ag); *c* nach ZWORYKIN und RAMBERG (Cs_3Sb) [*363a*].

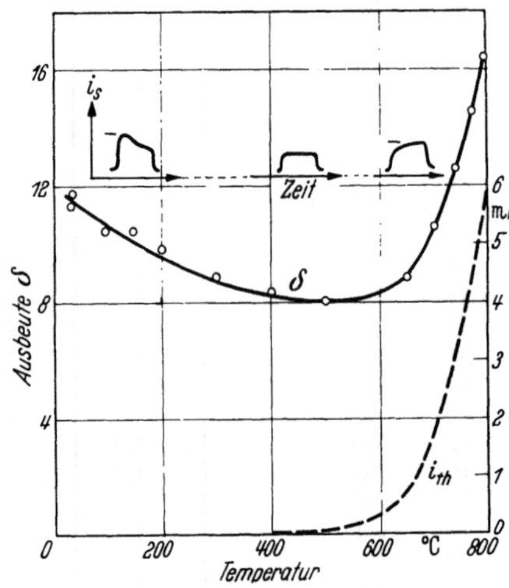

Fig. 36. Ausbeute an einer Oxydkathodenschicht als Funktion der Temperatur nach JOHNSON [*338*].

während TIMOFEEV der Ansicht ist, daß diese freien Metallpartikel selbst Elektronenquellen sind und daß die Emission der S durch positive Raumladungen unterstützt wird. Experimente, die schlüssig zeigen, welche Ansicht der Wirklichkeit am nächsten kommt, wurden bis jetzt nicht durchgeführt."

Der technischen Verwendung dieser Schichten in den SE-Vervielfachern tut die Unkenntnis des eigentlichen Emissionsmechanismus natürlich nicht den geringsten Abbruch. So sind heutzutage SE-Vervielfacher mit verschiedenen Alkalischichten käuflich (z. B. auch mit CsSb-Photoschichten). Wesentlich ist, daß mit solchen Schichten maximale Ausbeuten von der Größenordnung $\delta \approx 10$ erreichbar sind und daß die Ausbeutekurven bei Primärenergien $E_P = 100$ eV schon den Wert 4 erreichen (vgl. Fig. 35, wo die Ausbeutekurven für einige Cs-Photoschichten wiedergegeben sind).

Auch die üblicherweise für Glühemission benutzten Oxydkathoden wurden bezüglich ihrer SE-Ausbeute untersucht [*207*], [*328*], [*338*], [*341*], [*353*]. JOHNSON führte seine Ausbeutemessungen nach der Impulstechnik durch (vgl. den Schlußabsatz von Ziff. 10) und untersuchte eingehend die Temperaturabhängigkeit der Ausbeute (Fig. 36): Bei niedrigen Temperaturen lädt der Primärstrom im Verlauf des Impulses (einige μsec) die Oberfläche auf und vermindert dadurch die effektive Ausbeute; das wird am S-Impuls deutlich, dessen Form in Fig. 36 über den entsprechenden Temperaturgebieten angegeben ist[1]. Mit zunehmender

[1] Diese Erscheinung wurde von POMERANTZ [*341*] nicht beobachtet, der annimmt, daß die Leitfähigkeit der von ihm verwendeten Oxydkathodenschichten wesentlich größer war.

Temperatur der Schicht nimmt die Ausbeute merklich ab, die Leitfähigkeit der Schicht hat anscheinend genügend zugenommen, um Aufladungserscheinungen nicht mehr zuzulassen. Der folgende steile Anstieg der Ausbeute mit wachsender Temperatur ist nur im etwa gleichen Betrag wie für 500° C als SE zu deuten; er wird dadurch hervorgerufen, daß die Glühemission durch das Bombardement mit P angefacht wird („enhanced thermionic emission"), wie aus der Kurvenform des Impulses zu entnehmen ist. Die Temperaturabhängigkeit der Oxydkathodenschichten ist für die technische Verwendung dieser Schichten ebensowenig erfreulich wie die starke Empfindlichkeit der weiter oben behandelten zusammengesetzten Photoschichten gegen erhöhte Temperatur.

Die Alkali-Photoschichten sind wegen ihrer großen Ausbeute und wegen ihrer gleichzeitigen Photoempfindlichkeit technisch außerordentlich wichtig, für manche Zwecke hätte man aber gern Schichten mit ähnlich großer Ausbeute, die temperaturbeständiger und damit durch stärkere Ströme belastbar sind und unter Umständen auch *keine* Lichtempfindlichkeit besitzen sollen (Teilchenzähler).

19. Die Ausbeute bei oxydierten Metallen und Metallegierungen. Die Untersuchung der Ausbeute an reinen Metallen hat Resultate ergeben, die für die Aufklärung der gesamten SE-Erscheinung außerordentlich wichtig geworden sind, speziell konnte dadurch die ursprüngliche Auffassung richtig gestellt werden, daß die Alkali- und Erdalkalimetalle besonders große Ausbeutewerte besitzen (vgl. hierzu die Darstellung von KOLLATH [*4*]). Es zeigte sich, daß Alkali- und Erdalkalimetalle im *reinen* Zustand sehr niedrige Ausbeutewerte haben, meist $\lessapprox 1$, und daß die zunächst gemessenen großen Ausbeutewerte mit der Oxydationsschicht zusammenhängen, die sich bei Anwesenheit von Luft sofort auf diesen Metallen ausbildet. Darauf hat BRUINING als erster hingewiesen; vgl. [*2*].

Um SE-Schichten großer Ausbeute, aber geringer Empfindlichkeit gegenüber Temperaturänderungen (und auch gegenüber Einwirkung von Luft) zu finden, wurden auch oxydierte Metalle, Metallverbindungen und Metallegierungen genauer untersucht; dabei stieß man auf technisch brauchbare Schichten. Ausgehend von den relativ großen Ausbeutewerten kompakter Mg-Schichten (d. h. oxydierter Mg-Oberflächen) gelangte man zu den oxydierten („formierten") AgMg-Legierungsschichten [*268*], [*286*], [*310*], ausgehend von den Ausbeutewerten kompakter Be-Schichten (d. h. oxydierter Be-Oberflächen) zu den oxydierten („formierten") NiBe- und CuBe-Legierungsschichten [*287*], [*292*], [*296*], [*304*]. Schichten dieser Art wurden mit gutem Erfolg unter anderem von ALLEN [*222*], [*364*] sowie BAY [*188*], [*281*] zur Messung von sehr kleinen Teilchenströmen, sogar zum Einzelnachweis von Elektronen, Ionen und Neutralteilchen benutzt, ferner haben sie eine praktische Verwendung in der Massenspektroskopie gefunden.

BAY [*281*] konnte ferner zeigen, daß sich noch weitere in geeigneter Weise oxydierte („formierte") Legierungen für den gleichen Zweck eignen, unter anderem Messing. Wesentlich für den speziellen Verwendungszweck als *Teilchenzähler* ist die im allgemeinen große Austrittsarbeit dieser Schichten und damit ihre Lichtunempfindlichkeit bzw. ihr sehr geringer Dunkelstrom. In Fig. 37 sind Ausbeutekurven für einige „formierte" Metall- und Metallegierungsschichten zusammengestellt.

Soweit für Schichten dieser Art Ausbeuten gefunden wurden, die einen Wert von der Größenordnung 10 merklich überschreiten, handelt es sich nicht mehr um S, sondern um die *Feldemission dünner Schichten*[1], die durch SE nur

[1] Eine zusammenfassende Darstellung gibt TREY [*325*].

eingeleitet wird, den sog. Malter-Effekt. Malter hat als erster über solche Erscheinungen an Cs-Photoschichten berichtet, die auf oxydiertem Al als Unterlage hergestellt waren [*151*], später haben eine ganze Reihe von Autoren sich mit dem Malter-Effekt beschäftigt [*177*], [*180*], [*184*], [*205*], [*208*], [*223*], [*239*], [*332*], [*333*][1]. Feldemissionskathoden, die den Malter-Effekt zeigen, bestehen im Prinzip aus drei Schichten: einer Oberflächenschicht mit $\delta > 1$, einer gut isolierenden Zwischenschicht und einem Unterlagemetall. Der Emissionsmechanismus ist folgendermaßen erklärbar: Zunächst werden S durch die P in der Oberflächenschicht ausgelöst und bei genügend geringer Leitfähigkeit der Zwischenschicht lädt sich die Oberfläche positiv gegen die Metallunterlage auf. Dadurch entstehen in der isolierenden Zwischenschicht hohe Feldstärken, die schließlich dazu führen können, daß Elektronen aus dem Unterlagemetall durch Feldemission herausgezogen werden. Diese Anschauung wird gestützt durch Messung der Energieverteilung der austretenden „Malter-Elektronen"[2] [*180*], ferner durch die Abnahme des Effekts mit steigender Temperatur, d. h. erhöhter Leitfähigkeit der Zwischenschicht, und durch elektronenoptische Bilder der emittierenden Oberfläche [*205*]. Dieser Feldemissionsstrom unterscheidet sich von einem echten S-Strom durch seine *Trägheitserscheinungen*, denn bei der eigentlichen SE konnten bisher keinerlei Trägheitserscheinungen festgestellt werden, worauf in Ziff. 22 noch näher eingegangen wird. Die Feldemission dünner Schichten hält nämlich nach Abschalten des P-Stroms weiter an (exponentieller Abfall), wenn nur durch eine Absaugelektrode weiterhin aus der Schicht Strom gezogen wird. Der Malter-Effekt wurde außer an Al_2O_3 unter anderem auch an stark oxydiertem MgO und NiBe beobachtet. Nach den bisherigen Erklärungen zur Feldemission dünner Schichten[3] sollte man eigentlich einen stetigen Übergang der SE oxydierter Schichten in den Malter-Effekt mit wachsender Schichtdicke bzw. Isolationsfähigkeit der Schicht erwarten, doch ist über den Nachweis eines solchen kontinuierlichen Übergangs bisher noch nichts bekanntgeworden.

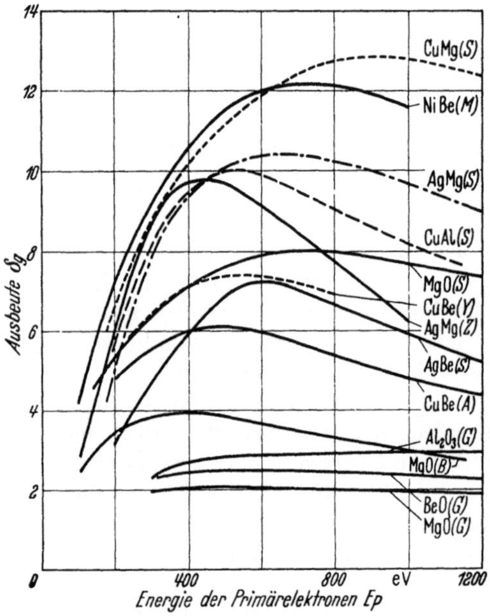

Fig. 37. Zusammenstellung einiger Ausbeutekurven für verschieden vorbehandelte Metallegierungen (vgl. Text); nach *A* Allen [*222*]; *B* Bruining, und de Boer [*225*]; *G* Geyer [*312*]; *M* Mathes [*296*]; *S* Salow [*367 a*]; *Y* Yoshida u. a. [*395 a*]; *Z* Zworykin u. a. [*310*].

20. Ausbeutemessungen an verschiedenen Halbleiter- und Isolatorschichten. Bei den Ausbeutemessungen an oxydierten Metallen lernten wir bereits Erscheinungen kennen, die in das Gebiet der Halbleiter bzw. Isolatoren gehören (z. B. Aufladungen der Schichten, Malter-Effekt). Diese Erscheinungen begrenzen die technische Brauchbarkeit der Schichten, ein Gesichtspunkt, der bisher im

[1] In engem Zusammenhang mit dem Malter-Effekt stehen auch die bereits früher von Güntherschulze beschriebenen „Spritzentladungen" [*110*], [*111*].

[2] Vgl. E. W. Müller [*394*].: Einfluß des Tunneleffektes auf die Energieverteilung bei der Feldemission.

[3] Vgl. z. B. die ausführliche Diskussion von McKay [*3*].

Vordergrund stand. In diesem Abschnitt sollen allgemein (auch dicke) Isolator- und Halbleiterschichten behandelt werden. Ausbeutemessungen begegnen hier erheblich gesteigerten Schwierigkeiten, weil die Anzahl der Variablen noch größer ist als bei den Metallen, vor allem aber, weil die *Aufladungserscheinungen* die Deutung der Resultate erschweren. Gerade deswegen wurden zur Ausbeutemessung an schlecht leitenden Schichten häufig Impulsverfahren verwendet (Ziff. 10), Messungen nach der üblichen „statischen" Methode müssen jedenfalls mit großer Sorgfalt und Kritik durchgeführt werden.

Ausbeutemessungen an Schichten mit großem Widerstand werden rein meßtechnisch zunächst durch zwei Erscheinungen beeinträchtigt:

a) bei konstantem P-Strom und positivem Auffängerpotential tritt keine Sättigung des S-Stroms ein oder jedenfalls erst bei großen Ziehspannungen [*209*], [*226*], [*336*], wofür in Fig. 38 ein Beispiel gegeben wird[1];

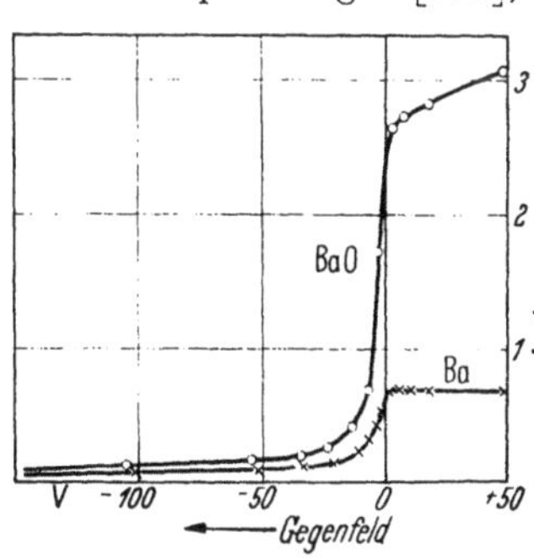

Fig. 38. Gegenfeldkurven der S aus BaO, zum Vergleich die entsprechende Kurve für Ba, nach BRUINING und DE BOER [*226*].

b) die Ausbeute δ ist in vielen Fällen von der Primärstromdichte und von der aufgestrahlten Menge der P und damit von der Beobachtungszeit abhängig.

Es sei daran erinnert, daß die Sättigung des S-Stroms und die Proportionalität zwischen i_K und i_P bei Metallen Voraussetzungen für die Definition und für einwandfreie Messung von δ waren (Ziff. 9). Es ist klar, daß die Erscheinungen a), b) durch Aufladungen an der Oberfläche oder im Inneren der Schicht bedingt sind; die quantitativen Zusammenhänge sind aber noch nicht restlos geklärt. Am klarsten treten Aufladungserscheinungen zutage in Arbeiten von HINTENBERGER [*231*] und NELSON [*209*], [*262*]. HINTENBERGER läßt einen festen P-Strom auf Glimmer, Al_2O_3 und NaCl auftreffen und mißt das Oberflächenpotential als Funktion der Zeit. Für E_P etwas größer als E_P^I, also δ-Werte etwas größer als 1, beobachtete er zunächst einen Potentialanstieg der Oberfläche bis nahe an das Potential des Auffängers heran (vgl. hierzu das Verhalten von Metallen, Ziff. 11). Dann aber fiel das Oberflächenpotential zunächst langsam, schließlich abrupt auf Kathodenpotential ab. Für E_P-Werte, die wesentlich größer waren als E_P^I, trat diese Erscheinung nicht auf. HINTENBERGER nimmt an, daß zunächst eine positive Aufladung der Oberfläche erfolgt wegen $\delta > 1$, daß aber die negativen Ladungen der im Isolator in einer gewissen Tiefe steckenbleibenden P dort allmählich eine negative Raumladung aufbauen. Diese unterstützt zunächst das Heraustreten der S aus der Schicht („field enhanced emission"), gleichzeitig bremst sie aber die neu hereinkommenden P ab: die negative Raumladung im Innern muß sich also zwangsläufig in Richtung auf die Oberfläche hin ausbreiten. Dadurch wird die Ausbeute immer kleiner (geringere Eindringtiefe der P), und wenn erst einmal $\delta < 1$ geworden ist, sinkt das Oberflächenpotential zwangsläufig auf das Kathodenpotential ab, den stabilen Punkt der Ausbeutekurve für $\delta < 1$. NELSON beobachtete eine plötzliche Zunahme der Ausbeute an MgO und NiCr bei Vergrößerung von E_P über E_P^I hinaus und konnte zeigen, daß diese Erscheinung mit einem plötzlichen Anwachsen des Oberflächenpotentials zusammenfiel[2].

[1] Bei sehr dünnen Schichten, z. B. von 1 μ Dicke und darunter scheint diese Schwierigkeit nicht aufzutreten [*329*].

[2] Ähnliche Beobachtungen machte COPELAND [*228*], vgl. hierzu auch YASNOPOLSKI [*280*].

Bei den meisten Ausbeuteuntersuchungen an Isolatoren müssen Aufladungserscheinungen zur Erklärung der Resultate herangezogen werden; die Versuche von HINTENBERGER sowie NELSON wurden nur deswegen eingangs etwas genauer diskutiert, weil diese Autoren durch gleichzeitige Messung des Oberflächenpotentials den *Beweis* für Aufladungen erbrachten. Die Versuche zeigen gleichzeitig die Bedeutung der Potentiale E_P^{I} und E_P^{II} im Bereich der SE von Isolatoren, insbesondere im Hinblick darauf, daß nahezu alle Isolatoren in einem gewissen E_P-Bereich den Ausbeutewert $\delta = 1$ überschreiten. Man sollte die Aufladungserscheinungen nicht vergessen, auch wenn im folgenden nicht mehr ausdrücklich darauf hingewiesen wird.

Der allgemeine Verlauf der Ausbeutekurve von Halbleitern und Isolatoren entspricht dem der Ausbeutekurve von Metallen mit der Einschränkung, daß die maximalen δ-Werte wesentlich größer sind; sie liegen nach Tabelle 3 zwischen 1 und 6. Der Grund für die wesentlich größere Ausbeute bei Isolatoren wird zur Zeit noch lebhaft diskutiert (z. B. BRUINING [*2*], S. 57ff). Der Gedanke der durch interne Felder angefachten Ausbeute („field enhanced emission") genügt jedenfalls nicht in allen Fällen zur Erklärung der großen Ausbeuten; denn zum Aufbau der notwendigen Raumladung ist eine endliche Zeit erforderlich. Ein zeitliches „Anklingen" der Ausbeute wird aber im allgemeinen nicht beobachtet. Man kommt wohl nicht um die Tatsache herum — und das ist für die theoretische Deutung der SE wichtig — daß im Durchschnitt Isolatoren a priori größere Ausbeutewerte besitzen als Metalle. Es seien hier nur einige Ansichten zu dieser Feststellung wiedergegeben: BRUINING und DE BOER [*229*], MAURER [*297*], MORGULIS [*259*], [*260*] glauben, daß große Ausbeuten dann auftreten, wenn das erste leere Leitungsband mit seinem unteren Rande über der Austrittsschwelle liegt oder jedenfalls, verglichen mit dem Bandabstand, zwischen oberstem besetzten und unterstem leeren Leitungsband, sehr dicht unter der Austrittsschwelle. TIMOFEEV und Mitarbeiter [*269*] bis [*275*] sowie FRIMER [*251*] u. a. nehmen an, daß ionisierte Atome in der Schicht den Austritt der S erleichtern und damit die eigentlichen Träger der großen Ausbeute darstellen.

Die Ausbeute bei Isolatoren hängt vom *Einfallswinkel der P* in gleicher Weise ab wie bei Metallen (Ziff. 13). SALOW [*266*] und SCHERER [*267*] konnten zeigen, daß die früher beobachteten „kritischen" Einfallswinkel (vgl. z. B. [*4*] S. 216), bei denen die Ausbeute mit steigendem Einfallswinkel plötzlich auf viel kleinere Werte abfällt, durch Aufladungen bedingt waren. Durch Vermeidung dieser Aufladungen (Impulsmethode) erhielt SALOW die gleiche Abhängigkeit der Ausbeute vom Einfallswinkel für Glimmer, Glas und ZnS wie MUELLER für Metalle (vgl. Fig. 26).

Über einen *Einfluß der Austrittsarbeit* auf die Ausbeute ist kaum etwas bekannt; mögliche Einflüsse der Austrittsarbeit diskutiert unter anderem SHOCKLEY [*242*].

Eine *Temperaturabhängigkeit der Ausbeute* ist nach HACHENBERG [*352*] auf Grund theoretischer Überlegungen für einen kristallinen Isolator mit großem Bandabstand zu erwarten, wenn er frei von Störstellen ist; die Ausbeute sollte dann mit steigender Temperatur proportional $1/T$ abnehmen. Dagegen sollte die Ausbeute weitgehend *un*abhängig von der Temperatur sein bei „Isolatoren" mit hinreichend vielen freien Elektronen im Leitungsband[1] ferner bei Halbleitern mit geringem Bandabstand und schließlich bei Körpern mit hoher Fehlordnung, z. B. Gläsern. Die Deutung der experimentellen Resultate stößt auf Schwierigkeiten, da die Leitfähigkeit im allgemeinen mit steigender Temperatur zunimmt

[1] Denn die Wechselwirkung der S auf dem Weg zur Oberfläche ist dann im wesentlichen durch diese freien Elektronen bedingt.

und weil mit steigender Temperatur auch Veränderungen in der Zusammensetzung der Schicht vor sich gehen können; beide Effekte werden in der Regel eine Temperaturabhängigkeit der Ausbeute zur Folge haben. Veränderungen in der Schicht führen z. B. bei NaCl und KCl zu einer starken Abnahme von δ

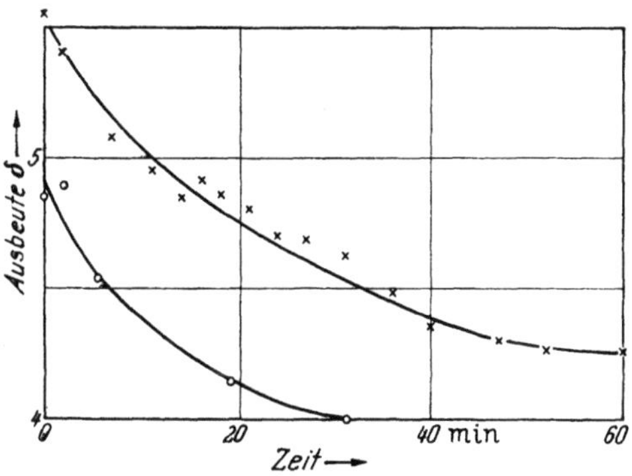

Fig. 39.

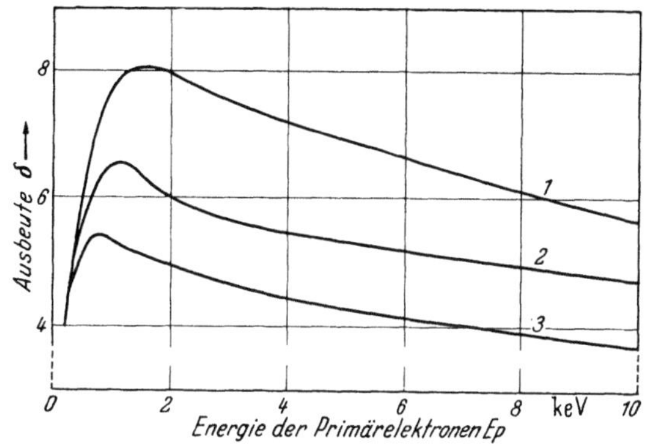

Fig. 40.

Fig. 39. Zeitliche Abnahme der Ausbeute bei Elektronenbestrahlung. Untere Kurve: NaCl, $E_P = 200$ eV, 10^{-7} A/cm² nach BRUINING und DE BOER [226]; obere Kurve: KCl, $E_P = 1000$ eV, 3×10^{-8} A/cm² nach KNOLL, HACHENBERG und RANDMER [329].

Fig. 40. Ausbeutekurven für KCl als Funktion der Temperatur *1*) 20° C; *2*) 200° C; *3*) 300° C nach KNOLL, HACHENBERG und RANDMER [329].

mit der Bestrahlungszeit, wie Fig. 39 zeigt. KNOLL, HACHENBERG und RANDMER [329] berichten, daß bei Messungen der Ausbeute als Funktion der Temperatur Zersetzungserscheinungen des KCl vermieden wurden und schreiben die Änderungen der Ausbeute (Fig. 40) einer echten Abnahme der SE mit steigender Temperatur zu[1]. Ganz entsprechende Resultate erhielten JOHNSON und McKAY [392] an kristallinen MgO-Schichten (Fig. 41). Die beiden letzten Figuren zeigen übereinstimmend, daß die Abnahme von δ mit steigender Temperatur nur für $E_P > 500$ eV auftritt und dann mit steigender E_P zunimmt, und ferner, daß sich das Maximum der Ausbeute mit zunehmender Temperatur nach kleineren E_P-Werten verschiebt. JOHNSON und McKAY geben dazu eine plausible Deutung: je größer E_P, desto größer wird die Entstehungstiefe der S; ihr Weg zur Oberfläche wächst und damit der Verlust an S infolge Wechselwirkung mit den Gitterschwingungen. Auch für andere Isolatoren ergab sich eine deutliche Abnahme der Ausbeute mit steigender Temperatur [370], [391][2]. Keine merkliche Temperaturabhängigkeit fanden MUELLER [336] für Pyrexglas und BLANKENFELD

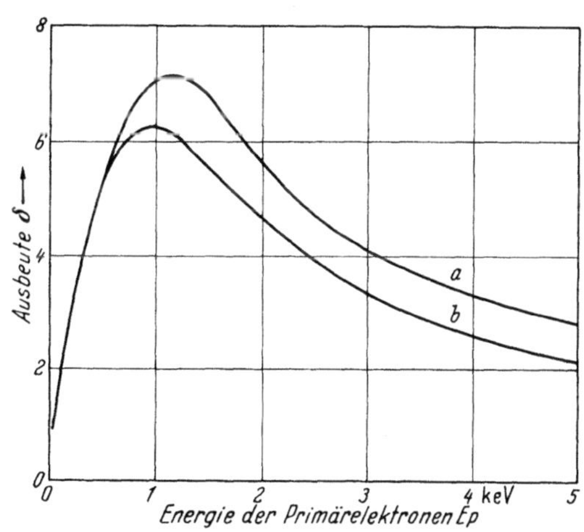

Fig. 41. Ausbeutekurven an MgO bei *a* 300° C; *b* 740° C nach JOHNSON und McKAY [392].

[1] VUDINSKY [245] stellte dagegen bei NaCl und KCl bis 250° C keine Temperaturabhängigkeit fest.

[2] Vgl. hierzu auch die Messungen von JOHNSON [338], [353] an Oxydkathoden, die in Ziff. 18 bereits diskutiert wurden.

[370] für verschiedene Gläser, in Übereinstimmung mit den theoretischen Überlegungen von HACHENBERG (s. oben). Übrigens zeigt auch Germanium nach JOHNSON und McKAY [401] eine kleine, aber deutliche Temperaturabhängigkeit

Tabelle 3.

Material	δ_{max}	$E_{P_{max}}$ in eV	E_P^{I} in eV	E_P^{II} in keV	Literatur
LiF	5,6		21		[104], [226]
NaF	5,7		20		[104], [226]
NaCl	6 bis 6,8	600	20	1,4	[104], [226], [231], [245], [253], [254]
NaBr	5,5 bis 6,2				[226]
NaJ	5,5				[226]
KCl	7,5		15		[104], [226]
KJ	5,5		12		[104], [226]
RbCl	5,8				[226]
CsCl	6,5				[226]
CaF_2	3,2				[226]
BaF_2	4,5				[226]
BeO	3,4	2000			[312]
MgO	2,4	1500			[312]
	4,0	400			[226]
CaO	2,2	500			[312]
SrO	2,6	500			[312]
BaO	2,3	1600			[312]
	4,8	400			[189]
BaO-SrO (Oxydkathode)	8	1500	60	3,5	[341]
	5 bis 12	1400	40		[353]
Al_2O_3	1,5 bis 4,8	350 bis 1300	20	1,2	[189], [231], [254], [312]
Ag_2O	1 bis 1,2				[137]
Cu_2O	1,2				[226]
MoO_2	1,1 bis 1,3				[137]
SiO_2	2,1 bis 2,9	400 bis 440	30 (< 50)	2,3	[265]
SnO_2	1,1	600			[345]
MoS_2	1,1				[226]
WS_2	1,0				[226]
Div. Gläser	2 bis 3	300 bis 420	< 60	0,9 bis 3,8	[210], [237], [254], [265]
Pyrexglas	2,3	340 bis 400	< 40	2,3	[265], [336]
Quarz	2,1 bis 2,9	400 bis 440	< 50	2,3	[265]
Glimmer	2,4	300 bis 380	20 bis 30	1 bis 3,5	[231], [254], [265], [267]
Zinksulfid				6 bis 9	[237]
Ca-Wolframat				3 bis 5	[237]
Willemit				3 bis 20	[183], [211], [237]

im gleichen Sinne: Ge besitzt eben schon Halbleitercharakter, bei reinen Metallen war ja keine Temperaturabhängigkeit nachweisbar (Ziff. 14).

Der *Einfluß von Störstellen* auf die Ausbeute von Halbleitern und Isolatoren ist besonders interessant, weil die Eigenschaften dieser Stoffe von Art und Anzahl der Störstellen wesentlich abhängen. Die Entstehung von F-Zentren (Verfärbung) bei KCl- und NaCl-Kristallen führt zu einer Verringerung der Ausbeute[1]. Andererseits erhöhen Metallzusätze die Ausbeute von Cs_2O-Schichten [*198*]; SCHNITGER [*268*] hält in MgO-Schichten freies Mg für notwendig, um große Ausbeuten zu erzielen. Auch bei den Alkalihalogeniden kann man die Ausbeute durch Zusätze erhöhen [*297*]. Nach GOBRECHT und SPEER [*390*] steigt die Ausbeute von Germanium mit zunehmender Beimischung von In, Sb, Ga und die Ausbeute von Selen mit zunehmender Beimischung von Cl und Br. Wieweit hierbei die Störstellen selbst als Emissionszentren anzusehen sind, ist noch ungeklärt.

Zum Abschluß dieses Abschnittes sind in Tabelle 3 Ausbeutemessungen an verschiedenen Halbleitern und Isolatoren zusammengestellt (Bezeichnungen wie in Tabelle 2). Vermutlich können die Daten der Tabelle 3 nicht die gleiche Sicherheit beanspruchen wie die Werte in Tabelle 2, man erhält aber einen gewissen Überblick über den Umfang der Messungen. Vielfach wurden nicht die gesamten Ausbeutekurven gemessen, sondern nur einzelne der charakteristischen Daten, worauf die Lücken in Tabelle 3 zurückzuführen sind.

D. Weitere Eigenschaften von Sekundärelektronen.

21. Die Winkelverteilung der Sekundärelektronen. Die Verteilung der Gesamtzahl der S auf die verschiedenen Austrittsrichtungen kann mit einer in Fig. 42 schematisch dargestellten Anordnung untersucht werden; der Austrittswinkel φ

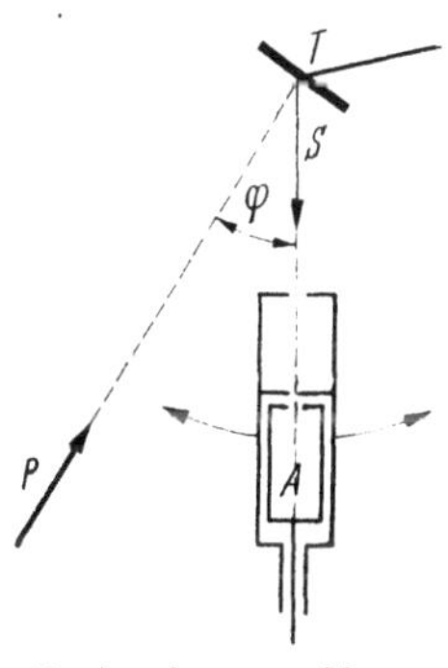

Fig. 42. Anordnung zur Messung der Winkelverteilung von S (schematisch).

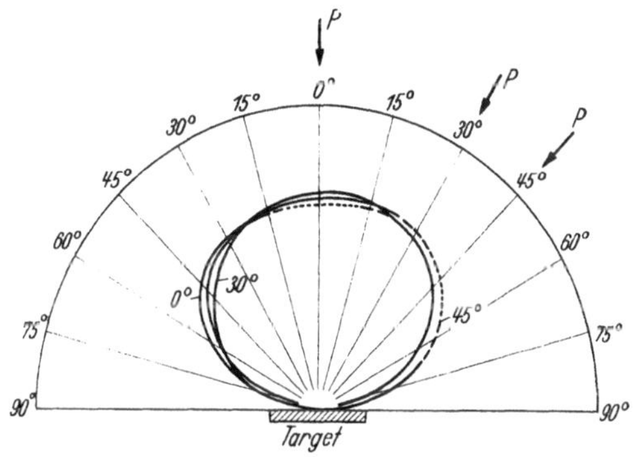

Fig. 43. Winkelverteilung der S für Einfallswinkel der P von 0°, 30°, 45°; $E_P = 100$ eV, nach JONKER [*374*].

wird dabei gegen die Flächennormale gemessen. Man fängt die S mit einem beweglichen Käfig A auf; der gemessene Käfigstrom als Funktion von φ stellt direkt die gesuchte Winkelverteilung der S dar (Anzahl der S je Raumwinkeleinheit). Bei $E_P = 2500$ eV hatten bereits DAENE und SCHMERWITZ [*81*] festgestellt, daß für die Winkelverteilung ein cos-Gesetz gilt, d. h. daß der Käfigstrom proportional zu $\cos\varphi$ ist, unabhängig vom Auftreffwinkel der P; TATE [*33*] fand bei kleineren Primärenergien eine Bevorzugung gewisser Winkel; diese Ergebnisse konnten von JONKER [*374*] nicht bestätigt werden. In Fig. 43 sind einige Messungen der Winkelverteilung nach JONKER für verschiedene Einfallswinkel der P dargestellt. Bei diesen Messungen wurden nur die „wahren" S mit Energien bis zu 15 eV zugelassen. Rückdiffundierende Elektronen mit Energien von etwa 50 eV (bei $E_P = 100$ eV) verhalten sich nach weiteren Messungen von JONKER bezüglich der Winkelverteilung wie die „wahren" S, während die elastisch

[1] BRUINING [*226*]; vgl. auch die ausführliche Diskussion der Vorgänge dabei durch BRUINING [*2*], S. 64 sowie KNOLL, HACHENBERG und RANDMER [*329*].

reflektierten P bekanntlich bevorzugte Austrittsrichtungen besitzen [*144*]. Die Resultate von JONKER gelten für polykristallines Nickel; es liegen noch keine Messungen darüber vor, ob das cos-Gesetz auch für Metall-Einkristalle und für Halbleiter und Isolatoren gültig ist.

22. Die Auslösezeit der Sekundärelektronen. Bei der Erzeugung der S übertragen die P irgendwie in Teilbeträgen ihre Energie an Elektronen des Metalls; bei der Diffusion der S zur Oberfläche hin wird die Energie der S in irgendeiner Form dissipiert, so daß nach einiger Zeit keine S mehr vorhanden sind, deren Energie zur Überwindung der Potentialschwelle an der Metalloberfläche ausreicht. Der durch die P hervorgerufene „Anregungszustand" im Metall muß also im Prinzip eine endliche Abklingzeit besitzen. Wir wollen mit „Auslösezeit" das Zeitintervall zwischen dem Eintritt eines P und dem Austritt des ersten S bezeichnen; das Zeitintervall zwischen dem Austritt des ersten und des letzten S soll als „Auslösebreite" definiert werden[1]. Auslösezeit und Auslösebreite sind beide sicher sehr klein, denn alle bisherigen Untersuchungen haben nur zur Angabe einer oberen Grenze für diese Größen geführt. HIRASHIMA [*367*] benutzte bei seinen Messungen eine Anordnung nach Fig. 16 und überlagerte der Ziehspannung zwischen T und K eine hochfrequente Wechselspannung (300 MHz); seine Resultate machen eine Auslösezeit $< 3 \cdot 10^{-9}$ sec wahrscheinlich. DIEMER und JONKER [*366*] folgerten aus Messungen an Dynatrons auf Auslösezeiten $< 10^{-11}$ sec. WANG [*337*] ersetzte den Reflektor in einem Reflexklystron durch eine SE-Schicht, trotzdem arbeitete das Klystron bei einer Frequenz von 4000 MHz noch zufriedenstellend. WANG schloß daraus, daß die Auslösebreite kleiner als $2 \cdot 10^{-10}$ sec sein muß. GREENBLATT und MILLER [*343*] schlossen aus Versuchen an Hochfrequenzvervielfachern auf eine Auslösezeit von weniger als $5 \cdot 10^{-11}$ sec. Auch Messungen des Schroteffekts lassen Schlüsse auf die Auslösezeit zu [*120*], [*125*], [*135*], [*161*], [*162*], [*173*] u. a., so fand HAYNER [*129*] die Auslösezeit $< 10^{-6}$ sec. Es sei erwähnt, daß die Summe der Laufzeiten von P und S im Material von der Größenordnung 10^{-14} sec ist, also wesentlich kleiner, als die bisher nachweisbare Grenze der Auslösezeit.

E. Daten über die Bewegung langsamer Elektronen in festen Körpern.

23. Die Eindringtiefe der primären und die Absorption der sekundären Elektronen. Für das Verständnis der SE-Erscheinungen wäre es wichtig, genaueres darüber zu wissen, in welcher Tiefe die S ausgelöst werden (Eindringtiefe der P), wie stark die S auf ihrem Weg zur Oberfläche absorbiert werden (Absorption der S) und in welcher Weise die P ihre Energie im Metall zur Erzeugung der S abgeben (Größe der Energieverluste der P). Leider sind darüber nicht viele quantitative Daten bekannt, da sie gerade im Bereich bis 1000 eV sehr schwer zu erhalten sind. In diesem und im nächsten Abschnitt wollen wir wenigstens etwas zu diesen Fragen sagen. Der Vorgang der SE wird dadurch kompliziert, daß die Elektronen nach dem Eintritt in das Metall einer starken Streuung unterliegen.

Auf eine („praktische") maximale Eindringtiefe der P kann aus Durchlässigkeitsmessungen an Folien geschlossen werden. Für eine Folie bekannter Dicke wird die durchgehende Elektronenmenge mit abnehmender P-Energie immer kleiner und die so gewonnene Intensitätskurve läßt sich mit genügender Sicherheit zur Intensität Null hin extrapolieren[2]. Messungen dieser Art führten z. B.

[1] Man kann natürlich auch das Zeitintervall zwischen dem Eintritt des P und dem Austritt des letzten S als Auslösezeit bezeichnen.

[2] Die „Grenzdicke" nach LENARD ist etwa doppelt so groß.

Wecker [*308*] sowie Lane und Zaffarano [*403*] an Folien verschiedener Dicke durch. Leider ist diese Methode im Bereich unterhalb einiger keV nicht mehr anwendbar, da wegen der geringen Eindringtiefe sehr dünne Folien erforderlich wären[1]. Über Absorptionsmessungen an Folien bei Elektronenenergien unter 100 eV vgl. den Schluß dieses Abschnitts. In Fig. 44 sind die Ergebnisse von

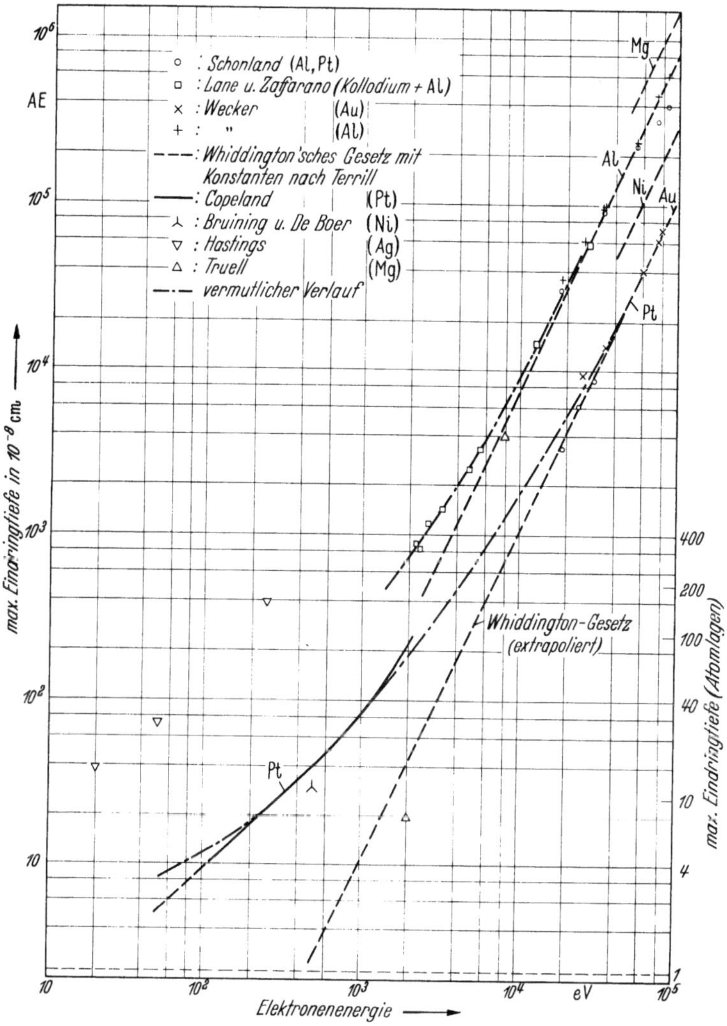

Fig. 44. Verschiedene Messungen zur Eindringtiefe der P; vgl. Text.

Wecker für Al und Au eingetragen, ferner die Messungen von Lane und Zaffarano für Al-bedampfte Collodiumfilme[2], schließlich noch einige Messungen von Schonland [*42*]. Im Energiegebiet zwischen etwa 40 und 80 keV werden die Messungen gut wiedergegeben durch das Whiddingtonsche Gesetz [*22*], das den mittleren Energieverlust der Elektronen und die durchlaufene Schichtdicke in Beziehung bringt:

$$E_P^2 - E_P(x)^2 = a \cdot x;$$

[1] Einwandfreie Folien, die dünner sind als einige Hundert Å, lassen sich nicht herstellen.

[2] Lane und Zaffarano geben an, daß (Z/A)eff für Collodium sich nicht wesentlich vom (Z/A)-Wert für Al unterscheidet; tatsächlich fügen sich auch ihre Werte der für Al zu denkenden Kurve in Fig. 44 zwanglos ein.

dabei ist E_P die Anfangsenergie des Elektrons, $E_P(x)$ seine Energie nach Durchlaufen der Strecke x und a eine Konstante, die z. B. von TERRILL [*43*] für verschiedene Materialien gemessen worden ist (Tabelle 4). Die einfach gestrichelten Linien mit der Neigung 2 in Fig. 44 geben, oben rechts beginnend, den Zusammenhang von x in Å (bzw. Atomlagen: rechte Ordinate) und E_P in eV für verschiedene Materialien an. Die Konstante a ist — jedenfalls in diesem Energiebereich — nahezu proportional der Dichte: $a/\varrho \approx 0{,}4 \cdot 10^{12}$ (eV)²/cm. Eine Extrapolation der gestrichelten Kurven nach kleineren Energien hin ist nicht sinnvoll, denn sie würde schon für Elektronen von 500 eV Energie eine Eindringtiefe von nur einigen Å ergeben, was sicher nicht zutrifft[1] (vgl. die Verlängerung der Linie für Au).

COPELAND [*250*] hat Versuche gemacht, um etwas über die Materialdicke zu erfahren, aus der maximal noch S herauskommen (maximale Entstehungstiefe der S). Er dampft auf eine Unterlage, deren Ausbeutekurve (δ-Kurve) bekannt ist, ein anderes Material mit ebenfalls bekannter δ-Kurve auf und mißt die Dicke der Aufdampfschicht, bei der die δ-Kurve eindeutig dem aufgedampften Material zugeordnet werden kann. Bei dieser Dicke der Aufdampfschicht stammen also alle S aus dem aufgedampften Material. Zwar ist die so gemessene maximale „Entstehungstiefe" der S nicht notwendig identisch mit der maximalen Eindringtiefe der P[2], es scheint aber, als ob die beiden Größen nicht sehr verschieden sind. Die Meßkurve von COPELAND für aufgedampftes Pt (in Fig. 44 stark ausgezogen) steigt im Energiebereich zwischen 100 und 1000 eV zunächst linear mit E_P, dann oberhalb 1000 eV steiler an. Extrapoliert man in Fig. 44 die Au-Kurve unter Berücksichtigung des Verlaufs der Al-Kurve, so lassen sich die Meßwerte von COPELAND (Pt) zwanglos dem Gesamtbild einfügen.

Tabelle 4. WHIDDINGTONsche *Konstanten* α *nach* TERRILL [*43*].

Material	$a \times 10^{-12}$ (eV)²/cm	$a/\varrho \times 10^{-12}$
Be	0,75	0,39
Al	1,1	0,41
Cu	3,6	0,40
Ag	4,2	0,40
Au	8,9	0,46

Schließlich kann man noch etwas über die Schichtdicke aussagen, die als untere Grenze der Eindringtiefe für Elektronen sehr kleiner Energie ($E_P \lesssim 10$ eV) anzunehmen ist: sie muß wohl immer noch einige Å betragen (bzw. eine oder einige Atomlagen); doch verliert in diesem Gebiet der Begriff der „Eindringtiefe" seinen Sinn, da mit „Eindringtiefe" im Vorangehenden der Abstand von der Oberfläche des Materials verstanden wurde, in welchem die P noch S erzeugen können[3].

Messungen der Entstehungstiefe haben HASTINGS [*253*] an Ag sowie TRUELL [*317*] an Mg als Aufdampfmaterial durchgeführt. HASTINGS glaubt aus seinen Messungen (20 eV $\lesssim E_P \lesssim$ 250 eV) schließen zu können, daß die langsameren S aus einer größeren Tiefe kommen als die rückdiffundierenden Elektronen. Die Werte von HASTINGS (20 eV $\lesssim E_P \lesssim$ 250 eV) und von TRUELL ($E_P = 2000$ eV bzw. 8000 eV) passen sich dem sonstigen Bild der Fig. 44 nicht an, ein Grund dafür ist zunächst nicht recht einzusehen.

[1] So ist es erklärlich, daß BRUINING [*2*] zu große $E_{P\max}$=Werte für Ni bzw. Au findet, wenn er die WHIDDINGTONsche Beziehung und eine plausible Größe des Absorptionskoeffizienten für die S den Rechnungen zugrunde legt. Eine gewisse Variation von a bringt dabei nicht viel, so lange die quadratische Beziehung zwischen E_P und x aufrechterhalten wird.

[2] Es könnte auch eine Grenzdicke für den Diffusionsvorgang der S zur Oberfläche sein, die aber grundsätzlich nicht größer sein könnte, als die Eindringtiefe der P.

[3] Bedenken gegen die Methode von COPELAND hat KADYSCHEWITSCH [*335*] geäußert. Sie hängen damit zusammen, daß bei gewisser Abhängigkeit der Ausbeute von der Austrittsarbeit und Änderung der Austrittsarbeit durch die aufgedampften Schichten Mißdeutungen entstehen können, ferner damit, daß eine nur wenige Atomlagen dicke Aufdampfschicht nicht ohne weiteres die Ausbeutekurve des gleichen Materials in kompakter Form zu zeigen braucht.

Wenn man nach diesem spärlichen Versuchsmaterial überhaupt einen ungefähren Verlauf der Eindringtiefe angeben will, so müßte er etwa den kurz-lang-gestrichelten Kurven für Au (Pt) und für Al in Fig. 44 entsprechen.

Auf einem anderen Wege hat BRUINING versucht, Aussagen über eine *mittlere Entstehungstiefe* $\bar{x}$ *der S* zu gewinnen. Die Abhängigkeit der Ausbeute vom Einfallswinkel Θ ist nach BRUINING durch Gl. (13.2) darstellbar. Die Auswertung zeigt, daß das Produkt $\alpha\bar{x}$ (α = Absorptionskoeffizient) im wesentlichen unabhängig von Θ ist. Mit wachsender Dichte nimmt $\alpha\bar{x}$ allgemein ab, für Ni wurde der Wert 0,5 gefunden. Daraus läßt sich $\bar{x}$ berechnen, wenn der Absorptionskoeffizient α für die S bekannt ist. Unter Benutzung des Wertes für Nickel $\alpha = 1{,}5 \cdot 10^6$/cm, der nach BECKER [*80*] zwischen 10 und 1000 eV gelten soll, erhält BRUINING für $\bar{x}$ den Wert 30 Å, d. h. etwa 14 Atomlagen. Dieser Wert von $\bar{x}$ ist in Fig. 44 als ⅄ eingetragen. Er hängt mit der maximalen Entstehungstiefe x_m in folgender Weise zusammen:

$$n_0 e^{-\alpha\bar{x}} = \int_0^{x_m} n(x) e^{-\alpha x} dx; \quad \text{mit} \quad n_0 = \int_0^{x_m} n(x)\, dx.$$

Über $n(x)$ müßte man spezielle Annahmen machen, daher wurde von einer Umrechnung abgesehen. Es ist aber sicher, daß x_m wesentlich größer ist als $\bar{x}$.

Bei der Lückenhaftigkeit der Daten muß man wohl zufrieden sein, daß die Mehrzahl der zusammengetragenen Messungen keine größenordnungsmäßigen Unterschiede ergeben haben. Alles weist darauf hin, daß in dem uns vornehmlich interessierenden Energiebereich zwischen 100 und 1000 eV je nach Primärenergie und Material 5 bis 100 Atomlagen an der Entstehung der S beteiligt sind[1].

Es bleibt allerdings ziemlich schwer verständlich, daß gerade die *sekundären* Elektronen, die ja relativ kleine Energien haben, noch aus einer Tiefe herauskommen können, in die nach dem vorigen allgemeinen Bild die *Primär*elektronen von z. B. 10facher Energie überhaupt nicht vordringen können (vgl. oben die Resultate von HASTINGS). Über die Absorption der langsamen S ist so gut wie nichts bekannt. BECKER [*80*] hat die Durchlässigkeit von Ni-Folien für langsame Elektronen gemessen und den Absorptionskoeffizienten α zu etwa $1{,}5 \cdot 10^6$/cm angegeben, der im Bereich $100 < E_P < 1000$ eV im wesentlichen konstant sein soll. Später hat KATZ [*175*] überraschend große freie Durchlässigkeiten von Ag-Folien für Elektronen von 2 bis 10 eV Energie gefunden, BERGER [*397*] konnte aber kürzlich zeigen, daß die Durchlässigkeit auf viele kleine Löcher von submikroskopischer Größe zurückzuführen ist, die nur im Elektronenmikroskop sichtbar werden. Nach BERGER sind einwandfreie Ag-Folien (Prüfung im Elektronenmikroskop) bei einer Dicke von 500 bis 1000 Å für langsame Elektronen völlig undurchlässig. BERGER schließt aus seinen Versuchen auf einen Absorptionskoeffizienten $\alpha \geq 10^7$/cm für Ag-Folien und hält auch eine Nachprüfung der Messungen von BECKER [*80*] an Ni-Folien für notwendig.

Zur Messung des Absorptionskoeffizienten von Metallen für langsame Elektronen müssen wahrscheinlich andere Wege beschritten werden. Der Absorptionskoeffizient von Isolatoren sollte um mindestens eine Zehnerpotenz kleiner sein als der von Metallen.

[1] Mit dieser Feststellung steht nicht im Widerspruch, daß in Versuchen von FARNSWORTH [*118*] durch Aufdampfen einer 2-atomaren Cu-Schicht auf einen Cu-Einkristall die selektiven Beugungsmaxima der elastisch reflektierten Elektronen bereits auf $^1/_3$ ihrer Anfangsintensität heruntergedrückt wurden. Für diese Intensitätsminderung genügt bereits *ein* elastischer oder unelastischer Streuvorgang; der dabei berechnete „Beeinflussungskoeffizient" (Größenordnung $2 \cdot 10^7$/cm) entspricht etwa dem „Gesamtwirkungsquerschnitt" nach RAMSAUER [*32*] in seiner *ursprünglichen* Bedeutung.

24. Energieverluste langsamer Elektronen beim Durchgang durch feste Körper. Die zweite wichtige Größe für den Entstehungsprozeß der S ist der Energiebetrag, der im Elementarakt von den P abgegeben wird. Vor allem würde interessieren, in welcher Weise dieser Energiebetrag zur Befreiung von einem oder mehreren Elektronen verwendet wird. Wird z. B. bei Metallen die Energie der P im wesentlichen an die Metallelektronen abgegeben oder an die fester gebundenen Elektronen in den unteren Niveaus? Sind bei den Isolatoren nur die Elektronen des obersten gefüllten Bandes am Emissionsprozeß beteiligt oder auch die fester gebundenen Elektronen in den tiefer gelegenen Niveaus und in welchem Verhältnis? Über diese Fragen ist noch weniger bekannt als über die Eindringtiefe der P. Immerhin hat es nicht an Versuchen gefehlt, wenigstens einige Anhaltspunkte über die Größe der Energieverluste der P im Elementarakt zu gewinnen. Diese Versuche sollen im folgenden kurz wiedergegeben werden.

Die erste Gruppe von Versuchen betrifft die *Feinstruktur der Ausbeutekurve* besonders im Bereich kleiner Primärenergien (kleine Maxima, Kurvenknicke, plötzliche Steilheitsänderungen). Auf die Untersuchung dieser Feinstrukturen ist viel Mühe verwendet worden, weil man grundlegende Zusammenhänge mit der Entstehung der S vermutete [*51*], [*53*], [*58*], [*63*], [*64*], [*65*], [*75*], [*76*], [*89*], [*90*], [*113*], [*114*]. Es zeigten sich sehr zahlreiche Knicke, aber FARNSWORTH [*70*], [*82*] und WARNECKE [*158*] konnten nachweisen, daß die meisten von ihnen beim Ausglühen der Proben verschwinden und daß sich einige als Maxima der elastisch reflektierten Elektronen erklären lassen (selektive Beugungsmaxima). Die Abszissenlage der nach WARNECKE *schließlich verbleibenden* plötzlichen Änderungen der Ausbeutekurve ist in Tabelle 5 zusammengestellt. Diese kritischen Potentiale versucht man mit der Anregung weicher Röntgenstrahlen in Verbindung zu bringen, vgl. z. B. [*52*], [*91*].

Tabelle 5. „*Kritische*" *Potentiale bei der SE-Ausbeutemessung nach* WARNECKE [*158*].

Material	Kritische Potentiale in Volt				
Wolfram	25±1	35±2	70±5		
Tantal	23±1	35±2	70		
Nickel	30±2	50±2	80±5	100±5	125±5

In ähnlicher Weise gingen HILSCH [*104*] und KRENZIEN [*178*] vor, die den Elektronenaustritt aus verschiedenen Alkalihalogeniden bei kleinen Primärenergien bis etwa 30 eV gemessen haben. Sie konnten zeigen, daß ein starker Abfall der elastischen Reflektion der auftreffenden P mit der langwelligen Seite des optischen Absorptionsbandes zusammenfällt (Fig. 45a). Ob an diesen Stellen bereits die Erzeugung von S einsetzt, bleibt offen.

Eine zweite Gruppe von Autoren beschäftigt sich mit der Analyse der Energieverteilung der rückdiffundierenden Elektronen in der Nähe der Primärenergie. Die Untersuchungen von RUDBERG [*152*] sowie RUDBERG und SLATER [*153*] bei $E_P = 150$ eV ergaben, daß auf die Lücke zwischen den elastisch reflektierten und den rückdiffundierenden P deutlich ausgeprägte Maxima und Minima folgen[1]. Die entsprechenden Energieverluste der P für Cu, Ag, Au haben nach RUDBERG folgende diskrete Werte (vgl. auch Fig. 45b):

Cu: 4,2 eV; Ag: 4,0 eV; Au: 2,8 eV.

Eine dritte Gruppe von Autoren analysiert die Energieverteilung von Elektronen ($2 < E_P < 20$ keV), die auf der Rückseite einer Metallfolie austreten (LANG

[1] Nach Abschluß dieses Berichtes erschien hierzu eine weitere Arbeit; D. GABOR and G. W. JULL: Experimental Evidence for the collective nature of the characteristic electron-energy losses in solids. Nature Lond. **175**, 718 (1955).

[354], RUTHEMANN [356], MÖLLENSTEDT [360]). Die Elektronen erleiden beim Durchgang durch die Folien diskrete Energieverluste, deren Größe von Material zu Material variiert. Wie Fig. 45c und Tabelle 6 zeigen, stimmen die Energieverluste nach LANG, RUTHEMANN nicht mit denen nach RUDBERG und nach WARNECKE überein. Zur Wahrscheinlichkeit der Energieverluste hat LANG beobachtet, daß 7,5 keV-Elektronen, die durch eine 350 Å dicke Al-Folie ohne wesentliche Streuung hindurchtraten, bis zu 6mal einen Energieverlust von je 14,8 eV erfahren haben können. HACHENBERG und BRAUER [400] glauben, daß

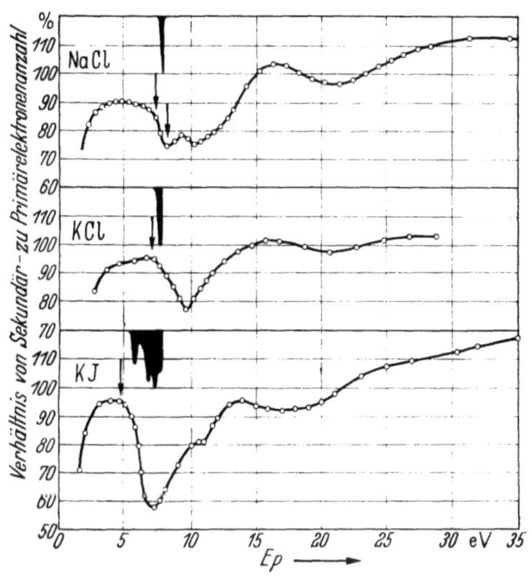

Fig. 45a. Abfall der Elektronenreflexion bei einer Primärelektronenenergie E_P entsprechend der optischen Absorptionsgrenze nach HILSCH [103].

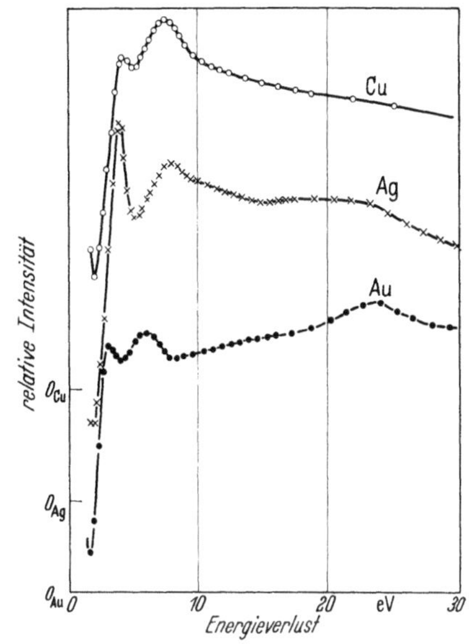

Fig. 45b. Energieverluste von Elektronen bei Streuung an Cu, Ag, Au nach RUDBERG [152].

die Anzahl der diskreten Energieverluste im Vergleich zu sonstigen Prozessen so klein sei, daß sie für die theoretische Behandlung der SE vorerst nicht ins Gewicht fallen.

Sehr interessant in diesem Zusammenhang ist eine kürzlich erschienene Arbeit von GERTHSEN und ALBERT [399], in welcher die Feinstruktur der Röntgenstrahl-Isochromaten dicht oberhalb der durch das DUANE-HUNTsche Verschiebungsgesetz gegebenen Grenze untersucht wird. Die Isochromaten steigen in ausgeprägter Form *stufenartig* an, wobei die Abstände der Stufen für Pt, W, Au mit den von LANG und RUTHEMANN angegebenen Größen der Energieverluste für diese Materialien quantitativ übereinstimmen[1]. GERTHSEN und ALBERT lassen die Frage offen, ob es sich bei ihren Versuchen um eine Energieübertragung an ein gebundenes Elektron handelt oder um eine Energieabgabe zur Anregung von Plasmaschwingungen der Metallelektronen nach PINES und BOHM [385].

Zu den hier behandelten Vorgängen im Festkörper gehört vielleicht auch der Nachweis von AUGER-Elektronen durch LANDER [393]. Ob die diskreten Energieverluste in irgendeinem Zusammenhang mit den Nebenmaximis in den Energieverteilungskurven der S stehen (vgl. Ziff. 7), müssen weitere Untersuchungen zeigen; die Maxima in den Energieverteilungskurven der S weisen allerdings im allgemeinen nur Energiedifferenzen von 10 bis 12 eV auf. In einem Vortrag vom 23. 3. 55 auf der 15. Jahrestagung über physikalische Elektronik im MIT,

[1] was in *diesem* Zusammenhang allein interessiert; über Einzelheiten und über die Durchführung der Versuche vgl. man die Originalarbeit.

Cambridge/Mass., berichtete G. A. HARROWER über „Feinstruktur der Energieverteilung von S aus Wolfram und Molybdän". Er kann die von HAWORTH und KOLLATH nachgewiesenen Nebenmaxima bestätigen und glaubt sie auf AUGER-Elektronen zurückführen zu können[1].

Eine vierte Gruppe von Arbeiten umfaßt Untersuchungen, die die Neigung der Ausbeutekurve mit der im Mittel zur Befreiung eines S aufgewendeten Energie $\overline{E}_0$ in Zusammenhang bringen. MAURER [*297*] und GEYER [*313*] bestimmen die *Anfangs*steilheit durch Extrapolation aus den gemessenen Ausbeutekurven. Der daraus berechnete Wert von $\overline{E}_0$ soll nach GEYER unabhängig sein von der Energie der P und von der Oberflächenbeschaffenheit des Materials, aber mit der Hauptquantenzahl n merklich ansteigen: im Bereich $2 < n < 6$ soll E_0 von 15 auf 125 eV anwachsen. Wesentlich andere Ergebnisse bezüglich der Größe von $\overline{E}_0$ und ihrer Abhängigkeit von der Ordnungszahl erhält STERNGLASS [*404*]. Entsprechend seiner Theorie (vgl. Abschnitt F) trägt STERNGLASS $\ln \frac{\delta - \eta}{E_P}$ als Funktion von $\sqrt{E_P}$ auf und erhält für $200 < E_P < 1500$ eV gerade Linien, deren Schnittpunkte mit der Ordinatenachse die Größe $\overline{E}_0$ (multipliziert mit einem Faktor von der Größenordnung 1) liefern. Die so gewonnenen $\overline{E}_0$-Werte liegen sämtlich zwischen 20 und 35 eV, die Stellung des betreffenden Materials im periodischen System kommt nur in einer gewissen periodischen Schwankung der $\overline{E}_0$-Werte zum Ausdruck. Wenn man die Ergebnisse von GEYER und von STERNGLASS als richtig voraussetzt, müssen die normierten Ausbeutekurven für verschiedene Materialien voneinander abweichen. Denn bei Annahme *einer* gemeinsamen normierten Ausbeutekurve für *alle* Materialien hängen, wie bereits in Ziff. 11 erwähnt, die Anfangssteilheit und die Neigung der Ausbeutekurve hinter dem Maximum unmittelbar zusammen.

Fig. 45c. 5fache Vergrößerung zweier Aufnahmen bei 6,4 keV Elektronenenergie (Streuwinkel $0 \pm 1{,}5°$, Schichtdicke der Folien etwa 300 Å), nach RUTHEMANN [*356*]; vgl. Text.

Tabelle 6. *Diskrete Energieverluste von Elektronen in Folien.*

Material	Energieverlust in eV	Literatur
Kollodium .	21,4	Nach RUTHEMANN [*356*]
Al_2O_3 . . .	22,3	Nach RUTHEMANN [*356*]
Ag	22,6	Nach RUTHEMANN [*356*]
Be	19,0	Nach RUTHEMANN [*356*]
Al.	14,7	Nach RUTHEMANN [*356*]
Ni	24,2	Nach LANG [*354*]
Cu	19,1	Nach LANG [*354*]

F. Theoretische Ansätze zur Deutung der SE.

25. O. HACHENBERG und W. BRAUER [*400*] haben kürzlich den gegenwärtigen Stand der Theorie in einer sehr übersichtlichen und umfassenden Darstellung behandelt, die aus zeitlichen Gründen nur die Theorie von STERNGLASS [*404*]

[1] Herr Dr. STERNGLASS hat mich freundlicherweise davon in Kenntnis gesetzt.

noch nicht einschließt. Wir begnügen uns daher mit einem Überblick über die Methodik, die Voraussetzungen und die Ergebnisse der verschiedenen theoretischen Ansätze (Tabelle 7) und gehen nur kurz auf Einzelheiten der Durchführung ein. Zur eingehenderen Orientierung sei auf die angegebene Originalliteratur verwiesen. Die Kolonnen in Tabelle 7 sind in 6 Abteilungen gegliedert: 1. Grundsätzliches zur Methode, 2. Entstehungsprozeß der S, 3. Vorgänge zwischen Entstehungsort und Außenraum, 4. Gültigkeitsbereich, 5. Ergebnisse, 6. Vergleich mit dem Experiment; die besonders wichtigen Kolonnen 2, 3, 5 sind noch weiter unterteilt. An Hand von Tabelle 7 sollen nun der Reihe nach die verschiedenen theoretischen Ansätze kurz diskutiert werden.

Als erster hat 1932 FRÖHLICH [*101*] eine Theorie der SE auf quantenmechanischer Grundlage gegeben (BLOCHsche Leitfähigkeitstheorie der Metalle). Diese Theorie wurde von WOOLDRIDGE [*248*] erweitert mit dem Resultat, daß der Verlauf der Ausbeutekurve befriedigend wiedergegeben werden konnte. Inzwischen haben aber BAROODY [*376*] und MARSHALL [*384*] darauf hingewiesen, daß die Winkel-Anisotropie bei der Entstehung der S in der Theorie von WOOLDRIDGE bei genauerer Rechnung verschwindet und damit auch das Ausbeutemaximum. Ein Haupteinwand gegen die Theorie von WOOLDRIDGE ist der temperatur*ab*hängige Erzeugungsmechanismus der S; um die experimentell gefundene *Un*abhängigkeit der Ausbeute bei Metallen von der Temperatur wiederzugeben, muß ein entsprechender entgegengesetzter Mechanismus für die Wechselwirkung der S auf ihrem Wege zur Oberfläche angenommen werden. Weiterhin sollten nach der Theorie von WOOLDRIDGE die S bei ihrer Entstehung vorzugsweise 20 eV Energie besitzen; es ist bisher nicht möglich gewesen, auf Grund experimenteller Ergebnisse Aussagen über einen solchen bevorzugten Energiebetrag zu machen.

DECKER und VAN DER ZIEL [*378*] haben eine allgemeine Theorie der SE auf quantenmechanischer Grundlage entwickelt, die WOOLDRIDGES Theorie als Spezialfall enthält. VAN DER ZIEL benutzt dabei ein abgeschirmtes COULOMB-Potential

$$U = \frac{e^2}{|\mathfrak{R} - \mathfrak{r}|} \exp(-\lambda |\mathfrak{R} - \mathfrak{r}|).$$

Für den Spezialfall $\lambda = 0$ erhält man das von WOOLDRIDGE verwendete reine COULOMB-Potential.

Es sei $\psi_{\mathfrak{k}}(\mathfrak{r})$ die Energieeigenfunktion des Gitterelektrons, $\psi_{\mathfrak{K}}(\mathfrak{R})$ die des P. Die Störungsrechnung für zeitabhängige Systeme nach DIRAC liefert im einfachsten Fall eines kubischen Gitters als Erhaltungssätze den Energiesatz in der Form

$$\frac{\hbar^2}{2m} \cdot K^2 + E(\mathfrak{k}) = \frac{\hbar^2}{2m} K'^2 + E(\mathfrak{k}')$$

und den Quasi-Impulssatz

$$\mathfrak{K} + \mathfrak{k} + \frac{2\pi}{a} \mathfrak{g} = \mathfrak{K}' + \mathfrak{k}',$$

wie bei WOOLDRIDGE. Dabei ist a die Gitterkonstante, der Vektor $\mathfrak{g}$ entspricht dem Einfluß des Gitters beim Entstehungsprozeß der S, $|\mathfrak{g}| = 0$ bedeutet, daß das Gitter keinen Einfluß auf den Prozeß hat. VAN DER ZIEL zeigt, daß den Erzeugungsprozessen ohne Mitwirkung des Gitters die weitaus größte Wahrscheinlichkeit zukommt, und daß daher die Wechselwirkung mit dem Gitter als Ganzem vernachlässigt werden kann. Hier liegt der wesentliche Unterschied gegenüber der Theorie von WOOLDRIDGE, der annimmt, daß das Gitter bei dem Erzeugungsprozeß der S von ausschlaggebender Bedeutung ist. Da sich aber die Untersuchungen von DECKER und VAN DER ZIEL nur auf Metalle beziehen, könnten die

Tabelle 7. *Zusammenstellung der theoretischen Ansätze (vgl. Text).*

Autor	Grundsätzliche Behandlung	Entstehungsprozess der S			Vorgänge zwischen Entstehungsort und Außenraum		
		Energieverlust der P	Winkelverteilung für S am Entstehungsort	Sonstiges	Arten der Beeinflussung der S	Absorptionsgesetz	Sonstiges
FRÖHLICH, WOOLDRIDGE	Quantenmechanik (BLOCH); Wechselwirkung mit Gitter als Ganzem; reines COULOMB-Potential		Anisotrop, mehr in Richtung Oberfläche, mit wachsender E_p Anisotropie abnehmend	Temperaturabhängigkeit der Zahl der entstehenden S			Temperaturabhängiger Bremsmechanismus, damit Ausbeute temperaturunabhängig wird
DECKER-V. D. ZIEL	Quantenmechanik Wechselwirkung mit Gitterelektronen, abgesch. COULOMB-Potential $U = \frac{e^2}{R}\exp(-\lambda R)$	$-\frac{dE_p}{dR} \sim \frac{1}{E_p}\ln(C E_p)$ für $E_p \gg E\lambda$ $\left(E\lambda = \frac{\hbar^2}{2m}\cdot\lambda^2\right)$		Energie der angeregten Gitterelektronen (Impuls $\mathfrak{k}'$) $E(\mathfrak{k}') \approx \frac{\hbar^2}{2m}\mathfrak{k}'^2$ (Näherung) Wellenfunktion des P unbeeinflußt vom Gitterpotential (Näherung)	Da quantenmechanische Behandlung der Diffusion der S nicht durchgeführt, Behandlungsmöglichkeit wie unter BAROODY		
KADYSCHEWITSCH	Klass. mit Bändermodell. P gegen freies Metallelektron, „da gebundene Elektronen nicht wesentlich beitragen“	Einführung von freien Weglängen für elastische und unelastische Streuung			Wechselwirkung mit Gitter und freien Elektronen, freie Wegelängen für elastische und unelastische Streuung		Berücksichtigung der Austrittsarbeit
BAROODY	Klass. Wechselwirkung mit einzelnem Metallelektron (2 Körper-Pr.) ohne Aussagen über relativen Anteil verschieden stark gebundener Elektronen (SOMMERFELDsches Elektronengas)	$-\frac{dE_p}{dR} \sim \frac{a}{E_p}$ (WHIDDINGTON-Gesetz)	Impuls parallel zur Oberfläche	Nur senkrechter Einfall der P, keine Richtungsänderung der P im Metall	Wechselwirkung mit Gitter: elastische Streuung, keine Energieverluste, Absorption im Elektronenplasma	e-Gesetz ($e^{-x\sigma}$); Einführung von $\bar{l}_{\mathrm{refl}}$ für elastische Streuung, $\bar{l}_{\mathrm{abs}}$ für Absorption: $\sigma = \frac{1}{\bar{l}_{\mathrm{refl}}} + \frac{1}{\bar{l}_{\mathrm{abs}}}$	Berücksichtigung der Austrittsarbeit; isotrope Winkelverteilung an der Oberfläche infolge Diffusion
STERNGLASS	Klass. Wechselwirkung mit gebundenen Elektronen. Zahl der erzeugten S gleich absorbierte Primärenergie dividiert durch mittleren Energ.-Aufwand $\bar{E}_0$ pro S.	$-\frac{dE_p}{dR} = \frac{\mathrm{const}}{\sqrt{E_p}}$ (ben. Theorie: BOHR [406] BETHE [407]) mittlere Energieeindringtiefe $\bar{d}$. Absorbierte Energie $E'_p = E_p(1-\eta\bar{K})$	Da in der Tiefe $\bar{d}$ bereits isotrope Richtungsverteilung der P (BETHE), auch isotrope Verteilung der S	$\bar{E}_0$ = mittlerer Energieaufwand pro S (vgl. Text)	Keine physikalische Aussage über Vorgang nur $\bar{\lambda}_s \sim 1/\sigma_s$ (σ_s vgl. Text). Größe von $\bar{\lambda}_s$ durch Vergleich mit experimenteller Ausbeutekurve	e-Gesetz; $A\,e^{-(\bar{d}/\bar{\lambda}_s)}$ $\bar{\lambda}_s$ = mittlere freie Weglänge der S für unelastische Streuung am Einzelatom. Temperaturabhängigkeit wegen Gitterschwingung	Vernachlässigung der Austrittsarbeit (keine Absorption durch Oberfläche $\Gamma = 1$)
BRUINING, JONKER	Halbempirisch	WHIDDINGTON-Gesetz	Annahme: Isotropie		Absorption	e-Gesetz: $e^{-\alpha x}$ (α relativ klein nach Versuchen von KATZ, vgl. S. 279)	Vernachlässigung der Austrittsarbeit

Tabelle 7. (Fortsetzung.)

Autor	Gültigkeitsbereich	Resultate: Ausbeute	Resultate: Energieverteilung	Resultate: Sonstiges	Vergleich mit Experimenten und Bemerkungen
Fröhlich, Wooldridge		Ausbeutemaximum nur bei Anisotropie der Winkelverteilung			Ausbeutekurve vgl. Text. Energieverteilung vgl. Text.
Decker-v. d. Ziel	$E_p \geqq 100$ eV		$P(k')\,dk' = \frac{8\pi m^2 e^4}{\hbar^4 K^2} \frac{dk'}{\lvert k'\rvert^3}$ $K = \lvert\mathfrak{K}\rvert$ = Impuls der P $\mathfrak{k}'$ = Impuls der S für $E(\mathfrak{k}') \gg E_\lambda$ bzw. $k'^2 \gg \lambda^2$	Wechselwirkung mit Gitterelektronen sehr viel wahrscheinlicher als Wechselwirkung mit Gitter als Ganzem	Energieverteilung in Übereinstimmung mit der klass. abgeleiteten Formel von Baroody, keine Aussagen für kleine E_s
Kadyschewitsch	$E_p \geqq 200$ eV	Ausbeute für bestimmte Materialien (Ag und Ni)	Energieverteilung der S für stimmte Materialien	Richtungsverteilung der S, Abhängigkeit vom Einfallswinkel der P für bestimmte Materialien	Keine *Vorher*sagen möglich, da Werte der Parameter (freie Weglänge usw.) erst durch Vergleich von Theorie und Experiment
Baroody	$E_p \geqq 100$ eV	Allgemeine Form für $\frac{\delta}{\delta_{\max}} = f\left(\frac{E_p}{E_{p\max}}\right)$; Formel für Ausbeutekurve (vgl. Text)	$-dN(E_s) = \frac{P E_m^{\frac{3}{2}}\, dE}{E_p (E_s - E_m)^2}$ (Energieverteilung der S am Entstehungsort)	Richtungsverteilung der S: Gute Übereinstimmung mit dem experimentellen cos-Gesetz	Qual. Übereinstimmung der Energieverteilung mit dem Experiment (vgl. Text) bzw. Ausbeute vgl. Fig. 46
Sternglass	$200 < E_p < 1500$ eV	$\Delta = \frac{E_p(1-\eta\overline{K})}{\overline{E}_0} \Gamma A\, e^{-(\overline{d}/\overline{\lambda}_s)}$	Energieverteilung keine Aussagen		Allgemeine Form der Ausbeutekurve vgl. Fig. 46
Bruining, Jonker		$\delta = k\,a \int_0^{x_{\max}} (E_p^2 - a\,x)^{-\frac{1}{2}} e^{-\alpha x}\, dx$ allg. Form f. $\frac{\delta}{\delta_{\max}} = f\left(\frac{E_p}{E_{p\max}}\right)$	keine Aussagen	Richtungsverteilung der S: cos-Gesetz	Ausbeutekurve vgl. Fig. 46

Wooldridgeschen Prozesse bei der SE der Halbleiter und Isolatoren von Bedeutung sein. Die unter den oben genannten Voraussetzungen von Decker und van der Ziel berechnete Energieverteilung der angeregten Elektronen stimmt für $E(\mathfrak{k}') \gg 3{,}4$ eV mit der von Baroody klassisch abgeleiteten überein. Da eine Behandlung der Diffusion der S zur Oberfläche auf quantenmechanischer Grundlage noch nicht befriedigend durchgeführt ist, könnte man diesen Teil der Rechnung z.B. wie bei Baroody ausführen (vgl. weiter unten).

Nach *klassischen* Methoden haben Kadyschewitz, Baroody und Sternglass gerechnet.

Kadyschewitz [*255*], [*290*], [*334*] führt in seiner Theorie eine größere Anzahl von Parametern ein, z.B. freie Weglängen für elastische und unelastische Streuung der P und der S. Durch Angleichung an das Experiment (geeignete Wahl dieser Parameter) kann er für Ag und Ni recht gute Übereinstimmung der Ausbeute und der Energieverteilung mit der Erfahrung erreichen. Es können aber — und das ist der Nachteil dieser Theorie — keine *Vorher*sagen über Ausbeute- und Energieverteilungskurven von Materialien gemacht werden, für die noch keine Messungen vorliegen. Es sei in diesem Zusammenhang auf die Diskussion der Theorien von Wooldridge und Kadyschewitz durch Massey und Burhop hingewiesen[1].

Baroody hat alle wichtigen experimentellen Ergebnisse der SE wenigstens qualitativ mit Hilfe des Sommerfeldschen Elektronengasmodells abgeleitet. Zur Behandlung des Erzeugungsprozesses der S nimmt Baroody an, daß der P-Strahl senkrecht auf die Metalloberfläche auffällt. Die Wechselwirkung der P mit dem Gitter wird vernachlässigt, die Energie der P soll groß gegenüber der Energie der Metallelektronen sein. Beim Zusammentreffen mit einem Metallelektron gibt das P einen Impulsbetrag Δp ab, dabei soll es seine Geschwindigkeit praktisch nicht ändern. Hierdurch entstehen S mit Impulsen μp_0, die dem Betrage nach größer sind als der maximale Impulsbetrag p_0 der Fermi-Kugel (d.h. $\mu > 1$). Die Zahl dieser S, die pro cm Weg in der Tiefe x (von der Oberfläche aus gerechnet) entstehen, ist nach Baroody

$$N(\mu\, x) = \frac{B \cdot E_m^{\frac{1}{2}}}{E_P(x) \cdot (\mu^2 - 1)}\,; \quad E_m = \frac{p_0^2}{2m}\,, \quad B = 2{,}95 \cdot 10^8\,(\mathrm{eV})^{\frac{1}{2}}\,\mathrm{cm}^{-1}. \qquad (25.1)$$

Aus dieser Formel hat Baroody abgeschätzt, daß für $E_P \approx 100$ eV auf je 10 Å Weg des P etwa ein S erzeugt wird mit genügend großer Energie um den Potentialwall an der Metalloberfläche zu überwinden. Wegen der senkrechten Inzidenz der P können S die Metalloberfläche nur erreichen, wenn sie wenigstens einmal nach ihrer Erzeugung gestreut werden. Abgesehen davon wird ein Teil der S auf dem Weg zur Oberfläche absorbiert. Baroody führt daher zwei mittlere freie Weglängen für Streuung am Gitter und Absorption im Elektroplasma ein $\bar{l}_{\mathrm{refl}}$ und $\bar{l}_{\mathrm{abs}}$, und zeigt — unter Benutzung des Whiddingtonschen Gesetzes $E_P^2(x) = E_P^2(0) - a\,x$ für den Energieverlust der P—, daß man für die Grenzfälle $\bar{l}_{\mathrm{refl}} \gg \bar{l}_{\mathrm{abs}}$ (Einzelstreuung) und $\bar{l}_{\mathrm{refl}} \ll \bar{l}_{\mathrm{abs}}$ (Vielfachstreuung) praktisch die gleiche Abhängigkeit der Ausbeute von der Austrittsarbeit erhält, und daß außerdem die Abhängigkeit der Ausbeute von der Primärenergie bei Vielfachstreuung dieselbe ist, wie bei der Näherung mit Einzelstreuung. Die Ausbeute ist für $\bar{l}_{\mathrm{refl}} \gg \bar{l}_{\mathrm{abs}}$ gegeben durch die Formel

$$\delta_1 = \frac{B\,E_m^{\frac{1}{2}}}{2\,\bar{l}_{\mathrm{refl}}\,\sigma} \int\limits_0^1 \int\limits_0^{E_P^2/a} \frac{q^2 \exp(-\sigma\,x/q)}{(E_P^2 - a\,x)^{\frac{1}{2}}\,(\mu_0^2 - q^2)}\,dq\,dx.$$

[1] H. S. W. Massey and E. H. S. Burhop: Electronic and Ionic Impact Phenomena. Oxford: Clarendon Press 1952. Speziell S. 295ff.; dort findet man auch eine kurze Gesamtdarstellung der SE-Erscheinungen.

Dabei ist:

x der Abstand von der Oberfläche,

$q = \cos\alpha$, α = Winkel zwischen Streurichtung und Einfallslot,

$B = 2{,}95 \cdot 10^8 (\mathrm{eV})^{\frac{1}{2}}\,\mathrm{cm}^{-1}$,

E_m die Energie, die zum maximalen Impuls der FERMI-Kugel gehört,

$\bar{l}_{\mathrm{refl}}$ die mittlere Weglänge für Streuung am Gitter,

a die Konstante im WHIDDINGTONschen Gesetz,

$\mu_0 = 1 + \frac{e\Phi}{E_m}$ ($e\Phi$ = effektive Austrittsarbeit).

Damit ergibt sich für alle Materialien die gleiche normierte Form für $\delta/\delta_{\max} = f(E_P/E_{P_{\max}})$, die in Fig. 46 dargestellt ist. Die BAROODYsche Theorie gibt

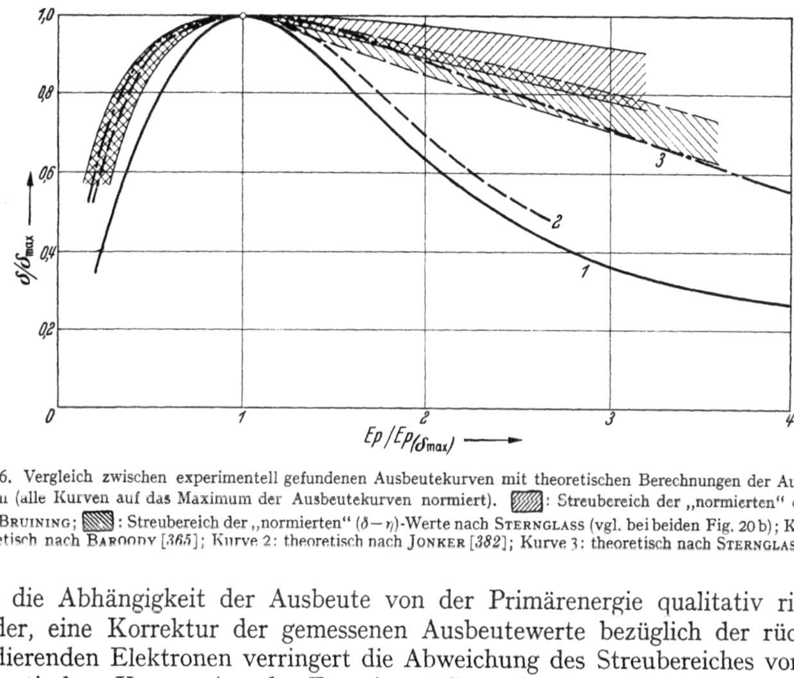

Fig. 46. Vergleich zwischen experimentell gefundenen Ausbeutekurven mit theoretischen Berechnungen der Ausbeutekurven (alle Kurven auf das Maximum der Ausbeutekurven normiert). ▨: Streubereich der „normierten" δ-Werte nach BRUINING; ▧: Streubereich der „normierten" $(\delta-\eta)$-Werte nach STERNGLASS (vgl. bei beiden Fig. 20b); Kurve 1: theoretisch nach BAROODY [365]; Kurve 2: theoretisch nach JONKER [382]; Kurve 3: theoretisch nach STERNGLASS [404].

also die Abhängigkeit der Ausbeute von der Primärenergie qualitativ richtig wieder, eine Korrektur der gemessenen Ausbeutewerte bezüglich der rückdiffundierenden Elektronen verringert die Abweichung des Streubereiches von der theoretischen Kurve. Aus der Energieverteilung der S am Entstehungsort, die man durch Ableitung der Formel (25.1) gewinnt (vgl. Tabelle 7), berechnet BAROODY die Energieverteilung der austretenden S

$$-\frac{dP}{d(E/e\Phi)} = \frac{\mu_1-\mu_0}{(\mu^2-1)^2}\cdot\frac{\mu_0^2-1}{2\mu_1}\Bigg/\int\limits_{\mu_0}^{\infty}\frac{\mu-\mu_0}{(\mu^2-1)^2}\,d\mu\,;\qquad \mu_1^2=\mu_0^2+(\mu_0^2-1)\,\frac{E}{e\Phi}$$

in guter Übereinstimmung mit den Messungen von HAWORTH an Mo[1] (vgl. Fig. 47). Über die Abhängigkeit der Ausbeute vom Einfallswinkel der P macht die Theorie von BAROODY keine Aussagen, da senkrechter Einfall der P zugrunde gelegt wird.

Dagegen berechnet BAROODY die Winkelverteilung der austretenden S und erhält unter der Annahme, daß die S isotrop verteilt an der Metalloberfläche ankommen, die in Fig. 48 dargestellte Funktion. Die experimentell festgestellte cos-Verteilung der emittierten S steht, wie Fig. 48 zeigt, mit dem theoretischen

[1] Die bei HAWORTH auftretenden Nebenmaxima (vgl. Ziff. 7) werden allerdings nicht wiedergegeben.

Ergebnis in guter Übereinstimmung. Trotz der weitgehenden Schematisierung liefert die Theorie von BAROODY eine Menge richtiger Ergebnisse. Bezüglich der Mängel, die in dieser Schematisierung begründet sind, vgl. man die ausführliche Diskussion von HACHENBERG und BRAUER ([*400*], speziell S. 461 ff.).

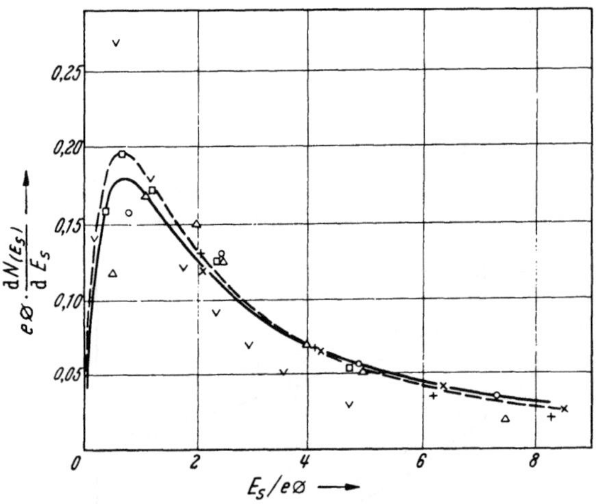

Fig. 47. Energieverteilung der S. (Energie E_s der S bezogen auf die Austrittsarbeit als Einheit) nach BAROODY (*365*), vgl. Text. Theorie von BAROODY: —— berechnet für $\mu_0^2 = 1{,}6$; ------ berechnet für $\mu_0^2 = 2{,}0$. Experimentelle Punkte von RUDBERG [*152*]: + (Au), × (Ag), ○ (Cu); HAWORTH [*128*, *145*]: □ (Mo), △ (Nb); KOLLATH [*293*]: ∨ (Mo).

Die Theorie von STERNGLASS [*404*] soll hier etwas ausführlicher besprochen werden, weil sie in den bisher erschienenen Darstellungen noch nicht behandelt wird und weil sie einige neue Gesichtspunkte enthält. STERNGLASS legt seiner Theorie folgendes physikalische Bild zugrunde:

Die P werden stark gestreut und diffundieren oberhalb einer gewissen Tiefe gleichmäßig nach allen Richtungen (vgl. Fig. 49). Die P verlieren ihre Energie bei der Erzeugung von S und von Strahlung; die Strahlung ihrerseits kann wieder zur Erzeugung von S Veranlassung geben. Dies führt schließlich zur Bildung einer mittleren Zahl von „eigentlichen" S, die proportional der absorbierten Primärenergie ist. Ein Teil aller S verliert seine Energie, bevor er die Oberfläche erreicht; der Rest tritt aus dem Material aus.

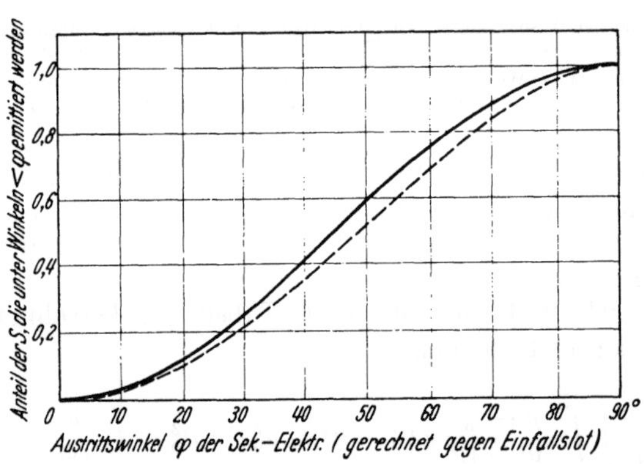

Fig. 48. Winkelverteilung der S nach BAROODY [*365*]. —— experimentelle Verteilung ($\sim \sin^2 \Theta$), — — — berechnet von BAROODY mit $\mu_0^2 = 1{,}6$ (vgl. Text).

Diesem Bild entsprechend führt STERNGLASS folgende Größen ein:

1. *Eine mittlere Eindringtiefe* $\bar{d}$ für die Primärenergie (und damit eine mittlere Entstehungstiefe der S); $\bar{d}$ ist der mittleren freien Weglänge für Streuung proportional. In der Tiefe $\bar{d}$ ist die Richtungsverteilung der P isotrop und damit auch die Richtungsverteilung der S. $\bar{d}$ ist sehr viel kleiner als die früher behandelte maximale Eindringtiefe in Fig. 44.

2. *Eine mittlere Energie* $\overline{E}_0$, die je S vom P-Strahl geliefert werden muß:

$$\overline{E}_0 = \frac{\text{Absorbierte Primärenergie}}{\text{mittlere Anzahl der entstehenden S}}\,.$$

3. *Die Zahl der austretenden „eigentlichen"* S, die je P erzeugt werden.

4. *Die im Material absorbierte Primärenergie* E'_P, die durch die Differenz zwischen ins Metall eintretender P-Energie und aus dem Material durch rückdiffundierende Elektronen weggeführter Energie gegeben ist: $E'_P = E_P(1 - \eta \overline{K})$.

wobei η der Teil der P ist, der rückdiffundiert und $\overline{K}$ die mittlere Energie dieser Elektronen relativ zu E_P.

5. *Eine mittlere freie Weglänge* $\bar{\lambda}_s$ für unelastische Stöße der S an Einzelatomen; $\bar{\lambda}_s$ entspricht einem mittleren Wirkungsquerschnitt pro Atom: $\bar{\lambda}_s = 1/N\bar{\sigma}_s$.

Die Endformel für die Ausbeute hat die Form

$$\Delta = \frac{E'_P}{\overline{E}_0} A\, e^{-\bar{d}/\bar{\lambda}_s},$$ [1]

wobei die e-Funktion der Absorption der S Rechnung trägt. Unter Benutzung der Theorien von BETHE und BOHR berechnet STERNGLASS $\bar{d}$ proportional zu $(E_P)^{\frac{1}{2}}$:

$$\bar{d} = b \cdot E_P^{\frac{1}{2}}.$$

Er setzt dabei voraus, daß die Bindungsenergie $I_{n,l}$ der Schale, die durch die Quantenzahlen n, l charakterisiert wird, viel kleiner ist als E_P. Genauer gesagt, verlangt STERNGLASS $2{,}4 < E_P/I_{n,l} < 47$; nur in diesem Bereich kann man die Formel für den Energieverlust

$$-\frac{dE_P}{dR} \sim \frac{1}{E_P} \ln(CE_P)$$

annähernd durch den Ausdruck

$$-\frac{dE_P}{dR} \sim \frac{1}{\sqrt{E_P}}$$

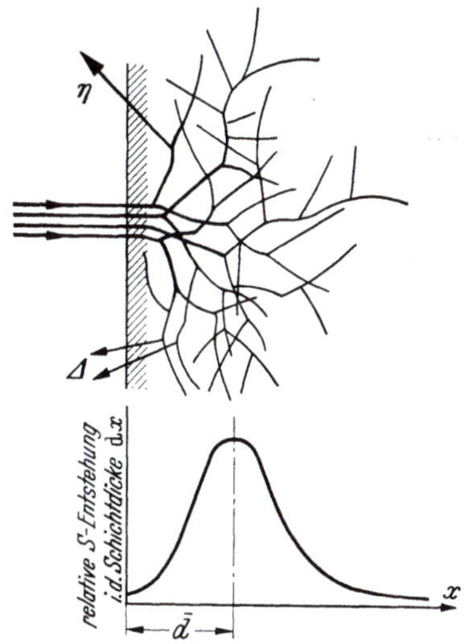

Fig. 49. Schematisches Bild der S-Entstehung nach STERNGLASS [404]. ⟵ bezeichnet rückdiffundierende Elektronen; ← bezeichnet austretende (eigentliche) S; $\bar{d}$ = mittlere Entstehungstiefe der S.

ersetzen. Die Annahme, daß nur Schalen mit $I_{n,l} < E_P$ für die Größe von $\bar{d}$ eine Rolle spielen, ist physikalisch vernünftig. $\bar{d}$ ist ferner abhängig von dem Aufbau der äußeren Schalen, der Fehler bei der Berechnung von b wird mit etwa 20% angegeben.

STERNGLASS erhält dann für die Ausbeute

$$\Delta = \frac{E'_P}{\overline{E}_0} A \cdot e^{-\beta E_P^{\frac{1}{2}}}$$

oder

$$\frac{\Delta}{E_P} = \frac{(1-\eta\overline{K})}{\overline{E}_0} A \cdot e^{-\beta E_P^{\frac{1}{2}}}. \tag{25.2}$$

Trägt man $\ln(\Delta/E_P)$ als Funktion von $\sqrt{E_P}$ auf, so erhält man gerade Linien, deren Neigung β ist und deren Ordinatenabschnitt eine Bestimmung von $\overline{E}_0$ gestattet, da η, $\overline{K}$ aus anderweitigen Messungen (Ziff. 15) bekannt sind. Wie STERNGLASS zeigt, lassen sich die experimentellen Ausbeutekurven in einem weiten Bereich ($200\,\text{eV} < E_p < 1500\,\text{eV}$) bei Auftragung über $\sqrt{E_P}$ als gerade Linien darstellen (vgl. Fig. 50). Die so gefundenen $\overline{E}_0$-Werte liegen zwischen 20 und 35 eV. Die $1/\beta$-Werte zeigen periodische Schwankungen als Funktion der Ordnungszahl (vgl. Fig. 33), die einem allgemeinen Anstieg überlagert sind. $1/\beta$ sollte

[1] Der Faktor A berücksichtigt die Richtungsverteilung der entstehenden S; wegen 1. wird $A \approx 0{,}5$. Mit einem weiteren Faktor Γ berücksichtigt STERNGLASS die Oberflächenbeschaffenheit, setzt aber $\Gamma = 1$ (glatte Oberflächen, Vernachlässigung der Austrittsarbeit).

aus theoretischen Gründen mit $Z^{\frac{1}{3}}$ ansteigen. Auf Grund der qualitativen Übereinstimmung dieser Voraussagen mit den experimentellen Daten schließt STERNGLASS, daß $\bar{\sigma}_s$ in engem Zusammenhang mit dem gaskinetischen Querschnitt steht. $\bar{\sigma}_s$ kann qualitativ durch die Formel $\bar{\sigma}_s = 0{,}1 \cdot (10 - n) \cdot Z^{\frac{1}{3}}$ dargestellt werden (n ist die Hauptquantenzahl).

Es ergibt sich damit folgendes Bild: Das Atom besitzt einen inneren Bereich („core"), dessen (abgeschlossene) Elektronenschalen keine Wechselwirkung mit den P ergeben, der Rest des Atoms (äußere Schalen) trägt allein zur Entstehung der S durch Wechselwirkung mit den P bei. Für die Absorption der S ist der gaskinetische Querschnitt maßgebend.

Verlängert man die Verbindungslinien zusammengehörender $1/\beta$-Werte wie in Fig. 33, so schneiden die Verlängerungen die Abszissenachse in den mit Pfeilen (bzw. He, Ne, Ar, ...) bezeichneten Punkten: Die zu diesen Punkten gehörenden Schalen (Edelgaskonfiguration) tragen zur SE nicht mehr bei. Daraus erhält man eine Abschätzung der Elektronenzahl in den „aktiven" Schalen.

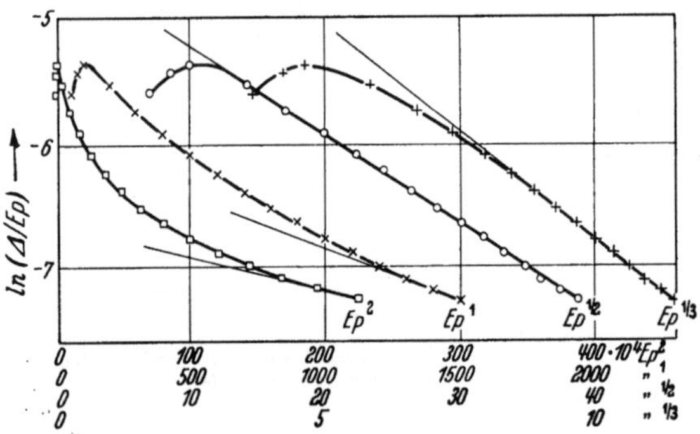

Fig. 50. Prüfung des Exponentialgesetzes für den Abfall der Ausbeutekurve von Ag durch STERNGLASS [404] nach Messungen von WARNECKE [158].

Ferner ergibt sich aus Gl. (25.2) eine gemeinsame „normierte" Ausbeutekurve für alle Materialien

$$\frac{\Delta}{\Delta_{\max}} = 2{,}72^2 \, \frac{E_P}{E_{P_{\max}}} \times$$

$$\times \exp\left[-2\left(\frac{E_P}{E_{P_{\max}}}\right)^{\frac{1}{2}}\right],$$

die in Fig. 46 strichpunktiert eingetragen ist.

BRUINING [2] geht von dem Gedanken aus, daß die Zahl der in Tiefe x erzeugten S direkt proportional ist dem Energieverlust der P (Proportionalitätsfaktor K). Auf ihrem Wege zur Oberfläche werden die S absorbiert [Absorptionsgesetz von der Form $\exp(-\alpha x)$]. Aus dem WHIDDINGTONschen Gesetz erhält BRUINING eine maximale Eindringtiefe der P: $x_{\max} = E_P^2/a$. Damit ergibt sich der S-Strom durch Integration von $x = 0$ bis $x = x_{\max}$:

$$i_s = \tfrac{1}{2} K a i_P \int_0^{E_P^2/a} (E_P^2 - a x)^{-\frac{1}{2}} \cdot e^{-\alpha x} \cdot dx. \tag{25.3}$$

Dabei ist a die WHIDDINGTONsche Konstante, α der Absorptionskoeffizient der S. Die Funktion (25.3) hat ein Maximum für $r = E_P \sqrt{\alpha/a} = 0{,}92$. Die reduzierte Ausbeute $\delta/\delta_{\max} = i_s/i_{s_{\max}}$ ist nur noch eine Funktion von E_P, d.h. es ergibt sich für alle Materialien ein und dieselbe reduzierte Ausbeutekurve. Eine Berücksichtigung der isotropen Winkelverteilung der S am Entstehungsort (JONKER [382]) ändert lediglich die Zahlenfaktoren der reduzierten Ausbeute. Die Werte von a und α müssen anderweitigen Messungen entnommen werden (vgl. hierzu Ziff. 23). Die reduzierte Ausbeutekurve nach JONKER ist in Fig. 46 eingetragen.

Bemerkung: Nach Abschluß der Korrektur erschien in den „Physikalischen Abhandlungen aus der Sowjetunion", Folge 1, S. 7 (1956) eine Arbeit von A. J. WJATSKIN zur Theorie der Sekundäremission, in der erstmalig die Nebenmaxima in der Energieverteilung der S aus Molybdän in Übereinstimmung mit den Messungen von HAWORTH (vgl. S. 245) berechnet werden.

Literatur.

Zusammenfassungen.

[1] BRUINING, H.: Die Sekundärelektronenemission fester Körper. Berlin: Springer 1942.
[2] BRUINING, H.: Physics and applications of secondary electron emission. London: Pergamon Press Ltd 1954.
[3] MCKAY, K. G.: Secondary electron emission. Adv. Electronics **1**, 66 (1948).
[4] KOLLATH, R.: Sekundärelektronenemission fester Körper. Phys. Z. **38**, 202 (1937).
[5] KOLLER, L. R.: Secondary emission. I. Gen. Electr. Rev. **51**, 33 (1948). – II. Gen. Electr. Rev. **51**, 50 (1948).
[6] LENARD, P.: Quantitatives über Kathodenstrahlen. Heidelberg 1918 u. 1925.
[7] LENARD, P., u. A. BECKER: Handbuch der Experimentalphysik, Bd. 14, Kathodenstrahlen.
[8] WARNECKE, R.: Lois principales de l'émission d'électrons secondaires par la surface des métaux. Onde électr. **16**, 509 (1937).

Ältere Literatur (bis 1920).

[9] AUSTIN, L., u. H. STARKE: Über die Reflexion der Kathodenstrahlen und eine damit verbundene neue Erscheinung sekundärer Emission. Ann. Phys., Lpz. **9**, 271 (1902).
[10] AUSTIN, L., u. H. STARKE: Über die Reflexion der Kathodenstrahlen und eine damit verbundene neue Erscheinung sekundärer Emission. Verh. dtsch. phys. Ges. **4**, 106 (1902).
[11] LENARD, P.: Über die Beobachtung langsamer Kathodenstrahlen mit Hilfe der Phosphoreszenz und Sekundärentstehung von Kathodenstrahlen. Ann. Phys., Lpz. **12**, 449 (1903).
[12] LENARD, P.: Über sekundäre Kathodenstrahlung in gasförmigen und festen Körpern. Ann. Phys., Lpz. **15**, 485 (1904).
[13] FÜCHTBAUER, CH.: Über die Geschwindigkeit der von Kanalstrahlen und von Kathodenstrahlen beim Auftreffen auf Metalle erzeugten negativen Strahlen. Phys. Z. **7**, 748 (1906).
[14] FÜCHTBAUER, CH.: Über die Geschwindigkeit der von Kanalstrahlen und Kathodenstrahlen beim Auftreffen auf Metalle erzeugten negativen Strahlen. Verh. dtsch. phys. Ges. **8**, 394 (1906).
[15] FÜCHTBAUER, CH.: Über Sekundärstrahlen. Ann. Phys., Lpz. **23**, 301 (1907).
[16] LAUB, J.: Über sekundäre Kathodenstrahlen. Ann. Phys., Lpz. **23**, 285 (1907).
[17] BAEYER, O. v.: Über langsame Kathodenstrahlen. Verh. dtsch. phys. Ges. **10**, 96 (1908).
[18] BAEYER, O. v.: Über langsame Kathodenstrahlen. Phys. Z. **10**, 168 (1909).
[19] BAEYER, O. v.: Reflexion und Sekundärstrahlung langsamer Kathodenstrahlen. Phys. Z. **10**, 176 (1909).
[20] GEHRTS, A.: Reflexion und Sekundärstrahlen lichtelektrisch ausgelöster Kathodenstrahlen. Ann. Phys., Lpz. **36**, 995 (1911).
[21] CAMPBELL, N. R.: Delta rays produced by beta rays. Phil. Mag. **24**, 783 (1912).
[22] WHIDDINGTON, R.: Proc. Roy. Soc. Lond. **86**, 360 (1912).
[23] CAMPBELL, N. R.: Ionization by charged particles. Phil. Mag. **25**, 803 (1913).
[24] CAMPBELL, N. R.: The ionization of platinum by cathode rays. Phil. Mag. **28**, 286 (1914).
[25] RAMSAUER, C.: Ann. Phys., Lpz. **45**, 961 (1914).
[26] CAMPBELL, N. R.: The ionization of metals by cathode rays. Phil. Mag. **29**, 369 (1915).
[27] HULL, A. W.: The reflection of slow moving electrons by copper. Phys. Rev. **7**, 1 (1916).
[28] DADOURIAN, H. M.: Reflection of electrons from metal surfaces. Phys. Rev. **14**, 434 (1919).
[29] SLEPIAN: USA-Patent Nr. 1 450 265 vom 3. 4. 1923 (eingeg. 1919).

1921

[30] BARBER, J. G.: Secondary electron emission from copper surfaces. Phys. Rev. **17**, 322.
[31] MILLIKAN, R. A., and J. G. BARBER: On the reflection and re-emission of electrons from metal surfaces and a method of measuring the ionizing potential of such surfaces. Proc. Nat. Acad. Sci., Wash. **7**, 13.
[32] RAMSAUER, C.: Über den Wirkungsquerschnitt der Gasmoleküle gegenüber langsamen Elektronen. Ann. Phys., Lpz. **64**, 513.
[33] TATE, J. T.: The effect of angle of incidence on the reflection and secondary emission of slow moving electrons from platinum. Phys. Rev. **17**, 394.

1922

[34] BALTRUSCHAT, M., u. H. STARKE: Über sekundäre Kathodenstrahlung. Phys. Z. **23**, 403.
[35] BUSCH, H.: Eine neue Methode zur e/m-Bestimmung. Phys. Z. **23**, 438.
[36] DAVISSON, C., and C. H. KUNSMAN: The secondary electron emission from nickel. Phys. Rev. **20**, 110.
[37] FARNSWORTH, H. E.: Bombardment of metal surfaces by slow moving electrons. Phys. Rev. **20**, 358.
[38] MCALLISTER, L. E.: The effect of ageing on the secondary electron emission from copper surfaces. Phys. Rev. **19**, 246.
[39] TANK, F.: Zur Kenntnis der Vorgänge in Elektronenröhren. Jb. drahtl. Telegr. **20**, 82.

1923

[40] MCALLISTER, L. E.: Secondary electron emission from copper and copper oxide surfaces. Phys. Rev. **21**, 122.
[41] LANGMUIR, I.: A new photoelectric effect; reflection of electrons induced by light. Science, Lancaster, Pa. **58**, 398.
[42] SCHONLAND, B. F. I.: The passage of cathode rays through matter. Proc. Roy. Soc. Lond., Ser. A **104**, 235.
[43] TERRILL, H. M.: Loss of velocity of cathode rays in matter. Phys. Rev. **22**, 101.
[44] VAN DER POL jr.: Over electronenbewegingen in trioden. Physica, Haag **3**, 108.

1924

[45] STUHLMANN, O.: The minimum velocity of impact to produce secondary electron emission from tungsten. Phys. Rev. **23**, 296.

1925

[46] BECKER, A.: Über die Geschwindigkeit der sekundären Kathodenstrahlung. Ann. Phys., Lpz. **78**, 228.
[47] BECKER, A.: Über die Rückdiffision, Reflexion und Sekundärstrahlerregung langsamer Kathodenstrahlen. Ann. Phys. Lpz. **78**, 253.
[48] FARNSWORTH, H. E.: Electronic bombardment of metal surfaces. Phys. Rev. **25**, 41.
[49] HULL, A. W., and N. H. WILLIAMS: Determination of elementary charge from measurements of shot-effect. Phys. Rev. **25**, 147.
[50] LANGE, H.: Über die Sekundärstrahlung in Elektronenröhren. Jb. drahtl. Telegr. **26**, 38.
[51] PETRY, R. L.: Critical potentials in secondary electron emission from iron, nickel and molybdenum. Phys. Rev. **26**, 346
[52] ROSE, E. L.: Soft X-rays and secondary electrons. Phys. Rev. **25**, 883.
[53] STUHLMANN, O.: The excitation of the M-series in iron by electronic impact. Phys. Rev. **25**, 234.
[54] TINGWALDT, C.: Über den Einfluß der Entgasung einer von Kathodenstrahlen getroffenen Metallplatte auf die ausgelöste Elektronenstrahlung. Z. Physik **34**, 280.
[55] VAN DER POL jr.: Über Elektronenbahnen in Trioden. Jb. drahtl. Telegr. **25**, 121.

1926

[56] FARNSWORTH, H. E.: Secondary electrons from iron; critical potentials. Phys. Rev. **27**, 243.
[57] FARNSWORTH, H. E.: Secondary electrons from iron. Phys. Rev. **27**, 413.
[58] PETRY, R. L.: Secondary electron emission from tungsten, copper and gold. Phys. Rev. **28**, 362.

1927

[59] BRINSMADE, J. B.: Reflected and secondary electrons from an aluminium target Phys. Rev. **30**, 494.
[60] BROWN, D., and R. WHIDDINGTON: Electron „reflection" in vacuo. Nature, Lond. **119**, 427.
[61] BROWN, D., and R. WHIDDINGTON: Electron „reflection" in vacuo. Proc. Leeds Phil. Lit. Soc. **1**, 162.
[62] FARNSWORTH, H. E.: J. Opt. Soc. Amer. **15**, 290.
[63] KREFFT, H. E.: O and N energy levels in the secondary emission of hot tungsten. Phys. Rev. **29**, 908.
[64] KREFFT, H. E.: Critical primary velocities in the secondary electron emission of tungsten. J. Franklin Inst. **204**, 537.
[65] KREFFT, H. E.: Kritische Geschwindigkeiten für die durch Elektronen an Wolfram ausgelöste sekundäre Kathodenstrahlung. Ann. Phys., Lpz. **84**, 639.

[66] SHARMAN, C. F.: The applications of the method of the magnetic spectrum to the study of secondary electron emission. Proc. Cambridge Phil. Soc. **23**, 523.
[67] SHARMAN, C. F.: A differential retarding potential method for the study of the energy distribution of slow electron emission. Proc. Cambridge Phil. Soc. **23**, 922.

1928

[68] FARNSWORTH, H. E.: Energy distribution of secondary electrons from copper, iron, nickel and silver. Phys. Rev. **31**, 405.
[69] FARNSWORTH, H. E.: Large angle scattering of low velocity electrons from copper, iron, nickel and silver. Phys. Rev. **31**, 414.
[70] FARNSWORTH, H. E.: Secondary electron current as a function of crystal structure. Phys. Rev. **31**, 419.
[71] HYATT, J. M.: Secondary electron emission produced by positive caesium ions. Phys. Rev. **32**, 922.
[72] HYATT, J. M., and H. A. SMITH: Secondary electron emission from molybdenum. Phys. Rev. **32**, 929.
[73] JOFFÉ, A.: Über den Brechungsquotienten der DE BROGLIE-Wellen des Elektrons. Z. Physik **48**, 737.
[74] KLEMPERER, O.: Über den Brechungsquotienten der DE BROGLIE-Wellen des Elektrons. Z. Physik **47**, 417.
[75] KREFFT, H. E.: Critical primary velocities in the secondary electron emission of tungsten. Phys. Rev. **31**, 199.
[76] RICHARDSON, O. W.: The emission of secondary electrons and the excitation of soft X-rays. Proc. Roy. Soc. Lond., Ser. A **119**, 531.
[77] STEHBERGER, K. H.: Über Rückdiffusion und Sekundärstrahlung mittelschneller Kathodenstrahlen an Metallen. Ann. Phys., Lpz. **86**, 825.
[78] SUHRMANN, R.: DRP Nr. 656525.
[79] WELLS, D. A.: Energy distribution among secondary electrons from nickel, aluminium and copper. Phil. Mag. **5**, 367.

1929

[80] BECKER, A.: Über den Durchgang langsamer Kathodenstrahlen durch Metalle. II. Ann. Phys. **2**, 249.
[81] DAENE, H., u. G. SCHMERWITZ: Prüfung der theoretischen Erklärungen der Sekundärelektronenemission von Isolatoren und damit zusammenhängender Erscheinungen. Z. Physik **53**, 404.
[82] FARNSWORTH, H. E.: Electron emission and diffraction by a copper crystal. Phys. Rev. **34**, 679.
[83] RAMSAUER, C., u. R. KOLLATH: Über den Wirkungsquerschnitt der Edelgasmoleküle gegenüber Elektronen unterhalb 1 Volt. Ann. Phys., Lpz. **3**, 536.
[84] SIXTUS, K.: Untersuchungen über Sekundäremission. Ann. Phys., Lpz. **3**, 1017.

1930

[85] BETHE, H. A.: Zur Theorie des Durchgangs schneller Korpuskularstrahlen durch Materie. Ann. Phys., Lpz. **5**, 325.
[86] COPELAND, P. L.: Secondary electrons from contaminated metal surfaces. Phys. Rev. **35**, 293.
[87] COPELAND, P. L.: Secondary electrons from contaminated surfaces. Phys. Rev. **35**, 982.
[88] OBOLENSKY, F.: Über die Wirkung der Sekundärelektronen auf den statischen Arbeitszustand der Eingitterröhren. Arch. Elektrotechn. **52**, 834.
[89] RAO, S. R.: Total secondary electron emission from polycrystalline nickel. Proc. Roy. Soc. Lond., Ser. A **128**, 41.
[90] RAO, S. R.: Total secondary electron emission from a single crystal face of nickel. Proc. Roy. Soc. Lond., Ser. A **128**, 57.
[91] RICHARDSON, O. W.: The emission of secondary electrons and the excitation of soft X-rays. Proc. Roy. Soc. Lond., Ser. A **128**, 63.
[92] RUDBERG, E.: Characteristic energy losses of electrons scattered from incandescent solids. Proc. Roy. Soc. Lond., Ser. A **127**, 111.
[93] SOLLER, T.: The velocity distribution of secondary electrons from molybdenum. Phys. Rev. **36**, 1212.
[94] WAGNER, P. B.: Secondary electrons of high velocity from metals bombarded with cathode rays. Phys. Rev. **35**, 98.

1931

[95] AHEARN, A. J.: The emission of secondary electrons from tungsten. Phys. Rev. **38**, 1858.
[96] COPELAND, P. L.: Thesis, University of Iowa.
[97] HAWORTH, L. J.: Secondary electrons from molybdenum. Phys. Rev. **37**, 93.
[98] KALASCHNIKOFF, S.: Über den Aufladungsprozeß an der Oberfläche der Dielektrika bei der Beschießung mit langsamen Elektronen und positiven Teilchen. Z. Physik **69**, 380.

1932

[99] CHYLINSKI, S.: Magnetic spectra of secondary electrons from silver. Phys. Rev. **42**, 393.
[100] COPELAND, P. L.: Total secondary emission of electrons from metals as a function of primary energy. Phys. Rev. **40**, 122.
[101] FRÖHLICH, H.: Theorie der Sekundärelektronenemission von Metallen. Ann. Phys., Lpz. **13**, 229.
[102] HAWORTH, L. J.: Energy distribution of secondary electrons from molybdenum. Phys. Rev. **42**, 906.
[103] HILSCH, R.: Die Reflexion langsamer Elektronen an Ionenkristallschichten zum Nachweis optischer Energiestufen. Göttinger Nachr. **2**, 203.
[104] HILSCH, R.: Der Elektronenstoß an Kristallschichten zum Nachweis optischer Energiestufen. Z. Physik **77**, 427.
[105] TARTAKOWSKY, P., u. W. KUDRJAWZEWA: Sekundäre Elektronenemission von Nickel und Ferromagnetismus. Z. Physik **75**, 137.

1933

[106] COPELAND, P. L.: The variation of secondary emission with heat treatment. J. Franklin Inst. **215**, 435.
[107] COPELAND, P. L.: Secondary emission of electrons from molybdenum. J. Franklin Inst. **215**, 593.
[108] COPELAND, P. L.: Surface condition and stability of characteristic in screengrid tubes. J. Franklin Inst. **216**, 417.
[109] DE LUSSANET DE LA SABLONIÈRE, C. J.: Die Sekundärelektronenemission in Elektronenröhren, namentlich Schirmgitterröhren. Hochfrequenztechn. **41**, 195.
[110] GÜNTHERSCHULZE, A., u. H. FRICKE: Eine neue Art von Glimmentladung ohne HITTORFFschen Dunkelraum und ohne Kathodenfall. Z. Physik **86**, 451.
[111] GÜNTHERSCHULZE, A.: Die Elektronengeschwindigkeit in Isolatoren bei hohen Feldstärken und ihre Beziehung zur Theorie des elektrischen Durchschlags. Z. Physik **86**, 778.
[112] HAYAKAWA, K.: Studies on the transformation of metals by secondary electron emission. Sci. Rep. Tôhoku Univ. (1) **22**, 934.
[113] RAO, S. R.: Total efficiencies of soft X-ray excitation and secondary electron emission from metal surfaces. Current. Sci., **2**, 93.
[114] RAO, S. R.: The efficiency of secondary electron emission. Proc. Roy. Soc. Lond., Ser. A **139**, 436.

1934

[115] COPELAND, P. L.: Secondary emission of electrons from complex targets. Bull. Amer Phys. Soc. **9**, 31.
[116] COPELAND, P. L.: Secondary emission of electrons from complex targets. Phys. Rev. **45**, 763.
[117] COPELAND, P. L.: Correlation between variation of secondary electron emission and atomic number. Phys. Rev. **46**, 167.
[118] FARNSWORTH, P. T.: Television by electron image scanning. J. Franklin Inst. **218**, 411.
[119] HUGHES, A. L., and V. ROJANSKY: On the analysis of electronic velocities by electrostatic means. Phys. Rev. **34**, 284.
[120] MOULLIN, E. B.: Measurements of shotvoltage used to deduce the magnitude of secondary thermionic emission. Proc. Roy. Soc. Lond., Ser. A **147**, 100.
[121] RUDBERG, E.: Inelastic scattering of electrons from metals. Phys. Rev. **45**, 764.
[122] SCHWARZENBACH, H.: Bestimmung der Primärelektronenströme in Trioden durch Energiemessungen. Helv. phys. Acta **7**, 108.
[123] TURNBULL, J. C., and P. L. COPELAND: Secondary emission of electrons from gold and aluminium. Phys. Rev. **45**, 763.
[124] WARNECKE, R.: Recherches expérimentales sur l'émission secondaire du tantale. J. Phys. Radium **5**, 267.

1935

[*125*] Aldous, W. A., and N. R. Campbell: The effect of secondary emission upon the fluctuations of the current in a triode (Shot-effect). Proc. Roy. Soc. Lond., Ser. A **151**, 694.

[*126*] Bruining, H.: Secundaire electronenemissie. Handel. 25e. Ned. natuuren geneesk. Congr., S. 119.

[*127*] Copeland, P. L.: Secondary emission of electrons of complex targets. Phys. Rev. **48**, 96.

[*128*] Haworth, L. J.: The energy distribution of secondary electrons from molybdenum. Phys. Rev. **48**, 88.

[*129*] Hayner, L. J.: Shot effects of secondary electron currents. Physics **6**, 323.

[*130*] Iams, H., and B. Salzberg: The secondary emission phototube. Proc. Inst. Radio Engrs., N. Y. **23**, 55.

[*131*] Knoll, M.: Aufladepotential und Sekundäremission elektronenbestrahlter Körper. Phys. Z. **36**, 861.

[*132*] Knoll, M.: Aufladepotential und Sekundäremission elektronenbestrahlter Körper. Z. techn. Phys. **16**, 467.

[*133*] Langenwalter, H. W.: Über Rückdiffusion und Sekundärstrahlerregung langsamer Kathodenstrahlen an dünnen Metallschichten. Ann. Phys., Lpz. **24**, 273.

[*134*] Penning, F. M., u. A. A. Kruithof: Verstärkung von Photoströmen durch Emission von Sekundärelektronen. Physica, Haag **2**, 793.

[*135*] Ziegler, M.: Shot effect of secondary emission. Physica, Haag **2**, 415.

1936

[*136*] Afanasjèwa, A., P. Timofeew u. A. Ignaton: Sekundäre Elektronenemission dünner auf Glas niedergeschlagener Metallfilme. Techn. Physics USSR. **3**, 1011.

[*137*] Afanasjèwa, A., P. Timofeew u. A. Ignaton: Die sekundäre Elektronenemission von oxydierten Silber- und Molybdänoberflächen. Phys. Z. Sowjet **10**, 831.

[*138*] Bruining, H.: The depth at which secondary electrons are liberated. Physica, Haag **3**, 1046.

[*139*] Chlebnikow u. Nalimow: Die Sekundärelektronenemission. Fortschr. phys. Wiss. **16**, 467.

[*140*] Coeterier, F., u. M. C. Teves: An apparatus for the transformation of light of long wave length into light of short wave length. III. Amplification by secondary emission. Physica, Haag **4**, 33.

[*141*] Dobroljubski, A.: Über die Wechselbeziehung zwischen der Sekundärelektronenemission mit der Fotoempfindlichkeit und dem Thermioneneffekt. Z. Physik **102**, 626.

[*142*] Dobroljubski, A.: On the secondary electron emission from composite surfaces. J. techn. Phys. USSR. **6**, 1489.

[*143*] Dobroljubski, A.: On the correlation of the secondary emission of electrodes, possessing photosensibility and the thermo-effect of ions. Phys. Z. Sowjet. **10**, 242.

[*144*] Farnsworth, H. E.: Penetration of Low Speed Diffracted Electrons. Phys. Rev. **49**, 605.

[*145*] Haworth, L. J.: The energy distribution of secondary electrons from columbium. Phys. Rev. **50**, 216.

[*146*] Knoll, M.: Änderung der sekundären Elektronenemission von Isolatoren und Halbleitern durch Elektronenbestrahlung. Naturwiss. **24**, 345.

[*147*] Kollath, R.: Zur Messung von Geschwindigkeits- und Energieverteilungen. Ann. Phys., Lpz. **27**, 721.

[*148*] Kwarzchawa, I. F.: Sekundäremission und Ermüdungserscheinungen bei lichtempfindlichen Sauerstoff-Caesium-Elektroden. Phys. Z. Sowjet. **10**, 809.

[*149*] Lukjanov, S., u. W. N. Bernatovitch: J. exp. theoret. Phys. USSR. **7**, 856.

[*150*] Malter, L.: Anomalous secondary electron emission, a new phenomenon. Phys. Rev. **49**, 478.

[*151*] Malter, L.: Thin film field emission. Phys. Rev. **50**, 48.

[*152*] Rudberg, E.: Inelastic scattering of electrons from solids. Phys. Rev. **50**, 138.

[*153*] Rudberg, E., and J. C. Slater: Theory of inelastic scattering of electrons from solids. Phys. Rev. **50**, 150.

[*154*] Strübig, H.: Das Potential eines im Hochvakuum isolierten Auffangschirmes bei Beschießung mit Elektronen. Phys. Z. **37**, 402.

[*155*] Timofeew, P. W., u. A. I. Pjatnitzki: Die sekundäre Elektronenemission einer Sauerstoff-Caesium-Elektrode. Phys. Z. Sowjet. **10**, 518.

[*156*] Treloar, L. R. G.: A method of measuring secondary emission of filaments. Proc. Phys. Soc. Lond. **48**, 488.

[157] TRELOAR, L. R. G.: Relation between secondary emission and workfunction. Nature, Lond. **137**, 579.

[158] WARNECKE, R.: Emission sécondaire des métaux purs. J. Phys. Radium **7**, 270.

[159] WEISS, G.: Über Sekundärelektronen-Vervielfacher. Fernsehen **7**, 41.

[160] WEISS, G.: Über Sekundärelektronen-Vervielfacher. Z. techn. Phys. **17**, 623.

[161] ZIEGLER, M.: Shot effect of secondary emission. I. Physica, Haag **3**, 1.

[162] ZIEGLER, M.: Shot effect of secondary emission. II. Physica, Haag **3**, 307.

[163] ZWORYKIN, V. K.: Elektronenoptische Systeme und ihre Anwendung. Z. techn. Phys. **17**, 170.

[164] ZWORYKIN, V. K., G. A. MORTON and L. MALTER: The secondary emission multiplier, a new electronic device. Proc. Inst. Radio Engrs., N. Y. **24**, 351.

1937

[165] AFANASJÈWA, A., u. P. W. TIMOFEEW: Sekundärelektronenemission von mit dünnen Alkalimetallschichten bedecktem Gold, Silber, Platin. Techn. Physics USSR. **4**, 953.

[166] BHAWALKAR, P. R.: An explanation of the maximum in secondary electron-emission of metals. Proc. Indian Sci. **6**, 74.

[167] BRUINING, H., J. H. DE BOER u. W. G. BURGERS: Secondary electron emission of soot in valves with oxidecathode. Physica, Haag **4**, 267.

[168] BRUINING, H., J. H. DE BOER u. W. G. BURGERS: Secondary electron emission of metals with a low work-function. Physica, Haag **4**, 473.

[169] COPELAND, P. L.: Secondary emission of electrons from sodium films on tantal. Bull. Amer. Phys. Soc. **12**, 12.

[170] DOBROLJUBSKI, A.: Einige Daten zur Frage des Verhältnisses von sekundärer Elektronenemission und Fotoempfindlichkeit. Phys. Z. Sowjet. **11**, 118.

[171] *General Electric Company:* Secondary emission photocell. J. Sci. Instrum. **14**, 250.

[172] HAGEN, A., u. A. BEY: Aufladepotential elektronenbestrahlter Körper. Z. Physik. **104**, 681.

[173] HAYNER, L. J., and B. KURRELMEYER: Shot effect of secondary electrons from nickel and beryllium. Phys. Rev. **52**, 952.

[174] KATZ, H.: Durchgang langsamer Elektronen durch Metallfolien. Wiss. techn. Ber. Osram **17**, 75.

[175] KATZ, H.: Durchgang langsamer Elektronen durch Metallfolien. Z. techn. Phys. **18**, 555.

[176] KOLLATH, R., u. B. MROWKA: Elektronenemission. Jb. AEG-Forsch. **5**, 53.

[177] KOLLER, L. R., and R. P. JOHNSON: Visual observation of the Malter effect. Phys. Rev. **52**, 519.

[178] KRENZIEN, O.: Auslösung von Sekundärelektronen in Adsorptionsschichten. Z. techn. Phys. **18**, 568.

[179] KWARZCHAWA, I. F.: Secondary emission and fatigue photoelectric surfaces. J. exp. theoret. Phys. USSR. **7**, 68.

[180] MAHL, H.: Feldemission der geschichteten Kathoden bei Elektronenbestrahlung. Z. techn. Phys. **18**, 559.

[181] MÜLLER, H. O.: Die Abhängigkeit der Sekundärelektronenemission einiger Metalle vom Einfallswinkel des primären Kathodenstrahls. Z. Physik **104**, 475.

[182] MYERS, D. M.: The division of the primary electron current between grid and anode of triode. Proc. Phys. Soc. Lond. **49**, 264.

[183] NOTTINGHAM, W. B.: Electrical and luminescent properties of willemite under electron bombardment. J. Appl. Phys. **8**, 762.

[184] PIORE, E. R.: Thin film field emission. Phys. Rev. **51**, 1111.

[185] SCHWITZKE, W.: Das Verhalten der Sekundärelektronenemission von Metallen in der Umgebung des Schmelzpunktes. Diss. Berlin

[186] TIMOFEEW, P. W., u. A. PJATNITZKI: Die Sekundärelektronenemission von Rb und K-Komplex-Kathoden. Techn. Physics USSR. **4**, 945.

[187] TRELOAR, L. R. G.: Secondary electron emission from complex surfaces. Proc. Phys. Soc. Lond. **49**, 392.

1938

[188] BAY, Z.: The electron multiplier as an electron counting device. Nature, Lond. **141**, 284, 1011.

[189] BRUINING, H., u. J. H. DE BOER: Secondary electron emission. Part I. Secondary electron emission of metals. Physica, Haag **5**, 17.

[190] BRUINING, H.: Secondary electron emission. Part II. Absorption of secondary electrons. Physica, Haag **5**, 901.

[191] Bruining, H.: Secondary electron emission. Part III. Secondary electron emission caused by bombardment with slow primary electrons. Physica, Haag **5**, 913.
[192] Bruining, H.: Philips techn. Rev. **3**, 80.
[193] Bruining, H.: Thesis, Leiden.
[194] Copeland, P. L.: Secondary emission of electrons from sodium on tantal. Phys. Rev. **53**, 328.
[195] Dobroljubski, A.: Photosensitivity and secondary electron emission from the oxygen caesium layers at diffusion. J. techn. Phys. USSR. **8**, 266. Ref. Phys. Ber. **20**, 206 (1939).
[196] Engbert, W.: Das Rauschen der Sekundäremission. Telefunkenröhre **13**, 127.
[197] Katz, H.: Elektronenoptische Versuche zum Durchgang langsamer Elektronen. Ann. Phys., Lpz. **33**, 160, 169.
[198] Khlebnikow, N., u. Korshunova: Secondary emission of composite surfaces. Techn. Physics USSR. **5**, 363.
[199] Khlebnikow, N., u. Korshunova: The influence of gases on the secondary emission of certain metals. Techn. Physics USSR. **5**, 593.
[200] Kollath, R.: Die Sekundärelektronenausbeute an Aufdampfschichten von Beryllium. Fernsehen **1**, 4.
[201] Kollath, R.: Der Einfluß der geometrischen Anordnung der Atome auf die Sekundärelektronenemission. Naturwiss. **4**, 60.
[202] Kollath, R.: Über die Sekundärelektronenemission des Berylliums. Ann. Phys., Lpz. **33**, 285.
[203] Kollath, R.: Einige Versuche zur Sekundärelektronenemission. Phys. Z. **39**, 916.
[204] Lukjanov, S.: On the dependence of the coefficient of the secondary emission on the angle of incidence of the primary electrons. Phys. Z. Sowjet. **13**, 123.
[205] Mahl, H.: Feldemission aus geschichteten Kathoden bei Elektronenbestrahlung. Z. techn. Phys. **19**, 313.
[206] Morgulis, N., u. A. Nagorsky: A conference on the photoelectric effect and secondary emission. Techn. Physics USSR. **5**, 244.
[207] Morgulis, N., u. A. Nagorsky: Secondary electron emission from oxide-coated cathodes. Techn. Physics USSR. **5**, 848.
[208] Mühlenpfordt, J.: Über die Feldelektronenemission an dünnen Isolatorschichten. Z. Physik **108**, 698.
[209] Nelson, H.: Phenomenon of secondary emission. Phys. Rev. **55**, 985.
[210] Nelson, H.: Method of measuring luminescent screen potential J. Appl. Phys. **9**, 592.
[211] Pjatnitzki, A.: Distribution of the energy of secondary electrons emitted by a composite caesium cathode. J. techn. Phys. USSR. **8**, 1014.
[212] Rajchman, J.: Le courant résiduel dans les multiplicateurs d'électrons électrostatiques. Thèse, Genève.
[213] Rann, W. H.: Electron-multiplier for measuring ionisation currents. Nature, Lond. **141**, 410.
[214] Rao, S. R., and P. S. Vadachari: Secondary electron emission of nickel at the Curie-Point. Ref. Chem. Abstr. **32**, 8916.
[215] Schneider, E. G.: Secondary emission of beryllium. Phys. Rev. **54**, 185.
[216] Shockley, W., and J. R. Pierce: Proc. Inst. Radio Engrs., N. Y. **26**, 321.
[217] Treloar, L. R. G., and D. H. Landon: Secondary electron emission from nickel, cobalt and iron as a function of temperature. Proc. Phys. Soc. Lond. **50**, 625.
[218] Treloar, L. R. G.: The measurements of secondary emission in valves. Wireless Engr. **15**, 535.
[219] Turnbull, J. C., and H. E. Farnsworth: The inelastic scattering of slow electrons from a silver single crystal. Phys. Rev. **54**, 509.
[220] Vudinsky, M. M.: The investigation of secondary electron emission from dielectrics by a thermal method. J. techn. Phys. USSR. **8**, 790. Ref. Phys. Ber. **20**, 437.
[221] Warnecke, R., et M. Lortie: Relation entre le coefficient d'émission secondaire et le travail d'extraction électronique de surfaces métalliques. Bull. Soc. Franc. Phys. **412**, 8.

1939

[222] Allen, J. S.: The detection of Single Positive Ions, Electrons and Protons by a Secondary Electron Multiplier. Phys. Rev. **55**, 966.
[223] Bojinesco, A.: Electronic field emission after bombardment of aluminium oxide by electrons or negative ions H-, N-, O-, O_2-. C. R. Acad. Sci. Paris **209**, 1800.
[224] Borzyak, P.: The emission from composite cathodes under simultaneons electron bombardment and illumination. J. techn. Phys. USSR. **15**, 1380.
[225] Bruining, H., u. J. H. de Boer: Secondary electron emission, Part IV. Compounds with a high capacity for secondary electron emission. Physica, Haag **6**, 823.

[226] BRUINING, H., u. J. H. DE BOER: Secondary electron emission. Part V. The mechanism of secondary electron emission. Physica, Haag **6**, 834.
[227] COOMES, E. A.: Total secondary electron emission from tungsten and thorium-coated tungsten. Phys. Rev. **55**, 519.
[228] COPELAND, P. L.: Secondary electrons from sodium films contaminated by gas. Phys. Rev. **55**, 1270.
[229] DE BOER, J. H., u. H. BRUINING: Secondary electron emission. VI. Influence of external adsorbed ions and atoms on the secondary electron emission of metals. Physica, Haag **6**, 941.
[230] HAGEN, C.: Aufladepotential, Sekundäremission und Ermüdungserscheinungen elektronenbestrahlter Metalle und Leuchtsubstanzen. Phys. Z. **40**, 621.
[231] HINTENBERGER, H.: Über Sekundärelektronenemission und Aufladungserscheinungen an Isolatoren. Z. Physik **114**, 98.
[232] KNOLL, M., u. R. THEILE: Elektronenabtaster zur Strukturabbildung von Oberflächen und dünnen Schichten. Z. Physik **113**, 260.
[233] KUSHNIR, YU. M.: Secondary electron emission under the action of two electron beams. J. techn. Phys. USSR. **9**, 267.
[234] KUSHNIR, YU. M., u. J. MILYUTIN: On the secondary electron emission from mercury. J. techn. Phys. USSR. **9**, 1589.
[235] MAHL, H.: Beobachtungen über Sekundärelektronenemission von Alkali-Aufdampfschichten mit einer oszillographischen Methode. Jb. AEG-Forsch. **6**, 33.
[236] MAJEWSKI, W.: Contribution à la technique des mesures de l'émission secondaire d'électrons. Acta phys. polon. **7**, 237.
[237] MARTIN, S. T., and L. B. HEADRICK: Light output and secondary emission characteristics of luminescent materials. J. Appl. Phys. **10**, 116.
[238] NELSON, H.: Phenomenon of secondary emission. Phys. Rev. **55**, 985.
[239] PAETOW, H.: Über die als Nachwirkung von Gasentladungen an den Elektroden auftretende spontane Elektronenemission und die Feldelektronenemission an dünnen Isolatorschichten. Z. Physik **111**, 770.
[240] PESYATSKI, J. F.: Secondary electron emission from thin films. J. techn. Phys. USSR. **9**, 188.
[241] PJANTNITZKI, A.: Distribution of the energy of secondary electrons emitted by a composite caesium cathode. Phys. Ber. **20**, 437.
[242] SHOCKLEY, W.: On the surface states associated with a periodic potential. Phys. Rev. **56**, 317.
[243] SUHRMANN, R., u. W. KUNDT: Die Sekundärelektronenemission reiner Metalle in ungeordnetem und geordnetem Zustand. Naturwiss. **27**, 548.
[244] SUHRMANN, R., u. W. KUNDT: Zum Mechanismus der Sekundärelektronenemission. Naturwiss. **27**, 707.
[245] VUDINSKY, M. M.: Mechanism of secondary electron emission from thin dielectric layers. J. techn. Phys. USSR. **9**, 271.
[246] VUDINSKY, M. M.: On the velocity distribution of secondary electrons emitted by sodium chloride. J. techn. Phys. USSR. **9**, 1583.
[247] WARNECKE, R., et M. LORTIE: Sur l'émission secondaire du beryllium. C. R. Acad. Sci. Paris **208**, 429.
[248] WOOLDRIDGE, D. E.: Theory of secondary emission. Phys. Rev. **56**, 562.
[249] WOOLDRIDGE, D. E.: Secondary emission from evaporated nickel and cobalt. Phys. Rev. **56**, 1062.

1940

[250] COPELAND, P. L.: Secondary emission from films of platinum on aluminium. Phys. Rev. **58**, 604.
[251] FRIMER, A. I.: A study of the secondary electron emission from copper oxide, pure or treated with alkali metals. J. techn. Phys. USSR. **10**, 394.
[252] GUBANOV, A.: The effect of a charge on an electron beam during secondary emission. J. exp. theoret. Phys. USSR. **2**, 161.
[253] HASTINGS, A. E.: Secondary emission from films of silver on platinum. Phys. Rev. **57**, 695.
[254] HEIMANN, W., u. K. GEYER: Ein Verfahren zur direkten Messung der Sekundärelektronenausbeute an Isolatoren. Elektr. Nachr.-Techn. **17**, 1.
[255] KADYSCHEVITSCH, A. E.: Theory of secondary electron emission from metals. J. Phys. USSR. **2**, 115.
[256] KAMOGAWA, H.: Secondary emission and electron diffraction on the glass surface. Phys. Rev. **58**, 660.

[257] Kirvalidze, I. D.: A method for determining the charging potential of dielectrics and the lower limit of secondary electron emission from a monocrystal of NaCl. C. R. Acad. Sci. USSR. **26**, 635.

[258] Kollath, R.: Eine bisher nicht verwendete Methode zur Messung der Energieverteilung von Sekundärelektronen. Z. techn. Phys. **21**, 328.

[259] Morgulis, N. D.: The mechanism of secondary electron emission from composite surfaces. J. techn. Phys. USSR. **10**, 79.

[260] Morgulis, N. D.: On the nature of secondary emission from composite cathodes. J. techn. Phys. USSR. **10**, 1710.

[261] Morgulis, N. D., u. B. I. Dyatlovitskaya: On the emission from antimony-caesium cathodes. J. techn. Phys. USSR. **10**, 657.

[262] Nelson, H.: Field enhanced secondary electron emission. Phys. Rev. **57**, 560.

[263] Piore, E. R., and G. A. Morton: The behaviour of willemite under electron bombardment. J. Appl. Phys. **11**, 153.

[264] Reichelt, W.: Über den Temperatureinfluß auf die Sekundärelektronenemission der Metalle. Ann. Phys., Lpz. **38**, 293.

[265] Salow, H.: Über den Sekundäremissionsfaktor elektronenbestrahlter Isolatoren. Z. techn. Phys. **21**, 8.

[266] Salow, H.: Über die Winkelabhängigkeit der Sekundärelektronenemission von Isolatoren. Phys. Z. **41**, 434.

[267] Scherer, K.: Aufladung und Sekundärelektronenemission elektronenbestrahlter Isolatoren. Arch. Elektrotechn. **34**, 143.

[268] Schnitger, H.: Über die Eigenschaften von Sekundäremissionsschichten aus Magnesiumoxyd. Z. techn. Phys. **21**, 376.

[269] Timofeew, P. W., u. A. V. Afanasjèwa: Secondary emission from oxides of metals. J. techn. Phys. USSR. **10**, 28.

[270] Timofeew, P. W., u. P. M. Aranovich: Barium oxide and magnesium oxide emitters of secondary electrons. J. techn. Phys. USSR. **10**, 32.

[271] Timofeew, P. W., u. J. Lunkova: Electron emission from caesiumo xide cathodes with gold particles in the intermediate layer. J. techn. Phys. USSR. **10**, 12.

[272] Timofeew, P. W., u. J. Lunkova: Antimony-caesium emitters. J. techn. Phys. USSR. **10**, 20.

[273] Timofeew, P. W., u. A. I. Pyatnitski: Secondary electron emission from oxygen-caesium emitters at different primary current densities. J. techn. Phys. USSR. **10**, 39.

[274] Timofeew, P. W., u. K. A. Yumatov: Secondary electron emission from oxygen-caesium emitters at low velocities of primary electrons. J. techn. Phys. USSR. **10**, 8.

[275] Timofeew, P. W., u. K. A. Yumatov: Secondary electron emission from sulphur-caesium emitters. J. techn. Phys. USSR. **10**, 91.

[276] Varadachari, P. S.: Secondary electron emission of nickel at the Curie point. Proc. Indian. Sci. A **12**, 381.

[277] Wooldridge, D. E.: Temperature effects on the secondary electron emission from pure metals. Phys. Rev. **57**, 1080.

[278] Wooldridge, D. E.: Temperature effects in secondary emission. Phys. Rev. **58**, 316.

[279] Wooldridge, D. E., and C. D. Hartmann: The effects of order and disorder on secondary electron emission. Phys. Rev. **58**, 381.

[280] Yasnopolski, N.: On the „jumps" observed in emitters of poor conductivity caused by the blocking effect. J. techn. Phys. USSR. **10**, 1813.

1941

[281] Bay, Z.: Elektronenvervielfacher als Elektronenzähler. Z. Physik **117**, 227.

[282] Bekow, G.: Secondary emission from copper single crystals at small primary velocities. Phys. Z. **42**, 144.

[283] Bethe, H. A.: On the theory of secondary emission. Phys. Rev. **59**, 940.

[284] Bruining, H.: Secondary electron emission from metals with low work function. Physica, Haag **8**, 1161.

[285] Chaudri, R. M., and A. W. Khan: Secondary electron emission from nickel. Phil. Mag. **31**, 382.

[286] Friedheim, J., u. J. G. Weiss: Sekundärelektronenausbeute von Silber-Magnesium-Legierungen. Naturwiss. **29**, 777.

[287] Gille, C.: Sekundärelektronenemission von Nickel-Berylliumlegierungen. Z. techn. Phys. **22**, 228.

[288] Görlich, P.: Beitrag zur Frage der Sekundäremission aufgedampfter Erdalkalimetallschichten. Phys. Z. **42**, 129.

[289] Joffe, M. S., u. I. V. Nechlaev: The secondary electron emission from potassium. J. exp. theoret. Phys. USSR. **11**, 93.

[290] Kadyschevitsch, A. E.: Theory of secondary electron emission from dielectrics and semiconductors. J. Phys. USSR. **4**, 341.
[291] Kollath, R.: Über den Temperatureinfluß auf die Sekundärelektronenemission der Metalle. Ann. Phys., Lpz. **39**, 19.
[292] Kollath, R.: Neuere Untersuchungen über die Sekundärelektronenemission von Legierungen. Fernsehen H. 4/5, S. 13.
[293] Kollath, R.: Zur Energieverteilung der Sekundärelektronen I. Ann. Phys., Lpz. **39**, 59.
[294] Kushnir, Yu. M., and M. I. Frumin: The dependence of the energy distribution function of secondary electrons on the angle of emergence. J. techn. Phys. USSR. **11**, 317.
[295] Malter, L.: The behaviour of electrostatic electron multiplier as a function of frequency. Proc. Inst. Radio Engr., N. Y. **29**, 587.
[296] Mathes, I.: Untersuchungen über die Sekundärelektronenemission von verschiedenen Legierungen. Z. techn. Phys. **22**, 232.
[297] Maurer, G.: Die Sekundärelektronenemission von Halbleitern bzw. Isolatoren. Z. Physik **118**, 122.
[298] Mishibori, E.: Secondary electron emission from magnesium oxide. Proc. Phys.-Math. Soc. Japan **23**, 570.
[299] Morgulis, N. D.: The emission of electrons by active semi-conducting surfaces. Bull. Acad. Sci. USSR. **5**, 536.
[300] Morozov, P. M.: The effect of temperature on secondary electron emission. J. exper. theoret. Phys. USSR. **11**, 402.
[301] Morozov, P. M.: Secondary electron emission from lead, tin and bismuth in the solid and liquid state. J. exp. theoret. Phys. USSR. **11**, 410.
[302] Nemilov, Yu. A.: A new method for studying secondary electron emission. J. techn. Phys. USSR. **11**, 854.
[303] Paetow, H.: Eine neue, bei sehr niedrigen Drucken auftretende Form der feldelektrischen Emission an Metalloberflächen, die mit isolierenden Stoffen bestaubt sind. Z. Physik **117**, 399.
[304] Raudenbusch, H.: Einige Untersuchungen zur technischen Verwendung sekundäremittierender Schichten. Z. techn. Phys. **22**, 237.
[305] Tanaka, M.: After-effect of metal bombarded by electrons. Proc. Phys.-Math. Soc. Japan **22**, 899.
[306] Teichmann, H., u. K. Geyer: The occurence of structure dependent selectivity in secondary electron emission. Z. Naturwiss. **7**, 313.
[307] Vudinsky, M. M.: The stability of secondary electron emission from alkali halide cathodes. J. techn. Phys. USSR. **11**, 1066.
[308] Wecker, F.: Neue Messungen der Absorption, Rückdiffusion und Sekundärstrahlerregung mittels schneller Kathodenstrahlen. Ann. Phys., Lpz. **40**, 405.
[309] Wolff, H.: Sekundärelektronen an lichtelektrisch ausgezeichneten Halbleitern. Ann. Phys., Lpz. **39**, 591.
[310] Zworykin, V. K., J. E. Ruedy and E. W. Pike: Silvermagnesium alloy as a secondary emitting material. J. Appl. Phys. **12**, 696.

1942

[311] Geyer, K. H.: Über Eigenschaften der Ausbeute und Energieverteilung von Sekundärelektronen aus Aufdampfschichten wachsender Dicke. Ann. Phys., Lpz. **41**, 117.
[312] Geyer, K. H.: Beobachtungen an der Sekundärelektronenstrahlung aus Nichtleitern. Ann. Phys., Lpz. **42**, 241.
[313] Geyer, K. H.: Zur Kenntnis des Grundvorgangs der sekundären Elektronenstrahlung. Ann. Phys., Lpz. **42**, 337.
[314] Görlich, P.: Über die Sekundäremission aufgedampfter Antimonschichten. Phys. Z. **43**, 121.
[315] Krenzien, O.: Die Elementarprozesse bei der Sekundärelektronenemission von polaren Kristallen. Wiss. Veröff. Siemens-Werk **20**, 91.
[316] McKay, K. G.: Total secondary emission from thin films of sodium on tungsten. Phys. Rev. **61**, 708.
[317] Truell, R.: Range of secondary electrons in magnesium. Phys. Rev. **62**, 340.

1943

[318] Bronstein, J.: Techn. Physics USSR. **13**, 176.
[319] Geyer, K. H.: Zur Kenntnis des Grundvorgangs der sekundären Elektronenstrahlung. Ann. Phys., Lpz. **42**, 337.
[320] Gimpel, I., and O. Richardson: The secondary electron emission from metals in the low primary energy region. Proc. Roy. Soc. Lond., Ser. A **182**, 17.

[321] KENNEDY, W. R., and P. L. COPELAND: The temperature coefficient of the secondary emission yield and the work function of molybdenum coated with beryllium and treated with HCl. Phys. Rev. **63**, 61.
[322] SCHLECHTWEG, H.: Zur Quantenmechanik der Sekundärelektronen-Emission von Übergangsmetallen. Naturwiss. **31**, 204.
[323] SUHRMANN, R., u. W. KUNDT: Über die Sekundäremission reiner Metallschichten im ungeordneten und geordneten Zustand und deren Durchlässigkeit für Sekundärelektronen. Z. Physik **120**, 363.
[324] SUHRMANN, R., u. W. KUNDT: Über den Einfluß absorbierten Sauerstoffs auf die Sekundäremission aufgedampfter Metallschichten bei 293° und 83° abs. Z. Physik **121**, 118.
[325] TREY, F.: Der Einfluß der Feldemission auf die Sekundäremission. Phys. Z. **44**, 38.

1944

[326] AFANASJÈWA, A. V.: Stable emitters of secondary electrons. Univ. M. V. Lomousova **74**, 114.
[327] HARRIES, J. H. O.: Secondary electron radiation. Electronics, N. Y. **10**, 100.
[328] JOHNSON, J. B.: Enhanced thermionic emission. Phys. Rev. **66**, 352.
[329] KNOLL, M., O. HACHENBERG u. J. RANDMER: Zum Mechanismus der Sekundäremission im Inneren von Ionenkristallen. Z. Physik **122**, 137.
[330] KWARZCHAWA, I. F.: Change of conductivity of aluminium oxide upon electron bombardment. Bull. Acad. Sci. USSR. **8**, 373.
[331] LUKJANOV, S. J.: Secondary electron emission of solids. Bull. Acad. Sci. USSR. **8**, 330.
[332] ZERNOV, D. V.: On the influence of strong electric fields on the secondary electron emission of dielectric films. Bull. Acad. Sci. USSR. **8**, 352.
[333] ZERNOV, D. V., M. I. ELINSON u. N. M. LEVIN: Investigation of autoelectronic emission of thin dielectric films. Bull. Acad. Sci. USSR., Cl. sci. techn. 166.

1945

[334] KADYSCHEVITSCH, A. E.: The velocity distribution of secondary electrons of various emitters. J. Phys. USSR. **9**, 341.
[335] KADYSCHEVITSCH, A. E.: On the measurement of the depth of generation of secondary electrons in metals. J. Phys. USSR. **9**, 436.
[336] MUELLER, C. W.: The secondary emission of „Pyrex" glass. J. Appl. Phys. **16**. 453.
[337] WANG, C. G.: Reflex oscillator utilizing secondary emission currents. Phys. Rev. **68**, 284.

1946

[338] JOHNSON, J. B.: Secondary emission of thermionic oxide cathodes. Phys. Rev. **69**, 693, 702.
[339] KOLLER, L. R., and J. S. BURGESS: Secondary emission from germanium, boron and silicon. Phys. Rev. **70**, 571.
[340] LALLEMAND, A.: Application of secondary emission of electrons to multiplier tubes. Rev. sci. Paris **84**, 131.
[341] POMERANTZ, M. A.: Secondary emission from oxide coated cathodes. J. Franklin Inst. **241**, 415; **242**, 41. — Phys. Rev. **70**, 33.

1947

[342] FRIMER, A. I.: Study of secondary emission at low temperatures. J. techn. Phys. **17**, 71.
[343] GREENBLATT, M. H., and P. A. MILLER jr.: A microwave secondary emission multiplier. Phys. Rev. **72**, 160.
[344] KOLLATH, R.: Zur Energieverteilung der Sekundärelektronen. II. Ann. Physik (6) **1**, 357.
[345] MENDENHALL, H. E.: Secondary emission from conducting films of tin oxide. Phys. Rev. **72**, 532.
[346] PALLUEL, P. C. R.: Rediffused component of secondary radiation from metals. C. R. Acad. Sci. Paris **224**, 1492.
[347] PALLUEL, P. C. R.: On the mechanism of rediffusion of electrons from metals. C. R. Acad. Sci. Paris **224**, 1551.
[348] PALLUEL, P. C. R.: The true secondary emission coefficient in metals. C. R. Acad. Sci. Paris **225**, 383.
[349] TRUMP, J. G., and R. J. VAN DE GRAAFF: High voltages in vacuum. J. Appl. Phys. **18**, 327.

[350] ZERNOV, D. V., u. B. S. KULJVARSKAYA: Investigations of the temperature dependence of the electronic emission of dielectric films under the influence of the field of positive surface charge. J. techn. Phys. USSR. **17**, 309.

1948

[351] BOHR, N.: The penetration of atomic particles through matter. Kgl. danske Vid. Selsk., math.-fys. Medd. **18**, 8.
[352] HACHENBERG, O.: Absorption der im Festkörper ausgelösten Sekundärelektronen auf ihrem Weg zur Oberfläche. Ann. Physik **2**, 403.
[353] JOHNSON, J. B.: Secondary emission from targets of Ba-Sr-oxide. Phys. Rev. **73**, 1058.
[354] LANG, W.: Geschwindigkeitsverluste mittelschneller Elektronen beim Durchgang durch dünne Metallfolien. Optik **3**, 233.
[355] MENDENHALL, H. E.: Secondary emission from thoria on platinum and tantalum. Phys. Rev. **74**, 1265.
[356] RUTHEMANN, G.: Diskrete Energieverluste mittelschneller Elektronen beim Durchgang durch dünne Folien. Ann. Physik **2**, 113.
[357] RUTHEMANN, G.: Elektronenbeugung an Röntgenniveaus. Ann. Physik **2**, 135.

1949

[357a] BOTHE, W.: Zur Rückdiffusion schneller Elektronen. Z. Naturforsch. **4**a, 542.
[358] HEES, G. W., and H. JACOBS: Secondary emission from barium oxide as a function of temperature. Phys. Rev. **75**, 1312.
[359] MENDENHALL, H. E.: Secondary emission from silicon and germanium oxides. Phys. Rev. **76**, 458.
[360] MÖLLENSTEDT, G.: Die elektrostatische Linse als hochauflösender Geschwindigkeitsanalysator. Optik **5**, 499.
[361] STERNGLASS, E. J.: An interpretation of secondary electron emission data for homogeneous solids. Phys. Rev. **76**, 189.
[362] TODD, F. C., and L. J. RUEGER: Secondary emission from oxides at high temperatures. Phys. Rev. **76**, 189.
[363] TRUMP, J. G., and R. J. VAN DE GRAAFF: The secondary emission of electrons by high energy electrons. Phys. Rev. **75**, 44.
[363a] ZWORYKIN, V. K., and E. G. RAMBERG: Photoelectricity and its application, p. 137. New York: J. Wiley & Sons 1949.

1950

[364] ALLEN, J. S.: Recent applications of electron multiplier tubes. Proc. Inst. Radio Engrs., N. Y. **38**, 346.
[365] BAROODY, E. M.: A theory of secondary electron emission from metals. Phys. Rev. **78**, 780.
[366] DIEMER, G., and J. L. H. JONKER: On the time delay of secondary emission. Philips Res. Rep. **5**, 161.
[367] HIRASHIMA, M.: Preliminary experiment for attempting to measure the exit time of secondary electrons. Applied Physics Japan **19**, 121.
[367a] SALOW, H.: Die Sekundärelektronenemission von Metallmischungen. Ann. Physik **5**, 417.
[368] STERNGLASS, E. J.: Secondary electron emission and atomic shell structure. Phys. Rev. **80**, 925.

1951

[369] BAROODY, E. M.: Application of Wooldridge theory of secondary emission. Phys. Rev. **83**, 857.
[370] BLANKENFELD, G.: Sekundärelektronenemission von Ni, Mo, MgO und von Glas. Ann. Phys. **9**, 48.
[371] BROPHY, J. J.: Comparison of theories of secondary emission. Phys. Rev. **82**. 757.
[372] BROPHY, J. J.: Secondary emission of electrons from liquid metal surface. Phys. Rev. **83**, 534.
[373] JACOBS, H.: Field dependent secondary emission. Phys. Rev. **84**, 755.
[374] JONKER, J. L. H.: The angular distribution of the secondary electrons of nickel. Philips Res. Rep. **6**, 372.
[375] SCHAETTI, N.: Sekundärelektronenvervielfacher. Z. angew. Math. Phys. **2**, 123.

1952

[376] Baroody, E. M.: A note on Wooldridge's theory of secondary emission. Phys. Rev. **86**, 915.
[377] Brand, F., H. Jocobs and J. Freely: Mechanism of field dependent secondary emission. Rep. 12th Annual Conf. Phys. Electronics, p. 74.
[378] Dekker, A. J., and A. van der Ziel: Theory of the production of secondary electrons in solids. Phys. Rev. **86**, 755.
[379] Hirashima, M.: Secondary emission valves. Wireless Engr. **29**, 246.
[380] Jacobs, H., J. Martin and F. Brand: Secondary emission from composite surfaces. Phys. Rev. **85**, 441.
[381] Johnson, J. B., and K. G. McKay: Secondary electron emission from germanium. Phys. Rev. **85**, 390.
[382] Jonker, J. L. H.: On the theory of secondary electron emission. Philips Res. Rep. **7**, 1.
[383] Law, R. R.: Decay time of secondary electron emission. Phys. Rev. **85**, 391.
[384] Marshall, J. F.: The theory of secondary emission. Phys. Rev. **88**, 416.
[385] Pines, D., and D. Bohm: A collective description of electron interactions II. Collective as individual particle aspects of the interactions. Phys. Rev. **85**, 338.
[386] Rappaport, P.: New methods of processing silver magnesium secondary emitters. Rep. 12th Annual Conf. Phys. Electronics, p. 18.
[387] Rittner, E. S., and R. H. Ahlert: On the I—V Characteristic of the „L-Cathode". Phys. Rev. **85**, 390.
[388] Woods, J., and D. A. Wright: Secondary electron emission of barium oxide. Brit. J. Appl. Phys. **3**, 323.

1953

[389] Berger, W.: Durchlässigkeit von Silberfolien für langsame Elektronen. Naturwiss. **41**, 59 (1954).
[390] Gobrecht, H., u. F. Speer: Ein Beitrag zur Sekundärelektronenemission von Störstellenhalbleitern. Z. Physik **135**, 602.
[391] Johnson, J. B.: Secondary electron emission from diamond. Phys. Rev. **92**, 843.
[392] Johnson, J. B., and K. G. McKay: Secondary electron emission of crystalline MgO. Phys. Rev. **91**, 582.
[393] Lander, J. J.: Auger peaks in the energy spectra of secondary electrons from various materials. Phys. Rev. **91**, 1382.
[394] Müller, E. W.: Feldelektronen-Mikroskopie. Ergebn. exakt. Naturwiss. **27**, 290.
[395] Sternglass, E J.: An experimental study of electron backscattering and secondary emission from solids. Westinghouse Res. Lab., Res. Rep. R-94416-3-J.
[395a] Yoshida, S. u. a.: Secondary electron emission of Copper-Beryllium Alloy. J. Phys. Soc. Japan **8**, 318.
[396] Ziel, A. van der: A modified theory of production of secondary electrons in solids. Phys. Rev. **92**, 35.

1954

[397] Berger, W.: Durchgang langsamer Elektronen bis zu 200 eV durch Silberfolien. Diplomarbeit Techn. Hochschule, Berlin.
[398] Brauer, W.: Bemerkungen zur Theorie der Sekundäremission der Metalle. Arbeitstagung Festkörperphysik, Dresden, S. 159.
[399] Gerthsen, Ch., u. L. Albert: Über den Einfluß diskreter Energieverluste der Elektronen auf die Struktur der Röntgenstrahlisochromaten in der Nähe der kurzwelligen Grenze. Z. angew. Phys. **6**, 196.
[400] Hachenberg, O., u. W. Brauer: Der gegenwärtige Stand der Theorie der Sekundärelektronenemission. Fortschr. Phys. **1**, 439.
[401] Johnson, J. B., and K. G. McKay: Secondary electron emission from germanium. Phys. Rev. **93**, 668.
[402] Kottgardt, K. H. J., W. Berthold u. H. Dietrich: Die Sekundäremission von Leuchtschirmen in Elektronenstrahlröhren. Z. angew. Phys. **6**, 560, Nr. 2.
[403] Lane, R. O., and D. J. Zaffarano: Transmission of 0—40 KeV electrons by thin films with applications to beta-ray spectroscopy. Phys. Rev. **94**, 960.
[404] Sternglass, E. J.: Theory of secondary electron emission . Westinghouse Res. Lab., Sci. Paper 1772.
[405] Sternglass, E. J.: Backscattering of Kilovolt electrons from solids. Phys. Rev. **95**, 345.
[406] Bohr, N.: The penetration of atomic particles through matter. Kgl. danske Vid. Selsk., math.-fys. Medd. **18**, 8 (1948).
[407] Bethe, H. A.: Ann. Phys. **5**, 325 (1930).

Photoionization in gases and photoelectric emission from solids.

By

G. L. Weissler.

With 92 figures.

1. Introduction, scope and objectives. Problems involving the interaction of radiation with matter occur in a wide variety of fields including among others astrophysics, atmospheric electricity, and the ionospheric layers at high altitudes. In addition laboratory studies on the conduction of electricity through gases frequently must concern themselves with the participation of photons in certain events. In each case the individual problems differ widely, and of all processes by which radiation may exchange energy with matter, this article will be limited to a discussion of the present state of our knowledge on A). the photoionization in gases and B). the phenomena of the external photoelectric effect.

The point of view taken in this presentation was to provide the reader with a general survey of observed phenomena, following essentially along lines of historical development but stressing contributions of the last twenty years, and to include an optimum amount of numerical material in such a form as to make it immediately useful to researchers of diverse specializations. Because of the fact that quantitative work on photoionization in gases is of rather recent origin, its treatment here is more detailed and comprehensive than that of the external photoelectric effect for which a large amount of literature together with several review articles and books are available. In the following section some important phenomena will be mentioned whose interpretation may depend critically on the information on photoionization in gases and/or the photoelectric effect. It is hoped that this will delineate further the purposes of this article.

2. Some photon-dependent processes in electrical discharges in gases. The primary process responsible for the building up of charges in a gas due to an applied electric field may be explained in terms of a Townsend[1] avalanche caused by electron collisions with molecules. However, such disruptive current transitions which lead to sparks, wire coronas, etc. necessitate for their explanation the introduction of additional, secondary mechanisms which have been summarized by Loeb[2] and others[3]. Among them are the contributions to the discharge current of photoelectrons from the cathode and photoionization of the gas, both caused by radiation originating in the Townsend avalanche [*1*]. The latter process my be visualized best for the case of air, a mixture of gases of different ionization potentials, where may occur single electron impact excitation of a

---

[1] J. S. Townsend: Electrons in Gases. Hutchinson's Scientific and Technical Publications. London 1948.

[2] L. B. Loeb: Fundamental Processes of Electrical Discharges in Gases, Chapt. 9. New York: John Wiley & Sons 1939.

[3] A. v. Engel und M. Steenbeck: Elektrische Gasentladungen, Vol. II. Berlin: Springer 1934.

nitrogen molecule to such a level that photons energetic enough to ionize oxygen are produced.

By invoking such a mechanism[1], it is for instance possible to account qualitatively for the filamentary character and the high speed of propagation of pre-breakdown streamers[2] and the stepped leader strokes[3] of lightning discharges. In pure gases, however, it is somewhat more difficult to see how photoionization can be active. Indeed LLEWELLYN JONES and his associates[4] measure the growth of current in a TOWNSEND gap to within 0.5% of the breakdown potential in both nitrogen and air under conditions of uniform fields, and they come to the conclusion that photoionization in these gases cannot be the secondary process responsible for breakdown. Rather they produce evidence of cathode dependent mechanisms[5], such as the ejection of electrons by positive ions or photons.

Formative time lags of spark breakdown in uniform fields have been investigated by FISHER, BEDERSON, and KACHICKAS[6] in air, nitrogen and argon. Their analysis indicates that the results in air and N_2 can be accounted for by a TOWNSEND buildup before breakdown together with a secondary mechanism of photoemission from either the cathode or from the gas near the cathode. In argon, however, none of the accepted secondary mechanisms seems to explain the large observed time lags, except perhaps delayed photoelectric emission from the cathode because of the slow diffusion of resonance radiation[7].

It is appropriate to mention at this point the work of SCHWIECKER[8] who finds that radiation capable of ionizing the gas is not exited in the pre-breakdown TOWNSEND discharge and is strongly excited in the "Kanal" or "streamer" discharge. Thus it seems still possible to have both a pre-breakdown TOWNSEND buildup and a non-TOWNSEND, channelled and streamerlike breakdown mechanism.

The discharge mechanism in argon counters has been studied by COLLI and FACCHINI[9] who observed the shape and duration of primary pulses, produced by a collimated beam of α particles, and secondary ones caused by photons from the first avalanche. Their results are ascribed to electron emission from the cathode caused by non-resonance photons of about 10 ev of energy [*1*].

The significant role of light quanta initiating electrical breakdown by photoelectric action on the container walls has been pointed out by GILL, HARRIS, and VON ENGEL[10] for cases where a gas-filled glass vessel with external electrodes

[1] Simultaneously and independently both LOEB and RAETHER made use of the concept of photoionization in gases in order to explain breakdown phenomena. For their most recent views on this subject see L. B. LOEB [Phys. Rev. **73**, 798 (1948); **81**, 287 (1951); **94**, 224 (1954); Brit. J. Appl. Phys. **3**, 341 (1952] and H. RAETHER [Ergebn. exakt. Naturw. **22**, 73 (1949); Z. angew. Phys. **5**, 211 (1953)].

[2] J. M. MEEK and R. F. SAXE: Nature, Lond. **162**, 263 (1948).

[3] B. F. J. SCHONLAND: Proc. Roy. Soc. Lond., Ser. A **164**, 132 (1938). — K. B. MCEACHRON: J. Franklin Inst. **227**, 149 (1939).

[4] F. LLEWELLYN JONES and A. B. PARKER: Proc. Roy. Soc., Lond., Ser. A **213**, 185 (1952). — J. DUTTON, S. C. HAYDON and F. LLEWELLYN JONES: Proc. Roy. Soc. Lond., Ser. A **213**, 203 (1952); **218**, 206 (1953).

[5] F. LLEWELLYN JONES and G. C. WILLIAMS: Proc. Phys. Soc. Lond. B **66**, 345 (1953). — F. LLEWELLYN JONES: Brit. J. Appl. Phys. **5**, 49 (1954).

[6] L. H. FISHER and B. BEDERSON: Phys. Rev. **81**, 109 (1951). — G. A. KACHICKAS and L. H. FISHER: Phys. Rev. **88**, 878 (1952); **91**, 775 (1953). See also J. DUTTON, S. C. HAYDON and F. LLEWELLYN JONES: Brit. J. Appl. Phys. **4**, 170 (1953).

[7] T. HOLSTEIN: Phys. Rev. **72**, 1212 (1947); **83**, 1159 (1951). — D. ALPERT, A. O. MC COUBREY, and T. HOLSTEIN: Phys. Rev. **76**, 1257 (1949); **85**, 985 (1952).

[8] W. SCHWIECKER: Z. Physik **116**, 562 (1940). — Naturwiss. **28**, 380 (1940).

[9] L. COLLI and U. FACCHINI: Phys. Rev. **88**, 987 (1952). — L. COLLI: Phys. Rev. **95'** 892 (1954).

[10] E. W. B. GILL and A. VON ENGEL: Proc. Roy. Soc. Lond., Ser. A **197**, 107 (1949). — W. L. HARRIS and A. VON ENGEL: Proc. Roy. Soc. Lond., Ser. A **222**, 490 (1954).

was placed in a uniform high-frequency electric field. In their interpretation a chance electron is accelerated by the field and excites neon atoms to about 16 ev. The emitted quanta fall on glass walls and release photoelectrons which repeat the process. This concept, when evaluated quantitatively, leads to good numerical agreement with their experimental data.

A recent contribution by LITTLE and VON ENGEL[1] stresses that the essential secondary mechanism in the theory of glow discharges is the photoelectric emission from the cathode by ultraviolet light originating in the glow.

Studies of the earth's upper atmosphere pose a similar variety of questions on photoionization and dissociation phenomena [2].

A. Photoionization in gases.

I. Historical survey.

3. Early work on air. Among the first to investigate the ionization of air by light were C. T. R. WILSON[2] and P. LENARD[3]. The first found that ultraviolet light, passing through his expansion chamber, produced condensation nuclei which initiated droplet formation upon low expansion, while the latter observed the discharge of an electroscope from the ultraviolet light of a spark. Since in both cases the air was moist and unfiltered a surface photoelectric effect from minute droplets or dust particles might have been present. Under very carefully selected experimental conditions HUGHES[4] was able to show unambiguously that positive ions were produced in air by ultraviolet light which had the same mobility as those produced by X-rays in the same apparatus. His ultraviolet light was generated in a hydrogen glow discharge and passed through a fluorite window with a short-wavelength cut-off of 1250 Å into the ionization chamber through which filtered air was streaming.—In a detailed investigation LENARD and RAMSAUER[5] essentially confirmed the work of HUGHES in air. It should be noted that the first ionization limits of both O_2 (at 1040 Å) and N_2 (at 796 Å) lie towards shorter wavelengths than the fluorite window cut-off (1250 Å), and that therefore direct photoionization of these molecules should not have been possible. However, photodissociation of O_2 undoubtedly occurred, and ionization of reaction products from atomic oxygen might explain these results.

MARTIN[6] extended the work of LENARD and RAMSAUER and obtained values of photoionization relative to air in oxygen, nitrogen and carbon dioxide by passing undispersed light from a spark through a fluorite window into an ionization chamber.

4. Cd, Zn, Hg. Photoionization by two-step processes was observed in mercury. In this case STEUBING and later HOUTERMANS[7] and others found that the vapor could be ionized by its own resonance radiation (2537 Å or 4.9 volts) even though its ionization potential is 10.43 volts or 1189 Å. HOUTERMANS' view was that illumination of the vapor with 2537 Å produced, by absorption, atoms in the $^3P_1$ state some of which could then be converted into the metastable one $^3P_0$ by collisions. This conversion was facilitated by the addition of argon or

[1] P. F. LITTLE and A. VON ENGEL: Proc. Roy. Soc. Lond., Ser. A **224**, 209 (1954).
[2] C. T. R. WILSON: Phil. Trans. Roy. Soc. Lond., Ser. A **192**, 403 (1899).
[3] P. LENARD: Ann. Phys., Lpz. **1**, 486 (1900); **3**, 298 (1900).
[4] A. L. HUGHES: Proc. Cambridge Phil. Soc. **15**, 483 (1910).
[5] P. LENARD and C. RAMSAUER: Sitzgsber. Heidelberg. Akad. Wiss. **1910/11**.
[6] A. MARTIN: Sitzgsber. Heidelberg. Akad. Wiss. **10** (1918).
[7] F. G. HOUTERMANS: Z. Physik **41**, 619 (1927).

suppressed in the presence of H_2 which is known to dissociate upon impact with Hg $^3P_1$. The observed photoionization was shown to depend on the square of the light intensity, and could be attributed to collisions between two Hg atoms, one in the $^3P_1$ (4.9 volts) and the other in the metastable $^3P_0$ state (4.7 volts), giving rise to the formation of Hg_2^+ at an expenditure of 9.6 volts (to form Hg^+ requires 10.43 volts).—A somewhat similar two-step mechanism could probably explain the absorption coefficients in Zn, Cd and Hg obtained by FRANCK and by MROZOWSKI[1] for wavelength longer than 2100 Å.

5. Alkali vapors. Some of the alkali vapors were particularly suitable for direct measurements of photoionization in the region from the series limit towards shorter wavelengths, because their low ionization potentials permitted the use of quartz optics. MOHLER and associates[2] were the first to obtain quantitative photoionization cross sections as a function of wavelength in Cs and Rb vapors and qualitative ones in K. At about the same time LAWRENCE and EDLEFSEN[3], WILLIAMSON[4], and others[5] obtained relative cross section date in these vapors. TRUMPY[6] measured relative absorption coefficients of Na on either side of its series limit (2411.87 Å) between 2300 and 2700 Å and compared the decrease of the continuous absorption with theoretical predictions based on calculations by SUGIURA[7]. In nearly all instances the above authors also found photoionization and/or absorption towards longer wavelengths from the series limit, and this may be explained, as in Hg, by two-step processes. In particular, the effect of photoionization in Cs by absorption of light in and between its resonance lines has been studied in detail by MOHLER and BOECKNER[8] who accounted for it in terms of collisions between excited and normal atoms to form molecular ions.

The above cited initial results in the alkali vapors were significantly improved through the efforts of DITCHBURN and his co-workers who applied spectrophotometric techniques to determine absolute absorption cross sections [3]. BRADDICK and DITCHBURN[9] presented data for Cs between 2000 and 3300 Å, DITCHBURN, TUNSTEAD and YATES[10] for K from 1600 to 3200 Å, and TUNSTEAD[11] for Li between 1800 and 2300 Å. New absorption measurements in Na between 1600 Å and 2600 Å have been published recently by DITCHBURN, JUTSUM and MARR together with an evaluation of previous work in the alkali vapors [3]. In this paper and in a separate one by SEATON[12] the experimental findings are compared with theoretical calculations.

---

[1] S. MROZOWSKI: Z. Physik **91**, 600 (1934). (This paper includes FRANCK's Hg work.)

[2] F. L. MOHLER and C. BOECKNER: J. Res. Nat. Bur. Stand. **3**, 303 (1929). — F. L. MOHLER, P. D. FOOTE and R. L. CHENAULT: Phys. Rev. **27**, 37 (1926).

[3] E. O. LAWRENCE and N. E. EDLEFSEN: Phys. Rev. **34**, 233, 1056 (1929).

[4] R. C. WILLIAMSON: Phys. Rev. **21**, 107 (1923). — Proc. Nat. Acad. Sci., Wash. **14**, 793 (1928).

[5] R. SAMUEL: Z. Physik **29**, 209 (1924).

[6] B. TRUMPY: Z. Physik **47**, 804 (1928); **71**, 720 (1931). (Li, theoretical.)

[7] M. Y. SUGIURA: J. Phys. Radium **8**, 113 (1927). — Phil. Mag. **4**, 495 (1927).

[8] F. L. MOHLER and C. BOECKNER: J. Res. Nat. Bur. Stand. **5**, 51, 399, 831 (1930).

[9] H. J. BRADDICK and R. W. DITCHBURN: Proc. Roy. Soc. Lond., Ser. A **143**, 472 (1934); (this includes a complete description of experimental techniques); see also Proc. Roy. Soc. Lond., Ser. A **150**, 478 (1935). — R. W. DITCHBURN and J. HARDING: Proc. Roy. Soc. Lond., Ser. A **157**, 66 (1936).

[10] R. W. DITCHBURN, J. TUNSTEAD and J. G. YATES: Proc. Roy. Soc. Lond., Ser. A **181**, 386 (1943).

[11] J. TUNSTEAD: Proc. Phys. Soc. Lond. A **66**, 304 (1953).

[12] M. J. SEATON: Proc. Roy. Soc. Lond., Ser. A **208**, 418 (1951). (This paper contains a summary of theoretical calculations in the alkali vapors with details on Na.)

DITCHBURN'S group also extended their work to atomic systems containing more than one electron in the outer shell, such as Mg[1], Tl[2], Ca[3], and In[4].

6. Absorption cross sections in air, O_2, and N_2. In the real gases, direct measurements of the number of ions produced per photon absorbed were not attempted until very recently. However, it is possible to make certain inferences as to the magnitude of photoionization from the knowledge of absorption cross sections as a function of wavelength.

KREUSLER[5] obtained many years ago the absorption coefficient of air at 1860 Å, while BUISSON, JAUSSERAN and ROUARD[6] made measurements at 1855, 1858, 1863 Å and longer wavelengths which were repeated by VASSY[7]. Absorption in this region is really due to O_2 since N_2 is essentially transparent here. Vacuum ultraviolet absorption coefficients in air were first investigated by SCHNEIDER[8] between 380 and 1600 Å.

In O_2, LADENBURG and VAN VOORHIS[9] and, somewhat later, SCHNEIDER[10] determined the absolute absorption in the SCHUMANN-RUNGE dissociation continuum between 1300 and 1750 Å. This was supported by the theoretical calculations of STUECKELBERG[11]. Recently, their work was repeated by WATANABE, INN and ZELIKOFF[12] and, most carefully, by DITCHBURN and HEDDLE[13]. The difficult region of the vacuum ultraviolet, below the transmission limit of LiF, was investigated by WEISSLER and LEE[14] from 300 to 1300 Å. They concerned themselves primarily with the absolute values of the strong ionization continua below 1000 Å. In this region CLARK[15] presented values in both O_2 and N_2 for about 27 wavelengths. With the idea of separating, whenever possible, the various continua in O_2, PO LEE[16] repeated these measurements and was able to delineate the separate absorption contours of several dissociation and ionization continua. In addition, he paid particular attention to the "oxygen window" at LYMAN α (1215 Å) because of its importance to the understanding of the formation of the ionospheric layers of the upper atmosphere. At this wavelength his results are in good agreement with those of DITCHBURN, BRADLEY, CANNON and MUNDAY ([2], p. 327), with WATANABE'S and associates[12] and with PRESTON'S[17]. In a recent paper ABOUD, CURTIS, MERCURE and RENSE[18] present preliminary absorption results in O_2 which are somewhat higher than those of previous observers.

---

[1] R. W. DITCHBURN and G. V. MARR: Proc. Phys. Soc. Lond., A **66**, 655 (1953).
[2] G. V. MARR: Proc. Roy. Soc. Lond., Ser. A **224**, 83 (1954).
[3] P. J. JUTSUM: Proc. Phys. Soc. Lond. A **67**, 190 (1954).
[4] G. V. MARR: Proc. Phys. Soc. Lond. A **67**, 196 (1954).
[5] H. KREUSLER: Ann. Phys., Lpz. **6**, 412 (1901).
[6] H. BUISSON, C. JAUSSERAN and P. ROUARD: Rev. Opt. **12**, 70 (1933).
[7] A. VASSY: Ann. Phys., Paris **16**, 145 (1941).
[8] E. G. SCHNEIDER: J. Opt. Soc. Amer. **30**, 128 (1940).
[9] R. LADENBURG and C. C. VAN VOORHIS: Phys. Rev. **43**, 315 (1933).
[10] E. G. SCHNEIDER: J. Chem. Phys. **5**, 106 (1937).
[11] E. C. G. STUECKELBERG: Phys. Rev. **42**, 518 (1932).
[12] K. WATANABE, E. C. Y. INN and M. ZELIKOFF: J. Chem. Phys. **21**, 1026 (1953). (This paper includes techniques with a vacuum monochromator and photomultiplier intensity measurements.)
[13] R. W. DITCHBURN and D. W. O. HEDDLE: Proc. Roy. Soc. Lond., Ser. A **220**, 61 (1953).
[14] G. L. WEISSLER and PO LEE: J. Opt. Soc. Amer. **42**, 200 (1952).
[15] K. C. CLARK: Phys. Rev. **87**, 271 (1952).
[16] PO LEE: J. Opt. Soc. Amer. **45**, 703 (1955).
[17] W. M. PRESTON: Phys. Rev. **57**, 887 (1940). (Measurements of absorption in atmospheric gases at HLy α.)
[18] A. A. ABOUD, J. P. CURTIS, R. MERCURE and W. E. RENSE: J. Opt. Soc. Amer. **45**, 767 (1955) (I.).

In N_2, absorption cross sections were obtained by WEISSLER, LEE and MOHR[1] between 300 and 1200 Å. A reasonable estimate of the magnitude and contour of the ionization continuum was also given. The region of low absorption between 900 and 1050 Å was remeasured by PO LEE and the results are contained in his O_2 paper[2]. In the above quoted work of DITCHBURN and associates ([2], p. 327) on absorption cross sections at and near LYMAN α, values are tabulated for N_2, O_2, and other, minor constituents of the atmosphere. The region from 150 to 1000 Å has been measured in both N_2 and air by CURTIS[3] using two photomultiplier tubes as radiation detectors. In the N_2 ionization continuum his cross sections are about 40% lower than those of WEISSLER, LEE, and MOHR, probably because of instrumental factors. The production of positive ions in a low pressure atmosphere of N_2 due to absorption of photons excited by electron impacts in another region of the same gas has been studied by MAUNSSELL[4].

7. Absorption cross section in molecular gases and vapors. The researches on gases other than O_2 or N_2 are less detailed and much work still remains. While the absorption spectrum of H_2 has been studied by HOPFIELD[5], BEUTLER, DEUBNER and JÜNGER[6] and, most extensively, by TANAKA[7], no absolute values were obtained except for the preliminary results of PO LEE and WEISSLER[8] between 470 Å and 880 Å, who also estimated the contour of its ionization continuum.

Absorption cross sections for CO and CO_2 have been obtained by WATANABA, INN and ZELIKOFF[9] between 1050 and 1800 Å and by SUN and WEISSLER[10] from 370 to 1300 Å.

NO has been investigated by J. MAYENCE[11] down to 1400 Å, by MARMO[12] between 1100 and 2300 Å, while SUN and WEISSLER[13] give preliminary data from 370 to 1300 Å.

Cross sections have been measured in N_2O by ROMAND and MAYENCE[14] from 1400 to 2200 Å and with different techniques, by ZELIKOFF, WATANABE and INN[15] between 1100 and 2100 Å, with acceptable agreement between them.

WILKINSON and JOHNSTON[16], MOE and DUNCAN[17], and WATANABE, INN and ZELIKOFF[18] investigated the absolute absorption of CH_4 at wavelengths longer than 1000 Å. The region of the main continuum lies towards shorter wavelengths and was measured by DITCHBURN[19] and by SUN and WEISSLER[20] who found good

[1] G. L. WEISSLER, PO LEE and E. I. MOHR: J. Opt. Soc. Amer. **42**, 84 (1952). (Includes experimental details.)

[2] PO LEE: J. Opt. Soc. Amer. **45**, 703 (1955).

[3] J. P. CURTIS: Phys. Rev. **94**, 908 (1954).

[4] C. D. MAUNSSELL: Phys. Rev. **98**, 1831 (1955).

[5] J. J. HOPFIELD: Nature, Lond. **125**, 927 (1930). — Phys. Rev. **47**, 788 (1935).

[6] H. BEUTLER, A. DEUBNER and H. O. JÜNGER: Z. Physik **98**, 181 (1936); **100**, 80 (1936).

[7] Y. TANAKA: Sci. Pap. Inst. Phys. Chem. Res., Tokyo **42**, 49 (1944). (Contains 53 references.)

[8] PO LEE and G. L. WEISSLER: Astrophys. J. **115**, 570 (1952).

[9] K. WATANABE, E. C. Y. INN and M. ZELIKOFF: J. Chem. Phys. **21**, 1648 (1953).

[10] H. SUN and G. L. WEISSLER: J. Chem. Phys. **23**, 1625 (1955).

[11] J. MAYENCE: Ann. Phys., Paris **7**, 453 (1952). — See also J. GRANIER and N. ASTOIN [C. R. Acad. Sci. Paris **242**, 1431 (1956)], for μ of NO between 150 and 1100 Å.

[12] F. F. MARMO: J. Opt. Soc. Amer. **43**, 1186 (1953); see also K. WATANABE: J. Chem. Phys. **22**, 1564 (1954).

[13] H. SUN and G. L. WEISSLER: J. Chem. Phys. **23**, 1372 (1955) (L).

[14] J. ROMAND and J. MAYENCE: C. R. Acad. Sci. Paris **228**, 998 (1949). — See also N. ASTOIN and J. GRANIER [C. R. Acad. Sci. Paris **241**, 1736 (1955) for μ of N_2O], between 150 and 1100 Å.

[15] M. ZELIKOFF, K. WATANABE and E. C. Y. INN: J. Chem. Phys. **21**, 1643 (1953).

[16] P. G. WILKINSON and H. L. JOHNSTON: J. Chem. Phys. **18**, 190 (1950).

[17] G. MOE and A. B. F. DUNCAN: J. Amer. Chem. Soc. **74**, 3140 (1952).

[18] Private communication to the author in 1953.

[19] R. W. DITCHBURN: Proc. Roy. Soc. Lond., Ser. A **229**, 44 (1955).

[20] H. SUN and G. L. WEISSLER: J. Chem. Phys. **23**, 1160 (1955).

agreement with each other. However, their results are about 50% lower than the theoretical calculations by DALGARNO[1]. All observers cited find the absorption of CH_4 to be continuous over the entire region which is undoubtedly due to its spherically symmetric structure. In contrast, NH_3 shows distinct series of resonance bands above 1150 Å which were analysed by DUNCAN[2]. In this region its absorption cross sections were determined by WATANABE[3] and from 400 to 1300 Å by SUN and WEISSLER[4].

Water vapor occupies a position of importance because of its nearly universal appearance, and yet surprisingly little work has been reported on it. From 1450 to 1850 Å its absorption has been measured by WILKINSON and JOHNSTON[5] and over a more extended range by WATANABE and ZELIKOFF[6] and by A. JOHANNIN-GILLES[7]. The latter found a pronounced pressure dependence of her absorption coefficients. ASTOIN, JOHANNIN and VODAR[8] reported cross sections between 200 Å and 500 Å which were independent of the pressure.

Before concluding this survey on ultraviolet absorption in molecular gases, the work of TANAKA, INN and WATANABE[9] on O_3 between 1050 and 2200 Å deserves mention because of its significance in upper atmospheric physics and in gas discharges. Finally, two examples of the behaviour of more complex molecules may be in order: C_2H_4 (Ethylene) studied in the region between 1065 and 2000 Å ZELIKOFF and WATANABE[10], and C_2H_2 (Acetylene) from 1050 to 1520 Å by MOE and DUNCAN[11].

8. Cross sections in the rare gases. Investigations on the rare gases are of particular significance because of the fact that it is much easier to interpret experimental findings. Their absorption cross sections have been measured by PO LEE and WEISSLER in the region of their respective ionization continua, namely in helium[12] from 240 to 600 Å, in neon[13,14] between 240 and 660 Å and in argon[12] from 350 to 800 Å.

While it is somewhat out of place here it seems advisable to mention at this point the results of EHLER and WEISSLER[15] on the absorption cross sections of atomic nitrogen between 400 and 800 Å. They produced a rather significant concentration of N atoms in a low pressure discharge through N_2 and measured the cross sections of N within milliseconds after this discharge had been turned off.

In all cases reported above, the experimental results are on the whole in good agreement with theoretical calculations to be cited later.

9. Direct photoionization measurements. There remain to be mentioned the direct measurements of ions produced per photon absorbed, the photoionization

[1] A. DALGARNO: Proc. Phys. Soc. Lond. A **65**, 663 (1952). (CH_4 and A, theoretical.)
[2] A. B. F. DUNCAN: Phys. Rev. **47**, 822 (1935); **50**, 700 (1936).
[3] K. WATANABE: J. Chem. Phys. **22**, 1564 (1954).
[4] H. SUN and G. L. WEISSLER: J. Chem. Phys. **23**, 1160 (1955).
[5] P. G. WILKINSON and H. L. JOHNSTON: J. Chem. Phys. **18**, 190 (1950).
[6] K. WATANABE and M. ZELIKOFF: J. Opt. Soc. Amer. **43**, 753 (1953); see also K. WATANABE and A. JURSA: To be published.
[7] A. JOHANNIN-GILLES: C. R. Acad. Sci. Paris **236**, 676 (1953).
[8] N. ASTOIN, A. JOHANNIN and B. VODAR: C. R. Acad. Sci. Paris **237**, 558 (1953). — See also N. ASTOIN [C. R. Acad. Sci. Paris **242**, 2327 (1956)], for μ of H_2O between 160 and 1100 Å.
[9] Y. TANAKA, E. C. Y. INN and K. WATANABE: J. Chem. Phys. **21**, 1651 (1953).
[10] M. ZELIKOFF and K. WATANABE: J. Opt. Soc. Amer. **43**, 756 (1953).
[11] G. MOE and A. B. F. DUNCAN: J. Amer. Chem. Soc. **74**, 3136 (1952).
[12] PO LEE and G. L. WEISSLER: Phys. Rev. **99**, 540 (1955).
[13] PO LEE and G. L. WEISSLER: Proc. Roy. Soc. Lond., Ser. A **219**, 71 (1953).
[14] R. W. DITCHBURN: Private communication to the author of unpublished results on Ne absorption.
[15] A. W. EHLER and G. L. WEISSLER: J. Opt. Soc. Amer. **45**, 1035 (1955).

efficiencies, from which the true photoionization cross sections may be determined by simultaneously measuring the total absorption. Such data have been published recently by WAINFAN, WALKER and WEISSLER[1] on O_2, N_2, CO_2, A, H_2O, H_2 and CH_4 together with ionization thresholds. The latter agreed well with the appearance potentials of ions formed by electron impacts. WATANABE, MARMO and INN[2] used similar techniques and determined the photoionization cross sections of NO and NH_3 between 1070 and 1343 Å, while INN[3] obtained the thresholds of a number of gases with ionization limits longer than 1000 Å. Preliminary results in NO, N_2O, NH_3, C_2H_4 and C_2H_2 have also been reported by WALKER and WEISSLER[4].

10. Summary on theoretical cross sections. Theoretical work on photoionization was of course necessarily limited to atoms or in some isolated instances to quasi-atomic molecules (CH_4). Rather than giving a detailed review it seems more appropriate to mention the more recent summaries which contain extensive references to previous detailed work. In 1946 BATES [*4*] discussed the "calculation of the cross section of neutral atoms and positive and negative ions towards the absorption of radiation in the continuum". This was extended by BATES and SEATON[5] to further details on O, N and C. While the systems treated were primarily of astrophysical interest, some of these methods are directly applicable to photoionization measurements.

In 1951 SEATON[6] presented calculations on the variation of the absorption cross sections in the Ne ionization continuum from the limit (575 Å) down to 400 Å. He was in part motivated by the knowledge that experimental work on this gas was in progress[7,8]. In addition he gave cross sections at the series limit for the elements from B to Ne. With the help of Na wave functions he was able to extend the Ne cross sections to just beyond the L_1-edge at 256 Å and found reasonable agreement with experiment[9].

With similar methods DALGARNO[10] arrived at a cross section at the series limit of argon which agreed well with the data of PO LEE and WEISSLER[11]. However, his calculations on CH_4 (without exchange) were about 50% higher than the experimental values[12,13]. — In He the calculations of VINTI[14] seemed to give a somewhat better fit to the measured cross sections[11] than those of WHEELER[15]. A recent paper of STEWART'S[16] deals with the photoionization of Li and improves upon earlier calculations.

[1] N. WAINFAN, W. C. WALKER and G. L. WEISSLER: J. Appl. Phys. **24**, 1318 (1953); a more complete account to be found in Phys. Rev. **99**, 542 (1955)..

[2] K. WATANABE, F. F. MARMO and E. C. Y. INN: Phys. Rev. **91**, 1155 (1953); see also footnote 3, p. 310.

[3] E. C. Y. INN: Phys. Rev. **91**, 1194 (1953).

[4] W. C. WALKER and G. L. WEISSLER: J. Chem. Phys. **23**, 1540 (NH_3) and 1547 (C_2H_2, C_2H_4) (1955) (L); **23**, 1962 (1955) (L) (NO, N_2O).

[5] D. R. BATES and M. J. SEATON: Mon. Not. Roy. Astronom. Soc. **109**, 698 (1949).

[6] M. J. SEATON: Proc. Roy. Soc. Lond., Ser. A **208**, 498 (1951).

[7] PO LEE and G. L. WEISSLER: Proc. Roy. Soc. Lond., Ser. A **219**, 71 (1953).

[8] R. W. DITCHBURN: Private communication to the author of unpublished results on Ne absorption.

[9] M. J. SEATON: Proc. Phys. Soc. Lond. A **67**, 927 (1954).

[10] See footnote 1, p. 310; also H. BEUTLER: Z. Physik **93**, 177 (1935). (Relative absorption in A, Kr, Xe between 1070 and 600 A, pre-ionization is discussed.)

[11] PO LEE and G. L. WEISSELR: Phys. Rev. **99**, 540 (1955).

[12] R. W. DITCHBURN et al.: Proc. Roy. Soc. Lond., Ser. A **229**, 44 (1955).

[13] H. SUN and G. L. WEISSLER: J. Chem. Phys. **23**, 1160 (1955).

[14] J. P. VINTI: Phys. Rev. **44**, 524 (1933).

[15] J. A. WHEELER: Phys. Rev. **43**, 258 (1933).

[16] A. L. STEWART: Proc. Phys. Soc. Lond. A **67**, 917 (1954).

SEATON[1] made a comparison between theory and experiment for the photoionization cross sections of the alkali metals and gave detailed calculations on Na. The agreement is good at and near the Na series limit [3]. For the other metals neither the theoretical nor the experimental data are sufficiently accurate to warrant a detailed discussion of the existing differences. Even though the theory is able to account for the general trend of photoionization it seems to be less accurate in the alkali vapors than in the rare gases. This lack of agreement in the case of the alkali vapors arises from the fact that great cancellation occurs within the transition matrix elements, and in consequence slight errors in the wave functions used may yield large errors in the photoionization cross sections. The effect of the polarizability, for example, (which can only be taken into account approximately) may be considerable.

II. Experimental methods.

11. Concept and representation of the cross section. One of the earliest and most useful quantitative descriptions of the interaction of light with matter is given in terms of LAMBERT's law. If the diminution of a parallel beam of monochromatic radiation through a homogeneous absorbing medium is considered, it is found that the intensity I is reduced by the same fractional amount in equal succeeding portions of its path. One may write for the decrease in intensity $-\Delta I$ of the incident beam I in going a distance Δx

$$-\Delta I = I \mu \Delta x. \tag{11.1}$$

Since the proportionality constant μ is a property of the medium only, this equation may be integrated to yield the intensity I of the light beam after traversing x centimeters of the absorbing medium,

$$I = I_0 e^{-\mu x}, \tag{11.2}$$

where I_0 is the incident intensity, μ is called the absorption coefficient of the material measured in units of cm^{-1} and is a function of the wavelength of light. For the purposes of this article which concerns itself with gases and vapors exclusively, it is convenient to reduce x to normal conditions of temperature (0° C) and pressure (760 mm Hg), abbreviated by N. T. P. For example, when an absorption chamber of length L cm. contains gas at a pressure of p mm. Hg, then the reduced distance may be calculated from the relationship

$$x = L \frac{p}{760} \frac{273}{T}. \tag{11.3}$$

All absorption coefficients presented here are based on these conditions. This in turn implies that measurements at different pressures and temperatures all reduce to the same value of μ, which is another statement of BEER's law of proportionality between μ and the concentration of the absorber. Thus at N. T. P. the absorption coefficient is proportional to LOSCHMIDT's number $n_0 = 2.687 \times 10^{19}$ molecules per $cm.^3$: $\mu = \sigma n_0$. The constant of proportionality σ has the units of $cm.^2$ and may be interpreted as the target area or cross section which one molecule presents with respect to the incident radiation. If molecules are sufficiently far apart so that disturbing influences of intermolecular forces are negligible, σ will be a characteristic constant of the molecule and, in line with BEER's law, independent of pressure and temperature.

[1] M. J. SEATON: Proc. Roy. Soc. Lond., Ser. A **208**, 418 (1951).

It is often convenient to express the probability of an absorption transition, such as for instance the excitation of an electron from the ground state of the O_2 molecule into the SCHUMANN-RUNGE dissociation continuum of the ${}^3\Sigma_u^-$ state, in terms of the classical oscillator strenth[1] or f-value. This can be shown to be related to the absorption coefficient by the equation

$$f = \frac{m c^2}{\pi e^2 n_0} \int_{\nu_1'}^{\nu_2'} \mu_{\nu'}\, d\nu', \tag{11.4}$$

where ν' is the wavenumber in reciprocal centimeters, m and e the electronic mass and charge, and n_0 the number of molecules per cm.³ at N. T. P. The sum rule states that for a single electron the integration above over all its possible transitions will lead to a value $\sum f = 1$. The f-values also appear together with their characteristic wavelengths in the dispersion equations for gases and vapors[2].

According to recent custom, μ and σ are both frequently used, and for the convenience of the reader all measurements will be simultaneously expressed in terms of absorption coefficients at N. T. P. and cross sections, with σ in units of Megabarn (Mb) or 10^{-18} cm.² All experimental work to determine these quantities may therefore be reduced to the problem of evaluating, at different wavelengths, the equation

$$\mu = n_0 \sigma = \frac{1}{x} {}^e\!\log \frac{I_0}{I}. \tag{11.2a}$$

It should be clear that this relationship only describes the attenuation of a monochromatic light beam. No assumption as to the mechanisms by which this may occur is involved. However, such detailed information may be inferred, as will be shown later, from the course of the curves of σ versus λ together with the energy levels of the molecules under consideration.

12. General methodology for the determination of cross sections. It will be assumed in the following sections that certain basic experimental conditions, such as the intensity calibration of photographic plates, the constancy of light sources, and others have been fulfilled adequately. Such detailed questions will be dealt with separately at the end of subdivision II.

In the previous section the exponential law of the attenuation of light on passing through a medium has been introduced, but it has been left open as to how the actual measurements were conducted. There are certain general features which all individual methods have in common and which may be illustrated best by discussing first an idealized experimental arrangement. Since the ultimate aim is the measurement of photoionization cross sections, it is necessary to disperse the light from an appropriate source by means of a spectrograph or a monochromator into its component wavelengths, then to select a beam of monochromatic radiation and pass it through a region which contains the absorbing gas. In this region provision must be made to measure several quantities: 1. the ratio of I_0/I and x to give the total absorption cross section μ, 2. the number of ions created per second and 3. the absolute value of I_0, that is at a given wavelength the number of photons per second impinging on a column of the absorbing gas. From a knowledge of the total absorption μ and the initial number of quanta available, one can calculate the number of photons which are removed each second from the light beam over a specific section of its path.

[1] E. U. CONDON and G. H. SHORTLEY: The Theory of Atomic Spectra. Cambrigde University Press 1935.

[2] For example: The Dispersion of Standard Air, B. EDLÉN: J. Opt. Soc. Amer. **43**, 339 (1953); for O_2: R. LADENBURG and G. WOLFSOHN: Z. Physik **79**, 42 (1932).

If the ions created along this section are collected by appropriate electric fields, then the ratio of the ion current to the number of photons absorbed yields the photoionization efficiency. This number need only be multiplied by the total absorption cross section σ to give the photoionization cross section σ_i directly and as a function of wavelength. However, because of technical difficulties this seemingly straight-forward procedure has only very recently been attempted, and most investigators confined themselves to one or the other sections of the idealized experiment. In order to facilitate the reading of the following section, a summary of the various methods seems in order:

(I) Measurements of absorption cross sections.

(Ia) The absorption chamber is placed between the light source and the spectrograph.

(Ib) The absorption gas fills the spectrograph.

(Ic) The absorption chamber is located between the exit slit of the monochromator and the light detector.

(II) Measurements of direct photoionization.

(IIa) Relative photoionization is measured with an ionization chamber and/or space-charge detector of ions.

(IIb) Relative photoionization is determined as in (IIa) but supplemented with total absorption data.

(IIc) Absorption cross sections and photoionization efficiencies are measured simultaneously, yielding σ_i directly.

13. Method (I), determination of absorption cross sections. In method (Ia) the absorption tube is placed between the constant intensity light source and the spectrograph as shown in Fig. 1, which has been taken from DITCHBURN'S work [3] on alkali vapors. Undispersed radiation from say a hydrogen glow discharge passes through the chamber filled with the absorbing gas or vapor and impinges on the slit of a spectrograph. The ratio of transmitted to incident light intensities, I/I_0, may be obtained at different wavelengths λ from the photographic densities d of successive exposures for which the gas pressure is changed from zero to higher values. Usually, the pressures and/or the length of the absorption path are chosen such as to give, at any one wavelength, values of densities $d_{p_1}, d_{p_2}, d_{p_3}, \ldots$ which lie on the linear portion of the plate calibration curves d versus $^{10}\mathrm{Log}\,(I\,t)$. The reduced distance x is calculated with the aid of eq. (11.3) from the optical path (length of absorption tube) and the gas pressure and temperature. This together with a knowledge of $(I/I_0)_\lambda$ leads to the cross section versus wavelength curve.

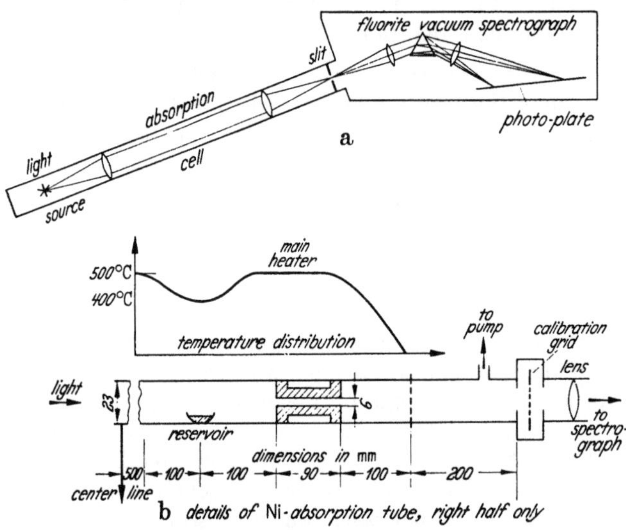

Fig. 1 a and b. Experimental method (Ia) to measure absorption coefficients in metallic vapors. a) Schematic; b) details of DITCHBURN Ni-absorption tube, right half only. Fluorite optics.

This method is particularly useful for work with alkali and other vapors which require higher temperatures to obtain appropriate pressures. However, it has also been used successfully with gases. A further advantage is that background radiation due to stray light may be kept to a minimum since the spectrograph is maintained at a high vacuum thus reducing the background from scattered radiation. Usually the absorption tube is closed off at its ends by either quartz or fluorite windows and can be pumped separately. Occasionally this method has been employed with a photomultiplier tube behind an exit slit placed at the focusing position instead of a photographic plate. This avoids the rather tedious intensity calibration of plates and facilitates the use of electromechanical recording of intensity data.

Method (Ib) is in most aspects similar to (Ia) with the main difference that the absorbing gas occupies the body of the spectrograph itself. Its principle

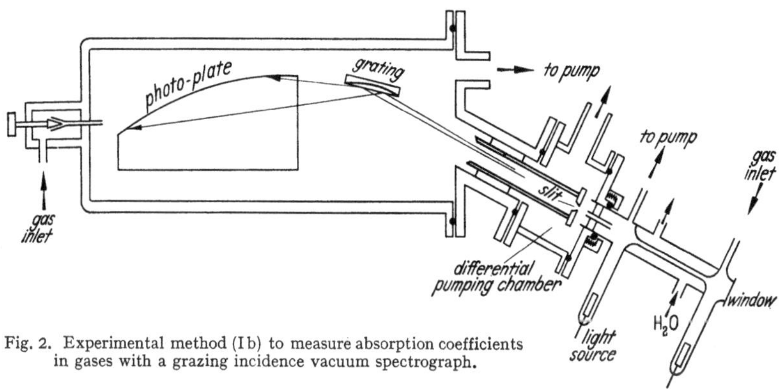

Fig. 2. Experimental method (Ib) to measure absorption coefficients in gases with a grazing incidence vacuum spectrograph.

features are shown in Fig. 2 as used by WEISSLER and associates[1] for work in gases below the lithium fluoride cut-off at 1050 Å. One advantage over (Ia) is that the light source, low-pressure sparks through a capillary, can be placed close to the slit of the spectrograph. This provides for sufficient intensities at reasonable exposure times for spark lines down to 200 Å or less. In order to separate the discharge carrier gas in the source from the absorbing gas in the spectrograph, a separately pumped chamber occupies the space between the slit and the source. Otherwise no windows are used. However, DITCHBURN has also been successful in separating these two regions with multiple layers of very thin collodion films and still obtaining sufficient intensities down to below 400 Å. It is of course possible to use method (Ib) equally well with fluorite optics.

Method (Ic) employs a vacuum spectrograph or monochromator with the absorption chamber located behind the exit slit. Fig. 3 shows a particularly versatile instrument designed by TOUSEY et al.[2] and adapted to absorption measurements by WATANABE and his group[3]. Behind the absorption cell is located a commercial photomultiplier tube whose glass envelope is coated with sodium salicylate, a material which fluoresces in the sensitive region of the multiplier when stimulated by radiation of the vacuum ultraviolet. With this method absorption cross sections may be obtained by recording the relative intensities I and I_0, as registered by the multiplier, when the absorption cell

[1] G. L. WEISSLER, PO LEE and E. I. MOHR: J. Opt. Soc. Amer. **42**, 84 (1952). — PO LEE: J. Opt. Soc. Amer., **45**, 703 (1955).
[2] R. TOUSEY, F. S. JOHNSON, J. RICHARDSON and N. TORAU: J. Opt. Soc. Amer. **41**, 669 (1951).
[3] K. WATANABE, E. C. Y. INN and M. ZELIKOFF: J. Chem. Phys. **21**, 1026 (1953).

contains a gas or is empty. The wavelength reaching the exit slit is changed continuously in TOUSEY's instrument by displacing the grating along the ROWLAND focusing circle.

Finally, it should be pointed out that this method is particularly useful for σ-measurements in gases which are sensitive to undesirable photochemical reactions such as $N_2O + h\nu \rightarrow N_2 + O$ and $O + N_2O \rightarrow 2NO$. The light flux in method (Ic) is by many orders of magnitude smaller than in either (Ia) or (Ib) and such reactions are far less disturbing.

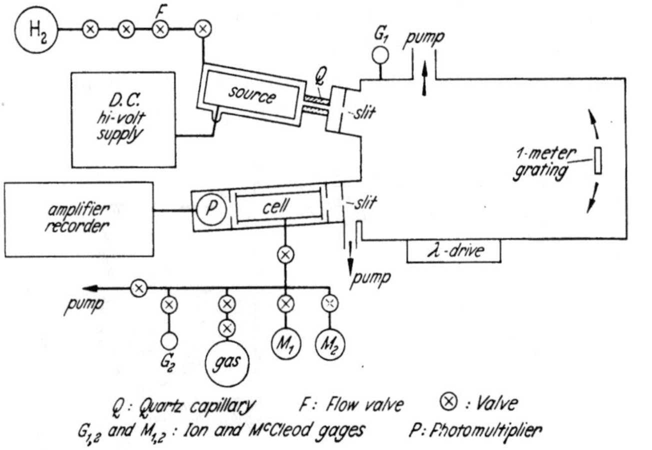

Fig. 3. Normal incidence vacuum monochromator, methods (Ic and IIc).

14. Method (II), direct determination of photoionization. In method (IIa) MOHLER[1], LAWRENCE and others measured relative photoionization in the alkali vapors as a function of wavelength. They employed either a tungsten filament lamp, a quartz mercury arc, or a hydrogen glow discharge as a light source and passed the radiation through a quartz monochromator into an ionization chamber or a positive ion space-charge detector both containing alkali vapors (Fig. 4a and b). The absolute photon flux was determined with a sensitive thermopile placed at the position of the ion detector. Thus, by collecting all ions formed in the chamber with a few volts across the electrodes, the relative ionization per unit light intensity could be measured.

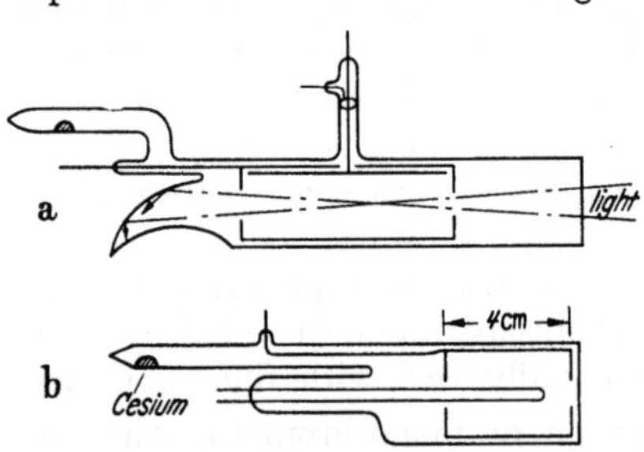

Fig. 4a and b. a) MOHLER's quartz-enclosed ionization chamber; b) his positive ion space-charge detector; methods (IIa, b).

While an ionization chamber operated under saturation conditions gives the absolute number of ions formed, its sensitivity is very much lower than that of a positive ion space-charge detector. This method was devised by HERTZ and independently by KINGDON[2], who measured its characteristics in some detail. The KINGDON cage, as it is also referred to, is essentially a diode operated in the space-charge limited region of its emission current with a plate potential of about

[1] F. L. MOHLER and C. BOECKNER: J. Res. Nat. But. Stand. **3**, 303 (1929).

[2] K. H. KINGDON: Phys. Rev. **21**, 408 (1923). — D. M. MILLEST: Phil. Mag. **37**, 323 (1946).

1 volt. With an alkali vapor present its ions may be detected by the change in emission current they produce by neutralization of the electron space-charge cloud. Since an individual ion on the average remains un-neutralized for a time interval many orders of magnitude larger than that of an electron, it may liberate from 10^4 to 10^6 electrons, a change in emission which is easily detected. Despite the great sensitivity of this instrument to positive ions and simultaneously its insensitivity to photoelectrons (because of the negligibly small area of its cathode filament), it has rather marked disadvantages for absolute work: difficulties in calibration, concentration gradients of the vapor in the vicinity of the hot filament, and other caprices. Nevertheless, much careful work on photoionization of alkali vapors was accomplished to a large extent with the aid of such a detector.

Even though little use has been made of method (IIb), MOHLER's absolute cross sections in Rb and Cs are still of value today. He employed eq. (11.1),

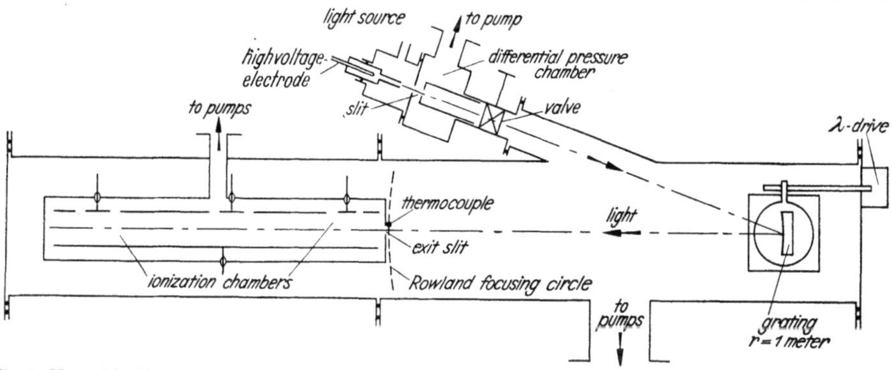

Fig. 5. Normal incidence vacuum monochromator with 3 ionization chambers in tandem behind the exit slit, methods (IIc and Ic).

$-\Delta I = \mu I \Delta x$, to evaluate the data obtained with the same procedure as in (Ic). By determining the light intensity I in ergs per second at the position of the ionization chamber with a thermocouple, and by assuming that the photon absorption in the chamber measured in terms of an ion current i could be completely accounted for by photoionization, namely $\frac{\Delta I}{h\nu} = \frac{i}{e}$ (ions produced per second), he wrote eq. (11.1) in the form $\frac{i h \nu}{e} = I \mu \Delta x$. Since furthermore $\mu = n_1 \sigma_i p$ where n_1 = number of atoms per cm.^{3} at a pressure of 1 mm. Hg and p was the vapor pressure, the photoionization cross section σ_i could be evaluated. This information was supplemented in a separate experiment in which an absorption tube containing Cs vapor was placed between the exit slit of the monochromator and the space-charge detector. By procedures of method (Ic) total absorption cross sections were obtained and found to be substantially in agreement with his values of σ_i.

Method (IIc) attempts the simultaneous measurements of total absorption cross sections σ and photoionization efficiencies to yield upon multiplication σ_i directly. The experimental arrangement shown in Fig. 5 is the one initiated by WEISSLER and associates[1] and is in its essential details similar to the one developed subsequently by WATANABE and his group. Light from a powerful spark through a ceramic capillary passes through a differential pumping chamber and illuminates the primary slit of a normal incidence 1 meter grating vacuum monochromator where it is dispersed and focused on an exit slit 0.3 mm. wide and 2 mm. long. The wavelength passing through this exit slit is changed by rotating the

[1] N. WAINFAN, W. C. WALKER and G. L. WEISSLER: J. Appl. Phys. **24**, 1318 (1953). — Phys. Rev. **99**, 542 (1955).

grating about its own vertical axis. Tests revealed that negligible defocusing occurred. Immediately adjacent to the exit slit and also in the focusing plane is placed a sensitive HILGER-SCHWARTZ thermocouple for absolute energy (photon flux) measurements. Radiation of known wavelength and intensity passes through the exit slit into a separately pumped chamber behind it, which contains the absorbing gas and three parallel plate ionization chambers in tandem separated by about 25 cm. each. [This chamber behind the exit slit can of course also be used for absorption work alone as in method (Ic) with a photomultiplier at the beginning and another one at the end of the optical path to yield simultaneously I_0 and I of eq. (11.2)]. From the respective ion currents measured separately by the three collectors, one in front, one in the middle, and one at the end of the chamber, it is possible to obtain the total absorption cross section σ of the gas (at a known pressure) for wavelentghs shorter than that corresponding to the ionization potential. This value of σ may then be used to calculate the number of photons which are removed each second from the beam under the first ionization chamber. The ratio of this number and the ions per second produced there yields immediately the photoionization efficiency. Some characteristic curves representing a number of different gases have been taken from the work of WEISSLER's group and assembled in Fig. 6. It should be realized that techniques have not as yet been perfected enough to provide sufficiently closely spaced measurements. Therefore the shape of the complex curves, such as the one for O_2, will undoubtedly exhibit still more structure in future investigations.

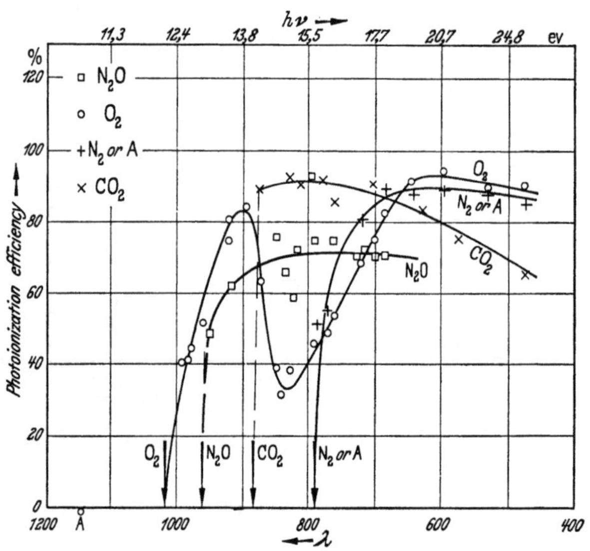

Fig. 6. Ionization efficiencies of various gases.

Falsification of these results by photoelectrons from walls and electrode surfaces has been looked for and is insignificant.

15. Some experimental details. An adequate source of light is of paramount importance in all methods and should be designed with particular attention to the following requirements: 1) It should have a high degree of constancy of intensity at all wavelengths, 2) it should emit a closely spaced line spectrum which can be resolved by the spectrograph or monochromator, and 3) it should furnish high intensities especially if method (IIc) is used. With respect to point 2) above it should be noted that a line spectrum is desirable for two reasons, first and most important, the interval between lines helps to establish a reference level to subtract any stray radiation and second to determine an accurate wavelength scale and single out higher order spectral lines. An emission continuum, such as that due to molecular helium[1] or the LYMAN continuum[2], is invaluable to deter-

[1] Y. TANAKA: Sci. Pap. Inst. Phys. Chem. Res., Tokyo **39**, 465 (1942). J. Opt. Soc. Amer. **45**, 710 (1955).

[2] R. E. WORLEY: Rev. Sci. Instrum. **13**, 67 (1942).

mine qualitatively the course of absorption, particularly for purposes of identification and classification to measure the exact positions and detailed structures of such regions as molecular resonance bands, pre-dissociated and pre-ionized bands, and pre-ionization lines [5]. However, these wavelength-continuous sources may produce quite deceptive results when applied to quantitative absorption and photoionization work.

There are two types of light sources which have proved to be most useful. For the region above 1000 A a glow discharge of the order of one ampere through H_2 yields a very intense many-lined spectrum[1] when passed through a water-cooled glass tube of about 15 cm. length and 5 mm. diameter. Below 1000 Å down to about 150 Å a low pressure spark through a ceramic capillary of high melting point seems to be best[2]. If impurity gases are carefully controlled and the pressure is kept steady, the breakdown potential and with it the intensity of emission will be constant to better than 5%. Its spectrum is mainly characterized by the discharge carrier gas flowing through the capillary and thus can be changed to cover different regions of the wavelength scale. At 700 Å and a band width of 5 Å it will produce an intensity of about 10 ergs per second behind an exit slit (method IIc) of 0.3×2 mm. dimensions.

The calibration of photographic plates in terms of curves plotting optical density versus ^{10}Log (intensity × time) may be accomplished by several methods of equal reproducibility. DITCHBURN[3] inserts into the light beam near the grating metal grids of known transmissions which is a technique similar to that of LADENBURG and VAN VOORHIS, who used sectored disks of different sizes in front of their fluorite lens. PO LEE[4] calibrated some of his plates in the spectrograph by varying the time of exposure within the limits of his absorption exposure times. He also refined a method originally developed by SCHNEIDER and only applicable to plates which are sensitized to the far ultraviolet by a thin lacquer film of fluorescent material. By irradiating a sample of this lacquer with 2537 Å light he excited and measured its fluorescent emission band which showed a peak at 3150 Å. Then the photographic plates could be calibrated in air with one curve for all wavelengths (1300 to 300 Å) using time exposures of 3150 Å light emitted from a mercury arc and selected by a quartz monochromator.

JOHNSON, WATANABE and TOUSEY[5] have made a detailed investigation of photomultipliers which were sensitized with various externally applied fluorescent materials for radiation of the far ultraviolet. The relative quantum efficiencies of fluorescence were studied for a number of materials, and among them only sodium salicylate and Eastman UV fluorescing lacquer exhibited nearly constant efficiencies between 2000 and 1000 Å with the salicylate being much less sensitive to stray light than the lacquer. Perhaps it is informative to mention

[1] E. O. LAWRENCE and N. E. EDLEFSEN: Rev. Sci. Instrum. **1**, 45 (1930). — F. S. JOHNSON, K. WATANABE and R. TOUSEY: J. Opt. Soc. Amer. **41**, 702 (1951).

[2] PO LEE and G. L. WEISSLER: J. Opt. Soc. Amer. **42**, 80 (1952). see also footnote 1, p. 317. In this connection, the "sliding spark", a source used successfully by VODAR and associates, should be mentioned. It is essentially a vacuum spark between metal electrodes (steel, Al, Cu, etc.) similar to MILLIKAN's hot spark, but sliding on either a semiconductor or an insulator placed between the electrodes. B. VODAR and J. ROMAND: Michrochim. Acta **1955**, 429. — J. ROMAND, G. BALLOFFET and B. VODAR: C. R. Acad. Sci., Paris **240**, 412 (1955).

[3] R. W. DITCHBURN, J. TUNSTEAD and J. G. YATES: Proc. Roy. Soc. Lond., Ser. A **181**, 386 (1943).

[4] PO LEE and G. L. WEISSLER: J. Opt. Soc. Amer. **43**, 512 (1953).

[5] F. S. JOHNSON, K. WATANABE and R. TOUSEY: J. Opt. Soc. Amer. **41**, 702 (1951); see also S. E. WILLIAMS, M. R. MEHARRY, V. W. MASLEN and R. L. FALCANER; J. Opt. Soc. Amer. **44**, 654 (1954); S. ROBIN and B. VODAR: J. Phys. Radium **12**, 634 (1951).

in this connection the constant γ region[1] for this lacquer between 300 and 1300 Å, where γ is the slope of the plate calibration curve d versus ${}^{10}\mathrm{Log}\,(It)$.

The use of thermocouples for absolute measurements of the intensities of spectral lines in the vacuum ultraviolet has been described by PACKER and LOCK[2].

In order to separate, in the alkali vapors, the atomic absorption cross section σ from the molecular one[3], it is assumed that σ varies linearly with the concentration C. Since the formation of molecules depends on the number of collisions per unit time, the molecular absorption σ_M will be proportional to C^2. Thus the total or measured absorption may be expressed as

$$\sigma_{\mathrm{tot}} = \sigma C + \sigma_M C^2 \quad \text{or} \quad \frac{\sigma_{\mathrm{tot}}}{C} = \sigma + \sigma_M C. \tag{15.1}$$

Here the concentration C is in turn proportional to the vapor pressure p divided by the mean temperature of the absorbing column. By plotting σ_{tot}/C against C, a straight line should result, and σ as well as σ_M can be determined from the intercept on the axis and the slope. The vapor pressure may be calculated from the temperature using an empirical equation of the form

$${}^{10}\mathrm{Log}\, p_{\mathrm{mm}} = -\frac{A}{T} - B\,{}^{10}\mathrm{Log}\,(T) - C, \tag{15.2}$$

where the constants A, B, and C may be obtained from DITCHBURN and GILMOUR[4], Table 2, for a number of monatomic vapors. Eq. (15.2) may be justified by integrating the CLAUSIUS-CLAPEYRON equation with the help of limiting assumptions.

When analysing experimental absorption data, it is of importance to keep in mind the resolution of the optical instrument employed, particularly in the range where the wavelength is longer than the first ionization limit. In molecular gases this region is usually profusely populated with absorption bands, sharp or diffuse, and as a consequence the cross section may change rapidly by many orders of magnitude depending upon whether it is measured on or between lines of a given band. Only in bands strongly diffuse because of predissociation may one expect reasonably accurate values of σ. Similar problems from preionized lines or bands arise in the regions near the ionization limits of both atoms and molecules. Even in the region of the ionization continuum itself lines or bands may be produced by the excitation of inner or more tightly bound electrons [5], [6].

This enables certain statements to be made with respect to gas purity. If measurements in the main ionization continuum are intended, impurities of the order of 1% may be tolerated since these continua are all of about the same magnitude. In contrast much greater care has to be exercised in regions of weak continua and strong resonance bands.

III. Experimental cross sections of photoabsorption and photoionization.

16. Introductory remarks. In the following the experimental results will be plotted uniformly with the ordinate in terms of both absorption coefficients μ and cross sections σ as defined by equations (11.2) and (11.2a) with μ in units

---

[1] Po LEE and G. L. WEISSLER: J. Opt. Soc. Amer. **43**, 512 (1953).

[2] D. M. PACKER and C. LOCK: J. Opt. Soc. Amer. **41**, 699 (1951).

[3] R. W. DITCHBURN, J. TUNSTEAD and J. G. YATES: Proc. Roy. Soc. Lond., Ser. A **181**, 386 (1943).

[4] R. W. DITCHBURN and J. C. GILMOUR: Rev. Mod. Phys. **13**, 310 (1941).

of cm^{-1} at N.T.P. and σ in Megabarn (Mb) or 10^{-18} cm^2. For greater ease of orientation the wavelength scale is in angstrom units and, simultaneously, in equivalent volts (ev.). Furthermore it should be noted that all absorption data have been shown to be independent of pressure and temperature except where specifically noted otherwise. Photoionization in the metal vapors at wavelengths longer than the first ionization limit, or other two-step processes, will not be discussed beyond what has been said about this subject in Sect. 4 and 5. A similar limitation applies to the presentation and discussion of absorption data in the more complex molecular gases and vapors, particularly the organic ones.

On the other hand, the absorption data of the molecular gases, shown in the following figures, usually include values for wavelengths longer than the first ionization limit. As already mentioned these regions exhibit strong absorption in resonance bands and dissociation continua and are of considerable importance to the understanding of, for instance, photoionization in gas mixtures [*1*], [*2*]. In addition, of course, an analysis of the position, structure, and magnitude of absorption in bands and continua leads to new information with respect to the electronic structure of molecules such as strong or weak bonding of electrons, oscillator strengths and transition probabilities. Details of this nature will not be discussed here except in specific cases which aid in the interpretation of the experimental results. In order to facilitate a more thorough understanding of the data presented in the following sections it might be well to keep in mind the different mechanisms of light absorption [*5*], [*6*] in atoms and molecules:

(a) Excitation to upper states (resonance transitions without pre-dissociation or pre-ionization; their appearance is that of sharp atomic absorption lines or of molecular bands with well delineated structure).

(b_1) Excitation to molecular states which are adjacent to a dissociation continuum (pre-dissociation; all or part of the band structure is missing, accompanied by a generally diffuse appearance. However, the absorption is not truly continuous).

(b_2) Excitation to upper states of atoms or molecules which are adhacent to an ionization continuum (pre-ionization, also called auto-ionization; atomic absorption lines are broad and diffuse, structure in molecular bands is wholly or partly missing and of diffuse appearance, but again the absorption is not truly continuous).

(c) Continuous absorption due to (1) direct excitation into an atomic ionization continuum; (2) dissociation of molecules into two fragments, each in the ground state; (3) dissociation of molecules into two fragments, at least one of which is excited; (4) direct excitation into one of the molecular ionization continua (without dissociation or pre-ionization); (5) dissociative ionization into an ion and an atom which may be in an excited state. (It is also possible to obtain two ions of opposite sign with one of them in an excited state.)

17. A general survey of cross sections in the main ionization continua. In the array of curves presented in Fig. 7 the absorption cross sections σ are plotted versus the wavelength, usually starting at the ionization limit. In some cases the curves are carried towards longer wavelengths and the limit is indicated by a vertical arrow marked I.P. Solid lines are meant to indicate more reliability than dotted or dashed ones. If only the photoionization cross section has been measured by method (IIc) the curve is marked with μ_i. Unfortunately a wide comparison of work from different observers is not possible because of their small number and the newness of these investigations. Curves [2], [3] and [4] are from DITCHBURN's group[1] with the exception of Rb in diagram [3] which is due to MOHLER and BOECKNER[2]. The other results originated in WEISSLER's laboratory[3] except for CH_4 in diagram [9] where his cross sections are averaged

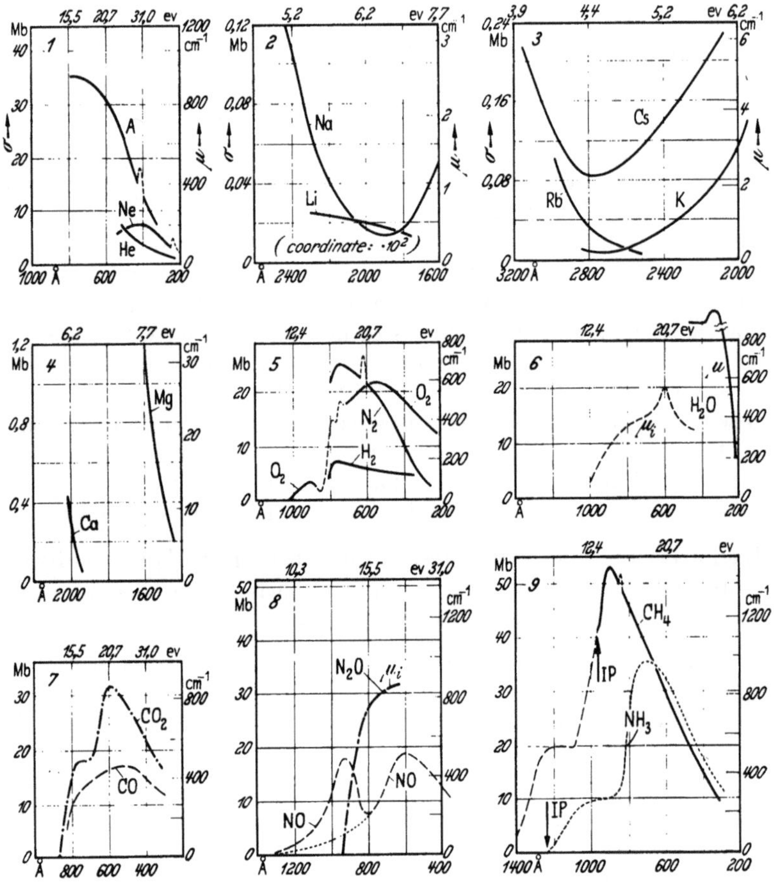

Fig. 7. A survey of absorption cross sections in the main ionization continua of various gases and vapors.

---

[1] Li: J. TUNSTEAD: Proc. Phys. Soc. Lond. A **66**, 304 (1953). — Na: R. W. DITCHBURN, P. J. JUTSUM and G. V. MARR: Proc. Roy. Soc. Lond., Ser. A **219**, 89 (1953). — Cs: H. J. BRADDICK and R. W. DITCHBURN: Proc. Roy. Soc. Lond., Ser. A **143**, 472 (1934). — K: R. W. DITCHBURN, J. TUNSTEAD and J. G. YATES: Proc. Roy. Soc. Lond., Ser. A **181**, 386 (1943). — Ca: P. J. JUTSUM: Proc. Phys. Soc. Lond. A **67**, 190 (1954). — Mg: R. W. DITCHBURN and G. V. MARR: Proc. Phys. Soc. Lond. A **66**, 655 (1953).

[2] F. L. MOHLER and C. BOECKNER: J. Res. Nat. Bur. Stand. **3**, 303 (1929).

[3] A and He: PO LEE and G. L. WEISSLER: Phys. Rev., **99**, 540 (1955). — Ne: PO LEE and G. L. WEISSLER: Proc. Roy. Soc. Lond., Ser. A **219**, 71 (1953). — N_2: G. L. WEISSLER, PO LEE and E. I. MOHR: J. Opt. Soc. Amer. **42**, 84 (1952). — O_2: G. L. WEISSLER and PO LEE: J. Opt. Soc. Amer. **42**, 200 (1952). — O_2 and N_2: PO LEE: J. Opt. Soc. Amer., **45**, 703 (1955). — H_2: PO LEE and G. L. WEISSLER: Astrophys. J. **115**, 570 (1952). — H_2O: N. WAINFAN, W. C. WALKER and G. L. WEISSLER: Phys. Rev. **99**, 542 (1955); also contains data on O_2, N_2, A, H_2, CO_2 and C_2H_4; see also J. Appl. Phys. **24**, 1318 (1953). — CO and CO_2: H. SUN and G. L. WEISSLER: J. Chem. Phys., **23**, 1625 (1955); see also references for H_2O. — NO: H. SUN and G. L. WEISSLER: J. Chem. Phys. **23**, 1372 (1955) (L). — W. C. WALKER and G. L. WEISSLER: J. Chem. Phys., **23**, 1962 (1955) (L). — N_2O: W. C. WALKER and G. L. WEISSLER: J. Chem. Phys., **23**, 1962 (1955) (L). — CH_4 and NH_3: H. SUN and G. L. WEISSLER: J. Chem. Phys., **23**, 1160 (1955); see also references for H_2O. — C_2H_4 and C_2H_2: W. C. WALKER and G. L. WEISSLER: J. Chem. Phys., **23**, 1547 (1955) (L).

with DITCHBURN'S[1] and for H_2O in diagram [6] which were obtained by VODAR and associates[2] between 250 and 460 Å.

The contours of the continua are determined by joining together the lowest absorption cross sections measured. In many instances intense resonance lines or bands, some of them strongly pre-ionized, are superimposed over these curves. For the sake of clarity they have been omitted in Fig. 7 and will be dealt with in the following sections where individual cases will be treated in detail. The two large peaks in NO, diagram [8], may not be characteristic of the ionization continuum even though they are rather broad, and the dotted line below may be more representative.

In general the accuracy of these curves and the ones to follow is about 10 to 15%, except in the case of σ_i-determinations (method IIc) where the error is about twice as large.

In the following Sect. 18 to 22, references to the original experimental work on μ or σ will not be quoted again, since they have been collected in tables 1 and 2 in Sect. 23 which summarize the results.

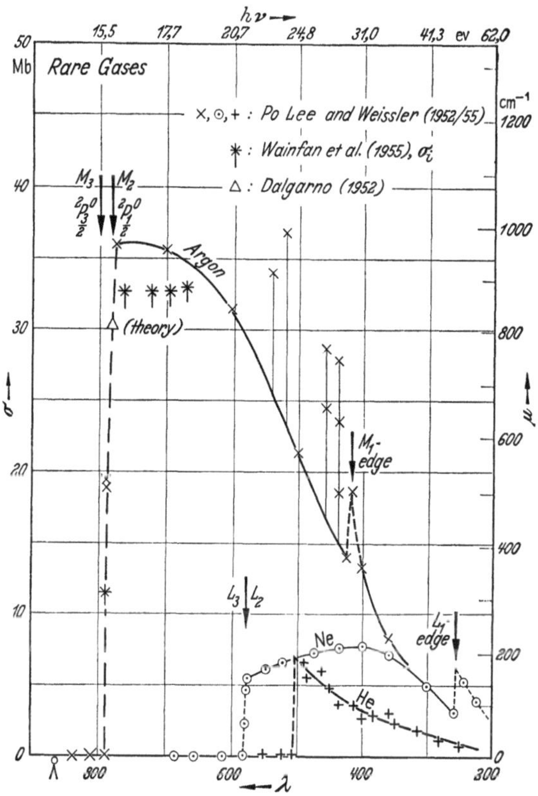

Fig. 8. Photoionization cross sections in Helium, Neon and Argon.

18. Photoionization in monatomic gases and vapors. The absorption mechanisms in the rare gases and in the alkali vapors are the least doubtful, and the experimental data as shown in Fig. 8 (method Ib) are easily interpreted in terms of the removal of an electron from the atom. This is verified by the fact that σ_i values of method (IIc) follow those of σ exactly. The curve in argon begins at its first ionization limits[3] which are obtained spectroscopically from the $3p \to ms$ and $3p \to md$ transitions (m integer) whose levels converge to the two states of the argon ion, namely ${}^2P^0_{\frac{3}{2}}$ and ${}^2P^0_{\frac{1}{2}}$ at 786.72 and 777.96 Å respectively. (This corresponds to the M_3- and M_2-edges in X-ray notation.) They are indicated by two arrows. The removal of a more tightly bound $3s$ electron occurs at the M_1-edge. The levels of the $3p \to md$ transitions which fall between the M_3- and M_2-edges are broadened because of the adjacent M_3-continuum and the additional probability of radiationless transitions from these levels into it [6]. Fig. 9 shows this region of pre-ionization in argon with its diffuse lines.

An entirely analogous situation prevails in Ne with the difference that the region of pre-ionization between the L_3- and L_2-edges is only 2 Å wide[4]. In

[1] R. W. DITCHBURN: Proc. Roy. Soc. Lond., Ser. A **229**, 44 (1955).
[2] N. ASTOIN, A. JOHANNIN-GILLES and B. VODAR: C. R. Acad. Sci. Paris **237**, 558 (1953).
[3] H. BEUTLER: Z. Physik **93**, 177 (1935).
[4] E. U. CONDON and G. H. SHORTLEY: The Theory of Atomic Structure, Cambridge University Press 1951. See also C. H. MOORE: Atomic Energy Levels, Vol. 1, Nat. Bur. Stand. Washington 1949, p. 76.

each case, only one or two experimental points are available in these regions, but in contrast to O_2 they are lower values than those to the short wavelength side of the limits. Helium shows no such pre-ionization. Between the M_1- and M_2-edges in argon a number of points lie above the curve. They may be assumed to fall into pre-ionized lines from levels converging on M_1. The theoretical calculations for these gases (see Sect. 22) agree well with the experimental values.

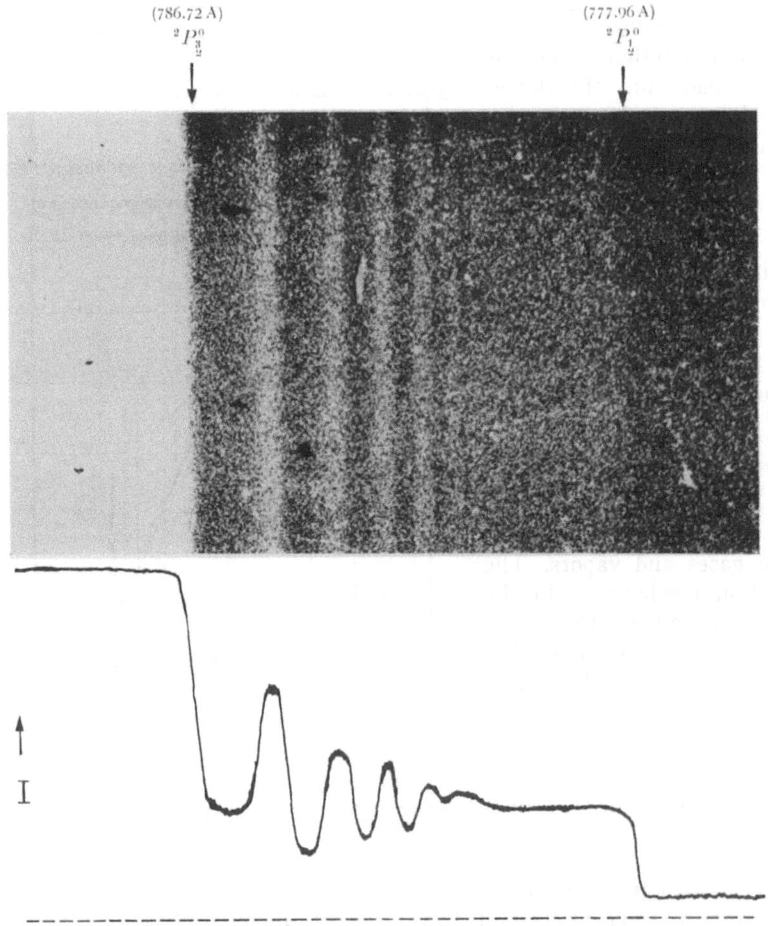

Fig. 9. Pre-ionization absorption spectrum of Argon between 786.72 and 777.96 A.

Representative data for the metallic vapors are shown for potassium in Fig. 10, for thallium in Fig. 11, and for the potassium molecule in Fig. 12, all from DITCHBURN's group (method Ia). If the relative photoionization data (method IIa) of MOHLER and BOECKNER, also indicated by four large circles in Fig. 10, are fitted to the absorption cross section curve, good agreement is obtained. The same is true to a much higher degree for the data of LAWRENCE and EDLEFSEN, which for the sake of greater clarity have been indicated with crosses on the upper, magnified ($\times 10$) curve of DITCHBURN.

In rubidium MOHLER and BOECKNER obtained one absolute measurement at 3150 Å (method IIb) and fitted their relative data to it as in diagram [3] of Fig. 7. Again their results compare well with those of LAWRENCE and EDLEFSEN.

While in cesium at the ionization limit the quantitative results from DITCHBURN and associates agree well with MOHLER's, the curve of the latter decreases more rapidly towards shorter wavelengths, and that of LAWRENCE is still steeper. However, in general the agreement between the results of direct photoionization (methods IIa, b) and absorption (method Ia) is such as to indicate without doubt that the absorption is entirely accounted for by photoionization.

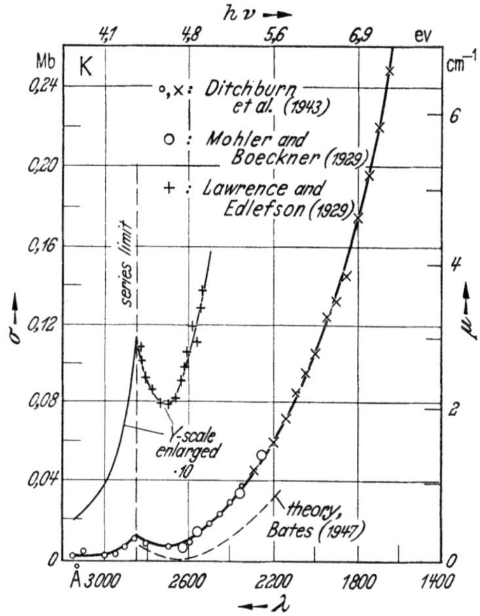

Fig. 10. Absorption cross sections in potassium.

MARR's results in thallium, Fig. 11, are of interest because of the extremely strong pre-ionization lines, giving rise to values far above those of the adjacent continuum. Somewhat similar results have been obtained for indium, but have not as yet been observed in other vapors such as calcium and magnesium.

The absorption cross sections for the potassium molecule[1] are shown in Fig. 12 and are probably due to the reaction[2] $K_2 + h\nu \rightarrow K^+ + K + e$. Other molecular absorption has been estimated by DITCHBURN et al. [3]. For Na_2 it is about 10 Mb near the series limit of Na, decreases to zero at 2100 Å, and then rises again to small values between 2000 and 1600 Å; in Cs_2 it is of the order of 50 Mb and in Li_2 about 500 Mb.

19. Absorption in O_2, N_2, and H_2. Absorption in these gases, particularly in oxygen, will be discussed in detail because they illustrate the various types of mechanisms which may come into play in molecular gases.

In Fig. 13 a survey is given for O_2 covering the entire range from 200 to 2000 Å with three continuous regions clearly in evidence. Immediately adjacent to the SCHUMANN-RUNGE bands the first dissociation continuum starts at 1750 Å, obtains a maximum of $\mu = 490$ cm.$^{-1}$ at 1450 Å and extends down to 1300 Å. It has been measured recently with great care by DITCHBURN and HEDDLE (method Ia) who compare their results with previous work. The second continuum appears at the first ionization limit at 1040 Å and reaches a maximum of $\mu = 100$ cm.$^{-1}$ at 910 Å. From this point on it decreases, according to PO LEE (method Ib), until at about 850 Å it is overlapped by a new region of strong continuous absorption.

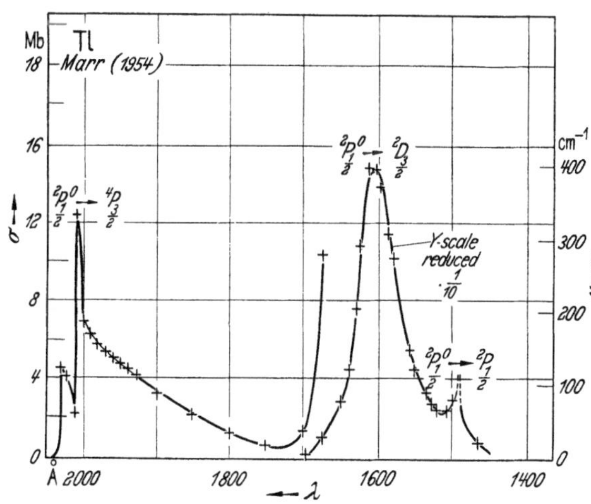

Fig. 11. Absorption cross sections in thallium, note pre-ionizations peaks.

[1] R. W. DITCHBURN, J. TUNSTEAD and J. G. YATES: Proc. Roy. Soc. Lond., Ser. A **181**, 386 (1943).

[2] R. W. DITCHBURN and F. L. ARNOT: Proc. Roy. Soc. Lond., Ser. A **123**, 516 (1929).

Between 850 and 740 Å it is difficult to assign an accurate absorption contour because of the presence of too many closely spaced and somewhat diffuse bands which have been photographed and analysed by PRICE and COLLINS[1]. However, below 740 Å and down to 200 Å PO LEE was again able to determine a smooth contour delineating the main ionization continuum with a maximum of $\mu = 590$ cm.$^{-1}$ at 510 Å. Even here bands are in evidence as may be noted from the experimental points above the contour line and many of them are diffuse in appearance. This together with the direct measurements (method IIc) of μ_i by WAINFAN et al., some of which also lie above the continuum, seems to indicate pre-ionization.

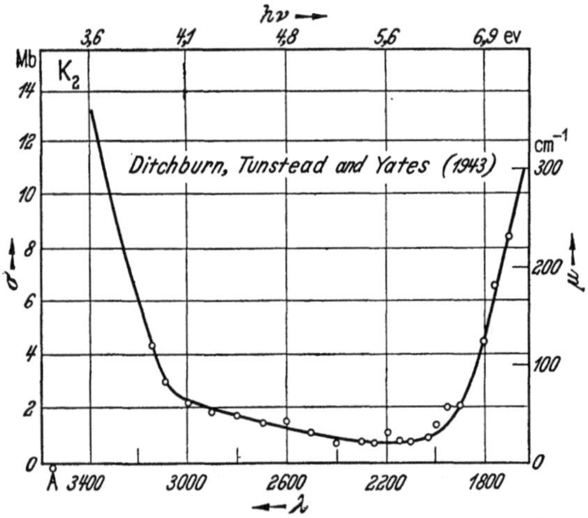

Fig. 12. Absorption cross sections in molecular potassium, K_2.

PRICE and COLLINS and separately TANAKA and TAKAMINE[2] have been able to group these bands into RYDBERG series progressions converging to four distinct series limits, which are indicated by arrows marked I.P.$_{1-4}$. The main continuum below 763 Å (I.P.$_2$) may therefore be interpreted

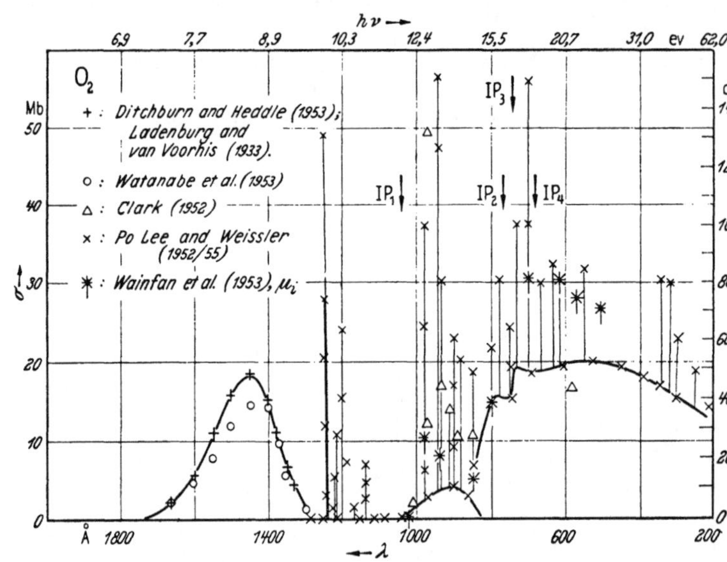

Fig. 13. A survey of absorption coefficients in oxygen. Note the strong resonance bands superimposed over regions of continuous absorption.

as a superposition of three ionization processes. However, the possible contribution of a dissociation continuum cannot be eliminated in view of the larger internuclear distances of these upper states [5].

[1] W. C. PRICE and G. COLLINS: Phys. Rev. **48**, 714 (1935).

[2] Y. TANAKA and T. TAKAMINE: Sci. Pap. Inst. Phys. Chem. Res., Tokyo **39**, 437 (1942). — Y. TANAKA: J. Chem. Phys. **20**, 1728 (1952).

The region between 1300 and 1040 Å is of particular interest to the physics of the upper atmosphere[1] [2]. If solar radiation, particularly its strong H LYMAN α line, can penetrate to lower levels through "windows" of low absorption, the photoionization of a constituent of the atmosphere such as NO is a distinct possibility. For this reason PO LEE measured the absorption for a large number of source emission lines, and the smallest ones found between 1200 and 1300 Å are shown in Fig. 14. One notices indeed rather low values for μ (less than 10 cm.$^{-1}$) in the immediate vicinity of the very strong sharp bands in regions F and G ($\mu_F = 1300$ cm.$^{-1}$ and $\mu_G = 600$ cm.$^{-1}$). At LYMAN α in Fig. 15 the lowest coefficient is 0.23 cm.$^{-1}$, in good agreement with DITCHBURN ([2], p. 327), PRESTON, and WATANABE who find 0.22_6, 0.28, and 0.30 cm.$^{-1}$ respectively.

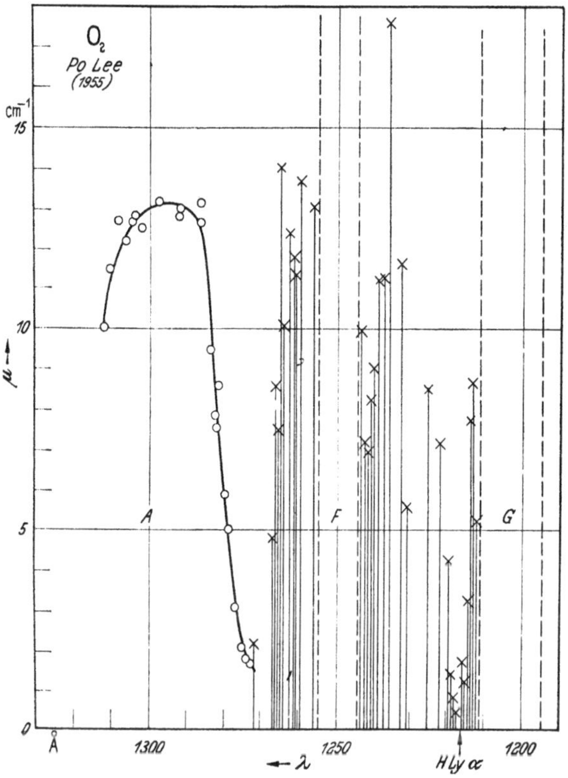

Fig. 14. Details of absorption in oxygen between two strong resonance bands, regions F and G.

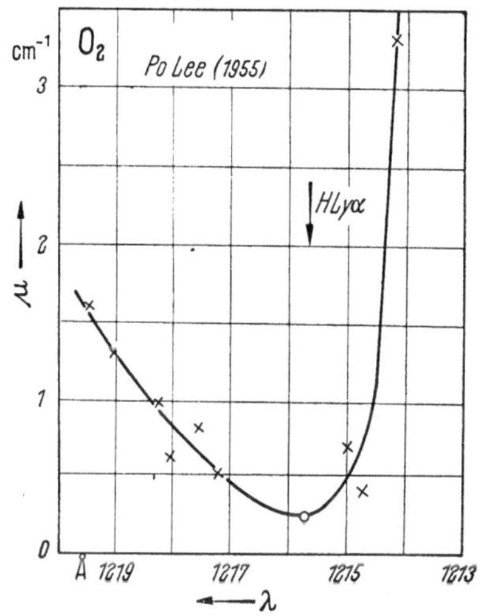

Fig. 15. Oxygen absorption near H LYMAN α.

In N_2 the picture is less complex as shown in Fig. 16. In a detailed spectroscopic analysis, WORLEY[2], TANAKA[3] and others were able to group most of the strong and sharp resonance bands between 600 and 1000 Å into RYDBERG series which converged on two limits at 796 and 661 Å indicated by arrows I.P.$_{1,2}$. It seems therefore obvious that the continuum below 800 Å is due to a photoionization process which is substantiated by the direct measurements (method II c) by WAINFAN et al. of μ_i which agree with the contour of WEISSLER et al. (method I b). Furthermore this seems to bear out the absence of pre-ionized bands in N_2 as might be expected since the upper states of the molecule and the molecular ion have essentially the same internuclear distance as the ground state [5]. The high absorption values which always lie in strong resonance bands should only be taken as an indication of the magnitude rather than as accurate

[1] D. R. BATES and M. J. SEATON: Proc. Phys. Soc. Lond. B **63**, 129 (1950).
[2] R. E. WORLEY: Phys. Rev. **64**, 207 (1943).
[3] Y. TANAKA and T. TAKAMINE: Sci. Pap. Inst. Phys. Chem. Res., Tokyo **39**, 427 (1942).

data because of the blending of the source lines with the resolved or unresolved rotational lines of a given band.

In order to supplement the O_2 information it should be pointed out that Po LEE found in N_2 coefficients of less than 1 cm.$^{-1}$ between bands in the region from 910 to 1040 Å, and at longer wavelengths μ was 0.05 cm.$^{-1}$ or less. This is in agreement with HOPFIELD's[1] estimate of an "atmospheric window" between 1000 and 1300 Å.

Only the preliminary data (method Ib) of Po LEE and WEISSLER on μ and of WAINFAN et al. on μ_i (method IIc) are available in H_2 and presented in Fig. 17. TANAKA[2] and others found both sharp and diffuse bands together with a dissociation continuum between 770 and 850 Å. Because of their overlapping with the ionization limit at 803.7 Å it is difficult to identify clearly the ionization continuum which is tentatively outlined by the dashed curve in Fig. 17.

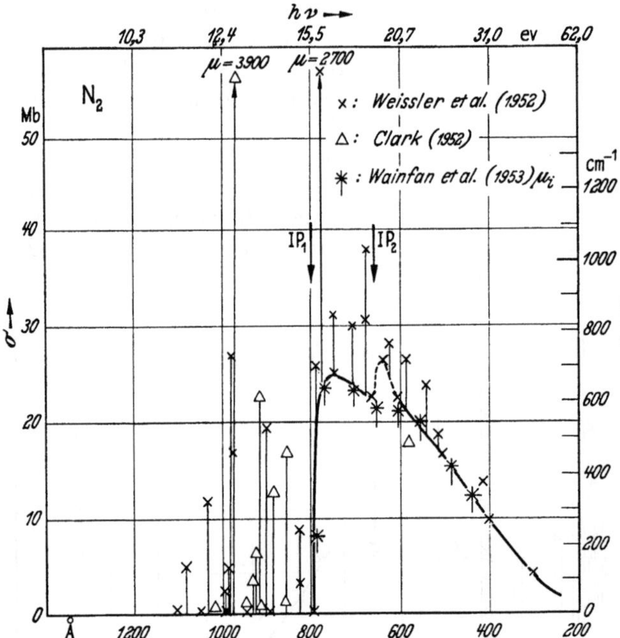

Fig. 16. Absorption coefficients in nitrogen.

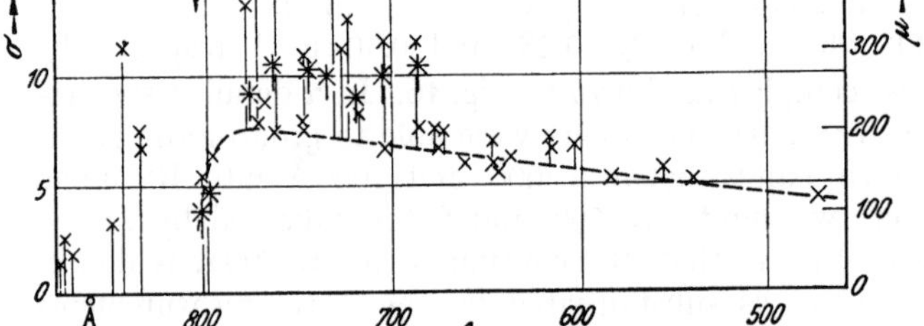

Fig. 17. Absorption in the hydrogen ionization continuum.

20. Absorption in CO, CO_2, and NO, N_2O. In CO, only the work of SUN and WEISSLER (method Ib) is available as shown in Fig. 18. A strong continuum, most likely due to photoionization, begins at 880 Å near the first ionization limit. The superimposed resonance bands have been analysed by TANAKA[3] and by HENNING[4] into RYDBERG progressions with the three limits, I.P.$_{1-3}$. The sharp bands in the region of wavelength

[1] J. J. HOPFIELD: Astrophys. J. **104**, 208 (1946).

[2] Y. TANAKA: Sci. Pap. Inst. Phys. Chem. Res., Tokyo **42**, 49 (1944); contains fifty-three references.

[3] Y. TANAKA: Sci. Pap. Inst. Phys. Chem. Res., Tokyo **40**, 371 (1943); **39**, 447 (1942).

[4] H. J. HENNING: Ann. Physik **13**, 599 (1932).

longer than 1000 Å, particularly the fourth positive system, have not as yet been measured quantitatively.

More complete data are available for CO_2 which was investigated by WATANABE et al. (method Ic) from 1100 to 1750 Å, by WILKINSON and JOHNSTON (method Ia) from 1450 to 1650 Å. and by WEISSLER'S group (methods Ib, IIc) below 1000 Å. A spectroscopic analysis of its band systems has been made by several investigators[1], and MULLIKEN[2] formulated the theory of its electronic structure.

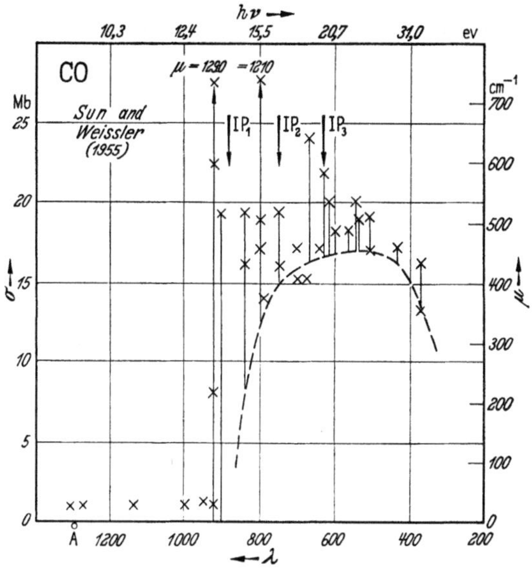

Fig. 18. Absorption coefficients in CO.

In Fig. 19 two weak continua are in evidence above 1200 Å, while below this wavelength some very strong and narrow absorption regions were observed. The ionization continuum begins at about 860 Å or 14.4 volts which is in agreement with electron impact data[3], but not with PRICE and SIMPSON'S first series limit at 903 Å. Strongly pre-ionized bands are in evidence between the two

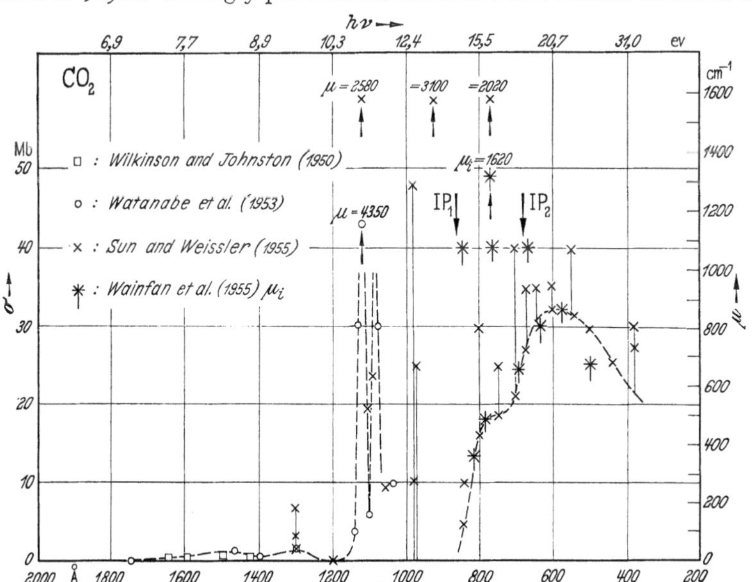

Fig. 19. Absorption coefficients in CO_2.

ionization limits $I.P._{1,2}$ as demonstrated by high values of μ_i there. The main continuum is probably due to photoionization of the molecule since no conspicuous bands due to dissociative ionization reactions were found.

[1] W. C. PRICE and D. M. SIMPSON: Proc. Roy. Soc. Lond., Ser. A **169**, 501 (1938); previous references are included.

[2] R. S. MULLIKEN: Rev. Mod. Phys. **14**, 204 (1942).

[3] H. D. SMYTH and E. C. G. STUECKELBERG: Phys. Rev. **36**, 472 (1930).

The available information on NO has been plotted in Fig. 20 from MARMO's absorption data (method Ic) above 1100 Å and from those of WEISSLER's group (methods Ib, IIc) between 1300 and 350 Å. Weak bands are found between

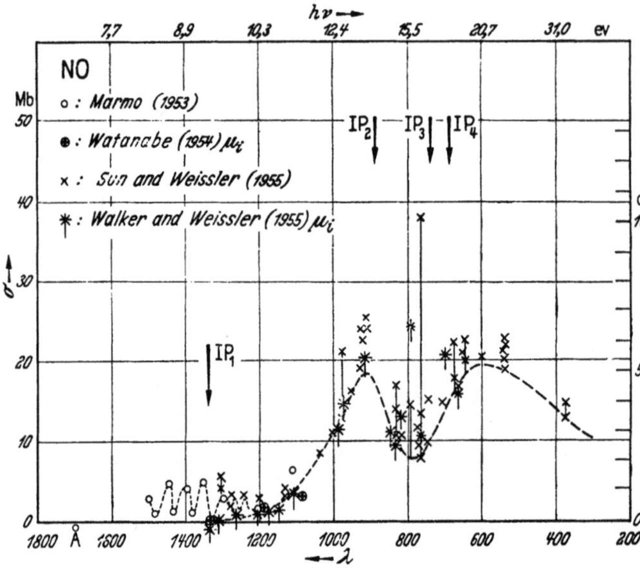

Fig. 20. Absorption coefficients in NO. For recent results between 150 and 1100 Å: J. GRANIER and N. ASTOIN: C. R. Acad. Sci. Paris 242, 1431 (1956).

1350 and 1600 Å. WATANABE also (method IIc) measured μ_i directly and found the onset of photoionization to occur at 1343 Å or 9.23 volts. His data stop at 1100 Å because of the cut-off of LiF windows. The absorption continuum drawn

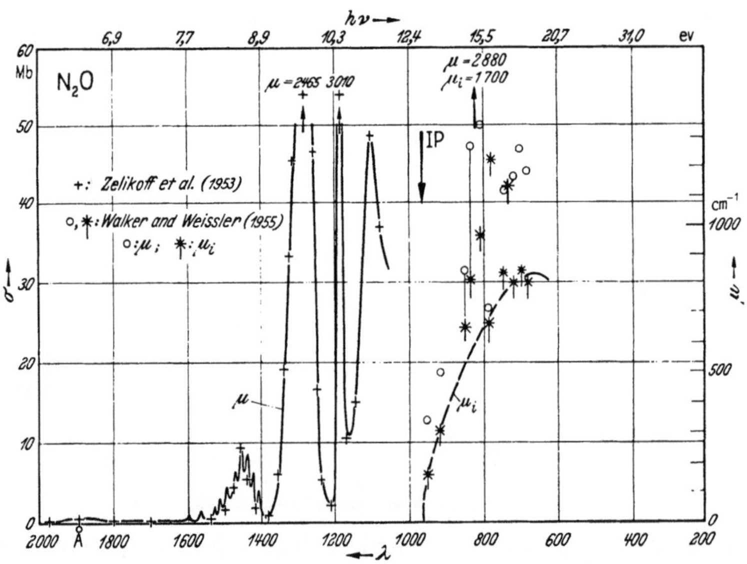

Fig. 21. Absorption coefficients in N_2O. For recent results between 150 and 1100 Å: N. ASTOIN and J. GRANIER: C. R. Acad. Sci. Paris 241, 1736 (1955).

from the preliminary work of SUN and WEISSLER may be in error because of N_2O impurities. ZELIKOFF's N_2O curve as shown in Fig. 21 reaches a value of $\mu = 2400$ cm.$^{-1}$ at about 1300 Å, and since SUN and WEISSLER's data at this wavelength in NO only amount to about $\mu = 130$ cm.$^{-1}$, the falsification of their NO results by N_2O impurities is probably not serious, at least over the region of the main continuum below 1000 Å. Whether the peak of $\mu = 490$ cm.$^{-1}$ at 920 Å is due to a dissociation and/or an ionization process can only be decided by more detailed measurements in this region. The preliminary photoionization data of μ_i by WALKER and WEISSLER are not spaced sufficiently closely to warrent a more complete analysis at this time. A similar speculation may be proposed for the maximum of $\mu = 500$ cm.$^{-1}$ at 620 Å or 20.0 volts. Since HAGSTRUM[1] observed the dissociative ionization reaction $NO \rightarrow N^+ + O^{-*}$ at 19.9 volts, it is possible that the peak is due to this process superimposed over the ionization continuum of TANAKA's[2] γ-series limit, I.P.$_4$.

The band system in N_2O between 800 and 2000 Å has been photographed and analysed by DUNCAN[3]. Its absorption coefficients have been measured by ZELIKOFF et al. (method Ic) from 1100 to 2000 Å and by ROMAND and MAYENCE (method Ia) between 1400 and 2200 Å whose values lie somewhat below ZELIKOFF's curve. The most characteristic feature in Fig. 21 is the very strong absorption region between 1215 and 1380 Å with a maximum of $\mu = 2465$ cm.$^{-1}$, and its symmetry suggests a dissociation continuum although the contribution of a RYDBERG series band cannot be ruled out. Below 1200 Å strong, diffuse bands with another underlying continuum make their appearance. Only preliminary values for ionization cross sections are available below 1100 Å from the measurements of WALKER and WEISSLER (method IIc).

21. Absorption in NH_3, CH_4, H_2O and briefly in O_3, C_2H_2, C_2H_4. The absorption spectrum of NH_3 has been thoroughly investigated by DUNCAN and HARRISON[4] down to 850 Å, and quantitative coefficients have been recently reported by WATANABE (method Ic) above 1100 Å, by SUN and WEISSLER (method Ib) between 350 and 1300 Å and preliminary μ_i results by WALKER and WEISSLER (method IIc) below 1000 Å. Their results are shown in Fig. 22. Strong and regularly spaced bands extend over the entire region of WATANABE's work and perhaps are superimposed over various weaker continua. He also measured directly (method IIc) the photoionization μ_i and found an onset at 1225 Å which could be corrected to 1210 Å or 10.25 volts by taking into account a BOLTZMANN distribution of the vibrational states of the molecular ground state. At shorter wavelengths a strong continuum with no major bands is in evidence.

CH_4 absorption has been measured by a number of investigators, all in reasonable agreement with each other (Fig. 23). WILKINSON and JOHNSTON (method Ia) worked between 1370 and 1450 Å, MOE and DUNCAN (method Ib) and also WATANABE et al. (method Ic) down to 1100 Å, DITCHBURN et al. (method Ib) and SUN and WEISSLER (method Ib) from 1300 Å to below 400 Å. WAINFAN et al. (method IIc) determined directly the photoionization efficiencies between the first ionization potential at 960 and 470 Å. As was to be expected from the symmetrical structure of this molecule, the absorption was continuous, rising from $\mu < 1$ cm.$^{-1}$ at 1450 Å to an essentially flat plateau between 1100 and 1300 Å and then continuing on to a peak at 960 Å. Towards shorter wavelengths there was good agreement between DITCHBURN and SUN and WEISSLER, while the preliminary results of WAINFAN et al. were somewhat higher. The slow rise of their efficiency curve from 960 Å to 100% at 800 Å is particularly interesting (see inset in Fig. 23 of P.I.E.-curve) in so far as the mechanism of the continuous

[1] H. D. HAGSTRUM: Rev. Mod. Phys. **23**, 185 (1951).
[2] Y. TANAKA: Sci. Pap. Inst. Phys. Chem. Res., Tokyo **39**, 456 (1942).
[3] A. B. F. DUNCAN: J. Chem. Phys. **4**, 638 (1936).
[4] A. B. F. DUNCAN and G. R. HARRISON: Phys. Rev. **49**, 211 (1936).

absorption is concerned. Quite clearly only below 800 Å can it be wholly accounted for by photoionization. It is significant that near the limit (I.P.) the efficiency and therefore σ_i is essentially zero. If the approximated theoretical photo-

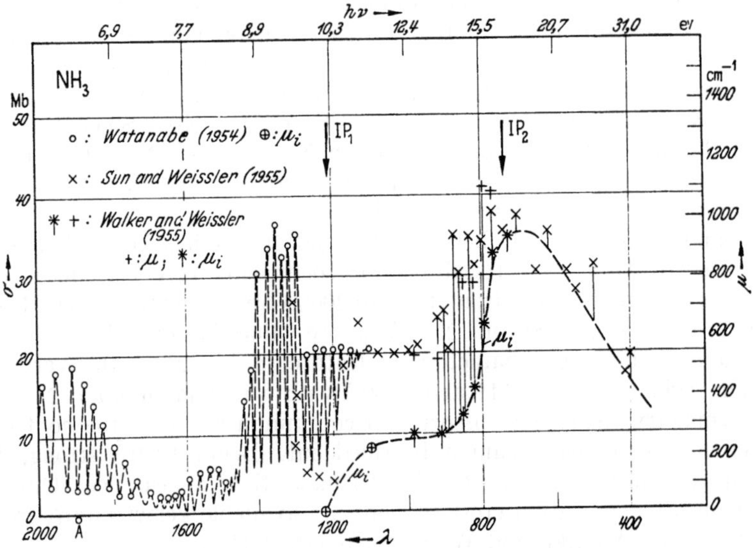

Fig. 22. Absorption coefficients in NH_3.

ionization cross section at the limit (DALGARNO, $\sigma_i = 94$ Mb) is compared with the total absorption work, the agreement is fair (factor of 2), but not so for the directly determined σ_i.

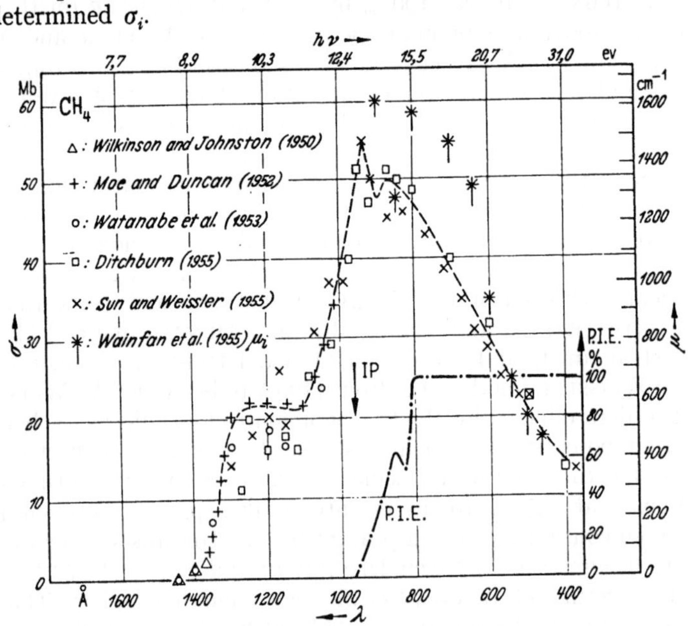

Fig. 23. Absorption coefficients in CH_4.

The results in water vapor are shown in Fig. 24 with reasonable agreement among various investigators in the region of longer wavelengths. Two continua, one between 1450 and 1850 Å and smooth in appearance and the other between

1250 Å and 1450 with a number of diffuse bands superimposed, are apparent. Below 1250 Å exist strong and diffuse bands which do not permit the indentification of a continuum. Only preliminary data are available towards still shorter

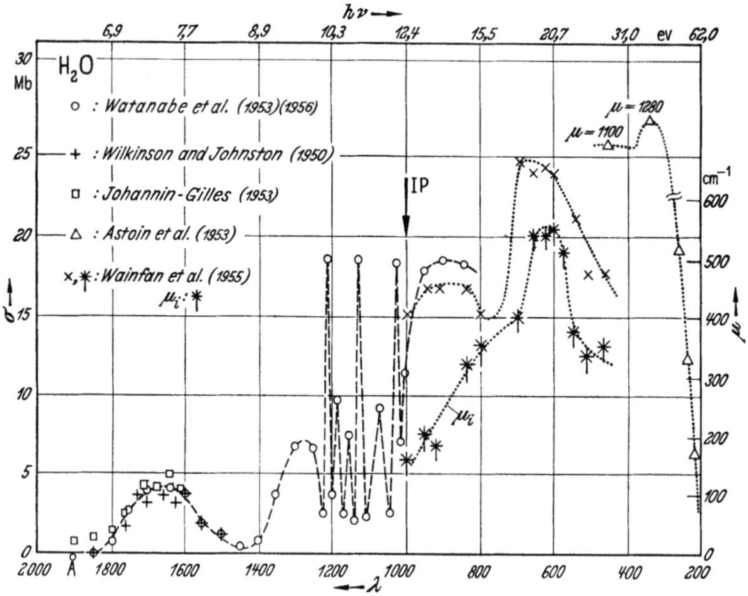

Fig. 24. Absorption coefficients in water vapor. For recent results between 160 and 1100 Å: N. Astoin: C. R. Acad. Sci. Paris **242**, 2327 (1956).

wavelengths. The results of Wainfan et al. (method IIc) between 980 and 470 Å are plotted both in terms of total absorption and photoionization cross sections marked μ_i, while the work of Astoin et al. (method Ib) is represented between 200 and 460 Å. In the overlapping region the agreement is not good, perhaps due to uncertainties of pressure values.

Because of the particular interest of atmospheric physicists in ozone absorption, the data of Tanaka et al. (method Ic) are shown in Fig. 25.

Finally, in order to give an idea as to the behavior of organic compounds the ethylene (C_2H_4) curve of Zelikoff and Watanabe (method Ic) and the one for acetylene (C_2H_2) by Moe and Duncan (method Ib) are shown in Fig. 26, together with the preliminary results of Walker and Weissler (method IIc) for the region of their onization continua.

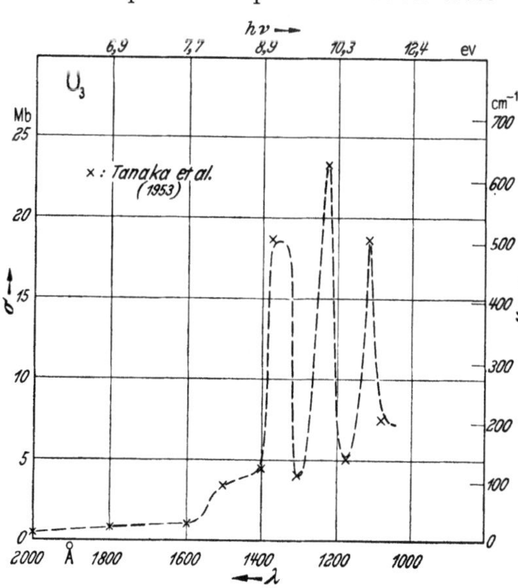

Fig. 25. Absorption coefficients in ozone (O_3).

22. Theoretical considerations. In the foregoing, cross sections from 2000 to 250 Å, corresponding to photon energies of 6 to 50 volts, have been presented. Nearly all photoionization and photodissociation processes occur in this range.

Because of their significance to the understanding of stellar and planetary atmospheres, including our own, a number of theorists have concerned themselves

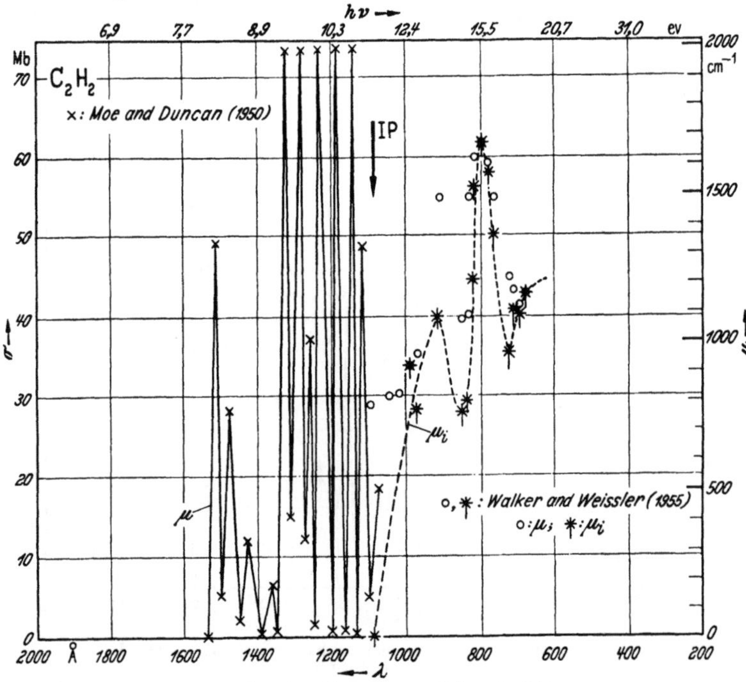

Fig. 26a. Absorption coefficients in C_2H_2.

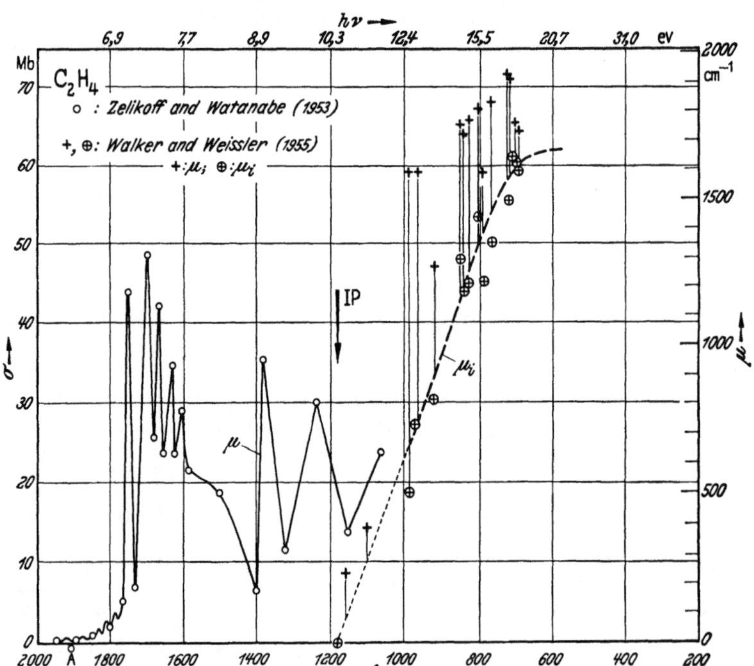

Fig. 26b. Absorption coefficients in C_2H_4.

with calculations on the absorption of radiation in the ionization continuum of atoms. Such transitions may in principle be calculated with the methods of quantum mechanics by employing an equation developed by BATES [4] for the ionization cross section which reads

$$\sigma_{i,\nu} = \frac{32\pi^4 m^2 e^2}{3h^3 c} \cdot \frac{1}{\omega_i} \sum_i \sum_f v\,\nu \left| \int \psi_i^* \left(\sum_j \boldsymbol{r}_j\right) \psi_f \, d\boldsymbol{r} \right|^2, \tag{22.1}$$

where $\boldsymbol{r}_j$ is the position vector of the jth electron with the $\sum_j \boldsymbol{r}_j$ characterizing the dipole moment, ω_i the statistical weight of the initial level and ν the frequency of the incident radiation. e, m, c and h have their customary meaning. ψ_i is the bound wave function for the initial state and ψ_f that for the ion together with the ejected electron having a velocity v. The integral is usually referred to as the dipole length matrix element, expressed in HARTREE units, and depends critically on these two wave functions. It is extremely sensitive to their exact form owing to the fact that the contributions from positive portions of the integrand may cancel almost completely. For instance in the case of potassium, the wave functions giving the experimental cross section at the ionization limit contributed positive and negative amounts to the transition integral which differed by only one part in fifty. Thus the problem at hand is the choice of the most appropriate approximate wave functions which may be obtained with the help of the variational principle giving the best fit for the determination of the energy. They are usually derived from HARTREE-FOCK equations of the self-consistent field including electron exchange.

In hydrogen-like systems (H, He^+, Li^{++}, etc.) exact calculations of σ_i were made by KRAMERS, SUGIURA, and others, but could not be compared directly to experimental work. Their application, even to non-penetrating orbits of a single electron outside a closed core as in the case of the alkalis, was not justified because it is necessary to use different values of the effective atomic numbers in their formulation of the bound and free wave functions. However, subsequent and improved calculations[1] by WHEELER and by VINTI on the photoionization of helium agree well with laboratory data.

The early theoretical work on lithium using simple wave functions and not including electron exchange has recently been refined by STEWART[2] with the aid of HARTREE-FOCK wave functions including exchange. He obtained $\sigma_i =$ 1.16 Mb at the series limit compared to 2.5 Mb from experiment, and he also found a gradual increase in σ_i towards shorter wavelengths, whereas the laboratory observations indicate a monotonic decrease (Fig. 7).

In order to account for the variation of σ_i with wavelength in sodium, potassium and cesium, BATES[3] suggested the introduction of a variable polarizability correction, as shown for the case of potassium in Fig. 27, which may be chosen to fit the observational trend. Sodium has been calculated recently by SEATON[4] using HARTREE-FOCK wave functions with exchange and, as may be seen from Fig. 28, the theory reproduces the main aspects of the experimental curve. He attributes the finite minima in the cross sections of the alkali metals to spin-orbit perturbations of the free electron wave functions and, on the basis of this effect, estimates minima which for cesium and rubidium compare well with

[1] He, theor.: J. A. WHEELER: Phys. Rev. **43**, 258 (1933). — J. P. VINTI: Phys. Rev. **44**, 524 (1933).

[2] Li, theor.: B. TRUMPY: Z. Physik **71**, 720 (1931). — J. HARGREAVES: Proc. Cambridg Phil. Soc. **35**, 75 (1929). — A. L. STEWART: Proc. Phys. Soc. Lond. A **67**, 917 (1954).

[3] K, theor.: D. R. BATES: Proc. Roy. Soc. Lond., Ser. A **188**, 350 (1947).

[4] Na and alkali metals, theor.: M. J. SEATON: Proc. Roy. Soc. Lond., Ser. A **208**, 418 (1951).

experiment. Entirely similar calculations have been performed for neon[1] and reasonable agreement was obtained with the experimental results (Fig. 29).

In their somewhat earlier work on calcium, BATES and MASSEY[2] obtained a cross section of 25 Mb at the series limit compared to 0.45 Mb from experiment. This has been attributed by BATES to the neglect of polarization which is likely

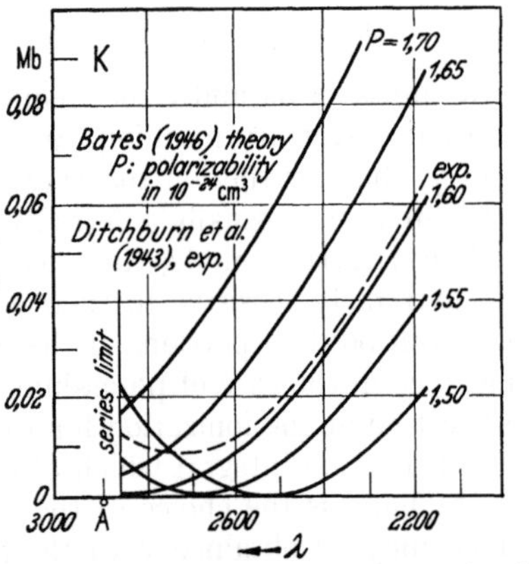

Fig. 27. Theoretical photoionization cross sections in potassium, showing the effect of a variable polarizability factor.

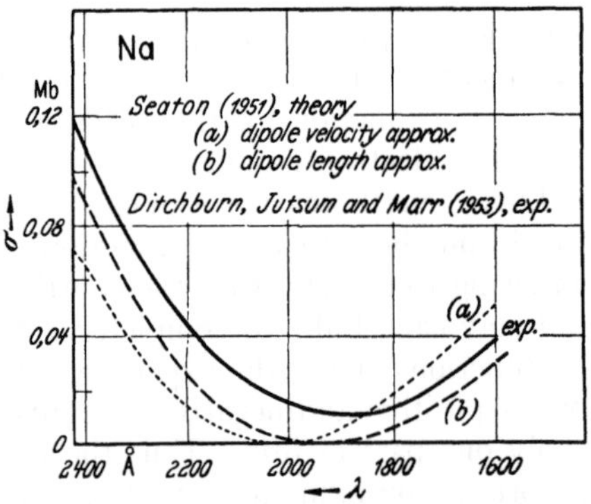

Fig. 28. Theoretical and experimental photoionization cross sections in sodium.

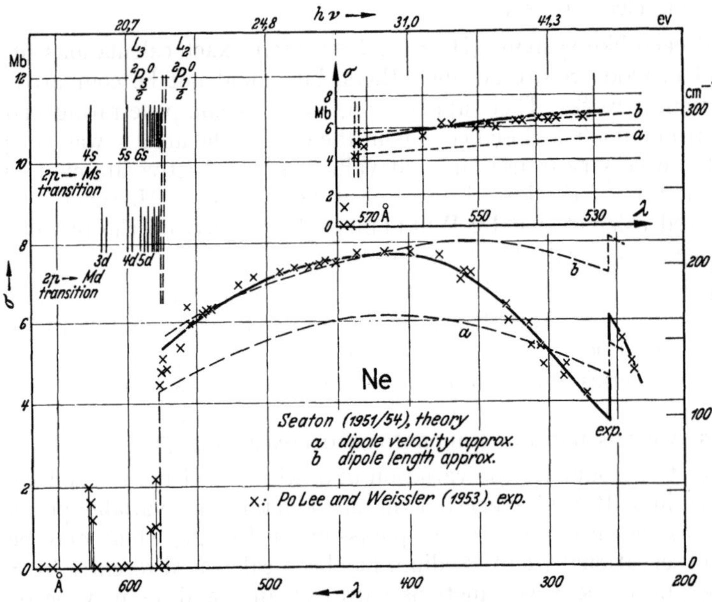

Fig. 29. Theoretical and experimental photoionization cross sections in neon.

to be very large. There is much better correspondence between the laboratory data of EHLER and WEISSLER (12.5 Mb) and the theoretical cross section (10.8 Mb) of BATES and SEATON[3] on atomic nitrogen, which lends considerable support to the latter's work on both atomic oxygen and carbon.

[1] Ne and elements from B to Ne, theor.: M. J. SEATON: Proc. Roy. Soc. Lond., Ser. A **208**, 408 (1951); see also Proc. Phys. Soc. Lond. A **67**, 927 (1954) on Ne.

[2] Ca, theor.: D. R. BATES and H. S. W. MASSEY: Proc. Roy. Soc. Lond., Ser. A **177**, 281 (1941).

[3] O, N, and C, theor.: D. R. BATES and M. J. SEATON: Mon. Not. Roy. Astronom. Soc. **109**, 698 (1949).

It has been suggested that photoionization cross sections of the highly symmetric methane molecule may be calculated on a quasi-atomic basis. This was done by DALGARNO[1] using self-consistent wave functions without exchange. At first sight his value at the limit (94 Mb) seems to be in the range of the experimental absorption (45 Mb), but it is much too large when compared to the photoionization efficiency curve in Fig. 23. When DALGARNO used the same approach with exchange included on argon, good agreement between theory and experiment prevailed.

23. Summary of results on photoabsorption and photoionization. For the convenience of the reader the previously discussed work on photoionization cross sections of monatomic gases and vapors has been summarized in Table 1 and that on absorption coefficients of molecular gases in Table 2.

Table 1. *Photoionization cross sections of monatomic gases and vapors.*

Gas	References	Meth.	Range (Å)	σ_{limit} (Mb)	σ_{min} (Mb)	Remarks
A	I.P.$_{-1}$: 786.72 Å or 15.75 volts					
	Po LEE and G. L. WEISSLER: Phys. Rev. **99**, 540 (1955).	Ib	850—350	36		σ decreases towards short λ.
	A. DALGARNO: Proc. Phys. Soc. Lond. A **65**, 663 (1952).	theor.		30		
	N. WAINFAN, W. C. WALKER and G. L. WEISSLER: Phys. Rev., **99**, 542 (1955).	IIc	900—500	35		as above.
Ne	I.P.$_{-1}$: 574.94 Å or 21.56 volts					
	R. W. DITCHBURN: Private note to the author.	Ib				as below.
	Po LEE and G. L. WEISSLER: Proc. Roy. Soc. Lond., Ser. A **219**, 71 (1953)	Ib	700—230	5.5		max. at 400 Å, then decrease.
	M. J. SEATON: Proc. Roy. Soc. Lond., Ser. A **208**, 408 (1951). Proc. Phys. Soc., Lond. A **67**, 927 (2954).	theor.		5.8 or 4.4		σ similar to exp.
He	I.P.$_{-1}$: 504.27 Å or 24.58 volts					
	Po LEE and G. L. WEISSLER: Phys. Rev. **99**, 540 (1955).	Ib	600—250	7.3		decrease, $\sigma \sim \nu^{-3}$.
	R. W. DITCHBURN: To be published.					
	J. A. WHEELER: Phys. Rev. **43**, 258 (1933).	theor.		7.4		decrease, $\sigma \sim \nu^{-3}$.
	J. P. VINTI: Phys. Rev. **44**, 524 (1933).	theor.		7.5		decrease, $\sigma \sim \nu^{-3}$.
Li	I.P.$_{-1}$: 2300 Å or 5.39 volts					
	J. TUNSTEAD: Proc. Phys. Soc. Lond. A **66**, 304 (1953).	Ia	2400—1700	2.5		σ decreases → short λ.
	B. TRUMPY: Z. Physik **71**, 720 (1931)	theor.		3.6		σ decreases.
	J. HARGREAVES: Proc. Cambridge Phil. Soc. **35**, 75 (1929).	theor.		2.9		max. at 1900 Å, then decrease.
	A. L. STEWART: Proc. Phys. Soc. Lond. A **67**, 917 (1954).	theor.		1.16		σ increases gradually.

[1] CH_4 and A, theor.: A. DALGARNO: Proc. Phys. Soc. Lond., Ser. A **65**, 663 (1952).

Table 1. (Continued).

Gas	References	Meth.	Range (Å)	σ_{limit} (Mb)	σ_{min} (Mb)	Remarks
Na	I.P.$_{\cdot 1}$: 2410 Å or 5.14 volts					
	B. TRUMPY: Z. Physik **47**, 804 (1928); semi-quantitative.	Ia	2500—2300			σ decreases.
	R. W. DITCHBURN, P. J. JUTSUM and G. V. MARR: Proc. Roy. Soc. Lond., Ser. A **219**, 89 (1953).	Ia	2500—1600	0.116	0.013	σ decreases to min. then rises towards short λ.
	M. J. SEATON: Proc. Roy. Soc. Lond., Ser. A **208**, 418 (1951).	theor.		0.10 or 0.07	10^{-5}	similar to exp.
K	I.P.$_{\cdot 1}$: 2860 Å or 4.34 volts					
	F. L. MOHLER and C. BOECKNER: J. Res. Nat. Bur. Stand. **3**, 303 (1929).	IIa	3000—2200	(0.12)	0.003	σ rises towards short λ.
	E. O. LAWRENCE and N. E. EDLEFSEN: Pnys. Rev. **34**, 233, 1056 (1929).	IIa	3000—2200		yes	σ rises towards short λ.
	R. W. DITCHBURN, J. TUNSTEAD and J. G. YATES: Proc. Roy. Soc. Lond., Ser. A **181**, 386 (1943).	Ia	3000—1600	0.12	0.008	σ rises towards short λ.
	D. R. BATES: Proc. Roy. Soc. Lond., Ser. A **188**, 350 (1947).	theor.			yes	σ rises towards short λ.
	M. J. SEATON: Proc. Roy. Soc. Lond., Ser. A **208**, 418 (1951).	theor.			3×10^{-4}	σ rises towards short λ.
Rb	I.P.$_{\cdot 1}$: 2972 Å or 4.18 volts					
	F. L. MOHLER and C. BOECKNER: J. Res. Nat. Bur. Stand. **3**, 303 (1929).	IIb	3000—2500	0.11	0.004	σ decreases towards min.
	E. O. LAWRENCE and N. E. EDLEFSEN: Phys. Rev. **34**, 233, 1056 (1929).	IIa	3000—2400			σ decreases towards min.
	M. J. SEATON: Proc. Roy. Soc. Lond., Ser. A **208**, 418 (1951).	theor.			0.004	
Cs	I.P.$_{\cdot 1}$: 3185 Å or 3,89 volts					
	F. L. MOHLER and C. BOECKNER: J. Res. Nat. Bur. Stand. **3**, 303 (1929).	IIb	3200—2400	0.23	0.043	σ rises towards short λ.
	E. O. LAWRENCE and N. E. EDLEFSEN: Phys. Rev. **34**, 233, 1056 (1929).	IIa	3200—2400			
	H. J. BRADDICK and R. W. DITCHBURN: Proc. Roy. Soc. Lond., Ser. A **143**, 472 (1934).	Ia	3300—2000	0.22	0.078	σ rises towards short λ.
	M. J. SEATON: Proc. Roy. Soc. Lond., Ser. A **208**, 418 (1951).	theor.			0.03	
Be, Mg, Ca, In, Tl						
Be	D. R. BATES: Mon. Not. Roy. Astronom. Soc. **106**, 423 (1946).	theor.		8,2		σ decrease $> \nu^{-3}$.
Mg	R. W. DITCHBURN and G. V. MARR: Proc. Phys. Soc. Lond. A **66**, 655 (1953).	Ia	1650—1450	1.2		σ decrease $\sim \lambda^{15}$.

Table 1. (Continued).

Gas	References	Meth.	Range (Å)	σ_{limit} (Mb)	σ_{min} (Mb)	Remarks
Be, Mg, Ca, In, Tl (Continued).						
Ca	P. J. JUTSUM: Proc. Phys. Soc. Lond. A **67**, 196 (1954) (L).	Ia	2100—1950	0.45		σ decrease $\sim\lambda^{47}$.
	D. R. BATES and H. S. W. MASSEY: Proc. Roy. Soc. Lond., Ser. A **177**, 281 (1941).	theor.		25		σ decrease $\sim\nu^{-3}$.
In	G. V. MARR: Proc. Phys. Soc. Lond. A **67**, 196 (1954).	Ia	2140—1758	$\sim$0.3		strong preionization.
Tl	G. V. MARR: Proc. Roy. Soc. Lond., Ser. A **224**, 83 (1954).	Ia	2100—1450	4.5		strong preionization.
N	I.P.$_1$: 852 Å or 14.55 volts; O I.P.$_1$: 910 Å or 13.62 volts					
N	A. W. EHLER and G. L. WEISSLER: J. Opt. Soc. Amer. **45**, 1035 (1955).	Ia	800—400	$\sim$10		nearly constant $\sigma_{max} = 12.5$ (650 Å).
	D. R. BATES and M. J. SEATON: Mon. Not. Roy. Astronom. Soc. **109**, 698 (1949).	theor.		10		$\sigma_i \sim 10.8$ (650) Å.
O	D. R. BATES and M. J. SEATON: As above.	theor.		2.8		$\sigma_i \sim 13$ (550 Å).

Table 2. *Absorption coefficients in molecular gases and vapors.*

Gas	References	Meth.	Range (Å)	Remarks (μ in units of cm^{-1} at N. T. P.)
O_2	I.P.$_1$: 1020 Å or 12.15 volts			
	R. LADENBURG and C. C. VAN VOORHIS: Phys. Rev. **43**, 315 (1933).	Ia	1720—1300	dissoc. cont. $\mu_{max} = 490$ (1450 Å).
	E. G. SCHNEIDER: J. Chem. Phys. **5**, 106 (1937).	Ia	1750—1300	dissoc. cont. $\mu_{max} = 460$ (1450 Å).
	R. W. DITCHBURN and D. W. O. HEDDLE: Proc. Roy. Soc. Lond., Ser. A **220**, 61 (1953).	Ia	1750—1300	dissoc. cont. $\mu_{max} = 490$ (1450 Å).
	K. WATANABE, E. C. Y. INN and M. ZELIKOFF: J. Chem. Phys. **21**, 1026 (1953)	Ic	2000—1050	dissoc. cont. $\mu_{max} = 380$ (1450 Å).
	G. L. WEISSLER and PO LEE: J. Opt. Soc. Amer. **42**, 200 (1952).	Ib	1300—300	bands, ioniz. cont. $\mu_{max} = 650$ ($\sim$550 Å).
	K. C. CLARK: Phys. Rev. **87**, 271 (1952).	Ib	1025—850	bands, $\mu_B = 53$ (1025 Å) to 1350 (972 Å).
	PO LEE: J. Opt. Soc. Amer. **45**, 703 (1955).	Ib	1320—200	bands, dissoc. and ioniz. continua, $\mu_{max} = 590$ (510 Å).
	A. A. ABOUD, J. P. CURTIS, R. MERCURE and W. A. RENSE: J. Opt. Soc. Amer. **45**, 767 (1955) (L).	Ib	1640—112	μ-values slightly higher than above data.
	N. WAINFAN, W. C. WALKER and G. L. WEISSLER: J. Appl. Phys. **24**, 1318 (1953). — Phys. Rev. **99**, 542 (1955).	IIc	1050—500	photoioniz.: $\mu_{i,max} = 750$ (600 Å).
N_2	I.P.$_1$: 796 Å or 15.58 volts			
	G. L. WEISSLER, PO LEE and E. I. MOHR: J. Opt. Soc. Amer. **42**, 84 (1952).	Ib	1300—300	bands, ioniz. cont. $\mu_{max} = 700$ (760 Å).
	K. C. CLARK: Phys. Rev. **87**, 271 (1952).	Ib	1025—850	bands, $\mu_B = 16$ (919 Å) to 3900 (972 Å).
	PO LEE: J. Opt. Soc. Amer. **45**, 703 (1955).	Ib	1300—950	bands, $\mu_{min} < 0.5$.

Table 2. (Continued.)

Gas	References	Meth.	Range (Å)	Remarks (μ in units of cm.$^{-1}$ at N. T. P.)
N_2	I.P.$_{-1}$: 796 Å or 15.58 volts (Continued).			
	N. WAINFAN, W. C. WALKER and G. L. WEISSLER: J. Appl. Phys. **24**, 1318 (1953). — Phys. Rev. **99**, 542 (1955).	IIc	800—500	photoioniz.: $\mu_{i,\max} = 700$ (750 Å).
	J. P. CURTIS: Phys. Rev. **94**, 908 (1954).	Ib	1000—150	μ-values are about 40% lower than above data.
	C. D. MAUNSSELL: Phys. Rev. **98**, 1831 (1955).	IIb	765?	photoioniz.: $\mu_{\max} = 11 \times 10^3$ (probably at 765 Å, preionized?).
H_2	I.P.$_{-1}$: 804 Å or 15.42 volts			
	PO LEE and G. L. WEISSLER: Astrophys. J. **115**, 570 (1952).	Ib	900—300	bands, ioniz. cont. $\mu_{\max} = 200$ (770 Å).
	N. WAINFAN, W. C. WALKER and G. L. WEISSLER: Phys. Rev. **99**, 542 (1955).	IIc	850—650	photoioniz.: $\mu_{i,\max} = 250$ (750 Å).
CO	I.P.$_{-1}$: 868.13 Å or 14.21 volts; CO_2 I.P.$_{-1}$: 903 Å (861 Å) or 13.73 volts (14.4 volts)			
CO	H. SUN and G. L. WEISSLER: J. Chem. Phys. **23**, 1625 (1955).	Ib	1300—400	bands, ioniz. cont. $\mu_{\max} = 460$ (600 Å).
CO_2	P. G. WILKINSON and H. L. JOHNSTON: J. Chem. Phys. **18**, 190 (1950).	Ia	1700—1300	dissoc. cont. $\mu_{\max} = 11$ (1500 Å).
	K. WATANABE, E. C. Y. INN and M. ZELIKOFF: J. Chem. Phys. **21**, 1648 (1953).	Ic	1800—1100	bands, dissoc. cont.
	H. SUN and G. L. WEISSLER: J. Chem. Phys. **23**, 1625 (1955).	Ib	1300—400	bands, ioniz. cont. $\mu_{\max} = 880$ (600 Å).
	N. WAINFAN, W. C. WALKER and G. L. WEISSLER: Phys. Rev. **99**, 542 (1955).	IIc	900—470	photoioniz.: $\mu_{i,\max} = 930$ (650 Å).
NO	I.P.$_{-1}$: 1340 Å or 9.25 volts			
	J. MAYENCE: Ann. Phys., Paris **7**, 453 (1952).	Ia	1400—1600	dissoc. cont. $\mu_{\max} = 10$ (1470 Å), bands.
	F. F. MARMO: J. Opt. Soc. Amer. **43**, 1186 (1953).	Ic	2300—1100	bands and continua.
	K. WATANABE: J. Chem. Phys. **22**, 1564 (1954).	IIc	1600—1050	bands, photoioniz. onset: 1340 Å.
	H. SUN and G. L. WEISSLER: J. Chem. Phys. **23**, 1372 (1955) (L).	Ib	1300—400	bands, ioniz. cont. $\mu_{\max} = 700$ to 1000 (600—900 Å).
	W. C. WALKER and G. L. WEISSLER: J. Chem. Phys. **23**, 1962 (1955) (L).	IIc	1350—680	as above.
	J. GRANIER and N. ASTOIN: C. R. Acad. Sci. Paris **242**, 1431 (1956).	Ib	1100—150	three ioniz. continua, $\mu_{1,\max} = 600$ (500 Å), $\mu_{2,\max} = 500$ (260 Å), $\mu_{3,\max} = 450$ (170 Å).
N_2O	I.P.$_{-1}$: 960 Å or 12.9 volts			
	J. ROMAND and J. MAYENCE: C. R. Acad. Sci. Paris **228**, 998 (1949).	Ia	2300—1400	two continua, $\mu_1 = 3.1$ (1840 Å) and $\mu_2 = 118$ (1450 Å).
	M. ZELIKOFF, K. WATANABE and E. C. Y. INN: J. Chem. Phys. **21**, 1643 (1953).	Ic	2100—1080	four continua, $\mu_3 = 2465$ (1285 Å) and $\mu_4 \sim 1000$ (1080 Å).
	W. C. WALKER and G. L. WEISSLER: J. Chem. Phys. **23**, 1962 (1955) (L).	IIc	970—680	ioniz. cont. $\mu_{i,\max} = 800$ (700 Å); pre-ioniz. $\mu_i = 1700$ (833 Å).
	N. ASTOIN and J. GRANIER: C. R. Acad. Sci. Paris **241**, 1736 (1955).	Ib	1100—150	three ioniz. continua, $\mu_{1,\max} = 850$ (700 Å), $\mu_{2,\max} = 800$ (357 Å), $\mu_{3,\max} = 800$ (225 Å).

Table 2. (Continued.)

Gas	References	Meth.	Range (Å)	Remarks (μ in units of $cm.^{-1}$ at N. T. P.)
NH_3	I.P.$_1$: 1210 Å or 10.25 volts			
	K. WATANABE: J. Chem. Phys. **22**, 1564 (1954).	Ic	2200—1050	strong bands, continua, photoioniz. onset: 1210 Å.
	H. SUN and G. L. WEISSLER: J. Chem. Phys. **23**, 1160 (1955).	Ib	1300—400	bands, ioniz. cont. $\mu_{max} = 950$ (700 Å).
	W. C. WALKER and G. L. WEISSLER: J. Chem. Phys. **23**, 1540 (1955) (L).	IIc	1240—680	as above.
CH_4	I.P.$_1$: 947 Å or 13.07 volts			
	P. G. WILKINSON and H. L. JOHNSTON: J. Chem. Phys. **18**, 190 (1950).	Ia	1460—1370	continuum $\mu_{max} = 70$ (1370 Å).
	G. MOE and A. B. F. DUNCAN: J. Amer. Chem. Soc. **74**, 3140 (1952).		1460—1050	two continua, $\mu_1 = 600$ (1200 Å) and $\mu_2 > 900$ (1050 Å).
	K. WATANABE, E. C. Y. INN and M. ZELIKOFF: Private communication.	Ic	1610—1065	continuous, $\mu_{max} = 500$ (1200 Å).
	R. W. DITCHBURN: Proc Roy. Soc., Lond., Ser. A **229**, 44 (1955).	Ib	1300—400	continuous, $\mu_{max} = 1400$ (940 Å).
	H. SUN and G. L. WEISSLER: J. Chem. Phys. **23**, 1160 (1955).	Ib	1300—400	continuous, $\mu_{max} = 1500$ (960 Å).
	N. WAINFAN, W. C. WALKER and G. L. WEISSLER: Phys. Rev. **99**, 542 (1955).	IIc	100—470	photoioniz.: $\mu_{i,max} = 1600$ (900 Å) or $\sigma_i = 60$ Mb.
	A. DALGARNO: Proc. Phys. Soc. Lond., A **65**, 663 (1952).	theor.		$\sigma_i = 94$ Mb. or $\mu_i = 2525$ (at limit).
H_2O	I.P.$_1$: 955 Å or 13.0 volts			
	P. G. WILKINSON and H. L. JOHNSTON: J. Chem. Phys. **18**, 190 (1950).	Ia	1850—1450	continiuum, $\mu_{max} = 100$ (1650 Å).
	K. WATANABE and M. ZELIKOFF: J. Opt. Soc. Amer. **43**, 753 (1953).	Ic	1860—1050	two continua, $\mu_1 = 120$ (1650Å), $\mu_2 = 180$ (1300 Å) bands.
	A. JOHANNIN-GILLES: C. R. Akad. Sci. Paris **236**, 676 (1953).	Ia	1750—1600	nearly continuous, pressure dependent.
	N. ASTOIN, A. JOHANNIN-GILLES and B. VODAR: C. R. Acad. Sci. Paris **237**, 558 (1953).	Ib	460—200	bands, ioniz. continuum $\mu_{max} \sim 1200$ (350 Å).
	N. WAINFAN, W. C. WALKER and G. L. WEISSLER: Phys. Rev. **99**, 542 (1955).	IIc	1000—470	photoioniz.: $\mu_{i,max} = 540$ (600 Å).
	K. WATANABE and A. JURSA: To be published.	Ic	1100—850	$\mu_{max} = 500$ (850—950 Å).
	N. ASTOIN: C. R. Acad. Sci. Paris **242**, 2327 (1956).	Ib	1100—160	four ioniz. continua, $\mu_{1,max} \sim 500$ (800 Å), $\mu_{2,max} = 550$ (625 Å), $\mu_{3,max} = 750$ (435 Å), $\mu_{4,max} = 630$ (355 Å).
O_3, C_2H_2, C_2H_4				
O_3	Y. TANAKA, E. C. Y. INN and K. WATANABE: J. Chem. Phys. **21**, 1651 (1953).	Ic	1200—1050	weak continuum, bands $\mu_B = 600$ (1220 Å).
C_2H_2	G. MOE and A. B. F. DUNCAN: J. Amer. Chem. Soc. **74**, 3136 (1952).	Ib	1540—1050	strong bands, $\mu_B = 2000$ (1160 Å).
	W. C. WALKER and G. L. WEISSLER: J. Chem. Phys. **23**, 1547 (1955) (L).	IIc	1100—680	photoioniz. onset: 1100 Å; $\mu_{i,max} = 1650$ (800 Å).
C_2H_4	M. ZELIKOFF and K. WATANABE: J. Opt. Soc. Amer. **43**, 756 (1953).	Ic	2000—1065	strong bands with continua, $\mu_B = 970$ (1380 A).
	P. G. WILKINSON and H. L. JOHNSTON: J. Chem. Phys. **18**, 190 (1950).	Ia	2000—1450	strong bands with continuum.
	J. R. PLATT, H. B. CLEVENS and W. C. PRICE: J. Chem. Phys. **17**, 466 (1949).	Ia	1750—1550	strong bands with continuum.
	W. C. WALKER and G. L. WEISSLER: J. Chem. Phys. **23**, 1547) 1955) (L).	IIc	1190—680	photoioniz. onset: 1185 Å; $\mu_{i,max} \sim 1600$ (~ 700 Å).

B. Photoelectric emission from solids.

I. A survey of photoelectric phenomena.

24. Introductory remarks. In order to discuss the various phenomena of the photoelectric effect, it is necessary to establish first a simple model of the electrons whithin solids. As is well known, the sharp electronic energy levels of isolated atoms may be broadened in the case of densely packed atoms into wide bands [7]. This holds true particularly for those states associated with outer electrons, which may be separated from each other by forbidden energy intervals. The most tightly bound innermost electrons are least disturbed. Another approach, and one more useful for theoretical calculations, is to consider the electrons in the periodic field of the lattice. This view also leads to certain allowed regions or bands of electronic energies as shown schematically in Fig. 30. At the boundary surface the potential energy of an electron approaches asymptotically zero value and represents a barrier to those inside a metal such as in an alkali, where some 10^{22} valence electrons per cm.^{3} in the upper conduction band, which is only half filled, are free to move in what amounts to an equipotential well. The bound electrons are contained in energy bands between and below the humps and are not permitted to leave their parent atoms.

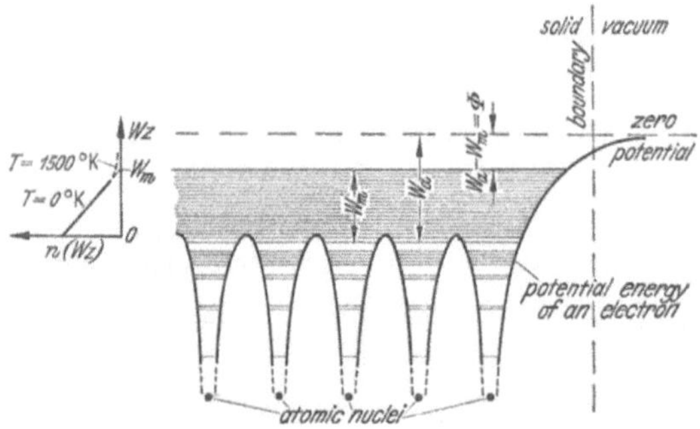

Fig. 30. Schematic diagram showing energy band structure in a solid.

SOMMERFELD and others[1] have applied FERMI-DIRAC statistics to the distribution of kinetic energy among the free electrons in a metal, on the assumption of a constant potential region, and have thus formulated the foundations of a new theory of metals which, because of its initial conspicuous success, has been extended into many directions. One of the startling predictions of the quantum-mechanical approach was the rather large width, W_m, of the free-electron or conduction band, of the order of 5 to 15 electron volts (ev), which has been born out in subsequent studies of the emission spectra of solids in the soft X-ray region[2]. With this technique, electron transitions from the valence band to lower, more sharply defined levels can be measured with considerable precision.

In order that an electron may escape from the metal it must secure an additional amount of energy over and above that which is associated with its velocity component perpendicular to the barrier surface. Such liberation may be accomplished by the processes of thermionic or field emission, by secondary emission from primary electrons, ions, metastables, or atoms, and of course by photons. Since the electrons of the conduction band have only a maximum energy W_m associated with a direction normal to the surface (see left insert in Fig. 30), light quanta must provide an additional amount, at least equal to that measured by the work function φ, namely $e\varphi = W_a - W_m$, in order to make an escape over the potential barrier W_a possible, such that

$$h\nu_0 = W_a - W_m, \tag{24.1}$$

[1] A. SOMMERFELD: Z. Physik **47**, 1 (1928); see also Sect. 31, and the article of J. C. SLATER in vol. XIX of this Encyclopedia.

[2] H. W. B. SKINNER: Phil. Trans. Roy. Soc. Lond., Ser. A **239**, 95 (1940); see also Vol. 30 of this Encyclopedia, article on Soft X-ray Spectroscopy, by D. H. TOMBOULIAN.

where ν_0 is the threshold frequency. For photons of wavelength shorter than c/ν_0 the excess appears as kinetic energy of the released electron, and the above equation reads

$$h\nu = (W_a - W_m) + \tfrac{1}{2} m v_{\max}^2, \tag{24.2}$$

which is the well known EINSTEIN relation. A complete experimental description of the external photoelectric effect then requires a knowledge of (a) its spectral distribution function, that is the number of electrons emitted per photon absorbed as a function of wavelength of the incident light, (b) its threshold frequency ν_0, and (c) the velocity distribution function of the ejected electrons for each wavelength. In addition it is necessary to determine how these quantities vary with the temperature, the applied field and the direction of the electric vector, if polarized light is used.

Within the limitations of this article a number of interesting photoelectric phenomena particularly those involving a detailed analysis in terms of the theory of the solid state or those investigations concerned with semiconductors or with technical applications of complex, multilayered cathodes, will be omitted or only mentioned in passing, and the reader must be referred to other articles of this Encyclopedia[1].

25. Early photoelectric investigations. It is of interest to note that within a very few years after the discovery of the photoelectric effect by H. HERTZ[2], quite a number of phenomena were observed by HALLWACHS[3], ELSTER and GEITEL[4], P. LENARD[5], J. J. THOMPSON[6] and others which are still of significance. Their observations indicated that light released electrons from metal surfaces and that the yield was greater for shorter wavelengths, particularly the ultraviolet. The alkali metals were shown to be more sensitive emitters in the visible region than other metals, and the presence of gases was found to have marked effects with respect to making a surface more or less sensitive to radiation. The first major theoretical interpretation, rewarded by a NOBEL prize, was put forth by EINSTEIN[7] in 1905 who suggested that light should be viewed as corpuscles or quanta of energy $h\nu$, which interacted with electrons in metals such that only a quantum of minimum energy specified by a threshold frequency ν_0 could liberate an electron. If $\nu > \nu_0$ then the excess would appear in form of kinetic energy of the electron as indicated in eq. (24.2).

This bold hypothesis differed radically from HUYGENS' and MAXWELL'S wave concept of light, in which the radiant energy is proportional to the square of the amplitude of the waves (regardless of frequency ν) and is distributed over the entire wave front. In the light of this classical picture it was difficult to understand how one electron could absorb without time lag[8] the large amount of energy necessary for its escape. When in later years BORN and others reconciled the wave-particle dualism of light by the statistical interpretation that the square of the wave amplitude was a measure of the local density of photons or quanta, it was easy to understand the three main characteristics of photoelectric emission, namely (1) that the yield is directly proportional to the intensity of light; (2) that the energy of emission of photoelectrons follows the EINSTEIN relationship by varying linearly with the frequency of the incident light; and (3) that the kinetic energy of photoelectrons is completely independent of the light intensity.

In the years subsequent to EINSTEIN'S work much effort was expended to verify his photoelectric equation by measurements of the energy of photoelectrons

---

[1] See e.g. Vols. XIX and XX, and the technical details in Vol. XXIII.
[2] H. HERTZ: Ann. Physik **31**, 983 (1887).
[3] W. HALLWACHS: Ann. Physik **33**, 301 (1888).
[4] J. ELSTER and H. GEITEL: Ann. Physik **38**, 40, 497 (1889).
[5] P. LENARD: Ann. Phys., Lpz. **2**, 359 (1900).
[6] J. J. THOMSON: Phil. Mag. **48**, 547 (1900).
[7] A. EINSTEIN: Ann. Phys., Lpz. **17**, 132 (1905).
[8] E. O. LAWRENCE and J. W. BEAMS: Phys. Rev. **32**, 478 (1928).

as a function of the frequency of the incident light [8]. This was accomplished by the application of retarding potentials between the photocathode and the electron collector, where the largest counter-potential was a measure of the maximum kinetic energy. The work of RICHARDSON and COMPTON[1] is of particular significance since they explained a number of previous discrepancies by calling attention to and correcting for the rather elusive contact potentials within every experimental tube. Their results were unambiguous and yielded a strictly linear relationship between energy and frequency as predicted by EINSTEIN.

MILLIKAN and co-workers[2] refined this approach in a very careful study with alkali metal cathodes which, when freshly scraped in the vacuum tube, exhibited photoelectric sensitivity in the visible region of the spectrum to about 6000 Å or 2 ev. With such surfaces they were able to test and verify the EINSTEIN relation over a much wider energy range and with sufficient precision to yield an independent value of the ratio h/e from

$$h(\nu - \nu_0) = \tfrac{1}{2} m v_{\max}^2 = e V, \tag{25.1}$$

where the retardation of electrons due to V has been corrected for contact potentials. It is interesting to note that their current-voltage curves plunged sharply into the axis at the cut-off potential, whereas later observers find an asymptotic approach which may be understood in terms of improved current measuring techniques.

LUKIRSKY and PRILEZAEV[3] investigated the current-voltage characteristics for a large number of metals, Al, Zn, Sn, Ni, Ag, Cd, Pb, Cu, Pt, by using a spherically symmetric electrode system: a small photocathode surrounded by a concentric silver sphere of large radius. This has the advantage that electrons emitted at any angle will always travel in directions parallel to the electric field. When retarding potentials are used the current-voltage curves reflect then the total kinetic energy of photoelectrons. This may be contrasted to a plane parallel geometry where only the normal component of the electron velocity is affected.

In all this work, so briefly discussed above, no evidence was apparent that the energy of photoelectrons was ever dependent on the light intensity, even though sources with emission lines of widely different strengths were used. In addition there is also very little evidence that the proportionality between light intensity and photocurrent does not hold, even though such deviations have been reported in the literature. In most instances such effects have only been noted at very high levels of illumination, which are known to alter the cathode surface, or in gas-filled tubes in which the initial photocurrent is amplified by electron impact ionization of the gas. In the latter cases either positive ion bombardment of the cathode or space-charge formation may cause serious deviations from the usual linear relationship. In 1916 ELSTER and GEITEL[4] were able to demonstrate that when proper precautions were taken strict proportionality prevailed even when the light intensities were changed by a factor of 10^8.

26. Threshold Determinations. It has been known from the very beginnings of photoelectric investigations that different surfaces varied widely, often in an unpredictable manner, in their response to light. With the EINSTEIN relationship generally accepted it was hoped that accurate measurements of threshold frequencies would contribute valuable information to the mechanism of the effect

[1] K. T. COMPTON: Phil. Mag. **23**, 579 (1912). — O. W. RICHARDSON and K. T. COMPTON: Phil. Mag. **24**, 575 (1912).
[2] R. A. MILLIKAN: Phys. Rev. **7**, 18, 355 (1916); **18**, 236 (1921).
[3] P. LUKIRSKY and S. PRILEZAEV: Z. Physik **49**, 236 (1928).
[4] J. ELSTER and H. GEITEL: Phys. Z. **14**, 741 (1913); **15**, 610 (1914); **17**, 268 (1916).

itself and help towards a better understanding of the structure of solids and their surfaces. In principle the threshold may be obtained by either varying the frequency of the incident light or by measuring the maximum kinetic energy of ejected photoelectrons for a given frequency ν. Then ν_0 may be obtained from eq. (25.1). Still another method was developed by SUHRMANN[1], and independently by ROY[2], which is based on the "complete" photoelectric emission integrated over all wavelengths. However, it will not be discussed in any detail here. In both of the first two methods the emission currents approach zero value asymptotically (see later figures). However, the first one, which depends on the variation of the frequency of the incident light until ν_0 is reached, has generally been accepted as more accurate, particularly since FOWLER'S[3] theoretical treatment of the problem facilitated the reduction of experimental data taken at room temperature or higher to zero degrees KELVIN [9]. The success of this method is illustrated in Fig. 31 by the work of DUBRIDGE and ROEHR[4] on palladium. Their results for eight different temperatures have been reduced by FOWLER'S method and are seen to fit nicely on a single theoretical curve.

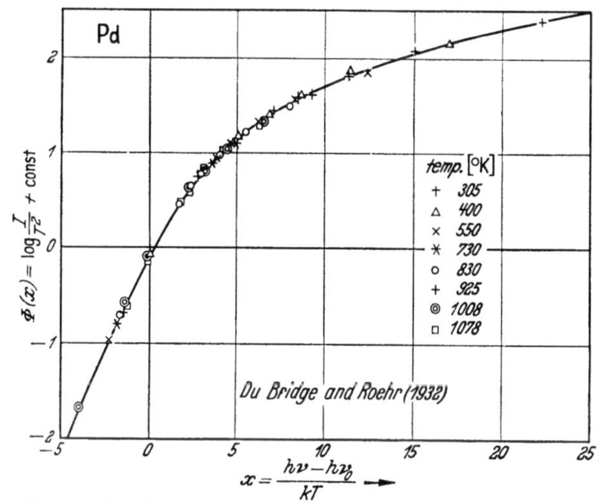

Fig. 31. Photoelectric emission from Pd at 8 different temperatures fitted to a FOWLER curve.

FOWLER showed that any temperature effects of ν_0 could be accounted for by changes in the FERMI-DIRAC distribution of energies of the electrons in the conduction band. On the basis of the initial assumption that all electrons with "normal components" of energy greater than W_a would escape, he obtained a relationship between the photoelectric yield, I, at or near ν_0 and the temperature, namely

$$I = \alpha A T^2 \varphi(x), \tag{26.1}$$

where $x = \frac{h(\nu - \nu_0)}{kT}$, α and A are assumed constant. He was able to demonstrate that within limiting assumptions the function φ was a series, whose numerical values could be tabulated, and that it was a universal function, namely one which did not depend on any of the other parameters. His calculated spectral distribution curve approached zero tangentially in agreement with experience and predicted the correct shape of the spectral yield curves, at least in the vicinity of ν_0. By plotting the experimental results in the form of ${}^{10}\text{Log}\,\frac{I}{T^2} + \text{Const.}$ against $h\nu/kT$ along the abscissa, the threshold ν_0 could be determined from the horizontal shift, $h\nu_0/kT$, of the experimental curve necessary to make it coincide with the

[1] R. SUHRMANN: Z. Physik **33**, 63 (1925); for a discussion of the "complete" photoelectric effect see also C. HERRING and M. H. NICHOLS: Rev. Mod. Phys. **21**, 185 (1949), Appendix VI.

[2] S. C. ROY: Proc. Roy. Soc. Lond., Ser. A **112**, 599 (1926).

[3] R. H. FOWLER: Phys. Rev. **38**, 45 (1931).

[4] L. A. DUBRIDGE and W. W. ROEHR: Phys. Rev. **39**, 99 (1932).

Table 3. *Photoelectric* (φ_{Phot}) *and thermionic* (φ_{Therm}) *workfunctions of elements.*

Element	φ_{Phot}	φ_{Therm}	References
Ag	3.67		58
	4.1–4.75		38
	4.74		94*
	4.56		94*
	4.75		36*
	4.81		36*
	4.3–4.8		36*
	6–3.7–4.6		94*
	3.9		47*
		3.09	91
		3.56	2
		3.08	40
Al	2.98		58
	3.43		44
	4.08		14
	4.20		79
	4.36		39*
	3.55		39*
	4.43		39*
As	5.11		75
	4.79		10
Au	4.73		64*
	4.82		64*
	4.86–4.92		38
	6–3.9		64*
		4.0–4.58	2
		4.32	40
B	4.4–4.6		9
Ba	2.48		60
	2.49		23
	2.52		50
	2.3		11
	1.9		75
		2.11	66
Be	3.17		75
	3.30		78
	3.92		59
Bi	4.14		44
	4.22–4.25		51
	4.31–4.32		71
	4.44		78
	4.46		90
	4.76–5.06		89
	4.26		10
	4.05		47
C	4.81		74
		4.34	68
		4.39	14a
Ca	2.42		75
	2.706–2.71		50
	2.76		92*
	3.20		70
	3.21		57
		2.24	34
Cd	3.68		75
	3.73		58
	3.94		44
	4.07		12
	4.099		79
	4.08		8a
Co	3.90		92*
	4.12–4.25		16
	4.55		75
		4.40	86
Cr	4.37		71
	3.76		75
		4.60	87
		4.70	56
Cs	1.9		49
	1.96		13
	1.38		75
		1.81	54
Cu	4.07		58
	4.18		92*
	4.86		80*
	5.61		80*
	4.16		47*
	4.8		75
		3.85	91
		4.26	2
		4.38	40
Fe	3.91		92*
	4.72		15
	4.77		40
	4.7		75
		4.04	28
		4.23	86*
		4.48	86*
		4.77	76
Ga	4.12		75
Ge	4.29		92*
	4.5		77
	4.73		75
	4.80		66
Hg	4.50		24
	4.52		73
	4.53		52
	4.52		43
	4.95		43*
K	2.0		48
	2.12		78
	2.24		49
	2.26		61
	2.20		26
	1.75		75
	2.24		13
Li	2.28		49
	2.42		75

Table 3. (Continued.)

Element	φ_{Phot}	φ_{Therm}	References
Mg	2.74		75
	$<$ 3.0		53
	3.59		20
	3.62		21
	3.68		59
	3.79		22
	2.4–2.65		20*
	2.15–3.75		20*
Mn	3.76		75
Mo	4.15		33
	4.34		71
	4.12		20
		4.15	33
		4.17	83
		4.20	95
		4.32	1
		4.38	96
		4.44	35
		4.37	95
Na	2.06		75
	2.28		60
	2.29		59
	2.47		49
	2.40		14
	2.27		26
Ni	4.06		92*
	4.87		71
	5.01		40
	5.06		19*
	5.05		19*
	5.20		19*
	3.96		47*
	4.47		75
		4.61	86
		4.63	28
		5.24	19
Pb	3.97		58
	4.14		44
Pd	4.97		32*
	5.2–3.7		32*
		4.99	32
Pt	4.09		58
	6.35		31*
	~6.2		71
		5.29	81
		5.32	93
		6.27	31
Rb	2.09		13
	2.16		49
Rh	4.57		29
		4.58	29
		5.7	29*
		4.80	85
Sb	4.01		63
	4.56		10
Se	4.62		44
	5.11		75
Si		3.59	14a
Sn	3.62		58
	3.87		44
	liq. 4.21		42
	γ 4.38		42
	β 4.50		42
Sr	2.06		30
	2.24		75
	2.74		5
Ta	4.12		70
	4.16		17*
	4.05		20
		4.07	35
		4.10	18
		4.19	37
Te	4.76		9
Th	3.38		70
	3.47		72
	3.57		44
	3.7		75
		3.35	96
Ti	3.95		75
	4.17		72
U	3.63		70
		3.27	46
V	3.77		75
W	4.35		62*
	4.50		62*
	4.54		88
	4.60		70
	4.49		9
		4.39	45*
		4.52	69
		4.52	65
		4.55	84
		4.56	45*
		4.58	1
		4.68	45*
		4.69	45*
		9.2	54*
		2.68	54
		0.715	54*
Zn	3.08		58
	3.32+3.57		27
	3.96		75
	3.60		44
	3.89		92*
	4.24		25*
	4.26		55*
	4.307		79
	3.6–4.35		25*
Zr	3.73		70
	3.91		46
	4.0		72
	3.73		72*
	3.67		72*
	3.85		75
		4.12	96

References to Table 3.

This table is a somewhat modified version of those complied by H. B. MICHAELSON, J. Appl. Phys. **21**, 536 (1950), and by C. HERRING and M. H. NICHOLS, Rev. Mod. Phys. **21**, 185 (1949). The starred (*) reference numbers indicate that comments may be found with the listing of the reference.

[1] A. J. AHEARN: Phys. Rev. **44**, 277 (1933).
[2] I. AMEISER: Z. Physik **69**, 111 (1931).
[3] P. A. ANDERSON: Phys. Rev. **47**, 958 (1935).
[4] P. A. ANDERSON: Phys. Rev. **49**, 320 (1936).
[5] P. A. ANDERSON: Phys. Rev. **54**, 753 (1938).
[6] P. A. ANDERSON: Phys. Rev. **59**, 1034 (1941).
[7] P. A. ANDERSON: Phys. Rev. **75**, 1205 (1949).
[8] P. A. ANDERSON: Phys. Rev. **76**, 388 (1949).
[8a] P. A. ANDERSON: Phys. Rev. **98**, 1739 (1955).
[9] L. APKER, E. TAFT and J. DICKEY: Phys. Rev. **74**, 1462 (1948).
[10] L. APKER, E. TAFT and J. DICKEY: Phys. Rev. **76**, 272 (1949).
[11] L. APKER, E. TAFT and J. DICKEY: Phys. Rev. **84**, 508 (1951).
[12] H. BOMKE: Ann. Phys., Lpz. **10**, 579 (1931).
[13] J. J. BRADY: Phys. Rev. **41**, 613 (1932).
[14] J. J. BRADY and P. JACOBSMEYER: Phys. Rev. **49**, 670 (1936).
[14a] A. BRAUN and G. BUSCH: Helv. phys. Acta **20**, 33 (1947).
[15] A. B. CARDWELL: Proc. Nat. Acad. Sci., Wash. **14**, 439 (1928).
[16] A. B. CARDWELL: Phys. Rev. **38**, 2033 (1931).
[17*] A. B. CARDWELL: Phys. Rev. **38**, 2041 (1931). $\varphi_{Ta} = 4.16$ ev after outgassing of 1000 hrs at 2200° K.
[18] A. B. CARDWELL: Phys. Rev. **47**, 628 (1935).
[19*] A. B. CARDWELL: Phys. Rev. **76**, 125 (1949); $\varphi_{Ni} = 5.06$ ev at 300° K, $\varphi_{Ni} = 5.05$ ev at the CURIE point of 623° K, $\varphi_{Ni} = 5.20$ ev at 1180° K.
[20*] R. J. CASHMAN and S. HUXFORD: Phys. Rev. **48**, 734 (1935); $\varphi_{Mg} = 2.4$ to 2.65 ev after exposure of Mg to H_2, $\varphi_{Mg} = 2.15$ to 3.75 ev after exposure of Mg to O_2, see figures in text.
[21] R. J. CASHMAN and E. BASSOE: Phys. Rev. **53**, 919 (1938) (A).
[22] R. J. CASHMAN: Phys. Rev. **54**, 753 (1938).
[23] R. J. CASHMAN and E. BASSOE: Phys. Rev. **55**, 63 (1939).
[24] H. CASSEL and A. SCHNEIDER: Naturwiss. **22**, 464 (1934).
[25*] C. F. DEVOE: Phys. Rev. **50**, 481 (1936); see figures in text for exposure of Zn to gases.
[26] J. DICKEY: Phys. Rev. **81**, 612 (1951).
[27] J. H. DILLON: Phys. Rev. **38**, 408 (1931).
[28] W. DISTLER and G. MÖNCH: Z. Physik **84**, 271 (1933).
[29*] E. H. DIXON: Phys. Rev. **37**, 60 (1931); $\varphi_{Rh} = 5.7$ ev after exposure of Rh to O_2.
[30] R. DOEPEL: Z. Physik **33**, 237 (1925).
[31*] L. A. DUBRIDGE: Phys. Rev. **31**, 236 (1928); **32**, 961 (1928); $\varphi_{Pt} = 6.35$ ev after outgassing of 300 hrs at 1250° C, see figures in text.
[32*] L. A. DUBRIDGE and W. W. ROEHR: Phys. Rev. **39**, 99 (1932); $\varphi_{Pd} = 4.97$ ev after outgassing of 1000 hrs at 900° K, $\varphi_{Pd} = 5.2$ to 3.7 ev during outgassing, see figure in text.
[33] L. A. DUBRIDGE and W. W. ROEHR: Phys. Rev. **42**, 52 (1932).
[34] S. DUSHMAN: Phys. Rev. **21**, 623 (1923).
[35] S. DUSHMAN, H. N. ROWE, J. EWALD and C. A. KIDNER: Phys. Rev. **25**, 338 (1925).
[36*] H. E. FARNSWORTH and R. P. WINCH: Phys. Rev. **58**, 812 (1940); crystal face (111): $\varphi_{Ag} = 4.75$, face (100): $\varphi_{Ag} = 4.81$ ev; φ_{Ag} changes during outgassing from 4.3 to 4.8 ev, see figure in text.
[37] M. D. FISKE: Phys. Rev. **61**, 513 (1942).
[38] R. H. FOWLER: Phys. Rev. **38**, 45 (1931).
[39*] E. GAVIOLA and J. STRONG: Phys. Rev. **49**, 441 (1936); $\varphi_{Al} = 4.36$ ev highly outgassed, 3.55 ev partially outgassed, 4.43 ev freshly distilled Al.
[40] G. N. GLASOE: Phys. Rev. **38**, 1490 (1931).
[41] A. GÖTZ: Z. Physik **43**, 531 (1927).
[42] A. GÖTZ: Phys. Rev. **33**, 373 (1929).
[43*] W. B. HALES: Phys. Rev. **32**, 950 (1928); $\varphi_{Hg} = 4.95$ ev after exposure of Hg to O_2.
[44] R. HAMER: J. Opt. Soc. Amer. **9**, 251 (1924).
[45] C. HERRING and M. H. NICHOLS: Rev. Mod. Phys. **21**, 185 (1949); for both crystal faces (111) and (116): $\varphi_W = 4.39$ ev, face (001): $\varphi_W = 4.56$ ev, face (110): $\varphi_W = 4.68$ ev, face (112): $\varphi_W = 4.69$ ev.

[46] W. L. HOLE and R. W. WRIGHT: Phys. Rev. **56**, 785 (1939).
[47]* J. S. HUNTER: Phil. Mag. **19**, 958 (1935); workfunctions for surfaces turned on a lathe.
[48] H. E. IVES: J. Opt. Soc. Amer. **8**, 551 (1924).
[49] H. E. IVES and A. R. OLPIN: Phys. Rev. **34**, 117 (1929).
[50] N. C. JAMISON and R. J. CASHMAN: Phys. Rev. **50**, 624 (1936).
[51] H. JUPNIK: Phys. Rev. **60**, 884 (1941).
[52] C. B. KAZDA: Phys. Rev. **26**, 643 (1925).
[53] C. KENTY: Phys. Rev. **43**, 776 (1933) (A).
[54]* K. H. KINGDON: Phys. Rev. **24**, 510 (1924); **25**, 892 (1925); for W plus oxygen: $\varphi_W = 9.2$ ev, for W plus thoria: $\varphi_W = 2.68$ ev, for W plus Cs: $\varphi_W = 0.715$ ev.
[55]* W. KLUGE and H. STEYSKAL: Z. Physik **116**, 415 (1940); crystal face (0001): $\varphi_{Zn} = 4.26$ ev.
[56] H. KÖSTERS: Z. Physik **66**, 807 (1930).
[57] I. LIBEN: Phys. Rev. **51**, 642 (1937).
[58] P. LUKIRSKY and S. PRILEZAEV: Z. Physik **49**, 236 (1928).
[59] M. M. MANN and L. A. DUBRIDGE: Phys. Rev. **51**, 120 (1937).
[60] R. J. MAURER: Phys. Rev. **57**, 653 (1940).
[61] H. MAYER: Ann. Phys., Lpz. **29**, 129 (1937).
[62]* C. E. MENDENHALL and C. F. DEVOE: Phys. Rev. **51**, 346 (1937); for crystal face (310); $\varphi_W = 4.35$ ev, face (210): $\varphi_W = 4.50$ ev.
[63] V. MIDDEL: Z. Physik **105**, 358 (1937).
[64]* L. W. MORRIS: Phys. Rev. **37**, 1263 (1931); $\varphi_{Au} = 4.73$ ev at 740° C, 4.82 ev at 20° C, 6 to 3.9 ev during outgassing.
[65] W. B. NOTTINGHAM: Phys. Rev. **47**, 806 (1935) (A).
[66] A. L. REIMANN: Thermionic emission, p. 37. London: Chapman & Hall 1934.
[67] A. L. REIMANN and C. K. GRANT: Phil. Mag. **22**, 34 (1936).
[68] A. L. REIMANN: Proc. Phys. Soc. Lond. **50**, 496 (1938).
[69] A. L. REIMANN: Phil. Mag. Suppl. **25**, 834 (1938).
[70] H. C. RENTSCHLER, D. E. HENRY and A. H. SMITH: Rev. Sci. Instrum. **3**, 794 (1932).
[71] H. C. RENTSCHLER and D. E. HENRY: J. Opt. Soc. Amer. **26**, 30 (1936).
[72]* H. C. RENTSCHLER and D. E. HENRY: Trans. Electrochem. Soc. **87**, 289 (1945/46); Zr plus dissolved N_2 or H_2: $\varphi_{Zr} = 3.73$ ev, Zr plus dissolved O_2: $\varphi_{Zr} = 3.67$ ev, see figure in text.
[73] D. ROLLER, W. H. JORDAN and C. S. WOODWARD: Phys. Rev. **38**, 396 (1931).
[74] S. C. ROY: Proc. Roy. Soc. Lond., Ser. A **112**, 599 (1926).
[75] R. SCHULZE: Z. Physik **92**, 212 (1934).
[76] G. SILJEHOLM: Ann. Phys., Lpz. **10**, 178 (1931).
[77] A. H. SMITH: Phys. Rev. **75**, 953 (1949).
[78] R. SUHRMANN and A. SCHALLAMACH: Z. Physik **91**, 775 (1934).
[79] R. SUHRMANN and J. PIETRZYK: Z. Physik **122**, 600 (1944).
[80]* N. UNDERWOOD: Phys. Rev. **47**, 502 (1935); crystal face (111): $\varphi_{Cu} = 4.86$ ev, face (100): $\varphi_{Cu} = 5.61$ ev.
[81] H. L. VAN VELZER: Phys. Rev. **44**, 831 (1933).
[82] H. B. WAHLIN and L. I. SORDAHL: Phys. Rev. **45**, 886 (1934).
[83] H. B. WAHLIN and J. A. REYNOLDS: Phys. Rev. **48**, 751 (1935).
[84] H. B. WAHLIN and L. V. WHITNEY: Phys. Rev. **50**, 735 (1936).
[85] H. B. WAHLIN and L. V. WHITNEY: J. Chem. Phys. **6**, 594 (1938).
[86]* H. B. WAHLIN: Phys. Rev. **61**, 509 (1942); $\varphi_{Fe} = 4.23$ ev for γ-phase, $\varphi_{Fe} = 4.48$ ev for β-phase.
[87] H. B. WAHLIN: Phys. Rev. **73**, 1458 (1948).
[88] A. H. WARNER: Phys. Rev. **38**, 1871 (1931).
[89] A. H. WEBER: Phys. Rev. **53**, 895 (1938).
[90] A. H. WEBER and C. J. EISELE: Phys. Rev. **59**, 473 (1941) (A).
[91] A. WEHNELT and S. SEELIGER: Z. Physik **38**, 443 (1926).
[92]* G. B. WELCH: Phys. Rev. **32**, 657 (1928); **40**, 470 (1932); workfunctions for surfaces which were not heat treated.
[93] L. V. WHITNEY: Phys. Rev. **50**, 1154 (1936).
[94]* R. P. WINCH: Phys. Rev. **37**, 1269 (1931); $\varphi_{Ag} = 4.74$ ev at 0° K, $\varphi_{Ag} = 4.56$ ev at 600° C after outgassing period of 1200 hrs, φ_{Ag} changes from 6 to 3.7 to 4.6 ev during outgassing, see figure in text.
[95] R. W. WRIGHT: Phys. Rev. **60**, 465 (1941).
[96] C. ZWIKKER: Proc. Roy. Acad., Amsterdam **29**, 792 (1926). — Phys. Z. **30**, 578 (1929).

theoretical one which can be plotted in terms of x as the independent variable versus $\varphi(x)$ [*9*]. This has been illustrated in Fig. 31.

Values of ν_0 for a number of surfaces are given in Table 3 together with references. Some of the thermionic workfunctions are included for the sake of comparison. After 1931 most of the threshold determinations were made with the help of FOWLER'S method. It should be noted that the spread of the observed values is indicative of the sensitivity of surfaces to contaminations even though most investigators did their work on clean and outgassed elements. The difficulties encountered with the removal of adsorbed gases and other surface impurities will be discussed separately in Sect. 32 and 33, where the profound influence of such layers on both the threshold and the spectral yield will be illustrated.

27. Spectral distributions of photoelectric yields. As in the case of the thresholds, so are quantitative yields of surfaces important to the understanding of the interaction of light with electrons in a solid, with many applications to other branches of physics and technology. Such measurements usually involve a light source together with a monochromator or filters, a sensitive thermopile which has been calibrated in absolute energy units, and the photocathode with its associated experimental apparatus. For details, such as surface preparations and other techniques which are intricate and vary widely, the reader must be referred to many excellent publications[1] on the subject [*8*], [*10*].

The large majority of all investigations cover the spectral range from the visible into the quartz ultraviolet, because of the convenience and ready availability of optical instruments for this region. Little work has been done below the quartz cut-off in the vacuum ultraviolet, requiring calcium- or lithium-fluoride optics, and even fewer data are available for the region of the extreme ultraviolet below 1100 Å.

In the following figures a variety of yield curves are represented, where in all instances the number of electrons per incident photon are plotted vertically versus the wavelength in units of both its electron volt equivalent (ev) and angstroms (Å). Some investigators who were particularly concerned with vectorial effects on thin films, using polarized light, have chosen to present yields in terms of photoelectrons per quanta *absorbed*. In such cases the curves will be marked accordingly.

When comparing the results of different investigators it is necessary to keep in mind a number of phenomena each one of which may well alter the yield curves in a significant way. Among them are the already mentioned impurity layers which may be important even when vacua of the order of 10^{-7} mm. Hg are used. In addition, the tickness of a vacuum-distilled surface is known to influence photoemission because of the fact that a very thin layer, of the order of 10^{-6} cm. thick, may produce patches of different sizes and workfunctions on its substratum (see Sect. 30). The deposited material may also form dipole layers or be adsorbed on the base to produce an ionic grid, thus representing a complex surface or at least seriously modifying the shape of the potential barrier.

It will be shown later that the optical constants of a photocathode play a rather dominant role in photoemission such that for thin layers the properties of both the underlying and the evaporated material have to be taken into account in order to understand the variations in the spectral yield. This is an integral part of the phenomenon called polarization selectivity or better the vectorial

[1] See also N. R. CAMPBELL and D. RITCHIE: Photoelectric cells, London: Sir Isaac Pittman & Sons 1934. — H. SIMON and R. SUHRMANN: Lichtelektrische Zellen und ihre Anwendungen. Berlin: Springer 1932. — B. GUDDEN: Lichtelektrische Erscheinungen. Berlin: Springer 1928.

effect (Sect. 29) which describes changes in yield by factors from 5 to 30 depending on the angle of incidence and the plane of polarization of the light. For mirror surfaces the yields usually reach a maximum when light polarized parallel to the plane of incidence is used at an angle of from 45 to 70° with the normal, whereas perpendicularly polarized light gives rise to very much smaller yields and usually does not exhibit a pronounced maximum. Thus even unpolarized light incident at such angles will produce markedly higher yields than at normal incidence. For rougher surfaces the dependence of the yield on both the angle of incidence and the plane of polarization may be obscured because of the multiplicity of orientations of the different normals associated with sub-microscopic

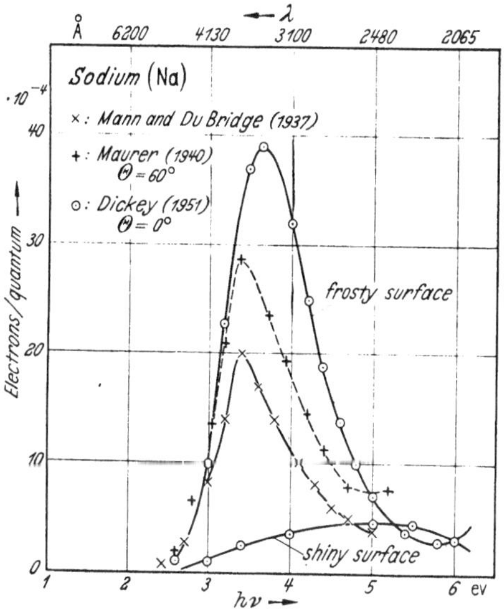

Fig. 32. Photoelectric yields from Na.

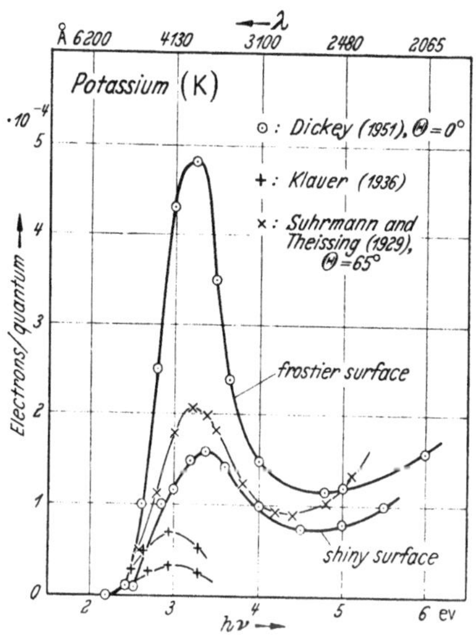

Fig. 33. Photoelectric yields from K.

hills and valleys of the surface. Emission will then depend on the degree of roughness, and the results of one observer will vary from those of another.

In Fig. 32 the spectral yield curves from thick sodium films obtained by MANN and DUBRIDGE[1], MAURER[2] and DICKEY[3] are reproduced. DICKEY's curve for the frosty or matt surface, resulting from slow deposition of the metal over a period of 15 minutes or more, shows a much more pronounced selective maximum than the one for a shiny surface which was formed by rapid evaporation. Both curves were taken with unpolarized light at normal incidence, and for the shiny surface a much smaller component of the electric vector was perpendicular to the emitting areas than for the frosty layer. When the angle of incidence was either 45 or 60° the shiny surface gave rise to yield maxima which were three or four times larger than those obtained at normal incidence. In addition this surface would then also exhibit a distinct minimum at $h\nu = 5.7$ ev. MAURER's curve taken on a matt surface at an angle of 60° and that of MANN and DUBRIDGE (curved cathode) thus compare well with DICKEY's work.

Her results[3] for potassium, Fig. 33, show similar characteristics to those for sodium and may be compared with the work of SUHRMANN and THEISSING[4]

[1] M. M. MANN and L. A. DUBRIDGE: Phys. Rev. **51**, 120 (1937).
[2] R. J. MAURER: Phys. Rev. **57**, 653 (1940).
[3] J. DICKEY: Phys. Rev. **81**, 612 (1951).
[4] R. SUHRMANN and H. THEISSING: Z. Physik **52**, 453 (1928/29).

and of KLAUER[1]. It is not possible to identify the causes for the differences in yields. DICKEY also mentions preliminary data on the photoelectric emission

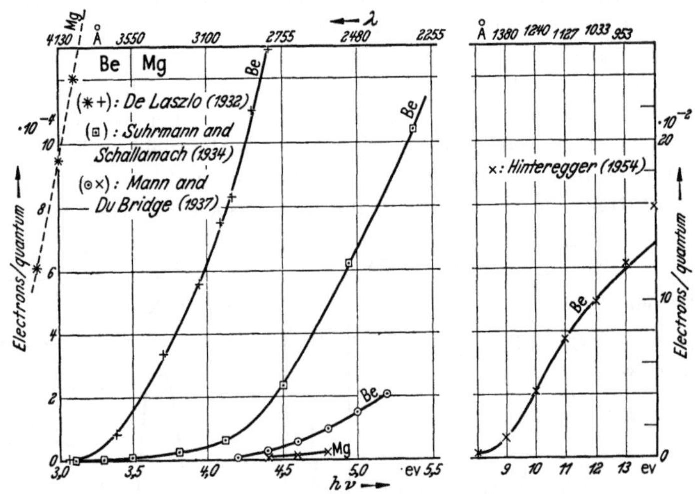

Fig. 34. Photoelectric yields from Be and Mg.

from rubidium which show the same type of spectral maximum and slight minimum as those for potassium. To what extent the yield increases in both sodium and potassium towards the short wavelength side ($h\nu > 5$ ev) signify the onset of a volume effect is still problematical and requires further work in the ultraviolet region of the spectrum.

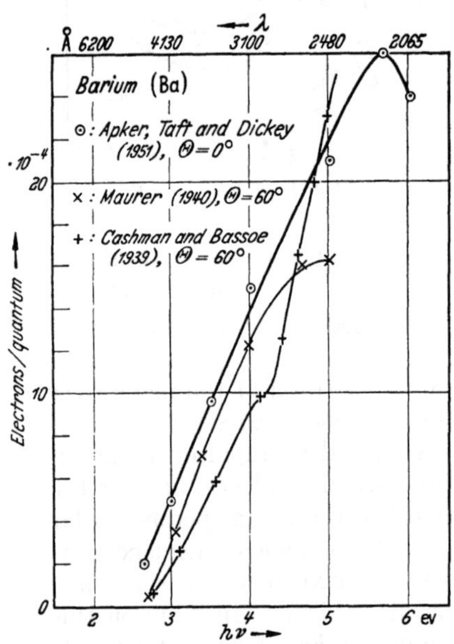

Fig. 35. Photoelectric yields from Ba.

For beryllium and magnesium the results of DE LASZLO[2], SUHRMANN and SCHALLAMACH[3], and MANN and DUBRIDGE[4] (Fig. 34) are rather more divergent, and particularly for magnesium it seems probable that slight contamination gave rise to the large observed differences. HINTEREGGER'S[5] recent work on Be in the far ultraviolet exhibits a steep rise in emission at about $h\nu = 9$ ev which is taken as the onset of the volume photoelectric effect to be discussed later. His yields increase to approximately 12% at 14 ev, a value which is probably too high and characteristic of a contaminated cathode. A magnitude of one percent seems more likely for pure Be in this spectral region as indicated by the results on other metals of WALKER et al., to be shown later.

[1] F. KLAUER: Ann. Phys., Lpz. **20**, 909 (1934).
[2] H. DE LASZLO: Phil. Mag. **13**, 1171 (1932).
[3] R. SUHRMANN and A. SCHALLAMACH: Z. Physik **91**, 775 (1934).
[4] M. M. MANN and L. A. DU BRIDGE: Phys. Rev. **51**, 120 (1937).
[5] H. E. HINTEREGGER: Phys. Rev. **96**, 538 (1954) (L).

Reasonably good agreement exists between the measurements of APKER, TAFT and DICKEY[1], of MAURER[2] and of CASHMAN and BASSOE[3] on barium (Fig. 35). However, the onset of the volume effect in Ba probably lies farther toward the ultraviolet than the value of $h\nu \sim 4.2$ ev claimed by CASHMAN and BASSOE.

In Fig. 36 are presented the photoelectric yields from Al, Cd, and Zn obtained by SUHRMANN and PIETRZYK[4], by DE LASZLO[5] and by BAKER[6]. In a reproducible fashion SUHRMANN and PIETRZYK were able to go from the state of greater order in Cd (at the temperature of liquid air, 83° K) to one of disorder (at 293° K). They also obtained entirely similar results for Al and Zn.

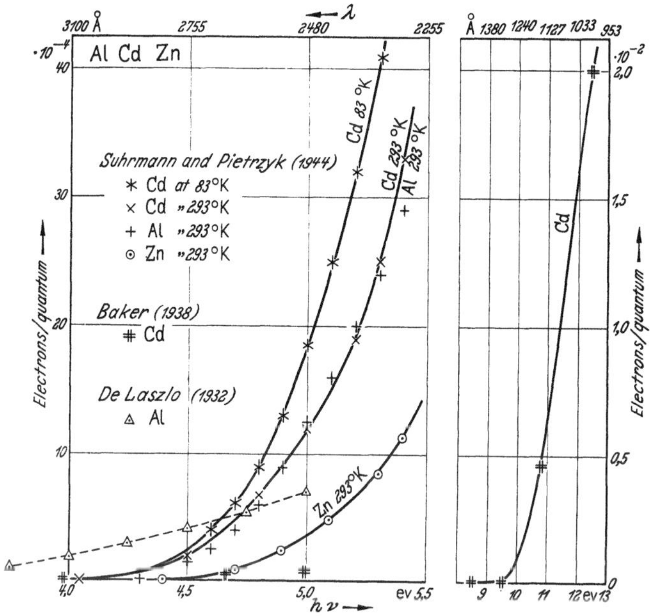

Fig. 36. Photoelectric yields from Al, Cd, and Zn.

Again the interesting phenomenon of the volume effect becomes apparent from BAKER'S Cd-curve which at 10 ev increases sharply to values of about two percent at 1000 A (12.4 ev). He does not claim high purity of the Cd-cathode, even though the metal was distilled several times, since his photo-tube could not be baked out because of the cement holding the LiF window.

Yield curves for Pd, W, Mo, and Ta have been reproduced in Fig. 37 from the work of WALKER, WAINFAN and WEISSLER[7]. The results for the untreated metals, washed only with an organic solvent, are shown by dotted lines and may be compared to the solid curves which were obtained after heating at about 1100° C until no further change in yield occurred. It is possible that higher temperatures and better vacuum conditions might further decrease the values for W, Mo, and Ta, since it is felt that the maximum between 16 and 22 ev is

[1] L. APKER, E. TAFT and J. DICHEY: Phys. Rev. **84**, 508 (1951).
[2] R. J. MAURER: Phys. Rev. **57**, 653 (1940).
[3] R. J. CASHMAN and E. BASSOE: Phys. Rev. **55**, 63 (1939).
[4] R. SUHRMANN and J. PIETRZYK: Z. Physik **122**, 600 (1944).
[5] H. DE LASZLO: Phil. Mag. **13**, 1171 (1932).
[6] R. F. BAKER: J. Opt. Soc. Amer. **28**, 55 (1938).
[7] W. C. WALKER, N. WAINFAN and G. L. WEISSLER: J. Appl. Phys. **24**, 1318 (1953); **26**, 1366 (1955).

probably caused by surface impurities. KENTY'S[1] values for degassed W are in reasonable agreement with those of WALKER et al., while the data of HINTEREGGER and WATANABE[2] join with the untreated W-curve at 12 ev and continue it to longer wavelengths. Again the onset of the volume effect appears to occur in the neighborhood of 9 ev.

Additional yield curves for degassed metals (Pt, Au, Ag, Ni, Cu) may be found in Sect. 32 and 33 where the influence of gas-contaminations on surfaces will be discussed.

APKER, TAFT and DICKEY[3] measured photoelectric emission in As, Sb, and Bi which may be compared in Fig. 38 with the Bi work of JUPNICK[4] and of

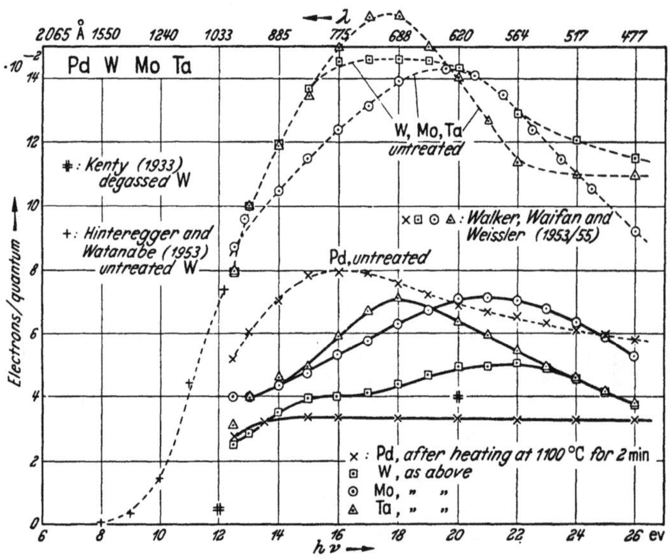

Fig. 37. Photoelectric yields from Pd, W, Mo, and Ta in the extreme ultraviolett. The effect from surfaces which were not outgassed is also indicated.

SUHRMANN and SCHALLAMACH[5] shown in the inset of Fig. 38 and again as a dotted line between 5.0 and 5.5 ev near the As-curve. One may only speculate that these large differences are caused by structural dissimilarity of the Bi films used by these observers.

The yields obtained by APKER, TAFT and DICKEY[6] for Te and Ge (Fig. 39) were found to be of the same order of magnitude as other elements such as Pt, while those from composite surfaces, such as Cs_3Sb or Cs_2Te (Fig. 40)[7], are by factors of 100 to 300 larger with a correspondingly pronounced excursion of the threshold towards longer wavelengths. All indications point to the volume effect as mainly responsible for the emission from these highly efficient complex cathodes.

The observational data may be summarized by saying that for most elements the surface photoelectric effect which occurs near the threshold gives rise to yields

[1] C. KENTY: Phys. Rev. **44**, 891 (1933).
[2] H. E. HINTEREGGER and K. WATANABE: J. Opt. Soc. Amer. **43**, 604 (1953).
[3] L. APKER, E. TAFT and J. DICKEY: Phys. Rev. **76**, 270 (1949).
[4] H. JUPNICK: Phys. Rev. **60**, 884 (1941).
[5] R. SUHRMANN and A. SCHALLAMACH: Z. Physik **91**, 775 (1934).
[6] L. APKER, E. TAFT and J. DICKEY: Phys. Rev. **74**, 1462 (1948).
[7] L. APKER, E. TAFT and J. DICKEY: J. Opt. Soc. Amer. **43**, 78 (1953) (on Cs_3Sb). — E. TAFT and L. APKTER: J. Opt. Soc. Amer. **43**, 81 (1953) (on Cs_2Te).

which increase initially in an exponential manner and then reach a maximum value of the order of 10^{-4} electrons per incident quantum. In some instances this maximum has not been observed because it probably lies between 2000 and 1000 Å, a region in which little work has as yet been done. In all instances the position of the threshold and the magnitude of the yields near the threshold are critically influenced by even the slightest changes in the structure of the cathode surface.

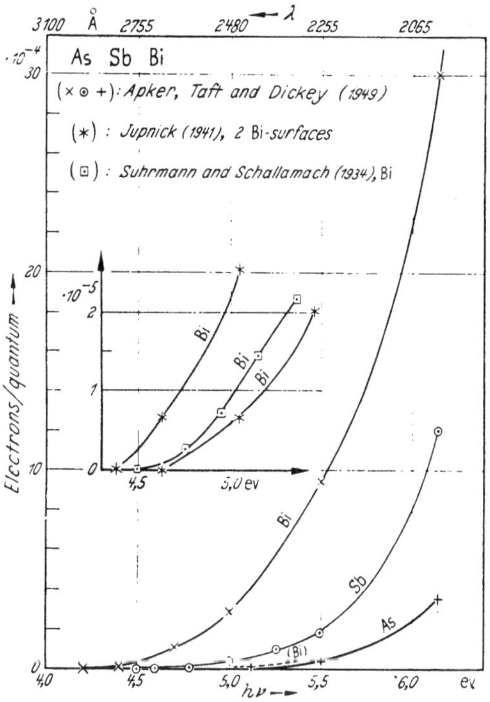

Fig. 38. Photoelectric yields from As, Sb, and Bi.

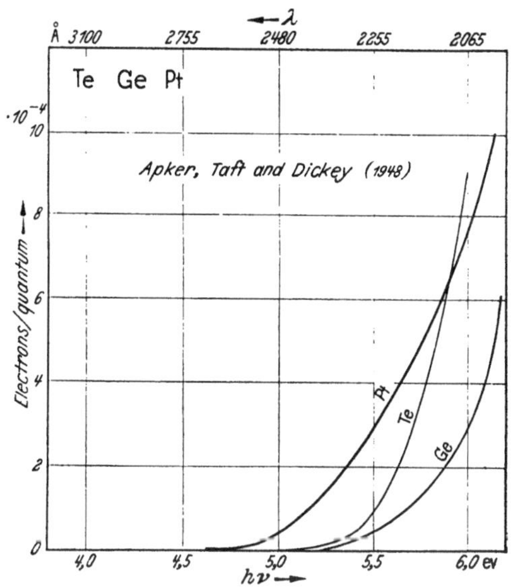

Fig. 39. Photoelectric yields from Te, Ge, and Pt.

Only sketchy information is available on the volume effect, and the preliminary experiments on a few metals point in those cases to an onset near 9 to 10 ev. Its largest value for pure elements seems to be of the order of 10^{-2} electrons per quantum, a factor of roughly 100 times larger than that of the surface effect, and is reached approximately at 12 to 14 ev. As is to be expected, surface changes are far less significant here than near the long wavelength threshold. Much careful work needs yet to be done in this little explored region of the spectrum.

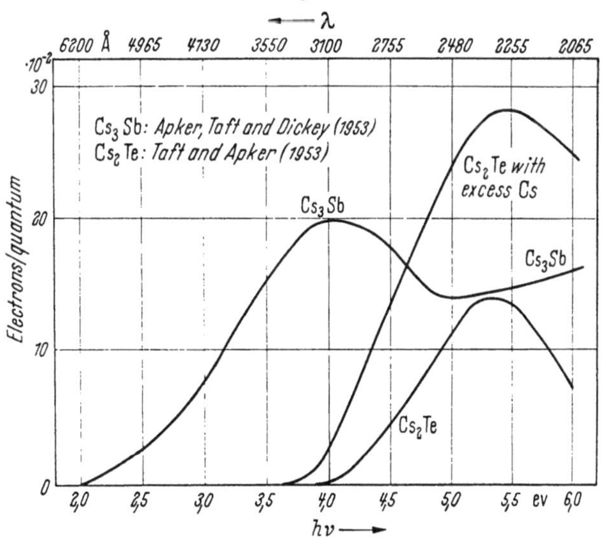

Fig. 40. Photoelectric yields from Cs_3Sb and Cs_2Te.

28. Energy distributions of photoelectrons. It has been pointed out before that the photoelectric properties of a surface are completely specified only when three quantities are known: (a) Its spectral distribution of yields, (b) its threshold frequency ν_0, and (c) the velocity or energy distribution of ejected electrons as a function of frequency. In principle there are two experimental approaches which have been used to measure electron energy distributions, the magnetic velocity selector and the stopping potential method. RAMSAUER[1] described one version of the magnetic selector with $\boldsymbol{H} \perp \boldsymbol{E}$ and obtained distribution curves for a Zn surface. His work was refined by others[2], but since difficulties with low energy electrons existed in both the magnetic and the stopping potential method, the bulk of research was done with the latter method which is considerably simpler. KOLLATH[3] suggested a modification with a longitudinal magnetic field (Fig. 41), but even though very promising this approach has not been used extensively.

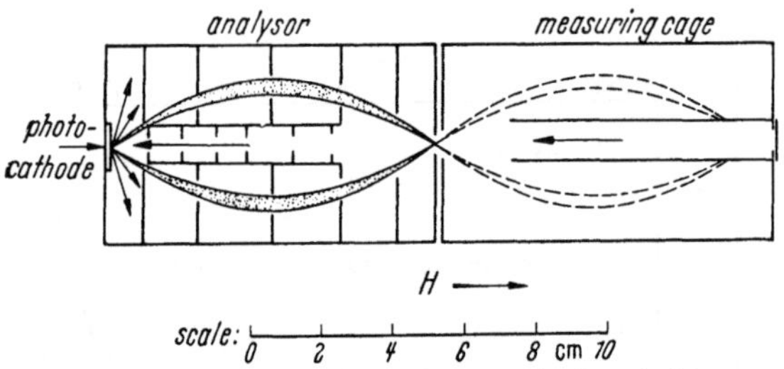

Fig. 41. KOLLATH's method for the determination of electron energy distributions with a longitudinal magnetic field.

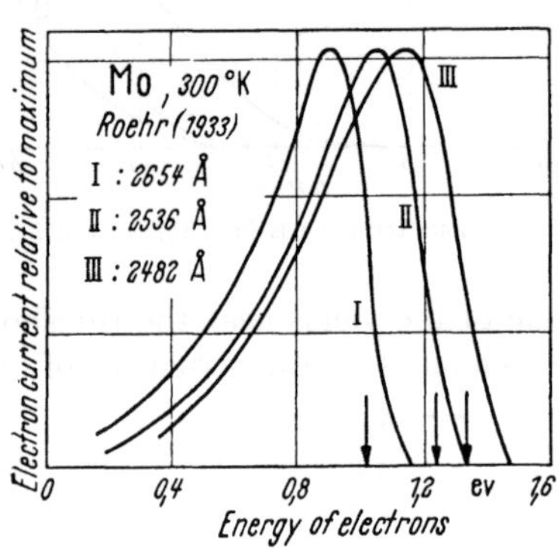

Fig. 42. Energy distributions of photoelectrons from Mo at 300° K. The arrows indicate the maximum energies at 0° K.

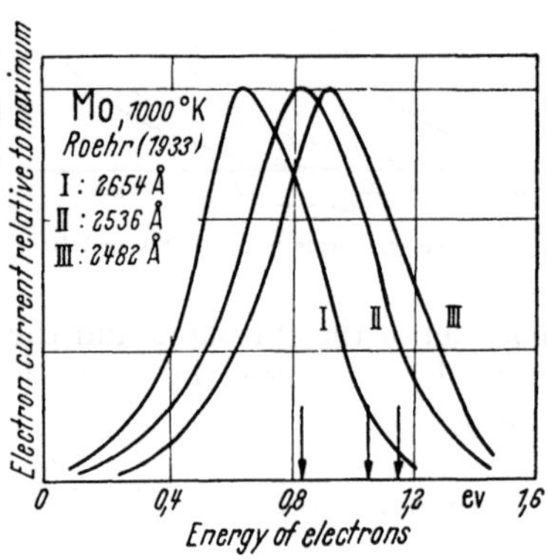

Fig. 43. Energy distributions of photoelectrons from Mo at 1000° K. The arrows indicate the maximum energies at 0° K.

The stopping potential method employing central fields in a spherically symmetric geometry for the retardation for electrons has the serious disadvantage in that it must rely on the differentiation of the electron current versus counter-voltage curve in order to obtain the energy distribution function. Any errors in the experimental current-voltage curve will be greatly enhanced by differentiating. In addition counter-currents of electron flow from the anode or grid to the photocathode due to scattered light or secondary electrons have to be avoided or compensated for [8][4]. Furthermore, in order to obtain absolute

---

[1] C. RAMSAUER: Ann. Physik **45**, 961, 1124 (1914); see also for experimental details on this method C. RAMSAUER and R. KOLLATH: Handbuch der Physik, Vol. XXII, part 2. Berlin: Springer 1933.

[2] O. KLEMPERER: Z. Physik **16**, 280 (1923). — F. WOLF: Ann. Physik **83**, 1001 (1927).

[3] R. KOLLATH: Ann. Phys., Lpz. **39**, 59 (1941).

[4] P. LUKIRSKY and S. PRILEZAEV: Z. Physik **49**, 236 (1928); see also A. L. HUGHES and L. A. DUBRIDGE: Photoelectric Phenomena, pp. 11–14, p. 25 and p. 116. New York: McGraw-Hill 1932.

values of the electron energies, any contact potential differences in the tube have to be taken into account[1,2].

In the following a selected number of distributions will be presented which characterize the influence of certain factors.

In the work of ROEHR[3] on Mo under conditions of careful outgassing (1600 hours at 2100° K) the effect of temperature was studied, and the results are shown in Figs. 42 and 43. It should be noted that when a FOWLER plot (see Sect. 31) was made of the yields at various temperatures, all points fell on just one curve giving rise to ν_0 at 0° K (as demonstrated for Pd, Fig. 31) and verified the theoretical postulate that the high-energy tail of the FERMI distribution of electrons in the metal is the sole cause for any observed temperature effect. This is also brought out when the two sets of curves in Figs. 42 and 43 are compared, where the

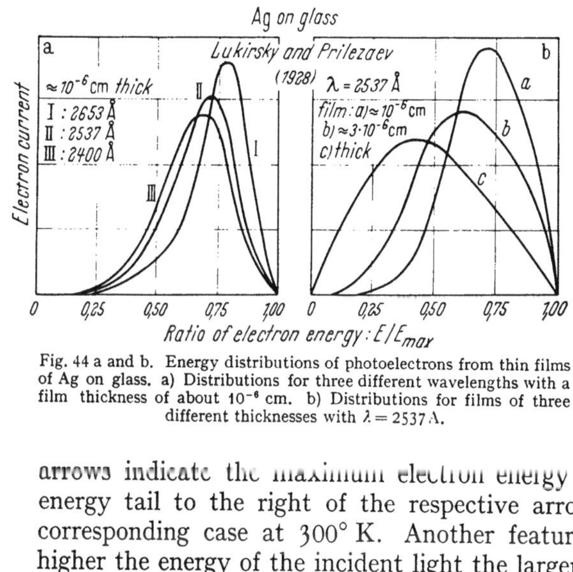

Fig. 44 a and b. Energy distributions of photoelectrons from thin films of Ag on glass. a) Distributions for three different wavelengths with a film thickness of about 10^{-6} cm. b) Distributions for films of three different thicknesses with $\lambda = 2537$ Å.

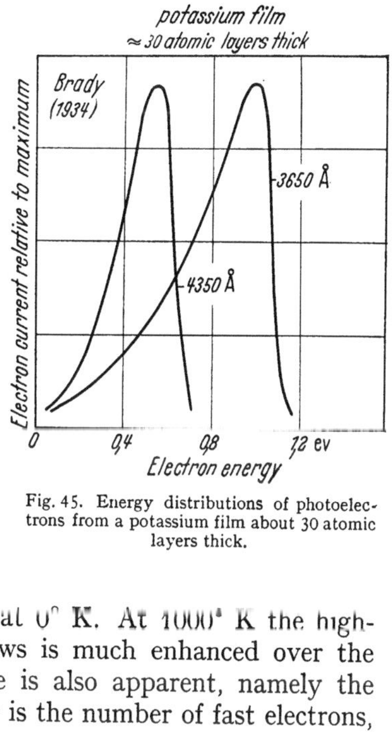

Fig. 45. Energy distributions of photoelectrons from a potassium film about 30 atomic layers thick.

arrows indicate the maximum electron energy at 0° K. At 1000° K the high-energy tail to the right of the respective arrows is much enhanced over the corresponding case at 300° K. Another feature is also apparent, namely the higher the energy of the incident light the larger is the number of fast electrons, at least for wavelengths near the threshold.

LUKIRSKY and PRILEZAEV[4] studied the distributions from thin films of Ag on glass and found that the relative number of slow electrons decreased materially with decreasing layer thickness, as shown in Fig. 44a, b. This was confirmed by similar studies on thin films of Pt, Al, Cu, and Pd[5].

In this connection BRADY'S[6] and HENSHAW'S[7] work, and particularly MAYER'S[8] investigations, are of significance in as much as they find for thin K-films on either a Ag-base[6], a Pt-base[7], or a quartz base[8], that one or two atomic layers of K behave like the bulk metal (thicker than 30 to 100 layers as in Fig. 45).

[1] See footnote 4, p. 356.

[2] J. DICKEY: Phys. Rev. **81**, 614 (1951).

[3] W. W. ROEHR: Phys. Rev. **44**, 866 (1933); see also L. A. DUBRIDGE and W. W. ROEHR: Phys. Rev. **42**, 52 (1932).

[4] P. LUKIRSKY and S. PRILEZAEV: Z. Physik **49**, 236 (1928).

[5] J. KLUGE: Ann. Physik **82**, 432 (1927). — W. BENNEWITZ: Ann. Physik **83**, 913 (1927). F. HEROLD: Ann. Physik **85**, 587 (1928), he obtains different but not systematically changing distributions with gas contaminations.

[6] J. J. BRADY: Phys. Rev. **46**, 772 (1934).

[7] C. L. HENSHAW: Phys. Rev. **52**, 854 (1937).

[8] H. MAYER: Z. Physik **124**, 326 (1948).

This seems to imply that the thickness will effect only the electron distribution in cases where less than one monolayer is deposited. Surface patches of differing work functions may cause serious changes, as will be indicated in a later section. MAYER also calculated from optical constants the amount of light *absorbed* in his thin films and arrived at quantum yields which could approach the optimum

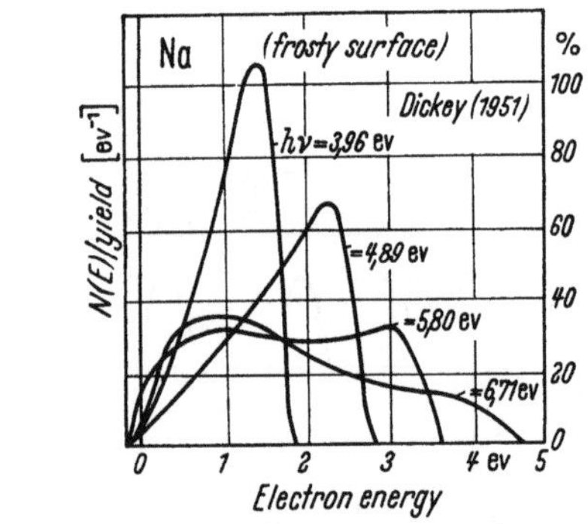

Fig. 46. Energy distributions of photoelectrons from a thick, frosty Na surface.

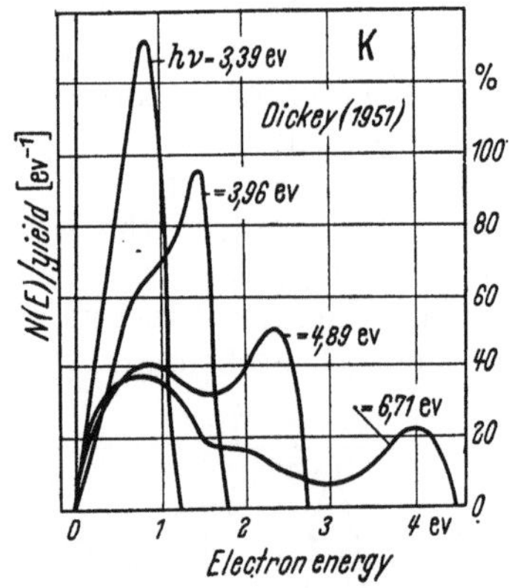

Fig. 47. Energy distributions of photoelectrons from a thick K surface.

of 50% (50% of the electrons travel backwards). This may have a bearing on the mechanisms in highly efficient complex cathodes.

DICKEY[1] recently re-measured the electron energy distributions from thick films of Na and K deposited on a Ta substratum. This gave the interesting results, shown in Figs. 46 and 47, where for the longer wavelengths there are relatively many more fast electrons than for shorter ones. It has been assumed that the appearance

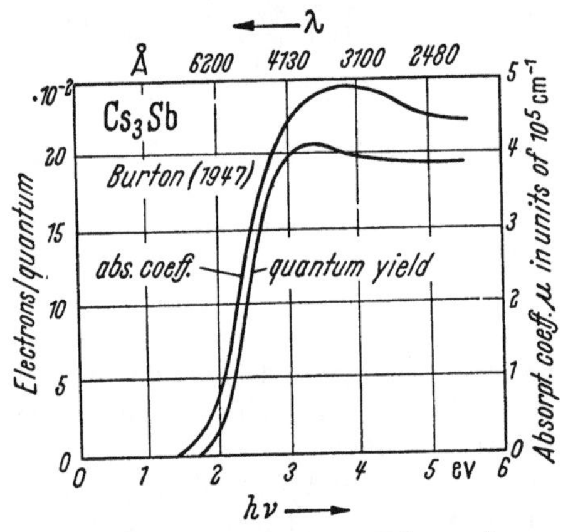

Fig. 48. Optical absorption coefficients and photoelectric yields from Cs_3Sb.

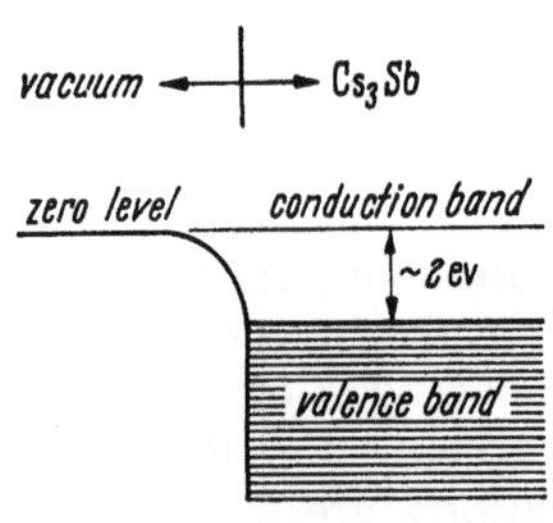

Fig. 49. Band structure model of Cs_3Sb.

of the low-energy hump in the potassium distributions, which makes its appearance at $h\nu \sim 4$ ev, may be connected with the observation[2] that the metal becomes more transparent to radiation of $h\nu \sim 3.9$ ev. This transparency occurs in Na in the vicinity of $h\nu \sim 5.9$ ev, and the low-energy peak becomes evident for $h\nu = 5.8$ ev, which seems to indicate some connection between absorbed radiation and low energy electrons.

[1] J. DICKEY: Phys. Rev. **81**, 614 (1951).

[2] H. E. IVES and H. B. BRIGGS: J. Opt. Soc. Amer. **28**, 337 (1938); **27**, 181, 395 (1937); **26**, 238 (1936). — R. W. WOOD: Phys. Rev. **44**, 353 (1933).

This receives further support from the work on cesium antimonide, Cs_3Sb, which is formed when a layer of evaporated antimony reacts with cesium vapor at 150° C. BURTON[1] has studied both the optical absorption coefficients and the quantum yields from such a surface [*10*] (p. 59), as shown in Fig. 48. He attributes the steep rise of the absorption coefficient at $h\nu = 2$ ev, accompanied with a similar one for the quantum yield, to the onset of absorption of light by valence electrons and proposes an energy band structure as in Fig. 49, with a

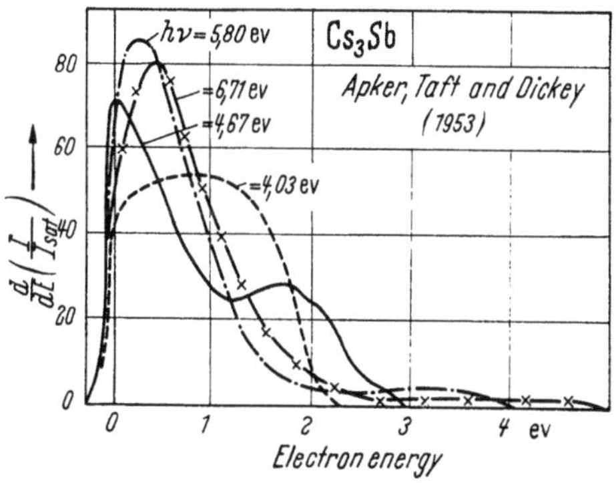

Fig. 50. Energy distributions of photoelectrons from Cs_3Sb.

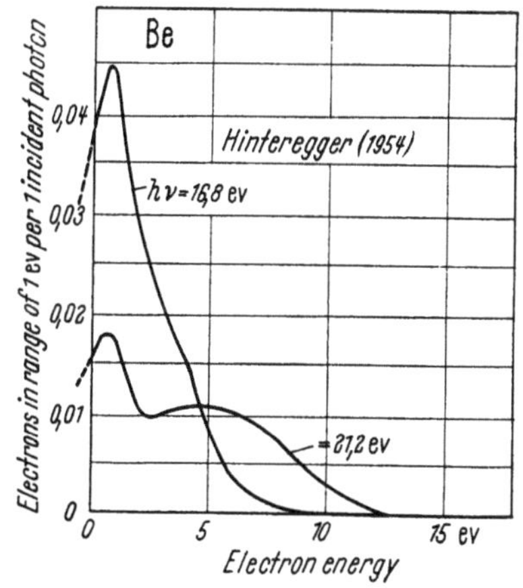

Fig. 51. Energy distributions of photoelectrons from Be with radiation of the extreme ultraviolet.

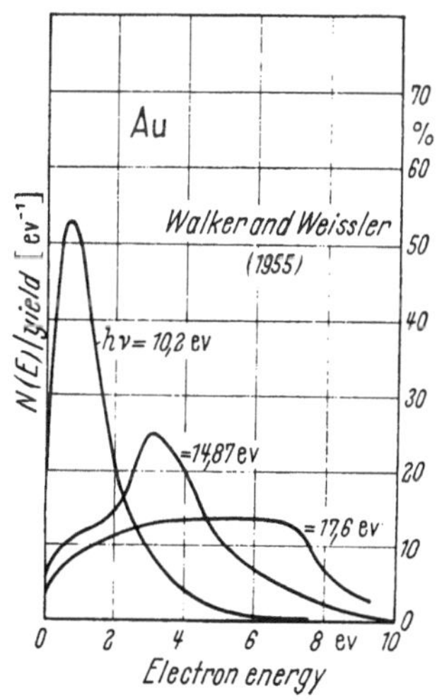

Fig. 52.* Energy distributions of photoelectrons from Au with radiation of the extreme ultraviolet.

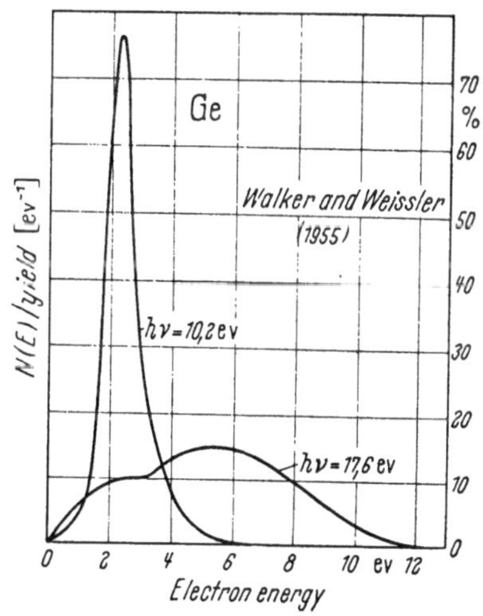

Fig. 53. Energy distributions of photoelectrons from Ge with radiation of the extreme ultraviolet.

forbidden region of about 2 ev width between the filled valence band and the empty conduction band. APKER, TAFT and DICKEY[2] have recently measured electron energy distributions in Cs_3Sb (Fig. 50) and find the same phenomenon of increased numbers of slow electrons for higher energy radiation as DICKEY in Na and K.

More specifically, they noticed a pronounced change in shape of the distribution curves for $h\nu > 4$ ev, where only relatively very few electrons have energies in excess of 2 ev (Fig. 50). This is interpreted with the help of BURTON's energy

[1] J. A. BURTON: Phys. Rev. **72**, 531 (1947) (A).
[2] L. APKER, E. TAFT and J. DICKEY: J. Opt. Soc. Amer. **43**, 78 (1953).

band structure model (Fig. 49) in terms of electron-electron scattering: A valence electron which has received less than 4 ev of energy from a photon will be lifted from the filled into the conduction band where it is subject only to lattice scattering. Thus it may migrate over large distances without significant loss in energy since this type of scattering is almost elastic. When on the other hand a valence electron receives energy greater than 4 ev, it may be scattered by another valence electron which receives a sufficient amount of energy in the process to cross the forbidden region of 2 ev, while the first one drops to near the bottom of the conduction band. Quite clearly such an interaction is not possible with an electron excited only to 3 ev, since it could not share its energy with another valence electron and itself remain in the conduction band. This is one of the rare examples where electron-electron scattering seems quite definite. However, other ways of accounting for the many low-energy electrons at $h\nu > 4$ ev, such as the effective excitation of two electrons per absorbed photon, can not be ruled out.

This trend for higher energy photons to produce a relatively larger number of slow electrons, perhaps caused by electron-electron scattering, might point to a volume effect, postulated by TAMM and SCHUBIN[1], namely that the light is not only interacting with the electrons in the non-periodic potential near the surface but also with those deeper lying ones within the periodic field of the lattice. Some recent measurements by HINTEREGGER[2] and by WALKER and WEISSLER[3] of electron distributions due to radiation of the extreme ultraviolet seem to support this view. Their results are shown in Figs. 51, 52 and 53 and continue the trend not only of DICKEY'S work quoted above but also that of APKER, TAFT and DICKEY[4] for the semiconductors Ge, Cs_3Sb, and Cs_2Te.

29. The vectorial effect with polarized light. It was already apparent in early investigations[5] that the effect of polarized light on photoelectric emission was critically dependent on the plane of polarization. For purposes of illustration, some of the work done on thin films of the alkali metals will be singled out, particularly since it leads to important correlations. For the sake of definition, $\boldsymbol{E}_{\parallel}$ shall indicate the direction of the electric vector when it is parallel to the plane of incidence defined by the normal of the surface and the oblique ray of light making an angle of incidence ϑ with the normal. Thus such a ray will always have a component of $\boldsymbol{E}_{\parallel}$ perpendicular to the surface, while $\boldsymbol{E}_{\perp}$ will be oriented at right angles to the plane of incidence and therefore will be parallel to the surface. The gross features of the effect are shown in Fig. 54 obtained by IVES[6] with polarized light of 5461 A falling on a thin film of potassium on a substratum of platinum. These and all other curves have been corrected for the fact that the cathode area covered by the beam is proportional to $\sec\vartheta$. The solid lines indicate the electric intensities of light, proportional to $|E|^2$, just at or slightly above the Pt-surface and were calculated on the basis of WIENER'S model of standing waves, created by interference between incident and reflected radiation, taking into account the optical constants of platinum. The thickness of the

---

[1] I. TAMM and S. SCHUBIN: Z. Physik **68**, 97 (1931).
[2] H. E. HINTEREGGER: Phys. Rev. **96**, 538 (1954).
[3] W. C. WALKER and G. L. WEISSLER: Phys. Rev. **97**, 1178 (1955) (L).
[4] L. APKER, E. TAFT and J. DICKEY: J. Opt. Soc. Amer. **43**, 78, 82 (1953). — Phys. Rev. **74**, 1471 (1948).
[5] J. ELSTER and H. GEITEL: Ann. Physik **52**, 433 (1894); **55**, 684 (1895); **61**, 445 (1895). — R. POHL and P. PRINGSHEIM: Verh. dtsch. phys. Ges. **12**, 682, 1039 (1910); **15**, 111 (1913). — H. GEITEL: Ann. Physik **67**, 420 (1922).
[6] H. E. IVES: Phys. Rev. **38**, 1209 (1931).

potassium film was very small compared to the wavelength of light, of the order of one atomic layer, and since it was not visible was not thought to affect the optical properties of the underlying platinum. The solid lines in Fig. 54 for $\boldsymbol{E}_{\parallel}$ and $\boldsymbol{E}_{\perp}$ were obtained as if no potassium were present. The $\boldsymbol{E}_{\perp}$ vector is parallel to the surface and therefore must be continuous. In contrast $\boldsymbol{E}_{\parallel}$ for oblique incidence has a component normal to the surface which must be multiplied by a factor depending on the absorption coefficient and the index of refraction of the K-film. When this was taken into account the dashed curve of Fig. 54 described the intensity variation of $\boldsymbol{E}'_{\parallel}$ at the Pt-surface with the K-film above it. Relative values of the emission current for $\boldsymbol{E}_{\parallel}$ are indicated by crosses and for $\boldsymbol{E}_{\perp}$ by circles. Similarly convincing results were obtained for thin rubidium films on glass.

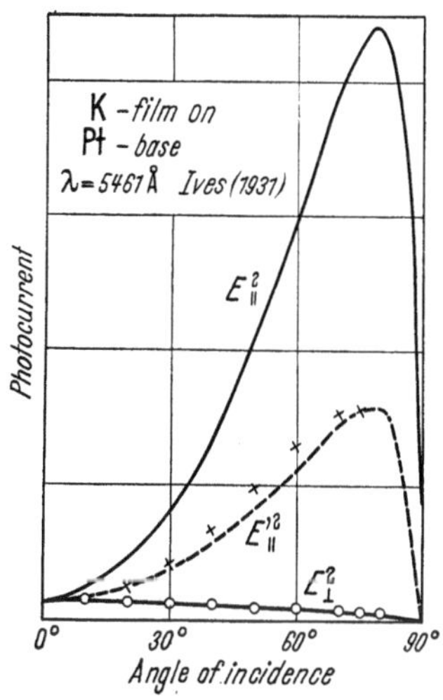

Fig. 54. The vectorial effect, showing the variation with angle of incidence of the radiation intensity and the emission current for both $E_{\perp}$ and $E_{\parallel}$.

The spectral response of polarization selectivity is characterized in Fig. 55 of

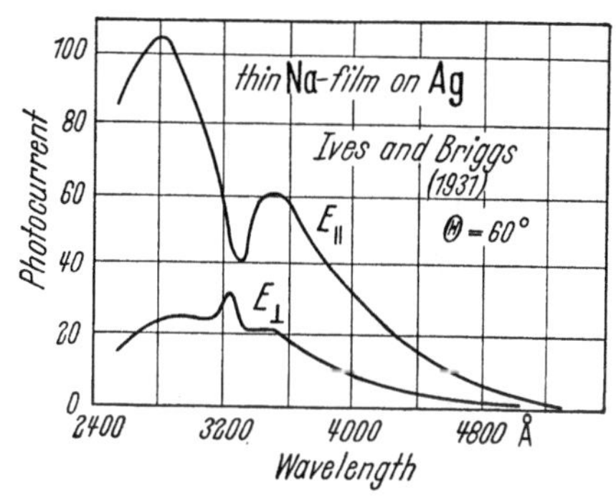

Fig. 55. The spectral distribution of photoemission from a thin film of Na on a Ag base for polarized light.

Ives and Briggs[1] for a thin sodium film on a silver base at an angle of incidence of 60°. The minimum at 3260 Å of the emission due to $\boldsymbol{E}_{\parallel}$ and the small maximum of the current at the same wavelength with $\boldsymbol{E}_{\perp}$ both show qualitatively the significance of the optical properties of the underlying material since silver has a pronounced transmission band in the immediate vicinity at 3160 Å. Their earlier work with platinum as a base did not show such abrupt changes since the optical constants of this metal vary slowly in this region of the spectrum.

In some later work Ives and Briggs[2] go considerably further in their investigations of thin alkali metal films (K, Rb, Cs) on a platinum-iridium mirror after some careful investigations of their optical properties[3]. As indicated in Fig. 56 for a K-film on a Pt-Ir base they were able to show a close correspondence between the photoemission current (full line) and the changes in intensity of the $\boldsymbol{E}_{\parallel}$ vector (dashed line) calculated on the basis of a 7×10^{-7} cm. film thickness and taking into account the known optical constants of Pt-Ir and K. The small side-maximum at 4400 Å was completely accounted for in terms of a potassium hydride curve (dotted line, see also Fig. 75). When this was subtracted from the main curve even closer agreement was achieved. The current due to $\boldsymbol{E}_{\perp}$

[1] H. E. Ives and H. B. Briggs: Phys. Rev. **38**, 1488 (1931).
[2] H. E. Ives and H. B. Briggs: J. Opt. Soc. Amer. **28**, 330 (1938).
[3] H. E. Ives and H. B. Briggs: J. Opt. Soc. Amer. **26**, 238 (1936); **27**, 181 and 395 (1937); see also R. W. Wood: Phys. Rev. **44**, 353 (1933); A. Smakula: Z. Physik **86**, 184 (1933); **88**, 114 (1934).

is not shown since it is nearly constant and too small to be of significance. Similarly close correspondence between the calculated intensity variation and measured emission was obtained with Rb- and Cs-films.

In this fashion a correlation between the energy density of the incident radiation at the surface and the vectorial selectivity seems firmly established. This view receives further support from the observations of FLEISCHMANN[1] on thin potassium films on glass who found them to be scarcely visible in light at normal incidence, but noticed strong colors at oblique incidence with parallel polarized light ($\boldsymbol{E}_{\parallel}$). These colors were complimentary to the regions of spectral absorption for which the photoemission of the films was a maximum.

HLUČLA[2] determined the optical properties of zink and was able to establish similar correlations for this metal. Further less direct support indicating the

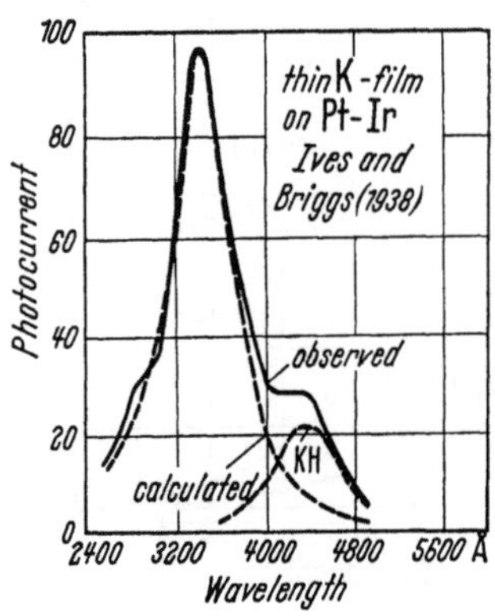

Fig. 56. The calculated variation of light intensity and of observed photocurrents with wavelength.

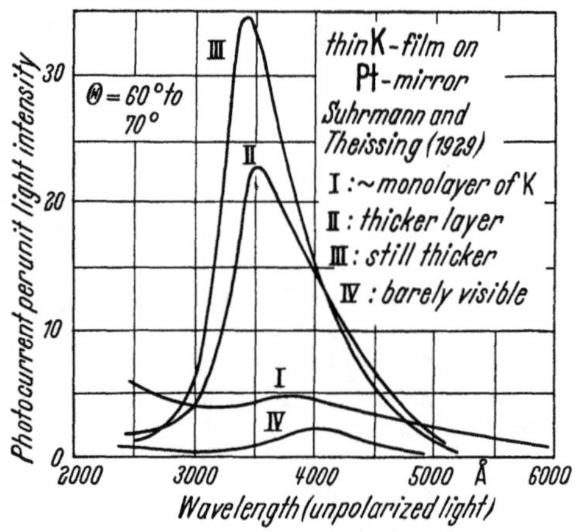

Fig. 57. Photoelectric yields from a thin K-film on a Pt mirror for different film thicknesses.

significance of optical constants in photoemission may be found in some work on total energy distributions as for example in the work at the end of Sect. 28.

SUHRMANN and THEISSING[3] (Fig. 57) and later BRADY and JACOBSMEYER[4] studied the photocurrents from thin potassium films on platinum and silver respectively as a function of film thickness. Both groups found that the thinnest layers, of the order of one atom thick, shifted the threshold frequency farthest toward the red in comparison to layers which were about 50 atoms thick and exhibited the properties of the bulk metal. For intermediate layer thicknesses they observed a maximum polarization selectivity which was 5 to 10 times higher than in either of the limiting cases. SUHRMANN'S and THEISSING'S curves agree with those of IVES in regions where overlapping occurs.

The shift of the threshold towards longer wavelengths for the thinnest layers has been linked to the observations of TAYLOR and LANGMUIR[5] and of BECKER[6] on the thermionic properties of tungsten surfaces covered with cesium. They

[1] R. FLEISCHMANN: Naturwiss. **19**, 826 (1931).

[2] F. HLUČLA: Z. Physik **103**, 237 (1936); **104**, 653 (1937).

[3] R. SUHRMANN and H. THEISSING: Z. Physik **55**, 701 (1929).

[4] J. J. BRADY and V. P. JACOBSMEYER: Phys. Rev. **49**, 670 (1936); see also J. J. BRADY: Phys. Rev. **41**, 613 (1932).

[5] J. B. TAYLOR and I. LANGMUIR: Phys. Rev. **44**, 423 (1933); also I. LANGMUIR and K. H. KINGDON, Proc. Roy. Soc. Lond., Ser. A **107**, 61 (1925).

[6] J. A. BECKER: Phys. Rev. **28**, 341 (1926); **29**, 364 (1927). — Rev. Mod. Phys. **7**, 95 (1935).

found that for incomplete surface coverage a much larger fraction could be evaporated in the form of Cs-ions, whereas for thicker layers an increased percentage of neutral Cs-atoms left the tungsten wire. This gave rise to the view that alkali atoms because of their low ionization potentials lost an electron to the tungsten which by virtue of its high workfunction has a strong electron affinity. Such an ion grid on a W-surface forms an electrical double layer which could explain a lower effective workfunction or photoelectric threshold than that of the bulk alkali metal.

30. Saturation phenomena, patch effect. Early results of IVES[1] showed that larger collecting potentials were needed to reach saturation of the photoelectric emission from thin alkali films in contrast to thicker ones. These findings were substantiated by SUHRMANN[2] who found in addition that higher fields were required if light frequencies near the threshold were used in contrast to shorter wavelength radiation. HUXFORD[3] (Fig. 58) observed the same effect only somewhat more pronounced in the case of ordinary oxide coated cathodes (BaO, SrO) and FLEISCHER and associates[4] for compound cesium cathodes (Fig. 59). The classic results of LAWRENCE and LINDFORD[5] for thin potassium films on a fine tungsten wire are reproduced in Fig. 60 where spectral distribution curves are plotted for different accelerating fields up to 63000 volts per cm.

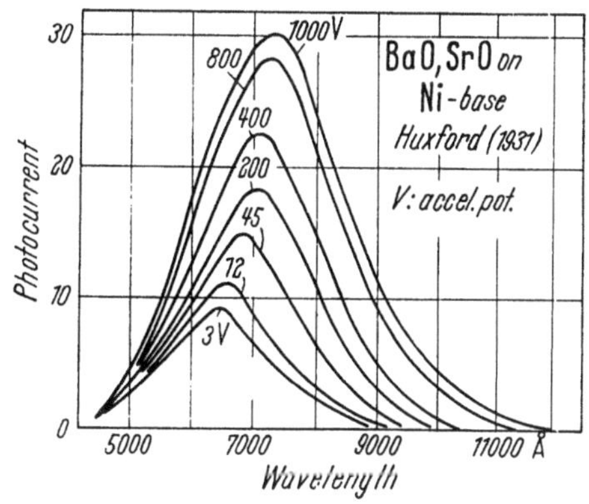

Fig. 58. Spectral yields as a function of different accelerating potentials.

This phenomenon may be qualitatively understood in terms of the lowering of the potential barrier at the surface of the metal by the action of the applied field as shown in Fig. 61. Here the simplest form of a surface barrier is indicated by the step function which should be modified by the fact that an escaping electron is subjected to its image force. Thus the corresponding image potential curve *O-C* represents more truly the actual situation. When an external electric field produces a potential near the metal surface as shown by the line *A-B*, the barrier will be modified to look like the solid curve *O-B* which shows a decrease $\Delta\varphi$ of the work function at a distance x_0 from the boundary. Thus

$$x_0 = \frac{1}{2}\sqrt{\frac{e}{E}} = \frac{1.90\times 10^{-4}}{\sqrt{E}}\ \text{cm} \tag{30.1}$$

and

$$\Delta\varphi = \sqrt{eE} = 3.79\times 10^{-4}\sqrt{E}\ \text{volts}, \tag{30.2}$$

which is essentially SCHOTTKY's method of accounting for the changes in thermionic emission brought about by accelerating fields. While the shifts in threshold measured by LAWRENCE and LINFORD follow approximately the above relationship, deviations from SCHOTTKY's equation [*12*] have been observed in both

[1] H. E. IVES: Astrophys. J. **60**, 209 (1924).
[2] R. SUHRMANN: Naturwiss. **16**, 336 (1928).
[3] W. S. HUXFORD: Phys. Rev. **38**, 379 (1931); see also Sect. 34, p. 379.
[4] R. FLEISCHER and P. GÖRLICH: Z. Physik **94**, 597 (1935). — R. FLEISCHER and H. PECH: Z. Physik **112**, 242 (1939).
[5] E. O. LAWRENCE and L. B. LINFORD: Phys. Rev. **36**, 482 (1930).

photoelectric and thermionic work among others by LINFORD[1], SUHRMANN and VON EICHBORN[2], NOTTINGHAM[3], and BECKER and MUELLER[4].

In order to provide an explanation for saturation phenomena in thermionic emission, LANGMUIR[5] suggested the model of a non-uniform or patchy surface

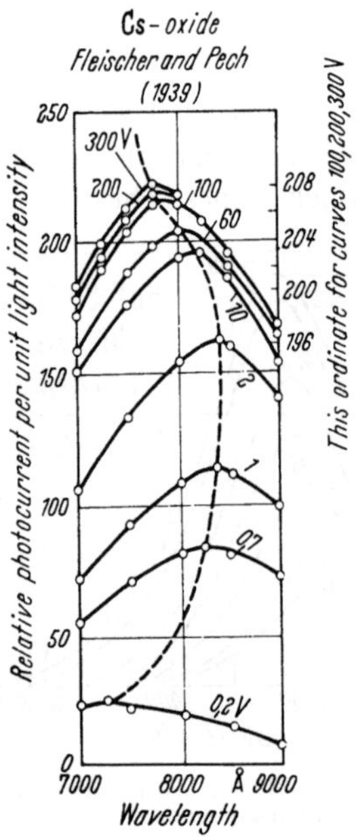

Fig. 59. The shift of the yield maximum caused by different accelerating potentials.

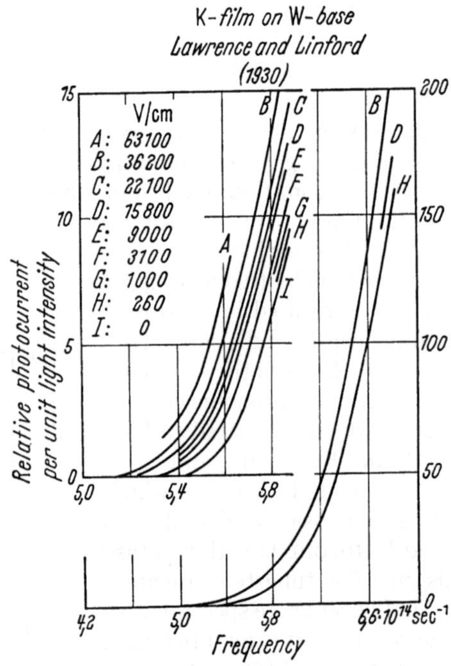

Fig. 60. Spectral yields as a function of different accelerating fields.

film with areas of different workfunction, which has since been used successfully by various workers in the field. Since HERRING and NICHOLS [*12*] treated this subject extensively in 1949, it will be sufficient to review it here rather briefly. (See also the article on thermionic emission by W. B. NOTTINGHAM in this volume.) From recent work with MÜLLER'S[6] field emission microscope, it has been convincingly demonstrated in terms of enlarged (magnifications of 10^5 to 10^6) electron emission maps from very fine points (radius about 10^{-5} to 10^{-4} cm.), that the workfunction is dependent on crystal orientation even for the cleanest specimens. Adsorbed gases or other surface films can be observed clearly by this technique, and they can be shown to occupy favorable surface sites more fre-

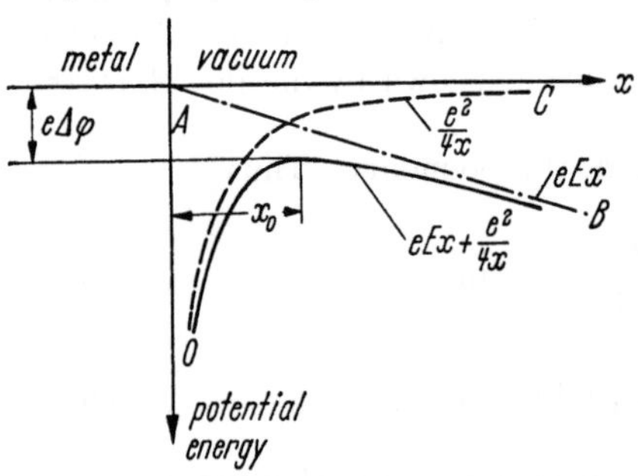

Fig. 61. The lowering of surface barriers by an accelerating field.

[1] L. B. LINFORD: Phys. Rev. **36**, 1100 (1930). — Phys. Rev. **5**, 34 (1933).
[2] R. SUHRMANN and J. VON EICHBORN: Z. Physik **107**, 523 (1937).
[3] W. B. NOTTINGHAM: Phys. Rev. **41**, 793 (1932); **49**, 78 (1936).
[4] J. A. BECKER and D. W. MUELLER: Phys. Rev. **31**, 341 (1928); see also J. A. BECKER: Rev. Mod. Phys. **7**, 95 (1935).
[5] I. LANGMUIR: Gen. Electr. Rev. **23**, 504 (1920). — S. DUSHMAN: Rev. Mod. Phys. **2**, 381 (1930).
[6] E. W. MÜLLER: Ergebn. exakt. Naturwiss. **27**, 290 (1953).

quently than others. Thus local dipole layers or areas of different workfunction exist because of unsymmetrical charge distribution in the vicinity of adsorbed surface atoms or ions, and even in the absence of applied voltages there will be electrostatic fields due to contact potential differences between patches which may enhance or impede electron emission. The resultant field outside the metal is called the patch field and dies off exponentially if the patches are small compared to the radius of curvature of the surface.

Some of the features encountered with patch fields are indicated in Fig. 62. The hight of the line *M-N* indicates the potential barrier of a clean surface, *A* and *B* represent patches of lower and higher workfunctions respectively, present in equal numbers and equal areas. Then the potential above the center of the patches outside the surface will fall off exponentially as shown by the dashed lines *A-N* and *B-N* where their slope at any point is a measure of the patch field. If an external collecting voltage is applied it may result in a potential distribution *M-C* which when added to the patch potentials for zero external field, *A-N* and *B-N*, gives rise to the potential barriers *A-C* and *B-C* for the low and high workfunction patches respectively. These barriers have been given a downward turn at both *A* and *B* in order to take account of the image potential, which in this case has been indicated as short-ranged compared to the patch fields. In addition these patch fields are of about the same magnitude as the applied field. It may be seen from Fig. 62 that the barrier of the high workfunction patch is much less affected than the low workfunction one when an external field is applied. As a consequence, emission will occur nearly exclusively from the electropositive patches *A-C*. This then explains SUHRMANN's early results of pronounced lack of saturation for patch surfaces with light near the threshold, since small fields will produce rather large changes in the workfunction of electro-positive patches and their yields (due to lowering of line *A-C*). In contrast, light of higher frequencies will produce emission from all patches and the effect of the low workfunction patches on the total yield will be materially decreased. In general, however, each case has to be carefully analysed since the patch fields depend critically on the patch size, their contact potential differences $\Delta\varphi$ and the percentage of the surface covered by each class of patches. From inspection of Fig. 62 it is also clear that great care has to be exercised in the interpretation of electron energy distributions from such surfaces measured by the stopping potential method.

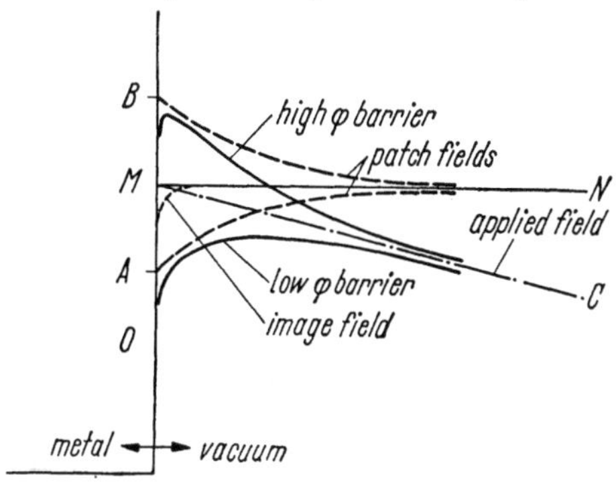

Fig. 62. The effect of surface patches on barriers.

31. Theoretical considerations. It has already been pointed out in the introduction that quantum-mechanical approaches to the theoretical treatment of the photoelectric effect were first indicated by SOMMERFELD[1] with the intro-

[1] A. SOMMERFELD: Z. Physik **47**, 1 (1928); see also R. H. FOWLER and L. NORDHEIM: Proc. Roy. Soc. Lond., Ser. A **119**, 173 (1928); G. WENTZEL: Probleme der Modernen Physik, SOMMERFELD-Festschrift, edided by P. DEBYE, Leipzig: S. Hirzel 1928. — H. FRÖHLICH: Ann. Phys., Lpz. **7**, 103 (1930); a modified version of FRÖHLICH's thin layer theory has been suggested by A. SOMMERFELD and H. BETHE: Geiger-Scheel Handbuch der Physik, second edition, Vol. **24**/2, p. 468.

duction of FERMI-DIRAC statistics for the energy distribution of electrons in a metal. Perhaps to the experimentalist one of the most useful versions of earlier theories is that due to FOWLER[1], who limited his considerations to the vicinity of the threshold [9].

In order to obtain the emission of photoelectrons per unit area of surface, it is necessary to calculate the number of electrons $n(W_z)\,dW_z$ with "normal" energies between W_z and W_z+dW_z which arrive at the surface in unit time. But this is the product of the FERMI distribution function (shown on the left of Fig. 30) and the normal velocity v_z. When this is multiplied by the transmission coefficient $D(W_z)$, which is a measure of the probability that an electron of energy W_z associated with a direction z normal to the surface will escape, then the current density j will be

$$j=\alpha e\int\limits_{W_a-h\nu}^{\infty} n(W_z)\,D(W_z)\,dW_z, \tag{31.1}$$

where α is a proportionality factor which involves the probability of an electron (charge e) absorbing a photon for unit intensity of incident radiation. The integration extends over all energies such that

$$W_z \geqq W_a - h\nu. \tag{31.2}$$

Since FOWLER was mainly interested in the spectral distribution function near the threshold ν_0, α could be considered as essentially constant. In addition the transmission function $D(W_z)$ was assumed to be zero for electron energies less than that necessary to surmount the barrier and one for greater values. Thus the solution of eq. (31.1) can be obtained with the help of the FERMI distribution of energies associated with a direction normal to the surface by integrating over $n(W_z)\,dW_z$ with the result that the photoelectric emission current per unit intensity of incident light becomes

$$j=\alpha A\,T^2\varphi(x), \tag{31.3}$$

where $x=\frac{(h\nu-h\nu_0)}{kT}$, $\varphi(x)$ is an exponential series of x which has been tabulated ([9], p. 21), and α and A are constants.

The usefulness of this relation is that it predicts the shape of the yield curve near the threshold with considerable accuracy and has been used extensively in determinations of ν_0 at 0° K mentioned in Sect. 26. Since light near the threshold will eject mostly those electrons whose energies lie close to the maximum value, W_m at 0° K, allowed by FERMI statistics, one would expect the photoelectric current in this case to be particularly strongly affected by changes in temperature. FOWLER has shown that there exists a sharply defined threshold ν_0 only at 0° K, while at elevated temperatures this will no longer be true because of the high-energy tail of the electron distribution. Thus measurements conducted at a temperature T can be reduced to 0° K with the help of eq. (31.3) to establish a definite value ν_0 characteristic of the surface (see Fig. 31). It must be born in mind, however, that experimental results will deviate from this theory for higher frequencies as shown in Fig. 63a, b.

At about the same time TAMM and SCHUBIN[2] made an important contribution towards a better understanding of the photoelectric effect by pointing out that photons cannot be absorbed by the free electrons in the SOMMERFELD equi-

[1] R. H. FOWLER: Phys. Rev. **39**, 45 (1931).
[2] I. TAMM and S. SCHUBIN: Z. Physik **68**, 97 (1931).

potential trough since energy and momentum cannot both be conserved in such a process. Light absorption can occur either when $h\nu$ interacts with an electron in the field of the surface barrier due to rapid changes of the potential there (surface effect) or with an electron bound to the periodic field of the lattice (volume effect). But since the barrier is of the order of 10^{-7} cm. thick only a small fraction of about 10^{-3} of the incident photons are absorbed there giving rise to the observed low yields of about 10^{-4} electrons per incident photon. In contrast, the majority of the photons is absorbed within a layer of the order of a wavelength thick without releasing photoelectrons. Then at a new threshold ν_0' the volume effect should become noticable leading to a second increase of the yield curve as perhaps in the earlier Figs. 34, 36 and 37. TAMM and SCHUBIN find that their calculations (shown by MITCHELL to be inaccurate, cf. below) for the surface effect alone, which they point out to be very similar to FRÖHLICH's[1] treatment of thin film emission, leads

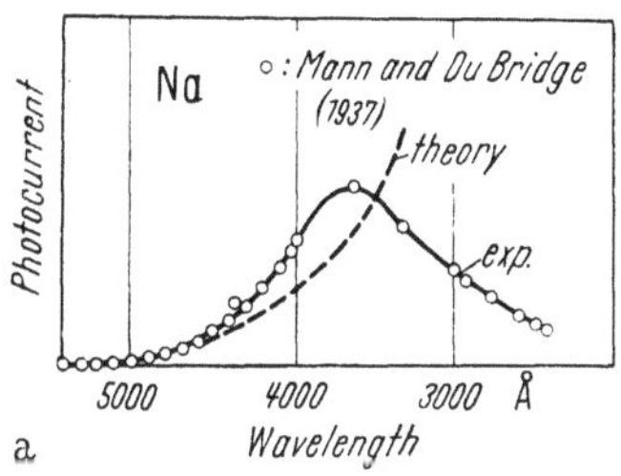

a

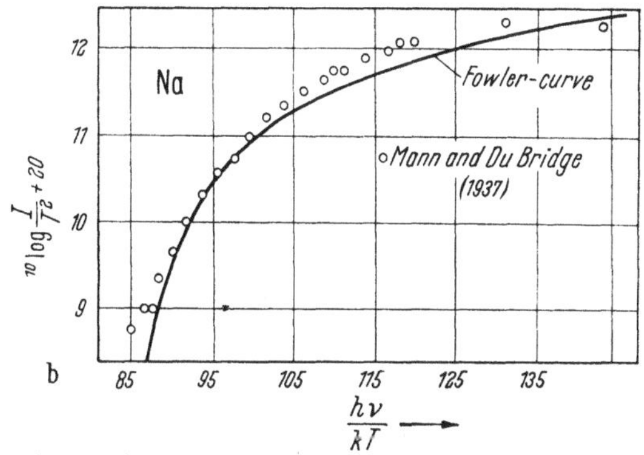

b

Fig. 63 a and b. An example of ill fit between theory and experiment. a) Linear and b) logarithmic plot.

to quantitative agreement with SUHRMANN's and THEISSING's experimental values for potassium (see also Fig. 33), except that the theoretical maximum is flatter and occurs at somewhat higher frequencies. However, they indicate that BLOCH's refinement of taking the periodic lattice field into account should steepen the yield maximum.

Thus TAMM and SCHUBIN's work on pure metals indicates a first yield maximum due to mainly the surface effect and a second one beyond the onset ν_0' of the volume effect which probably lies in the little explored region of the far ultraviolet. These latter predictions have recently received some support from the experiments of BAKER[2], HINTEREGGER and WATANABE[3], and of WALKER, WAINFAN and WEISSLER[4] (see also Figs. 72, 73, 74, 80, 81) who worked with radiation of the vacuum ultraviolet.

MITCHELL[5] reviews some of the earlier theories pointing out certain inadequacies and concerns himself with calculations of the surface effect. He assumes a simple potential step function at the surface and, neglecting variations of the potential in the interior due to lattice periodicity, establishes a complete system of electronic eigenfunctions to which he applies first order perturbation theory in order to account for photon interaction. In addition MITCHELL expands his treatment by including the orientation of the electric vectors of the incident, reflected and transmitted light on the basis of classical electromagnetic theory

[1] H. FRÖHLICH: Ann. Phys., Lpz. **7**, 103 (1930).
[2] R. F. BAKER: J. Opt. Soc. Amer. **28**, 55 (1938).
[3] H. E. HINTEREGGER and K. WATANABE: J. Opt. Soc. Amer. **43**, 604 (1953).
[4] N. WAINFAN, W. C. WALKER and G. L. WEISSLER: J. Appl. Phys. **24**, 1318 (1953). — W. C. WALKER, N. WAINFAN and G. L. WEISSLER: J. Appl. Phys. **26**, 1366 (1955).
[5] K. MITCHELL: Proc. Roy. Soc. Lond., Ser. A **146**, 442 (1934); **153**, 513 (1936).

which predicts an abrupt change at the surface. Furthermore, he considers the very important point of the roughness of a surface for both $\boldsymbol{E}_{\parallel}$ and $\boldsymbol{E}_{\perp}$ by averaging over the angles which the individual normals associated with submicroscopic hills and valleys make with the macroscopic surface normal. This yields a factor ω which will be larger the rougher the surface, and which has been adjusted to fit experimental data as shown in Fig. 64. Here MITCHELL compares his calculations with KLAUER's[1] results for unpolarized light on potassium which had been distilled five times for the sake of greater purity. ω values were chosen such as to give polarization selectivity. The divergence between experimental and theoretical curves is even more pronounced in the case of MAURER's[2] work on sodium. HILL[3] also used MITCHELL's theory on sodium, and while he finds good agreement with experiment near the threshold the maximum yields differ by a factor of ten.

It should be remembered at this point that experimental determinations of reproducible absolute yields are difficult to obtain for pure metals since even minute

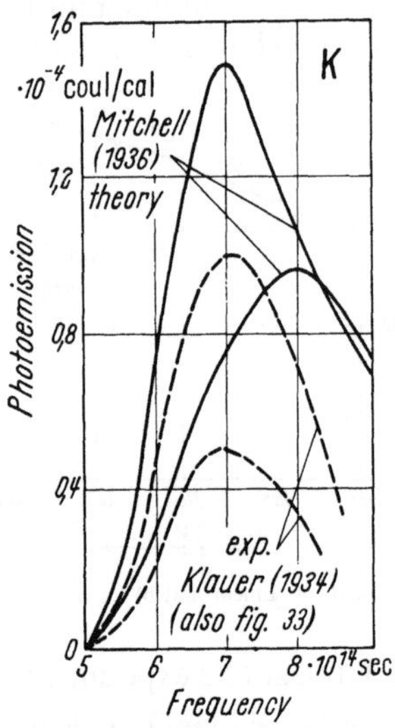

Fig. 64. A comparison between theoretical and experimental yields curves.

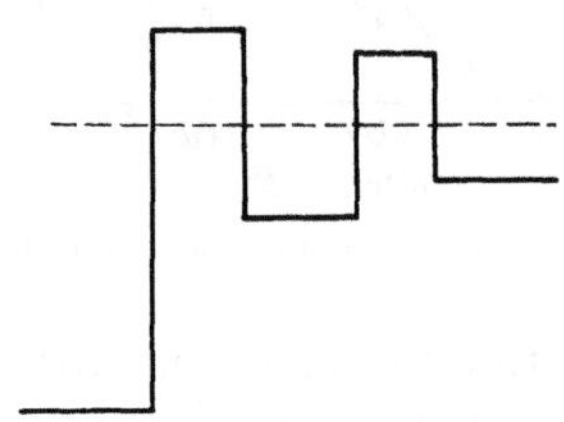

Fig. 65. FOWLER's complex barrier.

surface impurities will seriously affect the height of the spectral distribution curves (see Sect. 27). The seriousness of such effects was recognized among others by FOWLER[4] (Fig. 65) and by OPPENHEIMER[5] both of whom sought the solution in radically different potential functions at the surface of the metal in order to account for high electron transmission.

It was already realized by MITCHELL that his inclusion into the theory of optical constants which changed abruptly at the surface needed to be modified. MAKINSON[6] points out that the electron density does not fall sharply to zero but decreases from a high value inside the metal to a low one outside over a distance of 3×10^{-8} cm. It is in this region where the surface photoelectric effect takes place and where the light intensities are least accurately given by classical optical theory. SCHIFF and THOMAS[7] have worked out the quantum theory of metallic reflection, but because of its complexity did not attempt to find the photoelectric currents quantitatively. However, MAKINSON approximates their

[1] F. KLAUER: Ann. Phys., Lpz. **20**, 904 (1934).
[2] R. J. MAURER: Phys. Rev. **57**, 653 (1940).
[3] A. G. HILL: Phys. Rev. **53**, 184 (1938).
[4] R. H. FOWLER: Proc. Roy. Soc. Lond., Ser. A **128**, 123 (1930).
[5] Quoted by J. J. BRADY: Phys. Rev. **41**, 613 (1932).
[6] R. E. B. MAKINSON: Proc. Roy. Soc. Lond., Ser. A **162**, 367 (1937). — Phys. Rev. **75**, 1908 (1949).
[7] L. I. SCHIFF and L. H. THOMAS: Phys. Rev. **47**, 860 (1935).

approach by calculating first the electron density as a function of distance from the surface, and then he determines on a semi-classical basis the effect of this density distribution on the incident light by averaging over the reflected and refracted intensities calculated for different sections of this density distribution function. In spite of these important refinements MAKINSON obtains no better agreement with the height of experimental yield maxima than MITCHELL and ascribes it among other things to a lack of unambiguous experimental data. In general he concludes that photoemission arises as a surface effect from electrons of a given momentum in the metal as a product of their excitation function and barrier transmission coefficient as indicated earlier in eq. (31.1), and he furthermore proves that the excitation function does not vary rapidly near ν_0. However, in contrast to previous assumptions he finds no evidence that the excitation function is proportional to the component of the electronic momentum normal to the surface, which places some doubt on the FOWLER-DUBRIDGE interpretation of the temperature dependence of emission near ν_0. This also applies to the interpretation of MITCHELL and HILL of total energy distributions of photoelectrons, and indicates that further calculations are needed.

Recently, BUCKINGHAM[1] has refined MAKINSON'S approach by taking into account BARDEEN'S[2] suggestion that depending on its momentum each electron has its own effective surface barrier which is modified by momentum-dependent terms of exchange and correlation forces in the interior. This acts such as to reduce the theoretical magnitude for the absolute yields, thus bringing them somewhat more in line with experiment, at least for sodium. However, this does not help much with respect to the deviations from theory of Na near ν_0, since the shape of the spectral distribution curve remains essentially unchanged. BUCKINGHAM concludes by stressing the significance of the shape of the surface barrier on the form of the transmission coefficient function $D(W)$.

FAN[3] concerns himself extensively with the volume effect of electrons excited deep inside the metal, taking into account the influence of the periodicity of the lattice potential on the electron wave functions and the optical interactions with such electrons. The transition probability of an electron in the metal is computed by conventional time-dependent perturbation theory, the pertubation being due to the vector potential of a plane wave classical electromagnetic field. The procedure is formally the same as for the surface effect, but the unperturbed electron wave functions must now consist of a plane wave modulated by a factor which has the periodicity of the lattice. In consequence of this form, a selection rule is imposed which states that the transition probability is zero unless the initial and final momentum components in the plane wave portions of the electron wave functions differ by integral multiples of 2π times the reciprocal lattice constants. Taking this fact into account, as well as the number of electrons of various energies present as given by the FERMI distribution, and including the energy requirement that the surface barrier be overcome, an explicit expression for the emission current from the surface is obtained. The influence on the calculations of the finite size of the metal, electron reflection from the barrier and electron-electron scattering has not been assessed.

FAN argues that an estimate of the threshold frequency does not require a detailed knowledge of the form of the electronic wave functions, but rather is determined by the number of electrons of various energies available as given by the FERMI distribution, by the selection rule and by the dependence of energy

[1] M. J. BUCKINGHAM: Phys. Rev. **80**, 704 (1950).
[2] J. BARDEEN: Phys. Rev. **49**, 653 (1936); **58**, 727 (1940).
[3] H. Y. FAN: Phys. Rev. **68**, 43 (1945).

on momentum. Taking these facts into account by an ingenious geometrical argument, he obtains a volume threshold of the order of 10^{15} sec.$^{-1}$ (3000 Å) for Na and K, which indicates that the volume effect comes into play at frequencies somewhat higher than for the surface effect.

His rough estimate of the volume yield is comparable with that from the surface effect and of the same order as the experimental data. He concludes that although the volume effect does begin at a higher frequency than the surface threshold, it does contribute long before the first yield maximum due to the surface effect is reached, at least for the metals investigated, sodium and potassium.

HUNTINGTON[1] has extended the MAKINSON-MITCHELL treatment of transition probabilities in photoelectric emission from the free-electron model to the case where BLOCH wave functions in a periodic field are included. This is applied to the particular case of electrons near the top of a filled band in semiconductors.

II. Selected topics on complex surfaces.

32. The effect of unknown surface contaminations. The pronounced effect of surface impurities and gas layers on photoelectric emission was known even to the earliest investigators, and statements as to the degree of outgassing of the surface constitute a significant part of each report. The various techniques used to achieve this may be grouped into three main categories:

I. The deposition of new surfaces by a) evaporation in a high vacuum of substances of low melting point, or by b) sputtering of the more refractory elements through the action of glow discharges in rare gases, such as argon;

II. The cleaning of existing surfaces through heating them a) by electrical currents, conduction and induction, b) by radiation, and c) by electron bombardment;

III. The removal of undesirable surface layers by a) ion bombardment using rare gases (very effective), b) irradiation with ultraviolet light.

The action of the last effect (IIIb) is least understood and also least effective. MILLIKAN[2] found that illuminating various metals for 15 hours with light from a spark increased the total photocurrent by factors ranging from 1.5 to 10. This action was even more pronounced when WINCH[3] irradiated unbaked gold films, sputtered on quartz and about 2×10^{-6} cm. thick, with a mercury arc. He found increases of 136-fold together with a shift in the threshold by 576Å. A reversal of the applied potential such that photoelectrons were returned to the film was even more effective.

The strong influence of adsorbed or dissolved gases has been demonstrated by many investigators and is shown in Fig. 66 for a platinum filament, which DUBRIDGE[4] heated for 300 hours at 1250° C. Changes in the spectral distribution of Pt for successive stages of outgassing are shown in Fig. 67, where curve I represents the specimen after the initial baking of the tube. The final curve IV was obtained after 100 hours of heating at 1200 to 1500° C.

However, the trend during outgassing is not necessarily a regular one. For example the work of WINCH[5] on a silver ribbon shows an initial increase in the

---

[1] H. B. HUNTINGTON: Phys. Rev. **89**, 357 (1953). — H. B. HUNTINGTON and L. APKER: Phys. Rev. **89**, 352 (1953).

[2] R. A. MILLIKAN: Phys. Rev. **34**, 68 (1912).

[3] R. P. WINCH: Phys. Rev. **38**, 321 (1931).

[4] L. A. DUBRIDGE: Phys. Rev. **29**, 451 (1927); **31**, 236 (1928).

[5] R. P. WINCH: Phys. Rev. **37**, 1269 (1931); similar results were obtained for a Au-filament by L. W. MORRIS: Phys. Rev. **37**, 1263 (1931).

emission current (Fig. 68) with a corresponding shift to the red of the threshold and subsequently a decrease in both. The temperatures of the ribbon and the threshold wavelengths are indicated on the curve for various heating periods. In Fig. 69 DuBridge and Roehr[1] have plotted the change in threshold of palladium while heating it at a temperature of 900° K. During the interval *A*-*B* the tube was baked out again in an oven. A similar trend was observed by Farnsworth

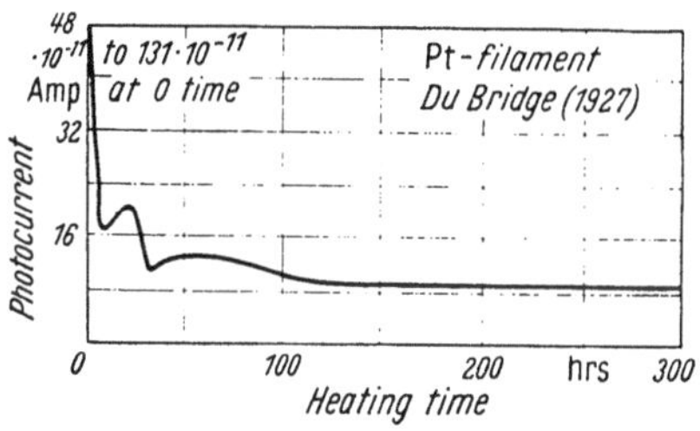

Fig. 66. Photocurrents from Pt during outgassing.

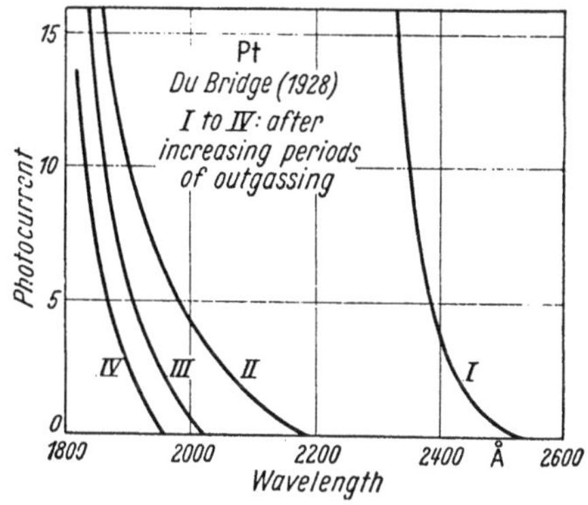

Fig. 67. Spectral yields from Pt during outgassing.

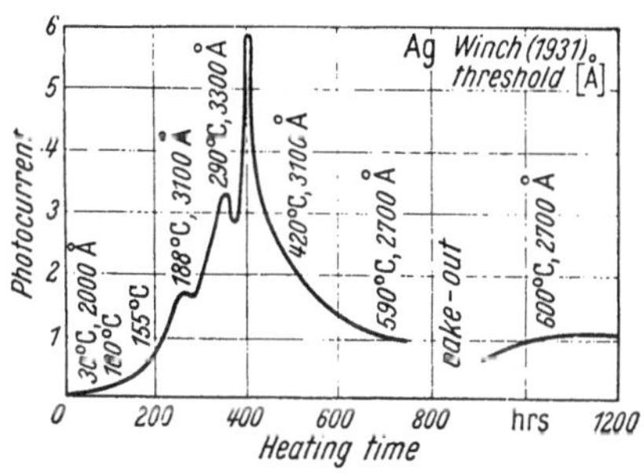

Fig. 68. Photocurrents and thresholds of Ag during outgassing.

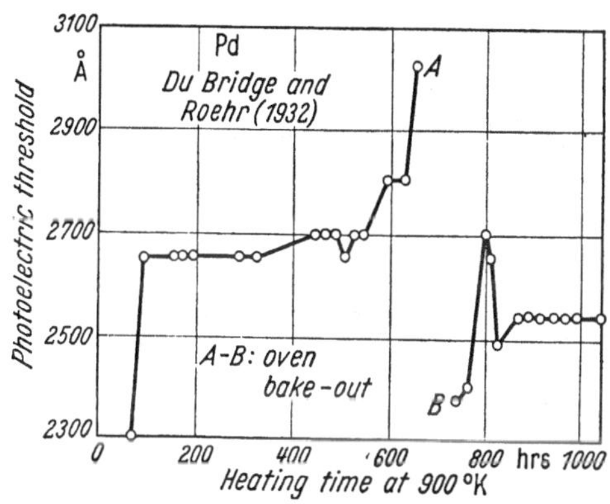

Fig. 69. Changes in the threshold of Pd during outgassing.

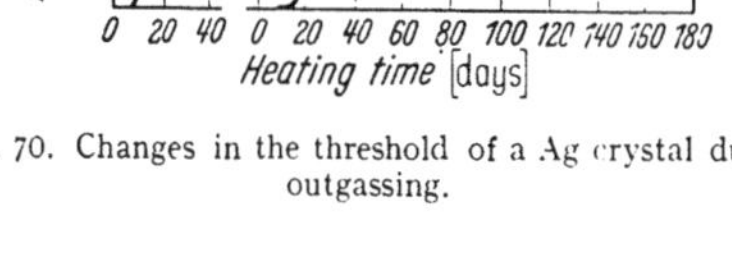

Fig. 70. Changes in the threshold of a Ag crystal during outgassing.

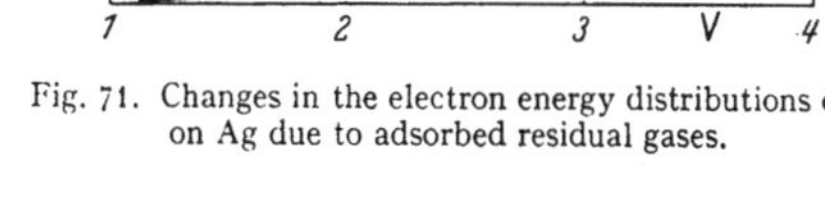

Fig. 71. Changes in the electron energy distributions of K on Ag due to adsorbed residual gases.

and Winch[2] in their careful work on the photoelectric workfunctions of different faces, [*111*] and [*100*], of single crystals of silver, Fig. 70, where their experiment *A* involved an outgassing period if 55 days and *B* 180 days! Further examples of threshold changes may also be found in Table 3.

[1] L. A. DuBridge and W. W. Roehr: Phys. Rev. **39**, 99 (1932); **42**, 52 (1932).
[2] H. E. Farnsworth and R. P. Winch: Phys. Rev. **58**, 812 (1940).

HENSHAW[1] measured the energy distributions of photoelectrons from thin potassium films, about three atoms thick, on a substrate of silver as shown in Fig. 71. Curve B' was taken immediately after evaporation of K, B'' two hours later while standing in a vacuum of 6×10^{-8} mm. Hg, and B''' $5\frac{1}{2}$ hours later. Probably the contamination was due to oxygen because of the shift of ν_0 towards shorter wavelengths. The saturation currents of B'' and B''' were 64 and 0.9%

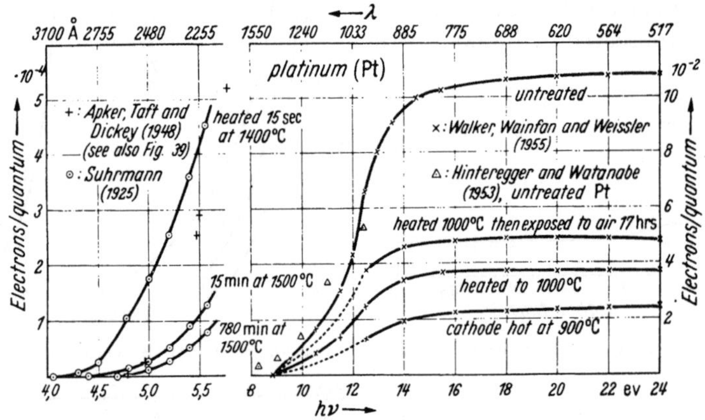

Fig. 72. The effect of heat treatments on the spectral yields of Pt both near and far from the threshold.

of B' respectively. Similar studies of impurity gases on electron energy distributions from Cu, AL, Pt, Pd, and soot on glass were made by HEROLD[2], BENNEWITZ[3], KLUGE[4] and others. Even though considerable variations were observed, they did not indicate unambiguous trends which could be interpreted.

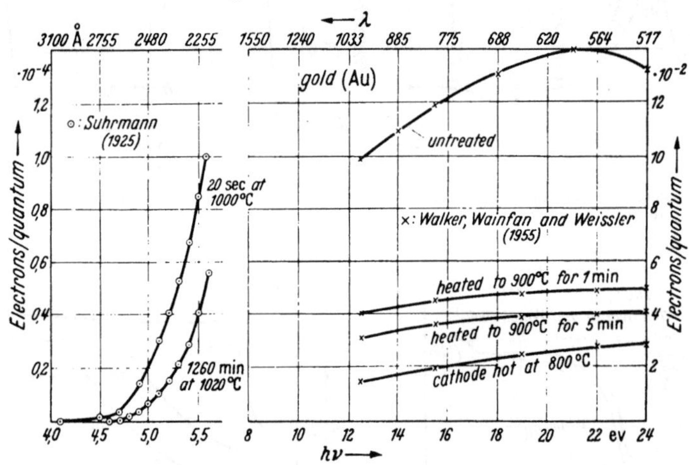

Fig. 73. The effect of heat treatments on the spectral yields of Au.

It is of interest to compare the absolute spectral yield curves of pure and impure Pt, Au, and Ag obtained by SUHRMANN[5] for the region near the threshold with those of WALKER, WAINFAN and WEISSLER[6] for the same metals using the short wavelength radiation of the vacuum ultraviolet, as shown in Figs. 72, 73

[1] C. L. HENSHAW: Phys. Rev. **52**, 854 (1937).
[2] F. HEROLD: Ann. Physik **85**, 587 (1928).
[3] W. BENNEWITZ: Ann. Physik **83**, 913 (1927).
[4] J. KLUGE: Ann. Physik **82**, 432 (1927).
[5] R. SUHRMANN: Z. Physik **33**, 63 (1925).
[6] W. C. WALKER, N. WAINFAN and G. L. WEISSLER: J. Appl. Phys. **26**, 1366 (1955).

and 74. As was to be expected the surface photoelectric effect which occurs predominantly near the threshold is much more sensitive to impurities giving rise to changes in yield by factors of the order of 100. In contrast, all indications point to a volume effect in the far ultraviolet which exhibits much smaller varia-

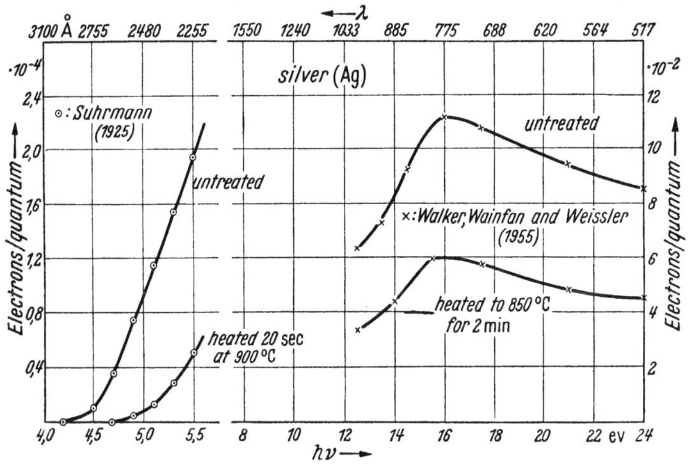

Fig. 74. The effect of heat treatments on the spectral yields of Ag.

tions during similar surface treatments. It should be noted that even though the yields increased after heating and subsequent exposure for several hours to room air at atmospheric pressure, they changed only by a factor of two or three in the vacuum ultraviolet and, most surprisingly, showed very good reproducibility, particularly for a gold surface.

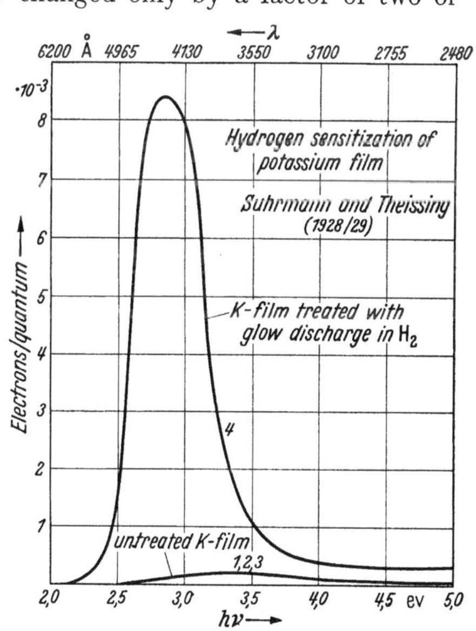

Fig. 75. Hydrogen sensitization of potassium.

33. Surface conditioning with known gases. The effect of certain gases on the work functions of metals was established early in various thermionic emission studies which have been very well summarized by DUSHMAN[1]. KINGDON[2], for instance, showed that whereas the workfunction of clean tungsten was 4.50 ev., an oxygen layer changed this to 9.2 ev., while tungsten plus thoria yielded a value of 2.68 ev. and tungsten plus cesium an even lower one of 0.715 ev.

Of all the gases the influence of hydrogen and oxygen on the photoelectric properties of surfaces has been most extensively studied. SUHRMANN and THEISSING[3] concerned themselves particularly with hydrogen and the changes it produced on a potassium surface freshly distilled on a silver base. Their results (Fig. 75) show that there is no significant difference between the yields from a new K-surface (curve 1), this same surface after 24 hours in a vacuum of better

[1] S. DUSHMAN: Rev. Mod. Phys. **2**, 415 (1930).
[2] K. H. KINGDON: Phys. Rev. **24**, 510 (1924).
[3] R. SUHRMANN and H. THEISSING: Z. Physik **52**, 453 (1928).

than 10^{-6} mm. Hg (curve 2) and after six hours of exposure to hydrogen gas at atmospheric pressure (curve 3). However, a glow discharge in H_2 produced a surface with greatly increased yield (curve 4). They conclude that therefore the effect is not due to the H_2 molecule but rather the atom or atomic ion. Fleischer and Teichmann[1] achieved the same results with hydrogen acting on a potassium layer, but without the use of a glow discharge. These investigators noted a decrease in yield of the freshly distilled K-surface to very nearly the same values due either to standing in a high vacuum for 24 hours or when exposed to hydrogen. However, when more potassium is evaporated onto the K-surface already saturated with hydrogen, a large increase in yield was noted. They obtained the same pronounced difference between a potassium surface distilled on platinum in a

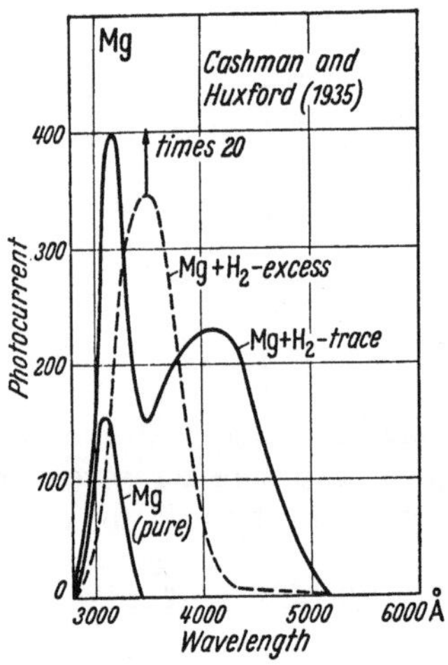

Fig. 76. Hydrogen sensitization of Mg.

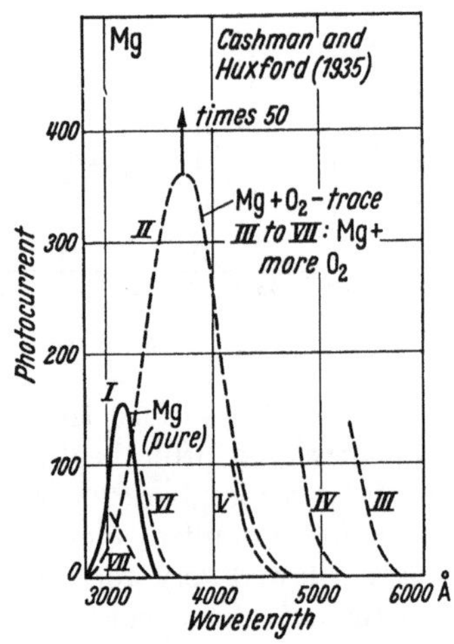

Fig. 77. The effect of oxygen on Mg.

high vacuum and one in an atmosphere of hydrogen at about 10^{-2} mm Hg. The most reasonable explanation of this effect seems to be the formation of a double layer, where first the hydrogen forms potassium hydride crystals onto which is deposited in the second step atomic potassium either due to sputtering in a glow discharge (which also forms the hydride, as in the work of Suhrmann and Theissing) or by subsequent evaporation (Fleischer and Teichmann). This idea receives further support from the studies of de Boer and Teves[2], who evaporated a thin layer first of calcium fluoride and then cesium, and those of Lukirsky and Rijanoff[3], who formed alternate layers of potassium hydride and potassium.

The work of Cashman and Huxford[4] on magnesium gave quite similar results as shown in Fig. 76, where curve I represents pure Mg, curve II Mg plus a trace of hydrogen, and curve III Mg with an excess of hydrogen. Their measurements also indicated changes in threshold from $\nu_0 = 3430$ Å for pure Mg to 5100 A for $Mg + H_2$. Pumping off the excess of H_2 caused a shift to 4700 Å which was quite reproducible. With traces of oxygen (about 10^{-7} mm. Hg) added, a sensitization similar to hydrogen was observed as indicated in Fig. 77, and again the threshold could be changed by alternately admitting a little O_2 and pumping

[1] R. Fleischer and H. Teichmann: Z. Physik **61**, 227 (1930).
[2] J. H. de Boer and M. C. Teves: Z. Physik **65**, 489 (1930).
[3] P. I. Lukirsky and S. Rijanoff: Z. Physik **75**, 249 (1932).
[4] R. J. Cashman and W. S. Huxford: Phys. Rev. **48**, 734 (1935).

it off. However, with larger amounts of oxygen the thresholds changed as in curves III, IV, V, VI, VII, and the yields decreased. For oxydized magnesium the threshold was 2000 Å or less, while for Mg distilled in the presence of H_2 it was about 7000 Å.

DeVoe[1] made a systematic study of a zink surface distilled in vacuum and exposed to various pressures of air, nitrogen, or helium. His findings are summarized in Figs. 78a, b, c, d. Both He and N_2 produced no change up to pressures of 3 and 10^{-3} mm. Hg respectively in contrast to air, which showed a marked increase in emission at about 10^{-7} mm. Hg with a corresponding shift of ν_0 to the red. Pressures greater than 10^{-5} mm. Hg reversed this completely. This is somewhat similar to Hales[2] work on mercury which showed that H_2, N_2, He, or A had no effect on its threshold, while O_2 and dry air both decreased the yield and increased the threshold frequency.

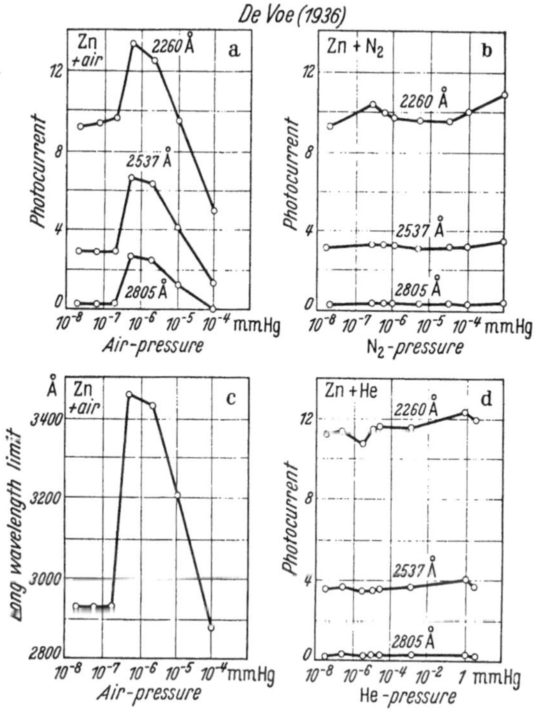

Fig. 78 a—d. The effect of air, N_2, and He on Zn.

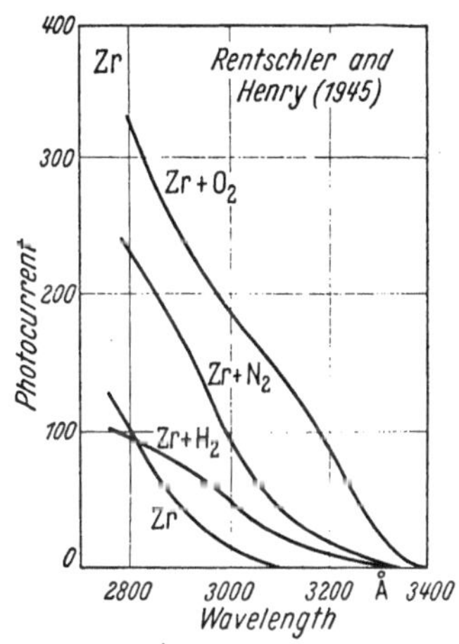

Fig. 79. The effect of dissolved O_2, N_2, and H_2 in Zr.

The relative photoelectric response of zirconium, titanium, thorium and others due to dissolved gases such as H_2, N_2, or O_2 has been investigated by Rentschler and Henry[3]. Fig. 79 shows their results for Zr. The Zr-filament was heated in 3 mm. Hg of O_2 until the pressure in the tube decreased to about 1 mm. Hg. After evacuation the filament was heated again and no gas was released. Then the measurements of curve IV were taken. Curve III and II represent a similar N_2 and H_2 treatment, while I is characteristic of pure Zr after heating it to 1500° C. In this case, as in many others, too little is known with respect to the structural changes produced by treatments to afford anything but speculations.

Photoemission of nickel and copper and changes due to gas treatment have been investigated by Walker, Wainfan and Weissler[4] in the region of the

[1] C. F. de Voe: Phys. Rev. **50**, 481 (1936).
[2] W. B. Hales: Phys. Rev. **32**, 950 (1928).
[3] H. C. Rentschler and D. E. Henry: Trans. Electrochem. Soc. **87**, 289 (1945).
[4] W. C. Walker, N. Wainfan and G. L. Weissler: J. Appl. Phys. **26**, 1366 (1955).

extreme ultraviolet and spectral yields are shown in Figs. 80 and 81. The lowest curve for each metal was obtained while simultaneously heating it, Ni to 900° C and Cu to 500° C. At these temperatures their thermionic emission currents were at least one order of magnitude lower than the photoelectric ones. Exposing a heated Ni-cathode to O_2 at 10^{-1} mm. Hg for one half hour gave rise to incraesed values, and heating it for one minute at 700° C in O_2 at a pressure of 5×10^{-5} mm. Hg resulted in still larger yields. Upon heating at about 1000° C in vacuum the clean state could be reproduced. The procedures were similar for copper. However, heating Cu in O_2 at 10^{-1} mm. Hg for one half hour gave results which were identical with those where the clean copper had been heated in a vacuum of the order of 5×10^{-5} mm. Hg before the measurements were taken. (A thick red Cu_2O-layer was present after heating in O_2, whereas the metal appeared clean after heating in vacuum.) This indicates clearly that oxides were present in the latter case too. Prolonged heating at 700° C of Cu cathodes, visibly oxidized or apparently clean, had very little effect.

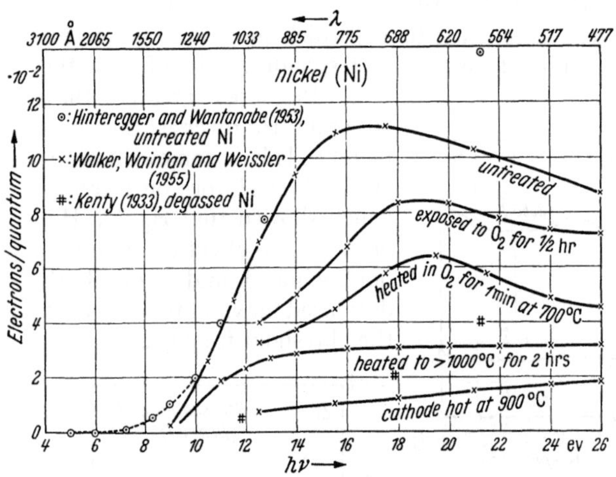

Fig. 80. Spectral yields of Ni for different surface treatments.

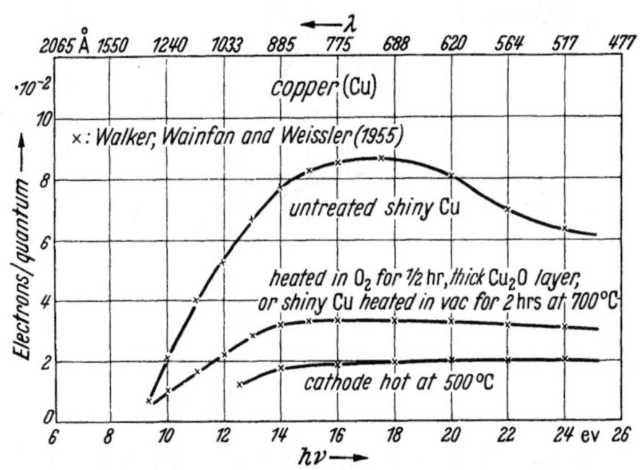

Fig. 81. Spectral yields of Cu for different surface treatments.

From what has been mentioned in this section[1] it seems clear that gas conditioning can not be explained in general and simple terms. Nearly each metal will behave differently, and only detailed information in individual cases will lead to a more complete understanding. There is, however, considerable evidence that oxygen treatment results in lower yields and in changes of the threshold towards the ultraviolet, although in the vacuum ultraviolet the reverse seems true. Oxygen in very small amounts, in contrast, may give rise to double layers, similar to the situation with complex cathodes, which may favor emission by increasing yields near ν_0 and shifting ν_0 to the red region of the spectrum.

34. Miscellaneous phenomena with composite surfaces. The trend of the photoelectric behaviour as revealed in the work summarized in the last section, particularly so far as hydrogen or oxygen treatments of surfaces are concerned,

[1] See also the review article on photoelectric emission from adsorbed films by R. SUHRMANN: Ergebn. exakt. Naturw. **13**, 148 (1934).

leads invariably to considerations of complex cathodes. Out of the large amount of research published [*15*] on this subject, only a limited number of papers can be discussed here in order to sketch some of the significant phenomena. As early as 1913 POHL and PRINGSHEIM[1] recognized the significance of sensitizing alkali surfaces with oxygen and found that this gas acted on a potassium layer such as to cause the formation of a coloured film accompanied by a strong rise in yields. Such a surface may be conveniently symbolized in the notation of DE BOER and TEVES by [K]—K_2O, K—K, where the base is indicated in brackets and the hyphens separate the constituents of individual layers. In this example, the intermediate layer of potassium oxide undoubtedly contains some colloidal potassium dispersed throughout its volume, while the outer layer is due to adsorbed K atoms, which have been sublimend to the surface. OLPIN[2] describes in a very extensive paper various techniques for sensitizing alkali metals by producing on their surfaces thin films of dielectrics through the action of oxygen, water vapor, sulphur vapor, sulphur dioxide, hydrogen sulphide, and a variety of organic compounds. For these different cases he gives corresponding yield curves which exhibit a number of different spectral maxima and thresholds.

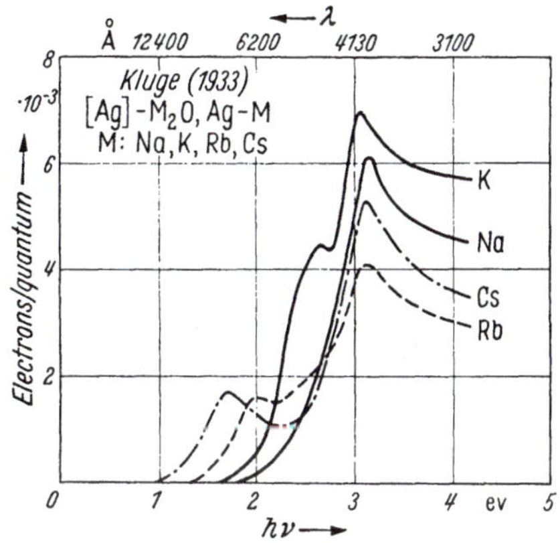

Fig. 82. Photocurrents from complex surfaces.

Instead of oxidizing the alkali metal by the addition of the gas, it has been found more effective to produce an oxide-coated base first, such as Ag_2O. In the second step, a thin layer of say cesium is evaporated onto it. The thin, porous Ag_2O-film facilitates the adsorption of Cs throughout its volume. During the third step of baking the cathode at about 180° C the cesium reacts with the silver oxide to form Cs_2O, leaving free Ag and excess Cs dispersed through the layer. Such a surface is symbolized by [Ag]—Cs_2O, Ag, Cs—Cs.

KLUGE[3] has made a comprehensive study of such cathodes, and the spectral yield curves for [Ag]—Cs_2O, Ag—Cs, [Ag]—K_2O, Ag—K, and [Ag]—Na_2O, Ag-Na have been reproduced in Fig. 82. In a series of systematic experiments on composite surfaces of the type Base-Alkali oxide-alkali or [T]—M_2O, T—M he varies both T and M, using for T: Ag, Cu, Ni, Au, and for M: Na, K, Rb, and Cs. The results are complicated and not open to simple interpretation. There are at least two and often more maxima for each surface in the region between 2000 Å and 4000 Å. Their relative heights change as either T or M is altered. For copper and silver he finds an additional long wavelength maximum at 6500 and 7500 Å respectively, which is strong in the case of Ag and weak for Cu. With silver as a base, this maximum was weak, when the alkali was rubidium, and strong, when it was cesium. KLUGE suggests that the band-like structure of the yield curves in the near ultraviolet may be considered as indicative of light absorption bands of the oxides whose lattices are disturbed by intersticial alkali or base atoms.

---

[1] R. POHL und P. PRINGSHEIM: Verh. dtsch. phys. Ges. **15**, 625 (1913).

[2] A. R. OLPIN: Phys. Rev. **36**, 251 (1930).

[3] W. KLUGE: Phys. Z. **34**, 115, 844 (1933); **39**, 911 (1938); — Z. Physik **67**, 497 (1931); **93**, 636 (1934/35); **95**, 734 (1935); **96**, 691 (1935).

A still further increase in yields by a factor of about ten was achieved by ASAO[1], who evaporated a number of metal atoms, particularly gold and silver, on a surface of [Ag]—Cs_2O, Ag, Cs—Cs and subjected it to a second heat treatment. The sensitivities of these surfaces are among the highest observed. They also exhibit reduced photoelectric fatigue, and both the long wavelength maximum and the threshold are shifted from the near infrared into the visible. ASAO's work with such layers on quartz as a base definitely established the semiconducting nature of these surfaces in view of their negative temperature coefficients of resistance [*14*].

The optical and photoelectric properties of semitransparent cesium oxide cathodes were investigated by SAYAMA[2] for surfaces deposited either on a base of silver, glass, or quartz and were of the usual type: [glass]—Cs_2O, Ag, Cs—Ag. From the optical constants of silver and glass, but neglecting those of Cs_2O, he calculated the light intensity at the surface as a function of wavelength and found that it corresponded to the variation of the relative yield curve 2 shown in Fig. 83. SAYAMA feels that the photoelectric properties of his surfaces are principally determined by the silver particles in the Cs_2O layer. The surfaces of these colloidal aggregates are assumed to be covered with a thin layer of Cs_2O with the result that their workfunction is decreased to about one volt. He ascribes the emission as due mainly to a volume effect.

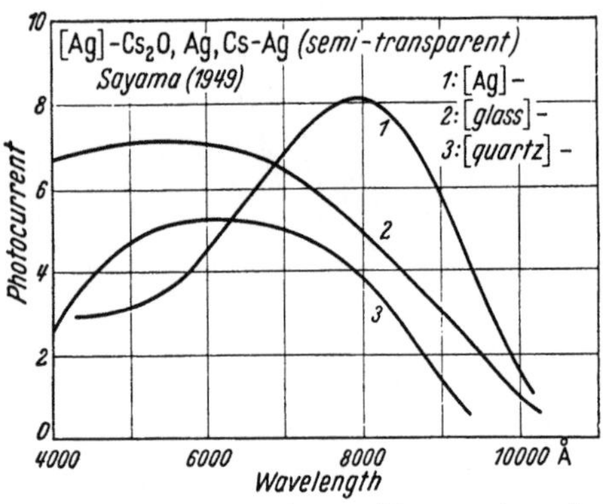

Fig. 83. Photocurrents from semi-transparent complex surfaces.

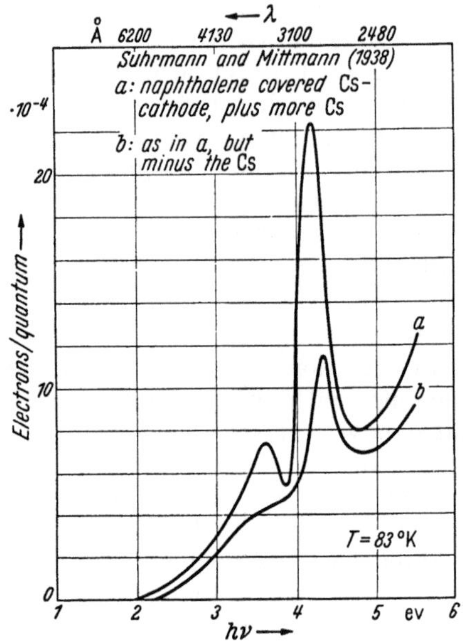

Fig. 84. Photocurrents from naphthalene covered Cs cathodes.

SUHRMANN and MITTMANN[3] made an effort to establish correlations between the yield maxima in the near ultraviolet for alkali surfaces whose intermediate dielectric layers were either hydrides, naphthalene, parafin, or anthrancene. For the case of a naphthalene surface symbolized by [K]—$C_{10}H_8$—K a strong selective maximum was observed in the near ultraviolet, but if instead of naphthalene paraffin was used this maximum disappeared. This is in agreement with the results of DE BOER and TEVES[4] who found that potassium was either not at

[1] S. ASAO and M. SUZUKI: Proc. Phys.-Math. Soc. Japan **12**, 247 (1930). — Proc. Inst. Radio Engrs. **19**, 655 (1931). — S. ASAO: Proc. Phys.-Math. Soc. Japan **22**, 448, (1940).

[2] Y. SAYAMA: J. Opt. Soc. Amer. **39**, 162 (1949).

[3] R. SUHRMANN and A. MITTMANN: Z. Physik **111**, 18 (1938); see also R. SUHRMANN: Z. Elektrochem. **37**, 678 (1931); Physik. Z. **32**, 929 (1931).

[4] J. H. DE BOER and M. C. TEVES: Z. Physik **73**, 192 (1931); **74**, 604 (1932).

all or only very slightly adsorbed on paraffin. Two curves of SUHRMANN and MITTMANN are shown in Fig. 84, which demonstrate a pronounced change in height of the selective maxima. When a naphthalene film covered the composite cesium cathode, the maximum at 3400 Å disappeared and the heigth of the one at 3000 Å was reduced. Subsequent evaporation of more Cs on the same naphthalene layer again brought out sharply these two maxima. These observers also noted marked fatigue of such surfaces when subjected for 20 minutes to intense illumination with ultraviolet light of $\lambda = 2970$ Å (Fig. 85).

Similar decreases in emission were measured by DE BOER and TEVES ([*13*], p. 335) and are reproduced in Fig. 86, where a [Ag]—Cs_2O, Cs, Ag—Cs cathode was irradiated with 1) infrared, 2) red, 3) green, and 4) blue light. Either allowing the surface to rest for 24 hours in darkness or heating it to 200° C would restore its initial emissivity. Such fatigue phenomena were explained on the basis of an insufficient supply of electrons from the base through the dielectric to those cesium atoms which had given up their electron and become ionized by the incident radiation. Thus the possible number of emitting centers was reduced, perhaps supplemented by two other hindering effects: 1) electrons may be trapped in the oxide lattice to form negative space charges and reduce further the flow of electrons from the metal, or 2) the alkali ions may migrate under the influence of external fields toward the base with the result that again the number of active surface centers is diminished.

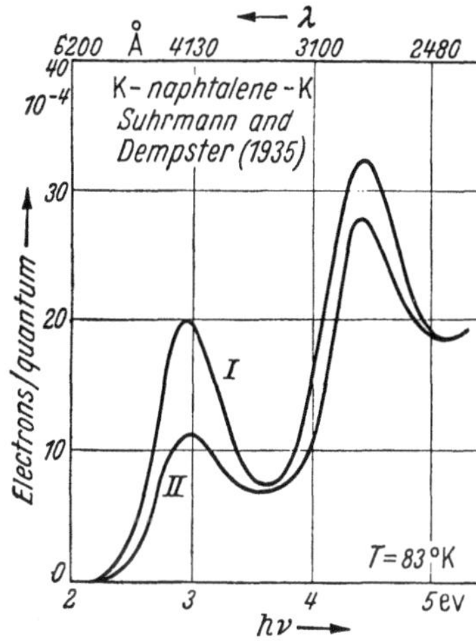

Fig. 85. Fatigue effect due to irradiation with $\lambda = 2970$ Å.

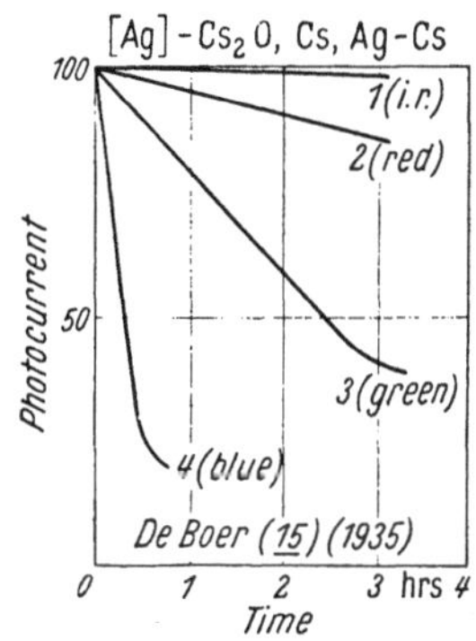

Fig. 86. Fatigue effect due to irradiation with a) infrared, 2) red, 3) green and 4) blue light.

Such a fatigue mechanism is supported indirectly by the previously cited research of ASAO, who worked with films containing an excess of metal atoms which provided an easier conduction path for electrons from the base to neutralize ionized adsorbed atoms, with the result that fatigue was absent or much reduced. This is further illustrated by HUXFORD'S observations (Fig. 58) mentioned in Sect. 30, who found for barium-strontium oxide surfaces that higher fields were required to furnish electrons from the base to the ionized barium and strontium atoms.

In order to account for the fact that the observed thresholds for most composite surfaces lie in the far red or infrared, DE BOER ([*13*], p. 194) has suggested an attractive model to explain this phenomenon. The potentials of a cesium atom and a cesium ion are shown in Fig. 87 as a function of the distance from the surface of a thin layer of calcium fluoride. In order to ionize a free atom (state a) an amount of energy equal to *I* has to be expended. However, in the adsorbed state b a much smaller amount will suffice for a FRANCK-CONDON transition from b to c, thus ionizing the atom with a photon energy $h\nu_0 < I$. Such potential curves will vary for spots of different adsorption activity on one crystal and of course for different atoms and surfaces, but in general a mechanism such as this provides an explanation for the observed facts.

The work of GÖRLICH[1] on semitransparent photocathodes is of interest. By using techniques such as his it is possible to obtain high yield maxima in the visible region in contrast to cesium oxide surfaces which exhibit their strongest emission in the vicinity of 8000 Å. Some of his results are shown in Fig. 88, where he evaporated bismuth and cesium on a conventional [Ag]—Cs_2O, Cs, Ag—Cs surface. Next to bismuth, antimony-cesium "alloys" were found to be very efficient. GÖRLICH explains these high sensitivities qualitatively in terms of

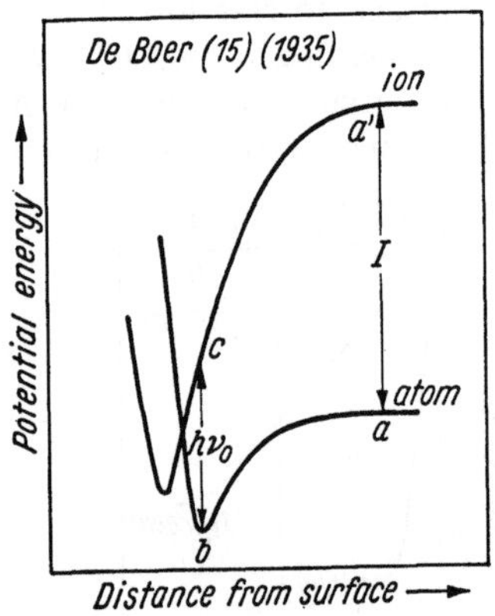

Fig. 87. Potential energy curves for a Cs ion and a Cs atom adsorbed on a surface.

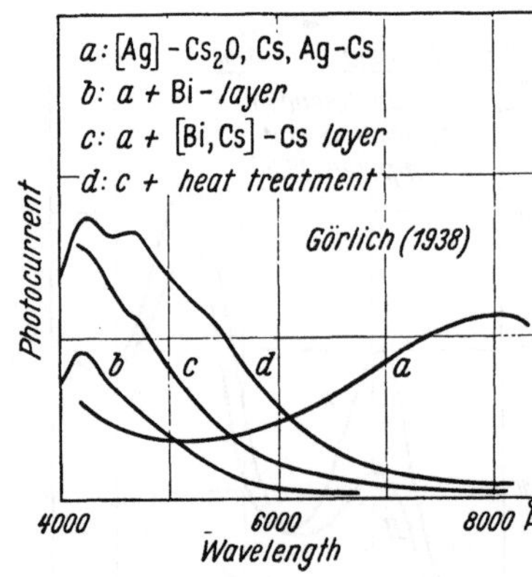

Fig. 88. Photocurrents from complex surfaces.

ionization of surface atoms, emission from metal atoms built into the Cs_2O crystal lattice, reduction of workfunctions due to dipole layers, and "loosening" of the lattice through heat treatment. It has been shown recently by BURTON[2] and others that these "alloy" surfaces are really intermetallic compounds, such as Cs_3Sb, with intrinsic semiconductor characteristics.

The promise held out towards a better understanding of complex surfaces in terms of solid state concepts is perhaps best illustrated by an example of some recent work. DE VORE and DEWDNEY[3] studied the optical properties, photoconductivities and photoelectric emission of barium oxide on a nickel base before, during and after activation by heat treatments. In Fig. 89 curves are shown to characterize the two states. The upper curve from $h\nu = 4.5$ to 6.5 ev is due to APKER, TAFT and DICKEY[4] and has been included to facilitate a more complete discussion in terms of the band structure of barium oxide represented in Fig. 90.

From the sharp rise of the emission current at $h\nu = 5$ ev, which occurred consistantly with all surfaces studied, it can be inferred that the top of the filled band lies below the zero level by this amount. DE VORE and DEWDNEY and TYLER[5] find that large optical absorption ($\mu \sim 10^5$ cm.$^{-1}$) begins at $h\nu \sim 3.7$ ev which is accompanied by a marked increase in photoconductivity at the same value (Fig. 91). This is interpreted as due to electron transitions from the filled band into the empty conduction band (or barely below it, from where they may reach the conduction band after receiving the additional small amount of energy

[1] P. GÖRLICH: Z. Physik **109**, 374 (1938).
[2] J. A. BURTON: Phys. Rev. **72**, 531 (1947) (A).
[3] H. B. DE VORE and J. W. DEWDNEY: Phys. Rev. **83**, 805 (1951).
[4] L. APKER, E. TAFT and J. DICKEY: Phys. Rev. **84**, 508 (1951).
[5] W. W. TYLER: Phys. Rev. **76**, 1887 (1949).

from the lattice vibrations). This is further substantiated by the proper temperature variation of the photoconductivity.

It will further be noticed from Fig. 89 that the photoelectric emission of unactivated BaO takes a sharp downward turn at $h\nu = 3.7$ ev. This is likely to be caused by a reduction in intensity of the electromagnetic radiation near the BaO surface. This intensity, which depends on both the incident and the reflected field vectors, suffers a considerable reduction in the vicinity of the principle absorption edge (Fig. 91) since the index of refraction rises there rapidly to large values.

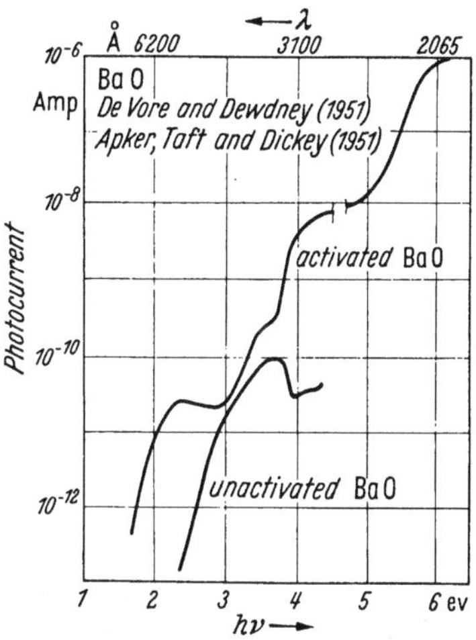

Fig. 89. Photocurrents from activated and unactivated BaO.

In order to attempt an understanding of the photoelectric emission from activated BaO below $h\nu = 5$ ev, it is necessary to assume that impurity levels exist in the forbidden region between the conduction and the filled band (Fig. 90). Those near the top, if occupied, may donate an electron to the normally empty conduction band and become an N-type impurity semiconductor. Others near the bottom may accept an electron from the filled band to form a negative ion or P-type impurity, giving rise to conduction in the filled band by virtue of the motion of positive holes [*14*]. The activation process may be regarded as forming a large number of donor sites.

It should be pointed out here that the emission curve for activated BaO in Fig. 89 was observed *after* irradiation of the cold oxide with light in the region of the principle absorption band ($h\nu > 3.7$ ev). The high photoemission thus obtained would decay slowly over a period of one day. These changes were reproducible, and increases in emission could be stimulated by short wavelength irradiation or suppressed by annealing at 700° C.

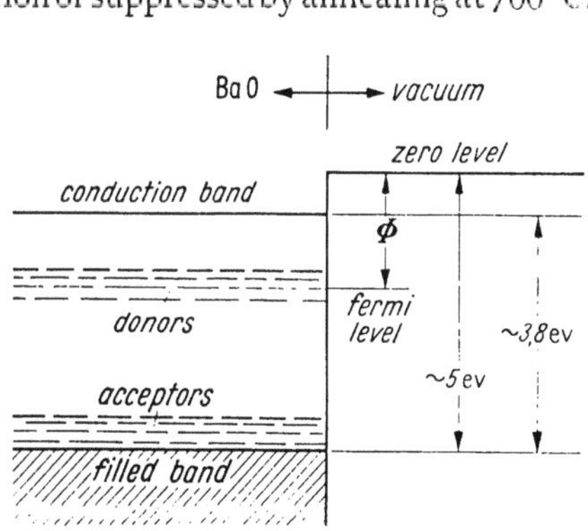

Fig. 90. Model of the energy band structure of BaO.

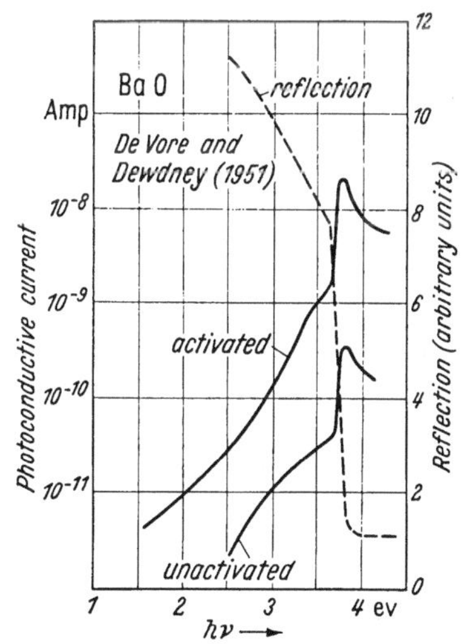

Fig. 91. Optical reflection and photoconductivity of BaO.

Thus both the marked rise in emission at $h\nu = 3.7$ ev and the peak at $h\nu = 2.5$ ev for activated and irradiated BaO probably do not represent intrinsic emission from BaO, but because of their pronounced dependence on temperature and irradiation they must rather be ascribed to photon interactions with electrons at lattice imperfections, some of which may be metastable. The concept of metastability is invoked here because of the observations of DICKEY and TAFT[1] on both SrO and BaO (Fig. 92). They found in BaO that photons with $h\nu > 3.8$ ev would cause emission which increased slowly with time towards a maximum value (full curve). After saturation was reached and irradiation stopped, the yields measured in subsequent time intervals decreased steadily (dashed curve).

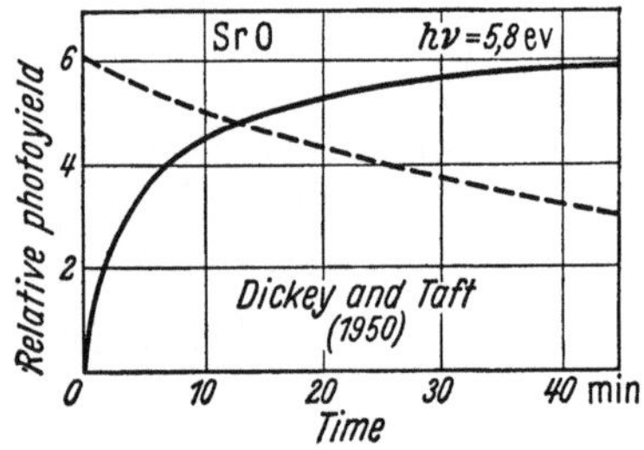

Fig. 92. Yield changes in SrO induced by ultraviolet irradiation.

It is clear that a detailed interpretation of such emission processes is still somewhat speculative because of lack of information on the band structure of composite layers as distinguished from those characteristics representative of surface conditions (such as patches). Nevertheless, all indications are that our understanding of these complicated photoelectric phenomena will be greatly enhanced with the help and guidance of modern solid state concepts.

Major References.

[1] LOEB, L. B.: Basic Processes in Gaseous Electronics. Berkeley: University of California Press 1955. Chapter 9 is particularly illuminating with respect to secondary processes in electrical discharges through gases.

A. Photoionization in gases.

[2] BOYD, R. L. F., and M. J. SEATON (Editors): Rocket Explorations of the Upper Atmosphere. London: Pergamon Press 1954. See also Vol. IL of this Encyclopedia, article on General Aeronomy, by M. NICOLET.

[3] DITCHBURN, R. W., P. J. JUTSUM and G. V. MARR: Proc. Roy. Soc. Lond., Ser. A **219**, 89 (1953). This article contains a complete review of experimental work on the photoionization in the alkali vapors together with references to pertinent theoretical research.

[4] BATES, D. R.: Mon. Not. Roy. Astronom. Soc. **106**, 423, 432 (1946). These papers outline the general theoretical approach towards calculations of photoionization cross sections for atomic systems. They include references to previous work and summarize all calculated cross sections for neutral atoms and positive and negative ions from Li to Ne.

[5] HERZBERG, G.: Molecular Spectra and Molecular Structure, second Edition. New York: D. Van Nostrand Co. 1950.

[6] HERZBERG, G.: Atomic Spectra and Atomic Structure, second Edition. Dover Publications. New York 1944.

B. Photoelectric emission from solids.

[7] SEITZ, F.: The Modern Theory of Solids. New York: McGraw-Hill Book Co. 1940. See also Vol. XIX of this Encyclopedia.

[8] HUGHES, A. L., and L. A. DUBRIDGE: Photoelectric Phenomena. New York: McGraw-Hill Book Co. 1932.

[9] DUBRIDGE, L. A.: New Theories of the Photoelectric Effect, Paris: Hermann & Cie. 1935.

[10] ZWORYKIN, V. K., and E. G. RAMBERG: Photoelectricity and its Applications. New York: John Wiley & Sons 1949.

[11] SUHRMANN, R.: Elektronenemission Metallischer Leiter, in MÜLLER-POUILLETS Lehrbuch der Physik, Bd. IV/4, Kap. 3. Braunschweig: F. Vieweg & Sohn 1934.

[12] HERRING, C., and M. H. NICHOLS: Thermionic Emission. Rev. Mod. Phys. **21**, 185 (1949); this extensive article contains among other things a detailed treatment of electron emission from non-uniform surfaces (patch effect).

[13] BOER, DE, J. H.: Electron Emission and Adsorption Phenomena. Cambridge: University Press 1935.

[14] WRIGHT, D. A.: Semiconductors. London and New York: Methuen & Co., and John Wiley & Sons 1950.

[15] NOTTINGHAM, W. B.: Bibliography on Physical Electronics. Cambridge: Addison-Wesley Publ. Co. 1954.

[1] J. E. DICKEY and E. A. TAFT: Phys. Rev. **80**, 308 (1950).

Motions of Ions and Electrons.

By

W. P. Allis.

With 15 figures.

Introduction.

This article is divided into five parts according to the mathematical method used, rather than according to the physical situation. In part I we attempt to follow the motion of an individual particle under a Lorentz force including the effects of gradients of the magnetic field. This is no longer possible with the introduction of collisions in part II, but one may yet follow the motion of an average particle. One is then tempted to expect that a swarm of particles will have the behavior of the average particle, and this is remarkably close to the truth for high values of E/p, the ratio of the continuous to the stochastic force, and has the advantage of being more intuitive than the distribution function methods which follow.

The method of orbits can be extended to give dispersion about the mean but this is unprofitable. If one cannot follow an individual particle experimentally it is best to let it lose its identity in the theory as well, and calculate the distribution function $f(\boldsymbol{r}, \boldsymbol{v}, t)$ of an assembly. The Boltzmann equation then replaces the Newtonian equations.

In part III the Boltzmann equation is applied to a "Lorentzian gas", a gas of light particles which do not collide with each other, but do collide with heavier particles; that is to free electrons in a gas.

The Boltzmann equation is difficult to solve when the restriction to light particles is removed, in part IV, so as to be applicable to ions. The equation is then transformed to an integral form, the Boltzmann transport equation, which gives directly the average value of any quantity $X(v)$ depending on the velocity of the ions. The transport equation contains the unknown $f(v)$, but as it is under an integral sign an approximation to f results in a higher approximation for $\bar{X}$. This property of the transport equation was shown by Maxwell to be particularly prominent if the law of force between particles varied inversely as the fifth power of their distance. In that case the equation for $\bar{X}$ is completely independent of f. Fortunately the inverse fifth power law is exactly right for a slow ion and a molecule, aside from the property of charge transfer.

The consideration of Coulomb interactions in part V introduces a force of so much longer range that an electron is interacting simultaneously with many others and the number N with which it interacts simultaneously enters explicitly in the equations. The basic assumption of Boltzmann theory is that the interaction time Δt is very small. With Coulomb interactions the fluctuation time τ is smaller than Δt, and the appropriate equation for this situation is that of Fokker-Planck.

When the interactions between electrons are large compared to their interactions with the gas, a new type of phenomenon appears: collective oscillations superposed on the random motions. These will not be considered here in part because the subject is new and not well understood.

MKS rationalized units will be used throughout this article. In this system the permeability of free space is $\mu_0 = 4\pi \times 10^{-7}$ henry/meter and the permittivity $\varepsilon_0 = 1/\mu_0 c^2$ farad/meter. To convert to c.g.s. substitute $\varepsilon_0 = 1$. $q = \pm Ze$ is the charge on the ion, and $e = 1.602 \times 10^{-19}$ coulombs is always positive. The word "ion" is generally used to distinguish the heavy particle from an electron, but is also occasionally used to mean either ion or electron, as in the preceding sentence, care being taken that the context is such that it is not misunderstood.

I. Electron and ion orbits.

a) Orbits in uniform fields.

1. Orbits in a uniform magnetic field only. In the absence of collisions a particle of charge q moving in a magnetic field $\boldsymbol{B}$ is subject to a force

$$\boldsymbol{F} = q\,\boldsymbol{v} \times \boldsymbol{B} = m\dot{\boldsymbol{v}}. \tag{1.1}$$

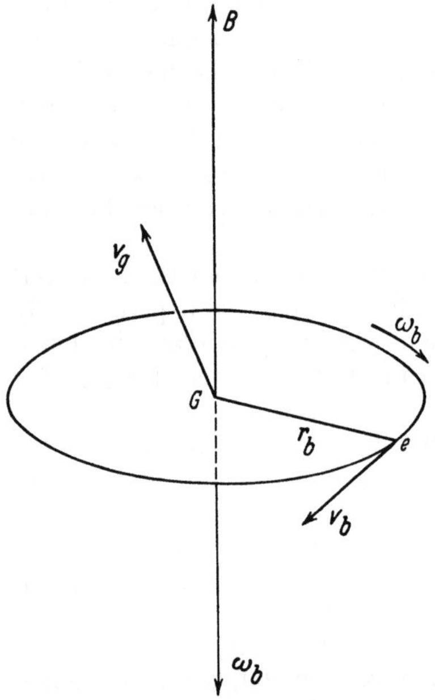

Fig. 1. Vector relations in the motion of an ion in a magnetic field.

We shall assume that the motion of the particle is the resultant of a circular motion $\boldsymbol{v}_b$ about a guiding center G, which itself has a uniform motion $\boldsymbol{v}_g$

$$\boldsymbol{v} = \boldsymbol{v}_g + \boldsymbol{v}_b = \boldsymbol{v}_g + \boldsymbol{\omega}_b \times \boldsymbol{r}_b. \tag{1.2}$$

The acceleration for this motion is

$$\dot{\boldsymbol{v}} = \boldsymbol{\omega}_b \times \boldsymbol{v}_b = -\omega_b^2\, \boldsymbol{r}_b. \tag{1.3}$$

Substituting in (1.1) gives

$$q\,\boldsymbol{v}_g \times \boldsymbol{B} + q\,\boldsymbol{v}_b \times \boldsymbol{B} = m\,\boldsymbol{\omega}_b \times \boldsymbol{v}_b$$

whence

$$\boldsymbol{\omega}_b = -q\boldsymbol{B}/m, \tag{1.4}$$

$$\boldsymbol{v}_g \times \boldsymbol{B} = 0. \tag{1.5}$$

Thus the motion is circular around $\boldsymbol{B}$ with the "cyclotron" frequency ν_b given by

$$\frac{\nu_{b1}}{B} = \frac{q}{2\pi m} = 2.799 \text{ Mc/sec. gauss for electrons}$$

$$= 1.525 \text{ Kc/sec. gauss for protons}$$

combined with an arbitrary uniform translation along $\boldsymbol{B}$. Equation (1.4) will be kept as the definition of the cyclotron vector $\boldsymbol{\omega}_b$. Equation (1.5) will be generalized in the following sections.

2. Orbits in uniform electric and magnetic fields. A particle in the presence of both an electric and a magnetic field experiences the LORENTZ force

$$\boldsymbol{F} = q(\boldsymbol{E} + \boldsymbol{v} \times \boldsymbol{B}) = m\dot{\boldsymbol{v}}. \tag{2.1}$$

If we assume a solution of this equation of the form (1.2), we find

$$\boldsymbol{E} + \boldsymbol{v}_g \times \boldsymbol{B} = 0 \qquad (2.2$$

which yields

$$\boldsymbol{v}_g = \boldsymbol{E} \times \boldsymbol{B}/B^2 + \boldsymbol{v}_\parallel \qquad (2.3$$

where the velocity $\boldsymbol{v}_{\parallel}$ parallel to the magnetic field is not determined by (2.2) but satisfies the usual equation

$$m\,\dot{\boldsymbol{v}}_{\parallel} = q\boldsymbol{E}_{\parallel} \tag{2.4}$$

and hence the conservation of energy principle

$$\tfrac{1}{2}\,m\,v_{\parallel}^2 + q\,V = \text{constant.} \tag{2.5}$$

The motion of the guiding center is therefore normally accelerated along the magnetic field and has a constant drift of magnitude $E_{\perp}/B$ across the magnetic field. If the medium is neutral, containing equal amounts of positive and negative charges, an electric field produces motion along $\boldsymbol{B}$ in the opposite direction for the two signs of charge, and hence a net electric current along the field. But the cross drift is the same for both kinds of charge, and hence in this direction $\boldsymbol{E}$ produces a mass motion with no electric current.

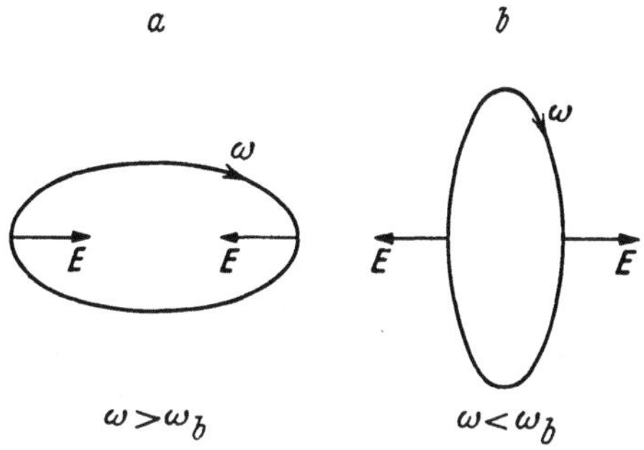

Fig. 2 a u. b. Elliptical motion in an alternating electric field. B is normal to the paper. *a* Displacement out of phase. *b* Displacement in phase.

3. Orbits in a uniform magnetic field and alternating electric field. If an alternating electric field

$$\boldsymbol{E} = \boldsymbol{E}_p\, e^{j\omega t} \tag{3.1}$$

is superposed on a magnetic field $\boldsymbol{B}$ the resultant velocity will have a random component with three arbitrary constants

$$\boldsymbol{v}_r = \boldsymbol{v}_{\parallel} + \boldsymbol{\omega}_b \times \boldsymbol{r}_b \tag{3.2}$$

in which $\boldsymbol{v}_{\parallel}$ is constant (zero frequency) and $\boldsymbol{\omega}_b \times \boldsymbol{r}_b$ has the cyclotron frequency, plus a forced motion, $\boldsymbol{v}_g e^{j\omega t}$, at the applied frequency, which has no arbitrary constants. Substituting (3.1), (3.2) and the forced motion into (2.1) gives

$$(j\omega - \boldsymbol{\omega}_b \times)\,\boldsymbol{v}_g = q\boldsymbol{E}/m = \boldsymbol{a} \tag{3.3}$$

whose solution is

$$\boldsymbol{r}_g = \frac{(j\omega + \boldsymbol{\omega}_b\times)\,\boldsymbol{a}_{\perp}}{\omega_b^2 - \omega^2} - \frac{j\,\boldsymbol{a}_{\parallel}}{\omega}. \tag{3.4}$$

This represents a three-dimensional elliptical motion, which is, of course, superposed on the helical random motion. At frequencies above the cyclotron frequency the component in a plane perpendicular to $\boldsymbol{B}$ has its major axis along $\boldsymbol{E}$, and is out of phase with $\boldsymbol{E}$. At lower frequencies the reverse is true.

At the cyclotron frequency (3.4) is singular and there is no steady state. The solution of the equation of motion is

$$\boldsymbol{r}_g = \left[\left(\omega t - \frac{j}{2}\right)\boldsymbol{E}_{\perp} + \left(j t - \frac{1}{2\omega}\right)\boldsymbol{E}\times\boldsymbol{\omega} - j\boldsymbol{E}_{\parallel}\right]\frac{e^{j\omega t}}{B}. \tag{3.5}$$

The vector solution (3.4) is somewhat easier to visualize as a tensor written with the z-axis along $\boldsymbol{B}$

$$\|\mathcal{K}_{kl}\| = \left\|\begin{matrix} \dfrac{\omega^2}{\omega^2-\omega_b^2} & \dfrac{j\,\omega_b}{\omega^2-\omega_b^2} & 0 \\[2ex] \dfrac{-j\,\omega_b}{\omega^2-\omega_b^2} & \dfrac{\omega^2}{\omega^2-\omega_b^2} & 0 \\[2ex] 0 & 0 & 1 \end{matrix}\right\|. \tag{3.6}$$

Then

$$v_{g,k} = \frac{-jq}{m\omega} \mathcal{K}_{kl} E_l \tag{3.7}$$

where the usual summation convention over the repeated index is implied.

The velocities of the charged particles determine the permittivity of a plasma, and hence its ability to propagate electromagnetic waves, in the following way. From the velocity of the guiding center one obtains the real current at the frequency ω

$$J_k = nqv_{g,k} = \frac{nq^2}{m} \frac{-j}{\omega} \mathcal{K}_{kl} E_l \tag{3.8}$$

and the total current is

$$J + \dot{D} = J + j\omega\varepsilon_0 E \tag{3.9}$$

which leads to the permittivity tensor

$$\varepsilon_{kl} = \varepsilon_0 \left(\delta_{kl} - \frac{\omega_p^2}{\omega^2} \mathcal{K}_{kl}\right) \tag{3.10}$$

where the symbol $\omega_p^2 = nq^2/\varepsilon_0 m$ defines the plasma frequency

$$\nu_p = \frac{\omega_p}{2\pi} = 8.984 \sqrt{n} \text{ in mks units for electrons.} \tag{3.11}$$

The ratio of plasma to cyclotron frequency has the interesting property

$$\frac{\omega_p^2}{\omega_b^2} = \frac{n m c^2}{B \cdot H} = \frac{\text{material energy density}}{2 \times \text{magnetic energy density}}. \tag{3.12}$$

In general we are not interested in the off-diagonal terms of the permittivity. For electric fields parallel to $\boldsymbol{B}$, the dielectric coefficient $\varepsilon/\varepsilon_0$ is less than 1, and is negative for frequencies below the plasma frequency. Electromagnetic waves of these frequencies therefore cannot propagate and the plasma frequency represents a lower frequency limit for propagation. For waves whose electric field is perpendicular to the magnetic field $\boldsymbol{B}$, there is a forbidden band[1]

$$\omega_b^2 < \omega^2 < \omega_b^2 + \omega_p^2. \tag{3.13}$$

Waves of frequency above this band propagate at a phase velocity faster than in free space while waves of frequency below the band propagate slower than light.

4. Orbits in a changing magnetic field. In a magnetic field changing at the rate $\dot{\boldsymbol{B}} \ll \boldsymbol{B}\omega_b$ there is an induced electric field $\boldsymbol{E}_i$ and applying the induction law to the circular orbit of the charged particle we have

$$2\pi r_b E_i = \oint \boldsymbol{E}_i \cdot d\boldsymbol{s} = -\dot{\Phi} = -\pi r_b^2 \dot{B}. \tag{4.1}$$

Consequently, the motion perpendicular to the magnetic field will be accelerated and we have

$$\dot{v}_\perp = qE_i/m = r_b \dot{\omega}_b/2. \tag{4.2}$$

From (1.2)

$$\dot{v}_\perp = r_b \dot{\omega}_b + \dot{r}_b \omega_b \tag{4.3}$$

whence

$$\omega_b \dot{r}_b = -r_b \dot{\omega}_b/2 = -\dot{v}_\perp \tag{4.4}$$

and

$$\frac{d}{dt}(\omega_b r_b^2) = \frac{d}{dt}(r_b v_\perp) = r_b(\omega_b \dot{r}_b + \dot{v}_\perp) = 0. \tag{4.5}$$

---

[1] R. JANCEL and T. KAHAN: Nuovo Cim. **12**, 573 (1954).

This equation expresses the conservation of angular momentum of the ion about the guiding center, or alternatively the conservation of magnetic flux linked by the circular motion. As the magnetic field B is increased, the orbits decrease in size and also move closer together, each guiding center remaining on a line of force and each orbit on a tube of force. The density of charged particles, as well as their energy, therefore increases proportionately to the magnetic field B.

$$\frac{d}{dt}\left(\frac{u_\perp}{B}\right) = \frac{d}{dt}\left(\frac{n}{B}\right) = 0. \tag{4.6}$$

where $u_\perp = \frac{1}{2} m v_\perp^2$ is the kinetic energy of the circular motion.

b) Orbits in inhomogeneous magnetic fields.

We will now consider certain peculiarities of ionic orbits which appear in inhomogeneous magnetic fields and which have been of particular interest to astronomers. They have been treated by H. ALFVEN[1]. We shall always assume small gradients, grad $B \ll B/r_b$, so that the circular motion is not sufficiently deformed so as to prevent the treatment in terms of the motion of a guiding center.

5. Longitudinal magnetic gradient. Consider first the effect of a gradient $(\boldsymbol{B}\cdot\text{grad}\,B)$ such that the magnitude of B changes along a straight line of force. The particle then observes a uniform magnetic field but one which changes with time as the guiding center moves along a line of force. Accordingly, the conditions of the last section hold in a coordinate system moving with the guiding center and the orbit of the particle will link a constant amount of magnetic flux (BUSCH's Theorem). The orbit is then a spiral described on the surface of a tube of force. The pitch of the spiral, $2\pi r_b v_\parallel / v_\perp = 2\pi r_b \tan\vartheta$, changes, however, as the conservation of energy requires $u = \text{constant}$, and conservation of angular momentum requires $u_\perp / B = \text{constant}$. Combining these two gives

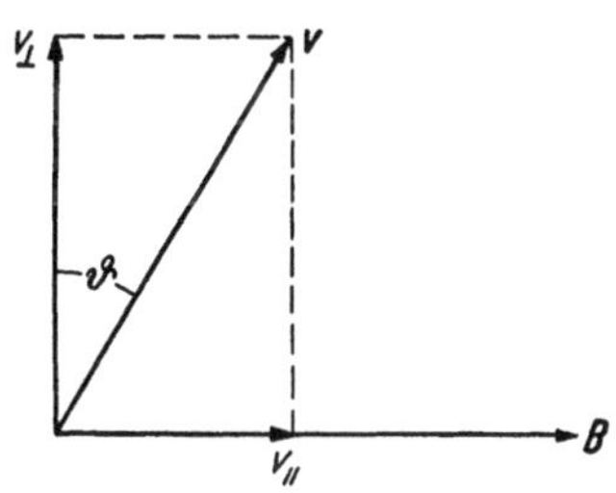

Fig. 3. The pitch angle ϑ.

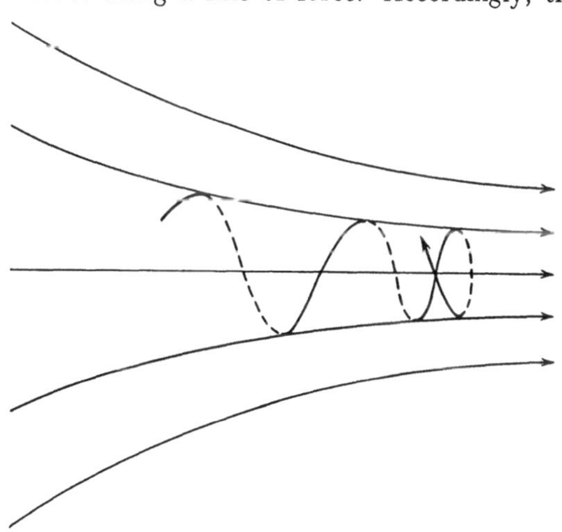

Fig. 4. Reflection of a spiral path in a converging magnetic field.

$$\frac{B u}{u_\perp} = \frac{B v^2}{v_\perp^2} = B \sec^2\vartheta = B_0 \tag{5.1}$$

where B_0 is a constant. An ion spiraling along a tube of force in the direction of increasing field has a decreasing pitch angle ϑ. B_0 is the value of B at which the angle ϑ goes through zero and therefore at which the motion of the particle

[1] H. ALFVEN: Cosmical Electrodynamics. Oxford: Clarendon Press 1950.

is reflected. Reciprocally, if there is a distribution of particles at B with various pitch angles, only particles for which

$$\cos^2\vartheta < \frac{B}{B_0} \tag{5.2}$$

will penetrate to fields greater than B_0.

From (5.1) it readily follows that

$$v_{\parallel}^2 = v^2\,\frac{B_0 - B}{B_0}, \tag{5.3}$$

whence the equation of motion of the guiding center along the line of force is

$$v t = \int \sqrt{\frac{B_0}{B_0 - B}}\,d z. \tag{5.4}$$

6. Transverse gradients. We must now consider two effects of magnetic gradients which produce motion of the guiding center at right angles to the magnetic field. If there is a magnetic gradient at right angles to the magnetic field, the erstwhile circular orbit will no longer be circular as the curvature will now be greater on one side than on the other and this will produce a crosswise motion, illustrated in Fig. 5. Similarly if the lines of force are curved, the vector curvature is given by

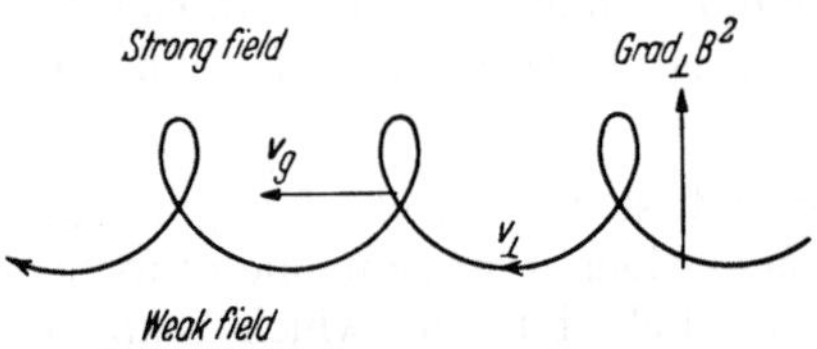

Fig. 5. Cross-drift in a magnetic gradient. B is perpendicular to the paper.

$$\left(\frac{1}{\boldsymbol{R}}\right) = \frac{(\boldsymbol{B}\cdot\mathrm{grad})\,\boldsymbol{B}_\perp}{\boldsymbol{B}\cdot\boldsymbol{B}} = \frac{1}{B}\,\frac{\partial \boldsymbol{B}_\perp}{\partial z}.\text{[1]} \tag{6.1}$$

The centrifugal force due to motion of the guiding center along the line will produce an additional motion of the guiding center at right angles to the magnetic field and to the curvature. We shall assume that the gradients are small so that we can expand the magnetic field about the center of the almost circular orbit

$$\boldsymbol{\omega}_b = \boldsymbol{\omega}_0 + (\boldsymbol{r}_b\cdot\mathrm{grad})\,\boldsymbol{\omega}_0 \tag{6.2}$$

and we shall assume

$$\boldsymbol{v} = \boldsymbol{v}_g + \boldsymbol{\omega}_0\times\boldsymbol{r}_b \tag{6.3}$$

as before. Substituting in the equation of motion (1.1) gives

$$\left.\begin{aligned} \boldsymbol{\omega}_b\times\boldsymbol{v} &= \boldsymbol{\omega}_0\times\boldsymbol{v}_g + \boldsymbol{\omega}_0\times(\boldsymbol{\omega}_0\times\boldsymbol{r}_b) + (\boldsymbol{r}_b\cdot\mathrm{grad})\,(\boldsymbol{\omega}_0\times\boldsymbol{v}_g) - \\ &\quad - (\boldsymbol{\omega}_0\times\boldsymbol{r}_b)\times(\boldsymbol{r}_b\cdot\mathrm{grad})\,\boldsymbol{\omega}_0 \\ &= \dot{\boldsymbol{v}} = -\,\omega_0^2\,\boldsymbol{r}_b + \left(\frac{v_\parallel}{B}\right)^2(\boldsymbol{B}\cdot\mathrm{grad})\,\boldsymbol{B}_\perp. \end{aligned}\right\} \tag{6.4}$$

Expanding the triple cross products and remembering that $\boldsymbol{r}_b\cdot\boldsymbol{\omega}_0 = 0$, we find

$$\boldsymbol{\omega}_0\times\boldsymbol{v}_g + (\boldsymbol{r}_b\cdot\mathrm{grad})\,(\boldsymbol{\omega}_0\times\boldsymbol{v}_g) - \frac{1}{2}\,\boldsymbol{r}_b(\boldsymbol{r}_b\cdot\mathrm{grad}\,\omega_0^2) = \left(\frac{v_\parallel}{B}\right)^2(\boldsymbol{B}\cdot\mathrm{grad})\,\boldsymbol{B}_\perp. \tag{6.5}$$

This equation contains terms of frequency ω_0 because that part of the motion which has been assumed circular in (6.3) is not truly circular. These periodic terms are eliminated by averaging over a cycle which gives

$$\overline{\boldsymbol{r}_b} = 0 \tag{6.6}$$

and

$$\overline{\boldsymbol{r}_b(\boldsymbol{r}_b\cdot\mathrm{grad}\,\omega_0^2)} = \left(\frac{r_b^2}{2}\right)\mathrm{grad}_\perp\,\omega_0^2. \tag{6.7}$$

[1] Note that although $\mathbf{B}_\perp$, the component of $\mathbf{B}$ perpendicular to itself, is zero, its derivative need not be zero.

When these are substituted in (6.5) one obtains

$$\left.\begin{aligned}\boldsymbol{\omega}_0\times\boldsymbol{v}_g &= \left(\frac{r_b^2}{4}\right)(\operatorname{grad}_\perp \omega_0^2) + \left(\frac{v_\parallel}{B}\right)^2 (\boldsymbol{B}\cdot\operatorname{grad})\boldsymbol{B}_\perp \\ &= \left(\frac{v_\perp}{2B}\right)^2 (\operatorname{grad}_\perp B^2) + \left(\frac{v_\parallel}{B}\right)^2 (\boldsymbol{B}\cdot\operatorname{grad})\boldsymbol{B}_\perp .\end{aligned}\right\} \tag{6.8}$$

The two terms on the right represent, respectively, a magnetic pressure and a centrifugal force. As the force is independent of the sign of q, the velocity of the guiding center v_g does depend on the sign of q; that is, in a neutral medium these forces will produce a current or a polarization of the medium.

Introducing the current density $\boldsymbol{J}$ in the medium by the relation

$$\boldsymbol{J}\times\boldsymbol{B} = (\boldsymbol{B}\cdot\operatorname{grad})\,\boldsymbol{H}_\perp - \tfrac{1}{2}\operatorname{grad}_\perp \boldsymbol{B}\cdot\boldsymbol{H} \tag{6.9}$$

equation (6.8) may be transformed to

$$\boldsymbol{\omega}_0\times\boldsymbol{v}_g = \frac{v_\perp^2 + 2v_\parallel^2}{4B^2}\operatorname{grad}_\perp B^2 + \frac{v_\parallel^2}{\boldsymbol{B}\cdot\boldsymbol{H}}\,\boldsymbol{J}\times\boldsymbol{B} \tag{6.10}$$

which relates the force on a particle $m\,\boldsymbol{\omega}_0\times\boldsymbol{v}_g$ to the force per unit volume $\boldsymbol{J}\times\boldsymbol{B}$ and in addition to the normal gradient of B^2. Solving for the velocity of the guiding center gives

$$\boldsymbol{v}_g = \frac{u_\perp + 2u_\parallel}{2q}\,\frac{\boldsymbol{B}\times\operatorname{grad} B^2}{B^4} + \frac{2u_\parallel}{q}\,\frac{\boldsymbol{B}\times(\boldsymbol{J}\times\boldsymbol{B})}{B^2\,\boldsymbol{H}\cdot\boldsymbol{B}} + \boldsymbol{v}_\parallel . \tag{6.11}$$

This equation simplifies considerably if there is no current ($J=0$) and if it is applied to an isotropic distribution such that

$$u_\perp = 2u_\parallel = kT \tag{6.12}$$

where the factor 2 comes from the two directions perpendicular to $\boldsymbol{B}$. The expression then reduces to

$$\boldsymbol{v}_g = \frac{kT}{q}\,\frac{\boldsymbol{B}\times\operatorname{grad} B^2}{B^4} + \boldsymbol{v}_\parallel . \tag{6.13}$$

These relations have been given by SPITZER[1].

The velocity of the guiding center at right angles to the magnetic field $\boldsymbol{B}$ is in such a direction that the magnetic field of the current carried by $\boldsymbol{v}_g$ reduces the inhomogeneity in $\boldsymbol{B}$. If the charges carried by $\boldsymbol{v}_g$ are allowed to accumulate they will produce an electric field which in turn will produce mass motion in the direction of lower magnetic fields.

7. The drift velocity. The drift velocity $\boldsymbol{v}_d$ is defined as the average velocity of the particles contained in a small volume V. More generally, it is the limit of this as the volume V becomes small, but not smaller than the radius r_b. The drift velocity $\boldsymbol{v}_d$ is to be distinguished very carefully from the velocity $\boldsymbol{v}_g$, which is the time average of the velocity of a single particle. These two velocity averages differ because the walls of the volume V cut through some orbits so that V contains only parts of the orbits of certain particles, and when there are magnetic gradients these fractional orbits correspond to velocities greater than $\boldsymbol{v}_g$ on one side of V and smaller than, or even opposed to, $\boldsymbol{v}_g$ on the other side. For a uniform density and magnetic field these fractional orbits on the two sides may be added to give complete orbits so that $\boldsymbol{v}_d = \boldsymbol{v}_g$, but in this case $\boldsymbol{v}_g = 0$ and both averages vanish. For an inhomogeneous field there are more fractional orbits on the

[1] L. SPITZER jr.: Astrophys. J. **116**, 299 (1952).

weak side, assuming uniform particle density, because these orbits are larger. But these fractional orbits correspond to motion *against* $\boldsymbol{v}_g$, and therefore $\boldsymbol{v}_d < \boldsymbol{v}_g$. A careful computation shows[1] that $\boldsymbol{v}_d = 0$ and we shall not repeat the argument, but merely point out that this result must be so, by considering a system in thermodynamic equilibrium. In such a system even when there are magnetic gradients present, provided there are no other forces, the density is uniform. Any drift velocities would disturb this uniformity and therefore must be zero. This cancellation of $\boldsymbol{v}_g$ by the intercepted orbits is exactly the same as the cancellation of the diamagnetism of free electrons due to surface effects[2].

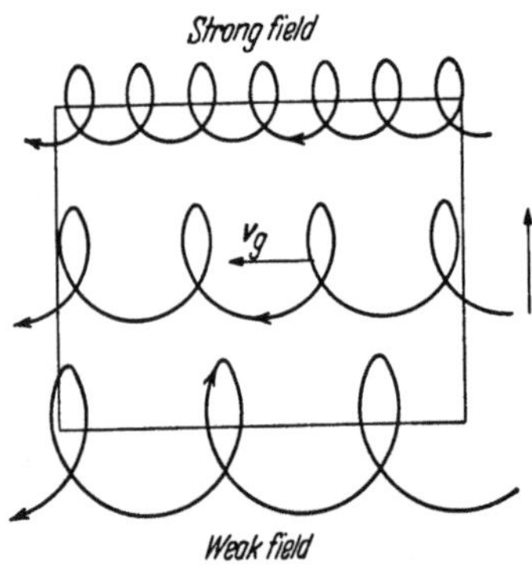

Fig. 6. A volume V contains fractional orbits for which the motion is opposed to the guiding center.

In a physical enclosure with real walls, the situation may be different and depends on the nature of the walls. If they are perfectly reflecting, so that equilibrium is possible, $\boldsymbol{v}_d = 0$ even though $\boldsymbol{v}_g$ is different from zero, as above. The particles intercepted by the wall hop along the wall in the reverse direction to $\boldsymbol{v}_g$ and this produces the cancellation. If charged particles striking the walls are absorbed by it, this eliminates the surface currents discussed above and then $\boldsymbol{v}_d \approx \boldsymbol{v}_g$, but the distribution of these velocities is different: the drift velocity is confined to the region near the walls until density gradients are generated, and then the problem becomes more complicated.

In the case of electric fields studied in Sect. 2, $\boldsymbol{v}_d$ and $\boldsymbol{v}_g$ are equal. Henceforth we will deal with assembly averages, and therefore with $\boldsymbol{v}_d$ and not with $\boldsymbol{v}_g$.

c) Pressure gradients.

8. Equation for the drift velocity. Pressure gradients are exactly the opposite of magnetic gradients, as they produce no motion of the individual particles ($\boldsymbol{v}_g = 0$), unless the particles collide with each other, but they do produce a drift velocity $\boldsymbol{v}_d$. Thus we have the following scheme:

Table 1.

Type of Force	Velocity Average
$\boldsymbol{E}$	$\boldsymbol{v}_g = \boldsymbol{v}_d$
grad B^2	$\boldsymbol{v}_g \neq 0, \quad \boldsymbol{v}_d = 0$
grad p	$\boldsymbol{v}_g = 0, \quad \boldsymbol{v}_d \neq 0$

In dealing with an assembly whose velocity distribution is isotropic, a pressure gradient is equivalent to a force, so that one can write the equation for the drift velocity

$$m\dot{\boldsymbol{v}}_d = q(\boldsymbol{E} + \boldsymbol{v}_d \times \boldsymbol{B}) - \frac{\operatorname{grad} p}{n}, \tag{8.1}$$

where

$$p = \tfrac{1}{3} n m \overline{v^2} = n k T. \tag{8.2}$$

Thus $\boldsymbol{v}_d$ is obtained from a force equation exactly like that for a single particle. $\boldsymbol{v}_d$ is the velocity of the average ion and determines the orbit of the average ion. This equation may be solved for $\boldsymbol{v}_d$ and gives

$$\boldsymbol{v}_d = \left(\boldsymbol{E} - \frac{\operatorname{grad} p}{n q}\right) \times \frac{\boldsymbol{B}}{B^2} + \boldsymbol{v}_{\parallel}, \tag{8.3}$$

[1] L. SPITZER jr.: Astrophys. J. **116**, 299 (1952).

[2] J. H. VAN VLECK: Theory of Electric and Magnetic Susceptibilities, p. 100. Oxford; Clarendon Press 1932

and

$$m \dot{\boldsymbol{v}}_{\|} = q E_{\|} - \frac{1}{n} \frac{\partial p}{\partial z}. \tag{8.4}$$

We have neglected here a term in $\boldsymbol{B} \times \dot{\boldsymbol{v}}_d$. This term would give the circular motion of Sect. 1 and would be applicable if all the ions of the assembly had the same starting conditions. If the starting conditions are random, the phases of the circular motions will be random, and the $\boldsymbol{B} \times \dot{\boldsymbol{v}}_d$ term averages out.

9. Equilibrium. In the case of thermodynamic equilibrium $\boldsymbol{v}_d = 0$ and (8.3) and (8.4) lead to

$$q\boldsymbol{E} = \frac{\operatorname{grad} p}{n} = kT \frac{\operatorname{grad} n}{n} \tag{9.1}$$

or

$$n = n_0 \exp(-qV/kT). \tag{9.2}$$

Equations (8.3) and (8.4) are therefore consistent with a Boltzmann distribution which is entirely independent of the magnetic field.

10. Stationary distributions. We shall now assume a stationary state in which $\boldsymbol{v}_d$ is constant, and

$$\boldsymbol{J} = n q \boldsymbol{v}_d = \operatorname{curl} \boldsymbol{H}, \tag{10.1}$$

that is that there are no currents other than those due to the particles in question. This is somewhat different than the assumption of Sect. 6 where the magnetic field was given and assumed due to external sources. Equation (8.1) then gives

$$\operatorname{grad} p - nq\boldsymbol{E} - \boldsymbol{B} \times \operatorname{curl} \boldsymbol{H} \tag{10.2}$$

or

$$\operatorname{grad}(p + \tfrac{1}{2}\boldsymbol{B}\cdot\boldsymbol{H}) = nq\boldsymbol{E} + (\boldsymbol{B}\cdot\operatorname{grad})\boldsymbol{H}. \tag{10.3}$$

Two special cases with $E = 0$ will be considered.

α) Straight parallel magnetic field lines. In this case, the right hand side of (10.3) vanishes and one has[1]

$$p + \tfrac{1}{2}\boldsymbol{B}\cdot\boldsymbol{H} = \text{constant}. \tag{10.4}$$

The charged particles responsible for the current concentrate in the regions of weak field, the total material plus magnetic pressure being a constant. The process by which this concentration is reached is that described in Sect. 6. Perhaps the important thing about the rather logical result (10.4) is that it is *not* correct unless $(\boldsymbol{B}\cdot\operatorname{grad})\boldsymbol{H}$ vanishes. This is illustrated by the following example.

β) Circular magnetic field lines. W. H. Bennett[2] first studied the dynamic equilibrium of an electrically neutral double stream of cylindrical symmetry. We assume equal numbers of positive and negative charges, $n_+ = n_- = n$, with the relative velocity $\boldsymbol{v}_d$, so that the current density is $J = nqv_d$. Bennett finds that the equilibrium current distributions is

$$J = \frac{J_0}{(1 + b r^2)^2} \tag{10.5}$$

where J_0 is the current density on the axis and b is a parameter determining the size of the stream. The field of this stream is

$$H = \frac{J_0 r/2}{1 + b r^2} \tag{10.6}$$

---

[1] L. Tonks: Phys. Rev. **97**, 1443 (1955).

[2] W. H. Bennett: Phys. Rev. **45**, 890 (1934); **98**, 1584 (1955).

and the force equation (10.2) is readily integrated to give

$$p = \frac{\mu_0 J_0^2/8b}{(1+br)^2} = \frac{\mu_0 J_0}{8b} J, \tag{10.7}$$

with no constant of integration if the pressure is to vanish at infinity. It is noted that $p + \frac{1}{2}\boldsymbol{B}\cdot\boldsymbol{H}$ is not constant. The proportionality between $p = 2nkT$ and J allows the relative drift velocity

$$v_d = 16b\,kT/\mu_0 q J_0 \tag{10.8}$$

to be constant over the cross section of the stream, which is essential if particles are to move freely.

The total current in the stream is found by integrating (10.5)

$$I = \frac{\pi J_0}{b} \tag{10.9}$$

and by integrating the pressure (10.7) one finds

$$\int_0^\infty p\,2\pi r\,dr = NkT = \frac{\mu_0 I^2}{8\pi} \tag{10.10}$$

where N is the total number of particles, positive and negative, per unit length of stream. Given N and T, (10.10) fixes a critical current I for equilibrium. If the current is less than this the magnetic force is insufficient to contain the thermal pressure and the stream expands. Conversely, a larger current is focused.

II. The LANGEVIN equation.

a) Drift.

11. Relaxation times. LANGEVIN[1] added to the steady forces a stochastic force $m\boldsymbol{A}(t)$ due to the collisions of an ion with the molecules of a gas, so that

$$\left.\begin{aligned} \dot{\boldsymbol{v}} &= \boldsymbol{a} + \boldsymbol{\omega}_b \times \boldsymbol{v} + \boldsymbol{A}(t), \\ \boldsymbol{a} &= q\boldsymbol{E}/m, \qquad \boldsymbol{\omega}_b = -q\boldsymbol{B}/m \end{aligned}\right\} \tag{11.1}$$

where $\boldsymbol{A}(t)$ cannot be specified because it is fluctuating randomly. However, certain average properties of $\boldsymbol{A}(t)$ may be known and we shall assume that these are

$$\overline{\boldsymbol{A}} = -\nu_c \boldsymbol{v}_g, \tag{11.2}$$

$$m\,\overline{\boldsymbol{v}\cdot\boldsymbol{A}} = \lambda \nu_c (\overline{U} - u). \tag{11.3}$$

ν_c will be shown [Eq. (27.6)] to be the collision frequency and λ [Eqs. (25.2) and (41.9)] an energy loss parameter which, for elastic collisions, is

$$\lambda = \frac{2m}{(M+m)} \tag{11.4}$$

where m is the mass of the ion and M that of a molecule of the gas. λ is therefore very small for an electron but of the order 1 for an ion. u is the energy of the ion and $\overline{U}$ the average energy of the gas molecules. The bar over $\overline{\boldsymbol{A}}$ and $\overline{\boldsymbol{v}\cdot\boldsymbol{A}}$ represents a time average during the motion of a single particle, over a time long

[1] P. LANGEVIN: C. R. Acad. Sci. Paris **146**, 530 (1908). Several basic papers in this field have been reprinted in "Noise and Stochastic Processes", edited by NELSON WAX. Dover 1954.

compared to $1/\nu_c$ but during which ν_c, and therefore u, does not change appreciably[1]. This is an impossible condition for ions, but may be satisfied for electrons.

The results of this section, using the LANGEVIN equation, will be functions of u, the instantaneous energy of the electron. They will not be directly applicable to an assembly having a distribution of energy, but relate to a "microcanonical" assembly with a known energy. These results will fit into the BOLTZMANN theory before integration over energies.

Omitting for the moment all external forces, the solution of (11.1) is found by first multiplying both sides by $m\boldsymbol{v}$ and averaging. This yields

$$\frac{du}{u-U} = -\lambda \nu_c dt. \tag{11.5}$$

$1/(\lambda\nu_c)$ is the relaxation time for energy and, when the energy dependence of $\nu_c(u)$ is known, (11.5) can be integrated for $u(t)$.

Averaging (11.1) directly yields

$$\frac{dv_g}{v_g} = -\nu_c dt \tag{11.6}$$

so $1/\nu_c$ is the relaxation time for momentum. Knowing $\nu_c(t)$ it can be integrated for $v_g(t)$.

As we shall not consider magnetic gradients the subscripts "g" or "d" are rather arbitrary. The energy associated with v_g will be denoted $u_d = \frac{1}{2} m v_g^2$ for convenience later. The total energy u is then conveniently divided into "drift" and "random" energy

$$u = u_d + u_r. \tag{11.7}$$

The drift energy disappears, in the average, at each collision, most of it being converted into random energy and a small fraction λ going to the gas.

12. Mobility. Reintroducing the electric and magnetic fields and averaging (11.1) over the time gives

$$\dot{\boldsymbol{v}}_g = \boldsymbol{a} + \boldsymbol{\omega}_b \times \boldsymbol{v}_g - \nu_c \boldsymbol{v}_g. \tag{12.1}$$

We shall consider the case of an alternating electric field $\boldsymbol{E} \sim e^{j\omega t}$ as the direct current case is readily derived from the a.c. solution. The velocity $\boldsymbol{v}_g$ then also varies as $e^{j\omega t}$ and (12.1) becomes

$$(\nu_c + j\omega - \boldsymbol{\omega}_b \times)\boldsymbol{v}_g = \boldsymbol{a} \tag{12.2}$$

whence

$$(\nu_c + j\omega)\, v_g^2 = \boldsymbol{a} \cdot \boldsymbol{v}_g. \tag{12.3}$$

For a given $\boldsymbol{v}_g$ the component of the field $\boldsymbol{E}$ in the direction of the drift is independent of the magnetic field, even though the drift may be partially across the magnetic field. This is TONKS' theorem[2]. Solving (12.2) for the velocity v_g gives

$$[\nu_c + j(\omega - \omega_b)]\,[\nu_c + j(\omega + \omega_b)]\,\boldsymbol{v}_g - (\boldsymbol{\omega}_b \cdot \boldsymbol{v}_g)\,\boldsymbol{\omega}_b = (\nu_c + j\omega + \boldsymbol{\omega}_b \times)\,\boldsymbol{a}. \tag{12.4}$$

This vector equation is somewhat better written in tensor notation

$$v_{g,i} = \mu_{ik} E_k. \tag{12.5}$$

---

[1] S. CHANDRASEKHAR: Stochastic Problems in Physics and Astronomy. Rev. Mod. Phys. **15**, 21 (1943).

[2] L. TONKS: Phys. Rev. **51**, 744; **52**, 710 (1937).

Introducing the symbols

$$l = \frac{1}{\nu_c + j(\omega + \omega_b)}, \quad r = \frac{1}{\nu_c + j(\omega - \omega_b)}^{\dagger}, \quad p = \frac{1}{\nu_c + j\omega} \tag{12.6}$$

to correspond to left and right circular polarizations of the field normal to $\boldsymbol{B}$, and $\boldsymbol{E}$ parallel to $\boldsymbol{B}$, the mobility tensor μ_{ik} is

$$\mu_{ik} = \frac{q}{2m} \begin{Vmatrix} l+r & -j(l-r) & 0 \\ j(l-r) & l+r & 0 \\ 0 & 0 & 2p \end{Vmatrix}. \tag{12.7}$$

It is convenient to refer to the three tensor components as the

parallel $\mu_{\parallel} = \frac{p\,q}{m}$,

transverse $\mu_{\mathrm{T}} = \frac{l+r}{2}\frac{q}{m}$

and perpendicular $\mu_{\perp} = j\frac{l-r}{2}\frac{q}{m}$

components.

The energy of the drift motion when it is at right angles to the magnetic field is obtained by replacing $e^{j\omega t}$ by $\cos\omega t$ and then squaring (12.5). It contains, of course, terms of double frequency.

$$u_d(t) = \frac{m\,a^2}{4} \frac{(\nu_c^2 + \omega^2 + \omega_b^2) + (\nu_c^2 - \omega^2 + \omega_b^2)\cos 2\omega t + 2\nu_c\omega\sin 2\omega t}{[\nu_c^2 + (\omega - \omega_b)^2]\,[\nu_c^2 + (\omega + \omega_b)^2]}. \tag{12.8}$$

Averaging over the time eliminates the periodic terms and gives for the average kinetic energy of the drift motion the simpler expression

$$u_d = \frac{m\,a^2}{8}\left[\frac{1}{\nu_c^2 + (\omega - \omega_b)^2} + \frac{1}{\nu_c^2 + (\omega + \omega_b)^2}\right]. \tag{12.9}$$

Simpler expressions yet are obtained in the special case of a constant field ($\omega = 0$) when the mobility tensor is

$$\mu_{ik} = \frac{q}{m} \begin{Vmatrix} \frac{\nu_c}{\nu_c^2 + \omega_b^2} & \frac{\omega_b}{\nu_c^2 + \omega_b^2} & 0 \\ \frac{-\omega_b}{\nu_c^2 + \omega_b^2} & \frac{\nu_c}{\nu_c^2 + \omega_b^2} & 0 \\ 0 & 0 & \frac{1}{\nu_c} \end{Vmatrix} \tag{12.10}$$

and the drift energy is

$$u_d = \frac{m\,a^2}{2(\nu_c^2 + \omega_b^2)}. \tag{12.11}$$

Equation (12.11) seems to differ from (12.9) by a factor 2 but this comes about because the acceleration $\boldsymbol{a}$ in (12.9) is defined in terms of the peak instead of the root-mean-square field.

Similarly, in the absence of a magnetic field the mobility becomes a scalar

$$\mu = \frac{q/m}{\nu_c + j\omega} \tag{12.12}$$

† Note that ω_b is negative for an ion but positive for an electron. Resonance occurs for a field rotating in the right hand direction around the magnetic field when electrons are responsible.

and the drift energy is

$$u_d = \frac{q^2 E_p^2/4m}{\nu_c^2 + \omega} = \frac{q^2 E_{\text{eff}}^2}{2m\nu_c^2}. \tag{12.13}$$

This last equation serves as the definition of an "effective" field E_{eff} which incorporates both the factor $\frac{1}{2}$ of an average cosine squared and the factor $\nu_c^2/(\nu_c^2+\omega^2)$ by which the drift energy in an alternating field is reduced from that in a constant field.

b) Diffusion.

13. The random flight. Diffusion currents are not obtained directly from the LANGEVIN equation because this equation treats the average motion of a single particle, and it has been shown that $\boldsymbol{v}_g = 0$ under a concentration gradient. A diffusion coefficient may, however, be defined in terms of the mean square displacement of a single particle through the theory of the *Random Flight*[1]. The procedure is as follows:

The motion of a free ion or electron in the absence of an electric field is along a helix, so that the displacement $\boldsymbol{r}$ is the chord of a helix. That is, $\boldsymbol{r}_\perp$ is the chord of a circle and $z = v_\parallel t$. In moving along the helix, collisions occur with the frequency ν_c so that the probability of having escaped a collision after a time t is $e^{-\nu_c t}$. One can then define a vector mean free path

$$\boldsymbol{l}_c = \int_0^\infty \boldsymbol{r}\, e^{-\nu_c t}\, \nu_c\, dt \tag{13.1}$$

and a particular mean square free path

$$l_{ik}^2 = \int_0^\infty r_i r_k\, e^{-\nu_c t}\, \nu_c\, dt. \tag{13.2}$$

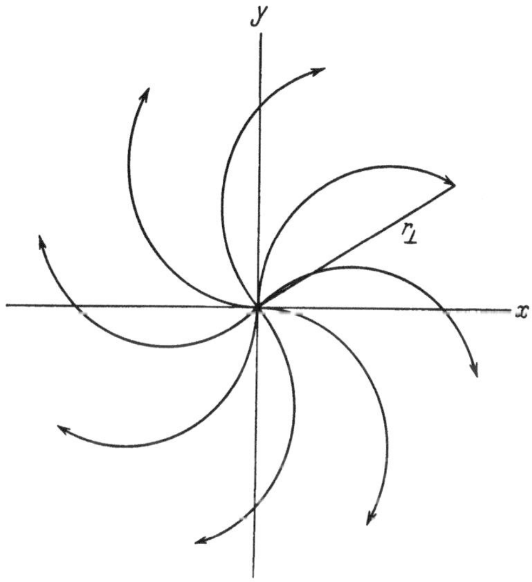

Fig. 7. Orbits considered in defining the mean square free path.

The scalar mean free path l_c will be defined in terms of particles moving along the magnetic field.

$$l_c = \frac{v}{\nu_c} \tag{13.3}$$

and a tensor mean square free path $\overline{l_{ik}^2}$ by averaging (13.2) over initial orientations assumed isotropic

$$\overline{l_{zz}^2} = 2\,\overline{\frac{v^2}{\nu_c^2}} = \frac{2}{3}\,\frac{v^2}{\nu_c^2} = \frac{2}{3}\, l_c^2, \tag{13.4}$$

$$\overline{l_{xx}^2} + \overline{l_{yy}^2} = \overline{\frac{2\, v_\perp^2}{\nu_c^2 + \omega_b^2}} = \frac{4}{3}\,\frac{v^2}{\nu_c^2 + \omega_b^2} = \frac{4}{3}\,\frac{\nu_c^2}{\nu_c^2 + \omega_b^2}\, l_c^2, \tag{13.5}$$

$$\overline{l_{xy}^2} = \overline{l_{yx}^2} = \overline{l_{zx}^2} = 0. \tag{13.6}$$

The tensor is, by its definition, symmetric and the components at right angles to the magnetic field are reduced in the same ratio as the effective field, $\nu_c^2/(\nu_c^2 + \omega_b^2)$,

[1] S. CHANDRASEKHAR: Rev. Mod. Phys. **15**, 1 (1943).

due to the curvature of the orbit. In the limit of high magnetic fields (or low pressures) $\overline{l_{xx}^2}+\overline{l_{yy}^2}=2r_b^2$. The mean square free path at right angles to the field is at most twice the orbit radius squared.

The mean square displacement tensor at the end of $N_c=\nu_c t$ collisions is

$$\overline{r_i r_k}=\nu_c t\,\overline{l_{ik}^2} \tag{13.7}$$

and the symmetric diffusion tensor may be defined[1] by

$$D_{ik}=\frac{\overline{r_i r_k}}{2t}=\frac{v^2}{3}\left\|\begin{matrix}\frac{\nu_c}{\nu_c^2+\omega_b^2} & 0 & 0\\ 0 & \frac{\nu_c}{\nu_c^2+\omega_b^2} & 0\\ 0 & 0 & \frac{1}{\nu}\end{matrix}\right\|. \tag{13.8}$$

It is frequently convenient to think of a magnetic field as increasing the effective size of the container in directions perpendicular to the field. Thus a cavity whose normal diffusion length Λ is given by

$$\frac{1}{\Lambda^2}=\frac{1}{\Lambda_x^2}+\frac{1}{\Lambda_y^2}+\frac{1}{\Lambda_z^2} \tag{13.9}$$

will, in the presence of a magnetic field along the z-axis, have an effective diffusion length Λ_b given by

$$\frac{1}{\Lambda_b^2}=\frac{\nu_c^2}{\nu_c^2+\omega_b^2}\left(\frac{1}{\Lambda_x^2}+\frac{1}{\Lambda_y^2}\right)+\frac{1}{\Lambda_z^2}. \tag{13.10}$$

The diffusion equation in such a cavity is then

$$\frac{\partial n}{\partial t}=\sum_1^3\frac{\partial^2}{\partial x_i^2}(D_{ii}n)=-\frac{Dn}{\Lambda_b^2} \tag{13.11}$$

where $D=v^2/3\nu_c$ is the field free diffusion coefficient and it has been assumed that n is distributed in the mode appropriate to Λ.

Equation (13.11) leads to the solution

$$n=n_0\,e^{-Dt/\Lambda_b^2} \tag{13.12}$$

whence one obtains the average number N_c of collisions suffered by an ion in diffusing out of the cavity

$$N_c=\frac{1}{n_0}\int_0^\infty n\,\nu_c\,dt=\frac{\nu_c\Lambda_b^2}{D}=\frac{3\Lambda_b^2}{l_c^2}. \tag{13.13}$$

This provides a useful random-flight definition of the diffusion length Λ_b.

14. The EINSTEIN relation. A better definition of the diffusion coefficient can be obtained by considering a pressure gradient as a force. Equation (8.1) shows that $nq\boldsymbol{E}$ and $-\operatorname{grad} p$ are equivalent in producing a drift velocity $\boldsymbol{v}_d$. Using the results of Sect. 12 it then follows that

$$v_{d,i}=-\frac{\mu_{ik}}{nq}\frac{\partial p}{\partial r_k}=-\frac{1}{n}\frac{\partial(D_{ik}n)}{\partial r_k} \tag{14.1}$$

whence

$$D_{ik}=\frac{p}{nq}\mu_{ik}=\frac{m}{q}\frac{v^2}{3}\mu_{ik} \tag{14.2}$$

[1] S. CHANDRASEKHAR: Rev. Mod. Phys. 15, 1 (1943), equation (120).

provided μ_{ik} is not a function of position, that is provided ν_c, and B do not depend on position. This is frequently a good assumption in electrical discharges and then (14.1), with D under the derivative sign, is a suitable definition of the diffusion coefficient. In the kinetic theory of gases, the gas pressure is frequently variable and then it is more suitable to put D outside the derivative sign. But then it must be remembered that grad p, and not grad n, is the driving force, so that the factor $\frac{1}{3} v^2$ must be retained under the derivative.

Equation (14.2) is known as the EINSTEIN relation, and by comparison with (12.10) gives the diffusion tensor

$$D = \frac{v^2}{3} \begin{Vmatrix} \frac{\nu_c}{\nu_c^2+\omega_b^2} & \frac{\omega_b}{\nu_c^2+\omega_b^2} & 0 \\ \frac{-\omega_b}{\nu_c^2+\omega_b^2} & \frac{\nu_c}{\nu_c^2+\omega_b^2} & 0 \\ 0 & 0 & \frac{1}{\nu_c} \end{Vmatrix} \tag{14.3}$$

which differs from (13.8) by the addition of antisymmetric terms. These terms are lost in the random flight definition of D_{ik} which is inherently symmetric.

Both this expression and (13.8) predict that diffusion transverse to a magnetic field decreases as $1/B^2$ at high fields. So far as the author is aware this result is verified experimentally. Claims to the contrary[1] refer to conditions in which space charge has a large effect and this situation must now be considered.

15. Ambipolar diffusion. In a plasma in which there are ions of both signs, the faster ions, generally electrons, tend to diffuse out of the plasma and leave an excess of positive charge, whose space charge field $\boldsymbol{E}_s$ then retards the electrons and accelerates the slower ions, thus linking their diffusion.

DEBYE and HÜCKEL[2] showed that the field of an ion in an electrolyte was reduced by a factor $\exp(-r/l_D)$ due to shielding by the ions of opposite sign, and the theory applies equally well to the positive ions in a plasma, where

$$l_D^2 = \frac{\varepsilon_0 D_-}{n_- e \mu_-}. \tag{15.1}$$

If the plasma is contained in a vessel of dimensions greater than l_D the ions will be able to hold their shielding electrons and the diffusion will be called ambipolar, whereas if the dimensions are much less than l_D the electrons will diffuse independently of the ions, and the diffusion will be called free. These limits correspond to high and low ion concentrations, respectively. One can also consider the two limits of high and low gas concentrations in which the vessel is large or small compared to the mean free path. ALLIS and ROSE[3] have considered the first situation, BERNSTEIN and HOLSTEIN the latter[4]. We shall consider the high pressure case as it is more closely related to the diffusion process. In this case the flow vectors $\boldsymbol{\Gamma}_+$ and $\boldsymbol{\Gamma}_-$ of the positive and negative ions are given by

$$\boldsymbol{\Gamma}_+ = -\operatorname{grad} D_+ n_+ + \boldsymbol{E}_s \mu_+ n_+, \tag{15.2}$$

$$\boldsymbol{\Gamma}_- = -\operatorname{grad} D_- n_- - \boldsymbol{E}_s \mu_- n_- \tag{15.3}$$

---

[1] A. GUTHRIE and R. K. WAKERLING: The Characteristics of Electrical Discharges in Magnetic Fields, p. 64. New York: McGraw Hill 1949. — ALBERT SIMON: Phys. Rev. **98**, 317 (1955).

[2] P. DEBYE and E. HÜCKEL: Phys. Z. **24**, 190 (1923).

[3] W. P. ALLIS and D. J. ROSE: Phys. Rev. **93**, 84 (1954).

[4] I. B. BERNSTEIN and T. HOLSTEIN: Phys. Rev. **94**, 1475 (1954).

and the space charge field $\boldsymbol{E}_s$ is given by POISSON's equation

$$\operatorname{div} \varepsilon_0 \boldsymbol{E}_s = (n_+ - n_-)\, e. \tag{15.4}$$

These equations are completed by the continuity equations

$$\frac{\partial n_+}{\partial t} = I - \operatorname{div} \boldsymbol{\Gamma}_+,$$

$$\frac{\partial n_-}{\partial t} = I - \operatorname{div} \boldsymbol{\Gamma}_-$$

where I is the ionization rate, and by boundary conditions. We shall consider in particular the stationary state in which

$$\operatorname{div} \boldsymbol{\Gamma}_+ = \operatorname{div} \boldsymbol{\Gamma}_-. \tag{15.5}$$

Equations (15.2) and (15.3) are quadratic due to the products $\boldsymbol{E}_s n$ and therefore methods of solution applicable to linear equations do not apply and machine computation is necessary to obtain good solutions. However, useful results can be obtained by making two simple, though inexact, assumptions: the assumption of congruence

$$\boldsymbol{\Gamma}_+ = \boldsymbol{\Gamma}_- \tag{15.6}$$

which is suggested, but not required, by (15.5) and the assumption of proportionality

$$\frac{\operatorname{grad} n_+}{n_+} = \frac{\operatorname{grad} n_-}{n_-} \tag{15.7}$$

which is true in the free limit because then the positive and negative ions satisfy independently the same diffusion equation, and is also true in the ambipolar limit because then the difference between n_+ and n_- is very small compared to either one of these. We shall assume that D_+ and D_- are constants so that they can be taken outside of the gradient operator. Equations (15.2) and (15.3) are then soveld by first eliminating the space charge field $\boldsymbol{E}_s$ between them, obtaining

$$\begin{aligned}(\mu_+ n_+ + \mu_- n_-)\, \boldsymbol{\Gamma} &= -\mu_+ n_+ \operatorname{grad}(D_- n_-) - \mu_- n_- \operatorname{grad} D_+ n_+) \\ &= -(\mu_+ D_- + \mu_- D_+)\, n_+ \operatorname{grad} n_-.\end{aligned}$$

Following SCHOTTKY[1] we define the ambipolar diffusion coefficient

$$D_a = \frac{\mu_+ D_- + \mu_- D_+}{\mu_+ + \mu_-} \tag{15.8}$$

and setting

$$\varrho = (n_+ - n_-)\, e, \tag{15.9}$$

$$\sigma = (\mu_+ n_+ + \mu_- n_-)\, e \tag{15.10}$$

for the space charge density and plasma conductivity, we obtain

$$\boldsymbol{\Gamma} = -D_a (1 + \mu_- \varrho/\sigma) \operatorname{grad} n_- = -D_s \operatorname{grad} n_- \tag{15.11}$$

where

$$D_s = D_a (1 + \mu_- \varrho/\sigma) \tag{15.12}$$

is an effective diffusion coefficient for electrons, and by (15.7) is not a function of position. D_a is obviously the minimum of D_s. The space charge field is obtained by substituting (15.11) in (15.3)

$$E_s = \frac{D_s - D_-}{\mu_-} \frac{\operatorname{grad} n_-}{n_-} \tag{15.13}$$

---

[1] W. SCHOTTKY: Phys. Z. **25**, 635 (1924).

and then POISSON's equation gives the charge density

$$\varrho = \varepsilon_0 \frac{D_- - D_s}{\mu_-} \left[\left(\frac{\operatorname{grad} n}{n} \right)^2 - \frac{\nabla^2 n}{n} \right]. \tag{15.14}$$

Defining the diffusion length Λ by

$$\Lambda^2 = - \frac{n}{\nabla^2 n} \tag{15.15}$$

and calling "center" that point in the plasma where grad $n = 0$, we find for the space charge at the center

$$\varrho_0 = \frac{D_- - D_s}{\mu_-} \frac{\varepsilon_0}{\Lambda^2}. \tag{15.16}$$

Obviously $D_s = D_-$ when $\varrho_0 = 0$. D_s decreases as the central space charge increases but the space charge must saturate when $D_s = D_a$. The relation between space charge and DEBYE length is found by substituting from (15.1) which gives, at the center,

$$\frac{n_+ - n_-}{n_-} = \frac{D_- - D_s}{D_-} \frac{l_D^2}{\Lambda^2}. \tag{15.17}$$

When the DEBYE length equals the diffusion length, the central ion density is close to twice the electron density. As the plasma conductivity increases, the DEBYE length decreases, and the central ratio n_+/n_- tends to one. The effective diffusion coefficient D_s can be expressed in terms of the central conductivity σ_0 by substituting from (15.16) in (15.12). This gives

$$D_s = D_a \frac{D_- + \Lambda^2 \sigma_0/\varepsilon_0}{D_a + \Lambda^2 \sigma_0/\varepsilon_0} \tag{15.18}$$

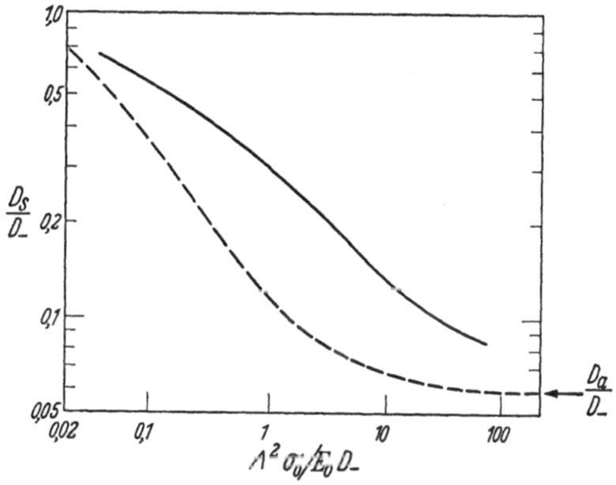

Fig. 8. Comparison of a machine solution of the combined diffusion equations with the approximate solution (15.18). The approximate solution is shown dotted.

which shows clearly the two limits, D_- and D_a, for the effective diffusion coefficient as the conductivity goes from 0 to ∞.

We must now discuss the errors of this theory. If the proportionality assumption held rigorously the charge density would be proportional to n and have a maximum at the center. However (15.14) indicates a minimum at the center. At low conductivities the error was made in obtaining the field $\boldsymbol{E}_s$ from a small difference of large quantities $\boldsymbol{\Gamma}_-$ and $-$grad $D_- n_-$, so (15.14) is false. At high conductivities one cannot conclude that ϱ is proportional to n in taking the small difference $n_+ - n_-$, so (15.14) is correct. In neither limit is the assumption of proportionality (15.7) itself very bad. Only in the transition is it bad, and the degree of error is shown in Fig. 8.

The assumption of congruence follows from (15.5) when the flow is irotational, as is true in the absence of a magnetic field. But in a magnetic field this is surely false, and hence the whole derivation, including the ambipolar coefficient (15.8) must be false. Observations of diffusion in high magnetic fields indicate coefficients of the order of the off-diagonal components of the diffusion tensor (14.3), and these must certainly enter an appropriate theory[1].

[1] A. SIMON: Phys. Rev. 98, 317 (1955).

In a manner similar to the case of a magnetic field, one can define an effective diffusion length to take account of space charge

$$\Lambda_s^2 = \Lambda^2 \frac{1 + \Lambda^2 \sigma_0/\varepsilon_0 D_a}{1 + \Lambda^2 \sigma_0/\varepsilon_0 D_-} \tag{15.19}$$

and say that, to the approximation treated here, the effect of space-charge on the electrons is to increase the effective size of the volume occupied by the plasma.

c) Energy and gain.

16. Elastic collisions only. In order to use the expressions derived in the preceding section, it is necessary to know the average energy of the ions. This is particularly true of the expression for the diffusion coefficient which contains v^2 explicitly, but may be true for the others when ν_c is a function of energy. We shall assume, however, in this section that ν_c and λ are not functions of energy in order to make the integrations easier. Nothing essential is changed if they are functions of energy and the corresponding integrals can generally be performed.

The power input to the ions is obtained by multiplying the LANGEVIN equation (11.1) by $m\boldsymbol{v}$, yielding

$$m\boldsymbol{v}\cdot\dot{\boldsymbol{v}} = m\boldsymbol{v}\cdot\boldsymbol{a} + m\boldsymbol{v}\cdot\boldsymbol{A}. \tag{16.1}$$

The term in ω_b has dropped out because magnetic fields do no work. Averaging over many collision times $1/\nu_c$, one has

$$\overline{m\boldsymbol{v}\cdot\boldsymbol{a}} = \boldsymbol{v}_g\cdot e\boldsymbol{E} = e\boldsymbol{v}_g\cdot\mu^{-1}\boldsymbol{v}_g = \nu_c m v_g^2 = 2\nu_c u_d$$

and making use of (11.3)

$$\frac{du}{dt} = \nu_c\left[2u_d + \lambda(\overline{U} - u)\right]. \tag{16.2}$$

The first term on the right is the gain of energy from the field and we note that the ions gain twice the drift energy per collision. The second term is the loss of energy to the gas where λ is the fractional loss per collision. The drift energy u_d is the rather complicated expression obtained in Sect. 12, Eq. (12.8). It can be written

$$u_d = \bar{u}_d\left[1 + \alpha\cos(2\omega t + \varphi)\right] \tag{16.3}$$

where

$$\bar{u}_d = \frac{e^2 E_p^2}{4m}\,\frac{\nu_c^2 + \omega^2 + \omega_b^2}{[\nu_c^2 + (\omega - \omega_b)^2]\,[\nu_c^2 + (\omega + \omega_b)^2]},$$

$$\alpha = \frac{\sqrt{[\nu_c^2 + (\omega - \omega_b)^2]\,[\nu_c^2 + (\omega + \omega_b)^2]}}{\nu_c^2 + \omega^2 + \omega_b^2}.$$

$$\tan\varphi = \frac{2\nu_c\omega}{-\nu_c^2\omega^2 + \omega_b^2}$$

$\alpha = 1$ when there is no magnetic field and in the d.c. case but is less than one when otherwise. Substituting in (16.2), and assuming that λ and ν_c are not functions of the energy, we obtain for the random energy

$$u = \overline{U} + \frac{2\bar{u}_d}{\lambda}\left[1 + \frac{\alpha\cos(2\omega t + \varphi')}{\sqrt{1 + (2\omega/\lambda\nu_c)^2}}\right] + A\,e^{-\lambda\nu_c t}. \tag{16.4}$$

The ion energy consists of three terms. The first is the mean gas energy as required for thermal equilibrium. The second depends on the field. At high frequencies $\omega/\nu_c \gg \lambda/2$ it is substantially constant with a slight alternating ripple.

At low frequencies $\omega/\nu_c \ll \lambda/2$, it follows E^2 and is completely modulated. The third term is a transient and gives the time constant $1/(\lambda\nu_c)$ at which the energy approaches its equilibrium value.

A difficulty might have been expected at low pressures when there are many oscillations per collision as the average taken in the LANGEVIN equation then covers many oscillations. However, in this case $2\omega/\lambda\nu_c$ is very large and the energy is not modulated so that it makes no difference.

In either the high frequency $(\omega \gg \lambda\nu_c/2)$ or the d.c. case the energy eventually reaches an equilibrium value

$$u_e = \overline{U} + 2\bar{u}_d/\lambda. \tag{16.5}$$

17. Energy gain and inelastic collisions. Whereas the energy transfers by an electron to recoil in an elastic collision, or to rotation and vibration of a molecule,

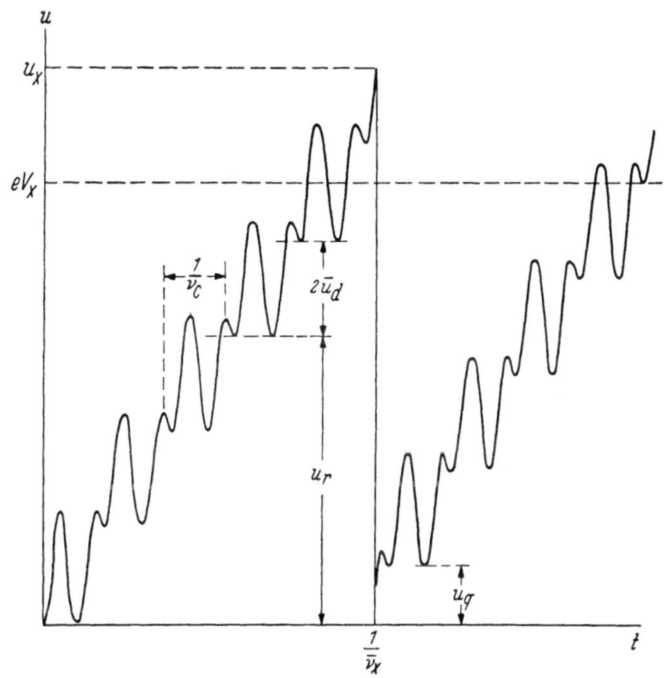

Fig. 9. Energy of the average electron as a function of time.

are always small compared to the electron's energy u, and therefore can be treated as in (16.2) as continuous, the losses to electronic excitation are quite otherwise. The probability of excitation of any level has a maximum not far above the excitation potential of that level, so that, no matter what the energy u of an electron may be, provided it is above the lowest excitation energy eV_x, the chances are it will excite a level not far below u and therefore be left with a small energy u_q. The life of an electron therefore consists of a succession of periods in which the energy rises from u_q to u_x, followed by sudden drops back to u_q. After N_x such rises the electron may ionize an atom, producing a new electron, or may disappear through diffusion to the walls or attachment to a molecule, the probabilities of ionization and disappearance being equal in the steady state. We shall consider here the case of disappearance by diffusion, and shall neglect the small energy losses to recoil, rotation, and vibration.

To render these ideas quantitative, consider a group of n electrons of the same energy u. Their energy is increasing at the rate

$$\frac{du}{dt} = 2\nu_c u_d. \tag{17.1}$$

At the same time the number n is being depleted by diffusion at the rate determined by Eq. (13.13)

$$\frac{dn}{dt} = -n\nu_c \frac{l_c^2}{3\Lambda^2}. \tag{17.2}$$

The time can be eliminated between these equations to give

$$\frac{dn}{n} = -\frac{l_c^2}{3\Lambda^2}\frac{du}{2u_d} = -\frac{\frac{4}{3}}{(eE_p\Lambda)^2}\frac{[\nu_c^2+(\omega-\omega_b)^2][\nu_c^2+(\omega+\omega_b)^2]}{\nu_c^2(\nu_c^2+\omega^2+\omega_b^2)}u\,du. \tag{17.3}$$

This expression can be integrated if ν_c is any known function of u. Assuming a constant collision frequency the integral is

$$\ln\frac{n_q}{n_x} = \frac{[\nu_c^2+(\omega-\omega_b)^2][\nu_c^2+(\omega+\omega_b)^2]}{\nu_c^2(\nu_c^2+\omega^2+\omega_b^2)}\frac{2}{3}\frac{u_x^2-u_q^2}{(eE_p\Lambda)^2}. \tag{17.4}$$

If every excitation were an ionization two electrons would start out at u_q for every one which reached u_x and one would have $n_q/n_x = 2$, but if N_x successive periods of gain are required to ionize

$$\frac{n_q}{n_x} = 1 + \frac{1}{N_x} \tag{17.5}$$

and if the number of excitations per ionization, N_x, is large, as it generally is, one has

$$\frac{(E_p\Lambda)^2}{2}\frac{\nu_c^2(\nu_c^2+\omega^2+\omega_b^2)}{[\nu_c^2+(\omega-\omega_b)^2][\nu_c^2+(\omega+\omega_b)^2]} = N_x\frac{u_x^2-u_q^2}{3e^2}. \tag{17.6}$$

This formula has a simple interpretation. $E_p\Lambda$ is in effect the voltage applied to the plasma. The factor $\frac{1}{2}$ reduces it to an r.m.s. value, and the factor containing ω is a power factor due to the phase of the electron's motion, but also contains a resonance denominator which enhances the field when the applied frequency equals the cyclotron frequency[1]. One can combine these to define an "effective" field E_{ef} by

$$E_{\text{ef}}^2 = \frac{E_q^2}{2}\frac{\nu_c^2(\nu_c^2+\omega^2+\omega_b^2)}{[\nu_c^2+(\omega^2+\omega_b)^2][\nu_c^2+(\omega+\omega_b)^2]} \tag{17.7}$$

or the corresponding integral expression if ν_c is a function of the energy. The power factor is 1 at higher pressures such that $\nu_c^2 \gg \omega^2+\omega_b^2$ no matter what the energy dependence of ν_c. The right hand side of (17.6) is a geometric mean between the total energy $N_x(u_x-u_q)/e$ (in electron volts) an electron must gain to ionize and a "mean" energy $(u_x+u_q)/3e$, which corresponds to the energy lost to diffusion.

Equation (17.6) can now be written

$$(E_{\text{ef}}\Lambda)^2 = N_x\frac{u_x^2-u_q^2}{3e^2}, \tag{17.8}$$

and determines the field at which ionization balances diffusion, a condition termed by LANGMUIR and TONKS[2] "plasma balance". It determines the high

[1] B. LAX, W. P. ALLIS and S. C. BROWN: J. Appl. Phys. 21, 1297 (1950).
[2] L. TONKS and I. LANGMUIR: Phys. Rev. 34, 876 (1929).

frequency breakdown field in a cavity if one uses the diffusion length Λ or Λ_b, or it determines the operating field in an active discharge, either a.c. or d.c., if one uses Λ_s. It is limited, however, to values of E/p sufficiently large that recoil losses are, in fact, negligible.

Present knowledge of the excitation functions of individual levels is not sufficient to determine u_q, but it is of the order of one or two volts and its square is negligible. The average excitation energy u_x is calculable if the total probability of excitation of all levels P_x is known. A reasonable assumption is that the ratio of inelastic to total collision probability varies linearly with the energy above $e V_x$

$$P_x = P_c h_x (u - e V_x). \tag{17.9}$$

Extending Eq. (17.2) to energies above excitation it becomes

$$\frac{dn}{dt} = -n v_c \left[\frac{l_c^2}{3\Lambda^2} + h_x (u - e V_x)\right] \tag{17.10}$$

and the diffusion term, $l_c^2/3\Lambda^2$, can generally be neglected compared to the excitation term in this range. Combining this with (17.1) and assuming u_d constant, yields for the number of electrons which reach the energy u

$$n = n_x \exp\left[-\frac{h_x}{4 u_d} (u - e V_x)^2\right] \tag{17.11}$$

and the average excitation energy is

$$u_x = \frac{1}{n_x} \int_{eV_x}^{\infty} u\, dn = e V_x + \sqrt{\frac{\pi u_d}{h_x}} = e V_x + c \frac{E}{p}. \tag{17.12}$$

The average electron will suffer an inelastic collision when it has exceeded the lowest excitation potential by an amount proportional to E/p. This overshoot is normally of the order of a volt, but becomes large at low pressures.

A similar calculation applied to the excitation number N_x yields

$$N_x = \frac{n_x}{n_i} = \frac{h_x}{n_i} 4 X^2 e^{X^2} \tag{17.13}$$

where

$$X = \sqrt{\frac{h_x}{u_d}} \frac{e V_i - e V_x}{2} \approx \frac{p}{E}$$

but as N_x is often very large the approximations of this section are not adequate and we must refer to (34.8).

The approximation made in this section of neglecting recoil losses (17.1) could be corrected, but it is useless to do so because of inherent errors in the method used. This is particularly obvious in calculating infrequent processes such as ionization. The equilibrium energy of an electron *without* inelastic collisions is

$$u_e = \overline{U} + 2\bar{u}_d/\lambda \tag{17.14}$$

and substitution of experimental values of E/p for high frequency breakdown at higher pressures often shows this to be below the ionization potential, and even below the excitation potential. In these cases the "average electron" never ionizes. It is only the exceptional electron which is able to ionize, and one must know the distribution of electrons about the average. The proper way to calculate this is by the BOLTZMANN equation, to which we now turn.

III. The BOLTZMANN equation.

a) The derivative terms.

18. Continuity in phase space. We now abandon the attempt to follow the motion of an individual electron and treat instead the changes in an assembly of electrons defined by giving the number d^6N of electrons in a volume d^3r of configuration space and d^3v of velocity space. This number defines a distribution function $f(\boldsymbol{r}, \boldsymbol{v}, t)$ by

$$d^6 N = f\, d^3 r\, d^3 v. \tag{18.1}$$

It also defines a particle current in phase space whose component in configuration space is given by

$$\gamma = \boldsymbol{v} f \tag{18.2}$$

and whose component in velocity space is determined by the applied forces

$$\boldsymbol{g} = (\boldsymbol{a} + \omega_b \times \boldsymbol{v})\, f. \tag{18.3}$$

For continuous forces one would have the continuity equation

$$\frac{\partial f}{\partial t} + \operatorname{div}_r \gamma + \operatorname{div}_v \boldsymbol{g} = 0$$

but collisions have the effect of removing an electron from one element of velocity space and replacing it in another, or even creating a new one in the case of ionization. One provides for this by writing

$$\frac{\partial f}{\partial t} + \operatorname{div}_r \gamma + \operatorname{div}_v \boldsymbol{g} = B(t) \tag{18.4}$$

where B is an integral to be discussed in the next section. This is the BOLTZMANN equation.

19. Expansion in spherical harmonics. Except for the singular case of a constant mean free time it is always necessary to expand the function f, and there are two convenient expansions. When the system considered is close to thermal equilibrium, departures from equilibrium being caused by some agent α, the distribution function can be expanded in powers of α

$$f = f^0 + \alpha f^1 + \alpha^2 f^2 \ldots. \tag{19.1}$$

This was the method by which ENSKOG solved BOLTZMANN's equation and is highly developed by CHAPMAN and COWLING[1]. It does not converge well for charged particles in an electric field. In all other cases the distribution function is expanded in spherical harmonics in velocity space

$$f(\boldsymbol{r}, \boldsymbol{v}) = \sum_l f^l(\boldsymbol{r}, \boldsymbol{v})\, P_l(\cos\vartheta) = f^0 + f^1 \cos\vartheta + \cdots \tag{19.2}$$

where, if necessary, one may also use the associated LEGENDRE functions P_l^m. We shall assume that this series converges sufficiently rapidly for the first two terms to suffice, but we shall not assume that f^0 is MAXWELLian, as this is generally far from true when any appreciable electric field is present.

Because of the orthogonality of the spherical harmonics one has, for any scalar function $s(v)$,

$$\oint s\, f\, d^2\Omega = 4\pi\, s\, f^0 \tag{19.3}$$

[1] S. CHAPMAN and T. G. COWLING: The Mathematical Theory of Non-uniform Gases, Cambridge University Press 1939.

and for the vector $\boldsymbol{v}$

$$\oint \boldsymbol{v} f \, d^2\Omega = \frac{4\pi}{3} v \boldsymbol{f}^1 \tag{19.4}$$

where the vector $\boldsymbol{f}^1$ is directed along the polar axis of P_1.

The two scalars of greatest interest are the electron concentration

$$n = \int_0^\infty f^0 \, 4\pi v^2 \, dv \tag{19.5}$$

and the energy density

$$n\bar{u} = \int_0^\infty u f^0 \, 4\pi v^2 \, dv \tag{19.6}$$

and there are two corresponding vector integrals: the particle flow in configuration space is

$$\boldsymbol{\Gamma} = n \boldsymbol{v}_d = \int \boldsymbol{f}^1 \frac{4\pi}{3} v^3 \, dv \tag{19.7}$$

and the total radial flow in velocity space is

$$G = \oint \boldsymbol{g} \cdot \boldsymbol{v} \, v \, d^2\Omega = \frac{4\pi}{3} v^2 \boldsymbol{a} \cdot \boldsymbol{f}^1. \tag{19.8}$$

The quantity G represents the rate of flow of electrons across any energy value in the direction of increasing energy. It deserves a convenient name and we shall call it the "gain".

We shall now introduce the expansion (19.2) in the BOLTZMANN equation and expand each term separately. Three relations between spherical harmonics are useful

$$(2l+1) P_l \cos\vartheta = (l+1) P_{l+1} + l P_{l-1}, \tag{19.9}$$

$$(2l+1) \sin^2\vartheta \frac{\partial P_l}{\partial \cos\vartheta} = l(l+1)(P_{l-1} - P_{l+1}), \tag{19.10}$$

$$(2l+1) \oint P_l P_m \, d^2\Omega = 4\pi \delta_{lm}. \tag{19.11}$$

20. The gradient term. Taking the polar axis along the direction of grad n and assuming that there is rotational symmetry about that axis so that the associated LEGENDRE polynomials are not required,

$$\operatorname{div}(\boldsymbol{v} f^l P_l) = \boldsymbol{v} \cdot (\operatorname{grad} f^l) P_l = v \frac{\partial f^l}{\partial z} \cos\vartheta \, P_l = v \frac{\partial f^l}{\partial z} \frac{l P_{l-1} + (l+1) P_{l+1}}{2l+1} \tag{20.1}$$

so that

$$\left.\begin{aligned} \operatorname{div}(\boldsymbol{v} f) &= v \sum_0^\infty \frac{\partial}{\partial z} \left[\frac{l}{2l-1} f^{l-1} + \frac{l+1}{2l+3} f^{l+1} \right] P_l \\ &= \frac{v}{3} \frac{\partial f^1}{\partial z} + v \cos\vartheta \frac{\partial}{\partial z} \left[f^0 + \frac{2}{5} f^2 \right] + \cdots \\ &= \frac{v}{3} \operatorname{div} \boldsymbol{f}^1 + \boldsymbol{v} \cdot \operatorname{grad} \left[f^0 + \frac{2}{5} f^2 \right] + \cdots. \end{aligned}\right\} \tag{20.2}$$

In the last line the scalar function f^1 has been replaced by the vector $\boldsymbol{f}^1$, and this notation removes the initial restriction on the direction of the polar axis of P_1.

21. Electric terms. In this section we take the polar axis of the spherical harmonics along the electric field $\boldsymbol{E}$. Using the expression for partial derivatives

$$\left(\frac{\partial}{\partial v_z}\right)_{v_x v_y} = \cos\vartheta \left(\frac{\partial}{\partial v}\right)_{\vartheta,\varphi} + \frac{\sin^2\vartheta}{v}\left(\frac{\partial}{\partial \cos\vartheta}\right)_{v,\varphi} \tag{21.1}$$

we find

$$\mathrm{div}_v\, \boldsymbol{a} f^l P_l = \boldsymbol{a}\cdot \mathrm{grad}_v f^l P_l = a\cos\vartheta \frac{\partial f^l}{\partial v} P_l + \frac{a f^l}{v}\sin^2\vartheta \frac{\partial P_l}{\partial \cos\vartheta}, \tag{21.2}$$

$$\left.\begin{aligned}\mathrm{div}_v \boldsymbol{a} f &= a\sum\left[\frac{l}{2l-1} v^{l-1}\frac{d}{dv}\frac{f^{l-1}}{v^{l-1}} + \frac{l+1}{2l+3}\frac{1}{v^{l+2}}\frac{d}{dv} v^{l+2} f^{l+1}\right]P_l\\ &= a\left[\frac{1}{3v^2}\frac{d}{dv} v^2 f^1 + \left(\frac{df^0}{dv} + \frac{2}{5}\frac{1}{v^3}\frac{d}{dv} v^3 f^2\right)\cos\vartheta + \cdots +\right.\\ &= \frac{1}{3v^2}\frac{d}{dv} v^2 \boldsymbol{a}\cdot\boldsymbol{f}^1 + \boldsymbol{a}\cdot\boldsymbol{v}\left(\frac{1}{v}\frac{df^0}{dv} + \frac{2}{5v^4}\frac{d}{dv} v^3 f^2\right) + \cdots\end{aligned}\right\} \tag{21.3}$$

where we have again introduced the vector notation to eliminate the requirement on the polar axis P_1.

22. Magnetic terms. As

$$\mathrm{div}_v\,(\boldsymbol{\omega}_b \times \boldsymbol{v}) = 0 \tag{22.1}$$

we have

$$\mathrm{div}_v\,(\boldsymbol{\omega}_b \times \boldsymbol{v} f) = (\boldsymbol{\omega}_b \times \boldsymbol{v})\cdot \mathrm{grad}_v f = \boldsymbol{v}\cdot(\boldsymbol{\omega}_b \times \mathrm{grad}_v f). \tag{22.2}$$

The expansion involves the associated LEGENDRE functions and the general term will not be given. The first two terms yield

$$\mathrm{div}_v\,(\boldsymbol{\omega}_b \times \boldsymbol{v} f) = -\boldsymbol{v}\cdot(\boldsymbol{\omega}_b \times \boldsymbol{f}^1)/v. \tag{22.3}$$

23. Expansion in FOURIER series. We shall assume an alternating electric field, $\boldsymbol{E} \sim e^{j\omega t}$, as we can always obtain the direct current formulae by setting $\omega = 0$ and replacing root-mean-square values by the d.c. value. The distribution function will then in general be a FOURIER series in ωt:

$$f = \sum\sum_{m} f^l_m P_l e^{mj\omega t} \tag{23.1}$$

and

$$\frac{\partial f}{\partial t} = \sum_l\sum_m m j\omega f^l_m P_l e^{mj\omega t}. \tag{23.2}$$

The exponential notation is very convenient for all linear equation. Unfortunately, we shall have to take products such as $\boldsymbol{a} f^l$ and the meaning of such a product is the product of the real parts and not the real part of the product. Performing this operation and then reintroducing the exponential notation, we have

$$2a f^l = a\left[f^l_{1r} + (2f^l_0 + f^l_2)\, e^{j\omega t} + \sum_{m=2}^{\infty}(f^l_{m-1} + f^l_{m+1})\, e^{mj\omega t}\right] \tag{23.3}$$

where f^l_{1r} indicates the real part of f^l_1.

24. The component equations. Substituting from Sect. 20 to 23 into (18.4) and decomposing the BOLTZMANN integral term B as well, one obtains an infinite set of equations of which the zero and first order ones are given below:

$$B^0_0 = \frac{v}{3}\,\mathrm{div}\,\boldsymbol{f}^1_0 + \frac{1}{6v^2}\frac{d}{dv}\left(v^2 \boldsymbol{a}\cdot\boldsymbol{f}^1_{1r}\right), \tag{24.1}$$

$$B^0_1 = j\omega f^0_1 + \frac{v}{3}\,\mathrm{div}\,\boldsymbol{f}^1_1 + \frac{1}{6v^2}\frac{d}{dv}\left[v^2 \boldsymbol{a}\cdot(2\boldsymbol{f}^1_0 + \boldsymbol{f}^1_2)\right], \tag{24.2}$$

$$\boldsymbol{B}_0^1 = v \operatorname{grad}\left(f_0^0 + \frac{2}{5} f_0^2\right) + \boldsymbol{a}\left[\frac{1}{2} \frac{d f_{1r}^0}{dv} + \frac{1}{5v^3} \frac{d}{dv}(v^3 f_{1r}^2)\right] - \boldsymbol{\omega}_b \times \boldsymbol{f}_0^1, \tag{24.3}$$

$$\left.\begin{aligned} \boldsymbol{B}_1^1 &= j\,\omega \boldsymbol{f}_1^1 + v \operatorname{grad}\left(f_1^0 + \frac{2}{5} f_1^2\right) + \\ &+ \boldsymbol{a}\left[\frac{1}{2} \frac{d}{dv}(2 f_0^0 + f_2^0) + \frac{1}{5v^3} \frac{d}{dv} v^3 (2 f_0^2 + f_2^2)\right] - \boldsymbol{\omega}_b \times \boldsymbol{f}_1^1. \end{aligned}\right\} \tag{24.4}$$

In the case of electrons (but not ions) the spherical harmonic expansion converges well and the terms in f^2 are negligible. The FOURIER series also converges well at high frequencies ($\omega/\nu_c > \lambda/2$) (Sect. 16). The transition to low frequencies where the second harmonic terms become large has been studied by MARGENAU[1] and this case will not be considered here. We therefore simplify the above set of equations to

$$B_0^0 = \frac{v}{3} \operatorname{div} \boldsymbol{f}_0^1 + \frac{1}{6v^2} \frac{d}{dv}(v^2 \boldsymbol{a} \cdot \boldsymbol{f}_{1r}^1), \tag{24.5}$$

$$B_1^0 = j\,\omega f_1^0 + \frac{v}{3} \operatorname{div} \boldsymbol{f}_1^1 + \frac{1}{3v^2} \frac{d}{dv}(v^2 \boldsymbol{a} \cdot \boldsymbol{f}_0^1). \tag{24.6}$$

$$\boldsymbol{B}_0^1 = v \operatorname{grad} f_0^0 + \frac{\boldsymbol{a}}{2} \frac{d f_{1r}^0}{dv} - \boldsymbol{\omega}_b \times \boldsymbol{f}_0^1, \tag{24.7}$$

$$\boldsymbol{B}_1^1 = j\,\omega \boldsymbol{f}_1^1 + v \operatorname{grad} f_1^0 + \boldsymbol{a} \frac{d f_0^0}{dv} - \boldsymbol{\omega}_b \times \boldsymbol{f}_1^1. \tag{24.8}$$

Some insight into these equations is gained by neglecting the collision terms and then multiplying (24.5) and (24.6) by $4\pi v^2 dv$ and integrating. Equation (24.5) gives

$$\operatorname{div} \boldsymbol{\Gamma}(u) = \operatorname{div} \int_v^\infty \boldsymbol{f}_0^1 \frac{4\pi}{3} v^3\, dv = \frac{2\pi}{3} v^2 \boldsymbol{a} \cdot \boldsymbol{f}_{1r}^1 = G. \tag{24.9}$$

From (19.7) we see that the left side is the divergence of the flow of particles whose velocity is above v. On the right we have, from (19.8) and noting that a factor $\frac{1}{2}$ comes from the average of $\cos^2 \omega t$, the gain through the sphere radius v. Equation (24.9) therefore expresses the conservation of particles whose speed is above v. Extending the integral down to zero yields

$$\operatorname{div} \boldsymbol{\Gamma}_0 = 0 \tag{24.10}$$

and the similar integration of (24.6) yields

$$\operatorname{div} \boldsymbol{\Gamma}_1 = -j\,\omega\, n_1 \tag{24.11}$$

The divergence of the alternating flow $\boldsymbol{\Gamma}_1$ produces an alternating electron density n_1. The alternating flow $\boldsymbol{\Gamma}_1$ is driven by $\boldsymbol{a} n_0$ and therefore n_1 exists in virtue of a divergence in $\boldsymbol{a} n_0$. As $\boldsymbol{a}$ is divergenceless, n_1 is produced by the interaction of an alternating electric field and a concentration gradient in the same direction. Such interactions may well occur, but we shall study situations designed to minimize them. It will be assumed that n_1 and f_1^0 may be neglected. The four equations then reduce to three simpler equations

$$B_0^0 = \frac{v}{3} \operatorname{div} \boldsymbol{f}_0^1 + \frac{1}{6v^2} \frac{d}{dv}(v^2 \boldsymbol{a} \cdot \boldsymbol{f}_{1r}^1), \tag{24.12}$$

$$\boldsymbol{B}_0^1 = v \operatorname{grad} f_0^0 + - \boldsymbol{\omega}_b \times \boldsymbol{f}_0^1, \tag{24.13}$$

$$\boldsymbol{B}_1^1 = j\,\omega \boldsymbol{f}_1^1 + \boldsymbol{a} \frac{d f_0^0}{dv} - \boldsymbol{\omega}_b \times \boldsymbol{f}_1^1. \tag{24.14}$$

---

[1] H. MARGENAU: Phys. Rev. **73**, 309 (1948).

We must now evaluate the collision integral B, and its components B_m^l defined by

$$B = \sum B_m^l \, P_l \, e^{m i \omega t}. \tag{24.15}$$

b) The collision integral.

25. Geometry of a collision. Consider a collision between an electron, denoted by lower case letters, and a neutral molecule, denoted by capital letters. In a collision, the positions of these particles do not change appreciably but their velocities change and we shall denote by primes the velocities of particles about to be scattered into $d^3v\, d^3V$ about the velocities $\boldsymbol{v}$ and $\boldsymbol{V}$. We do not consider the changing velocities during the collision but only the velocities before and after the particles are within range of their interaction forces. During the collision the velocity of the center of gravity, $\boldsymbol{v}_g$, remains fixed and if the collision was elastic, the relative velocity c has the same magnitude before and after but has changed in direction by an angle χ_g called the scattering angle in the center of gravity system. The center of gravity divides the relative velocity inversely as the masses so that the velocities of electron and molecule remain on two spheres in velocity space centered on the velocity of the center of gravity and of radii $Mc/(M+m)$ and $mc/(M+m)$. The radius of the molecule's sphere is very small compared with that of the electron so that it can almost be considered as a point. If the molecule was at rest before the collision, the origin of velocity coordinates should be taken at $\boldsymbol{V}'$ and the electronic velocities before and after the collision are given by the lines $V'v'$ and $V'v$ of Fig. 10. In the triangle, $\boldsymbol{V}', \boldsymbol{v}_g, \boldsymbol{v}$, we have

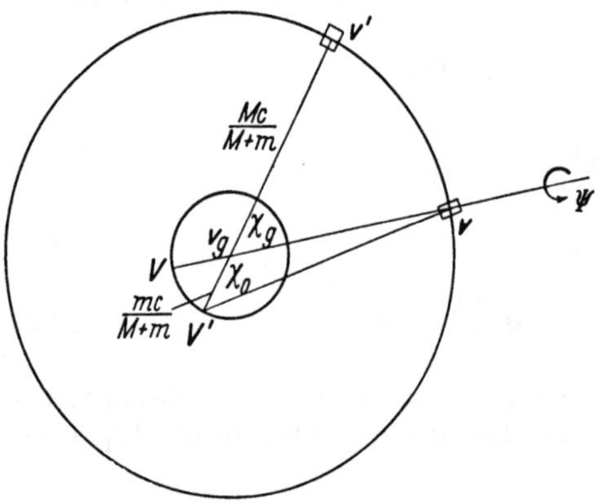

Fig. 10. Geometry of a collision in velocity space. The circles represent spheres. If the molecule is at rest before the collision the origin is at V'.

$$(\boldsymbol{v} - \boldsymbol{V}')^2 = c^2 - \frac{2Mm}{(M+m)^2}\, c^2 (1 - \cos\chi_g) \tag{25.1}$$

or if $V' = 0$

$$v'^2 - v^2 = \frac{2Mm}{(M+m)^2}\, v'^2 (1 - \cos\chi_g). \tag{25.2}$$

Equation (25.2) gives the energy loss by an electron in striking a molecule at rest. This energy goes into the recoil energy of the molecule. For electrons it is small and we can write it in the form

$$-\frac{\Delta v}{v} = \frac{Mm}{(M+m)^2} (1 - \cos\chi_g). \tag{25.3}$$

Another useful relation from the triangle $\boldsymbol{V}', \boldsymbol{v}_g, \boldsymbol{v}$ gives the scattering angle χ_0 in the laboratory system in terms of the scattering angle χ_g in the center of gravity system. It is

$$\sin^2\chi_0 = \frac{M^2 \sin^2\chi_g}{M^2 + m^2 + 2Mm\cos\chi_g}. \tag{25.4}$$

The scattering function $\sigma(\chi)$ is deduced from the force law as a function of χ_g and c, and is measured experimentally in terms of χ_0 and v', and it is always necessary to reduce one to the other. The dependence of σ on the azimuthal angle ψ

is never known and it will be assumed that σ does not depend on ψ. The combination $c\sigma$ always enters the equations and we shall use the symbol $\varrho(c) = c\sigma$ for it. ϱ has the dimensions of a volume per second per ster-radian.

26. Collision integrals. The rate of change of the number of particles in a velocity volume element d^3v is given by taking the difference between the number of particles scattered into that volume element and the number scattered out

$$B\,d^3v = \int F(\boldsymbol{V}')\,f(\boldsymbol{v}')\,\varrho\,d^2\Omega\,d^3V'\,d^3v' - \int F(\boldsymbol{V})\,f(\boldsymbol{v})\,\varrho\,d^2\Omega\,d^3V\,d^3v. \tag{26.1}$$

ϱ is the same in both these integrals and $d^2\Omega$ is the solid angle of the scattering sphere. $F(\boldsymbol{V}')\,d^3V'$ and $f(\boldsymbol{v}')\,d^3v'$ are the numbers of particles in position to scatter into the volume elements d^3V and d^3v, and the integrals are to be taken over all scattering angles and all molecular velocities. By LIOUVILLE's theorem

$$d^3v'\,d^3V' = d^3v\,d^3V \tag{26.2}$$

so that the two integrals may be combined into one

$$B = \int \left[F(\boldsymbol{V}')\,f(\boldsymbol{v}') - F(\boldsymbol{V})\,f(\boldsymbol{v})\right]\varrho\,d^2\Omega\,d^3V \tag{26.3}$$

which is the conventional way of writing BOLTZMANN's collision integral. Thit form is, however, rarely convenient as $F(V')$ contains V' and not V and it is nos easy to integrate it with respect to V. It is more convenient to keep dV' as in (26.1) and to convert d^3v' into d^3v. Referring to Fig. 10, we wish to determine the volume element d^3v' such that for given $\boldsymbol{V}'$ and χ_g, the final velocity after the collision ends up in d^3v. As all angles are held constant and $\boldsymbol{V}'$ is held fixed, the volume element d^3v' must be exactly similar in shape to d^3v, but all linear dimensions increased in the ratio $|(\boldsymbol{v}'-\boldsymbol{V}')/(\boldsymbol{v}-\boldsymbol{V}')|$, so that

$$d^3v' = \left(\frac{|\boldsymbol{v}'-\boldsymbol{V}'|}{|\boldsymbol{v}-\boldsymbol{V}'|}\right)^3 d^3v. \tag{26.4}$$

Substituting this in (26.1), we obtain

$$B = \int F(V')\,f(v')\left(\frac{|\boldsymbol{v}'-\boldsymbol{V}'|}{|\boldsymbol{v}-\boldsymbol{V}'|}\right)^3 \varrho\,d^2\Omega\,d^3V' - \int F(V)\,f(v)\,\varrho\,d^2\Omega\,d^3V. \tag{26.5}$$

27. Expansion in spherical harmonics. We now substitute for $f(\boldsymbol{v})$ its expansion in spherical harmonics and for $f(\boldsymbol{v}')$ the corresponding series in terms of ϑ'. Using the addition theorem for spherical harmonics, we have

$$f(\boldsymbol{v}') = \sum f^l\left[P_l(\vartheta)\,P_l(\chi_0) + 2\sum_1^l \frac{(l-m)!}{(l+m)!}\,P_l^m(\vartheta)\,P_l^m(\chi_0)\cos m(\varphi-\psi)\right]. \tag{27.1}$$

As ϱ has been assumed independent of the azimuthal angle ψ, the term in $\cos m(\vartheta-\psi)$ goes out on integrating, so that the BOLTZMANN integral for the coefficient of $P_l(\vartheta)$ becomes

$$B^l = \int\left(\frac{|\boldsymbol{v}'-\boldsymbol{V}'|}{|\boldsymbol{v}-\boldsymbol{V}'|}\right)^3 F(\boldsymbol{V}')\,f^l(v')\,P_l(\chi_0)\,\varrho\,d^2\Omega\,d^3V' - \int F(\boldsymbol{V})\,f^l(v)\,\varrho\,d^2\Omega\,d^3V. \tag{27.2}$$

So far the equations have been perfectly general, but we must now make assumptions which in practice restrict their application to electrons, or at most to light ions in a heavy gas. We will assume that the mass ratio m/M is zero, so that $v' = v$, and that $F(\boldsymbol{V})/n_g$ is a δ-function. Equation (27.2) then reduces to

$$l = 0 \qquad B^0 = 0 \tag{27.3}$$

$$l \neq 0 \qquad B^l = -f^l(v)\,n_g \int \varrho(v,\chi_0)\left[1 - P_l(\chi_0)\right]d^2\Omega = -\nu_{cl}\,f^l. \tag{27.4}$$

The "collision frequencies" ν_{cl} introduced above are defined in the laboratory reference system. Introducing an arbitrary definition for ν_{c0}, the first three collision frequencies are

$$\nu_{c0} = \int n_g \varrho \, d^2\Omega, \tag{27.5}$$

$$\nu_{c1} = \int n_g \varrho (1 - \cos\chi_0) \, d^2\Omega, \tag{27.6}$$

$$\nu_{c2} = \int n_g \varrho \tfrac{3}{2} \sin^2\chi_0 \, d^2\Omega. \tag{27.7}$$

ν_{c0} is the total collision frequency. The integral is generally improper as ϱ has a singularity in the forward direction. The apertures of experimental apparatus prevent observations from covering this singularity so that the quantities tabulated as experimental "probabilities of collision" are expressed in terms of (27.5) with an incomplete integral

$$P_c = \frac{\nu'_{c0}}{v p_0} \tag{27.8}$$

where p_0 is the "pressure" reduced to 0°C.

ν_{c1} is the collision frequency for "momentum transfer" and is of such importance in transport theory that the simpler symbol ν_c is used for it. It is a weighted collision frequency in which the backward scattering counts double, right angle scattering has the weight one, and forward scattering is not counted.

ν_{c2} might well be called the "scattering" frequency as right angle scattering has the weight $\frac{3}{2}$ and neither forward nor backward scattering is counted.

28. Recoil. The conclusion that $B^0 = 0$ means that collisions do not change the energy distribution, and follows from the assumption that the molecules were elastic and infinitely heavy. In fact it may take many thousand collisions to change the energy of an electron appreciably whereas exactly one "collision" defined by the frequency ν_{c1} will cancel its momentum. Nevertheless many thousand collisions do occur and we must introduce the higher order terms into (27.2) for $l=0$.

We first relax the restriction to infinitely heavy molecules so as to allow the molecule to recoil under electronic impact ($v' > v$) but continue to neglect thermal motion ($V'=0$). Then $c=v'$ and

$$v'^2 - v^2 = v'^2 \frac{2Mm}{(M+m)^2} (1 - \cos\chi_g) = \mu v'^2. \tag{28.1}$$

Let us expand in terms of μ which is assumed small

$$v'^3 \varrho(\boldsymbol{v}') f(\boldsymbol{v}') = v^3 \varrho f(\boldsymbol{v}) + \mu v^2 \frac{d(v^3 \varrho f)}{d(v^2)}. \tag{28.2}$$

Integrating the expression (27.2) for B^0 with respect to d^3V' for $T_g=0$ gives

$$\left.\begin{aligned} B^0 &= \int \left[\frac{v'^3}{v^3} n_g \varrho(\boldsymbol{v}') f^0(v') - n_g \varrho f^0(v)\right] d^2\Omega \\ &= \int \frac{\mu}{v} n_g \frac{d v^3 \varrho f^0}{d(v^2)} d^2\Omega = \frac{2Mm}{(M+m)^2} \frac{1}{v} \frac{d}{d(v^2)} (v^3 \nu_{g1} f^0) \end{aligned}\right\} \tag{28.3}$$

where ν_{g1} is the same as ν_{c1} of (27.6) but in the center of mass instead of the laboratory system. The relation between these two depends on how the cross-section ϱ

depends on χ_g and is related to the "persistence of velocity" factor[1]. It is given in first order by

$$\nu_c = \frac{M}{M+m}\,\nu_{g1} \tag{28.4}$$

so that we can write (28.3) as

$$B^0 = \frac{2m}{M+m}\,\frac{1}{v}\,\frac{d}{d(v^2)}\,(v^3\nu_c f^0)\,. \tag{28.5}$$

Next we take account of the MAXWELL distribution of the molecules $F(\boldsymbol{V}')$, which was replaced by a delta function in (28.3), and this will be done in a somewhat qualitative way. In integrating over d^3V' the point $\boldsymbol{V}'$ of Fig. 10 moves over a MAXWELL distribution centered about $V'=0$. For fixed v and scattering angle χ_g, this menas that $\boldsymbol{v}'$ moves about $\boldsymbol{v}_0'$, its value for $V'=0$, by amounts of the order $\boldsymbol{V}'$.

$$\boldsymbol{v}' = \boldsymbol{v}_0' + O(\boldsymbol{V}') \tag{28.6}$$

and

$$f(\boldsymbol{v}') = f(\boldsymbol{v}_0') + O(V'^2)\,\frac{\partial f}{\partial (v'^2)}\,. \tag{28.7}$$

This correction to $f(\boldsymbol{v})$ must enter (28.5) so that it becomes

$$B^0 = \frac{2m}{M+m}\,\frac{1}{v}\,\frac{d}{d(v^2)}\left[v^3\nu_c\left(f^0 + O(V'^2)\,\frac{d f^0}{d(v^2)}\right)\right]. \tag{28.8}$$

But we know that if the electrons have a MAXWELL distribution at the same temperature as that of the gas, collisions will not change this distribution and, therefore, (28.8) must give zero if f is a MAXWELL distribution. This determines the factor $O(V'^2)$ to be equal to $2kT_g/m$ and we have

$$B^0 = \frac{2m}{M+m}\,\frac{1}{v}\,\frac{d}{d(v^2)}\left[v^3\nu_c\left(f^0 + \frac{2kT_g}{m}\,\frac{d f^0}{d(v^2)}\right)\right] \tag{28.9}$$

for the rate of change of the energy distribution of electrons of mass m due to elastic collisions with molecules of finite mass M having a temperature T_g.

29. Inelastic collisions. If the gas is molecular, the electrons colliding with the molecules can excite rotational and vibrational states in them. As the spacings of the rotational and vibrational energy levels are generally small compared to the mean energy of the electrons, and therefore the losses are in small amounts, of the same order as the recoil losses, these losses can be treated in the same way as the recoil losses and incorporated in (28.9). However, the excitation probabilities of rotation and vibration are not well known either theoretically[2] or experimentally and therefore they can only be taken into account by introducing an arbitrary factor λ in place of the mass ratios $m/(M+m)$ where $2m/(M+m)\leq\lambda\leq 1$. ν_{c1} and λ were the two parameters introduced in the LANGEVIN equation in Sect. 11.

In the excitation of an electronic level, of excitation potential V_x, the electron loses the excitation energy eV_x and remains with very little energy left, generally of the order of one electron volt.

$$\tfrac{1}{2}m v^2 = \tfrac{1}{2}m v' - e V_x\,. \tag{29.1}$$

Introducing this in the collision integral gives

$$\int\left(\frac{v'}{v}\right)^3 \sum_x \varrho_x f(v')\,d^2\Omega = \varphi(v) = \xi(v)\,f(v) \tag{29.2}$$

[1] J. JEANS: The Dynamical Theory of Gases. 1925.

[2] E. GERJUOY and S. STEIN: Phys. Rev. 97, 1671 (1955).

where $\varphi(v)$ is the energy distribution of the electrons after the collision, and it is convenient to express this function as a product $\xi(v) f(v)$, which defines the function $\xi(v)$. Because the excitation probabilities are large immediately above the excitation potential and also because the distribution function is a rapidly decreasing function of v', the functions φ or ξ are both appreciable only for small velocities corresponding to the energies of the order of one electron volt. The function $\xi(v)$ corresponds to the excitation of all excitation levels, including ionization, and is therefore a sum of partial ξ's and has no simple relation to the velocity v' of the electron before the collision. For our purpose it will generally be sufficient to assume that the function ξ is a delta-function because, as noted in (17.8) it is the square of the remaining energy which enters the final formulae.

If attachment or recombination is occurring this can also be included in $\xi(v)$. Introducing ξ and also the total excitation frequency $\nu_x(v)$ gives the inelastic term

$$B^0 = [\xi(v) - \nu_x(v)] f^0. \tag{29.3}$$

The integral of $\xi(v)$ is related to the inelastic, ionization, and attachment frequencies, ν_x, ν_i, and ν_a because every excitation produces one slow electron, or two if it is an ionization, and attachment removes one. Therefore we have

$$\int \xi f^0 4\pi v^2 dv = \int (\nu_x + \nu_i - \nu_a) f^0 4\pi v^2 dv = n(\bar{\nu}_x + \bar{\nu}_i - \bar{\nu}_a). \tag{29.4}$$

The functions ξ and ν_x generally do not overlap, ξ being appreciable only for very low energies, ν_x only for high energies and there is a wide gap between these two functions.

c) Motions of electrons.

30. Diffusion. The results of Sect. a and b must now be put together to obtain equations for the component distribution functions f_0^0, $\boldsymbol{f}_0^1$, and $\boldsymbol{f}_1^1$. From (24.13) and (27.4) we have for the d.c. vector equation

$$(\nu_c - \boldsymbol{\omega}_b \times) \boldsymbol{f}_0^1 = - v \operatorname{grad} f_0^0. \tag{30.1}$$

This equation can be solved by vector methods to give

$$(\nu_c^2 + \omega_b^2) \boldsymbol{f}_0^1 - \boldsymbol{\omega}_b \cdot \boldsymbol{f}_0^1 \boldsymbol{\omega}_b = -(\nu_c - \boldsymbol{\omega}_b \times) \operatorname{grad}(v f_0^0) \tag{30.2}$$

or in terms of a matrix

$$\boldsymbol{f}_0^1 = - \left\| \begin{matrix} \dfrac{\nu_c}{\nu_c^2+\omega_b^2} & \dfrac{\omega_b}{\nu_c^2+\omega_b^2} & 0 \\ \dfrac{-\omega_b}{\nu_c^2+\omega_b^2} & \dfrac{\nu_c}{\nu_c^2+\omega_b^2} & 0 \\ 0 & 0 & \dfrac{1}{\nu_c} \end{matrix} \right\| \operatorname{grad} v f_0^0. \tag{30.3}$$

This is to be compared to the tensor in expression (14.3) of Part II. The diffusion coefficient D is defined in terms of the flow $\boldsymbol{\Gamma}$ by

$$\boldsymbol{\Gamma} = \int \frac{4\pi}{3} \boldsymbol{f}_0^1 v^3 dv = -\operatorname{grad}(nD). \tag{30.4}$$

D is a tensor whose parallel, transverse and perpendicular components are given by

$$D_\parallel = \frac{1}{n} \int \frac{v^2}{3\nu_c} f_0^0 4\pi v^2 dv = \frac{\overline{l_c v}}{3}, \tag{30.5}$$

$$D_\mathrm{T} = \frac{1}{n} \int \frac{\nu_c}{\nu_c^2+\omega_b^2} f_0^0 \frac{4\pi}{3} v^4 dv, \tag{30.6}$$

$$D_\perp = \frac{1}{n} \int \frac{\omega_b}{\nu_c^2+\omega_b^2} f_0^0 \frac{4\pi}{3} v^4 dv. \tag{30.7}$$

The diffusion coefficient D has been put inside the gradient operator in the definition (30.4) because, if the electron velocity distribution f_0^0 varies from point to point, this variation must be included in taking the gradient. On the other hand, if ν_c varies from point to point because of its gas pressure dependence, as distinct from its velocity dependence, or if the magnetic field varies from point to point so that ω_b is a variable, these variations must not be included in taking the gradient. The gradient is a partial derivative operating on the distribution function and through it on the velocity dependence of ν_c but not on its pressure dependence.

31. Mobility. The a.c. vector equation (24.14) is

$$(\nu_c + j\,\omega - \boldsymbol{\omega}_b \times)\,\boldsymbol{f}_1^1 = -\boldsymbol{a}\,\frac{d f_0^0}{d v} \tag{31.1}$$

which is solved as in Sect. 12 in terms of the quantities l, r, p defined in (12.6)

$$\boldsymbol{f}_1^1 = -\frac{\boldsymbol{a}}{2}\begin{Vmatrix} l+r & j(l-r) & 0 \\ j(r-l) & l+r & 0 \\ 0 & 0 & 2p \end{Vmatrix}\frac{d f_0^0}{d v}. \tag{31.2}$$

We now define the quantities L, R, and P by

$$L = \frac{-1}{n}\int \frac{4\pi}{3}\,l\,\frac{d f_0^0}{d v}\,v^3\,d v = \frac{1}{n}\int f_0^0\,\frac{d}{d v}\left[\frac{4\pi\,l\,v^3}{3}\right]d v \tag{31.3}$$

and similar equations for R and P in terms of r and p. It is noted that if ν_c is independent of v, so are l, p, and r, and $L=l$, $P=p$, $R=r$. The mobility of an electron is defined by

$$\mu\,\boldsymbol{E} = \boldsymbol{v}_d = \frac{1}{n}\int \frac{4\pi}{3}\,\boldsymbol{f}_1^1\,v^3\,d v \tag{31.4}$$

and leads to the tensor

$$\mu = \frac{q}{2m}\begin{Vmatrix} L+R & j(R-L) & 0 \\ j(L-R) & L+R & 0 \\ 0 & 0 & 2P \end{Vmatrix} \tag{31.5}$$

which reduces to (12.7) only when the mean free time, $\tau_c = 1/\nu_c$, is constant.

Expanding the expressions for the three mobility components gives

$$\mu_\parallel = \frac{q}{n\,m}\int f_0^0\,\frac{d}{d v}\left[\frac{\nu_c - j\,\omega}{\nu_c^2+\omega^2}\,\frac{4\pi\,v^3}{3}\right]d v\,, \tag{31.6}$$

$$\mu_\mathrm{T} = \frac{q}{n\,m}\int f_0^0\,\frac{d}{d v}\left\{\frac{\nu_c(\nu_c^2+\omega^2+\omega_b^2) - j\,\omega(\nu_c^2+\omega^2-\omega_b^2)}{[\nu_c^2+(\omega+\omega_b)^2]\,[\nu_c^2+(\omega-\omega_b)^2]}\,\frac{4\pi\,v^3}{3}\right\}d v\,, \tag{31.7}$$

$$\mu_\perp = \frac{q\,\omega_b}{n\,m}\int f_0^0\,\frac{d}{d v}\left\{\frac{\nu_c^2-\omega^2+\omega_b^2-2j\,w\,\nu_c}{[\nu_c^2+(\omega+\omega_b)^2]\,[\nu_c^2+(\omega-\omega_b)^2]}\,\frac{4\pi\,v^3}{3}\right\}d v. \tag{31.8}$$

These quantities simplify considerably in the case of a direct current ($\omega=0$) when

$$\mu_\parallel = \frac{q}{n m}\int f_0^0\,\frac{d}{d v}\left(\frac{4\pi}{3}\,\frac{v^3}{\nu_c}\right)d v\,, \tag{31.9}$$

$$\mu_\mathrm{T} = \frac{q}{n\,m}\int f_0^0\,\frac{d}{d v}\left(\frac{\nu_c}{\nu_c^2+\omega_b^2}\,\frac{4\pi\,v^3}{3}\right)d v\,, \tag{31.10}$$

$$\mu_\perp = \frac{q}{n\,m}\int f_0^0\,\frac{d}{d v}\left(\frac{\omega_b}{\nu_c^2+\omega_b^2}\,\frac{4\pi\,v^3}{3}\right)d v. \tag{31.11}$$

Particular interest is attached to the ratio D/μ for a direct current. From (30.6) and (31.10) it is seen that

$$\frac{D}{\mu}=\frac{kT}{e} \quad \text{when} \quad f_0^0=A\,e^{-\frac{mv^2}{2kT}}, \tag{31.12}$$

$$\frac{D}{\mu}=\frac{m\overline{v^2}}{3e} \quad \text{when} \quad \nu_c \text{ is constant}, \tag{31.13}$$

$$\frac{D}{\mu}=\frac{m}{2e}\frac{\bar{v}}{\overline{1/v}} \quad \text{when} \quad \nu_c=\frac{v}{l_c}>\omega_b. \tag{31.14}$$

The first of these is the well-known EINSTEIN relation and the second is generalization of it to non-MAXWELLian distributions. The third is a warning that it is not always true. Nevertheless (31.13) is very nearly true in all practical cases.

d) The energy distribution.

32. Plasma balance. Combining (24.12) with (28.9) and (29.3), we obtain the equation

$$(\nu_x-\xi)f_0^0-\frac{m}{M+m}\frac{1}{v^2}\frac{d}{dv}\left[v^3\nu_c\left(f_0^0+\frac{kT_g}{mv}\frac{df_0^0}{dv}\right)\right]=-\frac{v}{3}\operatorname{div}\boldsymbol{f}_0^1-\frac{1}{6v^2}\frac{d}{dv}(v^2\boldsymbol{a}\cdot\boldsymbol{f}_{1r}^1) \tag{32.1}$$

or

$$(\nu_x-\xi)f_0^0=-\frac{v}{3}\operatorname{div}\boldsymbol{f}_0^1-\frac{1}{4\pi v^2}\frac{dG}{dv} \tag{32.2}$$

in which we have introduced the gain

$$G=\frac{4\pi v^2}{3}\left[\frac{1}{2}\boldsymbol{a}\cdot\boldsymbol{f}_{1r}^1-\frac{3\nu_c}{M+m}\left(m v f_0^0+kT_g\frac{df_0^0}{dv}\right)\right]. \tag{32.3}$$

This expression combines that part of the gain which was obtained in (19.8) with an additional part which represents the energy interchanges in elastic collisions with the molecules, but does not include the electronic excitations. The recoil terms are included because they consist of small energy jumps by which the electrons gain or lose energy almost continuously, passing through all intermediate energy values, whereas in electronic excitations, the electron jumps discontinuously from an energy near the excitation value to an energy near zero.

Multiplying (32.2) by $4\pi v^2 dv$ and integrating from zero to infinity, the gain term disappears at both limits. The other two terms have been considered in (29.4) and (30.4) so that the net result of the integration is

$$\nabla^2 D n+n(\bar{\nu}_i-\bar{\nu}_a)=0. \tag{32.4}$$

In computing the LAPLACian in this equation, it must be remembered when there is a magnetic field present, that the diffusion coefficient has a different value along and at right angles to the magnetic field. On the other hand, the perpendicular component $D_\perp$ does not enter because the two off-diagonal terms of the diffusion tensor cancel each other.

Equation (32.4) is a diffusion equation and requires a characteristic value for $\bar{\nu}_i$ in order to satisfy prescribed boundary conditions, which are generally that $n=0$ at the walls. If D is independent of position and Λ_b is the diffusion length of the cavity as defined in (13.10)

$$\bar{\nu}_i=\bar{\nu}_a+D/\Lambda_b^2. \tag{32.5}$$

33. The second order equation. Substituting from (30.3) for $\boldsymbol{f}_0^1$ into the diffusion term in (32.2), gives

$$-\frac{v}{3}\operatorname{div}\boldsymbol{f}_0^1 = \frac{v^2}{3}\frac{\nu_c}{\nu_c^2+\omega_b^2}\nabla^2 f_0^0 \approx -\frac{v^2}{3\Lambda_b^2\,\nu_c} f_0^0 \tag{33.1}$$

where the replacement of Δ^2 by $-1/\Lambda_b^2$ is based on the assumption that the function f_0^0 is the product of a function n of position and a normalized function of velocity only. This may not always be valid but is generally approximately so.

Introducing the value (31.2) for $\boldsymbol{f}_1^1$ into the power input term, $\frac{1}{2}\,\boldsymbol{a}\cdot\boldsymbol{f}_{1r}^1$, and making use of the drift energy expression (12.9) it is seen that

$$\frac{1}{2}\boldsymbol{a}\cdot\boldsymbol{f}_{1r}^1 = -\frac{a^2}{4}(l+r)_r\frac{d f_0^0}{dv} = -\frac{\nu_c}{m}2u_d\frac{d f_0^0}{dv} = -2v\,\nu_c\,u_d\frac{d f_0^0}{du}. \tag{33.2}$$

Let us define the energy gain per collision u_c by

$$\left.\begin{aligned} u_c &= 2u_d + \frac{3mkT_g}{M+m}, \\ u_d &= \frac{e^2E_p^2}{8m}\left[\frac{1}{\nu_c^2+(\omega-\omega_b)^2}+\frac{1}{\nu_c^2+(\omega+\omega_b)^2}\right] \end{aligned}\right. \tag{33.3}$$

where ω_b is that component of the vector which is perpendicular to E_p, and in the case of a d.c. field the 8 is to be replaced by 4. Introducing all this in (32.2) and (32.3) gives the second order differential equation for f_0^0

$$\left.\begin{aligned} &\left(\nu_x - \xi + \frac{v^2}{3\Lambda_b^2\nu_c}\right) f_0^0 + \frac{1}{4\pi v^2}\frac{dG}{dv} = 0, \\ &G = -\frac{4\pi}{3}v^3\nu_c\left(u_c\frac{d f_0^0}{du} + \frac{3m}{M+m}f_0^0\right). \end{aligned}\right\} \tag{33.4}$$

When ν_c is a constant this equation can be solved in terms of confluent hypergeometric functions[1], but in general the solution is more difficult. In any case it is necessary to divide the solution at eV_x because ν_x is discontinuous there.

An approximation valid for high energies can be obtained by setting

$$f_0^0 = A\,e^{-s} \tag{33.5}$$

and discarding all but the largest terms of (33.4) which, in the excitation range, are the terms in ν_x and u_c. This leads to a Riccati equation for s':

$$\nu_x = \tfrac{2}{3}\nu_c u_c u\,(s'^2 - s''). \tag{33.6}$$

Numerical substitutions show that for energies near the ionization potentials $s'^2 \gg s''$ so that one has, approximately,

$$s = \int\sqrt{\frac{\nu_x}{\nu_c}\,\frac{\frac{3}{2}}{u_c u}}\,du \approx \frac{p}{E}. \tag{33.7}$$

From this one calculates the number of excitations per ionization

$$N_x = \frac{\int\limits_{v_x}^{\infty}\nu_x f_0^0\,4\pi v^2\,dv}{\int\limits_{v_i}^{\infty}\nu_i f_0^0\,4\pi v^2\,dv} \approx A\,e^{Bp/E}. \tag{33.8}$$

[1] A. D. Macdonald and S. C. Brown: Phys. Rev. **76**, 1634 (1949).

The calculation of the "constants" A and B requires a knowledge of the velocity dependence of ν_c and ν_x. They have been calculated for helium, hydrogen, and neon, and the reader is referred to the original papers for the details[1–4].

It is noted that the "average electron" approximation (17.13) corresponds to the smaller term, s'', in the Riccati equation.

34. The gain equation. Except for very large E/p most free electrons have energies between one volt and the lowest excitation potential, and in this range ν_x is zero and ξ is small. Equation (33.4) can be multiplied by $4\pi v^2 dv$ and integrated from a velocity v in this range to infinity. Introducing the "diffusion frequency"

$$n\,\nu_d(v) = \int\limits_v^\infty \frac{4\pi v^4}{3\Lambda^2 \nu_c} f_0^0\, dv = \operatorname{div} \boldsymbol{\Gamma}(u)\,, \tag{34.1}$$

$$n\left((\bar{\nu}_x + \nu_d(v)\right) = G = -\frac{4\pi}{3} v^3 \nu_c \left(u_c \frac{d f_0^0}{du} + \frac{3m}{M+m} f_0^0\right). \tag{34.2}$$

This is the equation which Smidt[5] and more recently Hartman[6] have used. It is in the form of a conservation equation for electrons of speed above v as indicated schematically in Fig. 11. The losses by inelastic collisions $n\,\bar{\nu}_x$ and diffusion $n\nu_d$ are balanced against the gain G by acceleration from lower energies.

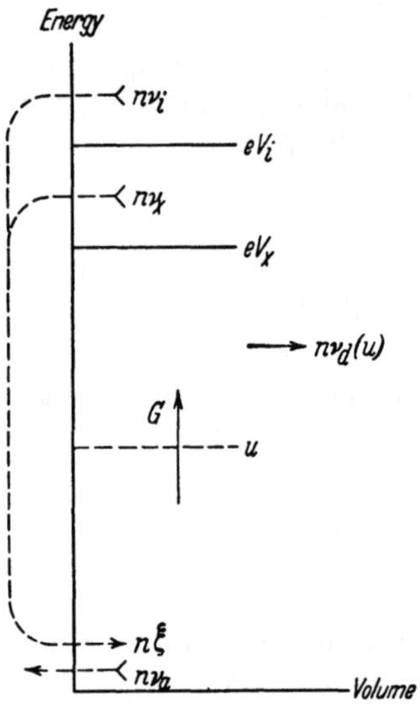

Fig. 11. Diagram to illustrate the terms in the gain equation.

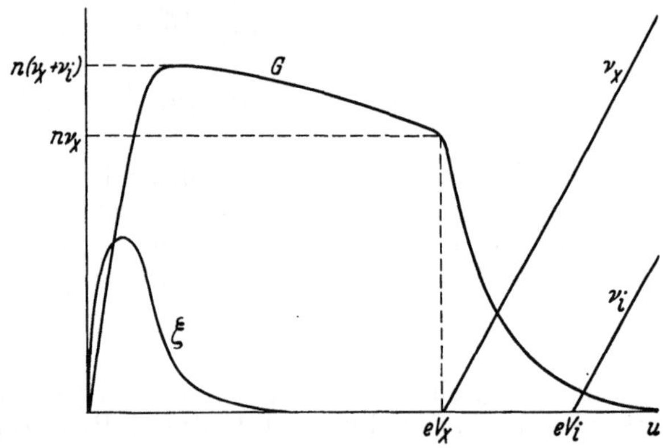

Fig. 12. The gain as a function of the energy.

The gain function G has the form shown in Fig. 12. It is constant over the range in question except for a slight drop due to diffusion. The "diffusion frequency" $\bar{\nu}_d(v)$ as defined varies from $\bar{\nu}_i - \bar{\nu}_a$ at zero, to near zero at u_x and, knowing these limits, it can be replaced by an approximation, or when the inelastic collision number $N_x = \bar{\nu}_x/\bar{\nu}_i =$ is large, as is the case except in Penning mixtures (mixtures of a rare gas and an impurity in which the rare gas metastable can ionize

[1] A. D. Macdonald and S. C. Brown: Phys. Rev. **75**, 411 (1949).
[2] W. P. Allis and S. C. Brown: Phys. Rev. **87**, 419 (1952).
[3] F. Reder and S. C. Brown: Phys. Rev. **95**, 885 (1954).
[4] A. D. Macdonald and D. D. Betts: Canad. J. Phys. **30**, 565 (1952).
[5] J. A. Smidt: Physica, Haag **3**, 543 (1937).
[6] L. M. Hartman: Phys. Rev. **37**, 316 (1948).

the impurity), it can be neglected entirely. We shall neglect it in the following. The gain equation then has a simple solution

$$f_0^0 = \frac{n\,\bar{v}_x}{4\pi}\,\frac{M+m}{m}\,e^{-w}\int\limits_{w}^{w_x}\frac{e^w\,dw}{v^3\,\nu_c} \tag{34.3}$$

where

$$w = \int\limits_0^u \frac{3m}{M+m}\,\frac{du}{u_c} \tag{34.4}$$

and w_x is the value of w at u_x which is slightly above $e\,V_x$. The overshoot $u_x - e\,V_x$ is found by joining (34.3) to (33.5) at $e\,V_x$

$$-\frac{f_0^{0\prime}}{f_0^0} = \frac{3m}{M+m}\,\frac{1}{u_c} + \frac{\dfrac{e^w}{v^3\,\nu_c\,u_c}}{\int\limits_{e\,V_x}^{u_x}\dfrac{e^w\,du}{v^3\,\nu_c\,u_c}} \approx \frac{3m}{M+m}\,\frac{1}{u_c} + \frac{1}{u_x - e\,V_x} = s_x'. \tag{34.5}$$

The evaluation of s_x' from (33.6) is difficult, but the approximation of (17.12) turns out to be quite good.

35. The distribution function. In Sect. 16 it was seen that the average electron, in the absence of inelastic collisions, reached an equillibrium energy

$$u_e = \frac{2u_d}{\lambda} = \frac{M+m}{2m}\,u_c. \tag{35.1}$$

The number w_x is, therefore, approximately $\frac{3\,u_x}{2\,u_e}$ It should be less than $\frac{3}{2}$ if the equilibrium energy is to be above the excitation potential but in practice ranges anywhere up to 2 even in discharges with active ionization. As examples $w_x = (5\,p/E_{\mathrm{ef}})^2$ in helium and $w_x = (12\,p/E_{\mathrm{ef}})^2$ in hydrogen, where p is in mm. and E in volts/cm.

The distribution function (34.3) has two limiting forms according to the value of w_x. For $w_x \gg 1$ the integral is dominated by its upper limit and is practically constant. Then

$$f_0^0 = A\,e^{-w}. \tag{35.2}$$

This is the distribution function introduced by DRUYVESTEYN[1] and extended by DAVYDOFF and MARGENAU[2]. It is applicable when excitation losses are negligible in comparison to recoil.

At the opposite limit $w_x \ll 1$, or when recoil losses are negligible

$$f_0^0 = \frac{3\,n\,\bar{v}_x}{4\pi}\int\limits_v^{v_x}\frac{m\,dv}{v^2\,\nu_c\,u_c} \to \frac{3\,m\,n\,\bar{v}_x}{4\pi\,\nu_c\,u_c}\left(\frac{1}{v} - \frac{1}{v_x}\right) \tag{35.3}$$

where the integral has been performed on the assumption that ν_c is a constant. This limit is surprisingly close to the actual distribution in all medium pressure discharges where neither diffusion, nor recoil, nor electron interaction, have taken control. It is nearly always much better than (35.2).

The exponential integral distribution (34.3) is not valid near the origin, where it has a singularity, but as all necessary integrals converge at the origin, and even

[1] M. J. DRUYVESTEYN and F. M. PENNING: Rev. Mod. Phys. **12**, 87 (1940).
[2] H. MARGENAU: Phys. Rev. **69**, 508 (1946).

contribute little from this region, the function can be used down to $v=0$. The distribution (34.3) is also invalid above eV_x where the expansion (33.5) should be used, but there are relatively few electrons in the range of inelastic collisions, and even these are largely included by extending the distribution (34,5) up to u_x where it vanishes.

The functions

$$\mathcal{D}_k = w_x^{-k-1} \int_0^{w_x} e^{-w}\, w^{k+\frac{1}{2}}\, dw \int_w^{w_x} e^t\, t^{-\frac{3}{2}}\, dt \tag{35.4}$$

supply all necessary integrals when ν_c is constant, which we shall now assume. They satisfy the recursion relation

$$w_x \mathcal{D}_k = \left(k + \frac{1}{2}\right) \mathcal{D}_{k-1} - \frac{1}{k} \tag{35.5}$$

and can be computed from the series

$$\mathcal{D}_0 = 4 \sum_0^{\infty} \frac{k!}{(2k+3)!} (4 w_x)^k. \tag{35.6}$$

Normalizing the distribution function yields the excitation frequency $\bar{\nu}_x$ in terms of E/p

$$\frac{\nu_c}{\bar{\nu}_x} = \frac{M+m}{2m} w_x \mathcal{D}_0 = \frac{u_x}{u_c}\left(1 + \frac{w_x}{5} + \frac{4 w_x^2}{105} \cdots\right). \tag{35.7}$$

The first term, which derives from the distribution (35.3), equates the power input $\nu_c u_c$ from the field to the power loss $\bar{\nu}_x u_x$ to excitation, and the series adds the power loss to recoil.

The ionization frequency follows from this by applying (33.8):

$$\bar{\nu}_i = \bar{\nu}_x / N_x. \tag{35.8}$$

The mean energy $\bar{u}$ is

$$\bar{u} = u_x \frac{\mathcal{D}_1}{\mathcal{D}_0} = \frac{3}{10} u_x \frac{1 + 4 w_x/21 + 2 w_x^2/63 \ldots}{1 + w_x/5 + 4 w_x^2/105 \ldots}. \tag{35.9}$$

It is generally quite reasonable to assume that the mean energy of the electrons in a plasma is $\frac{3}{10}$ of the lowest excitation potential, u_x being slightly larger than eV_x and the fraction being slightly less than one.

The diffusion constant, with no space-charge or magnetic field is

$$D_- = \frac{2}{3} \frac{\bar{u}}{m \nu_c} = \frac{2}{3} \frac{u_x}{m \nu_c} \frac{\mathcal{D}_1}{\mathcal{D}_0} \tag{35.10}$$

and introducing this and the ionization frequency in the plasma balance equation, with no attachment, yields the breakdown condition

$$u_c = \left(\frac{u_x}{\Lambda \nu_c}\right)^2 \mathcal{D}_1 N_x \tag{35.11}$$

and, in the absence of a magnetic field,

$$\frac{1}{2} (E_p \Lambda)^2 \frac{\nu_c^2}{\nu_c^2 + \omega^2} = \frac{N_x}{5} \left(\frac{u_x}{e}\right)^2 \left(1 + \frac{4 w_x}{21} + \frac{2 w_x^2}{63} \cdots\right). \tag{35.12}$$

$(E_p \Lambda)$ is the voltage "applied" to the plasma, the $\frac{1}{2}$ reduces it to root-mean-square value, $\frac{\nu_c^2}{\nu_c^2 + \omega^2}$ is a power factor because of the phase angle between field and current. The result is the "effective applied voltage" squared. The first term on the right comes from the distribution (35.3) and is quite close to the expression (17.8) found for the average electron. This shows that average electron calculations are reasonably good when $w_x < 1$ or when the "equilibrium" electron energy $u_e > 1.5\, eV_x$, that is for high E/p. The power series in (35.12) represents recoil losses.

e) Direct current.

36. The first TOWNSEND coefficient. The energy equations of the previous section can all be applied to direct current discharges by simply setting $\omega = 0$ and replacing the root-mean-square field by the d.c. field. Only the plasma balance equation must be reinterpreted because electrons may leave the plasma by mobility as well as by diffusion. For a long column the plasma balance equation remains as in (32.3) except that the diffusion length is computed for the two lateral directions only.

For a short column, such as is used in a TOWNSEND discharge, the lateral loss is generally negligible but one must consider the longitudinal diffusion. We shall consider the case of no magnetic field, space charge, or attachment, and shall define the first TOWNSEND coefficient α by assuming a simple exponential increase in the direction of the field

$$f(z, v) = f(v)\, e^{\alpha z}. \tag{36.1}$$

The vector equation, found by combining (30.1) and (31.1), is

$$\left.\begin{aligned} \nu_c \boldsymbol{f}^1 &= -v \operatorname{grad} f^0 - \boldsymbol{a}\frac{d f^0}{d v}, \\ &= -\left(\alpha \boldsymbol{v} + \boldsymbol{a}\frac{d}{d v}\right) f^0. \end{aligned}\right\} \tag{36.2}$$

Setting this in Eq. (32.2) gives

$$(\nu_x - \xi)\, f^0 = \frac{\alpha v}{3\nu_c}\left(\alpha v + a\frac{d}{d v}\right) f^0 - \frac{1}{4\pi v^2}\frac{d G}{d v}. \tag{36.3}$$

Multiplying by $4\pi v^2 d v$ and integrating from zero to infinity gives

$$\left.\begin{aligned} n \bar{\nu}_{i\alpha} &= -\int \frac{\alpha^2 v^2}{3\nu_c} f^0\, 4\pi v^2 d v - \int \frac{\alpha a}{3\nu_c}\frac{d f^0}{d v} 4\pi v^3 d v, \\ \bar{\nu}_{i\alpha} &= \alpha E \mu - \alpha^2 D = \alpha v_d. \end{aligned}\right\} \tag{36.4}$$

where v_d is the resultant drift velocity due to both field and concentration gradient.

This gives the relation between the ionization frequency $\bar{\nu}_{i\alpha}$ and the coefficient α in a TOWNSEND discharge, and it is easy to show that the diffusion term is small. Introducing $\eta = \alpha/E$, the number of ionizations per volt, the ratio of the two terms in $\bar{\nu}_{i\alpha}$ is

$$\frac{\alpha D}{\mu E} = \eta \frac{2\bar{u}}{3e} \leqq \frac{\eta u_x}{5e} < \frac{1}{5}. \tag{36.5}$$

Except for a PENNING mixture, where $\eta u_x/e$ may be close to 1, the diffusion term is always much less than one fifth of the mobility term.

However, we would like to compare α with $\bar{\nu}_i$ in a *uniform* discharge of the same E/p. Substituting (36.2) in the gain equation (32.3) gives

$$G = -\frac{4\pi}{3} v^3 \nu_c \left[\left(2 u_d + \frac{3m}{M+m} k T_g\right)\frac{d f^0}{d u} + \left(\frac{3m}{M+m} + \frac{\alpha a}{\nu_c^2}\right) f^0\right] \tag{36.6}$$

and it is seen that the non-uniformity is equivalent to an increase in the recoil loss. From (34.4) it follows that w_x is increased

$$w_{x\alpha} = w_x + \eta\, u_x/e \tag{36.7}$$

and introducing this in (35.7) gives the ratio

$$\frac{\bar{\nu}_i}{\bar{\nu}_{i\alpha}} = \frac{1 + \frac{w_{x\alpha}}{5} + \frac{4 w_{x\alpha}^2}{105} \cdots}{1 + \frac{w_x}{5} + \frac{4 w_x^2}{105} \cdots} = 1 + \frac{\eta\, u_x/e}{5 + w_x} \cdots. \tag{36.8}$$

The non-uniformity reduces $\bar{\nu}_i$ in just about the same ratio as v_d is reduced by diffusion. Unless $\eta u_x/e$ is unusually large one can write

$$\bar{\nu}_i = \alpha \mu E, \tag{36.9}$$

and

$$\frac{1}{\eta} = \frac{E}{\alpha} = N_x u_x \left(1 + \frac{w_x}{5} + \frac{4 w_x^2}{105} + \cdots\right). \tag{36.10}$$

The variation of η with E/p is due almost entirely, except for a PENNING mixture, to the variation of N_x which, as has been seen, varies as $e^{Bp/E}$. At low E/p this must be corrected by the series in (36.10) which is due to recoil. At high E/p the overshoot $u_x - e V_x$ causes a decrease in η. The transition between a decreasing N_x and an increasing u_x results in a maximum for η which corresponds to the PASCHEN minimum in d.c. breakdown.

IV. BOLTZMANN transport equation.

a) General theory.

37. Transport equation. In the case of ions whose mass m is comparable to the mass M of the molecules of the gas, the expansion (19.2) in spherical harmonics requires more than two terms, and it is no longer possible to obtain as simple an equation as (33.4) for the distribution function $f^0(v)$; but it is possible to transform the BOLTZMANN equation as given by (18.4) and (26.1) in such a way that it may be possible to calculate the average value of any quantity $X(v)$ without knowing too much about the distribution function f.

The BOLTZMANN equation is

$$\left.\begin{aligned} &\left[\frac{\partial f}{\partial t} + \operatorname{div}_r \boldsymbol{v} f + \operatorname{div}_v (\boldsymbol{a} + \boldsymbol{w}_b \times \boldsymbol{v})\, f\right] d^3 v = \\ &= \iint F(V')\, f(v')\, \varrho\, d^2\Omega\, d^3 V'\, d^3 v' - \iint F(V)\, f(v)\, \varrho\, d^2\Omega\, d^3 V\, d^3 v. \end{aligned}\right\} \tag{37.1}$$

Multiplying both sides of this equation by $X(v)$ and integrating over the velocities of the ions gives

$$\left.\begin{aligned} &\frac{\partial}{\partial t} (n \overline{X}) + \operatorname{div}_r (n \overline{\boldsymbol{v} X}) + \int X \operatorname{div}_v (\boldsymbol{a} f)\, d^3 v = \\ &= \iiint X F(V')\, f(v')\, \varrho\, d^2\Omega\, d^3 V'\, d^3 v' - \iiint X F(V)\, \varrho\, d^2\Omega\, d^3 V\, d^3 v \end{aligned}\right\} \tag{37.2}$$

where the bar indicates an average. The third term on the left can be transformed by GREEN's theorem, and in the first integral on the right it is allowable to interchange primed and unprimed variables because of the reversibility of elastic collisions. This leads to the transport equation

$$\left.\begin{aligned}&\frac{\partial}{\partial t}(n\overline{X}) + \operatorname{div}_r(n\overline{\boldsymbol{v}X}) - n\overline{(\boldsymbol{a}\cdot\operatorname{grad}_v)X} - n\overline{(\boldsymbol{w}_b\times\boldsymbol{v}\cdot\operatorname{grad}_v)X} = \\ &\qquad = \iiint (X'-X)\,F(V)\,f(v)\,\varrho\,d^2\Omega\,d^3V\,d^3v.\end{aligned}\right\}\tag{37.3}$$

The left hand terms of this equation determine the rate of change of $n\overline{X}$ due to flow in phase space. The right hand side gives the average rate of change of X due to collisions. We shall use $X=1$ and $X=\boldsymbol{v}$ as illustrations of this equation.

For $X=1$, $\overline{\boldsymbol{v}X}=\boldsymbol{v}_d$. On the right $X'=X$ for an elastic collision, but inelastic collisions can be taken into account by letting $X'=2$ in an ionizing collision, or $X'=0$ for an attachment or recombination. The the transport equation gives

$$\frac{\partial n}{\partial t} + \operatorname{div}(n\boldsymbol{v}_d) = n(\bar{\nu}_i - \bar{\nu}_a).\tag{37.4}$$

For $X=\boldsymbol{v}$, $\operatorname{grad}X=1$, and the transport equation becomes

$$\frac{\partial}{\partial t}(n\boldsymbol{v}_d) + \frac{\operatorname{div}\boldsymbol{p}}{m} - n(\boldsymbol{a} + \boldsymbol{w}_b\times\boldsymbol{v}_d) = -\,n\,\bar{\nu}_c\,\boldsymbol{v}_d\tag{37.5}$$

where the pressure $\boldsymbol{p}$ is a symmetric tensor

$$p_{ij} = n\,m\,\overline{v_i v_j}\tag{37.6}$$

and the right hand side is really a definition of $\bar{\nu}_c$.

The collision term will be discussed in the following sections, but the pressure term may be discussed immediately. For a nearly spherically symmetric distribution the tensor reduces to the usual scalar

$$p = n\,m\overline{v^2}/3 = n\,kT.\tag{37.7}$$

Neglecting the inelastic term $n(\bar{\nu}_i-\bar{\nu}_a)$ Eq. (37.4) and (37.5) combine to give

$$\frac{\partial\boldsymbol{v}_d}{\partial t} = \boldsymbol{a} + \boldsymbol{w}_b\times\boldsymbol{v}_d - \frac{\operatorname{grad}p}{n\,m} - \bar{\nu}_c\boldsymbol{v}_d + \boldsymbol{v}_d\frac{\operatorname{div}(n\,\boldsymbol{v}_d)}{n}.\tag{37.8}$$

This is to be compared with Eq. (8.1) obtained under simpler assumptions, but has, in addition to the collision term not considered in Sect. 8, a term quadratic in $\boldsymbol{v}_d$. This term is generally negligible if one is justified in replacing $\boldsymbol{p}$ by a scalar.

Let us now consider a group of ions whose velocities are all clustered about $\boldsymbol{v}_d$. The pressure tensor $\boldsymbol{p}$ then has a single component

$$p_{zz} = n\,m\,v_d^2.\tag{37.9}$$

The quadratic terms now add in such a way as to give

$$\frac{d\boldsymbol{v}_d}{dt} = \frac{\partial\boldsymbol{v}_d}{\partial t} + (\boldsymbol{v}_d\cdot\operatorname{grad})\,\boldsymbol{v}_d = \boldsymbol{a} + \boldsymbol{w}_b\times\boldsymbol{v}_d - \bar{\nu}_c\boldsymbol{v}_d.\tag{37.10}$$

This is the same as the LANGEVIN equation (12.1) for v_g. Besides referring to different situations ($\boldsymbol{v}_d$ to small drift motions, $\boldsymbol{v}_g$ to small random motions) the drift and guiding center methods are similar to the EULER and LAGRANGE methods in hydrodynamics.

38. Recursion equations. We now simplify the transport Eq. (37.3) by considering the steady state with no density gradient or magnetic field. Only the term in $\boldsymbol{a}$ remains on the left. The collision integral on the right is quite intractable without approximations except in the case of a constant mean free time, treated in section *b*. The usual assumption, treated by CHAPMAN and COWLING[1], is to consider small departures from thermal equilibrium. It turns out, as discussed in Sect. 43, that small departures from thermal equilibrium always fall under the mean free time assumption, which is treated exactly. We therefore make the opposite assumption, that the gas temperature is negligible compared to the random energy of the ions. The distribution function F then becomes n_g times a δ-function and the integration over $\boldsymbol{V}$ is immediate. In this way the transport equation becomes

$$\int f(\boldsymbol{a}\cdot\text{grad})\, X\, d^3v = \int\int (X - X')\, f\, n_g\, \varrho\, d^2\Omega\, d^3v. \tag{38.1}$$

We shall consider a set of quantities X of the form

$$X = v^k P_l(\vartheta) \tag{38.2}$$

where $\boldsymbol{a}$ is in the direction $\vartheta = 0$, and derive recursion relations between averages X_l^k of quantities of this form with different values of k and l.

Using the formulae for spherical harmonics (19.9) and (19.10), we then find for $\boldsymbol{a}\cdot\text{grad}\, X$

$$\int f(\boldsymbol{a}\cdot\text{grad})\, v^k P_l\, d^3v = n a\left[\frac{l(k+l+1)}{2l+1} X_{l-1}^{k-1} + \frac{(l+1)(k-l)}{2l+1} X_{l+1}^{k-1}\right]. \tag{38.3}$$

Using the rotation formula for spherical harmonics (27.1)

$$X' = v'^k\left[P_l(\vartheta)\, P_l(\chi_0) + 2\sum \frac{(l-m)!}{(l+m)!} P_l^m(\vartheta)\, P_l^m(\chi_0) \cos m(\varphi - \psi)\right] \tag{38.4}$$

and introducing this into the collision integral (38.1), the term in $\cos m(\varphi - \psi)$ integrates out because ϱ is not a function of φ. The spherical harmonic $P_l(\vartheta)$ then factors out and the only part of this integral depending on the scattering angle χ is

$$I_l^k = \int\left[1 - \left(\frac{v'}{v}\right)^k P_l(\chi_0)\right] n_g\, \varrho\, d^2\Omega. \tag{38.5}$$

The velocity ratio is obtained from the geometry of a collision

$$\frac{v'}{v} = \frac{\sqrt{M^2 + 2Mm\cos\chi_g + m^2}}{M+m} \tag{38.6}$$

and the laboratory scattering angle χ_0 is given in terms of the center of gravity angle χ_g by

$$\cos\chi_0 = \frac{m + M\cos\chi_g}{\sqrt{M^2 + 2Mm\cos\chi_g + m^2}}. \tag{38.7}$$

For the particular case of an ion of the same mass as the gas, $M = m$, these formulae reduce to

$$\frac{v'}{v} = \cos\chi_0 = \cos\frac{\chi_g}{2} \tag{38.8}$$

and the calculations are greatly simplified. This case has been worked out in detail by WANNIER[2] whose work we follow.

---

[1] S. S. CHAPMAN and T. G. COWLING: The Mathematical Theory of Non-uniform Gases. Cambridge University Press 1939.

[2] G. H. WANNIER: Phys. Rev. **83**, 281 (1951). — Bell Syst. Techn. J. **32**, 170 (1953).

The integrals I_l^k depend in unknown ways on the speed v, but we shall assume a power law, so that they can be written

$$I_l^k = N_l^k q v^\delta \tag{38.9}$$

where the N_l^k are pure numbers and qv^δ has the dimensions of a frequency. For a constant mean free time $\delta = 0$, $q = \nu_g$. For a constant mean free path $\delta = 1$, $q = 1/l_g$.

Introducing (38.4) and (38.9) in the transport equation leads to the recursion relations

$$\frac{a}{q}\left[l(k+l+1)\,X_{l-1}^{k-1} + (l+1)(k-l)\,X_{l+1}^{k-1}\right] = (2l+1)\,N_l^k X_l^{k+\delta}. \tag{38.10}$$

WANNIER has shown how to solve these equations in the cases $\delta = 0$ and $\delta = 1$.

b) Constant mean free time.

39. Drift velocity. MAXWELL[1] pointed out that in the case of the inverse fifth power law of force, because it gives a constant mean free time, all transport quantities can be calculated without any knowledge of the distribution function f. It is not even necessary to know the molecular function F and we can therefore use the general transport equation (37.3). Let

$$X = \boldsymbol{v}. \tag{39.1}$$

Then

$$(\boldsymbol{a}\cdot\operatorname{grad})\,\boldsymbol{v} = \boldsymbol{a} \tag{39.2}$$

and therefore

$$n\,\boldsymbol{a} = \iiint (\boldsymbol{v} - \boldsymbol{v}')\,F\,f\,\varrho\,d^2\Omega\,d^3V\,d^3v. \tag{39.3}$$

From the geometry of a collision, averaging over the azimuthal scattering angle ψ (see Fig. 10) gives

$$\langle \boldsymbol{v} - \boldsymbol{v}'\rangle_\psi = \frac{M}{M+m}\,(1 - \cos\chi_g)\,(\boldsymbol{v} - \boldsymbol{V}). \tag{39.4}$$

Integrating over the scattering angle χ_g, and defining collision frequencies ν_g in the center of gravity system in complete analogy to those, ν_c, in the laboratory system [Eqs. (27.5) to (27.7)], one obtains

$$n\,\boldsymbol{a} = \frac{M}{M+m}\iint F\,f\,\frac{\nu_{g1}}{n_g}\,(\boldsymbol{v} - \boldsymbol{V})\,d^3V d^3v. \tag{39.5}$$

The linear term in V disappears on integrating over $\boldsymbol{V}$ and one obtains, thanks to the independence of ν_g, on $|\boldsymbol{v} - \boldsymbol{V}|$

$$\boldsymbol{a} = \frac{M}{M+m}\,\nu_{g1}\boldsymbol{v}_d = \nu_{c1}\boldsymbol{v}_d. \tag{39.6}$$

The expression with ν_c is obtained from (31.9) and this shows the relation between ν_c and ν_g, the collision frequency in the laboratory and the center of gravity system. The factor $\frac{M+m}{M}$ is known as the persistence of velocity factor. If the scattering is isotropic in the center of gravity system, it is not isotropic in the laboratory system due to the recoil of the molecule, resulting in an average forward velocity of the scattered ion after a collision. As the frequency defined by (27.6) is the average frequency with which the forward motion is annulled by

[1] C. MAXWELL: Coll. Works 2, 36.

collisions, a "collision" in the center of gravity system is only a "fractional collision" in the laboratory system and the laboratory collision frequency is slightly less than the center of gravity collision frequency. Similarly, the laboratory mean free path l_c is slightly longer than the center of gravity mean free path l_g. Substituting for $\boldsymbol{a}$, we obtain the mobility

$$\mu = \frac{v_d}{E} = \frac{e}{m_r \nu_{g1}} = \frac{e}{m \nu_c} \tag{39.7}$$

where $m_r = Mm/(M+m)$ is the reduced mass.

40. Energy. Let

$$X = v^2 \tag{40.1}$$

whence

$$\int f(\boldsymbol{a}\cdot\text{grad})\, v^2\, d^3v = \int f\, 2\,\boldsymbol{a}\cdot\boldsymbol{v}\, d^3v = 2\,\boldsymbol{a}\cdot\boldsymbol{v}_d. \tag{40.2}$$

It is convenient to consider the velocity $\boldsymbol{v}$ as the sum of the velocity $\boldsymbol{v}_g$ of the center of gravity plus a fraction of the relative velocity c

$$\boldsymbol{v} = \boldsymbol{v}_g + \frac{M}{M+m}\,\mathbf{c}. \tag{40.3}$$

We then have

$$(M+m)^2 v^2 = (M+m)^2 v_g^2 + 2M(M+m)\,\boldsymbol{v}_g\cdot\mathbf{c} + M^2 c^2 \tag{40.4}$$

and as $\boldsymbol{v}_g$ and the magnitude c are conserved in a collision

$$(M+m)\,(v^2 - v'^2) = 2M\,\boldsymbol{v}_g\cdot(\mathbf{c}-\mathbf{c}'). \tag{40.5}$$

Averaging over the azimuthal angle gives

$$(M+m)\langle v^2 - v'^2\rangle_\varphi = 2M\boldsymbol{v}_g\cdot\mathbf{c}\,(1-\cos\chi_g). \tag{40.6}$$

Averaging over scattering angles gives

$$(M+m)\langle v^2 - v'^2\rangle_\Omega = 2M\,\boldsymbol{v}_g\cdot\mathbf{c}\,\nu_{g1}/n_g \tag{40.7}$$

or substituting back for $\boldsymbol{v}_g$ and $\mathbf{c}$

$$(M+m)^2\langle v^2 - v'^2\rangle_\Omega = 2M(m\boldsymbol{v} + M\boldsymbol{V})\cdot(\boldsymbol{v}-\boldsymbol{V})\,\nu_{g1}/n_g. \tag{40.8}$$

Substituting in the transport equation, the terms in $\boldsymbol{v}\cdot\boldsymbol{V}$ drop out on integrating over the distribution function F for the molecules. Integrating over the ion velocities gives

$$(M+m)^2\boldsymbol{a}\cdot\boldsymbol{v}_d = M\left(m\overline{v^2} - M\overline{V^2}\right)\nu_{g1}. \tag{40.9}$$

Substituting for $\boldsymbol{a}$ in terms of $\boldsymbol{v}_d$ from (39.6) gives

$$(M+m)\,v_d^2 = m\overline{v^2} - M\overline{V^2} \tag{40.10}$$

or for the random energy

$$m\overline{v_r^2} = M\left(\overline{V^2} + v_d^2\right). \tag{40.11}$$

It appears at first surprising that the drift velocity in this equation should be associated with the mass of the molecule rather than that of the ion. However, if we consider ourselves in the system moving with the drift velocity of the ions, (40.11) says that the mean energy of the ions is equal to the mean energy of the molecules, composed of their random energy and drift energy relative to the ions. This is therefore an application of the equipartition theorem and, viewed this way, exactly what one might expect.

The above calculations can be repeated with a component energy

$$X = v_x^2 \quad \text{and} \quad \boldsymbol{a} \cdot \operatorname{grad} v_x^2 = 0. \tag{40.12}$$

The collision integral leads, after somewhat more difficult manipulations, to the relation

$$6\nu_{g1}(m\, v_x^2 - M\, V_x^2) = \nu_{g2} M (v_z^2 - v_x^2) \tag{40.13}$$

where ν_{g2} is the collision frequency which emphasizes right-angle scattering [Eq. (27.7)]. Combining this with (40.11) gives the ellipticity in the distribution of the random energy

$$\varepsilon \equiv \frac{\overline{v_{rz}^2} - \overline{v_x^2}}{\overline{v_x^2}} = 3\,\frac{2\nu_{g1} - \nu_{g2}}{\nu_{g2}}\,\frac{m\overline{v_x^2} - M\overline{V_x^2}}{(m+M)\,\overline{v_x^2}}. \tag{40.14}$$

For low drift velocities, $m\overline{v_x^2} = M\overline{V_x^2}$, and in general for light ions such as electrons, the ellipticity is small and the distribution is almost spherical about the drift velocity $\boldsymbol{v}_d$. However, for heavy ions, $m > M$, the distribution about v_d is such that the random velocity in the direction of the drift is greater than the random velocity at right angles to the drift. If the scattering frequency ν_{g2} is small there may be very little random velocity at right angles to the direction of drift, or if $\nu_{g2} = 2\nu_{g1}$, which is unlikely, the ellipticity vanishes in all cases.

Similar calculations can be carried through for all higher moments of the distribution function and therefore the distribution function f can itself be determined. However, it is not necessary to do so in order to get the moments, and as the latter are the desired quantities there is no purpose in calculating the distribution.

c) Constant mean free path.

41. Heavy ions. It is seen from (40.11) that the random velocity of a heavy ion in a light gas ($m > M$) is small relative to the drift velocity, so that a heavy ion moves like a particle falling in a viscous medium, its distribution function resembling a delta function. As a better approximation, we can give the distribution function the form of a GAUSSian function

$$f = A\, e^{-\frac{m}{2M}\frac{3+\mathcal{E}}{V^2+v_d^2}\left[v_x^2 + v_y^2 + \frac{(v_z - v_d)^2}{1+\mathcal{E}}\right]}. \tag{41.1}$$

As the random velocity is small compared to the drift velocity, the actual velocity is nearly a constant and therefore it does not matter what form of variation the collision frequency has with velocity. The distribution (41.1) is good in general for a heavy ion, and so is the equation (39.6) for the drift velocity, provided ν_g is evaluated for the drift velocity.

Applying this to the case of a constant mean free path

$$l_g = \frac{v}{\nu_g} \tag{41.2}$$

and introducing this in (39.6), we find for the drift velocity of a heavy ion in a light gas

$$v_d = \sqrt{\frac{e E l_g}{m_r}} \tag{41.3}$$

A note on "mean free paths" is in order here. When the "probability" of collision P_c is measured by a monoenergetic ion beam method, and the angle scattering $I(\chi_0)$ is also measured, P_c can be "corrected" to the momentum transfer collision probability P_m by

$$P_m = P_c \frac{\int I(\chi_0)(1-\cos\chi_0)\,d\cos\chi_0}{\int I(\chi_0)\,d\cos\chi_0} \tag{41.4}$$

and then

$$l_c = 1/p_0 P_m \tag{41.5}$$

where p_0 is the pressure in millimeters reduced to 0° C. The observed scattering angles χ_0 can be corrected to the center of gravity angle χ_g by (38.7), and if these angles are used in (41.4) and (41.5) one obtains $l_g < l_c$. Neither of these is the MAXWELLian mean free path l_M, defined as the mean free path of the particles in a gas at a temperature T_g. If the mean free path is independent of velocity $l_M = l_g/\sqrt{2}$.

42. Light ions. The limit of a very light ion in a heavy gas is also known as this has been treated in Sect. 31 and 35 for electrons. Applying this to the case of a constant mean free path and with negligible thermal motion of the gas molecules, we know that the distribution function is of the form given by DRUYVESTEYN

$$f = A\,e^{-w}, \qquad w = \frac{3m}{M+m}\left(\frac{u}{e\,E\,l_c}\right)^2. \tag{42.1}$$

From this distribution, one can calculate the drift velocity by (31.9)

$$v_d = \frac{\sqrt{\pi}}{3\Gamma(\frac{3}{4})}\left(\frac{3m}{M+m}\right)^{\frac{1}{4}}\sqrt{\frac{2e\,E\,l_c}{m}} = \gamma\left(\frac{m}{M+m}\right)^{\frac{1}{4}}\sqrt{\frac{e\,E\,l_g}{m_r}} \tag{42.2}$$

where account has been taken of the persistence of velocity factor to convert mean free paths in the laboratory system to the mean free paths in the center of gravity system. In similar fashion, one can calculate the mean energy

$$m\,\overline{v^2} = \sqrt{\frac{M+m}{6m}}\,\pi\,e\,E\,l_c. \tag{42.3}$$

These expressions are not unlike those found above for heavy ions and one may therefore expect to be able to write for all mass ratios

$$v_d = \gamma\left(\frac{m}{M+m}\right)^{\frac{1}{4}}\sqrt{\frac{\pi\,e\,E\,l_g}{m_r}}, \tag{42.4}$$

$$m\,\overline{v_r^2} = M(V^2 + \alpha\,v_d^2), \tag{42.5}$$

$$\varepsilon = \beta\,\frac{3m}{M+m}\,\frac{m\,\overline{v_x^2} - M\,\overline{V_x^2}}{(M+m)\,\overline{v_x^2}}, \tag{42.6}$$

Table 2.

m/M	0	1	∞
α	1.0607	0.789	1
β		0.731	1
γ	0.8973	0.9643	1

where the parameters α, β and γ are slowly varying functions of the mass ratio. We have assumed that $\nu_{g2} = \nu_{g1}$ in the ellipticity formula and introduced the factor $\frac{m}{M+m}$ quite arbitrarily because it goes to zero and one at the two limits, as it should.

The work of WANNIER[1] gives the values of α, β, and γ for $m = M$ so that we can construct Table 2 above.

[1] G. H. WANNIER: Phys. Rev. **83**, 281 (1951). — Bell Syst. Techn. J. **32**, 170 (1953).

d) Polarizable molecules.

43. Force between ion and molecule. A molecule in an electric field E acquires a dipole moment

$$\boldsymbol{P} = \alpha \boldsymbol{E} \tag{43.1}$$

where α is the polarizability of the molecule, and this polarizability of the molecule results in a dielectric coefficient K for the gas given by

$$\frac{\alpha n_g}{\varepsilon_0} = 3 \frac{\varepsilon - \varepsilon_0}{\varepsilon + 2\varepsilon_0} \approx K - 1 \tag{43.2}$$

where n_g is the number of molecules per unit volume. The electric field of an ion will also polarize a molecule and the induced dipole moment of the molecule will produce an electric field at the ion resulting in an attractive force between ion and molecule

$$F = \left(\frac{e}{4\pi\varepsilon_0}\right)^2 \frac{2\alpha}{r^5} \tag{43.3}$$

whose potential is

$$\Phi = -\frac{\alpha}{2}\left(\frac{e}{4\pi\varepsilon_0 r^2}\right)^2. \tag{43.4}$$

This polarization force is exactly the type considered in Sect. 39 and 40 and is the major long range force between ion and molecule. The same force, of course, exists between an electron and a molecule. In the case of an electron at energy below the ionization energy, the electron's wavelength is comparable to the diameter of the atom, and therefore wave mechanics must be used, and diffraction effects occur, resulting in a rather complicated dependence of the cross section on the electron's velocity. The wavelengths of ions are very much smaller than those of electrons and therefore classical mechanics may be used. VOGT and WANNIER[1] have shown that quantum mechanics gives substantially the same orbits as classical mechanics in this case. The classical orbits are soluble in terms of elliptic functions[2]. If u is the relative kinetic energy of the particles at infinity and b the impact parameter, the orbits fall into two classes according to the value of $b^4 u$. If $b^4 u > 2\alpha(e/4\pi\varepsilon_0)^2$, the orbit is open, resembling a hyperbola. If $b^4 u < 2\alpha(e/4\pi\varepsilon_0)^2$, the orbit goes through the origin. In the singular case, $b^4 u = 2\alpha(e/4\pi\varepsilon_0)^2$, the orbit is a spiral asymptotic to a circle of radius $b/\sqrt{2}$. The central orbits certainly correspond to collisions and the frequency of such collisions is given by

$$\nu = n_g \pi b^2 v = \frac{e n_g}{2\varepsilon_0}\sqrt{\frac{\alpha}{m_r}} = \frac{e}{2}\sqrt{\frac{n_g(K-1)}{\varepsilon_0 m_r}}. \tag{43.5}$$

The open orbits also contribute to the collision frequency according to the scheme of (27.5), (27.6) and (27.7) and we have, in the center of gravity system,

$$\nu_{g0} = \infty, \tag{43.6}$$

$$\nu_{g1} = 1.1052\,\nu + \nu_t, \tag{43.7}$$

$$\nu_{g2} = 1.158\,\nu + \nu_t \tag{43.8}$$

where ν_t is the charge transfer frequency, which is effective when an ion moves in the parent gas, so that no energy loss is involved. The charge transfer cross sections for the rare gases have been computed by HOLSTEIN[3] and are large compared to the elastic cross sections. (43.7) and (43.8) are the appropriate frequencies to use in the equations of Sect. 39 and 40.

---

[1] E. VOGT and G. H. WANNIER Phys. Rev. **95**, 1190 (1954).

[2] W. D. MACMILLAN: Theoretical Mechanics, p. 297. New York: McGraw Hill 1927.

[3] T. HOLSTEIN: J. Phys. Chem. **56**, 832 (1952).

Besides the long range polarization force between ion and molecule, there is a short range repulsive force due to the exclusion principle which prevents the electron shells of ion and molecule from overlapping too much. This force increases very rapidly as molecule and ion approach and may be approximated by a rapidly rising potential, as in the collision of rigid spheres. This leads to a constant mean free path, and hence for energetic collisions the equations of Sect. 42 are appropriate. Measurements of P_c for ions[1] indicate that an energy above one electron-volt is necessary for the rigid sphere to begin to be effective in scattering. Ions of near thermal energy should therefore always show a constant mean free time.

Substituting from (43.7) and (43.5) in (39.6) gives for the mobility

$$\mu = 0.905 \sqrt{\frac{4\varepsilon_0}{n_g m_r (K-1)}}\,. \tag{43.9}$$

In comparing the mobilities at low E/p of different ions in the same gas (same value of K), the dependence on the $-\frac{1}{2}$ power of the reduced mass is well verified[2]. The dependence on the dielectric constant K, as measured macroscopically, is qualitatively correct but not exact. The difference can be due both to the non-uniformity of the electric field of an ion and to its time variation.

As E/p is increased there are two transitions: from thermal to non-thermal energies of the ions, and from polarization force to rigid sphere. Fortunately the first transition, which involves energies well below an electron-volt, occurs entirely in the polarization force range for which the equations include thermal motions and for which, in particular, the mobility is independent of thermal motions. We need not, therefore, include thermal energies in the constant mean free path formulae.

44. The LANGEVIN theory. The transition from a polarization force to a rigid sphere force was elegantly treated by LANGEVIN[3] and recalculated by HASSE[4]. The mobility was worked out as a function of a parameter λ given by

$$\lambda^2 = \frac{kT}{-\Phi_c} = \frac{2kT}{\alpha}\left(\frac{4\varepsilon_0 Q}{e}\right)^2 \tag{44.1}$$

where Q is the rigid sphere cross section, and Φ_c is the potential (43.4) of the ion when in contact with the rigid sphere part of the force. They found the mobility

$$\mu = A \sqrt{\frac{4\pi\varepsilon_0}{n_g m_r (K-1)}} \tag{44.2}$$

where A is a function of λ which goes to $0.9048/\sqrt{\pi}$ as λ goes to zero, corresponding to (43.9), and to $\frac{3}{4}\lambda$ for large λ, corresponding to (42.4). Unfortunately, their calculations were based on the assumption that the ions were in thermal equilibrium with the gas and had a random energy $\frac{3}{2}kT$. The LANGEVIN theory can be rescued, until a more correct theory is worked out, by replacing $3kT$ by the correct random energy $M(v_d^2 + V^2)$, or, as the thermal velocity of the gas does not matter for either low or high drift velocities, by simply replacing $3kT$ by Mv_d^2.

Let a and c be two constants defined by

$$a = \frac{4\varepsilon_0 Q}{e}\sqrt{\frac{2M}{3\alpha}} \tag{44.3}$$

[1] C. RAMSAUER and O. BEECK: Ann. Phys. **87**, 1 (1928).
[2] J. H. MITCHELL and K. E. W. RIDLER: Proc. Roy. Soc. Lond., Ser. A **146**, 911 (1934).
[3] P. LANGEVIN: Ann. Chim. Phys. **5**, 245 (1905).
[4] H. R. HASSE: Phil. Mag. **1**, 139 (1926).

where α is the polarizability and Q the rigid sphere cross section, and

$$c = \frac{n_g}{2\varepsilon_0}\sqrt{\frac{\alpha m_r}{\pi}}. \tag{44.4}$$

$$\frac{c}{a} = e\,\frac{K-1}{4\varepsilon_0 Q}\sqrt{\frac{3}{8\pi}\,\frac{m}{M+m}}\,. \tag{44.5}$$

We then have

$$v_d = \frac{\lambda}{a}, \tag{44.6}$$

$$\mu = \frac{A}{c}, \tag{44.7}$$

$$E = \frac{c\,\lambda}{a\,A}. \tag{44.8}$$

λ/A is therefore to be taken as the independent variable, proportional to E/p. A and λ are the dependent variables which give the mobility (as before) and the drift velocity. The ratio λ/A has been computed from HASSE's figures and is given below. A peculiarity of the LANGEVIN theory is that the mobility goes through a maximum during the transition. This has been observed by HERSHEY[1].

Table 3.

$\lambda/A = aE/c$	$A = c\mu$	$\lambda = a v_d$	$\lambda/A = aE/c$	$A = c\mu$	$\lambda = a v_d$	$\lambda/A = aE/c$	$A = c\mu$	$\lambda = a v_d$
0.00	0.5105	0.0	3.034	0.4614	1.4	10.773	0.2599	2.8
0.1822	0.5488	0.1	3.408	0.4402	1.5	11.531	0.2515	2.9
0.3541	0.5648	0.2	3.809	0.4201	1.6	12.315	0.2436	3.0
0.5212	0.5756	0.3	4.238	0.4011	1.7	13.124	0.2362	3.1
0.6854	0.5836	0.4	4.695	0.3834	1.8	13.962	0.2292	3.2
0.8495	0.5886	0.5	5.180	0.3668	1.9	14.825	0.2226	3.3
1.0163	0.5904	0.6	5.692	0.3514	2.0	15.719	0.2163	3.4
1.1909	0.5878	0.7	6.232	0.3370	2.1	16.635	0.2104	3.5
1.3803	0.5796	0.8	6.798	0.3236	2.2	17.578	0.2048	3.6
1.5895	0.5662	0.9	7.393	0.3111	2.3	18.556	0.1994	3.7
1.824	0.5483	1.0	8.016	0.2994	2.4	19.547	0.1944	3.8
2.084	0.5277	1.1	8.662	0.2886	2.5	20.580	0.1895	3.9
2.373	0.5057	1.2	9.339	0.2784	2.6	21.633	0.1849	4.0
2.689	0.4834	1.3	10.041	0.2689	2.7			

V. The FOKKER-PLANCK equation[2].

a) Flow in velocity space.

45. Relation of the FOKKER-PLANCK to the transport equation. The interactions of electrons with other electrons and with ions through their COULOMB fields are of very much longer range than even the inverse fifth power fields due to polarization which lead to a constant mean free path, and this introduces the difficulty that collisions are no longer distinct separate events, but an electron is in interaction with many other electrons and ions simultaneously. A less serious difficulty which enters in considering electron-electron interactions is that the two interacting distribution functions F and f are now identical, so that the BOLTZMANN equation is quadratic in f, and in order to solve it, it must somehow be linearized.

[1] A. V. HERSHEY: Phys. Rev. **56**, 908 (1939).

[2] This section is largely based on a Doctor's thesis (Mass. Inst. Tech. 1955) by HARRY DREICER, which, in turn, rested heavily on the work of L. SPITZER and R. HÄRM: Phys. Rev. **89**, 977 (1953) and of S. CHANDRASEKHAR: Astrophys. J. **97**, 255, 263 (1943); **98**, 54 (1943).

Both of these difficulties are alleviated by considering the FOKKER-PLANCK equation instead of the BOLTZMANN equation. This equation is readily derived from the transport equation (37.3). If the quantity $X' - X$ is expanded in powers of the small changes $\Delta \boldsymbol{v}$ of the velocity occurring in a collision

$$X' = X + \Delta v_i \frac{\partial X}{\partial v_i} + \frac{1}{2} \Delta v_i \Delta v_j \frac{\partial^2 X}{\partial v_i \partial v_j} + \cdots. \tag{45.1}$$

Introducing this in the transport equation without field terms

$$\int X \frac{\partial f}{\partial t} d^3 v = \iiint (X' - X) F(V) f(v) \varrho\, d^2\Omega\, d^3 V\, d^3 v \tag{45.2}$$

and integrating by parts yields

$$\left.\begin{aligned} &\int X \frac{\partial f}{\partial t} d^3 v \\ &= \iint \left[-X \frac{\partial}{\partial v_i} \int \Delta v_i f \varrho\, d^2\Omega + \frac{1}{2} X \frac{\partial^2}{\partial v_i \partial v_j} \int \Delta v_i \Delta v_j f \varrho\, d^2\Omega \ldots \right] F\, d^3 V d^3 v. \end{aligned}\right\} \tag{45.3}$$

We introduce a vector quantity

$$\gamma = \iint F \Delta \boldsymbol{v}\, \varrho\, d^2\Omega\, d^3 V \tag{45.4}$$

and a tensor quantity

$$\gamma_{ij} = \iint F \Delta v_i \Delta v_j \varrho\, d^2\Omega\, d^3 V. \tag{45.5}$$

We can then define a flow vector in velocity space by

$$g_i = \gamma_i f - \frac{1}{2} \frac{\partial \gamma_{ij} f}{\partial v_j} + \cdots. \tag{45.6}$$

The vector $-m\gamma$ has the dimensions of a force, enters the equations like a friction force, and has been termed the dynamical friction. In general it is directed oppositely to the velocity of an electron and is the force which will bring the velocity of a fast electron which is shot into a swarm of electrons and ions down to the average velocity of the swarm. The quantity γ_{ij} enters the equations like a diffusion coefficient but in velocity space instead of in configuration space. It is therefore called the diffusion-in-velocity tensor. $m\gamma_{ij}$ has the dimensions of power. The dynamical friction will reduce the average velocity of any electron in a swarm down to the drift velocity of the swarm. The diffusion in velocity will maintain the mean squared velocity of any electron near the average of the swarm and thus maintain equipartition. The higher terms in (45.6) have not been given names because they are not absolutely essential to thermal equilibrium, and also because random interactions tend to make them vanish. We note that they exist but will not calculate them.

Introducing $\boldsymbol{g}$ into equation (45.3) gives

$$\int X \frac{\partial f}{\partial t} d^3 v = \int -X \operatorname{div} \boldsymbol{g}\, d^3 v \tag{45.7}$$

and as this equation must hold for all functions X

$$\frac{\partial f}{\partial t} = -\operatorname{div} \boldsymbol{g}. \tag{45.8}$$

This is the FOKKER-PLANCK equation. As it is derived here, the FOKKER-PLANCK-equation is a series approximation for the BOLTZMANN equation in powers of the deflections Δv which occur in a collision. It will converge best when collisions result in many small deflections, that is, for long range forces, and therefore is appropriate for COULOMB forces. A different derivation based on the LANGEVIN equation will be given in Sect. d.

46. Conservation theorems. Because $\partial f/\partial t$ is given by the FOKKER-PLANCK equation as the divergence of a flow vector and as this flow vector vanishes at infinity with f, the conservation of particles is automatically assured. This is not automatic, however, for momentum or energy. The rate of change of total momentum and energy follows from the FOKKER-PLANCK equation by multiplying by $m\boldsymbol{v}$ and $\frac{1}{2} m v^2$ and integrating by parts.

$$\frac{\partial}{\partial t} m\bar{\boldsymbol{v}} = \int m\,\boldsymbol{\gamma} f\, d^3 v \tag{46.1}$$

and

$$\frac{\partial \bar{u}}{\partial t} = \int m\left(\boldsymbol{\gamma}\cdot\boldsymbol{v} + \frac{1}{2}\gamma\right) f\, d^3 v \tag{46.2}$$

where $\gamma = \Sigma\, \gamma_{ii}$ is the trace of the diffusion-in-velocity tensor. If the distribution F has spherical symmetry, the vector $\boldsymbol{\gamma}$ is radial, and if f also has spherical symmetry, the integral in (46.1) vanishes. In the theory of BROWNian motion the dynamical friction is taken from STOKES' law

$$\boldsymbol{\gamma} = -\beta\boldsymbol{v}. \tag{46.3}$$

It is then assumed that the diffusion-in-velocity coefficient is both constant and a scalar. In order for (46.2) to vanish it then follows that

$$\gamma = 2\beta\overline{v^2} = 6\beta k T/m. \tag{46.4}$$

This simple procedure is quite unsatisfactory in the present case where both γ and γ_{ij} turn out to be complicated functions of the velocity.

47. The flow vector in spherical coordinates. The distribution function f is generally expressed in polar coordinates and it is therefore convenient to have the flow vector $\boldsymbol{g}$ in polar coordinates also. The evaluation of the dynamical friction γ in curvalinear coordinates involves quantities equivalent to centrifugal and CORIOLIS forces and it is the purpose of this section to evaluate these quantities. Let $\boldsymbol{v}$ and $\boldsymbol{v}'$ be the velocities of an electron before and after a "collision", and the scalar difference $\Delta v = v' - v$. Consider the Cartesian coordinate system $\xi\eta\zeta$ such that the ξ axis is in the direction of v, η in the direction of increasing ϑ and ζ in the direction of increasing φ. The coordinates of the vector $\boldsymbol{v}'$ in this coordinates system are

$$\left.\begin{aligned} \xi' &= v'\left[\cos\vartheta'\cos\vartheta + \sin\vartheta'\sin\vartheta\cos(\varphi'-\varphi)\right],\\ \eta' &= v'\left[-\cos\vartheta', \sin\vartheta + \sin\vartheta'\cos\vartheta\cos(\varphi'-\varphi)\right],\\ \zeta' &= v'\sin\vartheta'\sin(\varphi'-\varphi). \end{aligned}\right\} \tag{47.1}$$

Applying TAYLOR's theorem and keeping terms of second order in Δv, $\Delta\vartheta$, $\Delta\varphi$, we obtain the components of $\Delta\boldsymbol{v}$

$$\left.\begin{aligned} \Delta\xi &= \Delta v - \frac{v}{2}(\Delta\vartheta)^2 - \frac{v}{2}\sin^2\vartheta(\Delta\varphi)^2 + \cdots,\\ \Delta\eta &= v\Delta\vartheta + \Delta v\Delta\vartheta - \frac{v}{2}\sin\vartheta\cos\vartheta(\Delta\varphi)^2 + \cdots,\\ \Delta\zeta &= v\sin\vartheta\Delta\varphi + \sin\vartheta\Delta v\Delta\varphi + v\cos\vartheta\Delta\vartheta\Delta\varphi + \cdots \end{aligned}\right\} \tag{47.2}$$

and the second order quantities

$$\left.\begin{aligned}(\Delta\xi)^2 &= (\Delta v)^2, & \Delta\eta\,\Delta\zeta &= v^2\sin\vartheta\,\Delta\vartheta\,\Delta\varphi,\\ (\Delta\eta)^2 &= (v\,\Delta\vartheta)^2, & \Delta\zeta\,\Delta\xi &= v\sin\vartheta\,\Delta v\,\Delta\varphi,\\ (\Delta\zeta)^2 &= (v\sin\vartheta\,\Delta\varphi)^2, & \Delta\xi\,\Delta\eta &= v\,\Delta v\,\Delta\vartheta.\end{aligned}\right\} \tag{47.3}$$

Substituting from Eqs. (47.3) back into Eqs. (47.2), we can solve these equations for the polar quantities in terms of the Cartesian ones

$$\left.\begin{aligned}\Delta v &= \Delta\xi + \frac{(\Delta\eta)^2 + (\Delta\zeta)^2}{2v},\\ v\,\Delta\vartheta &= \Delta\eta - \frac{\Delta\xi\,\Delta\eta}{v} + \frac{\cot\vartheta}{2v}(\Delta\zeta)^2,\\ v\sin\vartheta\,\Delta\varphi &= \Delta\zeta - \frac{\Delta\xi\,\Delta\zeta}{v} - \frac{\cot\vartheta}{v}\Delta\eta\,\Delta\zeta.\end{aligned}\right\} \tag{47.4}$$

The quantities γ_i and γ_{ij} which by Eqs. (45.4) and (45.5) are averages of Δv_i and $\Delta v_i \Delta v_j$ transform in exactly the same way as the Δv's. Remembering that the operation $\frac{\partial\gamma_{ij}}{\partial v_j}$ is a divergence, and after some regrouping of terms, one obtains

$$\begin{aligned}g_v &= \left(\gamma_\xi + \frac{\gamma_{\eta\eta}+\gamma_{\zeta\zeta}}{2v}\right)f - \frac{1}{2v^2}\frac{\partial}{\partial v}v^2\gamma_{\xi\xi}f - \frac{1}{2v\sin\vartheta}\frac{\partial}{\partial\vartheta}\sin\vartheta\,\gamma_{\xi\eta}f - \frac{1}{2v}\frac{\partial}{\partial\varphi}\gamma_{\xi\zeta}f,\\ g_\vartheta &= \left(\gamma_\eta + \frac{\gamma_{\zeta\zeta}-\gamma_{\eta\eta}}{2v}\cos\vartheta\right)f - \frac{1}{2v^4}\frac{\partial}{\partial v}v^4\gamma_{\xi\eta}f - \frac{1}{2v}\frac{\partial}{\partial\vartheta}\gamma_{\eta\eta}f - \frac{1}{2v}\frac{\partial}{\partial\varphi}\gamma_{\eta\zeta}f,\\ g_\varphi &= \gamma_\zeta f - \frac{1}{2v^4}\frac{\partial}{\partial v}v^4\gamma_{\xi\zeta}f - \frac{1}{2v}\frac{\partial}{\partial\vartheta}\gamma_{\eta\zeta}f - \frac{1}{2v}\frac{\partial}{\partial\varphi}\gamma_{\zeta\zeta}f.\end{aligned} \tag{47.5}$$

These are the general expressions for the flow in velocity space for which we must now calculate expressions for γ_i and γ_{ij} appropriate to COULOMB interactions.

b) RUTHERFORD scattering.

48. Mean deflections. Let us now consider two interacting particles and the coordinate system r, s, t attached to the center of gravity. The deflection of the particle of mass m in collision with a particle of mass M (Fig. 13) are given by

$$\left.\begin{aligned}\Delta v_r &= \frac{Mc}{M+m}(1-\cos\chi_g),\\ \Delta v_s &= \frac{Mc}{M+m}\sin\chi_g\cos\psi,\\ \Delta v_t &= \frac{Mc}{M+m}\sin\chi_g\sin\psi.\end{aligned}\right\} \tag{48.1}$$

Squaring, we obtain

$$\left.\begin{aligned}(\Delta v_r)^2 &= \left(\frac{Mc}{M+m}\right)^2(1-\cos\chi_g)^2,\\ (\Delta v_s)^2 &= \left(\frac{Mc}{M+m}\right)^2\sin^2\chi_g\cos^2\psi,\\ (\Delta v_t)^2 &= \left(\frac{Mc}{M+m}\right)^2\sin^2\chi_g\sin^2\psi.\end{aligned}\right\} \tag{48.2}$$

These quantities must be integrated over scattering angles and over the distribution F to give γ_i and γ_{ij}. Let us introduce the symbols δ_i and δ_{ij} for the results of the integration over scattering angles; and use the definitions (27.6) and (27.7) changed to the center of gravity system, and also use n_+ to indicate charged scatterers. Then

$$\delta_r = \int \Delta v_r \varrho \, d^2\Omega = -\frac{M c}{M+m} \frac{\nu_{g1}}{n_+} \tag{48.3}$$

$$\left.\begin{aligned} \delta_{rr} &= 2\left(\frac{M c}{M+m}\right)^2 \frac{3\nu_{g1}-\nu_{g2}}{3n_+}, \\ \delta_{ss} &= \delta_{tt} = \left(\frac{M c}{M+m}\right)^2 \frac{\nu_{g2}}{3n_+} \end{aligned}\right\} \tag{48.4}$$

and all other δ's vanish because ϱ does not depend on ψ. Thus the vector $\boldsymbol{\delta}$ is directed towards the center of gravity and the tensor δ_{ij} has the structure

$$\delta_{ij} = \left(\frac{M c}{M+m}\right)^2 \frac{1}{3n_+} \begin{Vmatrix} 6\nu_{g1} - 2\nu_{g2} & 0 & 0 \\ 0 & \nu_{g2} & 0 \\ 0 & 0 & \nu_{g2} \end{Vmatrix}. \tag{48.5}$$

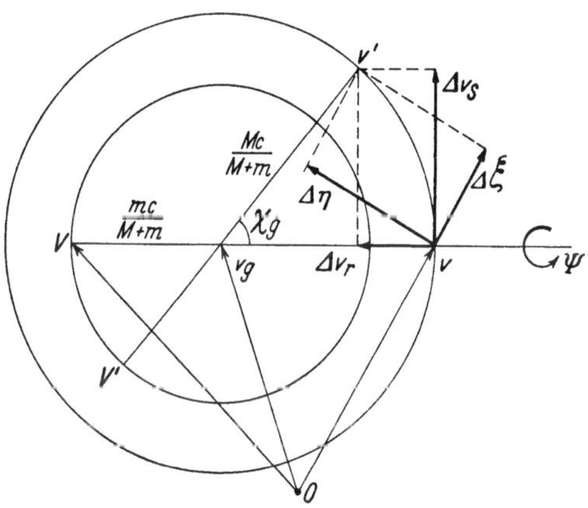

Fig. 13. Components of Δv in center of gravity coordinates r, s, t, and in laboratory coordinates ξ, η, ζ.

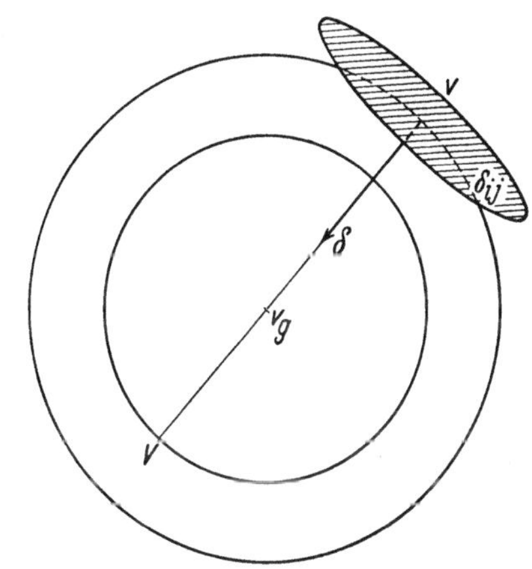

Fig. 14. Dynamical friction vector and diffusion tensor in relative coordinates.

This combination can be represented as in Fig. 14 by an arrow and an ellipsoid of revolution. For isotropic scattering the ellipsoid is elongated in the direction of the relative velocity, but if scattering is largely in the forward direction, that is, small angle scattering, the ellipsoid is very much flattened in the direction of the relative velocity. This situation comes from the factor $(1-\cos\chi_g)^2$ in (48.2) which strongly weights backward scattering. At an angle as large as 30°, this factor has only 4% of its value at 180°. The FOKKER-PLANCK equation converges rapidly only for small angle scattering and as $(1-\cos\chi_g)^2$ weights backward scattering very strongly, one may expect that δ_{rr} will appear in higher order terms of the FOKKER-PLANCK equation than δ_{ss}, δ_{tt} and δ_r. As we shall use only the first and second order terms of the FOKKER-PLANCK equation, we shall find it necessary eventually to neglect all terms in δ_{rr}, but in order to show how the difficulty enters we shall carry δ_{rr} along in the calculations until it is necessary to drop it.

It is noted that the four non-vanishing δ's contain only two ν_g's. There are therefore invariant relations between the δ's which do not depend on the scattering law. It is these relations which result in simpler expressions (52.5) for the flow vector $\boldsymbol{g}$ than are evident in (47.5). In particular the trace of the tensor δ_{ij} is a scalar

$$\delta = 2\left(\frac{Mc}{M+m}\right)^2 \frac{\nu_{g1}}{n_+} \tag{48.6}$$

and is simply related to the vector δ_r.

49. Shielded RUTHERFORD scattering. If the scattering particles of the last section are Z-fold ionized atoms ($n_- = Z n_+$), we can use for the scattering cross section σ the RUTHERFORD scattering formula

$$\sigma = \left[\frac{Z e^2}{4\pi \varepsilon_0 m_r c^2 (1 - \cos\chi_g)}\right]^2, \tag{49.1}$$

However, the integrals for both ν_{g1} and ν_{g2} diverge. Only the combination in δ_{rr} converges yielding

$$\delta_{rr} = \frac{\pi K}{c}, \qquad K = \left(\frac{Ze^2}{2\pi\varepsilon_0 m}\right)^2. \tag{49.2}$$

The other integrals diverge logarithmically for small scattering angles. COULOMB forces are too long ranged to consider the interaction to be simply between two particles at all distances. This problem was considered by CHANDRASEKHAR[1] in computing the time of relaxation of stellar systems and led him to cut off the RUTHERFORD cross Sect. (49.1), for an impact parameter equal to the average distance between stars. SPITZER[2] considers the same problem for electrons and ions and cuts off the RUTHERFORD cross section at a scattering angle χ_D for which the distance of closest approach is equal to the DEBYE distance l_D. The theory of DEBYE and HÜCKEL shows that the average potential of an ion is

$$V = \frac{Z e}{4\pi \varepsilon_0 r} \exp(-r/l_D), \qquad l_D^2 = \frac{\varepsilon_0 k T}{n_- e^2} \tag{49.3}$$

and therefore the average orbit of an electron going by an ion will be the classical orbit of a particle in such a field. This problem has not been solved, but cutting off the RUTHERFORD cross section at the angle χ_D is an approximation thereto. The use of an average potential is probably correct in calculating mean deflections, and its application to mean squared deflections is then required for consistency. In the interaction of an electron with an ion, the ion is hardly disturbed, and therefore will retain its electron cloud which shields its potential. The use of the DEBYE sphere in electron-electron interactions[3] is somewhat more difficult to justify. We shall assume the interaction (49.3) above to be good in both cases. Fortunately the cut off only appears in a logarithmic term and the difference between the assumptions of CHANDRASEKHAR and SPITZER is not as large as one might fear. The angle χ_D is given by

$$1 - \cos\chi_D = \frac{2}{1+q^2} \tag{49.4}$$

where

$$q = \frac{m_r c^2}{2} \frac{4\pi \varepsilon_0 l_D}{Z e^2}.$$

q is a variable as it contains c, but as q appears only in a logarithm it is convenient to treat it as a constant replacing $m_r c^2$ by its average value $3kT$. Then

$$\bar{q} = 6\pi n_- l_D^3/Z = 9 N_+/2 \tag{49.5}$$

[1] S. CHANDRASEKHAR: Principles of Stellar Dynamics, p. 58. University of Chicago Press 1942.

[2] L. SPITZER and R. HÄRM: Phys. Rev. **89**, 977 (1953).

[3] D. PINES and D. BOHM: Phys. Rev. **85**, 341 (1952).

where N_+ is the number of positive ions in the DEBYE sphere. It then follows from (48.6) and (48.3) that

$$\delta = 2\pi K L/c \tag{49.6}$$

and that

$$\delta_r = -\frac{M+m}{M}\frac{\pi K L}{c^2} \tag{49.7}$$

where

$$L = \ln\sqrt{1+q^2}. \tag{49.8}$$

Values of L for different values of $\log n_-\left(\frac{e}{kT}\right)^3$, where the argument is in (volt-cm.)$^{-3}$ are given below.

It is seen that for most discharges L is of the order 10 so that δ_{rr} is actually small compared to δ and the ellipsoid of Fig. 14 is flattened as indicated. The number N_+ varies as $1/\sqrt{n_-}$ and is down to one for $n_- = 3\times 10^{18}(kT/e)^3$ cm.$^{-3}$. As it seems unreasonable to extend the concept of DEBYE to less than one particle, the shielding at high concentrations has been set at $N_+ = 1$.

Table 4.

$\log n(e/kT)^3$	$L = \ln\sqrt{1+q^2}$
10	11.256
11	10.104
12	8.953
13	7.802
14	6.650
15	5.500
16	4.348
17	3.192
18	2.053
19	1.528
20	1.528

The deflection tensor δ_{ij} now has the form

$$\delta_{ij} = \frac{\pi K}{c}\begin{Vmatrix} 1 & 0 & 0 \\ 0 & L-\frac{1}{2} & 0 \\ 0 & 0 & L-\frac{1}{2} \end{Vmatrix}. \tag{49.9}$$

50. COULOMB's law in velocity space. The scalar diffusion-in-velocity δ and dynamical friction δ_r are quite remarkable in that they vary inversely as the relative velocity, and the square of the relative velocity, respectively, and thus obey, in velocity space, exactly the same laws as the electric potential and electric field in ordinary potential theory[1]. We can therefore apply our knowledge of potential theory to these quantities immediately. Consider, for instance, a spherical distribution of scattering particles. Then we know that the dynamical friction on a test particle in this distribution can be attributed entirely to particles whose speed is less than the speed of the test particle and these latter can be considered concentrated at the center of the spherical distribution.

α) Let us apply this to the conductivity of a completely ionized plasma. Let us take axes moving with the electron drift and assume that the electrons have a spherical distribution

$$f = n_-\left(\frac{m}{2\pi kT}\right)^{\frac{3}{2}} e^{-m v^2/2kT} \tag{50.1}$$

and let us suppose that the ions have a small drift velocity v_d relative to the electrons and have very little random velocity about this drift velocity. Then the dynamical friction acting on the ions is due entirely to the

$$n_d = \frac{2n_-}{3\sqrt{2\pi}}\left(\frac{m v_d^2}{kT}\right)^{\frac{3}{2}} \tag{50.2}$$

[1] M. ROSENBLUTH has pointed out that the part of δ_{ij} which contains L is also derivable from a "potential", $|c|$, by: $\delta_{ij} = \pi K L \frac{\partial^2 c}{\partial c_i\,\partial c_j}$.

electrons whose velocity is less than v_d, and the dynamical friction is equal to

$$M\boldsymbol{g} = -M\frac{M+m}{m}\frac{\pi K L}{v_d^3}\boldsymbol{v}_d n_+ n_d = -\frac{Z L \boldsymbol{v}_d}{3 m_r}\left(\frac{m}{2\pi k T}\right)^{\frac{3}{2}}\left(\frac{n_- e^2}{\varepsilon_0}\right)^2. \quad (50.3)$$

This dynamical friction must be balanced by the applied electric force

$$M\boldsymbol{g} + n_- e\boldsymbol{E} = 0 \quad (50.4)$$

from which we obtain the conductivity

$$\sigma = \frac{n_- e v_d}{E} = \frac{3 m_r}{Z L}\left(\frac{\varepsilon_0}{e}\right)^2\left(\frac{2\pi k T}{m}\right)^{\frac{3}{2}}. \quad (50.5)$$

This formula turns out to be too small by a factor 1.9747[1] mainly because of the neglect of diffusion-in-velocity which acts to reduce the dynamical friction, but it illustrates the application of potential theory.

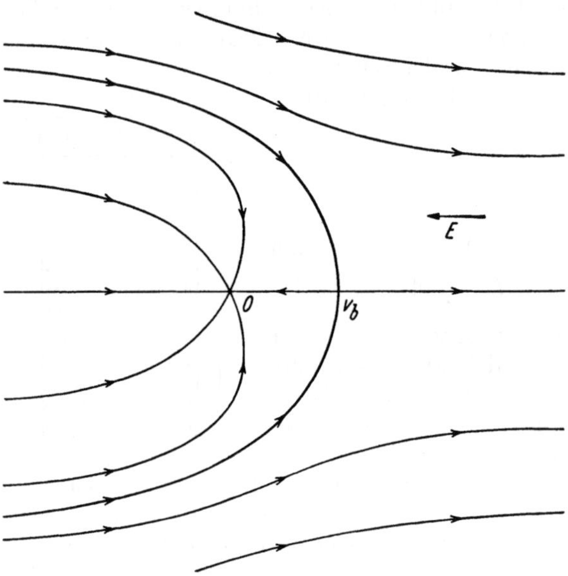

Fig. 15. Flow in velocity space due to external field and dynamical friction.

β) Consider now a fast electron, such that the dynamical friction acting on it is due to all the slower electrons and is $-2\pi n K L/v^2$. In the presence of an external electric field $\boldsymbol{E}$, the total force on such an electron is

$$\left.\begin{aligned}\boldsymbol{F} = &- e\boldsymbol{E} - \\ &-2\pi m K L n \boldsymbol{v}/v^3.\end{aligned}\right\} \quad (50.6)$$

This is the formula for the resultant of a uniform field plus a point charge, and the flow lines of such a field are given by

$$v^2 = \frac{2 v_b^2}{1+\cos\vartheta} + \frac{C}{\sin^2\vartheta}$$

where C is an arbitrary constant and the product

$$E\frac{m_r v_b^2}{2e} = 2\pi L\frac{e\, n^{\frac{2}{3}}}{4\pi\varepsilon_0}\frac{e\, n^{\frac{1}{3}}}{4\pi\varepsilon_0} = 13\times 10^{-18} n \text{ volt}^2/\text{meter} \quad (50.7)$$

is seen to be related to the product of the electric field and potential at the inter-electronic distance $n^{-\frac{1}{3}}$. The flow lines are of two kinds:

For negative C, the flow lines go through the origin. For positive C they go to infinity at both ends, the singular line separating these two categories being given by $C=0$. Neglecting the effects of diffusion-in-velocity, which give rise to transitions between lines, it is seen that electrons whose velocity is such that C is negative tend to stay with the electron distribution, whereas electrons of faster velocity for which C is positive will be accelerated indefinitely by the applied field and tend to escape. v_b is the escape velocity for an electron moving in the direction of the field.

[1] H. MAECKER, TH. PETERS and H. SCHENK: Z. Physik **140**, 119 (1955), give experimental values of σ requiring values for L below any given in Table 4.

51. Integration over the scattering field. We must now project from the r, s, t coordinate system which is attached to the relative velocity $\boldsymbol{c}$ onto the ξ, η, ζ coordinate system which is attached to the absolute velocity $\boldsymbol{v}$. Let ϑ_0, φ_0 be the angles between these two systems. Then

$$\left.\begin{aligned} \delta_\xi &= \delta_r \cos\vartheta_0 , \\ \delta_\eta &= \delta_r \sin\vartheta_0 \cos\varphi_0 , \\ \delta_\zeta &= \delta_r \sin\vartheta_0 \sin\varphi_0 . \end{aligned}\right\} \tag{51.1}$$

The scalar δ is, of course, invariant. Besides this we need the following tensor projections

$$\left.\begin{aligned} \delta_{\xi\xi} &= \delta_{rr}\left[\cos^2\vartheta_0 + (L-\tfrac{1}{2})\sin^2\vartheta_0\right] , \\ \delta_{\zeta\zeta} - \delta_{\eta\eta} &= \delta_{rr}(L-\tfrac{3}{2})\sin^2\vartheta_0\cos 2\varphi_0 , \\ \delta_{\xi\eta} = \delta_{\xi\zeta} &= \delta_{rr}(\tfrac{3}{2}-L)\sin\vartheta_0\cos\vartheta_0\cos\varphi_0 , \\ \delta_{\eta\zeta} &= \delta_{rr}(\tfrac{3}{2}-L)\sin^2\vartheta_0\sin\varphi_0\cos\varphi_0 . \end{aligned}\right\} \tag{51.2}$$

These quantities are then integrated over a distribution function

$$F = F^0 + F^1\cos\Theta. \tag{51.3}$$

Making use of the following relations between the variables

$$V^2 = v^2 + c^2 + 2cv\cos\vartheta_0 , \tag{51.4}$$

$$V\sin\psi = c\sin\vartheta_0 , \tag{51.5}$$

$$\cos\Theta = \cos\vartheta\cos\psi - \sin\vartheta\sin\psi\cos\varphi_0 \tag{51.6}$$

where the angle ψ is the angle between the two velocities $\boldsymbol{v}$ and $\boldsymbol{V}$. The integrals in Eqs. (45.4) and (45.5) can then all be evaluated using the differential

$$d^3V = dV\,d\cos\psi\,d\varphi_0 . \tag{51.7}$$

These tedious integrations have all been performed by SPITZER, and the results can be expressed in terms of the integrals

$$\left.\begin{aligned} I_j^l &= 4\pi v^{-j}\int_0^v F^l V^{j+2}\,dV , \\ J_j^l &= 4\pi v^{-j}\int_v^\infty F^l V^{j+2}\,dV . \end{aligned}\right\} \tag{51.8}$$

One finds that

$$\gamma_{\eta\eta} = \gamma_{\zeta\zeta} \quad \gamma_{\eta\zeta} = 0 \tag{51.9}$$

and

$$\left.\begin{aligned} \gamma_\xi &= -\frac{M+m}{M}\,\frac{\pi K L}{v^2}\left[I_0^0 + \frac{\cos\vartheta}{3}(2I_1^1 - J_{-2}^1)\right] , \\ \gamma_\eta &= -\frac{M+m}{M}\,\frac{\pi K L}{v^2}\left[\frac{\sin\vartheta}{3}(I_1^1 + J_{-2}^1)\right] , \end{aligned}\right\} \tag{51.10}$$

$$\left.\begin{aligned} \gamma &= \frac{2\pi K L}{v}\left[I_0^0 + J_{-1}^0 + \frac{\cos\vartheta}{3}(I_1^1 + J_{-2}^1)\right] , \\ \gamma_{\eta\eta} - \gamma_{\xi\xi} &= \frac{\pi K}{v}\left(L-\frac{3}{2}\right)\left[I_0^0 - I_2^0 + \frac{\cos\vartheta}{15}(5I_1^1 - 9I_3^1 - 4J_{-2}^1)\right] , \\ \gamma_{\xi\eta} &= -\frac{\pi K}{v}\left(L-\frac{3}{2}\right)\left[\frac{\sin\vartheta}{15}(5I_1^1 - 3I_3^1 + 2J_{-2}^1)\right] . \end{aligned}\right\} \tag{51.11}$$

The diffusion-in-velocity tensor having the general form

$$\gamma_{ij} = \begin{Vmatrix} A & C & 0 \\ C & B & 0 \\ 0 & 0 & B \end{Vmatrix}. \tag{51.12}$$

52. Reduction of the flow vector. It is noted that γ_ξ, γ_η, and γ are precisely the field and potential which would be obtained from a distribution of charge given by Eq. (51.3).

The integrals I and J satisfy the following differential relations

$$\left.\begin{aligned} \frac{\partial}{\partial v}\frac{I_j}{v} &= -(j+1)\frac{I_j}{v^2} + 4\pi v F\,, \\ \frac{\partial}{\partial v}\frac{J_j}{v} &= -(j+1)\frac{J_j}{v^2} - 4\pi v F \end{aligned}\right\} \tag{52.1}$$

and making use of these, one establishes the following three relations between the diffusion coefficients

$$\frac{\partial \gamma}{\partial v} = \frac{2M}{M+m}\gamma_\xi, \tag{52.2}$$

$$\left(\frac{3}{v} + \frac{\partial}{\partial v}\right)(\gamma_{\xi\xi} - \gamma_{\eta\eta}) + \frac{\frac{3}{2}}{v\sin\vartheta}\frac{\partial}{\partial\vartheta}(\sin\vartheta\,\gamma_{\xi\eta}) = \frac{M}{M+m}\frac{2L-3}{L}\gamma_\xi, \tag{52.3}$$

$$\frac{1}{v^4}\frac{\partial}{\partial v}(v^4\gamma_{\xi\eta}) = \frac{M}{M+m}\frac{2L-3}{L}\gamma_\eta. \tag{52.4}$$

Substituting these relations in the expressions for the flow vector, (47.5), the inertial terms in this vector cancel out and one obtains the much simpler expressions

$$\left.\begin{aligned} g_v &= \left(\frac{m}{M} + \frac{1}{L}\right)\frac{M\gamma_\xi}{M+m} f - \frac{1}{2}\left(\gamma_{\xi\xi}\frac{\partial f}{\partial v} + \gamma_{\xi\eta}\frac{\partial f}{v\,\partial\vartheta}\right), \\ g_\vartheta &= \left(\frac{m}{M} + \frac{2}{3L}\right)\frac{M\gamma_\eta}{M+m} f - \frac{1}{2}\left(\gamma_{\xi\eta}\frac{\partial f}{\partial v} + \frac{\partial(\gamma_{\eta\eta} f)}{v\,\partial\vartheta}\right), \\ g_\varphi &= 0. \end{aligned}\right\} \tag{52.5}$$

The inertial terms were somewhat artificial, coming as they did from the use curvilinear coordinates, and it is gratifying to see them drop out of these expressions. This has occurred by allowing part of the diffusion tensor to move into the position of dynamical friction and somewhat more than the inertial terms have been cancelled. This explains the small conductivity obtained in (50.5). It also leads one to expect that higher terms of the FOKKER-PLANCK equation may similarly move up, and might cancel the back-scattering terms coming from δ_{rr}. Two such terms are the $1/L$ in g_v and $2/3L$ in g_ϑ.

c) MAXWELLian distribution of scatterers.

53. Further reduction of the flow vector. We shall now first drop the back-scattering terms which are identified by not containing L. The dynamical friction remains as in (51.10), the diffusion tensor becomes

$$\left.\begin{aligned} \gamma_{\xi\xi} &= \frac{2\pi K L}{3v}\left[I_2^0 + J_{-1}^0 + 3\frac{\cos\vartheta}{5}(I_3^1 + J_{-2}^1)\right], \\ \gamma_{\eta\eta} &= \frac{\pi K L}{3v}\left[3I_0^0 - I_2^0 + 2J_{-1}^0 + \frac{\cos\vartheta}{5}(5I_1^1 - 3I_3^1 + 2J_{-1}^1)\right], \\ \gamma_{\xi\eta} &= \frac{\partial\gamma_{\eta\eta}}{\partial\vartheta}. \end{aligned}\right\} \tag{53.1}$$

and the flow vector is

$$\left.\begin{aligned} g_v &= \frac{m}{M+m}\gamma_\xi f - \frac{1}{2}\left(\gamma_{\xi\xi}\frac{\partial f}{\partial v} + \gamma_{\xi\eta}\frac{\partial f}{v\,\partial\vartheta}\right), \\ g_\vartheta &= \frac{m}{M+m}\gamma_\eta f - \frac{1}{2v}\left(\gamma_{\xi\eta}\frac{\partial v f}{\partial v} + \gamma_{\eta\eta}\frac{\partial f}{\partial\vartheta}\right). \end{aligned}\right\} \tag{53.2}$$

We then choose axes moving with the average scattering particle, so that $F^1=0$. The expressions then further reduce to

$$\gamma_\xi = -\frac{M+m}{M}\,\frac{\pi K L}{v^2}\,I_0^0. \tag{53.3}$$

$$\left.\begin{aligned} \gamma_{\xi\xi} &= \frac{2\pi K L}{3v}\,(I_2^0 + J_{-1}^0), \\ \gamma_{\eta\eta} &= \frac{\pi K L}{3v}\,(3I_0^0 - I_2^0 + 2J_{-1}^0), \end{aligned}\right\} \tag{53.4}$$

$$\left.\begin{aligned} g_v &= \frac{m\gamma_\xi}{M+m} f - \frac{1}{2}\gamma_{\xi\xi}\frac{\partial f}{\partial v}, \\ g_\vartheta &= -\frac{1}{2v}\gamma_{\eta\eta}\frac{\partial f}{\partial\vartheta}. \end{aligned}\right\} \tag{53.5}$$

If, furthermore, the scattering distribution is MAXWELLian

$$F = n_+\left(\frac{M}{2\pi kT}\right)^{\frac{3}{2}} e^{-MV^2/2kT}. \tag{53.6}$$

The integrations in (51.8) can all be performed and it is easy to show that

$$\gamma_\xi = -\frac{M+m}{M}\,\frac{\pi K L}{v^2}\,I_0^0 = -\frac{M+m}{2kT}\,v\gamma_{\xi\xi}, \tag{53.7}$$

$$\gamma_{\eta\eta} = \frac{\pi K L}{v}\left[\left(1 - \frac{kT}{M v^2}\right) I_0^0 - \frac{kT}{M}\,4\pi v F\right]. \tag{53.8}$$

Substituting in (53.5) gives

$$\left.\begin{aligned} g_v &= \frac{m\gamma_\xi}{M+m}\left(f + kT\frac{\partial f}{\partial u}\right), \\ g_\theta &= -\frac{\gamma_{\eta\eta}}{2v}\,\frac{\partial f}{\partial\vartheta} \end{aligned}\right\} \tag{53.9}$$

and it is evident that the flow is zero if the scattered particles have a MAXWELLian distribution at the same temperature as the scatterers. This relation does not follow if the terms not containing L had been retained, and this is the reason for believing that these terms must be cancelled by higher order terms in the FOKKER-PLANCK equation, for the principle of detailed balance requires that there be no flow in thermal equilibrium.

54. Electron interactions. Integrating (53.9) over a sphere in velocity space gives the gain due to interactions

$$G = -\frac{m}{M}\,4\pi^2 K L I_0\left(f^0 + kT\frac{\partial f^0}{\partial u}\right) \tag{54.1}$$

and this gives the rate at which a non-equilibrium distribution tends towards equilibrium. If f is MAXWELLian at a temperature above T the gain is everywhere negative and the distribution contracts. Reciprocally if f is too cold it expands. Ions are relatively inefficient in thermalizing an electron distribution because of the factor m/M. Of course if we are considering the interactions of electrons, I_0 itself is an integral over f and the factor "kT" should be a function depending on $I_2^0+J_{-1}^2$, but we shall disregard this error in the following. However, in solving the gain equation we must remember that kT is itself an unknown, representing the mean energy of the unknown f^0.

We would like to study the transition in electron distribution functions from those given in Sect. 35 as electron-electron interactions become important. To do this we merely add the gain (54.1) due to interactions to the gain (34.2) due to the field and collisions:

$$-\frac{G}{4\pi}=\left(\frac{\nu_c u_c v^3}{3}+\pi K L I_0 kT\right)\frac{\partial f}{\partial u}+\left(\frac{m}{M}\nu_c v^3+\pi K L I_0\right)f \tag{54.2}$$

and use this in the gain equation (34.2). Expanding I_0 in powers of v

$$I_0=\frac{n}{3}\sqrt{\frac{2}{\pi}}\left(\frac{m}{kT}\right)^{\frac{3}{2}} v^3\left(1-\frac{3mv^2}{10kT}\cdots\right), \tag{54.3}$$

and comparing this with the gain due to the electric field, it is seen that for slow electrons the interaction becomes dominant first at a concentration

$$n_-=\sqrt{\frac{mkT}{2\pi}}\left(\frac{2\pi\varepsilon_0}{e^2}\right)^2\frac{\nu_c u_c}{L} \tag{54.4}$$

or at a degree of ionization

$$\frac{n_-}{n_g}=\frac{u_c}{e}\sqrt{\frac{kT}{e}}\,\frac{\nu_c}{10^8 p_0}\,\frac{1.3}{10^3 L}\,. \tag{54.5}$$

Each factor in this product is generally less than one, particularly the last one which is about 10^{-4}. For degrees of ionization above 10^{-4} the distribution functions will look more MAXWELLian than those derived in Sect. 35.

Even though the electron concentration may considerably exceed the limit set by (54.4) there are two circumstances which keep the electron interactions from modifying the distribution functions too severely. Electron interactions modify neither their number nor their total energy. The plasma balance equation

$$\bar{\nu}_i=D/\Lambda^2 \tag{54.6}$$

and the energy balance equation

$$n e\mu E^2=\int_{v_x}^{\infty}\nu_x u f 4\pi v^2\,dv+\int_0^{\infty}\frac{3m}{M}\nu_c u f 4\pi v^2\,dv \tag{54.7}$$

do not contain the interaction terms explicitly.

As the electron concentration becomes high and the two interaction terms in the gain become large the temperature becomes indeterminate. We are then approaching thermal equilibrium and the temperature is determined by SAHA's equation

$$\frac{n_a}{n_+n_-}=\frac{g_a}{2g_i}\,\frac{h^3}{(2\pi m kT)^{\frac{3}{2}}}\,e^{eV_i/kT} \tag{54.8}$$

where n_a is the number of atoms remaining neutral, g_a and g_i the statistical weights of the ground states of atom and ion.

55. Electron-ion interactions. If the scattering particles are ions of mass $M \gg m$ all terms in $1/M$ in (53.7), (53.8), and (53.9) are negligible, leaving only

$$\gamma_{\eta\eta} = \frac{\pi K L}{v} I_0 = \frac{\pi L K n_+}{v}, \tag{55.1}$$

$$g_\vartheta = \frac{\pi K L n}{2v^2} + \frac{\partial f}{\partial \vartheta} = -\frac{\pi L K n_+}{2v^2} \sin\vartheta f^1 \tag{55.2}$$

all other terms being smaller than this one in the ratio m/M. Projecting on the Z-axis,

$$g_z = -g_\vartheta \sin\vartheta = \frac{\pi K L n_+}{2v^2} f^1 \sin^2\vartheta \tag{55.3}$$

and the total dynamical friction on the electrons is

$$m \iint g_z \, d^2\Omega \, v^2 \, dv = \frac{4\pi^2}{3} m K L n_+ \int f^1 \, dv \tag{55.4}$$

which must, in the absence of other collisions, be equal to the applied force $n_- e E$. The current is given by (19.7)

$$J = e \int \frac{4\pi}{3} f^1 v^3 \, dv = \sigma_i E \tag{55.5}$$

from which one obtains the conductivity σ

$$\sigma_i = \frac{n_- e^2}{\pi m K L n_+} \frac{\int f^1 v^3 \, dv}{\int f^1 \, dv} = \frac{\pi m}{Z L} \left(\frac{2\varepsilon_0}{e}\right)^2 \frac{\int f^1 v^3 \, dv}{\int f^1 \, dv}. \tag{55.6}$$

SPITZER shows that the distribution f^1 is given approximately by

$$f^1 = A v^2 e^{-m v^2/2kT} \tag{55.7}$$

whence the conductivity is

$$\sigma_i = \frac{m}{ZL} \left(\frac{4\pi\varepsilon_0}{e}\right)^2 \left(\frac{2kT}{\pi m}\right)^{\frac{3}{2}} \times 1.1632 \tag{55.8}$$

where the numerical factor was obtained by SPITZER by numerical integration of a distribution function slightly different than Eq. (55.7). The conductivity of a completely ionized plasma is independent of the concentration n_- because the number of resisting elements n_+ is proportional to the number of conducting elements n_-. However, if the atoms are more than singly ionized, the conductivity is reduced. The conductivity increases as the three-halves power of the temperature because c times the RUTHERFORD scattering cross section varies inversely as the cube of the relative velocity.

56. Conductivity of a partially ionized plasma. If the electrons collide with neutral gas as well as being deflected by the ions, the collision frequencies add and therefore the resistivities approximately add

$$\frac{1}{\sigma} = \frac{1}{\sigma_i} + \frac{1}{n_- e \mu}. \tag{56.1}$$

We are interested in the transition from "mobility" to "conductivity", that is in the ratio

$$X = \frac{n_- e \mu}{\sigma_i} = c w_i^{\frac{3}{2}} \frac{n_+}{n_g} \tag{56.2}$$

where

$$w_i = e V_i / k T \tag{56.3}$$

and

$$c = L(\mu n_g)\sqrt{\frac{m}{e}}\left(\frac{\pi}{2V_i}\right)^{\frac{3}{2}}\left(\frac{Ze}{4\pi\varepsilon_0}\right)^2 = \frac{Z^2 L}{4.2}\frac{(\mu p_0)}{V_i^{\frac{3}{2}}}. \tag{56.4}$$

In the expression for c, μ is in m.²/sec volt and p in mm. c is a number of the order of 2 so the value of n_g/n_+ for the transition depends mainly on w_i. As this is generally greater than 5, the conductivity transition takes place when the plasma is less than 5% ionized, a much higher percent than for the electron interaction transition. We are therefore justified in using SAHA's equation in this section.

The meaning of n_g is somewhat ambiguous. In principle it is the concentration of neutral atoms, $n_g = n_a$, but if the ions are not stripped, they have remaining electron shells and these contribute to the scattering. If they scatter equally with neutral atoms, $n_g = n_0$, the original concentration of gas atoms. We shall adopt this last assumption.

SAHA's equation (54.8) can be written

$$\frac{n_a}{n_+ n_-} = \frac{s}{Z n_0} w_i^{\frac{3}{2}} e^{w_i}, \tag{56.5}$$

$$s = \frac{g_a}{2g_i}\frac{Z n_0 h^3}{(2\pi m e V_i)^{\frac{3}{2}}} = \frac{g_a}{g_i}\frac{Z}{V_i^{\frac{3}{2}}}\frac{p}{17\times 10^4} \tag{56.6}$$

where p is in mm. s is therefore always very small.

Eliminating n_+ between (56.2) and (56.5) gives an equation for the ratio X in terms of the temperature:

$$s\, e^{w_i} = w_i^{\frac{3}{2}}\frac{c^2}{X^2} \pm \frac{c}{X} \tag{56.7}$$

$$(+ \text{ for stripped ions } \; n_g = n_a, \quad - \text{ for normal ions } \; n_g = n_0).$$

One can usually neglect the last term and the conductivity ratio is given by

$$X^2 = \frac{c^2}{s} w_i^{\frac{3}{2}} e^{-w}: \tag{56.8}$$

d) Stochastic processes.

57. The FOKKER-PLANCK equation. The derivation of the FOKKER-PLANCK equation given in Sect. 45 is not the conventional one. Following CHANDRASEKHAR[1] let $W(\boldsymbol{v}', \Delta\boldsymbol{v})$ be the probability that an electron at $\boldsymbol{v}'$ changes its velocity by $\Delta\boldsymbol{v}$ in a time Δt. Then

$$f(\boldsymbol{v}, t+\Delta t) = \int f(\boldsymbol{v}', t)\, W(\boldsymbol{v}', \Delta\boldsymbol{v})\, d^3\Delta v. \tag{57.1}$$

Expand both sides in TAYLOR series

$$f(\boldsymbol{v}, t+\Delta t) = f + \frac{\partial f}{\partial t}\Delta t + \cdots, \tag{57.2}$$

$$f(\boldsymbol{v}', t)\, W(\boldsymbol{v}', \Delta\boldsymbol{v}) = f(\boldsymbol{v}, t)\, W(\boldsymbol{v}, \Delta\boldsymbol{v}) - \frac{\partial(fW)}{\partial v_i}\Delta v_i + \frac{1}{2}\frac{\partial^2(fW)}{\partial v_i \partial v_j}\Delta v_i \Delta v_j \ldots \tag{57.3}$$

Substituting in (57.1) and introducing the notation

$$\int W\, d^3\Delta v = 1, \tag{57.4}$$

$$\int W\, \Delta\boldsymbol{v}\, d^3\Delta v = \gamma\,\Delta t, \tag{57.5}$$

$$\int W\, \Delta v_i \Delta v_j\, d^3\Delta v = \gamma_{ij}\,\Delta t \tag{57.6}$$

---

[1] S. CHANDRASEKHAR: Rev. Mod. Phys. **15**, 31 (1943).

there follows

$$\frac{\partial f}{\partial t} = -\operatorname{div}\gamma f + \frac{1}{2}\frac{\partial^2}{\partial v_i \partial v_j}\gamma_{ij} f + \cdots \tag{57.7}$$

which is formally identical to (45.8).

58. The LANGEVIN equation. The similarity between (45.8) and (57.7) is, however, purely formal. The γ's have different meanings in the two equations.

The LANGEVIN equation for an electron is

$$\frac{d\boldsymbol{v}}{dt} = \boldsymbol{A}(t) = -e\boldsymbol{E}(t)/m \tag{58.1}$$

where $\boldsymbol{E}(t)$ is the microfield acting on an electron due to its rapidly varying ionic environment. $\boldsymbol{A}(t)$ has an average value

$$\overline{\boldsymbol{A}(t)} = \gamma \tag{58.2}$$

and a correlation function

$$\overline{A_i(t+s)\,A_j(t)} = \psi_{ij}(s) \tag{58.3}$$

where the bar represents an assembly average.

Will we assume that

$$\int_0^\infty \psi_{ij}(s)\,ds = \gamma_{ij}. \tag{58.4}$$

From the LANGEVIN equation

$$\Delta\boldsymbol{v} = \int_0^{\Delta t} \boldsymbol{A}\,dt \tag{58.5}$$

and taking averages

$$\overline{\Delta\boldsymbol{v}} = \gamma\,\Delta t \tag{58.6}$$

which is the same as (57.5). Similarly[1]

$$\Delta v_i \Delta v_j = \int_0^{\Delta t}\int_0^{\Delta t} A_i(t)\,A_j(u)\,dt\,du, \tag{58.7}$$

$$\overline{\Delta v_i \Delta v_j} = \int_0^{\Delta t}\int_0^{\Delta t} \psi_{ij}(s)\,dt\,ds = \gamma_{ij}\Delta t, \tag{58.8}$$

provided $\Delta t \gg \tau$ where τ is the fluctuation time of $\boldsymbol{A}(t)$, or the mean width of the correlation function $\psi_{ij}(s)$. This is the same as (57.6).

The higher order correlations will generally be finite when integrated once, and therefore products such as $\overline{\Delta v_i \Delta v_j \Delta v_k}$ will be proportional to $(\Delta t)^2$ or higher powers. It follows then that if Δt is very small (but it must not go to zero), all terms after the first two in the FOKKER-PLANCK equation are also very small, and the two term equation we have used is correct.

The present derivation is not applicable when $\Delta t \to 0$, for if $\Delta t \ll \tau$, $\psi(s)$ in equation (58.8) is substantially constant and

$$\overline{\Delta v_i \Delta v_j} \approx \gamma_{ij}(\Delta t)^2/\tau. \tag{58.9}$$

Even the second order term in the FOKKER-PLANCK equation then vanishes[2].

The times Δt and τ are readily interpreted in the electron interaction problem. Δt is the time to produce Δv and is therefore the interaction time or the transit

[1] G. E. UHLENBECK and L. S. ORNSTEIN: Phys. Rev. **36**, 827 (1930).

[2] I am indebted to Dr. R. H. LYON for this observation.

time across a DEBYE sphere. The fluctuation time τ is the time to change nearest neighbors. $\tau \ll \Delta t$ if there are many particles in a DEBYE sphere. The unfavorable situation for BOLTZMANN theory is the favorable one for stochastic theory. But the coefficients γ_{ij} should be derived from the correlation of the microfield, not from a collision integral.

BOLTZMANN theory is ideally suited to rigid sphere collisions, in which $\Delta t \to 0$ but $\boldsymbol{A}(t) \to \infty$ so that $\Delta \boldsymbol{v}$ is not proportional to Δt. No terms in the FOKKER-PLANCK equation then vanish. The stochastic function $\boldsymbol{A}(t)$ is then a random series of sharp impulses so the fluctuation time τ also vanishes.

If, as in the theory of BROWNian motion, we are considering the motion of a very heavy particle in a light gas, so that $\Delta \boldsymbol{v}$ for one collision is very small, we can change our meanigs of Δt and $\Delta \boldsymbol{v}$ and let them extend over several collision. We are then back in the conditions of equation (58.8) and stochastic theory can be applied.

General References.

[1] ALFVEN, H.: Cosmical Electrodynamics. Oxford: Clarendon Press 1950. — Has a chapter on the motion of charged particles similar to part I of this article.

[2] CHAPMAN, S., and T. G. COWLING: The Mathematical Theory of Non-uniform Gases. Cambridge University Press 1939. — Deals exhaustively with small departures from thermal equilibrium.

[3] KIHARA, TARO: Mathematical Theory of Electrical Discharges in Gases. Rev. Mod. Phys **24**, 45 (1952); **25**, 844 (1953). — Covers the area of part III of this article.

[4] SPITZER, L., and R. HARM: Phys. Rev. **89**, 977 (1953). — Compute the conductivity of an ionized plasma by the method of part V.

[5] WANNIER, G. H.: Phys. Rev. **83**, 281 (1951). — This is the most recent work on ion mobilities.

[6] WAX, NELSON: Noise and Stochastic Processes. Dover Publ. Co. 1954. — Contains reprints of the basic articles by CHANDRASEKHAR and UHLENBECK.

The following references to electron distribution functions are given in historical order:

[7] DRUYVESTEYN, M. J.: Physica, Haag **10**, 69 (1930).

[8] MORSE, P. M., W. P. ALLIS and E. S. LAMAR: Phys. Rev. **82**, 412 (1935).

[9] DAVYDOV, B.: Phys. Z. Sowjet. **8**, 59 (1935).

[10] SMIT, J. A.: Physica, Haag **3**, 543 (1937).

[11] TONKS, L., and W. P. ALLIS: Phys. Rev. **52**, 710 (1937).

[12] HOLSTEIN, T.: Phys. Rev. **70**, 367 (1946).

[13] MARGENAU, H.: Phys. Rev. **73**, 297, 309 (1948).

[14] MACDONALD, A. D., and S. C. BROWN: Phys. Rev. **75**, 411 (1949).

[15] ALLIS, W. P., and S. C. BROWN: Phys. Rev. **87**, 419 (1952).

Formation of Negative Ions.

By

L. B. Loeb.

I. Fundamentals.

1. Introduction and historical. Negative ions appear in gases under two circumstances: (1) they may be created *in the gas* largely through attachment of free electrons to atoms, molecules and rarely by dissociation of molecules in a polar phase by electron impact, (2) they may be introduced into the gas by interaction of fast particles of atomic mass with surfaces or by liberation from hot surfaces. Attachment of electrons causes loss of the former as ionizing agents, it leads to delayed and undesirable electronic ionizing events in asymmetrical field breakdown, to negative ion space charge action limiting current flow and it may act to increase the rate of loss of carriers by recombination. Negative ions from surfaces or filaments are the cause of destruction of oscilloscope screens.

The presence of negative ions by process (2) was noted by J. J. Thomson [*1*] in his first mass spectrographs in 1910 and by him and many others in emission from heated metals at a still later date. The emission by ion impact was not intensively investigated for many years until the work of Arnot [*2*] and his associates in 1936 and by various investigators in connection with the destruction of oscilloscope screens. Later intensive study came from the work of Sloane and his associates [*3*].

Process (1) the attachment of electrons *in* gases was not clearly perceived until about 1910 when the vacuum techniques and gaseous purification of gases led Franck and Pohl [*4*] to study ion mobilities in inert gases and N_2 noting the presence of free electrons at higher pressures. Contemporaneously, the studies of Townsend [*5*] and his pupils around 1912 had led to the recognition of the existence of free electrons at lower pressures in gases. The experimental work leading to the ultimate discerning of electron attachment were studies of the variation of carrier mobilities in air as a function of pressure by A. F. Kovarick [*6*] in 1910 with the Rutherford A C method of mobility measurement using photo electrons and those of E. M. Wellisch [*7*] beginning in 1915 using the same method but producing ions by α particles from Po in an auxiliary field below a gauze following the method of Franck. The discrepancy between the results of these two methods led the writer [*8*] to study both techniques in 1919. In 1916 J. J. Thomson [*9*] had developed a theory for the formation of negative ions by attachment of electrons to molecules. On the basis of his own studies Wellisch [*7*] in 1917 put forth a rival theory to that of Thomson. In 1920 the writer adapted Thomson's theory to the data using techniques of Wellisch and Kovarick, proving that the Thomson theory represented the correct approach, and by means of this theory evaluating the attachment probability in air in order of magnitude. Since for exact evaluation electron drift velocities in air were needed at lower field strength to pressure ratios X/p, the writer proceeded to evaluate these in air and other gases and gave to Wahlin [*10*] in his laboratory the task of extended measurements of attachment to other gases. In 1923 the accurate theory of attachment as revealed by the Kovarick modification of the RUTHERFORD A C method [11] was developed by MOONEY and with the writer's data on drift velocities of electrons in air [12] led to a more exact evaluation of attachment probability h. The values apparently varied with field strength and pressure so that while applicable in any limited region the theory of THOMSON even if applied with an appropriate value of h led to poor overall agreement over a wide range of X/p. In 1925 V. A. BAILEY used a method involving the diffusion of electrons and ions to a series of electrodes out of an initially cylindrical electron beam in a field in the gas, assuming the THOMSON theory evaluated h at several values of X/p, or electron energy, as simultaneously derived from such measurements. He showed that h varied with average electron energy and thus X/p, thus clarifying the cause of the writer's deviations. The development of the writer's electron filter beginning in 1927 [14] and its application first by CRAVATH [15], and later in more perfected form by BRADBURY [16] in 1933, at once led to the evaluation of the attachment probability in a wide range of gases. Simultaneously the diffusion and drift velocity techniques of BAILEY were improved [17] and extended to yield basic data in a limited range of energy values for many gases. The simple more direct method of BRADBURY enabled him to delineate the mechanisms of attachment and to predict attachment mechanisms according to gas type. None of these techniques was capable of extending studies of attachment in high fields when electrons ionize by impact. This problem was solved for some gases by R. GEBALLE and his students [18] in 1951. A very beautiful method of study of attachment was developed by A. DOEHRING [19] 1944 using the four-gauze-shutter-time-of-flight method developed for ion mobilities by TYNDALL and POWELL. Because of the very low values of the attachment probability h in many gases, as well as the difficulty of achieving groups of electrons of narrow energy spread in gases of sufficient density for appreciable attachment, significant studies on the appearance potentials of ions and energy of ion formation with identification of ion species formed by mass spectrographs have not been successful, until very recently, when HICKAM and FOX [20] applied their retarding potential difference method to the study of attachment of electrons to SF_6 combined with mass spectrograph, perhaps ushering in a new era of investigation.

2. Basic relations. Following J. J. THOMSON [9] let it first be assumed that electrons do not readily attach to molecules to form negative ions and that there is a certain probability h that in a *kinetic theory type of impact* between an electron and molecule the electron will attach to the molecule thus forming a negative ion. To simplify the elementary analysis, assume that h is independent of the electron energy in such encounter. Let the average velocity of the electron be $\bar{c}_1$ and its kinetic free path λ. Its collision frequency $z=\bar{c}_1/\lambda$. In a field of strength X the electron has a drift velocity $v=k_e X$ in the field direction, k_e being the *mobility* of the electron. Thus, in advancing 1 cm. in the field direction along the X axis, the electron travels a time $t=1/v$ and makes $\alpha=zt=\bar{c}_1/\lambda v$ collisions. If the attachment process is a pure chance phenomenon of probability h then out of n electrons starting from a point x at a time $t=0$, the number of attachments dn occurring in a distance dx will be

$$dn=-\frac{h\bar{c}_1}{\lambda v}n\,dx=-\alpha n\,dx. \tag{2.1}$$

Integrating by setting $n=n_0$ at $x=0$ and $n=n$ at $x=x$, the number of electrons n, out of the initial n_0 starting, that *survive attachment* after a distance x are given by

$$n=n_0 e^{-\alpha x}=n_0 e^{-\frac{h\bar{c}_1}{\lambda v}x}=n_0 e^{-\frac{h\bar{c}_1 x}{\lambda k_e X}}. \tag{2.2}$$

It is seen that number of electrons *surviving* attachment after traversing a distance x in the field direction follows an exponential curve with an average distance $\bar{x}$ for attachment under the existing conditions given by

$$\bar{x} = \frac{\lambda v}{h \bar{c}_1} = \frac{\lambda k_e X}{h \bar{c}_1} = \frac{1}{\alpha}. \tag{2.3}$$

The quantity $\bar{x}$ thus represents the centroid of the survival curve.

Since $\alpha = \frac{h \bar{c}_1}{\lambda v}$ is the number of attachments per cm. advance in the field direction, $1/\alpha$ is the *average free path* for *attachment in the field direction.* In general it is inconvenient to introduce the quantities such as $\bar{c}_1/\lambda$ into relations; for though $\bar{c}_1$ is calculable from the energy distributions, and λ is the RAMSAUER free path for kinetic impact at $\bar{c}_1$, the two quantities are not directly obtained. However, the expression for electron drift velocity $v = k_e X$ is related to λ and $\bar{c}_1$ by a fundamental relation for drift velocity, viz.

$$v = k_e X = a \frac{\varepsilon X}{m} \frac{\lambda}{\bar{c}_1}. \tag{2.4}$$

Here a is a constant, near unity, which depends on the form of the energy distribution law having the value 0.815 for a MAXWELLian distribution, and 0.874 for the DRUYVESTEYN distribution. In consequence $\frac{\lambda}{\bar{c}_1} = \frac{m v}{a \varepsilon X}$, so that eq. (2.1) and (.2) become open to experimental exploitation as

$$d n = -\frac{a h \varepsilon X}{m v^2} n \, d x, \tag{2.5}$$

with

$$\alpha = \frac{a h \varepsilon X}{m v^2} = \frac{a h \varepsilon}{m k_e^2 X} = \frac{1}{\bar{x}}, \tag{2.6}$$

$$n = n_0 e^{-\frac{a h \varepsilon X x}{m v^2}} = n_0 e^{-\frac{a h \varepsilon x}{m k_e^2 X}}. \tag{2.7}$$

It is next also useful to regard the attachment process in terms of a collision cross section for capture. An electron travelling with average random velocity $\bar{c}_1$ among the relatively fixed gas molecules executes a random *path* of total lenth $\bar{c}_1$ cm. in one second. All molecules the centers of which lie within a distance σ_R collide with the electron, causing an *appreciable* deflection in the otherwise straight trajectory. Thus a volume $\pi \sigma_R^2 \bar{c}_1 = q_R \bar{c}_1$ is swept out in one second. Here $\pi \sigma_R^2 = q_R$ is the *collision cross section* for electron deflection which was first measured for many molecules and atoms by C. RAMSAUER, and later by others, and is known as the RAMSAUER, or gas kinetic scattering cross section for electrons in a gas. Since there are N molecules per cm.^{3} the $N q_R \bar{c}_1$ molecules lying within this cylinder will have collided with the electron in unit time. Since the total distance covered in unit time is $\bar{c}_1$ cm. the average distance between collisions, or the RAMSAUER free path is $\lambda = \frac{\bar{c}_1}{N q_R \bar{c}_1} = \frac{1}{N q_R}$. If p is the pressure in mm. and N_1 is LOSCHMIDT's number at 1 mm. pressure, $N = p N_1$, the survival equation for the n electrons out of n_0 that have traversed a random x cm. in the gas without a deflecting RAMSAUER collision with molecules is then

$$n = n_0 e^{-x/\lambda} = n_0 e^{-N_1 p q_R x}. \tag{2.8}$$

The probability of attachment, h, as defined by THOMSON [9] and used in eqs. (2.2) and (2.7) may now be expressed more precisely by means of an equation relating the number of attaching collisions per second, z_a, to the number of RAMSAUER collisions per second, z, i.e.,

$$h = \frac{z_a}{z} = \frac{\bar{c}_1}{\lambda_a} \Big/ \frac{\bar{c}_1}{\lambda} = \frac{q_a}{q_R}.$$

Thus

$$h = \frac{q_a}{q_R} = \frac{\lambda}{\lambda_a} \quad \text{or} \quad q_a = h\, q_R, \tag{2.9}$$

which enables q_a to be calculated if q_R and h are known from measurement. It is then possible to write the survival equation for attachment *in random motion in a gas* as

$$n = n_0\, e^{-x/\lambda_a} = n_0\, e^{-N_1 p\, q_a x}, \tag{2.10}$$

and the number attaching in a distance dx as

$$dn = -\frac{n_0\, e^{-x/\lambda_a}}{\lambda_a}\, dx = -n_0 N_1 p\, q_a\, e^{-N_1 p\, q_a x}\, dx, \tag{2.11}$$

with

$$\lambda_a = \frac{1}{N_1 p\, q_a}, \tag{2.12}$$

the *average random free path for attachment.* It should be noted that λ_a is the average random free path in the gas for attachment and has a value λ/h. In noting this, it must be emphasized that eqs. (2.2) and (2.7) which deal with distances x for attachment involving *motion in* the *field direction* are *not* to be *confused with eqs.* (2.10) and (2.11) dealing with distances x of *random motions involved in attachment.* Similar distinctions apply to $\bar{x} = 1/\alpha$ of eq. (2.3) and λ_a of eq. (2.12).

In consequence of the newer microwave techniques by which electron densities are measured as a function of time in a gas where loss by attachment occurs, eqs. (2.10) and (2.11) can be used to evaluate λ_a directly. Since the random path x is given by $x = \bar{c}_1 t$, the equations become

$$n = n_0\, e^{-\frac{\bar{c}_1}{\lambda_a} t} = n_0\, e^{-N_1 p\, q_a \bar{c}_1 t} = n_0\, e^{-\frac{t}{\tau_a}} \tag{2.13}$$

and

$$dn = -n \frac{\bar{c}_1}{\lambda_a}\, e^{-\frac{\bar{c}_1}{\lambda_a} t}\, dt = n N_1 p\, q_a c_1\, e^{-N_1 p\, q_a \bar{c}_1 t}\, dt = \frac{n}{\tau_a}\, e^{-\frac{t}{\tau_a}}\, dt. \tag{2.14}$$

Here, λ_a/c_1, the time of an attaching free path, is a characteristic time constant τ_a of the process. It is clear that from dn/dt, or from n as a function of t, if $\bar{c}_1$ is known λ_a and thus q_a can be evaluated.

With this information if now becomes possible to discuss the methods of evaluation of q_a or h.

II. Methods of evaluation of h and q_a.

3. General viewpoints. The choice of the method used must depend in some measure on the order of magnitude of h to be expected and on the nature of the gas. In principle nearly all methods in one fashion or another use the survival eq. (2.2), or its equivalent, in which measurements of relative numbers of electrons lost from the stream in a field are determined for a given distance of traverse in the field direction. Thus electrons starting from a certain point are lost to ion formation, and the number lost in a given distance x traversed is measured. Otherwise the evaluation of $\tau_a = \lambda_a/\bar{c}_1$ by temporal decline in electron concentration by attachment can be used. The attenuation-in-space method requires adequate numbers of electrons in the beam and an adequate density of attaching molecules in the path in the distance of measurement x so as to yield an accurately measured attenuation. The distance x is limited by geometry essential to insure uniform fields. Total gas pressures must be adequate to insure that electrons acquire steady drift velocities in distances much shorter than x. It is desired that insofar as possible the average energies of electrons in the field be known since h is a function of electron energy. Since h can vary from values as large as 10^{-1} or more down to 10^{-6} or less among the species there are definite limits imposed on measurement. Thus gases like the halogens require dilution in some inert gas, while gases like air require adequate pressures, or low field values, to insure adequate ratios of n/n_0. Measurements in any event are always limited in range by any single method. Except in one recent instance [*20*] owing to the relatively small value of q_a studies of direct attachment in beams of slow electrons of known energy moving through gases at low pressures using a mass spectrometer to identify ions have met with little success. As indicated, the loss of electrons as a function of t in the absence of fields through microwave techniques enables evaluation of τ_a, and thus of λ_a, directly. This method requires adequate densities of molecules and electrons, and study over appropriate time intervals. So far, while it appears promising, the method has not yielded satisfactory data since other loss mechanisms, such as by ambipolar diffusion and recombination, are also present. The temporal loss by ambipolar diffusion also shows an exponential electron decline. Fortunately the rate increases as pressure decreases, while the inverse holds for loss by attachment. At higher pressures recombination loss increases to compete with attachment. This loss fortunately obeys a different temporal law. However, *unless data* for *electron decline over a large range of t values are* at hand, the *two processes are hard to separate.* Most important in all these studies, however, is the need for the achievement of a high degree of gaseous purity since, if q_a is small, traces of impurities for which q_a is large will completely falsify data.

The various methods used will be described in the following Sects. 4 to 11.

4. Square wave alternating current method [*11*]. In this method electrons are photoelectrically liberated from the cathode in a plane parallel uniform field gap of length d and a square wave alternating potential of half period T of amplitude V is applied to the plates. In consequence, a current I of electrons and ions will arrive at the anode which is a function of the potential V, despite the retarding phase which sweeps the gap clean. The electron mobility is k_e and that of ions is k while $V_0 = d^2/Tk$ is the potential at which the ions without diffusion will just cross to the anode, while electrons cross at a much lower potential $V_{0e} = d^2/T k_e$. If the potential is steady, instead of alternating at a potential V, a steady current of ions and electrons flows from the cathode to the anode. Half this current to be compared with the on phase of the alternating potential yields a current which is designated as I_0. This quantity I_0 can be measured as a function of V and plotted. Thus having its measured values of I and I_0 as a function of V it has been shown by the writer and M. Mooney [*11*] that the ratio I/I_0 for various values of V with fixed d, T, etc., is given by:

$$\left.\begin{aligned}
\frac{I}{I_0} &= \frac{B k_e^2}{T k h}\left[e^{-\frac{B' h}{k_e^2}\left(\frac{d^2}{V} - T k\right)} - e^{-\frac{B' h}{k_e^2}\left(\frac{d^2}{V}\right)}\right] \quad \text{below } V_0,\\
\frac{I}{I_0} &= \frac{B k_e^2}{T k h}\left[1 - e^{-\frac{B' h}{k_e^2}\left(\frac{d^2}{V} - T k\right)}\right] \quad \text{at } V_0,\\
\frac{I}{I_0} &= \frac{V - V_0}{V} + \frac{B k_e^2}{T k h}\left[1 - e^{-\frac{B' h}{k_e^2}\left(\frac{d^2}{V}\right)}\right] \quad \text{above } V_0.
\end{aligned}\right\} \qquad (4.1)$$

Here B and B' are constants deriving from the quantity $\alpha\, e/m$ in eq. (2.7) and conversion from electrostatic units of charge to the potential V expressed in volts. It is seen that in measuring I/I_0 it is possible by means of eq. (4.1) to evaluate h at a given range of values of V about V_0 at a pressure p and hence at a given value of $V/pd = X/p$, if k_e is known for the same value of X/p. This was the first method used to evaluate h and is accurate if h does not vary too rapidly with X/p. It is limited in scope by the range of square wave potential frequencies and by values of d and p allowed.

5. The diffusion method. The method initiated by V. A. BAILEY in 1925 [*13*], [*17*] and later developed to a high degree of complexity is in principle a diffusion method in an electrical field. A rectangular beam of photoelectrons moves along the axis of a cylindrical tube in a uniform electrical field to an anode collecting plate E_0. It is defined by a longitudinal rectangular slit entrance E_5 in a diaphragm defining the cathode end of the field and sequentially passes through diaphragms E_3 and E_1 each having slits of the same size as those in E_5. All diaphragms are equipotential to their location in the field. The electron beam moves with a steady drift velocity v in the field direction with a radial diffusion of electrons and attachment of electrons to molecules to form ions underway. The ions also drift in the field direction and diffuse radially to the plates having of course different drift velocities and diffusion coefficients. The currents of negative carriers to E_3, E_1 and E_0 can be measured. In the original apparatus the distances between E_5 and E_3 and E_3 and E_1 were fixed and equal to c, but in later work these could be varied. In later work a magnetic field could also be imposed parallel to the electrical field. The ratio ζ of the currents to plates E_3 and E_1 and ξ to plates E_1 and E_0 could be measured for various values of the field X and pressure p. From these ratios, the values of S_1 and S_2 for plates E_3 and E_1 representing the ratio of current passing through the slit, to that caught by each plate, could be calculated to be

$$S_1 = \frac{\xi(1+\zeta)}{1+\xi(1+\zeta)} \quad \text{and} \quad S_2 = \frac{\zeta}{1+\zeta}. \tag{5.1}$$

Analysis using eq. (2.2) with n_0 electrons and N_0 ions passing through the plate E_5 allowed attenuation of electrons by attachment between E_5 and E_3 ... to be calculated. From solution of the diffusion equation in the field X in terms of a relation between the diffusion coefficient and mobility, the ratio R_e of electrons captured at E_3 and passing through the slit were computed as a function φ of $\frac{k_e}{D} = \frac{N\varepsilon X}{2\eta p}$. Here N is the density of molecules, X the field strength, η is the TOWNSEND multiplier into the average thermal energy of molecules at 15° C needed to yield the average electron energy $\frac{1}{2} m\, \bar{c}_1^2$, where the gas pressure p through $\eta p/N$ represents the kinetic energy of the electrons. In consequence it follows that

$$n_{E_3} = n_0 R_e e^{-\alpha c} \quad \text{and} \quad n_{E_1} = n_0 R_e^2 e^{-2\alpha c}. \tag{5.2}$$

The $n_0(1-e^{-\alpha c})$ electrons which attached to form ions between plates E_5 and E_3 arrived at E_3 with an unknown distribution, while N_0 ions arrived at E_3 with a ratio R which is a function φ of $\frac{N\varepsilon X}{2p}$ for ions, since the ions are in thermal equilibrium with the gas. With these considerations algebraic manipulation leads to the relation

$$R_e e^{-\alpha c} = \frac{S_1(R-S_2)}{R-S_1} = \varphi\left(\frac{N\varepsilon X}{2\eta p}\right) e^{-\alpha c} = a_1, \tag{5.3}$$

where a_1 is computed from values of $R=\varphi\left(\frac{N\varepsilon X}{2p}\right)$ and observed values of S_1 and S_2. As both α and $\frac{N\varepsilon X}{2\eta p}$ are unknown, it is necessary to change X and p both by the same factor b to values X/b and p/b. When this is done the quantity $\varphi\left(\frac{N\varepsilon\left(\frac{X}{b}\right)}{2\eta p}\right)e^{-\alpha c/b}$ is altered to a new value a_b, for α is changed to α/b by virtue of the pressure variation of λ while the values of $\bar{c}_1$ and v are unchanged in eq. (2.2) with X/p constant. Thus, with this change, two equations are obtained reading

$$\varphi\left(\frac{N\varepsilon\left(\frac{X}{b}\right)}{2\eta p}\right)e^{-\left(\frac{\alpha}{b}\right)c}=a_b, \tag{5.4}$$

$$\varphi\left(\frac{N\varepsilon X}{2\eta p}\right)e^{-\alpha c}=a_1. \tag{5.5}$$

These permit of simultaneous solution for $\alpha=\frac{h\bar{c}_1}{\lambda v}=\frac{h\bar{c}}{\lambda X k_e}$ and for $\frac{N\varepsilon X}{2\eta p}$. Thus, if $Xk_e=v$ and λ are known, η and h are evaluated. In 1925 there was at hand a method initiated by TOWNSEND in 1913 whereby through the lateral displacement of the electron beam by a magnetic field parallel to the slit length and its measurement on an appropriately sectored disc in place of E_0, the quantity v could be evaluated [*17*]. At the same time the lateral diffusion of the beam in a uniform electrical field X with no magnetic field could be determined from the ratio $\frac{N\varepsilon}{\eta p}$. Then, from eq. (2.4), assuming the constant a for a MAXWELLian distribution, $\bar{c}$, being evaluated from η the value of λ could be inferred. It is seen that through the use of magnetic fields and the relative amounts of lateral diffusion of electrons and ions in an electrical field, enough relations can be obtained to yield a consistent set of data for the values of α, η, $\bar{c}$, λ, v and h as a function of X/p [*17*]. The underlying theory is very involved. The various approximations and assumptions, introducing errors of varying magnitudes, especially with the varying forms of the distribution laws affecting the average values such as $\bar{c}_1$ and a, do not lead to high precision. In later years, BAILEY [*17*] and his students have broadened the analyses to include some variation in the form of the distribution laws and have so altered the form of the apparatus and techniques that the data leading to the evaluation of all of the variables enumerated including h can be taken in one of several ways on a given gas with a single piece of equipment. For details the reader should consult HEALEY and REED's "*Behaviour of Slow Electrons in Gases*" [*17*]. This method, as do all other loss methods, fails to yield proper values of h as soon as X/p reaches values where any appreciable ionization by collision occurs, for then generation of new electrons and diffusion of positive ions leads to an apparent reduced attachment loss and h appears to decrease. With this method, BAILEY [*13*] in 1925 first demonstrated the decrease of h in air with X/p and hence with energy and later with his students evaluated h for many gases.

6. The electron filter method. In 1926 the writer [*14*] developed the electron filter for removing electrons out of a mixed stream of electrons and ions in a gas so that those electrons that traversed a distance x in a gas without attachment could be removed from the stream, and the current of ions formed could directly be measured. The system was first tested by LUSK in 1926/27 [*14*] and used for measuring h in air and O_2 by CRAVATH [*15*] in 1927–1929. In that work, thermionic electrons were used with a single filter in a metal vessel using TOWNSEND and BAILEY'S values of drift velocity v. In 1931–1933 N. E. BRADBURY [*16*] used a set of two filters, an outgassed glass system, and measured the electron drift velocities directly in the same chamber [*21*]. The method is the following:

Two plane parallel grids G_1 and G_2 of fine wires about a mm. apart are mounted on their insulating frames having fairly large apertures through which streams of electrons and ions may pass. The grids are separated known distances x_1 and x_2 from a cathode A in uniform electrical field geometry. Beyond G_2 lies the collecting anode plane P connected to an electrometer. The grids G_1 and G_2 are equipotential to the uniform electrical field X along the axis, which is maintained uniform by means of surrounding equipotential vanes. The grids G_1 and G_2 are so mounted that alternately either G_1 or G_2 can be retracted, leaving G_2 or G_1 in the field, or both could be in place. The alternate wires of each of the grids are connected to the terminals of a high frequency oscillator giving from 10^6 to 10^7 oscillations per second. Photoelectrons are liberated from the cathode plate A by a stabilized ultraviolet arc source. In traversing the distances x_1 from A to G_1 and x_2 from A to G_2 electrons moving in the field X at a pressure p in the gas will attach to molecules, yielding ions according to eqs. (2.2) and (2.7). If then the potential of the high frequency applied between alternate grid wires and the frequency are appropriate to the electron drift velocities at the pressures used, the very nimble electrons will be picked up by the grids as they move from A to P in the field, while the relatively inert ions shudder slightly as they pass through the meshes but continue on, except for a few intercepted by the wires. The ions then continue on through to the collector where the current is measured. The electron currents surviving ion formation at G_1 and G_2 are then $I_1 = I_{01} - I_1'$ and $I_2' = I_{02} - I_2'$, I_{01} and I_1' being measured while G_1 is in place with alternating potential on and off, and I_{02} and I_2' representing similar currents to P with G_2 in place and alternating potential on and off. It then follows from eq. (2.2) that

$$\frac{I_2}{I_1} = e^{-\frac{a h}{k_e^2 X}(x_2 - x_1)}. \tag{6.1}$$

The action of the grids as filters is remarkably good if the operating variables are well chosen, though correction must be made for the ions intercepted by the grid wires. The use of two grids eliminates many errors. BRADBURY measured k_e *in situ* by means of an analysis of the current through the gas, at each X/p used, relative to that in vacuum using J. J. THOMSON'S back diffusion equation [*21*] and a measurement of the average energy of photoemission of the photoelectrons. The method of evaluating k_e used constitutes the least satisfactory aspect of this excellent study, but it appears to have generally yielded highly satisfactory data as indicated by later studies. Today the same two grids using variable frequency alternating potentials modulating a fixed steady bias potential between alternate wires, thus acting as electrical shutters, would on the same gas in the same apparatus yield values of $v = k_e X$ good to 1%. This method for v was developed by BRADBURY and NIELSEN [*22*] *after* the attachment studies were completed.

7. The four-gauze-shutter method of DOEHRING [*19*]. The four-gauze shutter method developed by TYNDALL and POWELL for ion mobility studies was improved and adapted by DOEHRING to attachment studies [*19*]. Electrons drawn from a heated filament advance in a uniform field to the first shutter composed of two fairly closely spaced gauzes lying on equipotential surfaces in the field. These gauzes are biassed by a small constant opposing potential difference preventing the electrons from traversing the gauzes except during instants when a very short repeated rectangular potential pulse is applied across the grid at regularly spaced intervals. Between the first two-gauze shutter and the second two-gauze shutter operated in the same fashion is the uniform field drift space of known and sometimes variable length. Beyond this second shutter the carriers move to a collecting electrode. The short square wave pulses opening the shutters for brief intervals are regularly and synchronously applied to both shutters. Thus carriers that pass the first shutter when open move across the uniform field drift space between shutters and if they arrive at the second shutter at a time when it is open they will pass through and register a current at a collecting eletrode. Thus carriers that arrive at the second shutter at any whole multiple of the pulse rate later will get through. While distance between shutters, and the drift field strength can be varied, in general these are held constant and the time between pulses is varied, the current to the anode collector being measured. The electrons, attaching at various points in the drift space, yield carriers of apparent drift velocity lying between those for electrons and ions. Drift velocities of electrons and ions can be directly measured by pulsing at very high and very low rates. Theory based on eq. (2.2) yields the value of α as

$$\alpha = \frac{1}{k X} \frac{\Delta \log I_i}{\Delta t},$$

where k is the ion mobility and I_i is the ion current received at various times. The method has a high resolving power and indicates clearly the presence of more than one negative carrier by its form. Results in pure O_2 showed just one carrier and are the best to date even the electron drift velocities closely approximating BRADBURY and NIELSEN's [*22*] accurate 1931 values, while those of HEALEY and KIRKPATRICK [*17*] involved in their evaluations of α lay 30% below.

8. Direct attachment studies with monoenergetic electrons and mass spectrographic analysis [*20*]. In the event that attachment cross sections are very high it should be possible to study the attachment directly at pressures so low that collisions do not appreciably alter energies. Thus very recently HICKAM and FOX [*20*] have applied their retarding potential difference method to study the attachment of monoenergetic electrons of very low energies (0.1 e.v.) in SF_6. The maximum cross sections for attachment were at very low energies for the formation of SF_6^- and at quite low energies for SF_5^- as revealed by mass spectrographic analysis. Their beautiful method is, however, applicable only to a relatively few gases of large attachment cross sections. Details of the techniques are not as yet available though reference should be made to their ionization potential methods [*20*] for the retarding potential technique.

9. Attachment studies in fields causing ionization by electron impact [*18*]. It has been shown by GEBALLE and HARRISON [*10*] that if the increase of a cathode liberated photoelectric current i_0 through ionization by electron impact to a current i in a gas be measured at constant uniform field strength with *gap lengths varied over a sufficient range of values*, following TOWNSEND's initial technique, the plots of log i/i_0 against x reveal curvatures near the origin in *electron attaching* gases. Theoretical analysis indicates that if the *first* TOWNSEND coefficient for the number of new electrons and positive ions created by an electron in traversing 1 cm. in the field direction is designated as α_i, and the number of attachments per cm. advance is designated as α as in eq. (2.2), then the i/i_0 curves are given by

$$\frac{i}{i_0} = \left(\frac{\alpha_i}{\alpha_i - \alpha}\right) e^{(\alpha_i - \alpha) x} - \frac{\alpha_i}{\alpha_i - \alpha}, \tag{9.1}$$

instead of TOWNSEND's original relation applicable to non attaching gases which reads:

$$\frac{i}{i_0} = e^{\alpha_i x}.$$

The method requires that x be varied over a range of values of the order of 100 for gases like air where α is small. The accuracy and resolving power of the method is not too good, and it has a limited range of values of X/p over which the necessary curving can be observed, extending from those where α_i begins to be measurable to regions where α_i so exceeds α as to obscure it. Results have been obtained in air, O_2 and certain spark suppressing gases. By this means α for dissociative attachment in O_2 was measured for the first time, and the results of earlier workers at high X/p thus clarified.

10. The microwave techniques [*23*]. In an outgassed quartz, or glass, flask of appropriate form in a microwave cavity the contained gas may be broken down by an adequately intense microwave discharge. The ionized gas is allowed to "cool" down for an appropriate time, usually of the order of milliseconds, and then scanned at regular time intervals by a very weak microwave searching beam. From the change in Q of the cavity, its resonant frequency f, or its complex and real conductivities, the average electron concentration can be evaluated as a function of time [*23*], [*24*]. The method is quite accurate for electron concentrations below 10^{11} per cm.^{3} and above 10^8 per cm.^{3}, provided the field distribution at breakdown is such as to insure relatively uniform ionization in the volume and that the diffusion constants are known and higher modes of diffusion are ruled out. The decline in electron concentration is due to ambipolar diffusion, recombination and electron attachment. By suitably varying pressures, the temporal variations of decline will permit of an evaluation of attachment coefficients using eq. (2.13), provided data are obtained over a long enough range in time with sufficient accuracy to yield an unambiguous functional relationship. Unless coupled with mass spectrographic study, the nature of the process and ions is not known. Gases of small h must be scrupulously clean. In one case, data in pure H_2 [*25*] were falsified by the reduction of the outgassed quartz container walls by atomic H yielding impurities. It has successfuly been applied by M. A. BIONDI [*24*] to attachment in I_2 of very slow electrons in 1953.

11. The method of HERRENG [*27*]. Among the newer techniques, using methods such as those of DOEHRING [*19*] and the writer [*11*] in which the current of attaching electrons and ions is measured as a function of time, thus leading to an evaluation of α or h, is that of HERRENG [*27*]. This was devised in LANGEVIN's laboratory during the occupation and well in advance of its time utilized a fast sweep cathode ray oscilloscope. In principle, HERRENG used a uniform field ionization chamber with electrode spacing some 12 cm. length that was traversed at any one of three accurately known points relative to the anode plane by a thin slab of X-ray ionization parallel to the anode. This ionization was flashed into the chamber in a pulse of 1 microsecond duration. The ions and electrons created then were drawn towards the anode by the constant field, the current to the anode being measured on the oscilloscope. In gases with little attachment and with *fast sweep* the current to the cathode as a function of time is produced by an element caused primarily by electron movement in the gap. This terminates in a sharp decline as electrons reach the anode. While the declining current pulse of electrons is diffused and broadened by the blurred X-ray ionization edges and electron diffusion, if the time elapsed between corresponding points on curves with the X-ray beam traversing the field at two points x_1 and x_2 cm.

from the anode is measured, accurate drift velocities can be determined. Values obtained this way by HERRENG in pure argon in an out-gassed system agreed well with those of BRADBURY and NIELSEN.

In electron attaching gases with fast sweep the current registered by the decline on arrival of electrons is just that component i_1 of the current contributed by the electrons that *survive* attachment in going the distance x_1 from the slab to the anode. If q is the number of electrons in the slab at x_1 and v their drift velocity, HERRENG shows that

$$i_1 = \frac{q\,v}{x_1}\, e^{-\frac{h\bar{c}_1}{\lambda} t_1}. \tag{11.1}$$

If for a second distance x_2 closer to the anode with the currents i_2 and time t_2 registered,

$$i_2 = \frac{q\,v}{x_2}\, e^{-\frac{h\bar{c}_1}{\lambda} t_2},$$

then it follows

$$\frac{i_1}{i_2} = e^{-\frac{h\bar{c}_1}{\lambda}(t_1-t_2)} = e^{-\frac{h\bar{c}_1}{\lambda}\left(\frac{x_1-x_2}{v}\right)} = e^{-\frac{a h e}{m k_e^2 X}(x_1-x_2)}, \tag{11.2}$$

using eqs. (2.2) and (2.7). The oscilloscope permits direct evaluation of $t_1 - t_2$. However, in any event, $v = k_e X$ must be evaluated to replace $\frac{\bar{c}_1}{\lambda}$ by $\frac{a\,e}{m\,k_e}$ so that the calculation of h in terms of x_1, x_2 and k_e is indicated. By this method, HERRENG, using O_2 and O_2 diluted with A, evaluated h in reasonable agreement with other workers.

III. Measurement of energy of ion formation.

12. Introduction. To date, relatively little data exist on the heat of negative ion formation and of its associated quantity, the appearance energy of a negative ion. Quantum mechanical calculation, or cyclic reasoning from known thermochemical reactions have given the heats of negative ion formation for many *atomic* species. Analogous data for negative molecular ions do not exist. If electron energies are known relative to X/p values, then some estimate can be made of the heat of formation from the X/p value for the appearance of the molecular ion in measurements such as BRADBURY'S in cases of dissociative attachment if the dissociation energy of the molecule is known, and if the nature of the ion formed is ascertained. In those gases such as O_2, SO_2, etc. in which the attachment is not dissociative, but in fact requires triple impact to remove the heat of formation the energy of ion formation cannot be determined in this fashion. In such instances, enough negative ions must be created so that the energy of electron *removal* by some external agency can be ascertained. Obviously electron bombardment of such ions will not serve. Thus electron detachment can only be achieved by collision processes of the negative ions with gas molecules in high electrical fields or by photodissociation.

13. The collision process. Observations of the capture of electrons by electron filters, as the potentials of the high frequency oscillations applied to the grid wires increased, led BRADBURY to the conclusion that detachment of electrons by negative ion impact might be occurring. The writer [*28*] undertook a study of this action in view of the urgent necessity of evaluating the heat of formation of the O_2^- ion. Toward this end, electrons from a high frequency glow discharge were allowed to form ions in O_2 of appropriate pressure and to drift through an electron filter of the conventional type which effectively removed most of the electrons.

The ions then passed through a second filter with grid wires of 2 mm. diameter spaced with surfaces 1 mm. apart. The high frequency fields on this filter, called the *smasher*, were raised in potential until the current of ions through the smasher to the collector plate beyond began to decline. The decline was associated with the *detachment* of electrons from the negative ions by impact with neutral gas molecules and capture by the field. Frequencies were too high for much capture of ions. Control measurements made this action plausible and the measurements with various frequencies, pressures and fields led to a linear relation between the value of the average field between wires and the pressure at which the very sharp decline was observed. The field-to-pressure ratio X/p was 90 volts/cm. per mm. Hg in O_2. Various ways of calculating the energy of the impacts at $X/p = 90$ based on modern mobility theory led to an energy of impact of some 0.7 e.v. or less, so that the available energy for detachment was definitely less than 0.35 e.v. The method is avowedly inaccurate but, as will be seen, probably furnishes the only experimental data of this sort to date.

14. The photodetachment process [*29*]. Many attempts have been made to create negative ions and to observe the electron detachment by means of light. Until recently techniques have not been adequate to the task. Very recently BRANSCOMBE [*29*] and his associates at the US Bureau of Standards have succeeded in the study for the case of H^- and O^- ions. In this process the H^- ions came from the anode end of the striations of a heavy glow discharge in H_2 contaminated by water vapor. The H^- entered a drift space with collecting electrodes and suitable transverse fields for sorting out free electrons from the ion beam. In this space they were subjected to intense illumination of light of various wave lengths passed through appropriate filters. Some of the surviving ions then passed through a slit in the anode to enter a mass spectrograph in which they were analyzed. In this fashion excellent photodetachment curves were obtained as a function of wave length using intermittent illumination to permit amplification of the weak currents of photodetached electrons. In consequence cross sections for photodetachment as a function of wavelength in good agreement with theoretical predictions of CHANDRASEKHAR for H^- were obtained. The threshold energy of photodetachment was found to agree with the calculated value of 0.74 e.v. Preliminary studies in O_2, using a mass spectrograph on the ion beam, revealed the formation of two ions in the glow discharge. These were O^- and O_2^-. The former was created by dissociative attachment of energetic electrons and the other by direct attachment of slow electrons. The O^- ion showed photodetachment at energies lying between 1.35 and 1.55 e.v., contrary to the previously accepted value 2.2 e.v. This conclusion is irrefutable. The O_2^- ion was observed and showed photodetachment *below* 1 e.v. However, the threshold for this action could not be determined owing to the presence of background impurities but more especially owing to the lack of suitable filters for passing extremely long wavelengths that were able to survive the heat of the high intensity source needed to produce measureable detachment. Further studies of this type are urgently needed.

IV. Experimental results and interpretation.

15. Evaluation and significance of data. Until the measurements of BRADBURY really no proper values of the attachment probabilities or cross sections were at hand. The earlier data of the author and of WAHLIN were of orienting magnitudes only, and even BAILEY's early data in air deviated from the later more precise measurements. The later values of BAILEY and his group as given by HEALEY and REED constitute data taken under better conditions. While BRADBURY's data cover a wide range of gases and X/p ratios and represent direct measurements, the legitimate criticism of his values of k_e which enter critically into the absolute evaluation of h, because they are squared, might cause some concern. It appears, however, that where later checks were made, as for example by DOEHRING, BRADBURY's values were not seriously in error. However, it must be equally stated of the work reported in HEALEY and REED that due to the indirect nature of the determinations, since complicated average values of important parameters, such as η, v, λ etc., with unknown distribution laws enter into the ultimate evaluation of h, the values are equally unsatisfactory. By and large, direct comparison of the data of BRADBURY and the BAILEY group for h as a function of X/p agrees in the order of magnitude of values assigned, but the variation of individual gases with X/p is not the same.

In discussing the data on h one more difference between the techniques of BRADBURY, DOEHRING, HERRENG, and those of the BAILEY type must be pointed out. That is, the former techniques give good values at low X/p and are thus well adapted for observing thresholds. Their validity should also apply to higher X/p. The BAILEY techniques are *not* very *sensitive* until higher values of X/p are reached. All measurements involving drift velocities are applicable to higher pressures in those gases in which h is relatively low. If h is very high, non-attaching diluting gases are required, since v loses its significance at low pressures and reasonable distances x. All methods except those of GEBALLE and HARRISON are inapplicable where excitation and ionization by collision occur in the gas. Attachment of electrons at near thermal energies and at really low energies can to date only be achieved by microwave or mono-energetic attachment techniques. The microwave techniques are applicable in general only where h is moderately high, as too high pressures are not convenient. The monoenergetic electron techniques, appearance potential studies and photodetachment techniques are possible only where h is really *very* high, as pressures must be low. All studies at lower pressures would do well to be accompanied by mass spectrographic analysis of the negative ions formed where ever possible.

All except the appearance potential and monoenergetic attachment studies suffer in that the energy of the electrons attaching are not known. In all gases the electron energies in a field are distributed according to some law. Where electron free paths vary in unknown ways with electron energy, the form of these laws can vary widely, ranging from those with a high energy cut off far steeper than even that of DRUYVESTEYN's law to one with far more electrons in the assymptotic high energy foot than exhibited by the MAXWELL law. If there is a minimum energy threshold, as is the case for most dissociative attachment processes, then even if the cross section for attachment were a maximum at his threshold, the observed curves for h or q_A as measured by other than low pressure techniques would rise gradually at *average* electron energies well below the appearance threshold, increase asymptotically and possibly reach a maximum value when the average passes the appearance threshold, falling beyond. Were distribution laws known, then study of the of h or q_a as a function of energy would in connection with the sensitivity of the method reveal the threshold and allow some interpretation to be made. The situation is here very bad since the variation of h, or q_a, as a function of X/p only is observed and the *average* electron energy as a function of X/p can only be evaluated by the indirect diffusion techniques of TOWNSEND and BAILEY, or by microwave techniques. The values of these quantities are not accurately determined and, even if they were, they are meaningless unless the form of the distribution law is known.

It is for this reason that the data from the BAILEY methods while yielding both values of h and the average energy at higher X/p have not lent themselves to any significant interpretation of the results obtained, and what is furthermore, correlate rather badly with deductions by BRADBURY and TATEL concerning processes at work. With his more sensitive method at lower X/p and his interest in interpretation of the processes along modern atomic theory lines, BRADBURY, whenever in doubt, studied h in mixtures of the active gas with several non-attaching gases in which the electron energy distributions were known in the pure state and relative to each other. The gases used were He, A and N_2. While the average energies were not known for the mixtures, at least the relative behaviors of the $X/p-h$ curves in the gases gave a much clearer understanding of the energies involved and as to what was occurring than the sole reliance on the $X/p-h$ data from diffusion studies.

16. Atomic and molecular structure and electron attachment in gases. In Table 1 are summarized data calculated by D. R. BATES [*29a*] concerning heats of formation of atomic negative ions expressed in electron volts. The calculations largely came from extrapolation down an isoelectronic series, though some other observed data are included. This table predicts in terms of the energies, positive or negative, whether a stabile negative ion can form and whether the reaction forming it is exothermic, which is, as will be seen, an important item. In Table 2 is presented an elaborate mass of data, initially as assembled by BRADBURY and, in its present form, as extended by the writer [*37*] when more data accumulated over the years. This table presents just the atoms,

Table 1. *Energy of atomic negative ion formation by extrapolation down an isoelectronic series, after* D. R. BATES [*29a*].

Atom	State	Energy	Probable value from experiment or exact theory
H	$1s^2$	+0.74	0.747 e.v., calculated wave—mechanically. Observed in mass spectrograph
He	$2s$	−0.53	
Li	$2s^2$	+0.47	0.54 e.v., calculated wave—mechanically. Observed in mass spectrograph
Be	$2p$	−0.8	
B	$2p^2$	−0.1	
C	$2p^3$	+0.9	
N	$2p^4$	−0.6	
O	$2p^5$	+1.0	2.2 ± 0.2, Experiment[1]. Observed in mass spectrograph
F	$2p^6$	+2.9	4.13 Calculated from cyclical process
Na	$3s^2$	+0.0	
Mg	$3p$	−0.9	
Al	$3p^2$	+0.2	
Si	$3p^3$	+1.7	
P	$3p^4$	+0.3	
S	$3p^5$	+1.5	Observed 3.6
Cl	$3p^6$	+3.1	3.79 Observed 3.78
Ne	—	−1.20	
A	—	−1.0	
Br	—	—	3.52
I	—	—	+3.14, 3.24, 3.15, 3.22

[1] This value has now been proven incorrect by BRANSCOMB [*29*], the correct value is 1.35 to 1.55 ev.

molecules or radicals so far observed in attachment studies of free electrons in the gaseous phase. In the second column it gives the ground state configuration of the system as far as known. The third column indicates whether the system forms an ion by direct electron attachment in the gas. The fourth column gives the threshold value of the electron energy at which attachment appears to set in judged from the electron energy distribution, or energy relations involved in dissociation and attachment where known. This applies as well to molecular types which do not of themselves attach electrons but may yield ions by dissociative attachment processes. Column five shows the reaction assumed to accompany the attachment, and column six classifies the process active as will be detailed below. This table omits some of the attachment processes occurring in the spark, suppressing gases studied by GEBALLE and his associates, viz. CCl_4, CF_3SF_2, CCl_2F_2, SF_4 and $SiCl_4$. Here the details of the attachment process are not sufficiently known. The data of HICKAM and FOX on the formation of SF_6^- and SF_5^- came too late to include.

A study of Table 2 quickly indicates that atoms in the $^1S_0$ states do not attach electrons, while those in the $^2S_{\frac{1}{2}}$, $^2P_{\frac{3}{2}}$ and $^3P_2$ states appear to have considerable affinity. Quite analogously, molecules in the $^1\Sigma$ states do *not* directly attach electrons, while sufficiently energetic electrons can attach dissociatively to fragments in the $^3\Sigma$, $^2\Sigma$, and $^2\Pi$ states. This behavior really presents no surprise in view of the character of these states.

That dissociative attachment in the gaseous phase might occur spontaneously in the case of the halogen molecules where the energy of negative atomic ion formation exceeds the energy of dissociation was first indicated to the writer by JAMES FRANCK in 1928. FRANCK also ventured the opinion that dissociative attachment to the halogen acids would occur at electron energies of some tenths of a volt. These predictions were predicated on the high observed attachment coefficients in these gases and were within four years directly verified by BRADBURY [16]. The dissociative attachment process is in fact the most frequently observed process for electrons in electrical fields in gases, when fields become adequate. Beyond this mechanism, BRADBURY however called attention to several other phenomena which may or may not properly be called separate processes.

The peculiar behaviour of O_2 at low values of X/p in which h rapidly falls as X/p and energy increase represents a distinctly different type of mechanism, shown only certainly by SO_2 and probably SF_6. That is, since the molecules of O_2, SO_2, and SF_6 have electron affinities, electrons may directly attach to form O_2^-, SO_2^- and SF_6^- provided a mechanism is present to carry off the excess electron energy and energy of ion formation within an appropriate time. The mechanism involved in these gases is one in which the energy of the electron plus the heat of attachment are temporarily absorbed into the vibrational energy of the system. If this energy is not dissipated in collision within a characteristic time τ in the excited state, the energy of vibration will be restored to the electron which will then detach itself and escape. Wave mechanical theory developed by BRADBURY and BLOCH [31] for this process in O_2 assuming a heat of formation under 0.2 e.v. required either a very large cross section for the triple impact, or else a long life time of the excited state to account for the pressure independence observed above 3 mm. Hg. No data are at hand for SO_2. In the case of SF_6^- the many vibrational states of this structure could share the very small attachment energy in such a fashion to give the state an inordinately long lifetime. In any event, theory predicts that the cross section, or h, should rise to a maximum with

Table 2. *Known negative ions created in gaseous phase giving atomic or molecular species and circumstances of ion formation, after* Bradbury *and* Loeb [24].

Molecule or atom	Ground state	Negative ions form directly	Electron energy at which attachment begins	Reaction accompanying attachment	Classification of process
He, Ne, A, Kr, Xe	$^1S_0$	No			
H_2, N_2, CO, CO_2	$^1\Sigma$	No			
Cl_2, Br_2, I_2	$^1\Sigma$	No	Large at thermal; decreases with energy	$Cl_2 + e \rightarrow Cl^- + Cl + (3.79–2.48)$ volts cross section for $I_2 \sim 4 \times 10^{-16}$ cm.^{2}	Dissociative attachment. Very large cross section for I_2, $h \sim 1$
HCl, HBr, HI	$^1\Sigma$	No	In HCl 0.7 e. v.	$HCl + e + (4.5 - 3.8)$ volts $\rightarrow H + Cl^-$	Dissociative attachment
NH_3	—	No	$> 5.8 \pm 0.3$ e. v. $> 6 \pm 0.5$ e. v.	$NH_3 + e + 3$ volts $\rightarrow H^- + NH_2$ $NH_3 + e +$? volts $\rightarrow NH_2^- + H$	Dissociative attachment
N_2O	$^1\Sigma$	No	1.8 e. v.	$N_2O + e + 1.8$ volts $\rightarrow O^- + N_2$ $N_2O + e + 3.84$ volts $= NO^- + N$	Dissociative attachment
CO_2	$^1\Sigma$	No	> 3.5 e. v. X/p 30	$CO_2 + e + 3.5$ volts $\rightarrow CO + O^-$	Dissociative. Has not been observed directly. X/p required too high
H_2S	—	No	$> 2.9 \pm 0.3$ e. v.	$H_2S + e + 2.9$ volts $\rightarrow S^- + H_2$	Dissociative attachment
H_2O	—	No	> 5.7	$H_2O + e + 5$ volts $\rightarrow OH + H^- + KE = 1.5$ e. v.	Dissociative attachment (both H_2O rows)
H_2O	—	No	$> 7.5 \pm 0.3$ e. v.	$H_2O + e + 7.5$ volts $\rightarrow O^- + 2H$	
O_2	$^3\Sigma$	Yes	Thermal to 0.5 e. v. h decreases as energy increases, but may decrease from 0.1 e. v. down	$O_2 + O_2 + e \rightarrow O_2^- + O_2$ Attachment energy between 0.2 and 0.5 e. v.	Direct attachment. Energy goes to vibration; lost in impact later
O_2	$^3\Sigma$	No	2.9 e. v. May set in at higher energy	$O_2 + e + 2.89$ volts $\rightarrow O^- + O$. Cross sect. $\sim 2.8 \times 10^{-19}$ cm.^{2} Attachment energy ~ 1.35 e. v.	Dissociative attachment with large cross section
NH_2	$^3\Sigma$	Yes	In $NH_3 > 6$ e. v.	$NH_3 + e + 6$ volts $\rightarrow NH_2^- + H$	Dissociative attachment
SO	$^3\Sigma$	May	In $SO_2 > 5.7$ volts	$SO_2 + e + 5.7$ volts $\rightarrow SO^- + O$	Dissociative attachment. Not surely established
SO_2	$^3\Sigma$	Yes	Thermal to 1 e. v. h decreases as energy increases	$2SO_2 + e \rightarrow SO_2^- + SO_2$.	Direct attachment. Energy goes to vibration; lost in impact later

SO_2	$^3\Sigma$	No	> 3.5	$SO_2 + e + 3.5$ volts $\rightarrow SO + O^-$	Dissociative attachment, alternative to formation of SO^-, and may predominate
NO	$^2\Pi$	Yes	All energies	$2NO \rightarrow (NO)_2$ $(NO)_2 + e \rightarrow NO^- + NO$	Pressure-dependent. Depends on complex $(NO)_2$ formation to take off energy
OH	$^2\Pi$	Yes	OH^- forms by collision process in H_2O vapor from O^-, H^-; possibly in excited state by electron attachment to OH at above 1 mm.	Not known	OH^{-*} in excited state optically verified. Also observed by mass spectrograph at higher pressures
H_2O	—	May	Low energies H_2O_2 aids	$(2H_2O) + e \rightarrow H_2O^- + H_2O$, or $(H_2O)_2 + O_2 + e \rightarrow O_2^- + (H_2O)_2$	Pressure-dependent. Requires formation of complexes to take off heat of formation
Cl	$^2P_{3/2}$	Yes	—	$2Cl + e \rightarrow Cl^- + Cl$ Attachment energy 3.79	Direct attachment if energy can be carried off by another atom
O	^3P	Yes	All energies	$2O + e \rightarrow O^- + O$ Attachment energy ~ 1.35 e.v.	Direct attachment if energy can be carried off by another atom

decreasing energy, pass a peak and fall to zero for very low energies in O_2. The peak has never been observed.

All three molecules at somewhat higher electron energies also yield dissociative attachment, that of O_2 occurring at 3 e.v. energy and SF_5^- occurring at very low energies. In O_2 the energy of 1.35 e.v. in the formation of O^- adds to effectively reduce the dissociation energy for O_2 of about 5 volts to near a 3 volt threshold.

A third process noted by BRADBURY was a pressure dependent attachment process which gave values of h decreasing as electron energy increased at any given pressure, with each curve for h rising to higher values as pressure increased. In NO the process was shown in essence to depend on triple impacts of a different sort from those in O_2 in which the effective stabilizing collisions of large cross sections could occur with any molecular species such as N_2. This was not the case with NO where N_2 molecules were ineffective.

It appears that at appropriate pressures of NO the NO polymerizes to $(NO)_2$ and the equilibrium concentration of $(NO)_2$ is a function of pressure and gas temperature. It thus seems likely that the negative ion formation in NO comes from an electron attachment to the $(NO)_2$ complexes with dissociation to give an NO^- ion and an NO molecule. That is, since NO cannot take up the considerable energy of attachment into its vibrational system as in O_2, permitting a later collision to remove the energy, attachment only occurs to the $(NO)_2$ complexes and attachment is a dissociative process at a very low threshold as the energy of formation of $(NO)_2$ is around 0.5 e.v. Attachment in H_2O at low X/p appears to follow the same pattern as NO. In this case BRADBURY was inclined to ascribe the pressure dependent process to the formation of H_2O complexes containing either O_2 or CO_2. Electrons then attach dissociatively forming perhaps O_2^- and CO_2^- ions. There is nothing that precludes H_2O^- ion formation dissociatively from x (H_2O) and an electron[1]. Little is known of the processes at work, and speculation is futile at this time. At higher energies, electrons attach dissociatively to H_2O as shown in Table 2.

So far dissociative attachment has been observed in impacts where electron energies are adequate to yield a stable atomic or molecular ion when account is taken of the dissociation energy, heat of ion formation and the electron energy. Very little correlation as yet exists between appearance potential studies of negative ions through electron impact and such processes as studied by BRADBURY and others. Unless attachment cross sections are relatively large, it is hard to detect much negative ion formation at the low pressures required of nearly monoenergetic electron beams. Today mass spectrometers revealing 1 part in 10^6, such as those of REYNOLDS, may help overcome this difficulty. Likewise as indicated, studies at higher pressures are falsified as to threshold by the unknown form of the energy distribution law and the meaning of the average energy in this regard. In some cases the expected behaviour of the gradual rise of h or q_a with increase in X/p above a certain value, the achievement of a peak followed by a decline has been observed. Likewise while values of h have been observed for many gases, lack of values of the RAMSAUER cross section as a function of electron energy in attaching gases do not permit of evaluation of *cross sections*. Data are fairly complete for the formation of O_2^- and of O^- above about 3 volts in oxygen. The most successful investigation so far is that of the dissociative attachment in I_2 from thermal energies up by BIONDI using microwave techniques. Here the cross section has a maximum of 4×10^{-16} cm.^{2} at thermal energies and drops very rapidly as energies increase to 0.4 e.v.

Probably it is sufficient to close this section with the comment that we are just now arriving at techniques which may in the future allow some understanding of the attachment mechanisms, cross sections and energetics. To date, however, aside from the development of new techniques it can be stated that *certain atomic* and *molecular* species do *not* form negative ions and that *others do*, the types so doing being consistent with established atomic-electronic behaviour. It has further been established that the predominating attachment mechanism and that showing the largest cross sections is the dissociative attachment process, very few molecules failing to show this at appropriate energies, though ions appearing may not fit presupposed patterns. The few pressure dependent reactions such as those in NO and H_2O probably again fall into the dissociative class. Thus far only three substances of low heat of attachment show a decline of attachment with increasing energy with no pressure dependence, indicating a direct attachment process with energy of attachment temporarily absorbed into the vibrational system and later dissipated in a molecular collision. In a sense these represent triple collision processes in which impact radius is decreased by increased life time of the vibrational state. New evidence by MUSCHLITZ and SIMONS (private communication) indicates that negative ions created by primary impact or other process may alter in subsequent collisions with parent molecule above 1 mm. pressure presenting complex results.

---

[1] Recent mass spectrographic data by MUSCHLITZ and SIMONS (private communication) indicate that H_2O^- forms by electron impact and more readily in $H_2O-H_2O_2$ mixtures probably in triple impact.

V. Formation of negative ions by processes other than attachment in the gaseous phase at low X/p.

17. Introduction. As early as 1912, J. J. THOMSON [*32*] in his first mass spectrograph observed negative ions of O^-, Cl^-, H^- and what he believed to be N^-. He at first ascribed these to possible dissociation of polar gaseous compounds by electron impact but control studies using ionization at low energies in glow discharges indicated that this was not the origin. O. W. RICHARDSON [*33*] in his book on emission of electricity from hot bodies reported negative ions to come from hot salts. From there on many experimental studies over the years indicated that negative ions could be formed by various processes. By the middle nineteen hundred and thirties the data fairly clearly identified several processes as being active and MASSEY and SMITH [*34*] developed the theory underlying some of them. More data are summarized in MASSEY's excellent little monograph on Negative Ions and in MASSEY and BURHOP'S recent book [*35*]. Since that period, stimulated by various investigations and certain industrial problems, very careful studies of the appearance of such ions by mass spectrograph have been carried out in the laboratory of K. G. EMELEUS in Belfast by SLOANE and his co-workers [*3*] that have clarified the questions and indicated what ions have been observed and something of the processes at work. Most important of all is the fact that they have shown that only those atomic ions listed in Table 1 with positive heats of ion formation are surely established as having been observed. Owing to lack of space, much of the earlier work cannot be detailed. The processes active may be listed as follows:

1. Dissociation of molecules in the polar phase by electron impact (usually rare except at high electron energies not usually achieved in gases), or other electron impact phenomena involving some sort of dissociative attachment not usually observed under ordinary values of electron energies in the gaseous phase.

2. Neutralization and creation of negative ions by impact of energetic positive ions on metal surfaces. With recent knowledge of what constitutes a clean surface it is not so far certainly established that this process is real, since process (3) below could account for most observations.

3. By impact ionization and/or sputtering of negative ions adsorbed as monolayers, or in thick layers, on metal surfaces by positive ion impact. Accumulation of data in recent times indicates that this process probably is the most common and important process.

4. Heating of salts, oxide coated cathodes, or metals contaminated by surface layers in cleansing, liberate negative thermions, if temperatures are high, applied fields favorable and heats of adsorption are low. Contaminated metal surfaces are usually cleaned of such ions by flashing at above 1000° C.

Since the initial studies of SLOANE and his co-workers, the remarkable vacuum techniques[1] of D. ALPERT [*36*] have made it possible to sustain background gas pressures in which it takes on the order of hours to form monolayers on an out-gassed surface. The studies of H. D. HAGSTRUM [*37*] and J. H. PARKER [*37*] on secondary emission by positive ion bombardment of surfaces using such techniques and the recent studies of sputtering of atoms by ion bombardment of WEHNER [*37*] and others indicate that no previous studies of negative ion formation at surfaces, could have been made on *clean* metal surfaces and that the proposed mechanism (2) must be critically reinvestigated with modern techniques.

In what follows the data on the four processes above accumulated to date will briefly be presented and discussed.

[1] Detailed description in D. ALPERT'S contribution to vol. XIII of this Encyclopedia.

18. Electron impact dissociation into negative and positive ions and high energy attachment in the gaseous phase. Such action as the electron impact dissociation of a polar molecule into positive and negative ions has been assumed to occur and looked for at various times over the years. Until recently the only case authenticated at relatively low energy was that of CO by TATE and LOZIER [*38*]. At well above 16 volts many molecules like H_2O etc. have been observed to do so. Relatively recently H. GUTBIER and H. NEUERT [*39*] have studied the two halogen acids HCl and HBr using mass spectrograph in relatively well controlled appearance potential studies. HCl dissociates at 4.43 volts and HBr at 3.72 volts. The heat of negative ion formation of Cl^- is 3.78 and of Br^- is 3.52 volts. Thus HCl and HBr should show dissociative attachment to yield H and Cl^- or Br^- respectively at 0.65 and 0.2 volts. The appearance potentials observed by these workers were (0.8 ± 0.3) volts and (0.6 ± 0.3) volts respectively which is in conformity with BRADBURY'S observation with HCl where attachment was observed above $X/p = 1$. The attachment rose to a maximum at 1.4 and 1.5 volts in HCl and HBr respectively in GUTBIER and NEUERT'S study. As electron energies rose to about 13.6 and 13.8 volts respectively for HCl and HBr these workers observed a very large current of Cl^- and Br^- ions which they ascribe to the dissociation of these acids to H^+ and Cl^- and Br^- respectively. The ionization potential of H is 13.5 volts and the energy needed to dissociate and form H^+ and Cl^- or Br^- is 0.65 and 0.2 volts. Thus appearance potentials of H^+ and Cl^- should be 14.2 and 13.7 volts respectively in agreement with observation. The yield of ions by this reaction rapidly increased with electron energy to values many times that for the dissociative attachment at peak and no decline was noted. The assumption as to mechanism that $e + HCl \rightarrow H^+ + Cl^- + e$ was *not firmly* established by finding H^+ ions in proportion to Cl^- at the threshold energy. However, H^- ions were looked for at this energy and found to be very few. Thus the evidence speaks very strongly for the reaction proposed as no alternatives exist. Owing to its high threshold this reaction will not occur in gases under ordinary discharge conditions. It might be suspected that dissociation of this sort in which negative ions are produced without electron attachment would apply more generally to polar molecules other than the few binary dipoles of which perhaps only the halogen acids are readily volatile at ordinary temperatures. This leads one at once to the study of water vapor which in turn involves interpretation of the contradictory data on the existence of the OH^- ion, including as well the studies of BRADBURY. Very extensive and careful analysis was made by M. M. MANN, A. HUSTRULID and J. T. TATE [*40*] in 1940 concerning appearance potentials of ions in H_2O vapor, and $NH^+_3 \cdot H_2O^+$ appeared at (13.0 ± 0.2) volts, OH^+ at (18.7 ± 0.2) volts and O^+ at 18.8 volts. Some H_3O^+ appeared at 13.8 volts and H^+ from H_2O *without* OH^- at 19.5 volts. H_2^+ appeared at 23 volts, H^- at (5.6 ± 0.5) volts and O^- appeared at (7.5 ± 0.3) volts. No OH^- was observed at any time[1]. The H^- appeared at (5.6 ± 0.5) volts and rose to a peak at about 7.1 volts. A second peak appeared at 8.9 volts. This confirms and agrees with the work of LOZIER [*41*] and of SMYTH and MUELLER [*41*]. Using retarding potentials the ions forming the first maximum have 1.5 volts kinetic energy and those from the second peak have 3.2 volts kinetic energy. To form H^- by dissociative attachment to H_2O giving OH and H^- requires

[1] All previous appearance potential studies were made at low pressures. Very recent work by BRANSCOMBE [*29*] and especially appearance potential studies at higher pressures with mass spectrograph of MUSCHLITZ and SIMONS [*51*] indicate the presence of a considerable number of OH^- ions above 4 mm. pressure H_2O. These presumably form by triple impact of O^- or ions with H_2O, and the reactions are highly pressure dependent.

4.2 volts, which if the heat of negative ion formation is 0.74 volts means that the heat of dissociation of H from H_2O is about 4.9 e.v. (calculated 5.1). With a heat of formation from H_2O of 4.2 e.v., the appearance of H^- ions with 1.5 e.v. would be at 5.7 volts in agreement with observation. The second peak caused by a higher excited state of H_2O^- with 3.2 volts of kinetic energy should be at 7.7 e.v. in rough agreement with the rise of the incompletely resolved electron peak. It may thus be concluded that with H_2O the action is the capture of an electron of 5.7 or of 7.7 volts giving an excited H_2O^- ion which subsequently dissociates into H^- and OH with excess kinetic energy going to the component parts. However, the appearance of a peak of O^- at $(7.5 \pm 0.3$ volts), ascribed to the process giving 2H and O^-, indicates that the H_2O^- complex may also dissociate into 2H and O^-. The heat of dissociation of H_2O into $2H + O$ is 9.5 volts, and the heat of formation of O^- is ~ 1.35 volts, thus the appearance potential of 7.3 volts for this peak is in agreement with observation. This mode of dissociation is more probable than the rough one giving H^- with 3 volts kinetic energy. Two higher peaks at 10 and 12 volts represent capture into higher excited states and dissociation into O^- with more kinetic energy. At 20.8 and 34.3 volts H_2O appears to dissociate into $H + H^+ + O^-$ and $H^+ + H^+ + O^-$ which might be interpreted as some sort of dissociation into a polar phase as the extra electron escapes. It certainly does not occur at the low energies characteristic of attachment studies.

It is thus seen that BRADBURY's data on the dissociative attachment of electrons to H_2O correctly indicated the electron energy to have been about 5 volts but misinterpreted the process ascribing the negative ion to OH^- when in fact it was H^-. This confirmation of BRADBURY's energy assignments using argon gas is gratifying in view of the conflict with conclusions from the BAILEY techniques.

Before leaving this study it must be indicated that an investigation of appearance potentials in O_2 indicated that in agreement with GEBALLE and HARRISON's conclusions O^- ions appear at 3 volts. This is also indicated by the rise of the curves of DOEHRING, BRADBURY and others.

It is next of interest to indicate that BENNETT and DARBY [*42*] utilized arcs in H_2O and H_2-H_2O mixtures to obtain copious sources of H^- ions. The most successful conditions were in a hot cathode arc at some 5×10^{-2} mm. Hg pressure at 88 volts and 50 mA current. The H^- appeared through a hole in the anode whenever pressure reduction brought the anode end of striation in contact with the anode. Currents of H^- ions of the order of 10^{-7} Amp could be extracted.

In regard to the absence of OH^- ions in all these studies, despite the importance of OH^- ions in solutions, the following may be said. Apparently the OH^- ion cannot form by electron impact on H_2O [*51*] but does form in secondary pressure dependent reactions of H^- or O^- with H_2O above 1 mm. Recently SCHÜLER, REINEBECK and MICHEL [*43*] in spectrographic investigations claim that OH spectra observed todate are not from OH and OH^+ radicals but from OH^-. The OH^- is assumed to form in the negative glow of an H_2O glow discharge by capture by OH of an electron into a highly excited vibrational state. This radiates the band spectrum observed.

H. NEUERT [*44*] has investigated H_2S and H_2Se. Energies of dissociation into H_2 and S or Se are about 6.5 and 5.3 volts respectively. Appearance potentials were at (2.9 ± 0.3) and (2.3 ± 0.3) volts with peaks at 4 and 3.4 volts yielding electron affinities of S^- and Se^- as 3.5 and 2.9 volts respectively. At 5.5 volts in Se a peak attributable to the formation of $Se^- + 2H$ requiring a total energy of 10.1 volts for separation into Se and H_2 and dissociation of H_2 appeared although the available energy observed was 8.4 volts with the 5.5 volt appearance potential. At 15 e.v. some sort of dissociation into H^+ and H or H_2^+ and Se^- was observed akin to the behaviour of water, the electron not attaching. In H_2S a similar general dissociation occurred at around 10 volts. H_2S however, yielded many fewer S^- ions and was difficult to work with. Unlike the case of H_2O no H^- was reported. BRADBURY reported a dissociative attachment in pure H_2S at an $X/p=6$ which he interpreted as yielding S^- or HS^- ions at 3.2 or 3.7 volts favoring the latter.

In a like manner negative [40] ions observed in NH^3 were H^- at (5.8 ± 0.3) volts and NH_2^- at (6 ± 0.5) volts. The curve for NH_2^- rises to a peak at (7.3 ± 0.7) volts with a small second peak rising to a peak at 11.5 volts starting near 9 volts. This peak disappears above 15 volts. The heat of dissociation giving NH_2 and H is 3.8 volts, thus kinetic energy of 2 volts plus the heat of formation of NH_2 of from 2 to 3 volts is released largely going to the H atom. The production of H^- at 5.8 volts with a maximum at 7.2 volts, and a second process giving H^- at 23 volts throw more light on the observations of BRADBURY. MANN, HUSTRULID and TATE [40] assume that NH_3 breaks into N, H and H_2 at 6.4 volts. If the heat of formation of 0.7 volts for H^- is included this accounts for the appearance potential of 5.8 volts. BRADBURY's estimate of an energy of 4 volts at X/p of about 10 in NH_3 and other gaseous mixtures is apparently low since 5.8 volts is required. His distribution may have had many fast electrons. The reaction suggested by the appearance potential studies are in agreement with the effective production of H_2 gas by NH_3 above $X/p=10$ observed by BRADBURY.

From this material certain conclusions may be drawn:

1. All future studies of attachment in the gas phase, whether dissociative or otherwise, require identification of the ions produced. This is evident from BRADBURY's erroneous assignments of ions formed as indicated from later correlative evidence.

2. It is also clear that there is no guarantee that in a homologous series of compounds, e.g., H_2O, H_2S and H_2Se the reactions will be the same. Thus although O^-, S^- and Se^- form with high energies the dissociative attachment in H_2O commonly observed leads to H^- ions and only at higher energies to O^- ions, while in the sulfides and selenides the ions are predominatingly S^- and Se^-.

3. It is finally clear that at sufficient energies electrons will dissociate polar molecules into appropriate ions of opposite sign *without* attachment. In general, conditions for these actions are such that under ordinary discharge conditions in the gaseous phase electron energies are not adequate and negative ion production by dissociative attachment is largely observed.

19. Negative ion formation by positive ion impact on clean metal surfaces. This process probably began to be invoked by O. TUXEN [45] in his careful mass spectroscopic study of ion formation in 1936, where he proved that the H^- he observed came directly from impact of H_2, H_2^+ or H on metal slits. This important investigation also definitely proved that in conformity to Table 1 N^- does not form, the ion observed by J. J. THOMSON [32] presumably having been CH_2^- or CH^- poorly resolved. The conclusion that H^- could be created by impact of H^+ ions on a metal surface was supported by the theoretical study of H. S. W. MASSEY and R. A. SMITH [34] in 1936 who had developed a suggestion by K. G. EMELEUS that positive ions of sufficient energy could leave a metal surface with a negative charge after impact provided the heat of negative ion formation was positive and adequate. Studies of F. L. ARNOT [2] and his associates beginning in 1936 confirmed and extended this concept. ARNOT's studies showed that negative ion production was proportional to the positive ion current incident and gave yields of negative ions relative to the number of positive ions incident. They further revealed that positive ions of above about 100 volts were capable of yielding negative ions and that the negative ions often emerged with a considerable portion of the energy of the incident positive ions. Unfortunately these studies made no pretense of using clean metal surfaces and a certain lack of control resulted in some erroneous conclusions. Thus the reported creation of Hg^- ions by impact of 200 volt Hg^+ ions on metal surfaces appears to have been incorrect. STILLE [*46*] in 1933 had not observed Hg^- though Hg^- had been accorded a heat of ion formation of $+1.74$ volts in MASSEY's table preceding that of BATES' [*29a*] recalculation shown as Table 1. SLOANE and PRESS [*3*], after very careful mass spectrographic study, found no trace of Hg^-. A similar very careful study not published but reported to the writer by letter of June 7, 1947 by W. H. BENNETT, using a very sensitive mass spectrograph, confirmed the non-existence of Hg^- ions. Thus while, in general, the early observations of ARNOT and his associates are correct, some caution should be used in accepting the details as too accurate. The work was in general repeated by SLOANE and PRESS [*3*] with much cleaner conditions, "outgassed" targets and using two mass spectrometers, the first to give ions of known character and the second to identify the ions knocked off. Analysis of yields of negative ions were made and much time was spent in studies of energy distributions of emitted negative ions. The probability of emission of a negative ion per impacting positive ion lay in the order of 10^{-4} to 10^{-5} in general conformity with magnitudes reported by ARNOT and MILLIGAN, but were in general less. The emitted negative ions had energies of many volts indicating that in the process of emission they had retained, or probably better received, a considerable fraction of the order of 20% and less of the energy of the incident positive ion.

At this point, it is not profitable to summarize the results in detail in connection with this section on the process of neutralization and negative ion creation from positive ions interacting with a *clean* metal surface. It is true that the surfaces of metals flashed say above 1000 to 1200° C are moderately clean though some, if not most, may retain monolayers of gas or impurity as H. D. HAGSTRUM [*37*] has amply shown. Under the conditions of operation by SLOANE, which in their time were excellent, the surfaces within seconds of time after they had cooled down were covered with monolayers of background gas, H_2O, O_2, N_2, H_2, perhaps some Hg, and organics from stopcock greases, etc. As will later be seen, there is considerable chance that some of these layers of adsorbed gases contain extra electrons or consist of negative ions of the adsorbed gas. Thus O and Cl adsorbed on surfaces are considered to be bound as O^- and Cl^- ions, since conditions are most propitious for such action. Less is known of the adsorbed molecular ions. In fact studies of secondary electron emission by energetic positive ion and neutral atom bombardment strongly indicate such action. Thus the blasting or sputtering off of pre-existent adsorbed negative ions on metal surfaces by energetic positive ions is to anticipated. The sputtering studies of WEHNER [*37*] and BRADLEY [*47*] make it plausible that the escaping atoms or ions should have an appreciable amount of momentum imparted to them by the incident positive ions, and that sputtering is much more propitious at grazing incidence. That such action occurs is attested by the fact that such fast positive ions as Hg^+ incident on the surface which do *not* create Hg^- have been observed to knock off a considerable number of lighter negative ions. A further indication that this is the action observed is that in the studies of all workers there is a *peculiar monotony* of the ions knocked off the surface. That is, irrespective of the bombarding ion, the ions observed are frequently those of the background gases, or else ions derived from molecular species with which the metal is being bombarded and its surface saturated. In the earlier studies of ions liberated from salts and soiled metals by bombardment with Li^+ ions by THOMPSON and WOODCOCK [*48*] in DEMPSTER's laboratory negative surface ions of the target material were observed. In the studies of SLOANE and LOVE [*3*] in which a clean Ni surface was bombarded by Li^+ ions not only were mass numbers 6 and 7 corresponding to the two Li isotopes observed but ions of mass 22 and 23 corresponding to LiO^- composed of O^{-16} adsorbed on the surface and the two Li isotopes were observed. Since Li^- ions are stable the formation of Li^- and LiO^- ions on the surface by the incident neutralized and condensed Li^+ ion and atom beam in the ratio of 1 in 10^{-5} to the ion beam is not surprising. These authors also attribute the negative ion of mass 23 earlier observed to be emitted from a Na containing treated KUNSMAN catalyst to LiO^- ions and not to Na^- as this ion should hardly be stable according to Table 1.

Thus at this writing it may safely be concluded that there is as of today no conclusive evidence that negative ions do directly form by impact of energetic positive ions on a *really clean metal surface*, except perhaps from the action of the Swindlotron where energetic H^+ ions of 1 M.e.v. passing through thin metal foils are in considerable number changed to H^- ions [*49*].

20. Negative ions produced by sputtering of negative surface ions on impact of energetic positive ions. This process has been accepted as occurring by most of the investigators of negative ion formation by fast positive ion impact generally since World War II and was observed previously by THOMPSON and WOODCOCK [*48*]. As indicated in Sect. 19 above, it now appears that this process can account for most, if not all, of the negative ion formation by positive ion impact observed to date. Under these circumstances, it is not believed that further discussion here is warranted as the matter has adequately been discussed above.

21. The thermal liberation of negative ions from surfaces. The earliest studies of emission of electricity from hot bodies [*33*] revealed the presence of positive and negative ions of common atoms liberated at temperatures below 1000° C from surface contaminated pure metals. In fact the early mass spectrographs of DEMPSTER derived positive ions from heated minerals and molten salts, as well as contaminated scintered metal combinations such as the KUNSMAN catalysts. Since if the energy of thermal agitation of negative ions on a metal, or in a molten salt film, exceeds the energy of adsorption on the surface the ions would be expected to leave the surface especially if a negative potential gradient aids their escape, it is not surprising that heated substances should emit negative ions. The destruction of cathode ray scope phosphors by what was apparently negative ion bombardment by ions created by the electron emitting filament led to renewed interest in the emission of such ions from heated surfaces. Beginning in 1935, VON ARDENNE [*50*], I. LEVY and D. W. WEST [*50*], H. SCHAEFER and W. WALCHER [*50*] in 1943, and particularly R. H. SLOANE and C. S. WATT [*3*] in 1948, investigated the problem. The study by SCHAEFER and WALCHER using a mass spectrograph on emission from oxide coated cathodes found that these yielded what they identified as H^-, CH^-, OH^-, O_2^- and Cl^- ions along with eleven other ions in smaller numbers. The sharpness of the focus indicated that all except some of the OH^- came from the cathode surface. The presence of some of the OH^- ions was attributed to the impact of H_2O molecules on the heated cathode surface. The peculiar behaviour of the OH^- ion, i.e., the diffusion of its image, should perhaps indicate that the nature and identification of this ion may not have been correct, and SLOANE and WATT have placed it on their list of ions whose identification is in doubt. SCHAEFER and WALCHER draw further conclusions concerning the mechanism of ion formation. They assume that the O^-, OH^- and Cl^- ions appear to be electrolyzed through the BaO layer while CH^- and C_2^- must wander along the crystalline interfaces of the surface. The O_2^- emission appears to be definitely related to the "forming" of the oxide cathode. The careful investigations of SLOANE and WATT extending over several years with various types of mass spectrographs including retarding potential measurements permit of more extensive information including the origin of the ions observed. Table 3 yields a summary of their findings as regards ion identification. Thermionic emission was studied as a function of cathode temperature, time, accelerating field and cathode surface condition. Their studies showed that some of the ions are liberated from the surface thermionically and that others result from positive ion impact on the surface, the latter having energies greater than that given by the extracting field. They indicate that certain atoms exist at the surface as negative ions. To what extent the ions are electrolyzed to the surface is not known but there is nothing to prevent Cl^- ions in a thin layer of molten NaCl from drifting to the outer surface of the melt in the applied field and then evaporating when they receive enough energy by thermal impact. Further work is needed and no doubt will add much light to these phenomena.

Table 3. *Summary of negative ions identified from oxide coated cathodes as given by* SLOANE *and* WATT *[24].*

Atomic mass number	Certain	In doubt	Atomic mass number	Certain	In doupt
1	H^-		30		NO^-
12	C^-				$HCHO^-$
13	CH^-		32	O_2^-	
14		CH_2^-	35	Cl^-	
16	O^-		37	Cl^-	
16		OH^-	39		K^-
18		H_2O^-	40		Ca^-
19	F^-		42	CNO^-	
23		Na^-	56		CaO^-
24	C_2		58		Ni^-
25	C_2H^-		79	Br^-	
26	CN^-	$C_2H_2^-$	81	Br^-	
27	Al^-				

References.

[1] THOMSON, J. J.: Phil. Mag. (VI) **13**, 561 (1906); **20**, 752 (1910).

[2] ARNOT, F. L., and J. C. MILLIGAN: Proc. Roy. Soc. Lond., Ser. A **156**, 359, 538 (1936). — ARNOT, F. L.: Proc. Roy. Soc. Lond., Ser. A **158**, 137 (1937). — ARNOT, F. L., and C. BECKETT: Proc. Roy. Soc. Lond., Ser. A **168**, 103 (1938).

[3] SLOANE, R. H., and R. PRESS: Proc. Roy. Soc. Lond., Ser. A **168**, 284 (1938). — SLOANE, R. H., and H. M. LOVE: Nature (Lond.) **159** (1947). — SLOANE, R. H., and C. S. WATT: Proc. Phys. Soc. Lond. A **61**, 217 (1948).

[4] FRANCK, J., u. W. POHL: Verh. dtsch. phys. Ges. **12**, 291, 613 (1910).

[5] TOWNSEND, J. S.: Electricity in Gases, p. 117ff. Oxford Press 1914. — LATTEY, R. T., and H. T. TIZZARD: Proc. Roy. Soc. Lond., Ser. A **86**, 349 (1912).

[6] KOVARICK, A. F.: Phys. Rev. **30**, 415 (1910).

[7] WELLISCH, E. M.: Amer. J. Sci. **39**, 583 (1915); **44**, 1 (1917). — Phil. Mag. **31**, 186 (1916); **34**, 33 (1917).

[8] LOEB, L. B.: Phys. Rev. **17**, 84 (1921). — Proc. Nat. Acad. Sci. **7**, 5 (1921).

[9] THOMSON, J. J.: Phil. Mag. **30**, 321 (1916).

[10] WAHLIN, H. B.: Phys. Rev. **19**, 173 (1922). — LOEB, L. B.: Phil. Mag. **43**, 229 (1922).

[11] LOEB, L. B.: J. Franklin Inst. **195**, 45 (1924).

[12] LOEB, L. B.: Proc. Nat. Acad. Sci. **9**, 335 (1923).

[13] BAILEY, V. A.: Phil. Mag. **50**, 825 (1925); **9**, 560, 625 (1930).

[14] LUSK, H. F.: Master's Thesis Univ. California, May 1927.

[15] CRAVATH, A. M.: Phys. Rev. **34**, 605 (1929).

[16] BRADBURY, N. E.: Phys. Rev. **44**, 885 (1933). — J. Chem. Phys. **2**, 827, 840 (1934). — BRADBURY, N. E., and H. E. TATEL: J. Chem. Phys. **2**, 835 (1934).
[17] For a complete discussion of later improvements of BAILEY's method see R. H. HEALEY and J. W. REED, The Behaviour of Slow Electrons in Gases, Amalgameted Wireless Press (Australasia), Ltd. Sydney, 1941, pp. 24ff.
[18] GEBALLE, R., and M. A. HARRISON: Phys. Rev. **85**, 372 (1952); **91**, 1 (1953).
[19] DOEHRING, A.: Z. Naturforsch. **7**a, 253 (1952).
[20] HICKAM, W. M., and R. E. FOX: Seventh Annual Conf. on Gaseous Electronics, New York, Oct. 14—16, 1954, paper A—1.
[21] BRADBURY, N. E.: Phys Rev. **40**, 980 (1932).
[22] NIELSEN, R. A., and N. E. BRADBURY: Phys. Rev. **49**, 338 (1936); **51**, 69 (1937). — NIELSEN, R. A.: Phys. Rev. **50**, 950 (1936).
[23] For microwave techniques in general see S. C. BROWN and D. J. ROSE: J. Appl. Phys. **23**, 711, 719, 1028 (1952). See also: Basic Processes of Gaseous Electronics, L. B. LOEB, chap. VI. Berkeley: Univ. of California Press 1955.
[24] BIONDI, M. A.: Phys. Rev. **84**, 1072 (1951); **89**, 337 (1953).
[25] VARNERIN, L. J.: Phys. Rev. **84**, 563 (1951).
[27] HERRENG, P.: Cahiers de Phys. **38**, 7 (1952).
[28] LOEB, L. B.: Phys. Rev. **48**, 684 (1935). See also: Basic Processes of Gaseous Electronics, loc. cit. chap. V.
[29] BRANSCOMB, L. M., and W. L. FITE: Phys. Rev. (A) **93**, 647 (1954). — BRANSCOMB, L. M., and S. J. SMITH: Nat. Bur. Stand. Technical Report No. 4073, 1 May 1955, to appear in Phys. Rev. **1955**. — Full report in Bur. Stand. J. Res., about Oct. **1955**.
[29a] BATES, D. R.: Proc. Roy. Irish. Acad. A **51**, 153 (1947).
[30] LOEB, L. B.: Fundamental Processes in Electrical Discharge in Gases, p 289. New. York: McGraw-Hill 1939.
[31] BLOCH, F., and N. E. BRADBURY: Phys. Rev. **48**, 689 (1935).
[32] THOMSON, J. J.: Rays of Positive Electricity, pp. 27, 70, 227. London: Longman's Green & Co. 1921.
[33] RICHARDSON, O. W.: Emission of Electricity from Hot Bodies, pp. 92—95, 244. London: Longman's Green & Co. 1916.
[34] MASSEY, H. S. W., and R. A. SMITH: Proc. Roy. Soc. Lond., Ser. A **155**, 473 (1936).
[35] MASSEY, H. S. W.: Negative Ions, second edit., p. 72. Cambridge: University Press 1950. See also MASSEY, H. S. W., and E. H. S. BURHOP: Electronic and Ionic Impact Phenomena, pp. 225ff., 249ff. Oxford: Clarendon Press 1952.
[36] ALPERT, D.: J. Appl. Phys. **24**, 810 (1953). — ALPERT, D., and R. S. BURITZ: J. Appl. Phys. **25**, 202 (1954).
[37] HAGSTRUM, H. D.: Phys. Rev. **89**, 338 (1953); **89**, 244 (1953); **91**, 541 (1953); **96**, 325 (1954). — PARKER, J. H.: Phys. Rev. **93**, 1148 (1954). — WEHNER, G. K.: Phys. Rev. **93**, 633 (1954). — J. Appl. Phys. **25**, 270 (1954). — WEHNER, G. K., and G. MEDICUS: J. Appl. Phys. **25**, 698 (1954).
[38] TATE, J. T., and W. W. LOZIER: Phys. Rev. **42**, 518 (1934).
[39] GUTBIER, H., u. H. NEUERT: Z. Naturforsch. **9**a, 335 (1954).
[40] MANN, M. M., A. HUSTRALID and J. T. TATE: Phys. Rev. **58**, 340 (1940).
[41] LOZIER, W. W.: Phys. Rev. **36**, 1417 (1930). — MUELLER, D. W., and H. D. SMYTH: Phys. Rev. **38**, 1920 (1931).
[42] BENNETT, W. H., and P. F. DARBY: Phys. Rev. **49**, 97 (1936).
[43] SCHÜLER, H., L. REINEBECK u. A. MICHEL: Z. Naturforsch. **9**a, 279 (1954).
[44] NEUERT, H.: Z. Naturforsch. **8**a, 459 (1953).
[45] TUXEN, O.: Z. Physik **103**, 463 (1936).
[46] STILLE, H.: Ann. Physik **17**, 635 (1933).
[47] BRADLEY, R. C.: Phys. Rev. **93**, 719 (1954).
[48] THOMPSON, J. S.: Phys. Rev. **38**, 1389 (1931). — WOODCOCK, K. S.: Phys. Rev. **38**, 1696 (1931).
[49] ALVAREZ, L. W.: Rev. Sci. Instrum. **22**, 705 (1951).
[50] ARDENNE, M. v.: Arch. Elektrotechn. **29**, 231 (1935). — LEVY, I., and D. H. WEST: J. Inst. Electr. Engrs. **79**, 11 (1936). — SCHAEFER, H., u. W. WALCHER: Z. Physik **121**, 679 (1943).
[51] MUSCHLITZ, E. E., J. H. SIMONS and coworkers, most of this very careful and significant work is in preparation for publication, presumably in J. Chem. Phys. **1955** and **1956**. Advance data for this article by private communication to the writer by letter of MUSCHLITZ, Aug. 1, 1955. The technique in part is described by E. E. MUSCHLITZ, T. L. BAILEY, J. M. MCGUIRE, J. Chem. Phys. **22**, 2088 (1954). For theory of formation of OH^- see LAIDLER, K. J.: J. Chem. Phys. **22**, 1740 (1954).

The Recombination of Ions.

By

L. B. Loeb.

With 4 figures.

I. Basic Relations.

1. Definition of recombination, ideal relations. Coincident with the study of the conductivity induced by X-rays in gases leading to the concept of gaseous conduction by ionic carriers at the Cavendish laboratory in 1896 it was observed that the conductivity in the *volume* of the gas declined with time after the ionizing X-rays were cut off. This decline was ascribed by J. J. Thomson [*1*] and his group to a neutralization of the positive ions by the negative ions in the volume of the gas and termed *recombination*. Assuming for simplicity that there were just one species each of negative and positive ions, that they were present in equal numbers, that at some initial time designated as zero, $t=0$, there were $n_+=n_-=n_0$ ions per cm.³ distributed uniformly and isotropically in space but with different signs at random relative to each other both on a macro and a micro scale, and that they by mutual "encounters", (unspecified), of ions of opposite sign experienced a neutralization by exchange of charge termed recombination, it was stated that: Loss of ions dn and thus of conductivity in a time dt would be proportional to n_+ the number density of positive ions, n_- the number density of negative ions and the time interval dt. Thus it follows that:

$$dn = -\alpha n_+ n_- \, dt = -\alpha n^2 dt \tag{1.1}$$

Here α is a *constant* of proportionality characteristic of the encounters, varying doubtless with the gas, its pressure, and temperature, the ions present, etc. This *constant* is called the *coefficient of recombination*. Since the number density of ions $n_+=q_+/V$, $n_-=q_-/V$ and $n=q/V$ where q are the numbers of ions corresponding to the volume V, the coefficient α has the dimensions:

$$\alpha = \frac{1}{n\,t} = \frac{V}{q\,t} = L^3 T^{-1} q^{-1},$$

the commonly used c.g.s. unit thus being cm.³ per ion second.

Eq. (1.1) can at once be integrated setting $n=n_0$ at $t=0$ and $n=n$ at $t=t$ as the boundary conditions to yield

$$n = \frac{n_0}{1+\alpha n_0 t} \tag{1.2}$$

and evaluating α from

$$\alpha = \frac{1}{t}\left(\frac{1}{n} - \frac{1}{n_0}\right). \tag{1.3}$$

It is noted that n/n_0 declines from unity to zero along a hyperbola with time and plots of $1/n$ against time should increase linearly with t yielding a slope of value α. Obviously this relation was put to immediate test and rough values of α were obtained. With existing techniques such determinations presented difficulties in measuring the small and ill controlled values of q and especially q_0, so that recourse was had to studying the number of carriers with time under conditions where Q ions were being created in unit volume by some source of ionization while the ions were being lost by recombination. The relations for change in the number of ions dn in time dt then become

$$dn = (Q - \alpha n^2)\, dt. \tag{1.4}$$

Setting $Q/\alpha = k^2$ integration from $n=0$ at $t=0$ to $n=n$ at $t=t$ leads to

$$n = k \frac{e^{2k\alpha t} - 1}{e^{2k\alpha t} + 1}. \tag{1.5}$$

Thus n rises along a complex transcendental curve which starts nearly linearly with t near the origin and approaches a steady state value of

$$n_\infty = \sqrt{\frac{Q}{\alpha}} \tag{1.6}$$

as t becomes indefinitely great. It is seen that eq. (1.3), (1.5) and (1.6) through measurement of n as a function of t, or of n_∞ and Q, lead to values of α. The three methods will be termed the methods of loss, growth and equilibrium. Initially all methods were used. Relatively early the disparate values of Q and n_∞ made it clear that great accuracy of simultaneous measurement of both quantities was impossible and it was abandoned. Even as late as 1937 the method of growth was used alternately with that of loss and usually abandoned since solution of the transcendental equation was not sensitive and since the important test of linearity of a $\frac{1}{n} - t$ plot revealing disturbing effects could not be applied. Today for evaluation of α the method of loss is generally used. In very recent times where secondary ionizating processes in plasmas yield a value of Q not amenable to direct evaluation, the method of growth with perhaps more complicated conditions and relations is indicated not for α but for Q.

2. Critique of ideal theory, and generalized relations. With the primitive techniques limited to longer time intervals, low ion concentrations, inadequate sweeping fields, stray potentials and especially impure gases with ill defined and changing carriers many of the departures from the simple idealized theory could not even be observed [*2*]. Thus despite great ingenuity and skill on the part of many able investigators there was not much progress possible, and the theory remained as primitive as given. It is also clear that the general character of the definitions and idealized equations could cover loss of carriers by recombination resulting from "encounters" between very many different types of carriers the very nature of the charge exchange effecting mechanism could differ radically following different pressure dependencies etc. Thus in fact until the controls of purity and measuring methods which followed the techniques of the post World War II period were brought into play, the theoretical extensions of the analysis to cases where $n_+ \neq n_-$ etc. were not needed.

However, rather early the anisotropy on a microscale incident to α particle ionization was recognized [*3*] and a proper theoretical approach was made by the studies of G. Jaffé ending with a rather complete analysis in 1923. As early as 1913 extending ranges of recombination times to lower values of t, the lack of linearity of the $\frac{1}{n} - t$ plots was recognized as being due to a microscale anisotropy in X-ray ionization by S. G. Plimpton [*4*] although it remained for the work of L. C. MARSHALL [4] and the interpretation of the writer to bring the significance of the anisotropies into relation with ionic motion and recombination theory in 1929. Thus by 1930, it was recognized that owing to initial anisotropies produced by the ionizing mechanisms then available to recombination studies the early stages would have recombination affected by changes in ion concentrations n with time by diffusive motion of the ions. This resulted in non linear $\frac{1}{n} - t$ plots, or expressed in terms of values of the recombination coefficients to values of an α' inferred from measurement that declined with time. That is as improving techniques reduced the values of t over which measurements of n could be made, it was found that values of α'_t inferred from a measurement of n_1 and n_2 between times t_1 and t_2 where $t_2 - t_1 < t$ by means of the relation,

$$\alpha'_t = \frac{1}{t_2 - t_1}\left(\frac{1}{n_2} - \frac{1}{n_1}\right) \tag{2.1}$$

fell continuously as $t = (t_1 + t_2)/2$ declined, tending, however, towards a lower limit Recognition of these difficulties led to proper data on α for ion-ion recombination only from 1938 on.

With the post world war techniques, dense isotropic plasmas could be achieved and ion densities measured in time intervals from 10^{-6} sec. on up [6], [7]. Use of pure gases and identification of ions present through ambipolar diffusion studies and mass spectrographs led to the discovery that recombinations observed aside from diffusive disturbances, especially where electronic carriers were involved, suffered from the effects of inequality in carrier densities n_+ and n_-, from the presence of two carriers of one sign recombining with an opposite carrier at different rates, with the two carriers in (a) dynamic equilibrium in the gas, (b) an initial fixed ratio to each other, and (c) one ion changing to the other with a rate commensurate with the recombination of one of the carriers [6], [7]. It was also discovered that long after the ionizing source was off secondary processes could produce ions that recombined, from, say, metastable states, at a rate commensurate with recombination [6]. While J. J. THOMSON [1] had contemplated such possibilities and set up the differential equations for them they had not been solved. Solution for two special sets of conditions in He were achieved numerically by HOLT and his associates [7]. It thus remained for KUNKEL [8] to achieve general solutions of the special cases for which solution of the complex relations were possible. Since shortage of space precludes giving the derivation of the relations in full, the initial differential equations will be given and special cases of the solution will be cited to be followed by certain general conclusions applying to the situations which arise in practice.

Let M represent the concentration of one type of positive ion, e.g., a molecular ion that recombines with the negative carriers, for example electrons, with a dissociative recombination coefficient α_1 of reasonably large magnitude. Let A represent the concentration of another type of positive ion, say an atomic species, that recombines with the negative carriers, electrons, of density n, with a radiative coefficient α_2 of very small magnitude compared to α_1. It is conceivable that in some sort of a reaction in the gas species A of positive ion changes to M with a probability β in unit time, β having such a value that rate of transformation of A into M can be commensurate with the rate of loss of M by recombination. These particular conditions were chosen as exemplary because they apply to actual situations occurring in microwave plasmas which furnish a fruitful source of data in recombination studies. However, nothing prevents assignment of other carriers to A, M, and n, A applying to free electrons,

M to negative ions, and n to one species of positive ion, or in fact any other combination to be encountered. The quantities M, A, and n vary with t while α_1, α_2 and β are constants. These in the first situation assumed, lead to the differential equations

$$\frac{dn}{dt} = -\alpha_1 M n - \alpha_2 A n, \tag{2.2}$$

$$\frac{dM}{dt} = -\alpha_1 M n + \beta A, \tag{2.3}$$

$$\frac{dA}{dt} = -\alpha_2 A n - \beta A. \tag{2.4}$$

Since

$$\frac{dn}{dt} = \frac{dM}{dt} + \frac{dA}{dt} \tag{2.5}$$

it is proper to set

$$n - M - A = \delta = n_0 - M_0 - A_0 \tag{2.6}$$

where n_0, M_0 and A_0 are the concentrations at $t=0$. Solutions are not possible for these non linear equations in all cases. One solution is possible if $\alpha_2 n \ll \beta$, that is, terms in α_2 can be neglected. Solution then yields $A = A_0 e^{-\beta t}$ giving

$$\frac{dn}{dt} = -\alpha_1 n^2 + [(\alpha_1 - \alpha_2) A_0 e^{-\beta t} + \alpha_1 \delta] n \tag{2.7}$$

of which the solution is

$$\frac{1}{n} = \frac{1}{g}\left(\frac{1}{n_0} + \alpha_1 \int_0^t g\, dt\right), \tag{2.8}$$

$$g = \exp\{\delta \alpha_1 t + A_0(\alpha_1 - \alpha_2)/\beta (1 - e^{-\beta t})\} \tag{2.9}$$

if $\alpha_1 = \alpha_2$ then either $A = 0$, or $\beta = 0$.

This is the equation for two carriers, say ions and electrons, one of which has an excess concentration δ, a case of some interest. The complete solution is

$$\frac{1}{n} - \frac{1}{\delta} = \left(\frac{1}{n_0} - \frac{1}{\delta}\right) e^{-\delta \alpha_1 t}. \tag{2.10}$$

If $\delta \alpha_2 t \ll 1$ this approximates to

$$\frac{1}{n} = \frac{1}{n_0} - \alpha_1 \left(1 - \frac{\delta}{n_0}\right) t. \tag{2.11}$$

This leads to a non linear $1/n - t$ plot with relatively little departure from linearity for small ranges of n. The slope inferred will lead to an erroneous value of α_2 which is diminished by the ratio of A_0 to n_0. For actual microwave operating conditions, where $n_0 = 2A_0$, α' can only vary 3% for changes in t by a factor of 5 yielding an $\alpha' = \frac{1}{2}\alpha_1$. If A_0 is very near n_0 but if α_1 is large, the α' inferred from the slope of the $1/n - t$ plot will vary with time and within the ranges of observation even differences between A_0 and n_0 of 1% will cause time variation of the coefficient inferred.

The integral of (2.8) cannot be evaluated in finite form but must be expressed as a series. If in the usual case $\delta = 0$

$$\int_0^t g\, dt = \left[t - \frac{1}{\beta} \sum_{k=1}^{\infty} \frac{A_0(\alpha_1 - \alpha_2)^K}{\beta} \left\{1 - \frac{e^{-K\beta t}}{K\,K!}\right\}\right] \exp\left[\frac{A_0(\alpha_1 - \alpha_2)}{\beta}\right]. \tag{2.12}$$

If $A_0(\alpha_1-\alpha_2)\leqq\beta$, which often occurs since $A_0\leqq n_0$ and cases are limited to those where $\alpha_2 n_0\ll\beta$ if $\alpha_2\ll\alpha_1$ then $\alpha_1 A_0\leqq\beta$ is more severe than the previous restriction. If, on the other hand, $A_0(\alpha_1-\alpha_2)\ll\beta$ terms in $K>1$ may be neglected and (2.8) becomes

$$\frac{1}{n}\approx\frac{1}{n_0}+\alpha_1 t-\frac{\alpha_0(\alpha_1-\alpha_2)}{\beta}\left[\left(\frac{1}{n_0}+\frac{\alpha_1}{\beta}\right)(1-e^{-\beta t}-\alpha_1 t\,e^{-\beta t})\right]. \tag{2.13}$$

This equation applies to conditions actually occurring in a microwave discharge investigated by HOLT and his group [7] and solution for which was achieved by numerical integration. The equation (2.13) has an asymptotic form

$$\frac{1}{n}\approx\frac{1}{n_0}-A_0\frac{\alpha_1-\alpha_2}{\beta\,n_0}+\alpha_1 t. \tag{2.14}$$

This yields an intercept at $t_1\approx A_0\frac{(\alpha_1-\alpha_2)}{N_0\alpha_1\beta}$.

If n_0 can be measured, $\frac{A_0}{\beta}\left(1-\frac{\alpha_2}{\alpha_1}\right)$ is found at once. Since α_1 is given by the slope of (2.14) it remains to evaluate A_0 or β separately. Since $A=A_0e^{-\beta t}$ and with the use of eq. (2.13), n, $M=n-A$ and thus $(Mn)^{\frac{1}{2}}$ can be plotted directly with the same constants. Since α_2 is negligible, the recombination radiation is proportional to Mn and maximum intensity should occur when $(Mn)^{\frac{1}{2}}$ is a maximum. The condition for the maximum is

$$e^{\beta t}=\left(1-\frac{\beta}{\alpha_1 M}\right)\frac{A_0}{2n}. \tag{2.15}$$

For the maximum M_0, α_1 and t can be readily measured, (2.15) yields a second relation between β and A_0. The case where n is changed by reactions of constituents within the plasma creating new carriers, e.g., by metastable interaction after the discharge is cut off was also solved by KUNKEL. The equation for this case neglecting diffusion loss is,

$$\frac{dn}{dt}=\frac{dA}{dt}=-\alpha\,n\,A+q(t) \tag{2.16}$$

where $q(t)$ is the production term. Again $n-A=\delta=n_0-A_0$, usually $\delta=0$ as the plasma as a whole is neutral. If the source decays exponentially as with the metastable productions and the metastable removal is not controlled by this process $q=b\,e^{-\beta t}$. If $\beta n_0\ll b$ the solution approximates

$$\frac{1}{n}\approx\frac{1}{n_0}-\alpha t-\frac{b}{\beta n_0}\left[-\frac{1}{n_0}(1-e^{-\beta t})-\alpha\,t\,e^{-\beta t}\right]. \tag{2.17}$$

If $n_0=0$ then the relation simplifies to

$$n\approx\frac{1-e^{-\beta t}}{\alpha t-(\beta/b)}. \tag{2.18}$$

The last equation approximates an actual case such as might be expected in the microwave studies of BIONDI [6].

Finally, in a study of conditions in the ionosphere owing to the photo dissociation of ions by sunlight and attachment to give ions there is a dynamic equilibrium ratio between electrons of concentration n_1, and negative ions of concentration n_2 such that the ratio $n_2/n_1=\gamma_p$ is constant. Here the changes from electron to ion and the reverse, photodetachment, occur at rates very high compared to that of recombination. Then if α_1 and α_2 represent the coefficients of recombination of electrons and ions with some single type of positive ion present in the concentration n_e, SAYERS [9] and BATES [9] showed that a single apparent coefficient of recombination will be observed of value $\alpha_1' = \alpha_1 + \gamma_p \alpha_2$. The general theory of such reactions was derived by KUNKEL [8] and the writer and found application in the analysis of pressure dependent recombination in microwave studies of Hg vapor by BIONDI in 1953.

If more generally concentrations of two species of carrier of one sign n_1 and n_2 undergo recombination with an oppositely charged carrier of concentration n_e and the two species are maintained in an equilibrium ratio, $n_2/n_1 = \gamma_p$, the coefficients for n_1 and n_2 with n_e being respectively α_1 and α_2 where γ_p may be pressure dependent, then it is possible to set

$$\frac{dn_1}{dt} = -\alpha_1 n_e n_1 - \left(\frac{dn}{dt}\right)_{\text{eq.}}, \tag{2.19}$$

$$\frac{dn_2}{dt} = -\alpha_2 n_e n_2 - \left(\frac{dn}{dt}\right)_{\text{eq.}} \tag{2.20}$$

where $(dn/dt)_{\text{eq.}}$ is the rate of transformation of one ion into the other and is eliminated in the treatment to follow. It is next possible to describe the loss of ions dn_1/dt as a recombination process with a coefficient α_1 which is experimentally observable.

$$\frac{dn_1}{dt} = -\alpha_1^1 n_1^2 = -\alpha_1 n_e n_1 - \left(\frac{dn}{dt}\right)_{\text{eq.}} \tag{2.21}$$

and likewise a coefficient α_2^1 applied for dn_2/dt such that

$$\frac{dn_2}{dt} = -\alpha_2^1 n_2^2 = -\alpha_2 n_2 n_e + \left(\frac{dn}{dt}\right)_{\text{eq.}}. \tag{2.22}$$

Since $n_e = n_1 + n_2$

$$\frac{dn_e}{dt} = \frac{dn_1}{dt} + \frac{dn_2}{dt} - \alpha n_1 n_e - \alpha_2 n_2 n_e = -\alpha_e^1 n_e^2. \tag{2.23}$$

Again

$$\gamma_p = \frac{n_2}{n_1}, \quad \frac{dn_2}{dt} = \gamma_p \frac{dn_1}{dt}$$

and

$$\alpha_2^1 n_2^2 = \gamma_p \alpha_1^1 n_1^2.$$

Since further:

$$\frac{n_2}{n_1} = \gamma_p, \tag{2.24}$$

$$\frac{n_e}{n_1} = 1 + \gamma_p, \tag{2.25}$$

$$\frac{n_e}{n_2} = \frac{1+\gamma_p}{\gamma_p} \tag{2.26}$$

it follows that:

$$\alpha_1^1 = \alpha_1 + \gamma_p \alpha_2, \tag{2.27}$$

$$\alpha_2^1 = \frac{\alpha_1 + \gamma_p \alpha_2}{\gamma_p}, \tag{2.28}$$

$$\alpha_e^1 = \frac{\alpha_1 + \gamma_p \alpha_2}{1 + \gamma_p}. \tag{2.29}$$

These relations have the following application:

Eq. (2.27) applies where n_1 is measured, e.g., as for electrons in the upper atmosphere. n_2 is a second negative carrier, e.g., an ion, and n_e represents the concentration of positive ions.

Eq. (2.28) applies when the concentration n_2 is measured, n_1 being a carrier of similar charge in dynamic equilibrium and n_e a negative carrier.

Eq. (2.29) applies to the more frequent laboratory case where n_e is an electron carrier in a plasma, the concentration of which is measured, which recombines with two different positive ions n_1 and n_2 in dynamic and perhaps pressure dependent equilibrium γ_p. The α_e^1 observed then appears pressure dependent while α_1 and α_2 are independent of pressure. Many such situations are likely to be encountered in the near future.

II. The Measurement of the coefficient.

3. Principles of measurement. As noted in Sect. I, basic theory indicates three general relations yielding values of the coefficient of recombination, viz., the method of loss, the method of growth, and the equilibrium method. Of these, most satisfactory data come from the method of loss. Again it might be suspected that conditions of measurement would depend in a measure on the nature of the carriers i.e., whether they are small, or normal ions, recombining with large slow, or LANGEVIN, ions, normal ions recombining with normal ions of opposite sign called ion-ion recombination or electrons recombining with positive ions. The theory and measurement of recombination of the LANGEVIN ions has not been extensively pursued. It requires long measuring intervals and the theoretical model is relatively simple. Practically all early studies of recombination involved the ion-ion process although satisfactory values were not achieved until in the late nineteen hundred and thirties. Attempts at measurement of electron-ion recombination were not begun until the later nineteen hundred and twenties and were largely confined to studies in arcs and plasmas. The coefficients observed in these studies had anomalous values, the interpretation of which is only becoming apparent now. Proper measurements of the coefficients have only been achieved with post World War II techniques and after 1945. These measurements involved electronic carriers as well as ions which required the extended theory developed by KUNKEL to take account of disparity in concentrations of electrons and ionic species as well as other complications. The ionospheric studies, it is true, first called attention to the multiple carrier recombination shortly prori to World War II and indicated a specialized analysis for one case.

The method of loss requires that the number density of ions

$$n = \frac{q}{V} = \frac{\left(\frac{e q}{V}\right)}{e}, \tag{3.1}$$

be measured at various times t and plotted as $1/n$ against t the slope of the curve yielding the unique value of α if linear, otherwise the data require more extensive study for evaluation of the coefficient. Methods of evaluating n are unfortunately limited and are often faulty. They involve the following procedures:

(1) Sweeping ions out of a known volume V at various time intervals t using high fields, in the hope of collecting all remaining ions and that microscopic and macroscopic anisotropies as altered in time by diffusion do not complicate the data, i.e., in the hope that n is given by eq. (3.1).

(2) By inserting probes into the gas and using standard probe techniques to measure electron concentrations at various times, with the hope that the probes do not alter concentrations in their neighborhood and that they yield correct absolute as well as relative values of the densities.

(3) Having identified certain spectral lines on a section of continuum as emanating from the act of recombination by measuring the decline of intensity as a function of time to yield the relative decrease of concentration of emitting carriers. Since in addition, the absolute concentration at some time is required, resort must be had to probes, STARK effect studies, or microwave measurement unless atomic theory permits calculation of the concentration of emitting carriers from the total energy radiated in a wave length region.

(4) If the gas is highly ionized, i.e., if ion densities are of the order of 10^{17} per cm.^{3}, then the STARK effect broadening of certain lines, e.g., H_α and H_β observed by introducing traces of H_2 into the gas, or by using other well known atomic lines and the theory of HOLTSMARK, permit ion densities to be measured.

(5) The use of a sensitive probing signal sent into a microwave cavity to evaluate the change in Q, or in resonant frequency f as the electron density declines with suitable precautions evaluates electron densities between 10^8 and 10^{11} electrons per cm.^{3} if the plasma is uniformly ionized.

(6) The analysis of the radio echoes from the ionosphere at different frequencies permits an evaluation of electron density in that region.

4. Limitations to the application of proposed techniques. These techniques suffer certain limitations to be noted, which are:

(a) The removal of ions and electrons by sweeping fields in order to insure collection of all carriers presents a problem as owing to geometrical, diffusive and other losses, *all* ions are *never* collected [*10*]. As carrier densities increase the polarization of the medium reduces fields within it and collection of even a reasonable portion of carriers becomes impossible. It is probable that the transition from sweepable to unsweepable condition in free electron gases would be in the region of densities marking the transition from free to ambipolar diffusion. Thus measurements might be extended up to 10^7 carriers per cm^3. Lower limits of densities would be about 10^5 for reasons to follow.

(b) Probes require plasmas with densities in excess of perhaps 10^{10} per cm^3. and only where negative carriers are exclusively electronic. Single probes should not be used in rapidly declining plasma densities.

(c) The decline of emitted light intensity requires comparison of the intensities of single lines, small areas of continua, etc. This means that carrier densities must be very high so that the light emitted by the carriers recombining in any very short time interval is sufficiently intense to be conveniently measured. This generally has implied high densities of ionization from 10^{10} to 10^{17} carriers/cm.^{3}. Very modern techniques such as used by MCCOUBREY and HOLSTEIN in the investigation of the decline of band fluorescence in resonance excited Hg vapor might reduce the limits cited.

(d) STARK effect studies have been confined to the order of 10^{15} to 10^{17} ions/cm.^{3} as the broadening of the lines is not sufficient at lower inter-ionic fields.

(e) The studies in microwave plasmas so far have had the densities limited to the range from 10^8 to 10^{11} electrons/cm.^{3}. Beyond these limits measurements are so far not reliable.

(f) It must again be indicated that ionization produced by particles like α particles or electrons are initially so anisotropic on macro and microscales as to make measurement of coefficients virtually impossible. Even ionization by microwave breakdown produces large scale anisotropy unless care is taken. Studies in arcs and glows also suffer from lack of adequate control. If relatively uniform breakdown is achieved random but isotropic distribution on all scales can usually be had by allowing the plasma to "cool down" for sufficient time. Under these conditions pressures must not be so low as to have ambipolar diffusion enter significantly. In general the higher the gas density the less diffusive loss enters and the more carrier loss can be ascribed to volume recombination, provided the gas is pure and electron attachment does not occur. In electron attaching gases ion formation must be assured before recombination is measured. Gas densities must then not be too low to prevent rapid formation nor must it be too high to induce a transition in recombination mechanism.

(g) One governing variable for all such studies is the time interval available to measurement. Regarding eq. (1.2) and obviously more so for the more involved relations, it is clear that the time interval $\Delta t = t_1 - t_2$ over which a recombination study is made must be such that n_2/n_1 can effectively be measured. This requires that in principle $n_1 \alpha t \sim 1$, or $t \sim \frac{1}{n_1 \alpha}$ otherwise Δt will extend over large ranges to change n_2/n_1 appreciably, and long ranges of Δt are undesirable because of intervention of other mechanisms or changes in carriers during study. If t is short compared to $\frac{1}{n_1 \alpha}$ then while large changes in $\frac{n_2}{n_1}$ are obtained the intervals Δt become too short for convenient measurement. Thus the values of t which are geared to the experimental techniques to be used require certain values of n_1 in order that a given range of α values may be studied. This places very distinct limits on the methods of measurement and ionization techniques to be used. Thus for radiative recombination studies where $\alpha \sim 10^{-12}$ it is clear that in order that Δt exceed 10^{-6} sec., which is a limit for convenient accurate measurement, t must be of the order of 10^{-5} sec. or less and n_1 must be of the order of 10^{17} electrons per cm.^{3}. If n_1 is less than 10^{15} electrons/cm.^{3} then time intervals of 10^{-3} sec. must be used which may well bring secondary mechanisms into the picture through molecular ion formation with higher recombination coefficients. In microwave techniques, intervals $\Delta t \sim 10^{-4}$ sec. are a lower limit as $t \sim 10^{-3}$ sec. are needed to homogenize the plasmas. Thus values of α from 10^{-8} on down can best be measured with concentrations from 10^{11} electrons per cm.^{3} on down. For ion-ion recombination, the time for formation of negative ions and some homogenizing of the plasma are required, thus with 10^{-2} to 10^{-1} sec. for Δt, with $t \sim 10^{-1}$ to 1 sec. and $\alpha \sim 10^{-6}$, concentrations of 10^7 to 10^6 ions/cm.^{3} can best be used.

With these considerations, it is possible to list the various methods used in the past. Needless to say, it is only the later methods developed around modern measuring techniques and controls that have yielded significant data.

5. Methods for measuring ion-ion recombination. (1) Air blast methods were among the earliest techniques resorted to! In this the decline of carrier densities was measured by probe currents at various downstream distances x in a non-turbulent stream of gas of velocity v along the axis of a cylindrical tube. The gas was ionized by X-rays or radiations from radioactive substances. The method suffered from countless inaccuracies, the worst being local inhomogeneities and lack of purity control as well as poor sampling [*2*]. The method was early used by RUTHERFORD, TOWNSEND and McCLELLAND [*1*], [*2*].

(2) The equilibrium method for growth and loss with mostly α ray ionization was used by McCLUNG, ERIKSON, and HENDREN [*1*], [*2*]. Needless to say, the difficulty in achieving significant saturation currents and various anisotropies rendered results of little significance.

(3) The method of growth was first tried by RUMELIN [13] with β and γ ray ionization with a rotating commutator. The same commutator techniques with X-ray ionization were later tried both by MARSHALL [4] and SAYERS [5]. It was abondoned by all in favor of the loss method [2].

(4) RUTHERFORD and THOMSON first measured the coefficient by the variation of an ionization current between parallel plates as the applied potential and the uniform field strength varied, ions being produced at a constant rate Q per cm.^{3} by an external source such as X-rays [1], [2]. Neglecting diffusion, the current flow and recombination loss must equal Q such that $Q=\alpha n^2+\frac{A}{V}\frac{i}{e}$, where A is the area of the plates, V is the ionized volume, i the current and e the electron charge. Since $i=ne(k_+ + k_-)X$ at a field strength X and since $V/A=l$ the interelectrode distance, if I is the "saturation" current with no recombination, $Q=\frac{A}{V}\frac{I}{e}$, thus evaluating Q when X is very large, then we obtain

$$\frac{\alpha}{e}=\frac{(I-i)\,(k_+ + k_-)^2 X^2}{i^2\, l}. \tag{5.1}$$

The method was later improved by RIECKE and MIE [10].

(5) The method of loss was first used by RUTHERFORD [1] who ionized air with a flash of X-rays and swept out ions at different times t using a pendulum interrupter. The method was later improved by LANGEVIN [15] using two identical ionization chambers at different potentials traversed by the same X-ray flash. It led to an evaluation of two charges Q_1 and Q_2 collected at the two different fields yielding a quantity

$$\alpha=\varepsilon\left[4\pi e(k_+ + k_-)\right] \tag{5.2}$$

where ε is a factor less than unity and the quantity $4\pi e(k_+ + k_-)$ was a theoretical expression deduced by LANGEVIN for the value of α.

(6) As indicated earlier, the only proper evaluation of α for ion-ion recombination came for O_2 and air using refinements of the commutator techniques and correcting for initial microscale anisotropies in the X-ray ionization used, as a result of the work of MARSHALL [4], LUHR [16] and GARDNER [5] in the writer's laboratory and SAYERS [5] at Cambridge in 1938. Analogous techniques were used for a study of columnar recombination by SCHEMEL [3] in collaboration with the theoretical studies of JAFFÉ on that subject in 1928.

6. Methods applicable to assumed radiative electron-ion recombination. (α) The work of KENTY [17], and MOHLER [17] and later by SAYERS [17] using arcs in A, Cs and Hg, sets the standards of investigation of what was supposed to be radiative electron-ion recombination before World War II. They interrupted the arcs by shorting the power supply and with various commutator devices measured the decline of electron densities by means of probes as a function of time. MOHLER also used the decline of the assumed recombination continuum in the yellow in Cs discharges measured with photometer. The data obtained optically by MOHLER agreed sufficiently well with probes so that probes were used in later work. Work was uniformly done at pressures below 1 mm. Hg and diffusion loss was never adequately corrected for. Data were sufficiently scattered and range of observations was limited so that the $\frac{1}{n}-t$ plots were never strictly linear. Values of α observed were of the order of 2×10^{-10} in all arcs which are too high for radiative recombination and far too small for other then known processes. Interpretation is just now possible.

(β) Using photomultipliers and fast sweep oscilloscopes, J. D. CRAGGS and J. M. MEEK [18] just after World War II, studied the hydrogen lines H_α, H_β etc., in spark channels in H_2 gas and argon lines in A at about atmospheric pressures. The potential and current flow were simultaneously oscillographically recorded. Calorimetric studies yielded the mean potential drop during current flow. Channel diameters were measured by photographs in the visible and ultraviolet. Filters isolated the BALMER lines in H. Ion densities were estimated by STARK effect broadening in H. The hydrogen was completely dissociated and after a microsecond thermal equilibrium among atoms could be assumed so that SAHA temperatures[1] of 20000 to 12000° K could be computed. Initial ion densities were of the order of 10^{17} electrons/cm.^{3} corresponding to complete ionization. Recombination yielded the BALMER series lines. Observed coefficients ranged from 2×10^{-12} to 1.4×10^{-11} cm.^{3}/ion sec. with declining electron temperatures and ion densities. It is probable that radiative recombination was being observed here, the values of α being enhanced by the influence of high interionic force fields. CRAGGS and HOPWOOD [18] later extended and confirmed the studies improving precision but adding nothing new. Data in A were not interpreted.

Somewhat similar studies were later carried out by OLSEN and HUXFORD [19] using oscillographic analysis of highly condensed discharges in A and Ne at lower pressures. They observed continua as did CRAGGS and MEEK. They associated the decline of electron energy with the continua ascribing the continua to free-free electron transitions. The STARK effect broadening of small amounts of H_2 introduced into the argon, which did not alter the discharge, gave them ion densities. Spark and arc lines argon showed different rates of growth and decline. The decay of the arc lines was correlated with decrease in electron density and from the decrease coefficients of recombination α of $\sim 2.5 \times 10^{-13}$ cm.^{3}/ion sec. in A and 5.3×10^{-14} cm.^{3}/ion sec. in Ne were estimated. The recombination was interpreted as being radiative electron-ion recombination at the elevated electron temperatures of around 8300° K and more, as molecular A_2 and Ne_2 ions probably are not stable at these temperatures. There are, however, many sources of complications in these measurements such as possible regeneration of excited states which could reduce the rate of decline of the lines and falsify values of α for which plots of $\frac{1}{n} - t$ were not properly linear over any extended range.

γ) Real advance came in the study of electron-ion recombination with microwave techniques [20], [6], [7]. In this method gas confined at suitable pressures in clean outgassed quartz or glass vessels of appropriate form, without electrodes is subjected to an intense ultra high frequency electrical field in a wave guide system by means of a pulsed magnetron. This causes the gas to be broken down by relatively low amplitude oscillations of the electrons appropriate to the pressures used. The breakdown fields act from around 1 to 3 milliseconds yielding uniform ionization, or at worst a known distribution of ionization in the vessel. The radiation from the vessel, if desired, can be observed but more usually the electron density existing at any time is determined either by measuring the resonant frequency of the cavity by means of a weak probing signal or by determining its Q. Modifications have been made with electrodes to determine the real and complex conductivities, but this variant may not be as good as the more conventional procedures [21]. The decline of electron density with time can

[1] See the article of FINKELNBURG and MAECKER in Vol. XXII.

be studied over time intervals extending down to several milliseconds. Electrons are lost by ambipolar diffusion to the walls, by attachment to molecules or atoms to form negative ions, and by recombination. Decline of electron concentration by either diffusion or electron attachment is indicated by an exponential law, giving linear $\log n - t$ plots while recombination processes with constant α should yield linear $\frac{1}{n} - t$ plots provided $n_+ = n_-$. Ambipolar diffusion can be calculated and corrected for since the shape and dimensions of the container permit of calculation of a characteristic diffusion length. The ambipolar diffusion coefficients observed can be of value in identifying the existing ions through their mobilities. The diffusion loss occurs prominently below 1 mm. and increases as pressure decreases. For most gases, attachment is a more effective loss mechanism at higher pressures since it depends on collision frequency. The pressure dependent ion-ion recombination cannot be measured by the microwave beam. Only if there are two recombining carrier species of different coefficients can the electron-ion recombination coefficients be pressure dependent. Thus as it is equally convenient in microwave studies to use higher pressures as well as low pressures for recombination studies, diffusion loss can be reduced and recombination loss emphasized relative to diffusion and perhaps attachment by proper choice of gas pressures. However it must be noted that the presence of higher modes of ambipolar diffusion can simulate recombination loss and that electron attaching impurities can lead to declines of electron concentration in time which, while yielding slightly non linear $\frac{1}{n} - t$ plots, can well be confused with recombination decline. This is especial y likely under the conditions so often existing in gases in which $n_+ \neq n_-$, as lmore than one species of positive or negative carriers may exist one of which recombines very little in the time of measurement. Furthermore so far techniques do not permit of studies with electron densities below 10^8 electrons/cm.^{3} nor above 10^{11} electrons/cm.^{3} It also appears necessary that the plasmas "cool" down for some time after ionization before measurement and for this reason recombination can be studied with electrons of near thermal velocities. With the first observations using this technique, beginning with those of M. A. BIONDI and S. C. BROWN [6], the coefficients observed in pure inert gases led to the establishment of dissociative molecular ion-electron recombination with large cross sections which had been inferred by D. R. BATES from ionospheric studies [6]. Studies were carried down to low temperatures. The investigations led to the establishment of the prevalence of molecular ions in inert gases [6], [7], under some conditions, to the existence of the more involved conditions underlying recombination indicated by eq. (2.2) through (2.29) and to a far better understanding of recombination mechanisms in general. Much more work remains to be done and the work using microwave technique will have to be done with the same meticulous care and avoidance of pit falls as the initial work of BIONDI and BROWN. It furthermore calls for accompanying mass spectrographic studies for ion identification.

Much further work remains to be done in the study of radiative electron-ion recombination by other techniques and ion-ion recombination requires further study unde better conditions. This is especially true for the high and low pressure regions where clean gases are used. It may properly be added at this point that while recombination processes are beginning to be understood in broad outline, progress in this field is less satisfactory than in any other area of discharge physics.

III. Experimental results and the theories of the recombination coefficient.

7. Presentation of tables and curves representing reliable data. The sum total of observed values for the coefficient of recombination that have any significance on the basis of reasonably reliable techniques, adequate gas purity, and similar considerations are summed up in Table 1 for ion-ion recombination in O_2 and Air, in Table 2 summarizing all processes and in Table 3 for electron-ion recombination together with Figs. 1 and 2 showing GARDNER'S [5] data for variation of α with pressure and temperature in O_2 and Figs. 3 and 4 showing SAYER'S [5] data for variation of α with pressure in air to two different scales, the last including MÄCHLER'S [22] data at higher pressures in air. Appropriate discussion of these data must await a consideration of the basic theory characteristic of the coefficient of recombination under varying conditions.

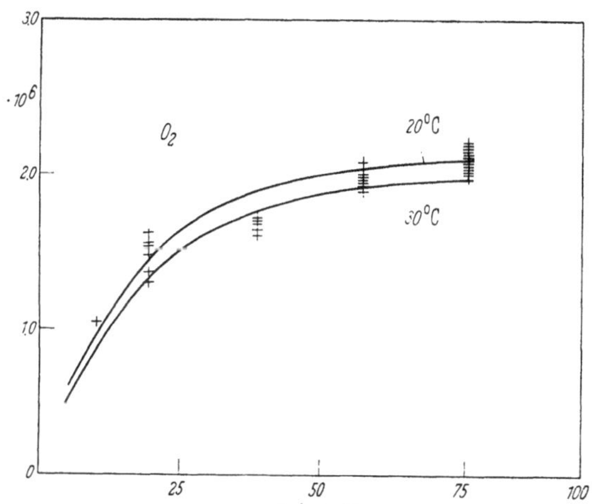

Fig. 1.

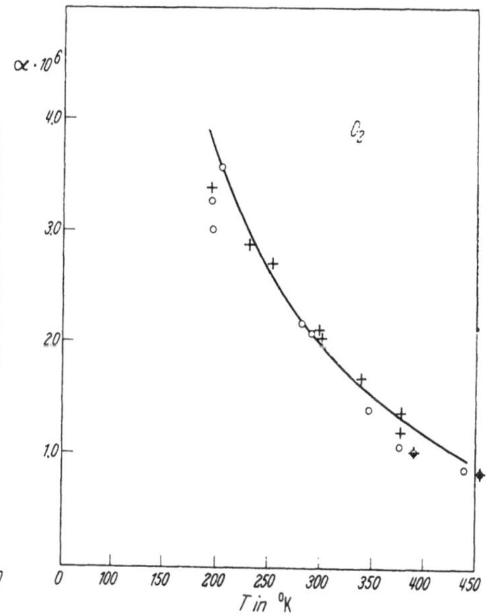

Fig. 2.

Fig. 1. GARDNER's data for the variation of α in O_2 as a function of pressure in mm of Hg. Temperatures varied between 20 and 30° C. Solid curves are calculated from J. J. THOMSON's theory.

Fig. 2. GARDNER's data for the variation of α in O_2 as a function of temperature in °K at constant density. Crosses are GARDNER's data, circles are *adjusted* data of H. A. ERIKSON and circled crosses of PHILIPPS in air. The solid curve is calculated from THOMSON's theory.

Table 1. *Only known reliable data for ion-ion recombination in O_2 and air after* GARDNER *and* SAYERS.

Gas	Observer	Ion concentration per cm.³	Temp. C°	Press mm. Hg	
O_2	GARDNER	2×10^6	25	760	$2.08 \pm 0.05 \times 10^{-6}$
O_2	GARDNER	10^5	25	760	2.5×10^{-6}
Air	SAYERS	1.5×10^6	20	760	2.65×10^{-6}
Air	SAYERS	4.4×10^6	20	760	2.3×10^{-6}

8. Basic models for recombination. The basic simplified assumptions leading to the definition of the coefficient of recombination given by eq. (1.1) indicate merely that ions of opposite sign must "come together" into some sort of an "encounter" leading to charge neutralization. How they do this and how charge exchange occur are not specified.

Very early LANGEVIN [*15*] formulated a model for the process of ion-ion recombination assuming essentially that owing to COULOMB forces the carriers, both ions of opposite sign drifted together as if moving in a continuous medium with a relative drift velocity $v = X(k_+ + k_-)$. This model led to the evaluation of α as

$$\alpha = \varepsilon\,[4\pi\, e\,(k_+ + k_-)] \quad (8.1)$$

with e the ionic charge, k_+ and $k-$ the mobilities of positive and negative ions and ε a factor less than unity introduced by LANGEVIN when α as evaluated by his method (5) (see p. 480) turned out to be less than unity. He ascribed ε to some sort of failure to recombine on "encounter". The assumption that carriers moved in a gas as if they were in a continuous resistant medium instead of as charged molecular or electronic configurations moving in a discontinuous medium of commensurate molecules in continuous heat motions can only apply at very low temperatures and/or in very dense gases. This subtle fallacy was not discovered until the measurements of MARSHALL [*4*] in 1928 convinced the writer that at 300° K and 760 mm. the ions were drifting apart rather than together. However, in 1924 J. J. THOMSON [*23*] had, quite independently of any experimental evidence, derived a theory applicable to ion-ion recombination below 1 atmosphere which was at once confirmed by the writer [*24*] and SOMMERFELD to represent the proper interpretation. The studies of the plasmas of arcs and glows in free electron gases in conjunction with the rapidly developing understanding of atomic structure led to quantum mechanical conclusions concerning radiative electron ion recombination. Confirmation of this theory was sought in the studies of KENTY and MOHLER [*17*] and, surprisingly, resulted in an apparent discrepancy which is only now receiving clarification.

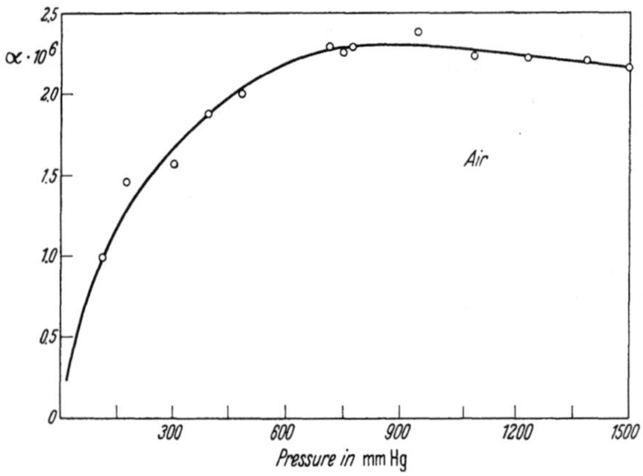

Fig. 3. SAYER's data for α in air as a function of pressure in mm Hg.

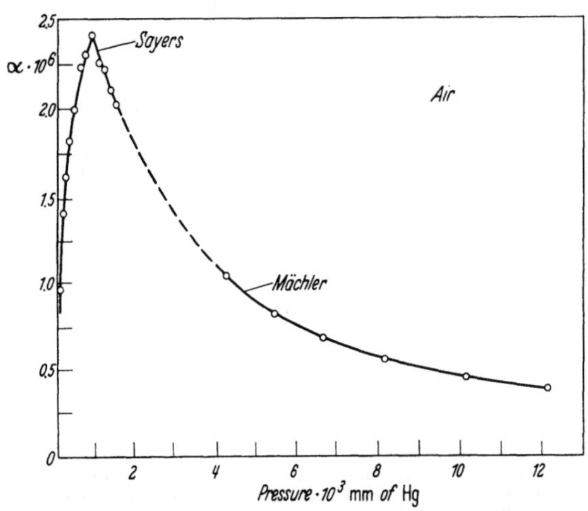

Fig. 4. SAYERS' data on α over an extended pressure range in air. Upper curve: with MÄCHLER's data at higher pressures, lower curve: showing the HARPER-LANGEVIN hyperbola beyond 2 atmospheres.

It was thus logical with the direct confirmation of the THOMSON theory by the measurements of GARDNER [*5*] in O_2 and SAYERS [*5*] in air and even with the somewhat peculiar electron-ion recombination data, that the writer was able to classify and in some measure to clarify existing recombination theory [*25*].

To round out this summary, there must be added the predictions and subsequent discovery as well as post war analysis of dissociative recombination by BATES [6] and BIONDI and BROWN [6]. In classifying the theoretical models, it must first be noted that recombination takes place between two different types of carriers arranged in two subgroups from the view point of the basic process of neutralization:

Class 1: Electron-Ion Recombination:

Atomic or molecular ion and electron by one of several processes.

Class 2: Ion-Ion Recombination:

(a) Between large ions, or LANGEVIN ions of radius about 10^{-6} cm. and ordinary small ions.

(b) Between ordinary small positive and negative ions.

Classes 1 and 2 can be analyzed most appropriately by different basic approaches. From what has been intimated in the discussion of the early LANGEVIN theory, the fundamental nature of the recombination process will differ depending on the density of the gas, the temperature and the number density of ions n. The role of these three quantities was first clearly indicated by J. J. THOMSON [23] in his model for ion-ion recombination in terms of the value of the ionic free path L, the radius of a sphere, best designated as the *sphere of active attraction*, of value $d_0 = \frac{2}{3}\frac{e^2}{kT}$ with e the charge, k the BOLTZMANN constant, T the absolute temperature, and an average distance between ions of $r_0 = \sqrt[3]{1/n_0}$ where n_0 is the initial number density of ions at $t=0$. The value of d_0 is defined as the distance at which the potential energy of one ion in the COULOMB field of the oppositely charged ions equals the kinetic energy of agitation of the ions. This is set by THOMSON as $e^2/d_0 = \frac{3}{2}kT$. Other theoretical deductions yield different values of the numerical coefficient $\frac{3}{2}$ depending on whether the energy of the ions is taken as that of a single ion or of both ions, etc. A relation deduced by HARPER [20] yields the factor as $3kT$ instead of $\frac{3}{2}$ while the BROWNIAN motion relation of SOMMERFELD gives $6kT$. In 1940 JAFFÉ [27] derived the generalized theory of recombination on the basis of statistical considerations and again emphasized the mean free path L and d_0 using a coefficient for the kT term of more nearly unity.

While the writer [25] in 1939 emphasized the relations of r_0, L and d_0 in relation to various models for recombination coefficients observed, it remained for the very important general paper of JAFFÉ to indicate the generalized appearance of r_0 in all but certain special oversimplified models of recombination such as those of THOMSON [23] and HARPER [26].

It is at once clear that the carrier behaviour will differ in principle if r_0 is greater or less than d_0, for in the former case, carriers will, in general, undergo random diffusion while in the latter case the COULOMB forces dominate and a more directed motion occurs. It is further to be suspected that the gas density will enter through the circumstance that $L \gtrless d_0$, for this situation governs movement within d_0.

If $r_0 > d_0$ both the electron-ion and ion-ion processes have the first feature in common. This is (1) a *diffusive approach period* in which either the electron diffuses to within a radius for recombination capture by the ion, or in which the two ions diffuse to within d_0 of each other. If $r_0 < d_0$ then the motion of oppositely charged carriers is a more restricted motion within d_0 and recombination assumes a different aspect which may for the present be ignored. From the point of diffusion of the *electron* to within a sphere representing the *capture cross section of the ion*, neutralization has occurred and the process of analysis differs from that for the ion-ion process in that the analysis involves the quantum theoretical evaluation of the *ionic capture cross section for the electron* by whatever mechanism is involved. In the case of ion-ion recombination, however, the diffusive approach period (1) is followed by three more periods. These are (2) the *period of active drift of the two ions towards each other* under their mutual COULOMB attractions within d_0, which will take the form of a drift velocity if $d_0 \gg L$, and a discontinuous series of few steps if $L \sim d_0$. (3) Once the ions are separated by a distance of the order of L they will execute orbits about the common center of mass, i.e., there is a *period of orbital encounter*. If the kinetic energy in the orbit is greater than the potential energy at perihelion, then the ions will separate along open hyperbolic orbits. If the kinetic energy is less, the orbits will be ellipses until the next impact with a neutral molecule. (4) If during the period of orbital encounter, the electron of the negative ion can tunnel through its barrier within the ion and reach the positive ion, charge transfer and neutralization results constituting the *period of charge transfer*. The steps 3 and 4 can also be regarded in terms of pseudo molecule formation. This approach was first used by BATES and MASSEY [28] in 1943 and was later presented by BATES [28]. If the ions A^+ and B^- approach each other very slowly along their mutual potential energy curve, they will remain in this curve during the remainder of their encounter and will finally separate as ions. However, if they approach with a finite velocity, there is a probability P of transition to a lower potential energy curve corresponding to that for the two neutralized initial carriers A and B occurring at the point of closest approach between the $A^+ - B^-$ and the $A - B$ curves. Since this point which lies at the high energy knee of the potential curve for $A - B$ is passed twice, the probability of separating as neutral atoms is $2P(1-P)$. The probability P can be calculated by quantal methods as shown by ZENER [28] and LANDAU [28]. It appears that the large coefficients do not arise when there is close energy balance, but rather when the process is exothermic by a few e.v. The reason for this is that when there is close energy balance, the crossing point on the zero order approximation is located at a greater radial distance where interaction is extremely weak. P is then almost unity and thus $2P(1-P)$ is very small. Thus for O_2^- and O_2^+ with electron binding energies of about 0.5 and 12.5 e.v. respectively, neutralization should be easy, but with Na^+ and Cl^- having energies of 5.1 and 3.84 e.v. respectively, the probability should be less.

In any event, the kinetic energies and/or orbit type, within step (3) will together with ionic species influence the chance of charge transfer and the alternative of subsequent separation as ions. Thus steps (2), (3) and (4) relative to values of L, d_0 and the nature of the ions, will determine the probability ε of capture once the ions diffuse to within d_0 of each other.

If, in addition to the relative values of r_0, L, and d_0, account is taken of other modifying conditions on the behaviour of carriers in gases such as action of the bounding walls, and microscopic or macroscopic anisotropies of initial distribution, it becomes possible to classify recombination processes so far observed under the following heads, assigning to each a characteristic coefficient α with proper subscript and a model theoretically evaluating it. These are:

1. Electron-ion recombination (see Sect. 9).

α) Volume recombination (1) By radiative capture, coefficient α_{er}. (2) By dissociative capture, coefficient α_{ed}. (3) By dielectronic capture, coefficient α_{ee}.

β) Possible preferential recombination where $d_0 > r_0$ with any capture process 1, 2 or 3 under α above, coefficient α_{ep}.

γ) Wall recombination, electrons, and ions diffusing to the walls and recombining there. While the recombination at the wall surface may have a coefficient α_ω of very large value because of two dimensional diffusion and the energy absorption by the wall, the decline of carriers from the volume will be *governed by ambipolar diffusion* and will follow the exponential law characteristic of such a process. Thus the decline is not temporally a true recombination process.

2. Ion-ion recombination.

a) Between normal ions and LANGEVIN ions, coefficient α_{SL} (see Sect. 10).

b) Between normal ions (see Sect. 11) under the headings:

α) Preferential recombination $d_0 > r_0 \gg L$, coefficient α_L, where the subscript stands for the LANGEVIN process.

β) Columnar recombination, coefficient α_c, where $r_0 > d_0 > L$, but where ions are distributed in high concentrations along the long, very narrow, straight columns produced by fast heavy particle ionization such as α particles, the columns being separated by many times their diameter, (macroscopic and microscopic anisotropy). The coefficient declines with time.

γ) Initial recombination, coefficient α_i, where generally $r_0 > d_0 > L$ and ion pairs are produced in close proximity but separate pairs are randomly distri buted in the gas as along beta ray tracks, (microscopic anisotropy). The coefficient declines with time due to diffusion.

δ) Volume recombination, involving three subheads: (1) Higher pressure recombination, (2 atmospheres and more), the LANGEVIN-HARPER process, $r_0 > d_0 \gg L$, coefficient α_H. (2) THOMSON process from around atmospheric on down to say 10^{-2} mm., where $r_0 > d_0$ and $L < d_0 < L$, in which the four steps outlined above are fully completed, which is strongly pressure and temperature dependent, coefficient α_T. (3) Low pressure recombination applicable to ionospheric and lower pressures, $r_0 > d_0$ and $L \gg d_0$, where recombination within d_0 is reduced to values such that $\pi d_0^2 \varepsilon < \pi \sigma^2$, with σ the ionic collision radius for charge transfer, coefficient α_σ. This varies only with temperature and ionic masses.

These processes, the conditions under which they appear, the ion densities involved, the techniques for study, the range of values of α as well as the dependence of α on time, pressure, temperature, concentration of ions, etc., are summarized in Table 2.

Unfortunately, lack of space precludes the detailed discussion of the various processes indicated. In what follows enough will be said about each process to orient the reader and yield adequate references to the literature. It must again be stated that in 1940 JAFFÉ [27] on generalized statistical theory derived relations governing the processes of recombination, previously obtained from the particular simplified models assumed under the coefficients listed under (1) and (2) particularly in regard to the factors L, d_0, and r_0. This analysis shows that except under the rather severe simplifying assumptions appertaining to specific models such as used by LANGEVIN, THOMSON, HARPER, etc., the relations are much involved and in many cases not amenable to calculation. The analysis of JAFFÉ improves the values of the constants involved and clarifies certain remarkable paradoxes deriving from earlier theory.

Table 2. *General table summarizing the various recombination*

1	2	3	4	5	6
Carrier	Type	Condition of Occurance	Gas Type	State Variable	Carrier Density
Electron-ion	Volume	$r_0 \gg d_0 \gtrless L$	Inert and pure nonelectron-attaching gases	10 ats. down 15000° K down	10^8 to 10^{17} per cm.^{3}
Electron-ion	Preferential	$d_0 > r_0 \gg L$	Inert and pure nonelectron-attaching gases	100 ats. 273° K	10^6 or less per cm.^{3}
Electron-ion	Wall	Carrier densities high enough to give plasma state, pressures low confined spaces $r_0 \gg L > d_0$	Inert and pure nonelectron-attaching gases	Usually $n_e > 10^8$ and below 1 mm. pressure in confined spaces	10^8 per cm.^{3} and up
Ion-ion	Preferential	$d_0 \gg r_0 \gg L$	Electron-attaching gases O_2, SO_2, air, Cl_2, etc. Impure inert gases	100 ats. ~273° K and less	Generally low (10^6 per cm.^{3}) Notable: 10^3
Ion-ion	Columnar	$r_0 \gg d_0 \lesseqgtr L$ Particle or heavy-particle ionization. Caused by macroscopic and microscopic anisotropy. Diffusion carries it to normal volume recombination in time	Electron-attaching gases O_2, SO_2, air, Cl_2, etc. Impure inert gases	10 ats. down to 0.01 ats.	Anisotropic. Roughly 10^9 to 10^3 about track axis
Ion-ion	Initial	Caused by microscale anisotropy reverts to normal in time by diffusion. $r_0 \gg d_0 \lesseqgtr L$ β ray or similar tracks, sparse ion pairs, with opposite ions close	Electron-attaching gases O_2, SO_2, air, Cl_2, etc. Impure inert gases	10 ats. to 0.01 ats.	Anisotropic. 10^8 to 10^3, in randomly spaced but close local pairs
Ion-ion	Volume	$r_0 \gg d_0 \gg L$	Electron-attaching gases O_2, SO_2, air, Cl_2, etc. Impure inert gases	20 ats. to 2 ats. air	10^{10} and down
Ion-ion	Volume	$r_0 \gg d_0 \lesseqgtr L$	Electron-attaching gases O_2, SO_2, air, Cl_2, etc. Impure inert gases	1 ats. to 10^{-2} mm.	10^{10} and down
Ion-ion	Volume. Low pressure	$L \gg r_0 \gg d_0$	Electron-attaching gases O_2, SO_2, air, Cl_2, etc. Impure inert gases	10^{-3} mm down	All, but usually small ($\ll 10^4$)

[1] All values of α may change with time if one or both partners change character with time.

processes, their character and the circumstances of their appearance.

7	8	9	10	11
Coefficient[1]	Theory	Value	Variation	How Studied
$\alpha_{er} = \pi\,\sigma_{er}^2\,\bar{c}_1$	Radiative capture quant. mech. for σ_{er}	$10^{-12}/V$ V energy in e.v.	Energy and species Indep. p, $\bar{c}$, $\propto \sqrt{T}$ Obs. so far H sparks	In arcs, in glows } high density
$\alpha_{ed} = \pi\,\sigma_{ed}^2\,\bar{c}_1$	Dissociative capture. BATES elementary theory	10^{-6} to 10^{-8}	Commonly observed energy and species. Indep. p, $\propto T^{-\frac{3}{2}}$ varies p if 2 ions same sign diff. α in equil. at p	In spark channels with microwave breakdown at appropriate pressures
$\alpha_{ee} = \pi\,\sigma_{ee}^2\,\bar{c}_1$	Capture to excite 2 electrons	10^{-9} to 10^{-12}		
	BRADBURY	Not known	Field-dependent	Unlikely to be observed due impurities
None. Declines exponentially with time, and thus cannot give an α	Ambipolar diffusion	Not known	Carrier concentration; pressure; ionic mobility; dimensions of chamber	At lower pressures in discharges, and in microwave breakdown
$\alpha_L = 4\pi\,e\,(k_+ + k_-)$	LANGEVIN; BRADBURY theory probably more accurate	10^{-8} and up	Proportional to $1/\varrho$ or $(1/p)_T$. Varies little with T at constant p	Not usually observed. Could have occurred in old cosmic ray studies
α_c	Special theory of JAFFÉ	10^{-3} on down to 10^{-6} from $t = 10^{-3}$ to $t = 1$ sec.	Proportional to $1/\varrho$ or as volume varies with time and density of ions	Appears with heavy-particle ionization along tracks. Goes to α_H or α_T in time
α_i	Special theory; rough- LOEB; exact- BRADBURY	10^{-4} to 10^{-6} $t = 10^{-3}$ to 1 sec.	Declines with time from 10^{-3} sec., or varies with ion density. Otherwise like α_H or α_T	Appears with X-ray or fast electron ionization. Goes to α_H or α_T in time
$\alpha_H = \alpha_L\,f_1 = 4\pi\,e\,(k_+ + k_-)\,f_1$	HARPER, or JAFFÉ theory	10^{-8} to 10^{-6}	Proportional to $1/\varrho$ or $(1/p)_T$. Varies little with T at constant p	In all gases forming ions above 2 ats. with isotropic ionization
$\alpha_T = \pi\,d_0^2\,\bar{c}\,(2\omega - \omega^2)$	THOMSON theory	10^{-6} to 10^{-8}	Varies $1/T^{\frac{3}{2}}$ and $\sqrt{p}$ roughly. Depends on concentration. Can change with t if ion changes	Isotropic ionization. All gases where negative ions form by electron attachment
$\alpha_\sigma = \pi\,\sigma_c^2\,\bar{c}$	σ_c collision cross-section for charge transfer	10^{-10}	Independent of p. Depends on $\sqrt{T}$	Occurs at low pressures with large volumes observed in ionosphere

Table 3. *The collected data for electron-ion recombination to date.*

Gas	Observers	Electron Temperature T, in °K	Pressure mm. Hg	Method Used	α_e in cm.³/ion sec.	Electron density per cm.³	Comments
A	Kenty	3100	8×10^{-1}	Probes in arc	2×10^{-10}	10^{13}	Probably rate production of A_2^+
Cs	Mohler	1300	$1–10\times10^{-2}$	Optical cont. spectrum Cs arc	3.6×10^{-10}	1.8×10^{12}	Could be ascribed to Dissoc. Recomb. and governed by creation Cs_2^+.
Cs	Mohler	1200	$1–10\times10^{-2}$	Probes in arc	3.4×10^{-10}	1×10^{12}	
Cs	Dandurand and Holt	1400	0.32–0.68	Microwave	3.5×10^{-7} 1.21×10^{-6}	10^{12}	Gas temperatures respectively 285° K and 270° K. Immature plasma
Hg	Mohler	2000	0.270	Probes in arc	2.3×10^{-10}	10^{13}	Dissoc. Recomb. influenced by creation of Hg_2^+
Hg	Dandurand and Holt	2000	0.5	Microwave	5×10^{-9}	10^{12}	Doubtful, plasma density too high
Hg	Biondi	300	1.4–46	Microwave	$5.1–14\times10^{-7}$	$\sim10^{9}$	Pressure-dependent. Hg pressure increases with T. Can involve equilibrium between Hg_2^+ and Hg_3^+. Exact.
He	Biondi and Brown	300	20–30 and below	Microwave breakdown-microwave analysis	1.7×10^{-8}	$10^{10}–10^{11}$	Probably exact. Dissociative Recomb. to He_2^+, proven.
He	Holt et al.	300	11–27	Microwave breakdown-microwave analysis	$9.5–9.8\times10^{-8}$	10^{11}	Perhaps influenced by immature plasma
Ne	Biondi and Brown	410 300 195	5–30	Microwave breakdown-microwave analysis	2.07×10^{-7}	$10^{10}–10^{11}$	Probably exact Dissociative Recomb. to Ne_2^+
Ne	Holt et al.	300	20	Microwave breakdown-microwave analysis	1.1×10^{-7}		
Ne	Biondi and Brown	77	5–25	Microwave breakdown-microwave analysis	2×10^{-7} 8×10^{-7} (depends on pressure)	$10^{10}–10^{11}$	Probably exact Dissociative to Ne_2^+ and some other Ne molecular ion in equilib.
Ne	Olsen and Huxford	8300	50–70	Pulsed, overvolted condenser discharge. Stark effect on H lines	5.3×10^{-14}	10^{17}	Possibly radiative electron-ion recombination

A	BIONDI and BROWN	300	20–30	Microwave	3.7×10^{-7}	10^{10}	A of doubtful purity
A	BIONDI	300	20–30	Microwave	8.8×10^{-7}	10^{11}	Probably pure
A	SAYERS	600 1 250	0.1–1	Probes in arc plasma	1.12×10^{-9} 4.2×10^{-10}	$>10^{11}$ $>10^{11}$	Constant with pressure decreases as $T^{-\frac{3}{2}}$. Possibly governed by creation of A_2^+.
A	HOLT et al.	300	2 30	Microwave analysis	5×10^{-7} 1.1×10^{-6}	10^{10}–10^{11}	Possibly ascribed to immature plasma
A	OLSEN and HUXFORD	8 300	50–70	Pulsed, overvolted condenser discharge. STARK effect on H lines	2.5×10^{-13}	10^{16}	Possibly radiative electron-ion recombination
H	HOPWOOD and CRAGGS	12000	760	Photomult. and oscilloscope in spark channel. Filter for H_α, H_β, H_γ	1.4×10^{-11}	10^{17}	Probably radiative capture to H^+. Influenced by high COULOMB fields.
H	HOPWOOD and CRAGGS	20000	760	Photomult. and oscilloscope in spark channel. Filter for H_α, H_β, H_γ	2×10^{-12}	2.5×10^{17}	Probably radiative capture to H^+. Influenced byhigh COULOMB fields
H_2	BIONDI and BROWN	300	10–15	Microwave breakdown-microwave analysis	2.5×10^{-6}	10^{10}	Study incomplete. H_2 possibly impure
H_2	VARNERIN[1]	300	3–60	Propagat. const. in microwave plasma	$\alpha_e=0.34\times10^{-6}$ to 0.34×10^{-6} 0.34–2.5×10^{-6}	10^9–2×10^{10}	Not a recombination but electron attachment with O or H_2O
H_2	PERSSON	300	3–60	Microwave breakdown	0	10^{10}	No dissociative recombination
N_2	BIONDI and BROWN	300	3–10	Microwave breakdown-microwave analysis	1.4×10^{-7} at about 5 mm. up to 1.9×10^{-7}	10^{10}	Study incomplete
N_2	BRYAN	300	6	Microwave breakdown-microwave analysis	2×10^{-6} 9×10^{-7} 2×10^{-6}	Pow power 10^{10}. High power 10^{11}	Immature plasma
O_2	BIONDI and BROWN	300	3–20	Microwave breakdown-microwave analysis	3×10^{-7} to 14×10^{-7}	10^9–10^{10}	Dissociative electron or ion-ion recombination. Study incomplete

[1] Now recognized as falsified by electron attachement to impurities. With pure H_2 no recombination was observed.

9. Electron-ion recombination. α) *The volume coefficients.* (1) The radiative capture, coefficient α_{er}. In all gases where electrons remain free and the ions are either atoms or molecules which do not dissociate on recombination with an electron, the recombination involves capture of an electron from the continuum of energy distribution, appropriate to the conditions, into an unoccupied level of the atom. For simple hydrogen like atoms quantum mechanics permits a calculation of the cross section for this process. Capture into the higher levels is more probable than directly to the ground state. This capture results in radiation corresponding to the ionization potential for electrons in that level plus the energy of the particular captured electron. Thus for each wave length corresponding to a transition from zero energy to the level in question, there will be a continuum of radiated light extending from that wave length into the violet. As capture is the more probable the less the energy to be radiated, the continuum will fade towards the violet. Most radiation of this sort will be into the infra red and visible as capture to higher states is more probable. Subsequent transitions from higher to lower levels result in radiation of the well known line spectra. Quantum mechanics has indicated that $\alpha_{er} \sim 10^{-12}/V$, where V is the energy in electron volts for the total cross section in hydrogen-like atoms the value in any specific computed case being in error by no more than perhaps 10%. Translated into cross sections if the velocity of thermal agitation of the electron be $\bar{c}$, then

$$\alpha_{er} = \pi \sigma_r^2 \bar{c}_1, \tag{9.1}$$

with $\pi\sigma_r^2$ the atomic cross section for radiative capture. Thus for electrons at roughly 1200° K $\alpha_{er} \sim 6 \times 10^{-12}$ cm.^{3}/ion sec. and $\pi\sigma_r^2 \sim 2.5 \times 10^{-19}$ cm.^{2}.

It is to be noted that the value of α_{er} is indeed very low and that if a significant range of changes in concentration is to be observed in $1/n$ then if t is of the order of 10^{-5} sec. n_0 will need to be around 10^{16} ions/cm.^{3}. It is therefore not surprising that the only one authenticated observed case of radiative recombination was achieved in a study of the channels of intense disruptive sparks in H_2 gas. Here only the BALMER series lines were observed by CRAGGS, MEEK and HOPWOOD [*18*]. Electron densities were estimated by STARK broadening. Temperatures ranged around 20000 to 12000° K in the channels, and values of α_{er} ranged from 10^{-12} to 10^{-11} cm.^{3}/ion sec. while theory at this energy would indicate that, a_{er} should be ~ 0.3 to 0.6×10^{-12} cm.^{3}/ion sec. The discrepancy observed could be attributed to the action of the strong ionic force fields in such plasmas. Note in this process the continua due to capture were not studied. Such intense breakdowns where plasmas at 12000° K are 100% ionized and the molecules are all dissociated will have much continuous radiation background and much radiation caused by the free-free electron transition as the gas cools, so that continua and the longer wave length lines due to capture cannot be discerned. Earlier studies in arcs in A, Cs and Hg where densities were of the order of 10^{14} ions/cm.^{3}, at most, by KENTY, MOHLER and later by J. SAYERS [*17*] yielding values of α of the order of 10^{10} when ion energies were 0.16 e.v. or more were ascribed to radiative recombination. As later indicated they probably are not due to this process.

(2) Dissociative recombination, coefficient α_{ed}. This process was one of the processes invoked before World War II by D. R. BATES [*9*] to account for large values of α inferred for electrons in the ionosphere by study of the radio echoes. In this, an electron of sufficient energy in interaction with a molecular ion causes the ion to dissociate into two neutral atoms, the difference between the energy of neutralization and the energy of dissociation going to kinetic energy of the separating atoms or to a combination of this energy and energy of excitation

of one or even perhaps both of the atoms. In this process, the energy radiated, would be the atomic line spectrum with DOPPLER broadening only, but yielding no continuum. In well ionized plasma of high energy much radiation in the visible and infra red including a continuous background can be expected through the high temperatures and free-free transitions [*19*] as electrons cool down. The inverse of this process had in previous years been postulated by MOHLER [*29*] to account for photoelectric ionization of gases like Cs and perhaps Hg at energies below the long wavelength limit. Thus MOHLER had indicated that Cs excited to 2.72 volts could in impacts with neutral atoms react as follows

$$h\nu + \mathrm{Cs} \rightarrow \mathrm{Cs}^* + \mathrm{Cs} \rightarrow \mathrm{Cs}_2^+ + e + \tfrac{1}{2} m v^2 .$$

The inverse reaction would be

$$e + \tfrac{1}{2} m v^2 + \mathrm{Cs}_2^+ \rightarrow \mathrm{Cs}^* + \mathrm{Cs} + \tfrac{1}{2} m v^2 \rightarrow 2\,\mathrm{Cs} + h\nu .$$

The yields from the ionization by molecular ion formation were of such size that the inverse reaction could be expected to show a large cross section. Following World War II, the new microwave techniques yielding uniform plasmas with 10^8 to 10^{12} electrons and ions under clean conditions, where electron loss could be measured in 10^{-3} sec. and loss by ambipolar diffusion could be corrected for, led to observation of values of α_{er} in pure *inert* gases ranging from 10^{-6} to 10^{-8}. Here *no negative ions* could form. BATES [*6*] postulated that these must be ascribed to dissociative recombination and calculated the rate of molecular ion formation from atomic inert gas ions in triple impact needed to account for the molecular ions. Even while BATES' calculations were in progress the work of MOLNAR, HORNBECK [*30*], HOLT and his associates [*7*] were proving the existence of the molecular ions in quantity in such plasmas and indicating the methods of creation. Subsequent studies of BIONDI and of PHELPS and BROWN [*6*] as well as BIONDI [*6*] and HOLSTEIN, have established the creation of the molecular ions and verified BATES' rough calculation as to rates with considerable success. BIONDI [*31*] finally established the dissociative process for He_2^+ ions by observing the DOPPLER broadening of the emitted lines.

D. R. BATES [*28*] has developed a rough theory of this process yielding α_{ed}. It has the form

$$\alpha_{ed} = 2.1 \times 10^{-16} \left\{ \frac{r}{T^{\frac{3}{2}}} (\tau_A + \tau_S) \right\} \int_G e^{-\frac{E}{kT}} f(E)\, dE \ \ \mathrm{cm.^3/ion\ sec.} \tag{9.2}$$

Here r is the ratio of statistical weights of the A_2^* to A_2^+ states where the A_2^* state is the excited molecular state that will dissociate, $f(E)$ is the FRANCK-CONDON factor measuring the degree of overlap between the nuclear wave functions concerned in the initial transitions; τ_A is the averaged characteristic time associated with autoionization, i.e., the reversal of $\mathrm{A}_2^+ + e \leftrightarrow \mathrm{A}_2^*$ and τ_S is the averaged time for effective separation to occur. Estimates of r by BATES place r relatively high if a number of excited states occur; he suggests 10. He estimates $\tau_A \sim 10^{-13}$ sec. and τ_S to be of the same order as velocities are 10^5 cm./sec. and distances less than 10^{-8} cm. are involved. The integral depends on details of the potential curves. It will be small unless $f(E)$ is appreciable for low E values, but in favorable circumstances it can equal the value of $2kT$ when expressed in e.v. With these possibilities at 250° K, $\alpha_{ed} \sim 10^{-7}$, which is good. It is to be noted that α_{ed} here varies as $T^{-\frac{3}{2}}$. It is too early as yet to say very much about values observed in specific cases. As indicated, the data for A, He, and Ne of BIONDI and BROWN and later BIONDI are doubtless good. As noted in Table 3, the values lie in the order of magnitude anticipated by theory. They apply to what were supposed to be thermal electrons, that is, they were in plasmas cooled down by aging for a millisecond or two. Values were independent of pressure except at 77° K in Ne where possibly equilibrium ratios of two positive ion species of different α_{ed} could occur. Strangely enough, the values of α_{ed} observed are insensitive to gas temperature contrary to the theory of BATES. This may indicate that the electrons were actually not in thermal equilibrium with the gas in each situation. This is not improbable in virtue of the nature of the plasma, which could elevate electron temperatures. These would depend on initial excitation energy and the time elapsed rather than on the thermal equilibrium temperature of the gas. BIONDI's later and higher value in argon is probably better than his earlier one, as the initial argon may not have been pure. BIONDI's value for Hg ions also indicates a dissociative process. The pressure variation must be ascribed to a second molecular ion in Hg in labile equilibrium following eq. (2.29) with different α values. Despite earlier work, the later study of K. B. PERSSON [21] in H_2, indicated that there is no dissociative recombination observable in H_2 at thermal energies. This observation is of great significance in that it indicates that while molecular ions can recombine dissociatively at thermal energies with large cross sections, *all molecular ions may not do so* even if the energy balance is in favor of this process as in H_2. BIONDI's data in N_2 and O_2 indicate dissociative recombination to be present in pressure dependent processes with large cross sections. BIONDI does not regard these results as more than preliminary and they are certainly not open to reasonable interpretation with the present data. It must be noted *à propos* of these studies that the $1/n-t$ curves were linear over a goodly range of values and that uniformity of density was assured while higher orders of ambipolar diffusion and other disturbances were eliminated. It will also be noted that pressures were relatively high so that the atomic species were able to form the molecular ions at a rate well exceeding loss by recombination.

It is now of interest to regard the early data of KENTY, MOHLER and later data of SAYERS in arcs [17]. All these data were taken with probes in the decaying plasmas of interrupted low voltage arcs. No adequate correction was made for ambipolar diffusion loss. Ranges of values in $1/n-t$ plots were not great and scatter of points was bad. There was no instance where really true linear plots of the nature of the later plots of BIONDI were observed. All data were taken at pressures of 1 mm. and usually much less. SAYERS in argon noted a temperature dependence of the observed α following a $T^{-\frac{3}{2}}$ law. In all cases the values of α were of the order of 2×10^{-10} and in no case were the values as low as 10^{-12} as for radiative capture, or as high as the established dissociative recombination for the same gases by BIONDI and others even corrected by the $T^{-\frac{3}{2}}$ law to room temperatures. As shown elsewhere, the explanation of these results comes from the fact that at the pressures indicated, *loss by dissociative recombination is faster than creation of the molecular ions from atomic ions in triple impacts* [32]. Analysis of this process by KUNKEL and the writer [32] shows that reasonably linear plots for loss of electrons by dissociative recombination under these conditions result, where the observed value of α is not α_{ed}, but a reduced value dependent on the rate of formation of molecular ions by triple impact. That is, the $1/n-t$ plots under these conditions would, over the limited ranges of observation, be no more curved than those observed, and the slope would yield values of a fictitious α of the order of 10^{-10}.

(3) Dielectronic recombination, coefficient α_{ee}. At the same time as BATES [9] suggested dissociative electronic recombination, he also proposed another process which he called dielectronic recombination. This phenomenon is the inverse of

an ionization process called auto-ionization. If, by electron impact, two electrons in an atom are raised to excited states, such that the sum of the energies exceeds the ionization energy, then by internal interaction akin to the AUGER effect, one electron returns to the ground state and the other leaves the atom. The inverse process to this is one in which an electron is captured into an atom, leading to the excitation of two electrons in lower states. These states can later radiate. GARTON [*33*], who has studied auto-ionization spectroscopically, used data on argon to calculate the value of the cross section and thus of the coefficient α_{ee}. This was done for one state of argon using BATES' and MASSEY'S relations. The value for α_{ee} so calculated was 3×10^{-12} at 3000° K, 10^{-10} at 300° K and 1.5×10^{-9} at 20° K. It is seen that this value is quite temperature sensitive and that it is only some order of magnitude larger than α_{er} for the radiative process. There is thus no justification in associating this process with any observed values to date, though the process no doubt occurs.

β) Preferential electron-ion recombination, coefficient α_{ep}. It might be expected that at very high pressures in free electron gases (requiring nearly fantastic degrees of purity), in which electrons lose their energy of liberation by the many impacts within d_0, with $d_0 \gg L$, that electrons and ions foregoing the diffusive phase would have a good chance of literally *re*-combining. Thus r_0 would be less than d_0 in the process of ionization so that a type of recombination with $d_0 > r_0 \gg L$ would occur called *preferential* recombination. The ionization and recombination can be detected only by the way in which currents drawn to properly designed and effective electrode systems vary with the applied field. In principle, nearly uniform high sweeping fields with values increasing above e/d_0^2 will remove electrons from increasingly greater regions within d_0. Early investigators, studying cosmic rays in high pressure inert gases, believed that they were observing this phenomenon. Actually their gases were not pure and they were observing preferential ion-ion recombination in improperly designed sweeping fields as BOWEN showed [*34*]. BRADBURY [*35*] discusses preferential recombination on more general grounds and concurs with the writer that with the required degree of purity of nonattaching gases it is unlikely that the phenomenon will be observed, since metal electrodes and containers and pressures in the hundreds of atmospheres are required.

γ) Electron-ion wall recombination. As earlier indicated, the actual recombination of electrons and ions at the wall with the high surface densities accruing from dense plasmas with ambipolar diffusion proceeds more rapidly than the diffusion of carriers to the walls. Thus the quantity leading to a loss of carriers from the volume of ionized gas in time is governed by ambipolar diffusion. Since this leads to an exponential decline of carriers with time, the measurement of loss does not lead to the evaluation of a proper recombination coefficient.

10. Small ion—large ion recombination, coefficient α_{SL}. This particular field has not been investigated very extensively owing to the experimental difficulty of creating enough large ions in controlled quantities and bringing them together in an appropriate mixture with normal or small ions. Computations have been made on disappearance of the intermediate and large ions from a volume of ionized air, presumably by loss to recombination with the normal ions. This is, in fact, the only way in which large ions can be lost. Since initially more heavily ionized moist dust-free air will quickly have half or more of the normal ions picked up to give intermediate and large ions leaving the residue of opposite ions as small ions, disappearance of ionization that can be collected by fields with time can be attributed to this phenomenon and subsequent neutralization of the charges on the large ions. Since the less permanent intermediate ions eventually rather quickly yield LANGEVIN ions, and these have radii σ_L in excess of 10^{-6} cm., $\sigma_L \gtrsim d_0$, the model for theoretical evaluation of α_{SL} is relatively simple. The concentration n of such ions is never much more than 10^6 per cm.^{3} so that $r_0 \gg d_0$. Thus the normal ions diffuse with their random thermal velocity $\bar{c}$ sweeping out a volume $\pi\sigma_L^2\bar{c}$ in each second. Every large ion of opposite sign of the n ions per cm.^{3} lying in $\pi\sigma_L^2\bar{c}$ will recombine in 1 sec. with the n opposite small ions. Thus,

$$\frac{dn}{dt} = -\pi\sigma_L^2\bar{c}\,n^2 = -\alpha_{SL}\,n^2$$

so that

$$\alpha_{SL} = \pi\sigma_L^2\bar{c}. \tag{10.1}$$

The diffusive approach is the only stage of the THOMSON process active since $\sigma_L > d_0$, so that COULOMB attractions are negligible. Once ions reach the surface at σ_L from the center of the large ion the charges neutralize with $\varepsilon = 1$ since the surface molecules take up the energy. The only complications present, aside from the difficulty of controlling, producing, and collecting the n ions, come from the fact that *the rate of charging of the* LANGEVIN *ions* is commensurate with α_{SL} since the same process governs, and that there is chance of creation of multiply-charged large ions as VASSAILS [*35*] and KUNKEL [*35*] have shown.

11. Ion-ion recombination. α) *Preferential ion-ion recombination, coefficient at very high pressures, the* LANGEVIN *law coefficient* α_L. In electron attaching gases at high pressures in which $d_0 > r_0 \gg L$, the ionizing process creates electrons which attach to form ions within d_0 irrespective of the values of r_0, provided these are not too large. Under such circumstances, the ions act on each other with a force eX, in which the field X is $X = e/r^2$ where r is the distance between ions. Then as the value of L is small compared to d_0, the ions can be considered as moving in a continuous medium with a drift velocity v given by $v = X(k_+ + k_-)$ in which k_+ and k_- are the mobilities of positive and negative ions respectively. Under such circumstances there is no period of diffusive movement, the *ions continuously drift together*. Charge transfer, when the ions approach to within a few Å, follows with $\varepsilon = 1$ since the neighboring molecules virtually insure impact between ions. This at once leads to a recombination coefficient evaluated as:

$$\alpha_L = 4\pi e(k_+ + k_-). \tag{11.1}$$

This model and eq. (11.1) were early derived by LANGEVIN from a more generalized approach and were accepted generally as applicable at all pressures for ion-ion recombination. They make α vary as $1/\varrho$, where ϱ is the gas density, and at constant temperature as $1/p$, where p is the pressure. The value of α_L at constant density is then as relatively insensitive to temperature as are ion mobilities. Applied by LANGEVIN to his measured value of α in gases, such as air at 760 mm. and below, the value of α observed for ions deviated from his theory, such that he was forced to write

$$\alpha_L = 4\pi e(k_+ + k_-)\,\varepsilon \tag{11.2}$$

with ε considerably less than unity and varying with pressure. It also follows from his theory that at all times the ions of opposite sign drift together. Thus, if ionization had initially created ions anisotropically on a microscale, it would have been observed that the anisotropy remained as time went on yielding an α constant with time. With improving techniques, all observations at 1 atmosphere and below showed just the contrary behaviour.

Actually, very little work has been done under the conditions applicable to this process. The process was observed and named by those workers making early cosmic ray measurements in high pressure chambers. It, as was indicated for preferential electron-ion recombination, involves sweeping more and more of the ions out of the volumes within d_0 of their partners as the external sweeping field X reduces the radius r in $e/r^2 \leqq X$ below d_0. In 1940, BRADBURY [34] set the outline of a general theory of preferential recombination in an external field X. The theory requires a suitable set of assumptions concerning electron attachment for the solution of a given problem and has not been further exploited.

β) Columnar recombination, coefficient α_C. This process was early noted when BRAGG and KLEEMANN [3] found that the currents drawn by a given sweep field with α ray tracks parallel to the field were markedly less than when they were perpendicular. Their tentative explanation was in terms of what would today be called preferential recombination. In 1910, MOULIN correctly showed that the deviations stemmed from an anisotropic distribution of ions. In 1913, JAFFÉ, who had studied conduction produced by ionization currents in liquids, confirmed MOULIN'S conclusions and developed the first part of his excellent theory of columnar ionization and recombination [27]. The α particles ionize intensively at the rate of 1000 ions per mm. length of straight trajectory, in paths that are perhaps 4 or more cm. long and can be made roughly parallel to each other. He assumed that, as a result of electron attachment, the ion density initially fell off radially from the axis of the track along a GAUSSian distribution with initial average radius from the axis of some tenth of a mm. or less. As time goes on, the ions diffuse radially and recombine at rates varying radially with the concentration. When the tracks expand radially by diffusion so that adjacent parallel tracks merge, the remaining ions may be thought of as having achieved essentially an isotropic distribution. If sweeping fields are applied along the paths, the ions are exposed to recombination in the dense regions much longer than at right angles to the tracks where the fields quickly separate them. The calculated currents at various angles with the α particle tracks gave quite satisfactory agreement with observation. In 1928, in conjunction with measurements of α as a function of time using α ray tracks by SCHEMEL, JAFFÉ [3] carried his theory as applied to impure H_2, CO_2, O_2 and air to the calculation of the observed coefficient as a function of time. The theory was in as good agreement with the experimental data as possible since SCHEMEL had to use finite flash times for adequate ionization. The α' was thus calculated to decline with time somewhat more rapidly than observation showed it to. Strangely, both the 1913 and 1928 theory assumed that the ions largely executed random diffusive motion and not the drift together that LANGEVIN'S theory required, but the inapplicability of LANGEVIN'S theory was not recognized.

γ) Initial recombination, coefficient α_i. This recombination process results from a microscopic anisotropy caused by rapid electron attachment to create adjacent pairs of positive and negative ions with $r_0 > d_0$ and $d_0 \gtrless L$ but pairs of adjacent opposite ions having $r_0 < n_0^{-\frac{1}{3}}$. This situation was first recognized by PLIMPTON [4] with X ray ionization who noted that the $1/n - t$ plots were not linear, and that, in fact, they were initially steeper, indicating that α appeared higher shortly after ionization. He interpreted this as being caused by the initially closer proximity of ion pairs which ultimately diffused to a random distribution as time went on. The phenomenon was also observed by RUMELIN [13] who did not grasp its significance. When, in 1928, L.C. MARSHALL [4] in the writer's laboratory observed the same phenomenon, the nature of the process was independently recognized, in the ignorance of PLIMPTON'S work of 15 years earlier. However, more important was the simultaneous recognition of the inconsistency between the diffusion of ions apart and LANGEVIN'S theory. This recognition permitted the writer to show that recombination at 1 atmosphere and below was consistent with J. J. THOMSON'S [*23*] model for the recombination process. With the help of A. SOMMERFELD [*4*] the nature of the recombination process at around atmospheric pressure in terms of BROWNian movement theory was established by the writer. The writer [*36*] later derived an elementary theory relating the observed time varying coefficient α to the true coefficient. This crude theory yielded reasonable agreement with GARDNER'S observations, the chief drawback again being ascribable to finite flash times needed to get adequate ions for measurement. Later an extensive general analysis of the problem was given by BRADBURY [*3*]. This assumes that the density of ions is a function of time given by

$$n_t = \frac{\mu}{\pi\left[r_r + \frac{4}{3} k X t + (\bar{c}\,\lambda t)^{\frac{1}{2}}\right]^2} \tag{11.3}$$

with L_β the length of the β or X ray secondary electron track, μ the number of ions per cm. of track, γ_β the number of tracks/cm.^{3}, k the drift velocity of the ions and the diffusion accounted for by $\bar{c}\lambda t$. It also involves r_r, the average distance between ions of a pair, as given by a distribution $\vartheta(r)$ of electrons attached to form negative ions relative to positive ions. From this BRADBURY evaluated n_0, the actual number of ions per cm.^{3} of gas as influenced by recombination, as

$$\frac{d n_0}{dt} = -\alpha\, n_t^2 L_\beta \gamma_\beta \pi \left[r_r + \frac{3}{4} k X t + (\bar{c}\,\lambda t)^{\frac{1}{2}}\right]^2 = \frac{\alpha \mu^2 \gamma_\beta L_\beta}{\pi\left[r_r + \frac{4}{3} k X t + (\bar{c}\,\lambda t)^{\frac{1}{2}}\right]^2}. \tag{11.4}$$

This must be integrated over a certain period of time to get the loss of ions during the initial period. The time can be found from the expression for n_t as that for which n_t becomes equal to the observed macroscopic density. If n_v is this value, then t_v can be found by setting $n_t = 8 n_v$. BRADBURY also considered the finite flash time in terms of very general relations which were not applied to existing data. In any event, the theoretical background is there for a more complete study when and if needed. However, agreement between more primitive theory and experiment is sufficient to leave no doubt about the interpretation of the phenomenon. With better ion generating and measuring techniques, its complications can be avoided and further study will be along lines of purely academic interest.

δ) Volume recombination. High pressure, or HARPER-LANGEVIN process, coefficient α_H. If $r_0 \gg d_0$ and $d_0 > L$, but not so many times greater than L that the gas can be considered a continuum, the chance ε of charge transfer, once the ions are within d_0, will be practically unity. Then the rate of recombination will, in principle, depend primarily on the rate of diffusion of the two carriers to within d_0 plus the time of approach from d_0 to neutralization of charge. The problem without imposing any pressure limits had been in principle solved in terms of the BROWNian motion of two particles interacting with COULOMB forces by SOMMERFELD in 1929 at the request of the writer [*4*]. It was published in a paper by the writer and MARSHALL [*4*]. Except for a numerical coefficient, it yielded results analogous to those to be described. In ignorance of the work of SOMMERFELD, HARPER [*26*] noted the problem posed by LANGEVIN'S theory and in 1932 published a solution to the problem of ion-ion recombination at intermediately high pressures. In this he set the radial movement of two ions of opposite charge as given by

$$\frac{dr}{dt} = \frac{3(D_+ + D_-)}{r} - \frac{(k_+ + k_-)}{r^2}, \tag{11.5}$$

with r the instantaneous separation, D_+ and D_-, and k_+ and k_- the diffusion and mobility coefficients of positive and negative ions respectively. At some value of $r = d_{0H}$ which is given by $d_{0H} = \frac{e^2}{3kT}$, $\frac{dr}{dt} = 0$. The value of d_{0B} from SOMMERFELD's BROWNian movement derivation is $d_{0B} = \frac{e^2}{6kT}$. If then only the diffuse motion outside d_{0H} is considered,

$$r^2 = 6(D_+ + D_-)\,t + r_0^2. \tag{11.6}$$

This gives

$$\frac{dr}{dt} = \frac{3(D_+ + D_-)\,t}{r}, \tag{11.7}$$

as the velocity of radial motion.

Thus (11.7) yields

$$\frac{dn_r}{dt} = -4\pi r^2 n^2 \frac{dr}{dt} = -4\pi r^2 n^2 \frac{3(D_+ + D_-)}{r} = -4\pi r \cdot 3(D_+ + D_-).$$

If $\varepsilon = 1$, and $r = d_{0H}$,

$$\frac{dn_r}{dt} = \frac{dn}{dt},$$

so that

$$\frac{dn}{dt} = -\alpha_H n^2 = -4\pi d_{0H} n^2 \cdot 3(D_+ + D_-), \tag{11.8}$$

whence

$$\alpha_H = 4\pi d_{0H} \cdot 3(D_+ + D_-). \tag{11.9}$$

Putting $d_{0H} = \frac{e^2}{3kT}$ above and recalling that $\frac{D_+}{k_+} = \frac{kT}{e} = \frac{D_-}{k_-}$, the value of α_H becomes

$$\alpha_H = 4\pi e(k_+ + k_-) = \alpha_L. \tag{11.10}$$

For the quantity α_H of eq. (11.10) JAFFÉ in a later paper on more general statistical grounds gets

$$\alpha_{HJ} = 4\pi d_{0J} f_1 (D_+ + D_-), \tag{11.11}$$

where $d_{0J} = e^2/kT$ and f_1 is a constant near unity so that

$$\alpha_{HJ} = 4\pi e f_1 (k_+ + k_-) = f_1 \alpha_L. \tag{11.12}$$

This leads to the surprising result that, irrespective of whether the ions randomly diffuse apart as in HARPER's and SOMMERFELD's treatment or are drawn together as in LANGEVIN's model, when $d_0 \gg L$ the expression of α derived even on the most general grounds is that initially deduced by LANGEVIN. That is, while in actual principle, the processes of LANGEVIN in which, due to a COULOMB attractive force, the ions drift together and that of HARPER and SOMMERFELD representing a random diffusion apart are diametrically opposed actions, the statistical laws operative lead to precisely the same expression for the coefficient, except for a constant numerical factor. This was first pointed out by JAFFÉ [27].

It is clear that as long as d_0 is sufficiently larger than L and $\varepsilon = 1$, the coefficient of ion-ion recombination should at constant density be independent of pressure, at constant temperature vary inversely as the pressure, at constant density vary relatively slowly with temperature where mobilities so vary, and vary with ion and gas masses following essentially the mass dispersion curve of TYNDALL and POWELL [73] where the nature of the carriers is known.

Unfortunately, there is little evidence to yield any accurate verification of this theory. A study of THOMSON's more general theory for ion-ion recombination would indicate ε as about unity near two atmospheres for air and O_2. Thus beyond two atmospheres pressure at about 300° K, the value of α should follow LANGEVIN's law. The data of MÄCHLER [22] on relatively contaminated air indicates that the value of α would extrapolate along the $1/p$ curve to join onto SAYERS' more accurate data at two atmospheres where, as shown in Fig. 4, SAYERS' values begin to decline. More study of this situation on reasonably clean air, or better O_2, could well be undertaken. At one atmosphere, in both air and O_2 α *declines* as pressure *decreases* following the THOMSON theory.

(2) The THOMSON Process, Coefficient α_T. With $r_0 > d_0$ and d_0 related to L by $d_0 \gtrsim L$ and $L > d_0$ the four step THOMSON process occurs. The first step is a random diffusive approach to d_0 between ions of opposite sign. Once within d_0 there is the period of attractive approach leading to an orbital encounter with a chance ε of effecting a charge transfer and recombining. Failing neutralization, the ions will separate ultimately to recombine with some other ions of opposite sign. If the relative thermal velocity of approach of the two ions in a MAXWELLian distribution of velocities is set as $\sqrt{\bar{c}_+^2 + \bar{c}_-^2}$, where $\bar{c}_+$ and $\bar{c}_-$ are the average thermal velocities of random motion of oppositely charged ions, and if the chance of recombination once within d_0 is ε, the coefficient of recombination α_T is given by

$$\frac{dn}{dt} = -n^2 \pi d_0^2 \varepsilon \sqrt{\bar{c}_+^2 + \bar{c}_-^2} = -\alpha_T n^2, \tag{11.13}$$

whence

$$\alpha_T = \pi d_0^2 \varepsilon \sqrt{\bar{c}_+^2 + \bar{c}_-^2}\,. \tag{11.14}$$

Under simplifying assumptions as to the trajectory of one ion relative to the other within d_0, it is possible to calculate the chance that the positive ion and the negative ion *each* have one collision with a neutral molecule within d_0. Through such collision, they lose enough of the kinetic energy gained by each ion from the potential fall within d_0, to make a closed orbit about the common center of mass possible and thus to insure charge exchange. This chance was earlier calculated by O. W. RICHARDSON [38] and later and quite independently derived by J. J. THOMSON [23]. These derivations contained a slight error that was later corrected by the writer [39]. The expressions for each ion read:

$$\omega_- = 1 - \frac{L_-^2}{2d_0^2}\left[1 - e^{-\frac{2d_0}{L_-}}\left\{\frac{2d_0}{L_-} + 1\right\}\right] \tag{11.15a}$$

and

$$\omega_+ = 1 - \frac{L_+^2}{2d_0^2}\left[1 - e^{-\frac{2d_0}{L_+}}\left\{\frac{2d_0}{L_+} + 1\right\}\right]. \tag{11.15b}$$

Here ω_- and ω_+ are the chances of encounter within d_0 for negative and positive ions respectively and L_- and L_+ are the free paths for negative and positive ions. Using THOMSON's concept,

$$\varepsilon = \omega_+ + \omega_- - \omega_+\omega_- \tag{11.16}$$

since on adding ω_+ and ω_- the chance of simultaneous impacts $\omega_+\omega_-$ are counted twice. L_+ and L_- can be inferred from the mobilities k which on a solid elastic theory have the form $k = 0.815\,\frac{e}{M}\,\frac{L}{C}\sqrt{\frac{M+m}{m}}$, where L is the free path, C is the root mean square molecular velocity, M is the mass of the gas molecules, and m

of the ions with e the electron charge. For most purposes L_+ can be set equal to L_- or to L, and $\bar{c}_+$ can be set equal to $\bar{c}_-$, or to $\bar{c}$, the mean molecular velocity. Then α_T becomes

$$\alpha_T = \pi \sqrt{2} d_0^2 \bar{c} (2\omega - \omega^2), \tag{11.17}$$

with ω given as above, replacing L_+ and L_- by L as a mean value. Since,

$$d_0 = \frac{e^2}{\frac{3}{2} k T} = 4.05 \times 10^{-6} \frac{273}{T}$$

and L can be set as $L = 10^{-5} \frac{760}{p} L_1$ where L_1 is the free path taken from k for 0°, 760 mm., divided by 10^{-5}, i.e., the mean free path at 0° and 760 mm. expressed in units of 10^{-5} mm. Then the parameters

$$x = 0.81 \frac{273}{T} \frac{p}{760} L_1, \tag{11.18}$$

at constant density, or

$$x = 0.81 \left(\frac{273}{T}\right)^2 \frac{p}{760} L_1, \tag{11.19}$$

at constant pressure and variable density, permit ω and ε to be evaluated in tabular, or curve, form as a function of x for use as needed. Calling m_m the molecular weight of the ion relative to the H-atom chosen as unity, the quantity $\pi d_0^2 \sqrt{2} \bar{c}$ becomes

$$0.922 \sqrt{2} \sqrt{\frac{2}{m_m}}\, 1.84 \times 10^5 \sqrt{\frac{T}{273}},$$

leading to the value of α_T as

$$\alpha_T = 1.73 \times 10^{-5} \left(\frac{273}{T}\right)^{\frac{3}{2}} \left(\frac{1}{m_m}\right)^{\frac{1}{2}} (2\omega - \omega^2). \tag{11.20}$$

Using such data with reasonable assumed values of m_m and L, GARDNER [5] computed values for α_T in O_2 in agreement with experiment. The results also varied both with p at constant T, and T at constant density, in agreement with GARDNER and SAYERS' data in O_2 and air. JAFFÉ, in his more complete study for $r_0 \gg d_0$ and $L_0 \gg d_0$, derived a modification in which he accepts

$$\varepsilon = 2\omega - \omega^2 - \varphi_2\left(\frac{d_0}{L}\right)$$

and finds

$$\alpha_{TJ} = \pi d_0^2 \sqrt{\bar{c}_+^2 + \bar{c}_-^2}\, \psi_1\left(\frac{d_0}{L}, \frac{d_0}{r_0}\right) \varphi_2\left(\frac{d_0}{L}\right). \tag{11.21}$$

This differs from the THOMSON relation in that α_{TJ} depends on d_0/r_0, i.e., on ion concentration yielding also a marked maximum at $r_0 = L$. Sufficiently above pressures for which $r_0 \gg L$ and for ionization so weak as to make $r_0 \gg d_0$, ψ_1 becomes unity, leading to the original THOMSON equation, if $L \gg d_0$. In discussing the general theory, JAFFÉ points out that only under the circumstances where LANGEVIN'S, HARPER'S or THOMSON'S model applies, as defined by the criteria given by him, is the coefficient independent of ion density. This accounts for variations of α observed with ion density by MARSHALL [4], GARDNER [5] and SAYERS [5]. When L and d_0 become comparable, neither THOMSON'S nor HARPER'S theory applies.

(3) Ion-ion recombination at very low pressure, coefficient α_σ. If $r_0 \gg d_0$ and $L \gg d_0$ as is the case in the ionosphere, the value of ε becomes vanishingly

small. Thus at 6.7×10^{-3} mm. pressure and 0° C α_T approaches 10^{-10}. At this value of pressure, the α_T becomes comparable with, or less than, a coefficient of recombination which makes $\varepsilon = 1$ when ions diffuse to within distances of the atomic separations in molecules. If the average distance between oppositely charged ions at which the chance of charge transfer approaches 1 becomes σ, then α_σ, the coefficient of recombination, can be estimated from

$$\frac{dn}{dt} = -\alpha_\sigma n^2 = -\pi \sigma^2 \sqrt{\bar{c}_+^2 + \bar{c}_-^2}\, n^2, \tag{11.22}$$

leading to

$$\alpha_\sigma = \pi \sigma^2 \sqrt{\bar{c}_+^2 + \bar{c}_-^2}. \tag{11.23}$$

If $\sigma \sim 2 \times 10^{-8}$ cm., as $c \sim 10^5$ cm./sec. or less, $\alpha_\sigma \sim 10^{-10}$. Thus the observed α, as p goes down below 10^{-3} mm., will change from α_T to α_σ. Under these conditions, it ceases to vary with p and varies only as $\sqrt{T}$.

References.

[1] Summary early studies and history see J. J. and G. P. Thomson: Conduction of Electricity Through Gases, Third Edit., vol. 1, p. 19ff. Cambridge: University Press 1928. — Loeb, L. B.: Fundamental Processes of Electrical Discharge in Gases, chap. II. New York: John Wiley & Sons 1939.

[2] Loeb, L. B.: Fundamental Processes, loc. cit. chap. II, and especially L. B. Loeb: Basic Processes of Gaseous Electronics, chap. VI. Berkeley: University of California Press 1955.

[3] Bragg, W. H., and R. D. Kleeman: Phil. Mag. **11**, 466 (1906). — Kleeman, R. D.: Phil. Mag. **12**, 273 (1906). — Moulin, M.: Ann. Chim. Phys. **21**, 550 (1910); **22**, 26 (1911). — Jaffé, G.: Radium **10**, 126 (1913). — Ann. Phys., Lpz. **42**, 303 (1913); **1**, 977 (1929). — Phys. Z. **30**, 849 (1929). — Schemel, J.: Ann Phys., Lpz. **85**, 137 (1928).

[4] Plimpton, S. G.: Phil. Mag. **25**, 65 (1913). — Marshall, L. C.: Phys. Rev. **34**, 618 (1929). — Loeb, L. B., and L. C. Marshall: J. Franklin Inst. **209**, 371 (1929).

[5] Gardner, M. E.: Phys. Rev. **53**, 75 (1938). — Sayers, J.: Proc. Roy. Soc. Lond., Ser. A **169**, 83 (1938). — Bradbury, N. E.: Terrest. Mag. a. Ats. Electricity **43**, 55 (1938).

[6] Biondi, M. A., and S. C. Brown: Phys. Rev. **75**, 1700 (1949); **76**, 1697 (1949). — Bates, D. R.: Phys. Rev. **77**, 718 (1950); **78**, 492 (1950). — Biondi, M. A., and T. Holstein Phys. Rev. **82**, 962 (1951). — Biondi, M. A.: Phys. Rev. **83**, 1078 (1951). — Phelps, A. V., and S. C. Brown: Phys. Rev. **86**, 102 (1952). — Biondi, M. A.: Phys. Rev. **88**, 660 (1952); **90**, 730 (1953).

[7] Holt, R. B., J. M. Richardson, B. Haviland and B. T. McClure: Phys. Rev. **77**, 239 (1950). — Johnson, R. A., B. T. McClure and R. B. Holt: Phys. Rev. **80**, 376 (1950). — Dandurand, P., and R. B. Holt: Phys. Rev. **82**, 278, 819 (1951). — Redfield, A., and R. B. Holt: Phys. Rev. **82**, 874 (1951).

[8] Kunkel, W. B.: Phys. Rev. **84**, 218 (1951). — Loeb, L. B., and W. B. Kunkel: Phys. Rev. **85**, 493 (1952).

[9] Appleton, E. V., and J. Sayers: J. U. R. S. I. **78**, 272 (1938). — Bates, D. R., R. H. Buckingham and H. S. W. Massey: Proc. Roy. Soc. Lond., Ser. A **170**, 322 (1939). — Bates, D. R., and H. S. W. Massey: Proc. Roy. Soc. Lond., Ser. A **187**, 261 (1946).

[10] Loeb, L. B.: Basic Processes of Gaseous Electronics, loc. cit. chap. VII.

[11] Loeb, L. B.: Basic Processes of Gaseous Electronics, loc. cit. chap. IV, part 2.

[12] Holstein, T., D. Alpert and A. O. McCoubrey: Phys. Rev. **76**, 1259 (1949); **85**, 985 (1952).

[13] Rumelin, G.: Phys. Z. **9**, 657 (1908). — Ann. Phys., Lpz. **43**, 84 (1914).

[14] Riecke, E.: Ann. Phys., Lpz. **12**, 814 (1903). — Mie, G.: Ann. Phys., Lpz. **13**, 857 (1904).

[15] Langevin, P.: Ann. Chim. Phys. **28**, 289, 443 (1903).

[16] Luhr, O.: Phys. Rev. 3**35**, 194 (1930); **36**, 24 (1930); **37**, 998, (1931); **44**, 459 (1933).

[17] Kenty, C.: Phys. Rev. **34**, 624 (1928). — Mohler, F. L., and C. Boeckner: Bur. Stand. J. Res. **2**, 489 (1929). — Boeckner, C.: Bur. Stand. J. Res. **6**, 277 (1931). — Mohler, F. L.: Bur. Stand. J. Res. **10**, 771 (1933); **19**, 447, 559 (1937). — Sayers, J.: Conf. Phys. Ionized Gases, Univ. Coll. Lon. Auspices Warren Research Fund, Royal Soc. April 1953, Report circulated Sept. 1953, p. 29.

[18] Craggs, J. D., and J. M. Meek: Proc. Roy. Soc. Lond., Ser. A **180**, 241 (1946). — Hopwood, W., and J. D. Craggs: Proc. Phys. Soc. Lond. **59**, 755, 771 (1947).
[19] Olsen, H. N., and W. S. Huxford: Phys. Rev. **87**, 922 (1952).
[20] Brown, S. C., and D. J. Rose: J. Appl. Phys. **23**, 711 719, 1028 (1952), as well as Biondi, M. A., and S. C. Brown: In Ref. [6].
[21] Varnerin, L. J.: Phys. Rev. **84**, 563 (1951). — Persson, K. B.: Sixth Conf. on Gaseous Electronics, Wash. D. C., Oct. 22—24, 1953. Paper added after program was printed.
[22] Mächler, W.: Z. Physik **104**, 1 (1936). — Kraus, P.: Ann. Physik **29**, 449 (1937).
[23] Thomson, J. J.: Phil. Mag. **47**, 337 (1924).
[24] Loeb, L. B., and L. C. Marshall: J. Franklin Inst. **209**, 371 (1929).
[25] Loeb, L. B.: Fundamental Processes of Electrical Discharge in Gases l. c. [1], Sect. 6 to 13 and especially p. 144.
[26] Harper, W. R.: Proc. Cambridge Phil. Soc. **28**, 219 (1937); **31**, 430 (1935). — Phil. Mag. **18**, 97 (1934); **20**, 740 (1935).
[27] Jaffé, G.: Phys. Rev. **58**, 968 (1940); **59**, 652 (1941).
[28] Bates, D. R., and H. S. W. Massey: Rep. Progr. Phys. **9**, 67 (1943). Also Phil. Trans. Roy. Soc. Lond., Ser. A **234**, 269 (1946). — Bates, D. R.: Conf. Phys. Ionized Gases Univ. Coll. Lon. April 1953. Loc. cit. p. 34. — Zener, C.: Proc. Roy. Soc. Lond., Ser. A **137**, 696 (1932). — Landau, L.: Phys. Z., Sowjetunion **2**, 46 (1932).
[29] Foote, P. D., and F. L. Mohler: Phys. Rev. **26**, 185 (1925); **27**, 37 (1926). — Mohler, F. L.: Phys. Rev. **31**, 187 (1928). — Mohler, F. L., and C. Boeckner: Bur. Stand. J. Res. **2**, 489 (1929).
[30] Hornbeck, J. A., and J. P. Molnar: Phys. Rev. **84**, 621 (1951). — Hornbeck, J. A.: Phys. Rev. **80**, 297 (1950); **83**, 374 (1951).
[31] Biondi, M. A.: Phys. Rev. **93**, 650A (1953).
[32] Loeb, L. B.: Basic Processes of Gaseous Electronics, loc. cit. chap. VI, p. 692ff.
[33] Garton, W. R. S.: Conf. Phys. Ionized Gases. Univ. Coll. Lon. April 1953, loc. cit. p. 90.
[34] Bowen, I. S.: Phys. Rev. **41**, 24 (1932). — Millikan, R. A.: Phys. Rev. **39**, 397 (1932). — Bradbury, N. E.: J. Appl. Phys. **11**, 267 (1940).
[35] Vassails, Gerard: Large Gaseous Ions, Thesis, Faculty of Sciences, Univ. of Paris Ser. H 2117, 1 May 1948. — Kunkel, W. B.: J. Appl. Phys. **21**, 833 (1950). See also Loeb, L. B.: Basic Processes of Gaseous Electronics, loc. cit. chap. I, Sect. 13.
[36] Loeb, L. B.: Fundamental Processes in Electrical Discharge in Gases, loc. cit. chap. II, Sect. 12, p. 131ff.
[37] Loeb, L. B.: Basic Processes of Gaseous Electronics, chap. I, part. 2, Sect. 5.
[38] Richardson, O. W.: Phil. Mag. **10**, 242 (1905).
[39] Loeb, L. B.: Fundamental Processes of Electrical Discharge in Gases, loc. cit. p. 116ff.

Ionization in Gases by Electrons in Electric Fields.

By

A. VON ENGEL.

With 69 figures.

A. Introduction.

I. Historical notes.

1. Prior to the 18th century it was thought that a gas was an ideal electrical insulator, that electric charges of either sign could not be made to pass through it and that electric currents could not flow in a circuit which contained a gap. The fact that charged conductors surrounded by air gradually lost their charge was explained by the finite surface conductivity of the insulators to which the conductors were fixed. COULOMB in 1750 was probably the first to prove conclusively that two metal spheres which carried opposite electric charges lost their charges by electric leakage through the gas and not by leakage along the surface of insulators in the torsion balance. Later on, his conclusion was confirmed by various experimenters [*1*]. The finite electric conductivity of ordinary gases was thus established.

At the end of the 19th century, and thus before the introduction of the quantum concept, a large number of investigations in electric discharges in gases had been carried out with the object of studying the origin of charges, the nature of the electric current flowing through rarefied or dense gases as well as the light emitted from various parts of the discharge. This knowledge is mainly due to researches by FARADAY [*2*], CROOKES [*3*], J. J. THOMPSON [*4*], STOLETOW [*5*], RIGHI [*8*] and many others.

It has to be remembered that the transport of charges through a gas has been treated for a long time like the flow of charges in electrolytes. In fact positive and negative charges in gases were called cations and anions, their motion was regarded as similar to that of ions in conducting liquids, and OHM's law was assumed to hold at least within a certain range of experimental conditions.

In order to show the dependence of the current in a gas on the various parameters, consider a gap consisting of two large plane electrodes parallel to one another to which a constant potential difference is applied. The space between the electrodes is filled with a gas of given pressure and temperature. The gas is extraneously irradiated, for example by a uniform beam of X-rays. The wide beam which ionizes the gas evenly is prevented by an aperture from falling on the electrodes. Fig. 1 shows the result when for a given inter-electrode distance and constant irradiation ($I=1$) the voltage between the two plates is raised. The current at first rises linearly and later less fast until it reaches constant (saturation) value. If the intensity of irradiation is increased ($I=2$) it is found that the current rises first more steeply and that it reaches a higher saturation value.

The explanation of these observations is that at low voltages or in weak electric fields the number of ions created by radiation is large compared with the number of ions dragged to the electrodes. Since the speed of the ions increases with increasing field, an increasing number of ions of both signs is driven to the electrodes and the current rises. However, since the number of ions produced in a gas in unit time is finite and proportional to the intensity of irradiation, the highest rate of removal of ions can just be equal to the rate of production. Therefore, a limiting value of current will be obtained when every ion pair which is produced reaches the electrodes without being able to recombine in the gas. Any further increase of the electric field within certain limits does not affect the saturation current. Its value can only be raised by an increase in irradiation, a principle which is used in ionization chambers.

2. When the voltage between the electrodes is increased still further the current begins to rise again, first slowly, then increasingly fast, until at a certain voltage the current suddenly becomes very large. Its final value is only restricted by the resistance of the circuit and the power of the potential source. The critical voltage at which this breakdown occurs does not depend any more on the irradiation nor is it easily possible to control the current in the gap by varying the irradiation once the current has reached values well above saturation value. The electric breakdown is not only accompanied by an abrupt rise in current but also by a fall in voltage and by emission of light. Depending on the pressure, the nature of the gas and of the electrodes, the luminous discharge will either fill a large volume of gas or be confined to a narrow path. In this chapter we shal deal mainly with those phenomena in the region of Fig. 1 where the current exceeds saturation current, but where it is still feeble enough to depend on the strength of the irradiating source.

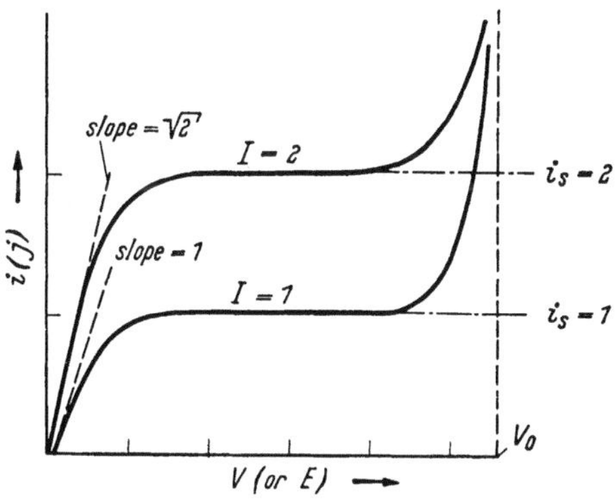

Fig. 1. Ionization current i (or its density j) as a function of the applied voltage V (or the electric field E) in a uniformly irradiated gas. The distance between the two large plane parallel electrodes and the gas pressure are kept constant (see Fig. 3). I intensity of irradiation, i_s saturation current. The initial slope of the curves is proportional to the electric conductivity of the gas. V_0 sparking potential. All magnitudes are in arbitrary units. Slope $\propto$ "conductivity" of ionized gas.

A more detailed investigation shows that—as for very weak fields—the current depends now on the electric field in the gas and also is more sensitive to the field the nearer its value is to the breakdown value. However, there is a marked difference between weak and strong fields. At low fields it is found that the current depends on two factors: the speed with which charges are lost in the gas by recombination and the speed of transport of charges to the electrodes. At high fields a new process sets in with the result that the number of charges in the gas increases. This "multiplication" of charges is caused mainly by collision of electrons (and more seldom of ions) with gas molecules under the action of the electric field.

II. Collisions between electrons, ions, quanta and atoms.

In order to understand the significance of the concepts used later on it is necessary to recall certain fundamental ideas based on the classical quantum theory which will be given now in brief outline. Electron impact phenomena are extensively treated in H. S. W. Massey's article in vol. XXXVI of this Encyclopedia.

3. Let electrons of known energy enter a field-free space which is filled with a gas. The gas pressure is asumed to be so low that the electron in its crossing has on the average not more than one collision. First we shall discuss the conditions in a gas which consists of *atoms* only (rare gas). As the energy of the electrons is slowly raised from zero, it is found that in a collision the direction of motion of the electrons is changed by a small angle and that the energy of the electrons after the collision is reduced by an amount of the order $2m/M$, that is in the ratio of the masses of electron to atom. On reaching a certain critical value of energy, however, some of the electrons lose upon collision a large fraction of their energy; at the same time they are scattered into a new direction. This critical energy is called the lowest excitation (resonance) energy. Thus kinetic energy of an electron is converted into potential energy of the atom by inducing one of its loosely bound electrons to jump from the ground level to the nearest higher one, the resonance level. After a short time interval of say 10^{-8} secs. it is possible to observe light quanta which indicates the return of the atom from its excited state into the ground state. The probability of such an excitation process to occur is of course zero when the energy of the electron is exactly equal to the critical energy. With increasing energy the probability is known to rise at first and after passing a maximum to decrease again. The maximum is sharp or extended according to whether the transition is a triplet-singlet or singlet-singlet transition, respectively. When the initial energy of the electrons is further increased it is found that excitations to higher levels take place. Since de-excitation is accompanied in general by the emission of light quanta of the corresponding energy, it is possible to investigate by means of electronic excitation the whole spectrum of an atom.

When the initial energy of the electrons entering the gas exceeds ionization energy, ionizing collisions can occur so that for each ionization two electrons and one ion will be finally present. Again the likelihood that a collision will lead to ionization is zero at ionization potential, rises steeply with increasing energy, passes a maximum, and decreases at large energy approximately proportional to the inverse of the energy (Fig. 2). The maximum probability lies for many gases between about 70 and 100 eV except for the alkalis which show several maxima between about 10 and 20 eV. With higher electron energies it is possible to set free more than one electron from the atom. If a collision occurs with an ion, an excited ion may be produced, etc. [*15*].

4. Let us now consider the collision between electrons and gas *molecules* and assume for simplicity diatomic molecules like hydrogen. Unfortunately the interpretation of the results obtained for electrons colliding with molecules is not perfectly clear. Besides, there is a lack of experimental data particularly on the probabilities of the various processes occurring which makes it necessary to make the following statements rather cautiously. Collisions between electrons and molecules are treated in detail by J. D. Craggs and H. S. W. Massey in vol. XXXVII of this Encyclopedia.

As the energy of electrons which enter the gas filled space (in zero electric field) is raised it is found that the average fraction of energy lost in a single collision is in general a great many times the value $2m/M$. Since this happens at relatively low energies (1 to 5 eV) it must be deduced that a fraction of this electron energy is used to induce vibrations and rotations in the molecule, though details of the latter process are not yet fully understood. With increasing electron energy the likelihood of these individual processes, which are quantized, will vary in a matter similar to those described above. At a certain still higher energy an electronic transition in the molecule will occur which brings the molecule into a higher state of excitation. Often these states are unstable with the result that the two atoms fly apart, the total energy being shared by the primary electron and the two atoms which usually acquire a certain amount of kinetic and potential energy. Following this dissociation one or both of the atoms can be in an excited state for a short time until they are able to radiate. At still higher electron energies an electron is removed from the molecule and a positive molecular ion is formed, which is stable. The likelihood of this process—its cross-section—as a function of the electron energy is again represented by a curve with a maximum.

5. The intensity of a beam of light which passes through a gas is known to decrease along its path because a certain fraction of the beam's energy is scattered sideways without heating the gas. Quantum theory expresses the classical (RAYLEIGH) scattering process by the number of quanta which through interaction with a sufficiently large number of molecules are deflected in various directions without loss of quantum energy (see vol. XXV of this Encyclopedia).

When the quantum energy is increased above a critical value, corresponding to the excitation energy of the atom or molecule, the energy can be absorbed by the particle and re-emitted a short time later in a general direction. In order that this process shall be effective the quantum energy must be almost exactly equal to the critical energy. Small deviations to higher or lower energies reduce the likelihood of excitation by absorption of quanta to negligibly small values. The data for the probability of this process—usually given in photo-absorption cross-sections—are mostly of recent date. All that can be said here is that in general for such a quantum the target represented by an unexcited atom appears as if it were of the order of 10^{-2} (or less) of the area of the first BOHR orbit.

When the energy of the quanta is further increased, it is found that at values corresponding to the ionization energy the quantum is absorbed and an electron ejected from the atom. The likelihood of this photo-ionization process is again of the order indicated above, which is equivalent to saying that out of 100 to 1000 quanta striking an atom one is able to ionize it. However, unlike the case of excitation, the likelihood of ionization does not vary much with the quantum energy over a considerable range of energies. It is to be noted that for most gases this region lies in the vacuum ultra violet. A further increase in the quantum energy brings us into the soft X-ray region in which (disregarding the discontinuities at the absorption edges) the probability of ionization decreases as the energyis increased. Quantitative results are to be found in G. L. WEISSLER's contribution part A on photo-ionization in this volume.

6. This short account would be incomplete if the following processes were not mentioned. First of all, excited and in particular metastable atoms and molecules seem to play in many cases a prominent rôle in the ionization of gases. Excited atoms can be ionized by electrons or quanta of low energy, and electrons can draw energy from excited atoms by collisions of the second kind. In certain cases collisions between ions and atoms, between ordinary atoms, and between excited atoms can lead to ionization. The cross sections for some of these processes are known [*15*].

7. A few numerical data should be given to illustrate the probability of ionization by electrons colliding with atoms and molecules. In Fig. 2 are shown curves for various representative gases and vapours giving s_e, the number of ion pairs produced when an electron of kinetic energy K passes through 1 cm. of the gas assumed to be at a density corresponding to 1 mm. Hg and 0° C.

This means that the ionization is as if the electron possesses an energy K along the whole length of 1 cm. Also the fact that instead of one, perhaps two or even three electrons are removed in a collision (an effect which occurs at higher energies) is allowed for. Thus the number of ion pairs given by the ordinate is the number of electrons liberated.

As can be seen from Fig. 2 many of the gases with higher ionization potential have a lower maximum value ($s_{e\,\max}$) though this rule is not applicable in all cases. It should be mentioned that the values given on the ordinate are numerically equal to the ionization cross sections in cm.²/cm.³. The cross section for a

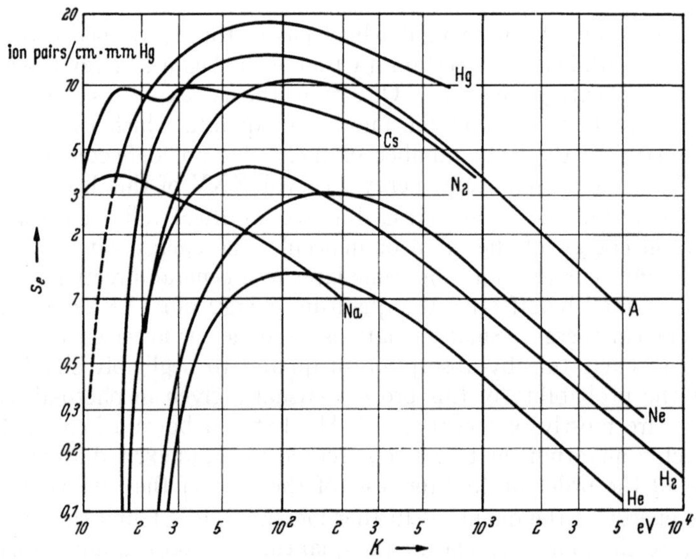

Fig. 2. Ionization efficiency s_e of electrons of uniform energy K as a function of their energy K in various gases. Note that the maxima of Cs and Na lie at considerably lower values of K than the maxima of the other gases.

single collision leading to ionization can be obtained from the values of s_e given here by dividing them by 3.6×10^{16}. In order to obtain the cross sections in units of the area of the first Bohr orbit the ordinate values are to be divided by π.

III. The problem.

8. From what has been said in I, it appears that in a gas containing free charges the application of an electric field does not only separate and remove charges of opposite signs to the respective electrodes but leads also to an increase in the number of charged particles by ionization by collision. The main ionization processes have been summarized in II. There is, however, a fundamental difficulty which has to be explained before the detailed treatments can be presented.

Consider electrons moving through a gas acted upon by a uniform electric field after having travelled a considerable distance. If the field intensity is not too high, a large number of electrons has suffered elastic collisions as a consequence of which they have been scattered, as explained above. This means that after a collision some of the electrons travel first against the field but soon their path is bent by the field until they finally travel parallel to the field again, unless their journey is interrupted by a collision with another molecule. Other electrons may have started with a strong side component which enables them to be moving sooner in field direction. The former group would pick up a smaller amount of energy between two elastic collisions than the latter. The result of all these scattering processes is that the electrons acquire a distribution of velocity or energy which under certain circumstances can be nearly of the same form as that of the molecules of a gas. For details, cf. W. P. ALLIS' contribution in this volume.

In the steady state therefore, there are present a large number of electrons of low and medium energy and a relatively small number of electrons of large energy, which may be loosely defined as the number which is able to ionize the gas. These fast electrons create new positive ions and electrons by collision thereby falling back into lower ranges of the energy distribution, but at the same time all charges are continuously accelerated by the electric field. Since equal numbers of positive and negative charges are produced in the gas the total net space charge is zero.

Provided the electric field is not too strong, one can easily visualize the steady state in which the losses by elastic scattering (and the inelastic losses) are balanced by the energy which the electrons take from the field. For strong fields the difficulty of developing a theory is not in the balancing of energies but rather of determining the energy distribution of electrons; in still higher fields the inelastic collisions of ions may have to be taken into consideration. But before entering into a discussion of the more complex cases some useful concepts associated with the motion of electrons in a gas will be introduced.

Again, a moderate electric field being assumed, the electrons moving in a gas be treated as a swarm of particles which have both a random and a drift velocity. The *random velocity* of the electron swarm—equivalent to the mean energy or electron temperature—is the velocity at or near the peak of the distribution curve, whereas the *drift velocity* is the average velocity with which the centre of the swarm moves in field direction. These two velocities are of different order of magnitude; the drift velocity is in few cases more, but often less than 10^{-3} of the random velocity. It is only because of the distribution of random velocities that there are electrons which have energies above that necessary for ionization. On the other hand only those electrons moving down the field and contributing to the drift velocity can drawn energy from the field. It has to be remembered that the distinction between random and drift velocity is only for mathematical convenience and has no physical significance. Thus it will serve no useful purpose when the drift velocity becomes comparable with the random velocity.

The electron swarm in a field can be regarded as a hot electron gas which is mixed with the cold neutral gas. This concept of a gas mixture with components at different temperatures is the most fruitful one which was introduced by TOWNSEND at on earlier stage. It may be asked why these two components do not exchange energy readily and so attain a common temperature. The answer is that as long as the average energy acquired by one electron between two collisions is sufficiently large compared with the average energy lost in an encounter, the exchange of energy between the electrons and the neutral gas is small and hence a large temperature difference can be maintained.

As the electric field is increased more and more, there will be not only a large increase of elastic collisions between electrons and gas molecules, but a directional effect sets in. So far the random velocity has been assumed to be the same in any direction because the drift velocity represented only a very small fraction of the former. In stronger fields however, the velocity of electrons acquired in field direction approaches gradually random velocity with the result that the electron energy distribution becomes anisotropic.

So far the positive ions have not been taken into consideration. The currents flowing through the gas have been assumed to be small enough to neglect space charge effects, electron-electron interaction or electron-ion interaction. When the electric field becomes strong enough, that is when the ion can pick up a sufficiently large amount of energy between two collisions, then inelastic collisions between ions and molecules can take place. Thus the ions can also ionize gas molecules, as has been shown in canal ray studies at the beginning of this century and this effect can contribute markedly to ionization [*20*]. Moreover fast ions have been shown to excite molecules by collisions.

It should be emphasised that the controlling parameter which determines ionization in the field is not simply the strength of the field but the average energy which an electron (or ion) acquires between two gas kinetic collisions. This concept led finally to the similarity rules which are discussed by G. FRANCIS in vol. 22 of this Encyclopedia. The dependent variable in nearly all the treatments which follow is, therefore, electric field per unit density of gas, or for constant temperature the field per unit gas pressure.

In the following sections the theory of the so-called ionization coefficients for electrons and ions in gases will be given as well as the experiments and their results. The ionization coefficient is the number of ion pairs which an electron produces on the average when moving unit length in field direction, in a gas of unit density. It depends on the average potential the electron has fallen through between two collisions and on the nature and state of the gas. The foundations of this approach and the first measurements are associated with J. S. TOWNSEND and his research school.

B. Ionization in uniform electric fields.

I. The electron multiplication.

9. It has been shown in Fig. 1 that a current flowing in a feebly ionised gas can be considerably increased when subjected to a sufficiently strong electric field. There are two kinds of experiments which have been devised to investigate

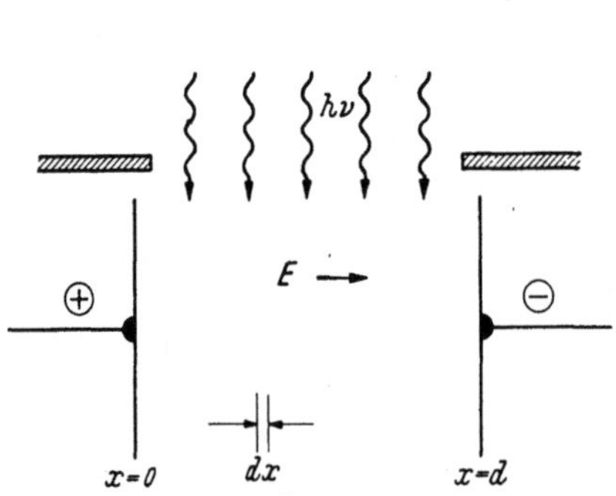

Fig. 3. Multiplication of charges when the gas is irradiated ($h\nu$) so that it is uniformly ionized.

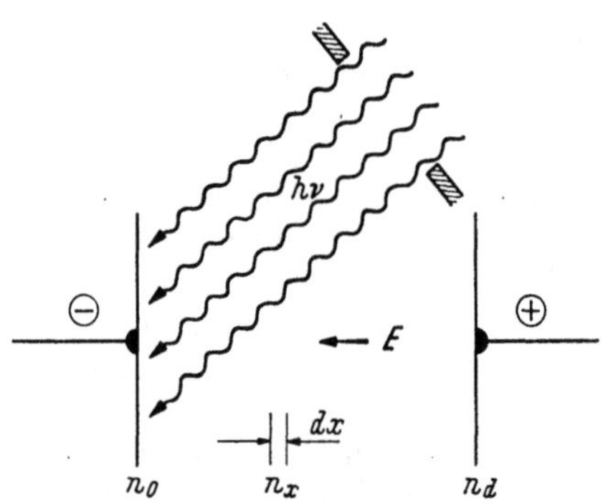

Fig. 4. Multiplication of charges when the cathode is irradiated ($h\nu$) to produce photoelectric emission. The radiation does not ionize directly the gas.

the increase of current with field. In Fig. 3 the gas between the electrodes is uniformly irradiated with X-rays or rays from another source, in Fig. 4 the cathode only is illuminated ultra violet light or other radiation and emits photoelectrons. This early work is connected with STOLETOW [*5*], KREUSLER [*6*], VON SCHWEIDLER [*7*], RIGHI [*8*] and TOWNSEND [*9*].

A most striking result emerged from STOLETOW's measurements with the arrangement of Fig. 4. At constant electric field and constant irradiation the current flowing in the circuit depended strongly on the distance between the electrodes. Since he used a potential difference of less than 100 V and a distance of only about of 0.1 cm. the accuracy of the measurements was inadequate. Thus the exponential relation between multiplication and inter-electrode distance was not discovered at that stage.

The multiplication in the steady state is conveniently defined as the ratio of the current flowing when the electric field is applied to the current when the field is practically absent, the irradiation being kept constant. (A small field is necessary to measure the photoelectron current.) The multiplication is also equal to the ratio of the number of electrons per second entering the anode with and without electric field respectively, or to the ratio of positive and negative charges crossing the cathode surface when the field is applied or absent respectively. To define the multiplication as a ratio of current densities is strictly speaking not correct since an illumination of only a part of the cathode releases photoelectrons which in the process of being multiplied move laterally because of their thermal motion. The anode area covered by the arriving electron swarm is thus larger than the emitting area at the cathode and this is the more so the lower the gas density and the field. (The spreading is amplified when certain secondary processes set in; see articles by P. F. LITTLE in this volume and by F. LL. JONES in vol. XXII.

Returning to the first experiments in which a multiplication of charges was observed depending on the distance between the electrodes, it should be mentioned in passing that this was first explained by J. J. THOMSON [*10*] as being due to ions generated by the light in a gas layer near the cathode. Though this would account for the growth of current with the field, it is more difficult to make the dependence on the electrode distance plausible. The hypothesis proved, however, to be untenable because a change in polarity would require that a small current remains flowing through the gas whereas in fact the current became zero when the field was reversed. STARK [*11*] suggested that the gas molecules are ionized by a "catalytic" effect of the electrodes; this too was disproved later by TOWNSEND [*14*].

The other hypothesis is based on an increase of the numbers of free charges in the gas by ionization. TOWNSEND suggested that negative ions are produced by light falling on the cathode or passing through the gas and that these ions would acquire speeds in the field which are sufficientla high to ionize gas molecules. Only at a later date he recognized that free electrons are the effective agents. On the other hand the ionization energy of "air" was found to be 175 eV according to RUTHERFORD and MCCLUNG [*13*] but TOWNSEND ([*14*], p. 263) states that by allowing for a numerical error the correct result should read 87 eV. Since potentials of 50 V were used in STOLETOW's measurements, ionization by ions was ruled out. As we know now the ionization potentials of gas atoms and molecules are much lower than first thought and the particles mainly responsible for ionization by collision are electrons. Furthermore, the applied potential between the electrodes is not a measure of the energy of the charged particles when a gas is present since a distribution in energy of the particles is obtained. Also, unless special experimental precautions are taken the energy distribution is cut off at higher energies at a value corresponding to the applied potential.

10. In order to obtain the simplest dependence of the current on the distance between the electrodes, the arrangement of Fig. 4 will be discussed first. If the number of photoelectrons released per second from the cathode is $n_0 = \text{constant}$, the electric field $\boldsymbol{E}$ and the position of cathode and anode at $x=0$ and $x=d$ respectively, then an electron moving in the gas produces α electrons and positive ions per cm. of its path in field direction. Thus with n_x electrons at x and $\alpha\,dx$ ion pairs produced in dx, we have an increase dn in number of ion pairs given by

$$dn = n_x \alpha\, dx. \tag{10.1}$$

Since $n=n_0$ at $x=0$, integration gives the multiplication

$$\frac{n_x}{n_0} = e^{\alpha x} \tag{10.2}$$

provided α is independent of x and n_x. As we shall see later this is not always the case. Since the number of electrons per sec. crossing the plane d is proportional to the total current, we have for $x=d$

$$\frac{i}{i_0} = e^{\alpha d} \tag{10.3}$$

where i_0 is the photoelectric current from the cathode. Eq. (10.3) states that the logarithm of the multiplication is proportional to the distance between the electrodes provided $\boldsymbol{E}$ and p are kept constant (Sect. 15a). The slope of this straight curve is a measure of α. It will be shown that (10.3) holds over a wide range of d and i (Sect. 20).

The above derivation assumes that ionization in the gas can take place at any x, provided $0<x<d$,and is as it were smeared out over the whole gas volume. A cruder picture assumes that all electrons start at the cathode with negligible initial speed and after having gained energy in the field ionize at the same distances x_1, x_2, etc. which are always δ cms apart. Hence for d cm. electrode distance the number of ionizations is

$$z = \frac{d}{\delta}. \tag{10.4}$$

Since each electron produces thereby one electron and one positive ion, i.e. the number of electrons is doubled, we obtain with n_0 electrons at the cathode the multiplication

$$\frac{n_d}{n_0} = 2^z. \tag{10.5}$$

Comparing this result with (10.3) it is seen that αd can be regarded as the average number of ionization along d. When ionization can take place at any x, which will happen because of the distribution of free path and for other reasons, we find in the result instead of the base 2 the base $e=2.718$.

Up to now only the ratio of the electron current at the anode to the photoelectric current has been discussed. What is the contribution of the ion current? To answer this question we have to digress into the theory of charge transport. When a molecule is ionized at x and the electron so produced subsequently enters the anode then a current flows in the circuit which is the result of the displacement of charge e from x to d. This displacement induces an image charge at the anode and cathode whose variation can be easily derived. Assuming that e follows the direction of the uniform field E, by equating the work done by e to the energy taken from the constant potential source, we have (Fig. 5)

$$V\, i_t\, dt = e\, E\, dx = V\, dq \tag{10.6}$$

or

$$\frac{\Delta q}{e} = \frac{\Delta V}{V} = \frac{d-x}{d}. \tag{10.7}$$

This means that the fraction of the charge (in units of e) which is induced at the anode corresponds to the fraction of the total distance travelled. The displacement of the positive ion from x to the cathode would bring the total charge conveyed to e. (This is not in contradiction with the indivisibility of e.) Further applications of this theorem see [*15*] and [*17*].

Applying this theorem to our problem it means that, by counting the number of electrons entering the anode, we have tacitly assumed that an equal number of positive ions (minus n_0) have entered the cathode during that time interval. Since (10.3) describes a steady state, we must obtain the same current i at any point x: for each electron which leaves the anode $e^{\alpha d}$ electrons arrive at d and $(e^{\alpha d} - 1)$ positive ions arrive at the cathode. Hence at the anode the current is a pure electron current $i = i_0 e^{\alpha d}$, whereas at the cathode

$$i = i^+ + i^- = [i_0(e^{\alpha d} - 1)] + i_0 = i_0 e^{\alpha d},$$

i.e. the cathode current has a large positive ion and a small electron component. Since the drift velocity of ions in a given field is several orders of magnitude smaller than that of electrons it follows that the numbers of ions per unit length dx must be very large at the cathode to carry practically the whole current.

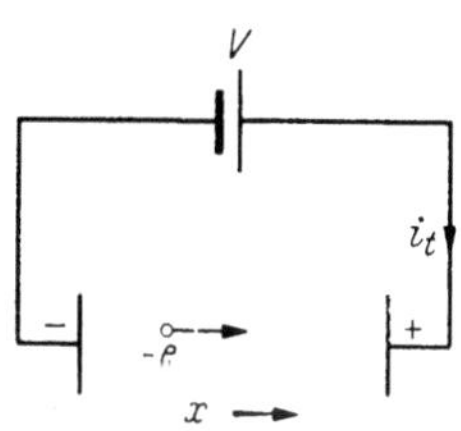

Fig. 5. To derive the instantaneous value i_t of the current caused by the motion of a charge $-e$.

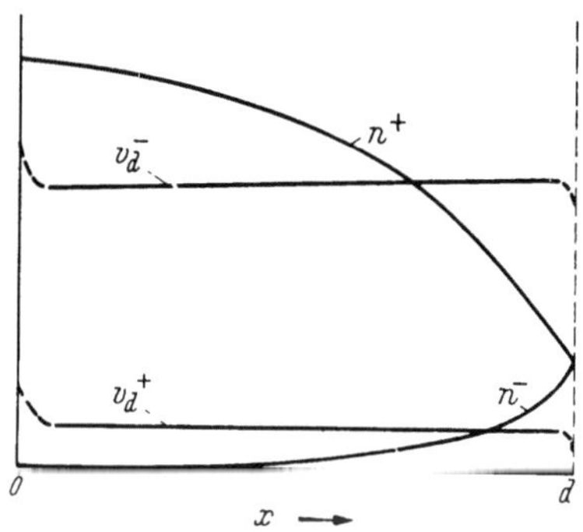

Fig. 6. Spatial distribution of positive and negative charges (n^+, n^-) and their respective drift velocities (v_d^+, v_d^-) in a gas through which a steady ionization current flows.

Fig. 6 shows schematically the spatial distribution n of ions and electrons (approximately equal to the concentration of charges) and their drift velocities which are to a first approximation independent of position.

11. At this point it is necessary to make one further observation. In order to obtain the number distribution of positive ions in space it is necessary to invoke the equation of continuity. In other words it is not possible to treat the positive ions as we did the electrons because whereas the electrons multiply and carry the current the ions are simply charge carriers—as long as their effect on the applied field or their ionizing power is neglected. In the steady state both ions and electrons are produced in the gas at the same rate and must enter the electrodes at this rate. Thus the current at the cathode and the anode must be the same, viz.

$$\frac{i}{e} = n_0^- v^- + n_0^+ v^+ = n_d^- v^- + n_d^+ v^+. \tag{11.1}$$

Also

$$n_d^- = n_0^- \cdot e^{\alpha d}, \tag{11.2}$$

and thus

$$\frac{n_0^+ - n^+}{n_0^-} = (e^{\alpha x} - 1)\,\frac{v^-}{v^+}. \tag{11.3}$$

Hence the ions increase considerably in number on the way from the anode to the cathode because both the velocity ratio and the multiplication are general large numbers. E.g. in H_2 at $\frac{E}{p} = 10\ \frac{\text{volts/cm.}}{\text{mm. Hg}}$, $v^+ = 10^5$ cm./sec. and $v^- = 4 \cdot 10^6$ cm./sec.; also when E/p large, $e^{\alpha d} \gg 1$. If the electrodes are small, the ion concentration may well become so large that a distortion of the field by positive space charges results. The value of n^+ very near to the anode is not easily determinable: $n_d^+=0$ at the anode surface but n^+ near the anode in the gas rises rapidly at a rate given by the electron distribution which itself is disturbed by the presence of the anode. Fig. 6 shows schematically $n^+=f(x)$ which follows from (11.3) viz.

$$\frac{n_0^+ - n^+}{n_0^-} = (e^{\alpha x} - 1)\frac{v^-}{v^+}. \tag{11.4}$$

It now remains to discuss the multiplication obtained in the case of Fig. 3. Here N_0 ion pairs per cm.³ are uniformly distributed throughout the gas. Consider the position of electrons which (for mathematical convenience) are here assumed to move from any x to $x=0$, the anode. Then, since each electron produces $e^{\alpha x}$ ion pairs on its way to the anode and $N_0\,dx$ electrons are produced by irradiation in the space dx, the total number of ion pairs between d and 0 (dx being negative) is

$$\int_d^0 N_0\, e^{\alpha x}\, dx = \frac{N_0}{\alpha}\left[e^{\alpha d} - 1\right], \tag{11.5}$$

and thus the multiplication defined by the ratio of (10.3) and the number of ion pairs caused by irradiation, i.e. $N_0\,d$, is

$$\frac{i}{i_0} = \frac{1}{\alpha d}\left[e^{\alpha d} - 1\right]. \tag{11.6}$$

Both (10.3) and (11.6) have been experimentally tested. This will be discussed in Sect. 15.

II. The electron avalanche.

12. Consider a single electron which is released from the cathode. When its motion through the gas is controlled by a uniform electric field of sufficient strength, then its advance is accompanied by ionization in the gas. Thus the number of electrons increases towards the anode, and if this process occurs so quickly that the positive ions remain essentially stationary, the number of electrons rises with position and time analogous to an avalanche. However, the number of electrons must not become so large that the electric field becomes distorted or other effects set in which invalidate the simple picture. Thus, provided the theory is applied within certain experimental limits, useful results can be obtained from such considerations concerning the rate of rise of current and the changes of other discharge parameters [*101*].

Consider the charge q in the circuit in Fig. 5 to vary with position as a result of ionization and field. The positive ions formed in the gas are assumed to remain at rest. In a field E the charge grows according to $q_x = e\cdot n_0\cdot e^{\alpha x}$, and since

$$V\,i\,dt = q_x E\,dx \tag{12.1}$$

and $x=vt$ the current i at any instant is given by

$$i = q_0\frac{v}{d}\,e^{\alpha v t} \propto i_0\, e^{t/\tau} \tag{12.2}$$

where q_0 is the charge at $x=0$ and v the drift speed of q_x. The time constant of the exponential rise is thus

$$\tau = \frac{1}{\alpha v}. \tag{12.3}$$

Applying this to electrons, in molecular gases $v^- \propto E/p$ approximately, and since for a given E/p a value of α/p follows, we find in agreement with experience that

$$\tau p = \text{const}. \tag{12.4}$$

Since, as will be seen later, $\alpha/p \propto e^{-\frac{p}{E}}$, we find that approximately

$$\tau p \propto e^{+p/E}. \tag{12.5}$$

In H_2 at $E/p = 200$ V/cm per mm. Hg, $v^- \approx 10^8$ cm./sec. and $\alpha/p \approx 2$. Hence $\tau \cdot p \approx 10^{-8}$ sec. Thus at $p = 1$ mm. Hg the current $i_0 = 10^{-12}$ A will rise to 10^{-6} A in several 10^{-7} secs.

A calculation can also be done about the time constant associated with the rise of current in a uniform field when secondary effects at the cathode are included ([*17*], p. 178). In this case besides α/p the drift velocity of positive ions is a decisive factor and because v^+ is much smaller than v^- larger values of τ ensue. This problem, however, is beyond the scope of this chapter.

III. The classical theory of ionization by collision.

13. Any theory of ionization in an electric field attempts to derive relations between the ionization coefficients, i.e. the number of ion pairs produced when an electron (or ion) has moved through unit length in the direction of the field and the pressure, the temperature, the nature and the state of the gas, the field strength, and other parameters as we shall see later. In what follows it is assumed that the electrons, which are regarded as the sole producers of ionization, have collided with gas molecules a great many times before an analysis of their average ionizing power in the gas is made. This means that any disturbances caused by electrodes and walls are excluded. Also only those gases are considered here which do not attach electrons.

It has been mentioned before that the controlling factor in ionization by collision is the average energy an electron has gained along a mean free path in field direction. If E is the field strength and λ_e the electron mean free path, $eE\lambda_e$ is the average energy which is proportional to E/p provided $T = \text{const}$. This can also be expressed in the form that the fraction of the number of encounters between an electron and the molecules leading to ionization will remain unchanged if a change in the density of the gas is accompanied by a corresponding change in field so that the average electron energy is the same. Since each electron ionizes (or collides) independent of the presence of the other charged particles and since the likelihood of a single ionization depends on the gas density, it can be assumed that the same holds here where a swarm of electrons moves through the gas. If α is the ionization coefficient for electrons—TOWNSEND's first ionization coefficient, as defined above—then $\alpha \propto p$ for a given E/p or

$$\frac{\alpha}{p} = f\left(\frac{E}{p}\right). \tag{13.1}$$

The classical treatment of $f(E/p)$ will now be presented [*14*].

It may be assumed that an electron can and will ionize a gas molecule when it has acquired an energy $\geq eV_i$ at the end of its journey. Supposing the electron has started with zero energy after the last collision—which in fact it is not likely to do—then in a field E only those electrons can ionize which have travelled a distance x in field direction given by

$$eEx \geq eV_i. \tag{13.2}$$

From kinetic theory it is known that the fraction $f(x)$ of the number of free paths exceeding x is given by

$$f(x) = \mathrm{e}^{-\frac{x}{\lambda}} \tag{13.3}$$

where λ is the mean free path. The number of (electron) mean free paths per unit length which exceed x is

$$\left|\frac{df(x)}{dx}\right| = \frac{1}{\lambda}\,\mathrm{e}^{-\frac{x}{\lambda}} \tag{13.4}$$

which is equal to the number of ionizing collisions per unit length, provided every such collision produces ionization. Assuming (13.1) to be an equality and inserting x from (13.2) in (13.4) and remembering that $\lambda = \frac{c_1}{p}$, we obtain

$$\frac{\alpha}{p} = c_1\,\mathrm{e}^{-\frac{c_1 V_i}{E/p}}. \tag{13.5}$$

Two points are striking: c_1 seems to follow straight from kinetic theory, and also the ionization potential V_i of a gas should be found from measurements of α and E at various p. c_1 is the number of molecules an electron encounters and ionizes when it moves 1 cm. distance parallel with the field. Hence, $1/c_1$ is the mean free path λ_e for ionization in that sense. The geometrical meaning of c_1 is the "saturation value" which α/p reaches when $E/p \to \infty$ (Fig. 7).

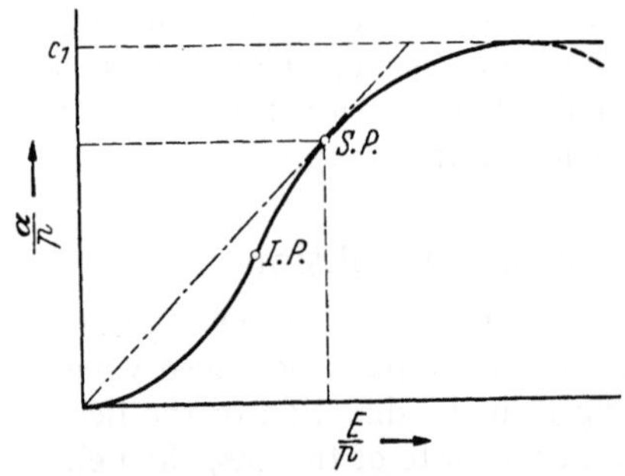

Fig. 7. Electron ionization coefficient α/p (Townsend's first coefficient) as a function of the electric field E/p. p is the equivalent pressure (the density of the gas). *I. P.* inflexion, *S. P.* Stoletow point, see Table 3, c_1 maximum of α/p.

Originally it was thought that $\lambda_e = 4\sqrt{2}\,\lambda$ where λ is the gas kinetic mean free path for molecules. For H_2 one expects $\lambda_e = 4\sqrt{2}\cdot 8.5\times 10^{-3}$ cm. at $p = 1$ mm. Hg so that $c_1 = 20$. From total cross section measurements $\lambda_e \approx 2.5\times 10^{-2}$ cm. so that $c_1 = 40$. This is to be compared with the value $c_1 \approx 5$ as is found from observations. However, the discrepancy between theory and experiment is not surprising: neither does the electron move parallel to the field as assumed nor does every collision give rise to ionization.

Moreover, attempts to use measurements of α and E to determine V_i were doomed to failure. Observations which make use of the medium range of E/p give values of V_i between 17 and 28 V for different gases i.e. up to 80% too high. But one has to remember that these measurements have been made around 1900 and that the results represented a great improvement compared with the data gained from older radioactivity experiments (Sect. 9). The reason is, as we shall see later, the neglect of the ionization probability in conjunction with the energy distribution of electrons i.e. the number of electrons having energies above ionization potential. The assumption of the classical theory are too crude to yield more precise results, and are not applicable to a wider range of variations.

It is interesting to find that Townsend was aware of certain restrictions of his theory which lead to eq. (13.5). In particular he thought it not applicable to values of $E/p < 100$ and states that at very high values of E/p (several 1000) α/p appears to decrease as E/p is raised. He has also found very early that the purity of the gas is an important factor particularly in rare gases [*14*].

14. One of the earliest observations on the variation of the ionization current with the gas pressure by Stoletow [*5*], Righi [*8*] and others gives a convincing proof of the correctness of the general ideas on which the classical theory of ionization by collision is based. In Fig. 8 STOLETOW's measurements are given of the current (in arbitrary units) which was flowing through air between two plane parallel plates when the pressure was varied between 10^{-2} and 10^3 mm. Hg. The cathode was irradiated with light, the potential difference V and inter-electrode distance d was kept constant. The main result of this experiment is that as the pressure is increased the current passes a maximum. This can be easily understood in terms of the theory presented above. A second result is that at the lower end of the pressure range the current is high whereas at the higher end it is low.

We write (13.5) in the form

$$\frac{\alpha}{p} = A\, e^{-\frac{B}{E/p}}. \qquad (14.1)$$

Since besides the irradiation, V and d are constant, E is constant and thus from (10.3) and (14.1) we have

$$\ln\frac{i}{i_0} = (A\,d)\,p\cdot e^{-\left(\frac{B}{E}\right)p}. \qquad (14.2)$$

This function has two factors, one increasing with p and one decreasing with p, and so i passes a maximum. Its abscissa can be obtained by differentiation giving

$$(p\,d)_m = \frac{V_m}{B}. \qquad (14.3)$$

Using the data given in Fig. 8 and the constant B for air, given later in Table 1, it is found that theory and experiment agree well. The value of p_m (about 1 mm. Hg at $d=0.4$ cm.) was found to remain unchanged when $E=V/d$ was kept constant but V varied between 66 and 236 V, again confirming the theory. The observed current at the maximum was about $(i/i_0)_m = 2.3$ whereas with the constant A and B for dry air (Table 1) a value of ≈ 4 is found.

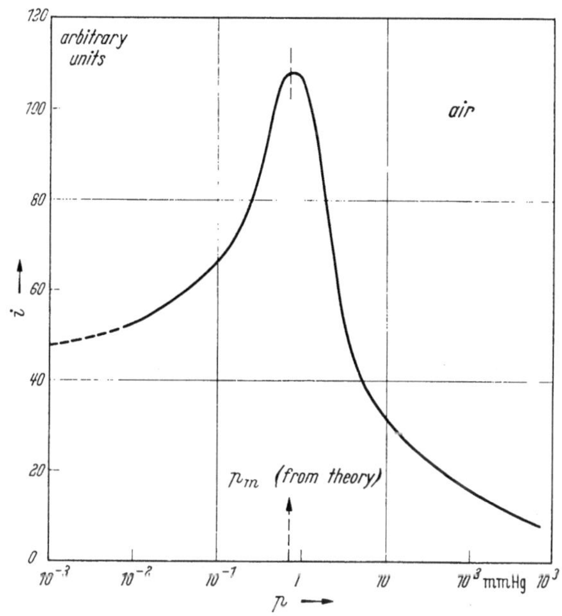

Fig. 8. Ionization current i in air as a function of the pressure p, the (uniform) field and the irradiation of the cathode being kept constant.

Having thus answered the formal part in a satisfactory way the physical picture has to be drawn. At low p an increase in p increases the chance of an ionizing collision though E/p gradually decreases. This is because for large values of E/p, α/p does not change appreciably and thus α and i rise as p rises. When p becomes sufficiently large E/p will decrease with rising p, thus α/p decreases rapidly and α and i fall. In this range the elastic losses which the electrons suffer by colliding with molecules, become larger. The energy an electron acquires between two collisions will become smaller as p rises and so fewer ionizing collisions will occur.

(14.3) can also be expressed in the form $B=(E/p)_m$ and by inserting it in (14.1) one obtains the maximum value of α from the expression $(\alpha/E)_m = 2.72\,A/B$. This ratio is found by drawing the tangent from the origin to the curve of Fig. 7. The inverse of this ratio is called the STOLETOW constant (Table 3); it represents the optimum value of E/α i.e. the least energy per ion pair produced.

Next the discrepancy between the larger calculated value of i/i_0 at $p=p_m$ and the observed current will be discussed. At 0.7 mm. Hg the mean free path of an electron is of the order $\frac{1}{2}$ mm. and with a gap width of 3.7 mm. about 8 collisions occur in the gap. Since an electron starts at the cathode usually with an energy of ≈ 1 eV and since it ionizes when its energy exceeds 15 eV (applied potential 93 V) it is clear that the effective number of collisions is reduced to about 6 (11.5 eV per collision). Some the of electrons will not produce ionization in the gas and will give up their energy to the anode. The majority may never acquire random velocities necessary for a possible application of the theory of ionization coefficients.

Another effect may contribute to the divergence between the maximum current as calculated be means of the ionization coefficient and the observed value. This effect is the back-diffusion. Assuming that the current observed at very low p is the photoelectric emission current i_0 in vacuum, one would expect the same current to flow at very high p where multiplication in the gas ceases. However, only $i_0/10$ or less is found at 1 atm (Fig. 8). According to J. J. THOMSON [18] the emission current from a cathode in a gas is greatly reduced as the pressure rises. The electrons which start with a small initial velocity are accelerated by the electric field but collide soon with gas molecules thereby acquiring a random velocity so that some of them are scattered back to the cathode. This is the more likely the higher the pressure of the gas and the lower the field. The emission current in the gas per unit charge and unit area is

Table 1. *Constants A and B in eq. (14.1).*

Gas	A ion pairs / cm. × mm. Hg	B V / cm. × mm. Hg	Range of validity E/p V / cm. × mm. Hg	V_i in V
H_2	5	130	150– 600	15.4
N_2	12	342	100– 600	15.5
O_2	—	—	—	12.2
CO_2	20	466	500–1000	13.7
air	15	365	100– 800	—
H_2O	13	290	150–1000	12.6
HCl	25	380	200–1000	—
He	3	34 (25)	20– 150 (3–10)	24.5
Ne	4	100	100– 400	21.5
A	14	180	100– 600	15.7
Kr	17	240	100–1000	14
Xe	26	350	200– 800	12.1
Hg	20	370	200– 600	10.4

$$\frac{j}{e} = n_0 - n_b \tag{14.4}$$

where n_0 is the number of electrons per cm.² and sec entering vacuum and n_b the number back-scattered. Further if v_r is the mean value of the electron velocity and since $j=eNv_d$, where N is the number of electrons per cm.³, the number of electrons diffusing back is

$$n_b = \frac{N v_r}{4} = \frac{j}{4e}\left(\frac{v_r}{v_d}\right) \tag{14.5}$$

where the bracket contains the ratio of random to drift velocity. The ratio of gas to vacuum current density is therefore

$$\frac{j}{j_0} = \frac{1}{1+\frac{1}{4}\frac{v_r}{v_d}} \tag{14.6}$$

where $j_0 = e n_0$.

Since in air at $E/p = 250/760 \approx 0.35$ V/cm., mm. Hg., $v_d = 7.3\cdot 10^5$ cm./sec. and $v_r = 2.1\cdot 10^7$ cm./sec., one obtains $j/j_0 \approx 0.12$ which is about right (see Fig. 8) particularly if the second effect above would have been allowed for. Summarizing it can be stated that the variations of the ionization current with pressure can be explained quantitatively with a sufficient degree of accuracy, particularly if the standard of experimental technique in 1890 is taken into consideration.

It has been pointed out earlier that (14.1) is not expected to hold at low E/p nor at very high E/p when drift and random velocities of electrons become comparable. However, it is convenient to obtain approximate values of α/p within a certain range and so the following Table 1 is given which contains the constants A and B in (14.1).

Table 1 shows that the coefficient A (c_1 in Fig. 7) which is the maximum value of α is small for the light rare gases and in H_2 and large for the heavy rare gases, Hg and the molecular gases. This is in accord with the ionization efficiencies (Fig. 2). The coefficient B is proportional to the average energy which an electron loses per ionizing collision [cf. (18.5) $B = (c_1/e)\, e V_i$] and is found to be small for the light rare gases and H_2 and large for example for the molecular gases. This again is qualitatively in agreement with observations.

IV. The methods of measuring ionization coefficients.

15. There exists a series of direct and indirect methods from which the electron ionization coefficient α in a uniform electric field can be obtained. The methods can also be divided into static and dynamic methods. The following list of them is probably not complete.

a) Measurement of α from the variation of current with the electrode separation d,

b) by measuring the breakdown potential as a function of p.d particularly the minimum sparking potential and the minimum value of p.d,

c) measuring the high frequency breakdown field at various pressures and electrode separations, and calculating α from it,

d) by measuring the longitudinal component of electric field in a positive column,

e) by calculating α from the observed ionization efficiency and a measured or calculated electron energy distribution,

f) by calculating the rate of ionizing collisions and measuring the electron drift velocities.

g) by analysis of the results of pulsed non-self-sustaining discharges.

h) by calculating α from the observed formative time lag of sparks.

Method a) will be discussed, including technical details, in Sects. 15 to 18. Of methods b) to h) a very brief outline is given in Sect. 19. Method g) gives instantaneous values of α. The evaluation of the observations, however, is difficult because under the conditions of experiment the primary and secondary ionization are not easily separable. Method h) gives only a rough average value of α.

a) Only one experimental method is known which gives directly absolute values of the ionization coefficient α. Its working principle is as follows: A cathode is illuminated so that photoelectrons are emitted from it, and a uniform electric field is established in the (often rarefied) gas between the cathode and an anode. The current i flowing in the circuit is measured for different distances x between cathode and anode, the field and gas density being kept constant. From the variation of current with distance α can be derived. This procedure is then repeated for other values of field and gas density.

The apparatus used by TOWNSEND is shown schematically in Fig. 9 [*14*]. Ultra violet light from a Hg discharge passes the clear quartz plate Q_1 and the silvered quartz plate Q_2—the anode—whose central part is ruled so as to allow radiation to fall on the central portion of the Zn plate which forms the cathode.

The cathode is kept at or near earth potential, the anode is connected to a battery of several 100 V; the exact value can be varied steadily. In order to prevent leakage currents, originating at Q_2 and flowing along the surface of it, from entering the electrometer circuit connected to the cathode, a "guard ring" at earth potential is provided. Since the electrode diameter is 4 cm. and $x \leq 1$ cm. the central part of the field between Q_2 and the Zn plate is uniform.

The currents (of order 10^{-13} to 10^{-10} A) can be measured either by means of an electrometer (or electrometer valve) or an electrostatic balance. If C is the

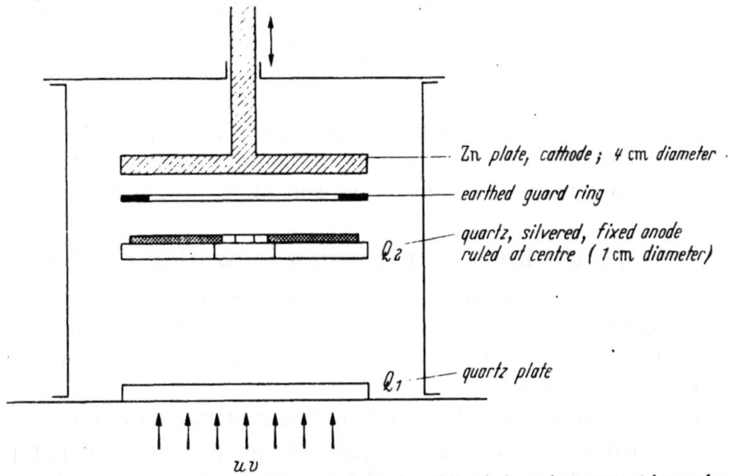

Fig. 9. Townsend's apparatus for measuring the electron ionization coefficient α in various gases at low and moderate gas pressure.

known capacity (Fig. 10) the current i is given by the time rate of change of potential difference $e_c (i = C \cdot d e_c/dt)$ measured by $V = - e_c$, the instantaneous potential for which El shows zero deflection, and the time t by a stop watch.

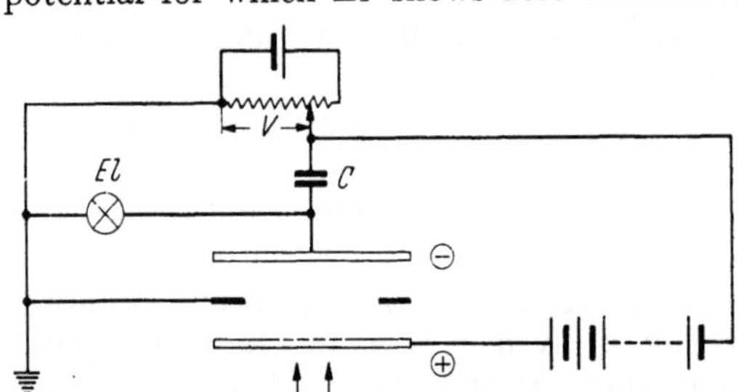

Fig. 10. Simple zero method (electrostatic balance) to measure ionization currents with an electrometer *El*.

When the current i is measured for different values of x, it is found that with a constant field and gas pressure

$$\frac{i_2}{i_1} = e^{\alpha (x_2 - x_1)} = e^{\alpha \Delta x} \qquad (15.1)$$

and the same ratio of currents is observed for any other pairs of x, provided Δx is the same. This procedure of measuring α by observing the ratio of currents which correspond to two inter-electrode distances x_1 and x_2 proved to be useful in that it avoids corrections which otherwise ought to be made. The electrons emitted from the cathode have a velocity distribution which is determined by the irradiating source and the properties of the emitting substances. Consequently the electrons must travel over a certain distance in the field until they can acquire (terminal) drift and random velocities in equilibrium with the field, and independent of position. Apart from allowing for a certain "relaxation distance" the results obtained by the above method are not free of spurious variations in emission of photoelectrons which may be due to changes on light intensity or chemical changes at the surface of the photo-cathode. Another source of trouble is the purity of the gas used; this will be discussed later.

A numerical example illustrates the theory of the apparatus. When filled with dry air at $p=1$ mm. Hg and when a constant field $E=350$ V/cm. was applied while x was varied, the following currents were observed (i in arbitrary units):

$x=0$	0.2	0.4	0.6	0.8	1.0	1.1 cm.
$i=(1)$	2.86	8.3	24.2	81	373	2250

Since α and $\Delta x (=0.2$ cm.) are constant, $e^{\alpha\Delta x}$ is constant and according to (15.1) i_2/i_1 should be constant. This is the case when x is not too large. Hence the primary process (α) is active for $x \lesssim 0.6$ cm. whereas at greater values of x a secondary process comes into play (cf. the following article by P. F. LITTLE). Instead of (15.1) we have then

$$\frac{i}{i_0} = \frac{e^{\alpha(x-x_0)}}{1-\gamma(e^{\alpha(x-x_0)}-1)} \tag{15.2}$$

where γ is the number of secondary electrons from the cathode liberated per incident positive ion plus other associated processes.

16. Another way of evaluating the measurements with an apparatus of Fig. 9 has been used by GOSSERIES [*19*]. It is known that a semi-logarithmic plot of $i/i_0 = e^{\alpha(x-x_0)}$ deviates from straight line when secondary processes set in (coefficient γ). This deviation becomes apparent at much lower d in rare gases than in molecular ones and also for lower E/p. Moreorer, there is a minimum distance x_0 which is the space needed for setting up a uniform steady state; x_0 is different under various conditions. In order to be able to find the values of these quantities the following procedure is adopted:

The complete equation on i/i_0 (15.2) is written in the form

$$\frac{1}{i_0} = \frac{1}{i}\,\frac{e^{\alpha(x-x_0)}}{1-\gamma(e^{\alpha(x-x_0)}-1)}\,. \tag{16.1}$$

Put $y=1/i$, $y_0=1/i_0$, and collect the constant terms containing x_0, γ and n_0. Then with $K - e^{\alpha x_0} y_0(1+\gamma)$ one obtains

$$y + \gamma y_0 = K\cdot e^{-\alpha x}. \tag{16.2}$$

Taking three values of x, namely x_1, x_2, x_3 corresponding to the inverse currents y_1, y_2, y_3 such that $x_2-x_1=x_3-x_2=\Delta$ then from (16.2)

$$\frac{y_1+\gamma y_0}{y_2+\gamma y_0} = \frac{y_2+\gamma y_0}{y_3+\gamma y_0} = e^{\alpha\Delta} = \frac{y_1-y_2}{y_2-y_3} \tag{16.3}$$

where the last expression follows from a theorem on ratios. Since the y's and Δ are known, α can be determined. Further K, γ, x_0 and y_0 can be found from (16.2). This method has been applied to measurements with H_2 for the medium range of E/p.

17. Up to this time all the apparatus for measuring α/p were sealed off against the atmosphere by grease, rubber or similar gaskets and wax. (Often oil vapour and Hg vapour from pumps and gauges was not frozen out.) Because of all these substances, it was not possible to clean and evacuate the system at elevated temperatures and so the purity of the gases used was not only unsatisfactory but changed during the measurements. Also it was not known earlier what kind and amount of impurities would cause in a particular case a change of the results. The introduction of glass to metal seals, of corrugated tubes for movable electrodes, of high frequency heating and of refractory methods was an important step in the development of the experimental technique.

An example of a modern tube (Fig. 11) built for the determination of α/p in pure argon is one developed by KRUITHOF and PENNING [20]. Cathode and anode consist of Cu. The cathode was movable by means of a screw and a corrugated bronze tube, the anode was highly insulated and contained a large number of holes to pass ultra-violet light from the 500 W quartz mercury lamp through a quartz window on to the cathode. Batteries were used throughout as potential sources. Cooling devices, shutters and gas inlet and purifier—the Ba coated vessel—can be seen in the diagram. The uniformity of the electric field was ensured by rounding off the edges of the electrodes and by a metal cylinder which prevented field distortion by wall charges. Leakage currents were collected internally and externally by a number of electrodes.

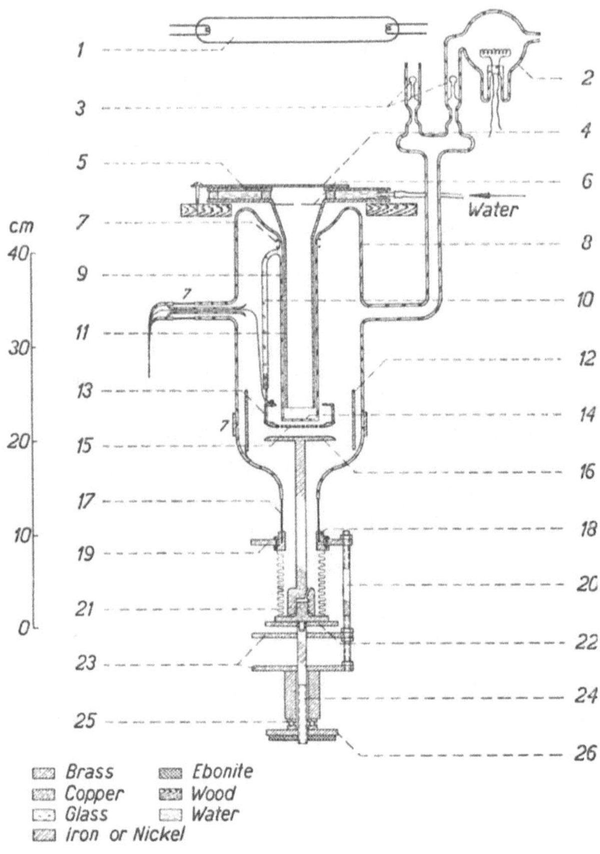

Fig. 11. KRUITHOF and PENNING's tube for measuring the ionization coefficient α in pure rare gases at $p \approx 1$ to 200 mm. Hg. *1* Quartz Hg lamp (500 W); *2* tube with Ba coated wall; *3* thin glass walls which can be broken magnetically; *4* cooled light shutter; *5* cooling device; *6* diaphragm; *7* electrodes for collecting leakage currents; *8* tube; *9* glass to quartz seal; *10* anode support; *11* Cu cylinder attached to 5 protecting 9; *12* electrostatic shield of Cu; *13* fixed anode; *14* quartz window; *15* holes in the anode (500, 1 mm. diameter); *16* movable cathode; *17* chrome-ion ring; *18* to *26* flexible parts and moving mechanism; liquid air traps and pressure gauges are not shown here.

A series of improvements in the measurements were made: Since the photoelectric emission varies during the readings due to the change of the work function and the photoelectric yield, allowance was made for it by repeating one of the measurements several times during a series of readings and correcting them. The contact potential between cathode and anode was measured by applying anode potentials of several volts and finding the potential at which the photo current decreased rapidly. This was observed to happen at about 1.6 V decreasing with time to about 0.4 V. The applied potential was then corrected by the contact potential.

The photoelectric current was found to depend on the inter-electrode distance. This is due to loss of radiation which is reflected from the cathode and passes through the holes in the anode. At larger distances radiation is scattered sideways and lost except for the part striking the cylindrical shield which emits photoelectrons; but this contribution accounts only to a few percent of that of the cathode. Other points are:

Electrons are reflected by the anode, but this makes itself felt only at very low values of E.

Back-diffusion of photoelectrons to the cathode in the presence of a gas is important at low E/p [18].

Field distortion becomes apparent at small distances when the current is found to decrease faster than expected as the distance is reduced. This is due to the field of the holes in the anode.

18. Whereas measurements in rare gases can be extended up to $E/p \approx 10^3$ V/cm. × mm. Hg and down to $E/p \approx 5$, there were until recently no reliable data on

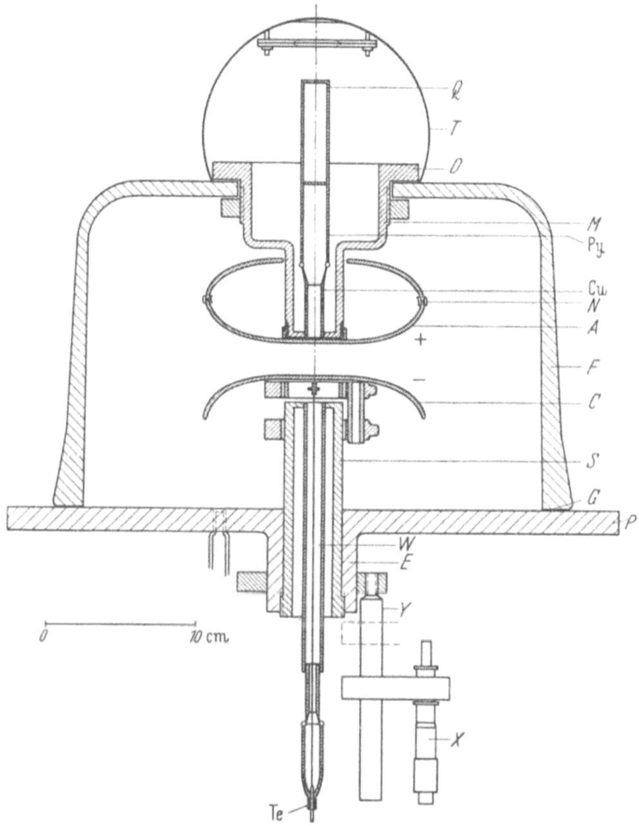

Fig. 12. Ionization chamber used by LLEWELLYN JONES *et al.* for measuring the ionization coefficient α in molecular gases at high pressures. *Q* quartz window; *T* positive high voltage terminal; *D* flange supporting *A*; *M* mild steel cylinder; *Py* pyrex glass tube with graded seal; *Cu* copper tube; *N* joint; *A* fixed nickel anode; *F* boro-silicate glass envelope; *C* movable nickel cathode mounted on quartz rods; *S* cylinder screwed into *P*; *G* rubber gasket; *P* base-plate; *W* cathode wire; *E* part of the base-plate; *Y* detachable mild steel framework; *X* micrometer; *Te* seal for wire *W*. Gas is admitted to the chamber through a hole in *P* on the left of the axis.

molecular gases at lower values of E/p say < 50. The ionization chamber used by LEWELLYN JONES, PARKER, DUTTON and HAYDON [21], [22] was designed so as to enable measurements of α/p at higher p (≤ 300 mm. Hg) electrode separations up to several cms. It is shown in Fig. 12. A glass bell jar was sealed on to a steel base with a rubber gasket coated with silicone grease. A well insulated nickel cathode of 15 cm. diameter was irradiated with ultra violet light through holes in the centre of the nickel anode of similar shape. A circle of 4 mm. radius was illuminated at the centre of the cathode where the field was uniform. Particular care has to be excercised with regard to the stability of the voltage which in these experiments was ≤ 40 kV with a ripple of $< 10^{-6}$ parts and slow fluctuations of $\approx 4 \cdot 10^{-4}$ parts. Corona discharges acting as fluctuating losses and sporadic ionization centres were avoided by graphite coating, plastic covers and spherical terminals.

The current was measured with a compensated electrostatic balance and electrometer (Fig. 13). With the switches S_1 and S_2 closed i.e. the quadrants earthed, and a certain electrode separation in C_1, a potential of say 100 V is applied. Then S_1 and S_2 are opened and C_3 changed until the least deflection of the electrometer needle is observed. Now with the switches closed the large potential is applied and again S_1 and S_2 opened and V_2 is varie so that zero electrometer deflection is obtained. The current i which flows in the ionization chamber after t secs is $i = V_2 C_2/t$. About 10^{-14} A could be measured to 10% with a normal Dolezalek electrometer. The variation of the leakage resistances of the screened wires through changes in humidity had to be eliminated by using heaters. It was also found that the photoelectric sensitivity could be stabilized by aging the cathode with a nitrogen glow discharge. The effect of sparks was to reduce the secondary coefficient over a part of the cathode surface. The results will be given in Sect. 21.

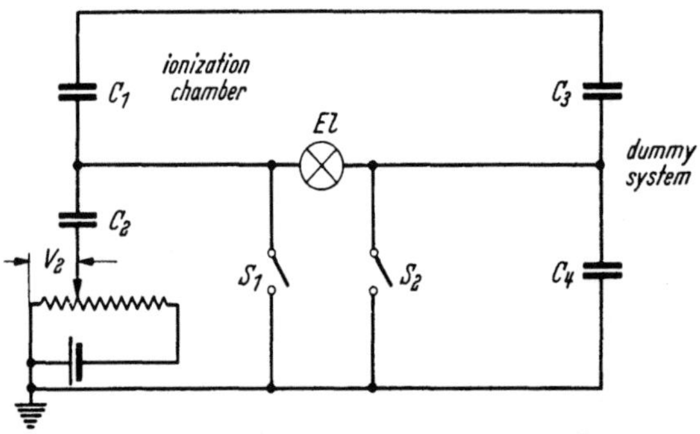

Fig. 13. Compensated electrostatic balance for measuring ionization currents with an electrometer *El* as a zero-instrument. The potential is applied between the upper electrode of C_1 and the earth. The electrostatic screens have been omitted in the diagram.

The apparatus described in the previous sections serve to test the validity of (10.3). It must be added that earlier work was also concerned with an experimental confirmation of (11.6). Satisfactory results were obtained [*14*].

Measurements of α/p in electro-negative gases require special precautions as regards impurities and interpretation. Geballe and Harrison [*45*], [*48*] measured α/p in O_2 at $p \leq 40$ mm. Hg and $d \leq 4$ cms. Assuming that the current is given by (28.5) and fitting this function to the experimental points, α and a i.e. the ionization and attachment coefficients for O_2 and other gases were found simultaneously (Sect. 21).

19. Let us now briefly outline methods b) to h) of Sect. 15.

b) When the voltage V between the electrodes of a gap is slowly increased the field and hence α and i will rise. Finally the denominator in (15.2) approaches zero and $i \to \infty$. (This state can also be reached by keeping E and α constant and increasing d up to the sparking distance.) If γ is constant and (14.1) inserted in $1 = \gamma(e^{\alpha d} - 1)$, the sparking potential V_s as a function of pd is found to be

$$V_s = \frac{B \cdot p d}{\ln\left\{\frac{A p d}{\ln(1 + (1/\gamma))}\right\}} \tag{19.1}$$

the minimum of this function corresponds to (e ≈ 2.72)

$$p d_m = \frac{\mathrm{e}}{A} \ln\left(1 + \frac{1}{\gamma}\right); \tag{19.2}$$

$$V_{sm} = \mathrm{e} \frac{B}{A} \ln\left(1 + \frac{1}{\gamma}\right). \tag{19.3}$$

By dividing, we find that with $\alpha/p = A \cdot e^{-\frac{Bpd}{V}}$

$$\left(\frac{V_s}{p\,d}\right)_m = B. \tag{19.4}$$

Assuming $A \approx s_{e\,\max}$ (Sect. 7) α/p is determined, provided that the various approximations hold.

c) The ionization coefficient α can be derived from a so-called a.c. ionization coefficient ζ (which unfortunately is defined differently and has another dimension). The latter is obtained from breakdown measurements of gases in cavities with ultra high frequencies.

Assuming that the loss of charge occurs by diffusion only, it can be shown (see S. C. BROWN's contribution in vol. XXII of this Encyclopedia) that ultra-high frequency breakdown occurs when

$$z_i = \frac{D^-}{\Lambda^2} \tag{19.5}$$

when z_i is the number of ionizing collisions per electron per sec. and Λ the "diffusion length", i.e. in the case of a cylinder approximately its radius. This relation is akin to the BROWNian displacement law $x^2 \approx 4Dt$ where D^- is the electron diffusion coefficient and t the time intervall which is here the inverse of the ionization frequency z_i. Eq. (19.5) simply means that an electron has to be replaced within its diffusion-life-time. From (19.9) we have $\alpha \cdot v_d = z_i$, where v_a is the drift velocity of electrons. Hence with $\eta = \alpha/E$ (in ion pairs/V), and the mobility μ^- defined by $v^- = \mu^- E$, E being the peak value of the electric field, we have

$$\eta = \frac{z_i}{v_d E} = \frac{z_i}{\mu^- E^2}. \tag{19.6}$$

In high frequency fields, instead of the loss of charge to the electrodes (controlled by mobility) losses occur through diffusion to the walls, and it was found convenient to define here an a.c. ionization coefficient (in ion pairs per [volt]^2)

$$\zeta = \frac{z_i}{D^- E^2} = \frac{1}{(\Lambda E)^2}. \tag{19.7}$$

Combining (19.5) and (19.6) and e.g. D^-/μ^- [cf. (19.8)], α can be found from the geometry of the discharge vessel (Λ) and the observed value of the breakdown field E. Also we find from (19.6) and (19.7)

$$\frac{\eta}{\zeta} = \frac{D^-}{\mu^-} \propto k T_e \tag{19.8}$$

where T_e is the electron temperature. For the evaluation of experimental results it is necessary to know the corresponding electron energy distributions [*23*] which are different in steady current and in a.c. measurements. The results [*24*] show good agreement with the theory and with older conventional observations.

d) If the nature of the multiplication process in the positive column of a glow discharge would be the same as that in an electron swarm experiment, the value of α could be derived from positive column measurements as follows: determine either from theory (Sect. 18) or experiment the electron temperature T_e and the drift velocity of electrons v_d. Assuming that the energy distribution function is known the rate of ionization z_i can be calculated (Sect. 24). Then

$$\alpha = \frac{z_i}{v_d}. \tag{19.9}$$

In many cases it is found that the values of α derived from the positive column are considerably larger than those from swarm experiments. This means either that the electron distribution is different or that other processes than ionization by simple collisions occur.

e) The dependence of α/p on E/p can be calculated from the rate of ionization, which contains the electron energy distribution, the ionization efficiency (Fig. 2) and the drift velocity of the electrons. This method is treated in Sect. 24.

f) A variation of the above method is to calculate the rate of ionization from T_e and to use the measured drift velocities of electrons to find α. This is done in more detail in Sect. 25.

g) A pulse of photoelectrons of duration 10^{-7} sec. at a repetition rate of order 50 cycles/sec. is released from a cathode by means of a spark-triggered ultraviolet light source [*25*]. The current flowing through the gas to the anode is measured as a function of the time by a cathode ray oscillograph. An analysis of the current oscillograms gives the value of α but at the present time this technique is not sufficiently developed [*58*].

h) The formative time lag τ of a discharge can be expressed [*26*], [*27*] by an equation of the form

$$\tau = \frac{C}{\alpha\, v_d} \tag{19.10}$$

where C is usually a factor of order 10. C can be evaluated from the theory and depends on the definition of τ. Hence α can be found from measured values of τ and v_d though not very accurate results can be expected.

V. The results of the measurements.

a) The observed multiplication.

20. In Sect. 15 the procedure for observing the multiplication i/i_0 has been explained. We shall give here several examples of the actual observations, of which only few are published, in order to illustrate the main disturbing factors which have to be considered in evaluating the results.

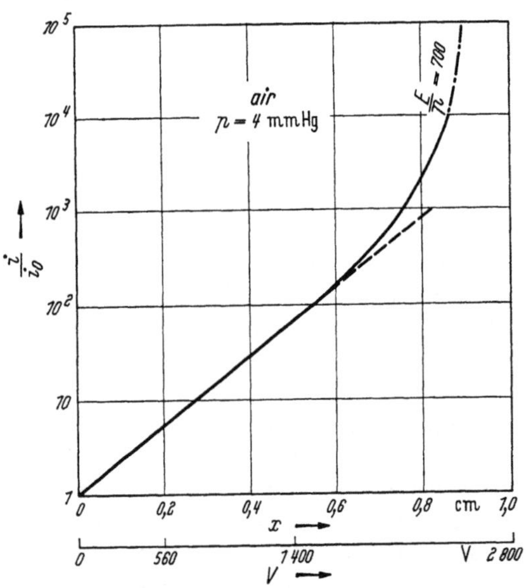

Fig. 14. Multiplication i/i_0 as a function of the electrode spacing d or the applied voltage V at a constant and high value of E/p (700 V/cm. mm. Hg) in air at 4 mm. Hg, obtained with the apparatus Fig. 9.

Fig. 14 gives i/i_0 on a logarithmic scale as a function of the electrode separation x or the applied voltage V for air at a constant, but high value of E/p [*28*]. i/i_0 is nearly unity at $x=0$; at larger x, ln i/i_0 increases more rapidly than linear because of secondary effects. Fig. 15 shows i (logarithmic scale) as a function of x in N_2 at high p [*22*], that is for low values of E/p, which has the same character as the curve in Fig. 14; earlier work suggested that breakdown (dash-dotted line) occurs without being preceded by upcurving. This would have meant that different secondary effects are operative in the two cases.

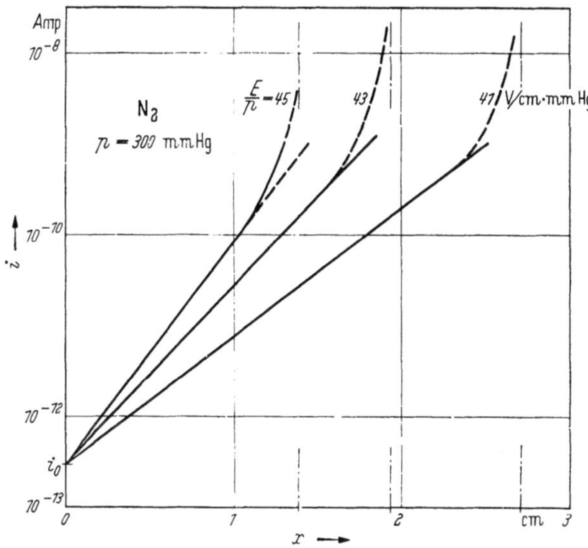

Fig. 15. Ionization current i (for a constant initial current $i_0 = 3 \cdot 10^{13}$ A) as a function of the electrode separation d for three low values of E/p in N_2 at 300 mm. Hg, obtained with the apparatus Fig. 12. Dash-dotted lines: sparking distances.

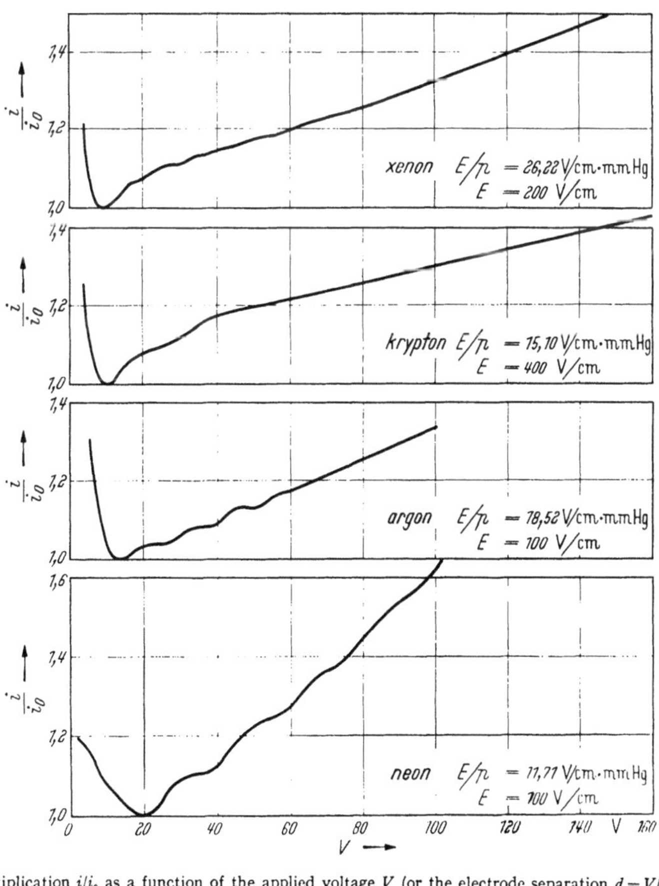

Fig. 16. Multiplication i/i_0 as a function of the applied voltage V (or the electrode separation $d = V/E$) in the range of low multiplication in four rare gases at low pressures $p = E/(E/p)$; see Fig. 11.

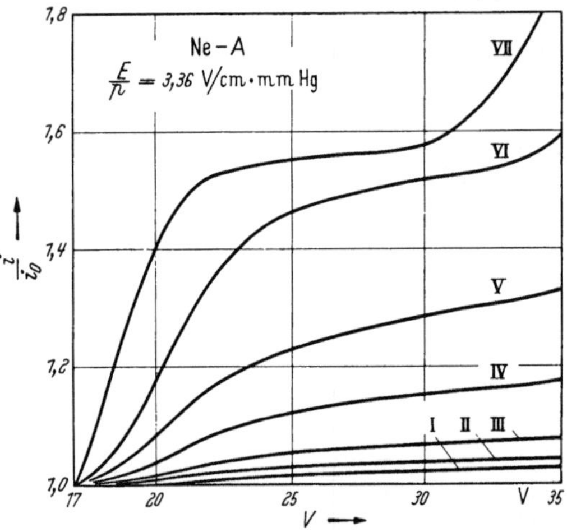

Fig. 17. Multiplication i/i_0 as a function of the applied voltage V for Ne with traces of A for a constant value of E/p and low values of V obtained with the apparatus Fig. 11. I pure Ne; II Ne + $8.2 \cdot 10^{-7}$ parts of A; III Ne + $2.9 \cdot 10^{-6}$; IV Ne + $9.6 \cdot 10^{-6}$; V Ne + $28 \cdot 10^{-6}$; VI Ne + $97 \cdot 10^{-6}$; VII Ne + $1 \cdot 10^{-2}$.

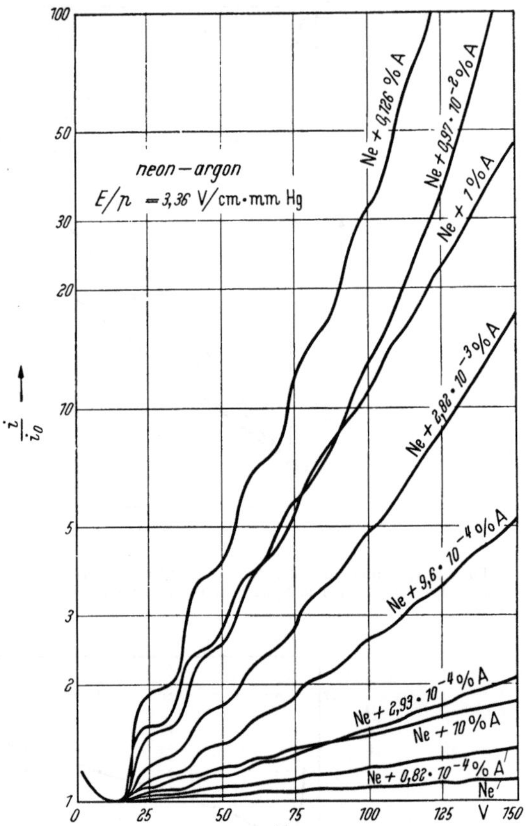

Fig. 18. Multiplication i/i_0 as a function of the voltage V as in Fig. 17 for higher values of V and d.

Fig. 18.

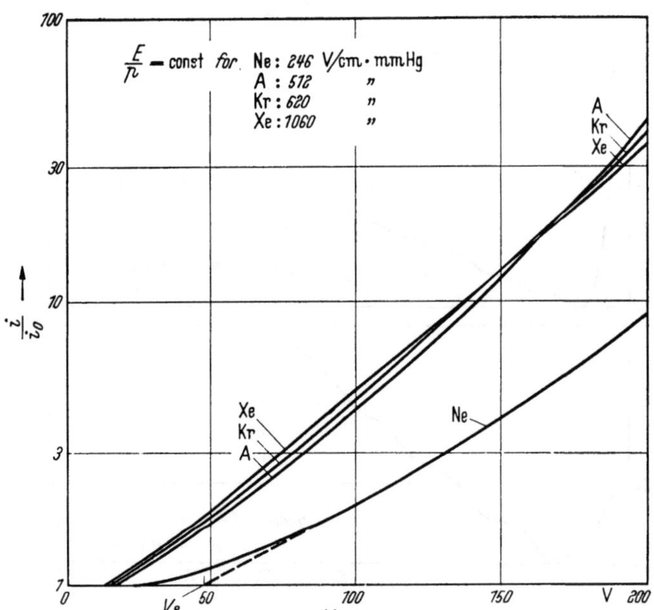

Fig 19. Multiplication i/i_0 as a function of the applied voltage V in the rare gases at large values of E/p.

A more complex picture presents itself with noble gases. Fig. 16 shows $i/i_0 = f(V)$ for Ne etc. again at constant but low E/p [29]. Two points are of main interest: for low V (several volts) as V increases i/i_0 falls. This is due to the perforations of the anode through which light reaches the cathode (Fig. 11): at low V, d is small and the electric field is concentrated at those parts of the cathode where there are no holes in the anode and i/i_0 is large. As V and x are increased, these non-uniformities gradually disappear and i/i_0 falls. At still larger x, photoelectrons are drawn from the whole cathode and i/i_0 rises steadilly.

The other effect is the undulatory character of the lower part of the curve. This is due to the non-equilibrium between electron energy and field near the

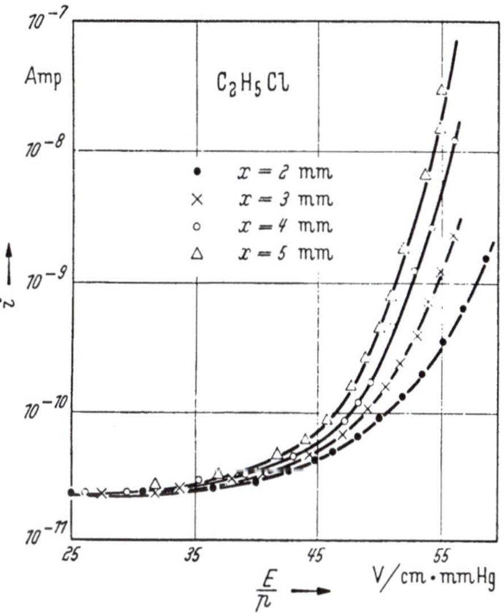

Fig. 20. Ionization current i ($i_0 \approx 2.5 \cdot 10^{-11}$ A) as a function of the applied field E/p for various electrode separations x in ethylalcohol.

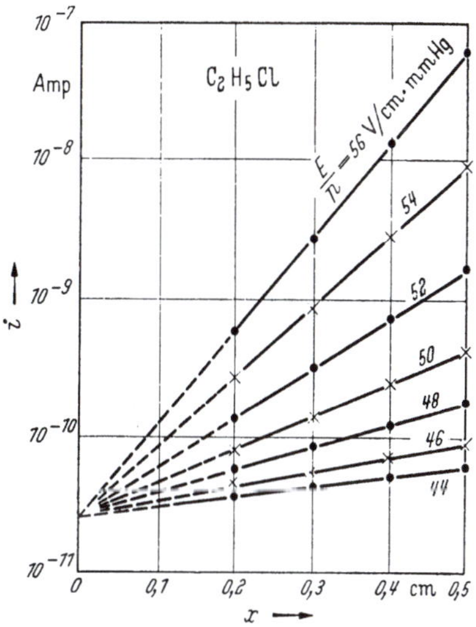

Fig. 21. Ionization current i as in Fig. 20 as a function of the separation x for various fields E/p.

cathode. Neglecting for the moment the distribution of their initial velocities, consider the photoelectrons moving through the gas in field direction without being scattered by collisions. When the electrons have fallen through a potential $V = V_{exc}$ (the principal excitation potential) some may have lost a larger part of their energy but no ionization or multiplication can yet occur. When $V = V_i$ multiplication starts and the current rises. From $V = V_i + V_{exc}$ onwards energy is lost again; hence this marks a minimum and the current starts to rise from here again, etc. In fact, the losses have to affect large numbers of electrons to be effective and so the arguments apply to the average energy of the electron swarm. After several such steps the steady state distribution has developed with the help of elastic collisions and the fine structure of the intial multiplication disappears.

In Ne with traces of A this effect is still more pronounced. This can be seen from Fig. 17 for low [30], and from Fig. 18 for high multiplication [31]. Both diagrams contain curves for various argon contents. Fig. 19 gives i/i_0 for pure gases at low p and high E/p in the range of strong multiplication [29]. The curve for Ne has a very short linear portion, the extension of which intersects the abscissa at V_δ. This is the potential corresponding to x_0 (15.2) and related to $\eta = \alpha/E$ by

$$\frac{i}{i_0} = \frac{e^{\alpha(x-x_0)}}{1-\gamma(e^{\alpha(x-x_0)}-1)} = \frac{e^{\eta(V-V_\delta)}}{1-\gamma(e^{\eta(V-V_\delta)}-1)}. \tag{20.1}$$

Finally i/i_0 for ethylalcohol is given as a function of E/p (Fig. 20) and of x (Fig. 21) from which it can be seen that further difficulties arise when x can only be varied by a factor 2 to 3 and when complex and easily dissociating molecules are to be investigated [32].

From the slope of the linear portion of $\ln (i/i_0) = f(x)$ the ionization coefficients α are found (15.1), from the slope $\ln i/i_0 = f(V)$ the coefficients η.

b) The ionization coefficients.

21. *α) Pure monatomic gases.* Measurements of the ionization coefficient α/p have been carried out in all the rare gases except radon [34]. The results are

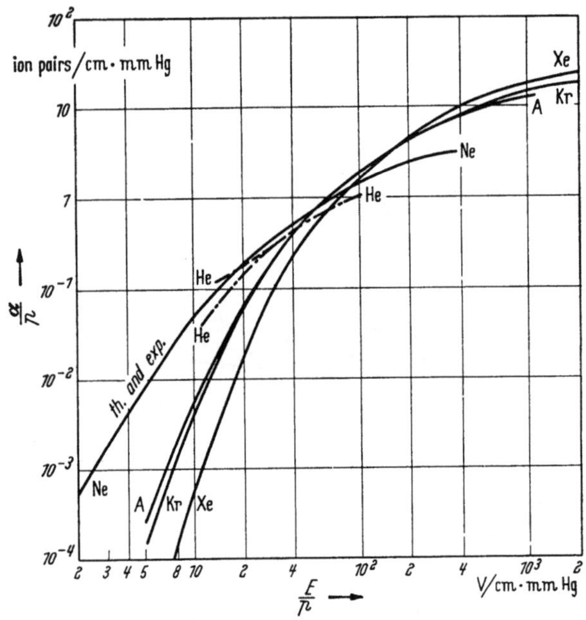

Fig. 22.

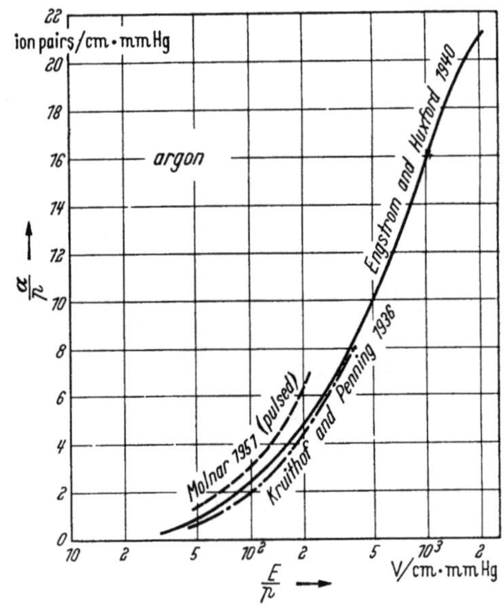

Fig. 23.

Fig. 22. Ionization coefficient α/p as a function of the electric field E/p for the rare gases. Dashed curve: He experimental, dashed-dotted curve: He calculated.

Fig. 23. Ionization coefficient α/p in pure A as a function of the electric field E/p.

given in Fig. 22 where α/p (in ion pairs per cm. in field direction per mm. Hg) as a function of E/p (in V/cm. mm. Hg) is plotted on a double logarithmic scale covering a range of three orders of magnitude in E/p. Except for helium [35] (Sect. 27) a certain regularity is found: for low values of E/p the values of α/p are larger the lighter the gas whereas for high values of E/p the opposite is the case. It is interesting to note that at low values of E/p the gas with the lowest ionization potential, namely Xe, has a much lower value of α/p than He or Ne. This is contrary to what one would expect to find if either the ionization potential and α/p would show a parallel trend or if the ionization probability (say its maximum) and α/p would vary in the same manner. Clearly some important factor must give rise to the reverse order as will be shown later (Sect. 26, 27). Fig. 23 shows results for pure argon from various investigators [20], [36] from which the degree of agreement can be seen. Fig. 24 is a linear graph of the rare gases up to large values of E/p.

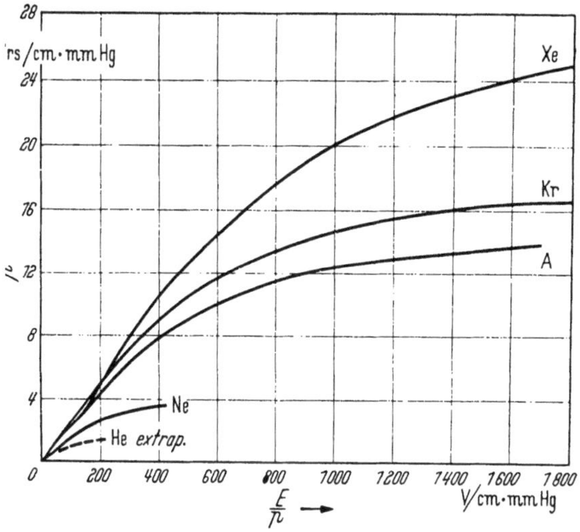

Fig. 24.

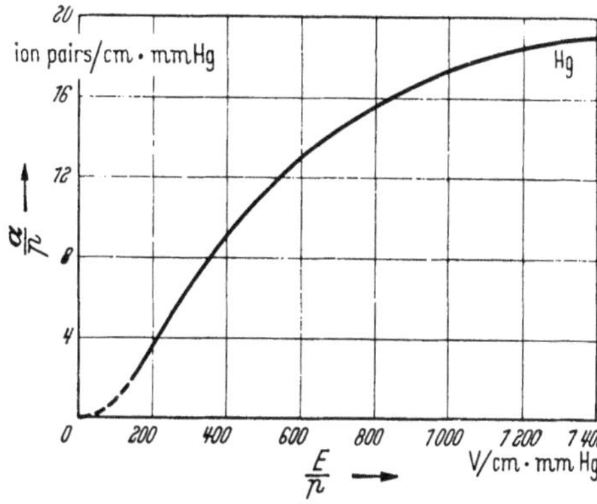

Fig. 25.

Fig. 24. Ionization coefficient α/p as a function of the electric field E/p for the rare gases at large values of the field.

Fig. 25. Ionization coefficient α/p as a function of the electric field E/p for Hg vapour.

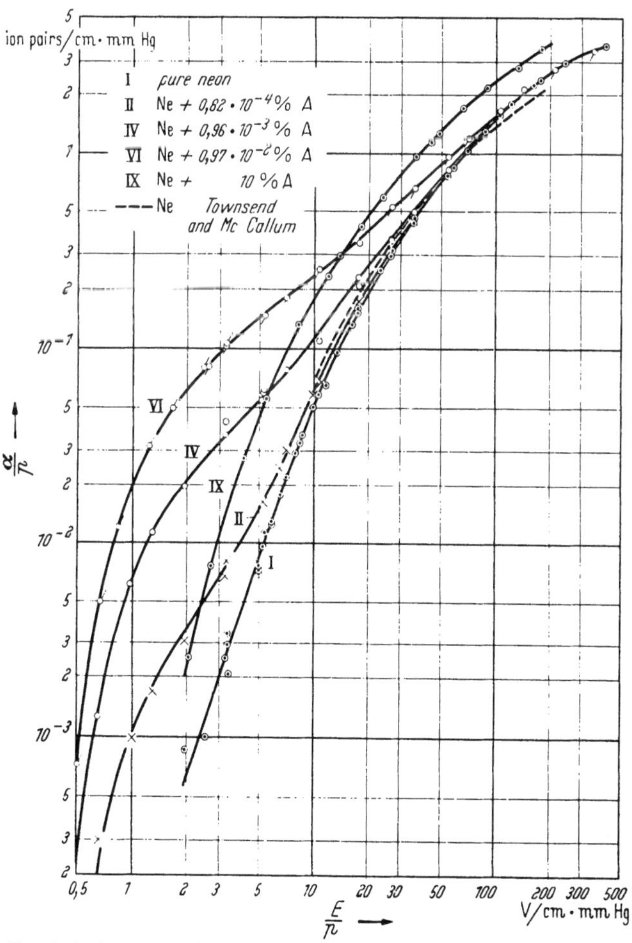

Fig. 26. Ionization coefficient α/p as a function of the electric field E/p for Ne with traces of A.

Observations in metal vapours are not available except in Hg [37]. Fig. 25 shows a linear plot of $\alpha/p = f(E/p)$ for mercury vapour for E/p up to 10^3 V/cm. × mm. Hg, all values of α/p being reduced to 0° C and 1 mm. Hg. A comparison of α/p in Hg with the values in the heavier rare gases shows that there is no great difference between them. Neither the rare gases nor Hg vapour show clearly whether α/p reaches a saturation value or has a maximum as E/p is increased to values above 10^3.

β) Mixtures of monatomic gases. The gas mixture best investigated is Ne + A [31]. This is because of the particular role which argon plays here when present in very small amounts. In short, whereas one would expect a variable mixture of two gases to give α/p varying steadily from the value of the pure Ne to that of pure A for a given value of E/p, it is found that α/p first rises and then falls as the amount of A added to Ne is increased. Fig. 26 shows the ionization coefficient for admixtures ranging from 10^{-6} parts of A in Ne to pure A. These measurements by KRUITHOF and PENNING in Ne with A of $< 10^{-6}$ parts compare favourably with TOWNSEND and MC CALLUM's observations [38] which have been obtained with differently treated samples of Ne.

There are unconfirmed reports that similar observations have been made with He with admixtures of A or Ne but no systematic measurements have been published as far as is known.

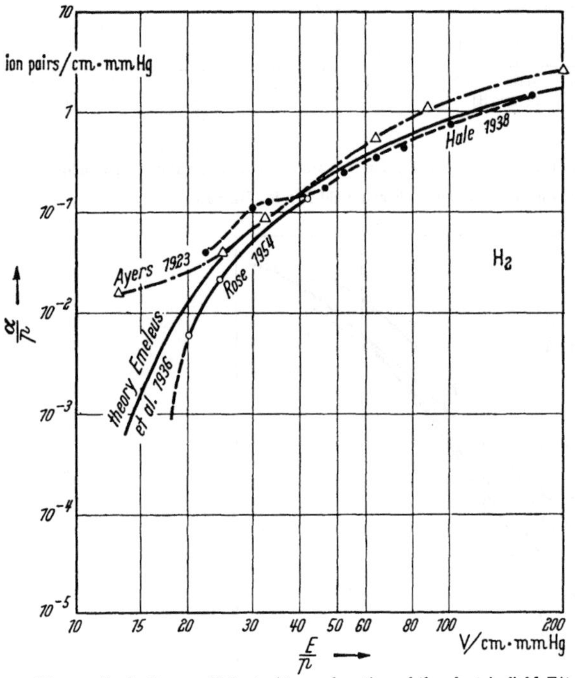

Fig. 27. Ionization coefficient α/p as a function of the electric field E/p for H_2. Measurements and theory.

γ) Diatomic gases. The various measurements on H_2 are shown in Fig. 27 [39], [40]. It can be seen that the points lie on a smooth curve for higher values of E/p but that humps and inflexions appear in earlier measurements at lower values of E/p which suggest that the experimental technique was not sufficiently advanced to ensure reliable results. It is certain that in molecular gases purity problems play an important role at low fields [41], [99], [100]. α/p in D_2 is probably slightly larger than in H_2.

A similar result is found in the case of α/p in N_2 (Fig. 28). Here old [39], [40] and new measurements [22] differ in the same way as has been shown for H_2. At low values of E/p earlier values of α/p have been found too high [43] which may be due to neglecting x_0 [see (20.1)] particularly at low V or due to traces of Hg (Fig. 29). E.g. at $E/p = 20$ the value of α/p for pure N_2 is probably of the order 10^{-4} but a value of about $3 \cdot 10^{-2}$ has been found earlier. Since in Hg $\alpha/p = 3.5$ at $E/p = 20$ it follows that with the same energy distribution of electrons 10^{-2} parts of Hg would explain the high value. If the distribution curve contained a larger number of fast electrons an even smaller admixture of Hg vapour would suffice to explain the differences in the observations. Fig. 30

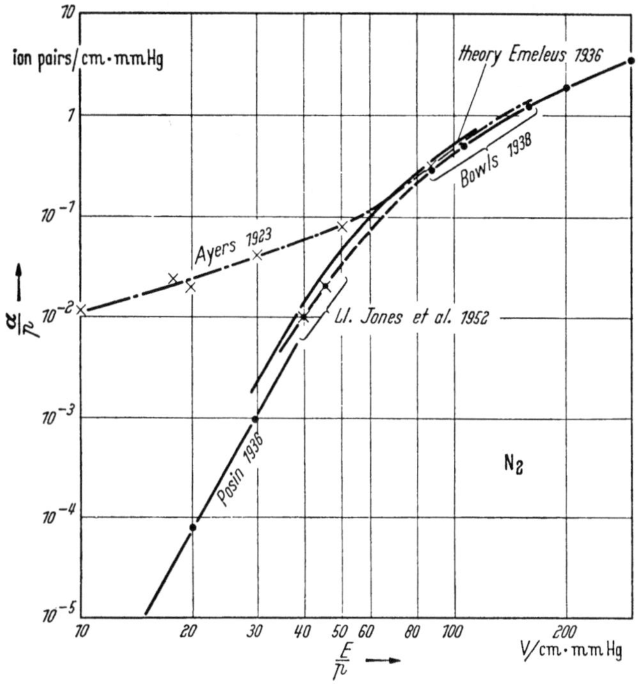

Fig. 28. Ionization coefficient α/p as a function of the electric field E/p for N_2. Measurements and theory.

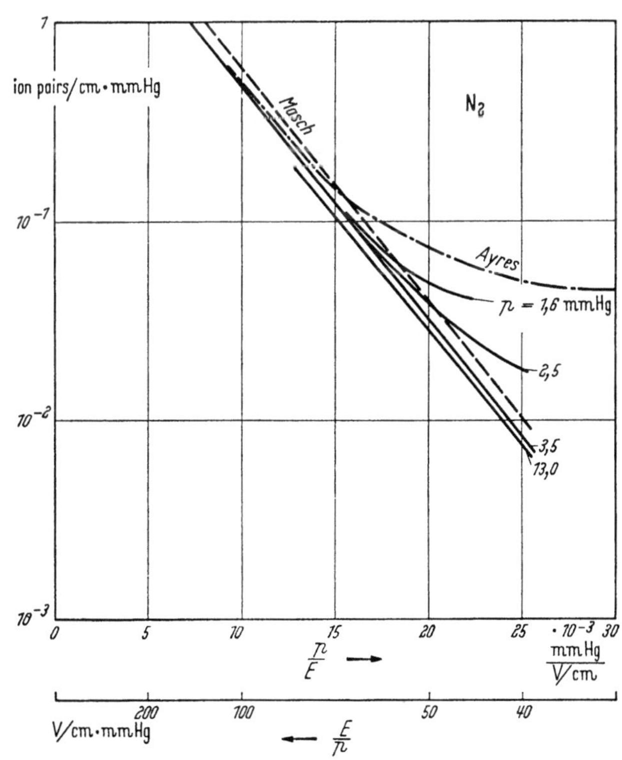

Fig. 29. Ionization coefficient α/p as a function of the inverse field p/E in N_2 at various values of p using a small electrode separation (probably $\leq$ 0.4 cm.).

gives α/p for N_2, H_2 and air up to high E/p. It shows that the ionization coefficient of H_2 can be larger or smaller than that of N_2 depending on E/p.

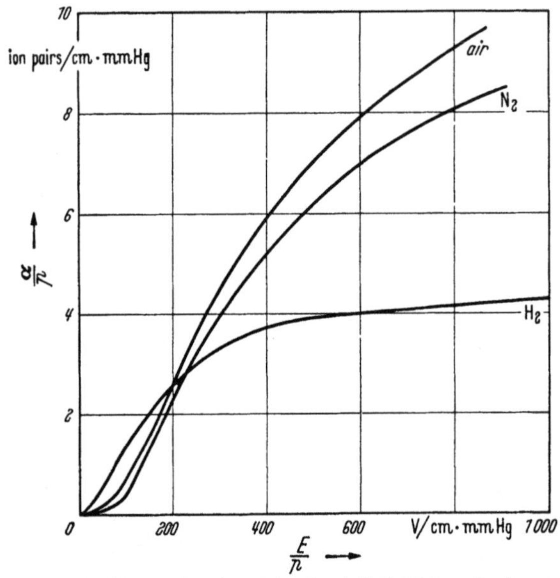

Fig. 30. Ionization coefficient α/p as a function of the electric field E/p in molecular gases for high values of E/p.

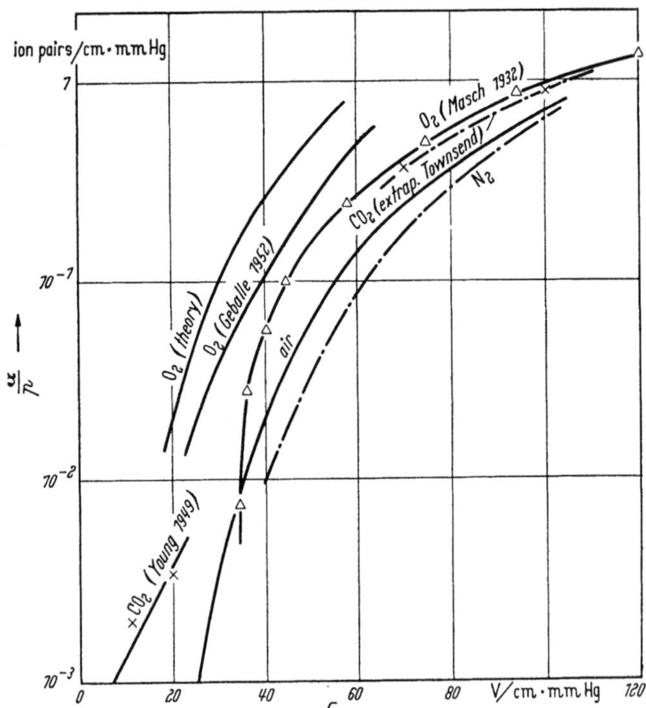

Fig. 31. Ionization coefficient α/p as a function of the electric field E/p in O_2, CO_2, air and N_2.

In the case of O_2 the measurements [44], [45] (Fig. 31) extend only over a very small range of values of E/p. The general shape of the curve is the same as for other diatomic gases and mixtures. Though O_2 attaches electrons its ionization coefficient is higher than that of the non-attaching gas N_2. Another diatomic gas with an electronegative component is HCl whose ionization coefficient is given in Fig. 32 [*14*]. Again it is apparent by comparison with air that the values of α/p are here particularly large. This seems to be so not in spite of but rather because of the presence of an electronegative element. The ionization coefficient of HCl at low values of E/p is given in [*98*].

δ) Mixtures of diatomic gases. The best known example is air [*14*], [*44*], [*46*]. The curve for low values of E/p is given in Fig. 31 and for high values in Fig. 32. It is seen that α/p for air is larger than for pure N_2 because of the larger ionization coefficient of O_2. This is instructive in the sense that the oft invoked electronegativity of O_2 leading to the formation of negative ions does not overshadow here the contribution of O_2 to the ionization. Though no results with other gas mixtures are available, it can be deduced by analogy with air and other evidence (e.g. sparking potential in mixtures) that α/p in mixtures of diatomic gases should go over continuously from the values for the pure component 1 to the values for the pure component 2 except when gaseous compounds of exceptional properties are formed (see ε).

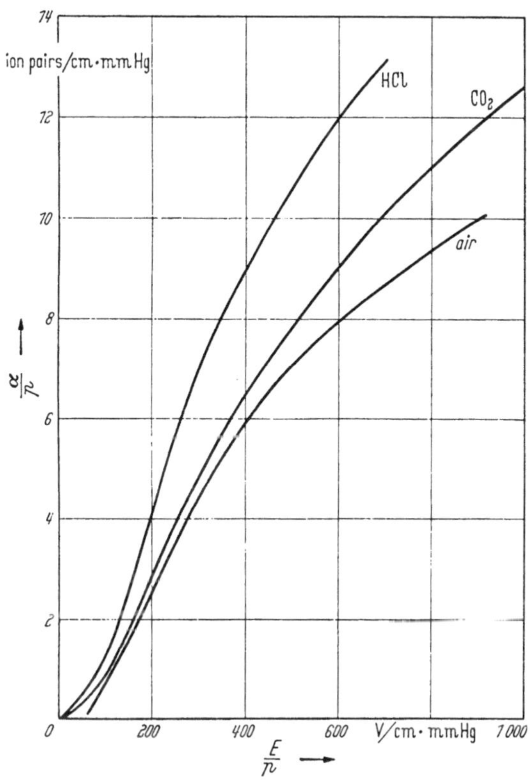

Fig. 32. Ionization coefficient α/p as a function of the electric field E/p in HCl and CO_2 (comparison with air) for strong fields.

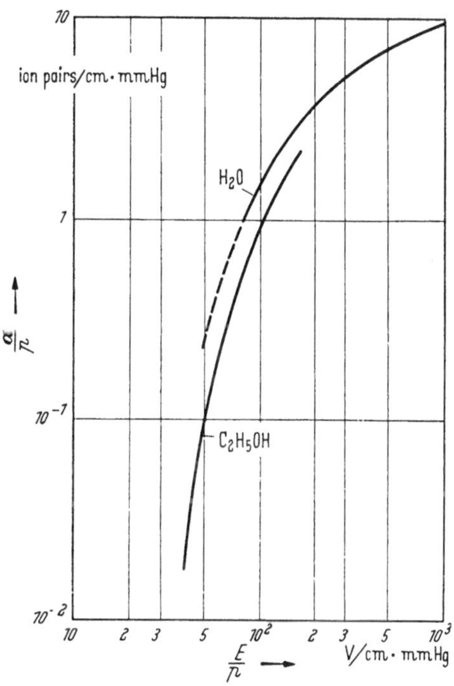

Fig. 33. Ionization coefficient α/p as a function of the electric field E/p for water vapour and ethylalcohol.

ε) Gases with complex molecules. Observations on triatomic molecules have been made. Results for CO_2 [*14*] H_2O [*14*], [*98*], and C_2H_5OH [*47*] are shown in Figs. 32 and 33. They do not differ much from those of other molecular gases.

There is, however, a marked difference for complex molecules with halogen atoms like CSF_8, freon-12 (CCl_2F_2) (Fig. 34) [*48*] and SF_6 (Fig. 35) [*49*]. These gases have much lower values of α/p at low values of E/p (say <100) than the other molecular gases, a property used for raising the sparking potential in electric devices [*33*]. Fig. 36 [*32*] and 37 [*50*] show the ionization coefficient for these complex substances. It may be remarked that C_6H_6 etc. have very large values of α/p compared with other gases. α/p for C_2H_4 and C_5H_{12} has been measured too [*97*]. For NH_3 see [*98*].

There is a certain amount of indirect evidence available from which the position of the curve $\alpha/p = f(E/p)$ of one gas with respect to the curves of other gases can be deduced, though not with great certainty. For example, comparing the sparking potential as a function of pd in H_2 with that of N_2 (or air) we find that the latter has a higher sparking potential at large pd, a higher minimum sparking potential and a lower value of $(pd)_m$ from which we conclude that α/p for N_2 is lower than for H_2 at high E/p. This of course is only so if the values of γ in the two cases compared do not vary in such a way as to cancel or reverse the effect of a change of α/p. A similar argument holds for the minimum breakdown potentials of Ba, Hg and Al cathodes in the rare gases with the corresponding curves for α/p. E.g. from Fig. 24 it is seen that α/p for argon is lower than for Ne. Hence the corresponding breakdown potential for argon should be higher than that for Ne. This has been found to be true. However, the curves for α/p often intersect at higher values of E/p (Fig. 22) and so the conclusions drawn are not always certain. There are no extensive observations of the variation of α with the gas temperature. (p is a measure of the density.)

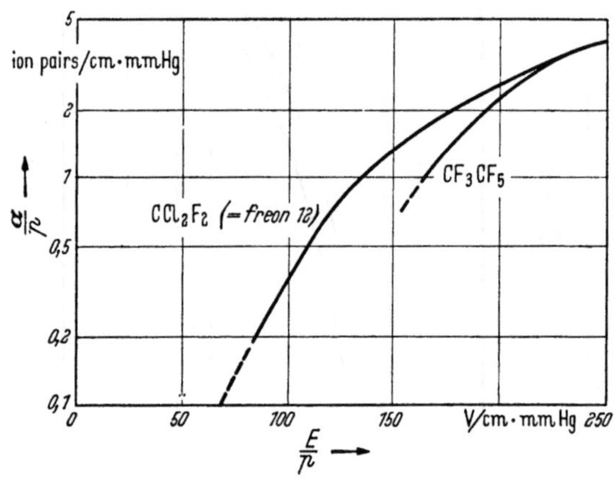

Fig. 34. Ionization coefficient α/p as a function of the electric field E/p for freon and other halogen containing gaseous compounds.

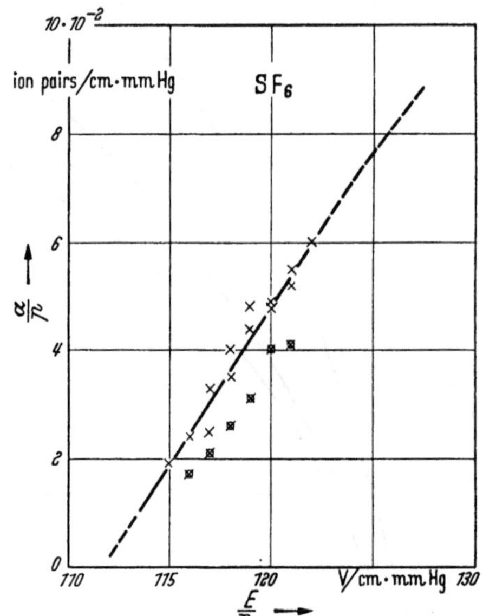

Fig. 35. Ionization coefficient α/p as a function of the electric field E/p in sulphur hexafluroide. Crosses: electrode separation $d = 0.6$ cm. and 0.8 cm. Encircled crosses: $d = 0.4$ cm.

In certain cases it has been more convenient to use the ionizations coefficient η (in ion pairs per V). Fig. 38 and 39 give $\eta = f(E/p)$ for rare gases (except He) [*29*] and for Ne—A mixtures [*34*] showing maxima between $E/p \approx 50$ and 300. $1/\eta$ corresponds to the ionizing power of a beam of mono-energetic electrons passing through a gas i.e. the average energy in eV necessary to produce one ion pair. The minimum value of $1/\eta = S$ is found in Table 3, see p. 539.

Table 2 gives a summary of the estimated maximum values of α/p and their corresponding values of E/p together with the initial slopes a, maximum ionization efficiencies s_e and the corresponding values of the electron energy K.

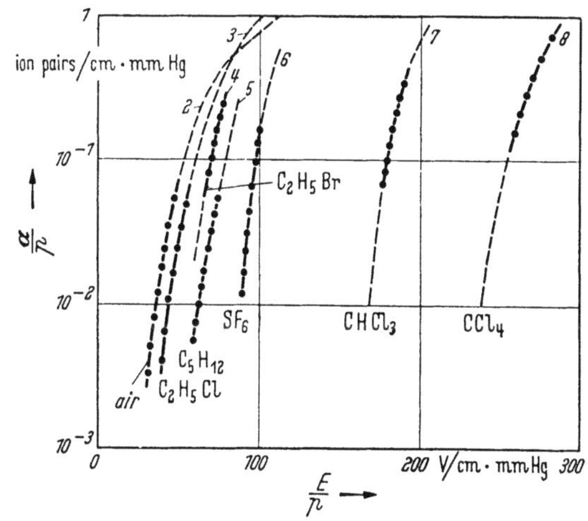

Fig. 36. Ionization coefficient α/p as a function of the electric field E/p for various complex molecular gases.

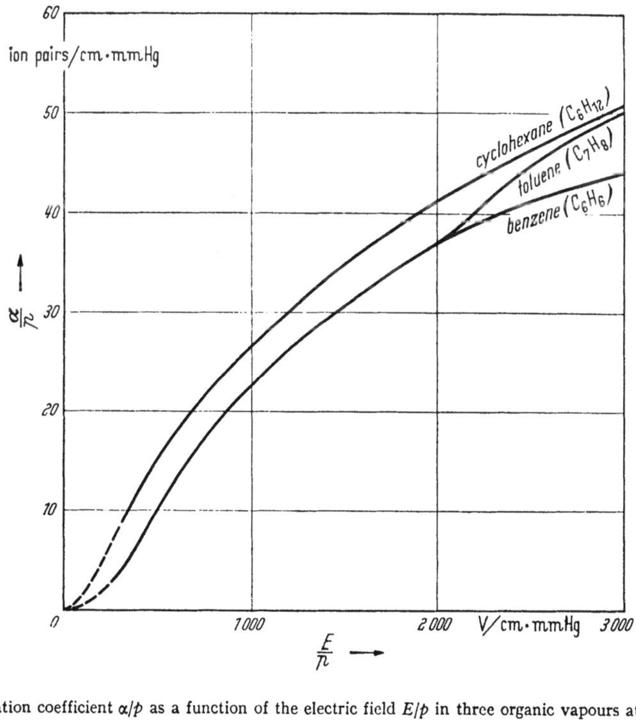

Fig. 37. Ionization coefficient α/p as a function of the electric field E/p in three organic vapours at high values of E/p.

Table 3 contains the STOLETOW constant S (Fig. 7) and the corresponding values of α/p and E/p for various gases as well as the values of the ionization coefficient and field at the point of inflexion.

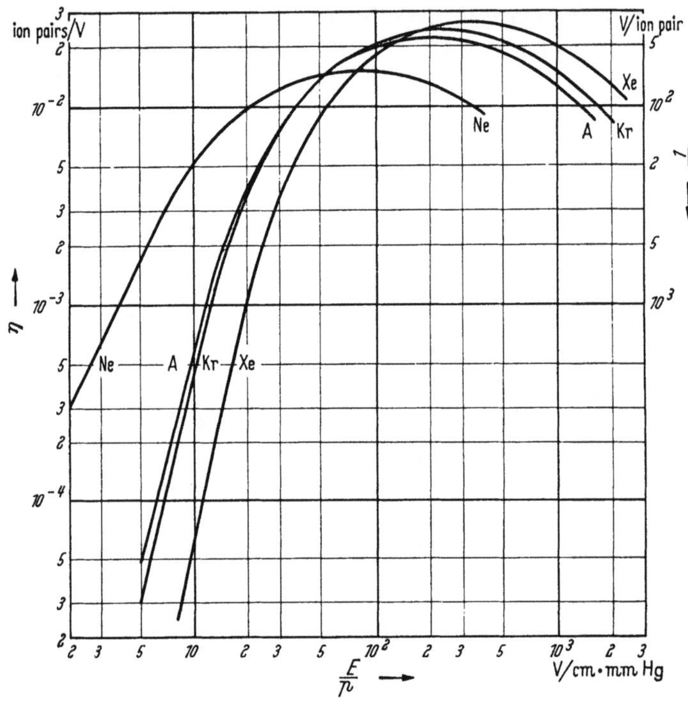

Fig. 38. Ionization coefficient η $(=\alpha/E)$ as a function of the electric field E/p in the rare gases (except He).

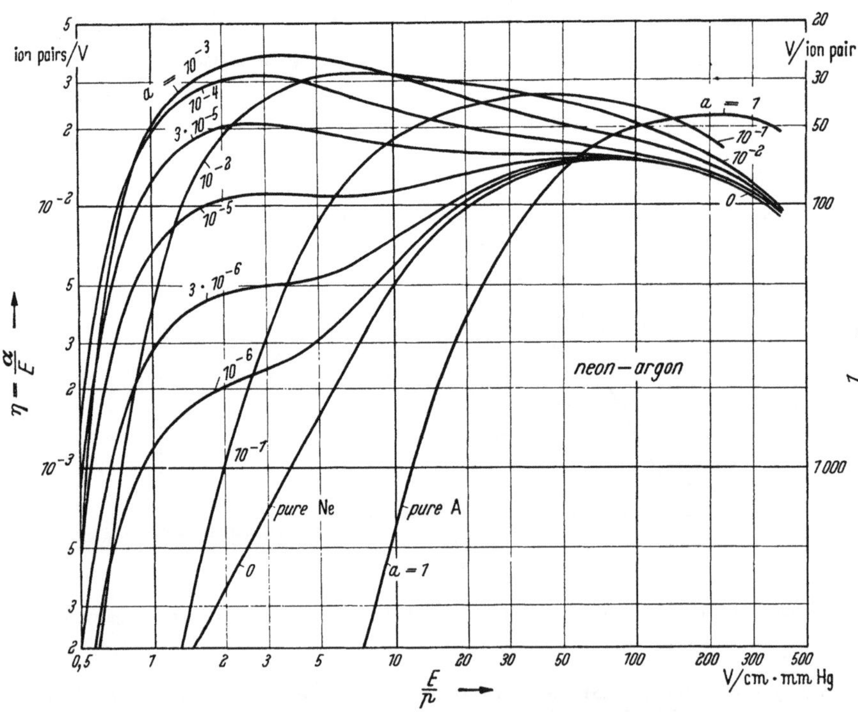

Fig. 39. Ionization coefficient η as a function of the electric field E/p in Ne-A mixtures. a is the number of parts of A in Ne. This diagram illustrates the Penning effect (increase in η in the presence of small traces of A in Ne).

Table 2. *Initial slope a, maximum ionization efficiency s_e and approximate values of $(\alpha/p)_{max}$ [15].* (The values s_e include multiply charged ions.)

Gas	a $\frac{1}{\text{V} \times \text{mm. Hg}}$	$(s_e)_{max}$ $\frac{\text{ion pairs}}{\text{cm.}}$	K_{max} eV	$(\alpha/p)_{max}$	$(E/p)_{max}$ $\frac{\text{V}}{\text{cm.} \times \text{mm. Hg}}$
H_2	0.21	3.6	70	4.5	~1000
N_2	0.26	10.2	100	9–10	~2000 (?)
O_2	0.24	10.5	120	>10	—
CO_2	—	—	—	16–18	2000
HCl	—	—	—	20	2000
H_2O	—	—	—	12	>1500
Hg	0.83	17	90	19	~1500
He	0.046	1.3	120	3	~ 400
Ne	0.05	3	70	4	~1000
A	0.7	13	60	14	~2000
Kr	—	—	—	17	>2000
Xe	—	—	—	27	~5000
C_6H_6	—	—	—	50	~5000
cyclohexane	—	—	—	60–70	—
C_2H_2	—	18	80	—	—

Table 3. Stoletow *constant S, optimum values of α/p and E/p, and values at the inflexion.*

Gas	$S = \frac{E}{\alpha}$ at $\frac{\text{V}}{\text{ion pair}}$	$\left[\left(\frac{\alpha}{p}\right);\left(\frac{E}{p}\right)\right]_{opt}$		$\left[\left(\frac{\alpha}{p}\right);\left(\frac{E}{p}\right)\right]_{infl}$	
H_2	73	2.2	160	1	80
N_2	76	4.2	320	2.3	200
O_2	~60	>4	270	—	—
CO_2	62	6.5	400	2.8	200
air	68	5	340	2.5	200
H_2O	58	4	220	2.2	150
HCl	43	7	300	3	170
Hg	45	9	400	5	250
He	~100	~1	~100	—	—
Ne	67	1.8	120	0.7	50
A	50	8	400	6	200
Kr	50	7	350	4	200
Xe	40	10	400	3	150
C_6H_6	44	20	900	~7	~400
cyclohexane	33	15	500	~7	~250

VI. Collision theory and ionization coefficient.

a) Elementary theories.

22. The first attempt to give the ionization coefficient α a theoretical basis has been treated in Sect. 13 and 14. There it was shown that it is possible to derive a relation between α/p and parameters of the gas by assuming that an electron which acquires an energy equal to or larger than ionization energy will ionize, that the electron mean free paths have a distribution according to kinetic theory and that they are essentially all parallel to the direction of the electric field. The eq. (13.5) so obtained contains, therefore, besides E/p two parameters only: the ionization potential and the mean free path of an electron at unit gas density.

These three assumptions which lead to a formula of great practical value are obviously not in accordance with experience. Ionization sets in when $V \geq V_i$ but the probability at $V = V_i$ is zero and rises slowly with V_i; it never exceeds 50%, the maximum lying at $V \approx 4-6 V_i$. Thus the result derived from (13.5) should be much too high. A second point is that the distribution of the mean free path was erroneous because the dependence of the mean free path on the electron energy was not included; the effect was not known at that time. The resulting error can have either sign. Thirdly, the electrons do not move parallel to the electric field nor do they ionize only when their "drift velocity" exceeds a certain critical value but they move at random with speeds exceeding the drift velocity usually by orders of magnitude and ionize by random collisions with gas molecules. Consequently their path is a zig-zag one and the neglect of this fact should make the result too small. The net result was, however, that over a certain range of E/p these errors balanced.

23. The next step in the development of the theory of the ionization coefficient for electrons was an attempt ([*17*], I p. 192) to show that TOWNSEND's eq. (13.5) is approximately of the same form as an equation which can be derived by invoking the fundamental processes of ionization in gases. The starting point is the postulate that a swarm of electrons which moves down a uniform electric field generates its ionization because its electrons have sufficiently high random velocities to ionize gas molecules in their encounters. Now provided that the electron distribution and the ionization probability are known, the rate of ionization can be found. Thus, an electron of that distribution which makes z_i ionizing collisions per sec. and moves in an electric field with the uniform drift speed v_d produces α ion pairs per cm. of path in field direction or

$$\frac{\alpha}{p} = \frac{z_i}{p} \frac{1}{v_d}. \tag{23.1}$$

The rate of ionization can be easily found if e.g. the distribution is assumed to be MAXWELLian. If dn is the number of collisions per sec. an electron with a speed in the range between v and $v+dv$ makes with molecules, and if $p(v)$ is the ionization probability (number of ionizing collisions divided by the total number of collisions), then the rate of ionization per electron

$$z_i = \int_{v_i}^{\infty} p(v)\, dn \tag{23.2}$$

where v_i is the velocity corresponding to the ionization energy eV_i. The ionization probability $p(v)$ is connected with the ionization efficiency s_e and the volt equivalent V of the electron energy by

$$p(v) = c_1 (v^2 - v_i^2) = s_e \lambda_e = \lambda_1 a\, (V - V_i). \tag{23.3}$$

Since $s_e \propto p$ (the values in Fig. 2 are reduced to 1 mm. Hg) and $\lambda_e \propto 1/p$, $p(v)$ is independent of p.

a (Table 2) and c_1 are constants; $\lambda_e \cdot p = \lambda_1$ is the electron mean free path at $p = 1$ mm. Hg and $eV = \frac{1}{2} m v^2$. With these substitutions and with a MAXWELLian distribution of velocities, viz.

$$dn = \frac{4}{\sqrt{\pi}} \frac{1}{\lambda_e} \left(\frac{v}{\bar{v}}\right)^2 e^{-\left(\frac{v}{\bar{v}}\right)^2} dv \tag{23.4}$$

the rate of ionization is obtained in terms of the electron temperature T_e (the BOLTZMANN constant $k = 1.37 \cdot 10^{-16}$ ergs/°K and $\frac{1}{2} m \bar{v}^2 = kT_e$):

$$z_i = \frac{2}{\sqrt{\pi}} \frac{a p}{e/m} \left(\frac{2kT_e}{m}\right)^{\frac{3}{2}} e^{-\frac{eV_i}{kT_e}} \left[1 + \frac{eV_i}{2kT_e}\right]. \tag{23.5}$$

For the evaluation it has to be remembered that electrostatic units have been used throughout. The relation between T_e and the field E is to be found next. In the steady state an electron picks up in unit time an amount of energy which it loses in collisions with molecules. Let $\varkappa$ be the average fraction of the energy $\frac{1}{2}mv^2$ lost per collision, then with $\bar{v}/\lambda_e$ collisions per second the power balance reads

$$e\,E\,v_d = \varkappa\,\frac{m\,\bar{v}^2}{2}\,\frac{\bar{v}}{\lambda_e}\,. \tag{23.6}$$

The electron drift velocity can be shown to be

$$v_d \approx \frac{e\,\lambda_e}{m\,\bar{v}}\,E\,. \tag{23.7}$$

From (23.6) and (23.7) expressions for v_d and $\bar{v}$ in terms of the field are found, viz.

$$v_d \approx \left(\frac{e}{m}\right)^{\frac{1}{2}} \varkappa^{\frac{1}{4}}\,(E\,\lambda_e)^{\frac{1}{2}}\,, \tag{23.8}$$

$$\bar{v} \approx \left(\frac{e}{m}\right)^{\frac{1}{2}} \frac{1}{\varkappa^{\frac{1}{4}}}\,(E\,\lambda_e)^{\frac{1}{2}}\,. \tag{23.9}$$

Since $kT_e = \frac{1}{2}m\bar{v}^2$, we find from (23.9) $T_e = f(E)$, viz.

$$kT_e = \frac{e}{2\sqrt{\varkappa}}\,E\cdot\lambda_e\,. \tag{23.10}$$

Substituting this for kT_e in (23.5) and combining with (23.9) and (23.1), we find

$$\frac{\alpha}{p} = \frac{2\,a\,V_i}{\sqrt{\varkappa}}\,\mathrm{e}^{-\frac{2\sqrt{\varkappa}\,V_i/\lambda_1}{E/p}} \tag{23.11}$$

provided that 1 is neglected against $eV_i/2kT_e$. Otherwise a small variation of the first factor with E/p occurs which counteracts the variation with E/p in the exponential term. However, it can be seen that (23.11) is of the same form as (14.1) viz.

$$\frac{\alpha}{p} = A\,\mathrm{e}^{-\frac{B}{E/p}} \quad \text{with} \quad A = \frac{2\,a\,V_i}{\sqrt{\varkappa}} \quad \text{and} \quad B = \frac{2\sqrt{\varkappa}\,V_i}{\lambda_1}$$

where λ_1 is the electron mean free path at $p = 1$ mm. Hg. Using suitable values for $\varkappa$, the constants A and B are found to be approximately of the right magnitude. This treatment, however, does not allow for the fact that $\varkappa = f(E/p)$ and that in many gases or in certain ranges of E/p eq. (23.5) does not apply because the MAXWELLian distribution fails.

24. A later treatment [*51*] avoids the pitfalls due to the use of the average collision loss. Again, eq. (23.1) is the starting point and also the MAXWELLian distribution is applied (alternatively as TOWNSEND's distribution which gives no better agreement with experiments). The rate of ionizing collisions is calculated thus: if v/λ_e is the number of collisions per sec. an electron makes in a gas at pressures p, if $P(\varepsilon)$ is the probability that a collision leads to ionization when the electron has the energy ε and if $f(\varepsilon)\,d\varepsilon$ is the energy distribution, then the ionization rate per electron is

$$z_i = \int\limits_{eV_i}^{\infty} \frac{v}{\lambda_e}\,P(\varepsilon)\,f(\varepsilon)\,d\varepsilon\,. \tag{24.1}$$

Table 4. *Drift velocities v_d of electrons (in 10^6 cm./sec.) in gases as a function of the relative electric field E/p (in V/cm. mm. Hg).*

E/p	He	Ne	A	H_2	N_2	O_2	Cl_2	Br_2	I_2	NO	CO	HCl	H_2O	N_2O	CO_2	NH_3	BF_3	C_2H_4	C_5H_{12}	Air	E/p
0.25	0.45	0.6	0.28	0.65	0.52	1.3				—	0.12			0.5		—	—	—	—	0.6	0.25
0.5	0.6	0.9	0.3	0.9	0.62	1.6				—	0.25			1.3		0.06	0.8	≈1	1	0.9	0.5
1	0.9	1.5	0.4	1.2	0.87	1.6				0.65	0.55			3		0.13	1.2	2	2	1.2	1
2	1.3	2.6	0.8	1.6	1.3	1.7				1	1.2			5.5	1.3	0.26	2	3.3	3.3	1.6	2
3	2	4	0.9	—	1.8	—				1.3	—			7	2.1	0.4	≈2.8	—	—	2	3
5	5	6.3	1.6	2.6	2.7	2.7			≈1	1.5	5				5	0.7		5.4	4.8	3	5
10		≈11		3.8	4.9	5	7	5	2.2	2.4	11	2.6	2.7		11	3		6	5.1	5	10
15				—	—	6.7	9	6	—		—	—	3.9		13.5	8		—	—	7	15
20				7	8.6	8.2	9	7.3	5.3		14	4.7	6.2			10		6.7	5.5	8	20
30				—	—	10	9	8	8		—	5.2	9.4					7.5	5.8	—	30
40				16	15	—	9.3	8.5	8.5		—	6						9	6.4	—	40
50				22	—	16	10	8.7	9		20									17	50
60				28	19		11													—	60
100				<50	<40		17													27	100

He: R. A. Nielsen: Phys. Rev. **50**, 950 (1936); J. A. Hornbeck: Phys. Rev. **83**, 374 (1951); older values J. S. Townsend and V. A. Bailey: Phil. Mag. **46**, 657 (1923).

Ne, A: see Nielsen.

H_2: J. S. Townsend and V. A. Bailey: Phil. Mag. **42**, 873 (1921); N. E. Bradbury and R. A. Nielsen: Phys. Rev. **49**, 388 (1936); approx. values for large E/p E. W. B. Gill and A. von Engel: Proc. Roy. Soc. Lond., Ser. A **197**, 107 (1949); contains notes on D_2.

N_2: see Nielsen; also J. S. Townsend: Motion of electrons in gases. Oxford: Clarendon Press 1925.

O_2: R. H. Healey and J. W. Reed: Slow electrons in gases. Sidney: Amalg. Wireless, Sidney 1941; see also J. S. Townsend above.

Cl_2, Br_2, I_2: see Healey and Reed above.

NO, CO: M. F. Skinker and J. V. White: Phil. Mag. **46**, 630 (1923).

HCl, H_2O: V. A. Bailey and W. E. Duncanson: Phil. Mag. **10**, 145 (1930).

N_2O: see NO; also R. A. Nielsen and N. E. Bradbury: Phys. Rev. **51**, 69 (1937).

CO_2: see H_2O.

NH_3: see Nielsen and Bradbury above ($p = 2 \ldots 10$ mm. Hg); Bailey and Duncanson loc. cit. ($p = 0.3 \ldots 1$ mm. Hg).

BF_3: J. A. Bistline: Rev. Sci. Instrum. **19**, 842 (1948); cf. B. B. Rossi and H. H. Staub: Ionisation chambers. New York: McGraw Hill 1949 and E. Segre: Experimental nuclear physics I. New York: J. Wiley & Sons 1953, but their values are probably 10 times too large.

C_2H_4, C_5H_{12}: J. D. McGee and J. C. Jaeger: Phil. Mag. **6**, 1107 (1928).

Air: J. S. Townsend, reference N_2; Nielsen and Bradbury reference N_2O.

v_d in Hg has been measured in the positive column at larger currents (order 1 A). At $E/p = 1$, 10, 100 V/cm. · mm. Hg $v_d = 0.5$, 2, $20 \cdot 10^6$ cm./sec. resp.; see W. L. Granowski: Der elektrische Strom I. Berlin: Akademie Verlag 1955.

v_d in gas mixtures see e.g. L. Colli and U. Facchini: Rev. Sci. Instrum. **23**, 39 (1952) [A—N_2]; W. A. English and G. C. Hanna: Canad. J. Phys. **31**, 768 (1953); J. S. Townsend, reference N_2; D. Taylor and J. Sharpe: Proc. Inst. Electr. Engr. **98**/II, 174 (1951); this paper and English and Hanna's also gives v_d for CH_4. v_d in C_2H_5OH see H. den Hartog, F. A. Müller and C. S. W. van Rooden: Physica, Haag **15**, 581 (1949).

But

$$v = \left(\frac{2e}{m}\right)^{\frac{1}{2}} \varepsilon^{\frac{1}{2}}, \qquad \lambda_e = \frac{\lambda_1}{p}, \tag{24.2}$$

$$P(\varepsilon) = s_e \lambda_e = \frac{s_e}{p} \lambda_1 = a \lambda_1 (\varepsilon - \varepsilon_i) \tag{24.3}$$

where s_e is the ionization efficiency in ion pairs per cm. and $\varepsilon_i = e V_i$. The last three equations give

$$\frac{z_i}{p} = \left(\frac{2e}{m}\right)^{\frac{1}{2}} a \int_{\varepsilon_i}^{\infty} \varepsilon^{\frac{1}{2}} (\varepsilon - \varepsilon_i) f(\varepsilon) \, d\varepsilon. \tag{24.4}$$

MAXWELL'S distribution in terms of energy reads [cf. (23.4)]

$$f(\varepsilon) \, d\varepsilon = C_1 \varepsilon^{\frac{1}{2}} e^{-b\varepsilon} d\varepsilon \quad \text{where} \quad b = \frac{1}{k T_e}, \quad C_1 = \frac{2}{\pi^{\frac{1}{2}} (k T_e)^{\frac{3}{2}}}. \tag{24.5}$$

Inserting (24.5) in (24.4) and integrating one obtains with (23.1):

$$\frac{\alpha}{p} = \frac{5.5 \times 10^7 a}{v_d} \overline{V}^{\frac{1}{2}} (V_i + 1.3 \overline{V}) e^{-\frac{V_i}{\overline{V}}}. \tag{24.6}$$

Here $\overline{V} = \frac{3}{2} k T_e / e$, a is the slope of the ionization efficiency curve (values in Table 2) and the numerical factor results from e/m in (24.2) the factors in (24.5), conversion of e.s.u. into practical units etc. Thus by using the measured values of a, V_i, $\overline{V}$ and v_d, α/p can be calculated. V_i is given in Table 1 and the drift velocities v_d in Table 4.

b) Comparison of the elementary theory with experiment.

25. Molecular gases. The theoretical result will now be compared with the experiments in molecular gases for which (24.6) is most likely to apply. In Fig. 27 were shown the measurements of α/p in H_2 by various investigators. Using TOWNSEND'S measurements of $\overline{V}$ and v_d and the slope of the ionization efficiency curve (s_e), corresponding to a value $a = 0.21$, it was shown [*51*] that the theory represents the observations quite satisfactorily for the range of $E/p > 40$. However, a very close fit in that range is obtained when the measured values of s_e are used for the computation instead of the linear approximation for s_e. The curve labelled "theory" represents the result. The agreement was unsatisfactory as far as low values of E/p are concerned. Here the result from the experiments by AYERS [*39*] as well as HALE'S data [*40*] deviate from the theoretical curve very strongly. Since all the measured values are above those theoretically expected, it looks as if impurities with low V_i were present which are either directly ionized by electron collisions or by a collision of the second kind between excited H_2 molecules and the impurity atoms. Moreover α/p does not show a maximum at large values of E/p as indicated in earlier measurements [*17*]. This maximum can not follow from any theory which omits the fact that the ionization efficiency has a maximum (Fig. 2).

Recent observations by ROSE [*41*] confirm the above conclusions. The measurements which were extended over a range $E/p \approx 20$ to 1000 show that the theory gives now rather larger values of α/p than the experiment, at low E/p. In the range of high E/p, AYERS' data have been confirmed (see also [*95*]).

Let us next enquire into α/p for N_2. Fig. 28 shows the comparison between theory [*51*] and the various experiments. We observe first that all the experiments by BOWLS [*42*] and DUTTON et al. [*22*] lie on a smooth curve except for the earlier results by Ayers [39]. His values at $E/p < 70$ are apparently too high as are the results of Posins [52], March [44] and Townsend [14]. This was shown to be mainly due to the presence of traces of Hg vapour which has al lower value of V_i than N_2 and can thus be ionized by excited N_2 (see Sect. 30).

However, comparing the recent measurements with the theory [51] leaves still a factor ≈ 2 unaccounted for. The usual explanation for the fact that the theory (which is based on a Maxwellian distribution) leads to too high values of α/p is that the number of electrons of energy above V_i is smaller than was assumed. The main cause of the shortage of fast electrons is said to be due to the large number of inelastic collisions which the electrons make. These collisions represent a serious drain in the upper range of the energy distribution. It reduces the number of electrons which climb up the energy scale by moving in field direction between collisions without reducing appreciably the average electron energy. This seems to be particularly so when the inelastic losses occur at energies just below V_i and if V_i is much higher than $\overline{V}$, and when the probability of excitation is high and the number of possible excitations large. The effect of a depletion of fast electrons in the tail of the energy distribution should be more pronounced at lower E/p. There is also the variation of the mean free path with the electron energy—the Ramsauer-Townsend effect—which influences the shape of the energy distribution in that it probably determines the rate of climb up on the energy scale assuming that the electron acquires its energy from the field.

In order to investigate the reasons for the good agreement in H_2 and the poorer agreement in N_2 between theory and experiment the theoretical curve was re-calculated by the author (Fig. 28). The agreement now obtained is very satisfactory. The difference between the above theoretical result and the one presented now is: $a = 0.26$ instead of $a = 0.3$, as assumed earlier; use of Bradbury and Nielsen's values of v_a in N_2 at low E/p, and Townsend's interpolated values of v_d and $\overline{V}$ for N_2 at higher E/p and (where not available), for air with small corrections. The remaining small difference between theory and experiment may be due to the reasons given earlier. The large deviations at low E/p between Ayers and recent observations are no doubt due to impurities in the gas.

The measurements of α/p in O_2 have been made originally in the manner given in Sect. 15 without allowing for an attachment of electrons to molecules. Fig. 31 gives older results and also recent measurements [45]. In the interpretation of the latter an equation has been used which contains besides α a coefficient of attachment a (Sect. 28). This means that the apparent values of α as found from the slope of the curves $i/i_0 = f(x)$ for constant E/p are lower than the real values, and this is the more so the smaller E/p. The practical difficulty in evaluating the observations is the accuracy with which the deviations from non-linearity can be obtained. It is for this reason that it was necessary to make provisions for large changes in the inter-electrode distance x, for potential sources with a high degree of constancy and for a constant initial current i_0 in order to be able to determine and separate α from a. These remarks do not apply to complex gases with electro-negative components. Here, instead of nearly straight graphs $\ln i/i_0 = f(x)$, the curves after an initial rise turn over to reach saturation provided E/p is large (Sect. 28).

From $s_e = f(\varepsilon)$ a theoretical curve for O_2 is found (Fig. 31) which has been calculated from (24.6) with $a = 0.21$ and Townsend's values of $\overline{V}$ and v_d. The use of Bradbury's results [53] of v_d would make the agreement closer but there can be no doubt that the agreement is at present not satisfactory. Neither can an opinion be voiced whether the measurements or the theory are at fault (probably both).

A comparison between theory and experiment is not always possible because of the lack of reliable data. For example in the case of CO_2 where measurements of $\overline{V}$ and v_d are available up to $E/p = 50$ the value of a is not known. Assuming it to be of the same order as for O_2 or N_2, a value for α/p is obtained which lies slightly above the observed value. However, no great significance can be attached to this result for at the time when these experiments were carried out no allowance was made for impurities and electron attachment to electronegative species produced simultaneously with ionization.

The dependence of α/p on E/p in air (Figs. 30, 31) has been calculated and found to give values which are some 20% below those observed. There are several reasons which may account for this result: e.g. both $\overline{V}$ and v_d have been found by extrapolation to high E/p, also the formation of NO with an ionization potential of 9.5 V and its probably large ionization efficiency has been neglected. However, it is well to note that adding O_2 to N_2 increases α/p (Fig. 31) and that α/p for HCl lies above the curve for CO_2 or air, a result which has to be attributed rather to the presence of Cl than of H.

The old belief that electronegative gases like O_2 or the halogens should produce fewer ion pairs per cm. because the electrons liberated in an electron collision attach themselves readily to the gas molecules, is based on a fallacy. Attachment to form negative ions is on the whole a rare process. It becomes quantitatively of importance only at low electron energies and is sometimes confined to a narrow energy range. There is therefore no attachment to be expected at large E/p where the average electron energy is large. However, this conclusion cannot be applied to complex molecules where an electron collision may produce dissociation into positive and negative ions and where the complexity of the molecule gives rise to a larger number of possible excitations and dissociations which reduce the average energy of the electrons.

It appears, therefore, that the naïve theory given in the previous section can explain satisfactorily the ionization processes in an electron swarm provided that the electric fields applied are neither very low nor very high, and that the gas molecules are not too complex in character. It follows that in these cases the assumed electron energy distribution, viz. the Maxwellian, is a good approximation. Finally it can be deduced that there are special physical reasons for the appearance of this type of distribution. It is suggested that these are caused by the energy distribution of the inelastic losses which occur when the electrons collide with the molecules. Though the detailed picture is not yet clear it seems that when the inelastic losses are spread out evenly over the greater part of the energy scale up to the ionization energy, then a Maxwellian distribution ensues. When, however, elastic collisions only occur over the larger part of the energy interval 0 to V_i and hence inelastic collisions below V_i are confined to a relatively narrow range, then appreciable deviations from the Maxwellian type are observed. The latter applies particularly to atomic gases. Thus the successful calculation of α/p in molecular gases based on a Maxwellian distribution may be simply fortuitous: the values of the mean energy of electrons were taken from diffusion experiments, their interpretation, however, makes it necessary to know the distribution. The number of electrons capable of ionizing the gas depends very critically of the form of the distribution. The drift velocity is not always found under the same experimental conditions. Alltogether the respective errors may just cancel. The most accurate calculations (see article by W. P. Allis) show that the distribution law varies with E/p in a given gas. As will be seen later at low E/p the scarcity of inelastic losses will favour more nearly a DRUYVESTEYN distribution and α/p should be smaller than expected whereas in the medium range of E/p more nearly MAXWELLian conditions should apply.

26. Monatomic gases. It has been known for a considerable time that the application of (24.6) to rare gases leads to erroneous results. A numerical example should illustrate the position. Neon is chosen as the gas because it shows a small RAMSAUER-TOWNSEND effect which was originally considered responsible for the failure. At $E/p = 20$ V/cm. × mm. Hg $\alpha/p = 0.2$ ion pairs/cm. × mm. Hg in pure Ne. From (24.6) with $\overline{V} \approx 13$ V, $v_d = 1.5 \times 10^7$ cm./sec. and $a = 5 \times 10^{-2}$ we obtain $\alpha/p \approx 2$. Although v_d and $\overline{V}$ are not accurately known, there seems to be no doubt that the data are sufficiently reliable to dismiss the explanation that the divergence by a factor 10 is the result of inaccuracy. Besides, similar calculations which have been made earlier [*51*] for argon show that at low E/p the naïve theory gives results which are 3 to 4 orders of magnitude too large. These facts are convincing evidence in the case against the MAXWELLian distribution, and so the question arises which type of electron distribution should applicable to atomic gases. A brief summary will be given now of one of the first succesful attempts to deal with this problem.

c) Theory of the ionization coefficient in monatomic gases.

27. Although the electron distribution function for the general case is treated in W. P. ALLIS' article it is necessary to show by way of example that it is possible to arrive at a solution which in spite of the numerous approximations gives satisfactory agreement with experiment. The treatment here is DRUYVESTEYN'S second attempt [*54*] to calculate the ionization coefficient α/p in pure neon.

Let $n(\varepsilon)$ be the flow density (number per cm.² and sec.) of electrons of energy ε which move in a uniform electric field E. (The electric current is assumed to be so small that space charge effects etc. are negligible.) Their energy gained along dx is $n(\varepsilon)eE\,dx$. The energy lost by elastic collisions is $N_0 \overline{V} \cdot e \cdot \varkappa \cdot dx$, where N_0 is the number of collisions with neutral atoms per cm.³ and sec., $\overline{V}$ the volt-equivalent of the mean energy, and $\varkappa$ the average fractional loss in a collision. If N_e and N_i are the number of collisions by the electrons per cm.³ and sec. leading to excited and ionized atoms respectively and V_e and V_i the corresponding critical potentials, then the inelastic losses are $N_e V_e e\,dx$ and $N_i V_i e\,dx$. In addition $dn = N_i\,dx = \alpha n\,dx$ electrons are produced along dx and these are assumed to be brought on the average from zero energy to $\overline{V}$ which represents a loss $\alpha n e \overline{V} dx$. The energy balance thus reads

$$n E = \varkappa N_0 \overline{V} + \alpha n \overline{V} + N_e V_e + N_i V_i. \tag{27.1}$$

This equation implies that recombination in the gas or on the walls is absent.

From the number of ionizations along dx above and neglecting diffusion we have

$$N_i = \alpha n \tag{27.2}$$

and calling the ratio of ionization losses to excitation losses

$$q = \frac{N_i V_i}{N_e V_e} \tag{27.3}$$

we find from (27.1) and (27.3), neglecting the elastic losses,

$$\frac{\alpha}{p} = \frac{1}{\left(\frac{1}{q} + 1 + \frac{\overline{V}}{V_i}\right) V_i} \frac{E}{p}. \tag{27.4}$$

(27.4) only holds when $N_e V_e / \varkappa N_0 \overline{V} \gg 1$. Since $\overline{V}/V_i < 1$ for moderate E/p, the main variation results from $q = f(E/p)$. In a certain range, q was found to vary approximately proportional to E/p so that $\alpha/p \propto (E/p)^2$. The meaning of (27.4) can be expressed in another way: α, the number of ion pairs per cm. in field direction, cannot attain the value E/V_i which is the inverse of the Stoletow constant (Table 3) and is smaller by a factor $(1/q + 1 + \overline{V}/V_i)^{-1}$. It now remains to show how $q = f(E/p)$ can be derived.

When a swarm of electrons moves simultaneously in an electric field and a concentration gradient of charge in the same direction, the number density of flow F of electrons in $d\varepsilon$ is given by

$$F = -\left(\frac{e\lambda}{mv}\right) E\varrho + \frac{v\lambda}{3}\frac{d\varrho}{dx} \tag{27.5}$$

where v is the electron velocity, ϱ the number of electrons/cm.^{3} and unit energy interval, and the first and second bracket the mobility and diffusion coefficient respectively. $\varrho = \varrho(\varepsilon, x)$ where $\varepsilon = \frac{1}{2} m v^2$ is the energy of the electrons whose steady distribution we seek, and v their (random) velocity

$$\frac{d\varrho}{dx} = \frac{\partial\varrho}{\partial x} + \frac{\partial\varrho}{\partial\varepsilon}\frac{d\varepsilon}{dx} = 0 + \frac{\partial\varrho}{\partial\varepsilon} eE.$$

Thus (27.5) can also be written

$$-F = \frac{3}{2}\frac{v\lambda}{3} E \frac{e\varrho}{\varepsilon} - \frac{v\lambda}{3} E e \frac{\partial\varrho}{\partial\varepsilon} \approx \frac{v\lambda}{3} E e \left[\frac{\varrho}{2\varepsilon} - \frac{\partial\varrho}{\partial\varepsilon}\right] = n(\varepsilon) \tag{27.6}$$

where ϱ is the energy distribution of the number density. The factor 3 has actually not been neglected since it has been shown [55] that under these circumstances field and diffusion are interdependent and thus the velocity vectors cannot simply by added with their full values.

The flow of electrons n takes from the field E an amount of energy which is assumed here to balance the losses by elastic collisions. Hence with $n(\varepsilon)$ electrons/cm.$^2 \times$ sec. and a density $\varrho(\varepsilon)$ both per unit energy range each making v/λ elastic collisions per sec. losing $\varkappa \cdot \varepsilon$ ergs on the average per collision, we have

$$n(\varepsilon)\, eE = \varrho(\varepsilon)\,(v/\lambda)\,\varkappa\,\varepsilon. \tag{27.7}$$

With $\varkappa$=const., and λ=const., from (27.6) and (27.7) by eliminating $n(\varepsilon)$ and integrating, the Druyvesteyn distribution is obtained, viz.

$$\left.\begin{aligned} \varrho(\varepsilon) &= C \cdot \varepsilon^{\frac{1}{2}}\, e^{-\left(\frac{\varepsilon}{\varepsilon_0}\right)^2}, \\ \varepsilon_0 &= \left(\frac{2}{3\varkappa}\right)^{\frac{1}{2}} eE\lambda, \\ C &= \frac{2}{\Gamma(\frac{3}{4})}\left(\frac{1}{\varepsilon_0}\right)^{\frac{3}{2}} \end{aligned}\right\} \tag{27.8}$$

where C is found by integrating $\varrho(\varepsilon)\, d\varepsilon$ from 0 to ∞ and equating it to unity. The mean energy is defined in Druyvesteyn's paper by $\bar{\varepsilon} = [\Gamma(\frac{5}{4})/\Gamma(\frac{3}{4})]\,\varepsilon_0$. ε_0 is a mean electron energy of this type of distribution, and is proportional to

$E/(p\sqrt{\varkappa})$. Fig. 40 shows the distribution (27.8) and the Maxwellian for the same mean energy $\bar{\varepsilon}$. It is apparent that in (27.8) there are fewer electrons of energy above eV_i provided the conditions of experiment are not such that the ionization energy of the gas lies below the point of intersection of the two curves. It will be shown that this result is markedly changed when inelastic collisions are included.

So far we have assumed that electrons of any energy ε in the distribution which have lost energy to the neutral gas (thereby increasing slightly its temperature) have taken up the energy loss from the electric field. That means that they move down and then up on the energy scale withing a range $d\varepsilon$ and thus $\partial\varrho/\partial x = 0$ and $\partial n/\partial x = 0$. This is not the case when inelastic collisions occur.

If the probabilities of ionization and excitation are $P_i(\varepsilon)$ and $P_e(\varepsilon)$, the number of electrons (per cm.[3] and sec.) which lose excitation or ionization energy and leave the range between ε and $\varepsilon + d\varepsilon$ is $\varrho(\varepsilon)\,(v/\lambda)\,(P_e + P_i)$. Let us neglect the few electrons coming from very large energy ranges which have suffered inelastic collisions but let us consider those which have collided elastically and those which have gained energy from the field. The corresponding number is

$$eE\frac{\partial n}{\partial\varepsilon} - \frac{\partial}{\partial\varepsilon}\left[\varrho\,\varkappa\,\varepsilon\,\frac{v}{\lambda}\right].$$

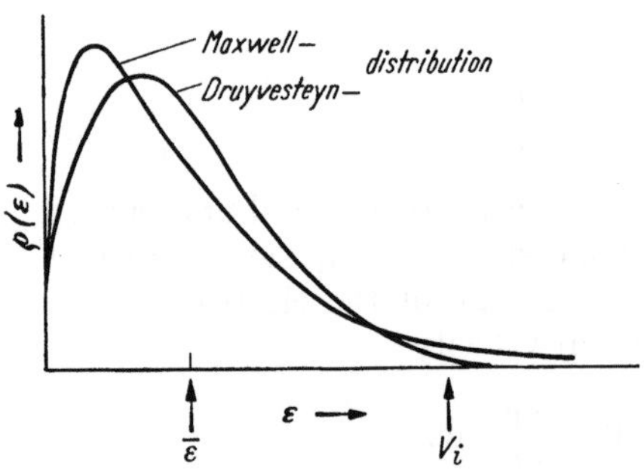

Fig. 40. Electron energy distribution ϱ as a function of the energy ε for the same mean energy $\bar{\varepsilon}$ and the same number of electrons (the same area covered by the curve).

Maxwell distribution: $\varrho(\varepsilon) = C'\varepsilon^{\frac{1}{2}-\frac{\varepsilon}{\bar{\varepsilon}}}$;

Druyvesteyn distribution: $\varrho(\varepsilon) = C\,\varepsilon^{\frac{1}{2}}\,e^{-0.55\left(\frac{\varepsilon}{\bar{\varepsilon}}\right)^2}$ identical with (27.8)

In the steady state and for energies $> eV_i$

$$-\frac{\partial n}{\partial x} = eE\frac{\partial n}{\partial\varepsilon} - \frac{\partial}{\partial\varepsilon}\left[\varrho\,\varkappa\,\varepsilon\,\frac{v}{\lambda}\right] + \varrho\left(\frac{v}{\lambda}\right)(P_e + P_i). \qquad (27.9)$$

A similar but more complex equation is obtained (54) for $eV_e < \varepsilon < eV_i$.

Assuming that ϱ and n vary as $e^{\alpha x}\cdot f(\varepsilon)$ and using the complete diffusion equation which is (27.6) plus a term $-\frac{v\lambda}{3}\frac{\partial\varrho}{\partial x}$, a differential equation is found of the form

$$\frac{\partial^2\varrho}{\partial\varepsilon^2} - \frac{3(P_e + P_i)}{(eE\lambda)^2}\,\varrho = 0. \qquad (27.10)$$

If the coefficient is constant, the general solution is the sum of two simple exponential terms. The solution is then $e^{-\frac{\varepsilon}{\varepsilon_0}}$ and not $e^{-\left(\frac{\varepsilon}{\varepsilon_0}\right)^2}$ as for elastic collisions. In fact P_e and P_i are proportional to the energy in excess of eV_e and eV_i respectively and thus the solution of (27.10) is involved and includes Hankel and Bessel functions. For example the distribution $\varrho(\varepsilon)$ which gives results not much different from more exact calculations is

$$\varrho = C\cdot e^{-\frac{(3P_e')^{\frac{1}{2}}\cdot\varepsilon}{eE\lambda}} \qquad (27.11)$$

where P_e' is the probability of excitation at an energy value $\frac{2}{3}$ up in the range between V_e and V_i and $C = (3P_e)^{\frac{1}{2}}/(eE\lambda)$.

The approximate solution for $\varepsilon > eV_i$ would contain $(P_e + P_i)$ instead of P_e. In Fig. 41 a numerical result is shown for He for various values of E/p obtained from lengthy calculations and using $P_e(\varepsilon)$ and $P_i(\varepsilon)$. If the value of ϱ at a certain ε is known, the value of ϱ at a larger value of ε can be found from the slope of ϱ from (27.11). But even this procedure does not give sufficiently accurate results unless the change in ε and E/p are kept small enough. Recent results [23] from high frequency measurements give ϱ for $E/p>10$.

Perhaps an explanation should be given for the curious result that when allowance is made for inelastic collisions the number of fast electrons decreases less steeply in the distribution than for elastic collisions. Instead of e^{-x^2} the variation is e^{-x} when x is the ratio of the energy to the mean energy of the electron distribution. First of all the onset of inelastic losses is roughly speaking equivalent to an increase in the fractional loss $\varkappa$. Secondly electrons which have collided inelastically jump down nearly to the bottom of the energy scale and have to be replaced by those moving slowly up through collisions whose component path is parallel to the direction of the electric field vector. This forces of course all the electrons (not only those at $\varepsilon - d\varepsilon$) to move up at much increased rate and so a large number of electrons, down to ε_0 and further, has to be moved up the scale to balance the losses. It is this point probably which reduces the slope $d\varrho/d\varepsilon$ and causes the change from e^{-x^2} to e^{-x}. The third point which is difficult to discuss qualitatively is the variation of the electron mean free path with energy. Here the reader is referred to the authoritative account which is given by W. P. ALLIS in this volume.

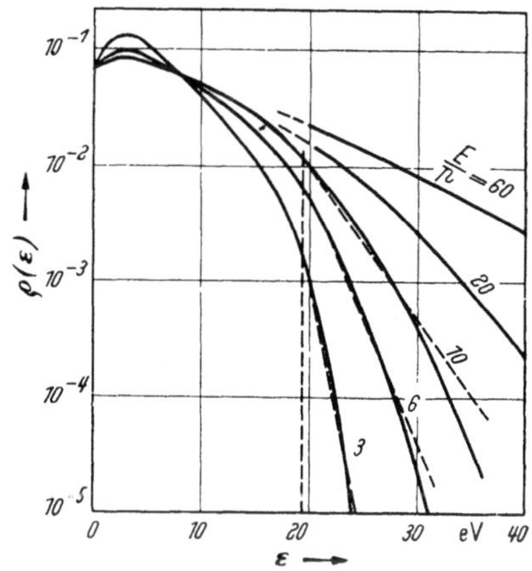

Fig. 41. Electron energy distribution $\varrho(\varepsilon)$ in He for different values of the electric field E/p. Full curves: solution of (27.10). Dashed curves: approximate solution (27.11).

It might be interesting to speculate about the reason why α/p at low values of E/p rises as the atomic number increases which is the reverse of what s_e of Fig. 2 suggests. The answer is probably that though s_e, the mean energy and the drift velocity is higher e.g. for A than for Ne (alltogether by a factor 10 or more) these are several parameters counter-balancing the increase; the collision cross section which is considerably smaller in A and the relative number of electrons in the distribution above V_i is much smaller in A than in Ne or He. Thus the magnitude of the tail of the distribution and the cross section for electron collisions seem to be the decisive factor in a rigorous theory of the ionization coefficient.

Having obtained the distribution function ϱ it is now possible to conclude our problem of finding the energy ratio of excitations to ionization $(1/q)$ in (27.4). If the number of electrons making ionizing and exciting collisions is N_i and N_e and their respective energies ε_i and ε_e, thus approximately

$$\frac{1}{q} = \frac{N_e\,\varepsilon_e}{N_i\,\varepsilon_i} \approx \frac{c_1 \int\limits_{\varepsilon_e}^{\infty} \varrho \cdot \varepsilon^{\frac{1}{2}} \cdot (\varepsilon - \varepsilon_e)\, d\varepsilon}{c_2 \int\limits_{\varepsilon_i}^{\infty} \varrho \cdot \varepsilon^{\frac{1}{2}} \cdot (\varepsilon - \varepsilon_i)\, d\varepsilon} \tag{27.12}$$

where c_1 and c_2 are the excitation and ionization efficiencies per unit energy change. Since ϱ is the number density of electrons of energy per unit energy and the collision frequency is $\propto \varepsilon^{\frac{1}{2}}$, the above integrals can be solved, usually by numerical methods. Fig. 42 shows $1/q=f(E/p)$ for Ne [29], [57]. Thus for fields below $E/p=3$ there are more than 100 exciting collision for one ionizing collision whereas the two types of collisions become equal at about $E/p=50$.

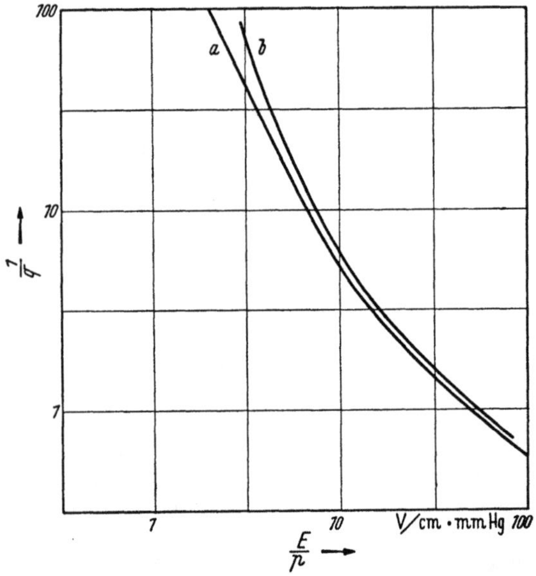

Fig. 42. Ratio $1/q$ of the rate of energy (per cm.³) of excitation to that of ionization as a function of the electric field E/p in Ne. Curve a: calculated; curve b measured.

Inserting the values q in (27.4) $\alpha/p=f(E/p)$ is found. It is not necessary to know accurately $\overline{V}=f(E/p)$ since $\overline{V}/V_i<1$. For large values of E/p (27.4) should fail because the function ϱ cannot be determined with sufficient accuracy. Fig. 22 shows the calculated curve for Ne together with the measured curve [54]. The corresponding curves for He do not fit so well; the method of calculation is, however, different from that given above [56]. Other approximate methods, see [34]. At large p the concentration of metastable He atoms can become so large that electrons of low energy may ionize metastable atoms. This would increase α, but $\alpha/p \neq f(E/p)$.

VII. Multiplication in the presence of negative ions.

28. The ionization of a gas by an electric field becomes somewhat modified when it is electro-negative. It is evident that in this case some of the electrons become attached (at least for a certain time) and because of their mass the negative ions so formed never attain sufficient energy to ionize another gas molecule. This effect is partly balanced by the greater ease with which electro-negative gases are ionized in large fields, a point which has been mentioned earlier (Sect. 25).

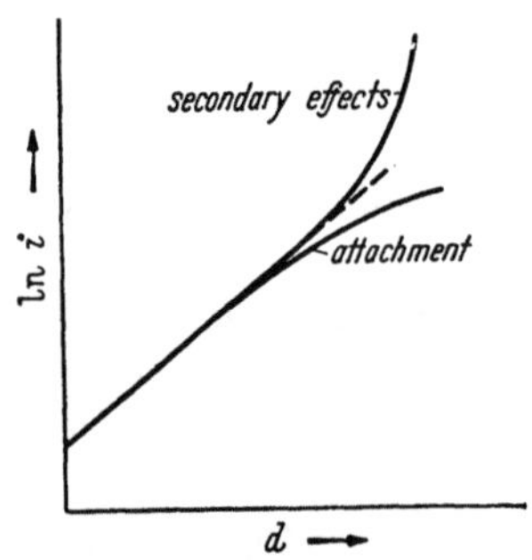

Fig. 43. Comparing the growth of ionization current i as a function of the electrode separation d at a constant field E/p for primary ionization by collision in the gas (dashed curve) or for ionization by collision and secondary effects (e.g. secondary electron emission from the cathode) with that by electron attachment associated with primary and secondary effects.

In general, when negative ions are present in an electron swarm which moves and multiplies in an electric field, the growth of current i/i_0 as a function of the distance d between the electrodes or as a function of potential difference V for a constant value of E/p varies as shown in Fig. 43. The diagram indicates that secondary effects may counteract effects due to electron attachment. In fact the logarithm of the multiplication i/i_0 may be only slightly curved and this makes it very difficult to measure with sufficient accuracy the attachment coefficient particularly when the ionization coefficient is relatively large.

The attachment coefficient a is defined as the number of negative ions which are formed by one electron moving along 1 cm. of path in field direction. The coefficient a/p depends on the strength of the electric field per unit pressure, the gas and its dissociation products. Attachment is in general favoured by slow relative velocities between electrons and molecules and thus becomes negligible in simple gases at high values of E/p.

In order to derive the variation of $a/p = f(E/p)$ the type of attachment has to be assumed. Only then would it be possible to obtain a relation between i/i_0, d, α/p and a/p. Let for example a neutral molecule attach an electron and the molecular ion dissociate into an atomic ion and an atom—a process called *dissociative attachment*. Then the increase in electrons dn_e along dx is proportional to the number of electrons n_e at x and to the difference between ionizations and attachment coefficients:

$$d n_e = (\alpha - a)\, n_e\, d x. \tag{28.1}$$

Integrating between 0 and d with $n_e = n_0$ at $x = 0$, we have

$$\frac{n_e}{n_0} = e^{(\alpha - a) d}. \tag{28.2}$$

The increase in negative ions along dx is similarly

$$d n^- = a\, n_e\, d x. \tag{28.3}$$

Substituting n_e from (28.2) and integrating, we find

$$\frac{n^-}{n_0} = \frac{a}{\alpha - a} \{e^{(\alpha - a) d} - a\}. \tag{28.4}$$

Thus the multiplication is

$$\frac{i}{i_0} = \frac{n_e + n^-}{n_0} = \frac{1}{\alpha - a} \{\alpha\, e^{(\alpha - a) d} - a\}. \tag{28.5}$$

It follows that for $E/p = \text{const.}$ or $\alpha = \text{const.} = c$ we have

$$\frac{i}{i_0} = A\, e^{c d} - (A - 1)$$

or

$$\ln \frac{i}{i_0} = \ln\{A\,(e^{c d} - 1) + 1\}$$

which for small values of d gives

$$\ln \frac{i}{i_0} = \ln\left\{A\left[\frac{c\,d}{1!} + \frac{(c\,d)^2}{2!} + \cdots\right] + 1\right\} \tag{28.6}$$

where $A = \alpha/(\alpha - a)$ and $c = \alpha - a$. Hence graphs of $\ln i/i_0$ as a function of α for constant E/p should have a concave curvature with respect to the abscissa. This in fact is confirmed by experiment. When $a \to 0$, from (28.5) $\ln i/i_0 = \alpha\, d$ which corresponds to (10.3).

Measurements have been carried out recently [*48*] which give more accurate results of a/p than obtained hitherto. This has been achieved by using gaps with a large variation of the distance between the electrodes, potential sources with small fluctuations, constant light sources for irradiation and chemically stable surfaces of the electrodes.

Figs. 44 to 47 show the current i as a function of the electrode distance d in a plane parallel gap for air, O_2, freon (CCl_2F_2), and CF_3SF_5 respectively at different values of E/p for $i_0 = 10^{-11}$ A. It is known that as E/p increases, α/p rises but a/p seems to fall, at least at lower values of E/p. Therefore, in this range $(\alpha - a)$ becomes small, i/i_0 appears to rise very slowly as d is increased and the evaluation is difficult. On the other hand some of the figures show that the deviation of the semi-logarithmic plot from the straight line is very small

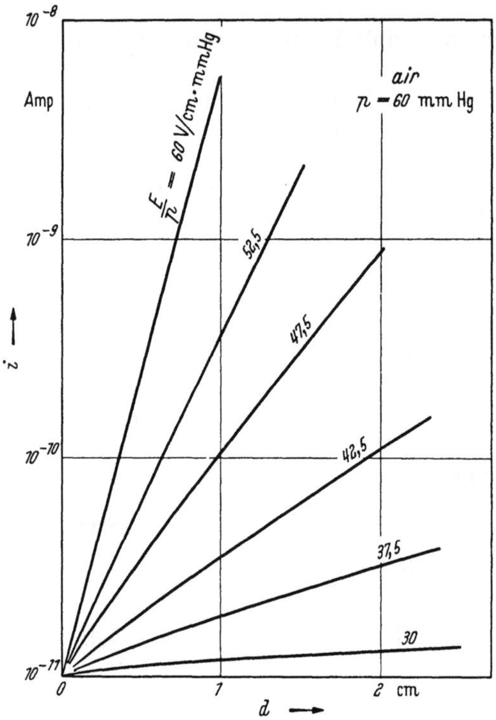

Fig. 44. Ionization current i ($i_0 = 1.10^{-11}$ A) as a function of the electrode separation d for various values of the electric field E/p in air at constant pressure p.

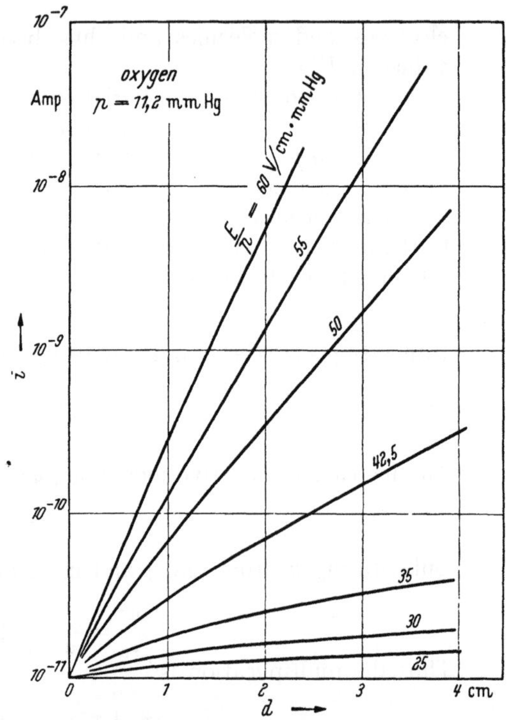

Fig. 45. Ionization current i ($i_0 = 10^{-11}$ A) as a function of the electrode separation d for various values of E/p in O_2 at constant pressure p.

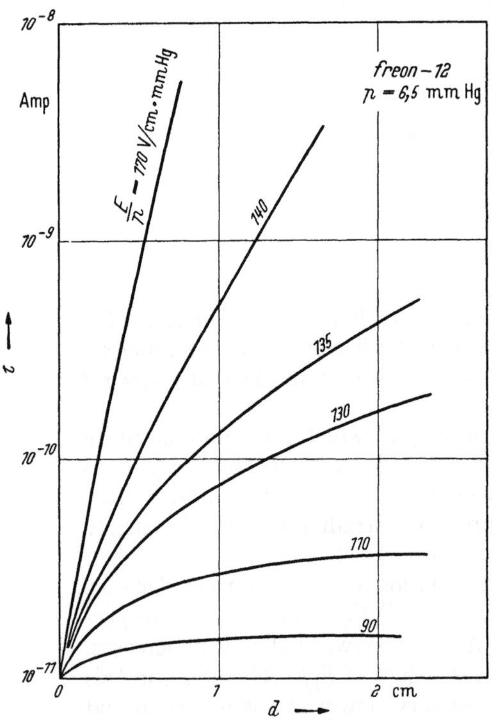

Fig. 46. Ionization current i ($i_0 = 10^{-11}$ A) as a function of the electrode separation d at different values of E/p in freon at constant pressure p.

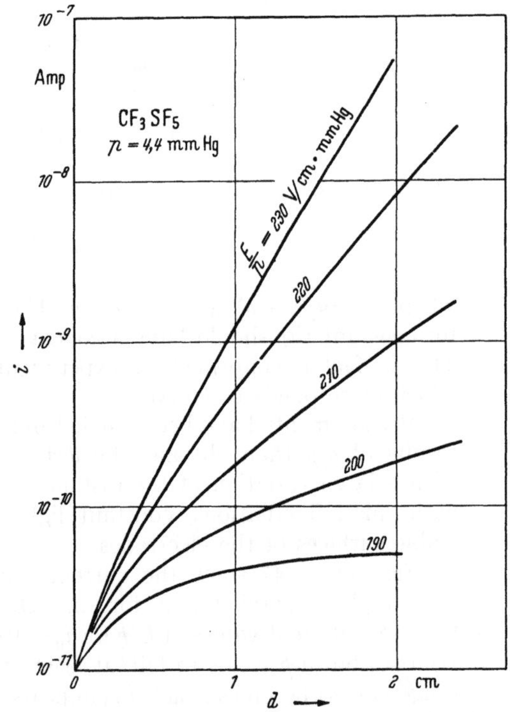

Fig. 47. Ionization current i ($i_0 = 10^{-11}$ A) as a function of the electrode separation d at different values of E/p in CF_3SF_5 at constant pressure.

and accordingly the accuracy of measurement of a/p not always satisfactory. The attachment coefficients a/p as derived from these measurements are shown in Figs. 48 and 49 for O_2 and freon respectively. The corresponding values of α/p have been given earlier (Sect. 21). We conclude that considerable errors may arise in measurements of α/p at low values of E/p in the presence of negative ions.

Knowledge on the formation of negative ions is still restricted, particularly as far as data on the probability of attachment are concerned and some of the available data are contradictory [*48*], [*59*], [*60*]. Also no attempt has yet been made to derive the attachment coefficients a/p in electric field from the electron distribution and the probability of attachment [*48*].

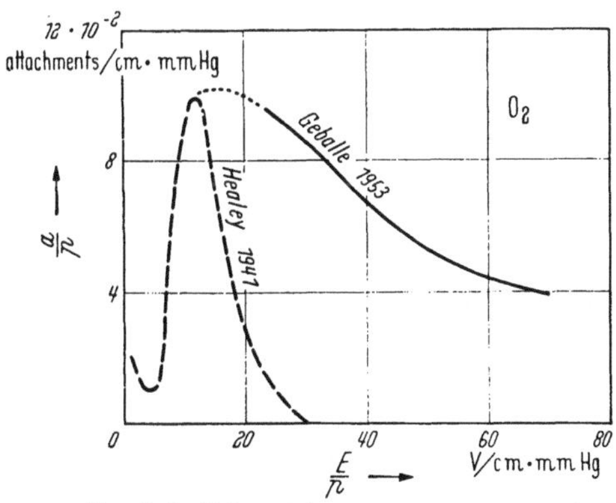

Fig. 48. Coefficient of electron attachment a/p as a function of the electric field E/p in O_2 derived from the measurements Fig. 45. $p = 10$ to 40 mm. Hg.

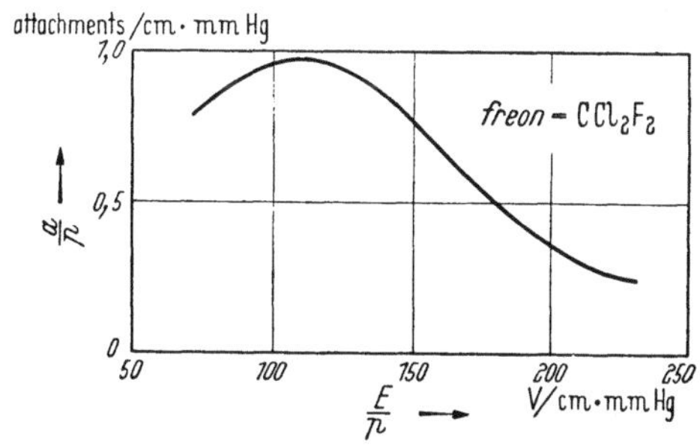

Fig. 49. Electron attachment coefficient a/p as a function of the electric field E/p in freon.

There are two other points which are to be considered here. One is that in deriving (28.5) we habe excluded secondary processes. Assuming a constant secondary coefficient γ (Sect. 15), the multiplication is found to be

$$\frac{i}{i_0} = \frac{\frac{\alpha}{\alpha - a} e^{(\alpha - a)d} - \frac{a}{\alpha - a}}{1 - \gamma \frac{\alpha}{\alpha - a}\{e^{(\alpha - a)d} - 1\}}. \qquad (28.7)$$

The second point is the different types of attachment. (28.5) was derived under the assumption that either direct attachment takes place, e.g. $O_2 + e \to O_2^-$ or dissociative attachment in which first an excited ion is thought to be formed which subsequently dissociates into an atomic ion and an atom, as e.g. $O_2 + e \to (O_2^-)^* \to O^* + O^-$. However, there is another type of collision viz. a dissociation into ions with the electron remaining free. If we apply this to O_2, we have $O_2 + e \to O^+ + O^- + e$. At present too little quantitative information is at hand to discuss the matter to advantage.

VIII. Multiplication in the presence of positive ions.

29. At first right it appears to be superfluous to discuss the presence of positive ions under a special heading. So far only the relation between the current and their distribution in space has been considered. We shall now list possible specific effects associated with positive ions:

Their space charge may distort the applied field; this will be treated in Sect. 35. The charge density depends to a certain extent on the charge transfer and the gas density (cf. H. S. W. Massey, vol. XXXVI of this Encyclopedia).

In very large fields (E/p) positive ions can excite and ionize gas molecules.

When the number density of the positive ions is large the electrons are scattered by ions rather than by gas molecules.

Positive ions may give rise to recombination in the presence of negative ions or slow electrons.

Collisions between positive ions and gas molecules have been regarded by TOWNSEND [14], [61] as an important secondary effect measured by the so-called second ionization coefficient β, giving the number of ion pairs produced by one positive ion along 1 cm. of path in field direction. To assess the relative importance of gas and electrode (secondary) effects, β/α is to be compared with the secondary coefficient γ. TOWNSEND and LL. JONES [62] conducted an experiment to discriminate quantitatively between β/α and γ. Though the result indicates that in H_2 for $E/p = 150$ to 300, β/α is effective and in He ($E/p = 30$ to 70) γ is in control, there is nevertheless some doubt whether the accuracy of the measurements was sufficient to arrive at a final conclusion. The few available data on the ionization efficiency of positive ions [15] do not support the above view. However, there is no doubt that the ionization by positive ions will play a part when E/p becomes sufficiently high.

If we assume that α/p retains its established meaning in highly ionized gases, then we have to allow for the fact that the mean free path of the electron at large charge densities is essentially determined by the density of positive ions and the electron temperature. It is thus considerably reduced because of the long-range field of the ions which acts as if the equivalent gas pressure were larger.

At sufficiently large ion densities the multiplication can be reduced by recombination in the gas between positive and negative charges. This magnitude can be easily estimated. From (23.1) $\alpha = z_i/v_d$ and since the rate of recombination/cm.^{3} sec $dN/dt = -\varrho N^2$ where N is the concentration in ions/cm.^{3} and ϱ the recombination coefficient in cm.^{3}/sec. we find the effective ionization coefficient

$$\alpha' = \frac{z_i - \varrho N}{v_d}. \tag{29.1}$$

Taking e.g. $z_i = 10^5$/sec., $\varrho \lesssim 10^{-7}$ cm.^{3}/sec., a change of $\gtrsim 10\%$ in α' would require $N \gtrsim 10^{11}$/cm.^{3}. This is much higher than the concentrations used in electron swarms.

IX. Multiplication by metastable atoms colliding with traces of a foreign gas.

30. From spectroscopic evidence FRERICHS [63] suggested that collisions of the second kind involving excited atoms, are responsible for the appearance of spark spectra, i.e. spectra of excited ions. The metastable atoms, like Ne_{met} and He_{met} were thought to transfer their excitation energy when colliding with Al or Cd respectively (in a hollow cathode discharge) and produce the spark spectra of these vapours. Further He_{met} with A added, induces the latter to emit spark spectra. Hence it was deduced that metastable atoms may ionize substances whose ionization energy is below the metastable energy level. At the same time PENNING [64], [65], [66] discovered that Ne containing traces of A has a lower sparking potential than pure Ne. For example a reduction by 50% is found for a plane gap of 3 cm. at $p = 40$ mm. Hg when 10^{-5} parts (in volume) of argon are added. He deduced from measurements of the ionization current that this variation is due to an increase in α/p—which at low E/p can amount to several orders of magnitude—and not to secondary effects at the cathode [67].

Also by irradiating e.g. Ne containing traces of A with visible Ne light [65], by irradiating with A light A with traces of Hg, or by adding a quenching gas, he showed that by destroying the ultimate cause, viz. the metastables, the effect disappears. Though the evidence appears to be conclusive, other explanations of the "PENNING effect" have been advanced [61], [68] but so far neither side has provided a convincing quantitative treatment. One argument is that though the concentration of "impurity atoms" is low, the number of electrons able to ionize them is very much larger than the number which can ionize the parent gas ([61] p. 143). This is certainly not true for A in Ne.

When a rare gas atom is excited by an electron colliding with it, then in general a normal excited atom with a short life (10^{-8} sec.) is formed. If the level of the excited atom is near a metastable level (say 1 to 15 times the kinetic energy of the atoms i.e. 10^{-2} to 0.15 eV) then a transition into a metastable state can occur by collisions with other gas atoms. The metastable atom so formed has usually a very much longer life τ which according to the conditions of experiment can last up to seconds. Its life is brought to an end either by collisions with other atoms (at high concentrations, which are excluded here, charged particles and quanta may be effective as "quenching" agents) or by collisions with the wall of the vessel. A calculation [69] of the diffusion problem gives τ as a function of the density of the gas and the radius R of the (long) cylindrical vessel and is expressed by the sum of the collision frequencies, viz.

$$\frac{1}{\tau} = \frac{C_1}{p R^2} + C_2 p \tag{30.1}$$

where the first term refers to wall and the second to gas collisions. The constant C_1 contains the relative velocity of the metastable and neutral atom and their atomic collision cross section, C_2 the probability of destruction in a collision (the quenching factor), the cross section and the velocity. Eq. (30.1) shows that $\tau = f(p)$ has a maximum which for some of the rare gases lies between 0.1 and 1 mm. Hg for values of R of order of several cms.

In order to transfer the potential energy of a metastable atom of type 1 to an atom of type 2 in the ground state to ionize it, the energies must be such that

$$V_{\text{met}\,(1)} > V_{i\,(2)}. \tag{30.2}$$

For example V_{met} of Ne is 16.5 V, V_i of argon 15.8 V and hence (30.2) is satisfied. Similarly He_{met} with 19.8 V should ionize A, further A_{met} and Ne_{met} should ionize Hg, Br_2 or Cl_2 etc. There is evidence that this is the case [65]. Furthermore Cd vapour is ionized by metastable A atoms [70] and K vapour by metastable Hg etc. [71], [63]. The probability of ionization for such collisions is high but not every collision leads to ionization. Recent results [72], [73] give the corresponding cross sections for $\text{He}_{\text{met}}-\text{A}$, $\text{He}_{\text{met}}-\text{Hg}$ and $\text{Ne}_{\text{met}}-\text{A}$ as 9.7×10^{-17}, 1.4×10^{-14} and 2.6×10^{-16} cm.² respectively. For $\text{Ne}_{\text{met}}-\text{A}$, say, every third collision which an A atom makes with any Ne_{met} ionizes A.

Before giving a more detailed account of the ionizing effects of metastable atoms, a number of elementary considerations should illustrate the problems. Let Ne* be the metastable neon atoms formed by electron collisions. These can be destroyed in two ways: either by colliding with Ne atoms in the ground state which leads to an emission of resonance light, viz.

$$\text{Ne}^* + \text{Ne} \rightarrow \text{Ne} + \text{Ne} + h\nu$$

or by a collision with an argon atom in the ground state thereby ionizing it, viz.

$$Ne^* + A \rightarrow Ne + A^+ + e.$$

Let us also assume that the pressure is high enough (say $p > 10$ mm. Hg) so that wall collisions can be neglected. Then the average life τ of Ne* is given by (30.1); at room temperature

$$\tau = \frac{10^{-2}}{p} \tag{30.3}$$

where τ is in seconds and p in mm. Hg [74], [76].

There is another process which destroys metastable atoms: red light of a certain wavelength is readily absorbed by Ne* and brings about a transition into a higher but normal excited state from which the atom returns into the ground state. Such light is produced by the discharge itself or can be obtained from an external source. This effect will be neglected here but discussed later on.

31. The presence of argon reveals itself only if the average time interval between the creation of a Ne* and the collision with an A atom is smaller than or equal to the life of a Ne* atom. This will depend of course on the fraction c_A of A in Ne. The number of collisions a Ne atom makes per sec. is v/λ and during this time it hits an A atom $c_A v/\lambda$ times where v and λ are the mean velocity and the mean free path of an atom. Since λ for Ne* is probably not much smaller, this result applies also to Ne*. The time interval between the birth of Ne* and its destruction by A (for every such collision is assumed here to lead to destruction) is with λ_1, the value of λ at $p=1$:

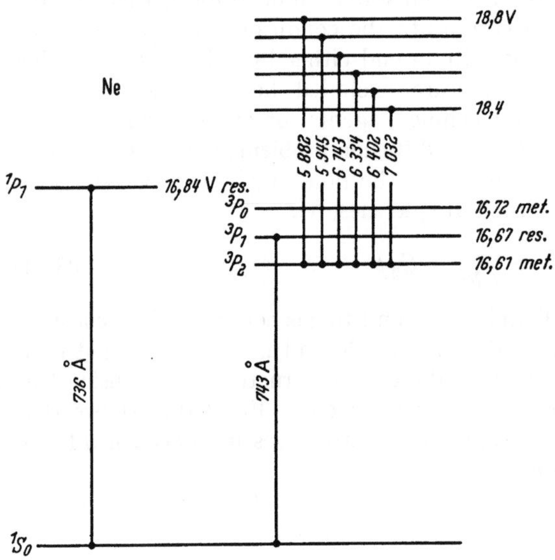

Fig. 50. Energy level diagram of Ne. (Only those transitions are given which have a bearing on the problem.)

$$\Delta t = \frac{\lambda}{c_A \cdot v} = \frac{\lambda_1}{c_A v \cdot p} . \tag{31.1}$$

Eq. (31.1) and (30.3) gives the fraction for which the effect is expected to be fully developed viz. $\Delta t = \tau$ or

$$c_A = \frac{10^2 \lambda_1}{v} . \tag{31.2}$$

Taking $\lambda_1 \approx 10^{-2}$ cm., $v = 5.10^4$ cm./sec., we find that $c_A \approx 2.10^{-5}$. In fact, both $\alpha/p = f(E/p)$, see Fig. 26, as well as the starting potential $V_s = f(c_A)$ show that with this concentration a pronounced effect is observed (cf. articles by L. B. LOEB and S. C. BROWN in vol. XXII).

Next it will be shown how an estimate of the intensity of light can be obtained which would destroy a greater portion of metastable atoms before any other of the effects mentioned earlier could operate. Again an electron is considered to have excited Ne into one of the two metastable states in which it has a life given by eq. (30.3). Thus we tacitly assume that traces of argon should not play a rôle in this case. According to Fig. 50 irradiation by any of the 6 red and yellow Ne lines would raise the atom from the metastable $^3P_2$ state into a higher level from which it could return directly into ground state thereby emitting an ultraviolet quantum. The most effective lines for irradiation have been recently isolated [77], [78].

If P quanta per cm.² and sec. pass through the excited gas, and q is the cross section in cm.² for absorption of a red quantum by a metastable Ne atom, then $P \cdot q$ is the number of quanta absorbed per sec. by one metastable. Equating this to the number of quenching collisions per sec., we have

$$P q = \frac{1}{\tau} \approx 10^2 \cdot p. \tag{31.3}$$

Hence, if a simple absorption process were to operate, with a light source producing $P \approx 10^{17}$ quanta/sec. cm.² at $p = 10$ mm. Hg, a photo-excitation cross section of metastables of about 10^{-14} cm.² would be needed. In fact, such a value of P has been estimated from the known efficiency of a Ne source (< 10%) and the geometry. The value of q is orders of magnitude higher than the known values of absorption of visible light by atoms in the ground state. However, one has to compare the absorption of quanta by metastables which excite them to a higher normal level with the absorption of resonance quanta by atoms in the ground state. Since large cross sections of the order quoted have been observed in the latter case [*69*], [*75*], we conclude that this could also apply to transitions from metastable states. The question now arises whether other processes may play a role. For example the destruction of metastables might reduce the population in the resonance level or the electron energy distribution might change and enhance the effect.

32. We shall now try to indicate how the ionization coefficient can be calculated in a mixture of Ne and A when $< 10^{-4}$ parts of A are present [*30*]. At higher concentrations, the excitation and ionization of argon by electron collisions etc. has to be considered too. At low concentrations the ionization in the gas is via Ne* — A. The treatment holds only for low E/p; at larger E/p direct ionization of Ne by electrons has to be included. The notation and arguments which follow are similar to those in Sect. 22 and 27.

From (23.1) we know that $\alpha = z_i/v_d$. The number of ionizing collisions/sec. is here proportional to the probability Π_1 that a metastable atom is created, to the probability Π_2 that during its life it will hit an A atom and ionize it, and to an integral corresponding to (23.3) which contains the collision frequency of electrons and their energy distribution and describes the number of exciting collisions per sec. and per electron. Hence we have instead of α

$$\alpha_{\text{met}} = \frac{\Pi_1 \Pi_2 \int\limits_{\varepsilon_1}^{\infty} \varrho(\varepsilon) \frac{v}{\lambda} d\varepsilon}{v_d} \tag{32.1}$$

where ε_1 is the excitation energy, v is the actual electron velocity given by $\frac{1}{2} m v^2 = \varepsilon$ and λ, Π_1, Π_2 are all assumed to be constant. For v_d we assume a relation of the form

$$v_d = \text{const} \frac{E}{p} \frac{1}{\overline{V}^{\frac{1}{2}}} \tag{32.2}$$

where $\overline{V}$ ist the mean energy of electrons. Using a distribution law for the inelastic case of the form (27.11), viz. $\varrho = e^{-\frac{\varepsilon}{\varepsilon_0}}$, we obtain finally with $\overline{V} \propto \varepsilon_0 \propto E/p$

$$\frac{\alpha_{\text{met}}}{p} \approx \text{const}\, \Pi_1 \Pi_2 \varepsilon_1^{\frac{1}{2}} \left(\frac{E}{p}\right)^{\frac{1}{2}} e^{-\frac{c \cdot \varepsilon_1}{E/p}}. \tag{32.3}$$

This shows that α_{met}/p rises steeply with E/p and that it is proportional to the fraction of argon present (through Π_2). In fact, a more refined calculation has been carried out [*30*] which should be consulted for future work. It should be pointed out that it is of course applicable to other examples of ionization by metastables. A comparison of (32.3) with experiment (Fig. 26) shows that the theory holds only for small concentration of A; see also [*96*].

If the concentration of metastables in a pure gas is high enough, slow electrons of the distribution will ionize metastables. It depends on the ratios of excitation— and ionization—to mean energy whether or not ionization of metastables is comparable to direct ionization. If the mean energy $\overline{V}$ is low and the concentration of metastables high then the many more slow electrons in the distribution may ionize metastables. The destruction of metastables by light in pure gases has been shown to cause an increase in the electric field of a positive colnmn [*78*].

X. Future work.

33. There seems to be plenty of scope for further work in this field. Theory has only been compared with experiment for a few gases. There are hardly any data of α/p for metal vapours. The extremely high values of α/p in organic vapours have not yet been explained and the combined ionization, dissociation and attachment processes for complex molecules are not really undestood. Simultaneous observations of ionization and excitation have just been started. The influenceof collisions other than by electrons has scarcely been studied. The recent results on photo-excitation and photo-ionization cross sections will further these investigations. There is a singular lack of experimental methods of measuring electron distributions or the fraction of electrons above a certain energy. Ionization rates in excited gases have not been studied extensively. Ionization involving metastable atoms has only been investigated for very small traces of a foreign gas. No work has been done to show that the emitted radiation distribution is changed by external irradiation nor has the absolute cross section for the wavelength of irradiation been measured. Another problem which needs clarification is the time and distance necessary to approach the steady state of the energy distribution of electrons and the corresponding relaxation values for small disturbances.

C. The non-uniform electric field.

The multiplication process in non-uniform fields is complicated by the fact that in many cases the current is not caused by a continuous flow of charge but that pulses or oscillations of different types develop. We are here concerned with simple multiplication processes only and thus we shall refrain from a discussion on all those effects which are known as burst pulses, streamers, counter pulses etc. These phenomena will not be treated here (cf. L. B. LOEB in Vol. XXII).

I. Non-uniformity by geometry and its effect on multiplication.

34. The distribution of the electric field in the gas is determined by the electrode field, provided the gas possesses a uniform dielectric constant and the net space charge is small. If the electrodes are e.g. two concentric cylinders of very different radii, then the ensuing field will vary rapidly from point to point. In order to find the ionization in the gas, it appears to be only necessary to calculate the field distribution as given by potential theory and to allow for a variation of α with $E(r)$. Thus, for example, with the cylinder radii r_0 and r_1 and a constant applied voltage V, the multiplication i/i_0, i and i_0 being currents per cm. of axial length, should follow from

$$\frac{i}{i_0} = e^{\int_{r_0}^{r_1} \alpha\, dr}. \tag{34.1}$$

The field E at the distance r from the axis is

$$\left.\begin{aligned} E &= \frac{C}{r} \\ \text{where}\quad C &= \frac{V}{\ln \frac{r_1}{r_0}}. \end{aligned}\right\} \tag{34.2}$$

The ionization coefficient α varies with E according to

$$\frac{\alpha}{p} = A\, e^{-\frac{Bp}{E}}. \tag{34.3}$$

Hence combining these equations, we have

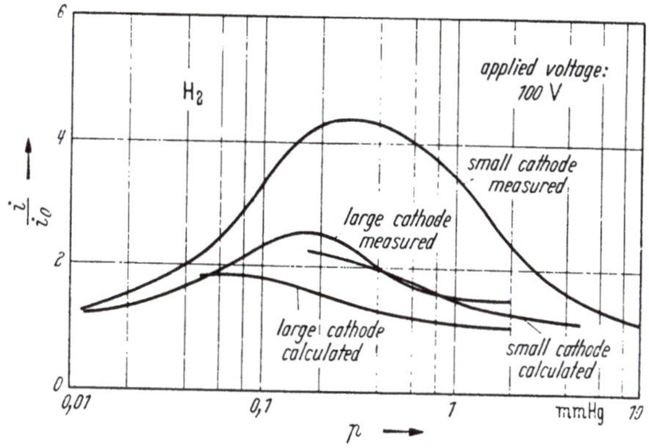

Fig. 51. Multiplication i/i_0 in H_2 as a function of the pressure p with α concentric cylinder electrodes at a constant applied voltage of 100 V. Small and large cathode wire radius $r = r_0 = 0.16$ and 0.56 cm. resp. Outer cylinder radius $r_1 = 4.4$ cm. Calculated curves from (34.5).

$$\ln\left(\frac{i}{i_0}\right) = \int_{r_0}^{r_1} \alpha\, dr = A \int_{r_0}^{r_1} e^{-\frac{Bpr}{C}}\, d(pr) = -\frac{AC}{B}\left[e^{-r_1 p \frac{B}{C}} - e^{-r_0 p \frac{B}{C}}\right]. \tag{34.4}$$

If $r_1 \gg r_0$, (34.4) reduces to

$$\ln\left(\frac{i}{i_0}\right) = \frac{AC}{B}\, e^{-r_0 p \frac{B}{C}} \propto e^{-\frac{\text{const}\, p}{V}}. \tag{34.5}$$

This relation as well as (34.4) shows that the multiplication i/i_0 falls as p increases and V decreases. A comparison with experiment [81], [79], [80] confirms the general dependence on V. However, it is found that, as p is raised from small to large values, i/i_0 first rises and after passing a maximum falls (as for the plane gap, Fig. 8), and in addition that the multiplication is always larger than expect from (34.5). Figs. 51 and 52 show the observations of $i/i_0 = f(p)$ in H_2 for negative wires in cylinders of about 40 times the wire radius for two values of V and Fig. 53 for air where the figures on the curves give the field E at r_0 in V/cm. The curves are taken therefore at constant voltage V, related to E by (34.2). It may be of interest to note that the multiplication for positive wires is found to be the same in air and slightly lower in H_2 than for negative wires.

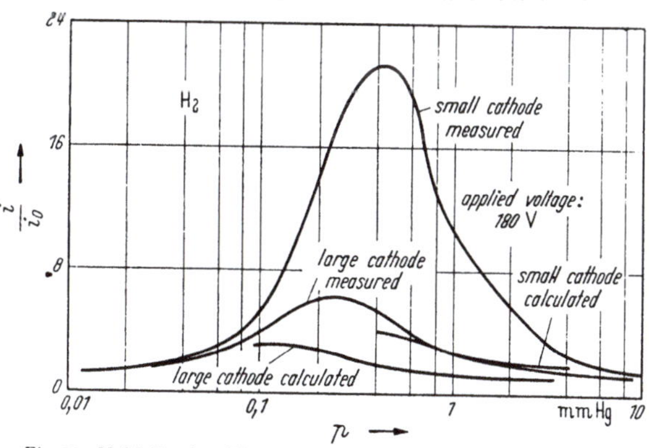

Fig. 52. Multiplication i/i_0 in H_2 as a function of the pressure p for an applied voltage of 180 V with the electrode system in Fig. 51. Calculated curves from (34.5).

The variation of the current i from a positive wire ($r_0 = 5 \cdot 10^{-3}$ cm.) in H_2 as a function of the applied voltage V for pressures between 10^{-3} and 70 mm. Hg is shown in Fig. 54. At low p and $V > 100$ V, i is the saturation current due to irradiation of the cathode with ultraviolet light. At $p \approx 0.1$ mm. ionization in the gas occurs and as the pressure increases a steep rise of current occurs at certain voltages whose values increase with rising p [81].

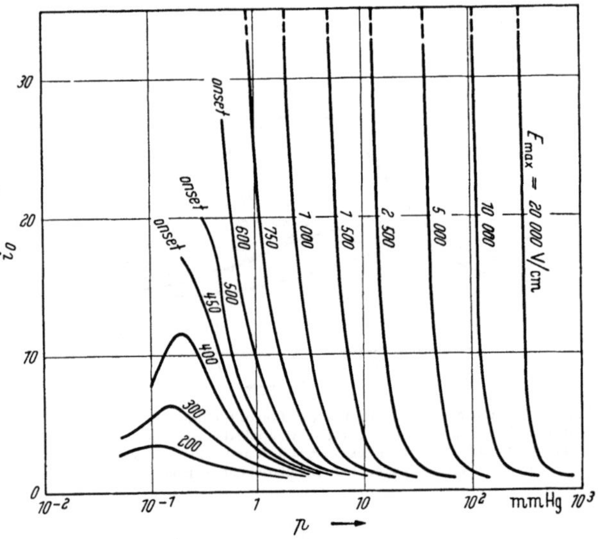

Fig. 53. Multiplication i/i_0 as a function of the pressure p in dry air for different values of the electric field at the central cathode wire E_{max} given by (34.2). Radius of the central wire $r = r_0 = 0.157$ cm. Radius of the positive cylinder $r_1 = 1.58$ cm. Below a certain pressure a self-containing discharge occurred.

The observed currents follow similarity relations. This is illustrated in Fig. 55 where the logarithm of the multiplication reduced to $p \cdot r$ is plotted as a function of E_0/p, the field at r at unit pressure, the wire being the cathode. The points represent results in air and H_2 at p between 0.4 and 200 mm. Hg and for radii r between 0.16 and 0.56 cm. The results are interesting in another respect: taking constant r and p, the curves represent simply the multiplication as a function of the voltage and show that i/i_0 in H_2 is larger than in air. As we have seen (Figs. 30 and 27) α/p is larger in H_2 than in air at large values of E/p and the reverse is true for low values ($E/p < 200$). Hence some deductions which are contained in previous work [79] are to be revised in the light of the new measurements of α/p.

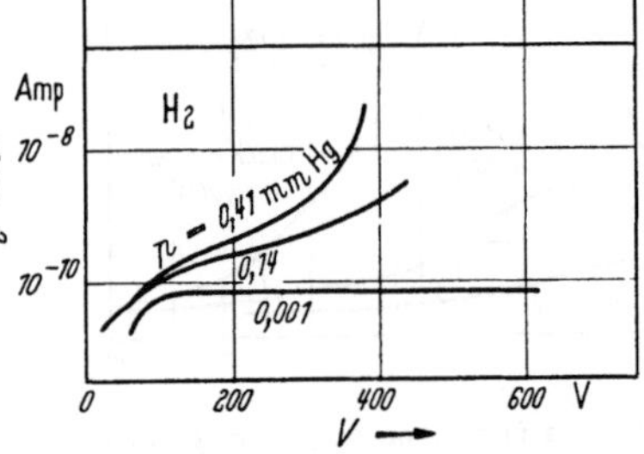

Fig. 54. Ionization current i between a positive wire at the axis of a negative cylinder in H_2 as a function of the applied voltage V for different pressures p. Wire radius: $5 \cdot 10^{-3}$ cm.; cylinder radius: ≈ 1 cm.

35. We shall now turn to the possible explanation of the multiplication being larger than anticipated. First we remark that since $(1/r\,p)$ $\ln i/i_0$ depends on E/p only, the multiplication is independent of the variation of dE/dr with r. Hence older theories [82] based on the assumption of a dependence are ruled out. Further, at sufficiently high p there is no difference in i/i_0 whether the field is convergent or divergent (wire positive or negative). At low p the difference becomes more pronounced at larger values of V and lower p.

Also from measurements with auxiliary anodes the distribution of ionization has been found. With wire radii of 0.24 and 0.56 cm. and cylinders 4.4 cm. radius the fraction f of ionization within 1.3 cm. radius in air and H_2 as a function of p is [80]:

p mm. Hg	f in air	p mm. Hg	f in H_2
0.1	0.2	0.2	0
0.3	0.6	0.3	0.6
0.8	0.9	0.7	0.8

This shows that for $p > 1$ mm. Hg ionization is confined to a cylinder less than 1 cm. in radius decreasing with increasing p. We also conclude that whereas on the basis of ionization by collision in a given field the bulk of the ions should be produced near the wire, it is clear that at low p a larger part is produced in the interspace and little ionization in the region of high field. The question of the influence of secondary processes has not yet been discussed satisfactorily. From experiments with different polarity and other indirect evidence it appears at first sight that secondary processes are not important here. At low V and high p, back-scattering of electrons to the cathode has to be allowed for in evaluating i_0.

From the foregoing it transpires that the ionization in non-uniform fields is higher than in uniform fields because in the former the overall efficiency is larger

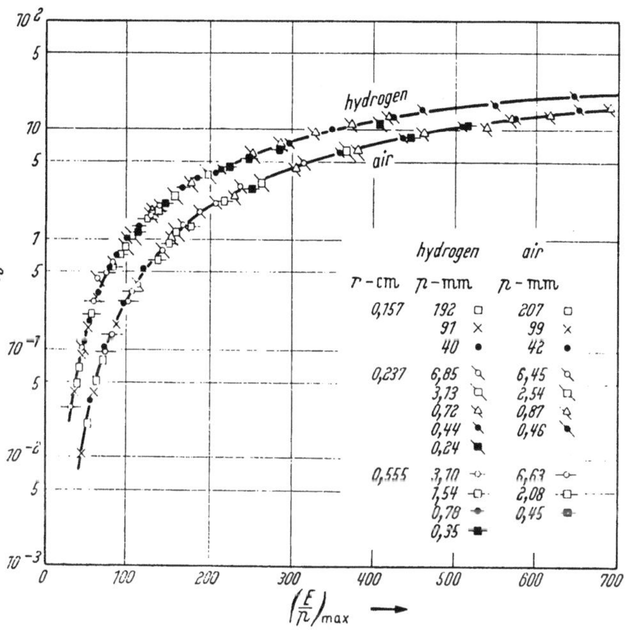

Fig. 55. Logarithm of the multiplication i/i_0 reduced to the radius of the cathode wire pr as a function of the field at the cathode wire $(E/p)_{max}$ in H_2 and air. The diagram shows that the law of similarity holds.

i.e. the losses smaller. Thus it has been suggested [80] that the inelastic losses of electrons leading to excitation are smaller—at least in molecular gases—because the electrons are here making fewer collisions when they pass an energy range over which the excitation probability is large. Consequently, in the case of electrons moving into lower fields, the electrons are entering regions of low fields with larger energy than they would posses if they were in equilibrium with the field. These are only qualitative speculations and a theory is still lacking. A first attempt [79] has led to a difference equation which had to be solved by numerical methods. The solution will not be given here because it can only explain the low pressure region in Fig. 51 up to the maximum, but cannot prove the existence of the latter. There are only few experiments on the multiplication in rare gases; they would be most useful for checking future theories.

There is more experimental evidence available on the multiplication in H_2 and A at larger p for a thin positive wire at the centre of a large cylinder [83]. Figs. 56 and 57 show i/i_0 as a function of V from which it can be seen the variation

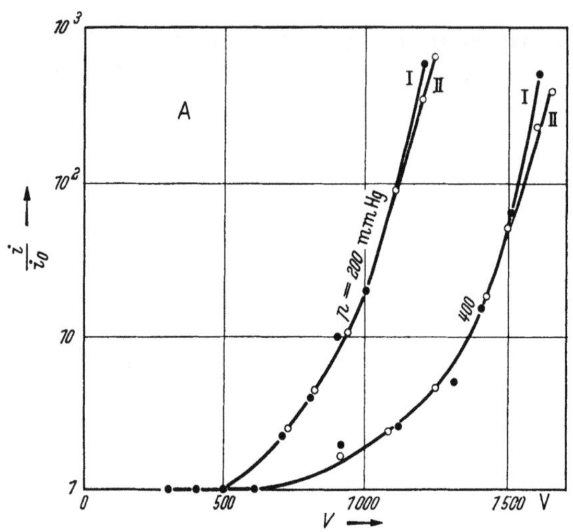

Fig. 56. Multiplication i/i_0 as a function of the applied voltage V in a gap with a positive wire in a cylinder in A at two different pressures.

$I\ldots\gamma$-ray source, $II\ldots\alpha$-particle } ionization

Radius of positive wire $= 6.4 \cdot 10^{-3}$ cm.; radius of cylinder $= 1.45$ cm.

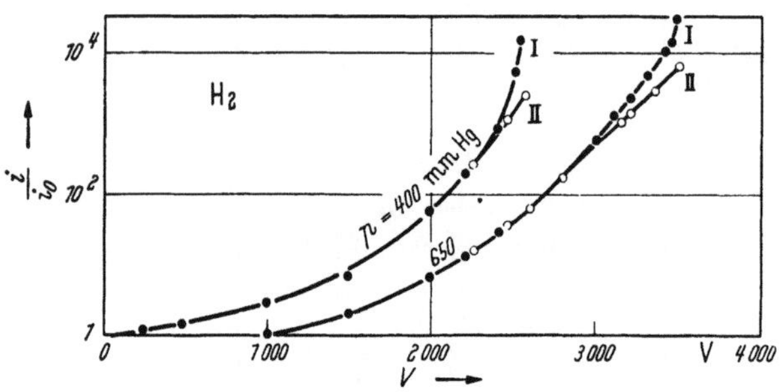

Fig. 57. Multiplication i/i_0 as a function of the applied voltage V in H_2 at two different pressures. Polarity, geometry of gap and irradiation see Fig. 56.

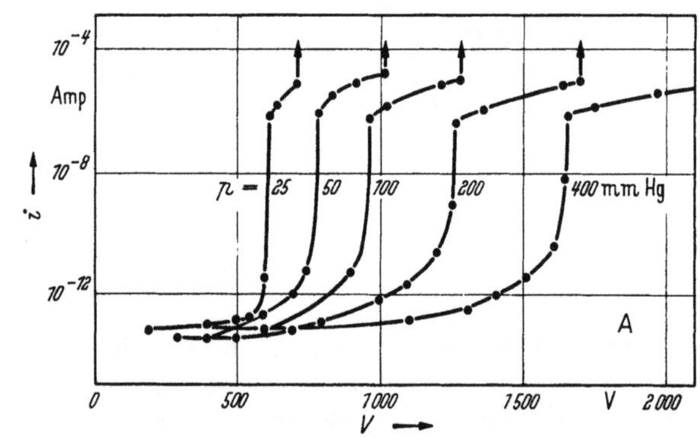

Fig. 58. Ionization current i as a function of the applied voltage V in A for various pressures. Polarity and geometry see Fig. 56. Irradiation: γ-rays.

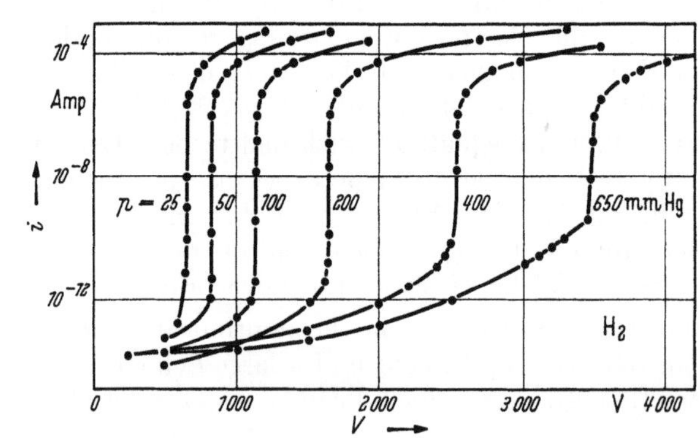

Fig. 59. Ionization current i as a function of the applied voltage V in H_2 at different pressures. Polarity and geometry see Fig. 56. Irradiation: γ-rays.

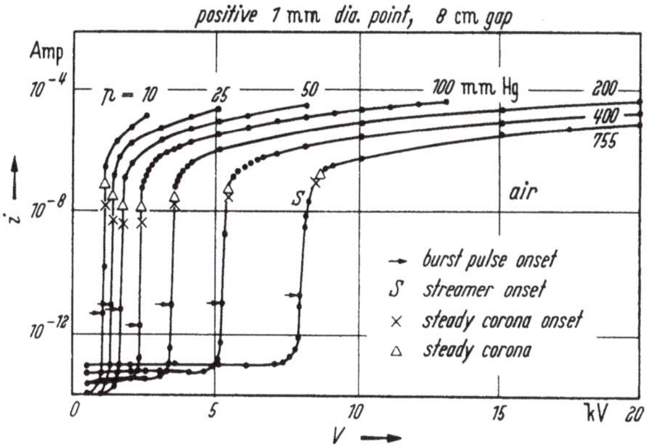

Fig. 60. Ionization current i as a function of the applied voltage V in air between a small positive point and a negative plane electrode at different pressures. Irradiation: γ-rays.

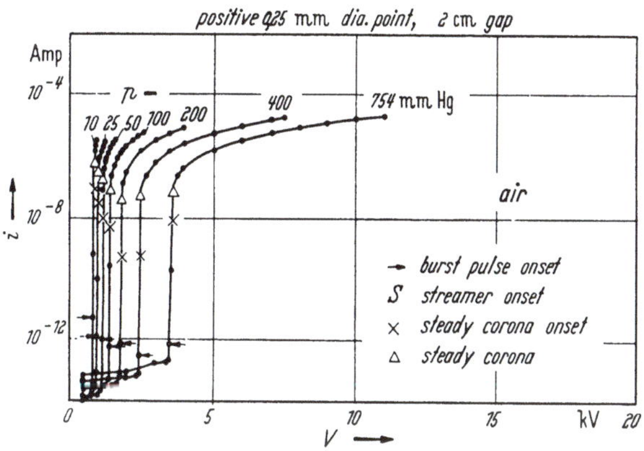

Fig. 61. Ionization current i as a function of the applied voltage V in air at different pressures for a positive point electrode and a negative plane, the radius of curvature of the point (hemispherically capped cylinder) being $\frac{1}{4}$ of that in Fig. 60.

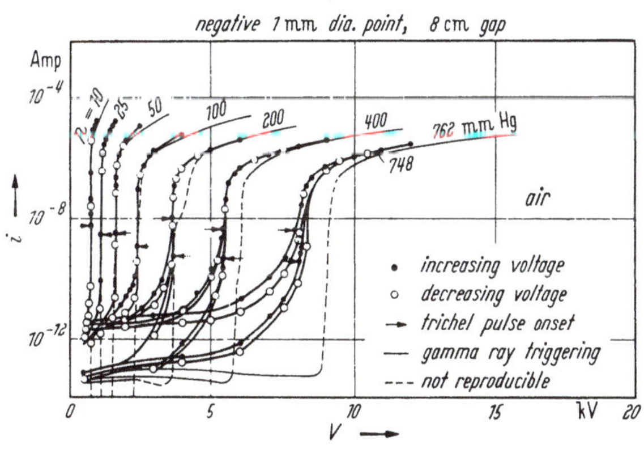

Fig. 62. Ionization current i as a function of the applied voltage V in air for a negative point and a positive plane electrode at different pressures. Geometry identical with Fig. 60. Irradiation: ultraviolet.

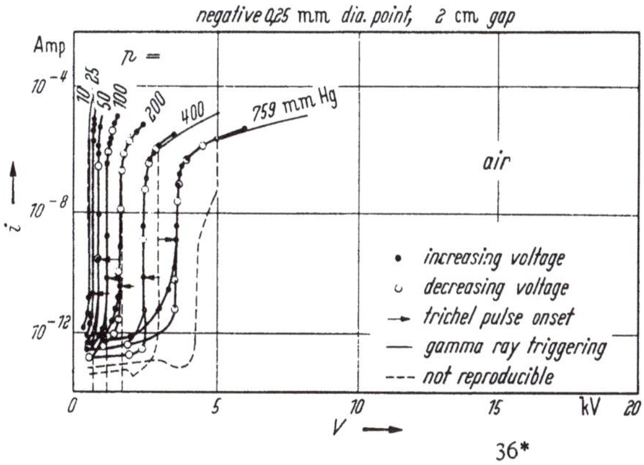

Fig. 63. Ionization current i as a function of the applied voltage V in air at different pressures for a negative point electrode. Geometry as in Fig. 61. Irradiation: ultraviolet.

with voltage is perhaps more rapid in argon than in H_2. Figs. 58 and 59 show the currents i as a function of V for p between 25 and 600 mm. Hg. The character of the families of curves is the same in both gases but they differ in details. For example sparking occurs in argon at lower values of V than in H_2 etc. Measurements have been made also in N_2, O_2 and their mixtures [86].

Finally, measurements are available in air [84] on positive and negative points, irradiated with ultra-violet light, the other electrode being a plane perpendicular to the point electrode, which is a wire 2 to 4 cm. long. In Figs. 60 and 61 are shown the currents from positive points of 1 mm. diameter 8 cm. distant from the plane, and $\frac{1}{4}$ mm. diameter 2 cm. from the plane, for pressures between 10 mm. and 760 mm. Hg. The results show that as V is increased the currents remain either constant or increase only slightly over a considerable range of voltage. These currents are due to natural ionization in the gas. Then a steep rise in current of several orders of magnitude follows which is caused by collision process in the gas. This is the range in which the current often flows in pulses. At still higher voltages a steady rise of current is recorded and a steady corona discharge is observed. The main difference between the current-voltage diagrams of the 1 mm. and $\frac{1}{4}$ mm. point is apparently in the scale on the abscissa: for larger p the curves belonging to the point of $^1/_4$ mm. diameter are simply reduced to values of about $\frac{1}{2}$ of those at 1 mm. diameter. Measurements in A, N_2 and H_2 have been made too [85].

The currents in air from negative point electrodes of diameters as before, again with irradiation, vary with the voltage as shown in Figs. 62 and 63. The main difference between these and the above results is that at higher pressure the steep rise of i becomes less pronounced the higher p and i_0. Again with a smaller point electrode the curves i *vs.* V are shifted in the direction of lower values of V to about the same degree as for the positive points electrodes.

II. Multiplication in space-charge distorted fields.

36. When a multiplication process occurs in a gas and the number of electrons and ions rises above 10^8 per cm.^{3}, or the current density exceeds $\approx 10^{-6}$ A/cm.^{2}, then the initially uniform field between plane parallel electrodes becomes distorted. The distortion will of course depend on whether the potential between the electrodes or the charges on them, i.e. the initial field strength is given. We shall discuss here first qualitatively the effect of the charges in the gas on the field and then the effect on the coefficient of ionization when the electrode potential is kept constant.

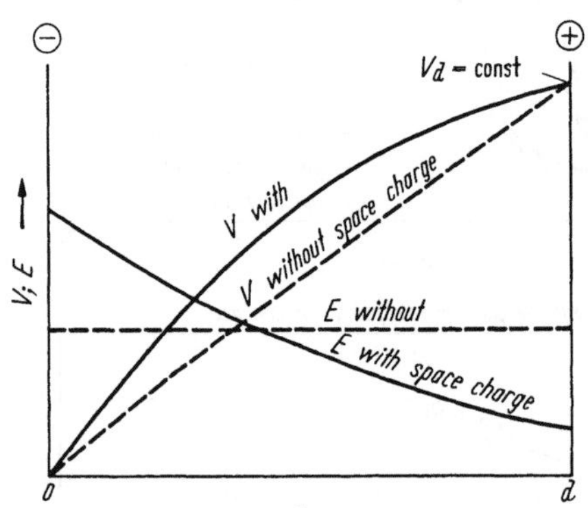

Fig. 64. Potential and field distribution between two plane parallel electrodes with a net space charge in the gas absent or present.

Fig. 64 shows the potential (V) and field distribution (E) in the gas when the net space charge $\varrho = 0$ and when $\varrho \neq 0$. It is seen that over a part of the space the field E is increased under the influence of the positive space charge associated with the current, and decreased over the remaining space so that the area under the curve is constant (the effect of the electron space charge is usually negligible). Consequently the ionization coefficient will rise near the cathode and fall near the anode. Thus the question arises whether the total ionization will rise or fall ([17], p. 51). We shall assume for the moment that the field distortion is such that it is permissible to maintain the concept of an ionization coefficient. For that to be true the field must not change by more than a few percent over an electron mean free path.

Let E_0 $(=V/d)$ be the field for $\varrho=0$. A small space charge will change it to $E_0+\delta E$ where $\delta E \lesseqgtr 0$ and the potential between the electrodes from V to $V+\delta V$. Expanding the ionization coefficient, we obtain from (14.1)

$$\left.\begin{aligned} \frac{\alpha}{p} &= A\,\mathrm{e}^{-\frac{Bp}{E_0+\delta E}} = A\,\mathrm{e}^{-\frac{Bp}{E_0}\left[\frac{1}{1+\delta E/E_0}\right]} \\ &= \frac{\alpha_0}{p}\left\{1+\frac{Bp\cdot\delta E}{E_0^2}+\frac{Bp}{E_0^4}\left(\frac{Bp}{2}-E_0\right)(\delta E)^2\right\} \end{aligned}\right\} \tag{36.1}$$

where $\alpha_0=\alpha$ for $\delta E=0$. Integrating (36.1), we find the change in the total ionization in the gap of length d caused an infinitesimal space charge is

$$\left(\int_0^d \alpha\,dx\right)-(\alpha_0 d)=\alpha_0\left\{\frac{Bp}{E_0^2}\,\delta V+\frac{Bp}{E_0^4}\left(\frac{Bp}{2}-E_0\right)\int_0^d(\delta E)^2 dx\right\}. \tag{36.2}$$

The integral on the right hand side is always positive; hence the last term is positive when $E_0/p<B/2$ and vice versa. If $V=\text{const}$, i.e. $\delta V=0$ (36.2), shows that for $E_0/p<B/2$ the total ionization in a space charge distorted field is larger than in a uniform field. $E_0/p=B/2$ corresponds to the point of inflexion (I. P.) of the curve $\alpha/p=f(E/p)$ (Fig. 7). Thus we conclude that space charges increase ionization when the field is in the section of the α/p curve below the point of inflexion (see also [*87*]) and vice versa. It is clear that space charge has no effect to a first approximation when measurements are carried out at the point of inflexion. The effect of the space charge distortion on the breakdown is discussed by L. B. Loeb in Vol. XXII.

From (36.2) we can also find the value of δV for which the ionization in the gap is the same whether space charge distortion is present or not. Thus equating the right hand side of (36.2) to zero, we have

$$\delta V=\frac{E_0-Bp/2}{E_0^2}\int_0^d(\delta E)^2 dx. \tag{36.3}$$

If we put $\delta E \propto \varrho_{\text{net}}$ and $\varrho_{\text{net}} \propto j$, j being the current density we find

$$V=V_0\pm c j^2 \tag{36.4}$$

the two signs applying to whether $E_0 \gtrless Bp/2$. Thus we find that when the field is below the point of inflexion, the voltage V decreases as the current density j increases. Since this result agrees with experiment, we conclude that it is permissible to use the ionization coefficient α to describe in a rough manner the transition from an electron swarm into a dark discharge.

37. For a description of the finer details, however, it is necessary to include the secondary processes and to allow for the spatial dependence of α [*88*]. Assuming the secondary coefficient γ (the number of secondary electrons released from the cathode per incident ion plus associated processes) to be constant, then we can write instead of (10.3) neglecting diffusion and spreading through secondary processes other than positive ions

$$\frac{j_x^-}{j_0}=\frac{\mathrm{e}^{u}}{1-\gamma\int_0^{\bar u}\mathrm{e}^{u}\,du}=\frac{\mathrm{e}^{u}}{1-\gamma(\mathrm{e}^{\bar u}-1)} \tag{37.1}$$

where $\bar{u}=\int_0^d \alpha\,dx$, $u=\int_0^x \alpha\,dx$ and $du=\alpha\,dx$. Since

$$\frac{dE}{dx}=4\pi e(N^+-N^-)\approx 4\pi\frac{j^+}{\mu^+E}, \tag{37.2}$$

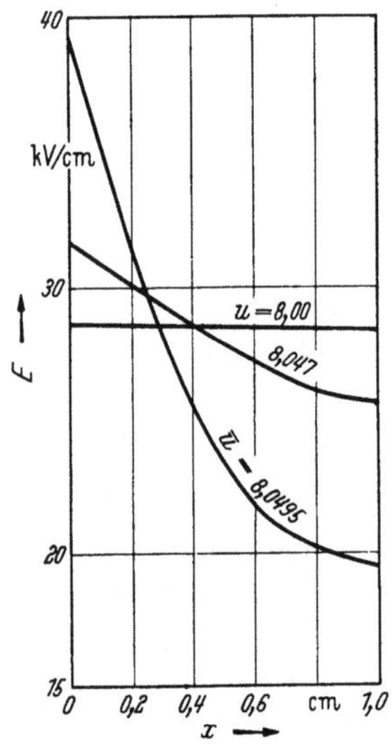

Fig. 65. Dependence of the electric field E between plane electrodes on the positive x for different values of the space charge parameter $\bar{u}$. Electrode separation 1 cm.; N_2, 700 mm. Hg.

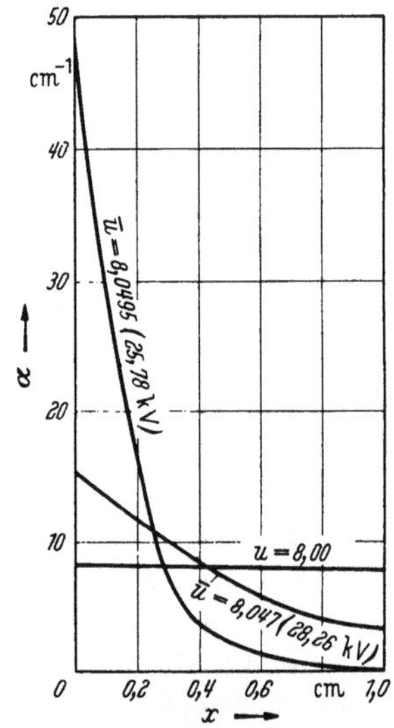

Fig. 66. Dependence of the ionization coefficient α on the position x in a plane parallel gap (Fig. 65) for different values of the space charge parameter $\bar{u}$.

and $j=j^+ + j^- = e(N^+v^+ + N^-v^-)$, and $\mu^-\gg\mu^+$, we have

$$\frac{\alpha E\,dE}{4\pi j_0/\mu^+}=\frac{(e^{\bar{u}}-e^{u})\,du}{1-\gamma(e^{\bar{u}}-1)}. \tag{37.3}$$

This equation can be integrated using

$$\frac{\alpha}{p}=A\,e^{-\frac{Bp}{E}} \tag{37.4}$$

by means of a function

$$2K(z)=\frac{e^{-z}}{z^2}-\frac{e^{-z}}{z}-Ei(-z) \tag{37.5}$$

where $z=Bp/E$ and Ei is the exponential integral. A numerical evaluation shows that in general the space charge effects are only to be considered for voltages $<\frac{1}{2}\%$ below the sparking potentials. Some of the results are presented in Figs. 65 and 66 which show the spatial distribution of E and α respectively for different values of the total ionization $\bar{u}$ in the gap. In Fig. 67 the resulting current *vs.* voltage relation is given which shows the transition from a Townsend into a self-sustaining dark discharge in N_2 at $p=700$ mm. Hg, $d=1$ cm. with $\gamma=3.2\cdot10^{-4}$, $\mu^+=3$ cm./sec. per V/cm. at 1 atm., $i_0=6.4\cdot10^{-11}$ A/cm.², $A=7$ and $B=260$ in practical units (see also Fig. 1 on p. 574).

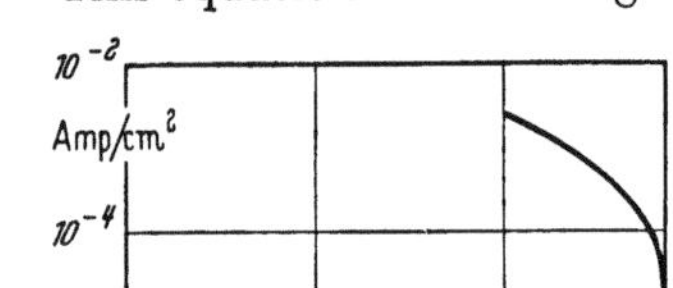

Fig. 67. Current density j as a function of the applied voltage V for a plane parallel gap of 1 cm. width in N_2 at 700 mm. Hg showing the transition from the Townsend discharge into a dark discharge.

38. So far only the steady state of a field distorted by space charges has been discussed. We shall now proceed to non-steady states and the effects of space charges on ionization in the gas. An idealized case of the multiplication process of a single avalanche of electrons has been treated recently [89]. Secondary process and radial diffusion have been neglected and the electron ionization coefficient α is assumed to apply even in non-uniform fields. The picture used is that of an avalanche which starts at the cathode supported by n_0 electrons which multiply in the field, initially of value E_0. The electron swarm leaves the positive ions behind which are considered stationary. These produce an ever increasing field, which retards the motion of electrons and the ionization in the gas. If the electrons form a thin layer (Fig. 68) free of ions with a field E_c at the centre and E_t and E_l at the trailing and leading edge respectively, and if the number of electrons at x is

$$n_x = n_0 e^{\int_0^x \alpha\, dx} \tag{38.1}$$

then by applying GAUSS' theorem we have

$$E_c = E_l - 2\pi n_0 e^{\int_0^x \alpha\, dx}. \tag{38.2}$$

Assuming the simplest dependence of α on E, viz.

$$\alpha = c_1 E_c \tag{38.3}$$

and a constant field $E_l = E_0$ at the electrodes, we find by differentiation of (38.2) with (38.3)

$$\frac{1}{c_1}\frac{d\alpha}{dx} = -2\pi n_0 e \alpha e^{\int_0^x \alpha\, dx}. \tag{38.4}$$

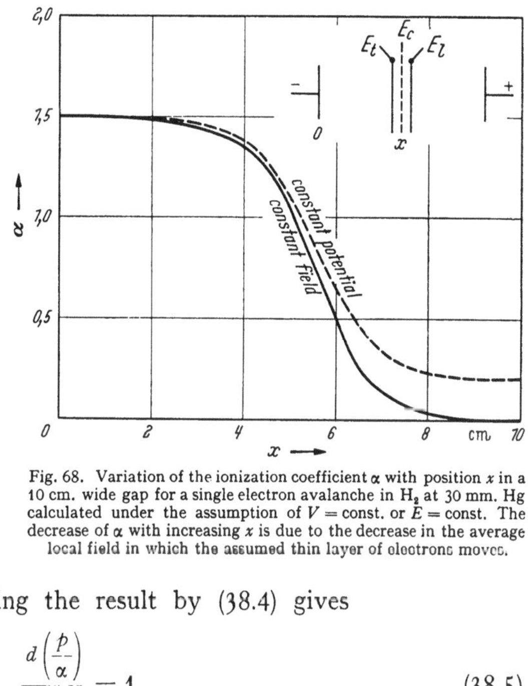

Fig. 68. Variation of the ionization coefficient α with position x in a 10 cm. wide gap for a single electron avalanche in H_2 at 30 mm. Hg calculated under the assumption of V = const. or E = const. The decrease of α with increasing x is due to the decrease in the average local field in which the assumed thin layer of electrons moves.

Differentiating again and dividing the result by (38.4) gives

$$\frac{d\left(\frac{p}{\alpha}\right)}{d\alpha} = 1 \tag{38.5}$$

where $p = d\alpha/dx$. With the boundary conditions

$$x = 0, \quad \alpha = \alpha_0, \quad \int \alpha\, dx = 0 \quad \text{and} \quad \left(\frac{d\alpha}{dx}\right)_0 = \alpha c_1 2\pi n_0 e \quad \text{from (38.4)}$$

the solution is found to be

$$\alpha = \frac{A}{1 + B e^{Ax}} \tag{38.6}$$

with

$$A = \alpha_0 + 2\pi n_0 e c_1,$$

$$B = \frac{2\pi n_0 e c_1}{\alpha_0}.$$

The multiplication is

$$M = e^{\int_0^x \alpha\, dx} = \frac{B+1}{B + e^{-Ax}}. \tag{38.7}$$

One concludes that α decreases with x and that this occurs the more rapidly the larger n_0, c_1 and α_0. A numerical example is shown in Fig. 68 with $\alpha_0 = 1.5$, $n_0 = 10^5$ and $c_1 = 2 \cdot 10^{-3}$ which illustrates the distribution of α in a gap of $d = 10$ cm. with H_2 at $p = 30$ mm. Hg for $E/p = 25$ V/cm. per mm. Hg.

The solution of the problem is more involved when the applied potential is assumed to be constant. This means that while the avalanche develops and the space charge grows and distorts the field, more charges are flowing from the constant potential source into the system. However, this strengthens the field and opposes the weakening effect of the space charge. The treatment of this case will not be given here but only the equation corresponding to (38.5) and the solution.

With a gap of length d and a constant applied voltage

$$\frac{d}{d\alpha}\left[\frac{p}{\alpha d - 2}\right] = \frac{\alpha}{\alpha d - 2}. \tag{38.8}$$

This can be integrated approximately when $\alpha_0 d > 2$ and α is not too small compared with α_0. In this case the solution of (38.8) is

$$\alpha = \alpha_0 - \left\{\alpha_0 - \frac{2}{d}\right\}\left\{\frac{e^{(1+B)\alpha_0 x} - 1}{e^{(1+B)\alpha_0 x} + 1/B}\right\}. \tag{38.9}$$

It is seen that whereas from (38.6) $\alpha \to 0$ as $x \to \infty$, we find from (38.9) that $\alpha \to 2/d$ as $x \to \infty$. This is shown in Fig. 68. If, however, $\alpha_0 d < 2$ then the initial field is low, the field distortion by space charges is less powerful than than the increase in field by supply of charges from the source, and hence α will increase with x. The treatment becomes invalid when the electrons become so numerous that the assumed thin layer spreads.

The time taken for a single avalanche to develop can be obtained from the drift velocity $v_d = f(E/p)$ where $E = E_c$. Since we found $\alpha = f(x)$ and assumed $\alpha \propto E_c$ the position of E_c at time t can be calculated. In the simple case where the field at the electrodes is given this leads to an equation of the form

$$t = C_1\left\{x + C_2 \ln \frac{C_3 + C_4 e^{Ax}}{C_3 + C_4}\right\} \tag{38.10}$$

where the C's are known constants. A similar relation is obtain when $V = \text{const.}$

39. Finally a treatment will be discussed which attempts to compute the ion and electron distribution along the axis of a single electron avalanche when a constant voltage is applied between the electrodes [*90*] From the results the formative time of a spark is derived assuming that ionization is caused by electron collision and determined by α. The electric fields is considered equal to the applied field up to a certain instant after the start of the avalanche. From the equations of the rate of growth of electrons which depends on ionization, diffusion and drift motion, and rate of growth of ions which depends on ionization alone (since the ions are practically at rest) the spatial distribution is found which is of the form

$$N^-(z) = f(D^-) \cdot e^{-\left\{\frac{(z-vt)^2}{4D^- t} + \alpha v t\right\}}, \tag{39.1}$$

$$N^+(z) = \alpha v \cdot \int_0^t f(D^- t') \cdot e^{-\left\{\frac{(z-vt')^2}{4D^- t'}\right\} + \alpha v t'} \tag{39.2}$$

where D^- is the electron diffusion coefficient, v the electron drift velocity in the z direction and N^+ and N^- the number of charges per cm.^{3}.

$N^-(z)$ represents a GAUSSian distribution when seen from a coordinate system which moves with the velocity v in the z direction. Putting $z_0 = vt$ and $a = 4D^-/v$, $(a z_0)^{\frac{1}{2}}$ is approximately the width or radius of the electron avalanche. Another parameter is $1/\alpha$, the distance (in field direction) between two ionizing events and hence a factor describing the spread of the positive ions. a is assumed to be independent of E. Using the dimensionless quantities

$$w = \frac{z - z_0}{(a\, z_0)^{\frac{1}{2}}} \tag{39.3}$$

which is a distance in terms of the width of the electron avalanche,

$$b = \alpha\,(a\, z_0)^{\frac{1}{2}} \tag{39.4}$$

the width of the avalanche in units of the ionizing distance, and the maximum electron density

$$N_0^- = (\pi\, a\, z_0)^{-\frac{3}{2}}\, e^{\alpha z_0}\,, \tag{39.5}$$

the spatial distribution of charges in the absence of space charge distortion is

$$\frac{N^+}{N_0^-} = e^{-w^2} \tag{39.6}$$

$$\frac{N^+}{N_0^-} = \left\{\begin{array}{lll} \frac{b}{2}\left(1 + \operatorname{erf}(-w)\right) e^{b w} & \text{when} & w < 0\,, \\ \frac{b}{2}\left(1 - \operatorname{erf}(-w)\right) & \text{when} & w > 0 \end{array}\right\} \tag{39.7}$$

where erf stands for the error function in (39.7). Fig. 69 shows (39.6) and (39.7), and it is seen that as b increases the electrons spread over a large portion of the ions. From this distribution a space charge field can be calculated and thus the concentration can be found at which the space charge field equals the applied field so that ionization would cease. A better approximation is to use the above results up to a certain instant t and to calculate the space charge field E_s by assuming that $E_s \propto N/r^2$, i.e. for a spherical distribution where r is the avalanche radius; N and N_t stands for the number of electrons (not their density) in the avalanche. With the applied field E it is found that

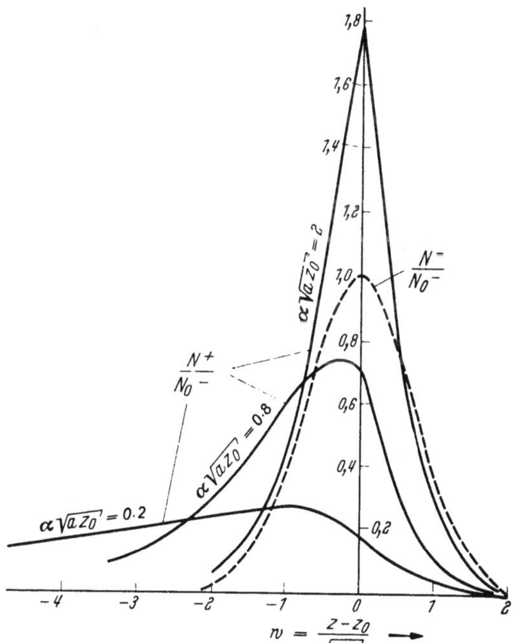

Fig. 69. Local density of positive ions N^+ and electrons N^- (in terms of the maximum electron density N^-) as a function of the distance from the centre of the electron cloud $(z - z_0)$ in terms of the radius of the avalanche $(a\, z_0)^{\frac{1}{2}}$ for different values of avalanche radius: ionization distance $1/\alpha$.

$$\frac{E_s}{E} \approx 0.2\, a\, \alpha \left(\frac{N}{N_t}\right)^{\frac{1}{2}} \tag{39.8}$$

where

$$\left.\begin{array}{l} N_t \propto E\, a^2\, (a\, \alpha \ln N_t)^{\frac{1}{2}} \quad \text{if} \\ \quad (\ldots)^{\frac{1}{2}} > 1\,. \end{array}\right\} \tag{39.9}$$

Further $E_s = E$ when

$$\frac{N}{N_t} \approx \frac{30}{(a\, \alpha)^2}\,. \tag{39.10}$$

If with the critical number N_c we have a spark, then the corresponding time lag is

$$\tau = \frac{1}{\alpha\, v} \ln N_c\,. \tag{39.11}$$

In this expression α, v and N_c depend on E, but the effect of E on α and v predominates. However, this analysis does not give the spatial distribution of α nor does it allow for the fact that the electrons may not have time to get into equilibrium with the field. Various authors [*82*], [*79*] have attempted to solve this problem, on the whole with little success.

III. Future work.

40. There are no relations describing in simple terms the multiplication of a concentric cylindrical system for the complete range of gas pressure which corresponds to the law for the system with plane electrodes. The reason for the higher multiplication in certain non-uniform fields compared with uniform fields are not fully understood. The multiplication in space charge-distorted fields is based on an ionization coefficient which is assumed independent of the current density and little is known about the range over which this concept applies.

D. Ionization in combined electric and magnetic fields.

I. Crossed uniform fields.

41. Though the motion of electrons through a vacuum is well understood when electric and magnetic fields act on them simultaneously, there is little known about the ionization in gases in combined fields. Experimental and theoretical investigations are mainly concerned with systems which are of interest in practical applications like low pressure gauges, electric breakdown problems, ion sources for beams of fast particles, etc.

However, there are certain laws of motion of charges in vacua to be remembered which are applicable to their motion in gases, since charges move between two successive collisions as if they were in vacuum. The force acting on a charge e moving with velocity $\boldsymbol{v}$ in *a* uniform magnetic field $\boldsymbol{H}$ is $\frac{e}{c}\,\boldsymbol{H}\times\boldsymbol{v}$, perpendicular to both $\boldsymbol{H}$ and $\boldsymbol{v}$, and thus corresponds to an electric force $e\boldsymbol{E}$ which does not do work. If this charge of mass m moves with velocity v normal to the field $\boldsymbol{H}$ it will travel along a circle of radius $r=\frac{1}{(e/m\,c)}\,\frac{v}{H}$ and require a time $t=\frac{2\pi}{(e/m\,c)\,H}$ to go round once. If this particle starts with zero velocity in an electric field and a magnetic field normal to it, it will move along a cycloid and it will thus drift normal to both $\boldsymbol{E}$ and $\boldsymbol{H}$; it has an angular velocity $\omega=\left(\frac{e}{m\,c}\right)\,H$ and a drift speed $\frac{c\,E}{H}$. Any initial velocity component parallel to $\boldsymbol{H}$ remains unchanged an a velocity component in a general direction changes the cycloid to a general trochoid but ω and the drift are not affected ([*17*]/I, [*91*]).

Consider now the motion of electrons in a gas when a magnetic field is applied. The path of the electrons between two collisions is the arc of a circle, its length equals λ, the mean free path, as with $H=0$. The distance between two successive collisions is the chord $2r\sin(\lambda/2r)$. The displacement of an electron is thus reduced by the magnetic field, the number of collisions per unit displacement is increased and the diffusion coefficient in that direction is reduced. However, there is no change either in gas pressure, velocity of the electrons nor in collision frequency.

In crossed magnetic and electric fields the effect of the cycloidal motion on α/p has been calculated [*92*]. Assuming that all electrons have the same free path λ and that they lose their energy in each collision, the distance x travelled in field direction in units of λ is

$$\frac{x}{\lambda} = 1 - \frac{\lambda}{a} \tag{41.1}$$

where a, the length of an arch of a cycloid, is given by

$$a = \frac{8 \cdot 10^8}{e/m} \frac{E}{H^2} \tag{41.2}$$

where e/m is 1.6×10^7 e.m.u., H is in Oe and E in V/cm. With (41.1) an equivalent gas pressure p_e can be defined by

$$\frac{p_e}{p} = \frac{\lambda}{x} = \frac{1}{1 - \lambda/a} \tag{41.3}$$

and an electron ionization coefficient in the fields $\boldsymbol{H}$ and $\boldsymbol{E}$ can be introduced which is given by

$$\frac{\alpha}{p} = A \cdot \mathrm{e}^{-\frac{B}{E/p_e}}. \tag{41.4}$$

Since $\lambda/a \propto (H/p)^2/(E/p)$, it is seen that, for constant E/p, p_e increases with rising H.

This theory has recently been improved [*73*] by using a distribution of free paths and allowing for recapture of electrons by the cathode. By including the distribution instead of (41.4) an expression for α is obtained, viz.

$$\frac{\alpha}{p} = A \frac{\lambda}{a} f\left(\frac{\lambda}{a}\right) \frac{\operatorname{Sin}\left\{\frac{a}{2\lambda}\left(1 - \frac{B}{(a/4\lambda)(E/p)}\right)^{\frac{1}{2}}\right\}}{\operatorname{Sin}\frac{a}{2\lambda}} \tag{41.5}$$

where $f(\lambda/a) = (\lambda/a)\,\{\operatorname{Cot}(a/2\lambda) - 2\lambda/a\}$ and a is given by (41.2). A and B are the constants in (41.4). Neither (41.4) nor (41.5) has been tested experimentally. If these results are applied to sparking potential measurements in simultaneous non-uniform electric and transverse uniform magnetic fields, no satisfactory agreement is obtained, which is not altogether surprising. The present theory does not include elastic collisions and does not use a distribution of the energy of electrons [*92*].

It is pertinent to enquire here about the effect of a magnetic field on the energy distribution [*94*]. Analysis of a deflection experiment using an electron swarm in a transverse magnetic field $\boldsymbol{H}$ shows that the distribution $\mathrm{e}^{-\left(\frac{\varepsilon}{\varepsilon_0}\right)^2}$ is changed into

$$f_0 = A_1 \mathrm{e}^{\frac{-\varepsilon^2 - \varepsilon H^2 \cdot C}{\varepsilon_0^2}} \tag{41.6}$$

where A_1 and C are constants. $C = 3\,(e\lambda)^2/M$ where λ is the electron mean free path and M the mass of a gas molecule; $\varepsilon_0 = e\lambda E$. Under experimental conditions the distribution f_0 changes slightly ($< 10\%$); it is seen from (41.6) that the application of a magnetic field reduces mainly the number of fast electrons and hence α.

II. Future work.

42. For experimental reasons nearly all observations have been made in non-uniform (radial symetrical) electric and uniform magnetic fields. To use the results obtained it is necessary to develop the theory of ionization in radial electric fields before the combined electric and magnetic effects can be reliably assured. The other possibility is to derive experiments in combined uniform fields. The study of this type of multiplication process has a bearing on certain problems in cosmic physics in particular in connexion with solar and auroral phenomena.

Acknowledgements.

I wish to thank Dr. A. E. Robson for reading and improving the manuscript. Professor L. B. Loeb has kindly sent me a list of references of the chapter on ionization coefficients of his book prior to publication.

References.

[1] Coulomb, C. A. de: Mem. l'Acad. Sci. **1785**, 612.
[2] Faraday, M.: Researches in Electricity. London 1844.
[3] Crookes, W.: Phil. Trans. Roy. Soc. Lond. **1**, 135 (1879).
[4] Thomson, J. J.: Conduction of Electricity through Gases. Cambridge 1903.
[5] Stoletow, A.: J. de phys. **9**, 468 (1890).
[6] Kreusler, H.: Verh. phys. Ges. **17**, 86 (1898). — Ann. Phys., Lpz. **6**, 398 (1901).
[7] Schweidler, E. v.: Wien. Ber. **18**, 273 (1899).
[8] Righi, A.: Mem. R. Acad. Bologna **10**, 85 (1890).
[9] Townsend, J. S.: Nature, Lond. **62**, 340 (1900).
[10] Thomson, J. J.: Phil. Mag. **48**, 552 (1899).
[11] Stark, J.: Phil. Mag. **6**, 116 (1903).
[12] Townsend, J. S.: Phil. Mag. **6**, 358 (1903).
[13] Rutherford, E., and R. K. Mc Clung: Phil. Trans. Roy. Soc. Lond., Ser. A **196**. 25 (1901).
[14] Townsend, J. S.: Electricity in Gases. Oxford: Clarendon Press 1915.
[15] Engel, A. von.: Ionized Gases. Oxford: Clarendon Press 1955.
[16] Wien, W.: Handbuch der Experimentalphysik, Bd. 14. Leipzig: Akademische Verlagsgesellschaft 1927.
[17] Engel, A. v., u. M. Steenbeck: Elektrische Gasentladungen, Bd. I u. II. Berlin: Springer 1932, 1934.
[18] Thomson, J. J.: Conduction of Electricity through Gases. 3rd edit., p. 466. 1928. See also C. W. Rice: Phys. Rev. **70**, 228 (1946) and J. K. Theobald: J. Appl. Phys. **24**, 123 (1953).
[19] Gosseries, A.: Physica, Haag **6**, 458 (1939).
[20] Kruithof, A. A., u. F. M. Penning: Physica, Haag **3**, 515 (1936).
[21] Jones, F. Ll., and A. B. Parker: Proc. Roy. Soc. Lond., Ser. A **213**, 185 (1952)
[22] Dutton, J., S. C. Haydon and F. Ll. Jones: Proc. Roy. Soc. Lond., Ser. A **213**, 203 (1952); see also Proc. Phys. Soc. Lond. B **69**, 2 (1956).
[23] Mac Donald, A. D., and D. D. Betts: Canad. J. Phys. **30**, 565 (1952) (Ne). — Reder, F. H., and S. C. Brown: Phys. Rev. **95**, 885 (1954) (He).
[24] Brown, S. C.: Proc. Inst. Radio Engrs. **39**, 1493 (1951).
[25] Hornbeck, J. A.: Phys. Rev. **83**, 374 (1951).
[26] Fletcher, R. C.: Phys. Rev. **76**, 1501 (1949)
[27] Kachickas, G. A., and L. H. Fisher: Phys. Rev. **91**, 775 (1953).
[28] Townsend, J. S.: In Handbuch der Radiologie, Bd. I. Leipzig: Akademische Verlagsgesellschaft 1920.
[29] Kruithof, A. A.: Physica, Haag **7**, 519 (1940).
[30] Kruithof, A. A., u. M. S. Druyvesteyn: Physica, Haag **4**, 450 (1937).
[31] Kruithof, A. A., u. F. M. Penning: Physica, Haag **4**, 430 (1937).
[32] Hochberg, B. M., and E. S. Sandberg: C. R. Acad. Sci. USSR. **53**, 511 (1946).
[33] Meek, J. M., and J. D. Craggs: Electrical breakdown of gases. Clarendon: Oxford Press 1953.
[34] Druyvesteyn, M. S., and F. M. Penning: Rev. Mod. Phys. **12**, 87 (1940).
[35] Dunlop, S. H.: Nature, Lond. **164**, 452 (1949).
[36] Engstrom, R. W., and W. S. Huxford: Phys. Rev. **55**, 754 (1939); **58**, 67 (1940).
[37] Badareu, E., et G. G. Bratescu: Bull. Soc. Rom. Fiz. **45**, 9 (1944).
[38] Townsend, J. S., and S. P. Mc Callum: Phil. Mag. **6**, 857 (1928).
[39] Ayers, T. L. R.: Phil. Mag. **45**, 353 (1923).
[40] Hale, D. H.: Phys. Rev. **54**, 241 (1938); **56**, 815 (1939).
[41] Rose, D. J.: (Bell Telephone Lab.) presented at Gaseons Electronics Conf. N. Y. 1954.
[42] Bowls, W. E.: Phys. Rev. **53**, 293 (1938).
[43] Hellmann, R.: Z. Physik **91**, 556 (1934).

[44] MASCH, K.: Arch. Elektrotechn. **26**, 582 (1932).
[45] GEBALLE, R., and M. A. HARRISON: Phys. Rev. **85**, 372 (1952).
[46] SANDERS, F. H.: Phys. Rev. **41**, 667 (1932); **44**, 1020 (1933).
[47] RAETHER, H.: Z. Physik **107**, 91 (1937).
[48] HARRISON, M. A., and R. GEBALLE: Phys. Rev. **91**, 1 (1953).
[49] HOCHBERG, B. M., i E. S. SANDBERG: J. techn. Phys. USSR. **12**, 65 (1942).
[50] BADAREU, E., et M. VALERIU: Bull. Soc. Rom. Fiz. **43**, No. 79 (1942).
[51] EMELEUS, K. G., R. W. LUNT and C. A. MEEK: Proc. Roy. Soc. Lond. **156**, 394 (1936).
[52] POSIN, D. A.: Phys. Rev. **50**, 650 (1936).
[53] BRADBURY, N. E., and R. A. NIELSON: Phys. Rev. **49**, 388 (1936).
[54] DRUYVESTEYN, M. J.: Physica, Haag **3**, 65 (1936).
[55] HERTZ, G.: Z. Physik **32**, 298 (1925).
[56] ABDELNABI, I., and H. S. W. MASSEY: Proc. Phys. Soc. Lond. A **66**, 288 (1935).
[57] ROGOWSKI, W.: Z. Physik **115**, 257 (1940).
[58] MOLNAR, J. P.: Phys. Rev. **83**, 933, 940 (1951).
[59] MASSEY, H. S. W.: Negative ions. Cambridge: University Press 1950.
[60] HAGSTRUM, H. D.: Rev. Mod. Phys. **23**, 185 (1951).
[61] TOWNSEND, J. S.: Electrons in Gases. London: Hutchinson 1947.
[62] TOWNSEND, J. S., and F. LL. JONES: Phil. Mag. **15**, 282 (1938).
[63] FRERICHS, R.: Ann. Phys., Lpz. **85**, 362 (1928).
[64] PENNING, F. M.: Physica, Haag **8**, 137 (1928).
[65] PENNING, F. M.: Proc. Acad. Sci. Amst. **32**, 341 (1929).
[66] PENNING, F. M.: Physica, Haag **10**, 47 (1930). See also E. W. PIKE: Phys. Rev. **49**, 513 (1936).
[67] PENNING, F. M.: Z. Physik **57**, 723 (1929).
[68] BIONDI, M. A.: Phys. Rev. **88**, 660 (1952).
[69] MITCHELL, A. C. G., and M. W. ZEMANSKY: Resonance radiation and excited atoms. Cambridge: University Press 1933.
[70] KESAEV, I.: Tech. Phys. USSR. **4**, 68 (1937).
[71] PURCARU, I. C.: Bull. Math. Phys. Polyt. Bucarest **7**, 45 (1937).
[72] SCHUT, T. G., u. J. A. SMIT: Physica, Haag **10**, 440 (1943).
[73] BIONDI, M. A.: Phys. Rev. **83**, 653 (1951); see [68].
[74] GRANT, F. A., and A. D. KRUMBEIN: Phys. Rev. **90**, 59 (1953).
[75] DORGELO, M. B.: Z. Physik **36**, 467 (1926).
[76] PHELPS, A. V., and J. P. MOLNAR: Phys. Rev. **89**, 1202 (1953).
[77] KENTY, C.: Phys. Rev. **80**, 95 (1950).
[78] MEISSNER, K. W., and W. F. MILLER: Phys. Rev. **92**, 896 (1953).
[79] MORTON, P. L.: Phys. Rev. **70**, 358 (1946).
[80] JOHNSON, G. W.: Phys. Rev. **73**, 284 (1948).
[81] WERNER, S.: Z. Physik **90**, 384 (1934); **92**, 705 (1934).
[82] COMPTON, K. T., and M. MORSE: Phys. Rev. **30**, 305 (1928).
[83] LAUER, E. S.: J. Appl. Phys. **23**, 300 (1952).
[84] BANDEL, H. W.: Phys. Rev. **84**, 92 (1951).
[85] WEISSLER, G. L.: Phys. Rev. **63**, 96 (1942).
[86] MILLER, C. G., and L. B. LOEB: J. Appl. Phys. **22**, 494 (1951).
[87] VARNEY, R. N. et al.: Phys. Rev. **48**, 818 (1935).
[88] CROWE, R. W. et al.: Phys. Rev. **96**, 10 (1954).
[89] FRANCIS, G.: Proc. Phys. Soc. Lond. B **68**, 369 (1955).
[90] FLETCHER, R. C.: Phys. Rev. **76**, 1501 (1949).
[91] ALFVÉN, H.: Cosmical Electrodynamics. Oxford: Clarendon Press 1950.
[92] VALLE, G.: Nuovo Cim. **7**, 174 (1950).
[93] SOMERVILLE, J. M.: Proc. Phys. Soc. Lond. B **65**, 620 (1952).
[94] ALLIS, W. P., and H. W. ALLEN: Phys. Rev. **52**, 703 (1937).
[95] CROWE, R. W. et al: J. Appl. Phys. **26**, 1121 (1955).
[96] GLOTOW, I. I.: Phys. Z. Sowjet. **12**, 257 (1937); **13**, 84 (1938).
[97] McGEE, J. D., and J. C. JAEGER: Phil. Mag. **6**, 1107 (1928).
[98] BAILEY, V. A., and W. E. DUNCANSON: Phil. Mag. **10**, 145 (1930).
[99] FROMHOLD, L.: Z. Physik **144**, 396 (1956).
[100] HOPWOOD, W., N. J. PEACOCK and A. WILKES: Proc. Roy. Soc. Lond., Ser. A **235**, 334 (1956).
[101] KOJIMA S. and K. KATO: J. Phys. Soc. Jap. **11** 322 (1956).

Secondary Effects.

By

P. F. Little.

With 47 Figures.

Introduction.

The nature of the secondary ionisation process in the conduction of electricity through a gas is considered in Part I of this article. We confine the discussion there to the TOWNSEND discharge, so that complications due to space charge may legitimately be ignored. This is the major part of the article.

Some remarks on discharges of special interest as far as secondary effects are concerned are given in Part II, but a full discussion of the secondary mechanisms in glow or arc discharges etc. is not given. Other authors in this Encyclopedia treat these discharges individually in detail, including the secondary ionisation processes.

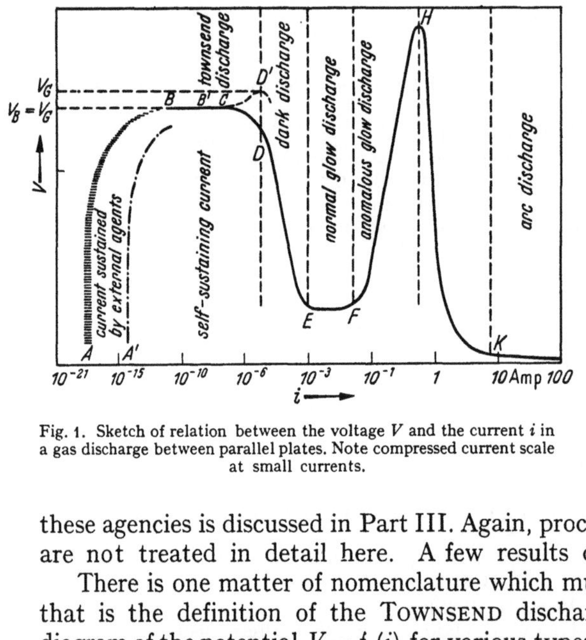

Fig. 1. Sketch of relation between the voltage V and the current i in a gas discharge between parallel plates. Note compressed current scale at small currents.

For comparison with values derived from discharges electron emission from surfaces bombarded with ions, photons etc., and volume ionisation by these agencies is discussed in Part III. Again, processes dealt with in other articles are not treated in detail here. A few results only are quoted in such cases.

There is one matter of nomenclature which must be made clear at this point: that is the definition of the TOWNSEND discharge. Fig. 1 shows a schematic diagram of the potential $V = f(i)$ for various types of discharge[1] between parallel plates: it is essentially that given by DRUYVESTEYN and PENNING [*10*].

When very minute currents are flowing, sustained by external agents (photoelectrons from the cathode or ionisation produced in the gap by X-rays, for example) then V increases as i increases. This is a very rapid increase at first and V at a given i is strongly dependent on i_0 the current provided by the external agency. Between the points B or B', however, depending on i_0, the current increases very rapidly while V does not increase much above V_B. Beyond C the characteristic is substantially independent of i_0 and usually V falls with increasing i (CD). The transition region DE in Fig. 1 represents the dark discharge, a self sustained discharge which is further discussed in Sect. 8. We

[1] See also E. PFENDER, Z. angew. Phys. 5, 450 (1953).

use the term TOWNSEND discharge for the whole region up to the point C which will be found to represent the point where space charge becomes important in determining the current. It should be added that the influence of i_0 in reducing V can be shown even in a glow discharge if i_0 is large enough[1].

If experiments are carried out at low pressures the curve CD' will be traced out. This is again an effect of space charge, causing V to rise to V_G.

The rest of the curve is self-explanatory and no difficulty arises in distinguishing between the various regions.

TOWNSEND discharges between wire and cylinder electrodes are briefly discussed in Sec. 18[2].

I. The TOWNSEND discharge.

a) Introduction.

1. The first idea of secondary ionisation. The early qualitative studies of ionisation in gases were first interpreted in a quantitative manner by the work of TOWNSEND and his school. The beginning of this work, the concept of the ionisation coefficient α, is dealt with in the previous article. TOWNSEND [*1*], [*2*] found that when a gap of length d between two plane parallel plates in air was irradiated, the current flowing was dependent on the electric field X in such a way that it seemed that ions were multiplying as they moved between the plates. Later work showed that "negative ions" were effective in multiplication, and that when the cathode only is irradiated, the current i flowing in the gap is given by

$$\frac{i}{i_0} = e^{\alpha(d - x_0)} \tag{1.1}$$

where i_0 is the photoelectric current from the cathode and α is the first TOWNSEND ionisation coefficient. Each "negative ion" produces α ion-pairs per cm. in direction of the electric field strength X, and α/p is a function of X/p for a given gas. x_0 [*12*] is the distance travelled before the electrons have sufficient energy to ionise. We shall put $x_0 = 0$ in this section.

It was found that if d is made large enough the graphs of $\log(i/i_0)$ against d curve up, *i.e.* the current becomes larger than expected from (1.1). The upward curvature indicates that another source of ionisation has become important.

It was natural to assume that positive ions are now producing ion-pairs, and TOWNSEND introduced a coefficient β to represent the number of ion-pairs formed by a positive ion in moving 1 cm in field direction.

Let n_0 be the number of electrons emitted from the cathode per second, n the number arriving at the anode per second and p, q the number of ion-pairs produced by collision in the volume on the cathode and anode side respectively of the plane at x. Then

$$n = n_0 + p + q$$

and $n_0 + p$ electrons cross the plane at x in unit time towards the cathode while q positive ions cross in the opposite direction. In the strip dx at x,

$$dp = -dq = (n_0 + p)\,\alpha\,dx + q\,\beta\,dx\,,$$

$$\frac{dp}{dx} = (n_0 + p)(\alpha - \beta) + n\beta$$

[1] H. DEMBER and G. GEHLHOFF: Verh. dtsch. phys. Ges. **8**, 264 (1906). — W. KLUGE: Z. angew. Phys. **5**, 173 (1953). — W. KLUGE and A. SCHULZE: Z. angew. Phys. **6**, 364 (1954), Naturwiss. **42**, 40 (1955). See G. FRANCIS, Vol. XXII.

[2] Corona discharges are discussed by L. B. LOEB in Vol. XXII.

or

$$\frac{d}{dx}(n_0 + p) - (\alpha - \beta)(n_0 + p) = n\beta.$$

The integrating factor is $e^{-(\alpha-\beta)x}$. Hence, since when $x=0$ $p=0$ and $n/n_0 = i/i_0$ we have

$$\frac{i}{i_0} = \frac{(\alpha-\beta)\,e^{(\alpha-\beta)d}}{\alpha-\beta\,e^{(\alpha-\beta)d}}. \tag{1.2}$$

If $\beta \ll \alpha$ then

$$\frac{i}{i_0} = \frac{e^{\alpha d}}{1-(\beta/\alpha)\,e^{\alpha d}}. \tag{1.2a}$$

The experimental results were formed to fit (1.2a) very well, for β was much smaller than α. For example, in He at $X/p = 400$ V/cm. per mm. Hg $\alpha/p = 1.1$ and $\beta/p = 0.16$. So long as gas conduction was regarded as analogous to liquid electrolytic conduction, this was felt to be a difficulty, for negative and positive ions should have similar charge and mass and hence should be roughly equal in their ability to ionise the parent gas. With the recognition that the negative carriers were electrons, this desire for symmetry was lost and the description given above seemed logical and attractive. β/p depends on X/p and the nature of the gas, as α/p does.

An alternative suggestion was put forward by TOWNSEND (though not favoured by him) and by HOLST and OOSTERHUIS[1] to explain the curvature of the $\log(i/i_0)$ *vs.* d graphs in terms of the emission of electrons from the cathode when bombarded by positive ions. We shall use the symbol γ_i to denote the average number of electrons released from a surface by an incident positive ion. γ_i would be expected to depend on the energy distribution and nature of the ions, on the nature of the surface, and also on the electric field at the surface. These are the same as for β, with the surface condition an extra parameter.

The influence of γ_i on i may be deduced as follows. For each electron leaving the cathode $\exp(\alpha d)$ electrons reach the anode and $[\exp(\alpha d)-1]$ ions are formed in the gap. These return to the cathode, releasing $\gamma_i[\exp(\alpha d)-1]$ electrons from the cathode: write this as f_i electrons. This "second generation" of electrons multiplies in the gas to give $f_i e^{\alpha d}$ electrons at the anode. They produce also $f_i[\exp(\alpha d)-1]$ ions in the gap as they cross, so releasing a third generation of electrons numbering $\gamma_i f_i[\exp(\alpha d)-1] = f_i^2$ electrons. Clearly the total number of electrons arriving at the anode due to one initial election is

$$e^{\alpha d}[1 + f_i + f_i^2 + \cdots] = \frac{e^{\alpha d}}{1-f_i}.$$

We have then in the steady state

$$\frac{i}{i_0} = \frac{e^{\alpha d}}{1-f_i} = \frac{e^{\alpha d}}{1-\gamma_i(e^{\alpha d}-1)} \tag{1.3}$$

which is of the same form as (1.2a) if $e^{\alpha d} \gg 1$ as is usually the case. Thus the experimental results could be explained in terms of cathode emission due to ions rather than ionisation in the gas by the positive ions.

2. The definition of a secondary effect. This discussion illustrates the meaning of primary and secondary ionisation processes. There is no doubt that the first consequence of the applied electric field is the ionisation by electrons, and only

[1] G. HOLST and E. OOSTERHUIS: Phil. Mag. **46**, 1117 (1923).

after ion-pairs have been formed in this manner do positive ions exist in the gap to cause ionisation or secondary electron emission from the cathode. When a source of charged particles is dependent on another source of ionisation or excitation within the discharge *it is termed a secondary effect.* Only if ionisation is a direct consequence of an external agency is it called a primary ionisation process.

Some physical phenomena maybe either primary or secondary processes depending on the conditions. Field emission, for example, may be directly due to an applied field (a primary process) or to a positive space charge (a secondary process). Similarly thermionic emission from an externally heated cathode is a primary source of electrons but if the heating is due to ion bombardment the thermionic emission is a secondary effect. Obviously the same physical phenomenon could be partly a primary and partly a secondary effect.

It is possible that some secondary mechanisms would not satisfy an equation of the type (1.2a). We shall examine the various types with this in mind: secondary sources of ionisation satisfying equations like (1.2a) are allowable processes.

b) Physical processes in the discharge.

3. Possible secondary mechanisms. As electrons move through any gas in the presence of a sufficiently strong electric field they ionise and excite the atoms of the gas. (We shall use the word "atoms" for both atoms and molecules in this article.) Possible causes of secondary ionisation are then

(i) Positive ions perhaps multiply charged, or formed by stepwise processes,
(ii) Photons,
(iii) Excited and metastable atoms.

Moreover, as the ions move in their own gas, charge exchange with slow neutral atoms is very likely. So we have also to consider

(iv) Fast neutral atoms, whose action depends on their kinetic energy [not potential energy as in (iii)].

In addition the impact of electrons on the anode may produce

(v) X rays [a special case of (ii)],
(vi) Positive ion emission.

All these agents may act in the gas or at the cathode, depending on the conditions. In some circumstances we may expect also at the cathode

(vii) Field emission,
(viii) Thermionic emission,
and in the gas,
(ix) Thermal ionisation.

In discharges in insulators[1] and in gases[2] when alternating potentials are applied the main secondary process will be emission due to (x) Electrons. We shall not however deal with this process. The article by R. Kollath in this volume should be consulted.

We shall use the symbols γ_i γ_m γ_p and γ_n to denote the number of electrons released from the cathode by one ion, metastable (or excited) atom, photon and fast neutral atom respectively. For ionisation in the gas, a distinction must be drawn between that produced by the positive ions and that due to any other agency. The ions multiply under the influence of the field and the other agents

[1] S. Whitehead: Dielectric Breakdown of Solids. Oxford 1951.

[2] E. W. B. Gill and A. von Engel: Proc. Roy. Soc. Lond., Ser. A **192**, 446 (1948), and references [*8*], [*13*].

do not. Also we may use the symbol β to denote the number of ion pairs formed per cm. path in *field* direction by one positive ion, but all other coefficients must be defined with reference to the *path* direction. Let us write μ_i and ζ_i for the number of ion pairs formed per cm. path in the gas by one photon and fast neutral atom respectively. The use of a symbol normally reserved for absorption coefficients will be a reminder that a photon which ionises an atom disappears completely and any excess energy available appears as kinetic energy of the electron.

It is not safe, however, to assume that all the electrons released from the cathode will assist in the maintenance of the discharge: the next section deals with this point.

4. The back-scattering of electrons to the cathode. In any discharge the number of electrons joining the main current stream will not be identical with the number released from the cathode by the various agencies acting there. Some of the electrons will be scattered back to the cathode by their collisions with gas molecules. The fraction lost in this way will vary with the energy distribution of the secondaries and with the electric field, or better with X/p. THOMSON [3] first discussed the matter theoretically.

The number of electrons striking the cathode per cm. per second is $(nC/4)$ where n is the electron density and C their average velocity. The current density in the gas is $j^- = nev^-$ where v^- is the electron drift velocity.

If the electron current emitted from the cathode per sq. cm. is j_0^- then

$$j_0^- - j^- = \frac{nC}{4}$$

and therefore

$$\frac{j^-}{j_0^-} = \frac{4v^-}{C + 4v^-}. \tag{4.1}$$

From this equation it will be seen that an increase in the average energy of emission C will reduce the current density in the gas, if all else is constant. If X/p increases, so that v^- increases, then the current density will increase if C is constant: the rate of increase will be more rapid the larger the value of C, unless C becomes so large that v^- is determined by C and not by X/p near the cathode.

BRADBURY[1] first verified this equation indirectly for photoelectrons from hydrogen and nitrogen assuming that $v^- = k^- X$ where k^- is the electron mobility. He found $K^- = \frac{k^- p}{760}$ (the mobility constant) to be a function of X/p only, using pressures from 9 to 70 mm. Hg and X/p between 0.05 and 20 V/cm. per mm. Hg, and his values were in good agreement with previous workers. This gives confidence in the theory and in the interpretation of C as the average energy of emission.

THEOBALD[2] has recently made a more direct test of equation (4.1) using photoelectrons whose energy distribution was measured in the apparatus. He found that i^-/i_0^- (assumed equal to j^-/j_0^-) depended only on X/p for A, H_2, and N_2 in the range $p = 100$ to $p = 760$ mm. Hg, with $X/p = 10$ V/cm. per mm. Hg. When the known values for v^- (due to NIELSEN) are used in equation (4.1), the predicted curves for $i^-/i_0^- = f(X/p)$ are in good agreement for the molecular gases, but fall below the curves observed in argon by a factor five.

[1] N. E. BRADBURY: Phys. Rev. **40**, 980 (1932). cf. [10] and W. HARRIES and G. HERTZ: Z. Physik **46**, 177 (1927) for earlier studies.
[2] J. K. THEOBALD: J. Appl. Phys. **24**, 123 (1953).

Now in nitrogen and hydrogen the electrons are emitted with roughly the equilibrium energy. The reason for the failure in argon is due to the large equilibrium energy of electrons there which is much greater than the initial energy of the photo electrons. Thus the drift velocity of the photoelectrons is larger initially than the value found by NIELSEN in measurements where the electrons were in equilibrium with the field. From (4.1) it can be seen that an increase in v^- increases j^- so the experimental values can be understood.

If direct measurements of C and v^- were made, a complete check of THOMSON's theory could be carried out. The theory assumes a constant energy distribution among the secondary electrons which will only be true if the initial velocity distribution is the same as that of the electrons when in equilibrium with the applied field. This means that an alteration in the value of X/p (which determines the equilibrium energy distribution) or in the agency responsible for emission will affect the validity of equation (4.1) in a given gas.

If the initial average energy of the secondary electrons is much higher than the equilibrium average, then the diffusion back to the cathode will be larger than predicted from THOMSON's theory using values for v^- derived from the usual mobility measurements in uniform field. Hence i^-/i_0^- will be smaller than the theoretical value. This situation is the reverse of that observed by THEOBALD for photoelectrons in argon.

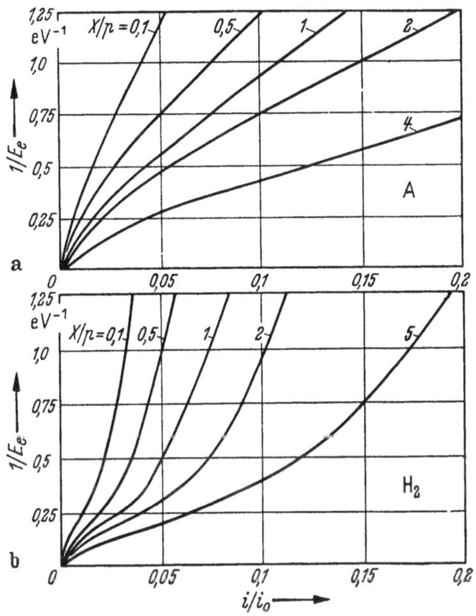

Fig. 2a and b. The influence of the mean secondary electron energy E_e on the electron current i flowing to the anode when i_0 is the electron current in vacuum (i.e. no back diffusion to the cathode). Shown as $1/E_e = f(i/i_0)$ for various X/p in argon and hydrogen, accurate to within a factor two.

Clearly, a complete solution of the problem involves a changing energy distribution among the electrons, and is concerned with the difficult intermediate region of scattering theory where neither single scattering nor true diffusion occurs. The THOMSON approach is oversimplified in that the rate at which electrons strike the cathode is assumed to be given by the kinetic theory of gases in equilibrium, but all other attempts to deal with the "plural scattering" which actually is taking place have failed. The reader is referred to LOEB [5] for a detailed account of the work.

On the basis of his experimental results for argon and molecular gases THEOBALD presents graphs of $1/E_e = f(i/i_0)$ for various X/p. E_e is the average energy of the electrons on emission and $i/i_0 = i^-/i_0^-$, the ratio of electrons moving off to the anode to the number released from the cathode. He considers that these represent the behavious of all rare gases and molecular gases respectively within a factor two. Figs. 2a and b show these curves, which may be used to correct observations of γ_i etc. for the effects of back scattering of electrons in the range of X/p covered in the figures (0.1 to 5 V/cm. per mm. Hg).

From the graphs it appears that i/i_0 is smaller in A than in H_2 if the secondary electrons have high energies ($1/E_e < 0.5$ approximately or $E_e > 2$). If the secondary electrons have lower energies $E_e \sim 2$ eV, i/i_0 is larger in A than in H_2. The influence of a change in X/p is greater the larger E_e for H_2 and the reverse is true in A. This is presumably caused by the difference in the average energy of the electrons in the two gases at the same X/p.

The graphs show that at very small X/p the back scattering is overwhelmingly important, and may change the apparent value of γ_i, for example, by two orders of magnitude if the initial electron energy is high. We shall write $\gamma_i' = \nu\gamma_i$ as the apparent secondary emission coefficient where $\nu = i/i_0$.

In equation (1.3) γ_i should be replaced by γ_i' for accuracy, but the discussion there is otherwise unaffected.

5. Electron emission from the cathode. A formal treatment showing the influence of the ions has been given above, and the same approach may be used for photons, metastable atoms or fast neutral atoms provided that the production rates for each replace the ion production rate. There is also a further complication arising from the scattering and absorption of the entity concerned which must be considered.

Let us first see the influence of *photons*. The energy of such photons must be large enough to overcome the work function of the cathode: assume that ϑ effective photons are produced by one election in moving 1 cm. in field direction. Then each initial election produces $\vartheta \exp(\alpha x)$ photons in dx at x as it multiplies in crossing the gap.

Assuming that the electrodes approximate to infinite plates, a fraction $g\,(g \sim \frac{1}{2})$ of these photons will move towards the cathode. For simplicity, we shall make further assumption that the photons move in field direction. This will give an upper limit to the number of photons arriving at the cathode, for the emission is isotropic. If the total absorption coefficient for the effective photons is μ, the number of photons reaching the cathode from dx at x is $\vartheta \exp(\alpha - \mu) x \cdot dx$.

By a simple analysis identical with that used in deriving (1.3) it can be seen that f_i in that work should be replaced by

$$f_p = \frac{\vartheta \gamma_p' g/(\alpha - \mu)}{e^{(\alpha-\mu)d} - 1}$$

so that finally in the steady state

$$\frac{i}{i_0} = \frac{e^{\alpha d}}{1 - \frac{\vartheta \gamma_p' g}{\alpha - \mu}\left(e^{(\alpha-\mu)d} - 1\right)}. \tag{5.1}$$

With similar simple assumptions (5.1) would hold for *metastable atoms*, if the symbols are interpreted in the appropriate way. We shall use ϑ_m instead of ϑ for the number of metastable atoms produced per cm. by an electron. Diffusion from the discharge space and into the anode is neglected, so μ expresses the probability of volume destruction of the metastables as they diffuse to the cathode.

There is one important distinction between the photons and the metastable atoms. The metastables formed near the anode will all diffuse to it, since their concentration at the anode is zero and the concentration gradient large. This means that g in (5.1) is a function of x, decreasing rapidly towards the anode. It is to be expected then that the effective γ will vary with d if metastable atoms are important.

Apart from this, (5.1) is similar to (1.2a) and (1.3) if $\mu \ll \alpha$, and with this limitation cathode emission due to photons or metastable atoms are satisfactory secondary processes. The assumption that these agents move parallel to the field is satisfactory if $\mu \ll \alpha$, provided that the electrodes are large.

If the electrode dimensions are comparable with the gap length, the value of g will vary with x very greatly for photons and metastables, so γ should depend very strongly on d. This has been demonstrated for metastables by a more refined analysis (Sect. 15) and experimentally (Sect. 16).

If $\mu \sim \alpha$

$$\frac{i}{i_0} = \frac{e^{\alpha d}}{1 - \vartheta \gamma_p' g d}, \tag{5.1 a}$$

and if $\mu \gg \alpha$

$$\frac{i}{i_0} = \frac{e^{\alpha d}}{1 - \vartheta \gamma_p' g/\mu}. \tag{5.1 b}$$

Obviously if μ is too great photo electric emission from the cathode will be no useful secondary process: the importance of the effect increases as μ tends to zero.

It must be remembered that μ in this derivation is an average absorption coefficient for all the quanta considered. The spectrum of the radiation from the discharge will depend on the gas and the value of X/p, so that μ and α depend on the same parameters. Similarly γ_p' is an average value, determined by the spectral energy distribution of the photons and back scattering i.e. by X/p. For metastable atoms the "absorption" is independent of X, being only a function of p. However, γ_m depends on the production of metastables and back-scattering as well as on μ, so it is a function of X/p.

Another point worthy of mention is that we have assumed that the process of absorption in the gas does not of itself produce secondary ionisation. For photons of not too high energy this will be true: μ includes losses due to scattering and the emission of one or more low energy quanta when one high energy quantum is absorbed. If the photon energy is sufficient to ionise the gas, then the absorption process will be largely due to photo ionisation, which itself produces secondary ionisation. This is considered separately (see below), but in practice may occur simultaneously with secondary emission at the cathode. Metastable atoms too may ionise in the volume if two collide or if gas mixtures are used (Sect. 37).

The influence of *fast neutral atoms* cannot be isolated from that of the positive ions from which they are formed by charge exchange collisions. Once formed, an atom cannot gain further energy from the field as an ion does. We assume it moves in field direction and does not lose much energy at a collision (see Sect. 19), if it has not sufficient energy to make inelastic collisions probable, but yet has energy large compared with thermal energies so that elastic scattering is partly forward scattering with consequent negligible energy transfer.

If we modify the argument leading to equation (1.3) then the upper limit for emission due to energetic atoms striking the cathode can easily be found. Instead of $(e^{\alpha d} - 1)$ ions returning to the cathode for one electron starting from the cathode we have $(k-1)(e^{\alpha d} - 1)$ ions where k is the fraction of ions which suffer charge exchange collisions in the gas. These ions are accompanied by $k(e^{\alpha d} - 1)$ energetic neutral atoms. The quantity f_i in the argument leading to (1.3) should clearly be replaced by

$$\gamma_i'(k-1)(e^{\alpha d} - 1) + \gamma_n' k (e^{\alpha d} - 1)$$

so finally we have

$$\frac{i}{i_0} = \frac{e^{\alpha d}}{1 - \{\gamma_i'(k-1)(e^{\alpha d} - 1) + \gamma_n' k (e^{\alpha d} - 1)\}}. \tag{5.2}$$

If $e^{\alpha d} \gg 1$

$$\frac{i}{i_0} = \frac{e^{\alpha d}}{1 - G e^{\alpha d}}$$

where

$$G = \gamma_i'(k-1) + \gamma_n' k.$$

This is of the same form as (1.2a) and (1.3), so that the action of fast neutral atoms at the cathode is an acceptable secondary ionisation process. The value of γ_n' depends on the energy distribution (which would be difficult to derive) and on the nature of the atoms, the nature and state of the surface, and possibly also the electric field at the surface. γ_n' is a function of X/p and other variables, as is γ_i'. A rigorous treatment would involve calculation of the number of atoms arriving at the cathode.

If *X-rays* are produced by the electrons striking the anode, they may release electrons from the cathode. If each electron at the anode produces ψ photons, a fraction g of these will move towards the cathode ($g \sim 1$), and using the same approximations as in dealing with the photons produced in the gas we have instead of f_p the quantity

$$f_x = \gamma_p' g \psi \, e^{(\alpha-\mu)d}$$

where μ is the average absorption coefficient. Then

$$\frac{i}{i_0} = \frac{e^{\alpha d}}{1 - \gamma_p' g \psi \, e^{(\alpha-\mu)d}}. \tag{5.3}$$

This again is a satisfactory form of secondary ionisation, process if the electrodes are effectively infinite in area and μ is small: the comments previously made as to the nature of μ and γ_p' apply here. ψ will depend on X/p and the nature of the anode.

The emission of positive ions from the anode has been suggested[1] as the important secondary process when high voltages are applied to low presures ($<10^{-4}$ mm. Hg) TOWNSEND discharges which precede the so-called vacuum breakdown. Under such conditions X-rays are certainly present, and have to be taken into account. The electron bombardment of the anode may cause local heating if the initial current density is high enough. If sufficiently high temperatures are reached positive ions may be emitted[2]. Alternatively, if a gas film is adhering to the anode this may be ionised and positive ions detached by electrostatic forces. This would seem to be more probable as absorbed gas is only completely absent if ultrahigh vacua is maintained and extreme heat treatment employed[3].

If such an effect is present, the formal treatment will be similar to that for X-rays, if ψ now is taken as the number of ions released per electron at the anode, γ_p' is replaced by γ_i', $g=1$ and $\mu=0$. Then (5.3) would become

$$\frac{i}{i_0} = \frac{e^{\alpha d}}{1 - \gamma_i' \psi \, e^{\alpha d}}. \tag{5.4}$$

This is an acceptable secondary ionisation process even if the electrode area is not effectively infinite.

If *field emission* in a TOWNSEND discharge of electrons is to be a secondary effect it must be due to a surface layer of positive ions on an insulating film which covers the cathode. Space charges are by definition negligible.

---

[1] J. G. TRUMP and R. J. VAN DER GRAAF: J. Appl. Phys. **18**, 327 (1947).
[2] M. HEALEA and C. HOUTERMANS: Phys. Rev. **58**, 608 (1940).
[3] J. A. BECKER and C. D. HARTMAN: J. Phys. Chem. **57**, 153 (1953).

From FOWLER and NORDHEIM'S work[1], it is known that the current density j due to an applied field X at a metal surface is

$$j = 6.2 \times 10^{-6} (\mu/\varphi)^{\frac{1}{2}} (\mu + \varphi)^{-1} X^2 \exp(-6.8 \times 10^7 . \varphi^{\frac{3}{2}}/X) \tag{5.5}$$

if j is in A/cm.², in V/cm., μ (the FERMI limit) and φ (the work function) in volts.

If, however, the surface is covered with an insulating film on which ions are deposited the potential barrier is not of the type treated by FOWLER and NORDHEIM (see Fig. 42a) but is of the form shown in Fig. 42b. The treatment given by STERN, GOSSLING, and FOWLER[2] applies here, and is quoted by LLEWELLYN JONES and MORGAN[3] to explain some observations made with aluminium cathodes. When the field intensification due to the surface charge is considered this is equivalent (Sect. 34) to emission from a metal of work function φ_1 covered with a thin layer of thickness d of another metal whose work function is φ_2: a potential gradient $(\varphi_1 - \varphi_2)/d$ is set up and equation (5.5) becomes

$$\left.\begin{aligned} j = 6.2 \times 10^{-6} (\mu/\varphi_1)^{\frac{1}{2}} (\mu + \varphi_1)^{-1} X^2 \exp(-6.8 \times 10^7 \varphi_2^{\frac{3}{2}}/X) \times \\ \times (\varphi_1/\varphi_2) \exp\left(-6.8 \times 10^7 d \left[\frac{\varphi_1 + \varphi_2 + (\varphi_1 \varphi_2)^{\frac{1}{2}}}{\varphi_1^{\frac{1}{2}} + \varphi_2^{\frac{1}{2}}}\right]\right) \end{aligned}\right\} \tag{5.6}$$

This is of the same form as far as $j = f(x)$ is concerned, but the constants are different numerically: lower applied fields are now sufficient to cause field emission (10^5 V/cm. instead of 10^7 V/cm.).

It is to be noted that the field must not only be large but extend over sufficient distance to make the potential drop V_d outside the barrier comparable with the work function φ. The wave mechanical treatment is based on a thin potential barrier at the levels where electrons exist in the metal, and the barrier is not changed if $V_d \ll \varphi$.

The nature of the film is also important. It must provide adequate insulation to withstand a high field across it: not all surface films can satisfy this condition. Yet in the steady state the surface is continually bombarded with positive ions, and the field will rise until the film breaks down or until the electrons are pulled out in sufficient numbers to recombine with the ions arriving. This may imply that the film is constantly being punctured and re-formed over the whole surface.

Supposing, however, that a film of the necessary type has been created, we have in a steady discharge an emission due to the ions arriving at the cathode and distinct from the secondary emission process described by the coefficient γ_i. The increased current flowing from the cathode could be formally included by increasing γ_i, though the electrons would not appear at the same time as the electrons extracted by the ions individually. There would be a time lag until the surface charge had built up to a sufficient intensity, but in a steady discharge the field emission due to an established surface layer of ions would be a satisfactory secondary process.

A similar argument may be used to include the effect of *thermionic emission* from the cathode, but this is not possible with the very small current density in a TOWNSEND discharge. For the same reason *thermal ionisation* in the volume need not be considered.

6. Ionisation in the volume. Turning now to the processes by which ionisation is created in the gas we again shall first deal with the *photons:* the influence of

[1] R. H. FOWLER and L. NORDHEIM: Proc. Roy. Soc. Lond., Ser. A **119**, 229 (1928).
[2] T. E. STERN, B. S. GOSSLING and R. H. FOWLER: Proc. Roy. Soc. Lond., Ser. A **124**, 699 (1929).
[3] F. LLEWELLYN JONES and C. G. MORGAN: Proc. Roy. Soc. Lond., Ser. A **218**, 88 (1953).

the ions has already been discussed in Sect. 1. Again we consider a gap of length d between infinite plane parallel electrodes, the cathode being irradiated to give an initial current i_0. An exact analysis is given by DUTTON *et al*[1], but here we give a simplified derivation.

Let ϑ photons be created by an election in moving 1 cm. in field direction. The total number of photons produced by one electron is

$$\int_0^d \vartheta\, e^{\alpha x}\, dx = \frac{\vartheta}{\alpha}\left(e^{\alpha d} - 1\right).$$

These photons are created throughout the gap, but most of them appear near the anode. Assuming that the photons are created at the anode and move parallel to the field so that a fraction g move towards the cathode, the number of photoionising events in dx at x from the cathode can easily be found as

$$\frac{\vartheta g \mu_i}{\alpha}\, e^{\alpha d - \mu (d - x)}\, dx$$

where μ is the total absorption coefficient for the photons and μ_i is the absorption leading to ionisation[2]. These photo-electrons in turn multiply as they move to the anode. The contribution to the current made by this second generation of electrons is by integration

$$\frac{\vartheta g \mu_i}{\alpha(\alpha - \mu)}\, e^{\alpha d}\left[e^{(\alpha - \mu) d} - 1\right].$$

At the same time more photons capable of ionising the gas are produced to create a third generation of electrons so that finally by summation we have

$$\frac{i}{i_0} = \frac{e^{\alpha d}}{1 - \varphi\left(e^{(\alpha - \mu) d} - 1\right)} \tag{6.1}$$

where

$$\varphi = \frac{\vartheta g \mu_i}{\alpha(\alpha - \mu)}.$$

If $\mu \ll \alpha$ (6.1) is of the same form as (1.2a) and (1.3); photoionisation is then an allowed type of secondary process. If, however, $\alpha \sim \mu$

$$\frac{i}{i_0} = \frac{e^{\alpha d}}{1 - \vartheta g d \mu_i/\alpha} \tag{6.1a}$$

and if $\mu \gg \alpha$:

$$\frac{i}{i_0} = \frac{e^{\alpha d}}{1 - \vartheta g/\alpha}. \tag{6.1b}$$

Clearly neither of these last two equations could account for the upcurving of the experimental log i/i_0 *vs.* d curves. Photo-ionisation then can only explain the results if the absorption coefficients μ and μ_i are not too large. On the other hand, there must be sufficient absorption to make the secondary process important, so that the range of allowed values of μ and μ_i is limited. The more refined analysis[1] confirms this. Again μ_i and γ_p are functions of X/p.

---

[1] J. DUTTON, S. C. HAYDON, F. LLEWELLYN JONES and P. M. DAVIDSON: Proc. Roy. Soc. Lond., Ser. A **218** 206 (1953). See Sect. 20.

[2] Stepwise ionisation is formally included in μ_i, making it a function of current at high current density.

No complete analysis of the action of *metastables* in the volume appears to be available. If a gas mixture is used, atoms of one gas may be ionised by metastable atoms of the other if $V_{i_1} \leqq V_{m_2}$ with obvious notation. It is not physically reasonable to apply the same approximate treatment as was used for the photons in a given mixture with given metastable half-life, for the point at which the metastables are destroyed depends on their diffusion velocity, which is determined by their distribution in space. Thus to assume that they are all produced near the anode and move with a constant speed to the cathode is unrealistic.

The differential equation describing the action of the metastables is very simply obtained, but in this summary of elementary derivations we shall not present it (see Sect. 21). This particular secondary mechanism, the action of metastable atoms of one gas ionising atoms of another, may lead always to an increased current that can be expressed as an increase in α. So far as at present appears, the standard forms (1.2a) or (1.3) do *not* describe the process of volume ionisation by metastables (see Sect. 21). This is definitely the case if the metastable atom has a short life. The effect on the curves of $\log i/i_0 = f(d)$ at constant X/p is to increase the *slope* of the curves at all d and not to introduce a *curvature* at large d.

Although the phenomenon is undoubtedly a secondary effect, the secondary electrons are produced so near the place of origin of the metastable that they need not be distinguished in the steady discharge from electrons directly produced by electron impact.

Let us now consider the ionisation to be attributed to *fast neutral atoms* moving through the gas. It has already been mentioned that apart from charge exchange the elastic scattering of ions and atoms of considerable energy is similar; approximately in fact it is half isotropic and half forward scattering in which there is little loss of energy. In the inelastic collisions leading to ionisation which concern us here ions and atoms again need not be treated differently. Theory indicates [*15*] that the curves for ionisation efficiently as a function of energy will be roughly the same for ions and atoms moving in their own gas, and the experimental values available confirm this.

If we assume then no energy loss except in inelastic collisions it is immaterial whether the energy is carried by ions or atoms, so that charge exchange need not be considered. As some energy loss will occur, and the energetic atoms once formed cannot gain further energy from the field we shall obtain an upper limit for effects due to fast neutral atoms in the gas if we ignore charge transfer and treat the case of positive ions moving through the gas. The original equation (1.2) then follows.

It appears then that a reduction of β by an unspecified amount would account formally for the influence of fast neutral atoms in producing ionisation in the gas. Hence this also is an allowed secondary process.

Consider finally the *X-rays* produced at the anode as an ionising agent in the gas. The number of photons produced at the anode by an electron from the cathode is $\psi\, e^{\alpha d}$, and of these a fraction g ($g \sim 1$) move towards the cathode. The number of photo ionising events in dx at x is

$$\mu_i\, g\, \psi\, e^{\alpha d - \mu(d-x)}\, dx$$

where μ is the total absorption coefficient of the X-ray quanta and μ_i is the coefficient for absorption leading to ionisation. The contribution made to the current is the second generation of electrons; by an argument similar to that leading to (6.1) we have

$$\frac{i}{i_0} = \frac{e^{\alpha d}}{1 - \varphi' \{e^{(\alpha-\mu)d} - 1\}} \tag{6.2}$$

where

$$\varphi' = \frac{\psi g \mu_i}{\alpha - \mu}.$$

This too would produce a graph similar to that experimentally observed, if $\mu_i, \mu \ll \alpha$: the comparison with (6.1) shows that μ and μ_i must lie within a certain range if the photo-ionisation in the gas due to X-rays from the anode is to be acceptable.

With the exception of ionisation by metastable atoms, then, all the effects listed at the beginning of this section are formally indistinguishable, if the constants are correctly chosen. In all instances, if absorption in the gas is small and $e^{\alpha d} \gg 1$ we may write

$$\frac{i}{i_0} = \frac{e^{\alpha d}}{1 - \gamma e^{\alpha d}} \tag{6.3}$$

where γ is a function depending on the physical basis of the secondary ionisation. We shall use γ as the general, non-commital secondary ionisation coefficient not presupposing the absence of volume ionisation.

7. The analysis of the secondary ionisation process. None of the experimental tests below (except *7 η*) are absolutely specific, but if two or three can be applied to a given discharge, the main cause or causes of secondary ionisation will probably appear.

α) Measurement of γ. This can be done most reliably by measuring i/i_0 as a function of pd for a given value of X/p. The value of X/p and γ can be calculated as in the original experiments of TOWNSEND. Values of γ might possibly be deduced from the energy balance at the cathode[1] but it has not been used. Careful account would have to be taken of the kinetic and potential energies of the incident particles, and the results would probably be unreliable.

Alternatively, if the breakdown potential $V_s = f(pd)$ is found, and if α/p is known, the approximate relation

$$\gamma (e^{\alpha d} - 1) = 1 \tag{7.1}$$

may be used to determine γ. This again is not a very satisfactory method, since V_s is not an accurately measurable quantity and α/p is not known for the particular apparatus used. It is often difficult to repeat exactly the conditions under which the values of α/p used were obtained, and α enters (7.1) in the exponential. Moreover, the influence of space charge at breakdown will invalidate (7.1) (see Sect. 8). Once the experimental value of γ is known it may be compared with that calculated for different agents. From experiments in vacuo the values of the coefficients e.g. γ_i, β, are known in many instances as a function of energy for ions, photons, etc. If the energy distribution of the ions, for example, can be calculated, the value of γ to be expected assuming secondary emission from the cathode may be found. If this agrees with the experimental value, this mechanism may be the one operating: if there is disagreement, some other process must be sought. This holds for every secondary effect, if the relevant energy distribution can be determined.

[1] A. GÜNTHERSCHULZE, W. BÄR and A. WINTER: Z. Physik **111**, 209 (1938) (cf Sect. 27 β).

The presence of gas in contact with the cathode will mean that values for the surface secondary coefficients determined with clean metals may not be accurate. Any chemical reactions in the discharge would further alter conditions. The secondary effects in the gas will change, too, if impurities are given off by the electrodes, so caution is needed in making use of the measurements under controlled conditions.

β) Variation of γ with X/p. γ is measured as above for various values of the uniform applied field and the pressure. It is known [12] that the ratio of ionising collisions by electrons in a gas to exciting collisions rises as the average energy rises, i.e. as X/p increases [10] if X/p is above a certain value. This follows from the shape of the curves for efficiency of ionisation and excitation. Effects due to photons or metastables therefore should decrease as X/p increases over part of the range. Therefore γ will decrease with increasing X/p just above the range where the secondary effect is due to photons or metastable atoms. If γ increases with X/p then the influence of the ions predominates.

γ) Variation of γ with X. If γ is found to be dependent on X and not on the reduced field X/p, then some process independent of the average electron energy is active. This points to field emission from the cathode: it is necessary to distinguish between a true secondary effect involving a surface layer of ions on an insulating film and field emission due directly to the applied field which is a primary source of ionisation. The constants in equation (5.6) could be evaluated, or the experiments repeated after all insulating films on the cathode surface have been removed. A simpler test, however, is to remove the source producing the initial current, and wait long enough to allow the surface charge to disperse. If a current flows on application of the field, then the field is acting directly as a primary source of ions (see also 7δ).

δ) Variation of γ with time. When with a given initial current i_0 the current i is measured as a function of time, then the differing speed of the various secondary effects enable them to be distinguished. Ions move in the gas at a speed determined by the field, photons (if not resonance photons) with the velocity of light and metastables with thermal velocities. When considering photons and metastable atoms, the life-time of the excited state in the gas must be taken into account. Resonance photons will be absorbed and re-emitted very frequently, so that the quanta are "imprisoned" in the gas and may move as if diffusing slowly [14].

This type of experiment may also serve to show whether field emission is a secondary or primary process: if the field acts directly the enhanced emission will appear immediately. If the field emission is a true secondary process a delay will appear, for the surface charge must build up. It is of course essential that the charge be dispersed before this experiment is attempted.

ε) Dependence of γ on the nature of the cathode. If a change in the cathode is found to change the value of γ, then obviously the agents of the secondary processes are acting at the cathode. The gas-dependent mechanisms cannot be excluded by this arguments of course.

ζ) Dependence of γ on the nature of the anode. If such a dependence is found it would indicate that the anode is emitting some entity which can produce ionisation in the gas or at the cathode. An obvious suggestion is that X-rays are being produced: these may be detected outside the vessel if they are energetic enough. It may, however, be that absorbed gas on the anode is being ionised by the incident electrons and the positive ion thus formed acts in the gas or at the cathode to produce ions. Outgassing the anode would reduce this effect.

These two mechanisms are to be anticipated in very low pressure discharges.

η) Dependence of γ on irradiation. If the discharge is irradiated with intense light of the frequency required to excite metastable atoms into a state from which they may return to the ground state without colliding, then effects due to the metastables will be greatly reduced. This is a specific test for secondary ionisation due to metastable atoms.

ϑ) Influence of a discharge on a neighbouring electrode system. When two similar electrode systems are close together, a discharge in one sometimes initiates a discharge or a flow of current in the other. If it is possible to determine the cause of such a second discharge then the same agency is probably responsible for the secondary process in the first.

The time interval between the two indicates the nature of the effect. Experiments may also be carried out with deflecting electric fields between the two gaps, or a window transparent to the photons expected: if some obstruction to the agent crossing is discovered, is may be possible to decide what the effective agent is.

These tests will prove useful in other types of discharge.

However, false values for the observed secondary coefficient γ will be obtained in TOWNSEND discharges if the space charges present are large: the next section shows what limitations exist.

8. The influence of space charge on the current. In the previous article of A. VON ENGEL on the first TOWNSEND coefficient α it has been shown that space charge in the gap increases the electron multiplication if the value of X/p is small, and decreases it at high values of X/p. If we may write [*1*], [*4*]

$$\frac{\alpha}{p} = A\,\mathrm{e}^{-B p/X} \tag{8.1}$$

then the value of X/p for which the change in the influence of the space charge occurs is given by

$$\frac{X}{p} = \frac{B}{2} \tag{8.2}$$

which is the point of inflexion in the curve of $\alpha/p = f(X/p)$. Near this point the influence of space charges will be small.

If the discharge conditions are such that $X/p < Bp/2$ or generally if α/p increases more rapidly than linearly with X/p [*7*], then the influence of space charge is to cause the discharge characteristic $V = f(i)$ to reach a maximum and fall. When V decreases with increasing i, the discharge becomes independent of i_0, the initial cathode current. The point at which this independency arises has been taken here to define the boundary between the TOWNSEND discharge and the dark discharge, as described in the introduction to this article. It depends to some extent on the value of i_0.

This result was derived by VON ENGEL and STEENBECK [*7*]. They also found that if the current density is j with applied potential V in the presence of space charge, and V_s is the greatest potential which can be maintained across the gap with an infinitesimal initial current i_0, then

$$V = V_s \pm c j^2 \quad \text{according to} \quad \frac{X_s}{p} \gtrless \frac{B}{2} \tag{8.3}$$

if (8.1) is assumed to hold.

As all work on the discharge is such that $X_s/p < B/2$ V falls off as j^2, the quantitative expression of the transition to the dark discharge (see Sect. 25).

ROGOWSKI and FUCKS [4] have worked on the derivation of $i/i_0 = f(V)$ for a given gap, and obtained a complicated analytical expression for the variation up to $V = V_s$, when (8.3) can be used. Although differing in detail, the curves obtained are similar in appearance to Fig. 4: the potential V_s is shown there, and the upper curve at high current corresponds to (8.3). The curves for a 1 cm. gap in atmospheric air is shown in Fig. 32 of Vol. II of VON ENGEL and STEENBECK, on the assumption that (8.1) holds and the secondary coefficient γ is constant.

Other expressions for i/i_0 as a function of d in the presence of space charge have been derived by VARNEY *et al*[1] and by CROWE, BRAGG and THOMAS[2] In both these papers the electron current density j^- is assumed proportional to i^- at any point i.e. the lateral spread of the discharge is neglected. It is also assumed that the ion drift velocity $v = k^+ X$, k^+ being the mobility, and that $v^+ \ll v^-$. All these assumptions are justified at fairly high pressures.

VARNEY *et al* deal with the case where no secondary effects occur and do not go beyond the point where $di/dV = \infty$ at V_s. This is discussed in the previous article. The work of CROWE *et al* includes secondary processes without specifying their nature and assumes γ constant. As X is not constant, $\alpha\, d$ in the previous work, must be replaced by $\int_0^d \alpha\, dx$.

From (6.3) with these assumptions we have

$$\frac{j_x^-}{j_0^-} = \frac{e^{u}}{1 - \gamma\,(e^{\bar{u}} - 1)} \tag{8.4}$$

where

$$\bar{u} = \int_0^d \alpha\, dx, \quad u = \int_0^x \alpha\, dx, \quad du = \alpha\, dx,$$

j_0^- is the current density of the initial photoelectrons from the cathode and j_x^- is the electron current density at x.

From POISSON's equation

$$\frac{dX}{dx} = 4\pi e (N^+ - N^-) = \frac{4\pi j^+}{k^+ X}\,. \tag{8.5}$$

Since $j = j^+ + j^-$ we have from (8.4) and (8.5)

$$\frac{k^+ \alpha X\, dX}{4\pi j_0^-} = \frac{(e^{\bar{u}} - e^{u})\, du}{1 - \gamma\,(e^{\bar{u}} - 1)}\,. \tag{8.6}$$

Using (8.1) this can be integrated by means of the function $K(z)$:

$$2K(z) = \frac{e^{-z}}{z^2} - \frac{e^{-z}}{z} - \mathrm{Ei}(-z) \tag{8.7}$$

where $Z = Bp/X$ and $\mathrm{Ei}(x)$ is the exponential integral.

Numerical integration is necessary. An example of the results is shown in Figs. 3a—c. The curves 1, 2, 3, 4 in each figure refer to the increasing values of $\bar{u}$, but the greatest increase is less than 1% as seen in the description of the figure. The very rapid increase in the space charge density ϱ^+ near the cathode

[1] R. N. VARNEY, H. J. WHITE, L. B. LOEB and D. Q. POSIN: Phys. Rev. **48**, 818 (1935). Cf. the article of A. VON ENGEL in this volume.

[2] R. W. CROWE, J. K. BRAGG and V. G. THOMAS: Phys. Rev. **96**, 10 (1954).

is clearly seen, accompanied by sharp rises in X and α. The effect of space charge is seen in Fig. 2b to produce an almost linear fall in X near the cathode, as in a glow discharge, though here a high pressure gap is under consideration.

The effect of the distortion of the distribution of ϱ^+, X and α on the current flowing in the gap is shown as a function of V for various initial currents in the range 10^{-14} to 10^{-8} A/cm.² in Fig. 4. Since the example is a gap with uniform field at constant distance, the shape of the curves $i/i_0 = f(V)$ is not exactly the same as $i/i_0 = f(d)$ at constant X/p, the curves discussed in Sect. 1. The graphs represent different values of X/p at constant d.

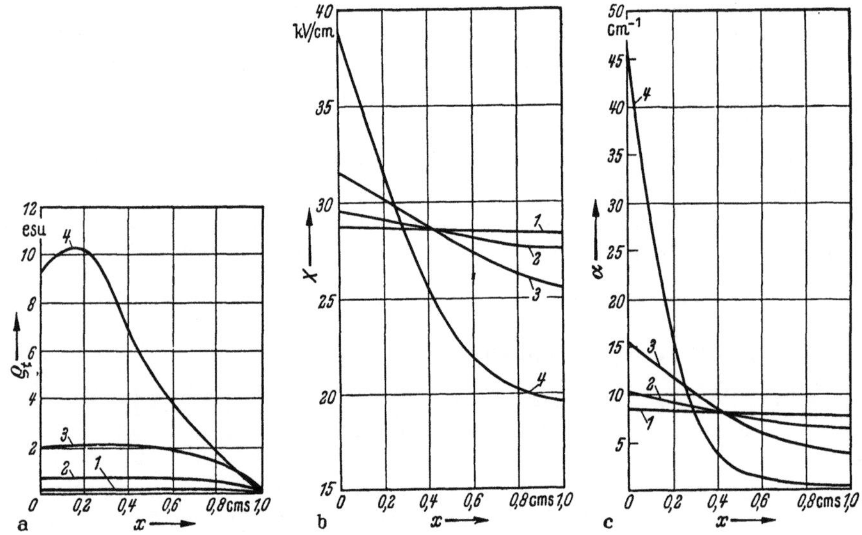

Fig. 3 a—c. Distribution of space charge ϱ^+, electric field X and ionisation coefficient α for a 1 cm. gap. 1. $\bar{u} = 8.00$, $V = 28.257$ kV; 2. $\bar{u} = 8.04$, $V = 28.348$ kV; 3. $\bar{u} = 8.047$, $V = 28.130$ kV; 4. $\bar{u} = 8.0495$, $V = 25.781$ kV.

From Fig. 4 it is seen that the curvature due to the value of γ assumed (3.2×10^{-4}) is identical in all the graphs except when V is very close to V_s. In fact, the effect of space charge is negligible up to within $\frac{1}{2}$% of V_s. This is not a direct proof that curves of log $(i/i_0) = f(d)$ at constant X/p will not be affected, but it is probably sufficient: it would be interesting to have a direct comparison with experimental curves.

CROWE *et al* did actually work out the initial current at which space charges might become important on producing a curvature of $i/i_0 = f(V)$ for one example of work done by LLEWLLYN JONES. He had tested the independence of γ and i_0 experimentally over a wide range.

However, this experimental test is only valid if the ions from one avalanche remain in the gap until a second photoelectron is released from the cathode by the ultra violet light. Otherwise, γ will be independent of i_0 although space charge may be distorting the field.

When the positive ions arrive at the cathode they will not release more than one electron at most[1], for breakdown would occur if the products of one avalanche could release an average of one electron. The time interval during which the ions remain in the gap is then of the order of (say) twice their transit time.

At $X/p \sim 40$ in air, this is $\sim 10^{-5}$ secs. for a 1 cm. gap at 760 mm. Hg. If one photoelectron is released every 10^{-5} secs., the value of i_0 is 10^{-14} amps: so for

[1] Much less than one if metastables or photons are prominent.,

all $i_0 > 10^{-14}$ the experimental test can be applied to a 1 cm. gap in atmospheric air legitimately. In general, the fact that γ is independent of i_0 means that space charge is negligible only if

$$i_0 > 5 \cdot 10^{-19}\, v^+/d \text{ amps.}$$

approximately, and γ_i alone is important.

This is satisfied for LLEWELLYN JONES' work in nitrogen (Table 1, Sect. 11) but in air (Table 2, Sect. 11) some of the curves do not comply with it. The general agreement in the values of γ among all the results suggests that no space charge distortion was present.

The theoretical work confirmed that an influence of i_0 was not to be expected until it was several orders of magnitude greater than any used in the experiments in nitrogen.

Space charge then can modify the form of the discharge. By making the field distribution violently non-uniform, it can cause also a change in the relative importance of the secondary processes, or introduce new ones (cf. Sect. 18).

If such large i_0, or to be exact j_0, are used that space charge effects occur, then the values of γ derived from the experimental curves are not consistent along the curve. POSIN observed this, and distortion due to space charge has probably not been present except[1] in the work of POSIN, PAAVOLA and SANDERS, and in some experiments of KRUITHOF and PENNING where very high i_0 was used [5].

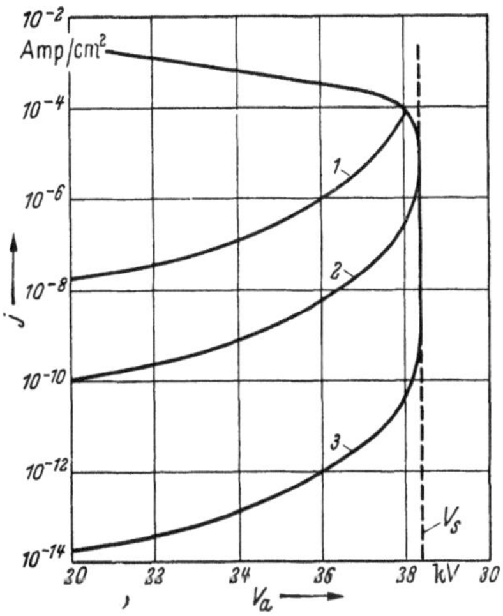

Fig. 4. Influence of initial current density j on the sparking threshold. Curves given for N_2, $p = 700$ mm. Hg, $d = 1$ cm., $\gamma = 3.19 \cdot 10^{-4}$ assuming $A = 7$ cm. per mm. Hg; $B = 260$ V/cm. per mm. Hg, $k^+ = 3.0(760/P)$ cm./sec. per V/cm. (see: $i_0 = 10^{-8}$ A/cm.^{2}, $6 \cdot 10^{-11}$ A/cm.^{2}, 10^{-14} A/cm.^{2}.

In any experimental work it is essential to test in a valid way that γ is not dependent on i_0. If this can be combined with a theoretical examination it will make assurance doubly sure.

c) Measurements in steady discharges.

The very early work on γ which TOWNSEND and his school carried out was exclusively in this field. This classical research was enough to establish the main principles of the physics of the low pressure discharges, but it was not conducted in clean systems with pure gases. Present day standards are not perfect, but there has been much improvement in the last fifty years.

The work reported here has been carried out in fairly clean conditions, and in particular mercury vapour has been excluded from the apparatus. The ionisation potential of mercury is 10.4 volts, and the presence of such an easily-ionised impurity with many excited states at still lower energies completely alters the electron energy distribution and so the discharge characteristic in gases with energy levels lying much higher.

A further effect of mercury is that it forms a film on the electrodes, and usually reacts with them chemically. At the cathode this has tremendous influence on the value of γ. Experiments carried out with such an impurity do not give much information about corresponding results in pure gases. This is clearly seen even

[1] Cf. B. HOCHBERG and E. SANDBERG: J. Techn. Phys. USSR. **12**, 65 (1942).

in some of the work with vapours and diatomic gases (Sect. 10 and 11), and has been shown to be strong in rare gases too[1].

We shall deal with rare gases, metal vapours, and diatomic and polyatomic gases in turn, reporting the experiments in which clean conditions prevailed.

9. Experiments in rare gases. Most of the work in this field has been done in *argon*. Both methods described in Sect. 7α have been used, though the determinations have been by measurement of V_s usually. ENGSTROM[2] obtained agreement between the two methods. He found that the measurement of log $i/i_0=f(d)$ gave a succession of small peaks in the curve for γ, while the use of equation (7.1) resulted in a line representing the average of the previous curve.

It has been suggested [7] that the sparking potential depends on the maximum value of γ to be found at any part of the cathode, while the values of i/i_0 depend on the average γ for the whole surface. This may be true at high pressures or if breakdown measurements are made at large overvoltages. At low pressures, however, if the voltage applied is not much larger than V_s the discharge will diffuse rapidly, especially if photoelectric effects or effects of metastable atoms are important. So it is to be expected that the same value of γ is appropriate to breakdown as to the measurement of i/i_0 as quite a large proportion of the cathode is active in the secondary process.

In order to determine whether or not it is important to consider variations in γ over the surface, it is sufficient in measurements at high X/p to consider the electron diffusion in crossing the gap. The main contribution to γ is γ_i under these conditions as will be seen later, and the ions will follow the field lines with little diffusion. If the lateral diffusion of the electrons is several orders of magnitude smaller than the electrode area then locally large values of γ will influence breakdown measurements.

At low X/p when photons and metastables contribute largely to γ, the whole area of the cathode is always concerned in breakdown and in current measurements. Although the discharge contracts in the breakdown process as the current increases, the difference in potential at which a steady state with the largest multiplication possible exists and the actual breakdown potential measured is negligible.

If the true values of α/p are used, it is not unreasonable then to expect both methods of Sect. 7α to agree.

Fig. 5 shows the results obtained by several authors with various cathodes in *argon*. The curves for Al and Mg and the dashed curve for Ba are derived by JACOBS and LA ROCQUE[3] from breakdown potential measurements, using α/p given by KRUITHOF and PENNING[4]. The full-line curves for Ba and Ni represent the results of SCHÖFER[5], again from V_s and KRUITHOF and PENNING's values for α/p. The average values of ENGSTROM[6] for Ba are on the dotted curve shown, in good agreement with SCHÖFER.

All the other curves are calculated by DRUYVESTEYN and PENNING from published curves of $V_s=f(pd)$. The curve for Fe also represents well the results of SCHÖFER for for this cathode material. The figure does not include this curve

[1] F. M. PENNING and C. C. J. ADDINK: Physica, Haag **1**, 1007, 1028 (1934).
[2] R. W. ENGSTROM: Phys. Rev. **55**, 239 (1939). — M. HEALEA and C. HOUTERMANS: Phys. Rev. **58**, 608 (1940).
[3] H. JACOBS and A. P. LA ROCQUE: J. Appl. Phys. **18**, 199 (1947).
[4] A. A. KRUITHOF and F. M. PENNING: Physica, Haag **3**, 515 (1936).
[5] R. SCHÖFER: Z. Physik **110**, 21 (1938).
[6] R. W. ENGSTROM: Phys. Rev. **55**, 239 (1939).

or SCHÖFER's curve for Na, which is 10 to 20% lower than his curve for Ba and of the same shape. This disagrees with the curve of JACOBS and LA ROCQUE in the same fashion as the curves for Ba [*10*].

It should be remembered that α/p was not measured in the experiments given above (ENGSTROM's work excepted). This introduces an uncertainty into the values of γ.

The most striking feature of the curves is that they do not lie together at all, indicating that the cathode is playing an important role in the secondary ionisation process. This is true even for the results of the same observers, so that it is most unlikely that impurities in the gas could be the cause of the variation. In fact, the differing results for (allegedly) the same cathode material e.g. Ba and Cu indicate that important unvoluntary variations of the cathode surface are hard to avoid. It is generally accepted that reproducible surfaces are very difficult to achieve in the physical and chemical violence of a gas discharge, and the cathode surface is of vital importance in uniform field discharges.

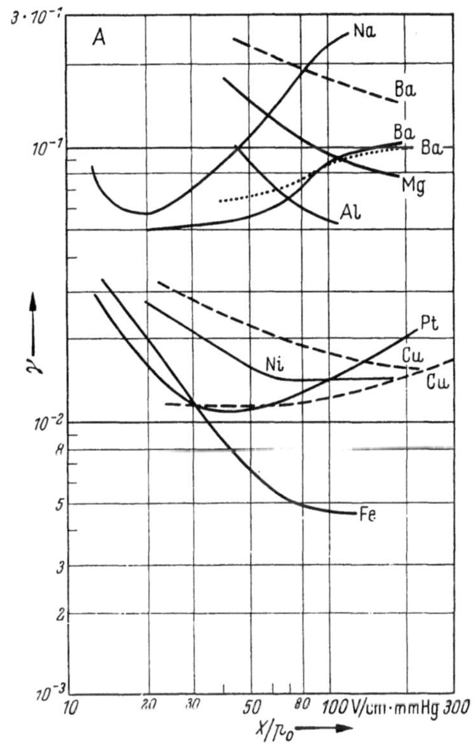

Fig. 5. The secondary ionisation coefficient γ in argon as a function of X/p with various cathodes, according JACOBS and LA ROCQUE (Al, Mg, Ba – – –); EHRENKKANZ (Na, Pt); SCHÖFER (Ba ———, Ni); ENGSTROM (Ba ······); PENNING and ADDINK (Fe); KRUITHOF and PENNING (Cu). p_0 is equivalent pressure at 0° C.

Another important feature of the curves is the rise in γ at low X/p displayed by each one. For Fe A γ increases by almost a factor 10 as X/p falls from 100 to 10.

It has been shown[1] that the number of excitations produced in the gas increased very rapidly in this range of X/p as X/p fell. From the nature of the excitation and ionisation functions it is to be expected that as the electron energy increases with increasing X/p the lowlying levels of the atom will be excited first, so that metastable atoms and light (including the ultra-violet) is produced before ionisation becomes important. When X/p is larger, the average electron energy rises above the peak of the excitation functions, so the ionisation becomes more prominent than the production of photons or metastable atoms. The shape of the curves indicates then that at low X/p photons or metastable atoms are the main agents in the secondary ionisation process in rare gases.

The fact that γ is so large at low X/p means that experimental curves of $\log i/i_0 = f(d)$ will show curvature at very low X/p: great care is needed in determining α in rare gases for this reason.

As X/p is increased, light and metastable atoms become less important and ions increasingly control the discharge. The fall in γ ceases when the influence of the ions is felt, and a minimum appears followed by a rising γ as X/p increases

[1] Cf. the article of A. VON ENGEL in this volume.

at large X/p. At extremely large X/p the electron energy should pass the peak of the ionisation curve i.e. $\alpha/p = f(X/p)$ reaches a maximum and begins to fall. Under these circumstances one might expect a second fall in γ, but the increased ion energy at high X/p more than compensates for the lower ion production rate, so γ continues to rise. This is the explanation of the peculiar breakdown curve $V_s = f(pd)$ in helium[1].

From Fig. 5 it also appears that surfaces of low work function tend to have a high γ, and vice versa. This is not absolutely true[2] but it is a reasonably good rule. There is some theoretical basis for it (Sect. 30).

The difference in the work function causes a difference in the photoelectric sensitivity and spectral response of the surface, which may account for the different position of the minimum on the X/p scale for different cathodes. A surface with a low photoelectric threshold would yield high values for γ_p at low X/p, where as a high threshold implies that γ_p will not contribute to γ until X/p is somewhat greater. The difference in the two curves for Cu and Ba is perhaps attributable to this cause, as both metals are very reactive chemically and surface films have a great influence on the photoelectric yield.

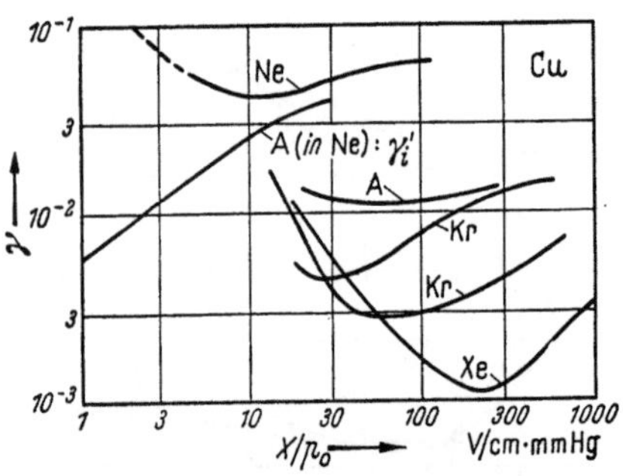

Fig. 6. The secondary ionisation coefficient γ for copper cathodes in different rare gases as a function of X/p_0, and the calculated γ'_i for A^+ in Ne–A. p_0 is equivalent pressure at 0° C.

Finally the order of magnitude of γ should be noted. It may be as high at 10^{-1}, and is never smaller than $3 \cdot 10^{-3}$. From Sect. 4 it seems that we may neglect the effects of back diffusion on γ (within 30%) at $X/p > 10$, for the secondaries will not have very great average energy; KENTY[3] estimated the mean energy of photoelectrons as only about 3 eV in argon. The true γ is certainly of the same order of magnitude as that measured, but is larger. The difference will be greatest at low X/p.

HUXFORD[4] measured α and γ from $i/i_0 = f(d)$ in *argon* for sealed-off tubes with a photosensitive Cs AgO electrode and Cs–Ni mesh electrode. Working at higher X/p, he found γ constant at 0.4 to 0.2 when X/p increased from approximately 100 to 700 V/cm. per mm. Hg for a Cs–Ni mesh cathode. For a Cs–AgO cathode γ was about 0.5 in this range. Above 700 V/cm. γ rose to 2 for the first cathode and 3 for the second at 1200 V/cm. per mm. Hg, approximately.

These larger values are to be expected with cathodes of low work function at high X/p: the rise to $\gamma > 1$ is noteworthy.

Values in other rare gases with a copper cathode calculated from $i/i_0 = f(d)$ are shown in Fig. 6, as given by KRUITHOF. The curve for Ne + A is γ'_i for A^+ calculated from γ in a mixture of a small percentage of A in Ne. The two curves for krypton refer to two different states of the copper cathode: after treatment with a glow discharge in argon the lower curve is obtained, and using a neon discharge the upper curve [*12*].

All the curves except that for Ne–A show the rise of γ at low X/p remarked before. Now the presence of argon, since

$$\mathrm{Ne}^* + \mathrm{A} \rightarrow \mathrm{A}^+ + \mathrm{Ne} + e \tag{9.1}$$

[1] F. M. PENNING: Proc. Kon. Ned. Akad. Wetensch. **34**, 1305 (1931) and [*5*].
[2] D. H. HALE and W. S. HUXFORD: J. Appl. Phys. **18**, 586 (1947).
[3] C. KENTY: Phys. Rev. **44**, 891 (1933).
[4] W. S. HUXFORD: Phys. Rev. **55**, 754 (1939).

replaces metastable atoms by equal numbers of ions. The low value of γ_i' implies that metastable atoms are far more efficient than ions of approximately *the same potential energy* in releasing electrons from surfaces i.e. $\gamma_m > \gamma_i$ under these conditions (Sect. 27, 32).

The figure also shows that γ is largest for the atoms of highest ionisation potential, and that the minimum in the curve occurs earliest in the X/p range for these atoms. A clue to the reason is found in the fact that α/p for Xe is much lower than for Ne at $X/p < 1000$, although the ionisation potential for Xe is 12.1 eV and for Ne it is 21.6 eV. The energy distribution of the electrons must be such that as the ion mass increases the relative number of high energy electrons falls. Thus in A, Kr, Xe we have progressively more excitation and less ionisation at any given X/p in the range studied. This explains why the rise due to ions is progressively more delayed: a smaller value of γ_i is also to be expected, for the potential energy of the ions is decreasing also.

One might expect that the photoelectric effect would be more marked in Xe than the other gases at low X/p. However it should be remembered that at any given X/p the larger numbers of photons from Xe will have a lower average energy, so the two effects may cancel out roughly. The measurements are not made with sufficiently well defined cathodes to make any deduction certain on this point, but it does seem that at given X/p Kr and Xe show the same γ below the minima of their curves.

From these results it is possible to determine the influence of γ_i and $(\gamma_p + \gamma_m)$ to some extent. KRUITHOF [*12*] was able to derive, with reasonable correction for back diffusion, values for γ_i and $(\gamma_p + \gamma_m)$, which he regarded as γ_p only. For *krypton* with a copper cathode, two of his curves could be represented if $\gamma_i = 4 \times 10^{-3}$ and $\gamma_p = 2.9 \times 10^{-3}$ for one and $\gamma_i = 2.2 \times 10^{-2}$ and $\gamma_p = 1.1 \times 10^{-3}$ for the other. The state of the cathode was changed by various treatments. The assumptions made are reasonable, and the values calculated are probably of the right order of magnitude (see Part III).

10. Experiments in vapours. The only direct measurements appear to be those of BADAREU and BRATESCU[1] in *mercury vapour*. They used an *iron* cathode, and measured α/p and V_s to obtain $\gamma = f(X/p)$ as shown in Fig. 7. The three points are obtained from other experiments on $i/i_0 = f(d)$ from which α/p and γ may be found more accurately. In the narrow range of X/p represented by these points the agreement between the two is good.

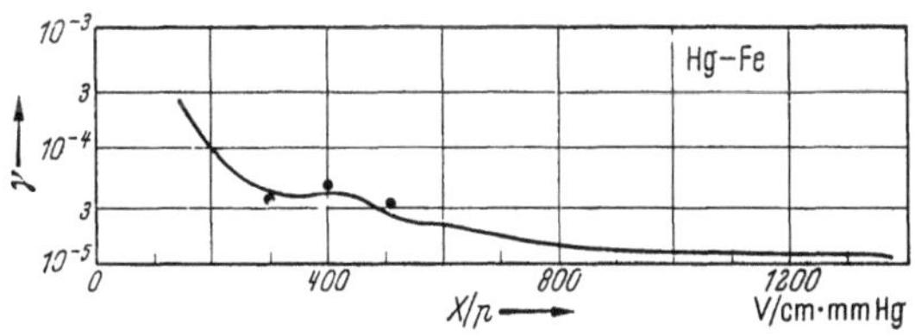

Fig. 7. Values of γ as a function of X/p for Hg vapour with Fe electrodes.

The values of X/p do not start at such a low level as those of Figs. 5 or 6, and γ is very much smaller in Hg than in any of the rare gases. From the low ionisation potential of Hg we would expect this, but Fig. 7 is remarkable because *no minimum* appears. There is a kink in the curve at $X/p \sim 400$, but after this γ decreases as X/p increases up to 1400 V/cm. per mm. Hg, where $\gamma \sim 10^{-5}$.

This indicates that, while metastable Hg atoms and the radiation from excited Hg atoms is as efficient a secondary mechanism as can be expected, Hg ions are very inefficient even when they possess large kinetic energies. It may be that the chemical action of mercury on the cathode or the formation of amalgam

[1] E. BADAREU and G. G. BRATESCU: Bull. Soc. roum. Fiz. **42**, 82 (1942).

with impurities in the iron has some bearing on this. The values of γ_i for Hg ions on Hg has been found to be very small indeed (Sect. 27β). The same is probably true of any Hg-contaminated surface.

11. Experiments in diatomic and polyatomic gases. The first experiments in these gases in which Hg vapour was deliberately excluded were those of BOWLS in *nitrogen*[1]. He used a platinum cathode, and varied the cathode surface by distilling in sodium or adding mercury. His results are shown in Fig. 8, and compared with those of TOWNSEND and POSIN in N_2 contaminated with Hg [5].

Under clean conditions with Pt cathode the curve shows a strong peak at $X/p \sim 70$, a minimum at $X/p \sim 250$ and then a continuous rise up to $X/p >$ 1000 V/cm. per mm. Hg. The portion $A\ B\ C$ is reminiscent of the curves observed for rare gases and γ is of the same order of magnitude. If the same interpretation holds, the peak is due to a photoelectric effect or electron emission due to metastables at the cathode, or photo-ionisation in the volume. This last is possible in a pure molecular gas, for two ionisation limits exist: see however Sect. 20γ.

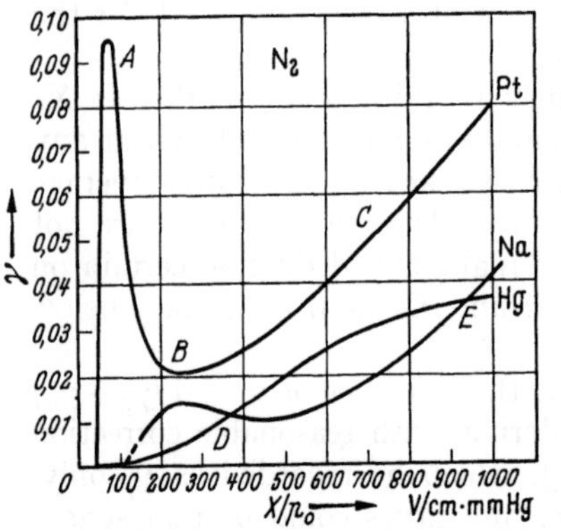

Fig. 8. Values of γ as a function of X/p_0 for N_2 contaminated with mercury (POSIN, BOWLS, TOWNSEND-Hg) and BOWLS' values for mercury-free N_2 with Pt cathode and Na cathode. p_0 is equivalent pressure at 22° C.

The sharp rise in γ at $X/p \sim 50$ may be due to several causes. There will be a definite photoelectric threshold for the cathode, and the spectrum of the radiation falling on it varies with X/p. The light too may be differently absorbed in the gas at different X/p, and the rate of production will increase rapidly with X/p at low X/p. Even so, the abrupt rise in γ is rather surprising.

When BOWLS distilled Na into his chamber, the peak became smaller and wider and moved out to $X/p \sim 270$. A minimum followed, and a rise as before, but γ was still lower than for Pt. The cathode was photoelectrically very sensitive, yet γ was small and the peak at low X/p reduced.

This strange result was interpreted in terms of a compound of sodium with the nitrogen or some impurity such as hydrogen. The presence of some surface compound is indicated by the great sensitivity to visible light (see BOWLS' paper). In the gas too some change must have occurred, for $\alpha/p = f(X/p)$ showed higher values with Na in the chamber, so the presence of a *volatile* compound (Na N_2) was suggested. It has not been identified.

When the chamber with a Pt cathode was contaminated with Hg, no trace of a peak in γ remained (Fig. 8). Above $X/p = 200$, α/p was increased, but not so much as when Na was added. The values *now* agreed closely with those of TOWNSEND. BOWLS does not mention any change in the illumination needed to produce i_0, so the contaminated cathode has probably about the same photosensitivity to the mercury lamp radiation as the clean cathode.

Since no marked increase in α/p occurs at low X/p, the agent responsible for the peak in clean Pt—N_2 must be absorbed in the Hg—contaminated gas without the production of ionisation, if we assume that no change in electron energy distribution occurs.

If the electron energy distribution is robbed of its high energy electrons by the Hg present, then the ionisation produced by the absorption of light by Hg

[1] W. E. BOWLS: Phys. Rev. **53**, 293 (1938).

atoms (or collisions between N_2 metastables and Hg atoms) must roughly compensate this to give the same α/p.

At high X/p, the ionisation of Hg in the volume more than compensates for loss of high energy electrons, for α/p is observed to increase. The absorption in the volume prevents the action of the photons or metastables at the cathode, and so γ falls below the values in pure N_2. Charge transfer collisions

$$N_2^+ + Hg \rightarrow Hg^+ + N_2$$

reduce γ_i, since Hg^+ is very inefficient at releasing electrons (Sect. 10 and 27β). The lowering of γ at high X/p is then easily understood.

Measurements of γ at low X/p in N_2 or any other gas are very difficult, as are all experiments at higher pressures where purity of gas and apparatus must be

Table 1. *Values of γ for $Ni-N_2$.*

X/p V/cm. per mm. Hg.	α/p (cm. per mm. Hg)$^{-1}$	γ
41	1.1×10^{-2}	1.31×10^{-4}
42	1.3	1.16
43	1.5	1.54
44	1.7	3.49
45	1.9	3.72

$i_0 = 3\cdot10^{-13}$ A
$p = 300$ mm. Hg.
$d = 0.3$ to 2.5 cms.

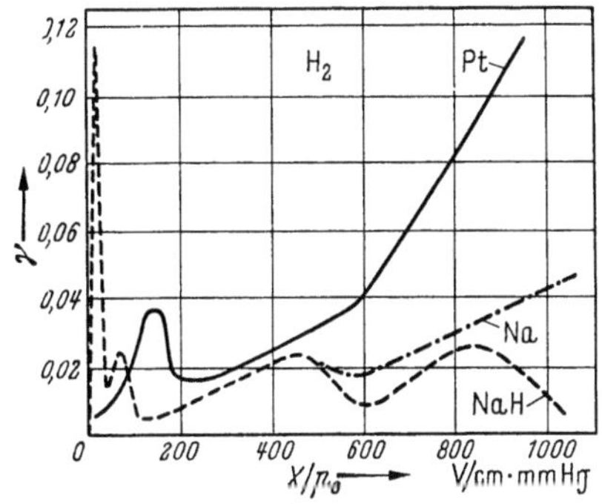

Fig. 9. Values of γ as a function of X/p_0 for H_2 with cathodes of Pt and Na (see text). p_0 is equivalent pressure at 22° C.

of the highest possible order. To obtain accurate results in this field is a great achievement, for high voltage sources of great stability are another essential, and means of measuring very small currents accurately yet another. The work of Sanders and Posin [5] first approached the required conditions, but in Hg-contaminated gas; the results of Llewellyn Jones and his school are the best available, and even these show much scattering.

Dutton *et al.* have measured γ in *nitrogen* for a nickel cathode over a very narrow range at extremely low X/p, by measuring $i/i_0=f(d)$ at constant X/p with a very stable voltage source[1]. The results are shown in Table 1.

The values are two orders of magnitude below the lowest observed by Bowls at pressures of order 1 mm. Hg. Their main interest, however, lies in the fact that γ can be measured at all at such high pressure and low X/p. It had been thought that breakdown could occur without any departure from linearity in the log (i/i_0) *vs.* d curves, as in the work of Sanders at the same pressure. It is suggested that Sanders did not observe curvatures in his experiments because he did not make measurements close enough to the sparking potential, being limited by the stability of his high voltage source[2], but see Sect. 8.

Hale carried out similar experiments[3] to those of Bowls for *hydrogen*. He found a series of peaks in the curve for $\gamma=f(X/p)$ in Na- and Hg-contaminated H_2 with a Pt cathode (Fig. 9). With clean conditions, a peak at $X/p=140$ appears, and a rise at high X/p which shows a sharp change in slope at $X/p\sim600$. Quali-

[1] J. Dutton, S. C. Haydon, F. Llewellyn Jones: Proc. Roy. Soc. Lond., Ser. A **213**, 203 (1952).
[2] F. H. Sanders: Phys. Rev. **41**, 667 (1932).
[3] D. H. Hale: Phys. Rev. **55**, 815 (1939).

tatively this is understandable, for the peak is due to influence of photons, and the rise to ions. The break in the curve may be due to a change in the nature of the ions, from H_2^+ to H^+.

When sodium was distilled into the chamber it was found impossible to obtain constant photosensitivity without striking a glow discharge or maintaining a TOWNSEND discharge for a long time. The surfaces formed by these treatments were taken to be NaH. They are labelled Na (1) and Na (2) respectively in Fig. 9, and both give the dashed line at $X/p < 500$.

The surface (2) shows several peaks, and a rise at high X/p; (1) displays no such rise. The interpretation of such curves is probably to be found in the absorption in the gas or transmission to the cathode of active agents, since α/p shows peaks for Na (? NaH)$-H_2$ which correspond to the minima in $\gamma = f(X/p)$.

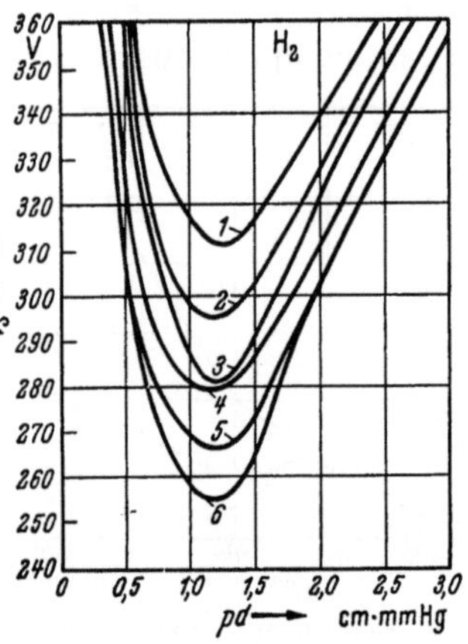

Fig. 10. Breakdown voltage V_s as a function of pd for various cathodes on hydrogen: *1* Steel cathode, *2* copper, *3* oxidised steel, *4* nickel, *5* pure aluminium, *6* commercial aluminium.

Results for the breakdown potential of H_2 obtained by LLEWELLYN JONES and HENDERSON were discussed by LLEWELLYN JONES[1]. Cathodes of various metals were used, and different values of V_s obtained (Fig. 10). The results for γ were of the type obtained by POSIN and BOWLS with Hg-contaminated apparatus in N_2; the values of γ rose monotonically to 0.12 approximately at $X/p \sim 700$. In clean H_2 HALE found a peak at low X/p with Pt, so that there was probably some contaminant which acted in H_2 as Hg did in N_2. The higher values than HALE's at $X/p > 200$ may be due to the difference in cathode material.

HALE[2] has also measured γ for H_2 with Al and Ni cathodes up to $X/p = 1700$ by measuring $i/i_0 = f(d)$. The surfaces were cleaned by running a discharge, as they were otherwise unstable: the aluminium was evaporated on to the nickel cathode in a layer a few hundred molecules thick.

The curve obtained for nickel was similar to that for Pt shown in Fig. 9, save that the peak at $X/p \sim 170$ was sharper and rose to 0.1 approx. Above $X/p = 1400$, however, a second break in the curve is drawn by HALE, so that three linear parts of the graph appear. It seems that a smooth curve would represent the results equally well. At $X/p = 1700$, $\gamma = 0.4$.

Results for aluminium show a peak $\gamma \sim 0.12$ at $X/p \approx 170$, and a slower rise afterwards. Between $X/p = 500$ (the minimum) where $\gamma \sim 0.02$, and $X/p = 1300$, γ only rises to 0.07. A very rapid rise then occurs, and at $X/p = 1600$, $\gamma \approx 0.2$.

These results are of the same form as those for platinum and are to be explained in the same fashion.

Other results for H_2 are given in the next section, as they show directly the dependence of γ on surface conditions.

Measurements of γ in *impure air* were made by SANDERS [*5*] at high X/p, and very much earlier by ROGOWSKI at low X/p [*4*], [*10*]. He found $\gamma < 10^{-10}$ for $X/p < 20$.

[1] F. LLEWELLYN JONES: Phil. Mag. **28**, 192, 328 (1939).

[2] D. H. HALE: Phys. Rev. **56**, 1199 (1939).

As back diffusion would not reduce γ by more than 60%, it is probable that absorption of photons is occurring in the mixture of gases contaminated with Hg. Possibly also the photoelectric threshold of the cathode lies in the region of the spectrum emitted from air at this value of X/p. It is not likely that γ_i would be important.

At low X/p in *clean air* the work of LLEWELLYN JONES and PARKER[1] with a nickel cathode is available, but no systematic study at high X/p with good conditions has been attempted. The work at low X/p actually preceded that in N_2, and the apparatus was not quite so refined. Their results are given in Table 2 below.

There seems to be a tendency for γ to fall as X/p decreases; though the variation of X/p is only 15% roughly, γ alters by a factor 10. A variation of almost

Table 2. *Values of γ for Ni—Air.*

X/p V/cm. per mm. Hg	α/p (cm. per mm. Hg)$^{-1}$	γ	i_0 (amps)
39	$1.61 \cdot 10^{-2}$	$8 \cdot 10^{-6}$	$1 \cdot 10^{-15}$
40	1.81	$2.3 \cdot 10^{-5}$	6
40	1.81	$1.5 \cdot 10^{-4}$	20
41	1.96	$4 \cdot 10^{-5}$	6
42	2.24	$4.6 \cdot 10^{-5}$	6
43	2.52	$1.13 \cdot 10^{-4}$	5
44	2.95	$1.05 \cdot 10^{-4}$	3
45	3.45	$8.4 \cdot 10^{-5}$	6

$p = 200$ mm. Hg.
$d = 1.1$ to 3.7 cms.

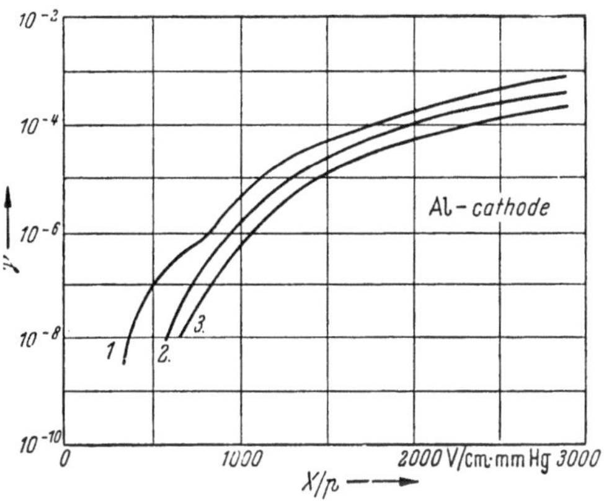

Fig. 11. Values of γ as a function of X/p for Al cathodes in various gases: *1* Benzene, *2* Cyclo-hexane, *3* Toluene.

this amount is observed by a change in the cathode at $X/p = 40$, so these values give tentative support only to ROGOWSKI's observations of very low γ for air when X/p is small.

It is interesting that when the cathode was more photosensitive, the value for γ increased at $X/p = 40$. This is the converse of BOWLS results in N_2; with Na present γ was reduced, though the sodium cathode was very sensitive. This supports the suggestion of gas contamination in BOWLS' work with Na.

Further values for γ in air at higher X/p with various cathodes are discussed in the next section (Table 3).

Information on polyatomic gases is very meagre. Results in *benzene, cyclohexane* and *toluene* obtained by VALERIU-PETRESCU[2] are shown in Fig. 11. These were obtained by measuring α/p and V_s, using aluminium cathodes.

The curves for all three show roughly the same shape, and γ is very small for $X/p < 400$ for all three.

There will be much dissociation of such molecules under electron bombardment and by radiation in a TOWNSEND discharge. The cathode will probably be covered with a film of a mixture of dissociation products, many of which will be electronegative. It is easy to understand why γ is small under such circumstances. It is not until high energy ions exist in large numbers that γ_i becomes important enough to make γ large, but even so values of 10^{-1} are never reached as they are in monatomic and diatomic gases. This is probably because large molecular ions

[1] F. LLEWELLYN JONES and A. B. PARKER: Proc. Roy. Soc. Lond., Ser. A **213**, 185 (1952).
[2] M. VALERIU-PETRESCU: Bull. Soc. roum. Fiz. **43**, 81 (1943).

tend to dissociate when neutralised at the cathode, rather than release electrons from the metal surface.

All these results have shown the importance of cathode conditions, but in order to establish the dependence of γ on the cathode material we shall include a short section reporting experiments designed to this end.

12. The influence of the cathode material. This was first emphasized by HOLST and OOSTERHUIS[1], and almost every experiment carried out in very clean conditions suggested a dependence of V_s on the cathode. The general references give the history of the subject.

The problem really lies in causing a change of the cathode without influencing the gas at all. The work already reported witnesses to the difficulty of this.

A simple and direct way of testing the influence of the cathode is to measure V_s for various cathodes in the same gas at the same time. This was done in H_2 by LLEWELLYN JONES and HENDERSON[2] who obtained the results shown in Fig. 10. At the minimum sparking potential V_{sm} there was a difference of 30% approximately between steel and aluminium cathodes. LLEWELLYN JONES deduced γ for these metals, using uncertain values for α/p [5]. He also showed from them that the cathode emission is important.

Suppose that there is a volume secondary ionisation process—say ionisation due to positive ions. Then, if suffixes refer to the cathodes used,

$$c_1 + \frac{\beta}{\alpha} = \gamma_1 \quad \text{and} \quad c_2 + \frac{\beta}{\alpha} = \gamma_2$$

where γ_1, γ_2 are the measured secondary coefficients at V_{sm} and $c_1\, c_2$ are the *cathode* secondary coefficients. β/α is the same for both, so

$$\gamma_1 - \gamma_2 = c_1 - c_2 .$$

That is, the difference in the coefficients is a measure of the difference in the cathode coefficients. With the values of γ deduced by LLEWELLYN JONES, c_1 for aluminium is 50% of the total emission if a steel cathode is completely inactive ($c_2 = 0$). If c_2 for steel is not zero, the percentage due to c_1 for Al rises. For Ni and Cu the figures are 41% and 28% respectively, at least, for the cathode contribution.

If c_2 for steel is half c_1 for aluminium, then the cathode emission accounts for the *whole* secondary effect at V_{sm}, where X/p lies between 190 to 250 V/cm. per mm. Hg.

In a later paper LLEWELLYN JONES and DAVIES [*13*] show the effect of the cathode surface on V_{sm} and γ at V_{sm}. The influence of thin films deposited by sputtering can be seen from the following table. γ is calculated from V_s and the best available α/p values. For air, these values are not satisfactory, as was mentioned in Sect. 11.

Very large effects on V_{sm} are observed here. It is to be noted that a *thin* film of one metal on another displays a value of γ characteristic of neither metal. Sometimes the value is higher, sometimes lower[3]. The same authors have measured $\gamma = f(X/p)$ for the cathode surface as it was changed by sputtering metal onto it. Fig. 12a shows $\gamma = f(X/p)$ for Ni on which a thin copper film has been deposited, and outgassed at 400% for 100 hrs. (curve a). A marked peak at 150 X/p is due to photons, and at $X/p \sim 450$ the effect of γ_i is appearing.

[1] G. HOLST and E. OOSTERHUIS: Phil. Mag. **46**, 1117 (1923).
[2] F. LLEWELLYN JONES: Phil. Mag. **28**, 192 (1939).
[3] F. M. PENNING and C. C. J. ADDINK: Physica (Haag) **1**, 1007 (1934) § 4.2 of chap. V.

Leaving the cathode in vacuo for a week resulted in curve b; a general reduction in γ is apparent. Increasing the film thickness resulted in curve c, where the peak is much reduced, and finally curve d, in which no peak appears.

It is to be noted that at $X/p \sim 150$ is reduced by 90% during this treatment. At $X/p \geqq 400$, γ is reduced by only 20% approximately. Thus γ_i is less sensitive

Table 3. *Influence of the cathode on γ.*

Gas	Cathode	V_{sm}	X/p	$\gamma \cdot 10^3$
Air + Hg	Cu amalgam	460	720	4
	Hg on Al	390	885	14
	Hg on Ni	390	585	6
Air	Oxidised Al	416	905	10
	Oxidised Ni	421	957	10
H_2 (cleaned by ion bombardment)	Al	243	200	100
	Al on Ni	212	200	140
	Ni	289	180	75
	Ni on Al	390	245	15
	Al (commercial)	225	200	125
	Al on staybrite steel	205	210	150
	Staybrite steel	274	190	75
	Steel on Al	282	190	75

to the surface than γ_p, when the radiation concerned is that emitted from a TOWNSEND discharge in hydrogen.

Other experiments on Ni and Ag surfaces were carried out, and also work on changes in the surface of given metals. Fig. 12b shows the effect in Cu. Curve a

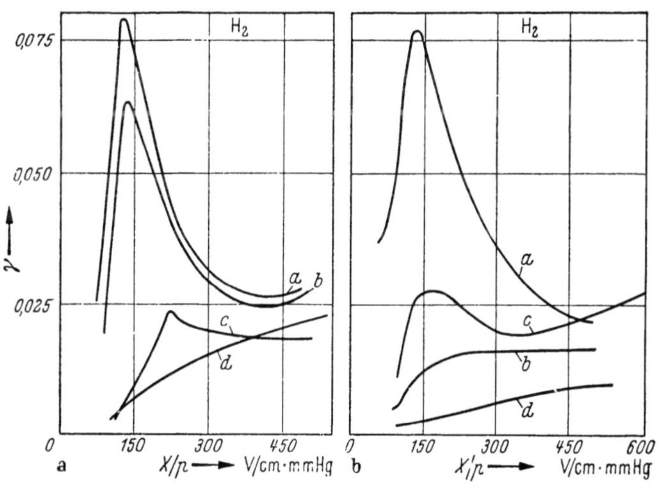

Fig. 12a and b. (a) The effect on $\gamma = f(X/p)$ in hydrogen of the deposition of nickel on a copper cathode (see text). (b) The effect on $\gamma = f(X/p)$ in hydrogen of the surface cleaning of a copper cathode (see text).

is for slightly oxidised copper, and appears very similar to curve a of Fig. 12a. When the cathode was baked for over 100 hrs. at 400° C in vacuum the curve b was obtained without a peak. Mere exposure to H_2 created a smaller peak at 150 V/cm. per mm. Hg (c) and this vanished under mild ion bombardment and outgassing (d).

It is obvious that thin films of oxides, foreign metal or possibly hydrides (curve c of Fig. 12b) greatly affect the emission. Cu and Ni have the same work function approximately, 4.6 eV. The changes in depositing Cu on Ni are interpreted by the authors in terms of a change in work function due to an electric double layer, which vanishes when the film deposited is sufficiently thick for conduction bands to appear. This idea of change in work function, giving a change in γ can explain all the results if low energy photons are important. The authors show that a rapid change in γ is connected with photo-effects, and a slow change with ions, as the work function alters. This agrees with our previous interpretation of the peak as due to photons and the slow rise at higher X/p to ions.

It is known (Sect. 32) that for high energy photons γ_p is not very dependent on work-function, in marked contrast to photons of visible light. This would conform to the explanation of LLEWELLYN JONES and DAVIES: the smaller changes in the intermediate region of X/p are probably the combined result of changes in γ_i and γ_p for energetic photons.

It is concluded that deposition of a thin film on a more electropositive metal raises the work function, while a thin film on a more electro-negative metal lowers the work function. Films of chemical compounds between metal and gas would have similar effects. Such changes in the cathode surface account for the great difficulty in obtaining consistent values of γ. There is no doubt, therefore, that the second TOWNSEND coefficient depends on the cathode surface.

In all this work it has only been possible to distinguish between the influence of ions and photons or metastable atoms by virtue of the dependence of γ on X/p. A more direct analysis is made possible by measurements of the time-dependence of γ in pulsed discharges, described in the next section. This has only been exploited fairly recently; it is a most promising study, though difficult.

d) Analysis of secondary effects by their time dependence.

13. Breakdown of a gap. The time interval between the application of a potential to a gap and the passage of a discharge is made up of two parts. There is the interval before an electron capable of starting a discharge appears in the gap, and the interval between its appearance and the establishment of a steady discharge. The first is called the statistical time lag, and the second the formative time lag which is itself a statistical average[1].

VON LAUE[2] first discussed the statistical lag, and the theory has been developed in detail[3]. Experiments have confirmed its correctness[4].

We shall not describe work on statistical lags here, for the formative time of the discharge only is determined by the primary and secondary mechanisms. Usually the statistical time lag is made negligible by irradiating the cathode with ultra-violet light to provide a copious supply of electrons in the gap.

For a description of work at high pressures and in long gaps, the articles on spark discharges should be consulted. The theory of the break-down of short gaps at fairly low pressures has been more fully developed, and there is general agreement about the process. Only in this region of low pd can the experiments be interpreted in detail, and even so accurate theories have so far never been applied because of the tremendous effort needed in calculation.

[1] α and γ are averages, so the multiplication is bound to show a distribution. See L. B. LOEB Vol XXII.

[2] M. v. LAUE: Ann. Phys., Lpz. **76**, 261 (1925).

[3] R. A. WIJSMAN: Phys. Rev. **75**, 833 (1949) and references.

[4] G. HERTZ: Z. Physik **106**, 102 (1937). — H. PAETOW: Z. Physik **111**, 770 (1939).

The first reproducible experiments seem to be those of TANK and GRAF and the first theoretical treatments are due to ROGOWSKI and STEENBECK[1]. This was amplified in VON ENGEL and STEENBECK's book [4]; they included only γ_i as the secondary coefficient.

SCHADE[2] too developed a simple theory in which only the γ_i mechanism was considered, and obtained a current $i(0, t)$ at the cathode in terms of τ_i, the ion transit time across the gap, and i_0, the externally sustained current from the cathode:

$$t = \frac{\tau_i}{\varepsilon} \ln\left(1 + \varepsilon\,\frac{i(0,t)}{i_0}\right) \tag{13.1}$$

where

$$\varepsilon = \gamma\,(e^{\alpha d} - 1) - 1 .$$

If breakdown is defined by the time t_b at which a given current i_b flows, then (13.1) defines the variation of formative time lag with α. Using TOWNSEND's expression for α [1], assuming that the field is uniform and that the drift velocities of ions and electrons are

$$\left.\begin{aligned} v^+ &= k^+ X \\ v^- &= k^- X . \end{aligned}\right\} \tag{13.2}$$

SCHADE found for small overvoltages, constant pd and small i_0

$$t_b = \frac{a}{V - V_s}\, e^{b/V} \tag{13.3}$$

where V is the applied potential and V_s the static breakdown potential, for which $\varepsilon = 0$.

He carried out experiments in neon which gave results in good accord with (13.3). In addition, the predicted dependence of t_b on i_0 (13.1) was confirmed for large values of i_0. HEYMANN[3] developed the theory for the case of a steadily rising voltage applied to the gap, and found that in argon experimental values of $\partial V/\partial t$ and the overvoltage at which the gap broke down were well explained by the theory.

BARTHOLOMEYCZYK[4] included γ_p and VON GUGELBERG[5] used the theory for γ_i and γ_p for comparison with experiment. We shall give these theories and the results of VON GUGELBERG first, then the development of BARTHOLOMEYCZYK's solution by DAVIDSON[6] and BANDEL[7], including their experiments.

In STEENBECK's more refined analysis the assumptions made are that (13.2) holds and that there is a characteristic time constant τ (see the contribution of A. VON ENGEL in this volume, for physical implications) so that the current of electron and ions may be written

$$\left.\begin{aligned} i^-(x,t) &= i_x^-\, e^{t/\tau} , \\ i^+(x,t) &= i_x^+\, e^{t/\tau} . \end{aligned}\right\} \tag{13.4}$$

---

[1] F. TANK and K. GRAF: Helv. phys. Acta **2**, 33 (1929), W. ROGOWSKI Arch. Elektrotech. **20** 99 (1928) and [20].
[2] R. SCHADE: Z. Physik **104**, 487 (1937).
[3] F. G. HEYMANN: Proc. Phys. Soc. Lond. **63**, 25 (1950).
[4] W. BARTHOLOMEYCZYK: Z. Physik **116**, 235 (1940).
[5] H. L. v. GUGELBERG: Helv. phys. Acta **20**, 307 (1947).
[6] J. DUTTON, S. C. HAYDON, F. LLEWELLYN JONES and P. M. DAVIDSON: Brit. J. Appl. Phys. **4**, 170 (1953).
[7] H. W. BANDEL: Phys. Rev. **95**, 1117 (1954).

Then if n^+, n^- are the ion and electron concentrations,

$$\left.\begin{aligned} \frac{\partial n^-}{\partial t} &= \alpha v^- n^- - \frac{\partial}{\partial x}(v^- n^-), \\ \frac{\partial n^+}{\partial t} &= \alpha v^- n^- - \frac{\partial}{\partial x}(v^+ n^+), \end{aligned}\right\} \tag{13.5}$$

we have

$$\left.\begin{aligned} \frac{\partial}{\partial t}\left(\frac{i^-}{v^-}\right) &= \alpha i^- - \frac{\partial i^-}{\partial x}, \\ \frac{\partial}{\partial t}\left(\frac{i^+}{v^+}\right) &= \alpha i^- + \frac{\partial i^+}{\partial x}. \end{aligned}\right\} \tag{13.6}$$

This neglects losses by diffusion to the walls, and assumes the discharge cross-section to be constant.

Using (13.2) these become on integration

$$\left.\begin{aligned} i^-(x) &= C_1 \exp\left(\int_0^x \left[\alpha - \frac{1}{\tau v}\right] dx\right), \\ i^+(x) &= \left[\exp\int_0^x \frac{dx}{\tau v^+}\right]\left[C_2 - C_1 \int_0^x \alpha\left\{\exp\int_0^x \left(\alpha - \frac{1}{\tau v}\right) dx\right\} dx\right] \end{aligned}\right\} \tag{13.7}$$

where

$$\frac{1}{v} = \frac{1}{v^+} + \frac{1}{v^-}.$$

Now the boundary conditions are

$$\left.\begin{aligned} j^-(0) &= \gamma_i' j^+(0), \\ j^+(d) &= 0 \end{aligned}\right\} \tag{13.8}$$

the external irradiation produces i_0 at the cathode, but $i_0 \ll i^-(0)$ and is neglected. We assume only γ_i, as STEENBECK did. Then the current rise is described by (13.4) if

$$1 = \gamma_i' \int_0^d \alpha \left\{\exp\left[\int_0^x \left(\alpha - \frac{1}{\tau v}\right) dx\right]\right\} dx = F(d). \tag{13.9}$$

BARTHOLOMEYCZYK derived this result on equivalent assumptions writing $\lambda = 1/\tau$ [see (13.17)] and also took into account diffusion to the walls. Then (13.5a) goes into

$$\frac{\partial n^-}{\partial t} = -\frac{\partial}{\partial x}(v^- n^-) + \alpha v^- n^- + D^-\left[\frac{1}{r}\frac{\partial n}{\partial r} + \frac{\partial^2 n}{\partial r^2}\right]$$

with a similar additional term in (13.5b) for the ions. Assuming

$$n^- = e^{t/\tau} f(x)\, g(r)$$

the variables may be separated and solutions for n^+, n^- found. Using the boundary conditions the final result is

$$1 = \gamma_i' \int_0^d \alpha \left\{\exp\left[\int_0^x \left(\alpha - \frac{1}{L} - \frac{1}{\tau v}\right) dx\right]\right\} dx \tag{13.10}$$

where

$$\frac{1}{L} = \left(\frac{2.41}{R}\right)^2 \left(\frac{D^+}{v^+} + \frac{D^-}{v^-}\right)$$

and R is the radius of the discharge tube. For large R, this is the same as (13.9). This is the only case we shall consider, but in the original paper the influence of R is discussed: τ is increased at small values of R.

BARTHOLOMEYCZYK also included the photoelectric effect at the cathode. Then (13.8) becomes

$$j^-(0) = \gamma_i' j^+(0) + \gamma_p' g \int_0^d j^-(x)\,\vartheta\, e^{-\mu x}\, dx \tag{13.11}$$

using the notation of that part of Sect. 5 dealing with photoelectric effects.

Since (13.8b) still holds we have the breakdown threshold relations with $j_0 \ll j^-(0)$ which replace (13.9) as

$$1 = F(d) + \gamma_p' g \int_0^d \vartheta \left\{\exp\left[\int_0^x \left(\alpha - \frac{1}{\tau v^-}\right) dx - \mu x\right]\right\} dx. \tag{13.12}$$

We now simplify by assuming that the field is everywhere uniform at breakdown i.e. space charge is neglected. Under such conditions (13.4) is acceptable. Then (13.9) becomes

$$1 = \gamma_i' \frac{e^{\left(\alpha - \frac{1}{\tau v}\right) d} - 1}{\alpha - \frac{1}{\tau v}} = F_1(d) \tag{13.13}$$

and (13.12) becomes

$$1 = F_1(d) + g \gamma_p' \vartheta \frac{e^{\left(\alpha - \mu - \frac{1}{\tau v}\right) d} - 1}{\alpha - \mu - \frac{1}{\tau v^-}}. \tag{13.14}$$

If τ is very large (13.13) goes into the familiar form

$$1 = \gamma_i' (e^{\alpha d} - 1)\,.$$

This is the condition that the current grows infinitely slowly—the static breakdown condition. We shall now obtain an expression for the dependence of the time constant of current growth on the discharge parameters, in particular d, in order that the theoretical basis of VON GUGELBERG's experiments may be understood. He did not modify the treatment of BARTHOLOMEYCZYK, but noted the omission of i_0.

On solving (13.13) for τ we have

$$\tau = \tau_i = \frac{d/v}{\alpha d - \log_e\left(\frac{1}{\gamma_i'} + 1\right)} \tag{13.15}$$

with the assumption that $\tau_i \gg 1/\alpha v$ i.e. this is near the threshold voltage.

From this it is seen that τ_i decreases as α, γ_i and d increase. The other extreme case is photon action alone, when $\gamma_i' = 0$ in (13.14). Then

$$\tau = \tau_p = \frac{d/v^-}{(\alpha - \mu)\, d - \log_e\left(1 + \frac{\alpha - \mu}{\vartheta g \gamma_p'}\right)} \tag{13.16}$$

again near the threshold.

Consider now the distance at which breakdown occurs, i.e. when (13.10) or (13.11) are satisfied for $\tau \to \infty$ (von GUGELBERG's definition). Let the solution of (13.13) be d_i, of (13.14) d_p, when $\gamma_i' = 0$, and of (13.14) as it stands d_{ip}.

Clearly, $d_{ip} < d_i$ and $d_{ip} < d_p$. If we keep X/p constant and increase d, then when

$$d_i,\, d_p > d > d_{ip}$$

breakdown occurs, with τ determined by the slowest agency, the ions. So $\tau = \tau_i$, although photons are acting.

When $d > d_p$ the photons can sustain breakdown alone, and $\tau = \tau_p$. So at $d = d_p$, the characteristic time of the discharge growth alters suddenly.

It follows that if a voltage pulse of length $t < \tau_p$ is used, no breakdown will be observed at voltages just above the threshold voltage for static breakdown. When $\tau_i > t > \tau_p$ breakdown will be observed at $d \geqq d_p$: when $t > \tau_i$ breakdown will be found at $d \geqq d_{ip}$.

If metastable atoms were present, it is obvious that breakdown would be observed at a still smaller distance with a still greater value for τ. This was not treated by VON GUGELBERG. He gave a theoretical curve for $1/\tau = f(d)$ assuming ion and photon action only.

In his experiments VON GUGELBERG used six sealed off tubes with plane parallel gaps, irradiating the cathode through the anode. Two tubes had variable gaps, and the fixed gaps were between 0.13 and 0.75 cm. long. Ni or Ni–Ba electrodes were used, and the gases were He, A, Kr, Xe, H_2 and N_2.

Purification and outgassing were good, and the time resolution of the oscilloscope used was of the order of 10^{-7} secs. Heavy initial currents were necessary and the measured currents ranged from 10^{-11} to 10^{-4} A as the discharge developed. The measurements could be reproduced to 10%.

The action of metastable atoms was noticed, being greatest in He and least in Xe, and absent altogether only in H_2. It was estimated that in the steady state the metastables provided 5 to 25% of the secondary coefficient. The values for τ_i were in good agreement with the best available data on α and v^+.

The transition from the combined action of ions and photons to photons only was not so sharp as expected, possibly because of the assumption of uniform field conditions at breakdown, but an estimate of the percentage q of total secondary emission due to photons was made. Table 4 shows the results: p is the pressure in mm, Hg at 0° C, d_s the breakdown distance ($d_s = d_{ip}$), V_s the breakdown potential and X_s/p the value of X/p at breakdown[1].

Table 4. *Influence of photons at breakdown.*

Gas	Cathode	p mm. Hg	d_s cm	V_s volts	X_s/p V/cm. per mm. Hg	q(%)
He	Ni	7.5	0.28	225	107	0
A	Ni	2.6	0.35	210	231	1.6—7
Kr	Ni	2.9	0.35	220	218	1.5—9
Xe	Ni	2.25	0.35	235	298	9—20
H_2	Ni	5.1	0.35	295	166	> 60
N_2	Ni	1.85	0.35	250	386	0
Ne + 1% A . .	Ni	7.1	0.28	145	73	0
He	Ba	6.35	0.35	150	68	0
A	Ba	2.4	0.35	110	131	0.3—4

Where no effect was measurable, then 0.1% is the upper limit for that particular gas-cathode combination. It is interesting that A–Ba shows less photo-effect than A–Ni: possibly the coated cathodes were contaminated in the processing with BaN_3 or other chemicals.

[1] R. SCHADE: Z. Phys. III, 437 (1938) also concluded that photons are important in H_2.

We now consider DAVIDSON's modification of BARTHOLOMEYCZYK's result. The original equation is (13.6) with boundary conditions (13.11). In the original treatment it is assumed that

$$i^-(x,t)\,e^{-\alpha x} = C\,e^{\lambda(t-x/v^-)} \tag{13.17}$$

where C is a constant and λ is the real solution of $f(d)=0$ where [cf. eq. (13.14)]

$$f(x) = 1 - \frac{\alpha\gamma_i'}{\varphi}\left(e^{\varphi x}-1\right) - \frac{g\gamma_p'\vartheta}{\psi}\left(e^{\psi x}-1\right) \tag{13.18}$$

in which

$$\varphi = \alpha - \frac{\lambda}{v},$$

$$\psi = \alpha - \mu - \frac{\lambda}{v^-},$$

$$\frac{1}{v} = \frac{1}{v^-} + \frac{1}{v^+}.$$

Now (13.17) satisfies the boundary conditions *but not the charge distribution at* $t=0$ i.e. when the final voltage is applied and a current is flowing due to the externally-sustained i_0 and the standing voltage before $t=0$.

In order to account for the externally sustained i_0, which was not included in VON GUGELBERG's analysis in a valid manner, we use the expression

$$P = 1 - \gamma_i'\left(e^{\alpha d}-1\right) - \frac{g\gamma_p'\vartheta}{\alpha-\mu}\left[e^{(\alpha-\mu)d}-1\right]. \tag{13.18'}$$

Then

$$i^-(x,t)\,e^{-\alpha x} = \frac{i_0}{P} + C\,e^{\lambda(t-x/v^-)} \tag{13.19}$$

satisfies (13.6) and the boundary conditions including i_0.

To obtain an *approximation* for the initial charge distribution if $i^-(0,t)=c$, put

$$C = c - \frac{i_0}{P}. \tag{13.20}$$

An *accurate* solution may be found by recognising that an infinite number of complex roots of $f(d)=0$ exist. Adding to (13.15) a sum of various complex λ's with appropriate C's will yield the true solution.

DAVIDSON derives this, but the solution is a very complicated one. DUTTON, HAYDON and LLEWELLYN JONES in the same paper used (13.19) with the assumption

$$i^-(x,0) = 0 \quad \text{or} \quad C = -\frac{i_0}{P} \tag{13.21}$$

to obtain theoretical graphs for the formative time of a TOWNSEND discharge in air leading to breakdown. Like other workers, they were unable to use the more complex theory because of the frightful labour of computation involved.

They defined breakdown in the same manner as SCHADE[1], by the attainment of a given current i_b: t_b was given by $i_b = i^-(0,t_b) = 10^{-9} - 10^{-1}$ A in various calculations. The practical definition of breakdown is the point where a fall in potential across the gap is observed, so these two definitions are fairly close if the choice of i_b is correct. The actual value of i_b chosen influences the curves to some extent, and a large i_b invalidates the assumption that the field is uniform. The physical process of breakdown when the current increases beyond i_b is not discussed.

[1] R. SCHADE: Z. Physik **104**, 487 (1937).

The example taken is a 1 cm. gap at 760 mm. Hg, where $V_s/pd = 41.6$ and α is taken from SANDERS' [5] values in Hg-contaminated air. Various values of the ratio $\vartheta g \gamma_p'/\gamma_i'$ were used, from 0 (only γ_i) to ∞ (only γ_p). Metastable atoms were not considered.

Small overvoltages ($<2\%$) were considered, γ, v^-, v^+ being assumed constant in this range. Graphical results derived from numerical calculations show:

(i) As γ_p/γ_i increases, t_b is less sensitive to i_0,

(ii) t_b depends more on γ_p/γ_i at low overvoltages.

(iii) The shape of $t_b = f(\Delta V/V_s)$ depends much on the value of γ_p/γ_i.

(iv) If $\vartheta g \gamma_p'/\gamma_i' > 1.0$, $t_b < 10^{-6}$ sec. at 2% overvoltage.

These conclusions are in general agreement with previous theories.

Direct comparison with FISHER and BEDERSON's[1] experimental values for air indicate that good agreement could be obtained by proper choice of γ_p/γ_i and i_b. Photoeffects appear to predominate almost to the exclusion of ion emission: 10% of the emission is due to ions approximately.

A more sensitive test of the theory has been made by BANDEL[2]. He measured i as a function of time for an air gap between plane brass electrodes. A constant i_0 of about 50 electrons/µ sec. was drawn from the cathode by irradiating it with a mercury lamp: a constant high voltage was maintained across the 1 cm gap and a pulse of 1000 volts applied to cause breakdown. Small overvoltages only were used.

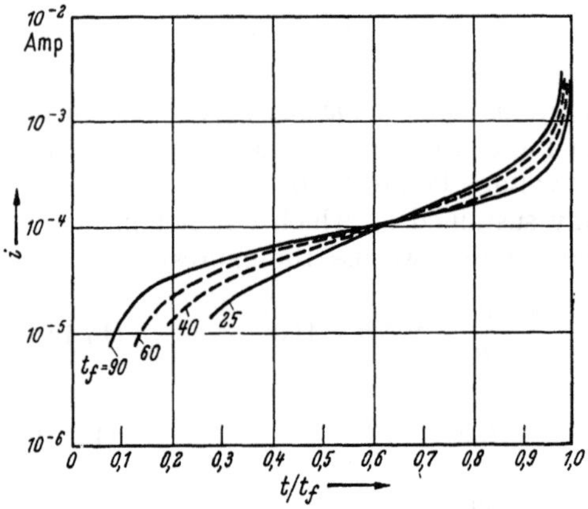

Fig. 13. The current i flowing between plane parallel plates in air ($p = 724$ mm. Hg, $d = 1$ cm.) when the initial current $i_0 = 60$ electrons/µsec. At $t = 0$ a voltage pulse is applied and at $t = t_f$ the gap has broken down. i is shown as a function of t/t_f for various values of t_f.

Fig. 13 shows some of the results, when statistical lag is negligible, for formative time lags of 25 to 90 µsecs. The first currents are recorded at about 8 µsecs after the pulse was applied, and were measured up to 1 µsec. of breakdown, when currents of order 10^{-3} amps were flowing. The initial rise is attributed to the very rapid increase in charge density in the gap due to primary multiplication and photoelectric effects.

If the multiplication was insufficient to cause breakdown, this would lead to a steady current in ~10 µsecs, judging from Fig. 13. With large multiplication, this steady state becomes the nearly-exponential rise in current displayed in the centre of the graph, as expected from theory, eqs. (13.2) and (13.15). When the current is large, ~100 µA, then a still more rapid rise appears which is due to space charge distortion of the field. This graph shows the time required to establish space charges, which have been discussed before only in steady state conditions.

BANDEL was able to show that a current of order 10^{-4} amps should lead to an apparent increase in α due to space charges, and he obtained curves having somewhat the same shape as the first half of Fig. 13 by solving the equations (13.6)

[1] L. H. FISHER and B. BEDERSON: Phys. Rev. **81**, 109 (1951). See F. LLEWELLYN JONES' article in Vol. XXII.

[2] H. W. BANDEL: Phys. Rev **95**, 1117 (1954).

with the boundary conditions

$$i^-(0,t) = i_0 + \gamma_i' i^+(0,t) + \gamma_p' g\vartheta \int_0^d i^-(x,t)\,dx$$
$$i^+(d,t) = 0$$

which ignores the absorption of light in the gas. He used a different approach to the problem. If $i^-(0,t)$ at the cathode is known, then

$$i^-(x,t) = i^-(0, t - x/v^-) \tag{13.22}$$

so that the ionisation in dx' at x' is

$$di^+ = \alpha i^-(x',t)\,dx' = \alpha i^-(0, t - x/v^-)\,\mathrm{e}^{\alpha x'}\,dx'.$$

The positive ions created in dx' will contribute to the current at x at a time $(x'-x)/v^+$ later, so the total positive ion current at x is

$$i^+(x,t) = \int_0^{v(t+x/v^+)\text{ or } d} \alpha i^-(0, t + x/v^+ - x/v)\,\mathrm{e}^{\alpha x'}\,dx'$$

where

$$\frac{1}{v} = \frac{1}{v^+} + \frac{1}{v^-}$$

and the upper limit of the integral expresses the fact that the ions from x' cannot contribute to i^+ at x until sufficient time has elapsed for electrons to reach x' and for ions to move from x' to x. If t is sufficiently great, however, the upper limit is given by d, the limit of the gap.

Taking into account the action of photons we may derive

$$\left.\begin{aligned} i^-(0,t) = i_0 + \gamma_i' \alpha \int_0^{vt\text{ or } d} i^-(0, t - x'/v)\,\mathrm{e}^{\alpha x'}\,dx' \\ + \gamma_p' g\vartheta \int_0^{v^- t\text{ or } d} i^-(0, t - x/v^-)\,\mathrm{e}^{\alpha x}\,dx. \end{aligned}\right\} \tag{13.23}$$

Trying

$$i^-(0,t) = i_0(A + B\,\mathrm{e}^{t/\tau}) \tag{13.24}$$

which is comparable to (13.20), BANDEL found that no solution is possible for $vt < d$, and for $vt > d$ Eq. (13.24) is the same as that of DAVIDSON (13.19)[1].

If $\gamma_i' \neq 0$, so that ions must be included, $v \sim v^+$ and the solution is of no value. If the ion contribution may be neglected then $v = v^-$ and for this extreme case

$$A = \frac{1}{1 - G(\mathrm{e}^{\alpha d} - 1)} \tag{13.25}$$

and

$$G\,\mathrm{e}^{\left(\alpha - \frac{1}{v^-\tau}\right)d} = 1 + G - \frac{1}{\alpha v^- \tau} \tag{13.26}$$

where

$$G = \gamma_p' g\vartheta/\alpha$$

and is assumed constant.

The other extreme case $\gamma_p' = G = 0$, may be solved if the assumption

$$\frac{di^-}{dt} \cdot \frac{d}{v^-} \ll i^-$$

[1] See however the end of this article.

is made i.e. that in one electron transit time the current increase is small compared with the current. This is justified if ions act alone, even near breakdown.

Assuming also

$$\gamma_i' \ll 1 \ll e^{\alpha d}$$

we have

$$B \sim \frac{\gamma_i' e^{\alpha d}\left(1 - \frac{1}{\alpha v \tau}\right)}{1 - \gamma_i' e^{\alpha d}}. \tag{13.27}$$

Now the integral equation (13.23) for $\gamma_i' = 0$ and $G = 0$ has the same form if v^- replaces v and G replaces γ_i'. If these changes are made (13.27) will give B for the case $\gamma_i' = 0$ if $t \leq d/v^-$: moreover, in this short interval the ions cannot reach the cathode, so the new expression for B will hold under all circumstances if $t \leq d/v^-$.

BANDEL discussed the matter further, but finally assumed that no great contribution due to ions could be expected in view of the short formative times observed, and he put $\gamma_i' = 0$. Then (13.25), (13.26) and (13.27) with G, v^- substituted for γ_i', v together define A, B and τ in (13.24).

The final value for $i(t)$ in the external circuit is[1]

$$i(t) = \frac{1}{d} \int_0^d \{i^-(x, t) + i^+(x, t)\}\, dx. \tag{13.28}$$

In the comparison with experiment, uniform field conditions were assumed, and SANDER's values for α/p used. The value of X at breakdown was adjusted, however, to obtain the best fit with experiment, but agreement was not good.

The main discrepancies are that the experimental curves seem to rise about 5 μsecs later than theory predicts, and the slope of the exponential part rises less rapidly than theory indicates with increasing overvoltage. These are considered to be real differences, not instrumental ones. Possibly they arise because α/p values measured in Hg-contaminated air were used or[2] because "delayed" photons contribute.

No theory was attempted for the space charge distorted region. RAETHER[3] has dealt with this in an approximate manner, but it is outside the scope of this article.

In spite of the unsatisfactory state of the theory, it is unanimously concluded by these workers in air that at higher pressures where X_s/p is low photo electric effects predominate. From all results of Sect. 11, this is to be expected.

Having then some experimental values for the contribution of photons and ions at breakdown in various gases, we shall consider the results of investigations using pulsed light sources for the photo-electrons.

14. The first studies with interrupted initial currents. This approach was first used by ENGSTROM and HUXFORD[4]. Many investigators had reported that the current in photo cells did not remain in phase with variable light incident on the cell, and KRUITHOF measured[5] the frequency response of a Ne discharge in this way. He found that the lag was reduced when argon was added, and deduced that metastable atoms were important in determing γ for pure neon.

[1] Cf. the introduction to VARNEY's work (Sect. 17).
[2] See Sects. 18 and 23.
[3] H. RAETHER: Z. angew. Phys. 5, 211 (1953).
[4] R. W. ENGSTROM and W. S. HUXFORD: Phys. Rev. 58, 67 (1940).
[5] A. A. KRUITHOF: Philips Techn. Rev. 4, 48 (1939).

ENGSTROM and HUXFORD made use of a variable gap d in *argon*, since the time lag for metastables diffusing to the cathode should increase with d and p contrary to the time lag due to other agencies [see equations (13.12) and (13.14)], so that the effects due to metastables could be isolated from these.

The theory assumes that the probability of arrival of metastable atoms at time t after their creation is

$$P(t) = \frac{1}{\tau_0} \cdot e^{-t/\tau_0}. \tag{14.1}$$

τ_0 being the average interval between creation and arrival. The value of τ_0 is obtained from diffusion theory, on the assumption that the metastables are created between x_m and d with a linear density m given by

$$m = C_i\, e^{\alpha(x-x_m)} \tag{14.2}$$

C_i being a constant. τ_0 is expressed as a series involving d, x_m and the coefficient of diffusion D. It can be shown that the current flowing at time t after a constant photocurrent i_0 began to flow is

$$i(t) = 1 - (1 - i_a)\, e^{-(i_a/\tau_0)t} \tag{14.3}$$

where the maximum equilibrium current is taken as unity and i_a is the fraction of the current *not* given by metastable atoms, and effectively in phase with the illumination. Similarly, if unit constant photocurrent is cut off at $t'=0$, then

$$i(t') = (1 - i_a - i_c)\, e^{-(i_a/\tau_0)t} \tag{14.4}$$

where i_c is the ratio between the equilibrium current and the actual current after the period $T=t-t'$ of the illumination.

Experimentally one may measure (τ_0/i_a), the decay constant, and i_a, as a function of distance and pressure. Knowing then τ_0 from the equation for $\tau_0=f(D)$ the value of $D=D_0$ at S.T.P. can be found.

The values of α and γ were measured by finding $i/i_0=f(X)$ at given d for various values of d and $i/i_0=f(d)$ for constant X/p found by interpolation. This was necessary to avoid photoelectric fatigue.

γ can also be measured from the time-lag curves from the estimated equilibrium values of the current, and γ_i can be obtained from the instantaneous amplification before any contribution from the metastables becomes important, according to ENGSTROM and HUXFORD. They consider that the instantaneous amplification gives γ_i only, but any photons which move rapidly through the gas will be included. Moreover, resonance radiation diffusing through the gas would be included in their values for γ_m; however, as they found no evidence of two time constants of comparable size this latter is probably not important[1].

Finally from the known total γ and the fast component (say γ_f) the slow component due to metastables may be evaluated. It is actually $\gamma'_m\,\vartheta_m\, g/\alpha$ in our notation, and thus it is necessary to estimate an average value for g (from the geometry of the system) and ϑ_m for the gas and X/p concerned. This latter quantity is not known, though the authors estimated the maximum value for ϑ_m. They assumed that all the energy the electron gained in the field went into this one metastable state of energy V_m: then comparing this with PENNING's work[2] a minimum value for γ_m is found.

The experiment consisted of chopping the light beam from a tungsten filament by a sectored disc. *Silver* electrodes were used, the cathode being activated with

[1] See however Sect. 18.

[2] F. M. PENNING: Physica, Haag **5**, 286 (1938).

Cs—O. It was illuminated through a mesh in the anode, 2 cm. in diameter in the 4 cm. diameter anode. The gas was purified by an arc discharge with Ba electrodes and left in contact with a Ba mirror. The initial photocurrent was 10^{-10} A/cm.^{2}; the range of pressures were within 0.1 to 1 mm. Hg.

The range of X/p was 50 to 200, and the values for α/p obtained scattered badly at different p, straddling the values of KRUITHOF and PENNING[1]. It seems that the electrons never reached equilibrium with the field at the lowest pressure.

The time constants for rise and fall were found to agree, but at low pressures there was a time constant of order 0.01 sec. present, affecting the current by a few percent. This was attributed to changes in the surface with bombardment.

Values for D had to be corrected for diffusion through the mesh in the anode, but gave (at 1 mm. Hg and room temperature) $p\,D_1 = 67.5$ cm.^{2} sec.$^{-1}$ mm. Hg,

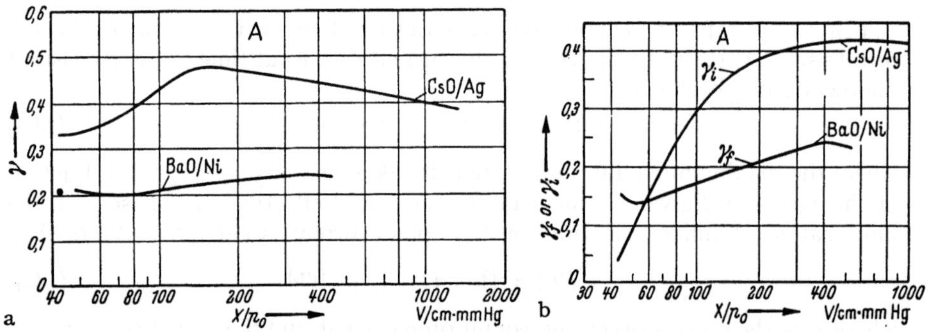

Fig. 14 a and b. Values of γ and γ_i as a function of X/p_0 for argon with cathodes of caesium oxide on silver (ENGSTROM and HUXFORD) and barium oxide on nickel (MOLNAR, Sect. 16).

and $D_0 = 0{,}084$ cm.^{2} sec.$^{-1}$ at S.T.P. For argon atoms in argon D_0 is calculated as 0.157. It is reasonable to expect lower diffusion with an excited atom of larger cross-section.

The values of γ and γ_i corrected for back diffusion obtained at higher pressures are shown in Fig. 14. From the remarks above γ_i is expected to be an upper limit.

The expected large contribution of metastables at low X/p and the fall in γ_i below $X/p = 200$ is observed. The minimum value for γ_m was found to be 0.4. The fraction of current due to metastable atoms was up to 67%.

The results obtained with a mesh cathode showed the same γ_i, but much less contribution from the metastables, probably because some were lost by diffusion.

When a *barium* cathode was used, no lagging current appeared: it is assumed that γ_m is very small for A—Ba. Contamination with H_2 in an effort to activate the cathode restored the lagging fraction to 62% with a time constant in fair agreement with the previous work. This shows the importance of surface layers on the cathode once more.

15. Development of the theory. Following the work of ENGSTROM and HUXFORD, NEWTON[2] derived the theory of the transients in TOWNSEND discharges more fully. We shall only quote results here. He considered a plane parallel gap with an externally sustained electron current i_0 emitted from the cathode while a constant X/p was maintained. He did not derive the form of the current while only photons could reach the cathode, but if all the radiation can be assumed to

[1] A. A. KRUITHOF and F. M. PENNING: Physica, Haag 3, 515 (1936).
[2] R. R. NEWTON: Phys. Rev. 73, 570 (1948).

cross the gap before ions arrive, the current flowing while γ_p acts alone rises to the constant value given by (5.1). NEWTON derives a similar equation and writes

$$i_p = i_0 M_p e^{\alpha(d-x_0)} \tag{15.1}$$

in our notation; M_p represents the multiplication factor given in (5.1), due to photons, and x_0 is defined in discussing (1.1). Then

$$M_p = (1 - \vartheta g \gamma_p' e^{\alpha(d-x_0)})^{-1} \tag{15.2}$$

and i_p is called the "photon steady state current". NEWTON ignores any "trapped" resonance radiation[1].

Similarly if all the ions produced by the first electrons in i_0 arrive before the metastables, then the "ion steady state" current is

$$i_i = i_0 M_p M_i e^{\alpha(d-x_0)} \tag{15.3}$$

where M_i is the multiplication due to ions

$$M_i = \{1 - \gamma_i' M_p [e^{\alpha(d-x_0)} - 1]\}^{-1}. \tag{15.4}$$

Finally, when the metastable atoms contribute the steady state current is

$$i = i_0 M_p M_i M_m e^{\alpha(d-x_0)} \tag{15.5}$$

where the multiplication due to metastables is

$$M_m = \frac{1}{1 - \frac{\gamma_m' M_p M_i \vartheta_m}{\alpha^2 d} f(N)}$$

where

$$f(N) = \frac{(1 + N\alpha d)(e^{\eta\alpha d} - 1) - \eta\alpha d}{1 + 2N}.$$

Here N is determined from the boundary conditions to be

$$N = \frac{4L}{3X}\left(\frac{1}{\gamma_m'} - \frac{1}{2}\right),$$

being the mean free path for metastable atoms moving in a gas of normal atoms, and the quantity $\eta = 1 - x_0/d$.

This is the true steady state current; it is seen that the contribution of γ_m no longer fits the general equation (6.3). The loss of metastable atoms by diffusion to the anode and out of the discharge volume is included in the analysis leading to (15.5) and this is the physical reason for the departure from the TOWNSEND-type equation. The diffusion to the anode increases with increasing d, since the production is more concentrated toward the anode[2].

This means that the measured γ in work on $i/i_0 = f(d)$ should *vary with d*. For large d,

$$\gamma_m' = \text{const}\left(N + \frac{1}{\alpha d}\right), \tag{15.6}$$

that is γ_m' falls as d increases even if the cathode is effectively infinite. This would also be qualitatively true for resonance and other radiation with a finite cathode.

[1] See Sect. 18.
[2] See Sect. 21.

If space charge distortion is occurring then (unless the pressure is so low that we are below the minimum of the PASCHEN curve) γ apparently increases as d increases. This is a fictitious change but there is *a real dependence of γ on d* if the agents concerned in γ may be lost from the discharge, i.e. if they are not charged particles.

Let us consider the solution of eqs. (15.1) to (15.5).

Now η and N may be determined from known constants, but the remaining quantities in these equations must be determined from the three currents i_p, i_i and i if measurements are confined to currents.

NEWTON shows that four parameters are to be found

$$\frac{\vartheta\gamma_p'}{\alpha}, \quad \gamma_i', \quad \frac{\vartheta_m\gamma_m'}{\alpha}, \quad \alpha.$$

These cannot be found from three equations.

They may be found if d is altered at constant X/p (the classical TOWNSEND method) of if the transients are measured. This brings in the new variables v^+ and D (see ENGSTROM and HUXFORD'S work). Using NEWTON'S analysis it is possible to study the positive ion transient. There are discontinuities in $i=f(t)$, from which (αd) and $(M_p\gamma_i')$ may be derived, and hence five equations now exist for four parameters. It is possible then to treat v^+ as an unknown quantity and measure the drift velocity of the ions and hence their mobility. These theoretical predictions were verified by later experimental work for the effects of ions and metastables. Experiments designed so that γ_m is important should demonstrate the variation of γ with d, but this had not been observed previously.

16. Later measurements of the influence of metastable atoms. The interpretation of the work to be described here is based on the theories of NEWTON, and ENGSTROM and HUXFORD. MOLNAR[1] and HORNBECK both developed the theory further in order to interpret the measurements they made, but we cannot present this for lack of space.

MOLNAR'S apparatus was essentially that of ENGSTROM and HUXFORD. A chopped light beam illuminated cathodes of *molybdenum*, *tantalum* and *coated tantalum* (BaO—Ta) through the anode of a plane parallel gap in *argon* for $5\cdot10^{-3}$ secs every 1/30 of a second. Difficulty in measuring α with surfaces of high γ led to this use of several surfaces in the same gas mixture, so that a surface of low γ could be used to measure α (Sect. 9). It was impossible to alter the gas pressure in the experiments with molybdenum and tantalum (clean or coated) without changing the cathode properties, in spite of great care in preparation. The electrode diameter was 2 inches, the maximum separation 12 cm.

Currents flowing were large up to the limit where space charge distortion occurs (10^{-7} amps). Time constants were measured by balancing out the current against an RC circuit of variable time constant.

Measurements of the diffusion coefficient for metastable atoms yielded $D_m=(45\pm4)$ cm.²/sec. at 1 mm. Hg; for normal argon atoms $D_a=139$ cm.²/sec., which indicates the metastable radius to be 2.5 times the normal. This is reasonables. The aceidental introduction of 0.1% of N_2 destroyed the argon metastables in 100 µsec., producing N_2 metastables for which $D_m=157$ cm.²/sec. These were not destroyed in the volume, but the argon metastables were so destroyed even in the purest gas. The probability of volume destruction was found to be

[1] J. P. MOLNAR: Phys. Rev. **83**, 933 (1951).

80 sec.$^{-1}$, corresponding to probability of destruction at a collision of $5 \cdot 10^{-6}$ i.e. $2 \cdot 10^5$ collisions are required to de-excite a metastable atom (cf. Sect. 37).

When radiation from an argon discharge fell on the gap, an increase in the fast component of the current and a decrease in the slow component was observed. If metastable atoms are raised to states from which they can radiate both these effects are readily understood, for some metastable atoms are made to deliver their potential energy to the cathode (as a photon) very quickly. This robs the slow component of the current.

An estimate (accurate to 20%) if γ_p/γ_m was made from this. For Ta it was 0.40, for Mo 0.08 and for BaO–Ta 0.10.

Table 5. *Variation of γ_s with discharge gap length.* Argon: $X/p = 72.6$

d	BaO–Ta	Mo	Ta
0.2	0.035	—	—
0.4	0.019	0.0055	0.0011
0.6	—	0.0032	0.0012
0.8	—	—	0.00095

In the evaluation of α, the effects of back diffusion and lack of equilibrium with the field were prominent at low d and the curves of $\log i/i_0 = f(d)$ did not display a straight portion of great length. It was difficult to assess α; finally the value which made the curves for γ_f (γ for the fast component of current) most nearly independent of d was assumed. As the electrode diameter was always greater than d by a factor five or more, few photons will be lost and γ_f should be constant. This was observed to be so for the "best" value of α for Ta and Mo, but not for BaO–Ta.

For all the surfaces however the value of γ_s, γ for the slow (lagging) component of current, was found to vary with d. For BaO–Ta and Mo γ_s fell to 1/3 of its original value when d increased by a factor two. These surfaces showed the larger values of γ_s: at $X/p = 72.6$, the values are was shown in Table 5 taken from a graph given by MOLNAR.

Table 6a. *Gas-dependent factors determining γ in A.*

X/p_0 V/cm. per mm. Hg	$\nu = \gamma_i'/\gamma_i$	α cm.$^{-1}$	$\vartheta_m = \vartheta$ cm^{-1}
195.4	0.97	6.63	2.6
117.2	0.91	3.72	2.4
72.6	0.86	5.10	7.0

It is seen then for Ta γ_s does not vary so much.

No complete graphs of $\gamma = f(X/p)$ could be obtained, since a change in p altered γ.

In another experiment, however, it was possible to determine γ (total coefficient) and γ_f' for a *coated nickel* surface BaO–Ni in argon; these curves are shown in

Table 6b. *Cathode dependent factors determining γ in A.*

p_0	X/p_0	Cathode	γ_f	γ_s (near V_s)	γ_i	γ_m	γ_p
		Ta	0.026	0.0025	0.026	0.023	0.009
1.53	195.4	Mo	0.076	0.0074	0.071	0.060	0.005
		BaO–Ta	0.23	0.050	0.23	0.31	0.031
		Ta	0.022	0.0026	0.022	0.023	0.009
1.53	117.2	Mo	0.065	0.011	0.071	0.065	0.005
		BaO–Ta	0.21	0.065	0.22	0.27	0.027
		Ta	0.0060	0.0009	0.0053	0.0035	0.0021[1]
4.13	72.6	Mo	0.030	0.0025	0.034	0.0020	0.001[1]
		BaO–Ta (lower efficiency cathodes)	0.073	0.013	0.085	0.078	—

[1] No light absorption measurements were used to obtain γ_m/γ_p for these values, so they are less reliable than the others.

Figs. 14a and b (Sect. 14); back diffusion, which was measured in this apparatus directly in other experiments, probably accounts for the variation in γ_f.

If the calculated values of $(\vartheta_m + \vartheta)/\alpha$ given by KRUITHOF [12] are used, then Table 6 may be drawn up, on the assumption $\vartheta_m = \vartheta$, and g (fraction of photons reaching cathode) $= 0.4$.

These values are not more accurate than 20%, so that $\gamma_m \approx \gamma_i$ always, and γ_p bears no simple relation to either. Clearly all the γ's are dependent on X/p.

17. The influence of the ions on very short current transients. HORNBECK[1] developed an apparatus for measuring the emission due to positive ions. It was similar to MOLNAR'S, using Ni cathode and anode, with BaO coating on the cathode. The currents used were of order 1 μA and lower than that needed to create space charge distortion when a 0.1 μsec pulse of photo electrons was released from the cathode. Impurities were found to have little effect on the results, perhaps because the time scale of observation is so short, when care was taken in outgassing and in the handling of the pure gases.

NEWTON'S theory for the current pulse due to ions assumes that the electrons cross the tube in negligible time, and that all ions have a uniform drift velocity v^+. When an exponential distribution of ions in the gap at $t=0$ is assumed, the ions are not formed until $x = x_0$ where the electrons have enough energy to ionise and the distribution ends at the anode, $x = d$. No emission is observed then, after the initial electrons have crossed, until $t = x_0/v^+$ (we include the emission due to photons from the first avalanche with the initial electrons). At $t = x_0/v^+$ a discontinuous upward jump in i is seen.

The form of the current at $t > x_0/v^+$ depends on the initial ion distribution ϱ_0^+ and on the products of the avalanches formed by the secondary electrons released by the ions. When $t = d/v^+$ all the initial ions have crossed and another discontinuity is observed as the current falls. The ions formed by the secondary avalanches continue to arrive at the cathode after $t = d/v^+$, so there is a "tail" in the current pulse, which can be shown to have a discontinuity in slope at $t = 2d/v^+$.

All these theoretical predictions were verified by HORNBECK though the distribution of v^+ smoothed out the discontinuities. He found that ϱ_0^+ was *not* exponential in form, for between $d/v^+ > t > x_0/v^+$ a ripple appeared in the pulse. This was probably because the electrons are not in equilibrium with the field. A. VON ENGEL'S contribution to this volume shows curves for $i/i_0 = f(d)$ displaying this time-lag to reach equilibrium in terms of distance.

Some indication of the diffusion of resonance quanta[2] was observed for $X/p < 75$ V/cm. per mm. Hg.

Because of the inaccuracies in theory and experiment HORNBECK gave only one example in his paper. For a BaO—Ni cathode in A, $p_0 = 0.82$ mm. Hg, $X/p_0 = 85$ V/cm. per mm. Hg, $\gamma_i' = 0.19$.

At V_s, $\gamma' \sim 0.35$ so $\gamma_i'/\gamma' = \gamma_i/\gamma = 0.54$.

The value of α determined from this work at this value of X/p_0 was 1.30, whereas KRUITHOF [12] quotes 1.25. The method seemed therefore reliable but the interpretation of the experimental results presents formidable difficulties.

VARNEY[3] used the same apparatus for measuring γ_i as a function of X/p. It has been applied in other ways to measure drift velocities. VARNEY'S work was made possible by a simpler and better theoretical approach.

---

[1] J. A. HORNBECK: Phys. Rev. **83**, 374 (1951).

[2] T. HOLSTEIN: Phys. Rev. **72**, 1212 (1947) and [14]. Cf. Sect. 18.

[3] R. N. VARNEY: Phys. Rev. **93**, 1156 (1954).

He made the same assumptions as NEWTON and HORNBECK, that:

1. $v^- \gg v^+$,
2. Diffusion and space charge can be neglected,
3. TOWNSEND's ionisation coefficient α may be used,
5. All ions have the same v^+,
6. Diffusion of resonance quanta may be neglected, and also
7. Electrons have always sufficient energy to ionise from $x=0$ i.e. $x_0=0$.

In order to calculate the current due to the electrons and positive ions in the gas, we must use the fact that the current due to a charge e moving with velocity v in a gap of length d is

$$i = \frac{e v}{d}$$

as may easily be shown by consideration of the work done by the external circuit as the charge moves.

After the initial electron pulse (photoelectrons due to γ_p included in it) is past, the initial ion distribution has a linear density s_1 where

$$s_1 = \alpha n_0 e^{\alpha x}.$$

The current i_1^+ due to such a line density is

$$i_1^+ = v^+ \int\limits_a^b \frac{s_1}{d} dx$$

if $s_1 = f(x)$ between a and b. Now at time t the ion cloud remaining in the gap is that which originally lay between $x = v^+ t$ and $x = d$. Thus

$$i_1^+ = \frac{e v^+}{d} (e^{\alpha d} - e^{\alpha v^+ t}), \quad 0 \leq t \leq \frac{d}{v^+} \tag{17.1}$$

where e is the electronic charge. All further equations refer to this time interval.

As the ions forming s_1 reach the cathode they release γ_i' electrons per ion on the average. These electrons produce new avalanches giving a second term in the ion current

$$i_2^+ = \frac{e v^+}{d} \int\limits_0^t \gamma_i' n^+(t') \left[e^{\alpha d} - e^{\alpha v^+ (t-t')}\right] dt' \tag{17.2}$$

where $n^+(t)$ is the rate of arrival of ions at the cathode. The integral represents the number of secondary ions remaining in the gap at time t.

One electron released from the cathode multiplies to $e^{\alpha x}$ electrons at x, so the electron current due to one initial electron at this instant is $e v e^{\alpha x}/d$. The total charge conveyed by all the electrons formed in the avalanche is then

$$\frac{e v}{d} \cdot \frac{e^{\alpha d} - 1}{\alpha d}$$

and this charge is conveyed in a time (d/v). Hence the *average* current flowing is $(e^{\alpha d} - 1)(e/\alpha d)$ per electron released at the cathode.

Since $\gamma_i' n^+(t)$ is the rate at which electrons are released by positive ions, the component of the current due to electron motion is

$$i^- = e \gamma_i' n^+(t) \frac{e^{\alpha d} - 1}{\alpha d}. \tag{17.3}$$

Another relation between i^+ and n^+ may be found by saying that the rate of change of i^+ is due to creation by ionisation in the gas and the sweeping out of the ions at the cathode. Then

$$\frac{di^+}{dt} = \frac{e\,v^+}{d}\left[n^+(t)\,\gamma_i'(e^{\alpha d}-1) - n^+(t)\right]. \tag{17.4}$$

Adding (17.1) and (17.2) and differentiating, we have another expression for di^+/dt; so using (17.4) we obtain a linear integral equation

$$n^+(t) = n_0\,\alpha\,v^+\,e^{\alpha v^+} + \gamma_i'\,\alpha\,v^+\int_0^t n^+(t')\,e^{\alpha v^+(t-t')}\,dt'$$

whose solution is

$$n^+(t) = n_0\,\alpha\,v^+\,e^{\alpha v^+(1+\gamma_i')t}. \tag{17.5}$$

Then substituting in (17.3) and (17.2) we obtain using (17.1) also

$$i^- = n_0\,e\,v^+\,\alpha\,\gamma_i'\exp\left(\alpha\,d + \alpha\,v^+(1+\gamma_i')\,t\right), \tag{17.6}$$

$$i^+ = n_0\,e\,v^+\left\{\frac{\gamma_i'}{1+\gamma_i'}\,(e^{\alpha d}-1)\,e^{\alpha v^+(1+\gamma)t} + \frac{1}{1+\gamma_i'}\,e^{\alpha d}\right\}. \tag{17.7}$$

The total current is the sum of (17.6) and (17.7).

This total current will be independent of time if the coefficient of

$$\exp\left[\alpha\,v^+(1+\gamma_i')\,t\right]$$

vanishes, i.e.

$$\gamma_i' = \left[\frac{e^{\alpha d}}{e^{\alpha d}-1}\right]^{\frac{1}{2}} - 1. \tag{17.8}$$

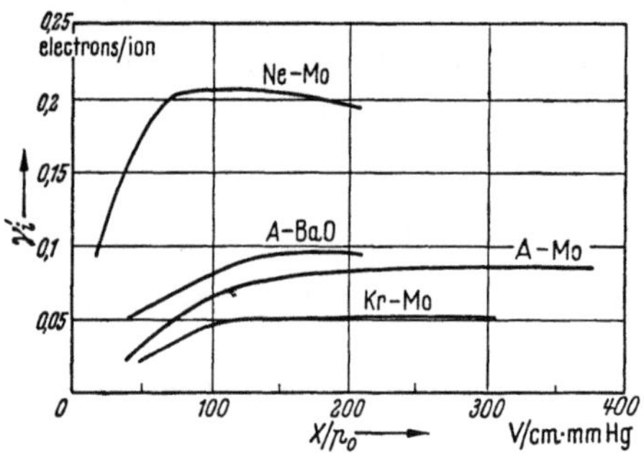

Fig. 15. Values of γ_i' as a function of X/p_0 for rare gases on molybdenum and argon on coated molybdenum.

If the voltage applied to the discharge gap is varied, the oscilloscope trace can be made horizontal so that (17.8) is satisfied for a given d. KRUITHOF's values for α/p were used to determine γ_i'.

VARNEY used Mo electrodes subjected to careful purifying by sputtering in heavy discharges in A, Kr and Xe, and BaO–Mo cathodes in A. His results for γ_i' are shown in Fig. 15. The fall at lower X/p_0 is probably to be attributed in part to back diffusion though a real change in γ_i is probable (Sect. 27). The energy of the ions above which $\gamma_i' =$ constant is estimated as 0.1 eV.

For the BaO cathode γ_i was very variable, and the curve shown is an intermediate one: γ_i was measured between 0.5 and 10^{-3} under various conditions, and was acutely sensitive to ion currents. The interest lies in the fact it does not approach the origin at low X/p as the others do. This may be due to a higher initial velocity for the electrons from a coated cathode [see Sect. 30, eq. (30.2)] which would reduce back diffusion (Sect. 4).

If radiation from de-excited metastable atoms reaches the cathode, the results for γ must be reduced: from MOLNAR's work this correction is expected to be $\sim 7\%$. It was noted that Mo surfaces too were easily contaminated and $\gamma \sim 10^{-4}$ was observed on occasions.

No effect indicating $x_0 \neq 0$ was observed, probably because the initial pulse concealed it.

e) Measurements in non-uniform fields.

18. Discharges between wire and cylinder. These may be TOWNSEND discharges, but in non-uniform fields the multiplication is not given by exp $(\int \alpha\, dr)$ for the electron energy distribution at a point where the field is X_1 is not the same as if they had been moving in a constant field X_1 for sufficient time to reach equilibrium with it. Hence the rate of production of ion-pairs is not α_1 per cm. per electron, if it is α_1 in a uniform field X_1. MORTON and JOHNSON (see overleaf) have demonstrated this. This immediately makes quantitative works more difficult. In addition, the geometry is such that space charges easily control the discharge by concentrating round the wire so that a wide range of currents in TOWNSEND discharges cannot be obtained.

We shall discuss some work which has shown the relative importance of various mechanisms in discharges where space charges were not too great. Some values of γ_i deduced from GEIGER counter work are also included for comparison.

Low pressure discharges in non uniform fields may be expected to emphasize volume ionisation at the expense of surface effects. When the wire is made negative the cathode surface is small, and when it is positive currents flow with very small fields at the cathode. However, LLEWELLYN JONES and WILLIAMS[1] have emphasized the importance of cathode effects in low pressure discharges between a cylinder and a wire. They deliberately altered the surface condition of the nickel cylinder and measured the breakdown potential with the wire negative and positive. *Air, hydrogen* and *helium* were used, at $p \sim 1$ mm. Hg.

Table 7.
Values of γ_i: A – Ni cylindrical cathode.

p mm. Hg	X_c/p [2] V/cm. per mm. Hg	γ_i'	γ_i
25	3.12	$7.8 \cdot 10^{-4}$	$3.5 \cdot 10^{-2}$
50	1.92	7.0	5.1
100	1.18	4.0	4.0
200	0.79	0.6	0.6
400	0.52	0.2	0.2

Values of γ were determined, using $\int \alpha\, dr$ to calculate the multiplication. This is not accurate, as MORTON and JOHNSON[3] have shown. Nevertheless, the investigations confirm that the cathode condition is important in non uniform fields at low pressure, as it is in uniform fields.

At high pressure photons will be absorbed still more strongly in the volume, and effects of metastables should be more prominent also. Yet in pure gases it has been shown that cathode effects predominate, and only in mixed gases has photoionisation been observed. The experimental results are given below. PENNING has shown [*10*] in mixed gases the action of metastables in co-axial discharges, which is explicable in the same way as in uniform fields. No analysis of the volume effects of photons and metastables seem to be available, probably because the rate of production is very difficult to calculate.

In *argon*, with *brass* and *nickel* cathodes no photoionisation occurred, according to COLLI and FACCHINI[4] but photoelectric effects at the cathode were very marked. LAUER, however[5], observed with very pure argon and a nickel cathode no contribution due to γ_p, but emission caused by positive ions. His results are shown in Table 7, with THEOBALD'S corrections for back diffusion[6], using MORTON'S estimate of the multiplication.

[1] F. LLEWELLYN JONES and G. C. WILLIAMS: Proc. Phys. Soc. Lond. B **66**, 345 (1953).
[2] X_c/p is the value of X/p at the cathode.
[3] P. L. MORTON: Phys. Rev. **70**, 358 (1946). — G. W. JOHNSON: Phys. Rev. **73**, 284 (1948).
[4] L. COLLI and U. FACCHINI: Phys. Rev. **88**, 987 (1952).
[5] E. J. LAUER: J. Appl. Phys. **23**, 300 (1952).
[6] J. K. THEOBALD: J. Appl. Phys. **24**, 123 (1953).

The reduction in yield due to back scattering is tremendous: only a few percent of the electrons escape at these low values of X/p.

The argon ions were A_2^+, as is to be expected (Sect. 19β). When corrected for back diffusion, the values at $p<100$ are comparable with the direct measurements of HAGSTRUM and PARKER (Sect. 27), if the effect of thin gas films and the lower potential energy of A_2^+ with respect to A^+ is borne in mind. These measurements show that at $p>100$ (low X/p) the value of γ_i falls sharply. Later work[1] showed that unless the gas is extremely pure γ_i at high pressures is not observed. This perhaps explains the lower values at high pressure in Table 7, since any impurity would have greater effects at higher pressure.

The reason why LAUER failed to observe photo-emission is that he used lower pressures and a smaller tube than the other workers; the energetic photons needed are produced by three-body collisions involving an argon metastable atom A_m^*:

$$A_m^* + 2A \rightarrow A_2^* + A$$

and after a time lag

$$A_2^* \rightarrow A_2 + h\nu.$$

This three-body collision is less likely at low p[2]. The photoelectric effect was demonstrated finally to be predominant at high pressure (> 100 mm. Hg) while below 50 mm. Hg γ_i was more important, in very pure argon. No effects due to photo ionisation were observed.

Table 8. *Photo emission: H_2^+—Ni-cylindrical cathode.*

p mm. Hg	$\frac{X_e}{p}$ $\frac{\text{V/cm.}}{\text{mm. Hg}}$	R'	R
100	1.43	$6.9\cdot10^{-4}$	$1.4\cdot10^{-2}$
200	1.04	1.5	0.4
400	0.79	0.74	0.2
650	0.46	0.49	0.15

In *hydrogen*, LAUER observed a strong photoelectric effect, and no emission due to ions. This agrees with other workers (Sect. 12 and 13). As the actual production rate of photons was unknown, he gave the value of the ratio R' of the number of electrons emitted from the cathode by photons to the number of ion pairs formed in the gap. In uniform fields $R'=\gamma_p' \vartheta g/\alpha$. Table 8 shows the results: R is the true emission corrected for back scattering. As there are so many unknown factors in R, it is impossible to discuss absolute values in detail. Comparison with Fig. 9 shows that at low X/p for Pt, which is similar to Ni (see text) the average value of γ is of the same order as R. The variation with X/p cannot be compared. No photoionisation in the volume was observed.

MILLER and LOEB[3] have observed TOWNSEND discharges (10^{-10} A) in positive wire cylindrical discharges in *nitrogen*, *oxygen* and mixtures of these gases. They found streamers at higher currents, especially in mixed gases, indicating that photo-ionisation was more prominent there. The later paper on negative wire discharges showed more complex behaviour. The various types of discharge are helpfully discussed[4].

HUBER[5] has done similar work to that of LAUER in pure *nitrogen* and *oxygen* and in mixtures. In pure N_2 no photoionisation or photoemission was found. The values of γ_i' and γ_i for N_4^+ ions are given in Table 9, using THEOBALD's approximate corrections. The absence of γ_p may be attributed to the low-energy photons which are common in N_2 emission spectrum.

[1] L. COLLI and U. FACCHINI: Phys. Rev. **96**, 1 (1954).
[2] A. V. PHELPS and J. P. MOLNAR: Phys. Rev. **89**, 1202 (1953).
[3] C. G. MILLER and L. B. LOEB: J. Appl. Phys. **22**, 494, 614 (1951).
[4] Some workers advocate β-process in cylinders, cf. H. MEDICUS: Z. angew. Phys. **1**, 316 (1949) and S. WERNER: Nature, Lond. **165**, 1018 (1950).
[5] E. HUBER: Phys. Rev. **97**, 267 (1955).

The rise and fall may be due to errors in the back scattering correction. The values of PARKER (Fig. 28b) for N^+ and N_2^+ on Pt are lower than these, but for Ta the value is roughly correct for very low energy ions.

The addition of ~1% O_2 reduced γ_i and caused a photoeffect at the cathode to appear. Oxidation of the cathode and a layer of electron-attaching molecules on it probably accounted for the reduction in γ_i; possibly γ_p arises from collisions between N_2^* and O_2 molecules, so that radiation from O_2 replaces that from N_2.

At higher concentrations effects due to photo-ionisation in the mixture appeared. These were found to lead to GEIGER counter action in some cases. This result of the use of cylindrical geometry is very important, but the subject is too vast to discuss here. Secondary processes are still less fully analysed in non uniform fields than in uniform ones, so no useful short summary can be given. Vol. XLV of this Encyclopedia should be consulted.

It is interesting to compare the results in extremely pure gases with measurements of γ_i' due to large molecular ions of quenching agents used in GEIGER counters, according to BERETTA and BIANCHI[1] and BERETTA and ROSTAGNI[2], based on PUTNAM'S work on the analysis of the time intervals between pulses. By isolating the counts due to preceding pulses (spurious counts) and attributing all these to secondary electrons released by ions, an estimate of γ_i' may be made. Any other cathode mechanisms are neglected, so the values obtained are possibly overestimated[3].

Table 9. *Values of γ_i: N_4^+–Ni-cylindrical cathode.*

p mm. Hg	γ_i'	γ_i
50	$1.4 \cdot 10^{-3}$	$2.0 \cdot 10^{-2}$
100	1.5	2.5
200	1.2	2.9
300	1.0	2.7
400	1.1	3.2
500	0.7	2.3
600	0.6	2.0

The counters are filled with 10 mm. Hg of quenching vapour and 90 mm. Hg of argon: it is assumed that only vapour ions arrive at the cathode. The cathode material is metal in Table 10A, and the effect of altering it is seen in 10B for ethyl alcohol ions.

Table 10 A. *γ_i' for complex ions on metal.*

Ion	$\gamma_i' \cdot 10^{10}$	Ion	$\gamma_i' \cdot 10^{10}$
CH_4	3.0 ± 0.5	C_3H_7 (so)	0.68 ± 0.09
$(C_2H_5)_2O$	1.7 ± 0.2	C_3H_7OH (n)	0.41 ± 0.09
$CH_2(OCH_3)_2$	0.57 ± 0.12	$C_2H_3O_2$—CH_3	1.32 ± 0.09
CH_3OH	0.97 ± 0.13	$C_2H_3O_2$—C_2H_5	1.01 ± 0.09
C_2H_5OH	0.76 ± 0.08	$C_2H_3O_2$—C_3H_7	0.76 ± 0.08

B. *γ_i' for $C_2H_5OH^+$.*

Cathode	$\gamma_i' \cdot 10^{10}$
Metal	0.33 ± 0.05
Glass	0.05 ± 0.03
Lampblack	0.12 ± 0.05

It is seen that the values of γ_i are very much lower than for the simpler ions. The correction for back diffusion would be expected to be little different than before, certainly not different by seven orders of magnitude. The reason is the likelihood that such molecules will dissociate after being neutralised, and because of the layer of dissociation products on the cathode surface. This will include electron-attaching substances. Note that increasingly complex molecules of similar structure show that γ_i falls as the size of the molecules increases.

[1] E. BERETTA and A. M. BIANCHI: Atti. Accad. Naz. Lincei (R. C. Cl. Sci. Fis. Mat. Nat.) **11**, 352 (1951).

[2] E. BERETTA and A. ROSTAGNI: Nuovo Cim. **6**, 391 (1949).

[3] See also S. P. PURI and P. S. GILL: Indian J. Phys. **29**, 95 (1955). — H. L. WISER and A. D. KRUMBEIN: Phys. Rev. **98**, 303 (1955).

f) The relative importance of various secondary mechanisms.

We shall now try to compare numerically the importance of ions, photons, and other agents producing ionisation in a discharge, making use of some experimental work other than that already reported where necessary.

19. Ions. *α) The production of ions in the discharge.* There is no doubt that ions are formed in every discharge. The rate of production in uniform fields in terms of α has long been studied; the results, electron energy distribution measurements and full details of the theory will be found in the preceding article.

The direct ionisation of an atom is not the only way in which ions may be formed, for molecular rare gas ions are observed (Sect. 19β), and at larger current densities, as in a positive column, metastables are sometimes responsible for ionisation (Sect. 25). In any particular experiment, especially with transient currents, it is essential that the values of α used should be appropriate to the conditions.

In non uniform fields, and before equilibrium with a uniform field is established, the ionisation cannot be described by α at all. This fact makes quantitative calculation under such conditions very difficult, for the electron energy distribution must be found. It is often possible, however, to determine the multiplication occurring before secondary processes are important, and then these last may be determined when they appear.

Table 11. *Molecular rare gas ions.*

Gas	Ion	V_i	B	X/p range
He	He^+ He_2^+ [1]	24.58 $23.18 + 0.2$ $- 0.7$	85	8 to 17
Ne	Ne^+ Ne_2^+ [1]	21.56 $20.85 + 0.3$ $- 0.7$	500	8 to 22
A	A^+ A_2^+ [1]	15.76 $15.06 + 0.2$ $- 0.7$	1200	24 to 60
Kr	Kr^+ Kr_2^+ [1]	13.96 $13.23 + 0.3$ $- 0.7$	20000	<75
Xe			40000	<75

β) The nature of the ion present. The work of HORNBECK and MOLNAR and VARNEY[2] has shown that a variety of ions exist in gas discharge work in pure rare and molecular gases. The apparatus of HORNBECK and VARNEY was that mentioned in Sect. 17, and mass spectrograph studies of HORNBECK and MOLNAR suggested the most likely process by which these molecular rare gas ions He_2^+, Ne_2^+ etc. are formed. This is the reaction[3]

$$\mathrm{He} + e \rightarrow \mathrm{He}^* + e$$

followed by

$$\mathrm{He}^* + \mathrm{He} \rightarrow \mathrm{He}_2^+ + e.$$

He* is probably a normal excited state. The production of atomic ions is proportional to the pressure: the production of molecular ions to the square of the pressure[4].

At such low pressures that scattering was small, the ratio B of the atomic ion peaks to the molecular ion peaks was as given in Table 11. The range

[1] Given by the authors as appearance potentials.

[2] J. A. HORNBECK and J. P. MOLNAR: Phys. Rev. **84**, 621 (1951). — R. N. VARNEY: Phys. Rev. **89**, 708 (1953). See also R. L. F. BOYD: Proc. Phys. Soc. Lond., Ser. A **63**, 543 (1950). — D. MORRIS: Proc. Phys. Soc. Lond., Ser. A **68**, 11 (1955).

[3] R. MEYEROTT: Phys. Rev. **70**, 671 (1946). He_2^+ detected by O. TUXEN: Z. Physik **103**, 463 (1936).

[4] For Hg_2^+ see F. G. HOUTERMANS: Z. Physik **41**, 699 (1937).

of X/p where molecular ions were observed is also shown, the pressure being 0.5 to 20 mm. Hg.

In *oxygen*, it is believed[1] that only O_2^+ exists.

In *nitrogen* at $X/p<60$ the ions are probably N_4^+. The intermediate range 60 to 100 shows a strange variation for $v^+=f(X/p)$; the ions probably change in nature several times as they cross the gap. Above $X/p=100$ the ions are N_2^+. In CO, a similar curve for $v^+=f(X/p)$ is observed, with an extra group of fast ions in the intermediate range which now lies at $100<X/p<150$. It is thought that CO^+ exists at high and low X/p, and a tentative explanation of the intermediate range is that CO_2^+ ions are formed and break up again, with C^+ ions (the fast group) present due to an irreversible reaction.

γ) The ion energy distribution. The drift velocity $v^+=f(X/p)$ measured in this work has provided a theoretical basis for the interpretation given above using the results of WANNIER[2]. He developed a method for finding averages without knowing the actual ion energy distribution, and the elastic sphere model for ion-atom collisions seems to provide results in agreement with experiment. The isotropic scattering of classical theory is attributed also to the special scattering when ions move in their own gas, e.g. A^+ in A (but not A_2^+ in A or Ne^+ in A). This "resonance" scattering is another term to be included as well as the polarisation attraction term: at high X/p polarisation forces are negligible, so WANNIER uses a hard sphere model of greater cross section than kinetic theory gives for atom—atom collisions[3].

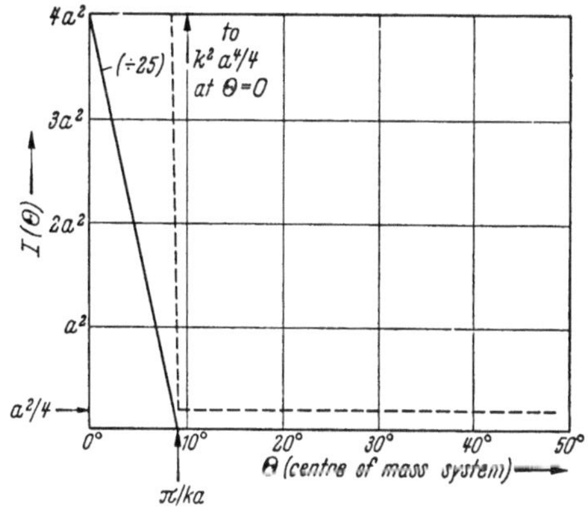

Fig. 16. Approximate angular distribution of scattered atoms for the rigid sphere model ($ka=20$) on quantum theory. – – – to scale; ——— $^1/_{25}$ scale. Classical theory predicts horizontal line at height $a^2/4$.

Concerning the ion energy distribution, it is important to remember the elastic scattering predicted by quantum theory for rigid spheres. Classically, isotropic scattering with the intensity $I(\vartheta)$ per unit solid angle independent of ϑ should be observed, ϑ being the angle of deflection in the centre of mass system. The quantum theory for the collision of two rigid spheres each of diameter a and mass M when the relative velocity is v predicts an angular distribution approximately as shown in Fig. 16 [*15*]: $k=\frac{1}{2}Mv/\hbar$ where $\hbar$ is PLANCK's constant.

The dashed line represents the intensity fullscale, and the full line the intensity divided by 25: $ka=20$ in the figure.

It can be shown that half of the ions are scattered within $\vartheta'=\pi/ka$ radians ($\vartheta'=9°$ for $ka=20$): only for $\vartheta>\vartheta'$ does isotropic scattering occur.

For ions moving in their own gas this treatment may be applied as an approximation.

In such collisions little loss of energy would occur. The value of ϑ' for ions of molecular weight M_w and energy E eV is roughly (in degrees)

$$\vartheta'=\frac{8}{\sqrt{M_w E}}$$

[1] O. LUHR: Phys. Rev. **44**, 459 (1933) also observed only O_2^+ in a mass spectrometer. O_3^+, O_4^+ were not found.

[2] J. A. HORNBECK and G. H. WANNIER: Phys. Rev. **82**, 458 (1951).

[3] Cf. T. KIHARA: Rev. Mod. Phys. **25**, 844 (1953).

assuming the diameter $a \sim 2.10^{-8}$ cm. If a is larger ϑ' is smaller proportionally. So for ions of more than thermal energies, the quantum theory of the elastic sphere model indicates that "glancing" collisions occur very frequently, superimposed on an isotropic distribution; for heavy ions this is true at low energies also.

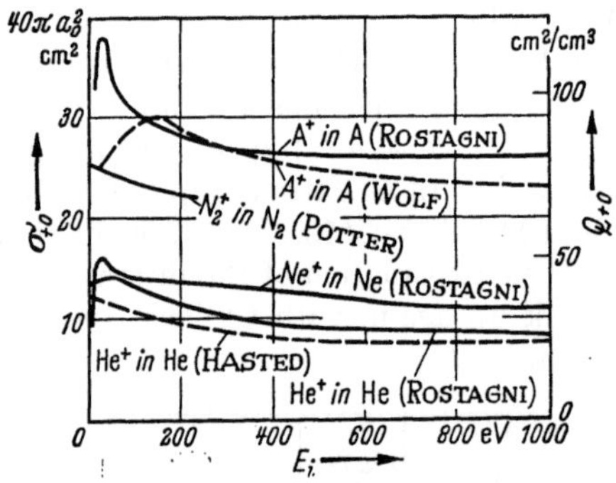

Fig. 17. Charge exchange neutralisation cross-section per atom σ_{+0} and per cm.³ at 1 mm. Hg (Q_{+0} for ions in their own gas measured by various authors.

If charge transfer is taken into account, then in some collisions the originally charged particle moves on almost undeflected as a neutral atom, and the atom smitten moves off almost at right angles as an ion of thermal energy. We may regard this as a collision in which an ion loses all its energy.

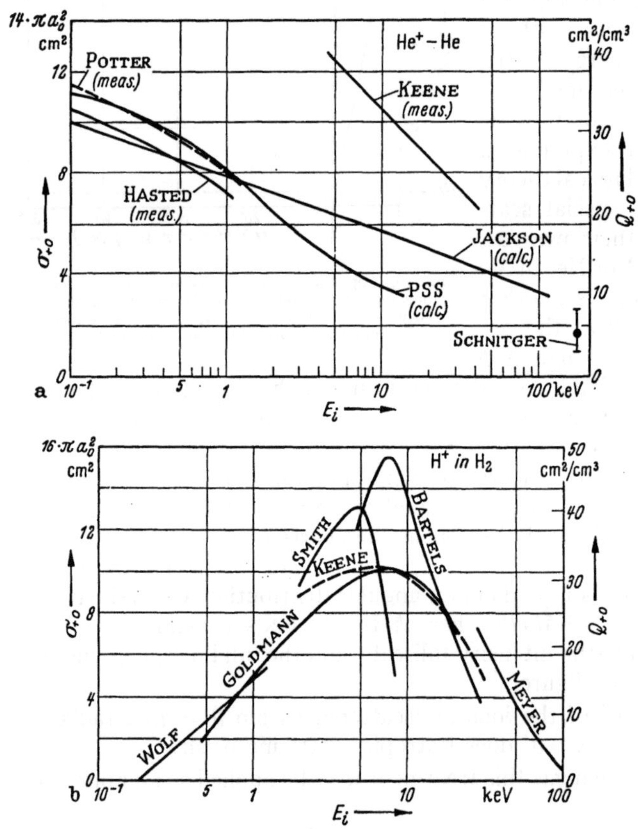

Fig. 18 a and b. Charge exchange neutralisation cross-section per atom (σ_{+0}) and per cm.³ at 1 mm. Hg Q_{+0} (a) for He^+- He: theories and experimental values, (b) for H^+–H_2 measured by various authors [15].

Fig. 17 shows the atomic cross section for neutralisation σ_{+0} in units of πa_0^2 (0.88×10^{-16} cm.², a_0 being the first BOHR orbit radius) and Q_{+0} the total effective charge transfer cross-section at 1 mm. Hg and 0° C [15]. Rare gases and N_2 are shown, and O_2 according to POTTER[1] lies between the two helium curves by ROSTAGNI and HASTED.

Fig. 18a shows the cross section for He^+ in He up to higher energies, and it is seen that various observers and the two theoretical calculations agree on the whole that this fall continues[2].

For ions moving in a different gas the picture is different. Fig. 18b shows that for H^+ in H_2 the cross-section is very small at zero energy and rises to a maximum at 5 to 10 keV [15]. This will always be true if the potential energy of the ion is different from the ionisation potential of the gas through which it moves, according to the theory.

The values for the total collision cross-section σ_i found by HORNBECK and WANNIER seem to be roughly five times larger than the charge transfer cross sections at low energy.

[1] R. F. POTTER: J. Chem. Phys. **22**, 974 (1954).
[2] J. D. JACKSON: Canad. J. Phys. **32**, 60 (1954).

δ) Volume ionisation by positive ions. This was the original proposal for the secondary effect in a gas, TOWNSEND's β-process. We shall make only a rough calculation, for it will appear that generally ionisation by positive ions may be neglected. TOWNSEND[1] calculated the number of free paths that exceed a certain length i.e. those in which the ion may gain more than a certain amount of energy. If we choose the minimum energy as $e V_i$, then in all free paths greater than λ where

$$2 V_i = X \lambda$$

in the gap of length d in which a uniform field X exists the ion has a chance of ionising (Sect. 35).

If we count only charge transfer collisions (those where the ion loses all its energy) then the number of free paths after which ionisation is possible is

$$c = (\sigma_{+0} N)(p\, d) \exp\left(\frac{2 V_i \sigma_{+0} N}{X/p}\right) \tag{19.1}$$

if the free paths are all in field direction and distributed at random in length. (19.1) is an overestimate, for we have neglected all other collisions than those where charge transfer occurs. If the total collision cross section σ_i were used, c would be much smaller, and would then probably be an underestimate.

Knowing then the number of collisions c in which ionisation is possible, and the value which β must have to maintain a discharge, the average probability of ionisation at a collision can be estimated.

For a 1 cm. gap in *air* at atmospheric pressure $V_s = 31600$ volts and $\beta = 7.1 \cdot 10^{-7}$, according to VARNEY, LOEB and HASELTINE[2]. These authors used an ion mean free path of $9 \cdot 10^{-6}$ cms. The value for $\sigma_{+0} N = Q_{+0}$ for nitrogen is roughly 80 cm.²/cm.³, giving a mean free path of $1.5 \cdot 10^{-5}$ cms. at 760 mm. Hg, so the value for c from (14.1) becomes $6 \cdot 10^{-21}$ if $V_i = 15$ volts. This would mean that at each collision of an ion of more than 30 eV energy some 10^{14} ion pairs must be formed to obtain $\beta = 7.1 \cdot 10^{7}$. This is of course impossible. VARNEY, LOEB and HASELTINE[2] obtained similar figures, leading to the same conclusion.

Similar results follow for rare gases: if we use the values for *neon* at low $p d$ so that X/p is large and conditions more favourable we have at $p d = 2$, $X/p = 130$ and $\gamma = 4 \cdot 10^{-3}$. Since $\alpha/p = 6 \cdot 10^{-4}$,

$$\beta = \alpha \gamma = 2.4 \cdot 10^{-6} p = 4.8 \cdot 10^{-6}/d.$$

With $V_i = 21.6$ volts and $Q_{+0} = 50$ cm.²/cm.³, (19.1) gives $c = 6.3 \cdot 10^{-6}$. Since the gap length is d, unit probability of ionisation at collision gives $\beta = 6.3 \cdot 10^{-6}/d$.

It is clear from Figs. 44 and 45, Sect. 35, that the probability is very small compared to unity, so that once more the β-process is unacceptable. It is, however, not so wildly inadequate as at low X/p in air at high pressure. Possibly at extremely low $p d$ the process may become important since X/p at breakdown increases rapidly as $p d$ falls: (19.1) is very sensitive to changes in X/p[3].

Note that Fig. 18b indicates small cross sections σ_{+0} for ions *not* in their own gas. If σ_i is also small, ions in gas mixtures may have higher average energies and higher β values will be found there.

---

[1] J. S. TOWNSEND: Phil. Mag. **28**, 111 (1939).

[2] R. N. VARNEY, L. B. LOEB and W. R. HASELTINE: Phil. Mag. **29**, 349 (1940). These authors also considered that the minimum ion energy for ionisation is much higher than $2 e V_i$ (see Sect. 35).

[3] Cf. H. FETZ: Brit. J. Appl. Phys. **6**, 288 (1955) and Sect. 18, footnotes.

It does not seem possible that β-processes can be important except perhaps in very low pressure breakdown where $pd<1$ and $X/p>1000$. Even then the ion contribution at the cathode may be more important still.

ε) Emission due to ions at the cathode. When the ions strike the cathode, secondary electrons are known to be emitted from direct study (Sect. 27). The best values for rare gas ions show the true value of β for singly charged ions to be of the order 10^{-1} electrons/ion on very clean surfaces and to be almost independent of kinetic energy (Fig. 26). The potential energy of the ion is very important in determining γ_i (Fig. 25a, b), and for multiply charged ions $\gamma_i>1$ (Fig. 27).

Gas layers may have the effect of lowering γ_i at very low energy (Fig. 28a). At higher energies γ_i always increased, and the yield is now almost proportional to the kinetic energy (Figs. 31a, c). This is always the surface conditions in any gas discharge.

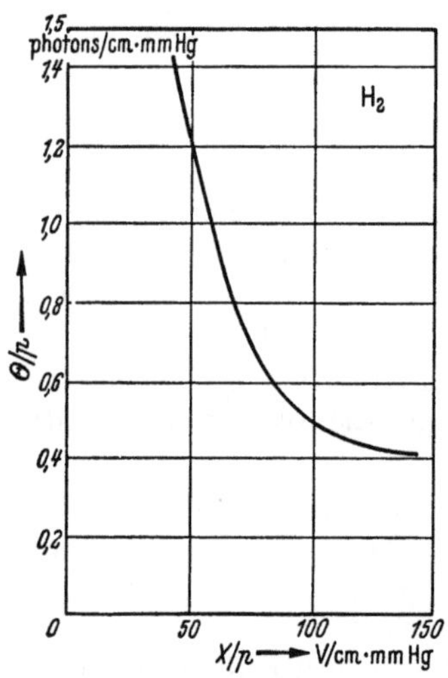

Fig. 19. Number of active photons ϑ produced per cm per electron in a uniform field in hydrogen: ϑ/p is shown as a function of X/p.

The direct measurement of γ_i for Hg has shown very low figures, and O_2 seems to have in general lower values than other molecular gases.

The gas discharge measurements have agreed with these general results very satisfactorily, although surface conditions cannot of course be controlled so well. The order of magnitude and the variation of γ_i with energy are as expected, so this mechanism is established as normally the most important one in the region of X/p where ions are formed. Mercury and hydrogen seem to be exceptions: photo effects or metastable atoms predominate in these gases.

In cylindrical geometry ions are less important, except apparently in very pure argon: no other rare gases have been investigated (Sect. 18).

20. Photons. *α) The production of photons.* Wiedemann[1] and later Thomson[2] and Dauvillier[3] showed that radiation in the Schumann ultra-violet range was emitted from *glow* discharges, and at low pressures ($p<1$ mm. Hg) soft X-rays appeared. The gas itself in the dark space emitted these penetrating radiations[4]. The photons from higher pressure discharge had then energies about the same as the ionisation energy of the gas (1000 Å ≡ 24.6 eV approx.).

Direct measurements of the photon production $\vartheta/p=f(X/p)$ for a Townsend discharge in hydrogen have been made by Geballe[5] (Fig. 19) who used a brass surface as a photoelectric detector of quanta at $p\sim1$ mm. Hg. The absorption coefficient of the effective radiation was ~0.5 cm.$^{-1}$ at 1 mm. Hg, and the relative yield is given in the figure. If $\gamma_p\sim0.1$, the ordinates must be divided by 10 to obtain absolute numbers of quanta/electron/cm.

There were some inconsistencies in this work, and the experimenter himself did not place reliance on the accuracy of the results. However, it is clear that ϑ/p is rising to more than 0.1 photon/cm./mm. Hg per electron at low X/p.

[1] E. Wiedemann: Z. Elektrochem. **2**, 159 (1895).
[2] J. J. Thomson: Proc. Cambridge Phil. Soc. **10**, 74 (1899). See also [3].
[3] A. Dauvillier: Phil. Mag. **7**, 1046 (1926).
[4] Air is transparent to some u.v. wavelengths. J. C. Boyce: Rev. Mod. Phys. **13**, 1 (1941).
[5] R. Geballe: Phys. Rev. **66**, 316 (1944).

No other direct quantitative work is available, but COSTA[1] has shown that radiation from a TOWNSEND discharge produces a photoelectric effect on a subsidiary silver cathode. He estimated the fraction of the total secondary electron emissions which were released by quanta. In hydrogen 50 to 100% was due to radiation, and the wavelength of most was < 1050 Å, for only 1% penetrated on LiF absorber. In air at 0.7 mm. Hg 50% of the secondaries were photoelectrons, but at 7 mm. Hg the proportion had fallen to 20%. This was probably due to absorption in the gas.

We do not know the mean value of γ'_p for the radiation concerned. COSTA gives graphs showing that $\gamma'_p \vartheta g/\alpha$ varies from 2 to $4 \cdot 10^{-3}$ in hydrogen as X/p increases from 100 to 400. For air the value rises from 2 to $2.5 \cdot 10^{-4}$ as X/p increases from 200 to 350.

He concluded that if $\gamma_p \sim 0.01$[2], then 1 effective photon per positive ion is produced in the discharge. This would mean $\vartheta_p \sim 0.1$ for the range of photons concerned, which agrees with GEBALLE'S figure: we do not know however the spectrum of the radiation which is being measured by the photo effect with two different cathode materials.

A little work has been done theoretically by using the excitation functions, the collision frequency and the electron energy distribution. This last is known for some molecular gases, over a certain range of X/p.

Calculations have been made by PENNING and KRUITHOF [*12*], which show qualitatively that the ratio ϑ/α of excitations to ionisations decreases as X/p increases. Extrapolation of known values was necessary, but results are in fair agreement with the measurements in hydrogen. It is difficult to make comparisons, as all photons are included in this theory.

No other work appears to have been done in this field.

β) The spectrum emitted by a TOWNSEND *discharge.* Little seems to be available on this subject, though the presence of ultraviolet radiation from such discharges has been established[3]. Future investigation should include a quantitative survey of the vacuum ultra violet, so that the number of photons capable of ionising in the volume or causing surface photoelectric effect can be estimated. An approximate knowledge of the number of resonance photons and their diffusion time of crossing the gas would also be of great use.

γ) The action of photons in the volume. The importance of this possibility was recognised early by KENTY, GRAVATH, DECHENE, GREINER, LOEB [*5*] and RAETHER[4] in sparks and GEIGER counters. Recent experiments (Sect. 18) show that with mixtures of gases photo ionisation is prominent, and a theoretical treatment of the plane parallel discharge shows some of the reasons why in pure gases this is not so.

DAVIDSON'S[5] treatment of volume photoionisation in a uniform field is as follows. In unit volume at x from the cathode let N/c photons of sufficient energy exist. A number

$$\varrho(x, \xi)\, d\Omega$$

move in directions occupying a solid angle $d\Omega$ between ξ and $\xi + d\xi$ with the positive x-direction.

[1] H. COSTA: Z. Physik **113**, 531 (1939); **116**, 508 (1940).
[2] C. KENTY: Phys. Rev. **44**, 891 (1933).
[3] See (Sect. 20α) W. SCHIECKER: Z. Physik **116**, 562 (1940) and footnote 3, p. 625.
[4] H. RAETHER: Z. Physik **110**, 611 (1938).
[5] J. DUTTON, S. C. HAYDON, F. LLEWELLYN JONES and P. M. DAVIDSON: Proc. Roy. Soc. Lond., Ser. A **218**, 206 (1953).

The rate of production of electrons is determined by TOWNSEND'S α and photo-ionisation, so

$$\frac{di^-}{dx} = \alpha i^- + \mu_i N \tag{20.1}$$

where $N\mu_i$ is the number of atoms ionised in unit distance by N photons.

The rate of production of energetic photons is subject to the condition that ϱ at given x is constant with time in the steady state. Photons are produced by direct excitation (ϑ photons/cm./electron) and lost either by ionising gas atoms or exciting them. The absorption coefficient μ_i for photoionisation has just been defined, and an analogous coefficient μ_e can be defined for excitation: a fraction μ_r/μ_e of the photons absorbed to excite atoms is re-emitted as resonance radiation. Then in the steady state,

$$c \cos\xi \frac{\partial \varrho}{\partial x} = \frac{\vartheta i^-}{4\pi} + \mu_r \frac{N}{4\pi} - (\mu_e + \mu_i)\, c\,\varrho . \tag{20.2}$$

Let

$$S = c \int \varrho \cos\xi \, d\Omega$$

and

$$\frac{1}{g} = \overline{\cos^2\xi} = \frac{c}{N} \int \varrho \cos^2\xi \, d\Omega .$$

Then from (20.2) by integration with respect to $d\Omega$

$$\frac{dS}{dx} = \vartheta i^- - \mu_d N . \tag{20.3}$$

Multiplying (20.2) by $\cos\xi$ and integrating

$$\frac{d}{dx}(N/g) = -\mu S \tag{20.4}$$

where

$$\mu = \mu_e + \mu_i$$

and

$$\mu_d = \mu - \mu_r .$$

It is evident that S represents the difference at the point x between the photon current towards the anode and the photon current towards the cathode, while g is the inverse of the R.M.S. value of $\cos\xi$ at x.

A general solution is obtained in the form

$$i = \mu_i \sum_{n=1}^{3} B_n e^{\alpha_n d}/(\alpha_n - \alpha) \tag{20.5}$$

where the α_n's are the roots of a cubic equation involving g, μ etc. which also fix the B_n's. The boundary conditions are given by the laws of reflection of the photons at the anode and cathode.

Under certain conditions, which must be worked out numerically for a given discharge, the coefficients B_n become infinite i.e. breakdown is possible. Several different combinations of values for μ_i μ_r etc. are worked out by DUTTON *et al.*, which would lead to breakdown for one particular value of d at the measured breakdown voltage. The theoretical curve for $i/i_0 = f(X/p)$ obtained with these values is in fair but not good agreement with the observed values.

The curve obtained by assuming that one of the "allowed" secondary processes acts, i.e. that (6.3) holds, is in excellent agreement with the experimental points. This is taken as a proof that even if the parameters of the discharge are adjusted so as to fit at breakdown, photo ionisation in the volume acting alone is not a satisfactory explanation of the multiplication occurring before breakdown.

Other serious objections to regarding this process as important in pure gases in plane parallel gaps may be raised. The value for μ_i has been measured (Sect. 36) and is two orders of magnitude larger than the values which must be *assumed* to explain the observations. This means that no breakdown is possible, since (20.5) becomes

$$i = i_0 \exp\left(\alpha + \frac{\mu_i}{\mu_r}\vartheta\right) \tag{20.6}$$

if $\mu \gg \alpha, \vartheta$. This is equivalent to an increase in α. Physically, a photon absorbed near its point of origin to form an ion-pair is equivalent to the direct creation of an ion pair.

Secondly, if μ_i were small, the emission due to photons reaching the cathode would contribute greatly as an efficient secondary process. For short gaps, this would be much more important than the volume ionisation.

Finally, the rate of production of the photons which can ionise the gas should be considered. They must have energies equal to or greater than the ionisation potential of the gas (Sect. 36) so direct excitation of the atom will not produce them, unless the electron energies are very high when an inner electron may be removed from the atom. Radiative recombination, excitation of ions or other two-stage processes would also give rise to energetic quanta, but very little radiation is likely in a TOWNSEND discharge from any of these causes.

In *gas mixtures*, radiation from one component may be capable of ionising the other, so this last objection does not hold. For air, however, the absorption of effective photons is still very great, so it still appears unlikely that photo ionisation is important in plane parallel gaps.

The experiments reported in Sect. 18 in cylindrical geometry show that even there where cathode effects are minimised no volume ionisation can be observed in pure gases. With gas mixtures however it is observed. This is because the very high field near the positive wire makes multiplication over small distances appreciable. A photon created at a point near the anode wire may move back only a short way before being absorbed by ionising an atom, but the photo electron created can multiply in that short distance.

Both theory and experiment agree then that in pure gases photo-ionisation is unlikely unless large current densities make two-stage processes common. In gas mixtures also it is not to be expected unless the cathode is remote and the field distribution non-uniform. In such instances as corona, GEIGER-MÜLLER counters or the long spark in air these conditions are fulfilled, so that here ionisation in the volume by light is all-important[1].

Now that measurements of the relevant cross-sections for absorption are available, it is to be hoped that experiment and theory will be developed further in non-uniform field conditions. Much work remains to be done.

δ) The surface photo-electric effect. Direct measurements of the yield (Sect. 32) show that γ_p may rise to more than 0.1 electrons/photons for radiation of wavelength 1000 to 500 Å. Different metals show roughly the same characteristics, but the point at which the $\gamma_p = f(\lambda)$ curve rises does depend on the metal somewhat.

---

[1] See appropriate articles in this encyclopedia, and L. B. LOEB: Brit. J. Appl. Phys. **3**, 341 (1952). — H. RAETHER: Z. angew. Phys. **7**, 50 (1955).

Experiments quoted in Sect. 20α and 20β above have shown that radiation from a discharge can cause photoemission. We can also see in what discharges this effect is pre-eminent.

The results of VON GUGELBERG, SCHADE (Sect. 13) and COSTA (Sect. 20α) agree that in *hydrogen* with Ag and Ni cathodes energetic non-resonance photons are the chief agents in the secondary process. LLEWELLYN JONES (Sect. 12) concluded that for H_2-Cu, when the cathode is covered with a thin film, low energy photons are important: with a clean cathode this is not so. Ions and high energy photons could not be distinguished in his work, so these results all agree that at low pressures energetic quanta control the hydrogen discharge.

In *nitrogen* and *light rare gases* quanta are not at all important (Table 4, Sect. 13) in low pressure breakdown.

In the *heavier rare gases* they contribute more, but always less than 20% of the secondary electrons. The increasing contribution in A, Kr, Xe agrees with the conclusion of Sect. 9 concerning the electron energy distributions. COSTA attributed up to 50% of the secondary emission to photons in air, but metastable atoms could not be isolated from the radiation in his apparatur.

At higher pressures little work on rare gases has been done, but from the measurements in air (Sect. 13) it is clear that these photons are important.

We expect photon action in other gases to be more marked at high pressures, since the value of X/p at breakdown falls as p increases.

In cylindrical geometry the photo effect at the cathode has been demonstrated at high and low pressure in pure gases, and in hydrogen a strong photoelectric effect again appears (Sect. 18). Since low fields prevail over most of the volume this is understandable also.

In general, there is good agreement between various observers about the role of photons under various conditions. Detailed calculation of the rate of production of quanta is still not available, however, and the importance of "delayed" resonance photons has not been estimated directly. However, the influence of radiation in discharges is now recognised to be very great.

21. Metastable atoms. Again the main hindrance to quantitative work is the uncertainty in the rate of production of metastables in the discharge.

α) The production of metastable states. DORRESTEIN[1] has used the secondary emission due to metastable atoms of He and Ne to measure the excitation functions for the metastable states. No values for ϑ_m/p measured (as ϑ/p has been measured) in a discharge seem to be available, apart from those estimated in transients (Table 6a).

DORRESTEIN's results for He showed that electrons of 22 eV energy produced metastable atoms and quanta at such a rate that four times as many electrons were emitted from a Pt electrode struck by metastable atoms as were released by photoelectric effects. At 30 eV the ratio rose to about five and the metastable current was at a peak here. It then fell almost to unity at about 100 eV. At this point the secondary electron current was about twice its value at 22 eV.

For Ne at 20 eV, the corresponding ratio was three, but while the metastable contribution rose to a peak at 30 eV and then fell at much the same rate as in He, the photoemission rose much more rapidly. It equalled the metastable secondary emission at 55 eV, and was twice as large at 100 eV, equal to the peak metastable yield.

[1] R. DORRESTEIN: Physica, Haag 9, 433 (1942).

This shows qualitatively what we may expect in a TOWNSEND discharge, but much hypothesis as to energy distribution is needed to obtain quantitative values for the production rates.

β) Ionisation in the volume. Metastable atoms will be destroyed in the volume by collision in pure gases without producing ionisation[1]. The excited atoms formed will radiate and contribute to the cathode emission (see MOLNAR's work) but no photoionisation in the gas will take place. With a mixture of gases, however, metastable atoms may be able to ionise directly and, when raised to a radiating state, by a photoionisation process.

We shall not deal with this latter complication, but consider the volume ionisation by collision. In a given element dx at x from the cathode, the condition that the number N of metastables per cc. be constant is

$$C\frac{d^2N}{dx^2} - \frac{N}{\tau} + n_e\vartheta = 0 \tag{21.1}$$

where C is the diffusion coefficient of the metastable atoms considered, τ is their half-life under the given conditions (gas mixture and pressure) and n_e is the number of electrons crossing the plane at x in unit time.

There is also a condition that the electron density in dx should not change

$$\frac{dn_e}{dx} - \frac{fN}{\tau} - n_e\alpha = 0 \tag{21.2}$$

where f is the fraction of metastables which are destroyed by ionising, and α is TOWNSEND's coefficient.

Then, from these two equations we find

$$\left[D^3 - \alpha D^2 - \frac{1}{\tau C}\{D + \alpha + \vartheta f\}\right] n_e = 0 \tag{21.3}$$

where $D = d/dx$. The solution of this depends on the solution of the cubic equation for D when the quantity in square brackets is made zero, and may have terms:

$$A_1 e^{\lambda x}, \quad (A_1 + A_2 x)\, e^{\lambda x}, \quad (A_1 + A_2 x + A_3 x^2)\, e^{\lambda x}$$

in the expression for $n_e = f(x)$, where λ is positive, negative, imaginary or complex. This type of equation cannot lead to breakdown, as n is always finite at finite x, unless the coefficients become infinite. With this proviso metastable atoms acting in the volume can only increase α, say to α', but do not act as a normal secondary process.

This result follows if one neglects the diffusion of the metastables of course, but it is strange that diffusion leaves the result unchanged. It might be expected that a particular choice of τ and C would lead to breakdown, as a particular choice of μ_i and μ does for photoionisation. There however breakdown occurs because the coefficients become infinite, and if this should happen here the action of metastables in the volume will also lead to breakdown.

For an electron avalanche without secondary effects, ROGOWSKI[2] has calculated the metastable distribution, neglecting ionisation by collision. Fig. 20 shows the result if $\vartheta_m = \alpha : n$ is the linear density of metastable atoms in a gap of length d such that $\alpha d = 3$ and

$$\eta = \frac{1}{\sqrt{\tau C}}\,.$$

[1] See, however, Sect. 22 for cumulative processes at higher current densities.

[2] W. ROGOWSKI: Z. Physik **115**, 257 (1940).

When ηd is very large the density is determined mainly by the production rate, for the metastables decay where they are produced effectively. When ηd is zero, the density distribution is determined by the rate of destruction at the anode and cathode because the volume decay is infinitely slow now. Hence a maximum appears at the centre.

This graph shows that when volume ionisation is strong (ηd large) the distribution of the metastables is such that they merely increase α. The only way in which the maximum density can be made to approach the cathode is by the assumption that the metastables decay slowly i.e. that there is no appreciable volume ionisation. This is the physical explanation of the mathematical result (13.3).

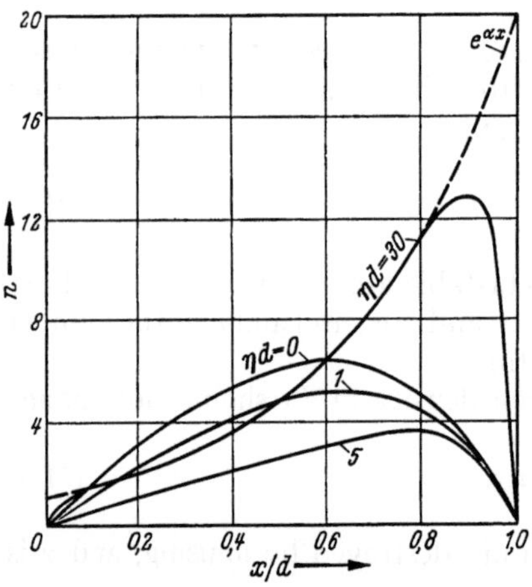

Fig. 20. Linear density distribution of metastables n per cm in a gasmultiplied electron current in uniform field for various values of ηd (see text).

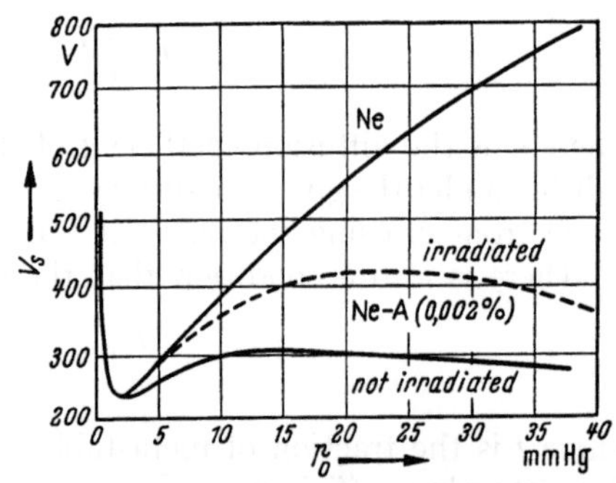

Fig. 21. Influence of light from a neon discharge on the breakdown voltage V_s as a function of p_0 (p at 0° C) of pure neon and neon-argon mixture.

The efficiency of ionisation of impurities by metastable atoms for rare gases has been estimated (Sect. 37) and proves to be between 0.1 and 1 for mixtures where the ionisation potential of one gas is equal to, or just lower than, a metastable level of the other.

PENNING[1] has shown the effect by irradiating pure Ne and Ne + A with neon light. Fig. 21 shows that in the first the breakdown potential V_s is unchanged. In the mixture, V_s is increased towards V_s for pure Ne. If all the metastables could have been destroyed before they collided inelastically with A atoms V_s should have been the same in the mixture as in Ne.

The importance of volume ionisation in coaxial discharges has also been demonstrated by PENNING [*10*] for rare gas mixtures.

γ) Surface emission due to metastable atoms. The measurements quoted in Sect. 33 show that metastable atoms can be more efficient than any other agency in releasing electrons from metal surfaces, unless ions or atoms of large kinetic energy are included. For the photons and low energy ions in a glow discharge $\gamma_i \sim 0.1$, but γ_m may be between 0.1 and 1. ENGSTROM and HUXFORD estimated $\gamma_m \sim 0.4$ for argon on silver (Sect. 14); the surface was activated there.

The emphasis on metastable action in rare gases (Sect. 14 to 16) is therefore readily understood. It appears that in *argon* at low $X/p \sim 60\%$ of the emission from activated Ba or Ag cathodes is due to metastable action, but lower values are found with unactivated cathodes, where γ_m is smaller (Sect. 16, Table 6b).

In gases where metastable states exist, then, we anticipate that in breakdown at high pd metastable action will be a main source of secondary electrons. This should be betrayed by a variation of γ with d.

[1] F. M. PENNING: Physica, Haag **12**, 65 (1932).

In coaxial discharges with a positive wire the γ_m-process should be prominent at low pressure and be less important at high pressure [*10*]. No wall effects should be noticeable with a negative wire, for electrons released from the cylindrical anode will immediately return to it.

22. Other agents. Neutral atoms accompanying the ions act as they do if energetic: generally they have low energies and are unimportant. At extremely low pressures soft X-rays will be emitted from the anode as in ionisation gauges[1]. These may be important also in cylindrical discharges with positive wires, but have never been discussed. Some volume photoionisation might be due to this process. Field emission has never been shown to act in the discharges so far considered[2]. Thermionic emission and thermal ionisation would not be anticipated in these low current discharges[3].

g) Conclusion.

23. Comments on the experiments. Here we list some points which arise from the work reported.

α) Purity of gases. The influence of impurities is seen most widely with mercury. This affects cathode and gas. Even if the cathode is unchanged, the ionisation coefficient may be different due to photoionisation or collisions involving metastable atoms.

β) Cleanliness of cathode. This can never be free from gas in discharge work. It should be cleaned by sputtering in heavy discharge or heating to 1400° C or above. The former method is usually easier: only non-reactive gases may be used.

γ) Values of α/p. Preferably α/p should be measured in the apparatus. If not, the conditions of experiment must be identical with those obtaining when the α/p values used were measured.

δ) Multiplication in time-lag studies. When current transients of very short time are studied, the multiplication may not be described by the coefficient α. It should be established that the electrons are in equilibrium with the field; allowance must be made for the time to reach equilibrium.

Apart from this, it is possible that α/p measured in steady discharges contains other ionisation processes than direct ionisation by electrons (i.e. collisions of the second kind leading to ionisation). It is desirable that the actual multiplication be measured in the transient.

ε) Imprisonment of radiation. When analysing the growth of current, it must be remembered that photons do not always move through the gas with the speed of light. Quanta emitted by an atom may be delayed by being absorbed and re-emitted by others of the same kind. For resonance quanta the effect is very great, but it may appear for other excited states to a lesser extent.

ζ) Variation of γ with the length of the gap. If photons or metastables are contributing to γ then the fraction of these which reach the cathode depends on the point of origin and hence on the anode-cathode distance d: γ decreases as d increases especially for metastable atoms.

η) Effect of space charge distortion. This has been widely recognised as a possible service of error. It is well to keep the initial current as low as convenient, and test the results by varying i_0 (by an order of magnitude if possible) after confirming

[1] D. ALPERT: J. Appl. Phys. **24**, 860 (1953).

[2] See Sect. 26.

[3] Delayed electron emission: H. NASSENSTEIN: Z. Naturforsch. **10**a, 944 (1955). — J. LOHFT and H. RAETHER: Z. Physik **142**, 310 (1955). — K. SEEGER- Naturwiss. **42**, 66 (1955).

that the succeeding avalanches can influence one another. Alternatively and perhaps better a theoretical analysis may be made.

ϑ) Rate of production of excited and metastable states. This is one of the most uncertain factors in the analysis of γ. Much work needs to be done in this field.

ι) The test of theories of breakdown. It is apparent that the most sensitive and detailed test of a theory may be made by measuring the current flowing in a discharge gap as a function of time. Measurements of potential are not so direct or satisfactory.

24. Future work. Some suggestions may be made on the basis of this survey.

1. The values of $\alpha/p = f(X/p)$ over a wide range of X/p in *clean* air are not known. This is of great practical importance and is needed in the measurement of γ also.
2. The electron multiplication in non-uniform fields and when the electrons are not in equilibrium with the field is known only to a small extent. The results would be valuable in theoretical work on high pressure breakdown, corona discharges and the low pressure glow discharge.
3. The rate of production of photons and metastable atoms in the discharge should be measured. The photons are specially important in hydrogen, and metastable atoms in rare gases.
4. Calculation of the ion energy distribution in uniform fields should be made, so that γ_i may be estimated more accurately.
5. The volume ionisation due to photons in pure gases and gas mixtures, particularly air, is of theoretical and practical value. The results already available have made an important contribution to the theory of secondary processes.
6. The study of current transients in argon should be extended to other gases so that a comparable analysis of γ may be made.

II. Other discharges.

25. The dark, the glow, and the arc discharge. These discharges show the influence of increasing current densities to a greater extent as we move from the TOWNSEND discharge to the arc. The first effect is the distortion of the field, seen in the dark discharge, as the space charge density increases. The effect is shown clearly in the reduction of V_s as i_0 is increased. The ratio $\Delta V_s/V_s$ is proportional to $\sqrt{i_0}$, and theory[1] shows that this is due to the increasep space charge. In the glow this becomes the over-riding factor in determining the field distribution, and cumulative processes appear. In arcs we have two types: in one the high current densities cause *thermionic emission* from the cathode, and in the other, the *low melting point metal* arc, there is no general agreement as to the mechanism.

Owing to the large fields at the cathode surface, back diffusion will be of minor importance unless the pressure is very large; we may write $\gamma_i' \equiv \gamma_i$ etc. in these discharges.

An interesting form of glow discharge is the "spray" discharge[2]. A cathode of carbon or aluminium covered by an insulating layer or powder e.g. Al_2O_3 can maintain a discharge at $V_c < 40$ volts, with current densities j thousands of times larger than in the glow discharge.

[1] W. ROGOWSKI and W. FUCKS: Arch. Elektrotechn. **29**, 362 (1935). — W. FUCKS and G. SCHUMACKER: Z. Physik **112**, 605 (1939). — J. M. MEEK: J. Inst. Electr. Engrs. **89**, 335 (1942).

[2] A. GÜNTHERSCHULZE and H. FRICKE: Z. Physik **86**, 451, 821 (1933). — F. HAWORTH: Phys. Rev. **80**, 223 (1950).

It was found that no cathode dark space appeared, and that the current density may be written

$$j = A V_c^2 e^{-b/V_c} \tag{25.1}$$

the form of the well-known equation for field emission. The high fields required for this process are due to the presence of positive ions on the insulating dust, causing emission from the underlying cathode.

A continuous transition from a glow to a spray discharge may be traced by varying the thickness of the insulating layer or the current density j[1]. The mechanism gradually alters from multiplication in a gas becoming pressure independent at large j. This indicates some form of multiplication in the insulator. The applied field strengths, estimated by assuming the whole discharge voltage to appear across the insulator whose thickness d is known, vary from $5 \cdot 10^6$ to $5 \cdot 10^5$ V/cm. as the thickness increases from 4 to 25 μ. If dust is used the grain diameter is used for d.

Although the applied field is low for a field emission process this seems from (25.1) to be the controlling feature. The ions which are certainly present on the insulator will increase the field (Fig. 42b), and it seems that the ion surface density is controlled by the applied field.

The emission from insulating layers is important in low pressure breakdown[2].

An example of the effect of increasing current densities is found in the smooth transition from the abnormal glow to a *thermionic* arc as shown in Fig. 1; a small spherical cathode on a wire is used to obtain a fairly large, uniformly hot emitting surface. The transition may be accounted for theoretically by including the heating effect of the ions at the cathode in γ (Sect. 5). References [*4*], [*6*], [*10*] give the treatment as follows, due to VON ENGEL and STEENBECK[3].

Let i, i^-, i^+, i_t be the current, electron current, positive ion current and thermionic current at the cathode when its temperature is T. Then

and
$$\left.\begin{aligned} i^- &= \gamma_i' \, i^+ + i_t \\ i_t &= f \cdot A T^2 e^{-B/T} \end{aligned}\right\} \tag{25.2}$$

where f is the cathode area emitting electrons and A, B are RICHARDSON'S constants. Now

$$\left.\begin{aligned} \gamma &= \frac{i^-}{i^+} = \frac{i - i^+}{i^+}, \\ \gamma &= \gamma_i' + \frac{1 + \gamma_i'}{\frac{i}{i_t} - 1}. \end{aligned}\right\} \tag{25.3}$$

Then, if we assume that all the cathode energy loss is lost by radiation, and that the whole energy of the discharge is dissipated at the cathode,

$$i V_c = f \sigma T^4 \tag{25.4}$$

where σ is STEFAN'S constant.

Fig. 22 compares the results of (25.3) and (25.4) with the observations of WEHRLI. The agreement is very good in view of the approximations in the theory, and calculations on the dark space agree also.

---

[1] G. DASCOLA: Nuovo Cim. **4**, 63 (1947).

[2] See Sect. 26.

[3] See however W. S. BOYLE and F. E. HAWORTH: Phys. Rev. **101**, 935 (1956).

The current densities are of the order 10^2 to 10^3 A/cm.^{2}, and the electric field at the cathode 10^4 to 10^6 V/cm. determined by the positive ion space charge in the very short dark space.

From Fig. 43 it is clear that field emission could never provide the observed current densities. There will be a contribution due to γ_i, but as $i^- > i^+$ and $\gamma_i \sim 0.1$ at the most this is again not the main contribution. It is very possible that the photoelectric emission at an arc cathode is important, but no work has been done on this problem.

In view of the success of the theory of thermionic emission due to heating by positive ions, it is unlikely that any other mechanism is as important as this. Much work has been done on the measurement of the cathode temperature T, and T is sufficiently high to account for the observed emission [*4*], [*6*], [*10*].

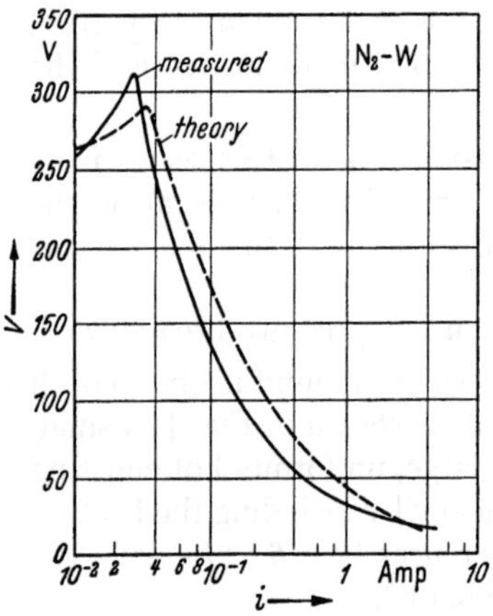

Fig. 22. WEHRLI's experimental values for discharge voltage as a function of current for tungsten electrodes in nitrogen compared with theory.

The arc on metals of low boiling point such as mercury or copper does not display a smooth transition region as thermionic arcs do. They cannot be accounted for satisfactorily by thermionic emission[1] since the metal boils far below that temperature at which emission is likely. The current densities in such arcs are of the order[2] of 10^6 A/cm.^{2}, their cathode falls may be lower than the ionisation potential of the cathode vapour[3], and it is known that such an arc may be extinguished if the potential is lowered for 10^{-8} secs.

ROBSON and VON ENGEL[4] recently proposed that the intense currents and the other main characteristics briefly listed above may be accounted for if the emission is due to excited atoms returned to the cathode by back scattering in the dense vapour immediately above the spot. In the short distance such atoms travel few will decay by radiating, so that they act as metastable atoms which are very efficient in releasing electrons.

The emitted electrons gain excitation energy only in the low cathode fall; ionisation occurs by collision between excited atoms.

Calculations of the rate of evaporation from the spot indicate that a sufficiently dense atmosphere may be maintained above the spot to make this process possible.

This seems to be the only mechanism for this type of arc which can account for all its properties in a simple manner. Field emission[5] could hardly supply the enormous current densities, except perhaps in very short arcs[6], and emission by positive ions would only be efficient enough at larger cathode falls of potential, to give two examples. These other effects may contribute in the high cathode field, but only as subsidiary effects. However, the matter is not yet settled.

In the *positive column* the discharge is maintained without satisfying an equation of the steady state such as (7.1). Electron multiplication must only be sufficient to replace losses, and at low currents single-stage ionisation described

---

[1] G. MIERDEL: Z. techn. Phys. **17**, 452 (1936). — C. G. SMITH: Phys. Rev. **62**, 48 (1942).
[2] K. D. FROOME: Proc. Phys. Soc. Lond., Ser. B **62**, 805 (1949).
[3] K. T. COMPTON and E. S. LAMAR: Phys. Rev. **37**, 1069 (1931).
[4] A. E. ROBSON and A. VON ENGEL: Nature, Lond. **175**, 646 (1955).
[5] C. G. SMITH: Phys. Rev. **69**, 96 (1946).
[6] P. KISLIUK: J. Appl. Phys. **25**, 897 (1954).

by TOWNSEND'S α occurs. Cumulative ionisation

$$A + e + KE \rightarrow A^* + e$$
$$A^* + e + KE \rightarrow A^+ + 2e$$

collisions of the second kind and ionisation of the type described by HORNBECK and MOLNAR[1] appear at higher current densities, as in the negative glow.

The effect of metastable atoms in the positive column of pure rare gases has recently been shown by MEISSNER and MILLER[2] who irradiated a low pressure positive column with radiation from the same gas. Fig. 23 shows the results for A and Kr: similar effects were observed in the other rare gases. In He it was shown that one particular infra-red line 20518 Å was effective in increasing the working voltage of the discharge. This is absorbed by the metastable $2\,^1S_0$ state and raises it to the $2\,^1P_1$ level; most atoms in this state return to the ground state with emission of 584 Å. Even if some fall back to the metastable state $2\,^1P_1$ the population of that level is diminished by the absorption of the 20518 Å line.

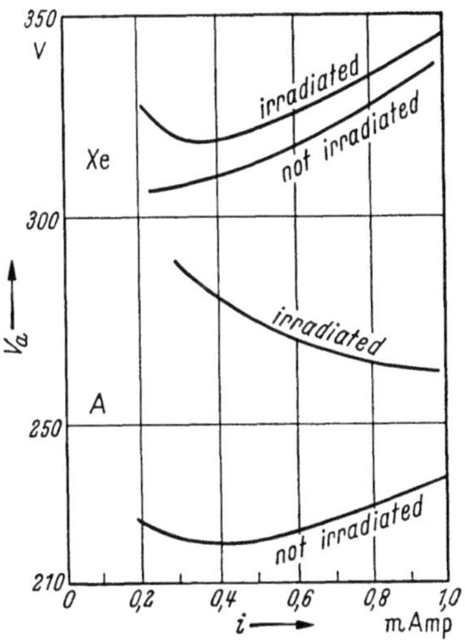

Fig. 23. Influence of irradiation from rare gas discharges on the working voltage V_a as a function of the current i for a discharge in the same pure gas.

In a pure gas the ionisation produced by metastables is due to collisions with other metastable and excited atoms or a cumulative ionisation process[3]. It is remarkable that this should be a strong contribution in the positive column even at low currents. De-excitation without ionisation would be expected to be more important (see Sect. 37). It was shown that the cathode regions were not affected greatly by irradiation, so that metastable atoms were not prominent in the maintenance of the discharge.

26. Low pressure breakdown. There is general agreement that field emission is important in such discharges, but QUARLES [6] and others found very low values for the field strength at the cathode due to the applied field. Fig. 24 shows his results for a mercury cathode and a spherical molybdenum anode in vacuum: he varied the work function by distillation.

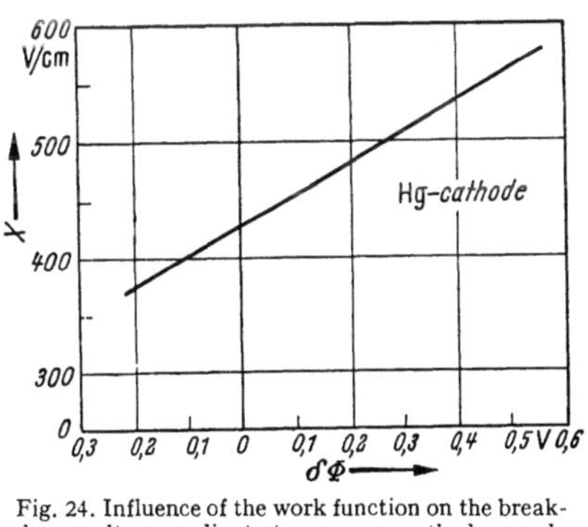

Fig. 24. Influence of the work function on the breakdown voltage gradient at a mercury cathode: anode is a molybdenum sphere.

The linear dependence on φ is stronger than the FOWLER-NORDHEIM theory gives, and X is very low for field emission currents. This suggests that a secondary mechanism is aiding the field emission, which is a *primary* source of ionisation directly due to the applied field.

Other evidence against pure field emission has been brought by TRUMP *et al.* [6]: X_s falls as the distance anode-cathode increases, and V_s rises[4]. However,

[1] J. A. HORNBECK and J. P. MOLNAR: Phys. Rev. **84**, 621 (1951).
[2] K. W. MEISSNER and W. F. MILLER: Phys. Rev. **92**, 896 (1953).
[3] Cf. G. FRANCIS and A. v. ENGEL: Z. Physik **145**, 560 (1956).
[4] J. G. TRUMP and R. J. VAN DER GRAAF: J. Appl. Phys. **18**, 327 (1947); see however W. S. BOYLE and P. KISLIUK: Phys. Rev. **97**, 255 (1955). — R. R. NEWTON: Phys. Rev. **73**, 1122 (1948).

WARMOLTZ[1] has discussed the effects of deformation of the cathode for mercury, and it is known that projections on the cathode (by increasing the field strength locally) reduce the breakdown potential [6]. If such projections become hot and evaporate, breakdown occurs in the vapour[2].

The influence of the anode cleanliness was shown by WARMOLTZ. It may be that an interchange of particles between anode and cathode contributes to breakdown[3] but the production of X ray quanta at the anode may be important. It has never been discussed.

Another factor to be considered is the presence of films on the cathode. Chemical compounds of the cathode material which are electrical insulators have been shown to be important in some instances[4], in the same way as insulating dust on the cathode of a glow discharge (Sect. 25 and 34).

III. Direct measurement of ionisation.

Some of the results we shall discuss here are reviewed in articles devoted exclusively to an individual interaction. These are only briefly mentioned, so that the values may be compared easily with the measurements in gas discharges.

We shall not discuss the behaviour of insulators, since it is important, normally, only in high frequency discharges or discharges in voids within dielectrics. The secondary effect then most prominent is often emission due to electrons, discussed in R. KOLLATH'S contribution in this volume.

a) Secondary electron emission from metals by positive ions.

The ejection of electrons from a metal by photons is related to the secondary emission due to fast electrons and ions. For the perturbing field of a moving charged particle may be represented by a pulse of electromagnetic radiation, and the emission due to a charged particle penetrating a metal is equivalent to the photoelectric emission due to this radiation[5].

The fact that an electron has only a small mass makes this analogy very good for electrons and photons, though it has not been developed in detail[6]. The kinetic energy transferred does not cause much secondary emission. In the case of an ion, however, much kinetic energy is delivered to the metal, and this alone can cause emission of secondary electrons. In addition, the ion carries potential energy within itself, and thus ions can eject electrons at zero speed, when electrons cannot. These factors explain the differences observed between the behaviour of ions and electrons, but there are still many resemblances (Sect. 27β and 30β). The effect of the transfer of potential energy is also seen in the behaviour of metastable atoms, which have no charge and yet are efficient at low speeds. Neutral unexcited atoms cannot cause secondary emission until they possess high kinetic energy, when they resemble fast ions in their effect.

Since the curve $\gamma_p = f(\lambda)$ shows a sharp rise (Sect. 32) and a slower fall as the wavelength λ decreases, the secondary emission of electrons by electrons (γ_e) would be expected to show a peak and a slower fall as the energy of the primary electron is increased. This is observed.

---

[1] N. WARMOLTZ: Philips Techn. Rev. **2**, 426 (1947).

[2] W. W. DOLAN, W. P. DYKE and J. K. TROLAN: Phys. Rev. **91**, 1043 (1953).

[3] See footnote 4, p. 637.

[4] C. G. MORGAN and D. HARCOMBE: Proc. Phys. Soc. Lond. B **66**, 665 (1953), and references.

[5] E. FERMI: Z. Physik **29**, 315 (1924). — E. J. WILLIAMS: Kgl. danske. Vid. Selsk., mat. fys. Medd. **13**, No. 4 (1935).

[6] Cf. H. FRÖHLICH: Proc. Phys. Soc. Lond. B **68**, 657 (1955).

It is to be expected that ions should show a similar peak in the curve $\gamma_i = f(E_i)$ at about the same ion velocity as the electron velocity at which γ_e was greatest. The potential energy and mass of the ion will cause γ_i to be greater than γ_e, and possibly the peak in γ_i will be shifted along the velocity scale relative to γ_e. We shall now see how far these expectations are justified.

Enough has been said already about the capricious nature of emission from contaminated surfaces. It is only recently that ultra-high vacua[1] ($< 10^{-9}$ mm. Hg) have been attained[2] and few experimenters have been able to avail themselves of these techniques by which we may obtain clean gas-free surfaces and observe the effects of gas layers. Knowing what must be done to achieve pure surface conditions it is possible to assess the surface state in previous work. The apparatus used by various workers is in principle simple. An ion source, usually a low pressure arc, produces ions which are accelerated to known energies. The charged particles are deflected in some way so that neutral atoms produced by charge exchange do not reach the target: often this deflection is made to provide beams homogeneous in energy. The target is capable of being degassed by heating, and surrounded by another electrode if retarding potential curves are to be taken of the secondary electron distribution. Reflected positive ions can be so measured also.

Early work was sometimes carried out with inhomogeneous beams of canal rays, and KUNSMAN sources have been used for alkali metal ions. An excellent source of ions for this work is the type where an electron stream accelerated to a known low voltage produces the ionisation in a low pressure gas, for the nature of the ions formed is more controllable[3].

We shall not describe individual apparatus in detail. The experiments discussed here were conducted with fairly clean surfaces in pressures of 10^{-5} to 10^{-6} mm. Hg, i.e. gas covered surfaces. Some workers attempted to keep the surface gas-free by heating (see later).

The very early work is described by RÜCHARDT and GEIGER [2] and some reviews [3], [9], [15] give the development of the subject.

We shall describe here some typical results first for the total emission γ_i, and the reflection of positive ions, then the work on energy distribution of secondary electrons and the theoretical ideas.

27. Measurement of the total emission. α) *Experiments on clean, gas-free surfaces.* The most complete investigation is due to HAGSTRUM who worked at pressures of 10^{-10} mm. Hg. Such extremes are necessary if monolayers of gas are to be avoided [18], for at 10^{-6} secs. a monolayer of N_2 forms on tungsten in approximately 1 second[4]. HAGSTRUM tested the state of the target after long outgassing at 250° C with metal parts at 1500° C by the rise in pressure occurring when it was heated. This varied with the length of time for which the target had been cold, and the pressure of the gas, but reached a constant level if a critical time t_c was exceeded. The value of t_c increased as the background pressure fell, and it was taken as the time for formation of an absorbed layer of gas: $t_c = 1.5$ hrs. at $p \sim 10^{-9}$ mm. Hg. The early work was done in this condition, but later results were obtained with $t_c \sim 40$ hrs. and $p \sim 10^{-11}$. Observations taken within a few minutes of the "flash" are on surfaces covered to only a few percent with gas atoms.

[1] Cf. D. ALPERT's article in vol. XIII of this Encyclopedia.

[2] D. ALPERT: J. Appl. Phys. **24**, 860 (1953).

[3] A full account of ion sources is given by D. KAMKE in vol. XXXIII of this Encyclopedia.

[4] J. A. BECKER and C. D. HARTMAN: J. Phys. Chem. **57**, 153 (1953).

The beam of ions was homogeneous to within one volt giving ions from 10 to 1000 eV: the apparatus has been described in detail [*18*]. Results for helium ions on molybdenum are shown in Fig. 25a. It is apparent that surfaces covered with approximately a monolayer of gas show more variation with energy in the range concerned. For He^{++} the curve is higher for the gas-covered surface, for He^{+} the curves cross at about 500 eV [*18*].

The potential energy, not the kinetic energy is obviously the most important variable. Doubling the potential energy (eV_i for $He^{++} = 79.0$, eV_i for $He^{+} = 24.6$) increases the value of γ_i by a factor three, at low kinetic energy. Doubling the kinetic energy is ineffective in raising γ_i. The energy after neutralisation is roughly proportional to γ_i.

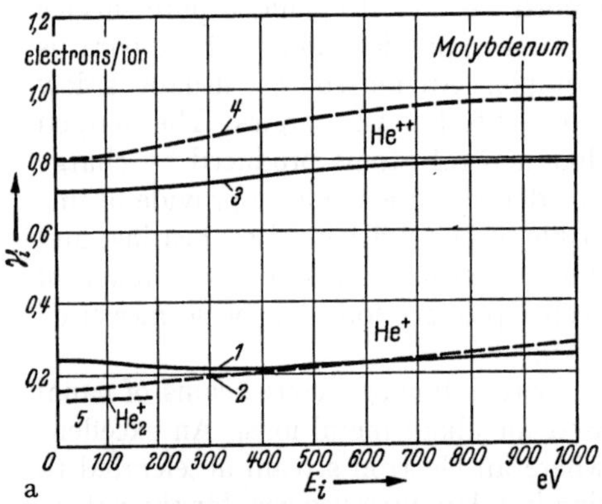

a

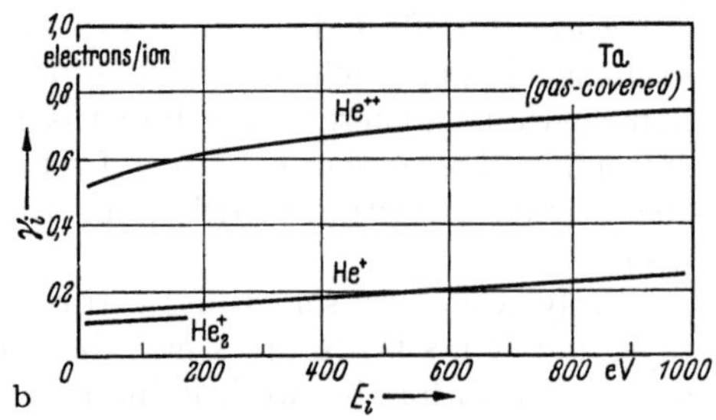

b

Fig. 25 a and b. The total electron yield γ_i as a function of ion energy for E_i for helium ions on clean molybdenum and tantalum, gas-free and gas-covered. Curves 1 and 3 in Fig. 25a atomically dean; 2, 4 and 5 covered with gas monolayer.

The reflection coefficient R was very small, being 5% at 1000 eV and decreasing at smaller energies.

The electron energy distributions were of the form shown in Fig. 37 and discussed there.

Fig. 25b shows the results with tantalum which apparently could *never* be freed from a gas film. Chemical reactions with gases and tantalum confirmed this, and the work function φ seemed to be about 0.8 eV greater on the cold surface, indicating the presence of a film. Hot tantalum (1750° K) apparently dissolved or absorbed the surface layer, which reappeared on cooling.

The results of Fig. 25b are to be compared with those for gas covered molybdenum in Fig. 25a. The value of γ_i for He_2^+ and He^+ is somewhat lower for tantalum, but for He^{++} the reduction is considerable, about 30%. Again, γ_i is roughly proportional to the energy available after neutralisation.

The reflection coefficient R was 20% at 1000 eV in this experiment, being smaller at smaller energies as before.

Electron energy distributions were similar to those for molybdenum, if the change in work function was 0.8 eV: otherwise the energy distribution shows no electrons of less than 0.8 eV, which is in disagreement with the results for molybdenum. This is the basis for the estimate of the change in φ, for theory does not suggest a lower limit (Sect. 30) in the electron energy distribution.

Similar curves for various rare gas ions are shown in Fig. 26 for a clean tungsten surface. No such effects as tantalum showed were observed. The curves for A^+, Kr^+ and Xe^+ were later found to be too high by 5% and 30% respectively as some metastable ions were formed.

The over-riding influence of potential energy is again apparent, though the curves for He^+ and Ne^+ require more detailed explanation (see Sect. 30).

No effect due to the difference of mass in Ne^{22} and Ne^{20} could be detected; this would be expected if potential energy alone is concerned.

Results for multiply charged ions of the rare gas atoms are given in Fig. 27. Again, the potential energy is seen to determine the yield in the main. The multiply charged ions and He^+ at low energies, show a fall in γ_i as energy increases. This is discussed in Sect. 30.

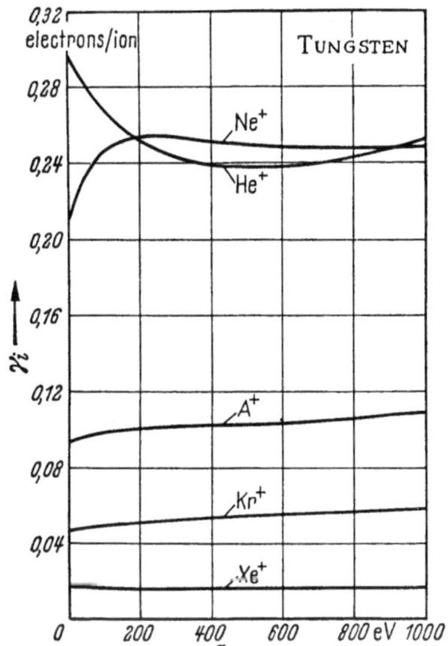

Fig. 26. The total electron yield γ_i as a function of ion energy for E_i rare gas ions on gas-free tungsten.

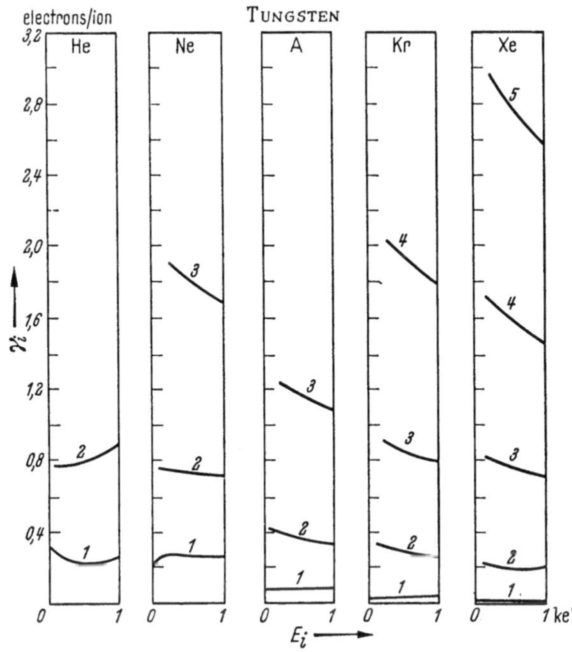

Fig. 27. The total electron yield γ_i as a function of the ion energy E_i applied to the ions for multiply charged ions on gas-free tungsten.

Reflection is again small, though increasing with energy slightly.

This work is the only extensive investigation under such good conditions.

Later HAGSTRUM[1] investigated the effect of gas layers. One typical result for He^+ may be summarised in the following table. The value for γ_i for ions of 0, 600 eV and 1000 eV are given for a W surface. (1) As put into the system, chemically clean; (2) after outgassing at 800° K; (3) after baking at 1200° K; (4) after flashing at 2400° K.

Surface state	Energy (eV) 0	600	1000
1	0.15	0.54	0.72
2	0.18	0.51	0.67
3	0.17	0.26	0.30
4	0.29	0.24	0.26

The reduction in γ_i at low energy and the increase at high energy agrees with the work next reported. In this the surfaces were gas-free for shorter periods, but they were surfaces of known cleanliness. PARKER[2] used also careful baking at 400° C with electrodes at 1400° C or above. These series of experiments were designed to show the effect of adsorbed gases.

Fig. 28a, curve 1, shows the emission from Pt due to A^+ as soon after heating as possible (25 sec.). Curve 2 shows the results 15 min. after the heating, the surface being in about $2 \cdot 10^{-5}$ mm. Hg of argon.

The independence of yield on energy is seen again and is maintained with slight fall down to 5 eV. The fall to lower values is due to the adsorption of gas on the surface. After 24 hrs. the value is 1/3 of curve 1.

[1] Private communication: To be published.

[2] J. H. PARKER: Phys. Rev. 93, 1148 (1954).

The other curves are the result of five minutes exposure to gases at presumably 10^{-5} mm. Hg. Oxygen is most effective in reducing the yield—it is an electronegative gas. A dependence on energy is apparent with such gas-covered surfaces.

Fig. 28b shows the emission for a mixture of ions of the molecular gases on platinum. A rise of γ_i with energy is found at more than 40 eV for all gases. In N_2^+ and N^+ mixtures a minimum is observed at 30 eV and a sharp rise below this, which is unexplained. The curve for N_2^+ and N^+ lies fairly close to A^+—Pt on N_2-covered surfaces above 30 eV and so for O_2^+, O^- and A^+—Pt on O_2-covered surface the energy available after neutralisation is about the same for these ions. Again O_2^+ and O^+ mixtures are the least efficient, probably because of the gas layer on the surface.

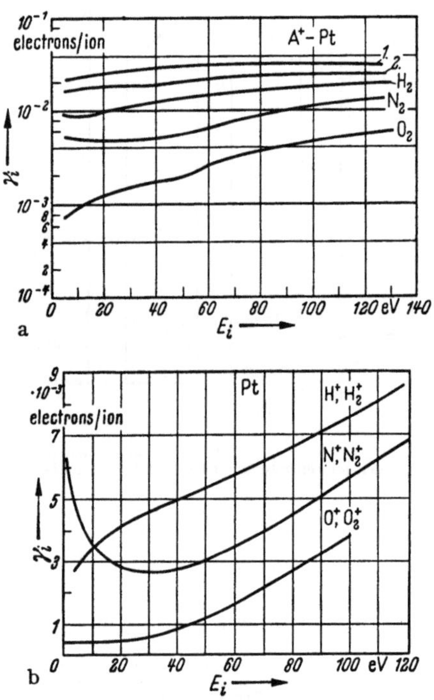

Fig. 28a and b. The total electron yield γ_i as a function of the ion energy E_i on gas-free and gas-covered platinum: *a* argon ions, *b* molecular gas ions.

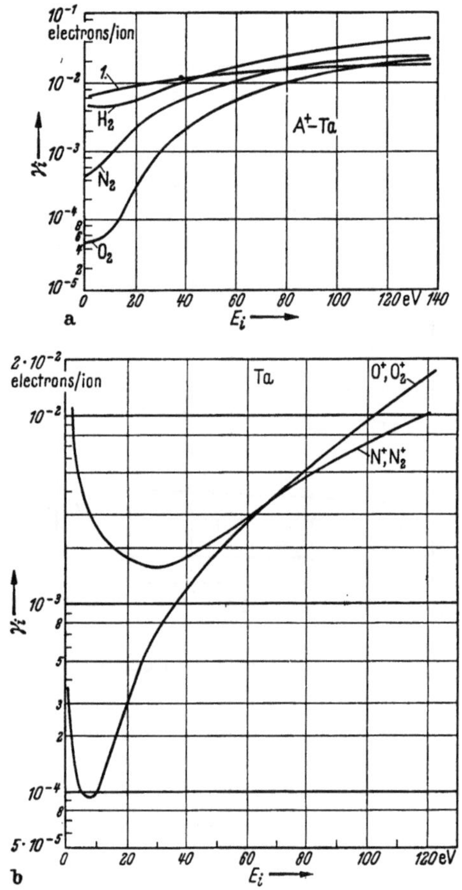

Fig. 29a and b. The total electron yield γ_i as a function of the ion energy E_i on gas-covered tantalum: *a* argon ions, *b* molecular gas ions.

Similar curves for tantalum are shown in Figs. 29a and b. Curve 1 lies below 1 of Fig. 28a. It represents results for the target clean after heating: When it has been standing in argon for up to 1 hr., the change in γ_i is only 20%, much less than for platinum. In long periods the difference between the two surfaces is less noticeable, but after 24 hrs. γ_i for Ta is 70% of curve 1) whereas in Pt γ_i was 30% of its initial value. This is explained by HAGSTRUM's observations of the tenacity of gas films on tantalum.

The values of γ_i rise slightly with energy, possibly due to the presence of the film. They are somewhat higher than MOLNAR's values for A^+—Ta (Sect. 16),

$7 \cdot 10^{-3}$ as against 2 to $6 \cdot 10^{-3}$. As MOLNAR's values depend on α, these direct values are to be preferred.

As with Pt, gas layers reduce γ_i, and the greatest reduction is again in O_2, by two orders of magnitude. The rise of γ_i as energy increases is more marked here: N_2 and O_2 surfaces tend to have the same γ_i as the "flashed" surface and H_2 rises well above it. This indicates perhaps that higher energy A^+ ions can interact with the metal in spite of the layer of gas. It may be that a chemical reaction occurs with hydrogen, so that the work function is reduced, though this is a very tentative suggestion.

No measurements of H_2^+, H^+ mixtures were made on tantalum surfaces, but oxygen and nitrogen ions were investigated (Fig. 29b). There is a pronounced minimum in the curve for N_2^+ and N^+ as before, but a smaller dip is observed for

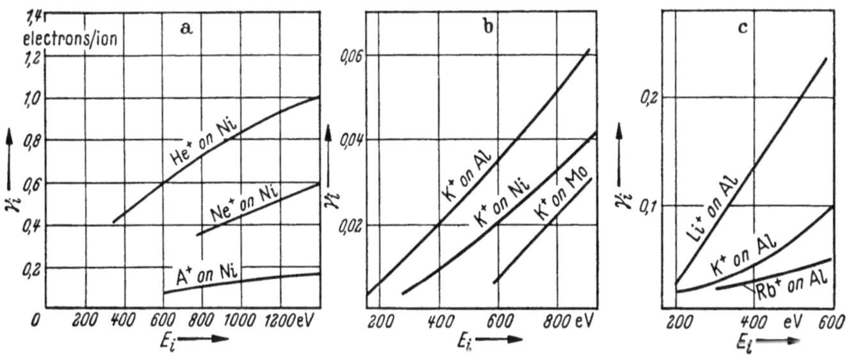

Fig. 30a—c. Total electron yield γ_i as a function of ion energy E_i: *a* rare gas ions on nickel, *b* potassium ions on various metals, *c* alkali ions on aluminium.

O_2^+, O^+ also. PARKER considers this may be an instrumental error, in which case the extent of the minimum in N_2^+, N^+ is suspect. Above 30 eV, as before N_2^+, N^+ lie close to A^+—Pt for N_2-covered surfaces, and similarly O_2^+, O^+ near to A^+—Pt on O_2-covered surfaces.

The effect of the gas layer on φ, the work function of a metal surface is variously reported as increasing or decreasing it. It is not only φ that determines γ_i, however, even on truly clean metals: for tantalum HAGSTRUM measured $\varphi = 4.9$ eV, and clean platinum has $\varphi = 5.3$. Nevertheless, γ_i is greater for platinum. The influence of a gas layer on γ_i was pointed out first by MCLENNAN and POUND[1]. We shall discuss this again when the energy distributions have been considered.

It is clear, however, that a gas film decreases γ_i at low energies, and becauses γ_i to increase with energy. Sometimes this means that γ_i increases when a gas layer is present, as PARKER found for H_2^+, H^+ on Ta at $E_i > 100$ eV and HAGSTRUM observed for He^+ on Mo at > 500 eV and for He^{++} on Mo from zero upwards.

β) Experiments on clean, gas covered surfaces. Most of the work on γ_i comes under this heading, as at pressures of 10^{-6} mm. Hg on cold surfaces a gas film is always present. Even a hot surface is not absolutely free; at $p = 10^{-6}$ mm. Hg a monolayer of gas clings to tungsten at 1400° C[2]. It is possible that many workers had much lower pressures than their ionisation gauges showed. The limit is set by the X rays emitted from the anode of the gauge: these produce photoelectrons from the cathode.

Rare gas ions of low energy. In Fig. 30, some measurements at low ion energies are shown. Fig. 30a is given by HEALEA and HOUTERMANS[3] for a hot nickel

[1] J. C. MCLENNAN and C. G. POUND: Phil. Mag. **30**, 491 (1915).
[2] J. A. BECKER and C. D. HARTMANN: J. Phys. Chem. **57**, 153 (1953).
[3] M. HEALEA and C. HOUTERMANS: Phys. Rev. **58**, 608 (1940).

target carefully cleaned and degassed as far as possible at 10^{-6} mm. Hg. They observed a rise in γ_i as the rare gases were admitted, and a fall as ion bombardment continued: they correctly attributed this to changes in the gas layer.

A comparison with Fig. 26 shows that some gas still clung to the target, for HAGSTRUM found much less variation with energy and lower values of γ_i. There may be differences in the surface condition apart from this.

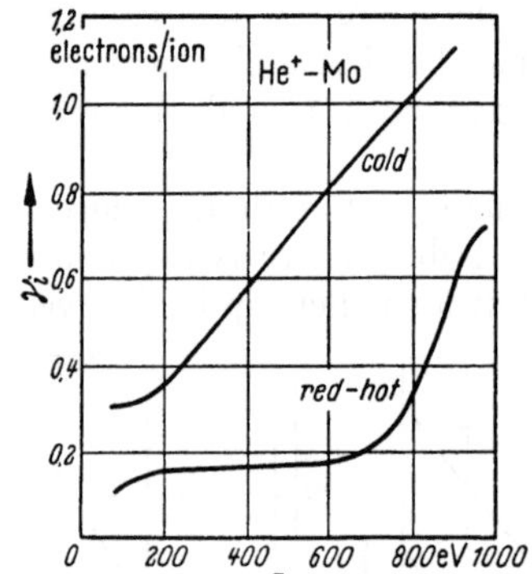

a

Fig. 30a shows that rare gas ions are more effective the higher the ionisation potential, in agreement with HAGSTRUM'S work.

Similar results, with variations due to different cleaning procedures, were found A^+—Al, Mo[1] and for rare gas ions on Fe, Ni, W[2], and Cu and brass[3]. Recently, OKANO[4] found that for rare gas ions on molybdenum γ_i increases with energy faster for gassy surfaces, and increases with the ionisation potential of the atom, or decreases with its mass (the two are inseparable). These conclusions agree with those of HAGSTRUM and PARKER, so at fairly low energies the experimental evidence for rare gas ions is fairly consistent.

OLIPHANT'S pioneer studies [*19*], Fig. 31a, showed the effect of heating very clearly. This work on He^+—Mo is the only evidence of the independence of

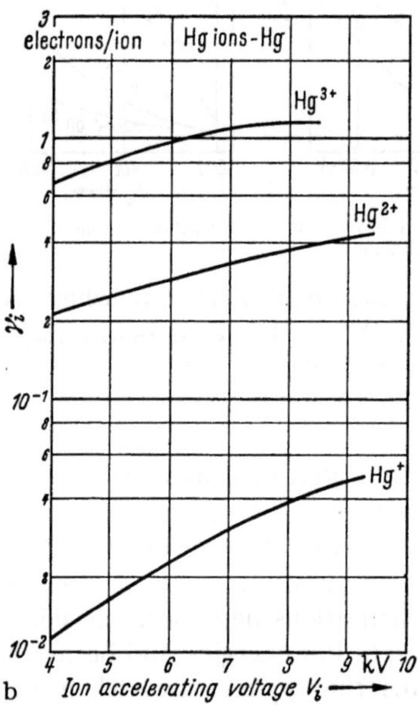

b

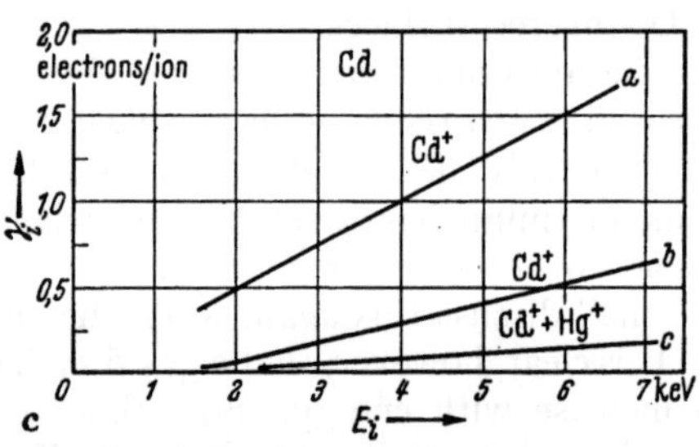

c

Fig. 31 a—c. Total electron yield γ_i as a function of ion energy E_i: *a* helium ions on cold and hot molybdenum, *b* mercury ions on mercury, *c* cadmium ions and mixed ions on cadmium (see text).

γ_i on energy previous to the results of HAGSTRUM and PARKER. The agreement with HAGSTRUM'S work (Fig. 25a) is best for gas-covered Mo, so that some gas probably remained on OLIPHANT'S cathode. The rise in γ_i at $E_i > 800$ eV is perhaps due to the release of gas from the body of the metal by local heating. The curve for cold Mo shows the almost linear rise characteristic of other results.

Alkali metal ions of low energy. Fig. 30b shows some of JACKSON'S[5] results. The yield decreases with increasing atomic number of the target: it is an order

[1] G. TIMOSHENKO: J. Appl. Phys. **12**, 69 (1941).
[2] F. M. PENNING: Proc. Acad. Sci. Amst. **33**, 841 (1930).
[3] A. ROSTAGNI: Z. Physik **88**, 55 (1934).
[4] J. OKANO: J. Phys. Soc. Japan **8**, 562 (1953).
[5] W. J. JACKSON: Phys. Rev. **28**, 524 (1926).

of magnitude below the results of Fig. 30a probably because V_i the ionisation potential of K^+ is about the same as the value of φ for the surface. If $V_i < \varphi$ then only the kinetic energy of the ion is available to release electrons, which is known from HAGSTRUM'S work to be inefficient on gas-free surfaces. The effect of the increasing ionisation potential is probably seen in Fig. 30c due to CHENEY[1].

PLOCH[2], however, found that even with ions of similar elements (alkali metals) the variation of γ_i with ionisation potential (or ion mass, as for rare gas ions) is not a simple one. The value of γ_i for 1.2 keV ions on Cu or Mo was lower for Na^+ than for Li^+ or K^+, though K^+, Rb^+, Cs^+ formed a descending series fairly smoothly.

Ions of moderate energy. Fig. 31 b shows the values[3] of γ_i for singly and multiply charged mercury ions on mercury. The values for the multiply charged ions are estimated from the results of mixtures of ions, so they are not accurate.

The most interesting feature is the remarkably low value of γ_i for all three ions. Even at 4 kV, γ_i for Hg^+ is only 10^{-2}: for K^+, which has a lower potential energy (4.3 eV as against 10.4 eV) γ_i on Mo is 10^{-2} when the kinetic energy of the ion is under 700 eV.

It is possible that the low value arises because the ion is of the same nature as the surface: but cadmium ions on cadmium (Fig. 31 c) have much higher efficiency. At 4 kV, $\gamma_i \sim 0.3$ after the surface had been outgassed (curve b). Before outgassing, the values were higher still (curve a).

When mercury was allowed to contaminate the cadmium vapour, much lower values for γ_i were observed (curve c). A similar reduction in γ_i was found by OLIPHANT [*19*]; when a mixture of $He^+ + Hg^+$ was incident on Mo, γ_i was much smaller than for pure He^+.

Now mercury has a lower ionisation potential than helium, so this last result might be expected, but it has a higher ionisation potential than cadmium. On the other hand, mercury is far heavier than either of the others, so the *ion mass* appears to be important when ion energies of a few keV are employed.

PLOCH[2] has studied lithium ions on molybdenum up to 8 keV energy. He found γ_i increasing up to 0.8 approximately at 8 keV, but the *rate* of increase with energy was falling as the energy increased. PLOCH also observed that the Li^6 isotope was about 10% more efficient in ejecting electrons as the Li^7 isotope—he worked only above 2 keV with both isotopes.

In the same energy range he investigated Li^+ on Be, Cu, Pt, and Ne^+ on Be and Pt, finding the same kind of dependence on energy. The mass of the ion alone does *not* determine γ_i, for ions of different elements.

The effect of degassing the target was always to reduce γ_i, as might be expected from the extrapolation of HAGSTRUM'S and PARKER'S results to higher energies.

COUCHET[4] observed γ_i for Li^+ on different surfaces up to 12 keV, where the rounding-off of γ_i was more pronounced. It appears as if a maximum for $\gamma_i \sim 1.5$ would be reached at higher energies.

Ions of high energy. Curves for high energy ions on Mo are shown in Figs. 32a[5]. These reveal the presence of a maximum in γ_i, between 5 and 15 electrons per ion for the various ions used. There is a sharp rise before the maximum and a slower fall after it. For heavier ions, the maximum appears later in the energy range.

---

1 W. L. CHENEY: Phys. Rev. **10**, 335 (1917).
2 W. PLOCH: Z. Physik **130**, 174 (1951).
3 M. SCHWARZ and P. L. COPELAND: Phys. Rev. **96**, 1467 (1954).
4 G. COUCHET: C. R. Acad. Sci. Paris **235**, 944 (1952).
5 A. G. HILL, W. W. BUECHNER, J. S. CLARK and J. B. FISK: Phys. Rev. **55**, 463 (1939).

This is still more clearly shown in Fig. 32b. The curves for H^+—Cu show strong differences[1], and both join well onto the curve for H^+—Au. This latter represents the results of AARET *et al.*[2] for H^+ on Mg, Al, Fe, Au, Pb within 10%: the same authors found γ_i for H_2^+-Al, Ni, Au to be represented by the curve shown to even greater accuracy. Similarly LINFORD[3] found Hg^+—Ni, Cu, Sn, Na, Mo, Al, W, Ag, Mg, Cd to lie closely on the curve drawn.

Apart from the independence of γ_i on the target surface, we see that H_2^+ still lies above H^+, as in Fig. 32a, and both are falling steadily. It is surprising to find Hg^+ now more efficient than the lighter ions, but in this energy range Hg^+ is still rising with energy and the others are falling.

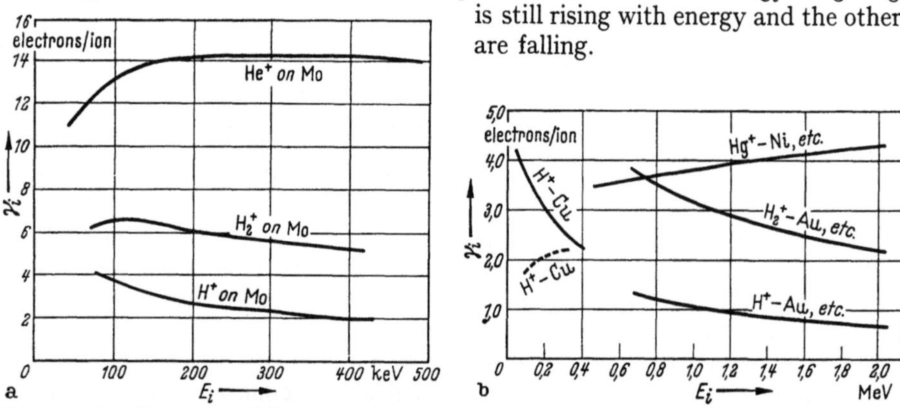

Fig. 32 a and b. Total electron yield γ_i as a function of ion energy E_i for high energy ions: *a* on molybdenum, *b* on various metals showing the same value of γ_i.

It is clear that the ion mass is very important at energies much greater than the potential energy of the ions, and $\gamma_i > 1$ under such circumstances.

The peak for H^+ and H_2^+ for various surfaces is shown clearly in Fig. 32a, b.

GÜNTHERSCHULZE'S *method for* γ_i. This method, based on heat transfer to the cathode (Sect. 7), has been applied to many ions and metals[4]. Fig. 33a shows his results for H_2^+—Fe compared with direct measurements of H_2^+-Ni and D_2^+—Ni[5]. The two curves for H_2^+—Ni are observed with different surface-conditions.

In spite of the uncertainties in the method, the values are as close as could be expected. It is probable that γ_i for H_2^+—Fe is lower than that shown by virtue of the other agents active in a discharge. Because of this and the uncertain nature of the ion energy distribution, the method is not so reliable as the direct one. The surface could be cleaned well by sputtering, but could never be gas-free.

Isotope effects. This has been mentioned for lithium and is very clearly shown in Fig. 33a for H_2 and D_2—on definitely the same surface. The difference in γ_i shows at lower energies for the lighter ions. Similar results for Ne^{20} and Ne^{22} are seen in Fig. 33b on a silver-magnesium alloy[6]. Again the lighter ion is more efficient at moderate energies. The other rare gas ions are seen to obey the same rule: Xe^+ is least efficient, He^+ most. The curve A^{++} for is interesting. At $E_i < 1000$ eV on gas-free surfaces γ_i for this ion is much higher than for A^+ (Fig. 27). Here it is slightly lower.

[1] J. S. ALLEN: Phys. Rev. **55**, 336 (1939). — HILL *et al.*: Phys. Rev. **55**, 463 (1939).
[2] B. AARET, R. W. CLOUD and J. G. TRUMP: J. Appl. Phys. **25**, 1365 (1954).
[3] L. H. LINFORD: Phys. Rev. **47**, 279 (1935).
[4] A. GÜNTHERSCHULZE, W. BÄR and A. WINTER: Z. Physik **111**, 209 (1938).
[5] M. HEALEA and E. L. CHAFFEE: Phys. Rev. **49**, 925 (1936). — C. HOUTERMANS: Phys. Rev. **58**, 608 (1940).
[6] M. J. HIGATSBERGER, H. L. DEMDREST and A. O. NIER: J. Appl. Phys. **25**, 883 (1954).

One tentative suggestion is that the emission on gas covered surfaces for $E_i > 1$ keV is due primarily to kinetic energy, and the doubly ionised argon atom serves as a more efficient trap for electrons so emitted than the singly charged ion. It would be worthwhile to trace the graph of γ_i for isotopic ions from 1 keV to 1 MeV on gas free and gas covered surfaces.

Complex ions. From the results just quoted, and those given in section [*18*] one might expect very low yields from heavy ions up to several MeV. PHILBERT[1] has obtained the results shown in Fig. 33 c for γ_i on graphite and molybdenum—both show the same γ_i. Ions of the form $C_nH_{n+1}^+$ lie together on a curve above that for CO_2^+, which itself is above A^+.

Comparing the curve for A^+—Mo with A^+—Ni at 1400 eV, PHILBERT'S values are higher so that his surface conditions were probably not so good. The unexpected behaviour of complex ions should be further confirmed.

Variation of γ_i with angle of incidence. This must be connected with the reflection of ions at different angles of incidence. OLIPHANT [*19*] obtained for 1000 eV He^+ on Ni the curve shown in Fig. 34. He expressed this as $\gamma_i = a - b \cos \zeta$, where a, b are constants, and ζ is the angle of incidence.

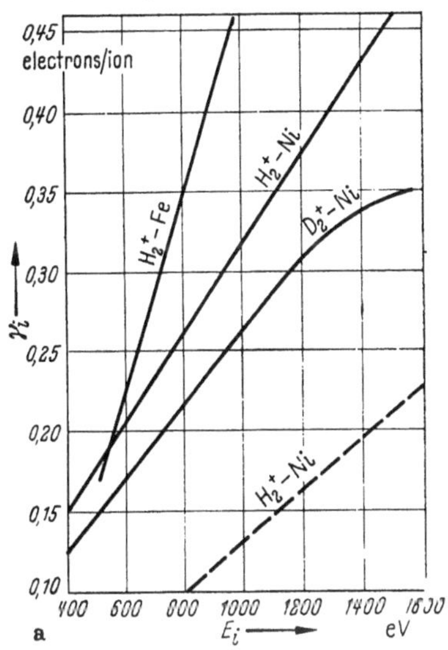

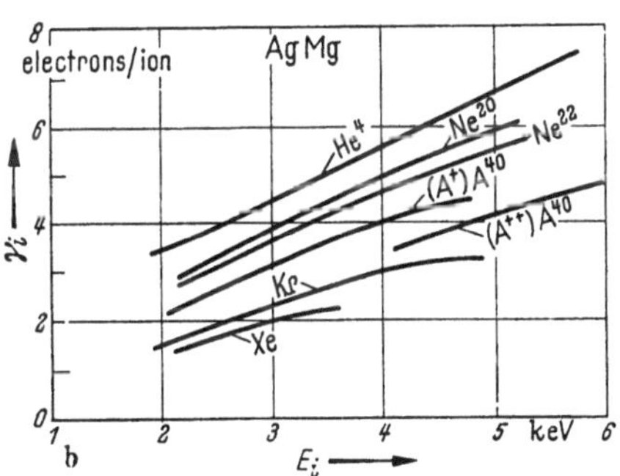

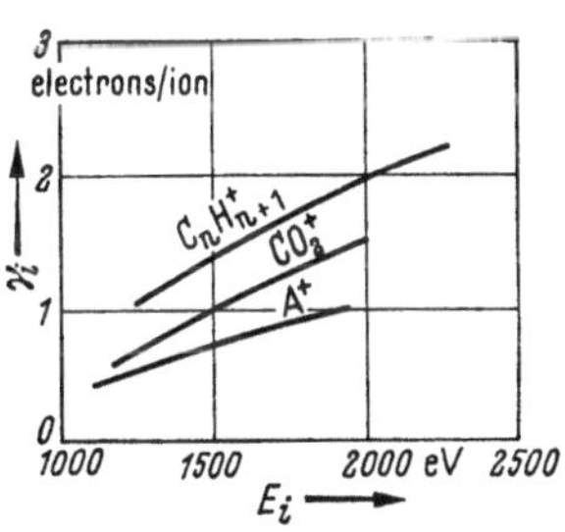

Fig. 33 a—c. Total electron yield γ_i as a function of ion energy E_i: *a* hydrogen and deuterium ions on iron and nickel, *b* various rare gas ions on silver-magnesium alloy (1.7% Mg), *c* complex ions on graphite and molybdenum, compared with argon ions.

ALLEN[2] for high energy protons on Be, C, Cu, Ni and Pt found $\gamma_i = c$ sec. ζ, c being a constant. OLIPHANT'S results fit this equation fairly well also.

Influence of adsorbed layers. This has been mentioned constantly in terms of gas adsorption.

It has been shown that monolayers of gas atoms sometimes increase and sometimes decrease the yield at low ion energy (10 to 100 eV). The dependence of γ_i on energy is more marked for gas-covered surfaces, so it appears that at higher energies (> 1000 eV) the gas film always increases γ_i.

No results of direct measurements with electronegative gases (O_2, Br_2 etc.) at high energies are known: these gases might not cause γ_i to be greater.

[1] G. PHILBERT: C. R. Acad. Sci. Paris **237**, 882 (1953).
[2] J. S. ALLEN: Phys. Rev. **55**, 336 (1939).

KOCH[1] and PAETOW and WALCHER[2] observed changes in γ_i as bombardment by alkali metal ions continued, which was interpreted in terms of formation of a layer on the target. Fig. 35 shows the results of γ_i for 500 eV $Cs^+ - W$.

The pure tungsten curve at the start of the experiment shows $\gamma_i \sim 0.01$. As the number of caesium atoms per cm.² increases γ_i rises until at M $\gamma_i \sim 0.047$. It now remains constant. M marks the points at which the surface is covered with a monolayer of atoms.

Curves for a tungsten surface partly covered with adsorbed oxygen lie above the pure tungsten graph. The curve labelled W—O has the most oxygen adsorbed on it, the others progressively less. It is seen that these curves rise to the constant value earlier: PAETOW and WALCHER suggest that Cs atoms form only over

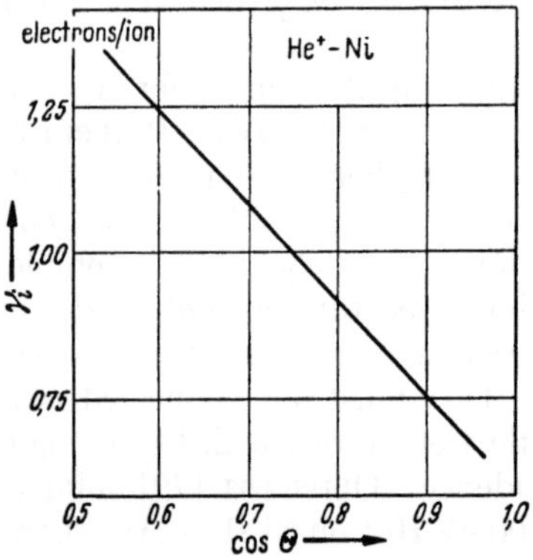

Fig. 34. Total electron yield γ_i as a function of cos Θ, the angle of incidence, for helium ions on nickel.

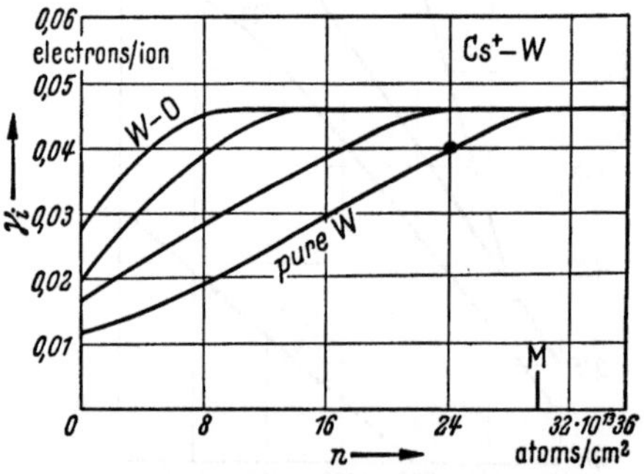

Fig. 35. Total electron yield γ_i as a function of density of adsorbed caesium atoms on tungsten covered with varying amounts of adsorbed oxygen. $M = 1$ complete monolayer.

the tungsten, and a composite film of Cs + O is the final result. The characteristic emission is that from the tungsten.

The same authors measured γ_i for tungsten covered with monolayers of hydrogen and potassium. The emission from W + Cs, W + K, W + H was 0.046, 0.049, and 0.018 respectively. The value of φ was least for W + H, so that φ alone does not define the effect of an adsorbed layer. It was suggested that the low electron density in the layer W + H may be significant.

28. Reflection of positive ions. *α) Gas-free surfaces.* HAGSTRUM found that the reflected fraction R for helium ions on molybdenum at $E_i = 1000$ eV was 5%, on tantalum 20%. Values for tungsten were about the same as for molybdenum. The value for R increased with energy.

β) Gas-covered surfaces. Fig. 36a shows results for caesium ions on various surfaces containing tungsten[2] and Fig. 36b shows typical results for rare gas ions and hydrogen ions on nickel[3].

There is a general agreement that R rises linearly as E_i increases. The probability that an ion will be reflected without neutralisation is large for rare gas ions, much smaller for H_2^+, and smallest for Cs^+. In the rare gas ions, the lighter ion suffers much more reflection.

The effect of the surface is clearly shown in Fig. 36a for Cs^+. PAETOW and WALCHER found that the ratio of probability of reflection as an ion to reflection as a neutral atom is $\exp[(\varphi - e V_i)/kT]$ where V_i is the ionisation potential of the

[1] J. KOCH: Z. Physik **100**, 685 (1936).
[2] H. PAETOW and W. WALCHER: Z. Physik **110**, 69 (1938).
[3] M. HEALEA and C. HOUTERMANS: Phys. Rev. **58**, 608 (1940).

ion and T the temperature. If $\varphi > e V_i$, as for oxidised tungsten and Cs^+, the ion reflection coefficient R will be large: if $\varphi < e V_i$, as for caesium monolayers on tungsten and Cs^+, R will be reduced. This is shown in the figure.

All these values refer to reflection at normal incidence, which is in practice the most important case. It would be expected that R should vary with angle of incidence ζ.

Early work with alkali metal ions[1] showed that R increased with ζ, and a diffuse beam with its centre at the specular reflection angle was observed[2]. In contrast to behaviour at normal incidence, a sharp peak in R occurred at 48 eV for Li^+ on Pt at $\zeta = 45°$. This has not been explained.

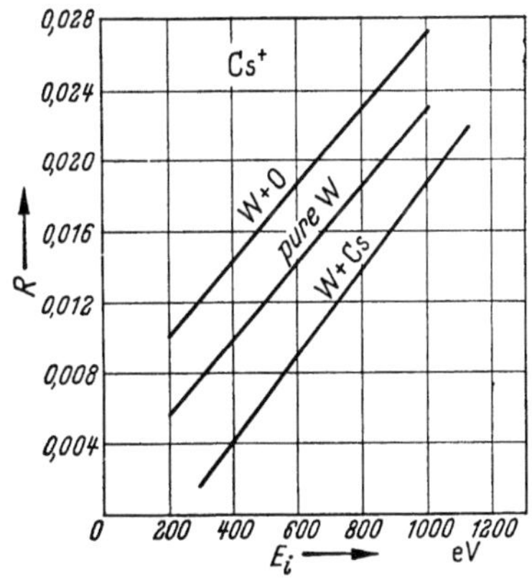

Fig. 36a. Reflection of caesium ions at surfaces of tungsten covered with adsorbed oxygen; clean tungsten; and tungsten covered with adsorbed caesium—Normal incidence.

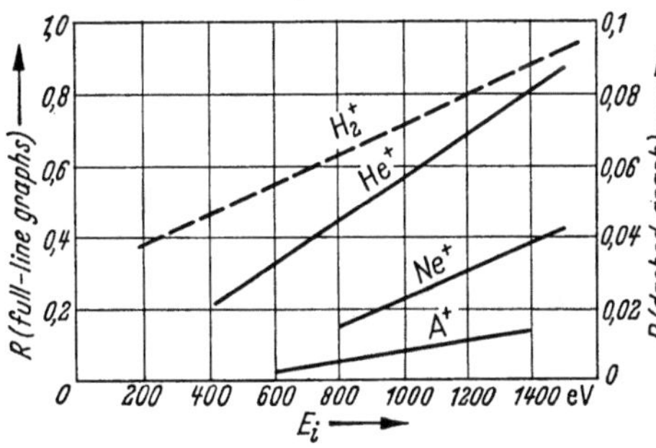

Fig. 36b. Reflection of hydrogen ions and rare gas ions at a nickel surface: normal incidence.

The energy of the reflected ions is only about 20% of the initial energy if they are deflected through 100°. As grazing incidence is approached, the ions retain more of their original energy, up to more than 90% at less than 10° deflection[3].

29. Energy distributions of secondary electrons. On clean and gas free surfaces HAGSTRUM has measured the distribution only for low energy, rare gas ions ($E_i < 1000$ eV). For comparison, Fig. 37a shows his theoretical curves for He^+ and Kr^+ on tungsten and the corresponding observations in Fig. 37b, c. The intervals marked with arrows are discussed in Sect. 30, and the extension below zero energy is due to spurious secondary effects [*18*].

It is clear that the form of the curves is correctly predicted except for He^+ at 1000 eV. The theoretical heights are rather larger than those measured, and the high energy tail goes to rather higher values and shows more dependence on energy than the experimental graphs.

In Fig. 38a the observations for 200 eV xenon ions multiply charged are shown. Each curve envelopes the curve for the ions of lower charge.

For comparison, OLIPHANT's curves of the distribution due to He^+ on molybdenum are given in Fig. 38b. The fine structure was repeatable, and appeared in magnetic deflection analysis as well as in retarding potential work: the target was red-hot during measurements. The great difference between Fig. 37c and Fig. 38b is probably due to surface differences: there is one similarity in that, at ion energies of 1000 eV and above, both graphs show a peak at 2 eV. The second peak at 20 eV in Fig. 38b is absent entirely in HAGSTRUM's work [*19*].

[1] G. E. READ: Phys. Rev. **31**, 629 (1928). — F. M. PENNING and C. C. J. ADDINK: Physica, Haag **1**, 1007 (1934).

[2] R. W. LONGACRE: Phys. Rev. **46**, 407 (1934).

[3] A. GURNEY: Phys. Rev. **32**, 467 (1928).

OLIPHANT observed that very gassy surfaces (i.e. cold surfaces) of molybdenum and other metals which showed higher values of γ_i showed more low energy end ons in the distribution.

The energy distribution of electrons ejected by ions of *low potential energy* has been found to be nearly MAXWELLian on surfaces clean but not gas-free. This suggested that a temperature could be assigned to the electrons and the emitting surface. Values of

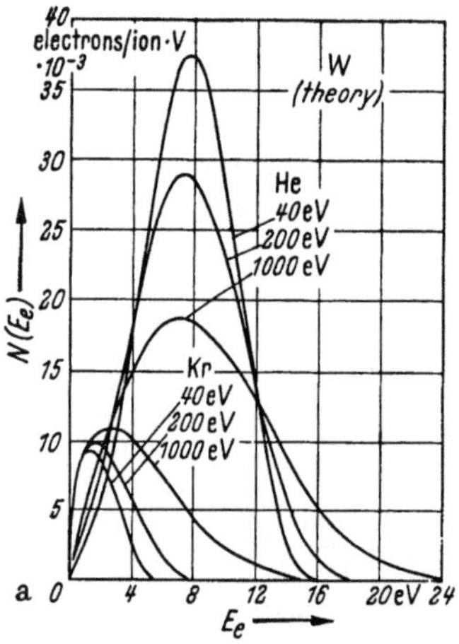

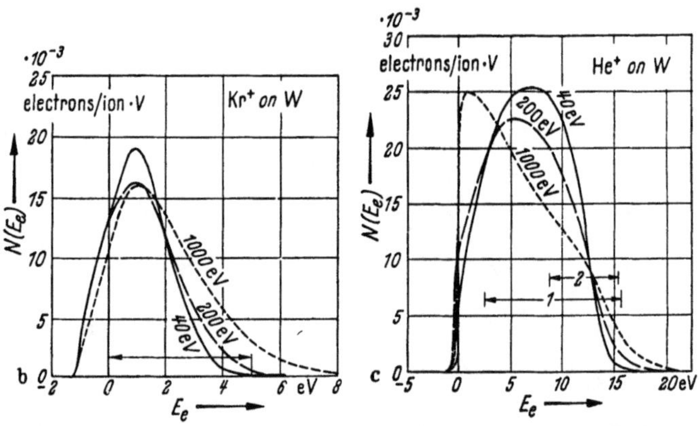

Fig. 37 a—c. Secondary electron energy distribution: $N(E_e)$ electrons/ion/eV as a function of electron energy E_e, for helium and krypton ions on tungsten: *a* theoretical, *b*, *c* experimental.

this temperature were from 10000 to 66000° C [*19*] and seem to be highest in metals of bad heat conductivity, which sputter greatly in glow discharges. The temperature so estimated rose slowly with ion energy.

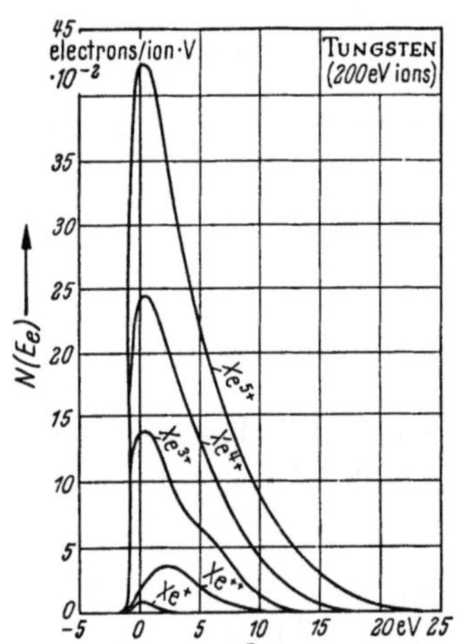

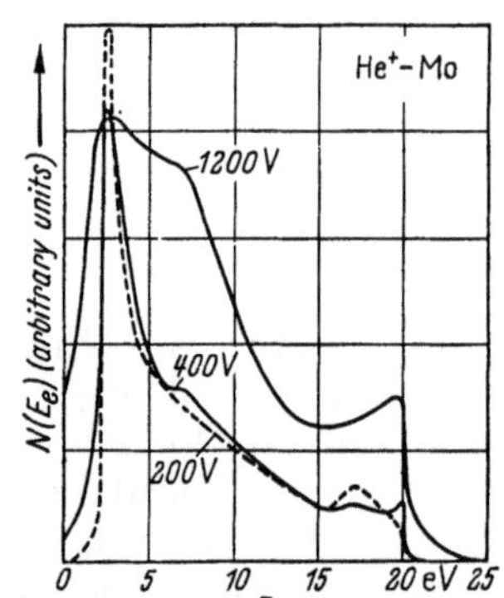

Fig. 38 a and b. Secondary electron energy distribution *a* for 200 eV xenon ions on tungsten, *b* for helium ions on molybdenum.

30. Theories of the secondary electron emission. When a positive ion approaches a metal surface it may interact with one metal electron or deliver its energy to a small volume of the metal. The first can occur if the ion has large potential energy and is not too fast. The second is the only possibility for ions of low potential energy, and predominates for fast ions of high potential energy.

α) Potential energy interaction. Let us first consider electron ejection by virtue of the potential energy of an ion. There is little likelihood of a radiative transition occurring in the electronic structure of the ion, for the time of strong interaction is too short ($\sim 10^{-12}$ secs). We have two alternatives; a two-stage interaction, and a single stage. Fig. 39b shows the energy levels for He^+ near Mo illustrating the two stage process of emission.

The first stage is the resonance capture of an electron from a level in the FERMI distribution (width $W - \varphi$) of the metal (assumed to have a smooth clean surface) into an unoccupied level of the atom of the same energy. The condition for this to be possible is

$$W < e(V_i - V_{ex}) > \varphi. \tag{30.1}$$

For He^+—Mo, this is clearly fulfilled for the 2^3S metastable state:

$$W = 14\,\text{eV}, \quad eV_i - eV_m = 24.47 - 19.77 = 4.7\,\text{eV}, \quad \text{and} \quad \varphi = 4.3\,\text{eV}.$$

The capture process at a level $e = 4.7$ eV is shown in the first part of Fig. 39b.

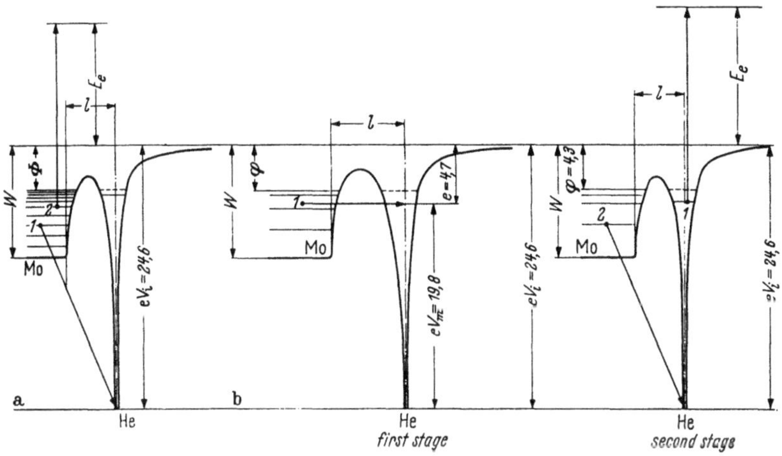

Fig. 39 a and b. Potential energy diagram and electronic transitions for helium ions on molybdenum *a* for direct AUGER neutralisation, *b* for resonance neutralisation followed by AUGER de-excitation.

The second stage is the ejection of an electron by the excited atom. This is illustrated showing electron exchange between metal and atom which is more likely for metastable helium [*18*]. Other excited states may decay without electron exchange and eject a metal electron. The kinetic energy of the emitted electron is E_e where

$$eV_{ex} - \varphi \geq E_e \geq eV_{ex} - W \tag{30.2}$$

since the energy of the electron in the metal is between φ eV and W eV below the zero line for energy.

The condition for this part of the process is

$$eV_{ex} > \varphi \tag{30.3}$$

which is to be satisfied at the same time as (30.1): helium metastable atoms clearly satisfy it.

The alternative single stage process is illustrated in Fig. 39a. It is, like the second part of the stage process, an AUGER transition — two electrons are involved at the same time. One is emitted and the other captured into the ground state of the ion.

The limits of the kinetic energy are

$$eV_i - 2\varphi \geq E_e \geq eV_i - 2W, \tag{30.4}$$

so the condition to be satisfied is

$$e V_i > 2\varphi. \tag{30.5}$$

This is again satisfied for He⁺—Mo.

Where both processes are energetically possible, the theory developed by VARNERIN [*19*] indicates that the two stage interaction can only occur if the distance s between ion and metal is beyond a critical value s_c. If $s < s_c$ only the second process can occur. The value of s_c depends on the ion and metal energy levels: for tungsten, only Ne^+ can be expected to have s_c small enough to make the two stage interaction at all likely.

The intervals 1 and 2 on Fig. 37c represent the limits of energy given by eqs. (30.2) and (30.4), and on Fig. 37b the arrow marks the limits of eq. (30.4). Obviously the two-stage process is not very important, but the single stage AUGER neutralisation is not a good description either.

The reason is that we have neglected the changes in the energy levels of the ion or atom near the surface, and the broading of the distribution because the initial and final states have only finite life times.

HAGSTRUM has developed the theory in detail for rare gas ions on tungsten [*19*]. He fitted the experimental results at the high energy end of the distribution for each ion to find the parameter A in the expression for the transition rate R:

$$R(s) = A\, e^{-as}$$

a was taken as $2 \cdot 10^{-8}\,cm^{-1}$ and $5 \cdot 10^{-8}\,cm.^{-1}$ in various calculations.

Then a parameter determining the ratio of escaping to reflected electrons was evaluated for He^+ of 40 eV by making γ_i for theory agree with γ_i for experiment.

The results are good for He^+ and within a factor two of the measured values for other low energy ions if for Ne^+ 10% or so of the transitions are two-stage processes. Transitions of other ions are pure AUGER neutralisation.

It would appear that kinetic energy cannot enter into the theory, but the changes in energy levels near the surface and the fact that faster ions approach closer before being neutralised make the ion speed important.

For higher energies the theory can account for all the features of Fig. 26 except the rise in γ_i for He^+ at higher energies. Probably other processes become important, for Fig. 37c shows that a different distribution is occurring.

It should be noted that the influence of adsorbed layers has not been treated specifically: it is likely that the change in φ will account for differences in emission for slow ions.

If the metal is hot, electrons will occupy levels above the normal FERMI limit. In such a case transitions may occur from such a higher occupied level and the conditions (30.1), (30.3) and (30.5) must be modified.

β) Kinetic energy exchange. For fast ions, interaction with the lattice is very important, and it should be similar for atoms of high energy. The relation is theoretically similar to ionisation of atoms by fast ions and neutral atoms in a gas [*15*].

The theory could be developed from the principles of normal collision theory.

An alternative suggestion that the kinetic energy can be regarded as causing very local heating has been developed on the assumption that thermionic emission from small areas occurs [*3*], [*9*], *10*], [*15*].

The variation of γ_i with ion mass for isotopes of the same element is in the right direction and even shows the right dependence ($\gamma_i \propto M^{\frac{1}{2}}$) according to

HIGATSBERGER *et al*[1]. Fig. 33a shows the values of γ_i for D_2^+ and H_2^+ on Ni to be tending toward the ratio 1.4 at high energies. PLOCH's results[2] for Li^6 and Li^7 also agree roughly, but he found that even for alkali metal ions of similar structure the mass *alone* did not decide γ_i.

It is difficult to conceive that a "temperature" can be assigned to a small volume containing 10 to 20 atoms, such as is postulated in this theory[3]. Nevertheless many observers have measured the energy distribution of the secondary electrons and observed a MAXWELLian distribution at higher ion energies or if the ions are such that the potential energy available cannot cause emission.

The sputtering of metals by ions and atoms should be identical on this theory, but ions are known to be much more effective [*10*], [*15*].

Another suggestion has been made that ions entering the metal from a "pass" in the potential barrier, so that many electrons can escape[3].

In order to explain the high yields from gas-covered surfaces, BERRY and ABBOTT[4] have suggested that high energy ions produce more ions in the gas layer, which then release electrons by the potential energy exchange described above. This would give a variation of γ_i with energy and a maximum of $\gamma_i > 1$. This process has obvious affinities with ionisation in a gas.

Much remains to be done in theory and experiment on this subject.

Table 12. *Comparison of the maximum yield due to ions and electrons (γ_i and γ_e).*

Particle	Surface	γ_i or γ_e	E_{max} eV	V_{max} cm/sec.
e	Mo	1.3	450	$1.2 \cdot 10^9$
e	Cu	1.3	550	$1.4 \cdot 10^9$
e	Ni	~ 1.3	500	$1.3 \cdot 10^9$
H^+	Mo	>4	$<5 \cdot 10^4$	$<3.1 \cdot 10^8$
H^+	Cu	>4	$<5 \cdot 10^4$	$<3.1 \cdot 10^8$
H^+	Cu	2.5	$4 \cdot 10^5$	$8.8 \cdot 10^8$
H_2^+	Mo	6.7	10^5	$3.1 \cdot 10^8$
Hg^+	Ni	>5	$<2 \cdot 10^6$	$<1.2 \cdot 10^8$

γ) Emission due to the velocity of the charge. There is a comparison also with emission due to electron bombardment at high energies, though not at low [*15*] (see Part III introduction). The curve of emission against speed for electrons shows a maximum, and the same form of curve appears for ions also. It is interesting to compare the maximum velocities of the ions and electrons and the values of γ_i and γ_e taken from [*15*], though the γ_i values known are few and not always consistent (see Table 12).

The ion velocities at the energy E_{max} for which the maximum yield is observed are lower by a factor between ten and two. The yields are much greater than those for electrons, as we might expect (see introduction to Part III). The comparison of velocities may be useful in predicting curves for other ions, since the peak in γ_i occurs at roughly the same velocity for all ions.

The emission due to fast multiply charged ions should be much greater than for simply charged ions, but it is not known.

31. Formation of negative ions[5]. If an ion has been neutralized to form an excited atom, or if an excited atom approaches sufficiently close to a metal surface, an electron may be captured to form a negative ion. The condition for this is

$$W > e V_{exc} + A > \varphi \tag{31.1}$$

where A is the electron affinity of the atom.

[1] M. J. HIGATSBERGER, H. L. DEMOREST and A. O. NIER: J. Appl. Phys. **25**, 883 (1954).
[2] W. PLOCH: Z. Physik **130**, 174 (1951).
[3] TOWNES: Phys. Rev. **65**, 319 (1944). — R. R. NEWTON: Phys. Rev. **73**, 1122 (1948).
[4] H. W. BERRY and R. C. ABBOTT: O.N.R. report, June 1954.
[5] The subject is fully treated by L. B. LOEB in his article on negative ion formation in this volume.

Occasionally if A is large and φ small an unexcited atom will form a negative ion if

$$W > A > \varphi. \tag{31.2}$$

Again, these relations must be modified for hot surfaces. For example, halogen atoms form negative ions when incident on cold alkali metal surfaces, but not on cold molybdenum. If they strike a hot tungsten or molybdenum surface, negative ions are formed.

Reflection of a positive ion as a negative ion can occur if (30.1) and (31.1) are satisfied or by a single process analagous to that given in Fig. 39b for electron emission. Reflection as a metastable may occur if (30.1) only satisfied.

b) Secondary electron emission by other means.

32. Photo emission from metals. This is reviewed in G. L. WEISSLER's contribution to this volume. We show here some of the work which indicates the rise in γ_p from very small values in the visible region to greater than 0.1 in the vacuum ultraviolet. Table 13 shows values for cadmium according to BAKER[1].

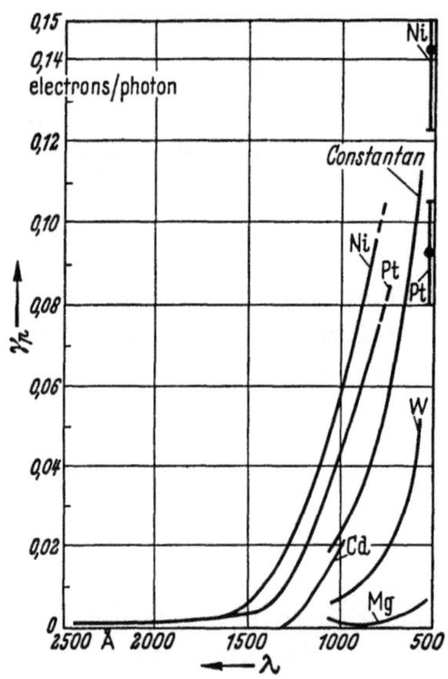

Fig. 40. Photoelectric yield γ_p as a function of wavelength for various metals.

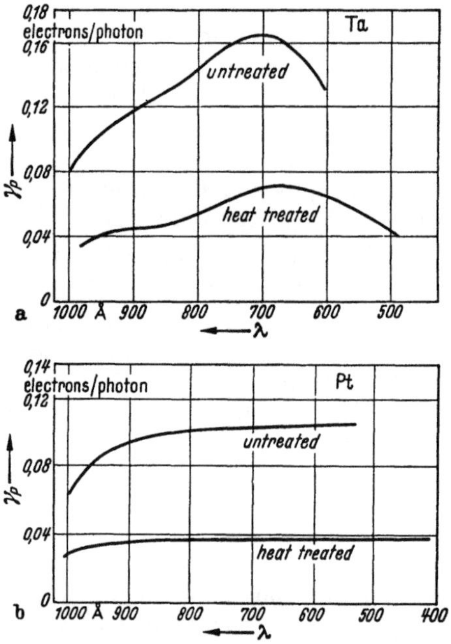

Fig. 41 a and b. Photoelectric yield γ_p as a function of wavelength by vacuum spectrometer measurements on gas-covered metals. *a* tantalum, *b* platinum.

These values are plotted in Fig. 40 with some of KENTY's results[2] and more recent ones of HINTEREGGER and WATANABE[3]. The latter authors obtained the

Table 13. γ_p *for Cd.*

λ(Å)	3130	2650	2480	1762	1608	1300	1149	1000
$\gamma_p \dfrac{\text{electrons}}{\text{photon}}$	$8 \cdot 10^{-7}$	$5 \cdot 10^{-5}$	$6.4 \cdot 10^{-5}$	$1.3 \cdot 10^{-4}$	$2.5 \cdot 10^{-4}$	$2.8 \cdot 10^{-4}$	$4.6 \cdot 10^{-3}$	$2 \cdot 10^{-2}$

[1] R. F. BAKER: J. Opt. Soc. Amer. **28**, 55 (1938).
[2] C. KENTY: Phys. Rev. **44**, 891 (1933).
[3] H. E. HINTEREGGER and K. WATANABE: J. Opt. Soc. Amer. **43**, 604 (1953).

two upper curves for Ni and Pt, together with the points at 500 Å for these metals. Their curve for tungsten lay between the Ni and Pt curves.

It is seen that KENTY'S values for Pt and W are lower than the more recent work, but all the curves show a sharp rise in γ_p below 1500 Å. Radiation of this wavelength is present in discharges.

Figs. 41a, b show the results of WAINFAN *et al*[1] for γ_p at the same wavelengths for Pt and Ta. The effect of outgassing is to reduce γ_p noticeably in these experiments, performed in a vacuum spectrograph. The curves show the maximum which is to be expected, for very penetrating radiation will release electrons at great depths in the metal. Few of these will be emitted from the surface. In fact, we move into the soft X ray region, where the absorption due to photoelectric effects is known to decrease with increasing wavelength.

Possibly the difference between Pt and Ta shown is associated with the persistent gas film present on tantalum (Sect. 27).

33. Emission due to neutral atoms. This may occur (Sect. 30) either by transfer of kinetic energy or potential energy of an impinging atom. Normally only metastable states can carry potential energy, but in some experiments ordinary excited states may be important[2].

WEBB[3] first detected emission by metastable mercury atoms on nickel. Values of γ_m for various atoms on several surfaces are now known.

Theoretical calculations by GUREVICH and YAVORSKY[4] for $6\,^3P_0$ Hg metastable atoms on mercury indicate that $\gamma_m \geqq 0.54$ if spin functions are ignored and $\gamma_m \geqq 0.75$ if spin functions are included. The process assumed is that the p-electron of the Hg atoms is excited to the continuum and becomes free while a metal electron falls to the s-state [Fig. 39b, the single stage process (modified) of Sect. 30α].

Some experimental values for the emission due to *metastable atoms* are given here.

Table 14. *Emission due to metastable atoms.*

Atom	State	Surface	Condition	γ_m
Hg	$6\,^3P_0$	W	Film of $Hg + O_2$	10^{-2} normally $1-10^{-5}$ extreme values[6]
He	—	Mo, Ni, Mg	Clean[5]	1[7]
He	$2\,^3S$	Pt	Gas film	0.24[8]
He	$2\,^1S$	Pt	Gas film	0.40[8]
Ne	—	Pt	Gas film	0.12[8]

The condition of the surface has been found to alter γ_m very greatly, as the first entry in Table 14 shows. It is clear that γ_m is very large if the excess potential energy $(eV_{ex} - \varphi)$ is large.

Measurements of emission due to *faster neutral atoms* have been made by ROSTAGNI[9]. It is possible that some metastable atoms were present, but not

[1] N. WAINFAN, W. C. WALKER and G. L. WEISSLER: J. Appl. Phys. **24**, 1318 (1953).
[2] A. E. ROBSON and A. VON ENGEL: Nature, Lond. **175**, 646 (1955). — D. GREENE: Proc. Phys. Soc. Lond. B **63**, 876 (1950).
[3] H. W. WEBB: Phys. Rev. **24**, 113 (1924).
[4] I. M. GUREVICH and B. M. YAVORSKY: C. R. Acad. Sci. URSS **53**, 789 (1946).
[5] Hot surface in vacuum, but not entirely gas free.
[6] S. SONKIN: Phys. Rev. **43**, 788 (1933).
[7] M. L. E. OLIPHANT: Proc. Roy. Soc. Lond., Ser. A **124**, 228 (1929).
[8] R. DORRESTEIN: Physica, Haag **9**, 447 (1942).
[9] A. ROSTAGNI: Z. Physik **88**, 55 (1934). — Nuovo Cim. **11**, 99 (1934).

in large numbers: the low values at low energies show that no potential energy exchange was active appreciably (Table 15).

The values are much below γ_i (which ROSTAGNI measured at the same time) at low energies, but at high energy the difference is greatly reduced.

A similar dependency was observed by BERRY and ABBOT[1] for helium ions and atoms on tungsten.

The secondary electron distribution has been studied by OLIPHANT, CHAUDHRI and KHAN, and GREENE[2,3]. The electrons ejected by ions of high potential energy show a distribution largely independent of velocity up to 2 keV, with an upper limit defined by eq. (30.2). The lower limit (30.2) was found by OLIPHANT for He—Mo, Ni, Mg, but GREENE found electrons of zero energy present for He, Ne, A—Mo.

On HAGSTRUM'S theory this can be understood, for metastable atoms may be ionised and subsequently neutralised by AUGER transitions as they approach a surface. This would lead to a zero limit for energy.

Table 15. *Emission due to energetic neutral unexcited atoms on copper and brass.*

Maximum kinetic energy (eV)	Values of γ_n		
	He	Ne	A
20	$8 \cdot 10^{-5}$	—	—
25	—	—	$8 \cdot 10^{-5}$
30	$6 \cdot 10^{-4}$	$2 \cdot 10^{-4}$	$2 \cdot 10^{-4}$
50	$8 \cdot 10^{-3}$	$3 \cdot 10^{-3}$	$4 \cdot 10^{-3}$
100	$8.1 \cdot 10^{-2}$	10^{-2}	$1.3 \cdot 10^{-2}$
200	$3.93 \cdot 10^{-1}$	$3.6 \cdot 10^{-2}$	$7 \cdot 10^{-2}$
400	$4.35 \cdot 10^{-1}$	$7.3 \cdot 10^{-2}$	$1.7 \cdot 10^{-1}$
600	—		$2.3 \cdot 10^{-1}$

OLIPHANT'S result may be due to an error in the zero of the retardation curve[4].

The presence of emission due to *kinetic energy* exchange was evident in GREENE'S work for argon. He also used H_2 atoms, and found there that the kinetic energy was important, as CHAUDHRI and KHAN found for K, Hg—Mo, Ni. These results showed a near-MAXWELLian distribution of electrons, indicating that some kind of thermionic emission was occurring. This is what OLIPHANT observed for K^+ ions on Pt, Ni, Al: the temperatures are of the same order as those he measured. For Hg—Ni the "temperature" of the emitting surface was 32000° C for 1500 to 2500 eV atoms and OLIPHANT found for K^+—Ni 30000° C for 600 eV ions—very good agreement.

The reflection of the neutral atoms was found to decrease slowly as the energy increased, but OLIPHANT found much smaller reflection coefficients than GREENE. At 800 eV for He—Mo their values were 0.15 and 0.54 respectively: again surface conditions were probably different.

34. Field and thermionic emission. Since it is known that a hot metal in an electric field emits electrons at a rate determined by the two parameters T and X, it is permissible to combine the two effects. The reviews [*16*], [*17*] and the two first articles in this volume deal with these processes in detail.

Fig. 42a shows the electron energy levels schematically for anode and cathode. If electrons are emitted from the cathode (on the left) by penetrating the barrier below the FERMI level (i.e. with the range of μ), pure field emission is occurring.

[1] H. W. BERRY and R. C. ABBOTT: O.N.R. report, June 1954.

[2] See footnote 7, p. 655.

[3] R. M. CHAUDHRI and A. W. KHAN: Proc. Phys. Soc. Lond. **61**, 526 (1948). — D. GREENE: Proc. Phys. Soc. Lond. B **63**, 876 (1950).

C. [4] H. PAETOW and WALCHER: Z. Physik **110**, 69 (1938). See also F. M. PENNING and C. J. ADDINK: Physica, Haag **1**, 1007 (1934).

If electrons have sufficient energy to be lifted above the top of the potential barrier, thermionic emission prevails. It is obvious that, if a high field strength $X = V/L$ exists, many electrons from a heated cathode will be emitted by penetrating the barrier above the FERMI limit (i.e. in the range of φ). This has been treated by many authors. DOLAN and DYKE[1] give the references for this, and derive a theory for a one-band electron model of the metal with a smooth clean surface. They assume a classical image force, and use SOMMERFELD and BETHE's transmission coefficient for electrons leaking through the barrier.

Numerical calculations for a metal of $\varphi = 4.5$ eV lead to the results in Fig. 43. F in the figure is the actual field allowing for irregularities in the metal surface.

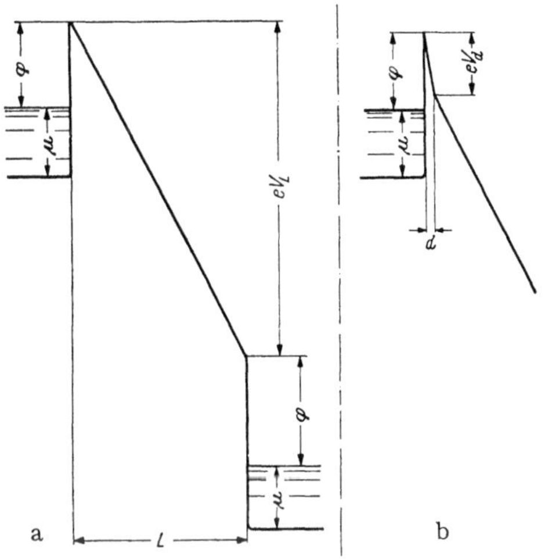

Fig. 42.

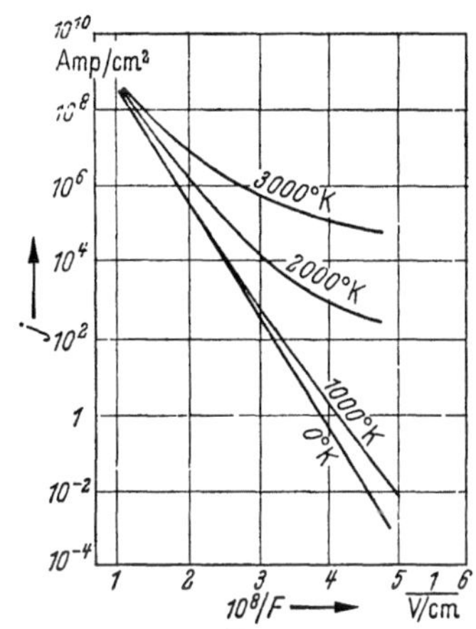

Fig. 43.

Fig. 42. *a* Potential energy diagram and electron energy levels for anode and cathode which have work function φ and a band of levels of width μ: applied potential V_L, separation L. *b* Modification of cathode potential diagram if a film of thickness d with a surface layer of charge on it exists.

Fig. 43. The current density j as a function of $10^8/F$, where F is the true surface field at various temperatures.

Clearly the contribution of thermal energy is greatest at low fields, as previous work suggested; it is negligible at temperatures below 1000° K.

If a layer of insulator of thickness d is present on the cathode and a positive surface charge exists on it, then the potential energy diagram of the cathode is modified (Fig. 42b). If the surface density is

$$\sigma = \frac{V_d - \frac{d}{L} V_L}{4\pi d} \sim \frac{V_d}{4\pi d}$$

then V_d is the potential difference across the layer.

The same type of barrier appears if a foreign metal (work function φ_2) is deposited on a metal of work function φ_1. Then (cf. Sect. 5)

$$V_d = \varphi_1 - \varphi_2 .$$

Now with low cathode temperatures a great increase in field emission can occur if $eV_d > \varphi$, and some influence will be evident if eV_d is a large fraction of φ. This is the probable explanation of field emission at low fields if the surface is smooth, or not rough enough to explain the observations, through of course it is possible for temperature—and field—emission to occur with a barrier of the form of Fig. 42b.

[1] W. W. DOLAN and W. P. DYKE: Phys. Rev. **95**, 327 (1954).

One limit for field emission which is independent of the value of X is obvious from Fig. 42a. If $eV_L < \varphi$, then no matter what $X = V_L/L$ may be, no thin barrier will be formed at the cathode. It is necessary to have $eV_L > \varphi$ for field emission to occur.

c) Ionisation in the gas.

35. Ionisation by positive ions and neutral unexcited atoms. There are two points of interest in this section. One is the lower limit at which the onset of ionisation begins, and the other is the actual magnitude of the cross-section. The theoretical and experimental work is thoroughly discussed in references [*1*] to [*6*], [*10*], [*15*] and in the article of R. KOLLATH in Vol. XXXIV of this Encyclopedia.

It is expected that for both the lower limits will be the same[1]. The classical picture of an inelastic collision between two equal masses shows that an energy eV_i may be handed over to a "stationary" atom if an ion or atom of energy $2eV_i$ strikes it. Some investigators attempted to detect the positive ions formed by their space charge[2]. Very high values for the onset potential were found, up to more than 400 eV for helium[3], where $2eV_i = 49.2$ eV. However, WAYLAND[4] showed that an electrometer was more sensitive than the space charge detector under the conditions of these experiments and ROSTAGNI[5] detected ionisation by helium atoms in helium at 60 eV. The fast atoms were obtained by charge exchange between fast He⁺ ions and helium atoms.

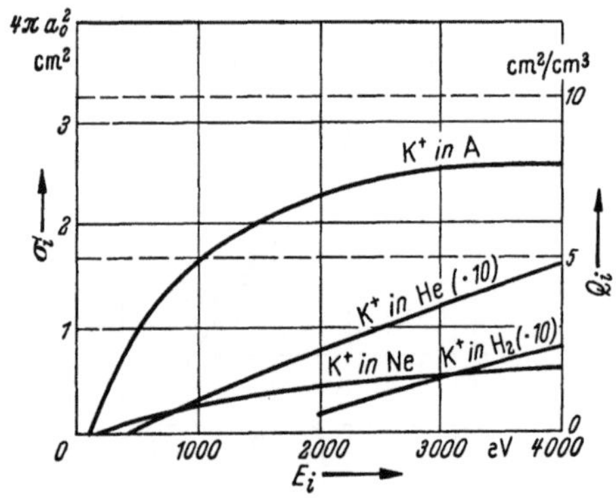

Fig. 44. The ionisation cross-section per atom (σ_i) and per cm³ at 1 mm. Hg (Q_i) for potassium in rare gases and hydrogen.

Recently, the difficult problem was solved by HORTON and MILLEST[6]. They used an electrometer to detect the tiny current of positive ions created in the gas by neutral He atoms, and found the onset of ionisation at 49.5 eV almost exactly the theoretical figure. It seems that the simple classical picture is correct.

The cross section however is very small at low energies, which accounts for the conflicting earlier results.

Fig. 44 shows some of the results of FRISCHE for the ionisation by K^+ ions in A, He, Ne and H_2. That for K^+–A is one of the largest cross-sections measured, and it is clear that a $E_i < 100$ eV the cross section is minute [*15*].

Figs. 45a, b shows the measured cross-sections for ions and atoms of A, Ne, Kr in their own gas, and of H_2 by A^+ ions. Except for BERRY'S results, the shape is the same as the curves in Fig. 44, i.e. a steady rise. There is considerable disagreement between the observations.

Quantum theory has not been applied in detail to the problem of ionisation by ions and atoms. It is known that if the relative velocity v is large compared

[1] A. ROSTAGNI: Nuovo Cim. **11**, 99 (1934) and reference [*15*].
[2] KINGDON: Phys. Rev. **21**, 408 (1923).
[3] C. J. BRASEFIELD: Phys. Rev. **43**, 785 (1933). — R. N. VARNEY: Phys. Rev. **50**, 159 (1936).
[4] H. WAYLAND: Phys. Rev. **52**, 31 (1937).
[5] A. ROSTAGNI: Nuovo Cim. **11**, 621 (1934).
[6] F. HORTON and D. M. MILLEST: Proc. Roy. Soc. Lond., Ser. A **185**, 381 (1946).

with the velocity of the atomic electrons u, then the behaviour should be similar to that of electrons of high energy. That is the cross-section decreases as the energy increases, the maximum value of the cross-section being of the order of the gas-kinetic value.

If $\frac{v^2}{u^2} \ll 1$, a parameter $\frac{a|\Delta E|}{\hbar v}$ is important, where $\frac{a}{v}$ is the time of the collision and ΔE is the energy change involved. If the parameter is greater than unity, the probability of the transition is small; the probability increases as the parameter approaches zero.

It is to be expected then that ionisation cross-sections would increase with energy until a $|\Delta E| \hbar v \sim 1$ and then decrease again. Charge transfer cross-sections (where $\Delta E = 0$) should a monotonic fall with energy.

These predictions are qualitatively correct, and for charge transfer some quantitative work has been carried out. Much remains to be done in both experiment and theory before ionisation cross-sections are properly understood.

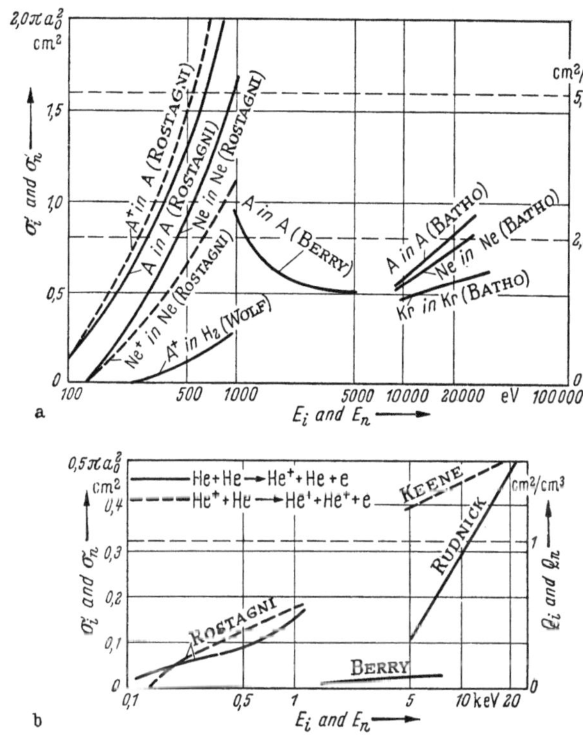

Fig. 45 a and b. The ionisation cross section per atom (σ_i) and per cm³ at 1 mm. Hg (Q_i), a for rare gas ions and atoms in their own gas, and for argon ions in hydrogen, b for helium atoms and ions in helium.

36. Photo-ionisation[1]. Many attempts have been made to measure the absorption of radiation capable of ionising a gas. The values obtained may be expressed by writing

$$\frac{I}{I_0} = e^{-kx} \tag{36.1}$$

where I is the intensity of the beam at a distance x when its intensity at the origin is I_0. Then k, the absorption coefficient was found to be 1 cm.$^{-1}$ at 760 mm. Hg and 0° C for some radiation in rare gases (or mixtures), and for other radiation $k \sim 100$ cm.$^{-1}$. Values intermediate between these limits were obtained also[2].

Calculations of the photoionisation cross section gave high values for the cross-section σ ($k = \sigma N$ where N is the number of atoms/cm.³ at N.T.P.). $\sigma \sim 10^{-17}$ cm.² for some gases according to BATES[3].

[1] Cf. the article of G. L. WEISSLER in this volume.

[2] J. DUTTON, S. C. HAYDON, F. LLEWELLYN JONES and P. M. DAVIDSON: Proc. Roy. Soc. Lond., Ser. A **218**, 206 (1953).

[3] D. R. BATES: Mon. Not. Roy. Astronom. Soc. **100**, 25 (1939).

Experimental confirmation of this was found by WEISSLER and his collaborators by vacuum-spectrograph techniques. Fig. 46a, b shows the number of ion-pairs per photon absorbed for nitrogen and oxygen[1]. The values are near unity for a range of wavelengths near the ionisation limits—the two limits for O_2 are very clear in the figure. Such photo-ionisation varies with energy in much the same manner as ionisation functions for electrons.

The absolute value of the measured absorption in the region $1000\,\text{Å} > \lambda > 500\,\text{Å}$ is shown for nitrogen[2] and neon[3] in Fig. 47a, b. The calculated value of σ for Ne is $5.8 \cdot 10^{-18}\,\text{cm.}^2$ in excellent agreement with the measured value of $5.4 \times 10^{-18}\,\text{cm.}^2$. The direct and unambiguous measurement of these quantities has been of great value in gas discharge research. Less strongly absorbed radiations cannot ionise gas atoms.

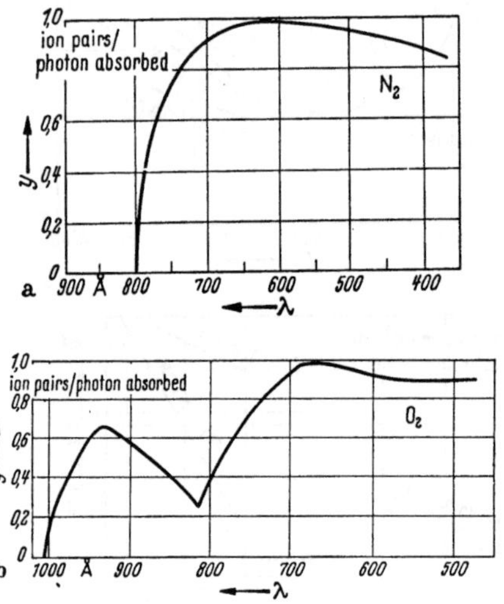

Fig. 46 a and b. The ratio y of ion-pairs produced to the total number of photons absorbed in gases, *a* nitrogen, *b* oxygen as a function of wavelength.

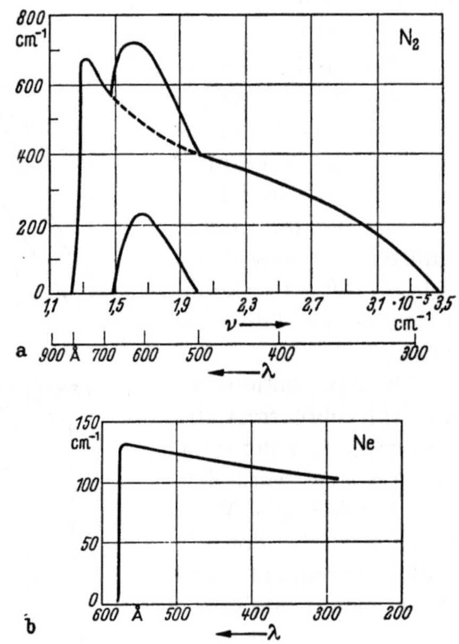

Fig. 47 a and b. Absorption coefficient k in the continuous absorption region where photo-ionisation is important *a* nitrogen, *b* neon.

37. Ionisation by metastable atoms. In gas mixtures such that one gas has a metastable level lying above the ionisation potential of the other, the reaction (taking helium and argon for an example)

$$\text{He}^* + \text{A} \rightarrow \text{He} + \text{A}^+ + e \tag{37.1}$$

is possible. He* means a metastable helium atom. This has long been known.

In a pure gas the reaction

$$\text{He}^* + \text{He}^* \rightarrow \text{He} + \text{He}^+ + e \tag{37.2}$$

can occur, but the reaction

$$\text{He}^* + \text{He} \rightarrow \text{He}_2^+ + e \tag{37.3}$$

seems infrequent compared to the similar one involving a normal excited state. This modification of (37.3) is important at pressures above 10^{-3} mm Hg, but little is known of the cross-section.

[1] N. WAINFAN, W. C. WALKER and G. L. WEISSLER: J. Appl. Phys. **24**, 1318 (1953)
[2] G. L. WEISSLER, PO LEE and E. I. MOHR: J. Opt. Soc. Amer. **42**, 84 (1952).
[3] PO LEE and G. L. WEISSLER: J. Opt. Soc. Amer. **42**, 214 (1952).

Let the cross section (37.1) be σ_i, and let it be σ_d for de-excitation of a metastable without producing ionisation. This normally[1] goes by collisions in which the metastable atoms is lifted to a radiating state just above it by collision with an unexcited atom. The energy level separation must not be large compared with thermal energies: for helium the separation is large, and then σ_d is small.

After much calculation, it is possible to obtain σ_i/σ_d from the electron multiplication in gas mixtures[2]. A value of σ_i has been estimated using mono-energetic electron beams[3] and computed from after-glow studies also[4]. Microwave investigations of afterglows[5] gave σ_i, σ_d and the diffusion coefficient D_m for metastable atoms. Optical studies of absorption can be interpreted to yield D_m[6] and σ_d, and the work on millisecond transients in glow discharges can be compared with these[7,8].

Table 16 gives the values known. There is general agreement as to the order of magnitude of the collision cross-sections, with the exception of one value of

Table 16. *Diffusion and collision of metastable atoms.*

Collision process	$D_m p$ (cm²/sec) mm Hg	σ_d cm²	
He*–He	520 ± 20	$9.6 \cdot 10^{-21}$	(5)
	535	$1.9 \cdot 10^{-20}$	(6)
Ne*–Ne	200 ± 20	$8.9 \cdot 10^{-20}$	(5)
	120 ± 10	—	(8)
	150 ± 30	—	(1)
A*–A	45 ± 4	$2.1 \cdot 10^{-20}$	(8)
	67.5	—	(7)
Xe*–Xe	13 ± 1	—	(8)
N_2^*–A	157	0	(8)
Hg_2^*–Hg	17	—	(9)

Collision process	σ_i cm²	
He*–A	$9.7 \cdot 10^{-17}$	(5)
He*–Hg	$1.4 \cdot 10^{-14}$	(5)
He*–H_2	$2.4 \cdot 10^{-16}$ †	(3)
Ne*–A	$2.6 \cdot 10^{-16}$	(5)
	$7.6 \cdot 10^{-16}$ †	(3)
	$7.5 \cdot 10^{-20}$ †	(4)
Ne*–H_2	$4.2 \cdot 10^{-16}$ †	(3)

σ_i for Ne*–A: this probably refers to A*–A, as under the conditions of this experiment many Ne* atoms would have been destroyed before observations began. The values marked by a dagger (†) of σ_i are calculated on the assumption that the total collision cross-section of Ne*–A, etc. is the gas-kinetic one. The figures in brackets give the authors (see footnotes below).

The cross-section for ionisation by metastable atoms is very large, almost as large as the gas-kinetic or even larger for He*–Hg. This is a most important mechanism in discharges containing impurities. The de-activation of metastables is very much less likely. The ratio σ_i/σ_d for Ne*–A is $2.5 \cdot 10^{-3}$ according to Biondi. Kruithof and Druyvesteyn estimated it as $4.1 \cdot 10^4$ earlier, from the multiplication in an electric field: Schut and Smit's preliminary results give $7.3 \cdot 10^{-3}$, so the first estimate is probably too high.

[1] A. V. Phelps and J. P. Molnar: Phys. Rev. **96**, 1 (1954).
[2] A. A. Kruithof and F. M. Penning: Physica, Haag **4**, 430 (1937).
[3] T. G. Schut and J. A. Smit: Physica, Haag **10**, 440 (1943).
[4] A. Hoffmann: Z. Physik **119**, 223 (1942).
[5] M. Biondi: Phys. Rev. **88**, 660 (1952).
[6] R. Ebbinghaus: Ann. Phys., Lpz. **7**, 267 (1930).
[7] R. W. Engstrom and W. S. Huxford: Phys. Rev. **58**, 67 (1940).
[8] J. P. Molnar: Phys. Rev. **83**, 933 (1951).
[9] A. O. McCoubrey and C. G. Matland: Phys. Rev. **101**, 603 (1956).

Note that ionisation in the volume by metastable atoms can occur in pure gases if (37.3) is prominent. This seems to happen only in the positive column (Sect. 25).

Conclusion.

The concept of a secondary ionisation coefficient links a wide range of facts. It only loses its value when charge densities and densities of excited atoms etc. become large. Primary and secondary coefficients then become functions of position and time, so they complicate the description of events and are better not used; rather, the elementary processes should be dealt with individually.

It is a pleasure to thank Dr. VON ENGEL for many helpful discussions and for an opportunity to see the text of his new book before publication. Professor LOEB also kindly allowed me an opportunity to study the relevant chapter of his book before it was in print. I am indebted to Professor LLEWELLYN JONES for a discussion of photo-ionisation in the gas, to Professor VARNEY for information concerning the theory of current transients, and to my colleagues for facts and suggestions throughout the preparation of this article.

References.

[1] TOWNSEND, J. S.: Electricity in gases. Oxford: O.U.P. 1915.
[2] GEIGER-SCHEEL's Handbuch der Physik, Vols XIV, XXIV. Berlin: Springer 1927.
[3] THOMSON, J. J., and G. P. THOMSON: Conduction of Electricity through Gases. Cambridge: CUP 1928—1933.
[4] ENGEL, A. v., and M. STEENBECK: Elektrische Gasentladungen. Berlin: Springer 1932 and 1934.
[5] LOEB, L. B.: Fundamental Processes of Electrical Discharge in Gases. New York: Wiley 1939.
[6] MEEK, J. M., and J. D. CRAGGS: Electrical Breakdown of Gases. Oxford: Clarendon Press 1953.
[7] LOEB, L. B.: Basic Processes of Electrical Discharge. 1955.
[8] ENGEL, A. v.: Ionised Gases. Oxford: Clarendon Press 1955.
[9] FRANCIS, V. J., and H. G. JENKINS: Electrical Discharges. Rep. Phys. Soc. Progr. Phys. **7**, 230 (1940).
[10] DRUYVESTEYN, M. J., and F. M. PENNING: Electrical Discharges. Rev. Mod. Phys. **12**, 87 (1940).
[11] LUNT, R. W., A. v. ENGEL and J. M. MEEK: Discharge phenomena. Rep. Phys. Soc. Progr. Phys. **8**, 338 (1941).
[12] KRUITHOF, A. A.: Physica, Haag **7**, 519 (1940).
[13] LLEWELLYN JONES, F.: Electrical Discharges. Rep. Phys. Soc. Progr. Phys. **16**, 216 (1953).
[14] MITCHELL, A. C. G., and M. W. ZEMANSKY: Resonance Radiation and Excited Atoms. Cambridge: C.U.P. 1934.
[15] MASSEY, H. S. W., and E. H. S. BURHOP: Electronic and Ionic Impact Phenomena. Oxford: Clarendon Press 1952.
[16] JENKINS, R. O.: Field emission of electrons. Rep. Phys. Soc. Progr. Phys. **9**, 177 (1942/43).
[17] HERRING, C., and M. H. NICHOLS: Thermionic emission. Rev. Mod. Phys. **21**, 185 (1949).
[18] HAGSTRUM, H. D.: Auger effects in electron emission by positive ions. Phys. Rev. **96**, 325, 336 (1954).
[19] OLIPHANT, M. L. E.: Proc. Roy. Soc. Lond., Ser. A **127**, 373 (1940). — MOON, P. B.: Proc. Roy. Soc. Lond., Ser. A **127**, 388 (1940).
[20] SEELIGER, R., and G. MIERDEL: Handbuch der Experimentalphysik, Vol. 13, p. 2. 1929.

Very recent work.

Studies of the fundamental processes have included absolute measurements of the rate of photon production in a TOWNSEND discharge in H_2 by CORRIGAN and VON ENGEL, and measurements of the light emitted from electron avalanches in air by LEGLER. Both included the vital ultraviolet radiation[1].

---

[1] A. v. ENGEL: Appl. Sci. Res. B **5**, 34 (1955). — W. LEGLER: Z. Physik **143**, 173 (1955).

A theoretical investigation of the movement of charge-carriers in non-homogeneous fields has been made by BEZ and HÖCKER[1].

The growth of current with time in a plane parallel gap has been derived rigorously by AUER[2], who simplified the problem by assuming that only the γ_p process was important and by neglecting the absorption of light in the gas and any "delayed" photons. The TOWNSEND expression for breakdown occurs if $t \to \infty$ and STEENBECK's assumption of an exponential current growth [Eq. (13.4)] can be justified without further approximations. A simplified solution is given in the second paper quoted. DAVIDSON[3] has published an alternative form for his solution of the problem, and pointed out errors in BANDEL's theory. AUER's result is equivalent to that of DAVIDSON, and the approximation he gives is accurate enough for all practical applications.

Experimental work on secondary ionisation coefficients in H_2 has been carefully carried out by two independent groups[4] up to high values of pd. GREEN[5] has measured conduction currents in hexane.

The importance of the γ_i process in low pressure breakdown has been emphasized by BOURNE *et al.*[6] though for very small gaps field emission enhanced by positive space-charge has been put forward as predominant[7]. Thin surface films can control the discharge in cylinders, according to HAEFER's results in methane at low pressure[8]. An investigation has been made of the elementary processes in the high voltage impulse spark at $p = 10^{-4}$ – 10^{-1} mm Hg[9].

The emission from a metal covered with a thin film has been discussed[10], and leads to an alternative explanation of the results of LLEWELLYN JONES (Sects. 11, 26) in terms of field dependent thermionic emission and patches of film of varying properties. Such surface irregularities may cause apparent space-charge field distortion. A further comparison of theory and experiment is also now available for clean surfaces[11].

---

[1] W. BEZ and K.-H. HÖCKER: Z. Naturforsch. **9**a, 64 (1954).

[2] P. L. AUER: Phys. Rev. **98**, 320 (1955); **101**, 1243 (1956).

[3] P. M. DAVIDSON: Phys. Rev. **99**, 1072 (1955).

[4] A. WILKES, W. HOPWOOD and N. J. PEACOCK: Nature, Lond. **176**, 837 (1955). — R.W. CROMPTON, J. DUTTON and S. C. HAYDON: Nature, Lond. **176**, 1079 (1955).

[5] W. B. GREEN: J. Appl. Phys. **26**, 1257 (1955).

[6] H. C. BOURNE, R. W. CLOUD and J. G. TRUMP: J. Appl. Phys. **26**, 596 (1955). — H. C. BOURNE: J. Appl. Phys. **26**, 625 (1955).

[7] W. S. BOYLE, P. KISLIUK and L. H. GERMER: J. Appl. Phys. **26**, 720 (1955).

[8] R. HAEFER: Acta phys. Austriaca **9**, 1 (1954).

[9] E. M. REIKHRUDEL, A. V. KUSTOVA and A. G. ZIMELEV: Z. tekh. Fiz. **24**, 1179 (1954).

[10] T. J. LEWIS: Proc. Phys. Soc. Lond. B **68**, 504, 938 (1955). — M. J. MORANT: Proc. Phys. Soc. Lond. B **68**, 513 (1955).

[11] W. P. DYKE, J. P. BARBOUR, E. E. MARTIN and J. K. TROLAN: Phys. Rev. **99**, 1192 (1955).

Sachverzeichnis.

(Deutsch-Englisch.)

Bei gleicher Schreibweise in beiden Sprachen sind die Stichwörter nur einmal aufgeführt.

Subject Index.

(English-German.)

Where English and German spelling of a word is identical the German version entry is omitted.

MIX
Papier aus verantwortungsvollen Quellen
Paper from responsible sources
FSC® C105338

If you have any concerns about our products,
you can contact us on
ProductSafety@springernature.com

In case Publisher is established outside the EU,
the EU authorized representative is:
Springer Nature Customer Service Center GmbH
Europaplatz 3, 69115 Heidelberg, Germany

Printed by Libri Plureos GmbH
in Hamburg, Germany